7-18 - Due friday
7-41 - Due Tuesday

midTerm
Ch 19, 7 8, 9 - 9.3   know exampy 9-1

Tuesday

                        7 89 19

MACHINE ELEMENTS IN MECHANICAL DESIGN

Fourth Edition

Robert L. Mott, P.E.

University of Dayton

PEARSON

Prentice
Hall

Upper Saddle River, New Jersey
Columbus, Ohio

To my wife, Marge
our children, Lynné, Robert, Jr., and Stephen
and my Mother and Father

Library of Congress Cataloging in Publication Data

Mott, Robert L.
    Machine elements in mechanical design / Robert L. Mott. — 4th ed.
        p. cm.
    ISBN 0-13-061885-3
    1. Machine design. 2. Mechanical movements. I. Title.

TJ230.M68 2004
621.8′15—dc21

2003042548

Editor in Chief: Stephen Helba
Executive Editor: Debbie Yarnell
Editorial Assistant: Jonathan Tenthoff
Production Editor: Louise N. Sette
Production Supervision: Carlisle Publishers Services
Design Coordinator: Diane Ernsberger
Cover Designer: Jason Moore
Production Manager: Brian Fox
Marketing Manager: Jimmy Stephens

This book was set in Times Roman and Helvetica by Carlisle Communications, Ltd. It was printed and bound by Courier Westford, Inc. The cover was printed by Phoenix Color Corp.

Pearson Education Ltd.
Pearson Education Singapore Pte. Ltd.
Pearson Education Canada, Ltd.
Pearson Education—Japan

Pearson Education Australia Pty. Limited
Pearson Education North Asia Ltd.
Pearson Educatión de Mexico, S.A. de C.V.
Pearson Education Malaysia Pte. Ltd.

10 9 8
ISBN 0-13-061885-3

Preface

The objective of this book is to provide the concepts, procedures, data, and decision analysis techniques necessary to design machine elements commonly found in mechanical devices and systems. Students completing a course of study using this book should be able to execute original designs for machine elements and integrate the elements into a system composed of several elements.

This process requires a consideration of the performance requirements of an individual element and of the interfaces between elements as they work together to form a system. For example, a gear must be designed to transmit power at a given speed. The design must specify the number of teeth, pitch, tooth form, face width, pitch diameter, material, and method of heat treatment. But the gear design also affects, and is affected by, the mating gear, the shaft carrying the gear, and the environment in which it is to operate. Furthermore, the shaft must be supported by bearings, which must be contained in a housing. Thus, the designer should keep the complete system in mind while designing each individual element. This book will help the student approach design problems in this way.

This text is designed for those interested in practical mechanical design. The emphasis is on the use of readily available materials and processes and appropriate design approaches to achieve a safe, efficient design. It is assumed that the person using the book will be the designer, that is, the person responsible for determining the configuration of a machine or a part of a machine. Where practical, all design equations, data, and procedures needed to make design decisions are specified.

It is expected that students using this book will have a good background in statics, strength of materials, college algebra, and trigonometry. Helpful, but not required, would be knowledge of kinematics, industrial mechanisms, dynamics, materials, and manufacturing processes.

Among the important features of this book are the following:

1. It is designed to be used at the undergraduate level in a first course in machine design.

2. The large list of topics allows the instructor some choice in the design of the course. The format is also appropriate for a two-course sequence and as a reference for mechanical design project courses.

3. Students should be able to extend their efforts into topics not covered in classroom instruction because explanations of principles are straightforward and include many example problems.

4. The practical presentation of the material leads to feasible design decisions and is useful to practicing designers.

5. The text advocates and demonstrates use of computer spreadsheets in cases requiring long, laborious solution procedures. Using spreadsheets allows the designer to make decisions and to modify data at several points within the problem while the computer performs all computations. See Chapter 6 on columns, Chapter 9 on spur gears, Chapter 12 on shafts, Chapter 13 on shrink fits, and Chapter 19 on spring design. Other computer-aided calculation software can also be used.

6. References to other books, standards, and technical papers assist the instructor in presenting alternate approaches or extending the depth of treatment.

7. Lists of Internet sites pertinent to topics in this book are included at the end of most chapters to assist readers in accessing additional information or data about commercial products.

8. In addition to the emphasis on original design of machine elements, much of the discussion covers commercially available machine elements and devices, since many design projects require an optimum combination of new, uniquely designed parts and purchased components.

9. For some topics the focus is on aiding the designer in selecting commercially available components, such as rolling contact bearings, flexible couplings, ball screws, electric motors, belt drives, chain drives, clutches, and brakes.

10. Computations and problem solutions use both the International System of Units (SI) and the U.S. Customary System (inch-pound-second) approximately equally. The basic reference for the usage of SI units is IEEE/ASTM-SI-10 *Standard for Use of the International System of Units (SI): The Modern Metric System*, which has replaced ASTM E380 and ANSI/IEEE Standard 268-1992.

11. Extensive appendices are included along with detailed tables in many chapters to help the reader to make real design decisions, using only this text.

MDESIGN–MECHANICAL DESIGN SOFTWARE INCLUDED IN THE BOOK

The design of machine elements inherently involves extensive procedures, complex calculations, and many design decisions. Data must be found from numerous charts and tables. Furthermore, design is typically iterative, requiring the designer to try several options for any given element, leading to the repetition of design calculations with new data or new design decisions. This is especially true for complete mechanical devices containing several components as the interfaces between components are considered. Changes to one component often require changes to mating elements. Use of computer-aided mechanical design software can facilitate the design process by performing many of the tasks while leaving the major design decisions to the creativity and judgment of the designer or engineer.

We emphasize that users of computer software must have a solid understanding of the principles of design and stress analysis to ensure that design decisions are based on reliable foundations. We recommend that the software be used only after mastering a given design methodology by careful study and using manual techniques.

Included in this book is the MDESIGN mechanical design software created by the TEDATA Company. Derived from the successful MDESIGN mec software produced for the European market, the U.S. version of MDESIGN employs standards and design methods that are typically in use in North America. Many of the textual aids and design procedures come directly from this book, *Machine Elements in Mechanical Design*.

Topics for which the MDESIGN software can be used as a supplement to this book include:

Beam stress analysis	Beam deflections	Mohr's circle	Columns
Belt drives	Chain drives	Spur gears	Helical gears
Shafts	Keys	Power screws	Springs
Rolling contact bearings	Plain surface bearings	Bolted connections	Fasteners
Clutches	Brakes		

Special icons as shown on the preceding page are placed in the margins at places in this book where use of the software is pertinent. Also, the Solutions Manual, available only to instructors using this book in scheduled classes, includes guidance for use of the software.

FEATURES OF THE FOURTH EDITION

The practical approach to designing machine elements in the context of complete mechanical designs is retained and refined in this edition. An extensive amount of updating has been accomplished through the inclusion of new photographs of commercially available machine components, new design data for some elements, new or revised standards, new end-of-chapter references, listings of Internet sites, and some completely new elements. The following list summarizes the primary features and the updates.

1. The three-part structure that was introduced in the third edition has been maintained.

   - Part I (Chapters 1–6) focuses on reviewing and upgrading readers' understanding of design philosophies, the principles of strength of materials, the design properties of materials, combined stresses, design for different types of loading, and the analysis and design of columns.

   - Part II (Chapters 7–15) is organized around the concept of the design of a complete power transmission system, covering some of the primary machine elements such as belt drives, chain drives, gears, shafts, keys, couplings, seals, and rolling contact bearings. These topics are tied together to emphasize both their interrelationships and their unique characteristics. Chapter 15, **Completion of the Design of a Power Transmission**, is a guide through detailed design decisions such as the overall layout, detail drawings, tolerances, and fits.

   - Part III (Chapters 16–22) presents methods of analysis and design of several important machine elements that were not pertinent to the design of a power transmission. These chapters can be covered in any order or can be used as reference material for general design projects. Covered here are plain surface bearings, linear motion elements, fasteners, springs, machine frames, bolted connections, welded joints, electric motors, controls, clutches, and brakes.

2. **The Big Picture, You Are the Designer**, and **Objectives** features introduced in earlier editions are maintained and refined. Feedback about these features from users, both students and instructors, has been enthusiastically favorable. They help readers to draw on their own experiences and to appreciate what competencies they will acquire from the study of each chapter. Constructivist theories of learning espouse this approach.

3. Some of the new or updated topics from individual chapters are summarized here.

   - In Chapter 1, the discussion of the mechanical design process is refined, and several new photographs are added. Internet sites for general mechanical design are included that are applicable to many later chapters. Some are for standards organizations, stress analysis software, and searchable databases for a wide variety of technical products and services.

   - Chapter 2, **Materials in Mechanical Design**, is refined, notably through added material on creep, austempered ductile iron (ADI), toughness, impact energy, and the special considerations for selecting plastics. An entirely new section on materials selection has been added. The extensive list of Internet sites provides readers access to industry data for virtually all types of materials discussed in the chapter with some tied to new practice problems.

- Chapter 3, a review of **Stress and Deformation Analysis**, has an added review of force analysis and refinement of the concepts of stress elements, combined normal stresses, and beams with concentrated bending moments.

- Chapter 5, **Design for Different Types of Loading**, is extensively upgraded and refined in the topics of endurance strength, design philosophy, design factors, predictions of failure, an overview of statistical approaches to design, finite life, and damage accumulation. The recommended approach to fatigue design has been changed from the *Soderberg criterion* to the *Goodman method*. The *modified Mohr method* is added for members made from brittle materials.

- In Chapter 7, synchronous belt drives are added and new design data for chain power ratings are included.

- Chapter 9, **Spur Gear Design**, is refined with new photographs of gear production machinery, new AGMA standards for gear quality, new discussion of functional measurement of gear quality, enhanced description of the geometry factor *I* for pitting resistance, more gear lubrication information, and a greatly expanded section on plastics gearing.

- In Chapter 11, new information is provided for keyless hub to shaft connections of the Ringfeder® and polygon types, and the Cornay™ universal joint. The extensive listing of Internet sites provides access to data for keys, couplings, universal joints, and seals.

- Critical speeds, other dynamic considerations, and flexible shafts are added to Chapter 12, **Shaft Design**.

- An all-new section, Tribology: Friction, Lubrication, and Wear, is added to Chapter 16, **Plain Surface Bearings**. More data on *pV factors* for boundary lubricated bearings are provided.

- Chapter 17 has been retitled **Linear Motion Elements** and includes power screws, ball screws, and linear actuators.

- Refinements to Chapter 18, **Fasteners**, include the shear strength of threads, components of torque applied to a fastener, and methods of bolt tightening.

Acknowledgments

My appreciation is extended to all who provided helpful suggestions for improvements to this book. I thank the editorial staff of Prentice Hall Publishing Company, those who provided illustrations, and the many users of the book, both instructors and students, with whom I have had discussions. Special appreciation goes to my colleagues at the University of Dayton, Professors David Myszka, James Penrod, Joseph Untener, Philip Doepker, and Robert Wolff. I also thank those who provided thoughtful reviews of the prior edition: Marian Barasch, Hudson Valley Community College; Ismail Fidan, Tennessee Tech University; Paul Unangst, Milwaukee School of Engineering; Richard Alexander, Texas A & M University; and Gary Qi, The University of Memphis. I especially thank my students—past and present—for their encouragement and their positive feedback about this book.

Robert L. Mott

Contents

**PART I Principles of Design and Stress
Analysis 1**

1 The Nature of Mechanical Design 2

The Big Picture 3
You Are the Designer 9
1–1 Objectives of This Chapter 9
1–2 The Mechanical Design Process 9
1–3 Skills Needed in Mechanical Design 11
1–4 Functions, Design Requirements, and
     Evaluation Criteria 11
1–5 Example of the Integration of Machine
     Elements into a Mechanical Design 14
1–6 Computational Aids in This Book 17
1–7 Design Calculations 17
1–8 Preferred Basic Sizes, Screw Threads, and
     Standard Shapes 18
1–9 Unit Systems 24
1–10 Distinction among Weight, Force, and Mass
     26
References 27
Internet Sites 27
Problems 28

2 Materials in Mechanical Design 29

The Big Picture 30
You Are the Designer 31
2–1 Objectives of This Chapter 32
2–2 Properties of Materials 32
2–3 Classification of Metals and Alloys 44
2–4 Variability of Material Properties Data 45
2–5 Carbon and Alloy Steel 46
2–6 Conditions for Steels and Heat Treatment 49
2–7 Stainless Steels 53
2–8 Structural Steel 54
2–9 Tool Steels 54
2–10 Cast Iron 54
2–11 Powdered Metals 56
2–12 Aluminum 57
2–13 Zinc Alloys 59
2–14 Titanium 60
2–15 Copper, Brass, and Bronze 60
2–16 Nickel-Based Alloys 61
2–17 Plastics 61
2–18 Composite Materials 65
2–19 Materials Selection 77
References 78
Internet Sites 79
Problems 80

3 Stress and Deformation Analysis 83

The Big Picture 84
You Are the Designer 85
3–1 Objectives of This Chapter 89
3–2 Philosophy of a Safe Design 89
3–3 Representing Stresses on a Stress Element 89
3–4 Direct Stresses: Tension and Compression 90
3–5 Deformation under Direct Axial Loading 92
3–6 Direct Shear Stress 92
3–7 Relationship among Torque, Power, and
     Rotational Speed 94
3–8 Torsional Shear Stress 95
3–9 Torsional Deformation 97
3–10 Torsion in Members Having Noncircular Cross
     Sections 98
3–11 Torsion in Closed, Thin-Walled Tubes 100
3–12 Open Tubes and a Comparison with Closed
     Tubes 100
3–13 Vertical Shearing Stress 102
3–14 Special Shearing Stress Formulas 104

3–15   Stress Due to Bending   105

3–16   Flexural Center for Beams   107

3–17   Beam Deflections   108

3–18   Equations for Deflected Beam Shape   110

3–19   Beams with Concentrated Bending Moments   112

3–20   Combined Normal Stresses: Superposition Principle   117

3–21   Stress Concentrations   119

3–22   Notch Sensitivity and Strength Reduction Factor   122

References   123

Internet Sites   123

Problems   123

4   Combined Stresses and Mohr's Circle   135

The Big Picture   136

You Are the Designer   136

4–1   Objectives of This Chapter   138

4–2   General Case of Combined Stress   138

4–3   Mohr's Circle   145

4–4   Mohr's Circle Practice Problems   151

4–5   Case When Both Principal Stresses Have the Same Sign   155

4–6   Mohr's Circle for Special Stress Conditions   158

4–7   Analysis of Complex Loading Conditions   161

References   162

Internet Site   162

Problems   162

5   Design for Different Types of Loading   163

The Big Picture   164

You Are the Designer   166

5–1   Objectives of This Chapter   166

5–2   Types of Loading and Stress Ratio   166

5–3   Endurance Strength   172

5–4   Estimated Actual Endurance Strength, s_n' 173

5–5   Example Problems for Estimating Actual Endurance Strength   181

5–6   Design Philosophy   182

5–7   Design Factors   185

5–8   Predictions of Failure   186

5–9   Design Analysis Methods   193

5–10   General Design Procedure   197

5–11   Design Examples   200

5–12   Statistical Approaches to Design   213

5–13   Finite Life and Damage Accumulation Method   214

References   218

Problems   219

6   Columns   229

The Big Picture   230

You Are the Designer   231

6–1   Objectives of This Chapter   231

6–2   Properties of the Cross Section of a Column   232

6–3   End Fixity and Effective Length   232

6–4   Slenderness Ratio   234

6–5   Transition Slenderness Ratio   234

6–6   Long Column Analysis: The Euler Formula   235

6–7   Short Column Analysis: The J. B. Johnson Formula   239

6–8   Column Analysis Spreadsheet   241

6–9   Efficient Shapes for Column Cross Sections   244

6–10   The Design of Columns   245

6–11   Crooked Columns   250

6–12   Eccentrically Loaded Columns   251

References   257

Problems   257

PART II   Design of a Mechanical Drive   261

7   Belt Drives and Chain Drives   264

The Big Picture   265

You Are the Designer   267

7–1   Objectives of This Chapter   267

7–2   Types of Belt Drives   268

7–3   V-Belt Drives   269

7–4   V-Belt Drive Design   272

7–5   Chain Drives   283

7–6   Design of Chain Drives   285

References   296

Internet Sites   298

Problems   298

8   Kinematics of Gears   300

The Big Picture   301

You Are the Designer   305

8–1   Objectives of This Chapter   306

8–2   Spur Gear Styles   306

8–3   Spur Gear Geometry: Involute-Tooth Form   307

8–4   Spur Gear Nomenclature and Gear-Tooth Features   308

8–5   Interference between Mating Spur Gear Teeth   320

8–6   Velocity Ratio and Gear Trains   322

8–7   Helical Gear Geometry   329

8–8   Bevel Gear Geometry   333

8–9   Types of Wormgearing   339

8–10   Geometry of Worms and Wormgears   341

8–11   Typical Geometry of Wormgear Sets   344

8–12   Train Value for Complex Gear Trains   347

8–13   Devising Gear Trains   350

References   357

Internet Sites   357

Problems   358

9   Spur Gear Design   363

The Big Picture   364

You Are the Designer   365

9–1   Objectives of This Chapter   365

9–2   Concepts from Previous Chapters   366

9–3   Forces, Torque, and Power in Gearing   367

9–4   Gear Manufacture   370

9–5   Gear Quality   372

9–6   Allowable Stress Numbers   378

9–7   Metallic Gear Materials   379

9–8   Stresses in Gear Teeth   385

9–9   Selection of Gear Material Based on Bending Stress   394

9–10   Pitting Resistance of Gear Teeth   399

9–11   Selection of Gear Material Based on Contact Stress   402

9–12   Design of Spur Gears   407

9–13   Gear Design for the Metric Module System   413

9–14   Computer-Aided Spur Gear Design and Analysis   415

9–15   Use of the Spur Gear Design Spreadsheet   419

9–16   Power-Transmitting Capacity   428

9–17   Practical Considerations for Gears and Interfaces with Other Elements   430

9–18   Plastics Gearing   434

References   442

Internet Sites   443

Problems   444

10   Helical Gears, Bevel Gears, and Wormgearing   449

The Big Picture   450

You Are the Designer   452

10–1   Objectives of This Chapter   452

10–2   Forces on Helical Gear Teeth   452

10–3   Stresses in Helical Gear Teeth   455

10–4   Pitting Resistance for Helical Gear Teeth   459

10–5   Design of Helical Gears   460

10–6   Forces on Straight Bevel Gears   463

10–7   Bearing Forces on Shafts Carrying Bevel Gears   465

10–8   Bending Moments on Shafts Carrying Bevel Gears   470

10–9   Stresses in Straight Bevel Gear Teeth   470

10–10 Design of Bevel Gears for Pitting Resistance   473

10–11 Forces, Friction, and Efficiency in Wormgear Sets   475

10–12 Stress in Wormgear Teeth   481

10–13 Surface Durability of Wormgear Drives   482

References   488

Internet Sites   488

Problems   489

11   Keys, Couplings, and Seals   491

The Big Picture   492
You Are the Designer   493
11–1   Objectives of This Chapter   493
11–2   Keys   494
11–3   Materials for Keys   498
11–4   Stress Analysis to Determine Key Length   499
11–5   Splines   503
11–6   Other Methods of Fastening Elements to Shafts   508
11–7   Couplings   513
11–8   Universal Joints   516
11–9   Retaining Rings and Other Means of Axial Location   518
11–10 Types of Seals   521
11–11 Seal Materials   525
References   526
Internet Sites   527
Problems   528

12   Shaft Design   530

The Big Picture   531
You Are the Designer   532
12–1   Objectives of This Chapter   532
12–2   Shaft Design Procedure   532
12–3   Forces Exerted on Shafts by Machine Elements   535
12–4   Stress Concentrations in Shafts   540
12–5   Design Stresses for Shafts   543
12–6   Shafts in Bending and Torsion Only   546
12–7   Shaft Design Example   548
12–8   Recommended Basic Sizes for Shafts   552
12–9   Additional Design Examples   553
12–10 Spreadsheet Aid for Shaft Design   561
12–11 Shaft Rigidity and Dynamic Considerations   562
12–12 Flexible Shafts   563
References   564
Internet Sites   564
Problems   565

13   Tolerances and Fits   575

The Big Picture   576
You Are the Designer   577
13–1   Objectives of This Chapter   577
13–2   Factors Affecting Tolerances and Fits   578
13–3   Tolerances, Production Processes, and Cost   578
13–4   Preferred Basic Sizes   581
13–5   Clearance Fits   581
13–6   Interference Fits   585
13–7   Transition Fits   586
13–8   Stresses for Force Fits   587
13–9   General Tolerancing Methods   591
13–10 Robust Product Design   592
References   594
Internet Sites   594
Problems   595

14   Rolling Contact Bearings   597

The Big Picture   598
You Are the Designer   599
14–1   Objectives of This Chapter   600
14–2   Types of Rolling Contact Bearings   600
14–3   Thrust Bearings   604
14–4   Mounted Bearings   604
14–5   Bearing Materials   606
14–6   Load/Life Relationship   606
14–7   Bearing Manufacturers' Data   606
14–8   Design Life   611
14–9   Bearing Selection: Radial Loads Only   613
14–10 Bearing Selection: Radial and Thrust Loads Combined   614
14–11 Mounting of Bearings   616
14–12 Tapered Roller Bearings   618
14–13 Practical Considerations in the Application of Bearings   621
14–14 Importance of Oil Film Thickness in Bearings   624
14–15 Life Prediction under Varying Loads   625
References   627
Internet Sites   627
Problems   628

15   Completion of the Design of a Power Transmission   630

The Big Picture   631

15–1   Objectives of This Chapter   631

15–2   Description of the Power Transmission to Be Designed   631

15–3   Design Alternatives and Selection of the Design Approach   633

15–4   Design Alternatives for the Gear-Type Reducer   635

15–5   General Layout and Design Details of the Reducer   635

15–6   Final Design Details for the Shafts   652

15–7   Assembly Drawing   655

References   657

Internet Sites   657

PART III   Design Details and Other Machine Elements   659

16   Plain Surface Bearings   660

The Big Picture   661

You Are the Designer   663

16–1   Objectives of This Chapter   663

16–2   The Bearing Design Task   663

16–3   Bearing Parameter, $\mu n/p$ 665

16–4   Bearing Materials   666

16–5   Design of Boundary-Lubricated Bearings   668

16–6   Full-Film Hydrodynamic Bearings   674

16–7   Design of Full-Film Hydrodynamically Lubricated Bearings   675

16–8   Practical Considerations for Plain Surface Bearings   682

16–9   Hydrostatic Bearings   683

16–10   Tribology: Friction, Lubrication, and Wear   687

References   691

Internet Sites   692

Problems   693

17   Linear Motion Elements   694

The Big Picture   695

You Are the Designer   698

17–1   Objectives of This Chapter   698

17–2   Power Screws   699

17–3   Ball Screws   704

17–4   Application Considerations for Power Screws and Ball Screws   707

References   709

Internet Sites   709

Problems   709

18   Fasteners   711

The Big Picture   713

You Are the Designer   714

18–1   Objectives of This Chapter   714

18–2   Bolt Materials and Strength   714

18–3   Thread Designations and Stress Area   717

18–4   Clamping Load and Tightening of Bolted Joints   719

18–5   Externally Applied Force on a Bolted Joint   722

18–6   Thread Stripping Strength   723

18–7   Other Types of Fasteners and Accessories   724

18–8   Other Means of Fastening and Joining   726

References   727

Internet Sites   727

Problems   728

19   Springs   729

The Big Picture   730

You Are the Designer   731

19–1   Objectives of This Chapter   732

19–2   Kinds of Springs   732

19–3   Helical Compression Springs   735

19–4   Stresses and Deflection for Helical Compression Springs   744

19–5   Analysis of Spring Characteristics   746

19–6   Design of Helical Compression Springs   749

19–7   Extension Springs   757

19–8   Helical Torsion Springs   762

19–9   Improving Spring Performance by Shot Peening   769

19–10   Spring Manufacturing   770

References   770

Internet Sites   770

Problems   771

20 Machine Frames, Bolted Connections, and Welded Joints   773

The Big Picture   774
You Are the Designer   775
20–1   Objectives of This Chapter   775
20–2   Machine Frames and Structures   776
20–3   Eccentrically Loaded Bolted Joints   780
20–4   Welded Joints   783
References   792
Internet Sites   792
Problems   793

21 Electric Motors and Controls   795

The Big Picture   796
You Are the Designer   797
21–1   Objectives of This Chapter   797
21–2   Motor Selection Factors   798
21–3   AC Power and General Information about AC Motors   799
21–4   Principles of Operation of AC Induction Motors   800
21–5   AC Motor Performance   802
21–6   Three-Phase, Squirrel-Cage Induction Motors   803
21–7   Single-Phase Motors   806
21–8   AC Motor Frame Types and Enclosures   808
21–9   Controls for AC Motors   811
21–10   DC Power   820
21–11   DC Motors   821
21–12   DC Motor Control   824
21–13   Other Types of Motors   824
References   826
Internet Sites   827
Problems   827

22 Motion Control: Clutches and Brakes   830

The Big Picture   831
You Are the Designer   833
22–1   Objectives of This Chapter   833
22–2   Descriptions of Clutches and Brakes   833

22–3   Types of Friction Clutches and Brakes   835
22–4   Performance Parameters   840
22–5   Time Required to Accelerate a Load   841
22–6   Inertia of a System Referred to the Clutch Shaft Speed   844
22–7   Effective Inertia for Bodies Moving Linearly   845
22–8   Energy Absorption: Heat-Dissipation Requirements   846
22–9   Response Time   847
22–10   Friction Materials and Coefficient of Friction   849
22–11   Plate-Type Clutch or Brake   851
22–12   Caliper Disc Brakes   854
22–13   Cone Clutch or Brake   854
22–14   Drum Brakes   855
22–15   Band Brakes   860
22–16   Other Types of Clutches and Brakes   862
References   864
Internet Sites   864
Problems   865

23 Design Projects   867

23–1   Objectives of This Chapter   868
23–2   Design Projects   868

Appendices   A–1

Appendix 1   Properties of Areas   A–1
Appendix 2   Preferred Basic Sizes and Screw Threads   A–3
Appendix 3   Design Properties of Carbon and Alloy Steels   A–6
Appendix 4   Properties of Heat-Treated Steels   A–8
Appendix 5   Properties of Carburized Steels   A–11
Appendix 6   Properties of Stainless Steels   A–12
Appendix 7   Properties of Structural Steels   A–13
Appendix 8   Design Properties of Cast Iron   A–14
Appendix 9   Typical Properties of Aluminum   A–15
Appendix 10   Typical Properties of Zinc Casting Alloys   A–16
Appendix 11   Properties of Titanium Alloys   A–16

Appendix 12   Properties of Bronzes   A–17

Appendix 13   Typical Properties of Selected Plastics   A–17

Appendix 14   Beam-Deflection Formulas   A–18

Appendix 15   Stress Concentration Factors   A–27

Appendix 16   Steel Structural Shapes   A–31

Appendix 17   Aluminum Structural Shapes   A–37

Appendix 18   Conversion Factors   A–39

Appendix 19   Hardness Conversion Table   A–40

Appendix 20   Geometry Factor *I* for Pitting for Spur Gears   A–41

Answers to Selected Problems   A–44

Index   I–1

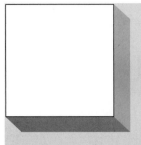

PART I

Principles of Design and Stress Analysis

OBJECTIVES AND CONTENT OF PART I

As you complete the first six chapters of this book, you will gain an understanding of design philosophies, and you will build on earlier-learned principles of strength of materials, materials science, and manufacturing processes. The competencies gained from these chapters are useful throughout the book and in general machine design or product design projects.

Chapter 1: The Nature of Mechanical Design helps you see the big picture of the process of mechanical design. Several examples are shown from different industry sectors: consumer products, manufacturing systems, construction equipment, agricultural equipment, transportation equipment, ships, and space systems. The responsibilities of designers are discussed, along with an illustration of the iterative nature of the design process. Units and conversions complete the chapter.

Chapter 2: Materials in Mechanical Design emphasizes the design properties of materials. Much of this chapter is probably review for you, but it is presented here to emphasize the importance of material selection to the design process and to explain the data for materials presented in the Appendices.

Chapter 3: Stress and Deformation Analysis is a review of the basic principles of stress and deflection analysis. It is essential that you understand the basic concepts summarized here before proceeding with later material. Reviewed are direct tensile, compressive, and shearing stresses; bending stresses; and torsional shear stresses.

Chapter 4: Combined Stresses and Mohr's Circle is important because many general design problems and the design of machine elements covered in later chapters of the book involve combined stresses. You may have covered these topics in a course in strength of materials.

Chapter 5: Design for Different Types of Loading is an in-depth discussion of design factors, fatigue, and many of the details of stress analysis as used in this book.

Chapter 6: Columns discusses the long, slender, axially loaded members that tend to fail by buckling rather than by exceeding the yield, ultimate, or shear stress of the material. Special design and analysis methods are reviewed here.

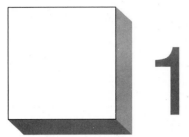

1

The Nature of Mechanical Design

The Big Picture

You Are the Designer

1–1   Objectives of This Chapter

1–2   The Mechanical Design Process

1–3   Skills Needed in Mechanical Design

1–4   Functions, Design Requirements, and Evaluation Criteria

1–5   Example of the Integration of Machine Elements into a Mechanical Design

1–6   Computational Aids in This Book

1–7   Design Calculations

1–8   Preferred Basic Sizes, Screw Threads, and Standard Shapes

1–9   Unit Systems

1–10   Distinction among Weight, Force, and Mass

The Nature of Mechanical Design

Discussion Map

☐ To design mechanical components and devices, you must be competent in the design of individual elements that comprise the system.

☐ But you must also be able to integrate several components and devices into a coordinated, robust system that meets your customer's needs.

Discover

Think, now, about the many fields in which you can use mechanical design:

What are some of the products of those fields?

What kinds of materials are used in the products?

What are some of the unique features of the products?

How were the components made?

How were the parts of the products assembled?

Consider consumer products, construction equipment, agricultural machinery, manufacturing systems, and transportation systems on the land, in the air, in space, and on and under water.

In this book, you will find the tools to learn the principles of *Machine Elements in Mechanical Design*.

Design of machine elements is an integral part of the larger and more general field of mechanical design. Designers and design engineers create devices or systems to satisfy specific needs. Mechanical devices typically involve moving parts that transmit power and accomplish specific patterns of motion. Mechanical systems are composed of several mechanical devices.

Therefore, to design mechanical devices and systems, you must be competent in the design of individual machine elements that comprise the system. But you must also be able to integrate several components and devices into a coordinated, robust system that meets your customer's needs. From this logic comes the name of this book, *Machine Elements in Mechanical Design.*

Think about the many fields in which you can use mechanical design. Discuss these fields with your instructor and with your colleagues who are studying with you. Talk with people who are doing mechanical design in local industries. Try to visit their companies if possible, or meet designers and design engineers at meetings of professional societies. Consider the following fields where mechanical products are designed and produced.

- *Consumer products:* Household appliances (can openers, food processors, mixers, toasters, vacuum cleaners, clothes washers), lawn mowers, chain saws, power tools, garage door openers, air conditioning systems, and many others. See Figures 1–1 and 1–2 for a few examples of commercially available products.

- *Manufacturing systems:* Material handling devices, conveyors, cranes, transfer devices, industrial robots, machine tools, automated assembly systems, special-purpose processing systems, forklift trucks, and packaging equipment. See Figures 1–3, 1–4, and 1–5.

- *Construction equipment:* Tractors with front-end loaders or backhoes, mobile cranes, power shovels, earthmovers, graders, dump trucks, road pavers, concrete mixers, powered nailers and staplers, compressors, and many others. See Figures 1–5 and 1–6.

FIGURE 1–1   Drill-powered band saw
[Courtesy of Black & Decker (U.S.) Inc.]

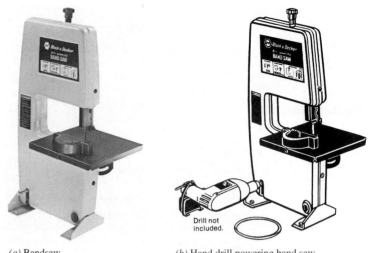

(a) Bandsaw   *(b)* Hand drill powering band saw

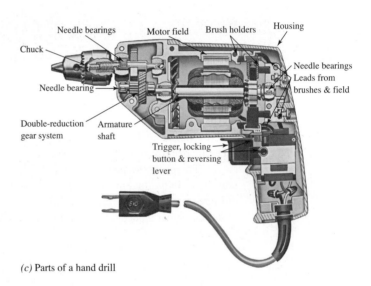

(c) Parts of a hand drill

FIGURE 1–2   Chain saw
(Copyright McCulloch
Corporation, Los Angeles, CA)

(a) Chain conveyor installation showing the drive system engaging the chain

(b) Chain and roller system
supported on an I-beam

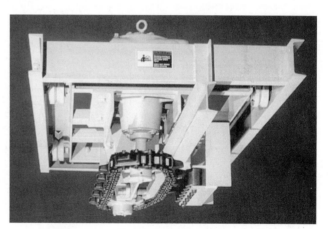

(c) Detail of the drive system
and its structure

FIGURE 1–3   Chain conveyor system (Richards-Wilcox, Inc., Aurora, IL)

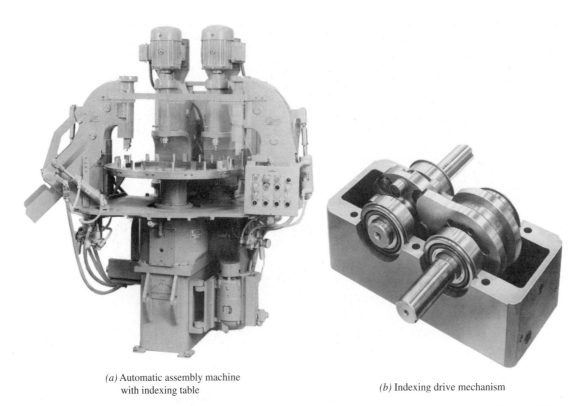

(a) Automatic assembly machine
with indexing table

(b) Indexing drive mechanism

FIGURE 1–4   Machinery to automatically assemble automotive components (Industrial
Motion Control, LLC, Wheeling, IL)

FIGURE 1–5   Industrial
crane (Air Technical
Industries, Mentor, OH)

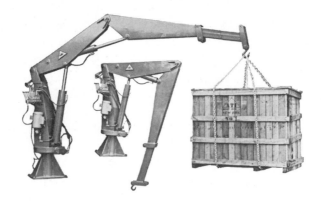

FIGURE 1–6   Tractor
with a front-end-loader
attachment (Case IH,
Racine, WI)

FIGURE 1–7 Tractor pulling an implement (Case IH, Racine, WI)

FIGURE 1–8 Cutaway of a tractor (Case IH, Racine, WI)

- *Agricultural equipment:* Tractors, harvesters (for corn, wheat, tomatoes, cotton, fruit, and many other crops), rakes, hay balers, plows, disc harrows, cultivators, and conveyors. See Figures 1–6, 1–7, and 1–8.

- *Transportation equipment:* (a) Automobiles, trucks, and buses, which include hundreds of mechanical devices such as suspension components (springs, shock absorbers, and struts); door and window operators; windshield wiper mechanisms; steering systems; hood and trunk latches and hinges; clutch and braking systems; transmissions; drive shafts; seat adjusters; and numerous parts of the engine systems. (b) Aircraft, which include retractable landing gear, flap and rudder actuators, cargo handling devices, seat reclining mechanisms, dozens of latches, structural components, and door operators. See Figures 1–9 and 1–10.

- *Ships:* Winches to haul up the anchor, cargo-handling cranes, rotating radar antennas, rudder steering gear, drive gearing and drive shafts, and the numerous sensors and controls for operating on-board systems.

- *Space systems:* Satellite systems, the space shuttle, the space station, and launch systems, which contain numerous mechanical systems such as devices to deploy antennas, hatches, docking systems, robotic arms, vibration control devices, devices to secure cargo, positioning devices for instruments, actuators for thrusters, and propulsion systems.

How many examples of mechanical devices and systems can you add to these lists?

What are some of the unique features of the products in these fields?

What kinds of mechanisms are included?

What kinds of materials are used in the products?

How were the components made?

How were the parts assembled into the complete products?

In this book, you will find the tools to learn the principles of *Machine Elements in Mechanical Design*. In the introduction to each chapter, we include a brief scenario called *You Are the Designer*. The purpose of these scenarios is to stimulate your thinking about the material presented in the chapter and to show examples of realistic situations in which you may apply it.

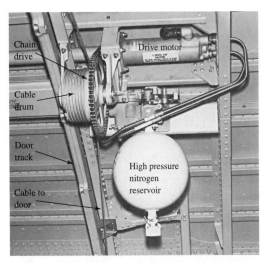

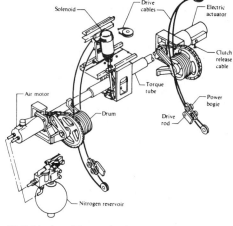

(a) Photograph of installed mechanism

(b) Cabin door drive mechanism

FIGURE 1–9   Aircraft door drive mechanism (The Boeing Company, Seattle, WA)

FIGURE 1–10
Aircraft landing gear
assembly (The Boeing
Company, Seattle, WA)

You Are the Designer

Consider, now, that you are the designer responsible for the design of a new consumer product, such as the band saw for a home workshop shown in Figure 1–1. What kind of technical preparation would you need to complete the design? What steps would you follow? What information would you need? How would you show, by calculation, that the design is safe and that the product will perform its desired function?

The general answers to these questions are presented in this chapter. As you complete the study of this book, you will learn about many design techniques that will aid in your design of a wide variety of machine elements. You will also learn how to integrate several machine elements into a mechanical system by considering the relationships between and among elements.

1–1
OBJECTIVES OF THIS CHAPTER

After completing this chapter, you will be able to:

1. Recognize examples of mechanical systems in which the application of the principles discussed in this book is necessary to complete their design.

2. List what design skills are required to perform competent mechanical design.

3. Describe the importance of integrating individual machine elements into a more comprehensive mechanical system.

4. Describe the main elements of the *product realization process*.

5. Write statements of *functions* and *design requirements* for mechanical devices.

6. Establish a set of criteria for evaluating proposed designs.

7. Work with appropriate units in mechanical design calculations both in the U.S. Customary Unit System and in SI metric units.

8. Distinguish between *force* and *mass*, and express them properly in both unit systems.

9. Present design calculations in a professional, neat, and orderly manner that can be understood and evaluated by others knowledgeable in the field of machine design.

1–2
THE MECHANICAL DESIGN PROCESS

The ultimate objective of mechanical design is to produce a useful product that satisfies the needs of a customer and that is safe, efficient, reliable, economical, and practical to manufacture. Think broadly when answering the question, "Who is the customer for the product or system I am about to design?" Consider the following scenarios:

- *You are designing a can opener for the home market.* The ultimate customer is the person who will purchase the can opener and use it in the kitchen of a home. Other customers may include the designer of the packaging for the opener, the manufacturing staff who must produce the opener economically, and service personnel who repair the unit.

- *You are designing a piece of production machinery for a manufacturing operation.* The customers include the manufacturing engineer who is responsible for the production operation, the operator of the machine, the staff who install the machine, and the maintenance personnel who must service the machine to keep it in good running order.

- *You are designing a powered system to open a large door on a passenger aircraft.* The customers include the person who must operate the door in normal service or in emergencies, the people who must pass through the door during use, the personnel who manufacture the opener, the installers, the aircraft structure designers who must accommodate the loads produced by the opener during flight and during operation, the service technicians who maintain the system, and the interior designers who must shield the opener during use while allowing access for installation and maintenance.

It is essential that you know the desires and expectations of all customers before beginning product design. Marketing professionals are often employed to manage the definition of customer expectations, but designers will likely work with them as a part of a product development team.

Many methods are used to determine what the customer wants. One popular method, called *quality function deployment* or *QFD*, seeks (1) to identify all of the features and performance factors that customers desire and (2) to assess the relative importance of these factors. The result of the QFD process is a detailed set of functions and design requirements for the product. (See Reference 8.)

It is also important to consider how the design process fits with all functions that must happen to deliver a satisfactory product to the customer and to service the product throughout its life cycle. In fact, it is important to consider how the product will be disposed of after it has served its useful life. The total of all such functions that affect the product is sometimes called the *product realization process* or *PRP*. (See References 3, 10.) Some of the factors included in PRP are as follows:

- Marketing functions to assess customer requirements
- Research to determine the available technology that can reasonably be used in the product
- Availability of materials and components that can be incorporated into the product
- Product design and development
- Performance testing
- Documentation of the design
- Vendor relationships and purchasing functions
- Consideration of global sourcing of materials and global marketing
- Work-force skills
- Physical plant and facilities available
- Capability of manufacturing systems
- Production planning and control of production systems
- Production support systems and personnel
- Quality systems requirements
- Operation and maintenance of the physical plant
- Distribution systems to get products to the customer
- Sales operations and time schedules
- Cost targets and other competitive issues
- Customer service requirements
- Environmental concerns during manufacture, operation, and disposal of the product
- Legal requirements
- Availability of financial capital

Can you add to this list?

You should be able to see that the design of a product is but one part of a comprehensive process. In this book, we will focus more carefully on the design process itself, but the producibility of your designs must always be considered. This simultaneous consideration of product design and manufacturing process design is often called *concurrent engineering.* Note that this process is a subset of the larger list given previously for the product realization process. Other major books discussing general approaches to mechanical design are listed as References 6, 7, and 12–16.

1–3 SKILLS NEEDED IN MECHANICAL DESIGN

Product engineers and mechanical designers use a wide range of skills and knowledge in their daily work, including the following:

1. Sketching, technical drawing, and computer-aided design
2. Properties of materials, materials processing, and manufacturing processes
3. Applications of chemistry such as corrosion protection, plating, and painting
4. Statics, dynamics, strength of materials, kinematics, and mechanisms
5. Oral communication, listening, technical writing, and teamwork skills
6. Fluid mechanics, thermodynamics, and heat transfer
7. Fluid power, the fundamentals of electrical phenomena, and industrial controls
8. Experimental design and performance testing of materials and mechanical systems
9. Creativity, problem solving, and project management
10. Stress analysis
11. Specialized knowledge of the behavior of machine elements such as gears, belt drives, chain drives, shafts, bearings, keys, splines, couplings, seals, springs, connections (bolted, riveted, welded, adhesive), electric motors, linear motion devices, clutches, and brakes

It is expected that you will have acquired a high level of competence in items 1–5 in this list prior to beginning the study of this text. The competencies in items 6–8 are typically acquired in other courses of study either before, concurrently, or after the study of design of machine elements. Item 9 represents skills that are developed continuously throughout your academic study and through experience. Studying this book will help you acquire significant knowledge and skills for the topics listed in items 10 and 11.

1–4 FUNCTIONS, DESIGN REQUIREMENTS, AND EVALUATION CRITERIA

Section 1–2 emphasized the importance of carefully identifying the needs and expectations of the customer prior to beginning the design of a mechanical device. You can formulate these by producing clear, complete statements of *functions*, *design requirements*, and *evaluation criteria*:

- ▪ *Functions* tell what the device must do, using general, nonquantitative statements that employ action phrases such as *to support a load*, *to lift a crate*, *to transmit power*, or *to hold two structural members together*.

- ▪ *Design requirements* are detailed, usually quantitative statements of *expected performance levels*, *environmental conditions in which the device must operate*, *limitations on space or weight*, or *available materials and components that may be used*.

- ▪ *Evaluation criteria* are statements of *desirable qualitative characteristics* of a design that assist the designer in deciding which alternative design is optimum—that is, the design that maximizes benefits while minimizing disadvantages.

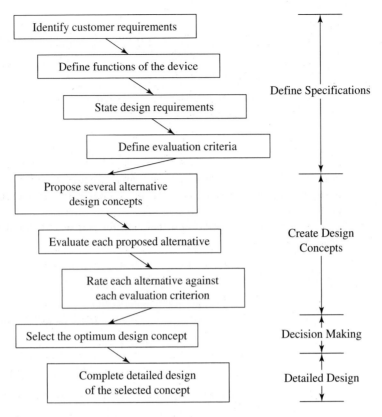

FIGURE 1–11   Steps in the design process

Together these elements can be called the *specifications* for the design.

Most designs progress through a cycle of activities as outlined in Figure 1–11. You should typically propose more than one possible alternative design concept. This is where creativity is exercised to produce truly novel designs. Each design concept must satisfy the functions and design requirements. A critical evaluation of the desirable features, advantages, and disadvantages of each design concept should be completed. Then a rational decision analysis technique should use the evaluation criteria to decide which design concept is the optimum and, therefore, should be produced.

The final block in the design flowchart is the detailed design, and the primary focus of this book is on that part of the overall design process. It is important to recognize that a significant amount of activity precedes the detailed design.

Example of Functions, Design Requirements, and Evaluation Criteria

Consider that you are the designer of a speed reducer that is part of the power transmission for a small tractor. The tractor's engine operates at a fairly high speed, while the drive for the wheels must rotate more slowly and transmit a higher torque than is available at the output of the engine.

To begin the design process, let us list the *functions* of the speed reducer. What is it supposed to do? Some answers to this question are as follows:

Functions

1. To receive power from the tractor's engine through a rotating shaft.
2. To transmit the power through machine elements that reduce the rotational speed to a desired value.
3. To deliver the power at the lower speed to an output shaft that ultimately drives the wheels of the tractor.

Now the *design requirements* should be stated. The following list is hypothetical, but if you were on the design team for the tractor, you would be able to identify such requirements from your own experience and ingenuity and/or by consultation with fellow designers, marketing staff, manufacturing engineers, service personnel, suppliers, and customers.

The product realization process calls for personnel from all of these functions to be involved from the earliest stages of design.

Design Requirements

1. The reducer must transmit 15.0 hp.
2. The input is from a two-cylinder gasoline engine with a rotational speed of 2000 rpm.
3. The output delivers the power at a rotational speed in the range of 290 to 295 rpm.
4. A mechanical efficiency of greater than 95% is desirable.
5. The minimum output torque capacity of the reducer should be 3050 pound-inches (lb·in).
6. The reducer output is connected to the drive shaft for the wheels of a farm tractor. Moderate shock will be encountered.
7. The input and output shafts must be in-line.
8. The reducer is to be fastened to a rigid steel frame of the tractor.
9. Small size is desirable. The reducer must fit in a space no larger than 20 in × 20 in, with a maximum height of 24 in.
10. The tractor is expected to operate 8 hours (h) per day, 5 days per week, with a design life of 10 years.
11. The reducer must be protected from the weather and must be capable of operating anywhere in the United States at temperatures ranging from 0 to 130°F.
12. Flexible couplings will be used on the input and output shafts to prohibit axial and bending loads from being transmitted to the reducer.
13. The production quantity is 10 000 units per year.
14. A moderate cost is critical to successful marketing.
15. All government and industry safety standards must be met.

Careful preparation of function statements and design requirements will ensure that the design effort is focused on the desired results. Much time and money can be wasted on designs that, although technically sound, do not meet design requirements. Design requirements should include everything that is needed, but at the same time they should offer ample opportunity for innovation.

Evaluation criteria should be developed by all members of a product development team to ensure that the interests of all concerned parties are considered. Often weights are assigned to the criteria to reflect their relative importance.

Safety must always be the paramount criterion. Different design concepts may have varying levels of inherent safety in addition to meeting stated safety requirements as noted in the design requirements list. Designers and engineers are legally liable if a person is injured because of a design error. You must consider any reasonably foreseeable uses of the device and ensure safety of those operating it or those who may be close by.

Achieving a high overall performance should also be a high priority. Certain design concepts may have desirable features not present on others.

The remaining criteria should reflect the special needs of a particular project. The following list gives examples of possible evaluation criteria for the small tractor.

Evaluation Criteria

1. Safety (the relative inherent safety over and above stated requirements)
2. Performance (the degree to which the design concept exceeds requirements)
3. Ease of manufacture
4. Ease of service or replacement of components
5. Ease of operation
6. Low initial cost
7. Low operating and maintenance costs
8. Small size and low weight
9. Low noise and vibration; smooth operation
10. Use of readily available materials and purchased components
11. Prudent use of both uniquely designed parts and commercially available components
12. Appearance that is attractive and appropriate to the application

**1–5
EXAMPLE OF THE
INTEGRATION
OF MACHINE
ELEMENTS INTO
A MECHANICAL
DESIGN**

Mechanical design is the process of designing and/or selecting mechanical components and putting them together to accomplish a desired function. Of course, machine elements must be compatible, must fit well together, and must perform safely and efficiently. The designer must consider not only the performance of the element being designed at a given time but also the elements with which it must interface.

To illustrate how the design of machine elements must be integrated with a larger mechanical design, let us consider the design of a speed reducer for the small tractor discussed in Section 1–4. Suppose that, to accomplish the speed reduction, you decide to design a double-reduction, spur gear speed reducer. You specify four gears, three shafts, six bearings, and a housing to hold the individual elements in proper relation to each other, as shown in Figure 1–12.

The primary elements of the speed reducer in Figure 1–12 are:

1. The input shaft (shaft 1) is to be connected to the power source, a gasoline engine whose output shaft rotates at 2000 rpm. A flexible coupling is to be employed to minimize difficulties with alignment.
2. The first pair of gears, A and B, causes a reduction in the speed of the intermediate shaft (shaft 2) proportional to the ratio of the numbers of teeth in the gears. Gears B and C are both mounted to shaft 2 and rotate at the same speed.
3. A key is used at the interface between the hub of each gear and the shaft on which it is mounted to transmit torque between the gear and the shaft.

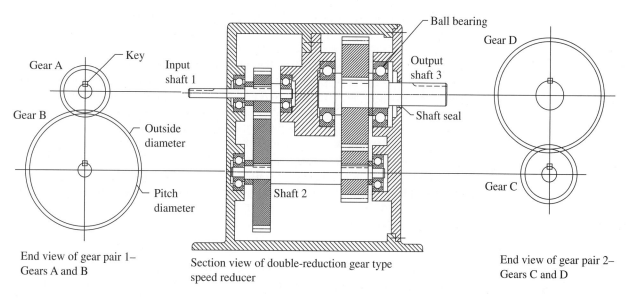

Gear A

Key

Input shaft 1

Ball bearing

Gear D

Output shaft 3

Gear B

Outside diameter

Shaft seal

Pitch diameter

Shaft 2

Gear C

End view of gear pair 1– Gears A and B

Section view of double-reduction gear type speed reducer

End view of gear pair 2– Gears C and D

FIGURE 1–12 Conceptual design for a speed reducer

4. The second pair of gears, C and D, further reduces the speed of gear D and the output shaft (shaft 3) to the range of 290 to 295 rpm.

5. The output shaft is to carry a chain sprocket (not shown). The chain drive ultimately is to be connected to the drive wheels of the tractor.

6. Each of the three shafts is supported by two ball bearings, making them statically determinate and allowing the analysis of forces and stresses using standard principles of mechanics.

7. The bearings are held in a housing that is to be attached to the frame of the tractor. Note the manner of holding each bearing so that the inner race rotates with the shaft while the outer race is held stationary.

8. Seals are shown on the input and output shafts to prohibit contaminants from entering the housing.

9. Other parts of the housing are shown schematically. Details of how the active elements are to be installed, lubricated, and aligned are only suggested at this stage of the design process to demonstrate feasibility. One possible assembly process could be as follows:

   ■ Start by placing the gears, keys, spacers, and bearings on their respective shafts.

   ■ Then insert shaft 1 into its bearing seat on the left side of the housing.

   ■ Insert the left end of shaft 2 into its bearing seat while engaging the teeth of gears A and B.

   ■ Install the center bearing support to provide support for the bearing at the right side of shaft 1.

   ■ Install shaft 3 by placing its left bearing into the seat on the center bearing support while engaging gears C and D.

   ■ Install the right side cover for the housing while placing the final two bearings in their seats.

   ■ Ensure careful alignment of the shafts.

   ■ Place gear lubricant in the lower part of the housing.

Figures 9–34 to 9–36 in Chapter 9 show three examples of commercially available double-reduction gear reducers where you can see these details.

The arrangement of the gears, the placement of the bearings so that they straddle the gears, and the general configuration of the housing are also design decisions. The design process cannot rationally proceed until these kinds of decisions are made. Notice that the sketch of Figure 1–12 is where *integration* of the elements into a whole design begins. When the overall design is conceptualized, the design of the individual machine elements in the speed reducer can proceed. As each element is discussed, scan the relevant chapters of the book. Part II of this book, including Chapters 7–15, provides details for the elements of the reducer. You should recognize that you have already made many design decisions by rendering such a sketch. First, you chose *spur gears* rather than helical gears, a worm and wormgear, or bevel gears. In fact, other types of speed reduction devices—belt drives, chain drives, or many others—could be appropriate.

Gears

For the gear pairs, you must specify the number of teeth in each gear, the pitch (size) of the teeth, the pitch diameters, the face width, and the material and its heat treatment. These specifications depend on considerations of strength and wear of the gear teeth and the motion requirements (kinematics). You must also recognize that the gears must be mounted on shafts in a manner that ensures proper location of the gears, adequate torque transmitting capability from the gears to the shafts (as through keys), and safe shaft design.

Shafts

Having designed the gear pairs, you next consider the shaft design (Chapter 12). The shaft is loaded in bending and torsion because of the forces acting at the gear teeth. Thus, its design must consider strength and rigidity, and it must permit the mounting of the gears and bearings. Shafts of varying diameters may be used to provide shoulders against which to seat the gears and bearings. There may be keyseats cut into the shaft (Chapter 11). The input and output shafts will extend beyond the housing to permit coupling with the engine and the drive axle. The type of coupling must be considered, as it can have a dramatic effect on the shaft stress analysis (Chapter 11). Seals on the input and output shafts protect internal components (Chapter 11).

Bearings

Design of the bearings (Chapter 14) is next. If rolling contact bearings are to be used, you will probably select commercially available bearings from a manufacturer's catalog, rather than design a unique one. You must first determine the magnitude of the loads on each bearing from the shaft analysis and the gear designs. The rotational speed and reasonable design life of the bearings and their compatibility with the shaft on which they are to be mounted must also be considered. For example, on the basis of the shaft analysis, you could specify the minimum allowable diameter at each bearing seat location to ensure safe stress levels. The bearing selected to support a particular part of the shaft, then, must have a bore (inside diameter) no smaller than the safe diameter of the shaft. Of course, the bearing should not be grossly larger than necessary. When a specific bearing is selected, the diameter of the shaft at the bearing seat location and allowable tolerances must be specified, according to the bearing manufacturer's recommendations, to achieve proper operation and life expectancy of the bearing.

Keys

Now the keys (Chapter 11) and the keyseats can be designed. The diameter of the shaft at the key determines the key's basic size (width and height). The torque that must be transmitted is used in strength calculations to specify key length and material. Once the working components are designed, the housing design can begin.

Housing

The housing design process must be both creative and practical. What provisions should be made to mount the bearings accurately and to transmit the bearing loads safely through the case to the structure on which the speed reducer is mounted? How will the various elements be assembled into the housing? How will the gears and bearings be lubricated? What housing material should be used? Should the housing be a casting, a weldment, or an assembly of machined parts?

The design process as outlined here implies that the design can progress in sequence: from the gears to the shafts, to the bearings, to the keys and couplings, and finally to the housing. It would be rare, however, to follow this logical path only once for a given design. Usually the designer must go back many times to adjust the design of certain components affected by changes in other components. This process, called *iteration,* continues until an acceptable overall design is achieved. Frequently prototypes are developed and tested during iteration.

Chapter 15 shows how all of the machine elements are finally integrated into a unit.

1–6 COMPUTATIONAL AIDS IN THIS BOOK

Because of the usual need for several iterations and because many of the design procedures require long, complex calculations, spreadsheets, mathematical analysis software, computer programs, or programmable calculators are often useful in performing the design analysis. Interactive spreadsheets or programs allow you, the designer, to make design decisions during the design process. In this way, many trials can be made in a short time, and the effects of changing various parameters can be investigated. Spreadsheets using Microsoft Excel are used most frequently as examples in this book for computer-aided design and analysis calculations.

1–7 DESIGN CALCULATIONS

As you study this book and as you progress in your career as a designer, you will make many design calculations. It is important to record the calculations neatly, completely, and in an orderly fashion. You may have to explain to others how you approached the design, which data you used, and which assumptions and judgments you made. In some cases, someone else will actually check your work when you are not there to comment on it or to answer questions. Also, an accurate record of your design calculations is often useful if changes in design are likely. In all of these situations, you are going to be asked to communicate your design to someone else in written and graphic form.

To prepare a careful design record, you will usually take the following steps:

1. Identify the machine element being designed and the nature of the design calculation.
2. Draw a sketch of the element, showing all features that affect performance or stress analysis.
3. Show in a sketch the forces acting on the element (the free-body diagram), and provide other drawings to clarify the actual physical situation.

4. Identify the kind of analysis to be performed, such as stress due to bending, deflection of a beam, buckling of a column, and so on.

5. List all given data and assumptions.

6. Write the formulas to be used in symbol form, and clearly indicate the values and units of the variables involved. If a formula is not well known to a potential reader of your work, give the source. The reader may want to refer to it to evaluate the appropriateness of the formula.

7. Solve each formula for the desired variable.

8. Insert data, check units, and perform computations.

9. Judge the reasonableness of the result.

10. If the result is not reasonable, change the design decisions and recompute. Perhaps a different geometry or material would be more appropriate.

11. When a reasonable, satisfactory result has been achieved, specify the final values for all important design parameters, using standard sizes, convenient dimensions, readily available materials, and so on.

Figure 1–13 shows a sample design calculation. A beam is to be designed to span a 60-in pit to support a large gear weighing 2050 pounds (lb). The design assumes that a rectangular shape is to be used for the cross section of the beam. Other practical shapes could have been used. The objective is to compute the required dimensions of the cross section, considering both stress and deflection. A material for the beam is also chosen. Refer to Chapter 3 for a review of stress due to bending.

1–8 PREFERRED BASIC SIZES, SCREW THREADS, AND STANDARD SHAPES

One responsibility of a designer is to specify the final dimensions for load-carrying members. After completing the analyses for stress and deformation (strain), the designer will know the minimum acceptable values for dimensions that will ensure that the member will meet performance requirements. The designer then typically specifies the final dimensions to be standard or convenient values that will facilitate the purchase of materials and the manufacture of the parts. This section presents some guides to aid in these decisions and specifications.

Preferred Basic Sizes

Table A2–1 lists preferred basic sizes for fractional-inch, decimal-inch, and metric sizes.[1] You should choose one of these preferred sizes as the final part of your design. An example is at the end of the sample design calculation shown in Figure 1–13. You may, of course, specify another size if there is a sound functional reason.

American Standard Screw Threads

Threaded fasteners and machine elements having threaded connections are manufactured according to standard dimensions to ensure interchangeability of parts and to permit convenient manufacture with standard machines and tooling. Table A2–2 gives the dimensions

[1] Throughout this book, some references to tables and figures have the letter A included in their numbers; these tables and figures are in the Appendices in the back of the book. For example, Table A2–1 is the first table in Appendix 2; Figure A15–4 is the fourth figure in Appendix 15. These tables and figures are clearly identified in their captions in the Appendices.

R. L. MOTT

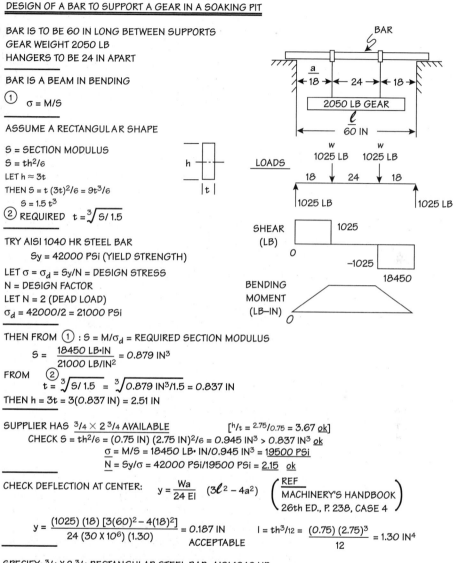

DESIGN OF A BAR TO SUPPORT A GEAR IN A SOAKING PIT

BAR IS TO BE 60 IN LONG BETWEEN SUPPORTS
GEAR WEIGHT 2050 LB
HANGERS TO BE 24 IN APART

BAR IS A BEAM IN BENDING

① $\sigma = M/S$

ASSUME A RECTANGULAR SHAPE

S = SECTION MODULUS
$S = th^2/6$
LET $h \approx 3t$
THEN $S = t(3t)^2/6 = 9t^3/6$
$S = 1.5\,t^3$
② REQUIRED $t = \sqrt[3]{S/1.5}$

TRY AISI 1040 HR STEEL BAR
$S_y = 42000$ PSI (YIELD STRENGTH)
LET $\sigma = \sigma_d = S_y/N$ = DESIGN STRESS
N = DESIGN FACTOR
LET $N = 2$ (DEAD LOAD)
$\sigma_d = 42000/2 = 21000$ PSI

THEN FROM ① : $S = M/\sigma_d$ = REQUIRED SECTION MODULUS

$$S = \frac{18450 \text{ LB·IN}}{21000 \text{ LB/IN}^2} = 0.879 \text{ IN}^3$$

FROM ②
$$t = \sqrt[3]{S/1.5} = \sqrt[3]{0.879 \text{ IN}^3/1.5} = 0.837 \text{ IN}$$
THEN $h = 3t = 3(0.837 \text{ IN}) = 2.51 \text{ IN}$

SUPPLIER HAS $3/4 \times 2\,3/4$ AVAILABLE   [$h/t = 2.75/0.75 = 3.67$ ok]
     CHECK $S = th^2/6 = (0.75 \text{ IN})(2.75 \text{ IN})^2/6 = 0.945 \text{ IN}^3 > 0.837 \text{ IN}^3$ ok
           $\sigma = M/S = 18450 \text{ LB·IN}/0.945 \text{ IN}^3 = \underline{19500 \text{ PSI}}$
           $N = S_y/\sigma = 42000 \text{ PSI}/19500 \text{ PSI} = \underline{2.15}$ ok

CHECK DEFLECTION AT CENTER: $y = \dfrac{Wa}{24\,EI}(3\ell^2 - 4a^2)$ $\left(\dfrac{\text{REF}}{\text{MACHINERY'S HANDBOOK}}\right.$
                                                                         $\left. \text{26th ED., P. 238, CASE 4}\right)$

$$y = \frac{(1025)(18)[3(60)^2 - 4(18)^2]}{24(30 \times 10^6)(1.30)} = 0.187 \text{ IN}$$ $I = th^3/12 = \dfrac{(0.75)(2.75)^3}{12} = 1.30 \text{ IN}^4$
                                                    ACCEPTABLE

SPECIFY: $3/4 \times 2\,3/4$ RECTANGULAR STEEL BAR. AISI 1040 HR

FIGURE 1–13   Sample design calculation

of American Standard Unified threads. Sizes smaller than 1/4 in are given numbers from 0 to 12, while fractional-inch sizes are specified for 1/4 in and larger sizes. Two series are listed: UNC is the designation for coarse threads, and UNF designates fine threads. Standard designations are as follows:

6–32 UNC   (number size 6, 32 threads per inch, coarse thread)

12–28 UNF   (number size 12, 28 threads per inch, fine thread)

$\frac{1}{2}$–13 UNC   (fractional size 1/2 in, 13 threads per inch, coarse thread)

$1\frac{1}{2}$–12 UNF   (fractional size $1\frac{1}{2}$ in, 12 threads per inch, fine thread)

Given in the tables are the basic major diameter (*D*), the number of threads per inch (*n*), and the tensile stress area (*A_t*), found from

⇨ **Tensile Stress
Area for Threads**

$$A_t = 0.7854 \left(D - \frac{0.9743}{n} \right)^2 \qquad \textbf{(1–1)}$$

When a threaded member is subjected to direct tension, the tensile stress area is used to compute the average tensile stress. It is based on a circular area computed from the mean of the pitch diameter and the minor diameter of the threaded member.

Metric Screw Threads

Table A2–3 gives similar dimensions for metric threads. Standard metric thread designations are of the form

$$M10 \times 1.5$$

where M stands for metric
    The following number is the basic major diameter, *D*, in mm
    The last number is the pitch, *P*, between adjacent threads in mm

The tensile stress area for metric threads is computed from the following equation and is based on a slightly different diameter. (See Reference 11, page 1483.)

$$A_t = 0.7854 \left(D - 0.9382P \right)^2 \qquad \textbf{(1–2)}$$

Thus, the designation above would denote a metric thread with a basic major diameter of $D = 10.0$ mm and a pitch of $P = 1.5$ mm. Note that pitch $= 1/n$. The tensile stress area for this thread is 58.0 mm^2.

Steel Structural Shapes

Steel manufacturers provide a large array of standard structural shapes that are efficient in the use of material and that are convenient for specification and installation into building structures or machine frames. Included, as shown in Table 1–1, are standard angles (L-shapes), channels (C-shapes), wide-flange beams (W-shapes), American Standard beams (S-shapes), structural tubing, and pipe. Note that the W-shapes and the S-shapes are often referred to in general conversation as "I-beams" because the shape of the cross section looks like the capital letter I.

Appendix 16 gives geometric properties of selected steel structural shapes that cover a fairly wide range of sizes. Note that many more sizes are available as presented in Reference 2. The tables in Appendix 16 give data for the area of the cross section (*A*), the weight per foot of length, the location of the centroid of the cross section, the moment of inertia (*I*), the section modulus (*S*), and the radius of gyration (*r*). The values of *I* and *S* are important in the analysis and design of beams. For column analysis, *I* and *r* are needed.

Materials used for structural shapes are typically called *structural steels,* and their characteristics and properties are described more fully in Chapter 2. Refer to Appendix 7 for typical strength data. Rolled W-shapes are most readily available in ASTM A992, A572 Grade 50, or A36. S-shapes and C-shapes are typically made from ASTM A572 Grade 50

TABLE 1–1 Designations for steel and aluminum shapes

Name of shape	Shape	Symbol	Example designation and Appendix table
Angle		L	$L4 \times 3 \times \frac{1}{2}$ Table A16–1
Channel		C	$C15 \times 50$ Table A16–2
Wide-flange beam		W	$W14 \times 43$ Table A16–3
American Standard beam		S	$S10 \times 35$ Table A16–4
Structural tubing—square			$4 \times 4 \times \frac{1}{4}$ Table A16–5
Structural tubing—rectangular			$6 \times 4 \times \frac{1}{4}$ Table A16–5
Pipe			4-inch standard weight 4-inch Schedule 40 Table A16–6
Aluminum Association channel		C	$C4 \times 1.738$ Table A17–1
Aluminum Association I-beam		I	$I8 \times 6.181$ Table A17–2

or A36. ASTM A36 should be specified for steel angles and plates. Hollow structural shapes (HSS) are most readily available in ASTM A500.

Steel Angles (L-Shapes)

Table A16–1 shows sketches of the typical shapes of steel angles having equal or unequal leg lengths. Called *L-shapes* because of the appearance of the cross section, angles are often used as tension members of trusses and towers, framing members for machine structures, lintels over windows and doors in construction, stiffeners for large plates used in housings and beams, brackets, and ledge-type supports for equipment. Some refer to these shapes as "angle iron." The standard designation takes the following form, using one example size:

$$L4 \times 3 \times \tfrac{1}{2}$$

where L refers to the L-shape
      4 is the length of the longer leg
      3 is the length of the shorter leg
      $\tfrac{1}{2}$ is the thickness of the legs
      Dimensions are in inches

American Standard Channels (C-Shapes)

See Table A16–2 for the appearance of channels and their geometric properties. Channels are used in applications similar to those described for angles. The flat web and the two flanges provide a generally stiffer shape than angles.

    The sketch at the top of the table shows that channels have tapered flanges and webs with constant thickness. The slope of the flange taper is approximately 2 inches in 12 inches, and this makes it difficult to attach other members to the flanges. Special tapered washers are available to facilitate fastening. Note the designation of the x- and y-axes in the sketch, defined with the web of the channel vertical which gives it the characteristic C-shape. This is most important when using channels as beams or columns. The x-axis is located on the horizontal axis of symmetry, while the dimension x, given in the table, locates the y-axis relative to the back of the web. The centroid is at the intersection of the x- and y-axes.

    The form of the standard designation for channels is

$$C15 \times 50$$

where C indicates that it is a standard C-shape
      15 is the nominal (and actual) depth in inches with the web vertical
      50 is the weight per unit length in lb/ft

Wide-Flange Shapes (W-Shapes)

Refer to Table A16–3, which illustrates the most common shape used for beams. W-shapes have relatively thin webs and somewhat thicker, flat flanges with constant thickness. Most of the area of the cross section is in the flanges, farthest away from the horizontal centroidal axis (x-axis), thus making the moment of inertia very high for a given amount of material.

Note that the properties of moment of inertia and section modulus are very much higher with respect to the x-axis than they are for the y-axis. Therefore, W-shapes are typically used in the orientation shown in the sketch in Table A16–3. Also, these shapes are best when used in pure bending without twisting because they are quite flexible in torsion.

The standard designation for W-shapes carries much information. Consider the following example:

$$W14 \times 43$$

where W indicates that it is a W-shape
       14 is the nominal depth in inches
       43 is the weight per unit length in lb/ft

The term depth is the standard designation for the vertical height of the cross section when placed in the orientation shown in Table A16–3. Note from the data in the table that the actual depth is often different from the nominal depth. For the W14 $\times$ 43, the actual depth is 13.66 in.

American Standard Beams (S-Shapes)

Table A16–4 shows the properties for S-shapes. Much of the discussion given for W-shapes applies to S-shapes as well. Note that, again, the weight per foot of length is included in the designation such as the S10 $\times$ 35, which weighs 35 lb/ft. For most, but not all, of the S-shapes, the actual depth is the same as the nominal depth. The flanges of the S-shapes are tapered at a slope of approximately 2 inches in 12 inches, similar to the flanges of the C-shapes. The x- and y-axes are defined as shown with the web vertical.

Often wide-flange shapes (W-shapes) are preferred over S-shapes because of their relatively wide flanges, the constant thickness of the flanges, and the generally higher section properties for a given weight and depth.

Hollow Structural Shapes (HSS Square and Rectangular)

See Table A16–5 for the appearance and properties for hollow structural shapes. These shapes are usually formed from flat sheet and welded along the length. The section properties account for the corner radii. Note the sketches showing the x- and y-axes. The standard designation takes the form

$$6 \times 4 \times \tfrac{1}{4}$$

where 6 is the depth of the longer side in inches
      4 is the width of the shorter side in inches
      $\frac{1}{4}$ is the wall thickness in inches

Square tubing and rectangular tubing are very useful in machine structures because they provide good section properties for members loaded as beams in bending and for torsional loading (twisting) because of the closed cross section. The flat sides often facilitate fastening of members together or the attachment of equipment to the structural members. Some frames are welded into an integral unit that functions as a stiff space-frame. Square tubing makes an efficient section for columns.

Pipe

Hollow circular sections, commonly called *pipe,* are very efficient for use as beams, torsion members, and columns. The placement of the material uniformly away from the center of the pipe enhances the moment of inertia for a given amount of material and gives the pipe uniform properties with respect to all axes through the center of the cross section. The closed cross-sectional shape gives it high strength and stiffness in torsion as well as in bending.

Table A16–6 gives the properties for American National Standard Schedule 40 welded and seamless wrought steel pipe. This type of pipe is often used to transport water and other fluids, but it also performs well in structural applications. Note that the actual inside and outside diameters are somewhat different from the nominal size, except for the very large sizes. Construction pipe is often called *Standard Weight Pipe,* and it has the same dimensions as the Schedule 40 pipe for sizes from 1/2 in to 10 in. Other "schedules" and "weights" of pipe are available with larger and smaller wall thicknesses.

Other hollow circular sections are commonly available that are referred to as *tubing*. These sections are available in carbon steel, alloy steel, stainless steel, aluminum, copper, brass, titanium, and other materials. See References 1, 2, 5, and 9 for a variety of types and sizes of pipe and tubing.

Aluminum Association Standard Channels and I-Beams

Tables A17–1 and A17–2 give the dimensions and section properties of channels and I-beams developed by the Aluminum Association (see Reference 1). These are extruded shapes having uniform thicknesses of the webs and flanges with generous radii where they meet. The proportions of these sections are somewhat different from those of the rolled steel sections described earlier. The extruded form offers advantages in the efficient use of material and in the joining of members. This book will use the following forms for the designation of aluminum sections:

$$C4 \times 1.738 \quad or \quad I8 \times 6.181$$

where C or I indicates the basic section shape
4 or 8 indicates the depth of the shape when in the orientation shown
1.738 or 6.181 indicates the weight per unit length in lb/ft

1–9 UNIT SYSTEMS

We will perform computations in this book by using either the U.S. Customary Unit System (inch-pound-second) or the International System (SI). Table 1–2 lists the typical units used in the study of machine design. *SI*, the abbreviation for "Le Système International d'Unités," is the standard for metric units throughout the world. (See Reference 4.) For convenience, the term *SI units* will be used instead of *metric units*.

Prefixes applied to the basic units indicate order of magnitude. Only those prefixes listed in Table 1–3, which differ by a factor of 1000, should be used in technical calculations. The final result for a quantity should be reported as a number between 0.1 and 10 000, times some multiple of 1000. Then the unit with the appropriate prefix should be specified. Table 1–4 lists examples of proper SI notation.

Sometimes you have to convert a unit from one system to another. Appendix 18 provides tables of conversion factors. Also, you should be familiar with the typical order of magnitude of the quantities encountered in machine design so that you can judge the reasonableness of design calculations (see Table 1–5 for several examples).

TABLE 1–2 Typical units used in machine design

Quantity	U.S. Customary unit	SI unit
Length or distance	inch (in)	meter (m)
	foot (ft)	millimeter (mm)
Area	square inch (in^2)	square meter (m^2) or square millimeter (mm^2)
Force	pound (lb)	newton (N)
	kip (K) (1000 lb)	($1\ N = 1\ kg \cdot m/s^2$)
Mass	slug ($lb \cdot s^2/ft$)	kilogram (kg)
Time	second (s)	second (s)
Angle	degree (°)	radian (rad) or degree (°)
Temperature	degrees Fahrenheit (°F)	degrees Celsius (°C)
Torque or moment	pound-inch (lb·in) or	newton-meter (N·m)
	pound-foot (lb·ft)	
Energy or work	pound-inch (lb·in)	joule (J)
		($1\ J = 1\ N \cdot m$)
Power	horsepower (hp)	watt (W) or kilowatts (kW)
	($1\ hp = 550\ lb \cdot ft/s$)	($1\ W = 1\ J/s = 1\ N \cdot m/s$)
Stress, pressure, or	pounds per square inch	pascal (Pa) ($1\ Pa = 1\ N/m^2$)
modulus of elasticity	(lb/in^2, or psi)	kilopascal (kPa) ($1\ kPa = 10^3\ Pa$)
	kips per square inch	megapascal (MPa) ($1\ MPa = 10^6\ Pa$)
	(K/in^2, or ksi)	gigapascal (GPa) ($1\ GPa = 10^9\ Pa$)
Section modulus	inches cubed (in^3)	meters cubed (m^3) or millimeters cubed (mm^3)
Moment of inertia	inches to the fourth	meters to the fourth power (m^4) or
	power (in^4)	millimeters to the fourth power (mm^4)
Rotational speed	revolutions per min (rpm)	radians per second (rad/s)

TABLE 1–3 Prefixes used with SI units

Prefix	SI symbol	Factor
micro-	μ	$10^{-6} = 0.000\ 001$
milli-	m	$10^{-3} = 0.001$
kilo-	k	$10^3\ \ = 1000$
mega-	M	$10^6\ \ = 1\ 000\ 000$
giga-	G	$10^9\ \ = 1\ 000\ 000\ 000$

TABLE 1–4 Quantities expressed in SI units

Computed result	Reported Result
0.001 65 m	1.65×10^{-3} m, or 1.65 mm
32 540 N	32.54×10^3 N, or 32.54 kN
1.583×10^5 W	158.3×10^3 W, or 158.3 kW;
	or $0.158\ 3 \times 10^6$ W; or 0.158 3 MW
2.07×10^{11} Pa	207×10^9 Pa, or 207 GPa

TABLE 1–5 Typical order of magnitude for commonly encountered quantities

Quantity	U.S. Customary unit	SI unit
Dimensions of a wood standard 2 × 4	1.50 in × 3.50 in	38 mm × 89 mm
Moment of inertia of a 2 × 4 (3.50-in side vertical)	5.36 in^4	$2.23 \times 10^6\ mm^4$, or $2.23 \times 10^{-6}\ m^4$
Section modulus of a 2 × 4 (3.50-in side vertical)	3.06 in^3	$5.02 \times 10^4\ mm^3$, or $5.02 \times 10^{-5}\ m^3$
Force required to lift 1.0 gal of gasoline	6.01 lb	26.7 N
Density of water	1.94 slugs/ft^3	1000 kg/m^3, or 1.0 Mg/m^3
Compressed air pressure in a factory	100 psi	690 kPa
Yield point of AISI 1040 hot-rolled steel	42 000 psi, or 42 ksi	290 MPa
Modulus of elasticity of steel	30 000 000 psi, or 30×10^6 psi	207 GPa

Example Problem 1–1   Express the diameter of a shaft in millimeters if it is measured to be 2.755 in.

Solution   Table A18 gives the conversion factor for length to be 1.00 in = 25.4 mm. Then

$$\text{Diameter} = 2.755 \text{ in} \frac{25.4 \text{ mm}}{1.00 \text{ in}} = 69.98 \text{ mm}$$

Example Problem 1–2   An electric motor is rotating at 1750 revolutions per minute (rpm). Express the speed in radians per second (rad/s).

Solution   A series of conversions is required.

$$\text{Rotational speed} = \frac{1750 \text{ rev}}{\text{min}} \frac{2\pi \text{ rad}}{\text{rev}} \frac{1 \text{ min}}{60 \text{ s}} = 183.3 \text{ rad/s}$$

1–10 DISTINCTION AMONG WEIGHT, FORCE, AND MASS

Distinction must be made among the terms *force*, *mass*, and *weight*. *Mass* is the quantity of matter in a body. A *force* is a push or pull applied to a body that results in a change in the body's motion or in some deformation of the body. Clearly these are two different physical phenomena, but the distinction is not always understood. The units for force and mass used in this text are listed in Table 1–2.

The term *weight*, as used in this book, refers to the amount of *force* required to support a body against the influence of gravity. Thus, in response to "What is the weight of 75 kg of steel?" we would use the relationship between force and mass from physics:

⇨ **Weight/Mass Relationship**

$$F = ma \quad \text{or} \quad w = mg$$

where F = force
$\quad m$ = mass
$\quad a$ = acceleration
$\quad w$ = weight
$\quad g$ = acceleration due to gravity

We will use

$$g = 32.2 \text{ ft/s}^2 \quad \text{or} \quad g = 9.81 \text{ m/s}^2$$

Then, to compute the weight,

$$w = mg = 75 \text{ kg}(9.81 \text{ m/s}^2)$$
$$w = 736 \text{ kg} \cdot \text{m/s}^2 = 736 \text{ N}$$

Remember that, as shown in Table 1-2, the newton (N) is equivalent to $1.0 \text{ kg} \cdot \text{m/s}^2$. In fact, the newton is defined as the force required to give a mass of 1.0 kg an acceleration of 1.0 m/s^2. In our example, then, we would say that the 75-kg mass of steel has a weight of 736 N.

REFERENCES

1. Aluminum Association. *Aluminum Standards and Data*. Washington, DC: Aluminum Association, 1997.

2. American Institute of Steel Construction. *Manual of Steel Construction, Load and Resistance Factor Design*. 3rd ed. Chicago: American Institute of Steel Construction, 2001.

3. American Society of Mechanical Engineers. *Integrating the Product Realization Process (PRP) into the Undergraduate Curriculum*. New York: American Society of Mechanical Engineers, 1995.

4. American Society for Testing and Materials. *IEEE/ASTM SI-10 Standard for Use of the International System of Units (SI): The Modern Metric System*. West Conshohocken, PA: American Society for Testing and Materials, 2000.

5. Avallone, Eugene A., and Theodore Baumeister III, eds. *Marks' Standard Handbook for Mechanical Engineers*. 10th ed. New York: McGraw-Hill, 1996.

6. Dym, Clive L., and Patrick Little. *Engineering Design: A Project-Based Introduction*. New York: John Wiley & Sons, 2000.

7. Ertas, Atila, and Jesse C. Jones. *The Engineering Design Process*. New York: John Wiley & Sons, 1993. Discussion of the design process from definition of design objectives through product certification and manufacture.

8. Hauser, J., and D. Clausing. "The House of Quality." *Harvard Business Review* (May–June 1988): 63–73. Discusses Quality Function Deployment.

9. Mott, Robert L. *Applied Fluid Mechanics*. 5th ed. Upper Saddle River, NJ: Prentice Hall, 2000.

10. National Research Council. *Improving Engineering Design: Designing for Competitive Advantage*. Washington, DC: National Academy Press, 1991. Describes the Product Realization Process (PRP).

11. Oberg, Erik, F. D. Jones, H. L. Horton, and H. H. Ryffell. *Machinery's Handbook*. 26th ed. New York: Industrial Press, 2000.

12. Pahl, G., and W. Beitz. *Engineering Design: A Systematic Approach*. 2nd ed. London: Springer-Verlag, 1996.

13. Pugh, Stuart. *Total Design: Integrated Methods for Successful Product Engineering*. Reading, MA: Addison-Wesley, 1991.

14. Suh, Nam Pyo. *Axiomatic Design: Advances and Applications*. New York: Oxford University Press, 2001.

15. Suh, Nam Pyo. *The Principles of Design*. New York: Oxford University Press, 1990.

16. Ullman, David G. *The Mechanical Design Process*. 2d ed. New York: McGraw-Hill, 1997.

INTERNET SITES FOR GENERAL MECHANICAL DESIGN

Included here are Internet sites that can be used in many of the chapters of this book and in general design practice to identify commercial suppliers of machine elements and standards for design or to perform stress analyses. Later chapters include sites specific to the topics covered there.

1. **American National Standards Institute (ANSI)** *www.ansi.org* A private, nonprofit organization that administers and coordinates the U.S. voluntary standardization and conformity assessment system.

2. **Global Engineering Documents** *http://global.ihs.com* A searchable database of standards and publications offered by many standards-developing organizations such as ASME, ASTM, and ISO.

3. **GlobalSpec** *www.globalspec.com* A searchable database of a wide variety of technical products and services that provides for searching by technical specifications, access to supplier information, and comparison of suppliers for a given product. The Mechanical Components category includes many of the topics addressed in this book.

4. **MDSOLIDS** *www.mdsolids.com* Educational software for strength of materials topics, including beams, flexure, torsion members, columns, axial structures, statically indeterminate structures, trusses, section properties, and Mohr's circle analysis. This software may serve as a review tool for the prerequisite knowledge needed in this book.

5. **StressAlyzer** *www.me.cmu.edu* A highly interactive problem-solving package for topics in strength of materials, including axial loading, torsional loading, shear force and bending moment diagrams, beam deflections, Mohr's circle (stress transformations), and load and stress calculations in three dimensions.

6. **Orand Systems–Beam 2D** *www.orandsystems.com* Stress and deflection analysis software package providing solutions for beams under static loading. Numerous beam cross sections, materials, loading patterns, and support conditions can be input. Output includes bending stress, shear stress, deflection, and slope for the beam.

7. **Power Transmission Home Page** *www.powertransmission.com* Clearinghouse on the Internet for buyers, users, and sellers of power transmission products and services. Included are gears, gear drives, belt drives, chain drives, bearings, clutches, brakes, and many other machine elements covered in this book.

PROBLEMS

Functions and Design Requirements

For the devices described in Problems 1–14, write a set of functions and design requirements in a similar manner to those in Section 1–4. You or your instructor may add more specific information to the general descriptions given.

1. The hood latch for an automobile

2. A hydraulic jack used for car repair

3. A portable crane to be used in small garages and homes

4. A machine to crush soft-drink or beer cans

5. An automatic transfer device for a production line

6. A device to raise a 55-gallon (gal) drum of bulk materials and dump the contents into a hopper

7. A paper feed device for a copier

8. A conveyor to elevate and load gravel into a truck

9. A crane to lift building materials from the ground to the top of a building during construction

10. A machine to insert toothpaste tubes into cartons

11. A machine to insert 24 cartons of toothpaste into a shipping container

12. A gripper for a robot to grasp a spare tire assembly and insert it into the trunk of an automobile on an assembly line

13. A table for positioning a weldment in relation to a robotic welder

14. A garage door opener

Units and Conversions

For Problems 15–28, perform the indicated conversion of units. (Refer to Appendix 18 for conversion factors.) Express the results with the appropriate prefix as illustrated in Tables 1–3 and 1–4.

15. Convert a shaft diameter of 1.75 in to mm.

16. Convert the length of a conveyor from 46 ft to meters.

17. Convert the torque developed by a motor of 12 550 lb·in to N·m.

18. A wide-flange steel-beam shape, W12 × 14, has a cross-sectional area of 4.12 in^2. Convert the area to mm^2.

19. The W12 × 14 beam shape has a section modulus of 14.8 in^3. Convert it to mm^3.

20. The W12 × 14 beam shape has a moment of inertia of 88.0 in^4. Convert it to mm^4.

21. What standard steel equal leg angle would have a cross-sectional area closest to (but greater than) 750 mm^2? See Table A16–1.

22. An electric motor is rated at 7.5 hp. What is its rating in watts (W)?

23. A vendor lists the ultimate tensile strength of a steel to be 127 000 psi. Compute the strength in MPa.

24. Compute the weight of a steel shaft, 35.0 mm in diameter and 675 mm long. (See Appendix 3 for the density of steel.)

25. A torsional spring requires a torque of 180 lb·in to rotate it 35°. Convert the torque to N·m and the rotation to radians. If the *scale of the spring* is defined as the applied torque per unit of angular rotation, compute the spring scale in both unit systems.

26. To compute the energy used by a motor, multiply the power that it draws by the time of operation. Consider a motor that draws 12.5 hp for 16 h/day, five days per week. Compute the energy used by the motor for one year. Express the result in ft·lb and W·h.

27. One unit used for fluid viscosity in Chapter 16 of this book is the *reyn*, defined as 1.0 lb·s/in^2. If a lubricating oil has a viscosity of 3.75 reyn, convert the viscosity to the standard units in the U.S. Customary System (lb·s/ft^2) and in the SI (N·s/m^2).

28. The life of a bearing supporting a rotating shaft is expressed in number of revolutions. Compute the life of a bearing that rotates 1750 rpm continuously for 24 h/day for five years.

2

Materials in Mechanical Design

The Big Picture

You Are the Designer

2–1 Objectives of This Chapter

2–2 Properties of Materials

2–3 Classification of Metals and Alloys

2–4 Variability of Material Properties Data

2–5 Carbon and Alloy Steel

2–6 Conditions for Steels and Heat Treatment

2–7 Stainless Steels

2–8 Structural Steel

2–9 Tool Steels

2–10 Cast Iron

2–11 Powdered Metals

2–12 Aluminum

2–13 Zinc Alloys

2–14 Titanium

2–15 Copper, Brass, and Bronze

2–16 Nickel-based Alloys

2–17 Plastics

2–18 Composite Materials

2–19 Materials Selection

Materials in Mechanical Design

Discussion Map

☐ You must understand the behavior of materials to make good design decisions and to communicate with suppliers and manufacturing staff.

Discover
Examine consumer products, industrial machinery, automobiles, and construction machinery.
What materials are used for the various parts?
Why do you think those materials were specified?
How were they processed?
What material properties were important to the decisions to use particular materials?
Examine the Appendices tables, and refer to them later as you read about specific materials.

This chapter summarizes the design properties of a variety of materials. The Appendices include data for many examples of these materials in many conditions.

It is the designer's responsibility to specify suitable materials for each component of a mechanical device. Your initial efforts in specifying a material for a particular component of a mechanical design should be directed to the basic kind of material to be used. Keep an open mind until you have specified the functions of the component, the kinds and magnitudes of loads it will carry, and the environment in which it must operate. Your selection of a material must consider its physical and mechanical properties and match them to the expectations placed on it. First consider the following classes of materials:

Metals and their alloys	Plastics	Composites
Elastomers	Woods	Ceramics and glasses

Each of these classes contains a large number of specific materials covering a wide range of actual properties. However, you probably know from your experience the general behavior of each kind and have some feel for the applications in which each is typically used. Most of the applications considered in the study of design of machine elements in this book use metal alloys, plastics, and composites.

Satisfactory performance of machine components and systems depends greatly on the materials that the designer specifies. As a designer, you must understand how materials behave, what properties of the material affect the performance of the parts, and how you should interpret the large amounts of data available on material properties. Your ability to effectively communicate your specifications for materials with suppliers, purchasing agents, metallurgists, manufacturing process personnel, heat treatment personnel, plastics molders, machinists, and quality assurance specialists often has a strong influence on the success of a design.

Explore what kinds of materials are used in consumer products, industrial machinery, automobiles, construction machinery, and other devices and systems that you come into contact with each day. Make judgments about why each material was specified for a particular application. Where do you see steel being used? Contrast that usage with where aluminum or other nonferrous materials are used. How are the products produced? Can you find different parts that are machined, cast, forged, roll-formed, and welded? Why do you think those processes were specified for those particular products?

Document several applications for plastics and describe the different forms that are available and that have been made by different manufacturing processes. Which are made by plastic molding processes, vacuum forming, blow molding, and others? Can you identify parts made from composite materials that have a significant amount of high-strength fibers embedded in a plastic matrix? Check out sporting goods and parts of cars, trucks, and airplanes.

From the products that you found from the exploration outlined previously, identify the basic properties of the materials that were important to the designers: strength, rigidity (stiffness), weight (density), corrosion resistance, appearance, machinability, weldability, ease of forming, cost, and others.

This chapter focuses on material selection and the use of material property data in design decisions, rather than on the metallurgy or chemistry of the materials. One of the uses of the information in this chapter is as a glossary of terms that you can use throughout the book; important terms are given in *italic* type. Also, there are numerous references to Appendices 3 through 13, where tables of data for material properties are given. Go there now and see what kinds of data are provided. Then you can study the tables in more depth as you read the text. Note that many of the problems that you will solve in this book and the design projects that you complete will use data from these tables.

Now apply some of what you have gained from **The Big Picture** exploration to a specific design situation as outlined in **You Are the Designer**, which follows.

 You Are the Designer

You are part of a team responsible for the design of an electric lawn mower for the household market. One of your tasks is to specify suitable materials for the various components. Consider your own experience with such lawn mowers and think what materials would be used for these key components: *wheels, axles, housing*, and *blade*. What are their functions? What conditions of service will each encounter? What is one reasonable type of material for each component and what general properties should it have? How could they be manufactured? Possible answers to these questions follow.

Wheels

Function: Support the weight of the mower. Permit easy, rolling movement. Provide for mounting on an axle. Ensure safe operation on flat or sloped lawn surfaces.

Conditions of service: Must operate on grass, hard surfaces, and soft earth. Exposed to water, lawn fertilizers, and general outdoor conditions. Will carry moderate loads. Requires an attractive appearance.

One reasonable material: One-piece plastic wheel incorporating the tire, rim, and hub. Must have good strength, stiffness, toughness, and wear resistance.

Manufacturing method: Plastic injection molding

Axles

Function: Transfer the weight of mower from the housing to the wheels. Allow rotation of the wheels. Maintain location of the wheels relative to the housing.

Conditions of service: Exposure to general outdoor conditions. Moderate loads.

One possible material: Steel rod with provisions for mounting wheels and attaching to housing. Requires moderate strength, stiffness, and corrosion resistance.

Manufacturing method: Commercially available cylindrical rod. Possibly machining.

Housing

Function: Support, safely enclose, and protect operating components, including the blade and motor. Accommodate the attachment of two axles and a handle. Permit cut grass to exit the cutting area.

Conditions of service: Moderate loads and vibration due to motor. Possible shock loads from wheels. Multiple attachment points for axles, handle, and motor. Exposed to wet grass and general outdoor conditions. Requires attractive appearance.

One possible material: Heavy-duty plastic with good strength, stiffness, impact resistance, toughness, and weather resistance.

Manufacturing method: Plastic injection molding. May require machining for holes and mounting points for the motor.

Blade

Function: Cut blades of grass and weeds while rotating at high speed. Facilitate connection to motor shaft. Operate safely when foreign objects are encountered, such as stones, sticks, or metal pieces.

Conditions of service: Normally moderate loads. Occasional shock and impact loads. Must be capable of sharpening a portion of the blade to ensure clean cutting of grass. Maintain sharpness for reasonable time during use.

One possible material: Steel with high strength, stiffness, impact resistance, toughness, and corrosion resistance.

Manufacturing method: Stamping from flat steel strip. Machining and/or grinding for cutting edge.

This simplified example of the material selection process should help you to understand the importance of the information provided in this chapter about the behavior of materials commonly used in the design of machine elements. A more comprehensive discussion of material selection occurs at the end of the chapter.

2–1 OBJECTIVES OF THIS CHAPTER

After completing this chapter, you will be able to:

1. State the types of material properties that are important to the design of mechanical devices and systems.

2. Define the following terms: *tensile strength, yield strength, proportional limit, elastic limit, modulus of elasticity in tension, ductility and percent elongation, shear strength, Poisson's ratio, modulus of elasticity in shear, hardness, machinability, impact strength, density, coefficient of thermal expansion, thermal conductivity,* and *electrical resistivity.*

3. Describe the nature of *carbon and alloy steels,* the number-designation system for steels, and the effect of several kinds of alloying elements on the properties of steels.

4. Describe the manner of designating the condition and heat treatment of steels, including *hot rolling, cold drawing, annealing, normalizing, through-hardening, tempering,* and *case hardening by flame hardening, induction hardening,* and *carburizing.*

5. Describe *stainless steels* and recognize many of the types that are commercially available.

6. Describe *structural steels* and recognize many of their designations and uses.

7. Describe *cast irons* and several kinds of *gray iron, ductile iron,* and *malleable iron.*

8. Describe *powdered metals* and their properties and uses.

9. Describe several types of *tool steels* and *carbides* and their typical uses.

10. Describe *aluminum alloys* and their conditions, such as *strain hardening* and *heat treatment.*

11. Describe the nature and typical properties of *zinc, titanium,* and *bronze.*

12. Describe several types of *plastics,* both *thermosetting* and *thermoplastic,* and their typical properties and uses.

13. Describe several kinds of *composite materials* and their typical properties and uses.

14. Implement a rational material selection process.

2–2 PROPERTIES OF MATERIALS

Machine elements are very often made from one of the metals or metal alloys such as steel, aluminum, cast iron, zinc, titanium, or bronze. This section describes the important properties of materials as they affect mechanical design.

Strength, elastic, and ductility properties for metals, plastics, and other types of materials are usually determined from a *tensile test* in which a sample of the material, typically in the form of a round or flat bar, is clamped between jaws and pulled slowly until it breaks in tension. The magnitude of the force on the bar and the corresponding change in length (strain) are monitored and recorded continuously during the test. Because the stress in the bar is equal to the applied force divided by the area, stress is proportional to the applied force. The data from such tensile tests are often shown on *stress-strain diagrams*, such as those shown in Figures 2–1 and 2–2. In the following paragraphs, several strength, elastic, and ductility properties of metals are defined.

Tensile Strength, s_u

The peak of the stress-strain curve is considered the *ultimate tensile strength* (s_u), sometimes called the *ultimate strength* or simply the *tensile strength*. At this point during the test, the highest *apparent stress* on a test bar of the material is measured. As shown in Figures 2–1 and 2–2, the curve appears to drop off after the peak. However, notice that the instrumentation used to create the diagrams is actually plotting *load versus deflection* rather than *true*

FIGURE 2–1 Typical stress-strain diagram for steel

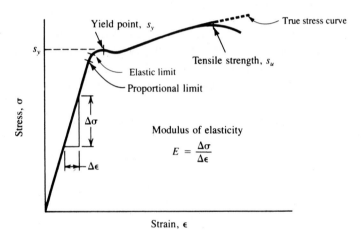

FIGURE 2–2 Typical stress-strain diagram for aluminum and other metals having no yield point

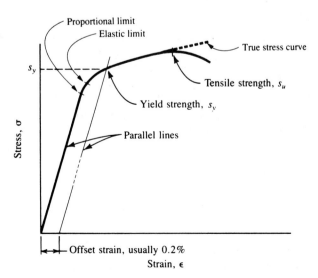

stress versus strain. The apparent stress is computed by dividing the load by the original cross-sectional area of the test bar. After the peak of the curve is reached, there is a pronounced decrease in the bar's diameter, referred to as *necking down.* Thus, the load acts over a smaller area, and the *actual stress* continues to increase until failure. It is very difficult to follow the reduction in diameter during the necking-down process, so it has become customary to use the peak of the curve as the tensile strength, although it is a more conservative value.

Yield Strength, s_y

That portion of the stress-strain diagram where there is a large increase in strain with little or no increase in stress is called the *yield strength* (s_y). This property indicates that the material has, in fact, yielded or elongated plastically, permanently, and to a large degree. If the point of yielding is quite noticeable, as it is in Figure 2–1, the property is called the *yield point* rather than the yield strength. This is typical of a plain carbon hot rolled steel.

Figure 2–2 shows the stress-strain diagram form that is typical of a nonferrous metal such as aluminum or titanium or of certain high-strength steels. Notice that there is no pronounced yield point, but the material has actually yielded at or near the stress level indicated as s_y. That point is determined by the *offset method,* in which a line is drawn parallel to the straight-line portion of the curve and is offset to the right by a set amount, usually 0.20% strain (0.002 in/in). The intersection of this line and the stress-strain curve defines the material's yield strength. In this book, the term *yield strength* will be used for s_y, regardless of whether the material exhibits a true yield point or whether the offset method is used.

Proportional Limit

That point on the stress-strain curve where it deviates from a straight line is called the *proportional limit.* That is, at or above that stress value, stress is no longer proportional to strain. Below the proportional limit, Hooke's law applies: Stress is proportional to strain. In mechanical design, materials are rarely used at stresses above the proportional limit.

Elastic Limit

At some point, called the *elastic limit,* a material experiences some amount of plastic strain and thus will not return to its original shape after release of the load. Below that level, the material behaves completely elastically. The proportional limit and the elastic limit lie quite close to the yield strength. Because they are difficult to determine, they are rarely reported.

Modulus of Elasticity in Tension, *E*

For the part of the stress-strain diagram that is straight, stress is proportional to strain, and the value of *E, the modulus of elasticity,* is the constant of proportionality. That is,

Modulus of Elasticity in Tension

$$E = \frac{\text{stress}}{\text{strain}} = \frac{\sigma}{\epsilon} \qquad (2\text{–}1)$$

This is the slope of the straight-line portion of the diagram. The modulus of elasticity indicates the stiffness of the material, or its resistance to deformation.

FIGURE 2–3
Measurement of
percent elongation

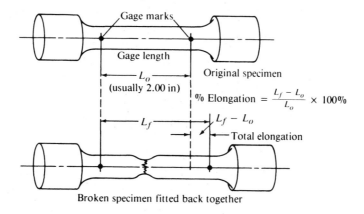

Ductility and Percent Elongation

Ductility is the degree to which a material will deform before ultimate fracture. The opposite of ductility is *brittleness*. When ductile materials are used in machine members, impending failure is detected easily, and sudden failure is unlikely. Also, ductile materials normally resist the repeated loads on machine elements better than brittle materials.

The usual measure of ductility is the *percent elongation* of the material after fracture in a standard tensile test. Figure 2–3 shows a typical standard tensile specimen before and after the test. Before the test, gage marks are placed on the bar, usually 2.00 in apart. Then, after the bar is broken, the two parts are fitted back together, and the final length between the gage marks is measured. The percent elongation is the difference between the final length and the original length divided by the original length, converted to a percentage. That is,

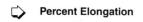 **Percent Elongation**

$$\text{percent elongation} = \frac{L_f - L_o}{L_o} \times 100\ \% \qquad (2\text{–}2)$$

The percent elongation is assumed to be based on a gage length of 2.00 in unless some other gage length is specifically indicated. Tests of structural steels often use a gage length of 8.00 in.

Theoretically, a material is considered ductile if its percent elongation is greater than 5% (lower values indicate brittleness). For practical reasons, it is advisable to use a material with a value of 12% or higher for machine members subject to repeated loads or shock or impact.

Percent reduction in area is another indication of ductility. To find this value, compare the original cross-sectional area with the final area at the break for the tensile test specimen.

Shear Strength, s_{ys} and s_{us}

Both the yield strength and the ultimate strength in shear (s_{ys} and s_{us}, respectively) are important properties of materials. Unfortunately, these values are seldom reported. We will use the following estimates:

 Estimates for
s_{ys} and s_{us}

$$s_{ys} = s_y/2 = 0.50\ s_y = \text{yield strength in shear} \qquad (2\text{–}3)$$

$$s_{us} = 0.75 s_u = \text{ultimate strength in shear} \qquad (2\text{–}4)$$

FIGURE 2–4
Illustration of Poisson's
ratio for an element in
tension

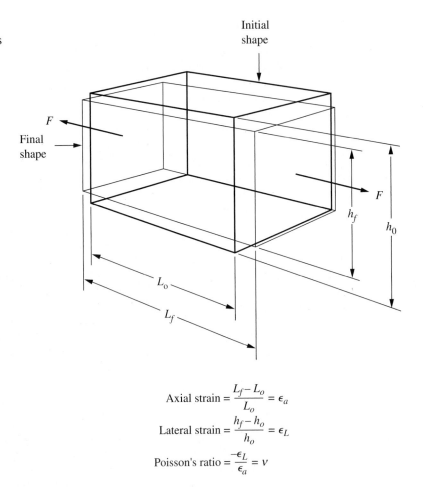

$$\text{Axial strain} = \frac{L_f - L_o}{L_o} = \epsilon_a$$

$$\text{Lateral strain} = \frac{h_f - h_o}{h_o} = \epsilon_L$$

$$\text{Poisson's ratio} = \frac{-\epsilon_L}{\epsilon_a} = \nu$$

Poisson's Ratio, ν

When a material is subjected to a tensile strain, there is a simultaneous shortening of the cross-sectional dimensions perpendicular to the direction of the tensile strain. The ratio of the shortening strain to the tensile strain is called *Poisson's ratio*, usually denoted by ν, the Greek letter nu. (The Greek letter mu, μ, is sometimes used for this ratio.) Poisson's ratio is illustrated in Figure 2–4. Typical ranges of values for Poisson's ratio are 0.25–0.27 for cast iron, 0.27–0.30 for steel, and 0.30–0.33 for aluminum and titanium.

Modulus of Elasticity in Shear, G

The *modulus of elasticity in shear* (G) is the ratio of shearing stress to shearing strain. This property indicates a material's stiffness under shear loading—that is, the resistance to shear deformation. There is a simple relationship between E, G, and Poisson's ratio:

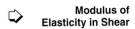 **Modulus of
Elasticity in Shear**

$$G = \frac{E}{2(1 + \nu)}$$

(2–5)

This equation is valid within the elastic range of the material.

Flexural Modulus

Another stiffness measure often reported, particularly for plastics, is called the *flexural modulus*, or *modulus of elasticity in flexure.* As the name implies, a specimen of the material is loaded as a beam in flexure (bending) with data taken and plotted for load versus deflection. From these data and from knowledge of the geometry of the specimen, stress and strain can be computed. The ratio of stress to strain is a measure of the flexural modulus. ASTM standard D 790[1] defines the complete method. Note that the values are significantly different from the tensile modulus because the stress pattern in the specimen is a combination of tension and compression. The data are useful for comparing the stiffness of different materials when a load-carrying part is subjected to bending in service.

Hardness

The resistance of a material to indentation by a penetrator is an indication of its *hardness.* Several types of devices, procedures, and penetrators measure hardness; the Brinell hardness tester and the Rockwell hardness tester are most frequently used for machine elements. For steels, the Brinell hardness tester employs a hardened steel ball 10 mm in diameter as the penetrator under a load of 3000-kg force. The load causes a permanent indentation in the test material, and the diameter of the indentation is related to the Brinell hardness number, which is abbreviated BHN or HB. The actual quantity being measured is the load divided by the contact area of the indentation. For steels, the value of HB ranges from approximately 100 for an annealed, low-carbon steel to more than 700 for high-strength, high-alloy steels in the as-quenched condition. In the high ranges, above HB 500, the penetrator is sometimes made of tungsten carbide rather than steel. For softer metals, a 500-kg load is used.

The Rockwell hardness tester uses a hardened steel ball with a 1/16-in diameter under a load of 100-kg force for softer metals, and the resulting hardness is listed as Rockwell B, R_B, or HRB. For harder metals, such as heat-treated alloy steels, the Rockwell C scale is used. A load of 150-kg force is placed on a diamond penetrator (a *brale* penetrator) made in a sphero-conical shape. Rockwell C hardness is sometimes referred to as R_C or HRC. Many other Rockwell scales are used.

The Brinell and Rockwell methods are based on different parameters and lead to quite different numbers. However, since they both measure hardness, there is a correlation between them, as noted in Appendix 19. It is also important to note that, especially for highly hardenable alloy steels, there is a nearly linear relationship between the Brinell hardness number and the tensile strength of the steel, according to the equation

Approximate Relationship between Hardness and Strength for Steel

⇨

$$0.50(\text{HB}) = \text{approximate tensile strength (ksi)} \tag{2–6}$$

This relationship is shown in Figure 2–5.

To compare the hardness scales with the tensile strength, consider Table 2–1. Note that there is some overlap between the HRB and HRC scales. Normally, HRB is used for the softer metals and ranges from approximately 60 to 100, whereas HRC is used for harder metals and ranges from 20 to 65. Using HRB numbers above 100 or HRC numbers below 20 is not recommended. Those shown in Table 2–1 are for comparison purposes only.

Hardness in a steel indicates wear resistance as well as strength. Wear resistance will be discussed in later chapters, particularly with regard to gear teeth.

---

[1] ASTM International. *Standard Test Method for Flexural Properties of Unreinforced and Reinforced Plastics and Electrical Insulating Materials, Standard D790.* West Conshohocken, PA: ASTM International, 2003.

FIGURE 2–5
Hardness conversions

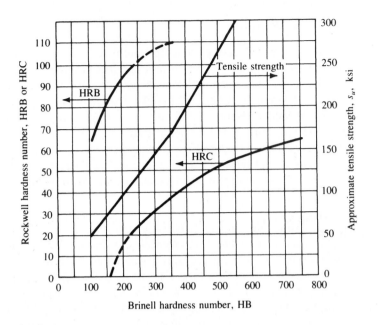

TABLE 2–1   Comparison of hardness scales with tensile strength

Material and condition	Hardness			Tensile strength	
	HB	HRB	HRC	ksi	MPa
1020 annealed	121	70		60	414
1040 hot-rolled	144	79		72	496
4140 annealed	197	93	13	95	655
4140 OQT 1000	341	109	37	168	1160
4140 OQT 700	461		49	231	1590

Machinability

Machinability is related to the ease with which a material can be machined to a good surface finish with reasonable tool life. Production rates are directly affected by machinability. It is difficult to define measurable properties related to machinability, so machinability is usually reported in comparative terms, relating the performance of a given material with some standard.

Toughness, Impact Energy

Toughness is the ability of a material to absorb applied energy without failure. Parts subjected to suddenly applied loads, shock, or impact need a high level of toughness. Several methods are used to measure the amount of energy required to break a particular specimen made from a material of interest. The energy absorption value from such tests is often called *impact energy* or *impact resistance*. However, it is important to note that the actual value is highly dependent on the nature of the test sample, particularly its geometry. It is not possible to use the test results in a quantitative way when making design calculations. Rather, the impact energy for several candidate materials for a particular application can

FIGURE 2–6 Impact testing using Charpy and Izod methods

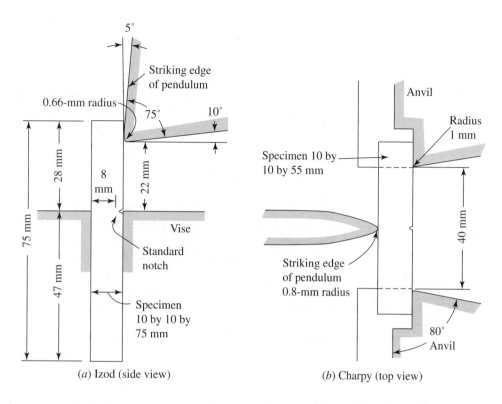

(a) Izod (side view)   (b) Charpy (top view)

be compared with each other as a qualitative indication of their toughness. The final design should be tested under real service conditions to verify its ability to survive safely during expected use.

For metals and plastics, two methods of determining impact energy, *Izod* and *Charpy*, are popular, with data often reported in the literature from vendors of the material. Figure 2–6 shows sketches of the dimensions of standard specimens and the manner of loading. In each method, a pendulum with a heavy mass carrying a specially designed striker is allowed to fall from a known height. The striker contacts the specimen with a high velocity at the bottom of the pendulum's arc; therefore, the pendulum possesses a known amount of kinetic energy. The specimen is typically broken during the test, taking some of the energy from the pendulum but allowing it to pass through the test area. The testing machine is configured to measure the final height to which the pendulum swings and to indicate the amount of energy removed. That value is reported in energy units of J (Joules or N·m) or ft·lb. Some highly ductile metals and many plastics do not break during the test, and the result is then reported as *No Break*.

The standard *Izod* test employs a square specimen with a V-shaped notch carefully machined 2.0 mm (0.079 in) deep according to specifications in ASTM standard D 256.[2] The specimen is clamped in a special vise with the notch aligned with the top edge of the vise. The striker contacts the specimen at a height of 22 mm above the notch, loading it as a cantilever in bending. When used for plastics, the width dimension can be different from that shown in Figure 2–6. This obviously changes the total amount of energy that the specimen will absorb during fracture. Therefore, the data for impact energy are divided by the

[2] ASTM International. *Standard Test Methods for Determining the Izod Pendulum Impact Resistance of Plastics, Standard D256.* West Conshohocken, PA: ASTM International, 2003.

actual width of the specimen, and the results are reported in units of N·m/m or ft·lb/in. Also, some vendors and customers may agree to test the material with the notch facing away from the striker rather than toward it as shown in Figure 2–6. This gives a measure of the material's impact energy with less influence from the notch.

The *Charpy* test also uses a square specimen with a 2.0 mm (0.079 in) deep notch, but it is centered along the length. The specimen is placed against a rigid anvil without being clamped. See ASTM standard A 370[3] for the specific geometry and testing procedure. The notch faces away from the place where the striker contacts the specimen. The loading can be described as the bending of a simply supported beam. The Charpy test is most often used for testing metals.

Another impact testing method used for some plastics, composites, and completed products is the *drop-weight* tester. Here a known mass is elevated vertically above the test specimen to a specified height. Thus, it has a known amount of potential energy. Allowing the mass to fall freely imparts a predictable amount of kinetic energy to the specimen clamped to a rigid base. The initial energy, the manner of support, the specimen geometry, and the shape of the striker (called a *tup*) are critical to the results found. One standard method, described in ASTM D 3763[4], employs a spherical tup with a diameter of 12.7 mm (0.50 in). The tup usually pierces the specimen. The apparatus is typically equipped with sensors that measure and plot the load versus deflection characteristics dynamically, giving the designer much information about how the material behaves during an impact event. Summary data reported typically include maximum load, deflection of the specimen at the point of maximum load, and the energy dissipated up to the maximum load point. The energy is calculated by determining the area under the load-deflection diagram. The appearance of the test specimen is also described, indicating whether fracture occurred and whether it was a ductile or brittle fracture.

Fatigue Strength or Endurance Strength

Parts subjected to repeated applications of loads or to stress conditions that vary with time over several thousands or millions of cycles fail because of the phenomenon of *fatigue*. Materials are tested under controlled cyclic loading to determine their ability to resist such repeated loads. The resulting data are reported as the *fatigue strength*, also called the *endurance strength* of the material. (See Chapter 5.)

Creep

When materials are subjected to high loads continuously, they may experience progressive elongation over time. This phenomenon, called *creep*, should be considered for metals operating at high temperatures. You should check for creep when the operating temperature of a loaded metal member exceeds approximately 0.3 (T_m) where T_m is the melting temperature expressed as an absolute temperature. (See Reference 22.) Creep can be important for critical members in internal combustion engines, furnaces, steam turbines, gas turbines, nuclear reactors, or rocket engines. The stress can be tension, compression, flexure, or shear. (See Reference 8.)

Figure 2–7 shows the typical behavior of metals that creep. The vertical axis is the creep strain, in units such as in/in or mm/mm, over that which occurs initially as the load

---

[3] ASTM International. *Standard Test Methods and Definitions for Mechanical Testing of Steel Products, Standard A370.* West Conshohocken, PA: ASTM International, 2003.

[4] ASTM International. *Standard Test Methods for High Speed Puncture of Plastics Using Load and Displacement Sensors, Standard D3763.* West Conshohocken, PA: ASTM International, 2003.

FIGURE 2–7
Typical creep behavior

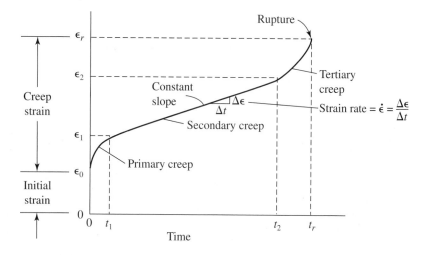

FIGURE 2–8

Example of stress versus strain as a function of time for nylon 66 plastic at 23°C (73°F) (DuPont Polymers, Wilmington, DE)

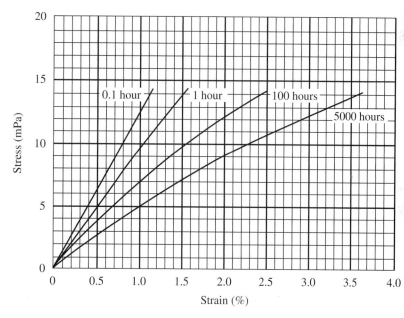

is applied. The horizontal axis is time, typically measured in hours because creep develops slowly over a long term. During the primary portion of the creep strain versus time curve, the rate of increase in strain initially rises with a rather steep slope that then decreases. The slope is constant (straight line) during the secondary portion of the curve. Then the slope increases in the tertiary portion that precedes the ultimate fracture of the material.

Creep is measured by subjecting a specimen to a known steady load, possibly through application of a dead weight, while the specimen is heated and maintained at a uniform temperature. Data for strain versus time are taken at least into the secondary creep stage and possibly all the way to fracture to determine the creep rupture strain. Testing over a range of temperatures gives a family of curves that are useful for design.

Creep can occur for many plastics even at or near room temperature. Figure 2–8 shows one way that creep data are displayed for plastic materials. (See Reference 8.) It is a graph of applied stress versus strain in the member with data shown for a specific temperature of the specimen. The curves show the amount of strain that would be developed within the specified times at increasing stress levels. For example, if this material were subjected to a constant stress of 5.0 MPa for 5000 hours, the total strain would be 1.0%. That is, the specimen would elongate by an amount 0.01 times the original length. If the stress were 10.0 MPa for 5000 hours, the total strain would be approximately 2.25%. The designer must take this creep strain into account to ensure that the product performs satisfactorily over time.

Example Problem 2–1   A solid circular bar has a diameter of 5.0 mm and a length of 250 mm. It is made from nylon 66 plastic and subjected to a steady tensile load of 240 N. Compute the elongation of the bar immediately after the load is applied and after 5000 hr (approximately seven months). See Appendix 13 and Figure 2–8 for properties of the nylon.

Solution   The stress and deflection immediately after loading will first be computed using fundamental equations of strength of materials:

$$\sigma = F/A \text{ and } \delta = FL/EA$$

See Chapter 3 for a review of strength of materials.

Then creep data from Figure 2–8 will be applied to determine the elongation after 5000 hr.

Results   *Stress:*
The cross-sectional area of the bar is

$$A = \pi D^2/4 = \pi(5.0 \text{ mm})^2/4 = 19.63 \text{ mm}^2$$

$$\sigma = \frac{F}{A} = \frac{240 \text{ N}}{19.63 \text{ mm}^2} = 12.2 \text{ N/mm}^2 = 12.2 \text{ MPa}$$

Appendix 13 lists the tensile strength for nylon 66 to be 83 MPa. Therefore, the rod is safe from fracture.

Elongation:
The tensile modulus of elasticity for nylon 66 is found from Appendix 13 to be $E = 2900$ MPa. Then the initial elongation is,

$$\delta = \frac{FL}{EA} = \frac{(240 \text{ N}) (250 \text{ mm})}{(2900 \text{ N/mm}^2) (19.63 \text{ mm}^2)} = 1.054 \text{ mm}$$

Creep:
Referring to Figure 2–8 we find that when a tensile stress of 12.2 MPa is applied to the nylon 66 plastic for 5000 hr, a total strain of approximately 2.95% occurs. This can be expressed as

$$\epsilon = 2.95\% = 0.0295 \text{ mm/mm} = \delta/L$$

Then,

$$\delta = \epsilon L = (0.0295 \text{ mm/mm}) (250 \text{mm}) = 7.375 \text{ mm}$$

Comment   This is approximately seven times as much deformation as originally experienced when the load was applied. So designing with the reported value of modulus of elasticity is not appropriate when stresses are applied continuously for a long time. We can now compute an apparent modulus of elasticity, E_{app}, for this material at the 5000 hr service life.

$$E_{app} = \sigma/\epsilon = (12.2 \text{ MPa})/(0.0295 \text{ mm/mm}) = 414 \text{ MPa}$$

Relaxation

A phenomenon related to creep occurs when a member under stress is captured under load, giving it a certain fixed length and a fixed strain. Over time, the stress in the member would decrease, exhibiting a behavior called *relaxation*. This is important in such applications as clamped joints, press-fit parts, and springs installed with a fixed deflection. Figure 2–9 shows the comparison between creep and relaxation. For stresses below approximately 1/3 of the ultimate tensile strength of the material at any temperature, the apparent modulus in either creep or relaxation at any time of loading may be considered similar for engineering purposes. Furthermore, values for apparent modulus are the same for tension, compression, or flexure. (See Reference 8.) Analysis of relaxation is complicated by the fact that as the stress decreases, the rate of creep also decreases. Additional material data beyond that typically reported would be required to accurately predict the amount of relaxation at any given time. Testing under realistic conditions is recommended.

Physical Properties

Here we will discuss density, coefficient of thermal expansion, thermal conductivity, and electrical resistivity.

Density.   *Density* is defined as the mass per unit volume of a material. Its usual units are kg/m^3 in the SI and lb/in^3 in the U.S. Customary Unit System, where the pound unit is taken to be pounds-mass. The Greek letter rho (ρ) is the symbol for density.

FIGURE 2–9
Comparison of creep and relaxation (DuPont Polymers, Wilmington, DE)

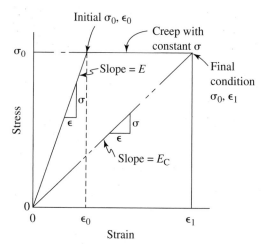

$E = \dfrac{\sigma_0}{\epsilon_0} = $ Tensile modulus

$E_C = \dfrac{\sigma_0}{\epsilon_1} = $ Creep modulus

(*a*) Creep behavior

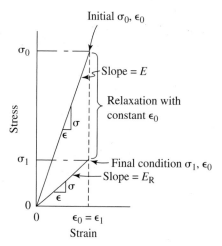

$E = \dfrac{\sigma_0}{\epsilon_0} = $ Tensile modulus

$E_R = \dfrac{\sigma_1}{\epsilon_0} = $ Relaxation modulus

(*b*) Relaxation behavior

In some applications, the term *specific weight* or *weight density* is used to indicate the weight per unit volume of a material. Typical units are N/m^3 in the SI and lb/in^3 in the U.S. Customary Unit System, where the pound is taken to be pounds-force. The Greek letter gamma (γ) is the symbol for specific weight.

Coefficient of Thermal Expansion.   The *coefficient of thermal expansion* is a measure of the change in length of a material subjected to a change in temperature. It is defined by the relation

⇨ Coefficient of Thermal Expansion

$$\alpha = \frac{\text{change in length}}{L_o\,(\Delta T)} = \frac{\text{strain}}{(\Delta T)} = \frac{\epsilon}{(\Delta T)} \qquad (2\text{–}7)$$

where L_o = original length
ΔT = change in temperature

Virtually all metals and plastics expand with increasing temperature, but different materials expand at different rates. For machines and structures containing parts of more than one material, the different rates can have a significant effect on the performance of the assembly and on the stresses produced.

Thermal Conductivity.   *Thermal conductivity* is the property of a material that indicates its ability to transfer heat. Where machine elements operate in hot environments or where significant internal heat is generated, the ability of the elements or of the machine's housing to transfer heat away can affect machine performance. For example, wormgear speed reducers typically generate frictional heat due to the rubbing contact between the worm and the wormgear teeth. If not adequately transferred, heat causes the lubricant to lose its effectiveness, allowing rapid gear-tooth wear.

Electrical Resistivity.   For machine elements that conduct electricity while carrying loads, the electrical resistivity of the material is as important as its strength. *Electrical resistivity* is a measure of the resistance offered by a given thickness of a material; it is measured in ohm-centimeters $(\Omega \cdot cm)$. *Electrical conductivity*, a measure of the capacity of a material to conduct electric current, is sometimes used instead of resistivity. It is often reported as a percentage of the conductivity of a reference material, usually the International Annealed Copper Standard.

2–3 CLASSIFICATION OF METALS AND ALLOYS

Various industry associations take responsibility for setting standards for the classification of metals and alloys. Each has its own numbering system, convenient to the particular metal covered by the standard. But this leads to confusion at times when there is overlap between two or more standards and when widely different schemes are used to denote the metals. Order has been brought to the classification of metals by the use of the Unified Numbering Systems (UNS) as defined in the Standard E 527-83 (Reapproved 1997), *Standard Practice for Numbering Metals and Alloys (UNS)*, by the American Society for Testing and Materials, or ASTM. (See References 12, 13) Besides listing materials under the control of ASTM itself, the UNS coordinates designations of the following:

The Aluminum Association (AA)

The American Iron and Steel Institute (AISI)

The Copper Development Association (CDA)

The Society of Automotive Engineers (SAE)

TABLE 2–2 Unified numbering system (UNS)

Number series	Types of metals and alloys	Responsible organization
Nonferrous metals and alloys		
A00001–A99999	Aluminum and aluminum alloys	AA
C00001–C99999	Copper and copper alloys	CDA
E00001–E99999	Rare earth metals and alloys	ASTM
L00001–L99999	Low-melting metals and alloys	ASTM
M00001–M99999	Miscellaneous nonferrous metals and alloys	ASTM
N00001–N99999	Nickel and nickel alloys	SAE
P00001–P99999	Precious metals and alloys	ASTM
R00001–R99999	Reactive and refractory metals and alloys	SAE
Z00001–Z99999	Zinc and zinc alloys	ASTM
Ferrous metals and alloys		
D00001–D99999	Steels; mechanical properties specified	SAE
F00001–F99999	Cast irons and cast steels	ASTM
G00001–G99999	Carbon and alloy steels (includes former SAE carbon and alloy steels)	AISI
H00001–H99999	H-steels; specified hardenability	AISI
J00001–J99999	Cast steels (except tool steels)	ASTM
K00001–K99999	Miscellaneous steels and ferrous alloys	ASTM
S00001–S99999	Heat- and corrosion-resistant (stainless) steels	ASTM
T00001–T99999	Tool steels	AISI

The primary series of numbers within UNS are listed in Table 2–2, along with the organization having responsibility for assigning numbers within each series.

Many alloys within the UNS retain the familiar numbers from the systems used for many years by the various associations as a *part* of the UNS number. Examples are shown in Section 2–5 for carbon and alloy steel. Also, the former designations remain widely used. For these reasons, this book will use the four-digit designation system of the AISI as described in Section 2–5 for most machine steels. Many of the designations of the SAE use the same four numbers. We will also use the designation systems of the ASTM when referring to structural steels and cast irons.

2–4 VARIABILITY OF MATERIAL PROPERTIES DATA

Tables of data such as those shown in Appendices 3 through 13 normally report single values for the strength, the modulus of elasticity (stiffness), or the percent elongation (ductility) of a particular material at a particular condition created by heat treatment or by the manner in which it was formed. It is important for you to understand the limitations of such data in making design decisions. You should seek information about the bases for the reported data.

Some tables of data report *guaranteed minimum values* for tensile strength, yield strength, and other values. This might be the case when you are using data obtained from a particular supplier. With such data, you should feel confident that the material that actually goes into your product has at least the reported strength. The supplier should be able to provide actual test data and statistical analyses used to determine the reported minimum strengths. Alternatively, you could arrange to have the actual materials to be used in a project tested to determine their minimum strength values. Such tests are costly, but they may be justified in critical designs.

Other tables of data report *typical values* for material properties. Thus, most batches of material (greater than 50%) delivered will have the stated values or greater. However,

FIGURE 2–10
Normal statistical
distribution of material
strength

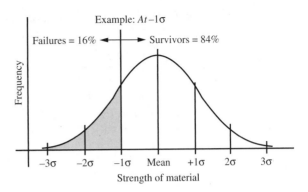

Assuming normal distribution of strength:

Stress level	% Surviving
Mean	50%
-1σ	84%
-2σ	97%
-3σ	99.8%
-4σ	99.99%

about 50% will have lower values, and this fact will affect your confidence in specifying a particular material and heat treatment if strength is critical. In such cases, you are advised to use higher than average design factors in your calculations of allowable (design) strength. (See Chapter 5.)

Using the guaranteed minimum values for strength in design decisions would be the safest approach. However, it is very conservative because most of the material actually delivered would have strengths significantly greater than the listed values.

One way to make the design more favorable is to acquire data for the statistical distribution of strength values taken for many samples. Then applications of probability theories can be used to specify suitable conditions for the material with a reasonable degree of confidence that the parts will perform according to specifications. Figure 2–10 illustrates some of the basic concepts of statistical distribution. The variation of strength over the entire population of samples is often assumed to have a normal distribution around some mean or average value. If you used a strength value that is one standard deviation (1σ) below the mean, 84% of the products would survive. At two standard deviations, greater than 97% would survive; at three standard deviations, more than 99.8%; and at four standard deviations, more than 99.99%.

As the designer, you must carefully judge the reliability of the data that you use. Ultimately, you should evaluate the reliability of the final product by considering the actual variations in material properties, the manufacturing considerations that may affect performance, and the interactions of various components with each other. There is more discussion on this point in Chapter 5.

**2–5
CARBON AND
ALLOY STEEL**

Steel is possibly the most widely used material for machine elements because of its properties of high strength, high stiffness, durability, and relative ease of fabrication. Many types of steels are available. This section will discuss the methods used for designating steels and will describe the most frequently used types.

The term *steel* refers to an alloy of iron, carbon, manganese, and one or more other significant elements. Carbon has a very strong effect on the strength, hardness, and ductility of any steel alloy. The other elements affect hardenability, toughness, corrosion resistance, machinability, and strength retention at high temperatures. The primary alloying

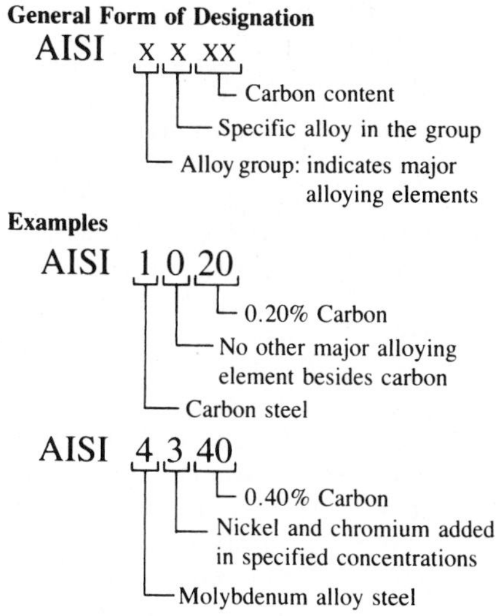

FIGURE 2–11 Steel designation system

General Form of Designation

AISI X X XX
- Carbon content
- Specific alloy in the group
- Alloy group: indicates major alloying elements

Examples

AISI 1 0 20
- 0.20% Carbon
- No other major alloying element besides carbon
- Carbon steel

AISI 4 3 40
- 0.40% Carbon
- Nickel and chromium added in specified concentrations
- Molybdenum alloy steel

elements present in the various alloy steels are sulfur, phosphorus, silicon, nickel, chromium, molybdenum, and vanadium.

Designation Systems

The AISI uses a four-digit designation system for carbon and alloy steel as shown in Figure 2–11. The first two digits indicate the specific alloy group that identifies the primary alloying elements other than carbon in the steel. (See Table 2–3.) The last two digits indicate the amount of carbon in the steel as described next.

Importance of Carbon

Although most steel alloys contain less than 1.0% carbon, it is included in the designation because of its effect on the properties of steel. As Figure 2–11 illustrates, the last two digits indicate carbon content in hundredths of a percent. For example, when the last two digits are 20, the alloy includes approximately 0.20% carbon. Some variation is allowed. The carbon content in a steel with *20 points of carbon* ranges from 0.18% to 0.23%.

As carbon content increases, strength and hardness also increase under the same conditions of processing and heat treatment. Since ductility decreases with increasing carbon content, selecting a suitable steel involves some compromise between strength and ductility.

As a rough classification scheme, a *low-carbon steel* is one having fewer than 30 points of carbon (0.30%). These steels have relatively low strength but good formability. In machine element applications where high strength is not required, low-carbon steels are frequently specified. If wear is a potential problem, low-carbon steels can be carburized (as discussed in Section 2–6) to increase the carbon content in the very outer surface of the part and to improve the combination of properties.

Medium-carbon steels contain 30 to 50 points of carbon (0.30%–0.50%). Most machine elements having moderate to high strength requirements with fairly good ductility and moderate hardness requirements come from this group.

High-carbon steels have 50 to 95 points of carbon (0.50%–0.95%). The high carbon content provides better wear properties suitable for applications requiring durable cutting edges and for applications where surfaces are subjected to constant abrasion. Tools, knives, chisels, and many agricultural implement components are among these uses.

TABLE 2–3   Alloy groups in the AISI numbering system

10xx	Plain carbon steel: No significant alloying element except carbon and manganese; less than 1.0% manganese. Also called *nonresulfurized*.
11xx	Free-cutting steel: Resulfurized. Sulfur content (typically 0.10%) improves machinability.
12xx	Free-cutting steel: Resulfurized and rephosphorized. Presence of increased sulfur and phosphorus improves machinability and surface finish.
12Lxx	Free-cutting steel: Lead added to 12xx steel further improves machinability.
13xx	Manganese steel: Nonresulfurized. Presence of approximately 1.75% manganese increases hardenability.
15xx	Carbon steel: Nonresulfurized; greater than 1.0% manganese.
23xx	Nickel steel: Nominally 3.5% nickel.
25xx	Nickel steel: Nominally 5.0% nickel.
31xx	Nickel-chromium steel: Nominally 1.25% Ni; 0.65% Cr.
33xx	Nickel-chromium steel: Nominally 3.5% Ni; 1.5% Cr.
40xx	Molybdenum steel: 0.25% Mo.
41xx	Chromium-molybdenum steel: 0.95% Cr; 0.2% Mo.
43xx	Nickel-chromium-molybdenum steel: 1.8% Ni; 0.5% or 0.8% Cr; 0.25% Mo.
44xx	Molybdenum steel: 0.5% Mo.
46xx	Nickel-molybdenum steel: 1.8% Ni; 0.25% Mo.
48xx	Nickel-molybdenum steel: 3.5% Ni; 0.25% Mo.
5xxx	Chromium steel: 0.4% Cr.
51xx	Chromium steel: Nominally 0.8% Cr.
51100	Chromium steel: Nominally 1.0% Cr; bearing steel, 1.0% C.
52100	Chromium steel: Nominally 1.45% Cr; bearing steel, 1.0% C.
61xx	Chromium-vanadium steel: 0.50%–1.10% Cr; 0.15% V.
86xx	Nickel-chromium-molybdenum steel: 0.55% Ni; 0.5% Cr; 0.20% Mo.
87xx	Nickel-chromium-molybdenum steel: 0.55% Ni; 0.5% Cr; 0.25% Mo.
92xx	Silicon steel: 2.0% silicon.
93xx	Nickel-chromium-molybdenum steel: 3.25% Ni; 1.2% Cr; 0.12% Mo.

A *bearing steel* nominally contains 1.0% carbon. Common grades are 50100, 51100, and 52100; the usual four-digit designation is replaced by five digits, indicating 100 points of carbon.

Alloy Groups

As indicated in Table 2–3, sulfur, phosphorus, and lead improve the machinability of steels and are added in significant amounts to the 11xx, 12xx, and 12Lxx grades. These grades are used for screw machine parts requiring high production rates where the resulting parts are not subjected to high stresses or wear conditions. In the other alloys, these elements are controlled to a very low level because of their adverse effects, such as increased brittleness.

Nickel improves the toughness, hardenability, and corrosion resistance of steel and is included in most of the alloy steels. Chromium improves hardenability, wear and abrasion resistance, and strength at elevated temperatures. In high concentrations, chromium provides significant corrosion resistance, as discussed in the section on stainless steels. Molybdenum also improves hardenability and high-temperature strength.

The steel selected for a particular application must be economical and must provide optimum properties of strength, ductility, toughness, machinability, and formability. Frequently, metallurgists, manufacturing engineers, and heat treatment specialists are consulted. (See also References 4, 14, 16, and 24.)

Table 2–4 lists some common steels used for machine parts, with typical applications listed for the alloys. You should benefit from the decisions of experienced designers when specifying materials.

TABLE 2–4 Uses of some steels

UNS number	AISI number	Applications
G10150	1015	Formed sheet-metal parts; machined parts (may be carburized)
G10300	1030	General-purpose, bar-shaped parts, levers, links, keys
G10400	1040	Shafts, gears
G10800	1080	Springs; agricultural equipment parts subjected to abrasion (rake teeth, disks, plowshares, mower teeth)
G11120	1112	Screw machine parts
G12144	12L14	Parts requiring good machinability
G41400	4140	Gears, shafts, forgings
G43400	4340	Gears, shafts, parts requiring good through-hardening
G46400	4640	Gears, shafts, cams
G51500	5150	Heavy-duty shafts, springs, gears
G51601	51B60	Shafts, springs, gears with improved hardenability
G52986	E52100	Bearing races, balls, rollers (bearing steel)
G61500	6150	Gears, forgings, shafts, springs
G86500	8650	Gears, shafts
G92600	9260	Springs

Examples of the Relationships between AISI and UNS Numbering Systems

Table 2–4 presents both the AISI and the UNS designations for the listed steels. Notice that for most carbon and alloy steels, the four-digit AISI number becomes the first four digits of the UNS number. The final digit in the UNS number is typically zero.

There are some exceptions, however. For high-carbon-bearing steels made in an electric furnace, such as AISI E52100, the UNS designation is G52986. Leaded steels have extra lead to improve machinability, and they have the letter L added between the second and third digits of the AISI number such as AISI 12L14, which becomes UNS G12144. Extra boron is added to some special alloys to improve hardenability. For example, alloy AISI 5160 is a chromium steel that carries the UNS designation G51600. But a similar alloy with added boron is AISI 51B60, and it carries the UNS designation G51601.

2–6 CONDITIONS FOR STEELS AND HEAT TREATMENT

The final properties of steels are dramatically affected by the way the steels are produced. Some processes involve mechanical working, such as rolling to a particular shape or drawing through dies. In machine design, many bar-shaped parts, shafts, wire, and structural members are produced in these ways. But most machine parts, particularly those carrying heavy loads, are heat-treated to produce high strength with acceptable toughness and ductility.

Carbon steel bar and sheet forms are usually delivered in the *as-rolled condition;* that is, they are rolled at an elevated temperature that eases the rolling process. The rolling can also be done cold to improve strength and surface finish. Cold-drawn bar and wire have the highest strength of the worked forms, along with a very good surface finish. However, when a material is designated to be *as-rolled*, it should be assumed that it was hot-rolled.

Heat Treating

Heat treating is any process in which steel is subjected to elevated temperatures to modify its properties. Of the several processes available, those most used for machine steels are annealing, normalizing, through-hardening (quench and temper), and case hardening. (See References 3 and 15.)

FIGURE 2–12 Heat
treatments for steel

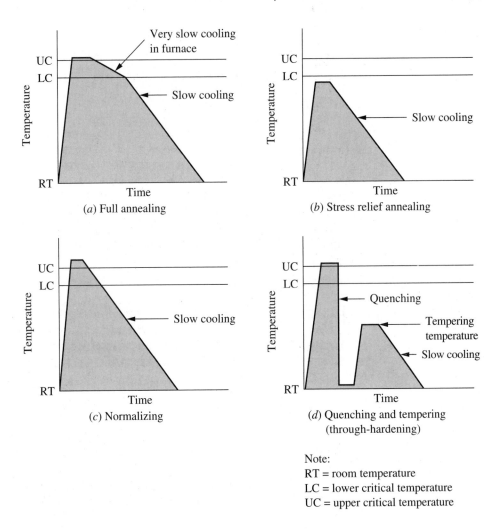

(a) Full annealing

(b) Stress relief annealing

(c) Normalizing

(d) Quenching and tempering
(through-hardening)

Note:
RT = room temperature
LC = lower critical temperature
UC = upper critical temperature

Figure 2–12 shows the temperature–time cycles for these heat treatment processes. The symbol RT indicates normal room temperature, and LC refers to the lower critical temperature at which the transformation of ferrite to austenite begins during the heating of the steel. At the upper critical temperature (UC), the transformation is complete. These temperatures vary with the composition of the steel. For most medium-carbon (0.30%–0.50% carbon) steels, UC is approximately 1500°F (822°C). References giving detailed heat treatment process data should be consulted.

Annealing. *Full annealing* [Figure 2–12(a)] is performed by heating the steel above the upper critical temperature and holding it until the composition is uniform. Then the steel is cooled very slowly in the furnace to below the lower critical temperature. Slow cooling to room temperature outside the furnace completes the process. This treatment produces a soft, low-strength form of the material, free of significant internal stresses. Parts are frequently cold-formed or machined in the annealed condition.

Stress relief annealing [Figure 2–12(b)] is often used following welding, machining, or cold forming to relieve residual stresses and thereby minimize subsequent distortion. The steel is heated to approximately 1000°F to 1200°F (540°C–650°C), held to achieve uniformity, and then slowly cooled in still air to room temperature.

Normalizing. *Normalizing* [Figure 2–12(c)] is performed in a similar manner to annealing, but at a higher temperature, above the transformation range where austenite is formed, approximately 1600°F (870°C). The result is a uniform internal structure in the steel and somewhat higher strength than annealing produces. Machinability and toughness are usually improved over the as-rolled condition.

Through-hardening and Quenching and Tempering. *Through-hardening* [Figure 2–12(d)] is accomplished by heating the steel to above the transformation range where austenite forms and then rapidly cooling it in a *quenching* medium. The rapid cooling causes the formation of martensite, the hard, strong form of steel. The degree to which martensite forms depends on the alloy's composition. An alloy containing a minimum of 80% of its structure in the martensite form over the entire cross section has *high hardenability*. This is an important property to look for when selecting a steel requiring high strength and hardness. The common quenching media are water, brine, and special mineral oils. The selection of a quenching medium depends on the rate at which cooling should proceed. Most machine steels use either oil or water quenching.

Tempering is usually performed immediately after quenching and involves reheating the steel to a temperature of 400°F to 1300°F (200°C–700°C) and then slowly cooling it in air back to room temperature. This process modifies the steel's properties: Tensile strength and yield strength decrease with increasing tempering temperature, whereas ductility improves, as indicated by an increase in the percent elongation. Thus, the designer can tailor the properties of the steel to meet specific requirements. Furthermore, the steel in its as-quenched condition has high internal stresses and is usually quite brittle. Machine parts should normally be tempered at 700°F (370°C) or higher after quenching.

To illustrate the effects of tempering on the properties of steels, several charts in Appendix 4 show graphs of strength versus tempering temperature. Included in these charts are tensile strength, yield point, percent elongation, percent reduction of area, and hardness number HB, all plotted in relation to tempering temperature. Note the difference in the shape of the curves and the absolute values of the strength and hardness when comparing the plain carbon AISI 1040 steel with the alloy steel AISI 4340. Although both have the same nominal carbon content, the alloy steel reaches a much higher strength and hardness. Note also the as-quenched hardness in the upper right part of the heading of the charts; it indicates the degree to which a given alloy can be hardened. When the case-hardening processes (described next) are used, the as-quenched hardness becomes very important.

Appendix 3 lists the range of properties that can be expected for several grades of carbon and alloy steels. The alloys are listed with their AISI numbers and conditions. For the heat-treated conditions, the designation reads, for example, AISI 4340 OQT 1000, which indicates that the alloy was oil-quenched and tempered at 1000°F. Expressing the properties at the 400°F and 1300°F tempering temperatures indicates the end-points of the possible range of properties that can be expected for that alloy. To specify a strength between these limits, you could refer to graphs such as those shown in Appendix 4, or you could determine the required heat treatment process from a specialist. For the purposes of material specification in this book, a rough interpolation between given values will be satisfactory. As noted before, you should seek more specific data for critical designs.

Case Hardening. In many cases, the bulk of the part requires only moderate strength although the surface must have a very high hardness. In gear teeth, for example, high surface hardness is necessary to resist wear as the mating teeth come into contact several million times during the expected life of the gears. At each contact, a high stress develops at the surface of the teeth. For applications such as this, *case hardening* is used; the surface

FIGURE 2–13

Typical case-hardened
gear-tooth section

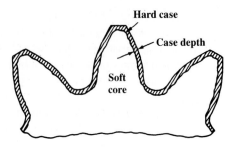

(or *case*) of the part is given a high hardness to a depth of perhaps 0.010 to 0.040 in (0.25–1.00 mm), although the interior of the part (the *core*) is affected only slightly, if at all. The advantage of surface hardening is that as the surface receives the required wear-resisting hardness, the core of the part remains in a more ductile form, resistant to impact and fatigue. The processes used most often for case hardening are flame hardening, induction hardening, carburizing, nitriding, cyaniding, and carbo-nitriding. (See Reference 17.)

Figure 2–13 shows a drawing of a typical case-hardened gear-tooth section, clearly showing the hard case surrounding the softer, more ductile core. Case hardening is used in applications requiring high wear and abrasion resistance in normal service (gear teeth, crane wheels, wire-rope sheaves, and heavy-duty shafts).

The most commonly used processes for case hardening are described in the following list.

1. *Flame hardening and induction hardening:* The processes of flame hardening and induction hardening involve the rapid heating of the surface of the part for a limited time so that a small, controlled depth of the material reaches the transformation range. Upon immediate quenching, only that part above the transformation range produces the high level of martensite required for high hardness.

   Flame hardening uses a concentrated flame impinging on a localized area for a controlled amount of time to heat the part, followed by quenching in a bath or by a stream of water or oil. *Induction hardening* is a process in which the part is surrounded by a coil through which high-frequency electric current is passed. Because of the electrical conductivity of the steel, current is *induced* primarily near the surface of the part. The resistance of the material to the flow of current results in a heating effect. Controlling the electrical power and the frequency of the induction system, and the time of exposure, determines the depth to which the material reaches the transformation temperature. Rapid quenching after heating hardens the surface. (See Reference 26.)

   Note that for flame or induction hardening to be effective, the material must have a good hardenability. Usually the goal of case hardening is to produce a case hardness in the range of Rockwell C hardness HRC 55 to 60 (Brinell hardness approximately HB 550 to 650). Therefore, the material must be capable of being hardened to the desired level. Carbon and alloy steels with fewer than 30 points of carbon typically cannot meet this requirement. Thus, the alloy steels with 40 points or more of carbon are the usual types given flame- or induction-hardening treatments.

2. *Carburizing, nitriding, cyaniding, and carbo-nitriding:* The remaining case-hardening processes—carburizing, nitriding, cyaniding, and carbo-nitriding—actually alter the composition of the surface of the material by exposing it to

carbon-bearing gases, liquids, or solids at high temperatures that produce carbon and diffuse it into the surface of the part. The concentration and the depth of penetration of carbon depend on the nature of the carbon-bearing material and the time of exposure. Nitriding and cyaniding typically result in very hard, thin cases that are good for general wear resistance. Where high load-carrying capability in addition to wear resistance is required, as with gear teeth, carburizing is preferred because of the thicker case.

Several steels are produced as carburizing grades. Among these are 1015, 1020, 1022, 1117, 1118, 4118, 4320, 4620, 4820, and 8620. Appendix 5 lists the expected properties of these carburized steels. Note when evaluating a material for use that the core properties determine its ability to withstand prevailing stresses, and the case hardness indicates its wear resistance. Carburizing, properly done, will virtually always produce a case hardness from HRC 55 to 64 (Rockwell C hardness) or from HB 550 to 700 (Brinell hardness).

Carburizing has several variations that allow the designer to tailor the properties to meet specific requirements. The exposure to the carbon atmosphere takes place at a temperature of approximately 1700°F (920°C) and usually takes 8 h. Immediate quenching achieves the highest strength, although the case is somewhat brittle. Normally, a part is allowed to cool slowly after carburizing. It is then reheated to approximately 1500°F (815°C) and then quenched. A tempering at the relatively low temperature of either 300°F or 450°F (150°C or 230°C) follows, to relieve stresses induced by quenching. As shown in Appendix 5, the higher tempering temperature lowers the core strength and the case hardness by a small amount, but in general it improves the part's toughness. The process just described is *single quenching and tempering*.

When a part is quenched in oil and tempered at 450°F, for example, the condition is *case hardening by carburizing, SOQT 450*. Reheating after the first quench and quenching again further refines the case and core properties; this process is *case hardening by carburizing, DOQT 450*. These conditions are listed in Appendix 5.

2–7
STAINLESS
STEELS

The term *stainless steel* characterizes the high level of corrosion resistance offered by alloys in this group. To be classified as a stainless steel, the alloy must have a chromium content of at least 10%. Most have 12% to 18% chromium. (See Reference 5.)

The AISI designates most stainless steels by its 200, 300, and 400 series. As mentioned previously (Section 2–3), another designation system is the unified numbering system (UNS) developed by the SAE and the ASTM. Appendix 6 lists the properties of several grades, giving both designations.

The three main groups of stainless steels are austenitic, ferritic, and martensitic. *Austenitic* stainless steels fall into the AISI 200 and 300 series. They are general-purpose grades with moderate strength. Most are not heat-treatable, and their final properties are determined by the amount of working, with the resulting temper referred to as 1/4 hard, 1/2 hard, 3/4 hard, and full hard. These alloys are nonmagnetic and are typically used in food processing equipment.

Ferritic stainless steels belong to the AISI 400 series, designated as 405, 409, 430, 446, and so on. They are magnetic and perform well at elevated temperatures, from 1300°F to 1900°F (700°C–1040°C), depending on the alloy. They are not heat-treatable, but they can be cold-worked to improve properties. Typical applications include heat exchanger tubing, petroleum refining equipment, automotive trim, furnace parts, and chemical equipment.

Martensitic stainless steels are also members of the AISI 400 series, including 403, 410, 414, 416, 420, 431, and 440 types. They are magnetic, can be heat-treated, and have higher strength than the 200 and 300 series, while retaining good toughness. Typical uses include turbine engine parts, cutlery, scissors, pump parts, valve parts, surgical instruments, aircraft fittings, and marine hardware.

There are many other grades of stainless steels, many of which are proprietary to particular manufacturers. A group used for high-strength applications in aerospace, marine, and vehicular applications is of the precipitation-hardening type. They develop very high strengths with heat treatments at relatively low temperatures, from 900°F to 1150°F (480°C–620°C). This characteristic helps to minimize distortion during treatment. Some examples are 17-4PH, 15-5PH, 17-7PH, PH15-7Mo, and AMS362 stainless steels.

2–8 STRUCTURAL STEEL

Most structural steels are designated by ASTM numbers established by the American Society for Testing and Materials. One common grade is ASTM A36, which has a minimum yield point of 36 000 psi (248 MPa) and is very ductile. It is basically a low-carbon, hot-rolled steel available in sheet, plate, bar, and structural shapes such as some wide-flange beams, American Standard beams, channels, and angles. The geometric properties of some of each of these sections are listed in Appendix 16.

Most wide-flange beams (W-shapes) are currently made using ASTM A992 structural steel, which has a yield point of 50 to 65 ksi (345 to 448 MPa) and a minimum tensile strength of 65 ksi (448 MPa). An additional requirement is that the maximum ratio of the yield point to the tensile strength is 0.85. This is a highly ductile steel, having a minimum of 21% elongation in a 2.00-inch gage length. Using this steel instead of the lower strength ASTM A36 steel typically allows smaller, lighter structural members at little or no additional cost.

Hollow structural sections (HSS) are typically made from ASTM A500 steel that is cold-formed and either welded or made seamless. Included are round tubes and square and rectangular shapes. Note in Appendix 7 that there are different strength values for round tubes as compared with the shaped forms. Also, several strength grades can be specified. Some of these HSS products are made from ASTM A501 hot-formed steel having properties similar to the ASTM A36 hot-rolled steel shapes.

Many higher-strength grades of structural steel are available for use in construction, vehicular, and machine applications. They provide yield points in the range from 42 000 to 100 000 psi (290–700 MPa). Some of these grades, referred to as *high-strength, low-alloy (HSLA) steels*, are ASTM A242, A440, A514, A572, and A588.

Appendix 7 lists the properties of several structural steels.

2–9 TOOL STEELS

The term *tool steels* refers to a group of steels typically used for cutting tools, punches, dies, shearing blades, chisels, and similar uses. The numerous varieties of tool steel materials have been classified into seven general types as shown in Table 2–5. Whereas most uses of tool steels are related to the field of manufacturing engineering, they are also pertinent to machine design where the ability to maintain a keen edge under abrasive conditions is required (Type H and F). Also, some tool steels have rather high shock resistance which may be desirable in machine components such as parts for mechanical clutches, pawls, blades, guides for moving materials, and clamps (Types S, L, F, and W). (See Reference 6 for a more extensive discussion of tool steels.)

2–10 CAST IRON

Large gears, machine structures, brackets, linkage parts, and other important machine parts are made from cast iron. The several types of grades available span wide ranges of strength, ductility, machinability, wear resistance, and cost. These features are attractive in many applications. The three most commonly used types of cast iron are gray iron,

TABLE 2–5 Examples of tool steel types

General type	Type symbol	Specific types — Major alloying elements	Examples — AISI No.	Examples — UNS No.	Typical uses (and other common alloys)
High-speed	M	Molybdenum	M2 M10 M42	T11302 T11310 T11342	General-purpose tool steels for cutting tools and dies for forging, extrusion, bending, drawing, and piercing (M1, M3, M4–M7, M30, M34, M36, M41–M47)
	T	Tungsten	T1 T15	T12001 T12015	Similar to uses for M-types (T2, T4, T5, T6, T8)
Hot-worked	H	Chromium	H10	T20810	Cold-heading dies, shearing knives, aircraft parts, low-temperature extrusion and die-casting dies (H1–H19)
		Tungsten	H21	T20821	Higher-temperature dies, hot shearing knives (H20–H39)
		Molybdenum	H42	T20842	Applications that tend to produce high wear (H40–H59)
Cold-worked	D	High-carbon, high-chromium	D2	T30402	Stamping dies, punches, gages (D3–D5, D7)
	A	Medium-alloy, air-hardening	A2	T30102	Punches, thread-rolling dies, die-casting dies (A3–A10)
	O	Oil-hardening	O1	T31501	Taps, reamers, broaches, gages, jigs and fixtures, bushings, machine tool arbors, tool shanks (O2, O6, O7)
Shock-resisting	S		S1	T41901	Chisels, pneumatic tools, heavy-duty punches, machine parts subject to shocks (S2, S4–S7)
Molded steels	P		P2	T51602	Plastic molding dies, zinc die-casting dies (P3–P6, P20, P21)
Special-purpose	L	Low-alloy types	L2	T61202	Tooling and machine parts requiring high toughness (L3, L6)
	F	Carbon-tungsten types	F1	T60601	Similar to L-types but with higher abrasion resistance (F2)
Water-hardened	W		W1	T72301	General-purpose tool and die uses, vise and chuck jaws, hand tools, jigs and fixtures, punches (W2, W5)

ductile iron, and malleable iron. Appendix 8 lists the properties of several cast irons. (See also Reference 9.)

Gray iron is available in grades having tensile strengths ranging from 20 000 to 60 000 psi (138–414 MPa). Its ultimate compressive strength is much higher, three to five times as high as the tensile strength. One disadvantage of gray iron is that it is brittle and therefore should not be used in applications where impact loading is likely. But it has excellent wear resistance, is relatively easy to machine, has good vibration damping ability, and can be surface-hardened. Applications include engine blocks, gears, brake parts, and machine bases. The gray irons are rated by the ASTM specification A48-94a in classes 20, 25, 30, 40, 50, and 60, where the number refers to the minimum tensile strength in kips/in^2(ksi). For example, class 40 gray iron has a minimum tensile strength of 40 ksi or 40 000 psi (276 MPa). Because it is brittle, gray iron does not exhibit the property of yield strength.

Malleable iron is a group of heat-treatable cast irons with moderate to high strength, high modulus of elasticity (stiffness), good machinability, and good wear resistance. The five-digit designation roughly indicates the yield strength and the expected percent elongation of the iron. For example, Grade 40010 has a yield strength of 40 ksi (276 MPa) and a 10% elongation. The strength properties listed in Appendix 8 are for the non-heat-treated condition. Higher strengths would result from heat treating. See ASTM specifications A 47-99 and A 220-99.

Ductile irons have higher strengths than the gray irons and, as the name implies, are more ductile. However, their ductility is still much lower than that of typical steels. A three-part grade designation is used for ductile iron in the ASTM A536-84 specification. The first number refers to the tensile strength in ksi, the second is the yield strength in ksi, and the third is the approximate percent elongation. For example, the grade 80-55-06 has a tensile strength of 80 ksi (552 MPa), a yield strength of 55 ksi (379 MPa), and a 6% elongation in 2.00 in. Higher-strength cast parts, such as crankshafts and gears, are made from ductile iron.

Austempered ductile iron (ADI) is an alloyed and heat-treated ductile iron. (See Reference 9.) It has attractive properties that lead to its use in transportation equipment, industrial machinery, and other applications where the low cost, good machinability, high damping characteristics, good wear resistance, and near-net-shape advantages of casting offer special benefits. Examples are drive train gears, parts for constant velocity joints, and suspension components. ASTM Standard 897-90 lists five grades of ADI ranging in tensile strength from 125 ksi (850 MPa) to 230 ksi (1600 MPa). Yield strengths range from 80 ksi (550 MPa) to 185 ksi (1300 MPa). Ductility decreases with increasing strength and hardness with percent elongation values in the range from approximately 10% to less than 1%. ADI begins as a conventional ductile iron with careful control of composition and the casting process to produce a sound, void-free casting. Small amounts of copper, nickel, and molybdenum are added to enhance the metal's response to the special heat treatment cycle shown in Figure 2–14. It is heated to the austenitizing temperature (1550° to 1750°F, or 843° to 954°C) depending on the composition. It is held at that temperature for one to three hours as the material becomes fully austenitic. A rapid quench follows in a medium at 460° to 750°F (238° to 400°C), and the casting is held at this temperature for one-half to four hours. This is the *austempering* part of the cycle during which all of the material is converted to a mixture of mostly austenite and ferrite, sometimes called *ausferrite*. It is important that neither pearlite nor bainite form during this cycle. The casting is then allowed to cool to room temperature.

2–11 POWDERED METALS

Making parts with intricate shapes by powder metallurgy can sometimes eliminate the need for extensive machining. Metal powders are available in many formulations whose properties approach those of the wrought form of the metal. The processing involves preparing a preform by compacting the powder in a die under high pressure. Sintering at a high temperature to fuse the powder into a uniform mass is the next step. Re-pressing is sometimes done to improve properties or dimensional accuracy of the part. Typical parts made by the powder metallurgy (PM) process are gears, gear segments, cams, eccentrics, and various machine parts having oddly shaped holes or projections. Dimensional tolerances of 0.001 to 0.005 in (0.025–0.125 mm) are typical.

One disadvantage of PM parts is that they are usually brittle and should not be used in applications where high-impact loading is expected. Another important application is in sintered bearings, which are made to a relatively low density with consequent high porosity. The bearing is impregnated with a lubricant that may be sufficient for the life of the part. This type of material is discussed further in Chapter 16.

Manufacturers of metal powders have many proprietary formulations and grades. However, the Metal Powder Industries Federation (MPIF) is promoting standardization of materials. Figure 2–15 shows photographs of some powder metal parts. (See Reference 3.)

FIGURE 2–14 Heat treatment cycle for austempered ductile iron (ADI)

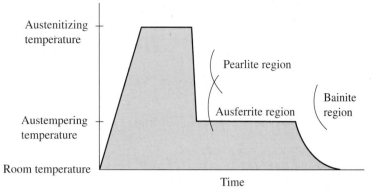

Austenitizing temperature

Pearlite region

Ausferrite region

Bainite region

Austempering temperature

Room temperature

Time

FIGURE 2–15 Examples of powder metal components (GKN Sinter Metals Auburn Hills, MI)

2–12 ALUMINUM

Aluminum is widely used for structural and mechanical applications. Chief among its attractive properties are light weight, good corrosion resistance, relative ease of forming and machining, and pleasing appearance. Its density is approximately one-third that of steel. However, its strength is somewhat lower, also. (See References 1, 8, and 12.) Table 2–6 lists the commonly used alloy groups.

The standard designations for aluminum alloys listed by the Aluminum Association use a four-digit system. The first digit indicates the alloy type according to the major alloying element. The second digit, if it is other than zero, indicates modifications of another alloy or limits placed on impurities in the alloy. The presence of impurities is particularly important for electrical conductors. Within each group are several specific alloys, indicated by the last two digits in the designation.

Table 2–7 lists several common alloys, along with the forms in which they are typically produced and some of their major applications. The table also lists several of the 50 or more available alloys that span the range of typical applications. The table should aid you in selecting a suitable alloy for a particular application.

TABLE 2–6 Aluminum alloy groups

Alloy designations (by major alloying element)
1xxx 99.00% or greater aluminum content
2xxx Copper
3xxx Manganese
4xxx Silicon
5xxx Magnesium
6xxx Magnesium and silicon
7xxx Zinc

TABLE 2–7   Common aluminum alloys and their uses

Alloy	Applications	Forms
1060	Chemical equipment and tanks	Sheet, plate, tube
1350	Electrical conductors	Sheet, plate, tube, rod, bar, wire, pipe, shapes
2014	Aircraft structures and vehicle frames	Sheet, plate, tube, rod, bar, wire, shapes, forgings
2024	Aircraft structures, wheels, machine parts	Sheet, plate, tube, rod, bar, wire, shapes, rivets
2219	Parts subjected to high temperatures (to 600°F)	Sheet, plate, tube, rod, bar, shapes, forgings
3003	Chemical equipment, tanks, cooking utensils, architectural parts	Sheet, plate, tube, rod, bar, wire, shapes, pipe, rivets, forgings
5052	Hydraulic tubes, appliances, sheet-metal fabrications	Sheet, plate, tube, rod, bar, wire, rivets
6061	Structures, vehicle frames and parts, marine uses	All forms
6063	Furniture, architectural hardware	Tube, pipe, extruded shapes
7001	High-strength structures	Tube, extruded shapes
7075	Aircraft and heavy-duty structures	All forms except pipe

The mechanical properties of the aluminum alloys are highly dependent on their condition. For this reason, the specification of an alloy is incomplete without a reference to its *temper*. The following list describes the usual tempers given to aluminum alloys. Note that some alloys respond to heat treating, and others are processed by strain hardening. *Strain hardening* is controlled cold working of the alloy, in which increased working increases hardness and strength while reducing ductility. Commonly available tempers are described next.

F (as-fabricated): No special control of properties is provided. Actual limits are unknown. This temper should be accepted only when the part can be thoroughly tested prior to service.

O (annealed): A thermal treatment that results in the softest and lowest strength condition. Sometimes specified to obtain the most workable form of the alloy. The resulting part can be heat-treated for improved properties if it is made from alloys in the 2xxx, 4xxx, 6xxx, or 7xxx series. Also, the working itself may provide some improvement in properties similar to that produced by strain hardening for alloys in the 1xxx, 3xxx, and 5xxx series.

H (strain-hardened): A process of cold working under controlled conditions that produces improved, predictable properties for alloys in the 1xxx, 3xxx, and 5xxx groups. The greater the amount of cold work, the higher the strength and hardness, although the ductility is decreased. The *H* designation is followed by two or more digits (usually 12, 14, 16, or 18) that indicate progressively higher strength. However, several other designations are used.

T (heat-treated): A series of controlled heating and cooling processes applied to alloys in the 2xxx, 4xxx, 6xxx, and 7xxx groups. The letter *T* is followed by one or more numbers to indicate specific processes. The more common designations for mechanical and structural products are T4 and T6.

Property data for aluminum alloys are included in Appendix 9. Because these data are typical values, not guaranteed values, the supplier should be consulted for data at the time of purchase.

For mechanical design applications, alloy 6061 is one of the most versatile types. Note that it is available in virtually all forms, has good strength and corrosion resistance, and is heat-treatable to obtain a wide variety of properties. It also has good weldability. In its softer forms, it is easily formed and worked. Then, if higher strength is required, it can be heat-treated after forming. However, it has low machinability.

2–13
ZINC ALLOYS

Zinc is the fourth most commonly used metal in the world. Much of it is in the form of zinc galvanizing used as a corrosion inhibitor for steels, but very large quantities of zinc alloys are used in castings and for bearing materials. Figure 2–16 shows examples of cast zinc parts. (See Reference 19.)

High production quantities are made using zinc pressure die casting, which results in very smooth surfaces and excellent dimensional accuracy. A variety of coating processes can be used to produce desirable finish appearance and to inhibit corrosion. Although the as-cast parts have inherently good corrosion resistance, the performance in some environments can be enhanced with chromate or phosphate treatments or anodizing. Painting and chrome plating are also used to produce a wide variety of attractive surface finishes.

In addition to die casting, zinc products are often made by permanent mold casting, graphite permanent mold casting, sand casting, and shell-mold casting. Other, less frequently used processes are investment casting, low-pressure permanent mold casting, centrifugal casting, continuous casting, and rubber-mold casting. Plaster-mold casting is often used for prototyping. Continuous casting is used to produce standard shapes (rod, bar, tube, and slabs). Prototypes or finished products can then be machined from these shapes.

Zinc alloys typically contain aluminum and a small amount of magnesium. Some alloys include copper or nickel. The performance of the final products can be very sensitive to small amounts of other elements, and maximum limits are placed on the content of iron, lead, cadmium, and tin in some alloys.

The most widely used zinc casting alloy is called *alloy No. 3*, sometimes referred to as *Zamak 3*. It has 4% aluminum and 0.035% magnesium. Another is called *Zamak 5*, and it

FIGURE 2–16 Cast zinc parts (INTERZINC, Washington, D.C.)

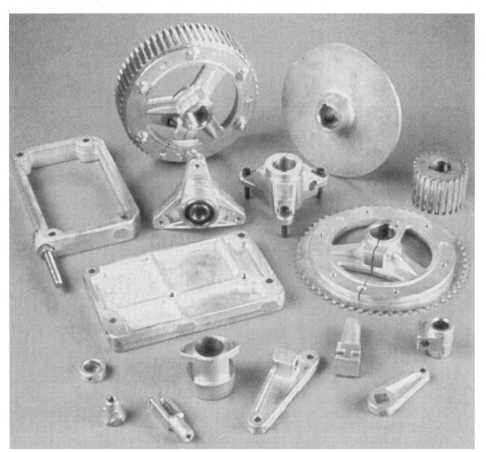

also contains 4% aluminum with 0.055% magnesium and 1.0% copper. A group of alloys having higher aluminum content are the ZA-alloys, with ZA-8, ZA-12, and ZA-27 being the most popular. Appendix 10 gives a summary of the composition and the typical properties of these alloys. As with most cast materials, some variations are to be expected with the size of the cast sections, the thermal treatment of the casting, the operating temperature of the product, and the quality assurance during the casting process.

2–14
TITANIUM

The applications of titanium include aerospace structures and components, chemical tanks and processing equipment, fluids-handling devices, and marine hardware. Titanium has very good corrosion resistance and a high strength-to-weight ratio. Its stiffness and density are between those of steel and aluminum; its modulus of elasticity is approximately 16×10^6 psi (110 GPa), and its density 0.160 lb/in^3 (4.429 kg/m^3). Typical yield strengths range from 25 to 175 ksi (172–1210 MPa). Disadvantages of titanium include relatively high cost and difficult machining.

The classification of titanium alloys usually falls into four types: commercially pure alpha titanium, alpha alloys, alpha-beta alloys, and beta alloys. Appendix 11 shows the properties of some of these grades. The term *alpha* refers to the hexagonal, close-packed, metallurgical structure that forms at low temperatures, and *beta* refers to the high-temperature, body-centered, cubic structure.

The grades of commercially pure titanium indicate the approximate expected yield strength of the material. For example, Ti-50A has an expected yield strength of 50 000 psi (345 MPa). As a class, these alloys exhibit only moderate strength but good ductility.

One popular grade of alpha alloy is titanium alloyed with 0.20% palladium (Pd), called *Ti-0.2Pd*. Its properties are listed in Appendix 11 for one heat-treat condition. Some alpha alloys have improved high-temperature strength and weldability.

Generally speaking, the alpha-beta alloys and the beta alloys are stronger forms of titanium. They are heat-treatable for close control of their properties. Since several alloys are available, a designer can tailor the properties to meet special needs for formability, machinability, forgeability, corrosion resistance, high-temperature strength, weldability, and creep resistance, as well as basic room-temperature strength and ductility. Alloy Ti-6Al-4V contains 6% aluminum and 4% vanadium and is used in a variety of aerospace applications.

2–15
COPPER, BRASS,
AND BRONZE

Copper is widely used in its nearly pure form for electrical and plumbing applications because of its high electrical conductivity and good corrosion resistance. It is rarely used for machine parts because of its relatively low strength compared with that of its alloys, *brass* and *bronze*. (See Reference 3.)

Brass is a family of alloys of copper and zinc, with the content of zinc ranging from about 5% to 40%. Brass is often used in marine applications because of its resistance to corrosion in salt water. Many brass alloys also have excellent machinability and are used as connectors, fittings, and other parts made on screw machines. *Yellow brass* contains about 30% or more of zinc and often contains a significant amount of lead to improve machinability. *Red brass* contains 5% to 15% zinc. Some alloys also contain tin, lead, nickel, or aluminum.

Bronze is a class of alloys of copper with several different elements, one of which is usually tin. They are useful in gears, bearings, and other applications where good strength and high wear resistance are desirable.

Wrought bronze alloys are available in four types:

Phosphor bronze: Copper-tin-phosphorus alloy

Leaded phosphor bronze: Copper-tin-lead-phosphorus alloy

Aluminum bronze: Copper-aluminum alloy

Silicon bronze: Copper-silicon alloy

Cast bronze alloys have four main types:

Tin bronze: Copper-tin alloy

Leaded tin bronze: Copper-tin-lead alloy

Nickel tin bronze: Copper-tin-nickel alloy

Aluminum bronze: Copper-aluminum alloy

The cast alloy called *manganese bronze* is actually a high-strength form of brass because it contains zinc, the characteristic alloying element of the brass family. Manganese bronze contains copper, zinc, tin, and manganese.

In the UNS, copper alloys are designated by the letter *C*, followed by a five-digit number. Numbers from 10000 to 79900 refer to wrought alloys; 80000 to 99900 refer to casting alloys. See Appendix 12 for typical properties.

2–16 NICKEL-BASED ALLOYS

Nickel alloys are often used in place of steel where operation at high temperatures and in certain corrosive environments is required. Examples are turbine engine components, furnace parts, chemical processing systems, and critical marine system components. (See Reference 7.) Some nickel alloys are called *superalloys*, and many of the commonly used alloys are proprietary. The following list gives some of the commercially available alloy types:

Inconel (International Nickel Co.): Nickel-chromium alloys

Monel (International Nickel Co.): Nickel-copper alloys

Ni-Resist (International Nickel Co.): Nickel-iron alloys

Hastelloy (Haynes International): Nickel-molybdenum alloys, sometimes with chromium, iron, or copper

2–17 PLASTICS

Plastics include a wide variety of materials formed of large molecules called *polymers*. The thousands of different plastics are created by combining different chemicals to form long molecular chains.

One method of classifying plastics is by the terms *thermoplastic* and *thermosetting*. In general, the *thermoplastic* materials can be formed repeatedly by heating or molding because their basic chemical structure is unchanged from its initial linear form. *Thermosetting* plastics do undergo some change during forming and result in a structure in which the molecules are cross-linked and form a network of interconnected molecules. Some designers recommend the terms *linear* and *cross-linked* in place of the more familiar *thermoplastic* and *thermosetting*.

Listed next are several thermoplastics and several thermosets that are used for load-carrying parts and that are therefore of interest to the designer of machine elements. These listings show the main advantages and uses of a sample of the many plastics available. Appendix 13 lists typical properties.

Thermoplastics

- *Nylon:* Good strength, wear resistance, and toughness; wide range of possible properties depending on fillers and formulations. Used for structural parts, mechanical devices such as gears and bearings, and parts needing wear resistance.

- *Acrylonitrile-butadiene-styrene (ABS):* Good impact resistance, rigidity, moderate strength. Used for housings, helmets, cases, appliance parts, pipe, and pipe fittings.

- *Polycarbonate:* Excellent toughness, impact resistance, and dimensional stability. Used for cams, gears, housings, electrical connectors, food processing products, helmets, and pump and meter parts.

- *Acrylic:* Good weather resistance and impact resistance; can be made with excellent transparency or translucent or opaque with color. Used for glazing, lenses, signs, and housings.

- *Polyvinyl chloride (PVC):* Good strength, weather resistance, and rigidity. Used for pipe, electrical conduit, small housings, ductwork, and moldings.

- *Polyimide:* Good strength and wear resistance; very good retention of properties at elevated temperatures up to 500°F. Used for bearings, seals, rotating vanes, and electrical parts.

- *Acetal:* High strength, stiffness, hardness, and wear resistance; low friction; good weather resistance and chemical resistance. Used for gears, bushings, sprockets, conveyor parts, and plumbing products.

- *Polyurethane elastomer:* A rubberlike material with exceptional toughness and abrasion resistance; good heat resistance and resistance to oils. Used for wheels, rollers, gears, sprockets, conveyor parts, and tubing.

- *Thermoplastic polyester resin (PET):* Polyethylene terephthalate (PET) resin with fibers of glass and/or mineral. Very high strength and stiffness, excellent resistance to chemicals and heat, excellent dimensional stability, and good electrical properties. Used for pump parts, housings, electrical parts, motor parts, auto parts, oven handles, gears, sprockets, and sporting goods.

- *Polyether-ester elastomer:* Flexible plastic with excellent toughness and resilience, high resistance to creep, impact, and fatigue under flexure, good chemical resistance. Remains flexible at low temperatures and retains good properties at moderately elevated temperatures. Used for seals, belts, pump diaphragms, protective boots, tubing, springs, and impact absorbing devices. High modulus grades can be used for gears and sprockets.

Thermosets

- *Phenolic:* High rigidity, good moldability and dimensional stability, very good electrical properties. Used for load-carrying parts in electrical equipment, switchgear, terminal strips, small housings, handles for appliances and cooking utensils, gears, and structural and mechanical parts. Alkyd, allyl, and amino thermosets have properties and uses similar to those of the phenolics.

- *Polyester:* Known as *fiber glass* when reinforced with glass fibers; high strength and stiffness, good weather resistance. Used for housings, structural shapes, and panels.

Special Considerations for Selecting Plastics

A particular plastic is often selected for a combination of properties, such as light weight, flexibility, color, strength, stiffness, chemical resistance, low friction characteristics, or transparency. Table 2–8 lists the primary plastic materials used for six different types of applications. References 11 and 23 provide an extensive comparative study of the design properties of plastics.

While most of the same definitions of design properties described in Section 2–2 of this chapter can be used for plastics as well as metals, a significant amount of additional information is typically needed to specify a suitable plastic material. Some of the special

TABLE 2–8 Applications of plastic materials

Applications	Desired properties	Suitable plastics
Housings, containers, ducts	High impact strength, stiffness, low cost, formability, environmental resistance, dimensional stability	ABS, polystyrene, polypropylene, PET, polyethylene, cellulose acetate, acrylics
Low friction—bearings, slides	Low coefficient of friction; resistance to abrasion, heat, corrosion	TFE fluorocarbons, nylon, acetals
High-strength components, gears, cams, rollers	High tensile and impact strength, stability at high temperatures, machinable	Nylon, phenolics, TFE-filled acetals, PET, polycarbonate
Chemical and thermal equipment	Chemical and thermal resistance, good strength, low moisture absorption	Fluorocarbons, polypropylene, polyethylene, epoxies, polyesters, phenolics
Electrostructural parts	Electrical resistance, heat resistance, high impact strength, dimensional stability, stiffness	Allyls, alkyds, aminos, epoxies, phenolics, polyesters, silicones, PET
Light-transmission components	Good light transmission in transparent and translucent colors, formability, shatter resistance	Acrylics, polystyrene, cellulose acetate, vinyls

characteristics of plastics follow. The charts shown in Figures 2–17 to 2–20 are examples only and are not meant to indicate the general nature of the performance of the given type of material. There is a wide range of properties among the many formulations of plastics even within a given class. Consult the extensive amount of design guidance available from vendors of the plastic materials.

1. Most properties of plastics are highly sensitive to temperature. In general, tensile strength, compressive strength, elastic modulus, and impact failure energy decrease significantly as the temperature increases. Figure 2–17 shows the tensile strength of nylon 66 at four temperatures. Note also the rather different shapes of the stress-strain curves. The slope of the curve at any point indicates the elastic modulus, and you can see a large variation for each curve.

2. Many plastics absorb a considerable amount of moisture from the environment and exhibit dimensional changes and degradation of strength and stiffness properties as a result. See Figure 2–18 that shows the flexural modulus versus temperature for a nylon in dry air, 50% relative humidity (RH), and 100% RH. A consumer product may well experience a major part of this range. At a temperature of 20°C (68°F, near room temperature), the flexural modulus would decrease dramatically from approximately 2900 MPa to about 500 MPa as humidity changes from dry air to 100% RH. The product may also see a temperature range from 0°C (32°F, freezing point of water) to 40°C (104°F). Over this range, the flexural modulus for the nylon at 50% RH would decrease from approximately 2300 MPa to 800 MPa.

3. Components that carry loads continuously must be designed to accommodate creep or relaxation. See Figures 2–17 to 2–19 and Example Problem 2–1.

4. Fatigue resistance data of a plastic must be acquired for the specific formulation used and at a representative temperature. Chapter 5 gives more information about fatigue. Figure 2–19 shows the fatigue stress versus number of cycles to failure for an acetal resin plastic. Curve 1 is at 23°C (73°F, near room temperature) with cyclic loading in tension only as when a tensile load is applied and

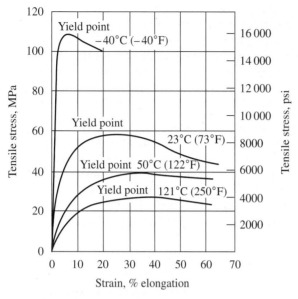

FIGURE 2–17   Stress-strain curves for nylon 66 at four temperatures (DuPont Polymers, Wilmington, DE)

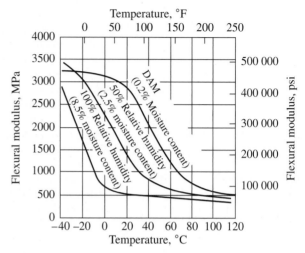

FIGURE 2–18   Effect of temperature and humidity on the flexural modulus of nylon 66 (DuPont Polymers, Wilmington, DE)

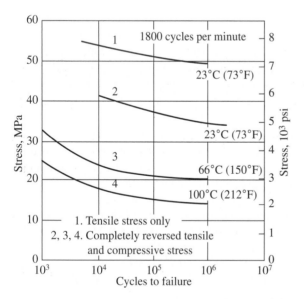

FIGURE 2–19   Fatigue stress vs. number of cycles to failure for an acetal resin plastic (DuPont Polymers, Wilmington, DE)

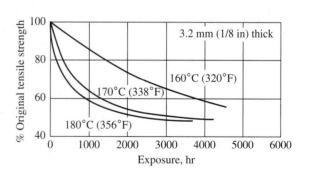

FIGURE 2–20   Effect of exposure to elevated temperature on a thermoplastic polyester resin (PET) (DuPont Polymers, Wilmington, DE)

removed many times. Curve 2 is at the same temperature, but the loading is completely reversed tension and compression as would be experienced with a rotating beam or shaft loaded in bending. Curve 3 is the reversed bending load at 66°C (150°F), and Curve 4 is the same loading at 100°C (212°F) to show the effect of temperature on fatigue data.

5. Processing methods can have large effects on the final dimensions and properties of parts made from plastics. Molded plastics shrink significantly during solidification and curing. Parting lines produced where mold halves meet may affect strength. The rate of solidification may be widely different in a given part depending on the section thicknesses, the complexity of the shape and the location of sprues that deliver molten plastic into the mold. The same material can produce different properties depending on whether it is processed by injection molding, extrusion, blow molding, or machining from a solid block or bar.

6. Resistance to chemicals, weather, and other environmental conditions must be checked.

7. Plastics may exhibit a change in properties as they age, particularly when subjected to elevated temperatures. Figure 2–20 shows the reduction in tensile strength for a thermoplastic polyester resin when subjected to temperatures from 160°C (320°F) to 180°C (356°F) for a given number of hours of exposure. The reduction can be as much as 50% in as little time as 2000 hours (12 weeks).

8. Flammability and electrical characteristics must be considered. Some plastics are specially formulated for high flammability ratings as called for by Underwriters Laboratory and other agencies.

9. Plastics used for food storage or processing must meet U.S. Food and Drug Administration standards.

2–18 COMPOSITE MATERIALS

Composite materials are composed of two or more different materials that act together to produce properties that are different from, and generally superior to, those of the individual components. Typical composites include a polymeric resin matrix material with fibrous reinforcing material dispersed within it. Some advanced composites have a metal matrix. (See References 10 and 20.)

Designers can tailor the properties of composite materials to meet the specific needs of a particular application by the selection of each of several variables that determine the performance of the final product. Among the factors under the designer's control are the following:

1. Matrix resin or metal
2. Type of reinforcing fibers
3. Amount of fiber contained in the composite
4. Orientation of the fibers
5. Number of individual layers used
6. Overall thickness of the material
7. Orientation of the layers relative to each other
8. Combination of two or more types of composites or other materials into a composite structure

Typically, the filler is a strong, stiff material, whereas the matrix has a relatively low density. When the two materials bond together, much of the load-carrying ability of the composite is produced by the filler material. The matrix serves to hold the filler in a favorable orientation relative to the manner of loading and to distribute the loads to the filler. The result is a somewhat optimized composite that has high strength and high stiffness with low weight. Table 2–9 lists some of the composites formed by combinations of resins and fibers and their general characteristics and uses.

A virtually unlimited variety of composite materials can be produced by combining different matrix materials with different fillers in different forms and in different orientations. Some typical materials are listed below.

Matrix Materials

The following are among the more frequently used matrix materials:

- Thermoplastic polymers: Polyethylene, nylon, polypropylene, polystyrene, polyamides
- Thermosetting polymers: Polyester, epoxy, phenolic polyimide
- Ceramics and glass
- Carbon and graphite
- Metals: Aluminum, magnesium, titanium

Forms of Filler Materials

Many forms of filler materials are used:

- Continuous fiber strand consisting of many individual filaments bound together
- Chopped strands in short lengths (0.75 to 50 mm or 0.03 to 2.00 in)
- Chopped strands randomly spread in the form of a mat
- Roving: A group of parallel strands
- Woven fabric made from roving or strands
- Metal filaments or wires
- Solid or hollow microspheres
- Metal, glass, or mica flakes
- Single-crystal whiskers of materials such as graphite, silicon carbide, and copper

TABLE 2–9 Examples of composite materials and their uses

Type of composite	Typical applications
Glass/epoxy	Automotive and aircraft parts, tanks, sporting goods, printed wiring boards
Boron/epoxy	Aircraft structures and stabilizers, sporting goods
Graphite/epoxy	Aircraft and spacecraft structures, sporting goods, agricultural equipment, material handling devices, medical devices
Aramid/epoxy	Filament-wound pressure vessels, aerospace structures and equipment, protective clothing, automotive components
Glass/polyester	Sheet-molding compound (SMC), body panels for trucks and cars, large housings

Types of Filler Materials

Fillers, also called *fibers*, come in many types based on both organic and inorganic materials. The following are some of the more popular fillers:

- Glass fibers in five different types:
  A-glass: Good chemical resistance because it contains alkalis such as sodium oxide
  C-glass: Special formulations for even higher chemical resistance than A-glass
  E-glass: Widely used glass with good electrical insulating ability and good strength
  S-glass: High-strength, high-temperature glass
  D-glass: Better electrical properties than E-glass
- Quartz fibers and high-silica glass: Good properties at high temperatures up to 2000°F (1095°C)
- Carbon fibers made from PAN-base carbon (PAN is polyacrylonitrile): Approximately 95% carbon with very high modulus of elasticity
- Graphite fibers: Greater than 99% carbon and an even higher modulus of elasticity than carbon; the stiffest fibers typically used in composites
- Boron coated onto tungsten fibers: Good strength and a higher modulus of elasticity than glass
- Silicon carbide coated onto tungsten fibers: Strength and stiffness similar to those of boron/tungsten, but with higher temperature capability
- Aramid fibers: A member of the polyamide family of polymers; higher strength and stiffness with lower density as compared with glass; very flexible. (Aramid fibers produced by the DuPont Company carry the name *Kevlar*™.)

Processing of Composites

One method that is frequently used to produce composite products is first to place layers of sheet-formed fabrics on a form having the desired shape and then to impregnate the fabric with wet resin. Each layer of fabric can be adjusted in its orientation to produce special properties of the finished article. After the lay-up and resin impregnation are completed, the entire system is subjected to heat and pressure while a curing agent reacts with the base resin to produce cross-linking that binds all of the elements into a three-dimensional, unified structure. The polymer binds to the fibers and holds them in their preferred position and orientation during use.

An alternative method of fabricating composite products starts with a process of preimpregnating the fibers with the resin material to produce strands, tape, braids, or sheets. The resulting form, called a *prepreg*, can then be stacked into layers or wound onto a form to produce the desired shape and thickness. The final step is the curing cycle as described for the wet process.

Polyester-based composites are often produced as *sheet-molding compounds (SMC)* in which preimpregnated fabric sheets are placed into a mold and shaped and cured simultaneously under heat and pressure. Large body panels for automotive applications can be produced in this manner.

Pultrusion is a process in which the fiber reinforcement is coated with resin as it is pulled through a heated die to produce a continuous form in the desired shape. This process is used to produce rod, tubing, structural shapes (I-beams, channels, angles, and so on), tees, and hat sections used as stiffeners in aircraft structures.

Filament winding is used to make pipe, pressure vessels, rocket motor cases, instrument enclosures, and odd-shaped containers. The continuous filament can be placed in a variety of patterns, including helical, axial, and circumferential, to produce desired strength and stiffness characteristics.

Advantages of Composites

Designers typically seek to produce products that are safe, strong, stiff, lightweight, and highly tolerant of the environment in which the product will operate. Composites often excel in meeting these objectives when compared with alternative materials such as metals, wood, and unfilled plastics. Two parameters that are used to compare materials are *specific strength* and *specific modulus*, defined as follows:

> **Specific strength** *is the ratio of the tensile strength of a material to its specific weight.*

> **Specific modulus** *is the ratio of the modulus of elasticity of a material to its specific weight.*

Because the modulus of elasticity is a measure of the stiffness of a material, the specific modulus is sometimes called *specific stiffness*.

Although obviously not a length, both of these quantities have the *unit* of length, derived from the ratio of the units for strength or modulus of elasticity and the units for specific weight. In the U.S. Customary System, the units for tensile strength and modulus of elasticity are lb/in^2, whereas specific weight (weight per unit volume) is in lb/in^3. Thus, the unit for specific strength or specific modulus is inches. In the SI, strength and modulus are expressed in N/m^2 (pascals), whereas specific weight is in N/m^3. Then the unit for specific strength or specific modulus is meters.

Table 2–10 gives comparisons of the specific strength and specific stiffness of selected composite materials with certain steel, aluminum, and titanium alloys. Figure 2–21 shows a comparison of these materials using bar charts. Figure 2–22 is a plot of these data with specific strength on the vertical axis and specific modulus on the horizontal axis. When weight is critical, the ideal material will lie in the upper-right part of this chart. Note that data in these charts and figures are for composites having the filler materials aligned in the most favorable direction to withstand the applied loads.

Advantages of composites can be summarized as follows:

1. Specific strengths for composite materials can range as high as five times those of high-strength steel alloys. See Table 2–10 and Figures 2–21 and 2–22.
2. Specific modulus values for composite materials can be as high as eight times those for steel, aluminum, or titanium alloys. See Table 2–10 and Figures 2–21 and 2–22.
3. Composite materials typically perform better than steel or aluminum in applications where cyclic loads can lead to the potential for fatigue failure.
4. Where impact loads and vibrations are expected, composites can be specially formulated with materials that provide high toughness and a high level of damping.
5. Some composites have much higher wear resistance than metals.
6. Careful selection of the matrix and filler materials can provide superior corrosion resistance.
7. Dimensional changes due to changes in temperature are typically much less for composites than for metals.

TABLE 2–10 Comparison of specific strength and specific modulus for selected materials

Material	Tensile strength, s_u (ksi)	Specific weight, γ (lb/in^3)	Specific strength (in)	Specific modulus (in)
Metals				
Steel ($E = 30 \times 10^6$ psi)				
AISI 1020 HR	55	0.283	0.194×10^6	1.06×10^8
AISI 5160 OQT 700	263	0.283	0.929×10^6	1.06×10^8
Aluminum ($E = 10.0 \times 10^6$ psi)				
6061-T6	45	0.098	0.459×10^6	1.02×10^8
7075-T6	83	0.101	0.822×10^6	0.99×10^8
Titanium ($E = 16.5 \times 10^6$ psi)				
Ti-6Al-4V, quenched and aged at 1000°F	160	0.160	1.00×10^6	1.03×10^8
Composites				
Glass/epoxy composite ($E = 4.0 \times 10^6$ psi)				
34% fiber content	114	0.061	1.87×10^6	0.66×10^8
Aramid/epoxy composite ($E = 11.0 \times 10^6$ psi)				
60% fiber content	200	0.050	4.0×10^6	2.20×10^8
Boron/epoxy composite ($E = 30.0 \times 10^6$ psi)				
60% fiber content	270	0.075	3.60×10^6	4.00×10^8
Graphite/epoxy composite ($E = 19.7 \times 10^6$ psi)				
62% fiber content	278	0.057	4.86×10^6	3.45×10^8
Graphite/epoxy composite ($E = 48 \times 10^6$ psi)				
Ultrahigh modulus	160	0.058	2.76×10^6	8.28×10^8

8. Because composite materials have properties that are highly directional, designers can tailor the placement of reinforcing fibers in directions that provide the required strength and stiffness under the specific loading conditions to be encountered.

9. Composite structures can often be made in complex shapes in one piece, thus reducing the number of parts in a product and the number of fastening operations required. The elimination of joints typically improves the reliability of such structures as well.

10. Composite structures are typically made in their final form directly or in a near-net shape, thus reducing the number of secondary operations required.

Limitations of Composites

Designers must balance many properties of materials in their designs while simultaneously considering manufacturing operations, costs, safety, life, and service of the product. Listed next are some of the major concerns when using composites.

1. Material costs for composites are typically higher than for many alternative materials.

2. Fabrication techniques are quite different from those used to shape metals. New manufacturing equipment may be required, along with additional training for production operators.

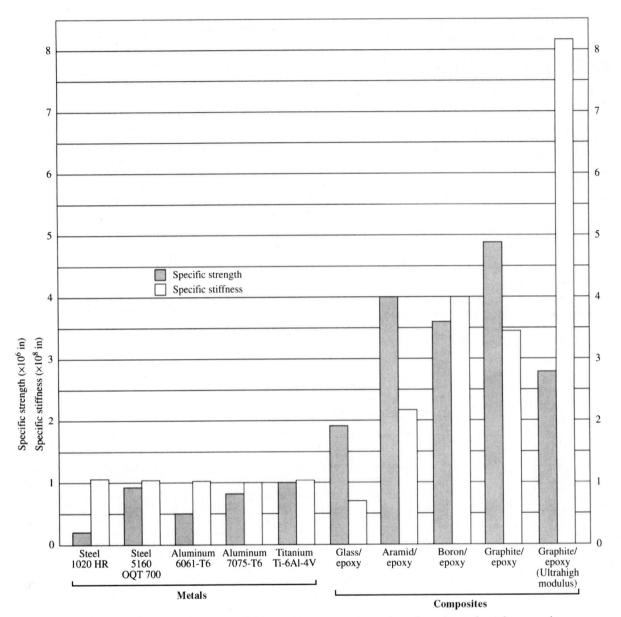

FIGURE 2–21   Comparison of specific strength and specific stiffness for selected metals and composites

3. The performance of products made from some composite production techniques is subject to a wider range of variability than the performance of products made from most metal fabrication techniques.

4. The operating temperature limits for composites having a polymeric matrix are typically 500°F (260°C). (But ceramic or metal matrix composites can be used at higher temperatures such as those found in engines.)

5. The properties of composite materials are not isotropic: Properties vary dramatically with the direction of the applied loads. Designers must account for these variations to ensure safety and satisfactory operation under all expected types of loading.

FIGURE 2–22
Specific strength versus specific modulus for selected metals and composites

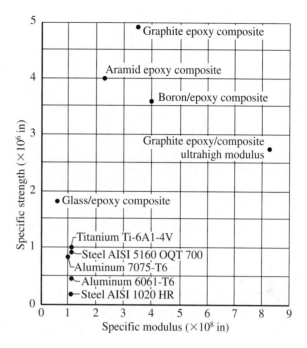

6. At this time, many designers lack understanding of the behavior of composite materials and the details of predicting failure modes. Whereas major advancements have been made in certain industries such as the aerospace and recreational equipment fields, there is a need for more general understanding about designing with composite materials.

7. The analysis of composite structures requires detailed knowledge of more properties of the materials than would be required for metals.

8. Inspection and testing of composite structures are typically more complicated and less precise than for metal structures. Special nondestructive techniques may be required to ensure that there are no major voids in the final product that could seriously weaken the structure. Testing of the complete structure may be required rather than testing of simply a sample of the material because of the interaction of different parts on each other and because of the directionality of the material properties.

9. Repair and maintenance of composite structures are serious concerns. Some of the initial production techniques require special environments of temperature and pressure that may be difficult to reproduce in the field when damage repair is required. Bonding of a repaired area to the parent structure may also be difficult.

Laminated Composite Construction

Many structures made from composite materials are made from several layers of the basic material containing both the matrix and the reinforcing fibers. The manner in which the layers are oriented relative to one another affects the final properties of the completed structure.

As an illustration, consider that each layer is made from a set of parallel strands of the reinforcing filler material, such as E-glass fibers, embedded in the resin matrix, such as polyester. As mentioned previously, in this form, the material is sometimes called a *prepreg*, indicating that the filler has been preimpregnated with the matrix prior to the forming of the structure and the curing of the assembly. To produce the maximum strength and stiffness in a particular direction, several layers or plies of the prepreg could be laid on top of one another with all of the fibers aligned in the direction of the expected tensile load. This is called a *unidirectional laminate*. After curing, the laminate would have a very high strength and stiffness when loaded in the direction of the strands, called the *longitudinal* direction. However, the resulting product would have a very low strength and stiffness in the direction perpendicular to the fiber direction, called the *transverse* direction. If any off-axis loads are encountered, the part may fail or deform significantly. Table 2–11 gives sample data for a unidirectional laminated, carbon/epoxy composite.

To overcome the lack of off-axis strength and stiffness, laminated structures should be made with a variety of orientations of the layers. One popular arrangement is shown in Figure 2–23. Naming the longitudinal direction of the surface layer the *0° ply*, this structure is referred to as

$$0°, 90°, +45°, -45°, -45°, +45°, 90°, 0°$$

The symmetry and the balance of this type of layering technique result in more nearly uniform properties in two directions. The term *quasi-isotropic* is sometimes used to describe such a structure. Note that the properties perpendicular to the faces of the layered

TABLE 2–11   Examples of the effect of laminate construction on strength and stiffness

	Tensile strength				Modulus of elasticity			
	Longitudinal		Transverse		Longitudinal		Transverse	
Laminate type	ksi	MPa	ksi	MPa	10^6 psi	GPa	10^6 psi	GPa
Unidirectional	200	1380	5	34	21	145	1.6	11
Quasi-isotropic	80	552	80	552	8	55	8	55

FIGURE 2–23
Multilayer, laminated, composite construction designed to produce quasi-isotropic properties

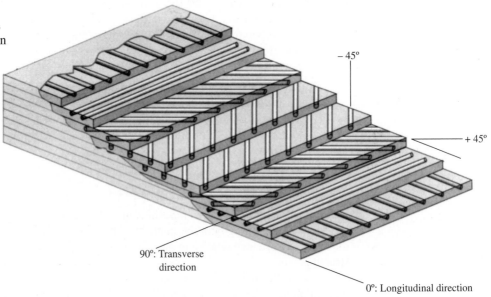

- 45°

+ 45°

90°: Transverse direction

0°: Longitudinal direction

structure (through the thickness) are still quite low because fibers do not extend in that direction. Also, the strength and the stiffness in the primary directions are somewhat lower than if the plies were aligned in the same direction. Table 2–11 also shows sample data for a quasi-isotropic laminate compared with one having unidirectional fibers in the same matrix.

Predicting Composite Properties

The following discussion summarizes some of the important variables needed to define the properties of a composite. The subscript c refers to the composite, m refers to the matrix, and f refers to the fibers. The strength and the stiffness of a composite material depend on the elastic properties of the fiber and matrix components. But another parameter is the relative volume of the composite composed of fibers, V_f, and that composed of the matrix material, V_m. That is,

$$V_f = \text{volume fraction of fiber in the composite}$$

$$V_m = \text{volume fraction of matrix in the composite}$$

Note that for a unit volume, $V_f + V_m = 1$; thus, $V_m = 1 - V_f$.

We will use an ideal case to illustrate the way in which the strength and the stiffness of a composite can be predicted. Consider a composite with unidirectional, continuous fibers aligned in the direction of the applied load. The fibers are typically much stronger and stiffer than the matrix material. Furthermore, the matrix will be able to undergo a larger strain before fracture than the fibers can. Figure 2–24 shows these phenomena on a plot of stress versus strain for the fibers and the matrix. We will use the following notation for key parameters from Figure 2–24:

$$s_{uf} = \text{ultimate strength of fiber}$$

$$\epsilon_{uf} = \text{strain in the fiber corresponding to its ultimate strength}$$

$$\sigma'_m = \text{stress in the matrix at the same strain as } \epsilon_{uf}$$

FIGURE 2–24
Stress versus strain for
fiber and matrix
materials

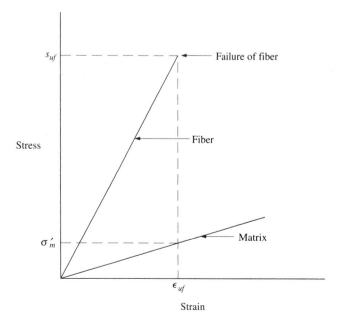

FIGURE 2–25
Relationship among
stresses and strains for
a composite and its
fiber and matrix
materials

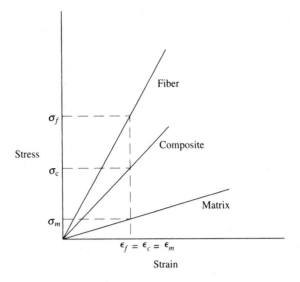

The ultimate strength of the composite, s_{uc}, is at some intermediate value between s_{uf} and σ'_m, depending on the volume fraction of fiber and matrix in the composite. That is,

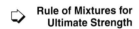 **Rule of Mixtures for Ultimate Strength**

$$s_{uc} = s_{uf} V_f + \sigma'_m V_m \qquad (2\text{–}8)$$

At any lower level of stress, the relationship among the overall stress in the composite, the stress in the fibers, and the stress in the matrix follows a similar pattern:

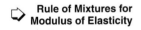 **Rule of Mixtures for Stress in a Composite**

$$\sigma_c = \sigma_f V_f + \sigma_m V_m \qquad (2\text{–}9)$$

Figure 2–25 illustrates this relationship on a stress-strain diagram.

Both sides of Equation (2–9) can be divided by the strain at which these stresses occur. Since for each material, $\sigma/\epsilon = E$, the modulus of elasticity for the composite can be shown as

 **Rule of Mixtures for Modulus of Elasticity**

$$E_c = E_f V_f + E_m V_m \qquad (2\text{–}10)$$

The density of a composite can be computed in a similar fashion:

Rule of Mixtures for Density of a Composite

$$\rho_c = \rho_f V_f + \rho_m V_m \qquad (2\text{–}11)$$

As mentioned previously (Section 2–2), density is defined as mass per unit volume. A related property, *specific weight,* is defined as weight per unit volume and is denoted by the symbol γ (the Greek letter gamma). The relationship between density and specific weight is simply $\gamma = \rho g$, where g is the acceleration due to gravity. Multiplying each term in Equation (2–11) by g gives the formula for the specific weight of a composite:

Rule of Mixtures for Specific Weight of a Composite

$$\gamma_c = \gamma_f V_f + \gamma_m V_m \qquad (2\text{–}12)$$

The forms of Equations (2–8) through (2–12) are examples of *the rules of mixtures.*

Table 2–12 lists example values for the properties of some matrix and filler materials. Remember that wide variations can occur in such properties, depending on the exact formulation and the condition of the materials.

TABLE 2–12   Example properties of matrix and filler materials

	Tensile strength		Tensile modulus		Specific weight	
	ksi	MPa	10^6 psi	GPa	lb/in^3	kN/m^3
Matrix materials:						
Polyester	10	69	0.40	2.76	0.047	12.7
Epoxy	18	124	0.56	3.86	0.047	12.7
Aluminum	45	310	10.0	69	0.100	27.1
Titanium	170	1170	16.5	114	0.160	43.4
Filler materials:						
S-glass	600	4140	12.5	86.2	0.09	24.4
Carbon-PAN	470	3240	33.5	231	0.064	17.4
Carbon-PAN (high-strength)	820	5650	40	276	0.065	17.7
Carbon (high-modulus)	325	2200	100	690	0.078	21.2
Aramid	500	3450	19.0	131	0.052	14.1

Example Problem 2–2   Compute the expected properties of ultimate tensile strength, modulus of elasticity, and specific weight of a composite made from unidirectional strands of carbon-PAN fibers in an epoxy matrix. The volume fraction of fibers is 30%. Use data from Table 2–12.

Solution   Objective   Compute the expected values of s_{uc}, E_c, and γ_c for the composite.

Given   Matrix-epoxy: $s_{um} = 18$ ksi; $E_m = 0.56 \times 10^6$ psi; $\gamma_m = 0.047$ lb/in^3.
Fiber-carbon-PAN: $s_{uf} = 470$ ksi; $E_f = 33.5 \times 10^6$ psi; $\gamma_f = 0.064$ lb/in^3.
Volume fraction of fiber: $V_f = 0.30$, and $V_m = 1.0 - 0.30 = 0.70$.

Analysis and Results   The ultimate tensile strength, s_{uc}, is computed from Equation (2–8):

$$s_{uc} = s_{uf}V_f + \sigma_m' V_m$$

To find σ_m', we first find the strain at which the fibers would fail at s_{uf}. Assume that the fibers are linearly elastic to failure. Then

$$\epsilon_f = s_{uf}/E_f = (470 \times 10^3 \text{ psi})/(33.5 \times 10^6 \text{ psi}) = 0.014$$

At this same strain, the stress in the matrix is

$$\sigma_m' = E_m\epsilon = (0.56 \times 10^6 \text{ psi})(0.014) = 7840 \text{ psi}$$

Then, in Equation (2–8),

$$s_{uc} = (470\,000 \text{ psi})(0.30) + (7840 \text{ psi})(0.70) = 146\,500 \text{ psi}$$

The modulus of elasticity computed from Equation (2–10):

$$E_c = E_f V_f + E_m V_m = (33.5 \times 10^6)(0.30) + (0.56 \times 10^6)(0.70)$$
$$E_c = 10.4 \times 10^6 \text{ psi}$$

The specific weight is computed from Equation (2–12):

$$\gamma_c = \gamma_f V_f + \gamma_m V_m = (0.064)(0.30) + (0.047)(0.70) = 0.052 \text{ lb/in}^3$$

Summary of Results

$$s_{uc} = 146\ 500 \text{ psi}$$
$$E_c = 10.4 \times 10^6 \text{ psi}$$
$$\gamma_c = 0.052 \text{ lb/in}^3$$

Comment   Note that the resulting properties for the composite are intermediate between those for the fibers and the matrix.

Design Guidelines for Members Made from Composites

The most important difference between designing with metals and designing with composites is that metals are typically taken to be homogeneous with isotropic strength and stiffness properties, whereas composites are decidedly *not* homogeneous or isotropic.

The failure modes of composite materials are complex. Tensile failure when the load is in-line with continuous fibers occurs when the individual fibers break. If the composite is made with shorter, chopped fibers, failure occurs when the fibers are pulled free from the matrix. Tensile failure when the load is perpendicular to continuous fibers occurs when the matrix itself fractures. If the fibers are in a woven form, or if a mat having shorter, randomly oriented fibers is used, other failure modes, such as fiber breakage or pullout, prevail. Such composites would have more nearly equal properties in any direction, or, as shown in Figure 2–23, multilayer laminate construction can be used.

Thus, an important design guideline to produce optimum strength is as follows:

Align the fibers with the direction of the load.

Another important failure mode is *interlaminar shear*, in which the plies of a multilayer composite separate under the action of shearing forces. The following is another design guideline:

Avoid shear loading, if possible.

Connections to composite materials are sometimes difficult to accomplish and provide places where fractures or fatigue failure could initiate. The manner of forming composites often allows the integration of several components into one part. Brackets, ribs, flanges, and the like, can be molded in along with the basic form of the part. The design guideline, then, is the following:

Combine several components into an integral structure.

When high panel stiffness is desired to resist flexure, as in beams or in broad panels such as floors, the designer can take advantage of the fact that the most effective material is near the outside surfaces of the panel or beam shape. Placing the high-strength fibers on these outer layers while filling the core of the shape with a light, yet rigid, material produces an efficient design in terms of weight for a given strength and stiffness. Figure 2–26 illustrates some examples of such designs. Another design guideline follows:

Use light core material covered with strong composite layers.

Because most composites use a polymeric material for the matrix, the temperatures that they can withstand are limited. Both strength and stiffness decrease as temperature in-

FIGURE 2–26
Laminated panels with
lightweight cores

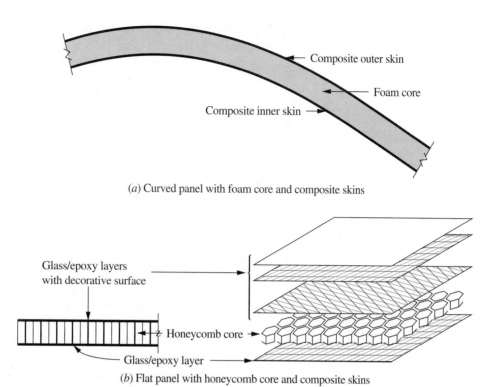

(*a*) Curved panel with foam core and composite skins

(*b*) Flat panel with honeycomb core and composite skins

creases. The polyimides provide better high-temperature properties [up to 600°F (318°C)] than most other polymer matrix materials. Epoxies are typically limited to from 250°F to 350°F (122°C–178°C). Any application above room temperature should be checked with material suppliers. The following is a design guideline:

Avoid high temperatures.

As described earlier in this section, many different fabrication techniques are used for composite materials. The shape can dictate a part's manufacturing technique. This is a good reason to implement the principles of concurrent engineering and adopt another design guideline:

Involve manufacturing considerations early in the design.

**2–19
MATERIALS
SELECTION**

One of the most important tasks for a designer is the specification of the material from which any individual component of a product is to be made. The decision must consider a huge number of factors, many of which have been discussed in this chapter.

The process of material selection must commence with a clear understanding of the functions and design requirements for the product and the individual component. Refer to Section 1–4 in Chapter 1 for a discussion of these concepts. Then, the designer should consider the interrelationships among the following:

- The functions of the component
- The component's shape
- The material from which the component is to be made
- The manufacturing process used to produce the component

Overall requirements for the performance of the component must be detailed. This includes, for example:

- The nature of the forces applied to the component
- The types and magnitudes of stresses created by the applied forces
- The allowable deformation of the component at critical points
- Interfaces with other components of the product
- The environment in which the component is to operate
- Physical size and weight of the component
- Aesthetics expected for the component and the overall product
- Cost targets for the product as a whole and this component in particular
- Anticipated manufacturing processes available

A much more detailed list may be made with more knowledge of specific conditions.

From the results of the exercises described previously, you should develop a list of key material properties that are important. Examples often include:

1. Strength as indicated by ultimate tensile strength, yield strength, compressive strength, fatigue strength, shear strength, and others
2. Stiffness as indicated by the tensile modulus of elasticity, shear modulus of elasticity, or flexural modulus
3. Weight and mass as indicated by specific weight or density
4. Ductility as indicated by the percent elongation
5. Toughness as indicated by the impact energy (Izod, Charpy, etc.)
6. Creep performance data
7. Corrosion resistance and compatibility with the environment
8. Cost for the material
9. Cost to process the material

A list of candidate materials should then be created using your knowledge of the behavior of several material types, successful similar applications, and emerging materials technologies. A rational decision analysis should be applied to determine the most suitable types of materials from the list of candidates. This could take the form of a matrix in which data for the properties just listed for each candidate material are entered and ranked. An analysis of the complete set of data will aid in making the final decision.

More comprehensive materials selection processes are described in References 2, 21, 22, and 25.

REFERENCES

1. Aluminum Association. *Aluminum Standards and Data.* Washington, DC: The Aluminum Association, 2000.

2. Ashby, M. F. *Materials Selection in Mechanical Design.* Oxford, England: Butterworth-Heinemann, 1999.

3. ASM International. *ASM Handbook.* Vol. 1, *Properties and Selection: Iron, Steels, and High-Performance Al-* *loys* (1990). Vol. 2, *Properties and Selection: Nonferrous Alloys and Special-Purpose Materials* (1991). Vol. 3, *Alloy Phase Diagrams* (1992). Vol. 4, *Heat Treating* (1991). Vol. 7, *Powder Metallurgy* (1998). Vol. 20, *Materials Selection and Design* (1997). Vol. 21, *Composites* (2001). Materials Park, OH: ASM International.

4. ASM International. *ASM Specialty Handbook: Carbon and Alloy Steels.* Edited by J. R. Davis. Materials Park, OH: ASM International, 1996.

5. ASM International. *ASM Specialty Handbook: Stainless Steels.* Edited by J. R. Davis. Materials Park, OH: ASM International, 1994.

6. ASM International. *ASM Specialty Handbook: Tool Materials.* Edited by J. R. Davis. Materials Park, OH: ASM International, 1995.

7. ASM International. *ASM Specialty Handbook: Heat-Resistant Materials.* Edited by J. R. Davis. Materials Park, OH: ASM International, 1997.

8. ASM International. *ASM Specialty Handbook: Aluminum and Aluminum Alloys.* Edited by J. R. Davis. Materials Park, OH: ASM International, 1993.

9. ASM International. *ASM Specialty Handbook: Cast Irons.* Edited by J. R. Davis. Materials Park, OH: ASM International, 1996.

10. ASM International. *ASM Engineered Materials Handbook: Composites.* Materials Park, OH: ASM International, 1987.

11. ASM International. *ASM Engineered Materials Handbook: Engineering Plastics.* Materials Park, OH: ASM International, 1988.

12. ASTM International. *Metals and Alloys in the Unified Numbering System.* 9th ed. West Conshohocken, PA: ASTM International. 2001. Jointly developed by ASTM and the Society of Automotive Engineers (SAE).

13. ASTM International. *Standard Practice for Numbering Metals and Alloys (UNS).* West Conshohocken, PA: ASTM International Standard E527-83 (1997), 2001.

14. Bethlehem Steel Corporation. *Modern Steels and Their Properties.* Bethlehem, PA: Bethlehem Steel Corporation, 1980.

15. Brooks, Charlie R. *Principles of the Heat Treatment of Plain Carbon and Low Alloy Steels.* Materials Park, OH: ASM International, 1996.

16. Budinski, Kenneth G. *Engineering Materials: Properties and Selection.* 6th ed. Upper Saddle River, NJ: Prentice Hall, 2001.

17. Budinski, Kenneth G. *Surface Engineering for Wear Resistance.* Upper Saddle River, NJ: Prentice Hall, 1988.

18. DuPont Engineering Polymers. *Design Handbook for DuPont Engineering Polymers: General Design Principles.* Wilmington, DE: The DuPont Company, 1992.

19. INTERZINC. *Zinc Casting: A Systems Approach.* Algonac, MI: INTERZINC.

20. Jang, Bor Z. *Advanced Polymer Composites: Principles and Applications.* Materials Park, OH: ASM International, 1994.

21. Lesko, J. *Industrial Design Materials and Manufacturing.* New York: John Wiley, 1999.

22. Mangonon, P. L. *The Principles of Materials Selection for Engineering Design.* Upper Saddle River, NJ: Prentice Hall, 1999.

23. Muccio, E. A. *Plastic Part Technology.* Materials Park, OH: ASM International, 1991.

24. Penton Publishing. *Machine Design Magazine,* Vol. 69. Cleveland, OH: Penton Publishing, 1997.

25. Shackelford, J. F., W. Alexander, and Jun S. Park. *CRC Practical Handbook of Materials Selection.* Boca Raton, FL: CRC Press, 1995.

26. Zinn, S., and S. L. Semiatin. *Elements of Induction Heating: Design, Control, and Applications.* Materials Park, OH: ASM International, 1988.

INTERNET SITES RELATED TO DESIGN PROPERTIES OF MATERIALS

1. **AZoM.com (The A to Z of Materials)** *www.azom.com* Materials information resource for the design community. No cost, searchable databases for metals, ceramics, polymers, and composites. Can also search by keyword, application, or industry type.

2. **Matweb** *www.matweb.com* Database of material properties for many metals, plastics, ceramics, and other engineering materials.

3. **ASM International** *www.asm-intl.org* The society for materials engineers and scientists, a worldwide network dedicated to advancing industry, technology, and applications of metals and other materials.

4. **TECHstreet** *www.techstreet.com* A store for purchasing standards for the metals industry.

5. **SAE International** *www.sae.org* The Society of Automotive Engineers, the engineering society for advancing mobility on land or sea, in air or space. A resource for technical information used in designing self-propelled vehicles. Offers standards on metals, plastics, and other materials along with components and subsystems of vehicles.

6. **ASTM International** *www.astm.org* Formerly known as the American Society for Testing and Materials. Develops and sells standards for material properties, testing procedures, and numerous other technical standards.

7. **American Iron and Steel Institute** *www.steel.org* AISI develops industry standards for steel materials and

products made from steel. Steel product manuals and industry standards are made available through the Iron & Steel Society (ISS), listed separately.

8. **Iron & Steel Society** *www.iss.org* Provides industry standards and other publications for advancing knowledge exchange in the global iron and steel industry.

9. **Aluminum Association** *www.aluminum.org* The association of the aluminum industry. Provides numerous publications that can be purchased.

10. **Alcoa, Inc.** *www.alcoa.com* A producer of aluminum and fabricated products. Website can be searched for properties of specific alloys.

11. **Copper Development Association** *www.copper.org* Provides a large searchable database of properties of wrought and cast copper, copper alloys, brasses, and bronzes. Allows searching for appropriate alloys for typical industrial uses based on several performance characteristics.

12. **Metal Powder Industries Federation** *www.mpif.org* The international trade association representing the powder metal producers. Standards and publications related to the design and production of products using powder metals.

13. **INTERZINC** *www.interzinc.com* A market development and technology transfer group dedicated to increasing awareness of zinc casting alloys. Provides

design assistance, alloy selection guide, alloy properties, and descriptions of casting alloys.

14. **RAPRA Technology Limited** *www.rapra.net* Comprehensive information source for the plastics and rubber industries. Formerly Rubber and Plastics Research Association. This site also hosts the Cambridge Engineering Selector, a computerized resource using the materials selection methodology of M. F. Ashby. See Reference 2.

15. **DuPont Plastics** *www.plastics.dupont.com* Information and data on DuPont plastics and their properties. Searchable database by type of plastic or application.

16. **PolymerPlace.com** *www.polymerplace.com* Information resource for the polymer industry.

17. **Plastics Technology Online** *www.plasticstechnology.com* Online resource of Plastics Technology magazine.

18. **PLASPEC Materials Selection Database** *www.plaspec.com* Affiliated with Plastics Technology Online. Provides current articles and information about plastics injection molding, extrusion, blow molding, materials, tooling, and auxiliary equipment.

19. **Society of Plastics Engineers** *www.4spe.org* SPE promotes scientific and engineering knowledge and education about plastics and polymers worldwide.

PROBLEMS

1. Define *ultimate tensile strength*.

2. Define *yield point*.

3. Define *yield strength* and tell how it is measured.

4. What types of materials would have a yield point?

5. What is the difference between proportional limit and elastic limit?

6. Define *Hooke's law*.

7. What property of a material is a measure of its stiffness?

8. What property of a material is a measure of its ductility?

9. If a material is reported to have a percent elongation in a 2.00-in gage length of 2%, is it ductile?

10. Define *Poisson's ratio*.

11. If a material has a tensile modulus of elasticity of 114 GPa and a Poisson's ratio of 0.33, what is its modulus of elasticity in shear?

12. A material is reported to have a Brinell hardness of 525. What is its approximate hardness on the Rockwell C scale?

13. A steel is reported to have a Brinell hardness of 450. What is its approximate tensile strength?

For Problems 14–17, describe what is wrong with each statement.

14. "After annealing, the steel bracket had a Brinell hardness of 750."

15. "The hardness of that steel shaft is HRB 120."

16. "The hardness of that bronze casting is HRC 12."

17. "Based on the fact that this aluminum plate has a hardness of HB 150, its approximate tensile strength is 75 ksi."

18. Name two tests used to measure impact energy.

19. What are the principal constituents in steels?

20. What are the principal alloying elements in AISI 4340 steel?

21. How much carbon is in AISI 4340 steel?

22. What is the typical carbon content of a low-carbon steel? Of a medium-carbon steel? Of a high-carbon steel?

23. How much carbon does a bearing steel typically contain?

24. What is the main difference between AISI 1213 steel and AISI 12L13 steel?

25. Name four materials that are commonly used for shafts.

26. Name four materials that are typically used for gears.

27. Describe the properties desirable for the auger blades of a post hole digger, and suggest a suitable material.

28. Appendix 3 lists AISI 5160 OQT 1000. Describe the basic composition of this material, how it was processed, and its properties in relation to other steels listed in that table.

29. If a shovel blade is made from AISI 1040 steel, would you recommend flame hardening to give its edge a surface hardness of HRC 40? Explain.

30. Describe the differences between through-hardening and carburizing.

31. Describe the process of induction hardening.

32. Name 10 steels used for carburizing. What is their approximate carbon content prior to carburizing?

33. What types of stainless steels are nonmagnetic?

34. What is the principal alloying element that gives a stainless steel corrosion resistance?

35. Of what material is a typical wide-flange beam made?

36. With regard to structural steels, what does the term *HSLA* mean? What strengths are available in HSLA steel?

37. Name three types of cast iron.

38. Describe the following cast iron materials according to type, tensile strength, yield strength, ductility, and stiffness:

    ASTM A48-83, Grade 30

    ASTM A536-84, Grade 100-70-03

    ASTM A47-84, Grade 35018

    ASTM A220-88, Grade 70003

39. Describe the process of making parts from powdered metals.

40. What properties are typical for parts made from Zamak 3 zinc casting alloy?

41. What are the typical uses for Group D tool steels?

42. What does the suffix *O* in aluminum 6061-O represent?

43. What does the suffix *H* in aluminum 3003-H14 represent?

44. What does the suffix *T* in aluminum 6061-T6 represent?

45. Name the aluminum alloy and condition that has the highest strength of those listed in Appendix 9.

46. Which is one of the most versatile aluminum alloys for mechanical and structural uses?

47. Name three typical uses for titanium alloys.

48. What is the principal constituent of bronze?

49. Describe the bronze having the UNS designation C86200.

50. Name two typical uses for bronze in machine design.

51. Describe the difference between thermosetting plastics and thermoplastics.

52. Suggest a suitable plastic material for each of the following uses:

    (a) Gears

    (b) Football helmets

    (c) Transparent shield

    (d) Structural housing

    (e) Pipe

    (f) Wheels

    (g) Electrical switch-gear, structural part

53. Name eight factors over which the designer has control when specifying a composite material.

54. Define the term *composite*.

55. Name four base resins often used for composite materials.

56. Name four types of reinforcement fibers used for composite materials.

57. Name three types of composite materials used for sporting equipment, such as tennis rackets, golf clubs, and skis.

58. Name three types of composite materials used for aircraft and aerospace structures.

59. What base resin and reinforcement are typically used for sheet-molding compound (SMC)?

60. For what applications are sheet-molding compounds used?

61. Describe six forms in which reinforcing fibers are produced.

62. Describe *wet processing* of composite materials.

63. Describe *preimpregnated materials*.

64. Describe the production processing of sheet-molding compounds.

65. Describe *pultrusion*, and list four shapes produced by this process.

66. Describe *filament winding* and four types of products made by this process.

67. Define the term *specific strength* as it is applied to structural materials.

68. Define the term *specific stiffness* as it is applied to structural materials.

69. Discuss the advantages of composite materials relative to metals with regard to specific strength and specific stiffness.

70. Compare the specific strength of AISI 1020 hot-rolled steel with that of AISI 5160 OQT 700 steel, the two aluminum alloys 6061-T6 and 7075-T6, and titanium Ti-6Al-4V.

71. Compare the specific stiffness of AISI 1020 hot-rolled steel with that of AISI 5160 OQT 700 steel, the two aluminum alloys 6061-T6 and 7075-T6, and titanium Ti-6Al-4V.

72. Compare the specific strengths of each of the five composite materials shown in Figure 2–21 with that of AISI 1020 hot-rolled steel.

73. Compare the specific stiffness of each of the five composite materials shown in Figure 2–21 with that of AISI 1020 hot-rolled steel.

74. Describe the general construction of a composite material identified as $[0/+30/-30/90]$.

75. List and discuss six design guidelines for the application of composite materials.

76. Why is it desirable to form a composite material in layers or plies with the angle of orientation of the different plies in different directions?

77. Why is it desirable to form a composite structural element with relatively thin skins of the stronger composite material over a core of light foam?

78. Describe why concurrent engineering and early manufacturing involvement are important when you are designing parts made from composite materials.

Internet-Based Assignments

79. Use the Matweb website to determine at least three appropriate materials for a shaft design. An alloy steel is preferred with a minimum yield strength of 150 ksi (1035 MPa) and a good ductility as represented by an elongation of 10% or greater.

80. Use the Matweb website to determine at least three appropriate plastic materials for use as a cam. The materials should have good strength properties and a high toughness.

81. Use the DuPont Plastics website to determine at least three appropriate plastic materials for use as a cam. The materials should have good strength properties and a high toughness.

82. Use the DuPont Plastics website to determine at least three appropriate plastic materials for use as a housing for an industrial product. Moderate strength, high rigidity, and high toughness are required.

83. Use the Alcoa website to determine at least three appropriate aluminum alloys for a mechanical component that requires moderate strength, good machinability, and good corrosion resistance.

84. Use the INTERZINC website to determine at least three appropriate zinc casting alloys for a structural component that requires good strength and that is recommended for die casting.

85. Use the Copper Development Association website to recommend at least three copper alloys for a wormgear. Good strength and ductility are desirable along with good wear properties.

86. Use the Copper Development Association website to recommend at least three copper alloys for a bearing application. Moderate strength and good friction and wear properties are required.

87. Locate the description of the ASTM Standard A992 structural steel that is commonly used for rolled steel beam shapes. Determine how to acquire a copy of the standard.

3

Stress and Deformation Analysis

The Big Picture

You Are the Designer

3–1 Objectives of This Chapter

3–2 Philosophy of a Safe Design

3–3 Representing Stresses on a Stress Element

3–4 Direct Stresses: Tension and Compression

3–5 Deformation under Direct Axial Loading

3–6 Direct Shear Stress

3–7 Relationship among Torque, Power, and Rotational Speed

3–8 Torsional Shear Stress

3–9 Torsional Deformation

3–10 Torsion in Members Having Noncircular Cross Sections

3–11 Torsion in Closed, Thin-Walled Tubes

3–12 Open Tubes and a Comparison with Closed Tubes

3–13 Vertical Shearing Stress

3–14 Special Shearing Stress Formulas

3–15 Stress Due to Bending

3–16 Flexural Center for Beams

3–17 Beam Deflections

3–18 Equations for Deflected Beam Shape

3–19 Beams with Concentrated Bending Moments

3–20 Combined Normal Stresses: Superposition Principle

3–21 Stress Concentrations

3–22 Notch Sensitivity and Strength Reduction Factor

The Big Picture

Stress and Deformation Analysis

Discussion Map

☐ As a designer you are responsible for ensuring the safety of the components and systems you design.

☐ You must apply your prior knowledge of the principles of strength of materials.

Discover

How could consumer products and machines fail?
Describe some product failures you have seen.

This chapter presents a brief review of the fundamentals of stress analysis. It will help you design products that do not fail, and it will prepare you for other topics later in this book.

A designer is responsible for ensuring the safety of the components and systems that he or she designs. Many factors affect safety, but one of the most critical aspects of design safety is that the level of stress to which a machine component is subjected must be safe under reasonably foreseeable conditions. This principle implies, of course, that nothing actually breaks. Safety may also be compromised if components are permitted to deflect excessively, even though nothing breaks.

You have already studied the principles of strength of materials to learn the fundamentals of stress analysis. Thus, at this point, you should be competent to analyze load-carrying members for stress and deflection due to direct tensile and compressive loads, direct shear, torsional shear, and bending.

Think, now, about consumer products and machines with which you are familiar, and try to explain how they *could fail.* Of course, we do not expect them to fail, because most such products are well designed. But some do fail. Can you recall any? How did they fail? What were the operating conditions when they failed? What was the material of the components that failed? Can you visualize and describe the kinds of loads that were placed on the components that failed? Were they subjected to bending, tension, compression, shear, or torsion? Could there have been more than one type of stress acting at the same time? Are there evidences of accidental overloads? Should such loads have been anticipated by the designer? Could the failure be due to the manufacture of the product rather than its design?

Talk about product and machine failures with your associates and your instructor. Consider parts of your car, home appliances, lawn maintenance equipment, or equipment where you have worked. If possible, bring failed components to the meetings with your associates, and discuss the components and their failure.

Most of this book emphasizes developing special methods to analyze and design machine elements. These methods are all based on the fundamentals of stress analysis, and it is assumed that you have completed a course in strength of materials. This chapter presents a brief review of the fundamentals. (See References 1, 3, 4, and 6.)

You Are the Designer

You are the designer of a utility crane that might be used in an automotive repair facility, in a manufacturing plant, or on a mobile unit such as a truck bed. Its function is to raise heavy loads. A schematic layout of one possible configuration of the crane is shown in Figure 3–1. It is comprised of four primary load-carrying members, labeled 1, 2, 3, and 4. These members are connected to each other with pin-type joints at A, B, C, D, E, and F. The load is applied to the end of the horizontal boom, member 4. Anchor points for the crane are provided at joints A and B that carry the loads from the crane to a rigid structure. Note that this is a simplified view of the crane showing only the primary structural components and the forces in the plane of the applied load. The crane would also need stabilizing members in the plane perpendicular to the drawing.

You will need to analyze the kinds of forces that are exerted on each of the load-carrying members before you can design them. This calls for the use of the principles of statics in which you should have already gained competence. The following discussion provides a review of some of the key principles you will need in this course.

Your work as a designer proceeds as follows:

1. Analyze the forces that are exerted on each load-carrying member using the principles of statics.
2. Identify the kinds of stresses that each member is subjected to by the applied forces.
3. Propose the general shape of each load-carrying member and the material from which each is to be made.
4. Complete the stress analysis for each member to determine its final dimensions.

Let's work through steps 1 and 2 now as a review of statics. You will improve your ability to do steps 3 and 4 as you perform several practice problems in this chapter and in Chapters 4 and 5 by reviewing strength of materials and adding competencies that build on that foundation.

Force Analysis:

One approach to the force analysis is outlined here.

1. Consider the entire crane structure as a free-body with the applied force acting at point G and the reactions acting at support points A and B. See Figure 3–2, which shows these forces and important dimensions of the crane structure.

FIGURE 3–1   Schematic layout of a crane

FIGURE 3–2 Free-body diagram of complete crane structure

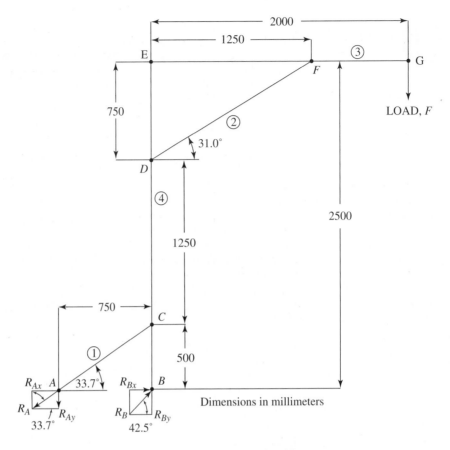

Reaction forces at supports A and B

2. Break the structure apart so that each member is represented as a free-body diagram, showing all forces acting at each joint. See the result in Figure 3–3.
3. Analyze the magnitudes and directions of all forces.

Comments are given here to summarize the methods used in the static analysis and to report results. You should work through the details of the analysis yourself or with colleagues to ensure that you can perform such calculations. All of the forces are directly proportional to the applied force F. We will show the results with an assumed value of $F = 10.0$ kN (approximately 2250 lb).

Step 1: The pin joints at A and B can provide support in any direction. We show the x and y components of the reactions in Figure 3–2. Then, proceed as follows:

1. Sum moments about B to find
   $R_{Ay} = 2.667\ F = 26.67$ kN
2. Sum forces in the vertical direction to find
   $R_{By} = 3.667\ F = 36.67$ kN.

At this point we need to recognize that the strut AC is pin-connected at each end and carries loads only at its ends. Therefore, it is a *two-force member,* and the direction of the total force, R_A, acts along the member itself. Then R_{Ay} and R_{Ax} are the rectangular components of R_A as shown in the lower left of Figure 3–2. We can then say that

$$\tan(33.7°) = R_{Ay}/R_{Ax}$$

and then

$$R_{Ax} = R_{Ay}/\tan(33.7°) = 26.67\ \text{kN}/\tan(33.7°) = 40.0\ \text{kN}$$

The total force, R_A, can be computed from the Pythagorean theorem,

$$R_A = \sqrt{R_{Ax}^2 + R_{Ay}^2} = \sqrt{(40.0)^2 + (26.67)^2} = 48.07\ \text{kN}$$

This force acts along the strut AC, at an angle of 33.7° above the horizontal, and it is the force that tends to shear the pin in joint A. The force at C on the strut AC is also 48.07 kN acting upward to the right to balance R_A on the two-force member as shown in Figure 3–3. Member AC is therefore in pure tension.

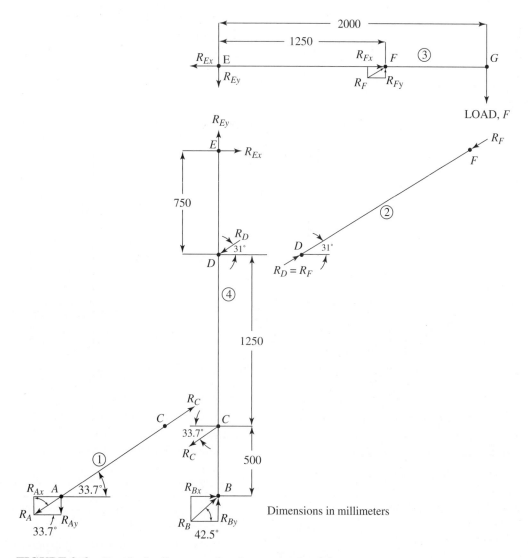

FIGURE 3–3 Free-body diagrams of each component of the crane

We can now use the sum of the forces in the horizontal direction on the entire structure to show that $R_{Ax} = R_{Bx} = 40.0$ kN. The resultant of R_{Bx} and R_{By} is 54.3 kN acting at an angle of 42.5° above the horizontal, and it is the total shearing force on the pin in joint B. See the diagram in the lower right of Figure 3–2.

Step 2: The set of free-body diagrams is shown in Figure 3–3.

Step 3: Now consider the free-body diagrams of all of the members in Figure 3–3. We have already discussed member 1, recognizing it as a two-force member in tension carrying forces R_A and

R_C equal to 48.07 kN. The reaction to R_C acts on the vertical member 4.

Now note that member 2 is also a two-force member, but it is in compression rather than tension. Therefore we know that the forces on points D and F are equal and that they act in line with member 2, 31.0° with respect to the horizontal. The reactions to these forces, then, act at point D on the vertical support, member 4, and at point F on the horizontal boom, member 3. We can find the value of R_F by considering the free-body diagram of member 3. You should be able to verify the following results using the methods already demonstrated.

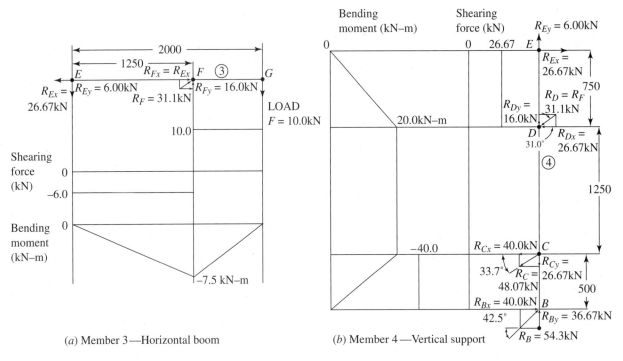

(a) Member 3—Horizontal boom   (b) Member 4—Vertical support

FIGURE 3–4   Shearing force and bending moment diagrams for members 3 and 4

R_{Fy} = 1.600 F = (1.600)(10.0 kN) = 16.00 kN

R_{Fx} = 2.667 F = (2.667)(10.0 kN) = 26.67 kN

R_F = 3.110 F = (3.110)(10.0 kN) = 31.10 kN

R_{Ey} = 0.600 F = (0.600)(10.0 kN) = 6.00 kN

R_{Ex} = 2.667 F = (2.667)(10.0 kN) = 26.67 kN

R_E = 2.733 F = (2.733)(10.0 kN) = 27.33 kN

Now all forces on the vertical member 4 are known from earlier analyses using the principle of action-reaction at each joint.

Types of Stresses on Each Member:

Consider again the free-body diagrams in Figure 3–3 to visualize the kinds of stresses that are created in each member. This will lead to the use of particular kinds of stress analysis as the design process is completed. Members 3 and 4 carry forces perpendicular to their long axes and, therefore, they act as beams in bending. Figure 3–4 shows these members with the additional shearing force and bending moment diagrams. You should have learned to prepare such diagrams in the prerequisite study of strength of materials. The following is a summary of the kinds of stresses in each member.

Member 1: The strut is in pure tension.

Member 2: The brace is in pure compression. Column buckling should be checked.

Member 3: The boom acts as a beam in bending. The right end between *F* and *G* is subjected to bending stress and vertical shear stress. Between *E* and *F* there is bending and shear combined with an axial tensile stress.

Member 4: The vertical support experiences a complex set of stresses depending on the segment being considered as described here.

Between *E* and *D*: Combined bending stress, vertical shear stress, and axial tension.
Between *D* and *C*: Combined bending stress and axial compression.
Between *C* and *B*: Combined bending stress, vertical shear stress, and axial compression.

Pin Joints: The connections between members at each joint must be designed to resist the total reaction force acting at each, computed in the earlier analysis. In general, each connection will likely include a cylindrical pin connecting two parts. The pin will typically be in direct shear.

<table>
<tr>
<td>

3–1
OBJECTIVES OF
THIS CHAPTER

</td>
<td>

After completing this chapter, you will:

1. Have reviewed the principles of stress and deformation analysis for several kinds of stresses, including the following:

   Direct tension and compression

   Direct shear

   Torsional shear for both circular and noncircular sections

   Vertical shearing stresses in beams

   Bending

2. Be able to interpret the nature of the stress at a point by drawing the *stress element* at any point in a load-carrying member for a variety of types of loads.

3. Have reviewed the importance of the *flexural center* of a beam cross section with regard to the alignment of loads on beams.

4. Have reviewed beam-deflection formulas.

5. Be able to analyze beam-loading patterns that produce abrupt changes in the magnitude of the bending moment in the beam.

6. Be able to use the principle of superposition to analyze machine elements that are subjected to loading patterns that produce combined stresses.

7. Be able to properly apply stress concentration factors in stress analyses.

</td>
</tr>
</table>

3–2
PHILOSOPHY OF
A SAFE DESIGN

In this book, every design approach will ensure that the stress level is below yield in ductile materials, automatically ensuring that the part will not break under a static load. For brittle materials, we will ensure that the stress levels are well below the ultimate tensile strength. We will also analyze deflection where it is critical to safety or performance of a part.

Two other failure modes that apply to machine members are fatigue and wear. *Fatigue* is the response of a part subjected to repeated loads (see Chapter 5). *Wear* is discussed within the chapters devoted to the machine elements, such as gears, bearings, and chains, for which it is a major concern.

3–3
REPRESENTING
STRESSES
ON A STRESS
ELEMENT

One major goal of stress analysis is to determine *the point* within a load-carrying member that is subjected to the highest stress level. You should develop the ability to visualize a *stress element*, a single, infinitesimally small cube from the member in a highly stressed area, and to show vectors that represent the kind of stresses that exist on that element. The orientation of the stress element is critical, and it must be aligned with specified axes on the member, typically called *x*, *y*, and *z*.

Figure 3–5 shows three examples of stress elements with three basic fundamental kinds of stress: tensile, compressive, and shear. Both the complete three-dimensional cube and the simplified, two-dimensional square forms for the stress elements are shown. The square is one face of the cube in a selected plane. The sides of the square represent the projections of the faces of the cube that are perpendicular to the selected plane. It is recommended that you visualize the cube form first and then represent a square stress element showing stresses on a particular plane of interest in a given problem. In some problems with more general states of stress, two or three square stress elements may be required to depict the complete stress condition.

Tensile and compressive stresses, called *normal stresses*, are shown acting perpendicular to opposite faces of the stress element. Tensile stresses tend to pull on the element, whereas compressive stresses tend to crush it.

FIGURE 3–5 Stress elements for three types of stresses

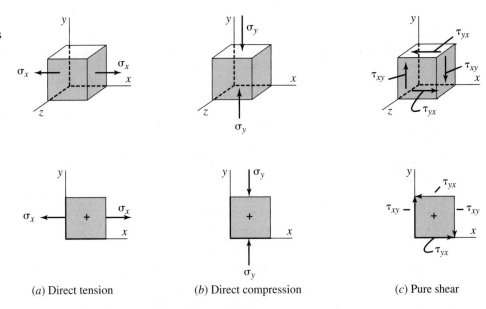

(a) Direct tension   (b) Direct compression   (c) Pure shear

Shear stresses are created by direct shear, vertical shear in beams, or torsion. In each case, the action on an element subjected to shear is a tendency to *cut* the element by exerting a stress downward on one face while simultaneously exerting a stress upward on the opposite, parallel face. This action is that of a simple pair of shears or scissors. But note that if only one pair of shear stresses acts on a stress element, it will not be in equilibrium. Rather, it will tend to spin because the pair of shear stresses forms a couple. To produce equilibrium, a second pair of shear stresses on the other two faces of the element must exist, acting in a direction that opposes the first pair.

In summary, shear stresses on an element will always be shown as two pairs of equal stresses acting on (parallel to) the four sides of the element. Figure 3–5(c) shows an example.

Sign Convention for Shear Stresses

This book adopts the following convention:

Positive shear stresses tend to rotate the element in a clockwise direction.

Negative shear stresses tend to rotate the element in a counterclockwise direction.

A double subscript notation is used to denote shear stresses in a plane. For example, in Figure 3–5(c), drawn for the *x-y* plane, the pair of shear stresses, τ_{xy}, indicates a shear stress acting on the element face that is perpendicular to the *x*-axis and parallel to the *y*-axis. Then τ_{yx} acts on the face that is perpendicular to the *y*-axis and parallel to the *x*-axis. In this example, τ_{xy} is positive and τ_{yx} is negative.

3–4 DIRECT STRESSES: TENSION AND COMPRESSION

Stress can be defined as the internal resistance offered by a unit area of a material to an externally applied load. *Normal stresses* (σ) are either *tensile* (positive) or *compressive* (negative).

For a load-carrying member in which the external load is uniformly distributed across the cross-sectional area of the member, the magnitude of the stress can be calculated from the direct stress formula:

⇨ **Direct Tensile or Compressive Stress**

$$\sigma = \text{force/area} = F/A \qquad\qquad (3\text{–}1)$$

The units for stress are always *force per unit area*, as is evident from Equation 3–1. Common units in the U.S. Customary system and the SI metric system follow.

U.S. Customary Units   *SI Metric Units*

lb/in^2 = psi N/m^2 = pascal = Pa

$kips/in^2$ = ksi N/mm^2 = megapascal = 10^6 Pa = MPa

Note: 1.0 kip = 1000 lb

1.0 ksi = 1000 psi

Example Problem 3–1   A tensile force of 9500 N is applied to a 12-mm-diameter round bar, as shown in Figure 3–6. Compute the direct tensile stress in the bar.

FIGURE 3–6   Tensile stress in a round bar

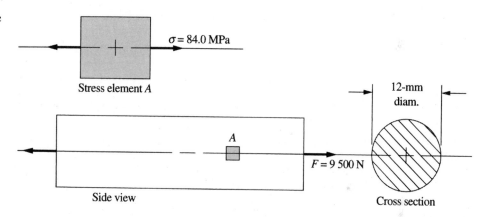

Solution   Objective   Compute the tensile stress in the round bar.

Given   Force = F = 9500 N; diameter = D = 12 mm.

Analysis   Use the direct tensile stress formula, Equation (3–1): $\sigma = F/A$. Compute the cross-sectional area from $A = \pi D^2/4$.

Results $A = \pi D^2/4 = \pi(12\ mm)^2/4 = 113\ mm^2$
$\sigma = F/A = (9500\ N)/(113\ mm^2) = 84.0\ N/mm^2 = 84.0\ MPa$

Comment   The results are shown on stress element A in Figure 3–6, which can be taken to be anywhere within the bar because, ideally, the stress is uniform on any cross section. The cube form of the element is as shown in Figure 3–5 (a).

The conditions on the use of Equation (3–1) are as follows:

1. The load-carrying member must be straight.

2. The line of action of the load must pass through the centroid of the cross section of the member.

3. The member must be of uniform cross section near where the stress is being computed.

4. The material must be homogeneous and isotropic.

5. In the case of compression members, the member must be short to prevent buckling. The conditions under which buckling is expected are discussed in Chapter 6.

3–5
DEFORMATION UNDER DIRECT AXIAL LOADING

The following formula computes the stretch due to a direct axial tensile load or the shortening due to a direct axial compressive load:

⇨ **Deformation Due to Direct Axial Load**

$$\delta = FL/EA \tag{3–2}$$

where δ = total deformation of the member carrying the axial load
  F = direct axial load
  L = original total length of the member
  E = modulus of elasticity of the material
  A = cross-sectional area of the member

Noting that $\sigma = F/A$, we can also compute the deformation from

$$\delta = \sigma L/E \tag{3–3}$$

Example Problem 3–2 For the round bar subjected to the tensile load shown in Figure 3–6, compute the total deformation if the original length of the bar is 3600 mm. The bar is made from a steel having a modulus of elasticity of 207 GPa.

Solution Objective Compute the deformation of the bar.

Given Force = F = 9500 N; diameter = D = 12 mm.
Length = L = 3600 mm; E = 207 GPa

Analysis From Example Problem 3–1, we found that σ = 84.0 MPa. Use Equation (3–3).

Results $$\delta = \frac{\sigma L}{E} = \frac{(84.0 \times 10^6 \text{N/m}^2)\,(3600 \text{ mm})}{(207 \times 10^9 \text{ N/m}^2)} = 1.46 \text{ mm}$$

3–6
DIRECT SHEAR STRESS

Direct shear stress occurs when the applied force tends to cut through the member as scissors or shears do or when a punch and a die are used to punch a slug of material from a sheet. Another important example of direct shear in machine design is the tendency for a key to be sheared off at the section between the shaft and the hub of a machine element when transmitting torque. Figure 3–7 shows the action.

The method of computing direct shear stress is similar to that used for computing direct tensile stress because the applied force is assumed to be uniformly distributed across the cross section of the part that is resisting the force. But the kind of stress is *shear stress* rather than *normal stress*. The symbol used for shear stress is the Greek letter tau (τ). The formula for direct shear stress can thus be written

⇨ **Direct Shear Stress** $$\tau = \text{shearing force/area in shear} = F/A_s \tag{3–4}$$

FIGURE 3–7   Direct
shear on a key

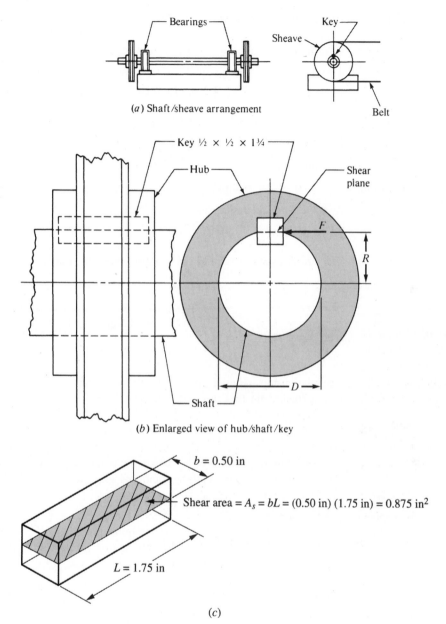

(a) Shaft/sheave arrangement

Key ½ × ½ × 1¾

(b) Enlarged view of hub/shaft/key

b = 0.50 in

Shear area = $A_s = bL$ = (0.50 in) (1.75 in) = 0.875 in^2

L = 1.75 in

(c)

This stress is more properly called the *average shearing* stress, but we will make the simplifying assumption that the stress is uniformly distributed across the shear area.

Example Problem 3–3   Figure 3–7 shows a shaft carrying two sheaves that are keyed to the shaft. Part (b) shows
that a force F is transmitted from the shaft to the hub of the sheave through a square key.
The shaft is 2.25 inches in diameter and transmits a torque of 14 063 lb.in. The key has a
square cross section, 0.50 in on a side, and a length of 1.75 in. Compute the force on the
key and the shear stress caused by this force.

Solution   Objective   Compute the force on the key and the shear stress.

Given   Layout of shaft, key, and hub shown in Figure 3–7.
Torque $= T = 14\,063$ lb·in; key dimensions $= 0.5 \times 0.5 \times 1.75$ in.
Shaft diameter $= D = 2.25$ in; radius $= R = D/2 = 1.125$ in.

Analysis   Torque $T =$ force $F \times$ radius R. Then $F = T/R$.
Use equation (3–4) to compute shearing stress: $\tau = F/A_s$.
Shear area is the cross section of the key at the interface between the shaft and the hub: $A_s = bL$.

Results $F = T/R = (14\,063 \text{ lb·in})/(1.125 \text{ in}) = 12\,500$ lb
$A_s = bL = (0.50 \text{ in})(1.75 \text{ in}) = 0.875 \text{ in}^2$
$\tau = F/A = (12\,500 \text{ lb})/(0.875 \text{ in}^2) = 14\,300 \text{ lb/in}^2$

Comment   This level of shearing stress will be uniform on all parts of the cross section of the key.

3–7 RELATIONSHIP AMONG TORQUE, POWER, AND ROTATIONAL SPEED

The relationship among the power *(P)*, the rotational speed *(n)*, and the torque *(T)* in a shaft is described by the equation

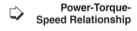
Power-Torque-Speed Relationship

$$T = P/n \tag{3–5}$$

In SI units, power is expressed in the unit of *watt* (W) or its equivalent, *newton meter per second* (N·m/s), and the rotational speed is in *radians per second* (rad/s).

Example Problem 3–4   Compute the amount of torque in a shaft transmitting 750 W of power while rotating at 183 rad/s. (*Note:* This is equivalent to the output of a 1.0-hp, 4-pole electric motor, operating at its rated speed of 1750 rpm. See Chapter 21.)

Solution   Objective   Compute the torque T in the shaft.

Given   Power $= P = 750$ W $= 750$ N·m/s.
Rotational speed $= n = 183$ rad/s.

Analysis   Use Equation (3–5).

Results $T = P/n = (750 \text{ N·m/s})/(183 \text{ rad/s})$
$T = 4.10$ N·m/rad $= 4.10$ N·m

Comments   In such calculations, the unit of N·m/rad is dimensionally correct, and some advocate its use. Most, however, consider the radian to be dimensionless, and thus torque is expressed in N·m or other familiar units of force times distance.

In the U.S. Customary Unit System, power is typically expressed as *horsepower*, equal to 550 ft · lb/s. The typical unit for rotational speed is rpm, or revolutions per minute. But the most convenient unit for torque is the pound-inch (lb · in). Considering all of these quantities and making the necessary conversions of units, we use the following formula to compute the torque (in lb · in) in a shaft carrying a certain power P (in hp) while rotating at a speed of n rpm.

***P-T-n*
Relationship for U.S.
Customary Units**

$$T = 63\ 000\ P/n \qquad\qquad (3\text{–}6)$$

The resulting torque will be in pound-inches. You should verify the value of the constant, 63 000.

Example Problem 3–5 | Compute the torque on a shaft transmitting 1.0 hp while rotating at 1750 rpm. Note that these conditions are approximately the same as those for which the torque was computed in Example Problem 3–4 using SI units.

Solution | Objective | Compute the torque in the shaft.

Given | $P = 1.0$ hp; $n = 1750$ rpm.

Analysis | Use Equation (3–6).

Results | $T = 63\ 000\ P/n = [63\ 000(1.0)]/1750 = 36.0\ \text{lb·in}$

**3–8
TORSIONAL
SHEAR STRESS** | When a *torque*, or twisting moment, is applied to a member, it tends to deform by twisting, causing a rotation of one part of the member relative to another. Such twisting causes a shear stress in the member. For a small element of the member, the nature of the stress is the same as that experienced under direct shear stress. However, in *torsional shear*, the distribution of stress is not uniform across the cross section.

The most frequent case of torsional shear in machine design is that of a round circular shaft transmitting power. Chapter 12 covers shaft design.

Torsional Shear Stress Formula

When subjected to a torque, the outer surface of a solid round shaft experiences the greatest shearing strain and therefore the largest torsional shear stress. See Figure 3–8. The value of the maximum torsional shear stress is found from

**Maximum Torsional
Shear Stress in a
Circular Shaft**

$$\tau_{\text{max}} = Tc/J \qquad\qquad (3\text{–}7)$$

where c = radius of the shaft to its outside surface
J = polar moment of inertia

See Appendix 1 for formulas for J.

FIGURE 3–8 Stress distribution in a solid shaft

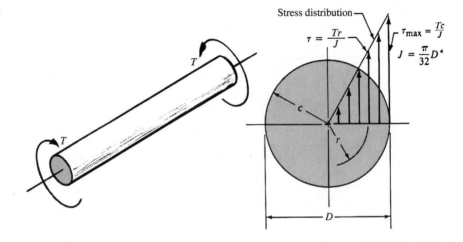

Example Problem 3–6	Compute the maximum torsional shear stress in a shaft having a diameter of 10 mm when it carries a torque of 4.10 N·m.
Solution Objective	Compute the torsional shear stress in the shaft.
Given	Torque = T = 4.10 N·m; shaft diameter = D = 10 mm. c = radius of the shaft = $D/2$ = 5.0 mm.
Analysis	Use Equation (3–7) to compute the torsional shear stress: $\tau_{max} = Tc/J$. J is the polar moment of inertia for the shaft: $J = \pi D^4/32$ (see Appendix 1).
Results	$J = \pi D^4/32 = [(\pi)(10 \text{ mm})^4]/32 = 982 \text{ mm}^4$ $\tau_{max} = \dfrac{(4.10 \text{ N·m})(5.0 \text{ mm})}{982 \text{ mm}^4} \dfrac{10^3 \text{ mm}}{\text{m}} = 20.9 \text{ N/mm}^2 = 20.9 \text{ MPa}$
Comment	The maximum torsional shear stress occurs at the outside surface of the shaft around its entire circumference.

If it is desired to compute the torsional shear stress at some point inside the shaft, the more general formula is used:

General Formula for Torsional Shear Stress

$$\tau = Tr/J \tag{3–8}$$

where r = radial distance from the center of the shaft to the point of interest

Figure 3–8 shows graphically that this equation is based on the linear variation of the torsional shear stress from zero at the center of the shaft to the maximum value at the outer surface.

Equations (3–7) and (3–8) apply also to hollow shafts (Figure 3–9 shows the distribution of shear stress). Again note that the maximum shear stress occurs at the outer surface. Also note that the entire cross section carries a relatively high stress level. As a result, the hollow shaft is more efficient. Notice that the material near the center of the solid shaft is not highly stressed.

FIGURE 3–9 Stress distribution in a hollow shaft

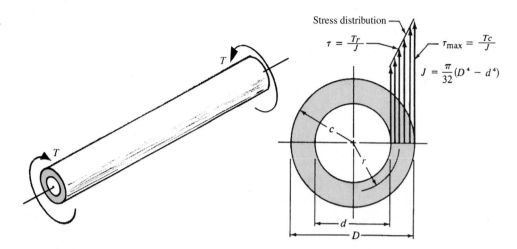

$$\tau = \frac{Tr}{J}$$

$$\tau_{max} = \frac{Tc}{J}$$

$$J = \frac{\pi}{32}(D^4 - d^4)$$

For design, it is convenient to define the *polar section modulus, Z_p*:

Polar Section Modulus

$$Z_p = J/c \qquad (3\text{–}9)$$

Then the equation for the maximum torsional shear stress is

$$\tau_{max} = T/Z_p \qquad (3\text{–}10)$$

Formulas for the polar section modulus are also given in Appendix 1. This form of the torsional shear stress equation is useful for design problems because the polar section modulus is the only term related to the geometry of the cross section.

3–9 TORSIONAL DEFORMATION

When a shaft is subjected to a torque, it undergoes a twisting in which one cross section is rotated relative to other cross sections in the shaft. The angle of twist is computed from

Torsional Deformation

$$\theta = TL/GJ \qquad (3\text{–}11)$$

where θ = angle of twist (radians)
$\quad L$ = length of the shaft over which the angle of twist is being computed
$\quad G$ = modulus of elasticity of the shaft material in *shear*

Example Problem 3–7

Compute the angle of twist of a 10-mm-diameter shaft carrying 4.10 N·m of torque if it is 250 mm long and made of steel with $G = 80$ GPa. Express the result in both radians and degrees.

Solution **Objective** Compute the angle of twist in the shaft.

Given Torque $= T = 4.10$ N·m; length $= L = 250$ mm.
Shaft diameter $= D = 10$ mm; $G = 80$ GPa.

Analysis Use Equation (3–11). For consistency, let $T = 4.10 \times 10^3$ N·mm and $G = 80 \times 10^3$ N/mm^2.
From Example Problem 3–6, $J = 982$ mm^4.

Results $\theta = \dfrac{TL}{GJ} = \dfrac{(4.10 \times 10^3 \text{ N·mm})(250 \text{ mm})}{(80 \times 10^3 \text{ N/mm}^2)(982 \text{ mm}^4)} = 0.013 \text{ rad}$

Using π rad $= 180°$,

$$\theta = (0.013 \text{ rad})(180 \text{ deg}/\pi \text{ rad}) = 0.75 \text{ deg}$$

Comment   Over the length of 250 mm, the shaft twists 0.75 deg.

3–10
TORSION IN
MEMBERS
HAVING
NONCIRCULAR
CROSS SECTIONS

The behavior of members having noncircular cross sections when subjected to torsion is radically different from that for members having circular cross sections. However, the factors of most use in machine design are the maximum stress and the total angle of twist for such members. The formulas for these factors can be expressed in similar forms to the formulas used for members of circular cross section (solid and hollow round shafts).
    The following two formulas can be used:

▷ **Torsional Shear Stress**

$$\tau_{max} = T/Q \tag{3–12}$$

▷ **Deflection for Noncircular Sections**

$$\theta = TL/GK \tag{3–13}$$

Note that Equations (3–12) and (3–13) are similar to Equations (3–10) and (3–11), with the substitution of Q for Z_p and K for J. Refer to Figure 3–10 for the methods of determining the values for K and Q for several types of cross sections useful in machine design. These values are appropriate only if the ends of the member are free to deform. If either end is fixed, as by welding to a solid structure, the resulting stress and angular twist are quite different. (See References 2, 4, and 6.)

Example Problem 3–8

A 2.50-in-diameter shaft carrying a chain sprocket has one end milled in the form of a square to permit the use of a hand crank. The square is 1.75 in on a side. Compute the maximum shear stress on the square part of the shaft when a torque of 15 000 lb·in is applied.
    Also, if the length of the square part is 8.00 in, compute the angle of twist over this part. The shaft material is steel with $G = 11.5 \times 10^6$ psi.

Solution   Objective   Compute the maximum shear stress and the angle of twist in the shaft.

Given   Torque $= T = 15\ 000$ lb·in; length $= L = 8.00$ in.
The shaft is square; thus, $a = 1.75$ in.
$G = 11.5 \times 10^6$ psi.

Analysis   Figure 3–10 shows the methods for calculating the values for Q and K for use in Equations (3–12) and (3–13).

Results $Q = 0.208a^3 = (0.208)(1.75 \text{ in})^3 = 1.115 \text{ in}^3$
          $K = 0.141a^4 = (0.141)(1.75 \text{ in})^4 = 1.322 \text{ in}^4$

Now the stress and the deflection can be computed.

FIGURE 3–10
Methods for determining values for K and Q for several types of cross sections

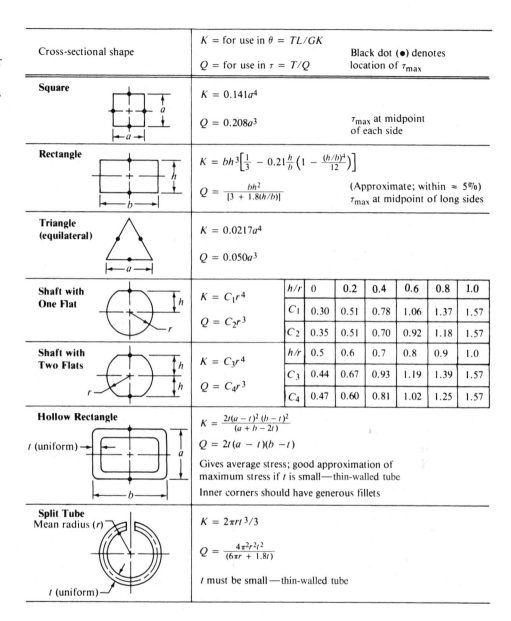

Cross-sectional shape	$K =$ for use in $\theta = TL/GK$ $Q =$ for use in $\tau = T/Q$	Black dot (●) denotes location of τ_{max}
Square	$K = 0.141a^4$ $Q = 0.208a^3$	τ_{max} at midpoint of each side
Rectangle	$K = bh^3\left[\frac{1}{3} - 0.21\frac{h}{b}\left(1 - \frac{(h/b)^4}{12}\right)\right]$ $Q = \dfrac{bh^2}{[3 + 1.8(h/b)]}$	(Approximate; within ≈ 5%) τ_{max} at midpoint of long sides
Triangle (equilateral)	$K = 0.0217a^4$ $Q = 0.050a^3$	

Shaft with One Flat $\quad K = C_1r^4 \quad Q = C_2r^3$

h/r	0	0.2	0.4	0.6	0.8	1.0
C_1	0.30	0.51	0.78	1.06	1.37	1.57
C_2	0.35	0.51	0.70	0.92	1.18	1.57

Shaft with Two Flats $\quad K = C_3r^4 \quad Q = C_4r^3$

h/r	0.5	0.6	0.7	0.8	0.9	1.0
C_3	0.44	0.67	0.93	1.19	1.39	1.57
C_4	0.47	0.60	0.81	1.02	1.25	1.57

Hollow Rectangle $\quad t$ (uniform)

$$K = \frac{2t(a - t)^2 (b - t)^2}{(a + b - 2t)}$$

$$Q = 2t(a - t)(b - t)$$

Gives average stress; good approximation of maximum stress if t is small—thin-walled tube

Inner corners should have generous fillets

Split Tube Mean radius (r) $\quad t$ (uniform)

$$K = 2\pi rt^3/3$$

$$Q = \frac{4\pi^2r^2t^2}{(6\pi r + 1.8t)}$$

t must be small—thin-walled tube

$$\tau_{max} = \frac{T}{Q} = \frac{15\,000\ \text{lb}\cdot\text{in}}{(1.115\ \text{in}^3)} = 13\,460\ \text{psi}$$

$$\theta = \frac{TL}{GK} = \frac{(15\,000\ \text{lb}\cdot\text{in})(8.00\ \text{in})}{(11.5 \times 10^6\ \text{lb/in}^2)(1.322\ \text{in}^4)} = 0.0079\ \text{rad}$$

Convert the angle of twist to degrees:

$$\theta = (0.0079\ \text{rad})(180\text{deg}/\pi\ \text{rad}) = 0.452\ \text{deg}$$

Comments   Over the length of 8.00 in, the square part of the shaft twists 0.452 deg. The maximum shear stress is 13 460 psi, and it occurs at the midpoint of each side as shown in Figure 3–10.

3–11
TORSION IN
CLOSED,
THIN-WALLED
TUBES

A general approach for closed, thin-walled tubes of virtually any shape uses Equations (3–12) and (3–13) with special methods of evaluating K and Q. Figure 3–11 shows such a tube having a constant wall thickness. The values of K and Q are

$$K = 4A^2t/U \tag{3–14}$$

$$Q = 2tA \tag{3–15}$$

where A = area enclosed by the median boundary (indicated by the dashed line in Figure 3–11)

t = wall thickness (which must be uniform and thin)

U = length of the median boundary

FIGURE 3–11
Closed, thin-walled
tube with a constant
wall thicknes

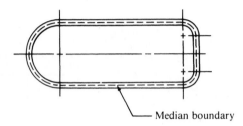

— Median boundary

The shear stress computed by this approach is the *average stress* in the tube wall. However, if the wall thickness t is small (a thin wall), the stress is nearly uniform throughout the wall, and this approach will yield a close approximation of the maximum stress. For the analysis of tubular sections having nonuniform wall thickness, see References 2, 4, and 7.

To design a member to resist torsion only, or torsion and bending combined, it is advisable to select hollow tubes, either round or rectangular, or some other closed shape. They possess good efficiency both in bending and in torsion.

3–12
OPEN TUBES
AND A
COMPARISON
WITH CLOSED
TUBES

The term *open tube* refers to a shape that appears to be tubular but is not completely closed. For example, some tubing is manufactured by starting with a thin, flat strip of steel that is roll-formed into the desired shape (circular, rectangular, square, and so on). Then the seam is welded along the entire length of the tube. It is interesting to compare the properties of the cross section of such a tube before and after it is welded. The following example problem illustrates the comparison for a particular size of circular tubing.

Example Problem 3–9   Figure 3–12 shows a tube before [Part (b)] and after [Part (a)] the seam is welded. Compare the stiffness and the strength of each shape.

Solution   **Objective**   Compare the torsional stiffness and the strength of the closed tube of Figure 3–12(a) with those of the open-seam (split) tube shown in Figure 3–12(b).

Given   The tube shapes are shown in Figure 3–12. Both have the same length, diameter, and wall thickness, and both are made from the same material.

Analysis   Equation (3–13) gives the angle of twist for a noncircular member and shows that the angle is inversely proportional to the value of K. Similarly, Equation (3–11) shows that the an-

FIGURE 3–12
Comparison of closed
and open tubes

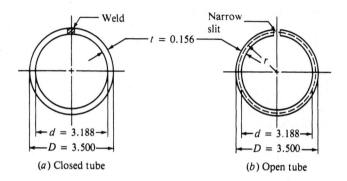

(a) Closed tube   (b) Open tube

gle of twist for a hollow circular tube is inversely proportional to the polar moment of inertia J. All other terms in the two equations are the same for each design. Therefore, the ratio of θ_{open} to θ_{closed} is equal to the ratio J/K. From Appendix 1, we find

$$J = \pi(D^4 - d^4)/32$$

From Figure 3–10, we find

$$K = 2\pi r t^3/3$$

Using similar logic, Equations (3–12) and (3–8) show that the maximum torsional shear stress is inversely proportional to Q and Z_p for the open and closed tubes, respectively. Then we can compare the strengths of the two forms by computing the ratio Z_p/Q. By Equation (3–9), we find that

$$Z_p = J/c = J/(D/2)$$

The equation for Q for the split tube is listed in Figure 3–10.

Results   We make the comparison of torsional stiffness by computing the ratio J/K. For the closed, hollow tube,

$$J = \pi(D^4 - d^4)/32$$
$$J = \pi(3.500^4 - 3.188^4)/32 = 4.592 \text{ in}^4$$

For the open tube before the slit is welded, from Figure 3–10,

$$K = 2\pi r t^3/3$$
$$K = [(2)(\pi)(1.672)(0.156)^3]/3 = 0.0133 \text{ in}^4$$
$$\text{Ratio} = J/K = 4.592/0.0133 = 345$$

Then we make the comparison of the strengths of the two forms by computing the ratio Z_p/Q. The value of J has already been computed to be 4.592 in^4. Then

$$Z_p = J/c = J/(D/2) = (4.592 \text{ in}^4)/[(3.500 \text{ in})/2] = 2.624 \text{ in}^3$$

For the open tube,

$$Q = \frac{4\pi^2 r^2 t^2}{(6\pi r + 1.8t)} = \frac{4\pi^2(1.672 \text{ in})^2(0.156 \text{ in})^2}{[6\pi(1.672 \text{ in}) + 1.8(0.156 \text{ in})]} = 0.0845 \text{ in}^3$$

Then the strength comparison is

$$\text{Ratio} = Z_p/Q = 2.624/0.0845 = 31.1$$

Comments   Thus, for a given applied torque, the slit tube would twist 345 times as much as the closed tube. The stress in the slit tube would be 31.1 times higher than in the closed tube. Also note that if the material for the tube is thin, it will likely buckle at a relatively low stress level, and the tube will collapse suddenly. This comparison shows the dramatic superiority of the closed form of a hollow section to an open form. A similar comparison could be made for shapes other than circular.

3–13
VERTICAL
SHEARING
STRESS

A beam carrying loads transverse to its axis will experience shearing forces, denoted by V. In the analysis of beams, it is usual to compute the variation in shearing force across the entire length of the beam and to draw the *shearing force diagram*. Then the resulting vertical shearing stress can be computed from

⇨ **Vertical Shearing**
Stress in Beams

$$\tau = VQ/It \tag{3–16}$$

where I = rectangular moment of inertia of the cross section of the beam
        t = thickness of the section at the place where the shearing stress is to be computed
        Q = *first moment*, with respect to the overall centroidal axis, *of the area* of that part of the cross section that lies away from the axis where the shearing stress is to be computed.

To calculate the value of Q, we define it by the following equation,

⇨ **First Moment of the**
Area

$$Q = A_p\bar{y} \tag{3–17}$$

where A_p = that part of the area of the section above the place where the stress is to be computed
      $\bar{y}$ = distance from the neutral axis of the section to the centroid of the area A_p

In some books or references, and in earlier editions of this book, Q was called the *statical moment*. Here we will use the term, *first moment of the area*.

For most section shapes, the maximum vertical shearing stress occurs at the centroidal axis. Specifically, if the thickness is not less at a place away from the centroidal axis, then it is assured that the maximum vertical shearing stress occurs at the centroidal axis.

Figure 3–13 shows three examples of how Q is computed in typical beam cross sections. In each, the maximum vertical shearing stress occurs at the neutral axis.

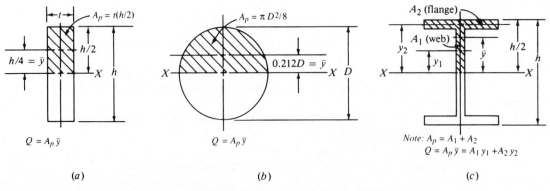

FIGURE 3–13   Illustrations of A_p and $\bar{y}$ used to compute Q for three shapes

FIGURE 3–14
Shearing force diagram
and vertical shearing
stress for beam

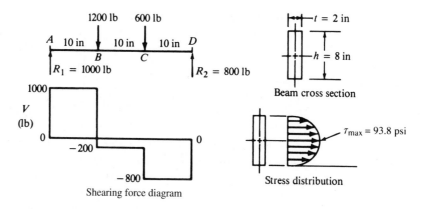

Shearing force diagram

Beam cross section

τ_{max} = 93.8 psi

Stress distribution

Example Problem 3–10   Figure 3–14 shows a simply supported beam carrying two concentrated loads. The shearing force diagram is shown, along with the rectangular shape and size of the cross section of the beam. The stress distribution is parabolic, with the maximum stress occurring at the neutral axis. Use Equation (3–16) to compute the maximum shearing stress in the beam.

Solution   Objective   Compute the maximum shearing stress τ in the beam in Figure 3–14.

Given   The beam shape is rectangular: h = 8.00 in; t = 2.00 in.
Maximum shearing force = V = 1000 lb at all points between A and B.

Analysis   Use Equation (3–16) to compute τ. V and t are given. From Appendix 1,

$$I = th^3/12$$

The value of the first moment of the area Q can be computed from Equation (3–17). For the rectangular cross section shown in Figure 3–13(a), $A_p = t(h/2)$ and $\bar{y} = h/4$. Then

$$Q = A_p\bar{y} = (th/2)(h/4) = th^2/8$$

Results $I = th^3/12 = (2.0 \text{ in})(8.0 \text{ in})^3/12 = 85.3 \text{ in}^4$
$Q = A_p\bar{y} = th^2/8 = (2.0 \text{ in})(8.0 \text{ in})^2/8 = 16.0 \text{ in}^3$

Then the maximum shearing stress is

$$\tau = \frac{VQ}{It} = \frac{(1000 \text{ lb})(16.0 \text{ in}^3)}{(85.3 \text{ in}^4)(2.0 \text{ in})} = 93.8 \text{ lb/in}^2 = 93.8 \text{ psi}$$

Comments   The maximum shearing stress of 93.8 psi occurs at the neutral axis of the rectangular section as shown in Figure 3–14. The stress distribution within the cross section is generally parabolic, ending with zero shearing stress at the top and bottom surfaces. This is the nature of the shearing stress everywhere between the left support at A and the point of application of the 1200-lb load at B. The maximum shearing stress at any other point in the beam is proportional to the magnitude of the vertical shearing force at the point of interest.

FIGURE 3–15 Shear stresses on an element

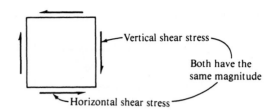

Note that the vertical shearing stress is equal to the *horizontal shearing stress* because any element of material subjected to a shear stress on one face must have a shear stress of the same magnitude on the adjacent face for the element to be in equilibrium. Figure 3–15 shows this phenomenon.

In most beams, the magnitude of the vertical shearing stress is quite small compared with the bending stress (see the following section). For this reason it is frequently not computed at all. Those cases where it is of importance include the following:

1. When the material of the beam has a relatively low shear strength (such as wood).
2. When the bending moment is zero or small (and thus the bending stress is small), for example, at the ends of simply supported beams and for short beams.
3. When the thickness of the section carrying the shearing force is small, as in sections made from rolled sheet, some extruded shapes, and the web of rolled structural shapes such as wide-flange beams.

3–14 SPECIAL SHEARING STRESS FORMULAS

Equation (3–16) can be cumbersome because of the need to evaluate the first moment of the area Q. Several commonly used cross sections have special, easy-to-use formulas for the maximum vertical shearing stress:

➪ τ_{max} **for Rectangle**

$$\tau_{max} = 3V/2A \text{ (exact)} \qquad (3\text{–}18)$$

where A = total cross-sectional area of the beam

➪ τ_{max} **for Circle**

$$\tau_{max} = 4V/3A \text{ (exact)} \qquad (3\text{–}19)$$

➪ τ_{max} **for I-Shape**

$$\tau_{max} \simeq V/th \text{ (approximate: about 15\% low)} \qquad (3\text{–}20)$$

where t = web thickness
h = height of the web (for example, a wide-flange beam)

➪ τ_{max} **for Thin-walled Tube**

$$\tau_{max} \simeq 2V/A \text{ (approximate: a little high)} \qquad (3\text{–}21)$$

In all of these cases, the maximum shearing stress occurs at the neutral axis.

Example Problem 3–11 Compute the maximum shearing stress in the beam described in Example Problem 3–10 using the special shearing stress formula for a rectangular section.

Solution Objective Compute the maximum shearing stress τ in the beam in Figure 3–14.

Given   The data are the same as stated in Example Problem 3–10 and as shown in Figure 3–14.

Analysis   Use Equation (3–18) to compute $\tau = 3V/2A$. For the rectangle, $A = th$.

Results $$\tau_{max} = \frac{3V}{2A} = \frac{3(1000\text{ lb})}{2[(2.0\text{ in})(8.0\text{ in})]} = 93.8\text{ psi}$$

Comment   This result is the same as that obtained for Example Problem 3–10, as expected.

**3–15
STRESS DUE TO
BENDING**

A *beam* is a member that carries loads transverse to its axis. Such loads produce bending moments in the beam, which result in the development of bending stresses. Bending stresses are *normal stresses*, that is, either tensile or compressive. The maximum bending stress in a beam cross section will occur in the part farthest from the neutral axis of the section. At that point, the *flexure formula* gives the stress:

⇨ **Flexure Formula
for Maximum
Bending Stress**

$$\sigma = Mc/I \qquad (3\text{–}22)$$

where M = magnitude of the bending moment at the section
I = moment of inertia of the cross section with respect to its neutral axis
c = distance from the neutral axis to the outermost fiber of the beam cross section

The magnitude of the bending stress varies linearly within the cross section from a value of zero at the neutral axis, to the maximum tensile stress on one side of the neutral axis, and to the maximum compressive stress on the other side. Figure 3–16 shows a typical stress distribution in a beam cross section. Note that the stress distribution is independent of the shape of the cross section.

Note that *positive bending* occurs when the deflected shape of the beam is concave upward, resulting in compression on the upper part of the cross section and tension on the lower part. Conversely, *negative bending* causes the beam to be concave downward.

The flexure formula was developed subject to the following conditions:

1. The beam must be in pure bending. Shearing stresses must be zero or negligible. No axial loads are present.
2. The beam must not twist or be subjected to a torsional load.
3. The material of the beam must obey Hooke's law.
4. The modulus of elasticity of the material must be the same in both tension and compression.
5. The beam is initially straight and has a constant cross section.
6. Any plane cross section of the beam remains plane during bending.
7. No part of the beam shape fails because of local buckling or wrinkling.

If condition 1 is not strictly met, you can continue the analysis by using the method of combined stresses presented in Chapter 4. In most practical beams, which are long relative to their height, shear stresses are sufficiently small as to be negligible. Furthermore, the maximum bending stress occurs at the outermost fibers of the beam section, where the shear stress is in fact zero. A beam with varying cross section, which would violate

FIGURE 3–16
Typical bending stress
distribution in a beam
cross section

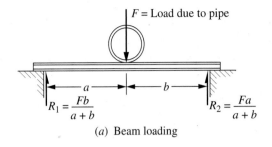

F = Load due to pipe

$R_1 = \dfrac{Fb}{a+b}$ $R_2 = \dfrac{Fa}{a+b}$

(a) Beam loading

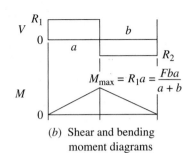

$M_{max} = R_1 a = \dfrac{Fba}{a+b}$

(b) Shear and bending
moment diagrams

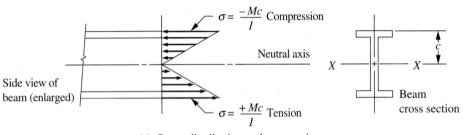

$\sigma = \dfrac{-Mc}{I}$ Compression

Neutral axis

$\sigma = \dfrac{+Mc}{I}$ Tension

Side view of
beam (enlarged)

Beam
cross section

(c) Stress distribution on beam section

(d) Stress element in compression
in top part of beam

(e) Stress element in tension
in bottom part of beam

condition 5, can be analyzed by the use of stress concentration factors discussed later in
this chapter.

For design, it is convenient to define the term *section modulus*, S, as

$$S = I/c \tag{3–23}$$

The flexure formula then becomes

▷ **Flexure Formula**

$$\sigma = M/S \tag{3–24}$$

Since I and c are geometrical properties of the cross section of the beam, S is also. Then, in
design, it is usual to define a design stress, σ_d, and, with the bending moment known, solve
for S:

▷ **Required Section
Modulus**

$$S = M/\sigma_d \tag{3–25}$$

This results in the required value of the section modulus. From it, the required dimensions
of the beam cross section can be determined.

Example Problem 3–12 For the beam shown in Figure 3–16, the load F due to the pipe is 12 000 lb. The distances are $a = 4$ ft and $b = 6$ ft. Determine the required section modulus for the beam to limit the stress due to bending to 30 000 psi, the recommended design stress for a typical structural steel in static bending.

Solution   Objective Compute the required section modulus S for the beam in Figure 3–16.

Given   The layout and the loading pattern are shown in Figure 3–16.

Lengths: Overall length $= L = 10$ ft; $a = 4$ ft; $b = 6$ ft.
Load $= F = 12\,000$ lb.
Design stress $= \sigma_d = 30\,000$ psi.

Analysis Use Equation (3–25) to compute the required section modulus S. Compute the maximum bending moment that occurs at the point of application of the load using the formula shown in Part (b) of Figure 3–16.

Results
$$M_{max} = R_1\, a = \frac{Fba}{a + b} = \frac{(12\,000\text{ lb})(6\text{ ft})(4\text{ ft})}{(6\text{ ft} + 4\text{ ft})} = 28\,800\text{ lb}\cdot\text{ft}$$

$$S = \frac{M}{\sigma_d} = \frac{28\,800\text{ lb}\cdot\text{ft}}{30\,000\text{ lb/in}^2}\frac{12\text{ in}}{\text{ft}} = 11.5\text{ in}^3$$

Comments A beam section can now be selected from Tables A16–3 and A16–4 that has at least this value for the section modulus. The lightest section, typically preferred, is the W8×15 wide-flange shape with $S = 11.8$ in^3.

3–16 FLEXURAL CENTER FOR BEAMS

A beam section must be loaded in a way that ensures symmetrical bending; that is, there must be no tendency for the section to twist under the load. Figure 3–17 shows several shapes that are typically used for beams having a vertical axis of symmetry. If the line of

FIGURE 3–17
Symmetrical sections.
A load applied through
the axis of symmetry
results in pure bending
in the beam.

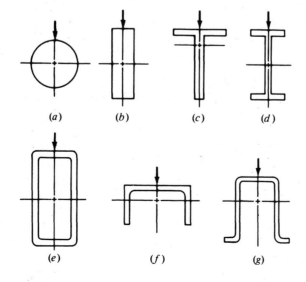

(a)   (b)   (c)   (d)

(e)   (f)   (g)

FIGURE 3–18
Nonsymmetrical
sections. A load applied
as at F_1 would cause
twisting; loads applied
as at F_2 through the
flexural center Q would
cause pure bending.

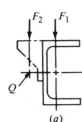
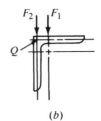
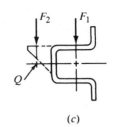

(a)   (b)   (c)

action of the loads on such sections passes through the axis of symmetry, then there is no tendency for the section to twist, and the flexure formula applies.

When there is no vertical axis of symmetry, as with the sections shown in Figure 3–18, care must be exercised in placement of the loads. If the line of action of the loads were shown as F_1 in the figure, the beam would twist and bend, so the flexure formula would not give accurate results for the stress in the section. For such sections, the load must be placed in line with the *flexural center,* sometimes called the *shear center.* Figure 3–18 shows the approximate location of the flexural center for these shapes (indicated by the symbol Q). Applying the load in line with Q, as shown with the forces labeled F_2, would result in pure bending. A table of formulas for the location of the flexural center is available (see Reference 7).

3–17 BEAM DEFLECTIONS

The bending loads applied to a beam cause it to deflect in a direction perpendicular to its axis. A beam that was originally straight will deform to a slightly curved shape. In most cases, the critical factor is either the maximum deflection of the beam or its deflection at specific locations.

Consider the double-reduction speed reducer shown in Figure 3–19. The four gears (A, B, C, and D) are mounted on three shafts, each of which is supported by two bearings. The action of the gears in transmitting power creates a set of forces that in turn act on the shafts to cause bending. One component of the total force on the gear teeth acts in a direction that tends to separate the two gears. Thus, gear A is forced upward while gear B is forced downward. For good gear performance, the net deflection of one gear relative to the other should not exceed 0.005 in (0.13 mm) for medium-sized industrial gearing.

To evaluate the design, there are many methods of computing shaft deflections. We will review briefly those methods using deflection formulas, superposition, and a general analytical approach.

A set of formulas for computing the deflection of beams at any point or at selected points is useful in many practical problems. Appendix 14 includes several cases.

For many additional cases, superposition is useful if the actual loading can be divided into parts that can be computed by available formulas. The deflection for each loading is computed separately, and then the individual deflections are summed at the points of interest.

Many commercially available computer software programs allow the modeling of beams having rather complex loading patterns and varying geometry. The results include reaction forces, shearing force and bending moment diagrams, and deflections at any point. It is important that you understand the principles of beam deflection, studied in strength of materials and reviewed here, so that you can apply such programs accurately and interpret the results carefully.

FIGURE 3–19 Shaft deflection analysis for a double-reduction speed reducer

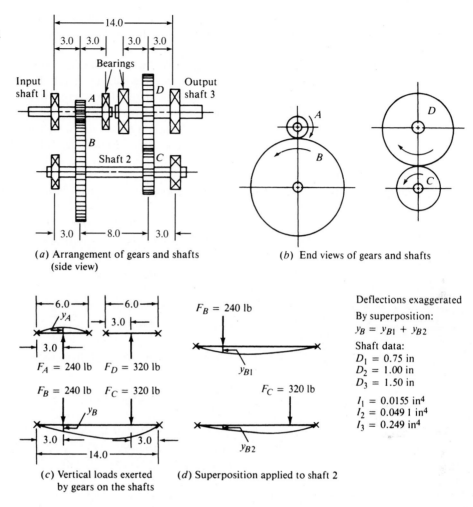

(a) Arrangement of gears and shafts
(side view)

(b) End views of gears and shafts

$F_B = 240$ lb

Deflections exaggerated

By superposition:
$y_B = y_{B1} + y_{B2}$

Shaft data:
$D_1 = 0.75$ in
$D_2 = 1.00$ in
$D_3 = 1.50$ in

$I_1 = 0.0155$ in^4
$I_2 = 0.049$ 1 in^4
$I_3 = 0.249$ in^4

$F_A = 240$ lb $F_D = 320$ lb

$F_B = 240$ lb $F_C = 320$ lb

$F_C = 320$ lb

(c) Vertical loads exerted
by gears on the shafts

(d) Superposition applied to shaft 2

Example Problem 3–13

For the two gears, A and B, in Figure 3–19, compute the relative deflection between them in the plane of the paper that is due to the forces shown in Part (c). These *separating forces*, or *normal forces*, are discussed in Chapters 9 and 10. It is customary to consider the loads at the gears and the reactions at the bearings to be concentrated. The shafts carrying the gears are steel and have uniform diameters as listed in the figure.

Solution   Objective   Compute the relative deflection between gears A and B in Figure 3–19.

Given   The layout and loading pattern are shown in Figure 3–19. The separating force between gears A and B is 240 lb. Gear A pushes downward on gear B, and the reaction force of gear B pushes upward on gear A. Shaft 1 has a diameter of 0.75 in and a moment of inertia of 0.0155 in^4. Shaft 2 has a diameter of 1.00 in and a moment of inertia of 0.0491 in^4. Both shafts are steel. Use $E = 30 \times 10^6$ psi.

Analysis   Use the deflection formulas from Appendix 14 to compute the upward deflection of shaft 1 at gear A and the downward deflection of shaft 2 at gear B. The sum of the two deflections is the total deflection of gear A with respect to gear B.

Case (a) from Table A14–1 applies to shaft 1 because there is a single concentrated force acting at the midpoint of the shaft between the supporting bearings. We will call that deflection y_A.

Shaft 2 is a simply supported beam carrying two nonsymmetrical loads. No single formula from Appendix 14 matches that loading pattern. But we can use superposition to compute the deflection of the shaft at gear B by considering the two forces separately as shown in Part (d) of Figure 3–19. Case (b) from Table A14–1 is used for each load.

We first compute the deflection at B due only to the 240-lb force, calling it y_{B1}. Then we compute the deflection at B due to the 320-lb force, calling it y_{B2}. The total deflection at B is $y_B = y_{B1} + y_{B2}$.

Results   The deflection of shaft 1 at gear A is

$$y_a = \frac{F_A L_1^3}{48\,EI} = \frac{(240)(6.0)^3}{48(30 \times 10^6)(0.0155)} = 0.0023 \text{ in}$$

The deflection of shaft 2 at B due only to the 240-lb force is

$$y_{B1} = -\frac{F_B\,a^2\,b^2}{3\,EI_2\,L_2} = -\frac{(240)(3.0)^2(11.0)^2}{3(30 \times 10^6)(0.0491)(14)} = -0.0042 \text{ in}$$

The deflection of shaft 2 at B due only to the 320-lb force at C is

$$y_{B2} = -\frac{F_c\,bx}{6\,EI_2\,L_2}(L_2^2 - b^2 - x^2)$$

$$y_{B2} = -\frac{(320)(3.0)(3.0)}{6(30 \times 10^6)(0.0491)(14)}\left[(14)^2 - (3.0)^2 - (3.0)^2\right]$$

$$y_{B2} = -0.0041 \text{ in}$$

Then the total deflection at gear B is

$$y_B = y_{B1} + y_{B2} = -0.0042 - 0.0041 = -0.0083 \text{ in}$$

Because shaft 1 deflects upward and shaft 2 deflects downward, the total relative deflection is the sum of y_A and y_B:

$$y_{\text{total}} = y_A + y_B = 0.0023 + 0.0083 = 0.0106 \text{ in}$$

Comment   This deflection is very large for this application. How could the deflection be reduced?

3–18
EQUATIONS FOR DEFLECTED BEAM SHAPE

The general principles relating the deflection of a beam to the loading on the beam and its manner of support are presented here. The result will be a set of relationships among the load, the vertical shearing force, the bending moment, the slope of the deflected beam shape, and the actual deflection curve for the beam. Figure 3–20 shows diagrams for these five factors, with θ as the slope and y indicating deflection of the beam from its initial straight position. The product of modulus of elasticity and the moment of inertia, EI, for the beam is a measure of its stiffness or resistance to bending deflection. It is convenient to

FIGURE 3–20

Relationships of load, vertical shearing force, bending moment, slope of deflected beam shape, and actual deflection curve of a beam

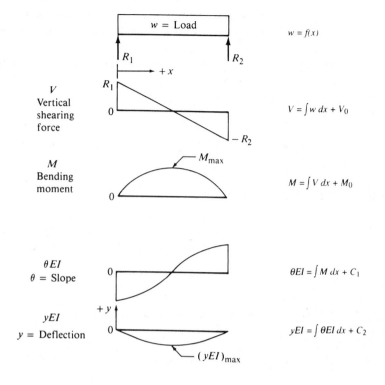

combine EI with the slope and deflection values to maintain a proper relationship, as discussed next.

One fundamental concept for beams in bending is

$$\frac{M}{EI} = \frac{d^2y}{dx^2}$$

where M = bending moment

x = position on the beam measured along its length

y = deflection

Thus, if it is desired to create an equation of the form $y = f(x)$ (that is, y as a function of x), it would be related to the other factors as follows:

$$y = f(x)$$

$$\theta = \frac{dy}{dx}$$

$$\frac{M}{EI} = \frac{d^2y}{dx^2}$$

$$\frac{V}{EI} = \frac{d^3y}{dx^3}$$

$$\frac{w}{EI} = \frac{d^4y}{dx^4}$$

where w = general term for the load distribution on the beam

The last two equations follow from the observation that there is a derivative (slope) relationship between shear and bending moment and between load and shear.

In practice, the fundamental equations just given are used in reverse. That is, the load distribution as a function of x is known, and the equations for the other factors are derived by successive integrations. The results are

$$w = f(x)$$
$$V = \int w \, dx + V_0$$
$$M = \int V \, dx + M_0$$

where V_0 and M_0 = constants of integration evaluated from the boundary conditions

In many cases, the load, shear, and bending moment diagrams can be drawn in the conventional manner, and the equations for shear or bending moment can be created directly by the principles of analytic geometry. With M as a function of x, the slope and deflection relations can be found:

$$\theta EI = \int M \, dx + C_1$$
$$yEI = \int \theta EI \, dx + C_2$$

The constants of integration must be evaluated from boundary conditions. Texts on strength of materials show the details. (See Reference 3.)

**3–19
BEAMS WITH
CONCENTRATED
BENDING
MOMENTS**

Figures 3–16 and 3–20 show beams loaded only with concentrated forces or distributed loads. For such loading in any combination, the moment diagram is continuous. That is, there are no points of abrupt change in the value of the bending moment. Many machine elements such as cranks, levers, helical gears, and brackets carry loads whose line of action is offset from the centroidal axis of the beam in such a way that a concentrated moment is exerted on the beam.

Figures 3–21, 3–22, and 3–23 show three different examples where concentrated moments are created on machine elements. The bell crank in Figure 3–21 pivots around point O and is used to transfer an applied force to a different line of action. Each arm behaves similar to a cantilever beam, bending with respect to an axis through the pivot. For analysis, we can isolate an arm by making an imaginary cut through the pivot and showing the reaction force at the pivot pin and the internal moment in the arm. The shearing force and bending moment diagrams included in Figure 3–21 show the results, and Example Problem 3–14 gives the details of the analysis. Note the similarity to a cantilever beam with the internal concentrated moment at the pivot reacting to the force, F_2, acting at the end of the arm.

Figure 3–22 shows a print head for an impact-type printer in which the applied force, F, is offset from the neutral axis of the print head itself. Thus the force creates a concentrated bending moment at the right end where the vertical lever arm attaches to the horizontal part. The freebody diagram shows the vertical arm cut off and an internal axial force and moment replacing the effect of the extended arm. The concentrated moment causes the abrupt change in the value of the bending moment at the right end of the arm as shown in the bending moment diagram. Example Problem 3–15 gives the details of the analysis.

Figure 3–23 shows an isometric view of a crankshaft that is actuated by the vertical force acting at the end of the crank. One result is an applied torque that tends to rotate the shaft ABC clockwise about its x-axis. The reaction torque is shown acting at the forward end of the crank. A second result is that the vertical force acting at the end of the crank cre-

FIGURE 3–21
Bending moment in a
bell crank

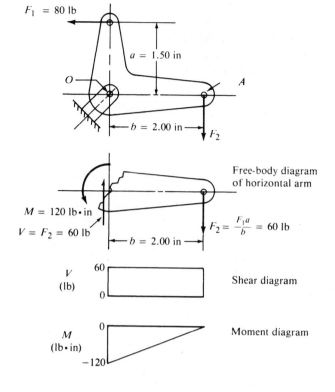

FIGURE 3–22
Bending moment on a
print head

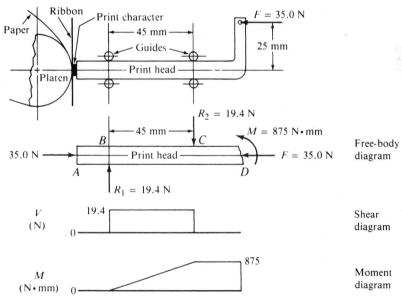

ates a twisting moment in the rod attached at B and thus tends to bend the shaft ABC in the x-z plane. The twisting moment is treated as a concentrated moment acting at B with the resulting abrupt change in the bending moment at that location as can be seen in the bending moment diagram. Example Problem 3–16 gives the details of the analysis.

When drawing the bending moment diagram for a member to which a concentrated moment is applied, the following sign convention will be used.

FIGURE 3–23

Bending moment on a
shaft carrying a crank

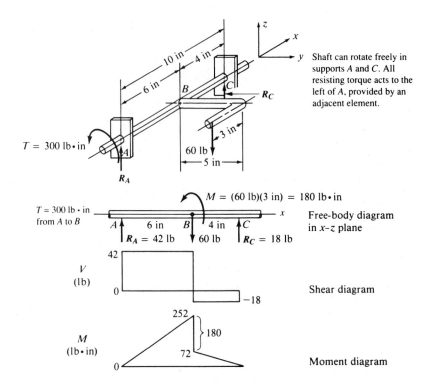

Shaft can rotate freely in
supports A and C. All
resisting torque acts to the
left of A, provided by an
adjacent element.

$T = 300$ lb·in

$T = 300$ lb · in
from A to B

$M = (60$ lb$)(3$ in$) = 180$ lb·in

Free-body diagram
in x–z plane

$R_A = 42$ lb   60 lb $R_C = 18$ lb

Shear diagram

Moment diagram

*When a concentrated bending moment acts on a beam in a counterclockwise
direction, the moment diagram drops; when a clockwise concentrated moment
acts, the moment diagram rises.*

Example Problem 3–14   The bell crank shown in Figure 3–21 is part of a linkage in which the 80-lb horizontal force
is transferred to F_2 acting vertically. The crank can pivot about the pin at O. Draw a free-
body diagram of the horizontal part of the crank from O to A. Then draw the shearing force
and bending moment diagrams that are necessary to complete the design of the horizontal
arm of the crank.

Solution   Objective   Draw the free-body diagram of the horizontal part of the crank in Figure 3–21. Draw the
shearing force and bending moment diagrams for that part.

Given   The layout from Figure 3–21.

Analysis   Use the entire crank first as a free body to determine the downward force F_2 that reacts to
the applied horizontal force F_1 of 80 lb by summing moments about the pin at O.
    Then create the free-body diagram for the horizontal part by breaking it through the
pin and replacing the removed part with the internal force and moment acting at the break.

Results   We can first find the value of F_2 by summing moments about the pin at O using the entire
crank:

$$F_1 \cdot a = F_2 \cdot b$$
$$F_2 = F_1(a/b) = 80 \text{ lb}(1.50/2.00) = 60 \text{ lb}$$

Below the drawing of the complete crank, we have drawn a sketch of the horizontal part, isolating it from the vertical part. The internal force and moment at the cut section are shown. The externally applied downward force F_2 is reacted by the upward reaction at the pin. Also, because F_2 causes a moment with respect to the section at the pin, an internal reaction moment exists, where

$$M = F_2 \cdot b = (60 \text{ lb})(2.00 \text{ in}) = 120 \text{ lb} \cdot \text{in}$$

The shear and moment diagrams can then be shown in the conventional manner. The result looks much like a cantilever that is built into a rigid support. The difference here is that the reaction moment at the section through the pin is developed in the vertical arm of the crank.

Comments   Note that the shape of the moment diagram for the horizontal part shows that the maximum moment occurs at the section through the pin and that the moment decreases linearly as we move out toward point A. As a result, the shape of the crank is optimized, having its largest cross section (and section modulus) at the section of highest bending moment. You could complete the design of the crank using the techniques reviewed in Section 3–15.

Example Problem 3–15   Figure 3–22 represents a print head for a computer printer. The force F moves the print head toward the left against the ribbon, imprinting the character on the paper that is backed up by the platen. Draw the free-body diagram for the horizontal portion of the print head, along with the shearing force and bending moment diagrams.

Solution   **Objective**   Draw the free-body diagram of the horizontal part of the print head in Figure 3–22. Draw the shearing force and bending moment diagrams for that part.

Given   The layout from Figure 3–22.

Analysis   The horizontal force of 35 N acting to the left is reacted by an equal 35 N horizontal force produced by the platen pushing back to the right on the print head. The guides provide simple supports in the vertical direction. The applied force also produces a moment at the base of the vertical arm where it joins the horizontal part of the print head.

We create the free-body diagram for the horizontal part by breaking it at its right end and replacing the removed part with the internal force and moment acting at the break. The shearing force and bending moment diagrams can then be drawn.

Results   The free-body diagram for the horizontal portion is shown below the complete sketch. Note that at the right end (section D) of the print head, the vertical arm has been removed and replaced with the internal horizontal force of 35.0 N and a moment of 875 N·mm caused by the 35.0 N force acting 25 mm above it. Also note that the 25 mm-moment arm for the force is taken from the line of action of the force *to the neutral axis of the horizontal part*. The 35.0 N reaction of the platen on the print head tends to place the head in compression over the entire length. The rotational tendency of the moment is reacted by the couple created by R_1 and R_2 acting 45 mm apart at B and C.

Below the free-body diagram is the vertical shearing force diagram in which a constant shear of 19.4 N occurs only between the two supports.

The bending moment diagram can be derived from either the left end or the right end. If we choose to start at the left end at A, there is no shearing force from A to B, and therefore there is no change in bending moment. From B to C, the positive shear causes an increase in bending moment from zero to 875 N·mm. Because there is no shear from C to D, there is no change in bending moment, and the value remains at 875 N·mm. The counterclockwise-directed concentrated moment at D causes the moment diagram to drop abruptly, closing the diagram.

Example Problem 3–16   Figure 3–23 shows a crank in which it is necessary to visualize the three-dimensional arrangement. The 60-lb downward force tends to rotate the shaft ABC around the x-axis. The reaction torque acts only at the end of the shaft outboard of the bearing support at A. Bearings A and C provide simple supports. Draw the complete free-body diagram for the shaft ABC, along with the shearing force and bending moment diagrams.

Solution   Objective   Draw the free-body diagram of the shaft ABC in Figure 3–23. Draw the shearing force and bending moment diagrams for that part.

Given   The layout from Figure 3–23.

Analysis   The analysis will take the following steps:

1. Determine the magnitude of the torque in the shaft between the left end and point B where the crank arm is attached.
2. Analyze the connection of the crank at point B to determine the force and moment transferred to the shaft ABC by the crank.
3. Compute the vertical reactions at supports A and C.
4. Draw the shearing force and bending moment diagrams considering the concentrated moment applied at point B, along with the familiar relationships between shearing force and bending moments.

Results   The free-body diagram is shown as viewed looking at the x-z plane. Note that the free body must be in equilibrium in all force and moment directions. Considering first the torque (rotating moment) about the x-axis, note that the crank force of 60 lb acts 5.0 in from the axis. The torque, then, is

$$T = (60\ \text{lb})(5.0\ \text{in}) = 300\ \text{lb} \cdot \text{in}$$

This level of torque acts from the left end of the shaft to section B, where the crank is attached to the shaft.

Now the loading at B should be described. One way to do so is to visualize that the crank itself is separated from the shaft and is replaced with a force and moment caused by the crank. First, the downward force of 60 lb pulls down at B. Also, because the 60-lb applied force acts 3.0 in to the left of B, it causes a concentrated moment in the *x-z plane* of 180 lb·in to be applied at B.

Both the downward force and the moment at B affect the magnitude and direction of the reaction forces at A and C. First, summing moments about A,

$$(60\ \text{lb})(6.0\ \text{in}) - 180\ \text{lb} \cdot \text{in} - R_C(10.0\ \text{in}) = 0$$
$$R_C = [(360 - 180)\text{lb} \cdot \text{in}]/(10.0\ \text{in}) = 18.0\ \text{lb upward}$$

Now, summing moments about C,

$$(60 \text{ lb})(4.0 \text{ in}) + 180 \text{ lb} \cdot \text{in} - R_A(10.0 \text{ in}) = 0$$
$$R_A = [(240 + 180) \text{ lb} \cdot \text{in}]/(10.0 \text{ in}) = 42.0 \text{ lb upward}$$

Now the shear and bending moment diagrams can be completed. The moment starts at zero at the simple support at *A,* rises to 252 lb·in at *B* under the influence of the 42-lb shear force, then drops by 180 lb·in due to the counterclockwise concentrated moment at *B,* and finally returns to zero at the simple support at *C.*

Comments   In summary, shaft *ABC* carries a torque of 300 lb·in from point *B* to its left end. The maximum bending moment of 252 lb·in occurs at point *B* where the crank is attached. The bending moment then suddenly drops to 72 lb·in under the influence of the concentrated moment of 180 lb·in applied by the crank.

**3–20
COMBINED
NORMAL
STRESSES:
SUPERPOSITION
PRINCIPLE**

When the same cross section of a load-carrying member is subjected to both a direct tensile or compressive stress and a stress due to bending, the resulting normal stress can be computed by the method of superposition. The formula is

$$\sigma = \pm Mc/I \pm F/A \qquad (3\text{–}26)$$

where tensile stresses are positive and compressive stresses are negative

An example of a load-carrying member subjected to combined bending and axial tension is shown in Figure 3–24. It shows a beam subjected to a load applied downward and to the right through a bracket below the beam. Resolving the load into horizontal and vertical components shows that its effect can be broken into three parts:

1. The vertical component tends to place the beam in bending with tension on the top and compression on the bottom.

2. The horizontal component, because it acts away from the neutral axis of the beam, causes bending with tension on the bottom and compression on the top.

3. The horizontal component causes direct tensile stress across the entire cross section.

We can proceed with the stress analysis by using the techniques from the previous section to prepare the shearing force and bending moment diagrams and then using Equation 3–26 to combine the effects of the bending stress and the direct tensile stress at any point. The details are shown within Example Problem 3–17.

Example Problem 3–17   The cantilever beam in Figure 3–24 is a steel American Standard beam, S6×12.5. The force *F* is 10 000 lb, and it acts at an angle of 30° below the horizontal, as shown. Use *a* = 24 in and *e* = 6.0 in. Draw the free-body diagram and the shearing force and bending moment diagrams for the beam. Then compute the maximum tensile and maximum compressive stresses in the beam and show where they occur.

Solution   Objective   Determine the maximum tensile and compressive stresses in the beam.

Given   The layout from Figure 3–24(a). Force = *F* = 10 000 lb; angle θ = 30°.
The beam shape: S6×12.5; length = *a* = 24 in.

FIGURE 3–24 Beam subjected to combined stresses

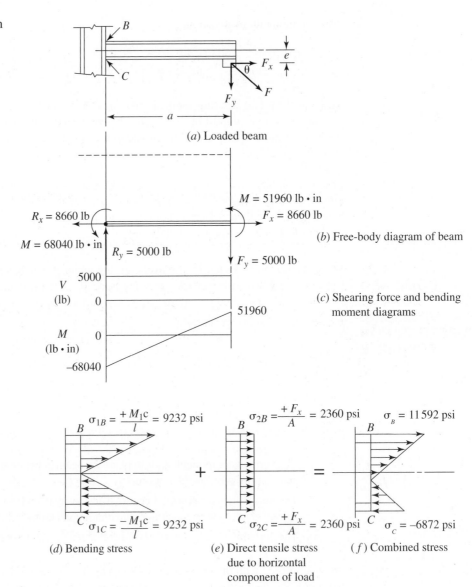

(a) Loaded beam

(b) Free-body diagram of beam

(c) Shearing force and bending moment diagrams

(d) Bending stress

(e) Direct tensile stress due to horizontal component of load

(f) Combined stress

Section modulus $= S = 7.37$ in^3; area $= A = 3.67$ in^2 (Table A16–4).
Eccentricity of the load $= e = 6.0$ in from the neutral axis of the beam to the line of action of the horizontal component of the applied load.

Analysis   The analysis takes the following steps:

1. Resolve the applied force into its vertical and horizontal components.
2. Transfer the horizontal component to an equivalent loading at the neutral axis having a direct tensile force and a moment due to the eccentric placement of the force.
3. Prepare the free-body diagram using the techniques from Section 3–19.

4. Draw the shearing force and bending moment diagrams and determine where the maximum bending moment occurs.

5. Complete the stress analysis at that section, computing both the maximum tensile and maximum compressive stresses.

Results   The components of the applied force are:

$$F_x = F\cos(30°) = (10\ 000\ \text{lb})[\cos(30°)] = 8660\ \text{lb acting to the right}$$
$$F_y = F\sin(30°) = (10\ 000\ \text{lb})[\sin(30°)] = 5000\ \text{lb acting downward}$$

The horizontal force produces a counterclockwise concentrated moment at the right end of the beam with a magnitude of:

$$M_1 = F_x(6.0\ \text{in}) = (8660\ \text{lb})(6.0\ \text{in}) = 51\ 960\ \text{lb} \cdot \text{in}$$

The free-body diagram of the beam is shown in Figure 3–24(b).

Figure 3–24(c) shows the shearing force and bending moment diagrams.

The maximum bending moment, 68 040 lb in, occurs at the left end of the beam where it is attached firmly to a column.

The bending moment, taken alone, produces a tensile stress (+) on the top surface at point B and a compressive stress (−) on the bottom surface at C. The magnitudes of these stresses are:

$$\sigma_1 = \pm M/S = \pm (68\ 040\ \text{lb in}) / (7.37\ \text{in}^3) = \pm 9232\ \text{psi}$$

Figure 3–24(d) shows the stress distribution due only to the bending stress.

Now we compute the tensile stress due to the axial force of 8660 lb.

$$\sigma_2 = F_x/A = (8660\ \text{lb})/(3.67\ \text{in}^2) = 2360\ \text{psi}$$

Figure 3–24(e) shows this stress distribution, uniform across the entire section.

Next, let's compute the combined stress at B on the top of the beam.

$$\sigma_B = +\sigma_1 + \sigma_2 = 9232\ \text{psi} + 2360\ \text{psi} = 11\ 592\ \text{psi Tensile}$$

At C on the bottom of the beam, the stress is:

$$\sigma_C = -\sigma_1 + \sigma_2 = -9232\ \text{psi} + 2360\ \text{psi} = -6872\ \text{psi Compressive}$$

Figure 3–24(f) shows the combined stress condition that exists on the cross section of the beam at its left end at the support. It is a superposition of the component stresses shown in Figure 3–24(d) and (e).

3–21
STRESS
CONCENTRATIONS

The formulas reviewed earlier for computing simple stresses due to direct tensile and compressive forces, bending moments, and torsional moments are applicable under certain conditions. One condition is that the geometry of the member is uniform throughout the section of interest.

In many typical machine design situations, inherent geometric discontinuities are necessary for the parts to perform their desired functions. For example, as shown in Figure 12–2 in Chapter 12, shafts carrying gears, chain sprockets, or belt sheaves usually have several diameters that create a series of shoulders that seat the power transmission members and support bearings. Grooves in the shaft allow the installation of retaining rings. Keyseats milled into the shaft enable keys to drive the elements. Similarly, tension members in linkages may be designed with retaining ring grooves, radial holes for pins, screw threads, or reduced sections.

Any of these geometric discontinuities will cause the actual maximum stress in the part to be higher than the simple formulas predict. Defining *stress concentration factors* as the factors by which the actual maximum stress exceeds the nominal stress, σ_{nom} or τ_{nom}, predicted from the simple equations allows the designer to analyze these situations. The symbol for these factors is K_t. In general, the K_t factors are used as follows:

$$\sigma_{max} = K_t \sigma_{nom} \quad \text{or} \quad \tau_{max} = K_t \tau_{nom} \tag{3–27}$$

depending on the kind of stress produced for the particular loading. The value of K_t depends on the shape of the discontinuity, the specific geometry, and the type of stress. Appendix 15 includes several charts for stress concentration factors. (See Reference 5.) Note that the charts indicate the method of computing the nominal stress. Usually, we compute the nominal stress by using the net section in the vicinity of the discontinuity. For example, for a flat plate with a hole in it subjected to a tensile force, the nominal stress is computed as the force divided by the minimum cross-sectional area through the location of the hole.

But there are other cases in which the gross area is used in calculating the nominal stress. For example, we analyze keyseats by applying the stress concentration factor to the computed stress in the full-diameter portion of the shaft.

Figure 3–25 shows an experimental device that demonstrates the phenomenon of stress concentrations. A model of a beam having several different cross-sectional heights is made from a special plastic that reacts to the presence of varying stresses at different points. When the model is viewed through a polarizing filter, several black *fringes* appear. Where there are many closely spaced fringes, the stress is changing rapidly. We can compute the actual magnitude of stress by knowing the optical characteristics of the plastic.

The beam in Figure 3–25 is simply supported near each end and is loaded vertically at its middle. The largest stress occurs to the left of the middle where the height of the cross section is reduced. Note that the fringes are very close together in the vicinity of the fillet connecting the smaller section to the larger part where the load is applied. This indicates

FIGURE 3–25
Illustration of stress concentrations
(Source: Measurements Group, Inc., Raleigh, North Carolina, U.S.A.)

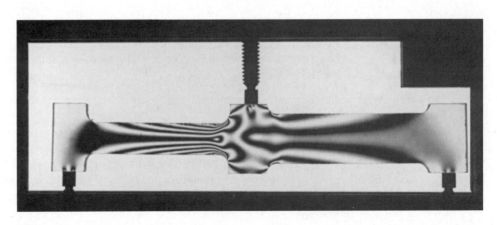

that the highest stress occurs in the fillet. Figure A15–2 gives data for the values of the stress concentration factor, K_t. The nominal stress, σ_{nom}, is computed from the classic flexure formula, and the section modulus is based on the smaller cross section near the fillet. These formulas are listed near the stress concentration chart.

An interesting observation can be made from Figure A15-3 showing the stress concentration factors for a flat plate with a central hole. Curves A and B refer to tension loading, while curve C is for bending. The nominal stress for each case is computed on the basis of the net cross section accounting for the material removed by the hole. Curve C indicates that the stress concentration factor is taken to be 1.0 for smaller holes, with the ratio of the hole diameter to the width of the plate <0.50.

Figure A15–4 covers the case of a round shaft that has a circular hole completely through it. The three curves cover tension, bending, and torsional loading, and each is based on the stress in the gross section, that is, the geometry of the shaft without the hole. Therefore, the value of K_t includes the effects of both the material removal and the discontinuity created by the presence of the hole. The values of K_t are relatively high, even for the smaller holes, indicating that you should be cautious when applying shafts with holes to ensure that the local stresses are small.

The following are guidelines on the use of stress concentration factors:

1. The worst case occurs for those areas in tension.

2. Always use stress concentration factors in analyzing members under fatigue loading because fatigue cracks usually initiate near points of high local tensile stress.

3. Stress concentrations can be ignored for static loading of ductile materials because if the local maximum stress exceeds the yield strength of the material, the load is redistributed. The resulting member is actually stronger after the local yielding occurs.

4. The stress concentration factors in Appendix 15 are empirical values based only on the geometry of the member and the manner of loading.

5. Use stress concentration factors when analyzing brittle materials under either static or fatigue loading. Because the material does not yield, the stress redistribution described in item 3 cannot occur.

6. Even scratches, nicks, corrosion, excessive surface roughness, and plating can cause stress concentrations. Chapter 5 discusses the care essential to manufacturing, handling, and assembling components subjected to fatigue loading.

Example Problem 3–18   Compute the maximum stress in a round bar subjected to an axial tensile force of 9800 N. The geometry is shown in Figure 3–26.

Solution   Objective   Compute the maximum stress in the stepped bar shown in Figure 3–26.

Given   The layout from Figure 3–26. Force $= F = 9800$ N.
The shaft has two diameters joined by a fillet with a radius of 1.5 mm.
Larger diameter $= D = 12$ mm; smaller diameter $= d = 10$ mm.

Analysis   The presence of the change in diameter at the step causes a stress concentration to occur. The general situation is a round bar subjected to an axial tensile load. We will use the top

FIGURE 3–26
Stepped round bar
subjected to axial
tensile force

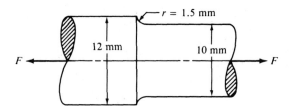

graph of Figure A15–1 to determine the stress concentration factor. That value is used in Equation (3–27) to determine the maximum stress.

Results   Figure A15–1 indicates that the nominal stress is computed for the smaller of the two diameters of the bar. The stress concentration factor depends on the ratio of the two diameters and the ratio of the fillet radius to the smaller diameter.

$$D/d = 12 \text{ mm}/10 \text{ mm} = 1.20$$
$$r/d = 1.5 \text{ mm}/10 \text{ mm} = 0.15$$

From these values, we can find that $K_t = 1.60$. The stress is

$$\sigma_{nom} = F/A = (9800 \text{ N})/[\pi(10 \text{ mm})^2/4] = 124.8 \text{ MPa}$$
$$\sigma_{max} = K_t\sigma_{nom} = (1.60)(124.8 \text{ MPa}) = 199.6 \text{ MPa}$$

Comments   The maximum tensile stress of 199.6 MPa occurs in the fillet near the smaller diameter. This value is 1.60 times higher than the nominal stress that occurs in the 10-mm-diameter shaft. To the left of the shoulder, the stress reduces dramatically as the effect of the stress concentration diminishes and because the area is larger.

**3–22
NOTCH
SENSITIVITY AND
STRENGTH
REDUCTION
FACTOR**

The amount by which a load-carrying member is weakened by the presence of a stress concentration (notch), considering both the material and the sharpness of the notch, is defined as

K_f = fatigue strength reduction factor

$$K_f = \frac{\text{endurance limit of a notch-free specimen}}{\text{endurance limit of a notched specimen}}$$

This factor could be determined by actual test. However, it is typically found by combining the stress concentration factor, K_t, defined in the previous section, and a material factor called the *notch sensitivity*, q. We define

$$q = (K_f - 1)/(K_t - 1) \qquad \qquad \textbf{(3–28)}$$

When q is known, K_f can be computed from

$$K_f = 1 + q(K_t - 1) \qquad \qquad \textbf{(3–29)}$$

Values of q range from zero to 1.0, and therefore K_f varies from 1.0 to K_t. Under repeated bending loads, very ductile steels typically exhibit values of q from 0.5 to 0.7. High-strength steels with hardness approximately HB 400 ($s_u \cong 200$ ksi or 1400 MPa) have values of q from 0.90 to 0.95. (See Reference 2 for further discussion of values of q.)

Because reliable values of q are difficult to obtain, the problems in this book will assume that $q = 1.0$ and $K_f = K_t$, the safest, most conservative value.

REFERENCES

1. Blake, Alexander. *Practical Stress Analysis in Engineering Design*. New York: Marcel Dekker, 2nd ed. 1990.

2. Boresi, A. P., O. M. Sidebottom, and R. J. Schmidt. *Advanced Mechanics of Materials*. 5th ed. New York: John Wisey, 1992.

3. Mott, R. L. *Applied Strength of Materials*. 4th ed. Upper Saddle River, NJ: Prentice Hall, 2002.

4. Muvdi, B. B., and J. W. McNabb. *Engineering Mechanics of Materials*. 2d ed. New York: Macmillan, 1984.

5. Pilkey, Walter D. *Peterson's Stress Concentration Factors*. 2d ed. New York: John Wiley, 1997.

6. Popov, E. P. *Engineering Mechanics of Solids*. 2nd ed. Upper Saddle River, NJ: Prentice Hall, 1998.

7. Young, W. C. and R. G. Budynas. *Roark's Formulas for Stress and Strain*. 7th ed. New York: McGraw-Hill, 2002.

INTERNET SITES RELATED TO STRESS AND DEFORMATION ANALYSIS

1. **BEAM 2D-Stress Analysis 3.1**
   www.orandsystems.com Software for mechanical, structural, civil, and architectural designers providing detailed analysis of statically indeterminate and determinate beams.

2. **MDSolids** *www.mdsolids.com* Educational software devoted to introductory mechanics of materials. Includes modules on basic stress and strain; beam and strut axial problems; trusses; statically indeterminate axial structures; torsion; determinate beams; section properties; general analysis of axial, torsion, and beam members; column buckling; pressure vessels; and Mohr's circle transformations.

3. **StressAlyzer** *http://hpme16.me.cmu.edu/stressalyzer* Interactive courseware for mechanics of materials including modules on axial loading, torsion loading, shear force and bending moment diagrams, load and stress calculations in 3D, beam deflections, and stress transformations.

PROBLEMS

Direct Tension and Compression

1. A tensile member in a machine structure is subjected to a steady load of 4.50 kN. It has a length of 750 mm and is made from a steel tube having an outside diameter of 18 mm and an inside diameter of 12 mm. Compute the tensile stress in the tube and the axial deformation.

2. Compute the stress in a round bar having a diameter of 10.0 mm and subjected to a direct tensile force of 3500 N.

3. Compute the stress in a rectangular bar having cross-sectional dimensions of 10.0 mm by 30.0 mm when a direct tensile force of 20.0 kN is applied.

4. A link in a packaging machine mechanism has a square cross section 0.40 in on a side. It is subjected to a tensile force of 860 lb. Compute the stress in the link.

5. Two circular rods support the 3800 lb weight of a space heater in a warehouse. Each rod has a diameter of 0.375 in and carries 1/2 of the total load. Compute the stress in the rods.

6. A tensile load of 5.00 kN is applied to a square bar, 12 mm on a side and having a length of 1.65 m. Compute the stress and the axial deformation in the bar if it is made from (a) AISI 1020 hot-rolled steel, (b) AISI 8650 OQT 1000 steel, (c) ductile iron A536-88 (60-40-18), (d) aluminum 6061-T6, (e) titanium Ti-6Al-4V, (f) rigid PVC plastic, and (g) phenolic plastic.

7. An aluminum rod is made in the form of a hollow square tube, 2.25 in outside, with a wall thickness of 0.120 in. Its length is 16.0 in. What axial compressive force would cause the tube to shorten by 0.004 in? Compute the resulting compressive stress in the aluminum.

8. Compute the stress in the middle portion of rod *AC* in Figure P3–8 if the vertical force on the boom is 2500 lb. The rod is rectangular, 1.50 in by 3.50 in.

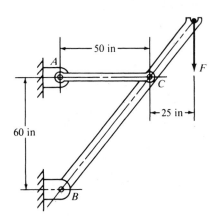

FIGURE P3–8   (Problems 8, 16, and 56)

9. Compute the forces in the two angled rods in Figure P3–9 for an applied force, $F = 1500$ lb, if the angle θ is 45°.

10. If the rods from Problem 9 are circular, determine their required diameter if the load is static and the allowable stress is 18 000 psi.

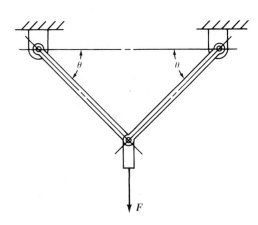

FIGURE P3–9   (Problems 9, 10, 11, 17, and 18)

11. Repeat Problems 9 and 10 if the angle θ is 15°.

12. Figure P3–12 shows a small truss spanning between solid supports and suspending a 10.5 kN load. The cross sections for the three main types of truss members are shown. Compute the stresses in all of the members of the truss near their midpoints away from the connections. Consider all joints to be pinned.

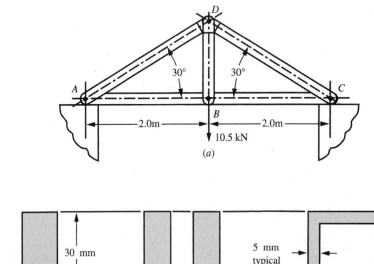

(a)

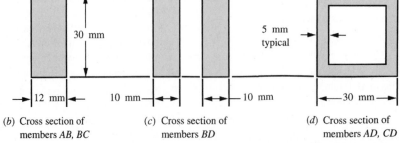

(b) Cross section of members *AB, BC*

(c) Cross section of members *BD*

(d) Cross section of members *AD, CD*

FIGURE P3–12   (Problem 12)

13. The truss shown in Figure P3–13 spans a total space of 18.0 ft and carries two concentrated loads on its top chord. The members are made from standard steel angle and channel shapes as indicated in the figure. Consider all joints to be pinned. Compute the stresses in all members near their midpoints away from the connections.

14. Figure P3–14 shows a short leg for a machine that carries a direct compression load. Compute the compressive

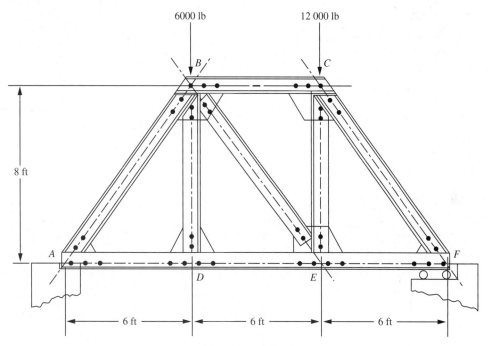

Member specifications:

AD, DE, EF L2 × 2 × 1/8 – doubled ——— ⌐⌐
BD, CE, BE L2 × 2 × 1/8 – single ——— ⌐
AB, BC, CF C3 × 4.1 – doubled ——— ⊐⊏

FIGURE P3–13   (Problem 13)

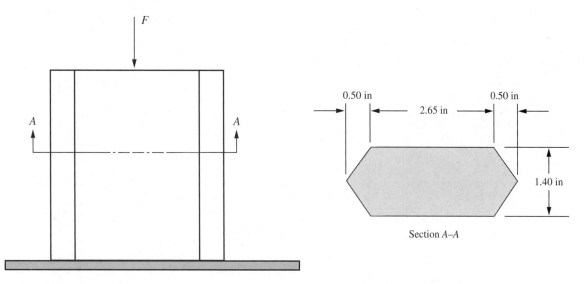

Section A–A

FIGURE P3–14   (Problem 14)

stress if the cross section has the shape shown and the applied force is $F = 52\,000$ lb.

15. Consider the short compression member shown in Figure P3–15. Compute the compressive stress if the cross section has the shape shown and the applied load is 640 kN.

Direct Shear Stress

16. Refer to Figure P3–8. Each of the pins at A, B, and C has a diameter of 0.50 in and is loaded in double shear. Compute the shear stress in each pin.

17. Compute the shear stress in the pins connecting the rods shown in Figure P3–9 when a load of $F = 1500$ lb is carried. The pins have a diameter of 0.75 in. The angle $\theta = 40°$.

18. Repeat Problem 17, but change the angle to $\theta = 15°$.

19. Refer to Figure 3–7. Compute the shear stress in the key if the shaft transmits a torque of 1600 N·m. The shaft diameter is 60 mm. The key is square with $b = 12$ mm, and it has a length of 45 mm.

20. A punch is attempting to cut a slug having the shape shown in Figure P3–20 from a sheet of aluminum having a thickness of 0.060 in. Compute the shearing stress in the aluminum when a force of 52 000 lb is applied by the punch.

21. Figure P3–21 shows the shape of a slug that is to be cut from a sheet of steel having a thickness of 2.0 mm. If the punch exerts a force of 225 kN, compute the shearing stress in the steel.

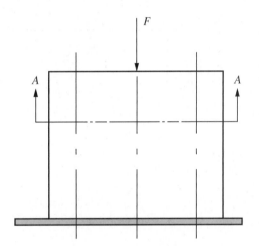

FIGURE P3–15 (Problem 15)

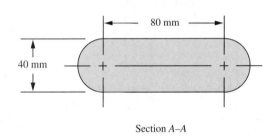

Section A–A

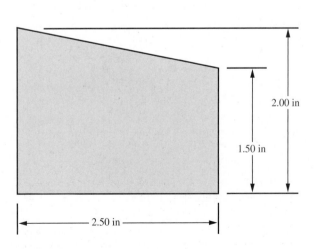

FIGURE P3–20 (Problem 20)

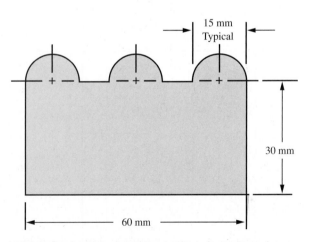

FIGURE P3–21 (Problem 21)

Torsion

22. Compute the torsional shear stress in a circular shaft with a diameter of 50 mm that is subjected to a torque of 800 N·m.

23. If the shaft of Problem 22 is 850 mm long and is made of steel, compute the angle of twist of one end in relation to the other.

24. Compute the torsional shear stress due to a torque of 88.0 lb·in in a circular shaft having a 0.40-in diameter.

25. Compute the torsional shear stress in a solid circular shaft having a diameter of 1.25 in that is transmitting 110 hp at a speed of 560 rpm.

26. Compute the torsional shear stress in a hollow shaft with an outside diameter of 40 mm and an inside diameter of 30 mm when transmitting 28 kilowatts (kW) of power at a speed of 45 rad/s.

27. Compute the angle of twist for the hollow shaft of Problem 26 over a length of 400 mm. The shaft is steel.

Noncircular Members in Torsion

28. A square steel bar, 25 mm on a side and 650 mm long, is subjected to a torque of 230 N·m. Compute the shear stress and the angle of twist for the bar.

29. A 3.00 in-diameter steel bar has a flat milled on one side, as shown in Figure P3–29. If the shaft is 44.0 in long and carries a torque of 10 600 lb·in, compute the stress and the angle of twist.

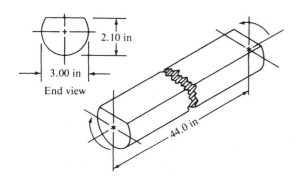

FIGURE P3–29 (Problem 29)

30. A commercial steel supplier lists rectangular steel tubing having outside dimensions of 4.00 by 2.00 in and a wall thickness of 0.109 in. Compute the maximum torque that can be applied to such a tube if the shear stress is to be limited to 6000 psi. For this torque, compute the angle of twist of the tube over a length of 6.5 ft.

Beams

31. A beam is simply supported and carries the load shown in Figure P3–31. Specify suitable dimensions for the beam if it is steel and the stress is limited to 18 000 psi, for the following shapes:

(a) Square

(b) Rectangle with height three times the width

(c) Rectangle with height one-third the width

(d) Solid circular section

(e) American Standard beam section

(f) American Standard channel with the legs down

(g) Standard steel pipe

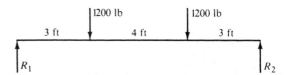

FIGURE P3–31 (Problems 31, 32, and 33)

32. For each beam of Problem 31, compute its weight if the steel weighs 0.283 lb/in³.

33. For each beam of Problem 31, compute the maximum deflection and the deflection at the loads.

34. For the beam loading of Figure P3–34, draw the complete shearing force and bending moment diagrams, and determine the bending moments at points A, B, and C.

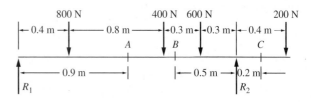

FIGURE P3–34 (Problems 34 and 35)

35. For the beam loading of Figure P3–34, design the beam, choosing a shape that will be reasonably efficient and will limit the stress to 100 MPa.

36. Figure P3–36 shows a beam made from 4 in steel pipe. Compute the deflection at points A and B for two cases: (a) the simple cantilever and (b) the supported cantilever.

37. Select an aluminum I-beam shape to carry the load shown in Figure P3–37 with a maximum stress of 12 000 psi. Then compute the deflection at each load.

38. Figure P3–38 represents a wood joist for a platform, carrying a uniformly distributed load of 120 lb/ft and two concentrated loads applied by some machinery. Compute

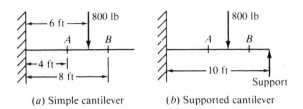

(a) Simple cantilever   (b) Supported cantilever

FIGURE P3–36   (Problem 36)

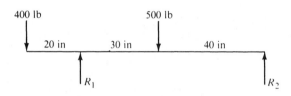

FIGURE P3–37   (Problem 37)

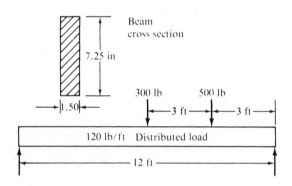

FIGURE P3–38   (Problem 38)

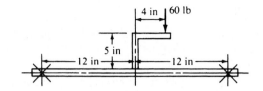

FIGURE P3–39   (Problems 39 and 57)

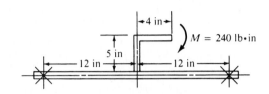

FIGURE P3–40   (Problem 40)

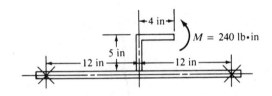

FIGURE P3–41   (Problem 41)

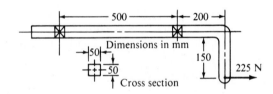

FIGURE P3–42   (Problems 42 and 58)

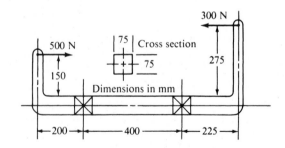

FIGURE P3–43   (Problems 43 and 59)

the maximum stress due to bending in the joist and the maximum vertical shear stress.

Beams with Concentrated Bending Moments

For Problems 39 through 50, draw the free-body diagram of only the horizontal beam portion of the given figures. Then draw the complete shear and bending moment diagrams. Where used, the symbol X indicates a simple support capable of exerting a reaction force in any direction but having no moment resistance. For beams having unbalanced axial loads, you may specify which support offers the reaction.

39. Use Figure P3–39.
40. Use Figure P3–40.
41. Use Figure P3–41.
42. Use Figure P3–42.
43. Use Figure P3–43.
44. Use Figure P3–44.

45. Use Figure P3–45.
46. Use Figure P3–46.
47. Use Figure P3–47.
48. Use Figure P3–48.
49. Use Figure P3–49.
50. Use Figure P3–50.

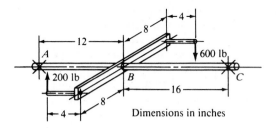

FIGURE P3–44 (Problem 44)

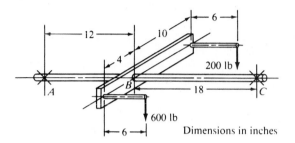

FIGURE P3–45 (Problem 45)

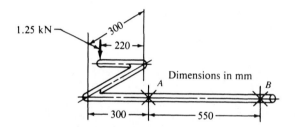

FIGURE P3–46 (Problem 46)

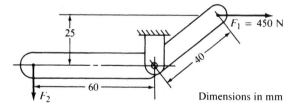

FIGURE P3–47 (Problem 47)

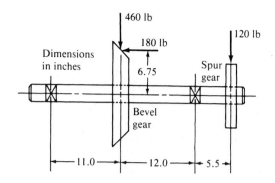

FIGURE P3–48 (Problem 48)

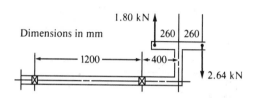

FIGURE P3–49 (Problem 49)

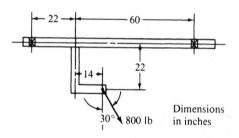

FIGURE P3–50 (Problems 50 and 60)

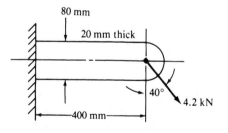

FIGURE P3–51 (Problem 51)

Combined Normal Stresses

51. Compute the maximum tensile stress in the bracket shown in Figure P3–51.

52. Compute the maximum tensile and compressive stresses in the horizontal beam shown in Figure P3–52.

53. For the lever shown in Figure P3–53 (a), compute the stress at section A near the fixed end. Then redesign the lever to the tapered form shown in Part (b) of the figure by adjusting only the height of the cross section at sec-

tions B and C so that they have no greater stress than section A.

54. Compute the maximum tensile stress at sections A and B on the crane boom shown in Figure P3–54.

55. Refer to Figure 3–22. Compute the maximum tensile stress in the print head just to the right of the right guide. The head has a rectangular cross section, 5.0 mm high in the plane of the paper and 2.4 mm thick.

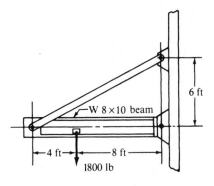

FIGURE P3–52   (Problem 52)

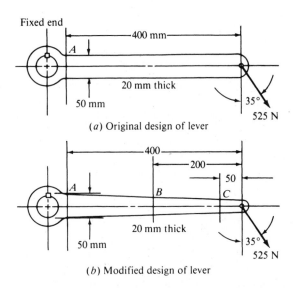

(*a*) Original design of lever

(*b*) Modified design of lever

FIGURE P3–53   (Problem 53)

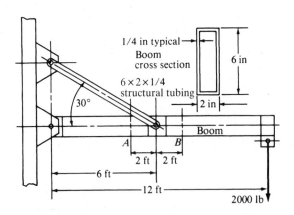

FIGURE P3–54   (Problem 54)

56. Refer to Figure P3–8. Compute the maximum tensile and compressive stresses in the member *B-C* if the load *F* is 1800 lb. The cross section of *B-C* is a 6 × 4 × 1/4 rectangular tube.

57. Refer to P3–39. The vertical member is to be made from steel with a maximum allowable stress of 12 000 psi. Specify the required size of a standard square cross section if sizes are available in increments of 1/16 in.

58. Refer to P3–42. Compute the maximum stress in the horizontal portion of the bar, and tell where it occurs on the cross section. The left support resists the axial force.

59. Refer to P3–43. Compute the maximum stress in the horizontal portion of the bar, and indicate where it occurs on the cross section. The right support resists the unbalanced axial force.

60. Refer to P3–50. Specify a suitable diameter for a solid circular bar to be used for the top horizontal member, which is supported in the bearings. The left bearing resists the axial load. The allowable normal stress is 25 000 psi.

Stress Concentrations

61. Figure P3–61 shows a valve stem from an engine subjected to an axial tensile load applied by the valve spring. For a force of 1.25 kN, compute the maximum stress at the fillet under the shoulder.

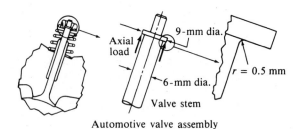

Automotive valve assembly

FIGURE P3–61   (Problem 61)

62. The conveyor fixture shown in Figure P3–62 carries three heavy assemblies (1200 lb each). Compute the maximum stress in the fixture, considering stress concentrations at the fillets and assuming that the load acts axially.

63. For the flat plate in tension in Figure P3–63, compute the stress at each hole, assuming that the holes are sufficiently far apart that their effects do not interact.

For Problems 64 through 68, compute the maximum stress in the member, considering stress concentrations.

64. Use Figure P3–64.

65. Use Figure P3–65.

66. Use Figure P3–66.

67. Use Figure P3–67.

68. Use Figure P3–68.

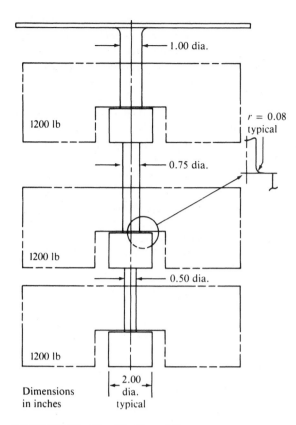

FIGURE P3–62 (Problem 62)

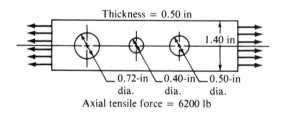

FIGURE P3–63 (Problem 63)

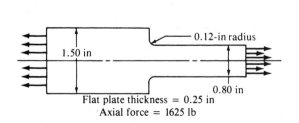

FIGURE P3–64 (Problem 64)

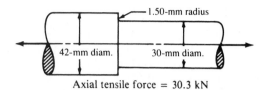

FIGURE P3–65 (Problem 65)

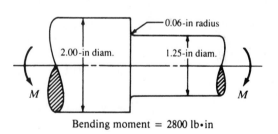

FIGURE P3–66 (Problem 66)

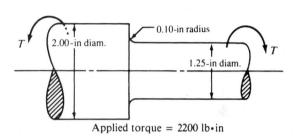

FIGURE P3–67 (Problem 67)

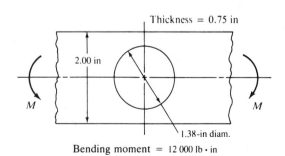

FIGURE P3–68 (Problem 68)

Problems of a General Nature

69. Figure P3–69 shows a horizontal beam supported by a vertical tension link. The cross sections of both the beam and the link are 20 mm square. All connections use 8.00 mm-diameter cylindrical pins in double shear. Compute the tensile stress in member *A-B*, the stress due to bending in *C-D*, and the shearing stress in the pins *A* and *C*.

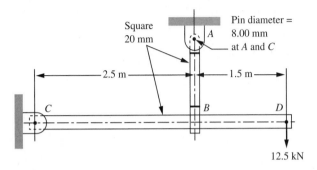

FIGURE P3–69   (Problem 69)

70. Figure P3–70 shows a tapered flat bar that has a uniform thickness of 20 mm. The depth tapers from $h_1 = 40$ mm near the load to $h_2 = 20$ mm at each support. Compute the stress due to bending in the bar at points spaced 40 mm apart from the support to the load. Let the load $P = 5.0$ kN.

71. For the flat bar shown in Figure P3–70, compute the stress in the middle of the bar if a hole of 25 mm diameter is drilled directly under the load on the horizontal centerline. The load is $P = 5.0$ kN. See data in Problem 70.

72. The beam shown in Figure P3–72 is a stepped, flat bar having a constant thickness of 1.20 in. It carries a single concentrated load at *C* of 1500 lb. Compare the stresses at the following locations:

(a) In the vicinity of the load

(b) At the section through the smaller hole to the right of section *C*

(c) At the section through the larger hole to the right of section *C*

(d) Near section *B* where the bar changes height

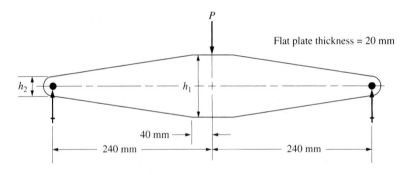

FIGURE P3–70   Tapered flat bar for Problems 70 and 71

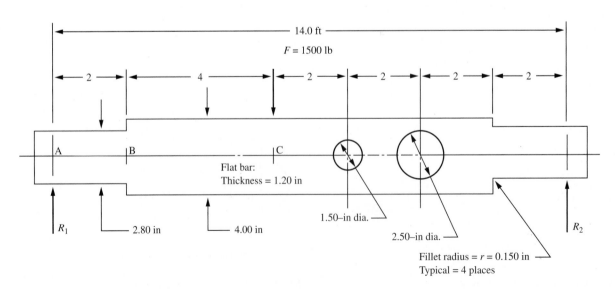

FIGURE P3–72   (Problem 72)

73. Figure P3–73 shows a stepped, flat bar having a constant thickness of 8.0 mm. It carries three concentrated loads as shown. Let $P = 200$ N, $L_1 = 180$ mm, $L_2 = 80$ mm, and $L_3 = 40$ mm. Compute the maximum stress due to bending, and state where it occurs. The bar is braced against lateral bending and twisting. Note that the dimensions in the figure are not drawn to scale.

74. Figure P3–74 shows a bracket carrying opposing forces of $F = 2500$ N. Compute the stress in the upper horizontal part through one of the holes as at B. Use $d = 15.0$ mm for the diameter of the holes.

75. Repeat Problem 74, but use a hole diameter of $d = 12.0$ mm.

76. Figure P3–76 shows a lever made from a rectangular bar of steel. Compute the stress due to bending at the fulcrum (20 in from the pivot) and at the section through the bottom hole. The diameter of each hole is 1.25 in.

77. For the lever in P3–76, determine the maximum stress if the attachment point is moved to each of the other two holes.

78. Figure P3–78 shows a shaft that is loaded only in bending. Bearings are located at points B and D to allow the shaft to rotate. Pulleys at A, C, and E carry cables that support loads from below while allowing the shaft to rotate. Compute the maximum stress due to bending in the shaft considering stress concentrations.

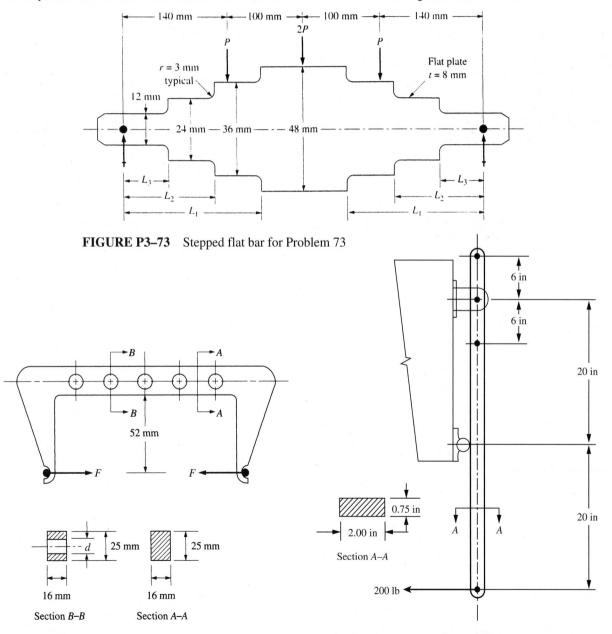

FIGURE P3–73   Stepped flat bar for Problem 73

FIGURE P3–74   Bracket for Problems 74 and 75

FIGURE P3–76   Lever for Problems 76 and 77

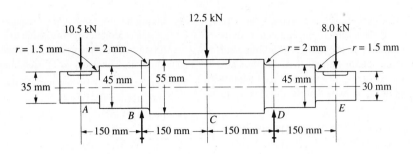

FIGURE P3–78   Data for Problem 78

Internet-Based Assignments

79. Use the MDSolids software to analyze the forces in all members of the truss shown in P3–12.

80. Use the MDSolids software to analyze the forces in all members of the truss shown in P3–13.

81. With the results from Problem 79, use the MDSolids software to analyze the axial tensile or compressive stresses in all members of the truss shown in P3–12.

82. With the results from Problem 80, use the MDSolids software to analyze the axial tensile or compressive stresses in all members of the truss shown in P3–13.

83. Use either the BEAM 2D, MDSolids, or StressAlyzer software to solve Problem 3–34.

84. For the beam shown in P3–37, use either the BEAM 2D, MDSolids, or StressAlyzer software to produce the shearing force and bending moment diagrams.

85. For the beam shown in P3–38, use either the BEAM 2D, MDSolids, or StressAlyzer software to produce the shearing force and bending moment diagrams.

86. For the shaft shown in P3–78, use either the BEAM 2D, MDSolids, or StressAlyzer software to produce the shearing force and bending moment diagrams. Determine the bending moment at point C and at each step in the diameter of the shaft.

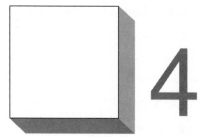

4

Combined Stresses and Mohr's Circle

The Big Picture

You Are the Designer

4–1 Objectives of This Chapter

4–2 General Case of Combined Stress

4–3 Mohr's Circle

4–4 Mohr's Circle Practice Problems

4–5 Case When Both Principal Stresses Have the Same Sign

4–6 Mohr's Circle for Special Stress Conditions

4–7 Analysis of Complex Loading Conditions

The Big Picture

Combined Stresses and Mohr's Circle

Discussion Map

☐ You must build your ability to analyze more complex parts and loading patterns.

Discover
Find products around you that have complex geometries or loading patterns.
Discuss these products with your colleagues.

This chapter helps you analyze complex objects to determine maximum stresses. We will use *Mohr's circle*, a graphical tool for stress analysis, as an aid in understanding how stresses vary within a load-carrying member.

In Chapter 3, you reviewed the basic principles of stress and deformation analysis, practiced the application of those principles to machine design problems, and solved some problems by superposition when two or more types of loads caused normal, either tensile or compressive, stresses.

But what happens when the loading pattern is more complex?

Many practical machine components experience combinations of normal and shear stresses. Sometimes the pattern of loading or the geometry of the component causes the analysis to be very difficult to solve directly using the methods of basic stress analysis.

Look around you and identify products, parts of structures, or machine components that have a more complex loading or geometry. Perhaps some of those identified in **The Big Picture** for Chapter 3 have this characteristic.

Discuss how the selected items are loaded, where the maximum stresses are likely to occur, and how the loads and the geometry are related. Did the designer tailor the shape of the object to be able to carry the applied loads in an efficient manner? How are the shape and the size of critical parts of the item related to the expected stresses?

When we move on to **Chapter 5: Design for Different Types of Loading,** we will need tools to determine the magnitude and the direction of maximum shear stresses or maximum principal (normal) stresses.

Completing this chapter will help you develop a clear understanding of the distribution of stress in a load-carrying member, and it will help you determine the maximum stresses, either normal or shear, so that you can complete a reliable design or analysis.

Some of the techniques of combining stresses require the application of fairly involved equations. A graphical tool, *Mohr's circle*, can be used as an aid in completion of the analysis. Applied properly, the method is precise and should aid you in understanding how the stresses vary within a complex load-carrying member. It should also help you correctly use commercially available stress analysis software.

 You Are the Designer

Your company is designing a special machine to test a high-strength fabric under prolonged exposure to a static load to determine whether it continues to deform a greater amount with time. The tests will be run at a variety of temperatures requiring a controlled environment around the test specimen. Figure 4–1 shows the general layout of one proposed design. Two rigid supports are available at the rear of the machine with a 24-in gap between them. The line of action of the load on the test fabric is centered on this gap and 15.0 in out from the middle of the supports. You are asked to design a bracket to hold the upper end of the load frame.

Assume that one of your design concepts uses the arrangement shown in Figure 4–2. Two circular bars are bent 90°. One end of each bar is securely welded to the vertical support surface. A flat bar is attached across the outboard end of each bar so that the load is shared evenly by the two bars.

One of your design problems is to determine the maximum stress that exists in the bent bars to ensure that

they are safe. What kinds of stress are developed in the bars? Where are the stresses likely to be the greatest? How can the magnitude of the stresses be computed? Note that the part of the bar near its attachment to the support has a combination of stresses exerted on it.

Consider the element on the top surface of the bar, labeled element A in Figure 4–2. The moment caused by the force acting at an extension of 6.0 in from the support places element A in tension due to the bending action. The torque caused by the force acting 15.0 in out from the axis of the bar at its point of support creates a torsional

FIGURE 4–1 Layout of the load frame supports—top view

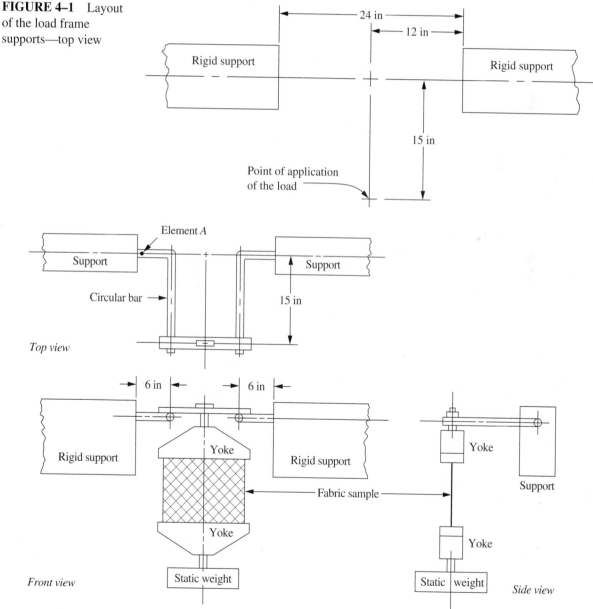

FIGURE 4–2 Proposed bracket design

shear stress on element A. Both of these stresses act in the x-y plane, subjecting element A to a combined normal and shear stress. How do you analyze such a stress condition? How do the tensile and shear stresses act together? What are the maximum normal stress and the

maximum shear stress on element A, and where do they occur?

You would need such answers to complete the design of the bars. The material in this chapter will enable you to complete the necessary analyses.

4–1
OBJECTIVES OF THIS CHAPTER

After completing this chapter, you will be able to:

1. Illustrate a variety of combined stresses on stress elements.

2. Analyze a load-carrying member subjected to combined stress to determine the maximum normal stress and the maximum shear stress on any given element.

3. Determine the directions in which the maximum stresses are aligned.

4. Determine the state of stress on an element in any specified direction.

5. Draw the complete Mohr's circle as an aid in completing the analyses for the maximum stresses.

4–2
GENERAL CASE OF COMBINED STRESS

To visualize the general case of combined stress, it is helpful to consider a small element of the load-carrying member on which combined normal and shear stresses act. For this discussion we will consider a two-dimensional stress condition, as illustrated in Figure 4–3. The x- and y-axes are aligned with corresponding axes on the member being analyzed.

The normal stresses, σ_x and σ_y, could be due to a direct tensile force or to bending. If the normal stresses were compressive (negative), the vectors would be pointing in the opposite sense, into the stress element.

The shear stress could be due to direct shear, torsional shear, or vertical shear stress. The double-subscript notation helps to orient the direction of shear stresses. For example, τ_{xy} indicates the shear stress acting on the element face that is perpendicular to the x-axis and parallel to the y-axis.

A positive shear stress is one that tends to rotate the stress element clockwise.

In Figure 4–3, τ_{xy} is positive, and τ_{yx} is negative. Their magnitudes must be equal to maintain the element in equilibrium.

It is necessary to determine the magnitudes and the signs of each of these stresses in order to show them properly on the stress element. Example Problem 4–1, which follows the definition of principal stresses, illustrates the process.

FIGURE 4–3
General two-dimensional stress element

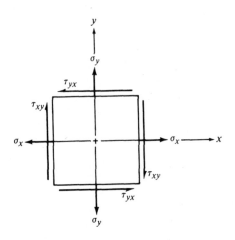

With the stress element defined, the objectives of the remaining analysis are to determine the maximum normal stress, the maximum shear stress, and the planes on which these stresses occur. The governing formulas follow. (See Reference 1 for the derivations.)

Maximum Normal Stresses: Principal Stresses

The combination of the applied normal and shear stresses that produces the maximum normal stress is called the *maximum principal stress*, σ_1. The magnitude of σ_1 can be computed from the following equation:

⇨ **Maximum Principal Stress**

$$\sigma_1 = \frac{\sigma_x + \sigma_y}{2} + \sqrt{\left(\frac{\sigma_x - \sigma_y}{2}\right)^2 + \tau_{xy}^2} \qquad (4\text{–}1)$$

The combination of the applied stresses that produces the minimum normal stress is called the *minimum principal stress,* σ_2. Its magnitude can be computed from

⇨ **Minimum Principal Stress**

$$\sigma_2 = \frac{\sigma_x + \sigma_y}{2} - \sqrt{\left(\frac{\sigma_x - \sigma_y}{2}\right)^2 + \tau_{xy}^2} \qquad (4\text{–}2)$$

Particularly in experimental stress analysis, it is important to know the orientation of the principal stresses. The angle of inclination of the planes on which the principal stresses act, called the *principal planes,* can be found from

⇨ **Angle for Principal Stress Element**

$$\phi_\sigma = \frac{1}{2}\arctan\left[2\tau_{xy}/(\sigma_x - \sigma_y)\right] \qquad (4\text{–}3)$$

The angle ϕ_σ is measured from the positive x-axis of the original stress element to the maximum principal stress, σ_1. Then the minimum principal stress, σ_2, is on the plane 90° from σ_1.

When the stress element is oriented as discussed so that the principal stresses are acting on it, the shear stress is zero. The resulting stress element is shown in Figure 4–4.

Maximum Shear Stress

On a different orientation of the stress element, the maximum shear stress will occur. Its magnitude can be computed from

⇨ **Maximum shear stress**

$$\tau_{max} = \sqrt{\left(\frac{\sigma_x - \sigma_y}{2}\right)^2 + \tau_{xy}^2} \qquad (4\text{–}4)$$

FIGURE 4–4
Principal stress element

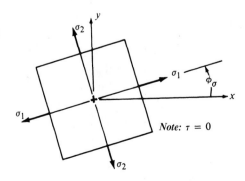

Note: $\tau = 0$

FIGURE 4–5
Maximum shear stress
element

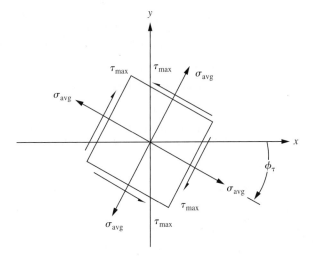

The angle of inclination of the element on which the maximum shear stress occurs is computed as follows:

**Angle for
Maximum Shear
Stress Element**

$$\phi_\tau = \tfrac{1}{2} \arctan\left[-(\sigma_x - \sigma_y)/2\tau_{xy}\right] \qquad (4\text{–}5)$$

The angle between the principal stress element and the maximum shear stress element is always 45°.

On the maximum shear stress element, there will be normal stresses of equal magnitude acting perpendicular to the planes on which the maximum shear stresses are acting. These normal stresses have the value

**Average
Normal Stress**

$$\sigma_{avg} = (\sigma_x + \sigma_y)/2 \qquad (4\text{–}6)$$

Note that this is the *average* of the two applied normal stresses. The resulting maximum shear stress element is shown in Figure 4–5. Note, as stated above, that the angle between the principal stress element and the maximum shear stress element is always 45°.

Summary and General Procedure for Analyzing Combined Stresses

The following list gives a summary of the techniques presented in this section; it also outlines the general procedure for applying the techniques to a given stress analysis problem.

**General
Procedure for
Computing
Principal Stresses
and Maximum
Shear Stresses**

1. Decide for which point you want to compute the stresses.
2. Clearly specify the coordinate system for the object, the free-body diagram, and the magnitude and direction of forces.
3. Compute the stresses on the selected point due to the applied forces, and show the stresses acting on a stress element at the desired point with careful attention to directions. Figure 4–3 is a model for how to show these stresses.
4. Compute the principal stresses on the point and the directions in which they act. Use Equations (4–1), (4–2), and (4–3).

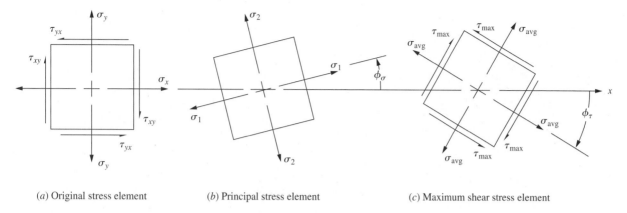

(a) Original stress element   (b) Principal stress element   (c) Maximum shear stress element

FIGURE 4–6 Relationships among original stress element, principal stress element, and maximum shear stress element for a given loading

> **5.** Draw the stress element on which the principal stresses act, and show its orientation relative to the original x-axis. It is recommended that the principal stress element be drawn beside the original stress element to illustrate the relationship between them.
>
> **6.** Compute the maximum shear stress on the element and the orientation of the plane on which it acts. Also, compute the normal stress that acts on the maximum shear stress element. Use Equations (4–4), (4–5), and (4–6).
>
> **7.** Draw the stress element on which the maximum shear stress acts, and show its orientation to the original x-axis. It is recommended that the maximum shear stress element be drawn beside the maximum principal stress element to illustrate the relationship between them.
>
> **8.** The resulting set of three stress elements will appear as shown in Figure 4–6.

The following example problem illustrates the use of this procedure.

Example Problem 4–1   The shaft shown in Figure 4–7 is supported by two bearings and carries two V-belt sheaves. The tensions in the belts exert horizontal forces on the shaft, tending to bend it in the x-z plane. Sheave B exerts a clockwise torque on the shaft when viewed toward the origin of the coordinate system along the x-axis. Sheave C exerts an equal but opposite torque on the shaft. For the loading condition shown, determine the principal stresses and the maximum shear stress on element K on the front surface of the shaft (on the positive z-side) just to the right of sheave B. Follow the general procedure for analyzing combined stresses given in this section.

Solution   Objective   Compute the principal stresses and the maximum shear stresses on element K.

Given   Shaft and loading pattern shown in Figure 4–7.

Analysis   Use the general procedure for analyzing combined stresses.

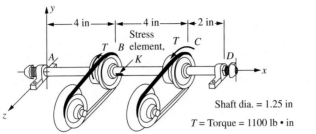

Shaft dia. = 1.25 in

T = Torque = 1100 lb • in

(a) Pictorial view of shaft

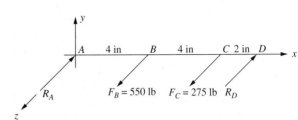

(b) Forces acting on shaft at B and C caused by belt drives

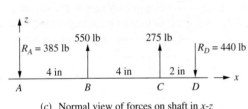

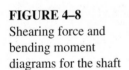

(c) Normal view of forces on shaft in x-z plane with reactions at bearings

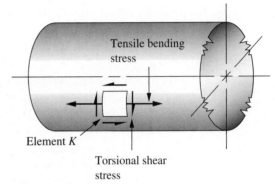

(d) Enlarged view of element K on front of shaft

FIGURE 4–7   Shaft supported by two bearings and carrying two V-belt sheaves

FIGURE 4–8
Shearing force and bending moment diagrams for the shaft

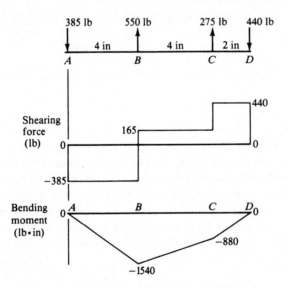

Results   Element K is subjected to bending that produces a tensile stress acting in the x-direction. Also, there is a torsional shear stress acting at K. Figure 4–8 shows the shearing force and bending moment diagrams for the shaft and indicates that the bending moment at K is 1540 lb · in. The bending stress is therefore

FIGURE 4–9

Stresses on element K

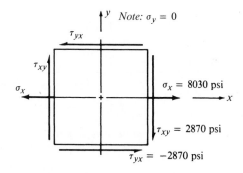

$$\sigma_x = M/S$$
$$S = \pi D^3/32 = [\pi(1.25 \text{ in})^3]/32 = 0.192 \text{ in}^3$$
$$\sigma_x = (1540 \text{ lb} \cdot \text{in})/(0.192 \text{ in}^3) = 8030 \text{ psi}$$

The torsional shear stress acts on element K in a way that causes a downward shear stress on the right side of the element and an upward shear stress on the left side. This action results in a tendency to rotate the element in a *clockwise* direction, which is the *positive* direction for shear stresses according to the standard convention. Also, the notation for shear stresses uses double subscripts. For example, τ_{xy} indicates the shear stress acting on the face of an element that is perpendicular to the x-axis and parallel to the y-axis. Thus, for element K,

$$\tau_{xy} = T/Z_p$$
$$Z_p = \pi D^3/16 = \pi(1.25 \text{ in})^3/16 = 0.383 \text{ in}^3$$
$$\tau_{xy} = (1100 \text{ lb} \cdot \text{in})/(0.383 \text{ in}^3) = 2870 \text{ psi}$$

The values of the normal stress, σ_x, and the shear stress, τ_{xy}, are shown on the stress element K in Figure 4–9. Note that the stress in the y-direction is zero for this loading. Also, the value of the shear stress, τ_{yx}, must be equal to τ_{xy}, and it must act as shown in order for the element to be in equilibrium.

We can now compute the principal stresses on the element, using Equations (4–1) through (4–3). The maximum principal stress is

$$\sigma_1 = \frac{\sigma_x + \sigma_y}{2} + \sqrt{\left(\frac{\sigma_x - \sigma_y}{2}\right)^2 + \tau_{xy}^2} \tag{4–1}$$
$$\sigma_1 = (8030/2) + \sqrt{(8030/2)^2 + (2870)^2}$$
$$\sigma_1 = 4015 + 4935 = 8950 \text{ psi}$$

The minimum principal stress is

$$\sigma_2 = \frac{\sigma_x + \sigma_y}{2} - \sqrt{\left(\frac{\sigma_x - \sigma_y}{2}\right)^2 + \tau_{xy}^2} \tag{4–2}$$
$$\sigma_2 = (8030/2) - \sqrt{(8030/2)^2 + (2870)^2}$$
$$\sigma_2 = 4015 - 4935 = -920 \text{ psi (compression)}$$

The direction in which the maximum principal stress acts is

$$\phi_\sigma = \tfrac{1}{2}\arctan\left[2\tau_{xy}/(\sigma_x - \sigma_y)\right] \tag{4–3}$$
$$\phi_\sigma = \tfrac{1}{2}\arctan\left[(2)(2870)/(8030)\right] = 17.8°$$

The positive sign calls for a *clockwise* rotation of the element.

FIGURE 4–10

Principal stress element

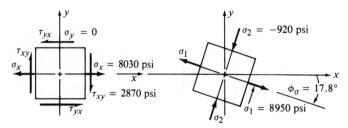

(*a*) Original stress element   (*b*) Principal stress element

The principal stresses can be shown on a stress element as illustrated in Figure 4–10. Note that the element is shown in relation to the original element to emphasize the direction of the principal stresses in relation to the original *x*-axis. The positive sign for ϕ_σ indicates that the principal stress element is rotated *clockwise* from its original position.

Now the maximum shear stress element can be defined, using Equations (4–4) through (4–6):

$$\tau_{max} = \sqrt{\left(\frac{\sigma_x - \sigma_y}{2}\right)^2 + \tau_{xy}^2} \qquad (4\text{–}4)$$

$$\tau_{max} = \sqrt{(8030/2)^2 + (2870)^2}$$

$$\tau_{max} = \pm\, 4935 \text{ psi}$$

The two pairs of shear stresses, $+\tau_{max}$ and $-\tau_{max}$, are equal in magnitude but opposite in direction.

The orientation of the element on which the maximum shear stress acts is found from Equation (4–5):

$$\phi_\tau = \tfrac{1}{2}\arctan\left[-(\sigma_x - \sigma_y)/2\tau_{xy}\right] \qquad (4\text{–}5)$$

$$\phi_\tau = \tfrac{1}{2}\arctan\left(-8030/[(2)(2870)]\right) = -27.2°$$

The negative sign calls for a *counterclockwise* rotation of the element.

There are equal normal stresses acting on the faces of this stress element, which have the value of

$$\sigma_{avg} = (\sigma_x + \sigma_y)/2 \qquad (4\text{–}6)$$

$$\sigma_{avg} = 8030/2 = 4015 \text{ psi}$$

Comments   Figure 4–11 shows the stress element on which the maximum shear stress acts in relation to the original stress element. Note that the angle between this element and the principal stress element is 45°.

Examine the results of Example Problem 4–1. The maximum principal stress, $\sigma_1 =$ 8950 psi, is 11 percent greater than the value of $\sigma_x = 8030$ psi computed for the bending stress in the shaft acting in the *x*-direction. The maximum shear stress, $\tau_{max} = 4935$ psi, is 72 percent greater than the computed applied torsional shear stress of $\tau_{xy} = 2870$ psi. You will see in Chapter 5 that either the maximum normal stress or the maximum shear stress is often required for accurate failure prediction and for safe design decisions. The angles

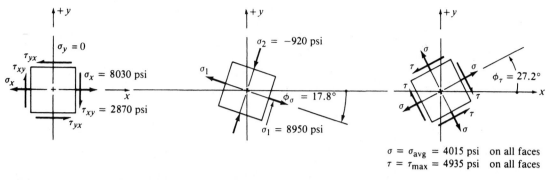

(a) Original stress element, K (b) Principal stress element   (c) Maximum shear stress element

FIGURE 4–11   Relation of maximum shear stress element to the original stress element and the principal stress element

of the final stress elements also predict the alignment of the most damaging stresses that can be an aid in experimental stress analysis and the analysis of actual failed components.

Another concept, called the *von Mises stress*, is used in the distortion energy theory of failure described in Chapter 5. The von Mises stress is a unique combination of the maximum principal stress, σ_1, and the minimum principal stress, σ_2, which can be compared directly with the yield strength of the material to predict failure by yielding.

The process of computing the principal stresses and the maximum shear stress shown in Example Problem 4–1 may seem somewhat abstract. These same results can be obtained using a method called *Mohr's circle*, which is discussed next. This method uses a combination of a graphical aid and simple calculations. With practice, the use of Mohr's circle should provide you with a more intuitive feel for the variations in stress that exist at a point in relation to the angle of orientation of the stress element. In addition, it provides a streamlined approach to determining the stress condition on any plane of interest.

**4–3
MOHR'S CIRCLE**

Because of the many terms and signs involved, and the many calculations required in the computation of the principal stresses and the maximum shear stress, there is a rather high probability of error. Using the graphic aid Mohr's circle helps to minimize errors and gives a better "feel" for the stress condition at the point of interest.

After Mohr's circle is constructed, it can be used for the following:

1. Finding the maximum and minimum principal stresses and the directions in which they act.

2. Finding the maximum shear stresses and the orientation of the planes on which they act.

3. Finding the value of the normal stresses that act on the planes where the maximum shear stresses act.

4. Finding the values of the normal and shear stresses that act on an element with any orientation.

The data needed to construct Mohr's circle are, of course, the same as those needed to compute the preceding values, because the graphical approach is an exact analogy to the computations.

If the normal and shear stresses that act on any two mutually perpendicular planes of an element are known, the circle can be constructed and any of items 1 through 4 can be found.

Mohr's circle is actually a plot of the combinations of normal and shearing stresses that exist on a stress element for all possible angles of orientation of the element. This method is particularly valuable in experimental stress analysis work because the results obtained from many types of standard strain gage instrumentation techniques give the necessary inputs for the creation of Mohr's circle. (See Reference 1.) When the principal stresses and the maximum shear stress are known, the complete design and analysis can be done, using the various theories of failure discussed in Chapter 5.

Procedure for Constructing Mohr's Circle

1. Perform the stress analysis to determine the magnitudes and directions of the normal and shear stresses acting at the point of interest.

2. Draw the stress element at the point of interest as shown in Figure 4–12(a). Normal stresses on any two mutually perpendicular planes are drawn with tensile stresses positive—projecting outward from the element. Compressive stresses are negative—directed inward on the face. Note that the *resultants* of all normal stresses acting in the chosen directions are plotted. Shear stresses are considered to be positive if they tend to rotate the element in a *clockwise* (cw) direction, and negative otherwise.

   Note that on the stress element illustrated, σ_x is positive, σ_y is negative, τ_{xy} is positive, and τ_{yx} is negative. This assignment is arbitrary for the purpose of illustration. In general, any combination of positive and negative values could exist.

3. Refer to Figure 4–12(b). Set up a rectangular coordinate system in which the positive horizontal axis represents positive (tensile) normal stresses, and the positive vertical axis represents positive (clockwise) shear stresses. Thus, the plane created will be referred to as the σ-τ *plane*.

4. Plot points on the σ-τ plane corresponding to the stresses acting on the faces of the stress element. If the element is drawn in the *x*-*y* plane, the two points to be plotted are σ_x, τ_{xy} and σ_y, τ_{yx}.

5. Draw the line connecting the two points.

6. The resulting line crosses the σ-axis at the center of Mohr's circle at the average of the two applied normal stresses, where

$$\sigma_{avg} = (\sigma_x + \sigma_y)/2$$

   The center of Mohr's circle is called O in Figure 4–12.

7. Note in Figure 4–12 that a right triangle has been formed, having the sides a, b, and R, where

$$R = \sqrt{a^2 + b^2}$$

   By inspection, we can see that

$$a = (\sigma_x - \sigma_y)/2$$
$$b = \tau_{xy}$$

   The point labeled O is at a distance of $\sigma_x - a$ from the origin of the coordinate system. We can now proceed with the construction of the circle.

8. Draw the complete circle with the center at O and a radius of R, as shown in Figure 4–13.

9. The point where the circle crosses the σ-axis at the right gives the value of the maximum principal stress, σ_1. Note that $\sigma_1 = \sigma_{avg} + R$.

10. The point where the circle crosses the σ-axis at the left gives the minimum principal stress, σ_2. Note that $\sigma_2 = \sigma_{avg} - R$.

FIGURE 4–12
Partially completed
Mohr's circle, Steps 1–7

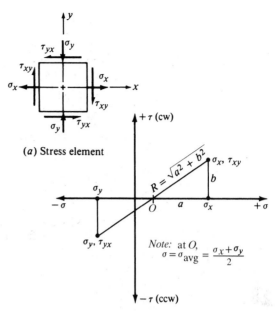

(a) Stress element

$$\text{Note: at } O,$$
$$\sigma = \sigma_{\text{avg}} = \frac{\sigma_x + \sigma_y}{2}$$

(b) Partially completed
Mohr's circle

FIGURE 4–13
Completed Mohr's
circle, Steps 8–14

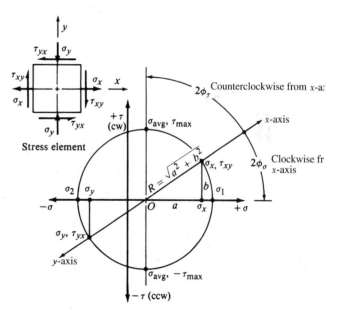

Stress element

11. The coordinates of the top of the circle give the maximum shear stress and the average normal stress that act on the element having the maximum shear stress. Note that $\tau_{\text{max}} = R$.

   Note: The following steps relate to determining the angles of inclination of the principal stress element and the maximum shear stress element in relation to the original x-axis. It is important to realize that angles on Mohr's circle are actually *double* the true angles. Refer to Figure 4–13; the line from O through the first point plotted, σ_x, τ_{xy}, represents the original x-axis, as noted in the figure. The line from O through the point σ_y, τ_{yx} represents the original y-axis. Of course, on the original element, these axes are 90° apart, not 180°, illustrating the double-angle feature of Mohr's circle. Having made this observation, we can continue with the development of the process.

12. The angle $2\phi_\sigma$ is measured from the x-axis on the circle to the σ-axis. Note that

$$2\phi_\sigma = \arctan (b/a)$$

It is also important to note the direction *from the x-axis to the σ-axis* (clockwise or counterclockwise). This is necessary for representing the relation of the principal stress element to the original stress element properly.

13. The angle from the x-axis on the circle to the vertical line through τ_{max} gives $2\phi_\tau$. From the geometry of the circle, in the example shown, we can see that

$$2\phi_\tau = 90° - 2\phi_\sigma$$

Other combinations of the initial stresses will result in different relationships between $2\phi_\sigma$ and $2\phi_\tau$. The specific geometry on the circle should be used each time. See Example Problems 4–3 to 4–8 that follow this section.

Again it is important to note the direction *from the x-axis to the τ_{max}-axis* for use in orienting the maximum shear stress element. You should also note that the σ-axis and the τ_{max}-axis are always 90° apart on the circle and therefore 45° apart on the actual element.

14. The final step in the process of using Mohr's circle is to draw the resulting stress elements in their proper relation to the original element, as shown in Figure 4–14.

We will now illustrate the construction of Mohr's circle by using the same data as in Example Problem 4–1, in which the principal stresses and the maximum shear stress were computed directly from the equations.

Example Problem 4–2   The shaft shown in Figure 4–7 is supported by two bearings and carries two V-belt sheaves. The tensions in the belts exert horizontal forces on the shaft, tending to bend it in the x-z plane. Sheave B exerts a clockwise torque on the shaft when viewed toward the origin of the coordinate system along the x-axis. Sheave C exerts an equal but opposite torque on the shaft. For the loading condition shown, determine the principal stresses and the maximum shear stress on element K on the front surface of the shaft (on the positive z-side) just to the right of sheave B. Use the procedure for constructing Mohr's circle in this section.

FIGURE 4–14
Display of results from
Mohr's circle

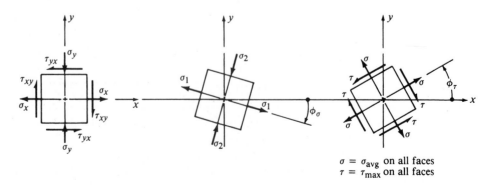

$\sigma = \sigma_{avg}$ on all faces
$\tau = \tau_{max}$ on all faces

(a) Original stress element   *(b)* Principal stress element   *(c)* Maximum shear stress element

Solution	Objective	Determine the principal stresses and the maximum shear stresses on element K.
	Given	Shaft and loading pattern shown in Figure 4–7.
	Analysis	Use the *Procedure for Constructing Mohr's Circle*. Some intermediate results will be taken from the solution to Example Problem 4–1 and from Figures 4–7, 4–8, and 4–9.
	Results	***Steps 1 and 2.*** The stress analysis for the given loading was completed in Example Problem 4–1. Figure 4–15 is identical to Figure 4–9 and represents the results of Step 2 of the Mohr's circle procedure.

Steps 3–6. Figure 4–16 shows the results. The first point plotted was

$$\sigma_x = 8030 \text{ psi}, \tau_{xy} = 2870 \text{ psi}$$

The second point was plotted at

$$\sigma_y = 0 \text{ psi}, \tau_{yx} = -2870 \text{ psi}$$

Then a line was drawn between them, crossing the σ-axis at O. The value of the stress at O is

FIGURE 4–15
Stresses on element K

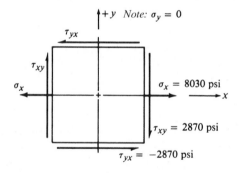

FIGURE 4–16
Partially completed
Mohr's circle

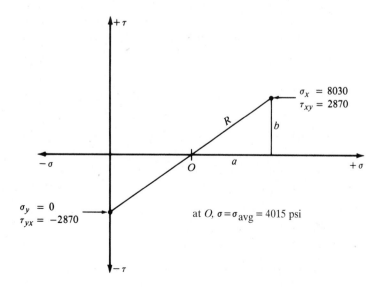

$$\sigma_{avg} = (\sigma_x + \sigma_y)/2 = (8030 + 0)/2 = 4015 \text{ psi}$$

Step 7. We compute the values for a, b, and R from

$$a = (\sigma_x - \sigma_y)/2 = (8030 - 0)/2 = 4015 \text{ psi}$$

$$b = \tau_{xy} = 2870 \text{ psi}$$

$$R = \sqrt{a^2 + b^2} = \sqrt{(4015)^2 + (2870)^2} = 4935 \text{ psi}$$

Step 8. Figure 4–17 shows the completed Mohr's circle. The circle has its center at O and the radius R. Note that the circle passes through the two points originally plotted. It must do so because the circle represents all possible states of stress on the element K.

Step 9. The maximum principal stress is at the right side of the circle.

$$\sigma_1 = \sigma_{avg} + R$$
$$\sigma_1 = 4015 + 4935 = 8950 \text{ psi}$$

Step 10. The minimum principal stress is at the left side of the circle.

$$\sigma_2 = \sigma_{avg} - R$$
$$\sigma_2 = 4015 - 4935 = -920 \text{ psi}$$

Step 11. At the top of the circle,

$$\sigma = \sigma_{avg} = 4015 \text{ psi}$$
$$\tau = \tau_{max} = R = 4935 \text{ psi}$$

FIGURE 4–17
Completed Mohr's circle

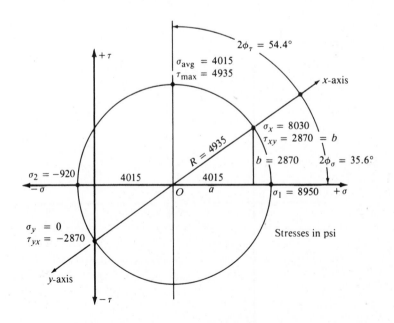

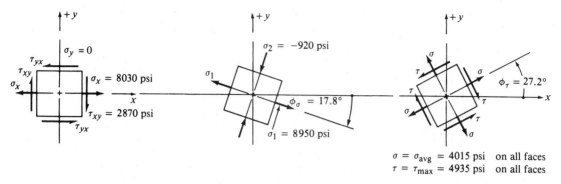

(a) Original stress element, K, from Figure 4–7

(b) Principal stress element

(c) Maximum shear stress element

FIGURE 4–18 Results from Mohr's circle analysis

The value of the normal stress on the element that carries the maximum shear stress is the same as the coordinate of O, the center of the circle.

Step 12. Compute the angle $2\phi_\sigma$ and then ϕ_σ. Use the circle as a guide.

$$2\phi_\sigma = \arctan(b/a) = \arctan(2870/4015) = 35.6°$$
$$\phi_\sigma = 35.6°/2 = 17.8°$$

Note that ϕ_σ must be measured *clockwise* from the original x-axis to the direction of the line of action of σ_1 for this set of data. The principal stress element will be rotated in the same direction as part of step 14.

Step 13. Compute the angle $2\phi_\tau$ and then ϕ_τ. From the circle we see that

$$2\phi_\tau = 90° - 2\phi_\sigma = 90° - 35.6° = 54.4°$$
$$\phi_\tau = 54.4°/2 = 27.2°$$

Note that the stress element on which the maximum shear stress acts must be rotated *counterclockwise* from the orientation of the original element for this set of data.

Step 14. Figure 4–18 shows the required stress elements. They are identical to those shown in Figure 4–11.

4–4
MOHR'S CIRCLE
PRACTICE
PROBLEMS

To a person seeing Mohr's circle for the first time, it may seem long and involved. But with practice under a variety of combinations of normal and shear stresses, you should be able to execute the 14 steps quickly and accurately.

Table 4–1 gives six sets of data (Example Problems 4–3 through 4–8) for normal and shear stresses in the x-y plane. You are advised to complete the Mohr's circle for each before looking at the solutions in Figures 4–19 through 4–24. From the circle, determine the two principal stresses, the maximum shear stress, and the planes on which these stresses act. Then draw the given stress element, the principal stress element, and the maximum shear stress element, all oriented properly with respect to the x- and y-directions.

TABLE 4–1   Practice problems for Mohr's circle

Example Problem	σ_x	σ_y	τ_{xy}	Fig. No.
4–3	+10.0 ksi	−4.0 ksi	+5.0 ksi	4–19
4–4	+10.0 ksi	−2.0 ksi	−4.0 ksi	4–20
4–5	+4.0 ksi	−10.0 ksi	+4.0 ksi	4–21
4–6	+120 MPa	−30 MPa	+60 MPa	4–22
4–7	−80 MPa	+20 MPa	−50 MPa	4–23
4–8	−80 MPa	+20 MPa	+50 Mpa	4–24

Example Problem 4–3

FIGURE 4–19
Solution for Example
Problem 4–3

Given:
$\sigma_x = +10.0$ ksi
$\sigma_y = -4.0$ ksi
$\tau_{xy} = +5.0$ ksi (cw)

Results:
$\sigma_1 = +11.60$ ksi
$\sigma_2 = -5.60$ ksi
$\phi_\sigma = 17.8°$ cw
$\tau_{max} = 8.60$ ksi
$\phi_\tau = 27.2°$ ccw
$\sigma_{avg} = +3.0$ ksi
x-axis in quadrant I

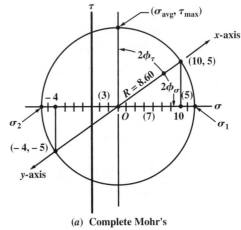

(a) Complete Mohr's circle

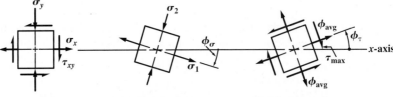

(b) Original stress element   (c) Principal stress element   (d) Maximum shear stress element

Example Problem 4–4

FIGURE 4–20
Solution for Example
Problem 4–4

Given:
$\sigma_x = +10.0$ ksi
$\sigma_y = -2.0$ ksi
$\tau_{xy} = -4.0$ ksi (ccw)

Results:
$\sigma_1 = +11.21$ ksi
$\sigma_2 = -3.21$ ksi
$\phi_\sigma = 16.8°$ ccw
$\tau_{max} = 7.21$ ksi
$\phi_\tau = 28.2°$ cw to $-\tau_{max}$
$\sigma_{avg} = +4.0$ ksi
x-axis in quadrant IV

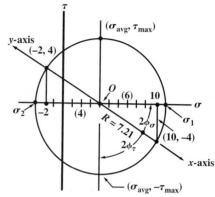

(a) **Complete Mohr's circle**

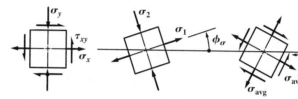

(b) **Original stress element** *(c)* **Principal stress element** *(d)* **Maximum shear stress element**

Example Problem 4–5

FIGURE 4–21
Solution for Example
Problem 4–5

Given:
$\sigma_x = +4.0$ ksi
$\sigma_y = -10.0$ ksi
$\tau_{xy} = +4.0$ ksi

Results:
$\sigma_1 = +5.06$ ksi
$\sigma_2 = -11.06$ ksi
$\phi_\sigma = 14.9°$ cw
$\tau_{max} = 8.06$ ksi
$\phi_\tau = 30.1°$ ccw
$\sigma_{avg} = -3.0$ ksi
x-axis in quadrant I

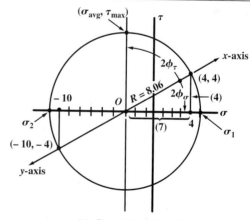

(a) **Complete Mohr's circle**

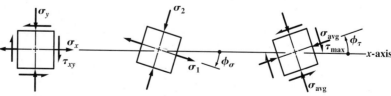

(b) **Original stress element** *(c)* **Principal stress element** *(d)* **Maximum shear stress element**

Example Problem 4–6

FIGURE 4–22
Solution for Example
Problem 4–6

Given:
σ_x = +120 MPa
σ_y = −30 MPa
τ_{xy} = +60 MPa

Results:
σ_1 = +141 MPa
σ_2 = −51 MPa
ϕ_σ = 19.3° cw
τ_{max} = 96 MPa
ϕ_τ = 25.7° ccw
σ_{avg} = +45 MPa
x-axis in quadrant I

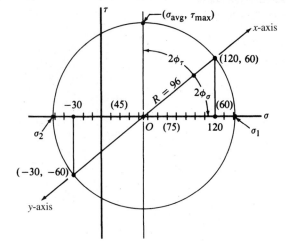

(a) Complete Mohr's circle

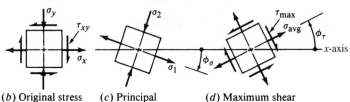

(b) Original stress   (c) Principal   (d) Maximum shear
element   stress element   stress element

Example Problem 4–7

FIGURE 4–23
Solution for Example
Problem 4–7

Given:
σ_x = −80 MPa
σ_y = +20 MPa
τ_{xy} = −50 MPa

Results:
σ_1 = +40.7 MPa
σ_2 = −100.7 MPa
ϕ_σ = 67.5° ccw
τ_{max} = 70.7 MPa
ϕ_τ = 22.5° ccw to $-\tau_{max}$
σ_{avg} = −30 MPa
x-axis in quadrant III

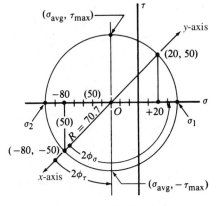

(a) Complete Mohr's circle

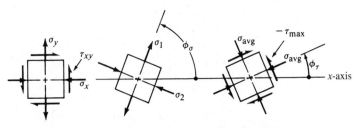

(b) Original stress   (c) Principal   (d) Maximum shear
element   stress element   stress element

Example Problem 4–8

FIGURE 4–24
Solution for Example
Problem 4–8

Given:
$\sigma_x = -80$ MPa
$\sigma_y = +20$ MPa
$\tau_{xy} = +50$ MPa

Results:
$\sigma_1 = +40.7$ MPa
$\sigma_2 = -100.7$ MPa
$\phi_\sigma = 67.5°$cw
$\tau_{max} = 70.7$ MPa
$\phi_\tau = 22.5°$cw
$\sigma_{avg} = -30$ MPa
x-axis in quadrant II

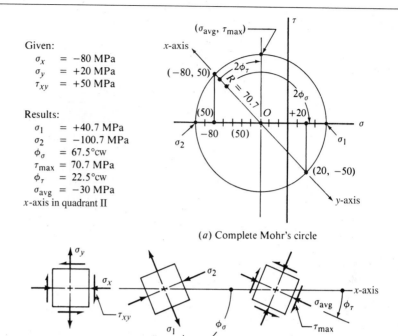

(a) Complete Mohr's circle

(b) Original stress element   (c) Principal stress element   (d) Maximum shear stress element

---

**4–5
CASE WHEN
BOTH PRINCIPAL
STRESSES HAVE
THE SAME SIGN**

Remember that all of the problems presented thus far have been plane stress problems, also called *biaxial stress* problems because stresses are acting in only two directions within one plane. Obviously, real load-carrying members are three-dimensional objects. The assumption here is that if no stress is given for the third direction, it is zero. In most cases, the solutions as given will produce the true maximum shear stress, along with the two principal stresses for the given plane. This will always be true if the two principal stresses have opposite signs—that is, if one is tensile and the other is compressive.

But the true maximum shear stress on the element will not be found if the two principal stresses are of the same sign. In such cases, you must consider the three-dimensional case.

Familiar examples of real products in which two principal stresses have the same sign are various forms of pressure vessels. A hydraulic cylinder with closed ends contains fluids under high pressure that tend to burst the walls of the cylinder. In strength of materials, you learned that the outer surfaces of the walls of such cylinders are subjected to tensile stresses in two directions: (1) tangent to its circumference and (2) axially, parallel to the axis of the cylinder. The stress perpendicular to the wall at the outer surface is zero.

Figure 4–25 shows the stress condition on an element of the surface of the cylinder. The tangential stress, also called *hoop stress*, is aligned with the x-direction and is labeled σ_x. The axially directed stress, also called *longitudinal stress*, acts in line with the y-direction and is labeled σ_y.

In strength of materials, you learned that if the wall of the cylinder is relatively thin, the maximum hoop stress is

$$\sigma_x = pD/2t$$

FIGURE 4–25
Thin-walled cylinder
subjected to pressure
with its ends closed

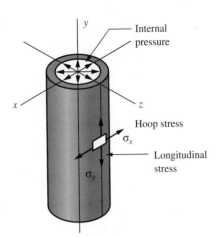

where p = internal pressure in the cylinder

D = diameter of the cylinder

t = thickness of the cylinder wall

Also, the longitudinal stress is

$$\sigma_y = pD/4t$$

Both stresses are tensile, and the hoop stress is twice as large as the longitudinal stress.

The analysis would be similar for any kind of thin-walled cylindrical vessel carrying an internal pressure. Examples are storage tanks for compressed gases, pipes carrying moving fluids under pressure, and the familiar beverage can that releases internal pressure when the top is popped open.

Let's use the hydraulic cylinder as an example for illustrating the special use of Mohr's circle when both principal stresses have the same sign. Consider that Figure 4–25 shows a cylinder with closed ends carrying an internal pressure of 500 psi. The wall thickness is $t = 0.080$ in, and the diameter of the cylinder is $D = 4.00$ in. The ratio of $D/t = 50$ indicates that the cylinder can be considered thin-walled. Any ratio over 20 is typically considered to be thin-walled.

The computed hoop and longitudinal stresses in the wall are

⇨ **Hoop Stress** $$\sigma_x = \frac{pD}{2t} = \frac{(500 \text{ psi})(4.0 \text{ in})}{(2)(0.080 \text{ in})} = 12\,500 \text{ psi (tension)}$$

⇨ **Longitudinal Stress** $$\sigma_y = \frac{pD}{4t} = \frac{(500 \text{ psi})(4.0 \text{ in})}{(4)(0.080 \text{ in})} = 6250 \text{ psi (tension)}$$

There are no shear stresses applied in the x- and y-directions.

Figure 4–26(a) shows the stress element for the x-y plane, and Part (b) shows the corresponding Mohr's circle. Because there are no applied shear stresses, σ_x and σ_y are the principal stresses for the plane. The circle would predict the maximum shear stress to be equal to the radius of the circle, 3125 psi.

But notice Part (c) of the figure. We could have chosen the x-z plane for analysis, instead of the x-y plane. The stress in the z-direction is zero because it is perpendicular to the

FIGURE 4–26 Stress analysis for a thin-walled cylinder

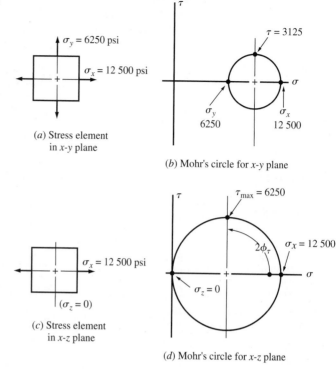

(a) Stress element in x-y plane

(b) Mohr's circle for x-y plane

(c) Stress element in x-z plane

(d) Mohr's circle for x-z plane

free face of the element. Likewise, there are no shear stresses on this face. The Mohr's circle for this plane is shown in Part (d) of the figure. The maximum shear stress is equal to the radius of the circle, 6250 psi, or *twice* as much as would be predicted from the x-y plane. This approach should be used any time the two principal stresses in a biaxial stress problem have the same sign.

In summary, on a general three-dimensional stress element, there will be one orientation of the element in which there are no shear stresses acting. The normal stresses on the three perpendicular faces are then the three principal stresses. If we call these stresses σ_1, σ_2, and σ_3, taking care to order them such that $\sigma_1 > \sigma_2 > \sigma_3$, then the maximum shear stress on the element will always be

$$\tau_{max} = \frac{\sigma_1 - \sigma_3}{2}$$

Figure 4–27 shows the three-dimensional element.

For the cylinder of Figure 4–25, we can conclude that

$$\sigma_1 = \sigma_x = 12\ 500 \text{ psi}$$
$$\sigma_2 = \sigma_y = 6250 \text{ psi}$$
$$\sigma_3 = \sigma_z = 0$$
$$\tau_{max} = (\sigma_1 - \sigma_3)/2 = (12\ 500 - 0)/2 = 6250 \text{ psi}$$

Figure 4–28 shows two additional examples in which the two principal stresses in the given plane have the same sign. Then the zero stress in the third direction is added to the diagram, and the new Mohr's circle is superimposed on the original one. This serves to illustrate that the maximum shear stress will occur on the Mohr's circle having the largest radius.

FIGURE 4–27
Three-dimensional
stress element

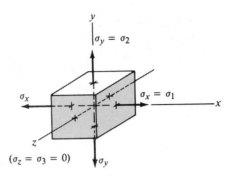

FIGURE 4–28
Mohr's circle for cases
in which two principal
stresses have the same
sign

Given:
σ_x = +150 MPa
σ_y = +30 MPa
σ_z = 0
τ_{xy} = +20 MPa

Results:
σ_1 = 153.2 MPa
σ_2 = 26.8 MPa
σ_3 = 0
τ_{max} = 76.6 MPa

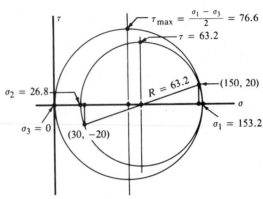

(a) σ_x and σ_y both positive

Given:
σ_x = −50 MPa
σ_y = −130 MPa
σ_z = 0
τ_{xy} = 40 MPa

Results:
σ_1 = 0
σ_2 = −33.4 MPa
σ_3 = −146.6 MPa
τ_{max} = 73.3 MPa

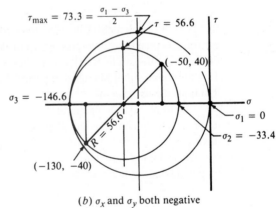

(b) σ_x and σ_y both negative

**4–6
MOHR'S CIRCLE
FOR SPECIAL
STRESS
CONDITIONS**

Mohr's circle is used here to demonstrate the relationship among the applied stresses, the principal stresses, and the maximum shear stress for the following special cases:

Pure uniaxial tension

Pure uniaxial compression

Pure torsional shear

Uniaxial tension combined with torsional shear

These are important, frequently encountered stress conditions, and they will be used in later chapters to illustrate failure theories and design methods. These failure theories are based on the values of the principal stresses and the maximum shear stress.

FIGURE 4–29
Mohr's circle for pure
uniaxial tension

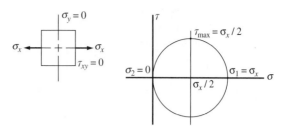

FIGURE 4–30
Mohr's circle for pure
uniaxial compression

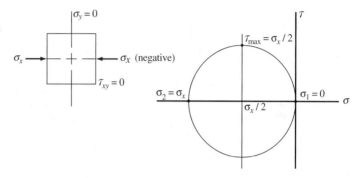

Pure Uniaxial Tension

The stress condition produced in all parts of a standard tensile test specimen is pure uniaxial tension. Figure 4–29 shows the stress element and the corresponding Mohr's circle. Note that the maximum principal stress, σ_1, is equal to the applied stress, σ_x; the minimum principal stress, σ_2, is zero; and the maximum shear stress, τ_{max}, is equal to $\sigma_x/2$.

Pure Uniaxial Compression

Figure 4–30 shows pure uniaxial compression as it would be produced by a standard compression test. Mohr's circle shows that $\sigma_1 = 0$; $\sigma_2 = \sigma_x$ (a negative value); and the magnitude of the maximum shear stress is $\tau_{max} = \sigma_x/2$.

Pure Torsion

Figure 4–31 shows that the Mohr's circle for this special case has its center at the origin of the σ-τ axes and that the radius of the circle is equal to the value of the applied shear stress, τ_{xy}. Therefore, $\tau_{max} = \tau_{xy}$; $\sigma_1 = \tau_{xy}$; and $\sigma_2 = -\tau_{xy}$.

Uniaxial Tension Combined with Torsional Shear

This is an important special case because it describes the stress condition in a rotating shaft carrying bending loads while simultaneously transmitting torque. This is the type of stress condition on which the procedure for designing shafts, presented in Chapter 12, is based. If the applied stresses are called σ_x and τ_{xy}, the Mohr's circle in Figure 4–32 shows that

$$\tau_{max} = R = \text{radius of circle} = \sqrt{(\sigma_x/2)^2 + \tau_{xy}^2} \qquad (4\text{--}7)$$

$$\sigma_1 = \sigma_x/2 + R = \sigma_x/2 + \sqrt{(\sigma_x/2)^2 + \tau_{xy}^2} \qquad (4\text{--}8)$$

$$\sigma_2 = \sigma_x/2 - R = \sigma_x/2 - \sqrt{(\sigma_x/2)^2 + \tau_{xy}^2} \qquad (4\text{--}9)$$

FIGURE 4–31
Mohr's circle for
pure torsional
shear

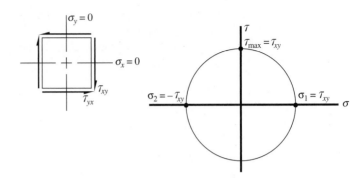

FIGURE 4–32
Mohr's circle for
uniaxial tension
combined with
torsional shear

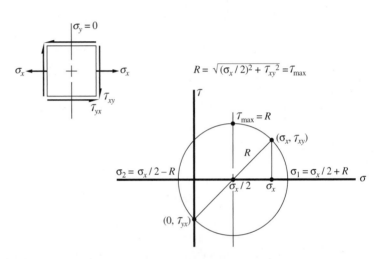

FIGURE 4–33
Circular bar in bending
and torsion

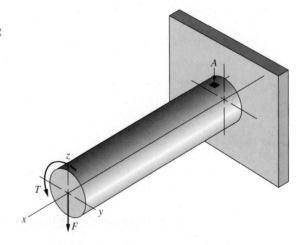

A convenient and useful concept called *equivalent torque* can be developed from Equation (4–7) for the special case of a body subjected to only bending and torsion.

An example is shown in Figure 4–33, where a circular bar is loaded at one end by a downward force and a torsional moment. The force causes bending in the bar with the maximum moment at the point where the bar is attached to the support. The moment causes a tensile stress on the top of the bar in the x-direction at the point called A, where the magnitude of the stress is

$$\sigma_x = M/S \qquad (4\text{–}10)$$

where S = section modulus of the round bar

Now the torsional moment causes torsional shear stress in the x-y plane at point A having a magnitude of

$$\tau_{xy} = T/Z_p \tag{4–11}$$

where Z_p = polar section modulus of the bar

Point A then is subjected to a tensile stress combined with shear, the special case shown in the Mohr's circle of Figure 4–32. The maximum shear stress can be computed from Equation (4–7). If we substitute Equations (4–10) and (4–11) in Equation (4–7), we get

$$\tau_{max} = \sqrt{(M/2S)^2 + (T/Z_p)^2} \tag{4–12}$$

Note from Appendix 1 that $Z_p = 2S$. Equation (4–12) can then be written as

$$\tau_{max} = \frac{\sqrt{M^2 + T^2}}{Z_p} \tag{4–13}$$

It is convenient to define the quantity in the numerator of this equation to be the *equivalent torque*, T_e. Then the equation becomes

$$\tau_{max} = T_e/Z_p \tag{4–14}$$

**4–7
ANALYSIS OF
COMPLEX
LOADING
CONDITIONS**

The examples shown in this chapter involved relatively simple part geometries and loading conditions for which the necessary stress analysis can be performed using familiar methods of statics and strength of materials. If more complex geometries or loading conditions are involved, you may not be able to complete the required analysis to create the original stress element from which the Mohr's circle is derived.

Consider, for example, a cast wheel for a high-performance racing car. The geometry would likely involve webs or spokes of a unique design connecting the hub to the rim to create a lightweight wheel. The loading would be a complex combination of torsion, bending, and compression generated by the cornering action of the wheel.

One method of analysis of such a load-carrying member would be accomplished by experimental stress analysis using strain gages or photoelastic techniques. The results would identify the stress levels at selected points in certain specified directions that could be used as the input to the construction of the Mohr's circle for critical points in the structure.

Another method of analysis would involve the modeling of the geometry of the wheel as a *finite-element model*. The three-dimensional model would be divided into several hundred small-volume elements. Points of support and restraint would be defined on the model, and then external loads would be applied at appropriate points. The complete data set would be input to a special type of computer analysis program called *finite-element analysis*. The output from the program lists the stresses and the deflection for each of the elements. These data can be plotted on the computer model so that the designer can visualize the stress distribution within the model. Most such programs list the principal stresses and the maximum shear stress for each element, eliminating the need to actually draw the Mohr's circle. A special form of stress, called the *von Mises stress*, is often computed by combining the principal stresses. (See Section 5–8 for a more complete discussion of the von Mises stress and its use.) Several different finite-element analysis programs are commercially available for use on personal computers, on engineering work stations, or on mainframe computers.

REFERENCE

Mott, Robert L. *Applied Strength of Materials*. 4th ed. Upper
    Saddle River, NJ: Prentice Hall, 2002.

INTERNET SITE RELATED TO MOHR'S CIRCLE ANALYSIS

1. **MDSolids** *www.mdsolids.com* Educational software devoted to introductory mechanics of materials. Includes modules on basic stress and strain, beam and strut axial problems, trusses, statically indeterminate axial structures, torsion, determinate beams, section properties, general analysis of axial, torsion, and beam members, column buckling, pressure vessels, and Mohr's circle transformations.

PROBLEMS

For the sets of given stresses on an element given in Table 4–2, draw a complete Mohr's circle, find the principal stresses and the maximum shear stress, and draw the principal stress element and the maximum shear stress element. Any stress components not shown are assumed to be zero.

31. Refer to Figure 3–23 in Chapter 3. For the shaft aligned with the x-axis, create a stress element on the bottom of the shaft just to the left of section B. Then draw the Mohr's circle for that element. Use $D = 0.50$ in.

32. Refer to Figure P3–44 in Chapter 3. For the shaft ABC, create a stress element on the bottom of the shaft just to the right of section B. The torque applied to the shaft at B is resisted at support C only. Draw the Mohr's circle for the stress element. Use $D = 1.50$ in.

33. Repeat Problem 32 for the shaft in Figure P3–45 $D = 0.50$ in.

34. Refer to Figure P3–46 in Chapter 3. For the shaft AB, create a stress element on the bottom of the shaft just to the right of section A. The torque applied to the shaft by the crank is resisted at support B only. Draw the Mohr's circle for the stress element. Use $D = 50$ mm.

35. A short cylindrical bar having a diameter of 4.00 in is subjected to an axial compressive force of 75 000 lb and a torsional moment of 20 000 lb·in. Draw a stress element on the surface of the bar. Then draw the Mohr's circle for the element.

36. A torsion bar is used as a suspension element for a vehicle. The bar has a diameter of 20 mm. It is subjected to a torsional moment of 450 N·m and an axial tensile force of 36.0 kN. Draw a stress element on the surface of the bar, and then draw the Mohr's circle for the element.

37. Use the Mohr's Circle module from the MDSolids software to complete any of Problems 1 to 30 in this chapter.

TABLE 4–2 Stresses for Problems 1–30

Problem	σ_x	σ_y	τ_{xy}
1.	20 ksi	0	10 ksi
2.	85 ksi	−40 ksi	30 ksi
3.	40 ksi	−40 ksi	−30 ksi
4.	−80 ksi	−40 ksi	−30 ksi
5.	120 ksi	40 ksi	20 ksi
6.	20 ksi	140 ksi	20 ksi
7.	20 ksi	−40 ksi	0
8.	120 ksi	−40 ksi	100 ksi
9.	100 MPa	0	80 MPa
10.	250 MPa	−80 MPa	110 MPa
11.	50 MPa	−80 MPa	40 MPa
12.	−150 MPa	−80 MPa	−40 MPa
13.	150 MPa	80 MPa	−40 MPa
14.	50 MPa	180 MPa	40 MPa
15.	250 MPa	−80 MPa	0
16.	50 MPa	−80 MPa	−30 MPa
17.	400 MPa	−300 MPa	200 MPa
18.	−120 MPa	180 MPa	−80 MPa
19.	−30 MPa	20 MPa	40 MPa
20.	220 MPa	−120 MPa	0 MPa
21.	40 ksi	0 ksi	0 ksi
22.	0 ksi	0 ksi	40 ksi
23.	38 ksi	−25 ksi	−18 ksi
24.	55 ksi	15 ksi	−40 ksi
25.	22 ksi	0 ksi	6.8 ksi
26.	−4250 psi	3250 psi	2800 psi
27.	300 MPa	100 MPa	80 MPa
28.	250 MPa	150 MPa	40 MPa
29.	−840 kPa	−335 kPa	−120 kPa
30.	−325 kPa	−50 kPa	−60 kPa

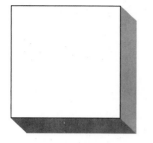

5

Design for Different Types of Loading

The Big Picture

You Are the Designer

5–1   Objectives of This Chapter

5–2   Types of Loading and Stress Ratio

5–3   Endurance Strength

5–4   Estimated Actual Endurance Strength, s'_n

5–5   Example Problems for Estimating Actual Endurance Strength

5–6   Design Philosophy

5–7   Design Factors

5–8   Predictions of Failure

5–9   Design Analysis Methods

5–10   General Design Procedure

5–11   Design Examples

5–12   Statistical Approaches to Design

5–13   Finite Life and Damage Accumulation Method

Design for Different Types of Loading

Discussion Map

☐ This chapter provides additional tools you can use to design load-carrying components that are safe and reasonably efficient in their use of materials.

☐ You must learn how to classify the kind of loading the component is subjected to: *static*, *repeated and reversed*, *fluctuating*, *shock*, and *impact*.

☐ You will learn to identify the appropriate analysis techniques based on the type of load and the type of material.

Discover

Identify components of real products or structures that are subjected to static loads.

Identify components that are subjected to equal, repeated loads that reverse directions.

Identify components that experience fluctuating loads that vary with time.

Identify components that are loaded with shock or impact, such as being struck by a hammer or dropped onto a hard surface.

Using the techniques you learn in this chapter will help you to complete a wide variety of design tasks

For the concepts considered in this chapter, the big picture encompasses a huge array of examples in which you will build on the principles of strength of materials that you reviewed in Chapters 3 and 4 and extend them from the analysis mode to the design mode. Several steps are involved, and you must learn to make rational judgments about the appropriate method to apply to complete the design.

In this chapter you will learn how to do the following:

1. Recognize the manner of loading for a part: Is it static, repeated and reversed, fluctuating, shock, or impact?

2. Select the appropriate method to analyze the stresses produced.

3. Determine the strength property for the material that is appropriate to the kind of loading and to the kind of material: Is the material a metal or a nonmetal? Is it brittle or ductile? Should the design be based on the yield strength, ultimate tensile strength, compressive strength, endurance strength, or some other material property?

4. Specify a suitable *design factor*, often called a *factor of safety*.

5. Design a wide variety of load-carrying members to be safe under their particular expected loading patterns.

The following paragraphs show by example some of the situations to be studied in this chapter.

An ideal *static load* is one that is applied slowly and is never removed. Some loads that are applied slowly and removed and replaced very infrequently can also be considered to be static. What examples can you think of for products or their components that are subjected to static loads? Consider load-carrying members of structures, parts of furniture pieces, and brackets or support rods holding equipment in your home or in a business or factory. Try to identify specific examples, and describe them to your colleagues. Discuss how the load is applied and which parts of the load-carrying member are subjected to the higher stress levels. Some of the examples that you discovered during **The Big Picture** discussion for Chapter 3 could be used again here.

Fluctuating loads are those that vary during the normal service of the product. They typically are applied for quite a long time so the part experiences many thousands or millions of cycles of stress during its expected life. There are many examples in consumer products around your home, in your car, in commercial buildings, and in manufacturing facilities. Consider virtually anything that has moving parts. Again, try to identify specific examples, and describe them to your colleagues. How does the load fluctuate? Is it applied and then completely removed each cycle? Or is there always some level of mean or average load with an alternating load superimposed on it? Does the load swing from a positive maximum value to a negative minimum value of equal magnitude during each cycle of loading? Consider parts with rotating shafts, such as engines or agricultural, production, and construction machinery.

Consider products that have failed. You may have identified some from **The Big Picture** discussion for Chapter 3. Did they fail the first time they were used? Or did they fail after some fairly long service? Why do you think they were able to operate for some time before failure?

Can you find components that failed suddenly because the material was brittle, such as cast iron, some ceramics, or some plastics? Can you find others that failed only after some considerable deformation? Such failures are called *ductile fractures*.

What were the consequences of the failures that you have found? Was anyone hurt? Was there damage to some other valuable component or property? Or was the failure simply an inconvenience? What was the order of magnitude of cost related to the failure? The answer to some of these questions can help you make rational decisions about design factors to be used in your designs.

It is the designer's responsibility to ensure that a machine part is safe for operation under reasonably foreseeable conditions. This requires that a stress analysis be performed in which the predicted stress levels in the part are compared with the *design stress*, or that level of stress permitted under the operating conditions.

The stress analysis can be performed either analytically or experimentally, depending on the degree of complexity of the part, the knowledge about the loading conditions, and the material properties. The designer must be able to verify that the stress to which a part is subjected is safe.

The manner of computing the design stress depends on the manner of loading and on the type of material. Loading types include the following:

Static

Repeated and reversed

Fluctuating

Shock or impact

Random

Material types are many and varied. Among the metallic materials, the chief classification is between *ductile* and *brittle* materials. Other considerations include the manner of forming the material (casting, forging, rolling, machining, and so on), the type of heat treatment, the surface finish, the physical size, the environment in which it is to operate, and the geometry of the part. Different factors must be considered for plastics, composites, ceramics, wood, and others.

This chapter outlines methods of analyzing load-carrying machine parts to ensure that they are safe. Several different cases are described in which knowledge of the combinations of material types and loading patterns leads to the determination of the appropriate method of analysis. It will then be your job to apply these tools correctly and judiciously as you continue your career.

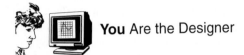

You Are the Designer

Recall the task presented at the start of Chapter 4, in which you were the designer of a bracket to hold a fabric sample during a test to determine its long-term stretch characteristics. Figure 4–2 showed a proposed design.

Now you are asked to continue this design exercise by selecting a material from which to make the two bent circular bars that are welded to the rigid support. Also, you must specify a suitable diameter for the bars when a certain load is applied to the test material.

**5–1
OBJECTIVES OF
THIS CHAPTER**

After completing this chapter, you will be able to:

1. Identify various kinds of loading commonly encountered by machine parts, including *static, repeated and reversed, fluctuating, shock or impact,* and *random.*

2. Define the term *stress ratio* and compute its value for the various kinds of loading.

3. Define the concept of *fatigue.*

4. Define the material property of *endurance strength* and determine estimates of its magnitude for different materials.

5. Recognize the factors that affect the magnitude of endurance strength.

6. Define the term *design factor.*

7. Specify a suitable value for the design factor.

8. Define the *maximum normal stress theory of failure* and the *modified Mohr method* for design with brittle materials.

9. Define the *maximum shear stress theory of failure.*

10. Define the *distortion energy theory*, also called the *von Mises theory* or the *Mises-Hencky theory.*

11. Describe the *Goodman method* and apply it to the design of parts subjected to fluctuating stresses.

12. Consider *statistical approaches, finite life,* and *damage accumulation methods* for design.

**5–2
TYPES OF
LOADING AND
STRESS RATIO**

The primary factors to consider when specifying the type of loading to which a machine part is subjected are the manner of variation of the load and the resulting variation of stress with time. Stress variations are characterized by four key values:

1. Maximum stress, σ_{max}

2. Minimum stress, σ_{min}

3. Mean (average) stress, σ_m

4. Alternating stress, σ_a (*stress amplitude*)

The maximum and minimum stresses are usually computed from known information by stress analysis or finite-element methods, or they are measured using experimental stress analysis techniques. Then the mean and alternating stresses can be computed from

$$\sigma_m = (\sigma_{max} + \sigma_{min})/2 \qquad\qquad \textbf{(5–1)}$$

$$\sigma_a = (\sigma_{max} - \sigma_{min})/2 \qquad\qquad \textbf{(5–2)}$$

FIGURE 5–1 Static
stress

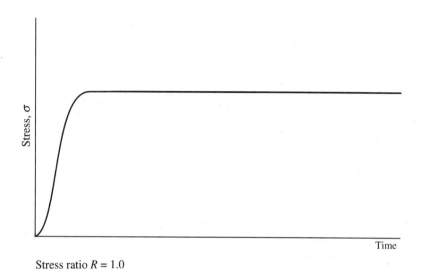

Stress ratio $R = 1.0$

The behavior of a material under varying stresses is dependent on the manner of the variation. One method used to characterize the variation is called *stress ratio*. Two types of stress ratios are commonly used, defined as

$$\text{Stress ratio } R = \frac{\text{minimum stress}}{\text{maximum stress}} = \frac{\sigma_{min}}{\sigma_{max}}$$

$$\text{Stress ratio } A = \frac{\text{alternating stress}}{\text{mean stress}} = \frac{\sigma_a}{\sigma_m}$$

(5–3)

Static Stress

When a part is subjected to a load that is applied slowly, without shock, and is held at a constant value, the resulting stress in the part is called *static stress*. An example is the load on a structure due to the dead weight of the building materials. Figure 5–1 shows a diagram of stress versus time for static loading. Because $\sigma_{max} = \sigma_{min}$, the stress ratio for static stress is $R = 1.0$.

Static loading can also be assumed when a load is applied and is removed slowly and then reapplied, if the number of load applications is small, that is, under a few thousand cycles of loading.

Repeated and Reversed Stress

A stress reversal occurs when a given element of a load-carrying member is subjected to a certain level of tensile stress followed by the *same level* of compressive stress. If this stress cycle is repeated many thousands of times, the stress is called *repeated and reversed*. Figure 5–2 shows the diagram of stress versus time for repeated and reversed stress. Because $\sigma_{min} = -\sigma_{max}$, the stress ratio is $R = -1.0$, and the mean stress is zero.

An important example in machine design is a rotating circular shaft loaded in bending such as that shown in Figure 5–3. In the position shown, an element on the bottom of the shaft experiences tensile stress while an element on the top of the shaft sees a compressive stress of equal magnitude. As the shaft is rotated 180° from the given position, these two elements experience a complete reversal of stress. Now if the shaft continues to rotate, all parts of the shaft that are in bending see repeated, reversed stress. This is a description of the classical loading case of *reversed bending*.

FIGURE 5–2
Repeated, reversed
stress

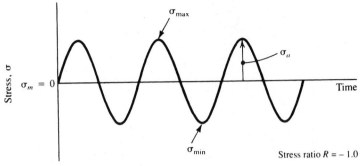

FIGURE 5–3
R. R. Moore fatigue
test device

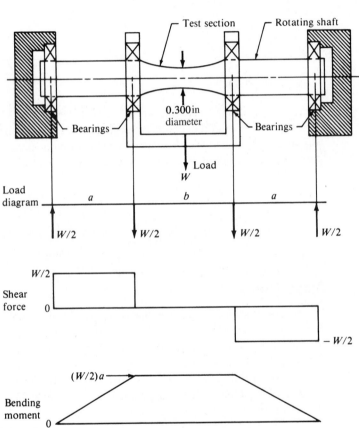

This type of loading is often called *fatigue loading*, and a machine of the type shown in Figure 5–3 is called a *standard R. R. Moore fatigue test device*. Such machines are used to test materials for their ability to resist repeated loads. The material property called *endurance strength* is measured in this manner. More is said later in this chapter about endurance strength. Actually, reversed bending is only a special case of fatigue loading, since any stress that varies with time can lead to fatigue failure of a part.

Fluctuating Stress

When a load-carrying member is subjected to an alternating stress with a nonzero mean, the loading produces *fluctuating stress*. Figure 5–4 shows four diagrams of stress versus time for this type of stress. Differences among the four diagrams occur in whether the various stress levels are positive (tensile) or negative (compressive). *Any varying stress with a*

FIGURE 5–4
Fluctuating stresses

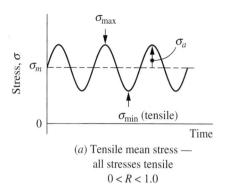

(a) Tensile mean stress —
all stresses tensile
$0 < R < 1.0$

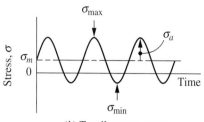

(b) Tensile mean stress —
σ_{max} tensile
σ_{min} compressive
$-1.0 < R < 0$

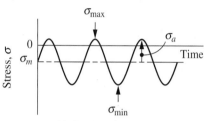

(c) Compressive mean stress —
σ_{max} tensile
σ_{min} compressive
$-\infty < R < -1.0$

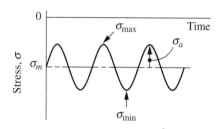

(d) Compressive mean stress —
all stresses compressive
$1.0 < R < \infty$

FIGURE 5–5
Repeated, one-direction
stress, a special case of
fluctuating stress

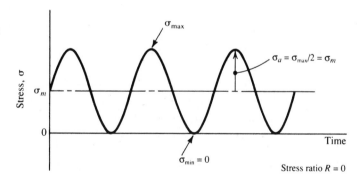

nonzero mean is considered a fluctuating stress. Figure 5–4 also shows the possible ranges of values for the stress ratio R for the given loading patterns.

A special, frequently encountered case of fluctuating stress is *repeated, one-direction stress*, in which the load is applied and removed many times. As shown in Figure 5–5, the stress varies from zero to a maximum with each cycle. Then, by observation,

$$\sigma_{min} = 0$$
$$\sigma_m = \sigma_a = \sigma_{max}/2$$
$$R = \sigma_{min}/\sigma_{max} = 0$$

An example of a machine part subjected to fluctuating stress of the type shown in Figure 5–4(a) is shown in Figure 5–6, in which a reciprocating cam follower feeds

FIGURE 5–6
Example of cyclic
loading in which the
flat spring is subjected
to fluctuating stress

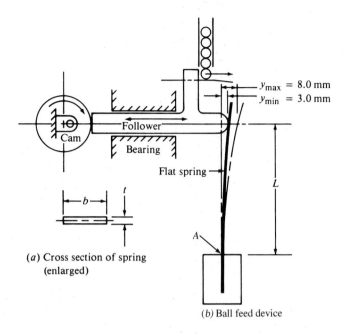

(a) Cross section of spring
(enlarged)

(b) Ball feed device

spherical balls one at a time from a chute. The follower is held against the eccentric cam
by a flat spring loaded as a cantilever. When the follower is farthest to the left, the spring
is deflected from its free (straight) position by an amount $y_{min} = 3.0$ mm. When the follower is farthest to the right, the spring is deflected to $y_{max} = 8.0$ mm. Then, as the cam
continues to rotate, the spring sees the cyclic loading between the minimum and maximum values. Point A at the base of the spring on the convex side experiences the varying tensile stresses of the type shown in Figure 5–4(a). Example Problem 5–1 completes
the analysis of the stress in the spring at point A.

Example Problem 5–1 For the flat steel spring shown in Figure 5–6, compute the maximum stress, the minimum
stress, the mean stress, and the alternating stress. Also compute the stress ratio, R. The length
L is 65 mm. The dimensions of the spring cross section are $t = 0.80$ mm and $b = 6.0$ mm.

Solution   Objective   Compute the maximum, minimum, mean, and alternating tensile stresses in the flat spring.
Compute the stress ratio, R.

Given   Layout shown in Figure 5–6. The spring is steel: $L = 65$ mm.

Spring cross section dimensions: $t = 0.80$ mm and $b = 6.0$ mm.

Maximum deflection of the spring at the follower $= 8.0$ mm.

Minimum deflection of the spring at the follower $= 3.0$ mm.

Analysis   Point A at the base of the spring experiences the maximum tensile stress. Determine the
force exerted on the spring by the follower for each level of deflection using the formulas
from Table A14–2, Case (a). Compute the bending moment at the base of the spring for each
deflection. Then compute the stresses at point A using the bending stress formula, $\sigma = Mc/I$.
Use Equations (5–1), (5–2), and (5–3) for the mean and alternating stresses and R.

Results   Case (a) of Table A14–2 gives the following formula for the amount of deflection of a cantilever for a given applied force:

$$y = PL^3/3EI$$

Solve for the force as a function of deflection:

$$P = 3EIy/L^3$$

Appendix 3 gives the modulus of elasticity for steel to be $E = 207$ GPa. The moment of inertia, I, for the spring cross section is found from

$$I = bt^3/12 = (6.00\text{mm})(0.80\text{mm})^3/12 = 0.256 \text{ mm}^4$$

Then the force on the spring when the deflection y is 3.0 mm is

$$P = \frac{3(207 \times 10^9 \text{ N/m}^2)(0.256 \text{ mm}^4)(3.0 \text{ mm})}{(65 \text{ mm})^3} \frac{(1.0 \text{ m}^2)}{(10^6 \text{ mm}^2)} = 1.74 \text{ N}$$

The bending moment at the support is

$$M = P \cdot L = (1.74 \text{ N})(65 \text{ mm}) = 113 \text{ N} \cdot \text{mm}$$

The bending stress at point A caused by this moment is

$$\sigma = \frac{Mc}{I} = \frac{(113 \text{ N} \cdot \text{mm})(0.40 \text{ mm})}{0.256 \text{ mm}^4} = 176 \text{ N/mm}^2 = 176 \text{ MPa}$$

This is the lowest stress that the spring sees in service, and therefore $\sigma_{min} = 176$ MPa.

Because the force on the spring is proportional to the deflection, the force exerted when the deflection is 8.0 mm is

$$P = (1.74 \text{ N})(8.0 \text{ mm})/(3.0 \text{ mm}) = 4.63 \text{ N}$$

The bending moment is

$$M = P \cdot L = (4.63 \text{ N})(65 \text{ mm}) = 301 \text{ N} \cdot \text{mm}$$

The bending stress at point A is

$$\sigma = \frac{Mc}{I} = \frac{(301 \text{ N} \cdot \text{mm})(0.40 \text{ mm})}{0.256 \text{ mm}^4} = 470 \text{ N/mm}^2 = 470 \text{ MPa}$$

This is the maximum stress that the spring sees, and therefore $\sigma_{max} = 470$ MPa.

Now the mean stress can be computed:

$$\sigma_m = (\sigma_{max} + \sigma_{min})/2 = (470 + 176)/2 = 323 \text{ MPa}$$

Finally, the alternating stress is

$$\sigma_a = (\sigma_{max} - \sigma_{min})/2 = (470 - 176)/2 = 147 \text{ MPa}$$

The stress ratio is found using Equation (5–3):

$$\text{Stress ratio } R = \frac{\text{minimum stress}}{\text{maximum stress}} = \frac{\sigma_{min}}{\sigma_{max}} = \frac{176 \text{ MPa}}{470 \text{ MPa}} = 0.37$$

Comments The sketch of stress versus time shown in Figure 5–4(a) illustrates the form of the fluctuating stress on the spring. In Section 5–9, you will see how to design parts subjected to this kind of stress.

Shock or Impact Loading

Loads applied suddenly and rapidly cause shock or impact. Examples include a hammer blow, a weight falling onto a structure, and the action inside a rock crusher. The design of machine members to withstand shock or impact involves an analysis of their energy-absorption capability, a topic not considered in this book. (See References 8 to 13).

Random Loading

When varying loads are applied that are not regular in their amplitude, the loading is called *random*. Statistical analysis is used to characterize random loading for purposes of design and analysis. This topic is not covered in this book. See Reference 14.

**5–3
ENDURANCE
STRENGTH**

The *endurance strength* of a material is its ability to withstand fatigue loads. In general, it is the stress level that a material can survive for a given number of cycles of loading. If the number of cycles is infinite, the stress level is called the *endurance limit*.

Endurance strengths are usually charted on a graph like that shown in Figure 5–7, called an *S-N diagram*. Curves *A*, *B*, and *D* are representative of a material that does exhibit an endurance limit, such as a plain carbon steel. Curve *C* is typical of most nonferrous metals, such as aluminum, which do not exhibit an endurance limit. For such materials, the number of cycles to failure should be reported for the given endurance strength.

Data for the endurance strength of the specific material for a part should be used whenever it is available, either from test results or from reliable published data. However, such data are not always readily available. Reference 13 suggests the following approximations for the basic endurance strength for wrought steel:

$$\text{Endurance strength} = 0.50(\text{ultimate tensile strength}) = 0.50(s_u)$$

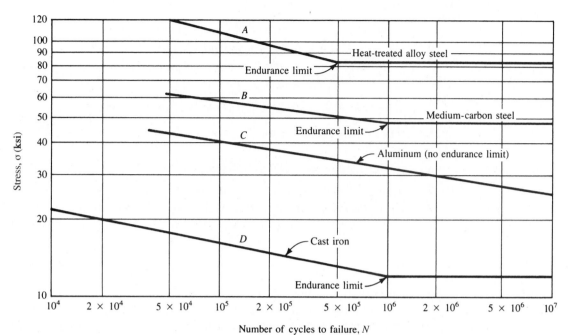

FIGURE 5–7 Representative endurance strengths

This approximation, along with published data, refers to the special case of repeated and reversed bending stress in a polished steel specimen having a diameter of 0.300 in (7.62 mm) as used in the R. R. Moore fatigue testing device shown in Figure 5–3. The next section discusses adjustments required when other, more realistic conditions exist.

**5–4
ESTIMATED
ACTUAL
ENDURANCE
STRENGTH, S_n'**

If the actual material characteristics or operating conditions for a machine part are different from those for which the basic endurance strength was determined, the fatigue strength must be reduced from the reported value. Some of the factors that decrease the endurance strength are discussed in this section. The discussion relates only to the endurance strength for materials subjected to normal tensile stresses such as bending and direct axial tensile stress. Cases involving endurance strength in shear are discussed separately in Section 5–9.

    We begin by presenting a procedure for estimating the *actual endurance strength*, s_n', for the material for the part being designed. It involves applying several factors to the basic endurance strength for the material. Additional elaboration on the factors follows.

**Procedure for
Estimating Actual
Endurance
Strength, s_n'**

1. Specify the material for the part and determine its ultimate tensile strength, s_u, considering its condition, as it will be used in service.
2. Specify the manufacturing process used to produce the part with special attention to the condition of the surface in the most highly stressed area.

3. Use Figure 5–8 to estimate the modified endurance strength, s_n.

4. Apply a material factor, C_m, from the following list.

     Wrought steel: $C_m = 1.00$ Malleable cast iron: $C_m = 0.80$

     Cast steel: $C_m = 0.80$ Gray cast iron: $C_m = 0.70$

     Powdered steel: $C_m = 0.76$ Ductile cast iron: $C_m = 0.66$

5. Apply a type-of-stress factor: $C_{st} = 1.0$ for bending stress; $C_{st} = 0.80$ for axial tension.

6. Apply a reliability factor, C_R, from Table 5–1.

7. Apply a size factor, C_s, using Figure 5–9 and Table 5–2 as guides.

8. Compute the estimated actual endurance strength, s_n', from

$$s_n' = s_n \, (C_m)(C_{st})(C_R)(C_s) \tag{5-4}$$

These are the only factors that will be used consistently in this book. If data for other factors can be determined from additional research, they should be multiplied as additional terms in Equation 5–4. In most cases, we suggest accounting for other factors for which reasonable data cannot be found by adjusting the value of the design factor as discussed in Section 5–8.

Stress concentrations caused by sudden changes in geometry are, indeed, likely places for fatigue failures to occur. Care should be taken in the design and manufacture of cyclically loaded parts to keep stress concentration factors to a low value. We will apply stress concentration factors to the computed stress rather than to the endurance strength. See Section 5–9.

While 12 factors affecting endurance strength are discussed in the following section, note that the procedure just given includes only the first five. They are *surface finish*, *material factor*, *type-of-stress factor*, *reliability factor*, and *size factor*. The others are mentioned to alert you to the variety of conditions you should investigate as you complete a design. However, generalized data are difficult to acquire for all factors. Special testing or additional literature searching should be done when conditions exist for which no data are provided in this book. The end-of-chapter references contain a huge amount of such information.

Surface Finish

Any deviation from a polished surface reduces endurance strength because the rougher surface provides sites where locally increased stresses or irregularities in the material structure promote the initiation of microscopic cracks that can progress to fatigue failures. Manufacturing processes, corrosion, and careless handling produce detrimental surface roughening.

Figure 5–8, adapted from data in Reference 11, shows estimates for the endurance strength s_n compared with the ultimate tensile strength of steels for several practical surface conditions. The data first estimate the endurance strength for the polished specimen to be 0.50 times the ultimate strength and then apply a factor related to the surface condition. U.S. Customary units are used for the bottom and left axes while SI units are shown on the top and right axes. Project vertically from the s_u axis to the appropriate curve and then horizontally to the endurance strength axis.

The data from Figure 5–8 should not be extrapolated for su > 220 ksi (1520 MPa) without specific testing as empirical data reported in Reference 6 are inconsistent at higher strength levels.

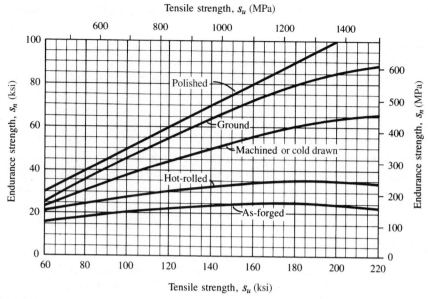

FIGURE 5–8 Endurance strength s_n versus tensile strength for wrought steel for various surface conditions

TABLE 5–1
Approximate reliability factors, C_R

Desired reliability	C_R
0.50	1.0
0.90	0.90
0.99	0.81
0.999	0.75

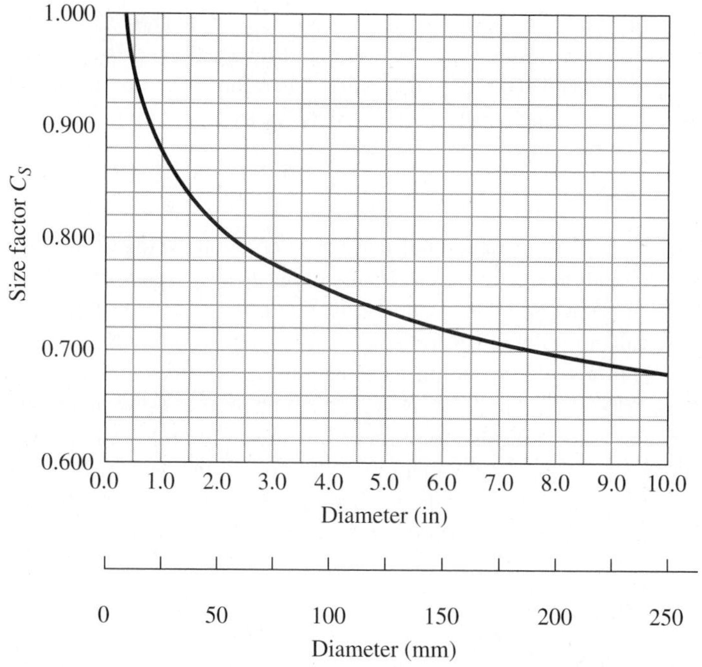

TABLE 5–2 Size factors

U.S. customary units	
Size Range	For D in inches
$D \leq 0.30$	$C_S = 1.0$
$0.30 < D \leq 2.0$	$C_S = (D/0.3)^{-0.11}$
$2.0 < D < 10.0$	$C_S = 0.859 - 0.02125D$

SI units	
Size Range	For D in mm
$D \leq 7.62$	$C_S = 1.0$
$7.62 < D \leq 50$	$C_S = (D/7.62)^{-0.11}$
$50 < D < 250$	$C_S = 0.859 - 0.000837D$

FIGURE 5–9 Size factor

Ground surfaces are fairly smooth and reduce the endurance strength by a factor of approximately 0.90 for $s_u < 160$ ksi (1 100 MPa), decreasing to about 0.80 for $s_u = 220$ ksi (1520 MPa). Machining or cold drawing produce a somewhat rougher surface because of tooling marks resulting in a reduction factor in the range of 0.80 to 0.60 over the range of strengths shown. The outer part of a hot rolled steel has a roughened oxidized scale that produces a reduction factor from 0.72 to 0.30. If a part is forged and not subsequently machined, the reduction factor ranges from 0.57 to 0.20.

From these data it should be obvious that you must give special attention to surface finish for critical surfaces exposed to fatigue loading in order to benefit from the steel's basic strength. Also, critical surfaces of fatigue-loaded parts must be protected from nicks, scratches, and corrosion because they drastically reduce fatigue strength.

Material Factors

Metal alloys having similar chemical composition can be wrought, cast, or made by powder metallurgy to produce the final form. Wrought materials are usually rolled or drawn, and they typically have higher endurance strength than cast materials. The grain structure of many cast materials or powder metals and the likelihood of internal flaws and inclusions tend to reduce their endurance strength. Reference 13 provides data from which the *material factors* listed in step 4 of the procedure outlined previously are taken.

Type-of-Stress Factor

Most endurance strength data are obtained from tests using a rotating cylindrical bar subjected to repeated and reversed bending in which the outer part experiences the highest stress. Stress levels decrease linearly to zero at the center of the bar. Because fatigue cracks usually initiate in regions of high tensile stress, a relatively small proportion of the material experiences such stresses. Contrast this with the case of a bar subjected to direct axial tensile stress for which *all* of the material experiences the maximum stress. There is a higher statistical probability that local flaws anywhere in the bar may start fatigue cracks. The result is that the endurance strength of a material subjected to repeated and reversed axial stress is approximately 80% of that from repeated and reversed bending. Therefore, we recommend that a factor $C_{st} = 1.0$ be applied for bending stress and $C_{st} = 0.80$ for axial loading.

Reliability Factor

The data for endurance strength for steel shown in Figure 5–8 represent average values derived from many tests of specimens having the appropriate ultimate strength and surface conditions. Naturally, there is variation among the data points; that is, half are higher and half are lower than the reported values on the given curve. The curve, then, represents a reliability of 50%, indicating that half of the parts would fail. Obviously, it is advisable to design for a higher reliability, say, 90%, 99%, or 99.9%. A factor can be used to estimate a lower endurance strength that can be used for design to produce the higher reliability values. Ideally, a statistical analysis of actual data for the material to be used in the design should be obtained. By making certain assumptions about the form of the distribution of strength data, Reference 11 reports the values in Table 5–1 as approximate reliability factors, C_R.

Size Factor—Circular Sections in Rotating Bending

Recall that the basic endurance strength data were taken for a specimen with a circular cross section that has a diameter of 0.30 in (7.6 mm) and that it was subjected to repeated and reversed bending while rotating. Therefore, each part of the surface is subjected to the maximum tensile bending stress with each revolution. Furthermore, the most likely place for fatigue failure to initiate is in the zone of maximum tensile stress within a small distance of the outer surface.

Data from References 2, 11, and 13 show that as the diameter of a rotating circular bending specimen increases, the endurance strength decreases because the stress gradient (change in stress as a function of radius) places a greater proportion of the material in the highly stressed region. Figure 5–9 and Table 5–2 show the size factor to be used in this book, adapted from Reference 13. These data can be used for either solid or hollow circular sections.

Size Factor—Other Conditions

We need different approaches to determining the size factor when a part with a circular section is subjected to repeated and reversed bending but it is *not rotating*, or if the part has a noncircular cross section. Here we show a procedure adapted from Reference 13 that focuses on the volume of the part that experiences 95% or more of the maximum stress. It is in this volume that fatigue failure is most likely to be initiated. Furthermore, in order to relate the physical size of such sections to the size factor data in Figure 5–9, we develop an equivalent diameter, D_e.

When the parts in question have a uniform geometry over the length of interest, the volume is the product of the length and the cross sectional area. We can compare different shapes by considering a unit length for each and looking only at the areas. As a base, let's begin by determining an expression for that part of a circular section subjected to 95% or more of the maximum bending stress, calling this area, A_{95}. Because the stress is directly proportional to the radius, we need the area of the thin ring between the outside surface with the full diameter D and a circle whose diameter is $0.95D$, as shown in Figure 5–10(a). Then,

$$A_{95} = (\pi/4)[D^2 - (0.95D)^2] = 0.0766D^2 \qquad \textbf{(5–5)}$$

You should demonstrate that this same equation applies to a hollow circular section as shown in Figure 5–10(b). This verifies that the data for size factor shown in Figure 5–9 and Table 5–2 apply directly to either the solid or hollow circular sections when they experience rotating bending.

Nonrotating Circular Section in Repeated and Reversed Flexure. Now consider a solid circular section that does not rotate but that is flexed back and forth in repeated and reversed bending. Only the top and bottom segments beyond a radius of $0.475D$ experience 95% or higher of the maximum bending stress as shown in Figure 5–10(c). By using properties of a segment of a circle, it can be shown that

$$A_{95} = 0.0105D^2 \qquad \textbf{(5–6)}$$

Now we can determine the *equivalent diameter*, D_e, for this area by equating equations (5–5) and (5–6) while designating the diameter in equation (5–5) as D_e and then solving for D_e.

$$0.0766D_e^2 = 0.0105D^2$$
$$D_e = 0.370D \qquad \textbf{(5–7)}$$

FIGURE 5–10
Geometry of sections
for computing A_{95} area

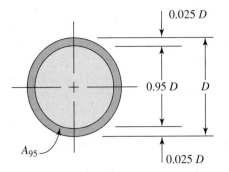

(a) Solid circular section-rotating

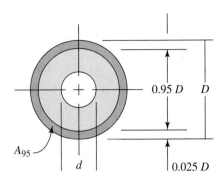

(b) Hollow circular section-rotating

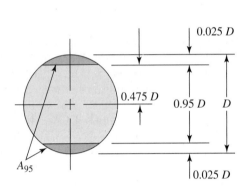

(c) Solid circular section not rotating

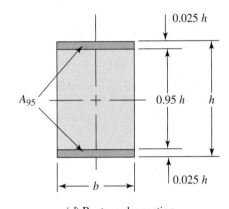

(d) Rectangular section

This same equation applies to a hollow circular section. The diameter, D_e, can be used in Figure 5–9 or in Table 5–2 to find the size factor.

Rectangular Section in Repeated and Reversed Flexure.   The A_{95} area is shown in Figure 5–10(d) as the two strips having a thickness of $0.025h$ at the top and bottom of the section. Therefore,

$$A_{95} = 0.05\ hb$$

Equating this to A_{95} for a circular section gives,

$$0.0766D_e^2 = 0.05\ hb \qquad\qquad \textbf{(5–8)}$$
$$D_e = 0.808\sqrt{hb}$$

This diameter can be used in Figure 5–9 or in Table 5–2 to find the size factor.
　　Other shapes can be analyzed in a similar manner.

Other Factors

The following factors are not included quantitatively in problem solutions in this book because of the difficulty of finding generalized data. However, you should consider each one as you engage in future designs and seek additional data as appropriate.

Flaws. Internal flaws of the material, especially likely in cast parts, are places in which fatigue cracks initiate. Critical parts can be inspected by x-ray techniques for internal flaws. If they are not inspected, a higher-than-average design factor should be specified for cast parts, and a lower endurance strength should be used.

Temperature. Most materials have a lower endurance strength at high temperatures. The reported values are typically for room temperatures. Operation above 500°F (260°C) will reduce the endurance strength of most steels. See Reference 13.

Nonuniform Material Properties. Many materials have different strength properties in different directions because of the manner in which the material was processed. Rolled sheet or bar products are typically stronger in the direction of rolling than they are in the transverse direction. Fatigue tests are likely to have been run on test bars oriented in the stronger direction. Stressing of such materials in the transverse direction may result in lower endurance strength.

Nonuniform properties are also likely to exist in the vicinity of welds because of incomplete weld penetration, slag inclusions, and variations in the geometry of the part at the weld. Also, welding of heat-treated materials may alter the strength of the material because of local annealing near the weld. Some welding processes may result in the production of residual tensile stresses that decrease the effective endurance strength of the material. Annealing or normalizing after welding is often used to relieve these stresses, but the effect of such treatments on the strength of the base material must be considered.

Residual Stresses. Fatigue failures typically initiate at locations of relatively high tensile stress. Any manufacturing process that tends to produce residual tensile stress will decrease the endurance strength of the component. Welding has already been mentioned as a process that may produce residual tensile stress. Grinding and machining, especially with high material removal rates, also cause undesirable residual tensile stresses. Critical areas of cyclically loaded components should be machined or ground in a gentle fashion.

Processes that produce residual *compressive* stresses can prove to be beneficial. Shot blasting and peening are two such methods. *Shot blasting* is performed by directing a high-velocity stream of hardened balls or pellets at the surface to be treated. *Peening* uses a series of hammer blows on the surface. Crankshafts, springs, and other cyclically loaded machine parts can benefit from these methods.

Corrosion and Environmental Factors. Endurance strength data are typically measured with the specimen in air. Operating conditions that expose a component to water, salt solutions, or other corrosive environments can significantly reduce the effective endurance strength. Corrosion may cause harmful local surface roughness and may also alter the internal grain structure and chemistry of the material. Steels exposed to hydrogen are especially affected adversely.

Nitriding. Nitriding is a surface-hardening process for alloy steels in which the material is heated to 950°F (514°C) in a nitrogen atmosphere, typically ammonia gas, followed by slow cooling. Improvement of endurance strength of 50% or more can be achieved with nitriding.

FIGURE 5–11 Effect of stress ratio R on endurance strength of a material

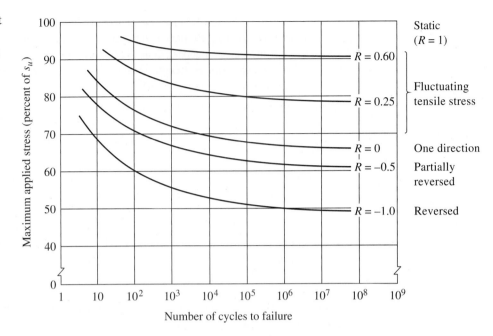

Effect of Stress Ratio on Endurance Strength. Figure 5–11 shows the general variation of endurance-strength data for a given material when the stress ratio R varies from -1.0 to $+1.0$, covering the range of cases including the following:

- Repeated, reversed stress (Figure 5–3); $R = -1.0$
- Partially reversed fluctuating stress with a tensile mean stress [Figure 5–4(b)]; $-1.0 < R < 0$
- Repeated, one-direction tensile stress (Figure 5–6); $R = 0$
- Fluctuating tensile stress [Figure 5–4(a)]; $0 < R < 1.0$
- Static stress (Figure 5–1); $R = 1$

Note that Figure 5–11 is only an example, and it should not be used to determine actual data points. If such data are desired for a particular material, specific data for that material must be found either experimentally or in published literature.

The most damaging kind of stress among those listed is the repeated, reversed stress with $R = -1$. (See Reference 6.) Recall that the rotating shaft in bending as shown in Figure 5–3 is an example of a load-carrying member subjected to a stress ratio $R = -1$.

Fluctuating stresses with a compressive mean stress as shown in Parts (c) and (d) of Figure 5–4 do not significantly affect the endurance strength of the material because fatigue failures tend to originate in regions of tensile stress.

Note that the curves of Figure 5–11 show estimates of the endurance strength, s_n, as a function of the ultimate tensile strength for steel. These data apply to ideal polished specimens and do not include any of the other factors discussed in this section. For example, the curve for $R = -1.0$ (reversed bending) shows that the endurance strength for steel is approximately 0.5 times the ultimate strength ($0.50 \times s_n$) for large numbers of cycles of loading (approximately 10^5 or higher). This is a good general estimate for steels. The chart also shows that types of loads producing R greater than -1.0 but less than 1.0 have less of an effect on the endurance strength. This illustrates that using data from the reversed bending test is the most conservative.

We will not use Figure 5–11 directly for problems in this book because our procedure for estimating the actual endurance strength starts with the use of Figure 5–8, which presents data from reversed bending tests. Therefore, the effect of stress ratio is already included. Section 5–9 includes methods of analysis for loading cases in which the fluctuating stress produces a stress ratio different from $R = -1.0$.

5–5 EXAMPLE PROBLEMS FOR ESTIMATING ACTUAL ENDURANCE STRENGTH

This section shows two examples that demonstrate the application of the *Procedure for Estimating Actual Endurance Strength*, s_n' that is presented in the previous section.

Example Problem 5–2

Estimate the actual endurance strength of AISI 1050 cold-drawn steel when used in a circular shaft subjected to rotating bending only. The shaft will be machined to a diameter of approximately 1.75 in.

Solution   Objective

Compute the estimated actual endurance strength of the shaft material.

Given

AISI 1050 cold-drawn steel, machined.

Size of section: $D = 1.75$ in.

Type of stress: Reversed, repeated bending.

Analysis

Use the Procedure for Estimating Actual Endurance Strength, s_n'

Step 1: The ultimate tensile strength: $s_u = 100$ ksi from Appendix 3.

Step 2: Diameter is machined.

Step 3: From Figure 5–8, $s_n = 38$ ksi

Step 4: Material factor for wrought steel: $C_m = 1.0$

Step 5: Type-of-stress factor for reversed bending: $C_{st} = 1.0$

Step 6: Specify a desired reliability of 0.99. Then $C_R = 0.81$ (Design decision)

Step 7: Size factor for circular section with $D = 1.75$ in.
From Figure 5–9, $C_s = 0.83$.

Step 8: Use Equation 5–4 to compute the estimated actual endurance strength.

$$s_n' = s_n(C_m)(C_{st})(C_R)(C_s) = 38 \text{ ksi}(1.0)(1.0)(0.81)(0.83) = 25.5 \text{ ksi}$$

Comments

This is the level of stress that would be expected to produce fatigue failure in a rotating shaft due to the action of reversed bending. It accounts for the basic endurance strength of the wrought AISI 1050 cold-drawn material, the effect of the machined surface, the size of the section, and the desired reliability.

Example Problem 5–3

Estimate the actual endurance strength of cast steel having an ultimate strength of 120 ksi when used in a bar subjected to a reversed, repeated, bending load. The bar will be machined to a rectangular cross section, 1.50 in wide × 2.00 in high.

Solution   Objective   Compute the estimated actual endurance strength of the bar material.

Given   Cast steel, machined: $s_u = 120$ ksi.
Size of section: $b = 1.50$ in., $h = 2.00$ in rectangular
Type of stress: Repeated, reversed bending.

Analysis   Use the Procedure for Estimating Actual Endurance Strength s_n'.

Step 1: The ultimate tensile strength is given to be $s_u = 120$ ksi.

Step 2: Surfaces are machined.

Step 3: From Figure 5–8, $s_n = 44$ ksi

Step 4: Material factor for cast steel: $C_m = 0.80$

Step 5: Type-of-stress factor for bending: $C_{st} = 1.00$

Step 6: Specify a desired reliability of 0.99. Then $C_R = 0.81$ (Design decision)

Step 7: Size factor for rectangular section: First use Equation 5–8 to determine the equivalent diameter,

$$D_e = 0.808\sqrt{hb} = 0.808\sqrt{(2.00\ \text{in})(1.50\ \text{in})} = 1.40\ \text{in}$$

Then from Figure 5–9, $C_s = 0.85$.

Step 8: Use Equation 5–4 to compute the estimated actual endurance strength.

$$s_n' = s_n(C_m)(C_{st})(C_R)(C_s) = 44\ \text{ksi}(0.80)(1.00)(0.81)(0.85) = 24.2\ \text{ksi}$$

**5–6
DESIGN
PHILOSOPHY**

It is the designer's responsibility to ensure that a machine part is safe for operation under reasonably foreseeable conditions. You should evaluate carefully the application in which the component is to be used, the environment in which it will operate, the nature of applied loads, the types of stresses to which the component will be exposed, the type of material to be used, and the degree of confidence you have in your knowledge about the application. Some general considerations are:

1. *Application.* Is the component to be produced in large or small quantities? What manufacturing techniques will be used to make the component? What are the consequences of failure in terms of danger to people and economic cost? How cost-sensitive is the design? Are small physical size or low weight important? With what other parts or devices will the component interface? For what life is the component being designed? Will the component be inspected and serviced periodically? How much time and expense for the design effort can be justified?

2. *Environment.* To what temperature range will the component be exposed? Will the component be exposed to electrical voltage or current? What is the potential for corrosion? Will the component be inside a housing? Will guarding protect access to the component? Is low noise important? What is the vibration environment?

3. *Loads.* Identify the nature of loads applied to the component being designed in as much detail as practical. Consider all modes of operation, including startup, shut down, normal operation, and foreseeable overloads. The loads should be characterized as *static, repeated and reversed, fluctuating, shock,* or *impact* as discussed in Section 5–2. Key magnitudes of loads are the *maximum, minimum,* and *mean.*

Variations of loads over time should be documented as completely as practical. Will high mean loads be applied for extended periods of time, particularly at high temperatures, for which creep must be considered? This information will influence the details of the design process.

4. ***Types of Stresses.*** Considering the nature of the loads and the manner of supporting the component, what kinds of stresses will be created: direct tension, direct compression, direct shear, bending, or torsional shear? Will two or more kinds of stresses be applied simultaneously? Are stresses developed in one direction (*uniaxially*), two directions (*biaxially*), or three directions (*triaxially*)? Is buckling likely to occur?

5. ***Material.*** Consider the required material properties of yield strength, ultimate tensile strength, ultimate compressive strength, endurance strength, stiffness, ductility, toughness, creep resistance, corrosion resistance, and others in relation to the application, loads, stresses, and the environment. Will the component be made from a ferrous metal such as plain carbon, alloy, stainless, or structural steel, or cast iron? Or will a nonferrous metal such as aluminum, brass, bronze, titanium, magnesium, or zinc be used? Is the material brittle (percent elongation < 5%) or ductile (percent elongation > 5%)? Ductile materials are highly preferred for components subjected to fatigue, shock, or impact loads. Will plastics be used? Is the application suitable for a composite material? Should you consider other nonmetals such as ceramics or wood? Are thermal or electrical properties of the material important?

6. ***Confidence.*** How reliable are the data for loads, material properties, and stress calculations? Are controls for manufacturing processes adequate to ensure that the component will be produced as designed with regard to dimensional accuracy, surface finish, and final as-made material properties? Will subsequent handling, use, or environmental exposure create damage that can affect the safety or life of the component? These considerations will affect your decision for the design factor, N, to be discussed in the next section.

All design approaches must define the relationship between the applied stresses on a component and the strength of the material from which it is to be made, considering the conditions of service. The strength basis for design can be yield strength in tension, compression, or shear; ultimate strength in tension, compression, or shear; endurance strength; or some combination of these. The goal of the design process is to achieve a suitable *design factor*, N, (sometimes called a *factor of safety*) that ensures the component is safe. That is, the strength of the material must be greater than the applied stresses. Design factors are discussed in the next section.

The sequence of design analysis will be different depending on what has already been specified and what is left to be determined. For example,

1. ***Geometry of the component and the loading are known:*** We apply the desired design factor, N, to the actual expected stress to determine the required strength of the material. Then a suitable material can be specified.

2. ***Loading is known and the material for the component has been specified:*** We compute a *design stress* by applying the desired design factor, N, to the appropriate strength of the material. This is the maximum allowable stress to which any part of the component can be exposed. We can then complete the stress analysis to determine what shape and size of the component will ensure that stresses are safe.

3. ***Loading is known, and the material and the complete geometry of the component have been specified:*** We compute both the expected maximum applied stress and the design stress. By comparing these stresses, we can determine the resulting

design factor, N, for the proposed design and judge its acceptability. A redesign may be called for if the design factor is either too low (unsafe) or too high (over designed).

Practical Considerations.   While ensuring that a component is safe, the designer is expected to also make the design practical to produce, considering several factors.

- Each design decision should be tested against the cost of achieving it.
- Material availability must be checked.
- Manufacturing considerations may affect final specifications for overall geometry, dimensions, tolerances, or surface finish.
- In general, components should be as small as practical unless operating conditions call for larger size or weight.
- After computing the minimum acceptable dimension for a feature of a component, standard or preferred sizes should be specified using normal company practice or tables of preferred sizes such as those listed in Appendix 2.
- Before a design is committed to production, tolerances on all dimensions and acceptable surface finishes must be specified so the manufacturing engineer and the production technician can specify suitable manufacturing processes.
- Surface finishes should only be as smooth as required for the function of a particular area of a component, considering appearance, effects on fatigue strength, and whether or not the area mates with another component. Producing smoother surfaces increases cost dramatically. See Chapter 13.
- Tolerances should be as large as possible while maintaining acceptable performance of the component. The cost to produce smaller tolerances rises dramatically. See Chapter 13.
- The final dimensions and tolerances for some features may be affected by the need to mate with other components. Proper clearances and fits must be defined, as discussed in Chapter 13. Another example is the mounting of a commercially available bearing on a shaft for which the bearing manufacturer specifies the nominal size and tolerances for the bearing seat on the shaft. Chapter 16 gives guidelines for clearances between the moving and stationary parts where either boundary or hydrodynamic lubrication is used.
- Will any feature of the component be subsequently painted or plated, affecting the final dimensions?

Deformations.   Machine elements can also fail because of excessive deformation or vibration. From your study of strength of materials, you should be able to compute deformations due to axial tensile or compressive loads, bending, torsion, or changes in temperature. Some of the basic concepts are reviewed in Chapter 3. For more complex shapes or loading patterns, computer-based analysis techniques such as finite element analysis (FEA) or beam analysis software are important aids.

Criteria for failure due to deformation are often highly dependent on the machine's use. Will excessive deformation cause two or more members to touch when they should not? Will the desired precision of the machine be compromised? Will the part look or feel too flexible (flimsy)? Will parts vibrate excessively or resonate at the frequencies experienced during operation? Will rotating shafts exhibit a critical speed during operation, resulting in wild oscillations of parts carried by the shaft?

This chapter will not pursue the quantitative analysis of deformation, leaving that to be your responsibility as the design of a machine evolves. Later chapters do address some critical cases such as the interference fit between two mating parts (Chapter 13), the position of the teeth of one gear relative to its mating gear (Chapter 9), the radial clearance between a journal bearing and the shaft rotating within it (Chapter 16), and the deformation of springs (Chapter 19). Also, Section 5–10, as a part of the general design procedure, suggests some guidelines for limiting deflections.

5–7 DESIGN FACTORS

The term *design factor*, N, is a measure of the relative safety of a load-carrying component. In most cases, the strength of the material from which the component is to be made is divided by the design factor to determine a *design stress*, σ_d, sometimes called the *allowable stress*. Then the actual stress to which the component is subjected should be less than the design stress. For some kinds of loading, it is more convenient to set up a relationship from which the design factor, N, can be computed from the actual applied stresses and the strength of the material. Still in other cases, particularly for the case of the buckling of columns, as discussed in Chapter 6, the design factor is applied to the *load* on the column rather than the strength of the material.

Section 5–9 presents methods for computing the design stress or design factor for several different kinds of loading and materials.

The designer must determine what a reasonable value for the design factor should be in any given situation. Often the value of the design factor or the design stress is governed by codes established by standards-setting organizations such as the American Society of Mechanical Engineers, the American Gear Manufacturers Association, the U.S. Department of Defense, the Aluminum Association, or the American Institute of Steel Construction. For structures, local or state building codes often prescribe design factors or design stresses. Some companies have adopted their own policies specifying design factors based on past experience with similar conditions.

In the absence of codes or standards, the designer must use judgment to specify the desired design factor. Part of the design philosophy, discussed in Section 5–6, discussed issues such as the nature of the application, environment, nature of the loads on the component to be designed, stress analysis, material properties, and the degree of confidence in data used in the design processes. All of these considerations affect the decision about what value for the design factor is appropriate. This book will use the following guidelines.

Ductile Materials

1. **N = 1.25 to 2.0.** Design of structures under static loads for which there is a high level of confidence in all design data.

2. **N = 2.0 to 2.5.** Design of machine elements under dynamic loading with average confidence in all design data. (Typically used in problem solutions in this book.)

3. **N = 2.5 to 4.0.** Design of static structures or machine elements under dynamic loading with uncertainty about loads, material properties, stress analysis, or the environment.

4. **N = 4.0 or higher.** Design of static structures or machine elements under dynamic loading with uncertainty about some combination of loads, material properties, stress analysis, or the environment. The desire to provide extra safety to critical components may also justify these values.

Brittle Materials

5. **N = 3.0 to 4.0.** Design of structures under static loads for which there is a high level of confidence in all design data.

6. **N = 4.0 to 8.0.** Design of static structures or machine elements under dynamic loading with uncertainty about loads, material properties, stress analysis, or the environment.

The following Sections 5–8 and 5–9 provide guidance on the introduction of the design factor into the design process with particular attention to the selection of the strength basis for the design and the computation of the design stress. In general, design for static loading involves applying the design factor to the yield strength or ultimate strength of the material. Dynamic loading requires the application of the design factor to the endurance strength using the methods described in Section 5–5 to estimate the actual endurance strength for the conditions under which the component is operating.

5–8 PREDICTIONS OF FAILURE

Designers should understand the various ways that load-carrying components can fail in order to complete a design that ensures that failure *does not occur*. Several different methods of predicting failure are available, and it is the designer's responsibility to select the one most appropriate to the conditions of the project. In this section we describe the methods that have found a high level of use in the field and discuss the situations in which each is applicable. The factors involved are the nature of the load (static, repeated and reversed, or fluctuating), the type of material involved (ductile or brittle), and the amount of design effort and analysis that can be justified by the nature of the component or product being designed.

The design analysis methods described in the following Section 5–9 define the relationship between the applied stresses on a component and the strength of the material from which it is to be made that is most relevant to the conditions of service. The strength basis for design can be yield strength, ultimate strength, endurance strength, or some combination of these. The goal of the design process is to achieve a suitable design factor, N, that ensures that the component is safe. That is, the strength of the material must be greater than the applied stresses.

The following types of failure prediction are described in this section. Reference 12 gives an excellent historical review of failure prediction and complete derivations of the fundamentals underlying the methods discussed here.

Failure Prediction Method	*Uses*
1. Maximum normal stress	Uniaxial static stress on brittle materials
2. Modified Mohr	Biaxial static stress on brittle materials
3. Yield strength	Uniaxial static stress on ductile materials
4. Maximum shear stress	Biaxial static stress on ductile materials [Moderately conservative]
5. Distortion energy	Biaxial or triaxial stress on ductile materials [Good predictor]
6. Goodman	Fluctuating stress on ductile materials [Slightly conservative]
7. Gerber	Fluctuating stress on ductile materials [Good predictor]
8. Soderberg	Fluctuating stress on ductile materials [Moderately conservative]

Maximum Normal Stress Method for Uniaxial Static Stress on Brittle Materials

The maximum normal stress theory states that a material will fracture when the maximum normal stress (either tension or compression) exceeds the ultimate strength of the material as obtained from a standard tensile or compressive test. Its use is limited, namely for brittle materials under pure uniaxial static tension or compression. When applying this theory, any stress concentration factor at the region of interest should be applied to the computed stress because brittle materials do not yield and therefore cannot redistribute the increased stress.

The following equations apply the maximum normal stress theory to design.

For tensile stress: $\qquad K_t\sigma < \sigma_d = s_{ut}/N$ (5–9)

For compressive stress: $\qquad K_t\sigma < \sigma_d = s_{uc}/N$ (5–10)

Note that many brittle materials such as gray cast iron have a significantly higher compressive strength than tensile strength.

Modified Mohr Method for Biaxial Static Stress on Brittle Materials

When stresses are applied in more that one direction or when normal stress and shear stress are applied simultaneously, it is necessary to compute the principal stresses, σ_1 and σ_2, using Mohr's circle or the equations in Chapter 4. *Stress concentrations should be included in the applied stresses before preparing Mohr's circle for brittle materials.*

For safety, the *combination* of the two principal stresses must lie within the area shown in Figure 5–12 that graphically depicts the *modified Mohr theory*. The graph is a plot of the maximum principal stress, σ_1, on the horizontal axis (abscissa) and the minimum principal stress, σ_2, on the vertical axis (ordinate).

Note that the failure criteria depend on the quadrant in which the principal stresses lie. In the first quadrant, both principal stresses are tensile, and failure is predicted when either one exceeds the ultimate tensile strength, s_{ut}, of the material. Similarly, in the third quadrant, both principal stresses are compressive, and failure is predicted when either one exceeds the ultimate compressive strength, s_{uc}, of the material. The failure lines for the second and fourth quadrants are more complex and have been derived semi-empirically to correlate with test data. The

FIGURE 5–12
Modified Mohr diagram with example data and a load line plotted

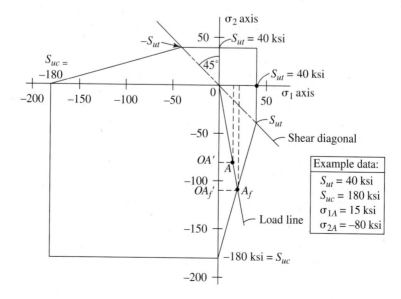

ultimate tensile strength lines are extended from the first quadrant into the second and fourth quadrants to the point where each intersects the *shear diagonal*, drawn at 45 degrees through the origin. Then the failure line proceeds at an angle to the ultimate compressive strength.

For design, because of the many different shapes and dimensions of safe-stress zones in Figure 5–12, it is suggested that a rough plot be made of the pertinent part of the modified Mohr diagram from actual material strength data. Then the actual values of σ_1 and σ_2 can be plotted to ensure that they lie within the safe zone of the diagram.

A *load line* can be an aid in determining the design factor, N, using the modified Mohr diagram. The assumption is made that stresses increase proportionally as loads increase. Apply the following steps for an example stress state, A, for which $\sigma_{1A} = 15$ ksi and $\sigma_{2A} = -80$ ksi. The material is Grade 40 gray cast iron having $s_{ut} = 40$ ksi and $s_{uc} = 180$ ksi.

1. Draw the modified Mohr diagram as shown in Figure 5–12.
2. Plot point A at $(15, -80)$.
3. Draw the load line from the origin through point A until it intersects the failure line on the diagram at the point labeled A_f.
4. Determine the distances $OA = 81.4$ ksi and $OA_f = 112$ ksi by scaling the diagram.
5. Compute the design factor from $N = OA_f/OA = 112/81.4 = 1.38$.
6. Alternatively, the projections of points A and A_f on the σ_1 or σ_2 axes could be used because the value of N is a ratio and similar triangles are formed as shown in Figure 5–12.
7. In this example, the projections onto the σ_2 axis are: $OA' = -80$ ksi, $OA'_f = -110$ ksi. Then,

$$N = OA'_f/OA' = -110/-80.0 = 1.38.$$

Yield Strength Method for Uniaxial Static Normal Stresses on Ductile Materials

This is a simple application of the principle of yielding in which a component is carrying a direct tensile or compressive load in the manner similar to the conditions of the standard tensile or compressive test for the material. Failure is predicted when the actual applied stress exceeds the yield strength. Stress concentrations can normally be neglected for static stresses on ductile materials because the higher stresses near the stress concentrations are highly localized. When the local stress on a small part of the component reaches the yield strength of the material, it does in fact yield. In the process, the stress is redistributed to other areas and the component is still safe.

The following equations apply the yield strength principle to design.

For tensile stress:	$\sigma < \sigma_d = s_{yt}/N$	(5–11)
For compressive stress:	$\sigma < \sigma_d = s_{yc}/N$	(5–12)

For most wrought ductile metals, $s_{yt} = s_{yc}$.

Maximum Shear Stress Method for Biaxial Static Stress on Ductile Materials

The maximum shear stress method of failure prediction states that a ductile material begins to yield when the maximum shear stress in a load-carrying component exceeds that in a tensile-test specimen when yielding begins. A Mohr's circle analysis for the uniaxial tension test, discussed in Section 4–6, shows that the maximum shear stress is one-half of the applied tensile stress. At yield, then, $s_{sy} = s_y/2$. We use this approach in this book to estimate s_{sy}. Then, for design, use

$$\tau_{max} < \tau_d = s_{sy}/N = 0.5\, s_y/N \qquad (5\text{–}13)$$

The maximum shear stress method of failure prediction has been shown by experimentation to be somewhat conservative for ductile materials subjected to a combination of normal and shear stresses. It is relatively easy to use and is often chosen by designers. For more precise analysis, the distortion energy method is preferred.

Distortion Energy Method for Static Biaxial or Triaxial Stress on Ductile Materials

The distortion energy method has been shown to be the best predictor of failure for ductile materials under static loads or completely reversed normal, shear, or combined stresses. It requires the definition of the new term, *von Mises stress*, indicated by the symbol, σ', that can be calculated for biaxial stresses, given the maximum and minimum principal stresses, σ_1 and σ_2, from

$$\sigma' = \sqrt{\sigma_1^2 + \sigma_2^2 - \sigma_1\sigma_2} \qquad (5\text{–}14)$$

Failure is predicted when $\sigma' > s_y$. The biaxial stress approach requires that the applied stress in the third orthogonal direction, typically σ_z, is zero.

Credit is given to R. von Mises for the development of Equation 5–14 in 1913. Because of additional contributions by H. Hencky in 1925, the method is sometimes called the *von Mises-Hencky method*. Be aware that the results from many finite element analysis software packages include the von Mises stress. Another term used is the *octahedral shear stress*.

It is helpful to visualize the distortion energy failure prediction method by plotting a failure line on a graph with σ_1 on the horizontal axis and σ_2 on the vertical axis as shown in Figure 5–13. The failure line is an ellipse centered at the origin and passing through the yield strength on each axis, in both the tensile and compressive regions. It is necessary that the material has equal values for yield strength in tension and compression for direct use of this method. The numerical scales on the graph are normalized to the yield strength so the ellipse passes through $s_y/\sigma_1 = 1.0$ on the σ_1 axis and similarly on the other axes. *Combinations of principal stresses that lie within the distortion energy ellipse are predicted to be safe, while those outside would predict failure.*

For design, the design factor, N, can be applied to the yield strength. Then use

$$\sigma' < \sigma_d = s_y/N \qquad (5\text{–}15)$$

For comparison, the failure prediction lines for the maximum shear stress method are shown also in Figure 5–13. With data showing that the distortion energy method is the best predictor, it can be seen that the maximum shear stress method is generally conservative and that it coincides with the distortion energy ellipse at six points. In other regions, it is as much as 16% low. Note the 45° diagonal line through the second and fourth quadrants, called the *shear diagonal*. It is the locus of points where $\sigma_1 = \sigma_2$ and its intersection with the failure ellipse is at the point $(-0.577, 0.577)$ in the second quadrant. This predicts failure when the shear stress is $0.577s_y$. The maximum shear stress method predicts failure at $0.50s_y$, thus quantifying the conservatism of the maximum shear stress method.

Also shown in Figure 5–13 are the failure prediction lines for the maximum principal stress method. It is coincident with the maximum shear stress lines in the first and third quadrants for which the two principal stresses have the same sign, either tensile (+) or

FIGURE 5–13
Distortion energy
method compared with
maximum shear stress
and maximum principal
stress methods

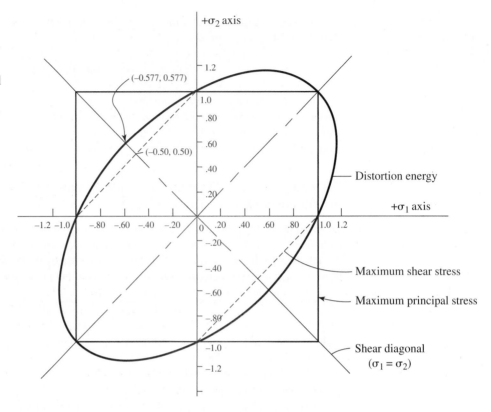

compressive ($-$). Therefore, it, too, is conservative in these regions. But note that it is dangerously nonconservative in the second and fourth quadrants.

Alternate Form for the von Mises Stress.   Equation 5–14 requires that the two principal stresses be determined from Mohr's circle, equations 4–1 and 4–2, or from a finite element analysis. Often you will first determine the stresses in some convenient orthogonal directions, x and y, namely σ_x, σ_y, and τ_{xy}. The von Mises stress can then be calculated directly from

$$\sigma' = \sqrt{\sigma_x^2 + \sigma_y^2 - \sigma_x\sigma_y + 3\tau_{xy}^2} \qquad \textbf{(5–16)}$$

For uniaxial stress with shear, $\sigma_y = 0$, Equation (5–16) reduces to

$$\sigma' = \sqrt{\sigma_x^2 + 3\tau_{xy}^2} \qquad \textbf{(5–17)}$$

Triaxial Distortion Energy Method.   A more general expression of the von Mises (distortion energy) stress is required when principal stresses occur in all three directions, σ_1, σ_2, and σ_3. We normally order these stresses such that $\sigma_1 > \sigma_2 > \sigma_3$. Then,

$$\sigma' = \left(\sqrt{2}/2\right)\left[\sqrt{(\sigma_2 - \sigma_1)^2 + (\sigma_3 - \sigma_1)^2 + (\sigma_3 - \sigma_2)^2}\,\right] \qquad \textbf{(5–18)}$$

Goodman Method for Fatigue Under Fluctuating Stress on Ductile Materials

Recall from Section 5–2 that the term *fluctuating stress* refers to the condition in which a load-carrying component is subjected to a nonzero mean stress with an alternating stress superimposed on the mean stress (see Figure 5–4). The Goodman method of failure pre-

FIGURE 5–14
Modified Goodman diagram for fatigue of ductile materials

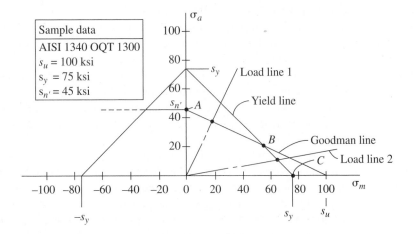

diction, sketched in Figure 5–14, has been shown to provide a good correlation with experimental data, falling just slightly below the scatter of data points.

The Goodman diagram plots the mean stresses on the horizontal axis and the alternating stresses on the vertical axis. Look first at the right part of the diagram representing fluctuating stresses with a tensile (+) mean stress. A straight line is drawn from the estimated actual endurance strength of the material, s_n', on the vertical axis to the ultimate tensile strength, s_u, on the horizontal axis. Combinations of mean stress, σ_m, and alternating stress, σ_a, above the line predict failure, while those below the line predict no failure from fatigue. The equation for the Goodman line is,

$$\frac{\sigma_a}{s_n'} + \frac{\sigma_m}{s_u} = 1 \qquad (5\text{–}19)$$

Design Equation. We can modify Equation 5–`th the ultimate and endurance strength values, as shown in Figure 5–15, to depict a "safe stress" line. Furthermore, any stress concentration factor in the region of interest should be applied to the alternating component but not to the mean stress component, because experimental evidence shows that the presence of a stress concentration does not affect the contribution of the mean stress of fatigue failure. Making these adjustments to the equation for the Goodman line gives,

$$\frac{K_t\sigma_a}{s_n'} + \frac{\sigma_m}{s_u} = \frac{1}{N} \qquad (5\text{–}20)$$

This is the design equation we will use in this book for fluctuating stresses.

FIGURE 5–15
Modified Goodman diagram showing safe stress line

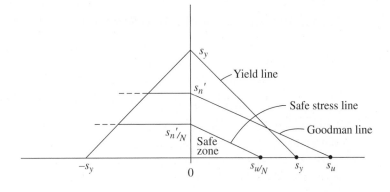

Checking for Early Cycle Yielding.   The Goodman line presents a difficulty near the right end because it seems to allow a pure mean stress greater than the yield strength of the material. Furthermore, as some value of alternating stress is added to the mean stress, the actual maximum stress ranges above the mean and may cause yielding. From a pure fatigue consideration, this condition is acceptable, provided the application can tolerate some local yielding in areas of high maximum stress. Any yielding would occur within the early cycles of loading, perhaps from the first cycle and certainly at less than 1000 cycles. After yielding, the stresses would be redistributed and the component would continue to be safe.

However, most designers choose to *not* permit yielding anywhere. To accomplish this, the *yield line* is added to the Goodman diagram, drawn between the yield strength plotted on both axes. Now the line segments between points labeled *A*, *B*, and *C* define the failure line. Consider two load lines drawn from the origin and extended through intersections with all of the failure lines on the diagram. Load line 1 intersects the Goodman line first, indicating that fatigue failure governs. Load line 2 intersects the yield line first and failure would commence as yielding.

We recommend completing first the design based on fatigue using Equation 5–20 and then checking for yielding separately. The design equation for the yield line is,

$$\frac{K_t \sigma_a}{s_y} + \frac{K_t \sigma_m}{s_y} = \frac{1}{N} \tag{5-21}$$

Here we do apply the stress concentration factor to the mean stress to ensure that yielding does not occur. In many cases, the safe stress line for fatigue actually falls completely below the yield strength line, indicating that no yielding is expected. See Figure 5–15. However, there may be a lower effective design factor for yielding than on fatigue failure, and you will need to judge whether this is acceptable or not. Equation 5–21 can be solved for *N* based on yielding, giving,

$$N = \frac{s_y}{K_t \left(\sigma_a + \sigma_m \right)} \tag{5-22}$$

Fluctuating Stresses with Compressive Mean Stress.   The left part of the Goodman diagram represents fluctuating stresses with compressive ($-$) stresses. Experimental data show that the presence of compressive mean stress does not significantly degrade the fatigue life beyond that predicted by the alternating stress only. So the failure line extends horizontally to the left from the s_n' point on the alternating stress axis. Its limit is the yield line for compressive yielding.

Gerber Method for Fluctuating Stress on Ductile Materials

Those interested in a more precise predictor of fatigue failure propose the Gerber method, shown in Figure 5–16. The Goodman line is shown for comparison. The end points of each are the same, but the Gerber line is parabolic and follows generally among the experimentally determined failure points, whereas the Goodman line lies below them. (See References 11 to 13.) This means that some failure points will lie below the Gerber line, an undesirable result. For this reason, we will use the Goodman line for problem solutions in this book.

FIGURE 5–16
Comparison of Gerber,
Goodman, and
Soderberg methods for
fluctuating stresses on
ductile materials

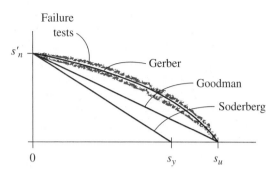

The equation of the Gerber line is,

$$\frac{\sigma_a}{s_n'} + \left[\frac{\sigma_m}{s_u}\right]^2 = 1 \qquad (5\text{--}23)$$

Soderberg Method for Fluctuating Stress on Ductile Materials

Another approach that has found significant use, and that was featured in earlier editions of this book, is called the *Soderberg method*. Figure 5–16 shows the Soderberg failure line in comparison with the Goodman and Gerber lines. The equation of the Soderberg line is,

$$\frac{K_t \sigma_a}{s_n'} + \frac{\sigma_m}{s_y} = 1 \qquad (5\text{--}24)$$

Drawn between the endurance strength and the yield strength, the Soderberg line is the most conservative of the three. One advantage of the Soderberg line is that it protects directly against early cycle yielding, whereas the Goodman and Gerber methods require the secondary consideration of the yield line as discussed previously. However, the degree of conservatism is considered too great for competitive efficient design.

   In summary, problem solving in this book will use the Goodman method for fluctuating stresses on ductile materials. It is only slightly conservative and its failure prediction line lies completely below the array of experimental failure data points.

**5–9
DESIGN
ANALYSIS
METHODS**

Here we summarize the recommended methods for design analysis based on the type of material (brittle or ductile), the nature of the loading (static or cyclical), and the type of stress (uniaxial or biaxial). The fact that 16 different cases are discussed is an indication of the large variety of approaches available. As you read about each case, refer to Figure 5–17 to follow the relationships among the factors to be considered.

   For cases C, E, F, and I, which involve ductile materials under four different types of loading, both the maximum shear stress and distortion energy methods are included. Recall from the discussions in the previous section that the maximum shear stress method is the simpler to use but somewhat conservative. The distortion energy method is the most accurate predictor of failure, but it requires the additional step of computing the von Mises stress. Both methods will be illustrated in the example problems in this book; the distortion energy method is recommended.

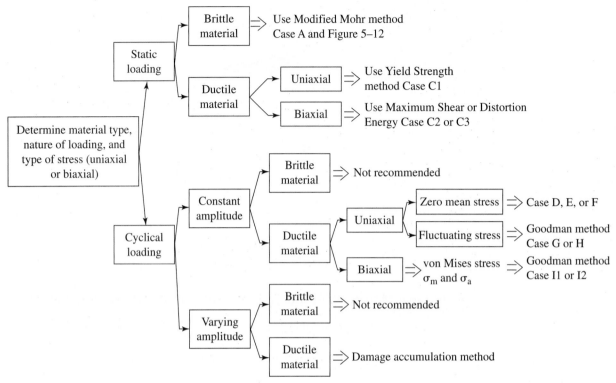

FIGURE 5–17   Logic diagram for visualizing methods of design analysis

Figure 5–17 includes a listing for the *damage accumulation method* for when ductile materials are subjected to cyclical loading with varying amplitude. This topic is discussed in Section 5–13.

The following symbols are used in the various cases.

$$s_u \text{ or } s_{ut} = \text{ultimate tensile strength}$$
$$s_{uc} = \text{ultimate compressive strength}$$
$$s_y = \text{yield strength or yield point}$$
$$s_{sy} = \text{yield strength in shear}$$
$$s'_n = \text{endurance strength of material under actual conditions}$$
$$s'_{sn} = \text{endurance strength in shear under actual conditions}$$
$$\sigma = \text{nominal applied stress, without } K_t$$

Case A: Brittle Materials under Static Loads

When the actual applied stress, σ, is simple tension or compression in only one direction, use the maximum normal stress theory of failure. Because brittle materials do not yield, you should always apply stress concentration factors when computing the applied stress.

Case A1: Uniaxial Tensile Stress

$$K_t\sigma < \sigma_d = s_{ut}/N \tag{5–9}$$

Case A2: Uniaxial Compressive Stress

$$K_t\sigma < \sigma_d = s_{uc}/N \qquad (5\text{--}10)$$

Case A3: Biaxial Stress. Use Mohr's circle to determine the principal stresses, σ_1 and σ_2. If both principal stresses are of the same sign, either tensile or compressive, use Case A1 or A2. If they are of different signs, use the modified Mohr method described in the preceeding section and illustrated in Figure 5–12. Any stress concentration factors should be applied to the computed nominal stresses.

Case B: Brittle Materials under Fatigue Loads

No specific recommendation will be given for brittle materials under fatigue loads because it is usually not desirable to use a brittle material in such cases. When it is necessary to do so, testing should be done to ensure safety under actual conditions of service.

Case C: Ductile Materials under Static Loads

Three failure methods are listed. The yield strength method is only for uniaxial normal stresses. For shear or biaxial loads, the maximum shear stress method is simpler but somewhat conservative. The distortion energy method is the best failure predictor.

C1: Yield Strength Method for Uniaxial Static Normal Stresses

For tensile stress: $\qquad \sigma < \sigma_d = s_{yt}/N \qquad\qquad (5\text{--}11)$

For compressive stress: $\qquad \sigma < \sigma_d = s_{yc}/N \qquad\qquad (5\text{--}12)$

C2: Maximum Shear Stress Method. Used for shear stresses and combined stresses. Determine the maximum shear stress from Mohr's circle. Then the design equation is,

$$\tau_{max} < \tau_d = s_{sy}/N = 0.50\, s_y/N \qquad (5\text{--}13)$$

C3: Distortion Energy Method. Used for shear stresses and combined stresses. Determine the maximum shear stress from Mohr's circle. Then compute the von Mises stress from,

$$\sigma' = \sqrt{\sigma_1^2 + \sigma_2^2 - \sigma_1\sigma_2} \qquad (5\text{--}14)$$

The alternate equations (5–16), (5–17), or (5–18) from the preceding section can also be used. For design use,

$$\sigma' < \sigma_d = s_y/N \qquad (5\text{--}15)$$

Stress concentrations are not needed for static loading if local yielding can be tolerated.

Case D: Reversed, Repeated Normal Stress

Figure 5–2 shows the general form of reversed, repeated normal stress. Note that the mean stress, σ_m, is zero and that the alternating stress, σ_a, is equal to the maximum stress, σ_{max}.

This case follows directly from the definition of the estimated actual endurance strength because the rotating beam testing method is used to acquire the strength data. Also, it is a special case of fluctuating stress, covered by Equation 5–20 in the preceding section. With a zero mean stress, the design equation becomes,

$$K_t \sigma_{\max} < \sigma_d = s'_n / N \tag{5–25}$$

Case E: Reversed, Repeated Shear Stress

Again the maximum shear stress theory or the distortion energy theory can be used. First compute the maximum repeated shear stress, $\tau_{\max}$, including any stress concentration factor. The discussion for Case D applies for shear stress as well.

Case E1: Maximum Shear Stress Theory

$$s'_{sn} = 0.5 \, s'_n \; (\text{estimate for endurance strength in shear})$$
$$K_t \tau_{\max} < \tau_d = s'_{sn}/N = 0.5 \, s'_n/N \tag{5–26}$$

Case E2: Distortion Energy Theory

$$s'_{sn} = 0.577 \, s'_n \; (\text{estimate for endurance strength in shear})$$
$$K_t \tau_{\max} < \tau_d = s'_{sn}/N = 0.577 \, s_n/N \tag{5–27}$$

Case F: Reversed Combined Stress

Use Mohr's circle to find the maximum shear stress and the two principal stresses by using the maximum values of the applied stresses.

Case F1: Maximum Shear Stress Theory.   Use Equation 5–26.

Case F2: Distortion Energy Theory.   Use Equation 5–27.

Case G: Fluctuating Normal Stresses: The Goodman Method

Use the Goodman method that was described in Section 5–8 and illustrated in Figure 5–15. A satisfactory design results if the combination of the mean stress and the alternating stress produces a point in the *safe zone* shown in Figure 5–15. Then you can use the Equation (5–20) to evaluate the design factor for fluctuating loads:

$$\frac{K_t \sigma_a}{s'_n} + \frac{\sigma_m}{s_u} = \frac{1}{N} \tag{5–20}$$

Case H: Fluctuating Shear Stresses

The preceding development of the Goodman method can also be done for fluctuating shear stresses instead of normal stresses. The design factor equation would then be

$$\frac{K_t \tau_a}{s'_{sn}} + \frac{\tau_m}{s_{su}} = \frac{1}{N} \tag{5–28}$$

In the absence of shear strength data, use the estimates, $s'_{sn} = 0.577 \, s'_n$ and $s_{su} = 0.75 \, s_u$.

Case I: Fluctuating Combined Stresses

The approach presented here is similar to the Goodman method described previously, but the effect of the combined stresses is first determined by using Mohr's circle.

Case I1.   For the maximum shear stress theory, draw two Mohr's circles, one for the mean stresses and one for the alternating stresses. From the first circle, determine maximum mean shear stress, $(\tau_m)_{max}$. From the second circle, determine the maximum alternating shear stress, $(\tau_a)_{max}$. Then use these values in the design equation

$$\frac{K_t\,(\tau_a)_{max}}{s'_{sn}} + \frac{(\tau_m)_{max}}{s_{su}} = \frac{1}{N} \qquad (5\text{–}29)$$

In the absence of shear strength data, use the estimates, $s'_{sn} = 0.577\,s'_n$ and $s_{su} = 0.75\,s_u$.

Case I2.   For the distortion energy theory, draw two Mohr's circles, one for the mean stresses and one for the alternating stresses. From these circles, determine the maximum and minimum principal stresses. Then compute the von Mises stresses for both the mean and the alternating components from

$$\sigma'_m = \sqrt{\sigma_{1m}^2 + \sigma_{2m}^2 - \sigma_{1m}\sigma_{2m}}$$
$$\sigma'_a = \sqrt{\sigma_{1a}^2 + \sigma_{2a}^2 - \sigma_{1a}\sigma_{2a}}$$

The Goodman equation then becomes

$$\frac{K_t\sigma'_a}{s'_n} + \frac{\sigma'_m}{s_u} = \frac{1}{N} \qquad (5\text{–}30)$$

**5–10
GENERAL
DESIGN
PROCEDURE**

The earlier parts of this chapter have provided guidance related to the many factors involved in design of machine elements that must be safe when carrying the applied loads. This section brings these factors together so that you can complete the design. The general design procedure described here is meant to give you a feel for the process. It is not practical to provide a completely general procedure. You will have to adapt it to the specific situations that you encounter.

The procedure is set up assuming that the following factors are known or can be specified or estimated:

- General design requirements: Objectives and limitations on size, shape, weight, desired precision, and so forth.
- Nature of the loads to be carried.
- Types of stresses produced by the loads.
- Type of material from which the element is to be made.
- General description of the manufacturing process to be used, particularly with regard to the surface finish that will be produced.
- Desired reliability.

General Design Procedure

1. Specify the objectives and limitations, if any, of the design, including desired life, size, shape, and appearance.

2. Determine the environment in which the element will be placed, considering such factors as corrosion potential and temperature.

3. Determine the nature and characteristics of the loads to be carried by the element, such as

   Static, dead, slowly applied loads.

   Dynamic, live, varying, repeated loads that may potentially cause fatigue failure.

   Shock or impact loads.

4. Determine the magnitudes for the loads and the operating conditions, such as

   Maximum expected load.

   Minimum expected load.

   Mean and alternating levels for fluctuating loads.

   Frequency of load application and repetition.

   Expected number of cycles of loading.

5. Analyze how loads are to be applied to determine the type of stresses produced, such as

   Direct normal stress, bending stress, direct shear stress, torsional shear stress, or some combination of stresses.

6. Propose the basic geometry for the element, paying particular attention to

   Its ability to carry the applied loads safely.

   Its ability to transmit loads to appropriate support points. Consider the *load paths*.

   The use of efficient shapes according to the nature of the loads and the types of stresses encountered. This applies to the general shape of the element and to each of its cross sections. Achieving efficiency involves optimizing the amount and the type of material involved. In Chapter 20, Section 20–2 gives some suggestions for efficient design of frames and members in bending and torsion.

   Providing appropriate attachments to supports and to other elements in the machine or structure.

   Providing for the positive location of other components that may be installed on the element being designed. This may call for shoulders, grooves, holes, retaining rings, keys and keyseats, pins, or other means of fastening or holding parts.

7. Propose the method of manufacturing the element with particular attention to the precision required for various features and the surface finish that is desired. Will it be cast, machined, ground, or polished, or produced by some other process? These design decisions have important impacts on the performance of the element, its ability to withstand fatigue loading, and the cost to produce it.

8. Specify the material from which the element is to be made, along with its condition. For metals the specific alloy should be specified, and the condition could include such processing factors as hot rolling, cold drawing, and a specific heat treatment. For nonmetals, it is often necessary to consult with vendors to specify the composition and the mechanical and physical properties of the desired material. Consult Chapter 2 and Section 20–2 in Chapter 20 for additional guidance.

9. Determine the expected properties of the selected material, for example

    Ultimate tensile strength, s_u.

    Ultimate compressive strength, s_{uc}, if appropriate.

    Yield strength, s_y.

    Ductility as represented by percent elongation.

    Stiffness as represented by modulus of elasticity, E or G.

10. Specify an appropriate design factor, N, for the stress analysis using the guidelines discussed in Section 5–7.

11. Determine which stress analysis method outlined in Section 5–9 applies to the design being completed.

12. Compute the appropriate design stress for use in the stress analysis. If fatigue loading is involved, the actual expected endurance strength of the material should be computed as outlined in Section 5–4. This requires the consideration of the expected size of the section, the type of material to be used, the nature of the stress, and the desired reliability. Because the size of the section is typically unknown at the start of the design process, an estimate must be made to allow the inclusion of a reasonable size factor, C_s. You should check the estimate at the end of the design process to verify that reasonable values were assumed at this stage of the design.

13. Determine the nature of any stress concentrations that may exist in the design at places where geometry changes occur. Stress analysis should be considered at all such places because of the likelihood of localized high tensile stresses that may produce fatigue failure. If the geometry of the element in these areas is known, determine the appropriate stress concentration factor, K_t. If the geometry is not yet known, it is advisable to estimate the expected magnitude of K_t. The estimate must then be checked at the end of the design process.

14. Complete the required stress analyses at all points where the stress may be high and at changes of cross section to determine the minimum acceptable dimensions for critical areas.

15. Specify suitable, convenient dimensions for all features of the element. Many design decisions are required, such as

    The use of preferred basic sizes as listed in Table A2–1.

    The size of any part that will be installed on or attached to the element being analyzed. Examples of this are shown in Chapter 12 on shaft design where gears, chain sprockets, bearings, and other elements are to be installed on the shafts. But many machine elements have similar needs to accommodate mating elements.

    Elements should not be significantly oversized without good reason in order to achieve an efficient overall design.

    Sometimes the manufacturing process to be used has an effect on the dimensions. For example, a company may have a preferred set of cutting tools for use in producing the elements. Casting, rolling, or molding processes often have limitations on the dimensions of certain features such as the thickness of ribs, radii produced by machining or bending, variation in cross section within different parts of the element, and convenient handling of the element during manufacture.

Consideration should be given to the sizes and shapes that are commercially available in the desired material. This could result in significant cost reductions both in material and in processing.

Sizes should be compatible with standard company practices if practical.

16. After completing all necessary stress analyses and proposing the basic sizes for all features, check all assumptions made earlier in the design to ensure that the element is still safe and reasonably efficient. See Steps 7, 12, and 13.

17. Specify suitable tolerances for all dimensions, considering the performance of the element, its fit with mating elements, the capability of the manufacturing process, and cost. Chapter 13 may be consulted. The use of computer-based tolerance-analysis techniques may be appropriate.

18. Check to determine whether some part of the component may deflect excessively. If that is an issue, complete an analysis of the deflection of the element as designed to this point. Sometimes there are known limits for deflection based on the operation of the machine of which the element being designed is a part. In the absence of such limits, the following guidelines may be applied based on the degree of precision desired:

Deflection of a Beam Due to Bending

General machine part:	0.000 5 to 0.003 in/in of beam length
Moderate precision:	0.000 01 to 0.000 5 in/in
High precision:	0.000 001 to 0.000 01 in/in

Deflection (Rotation) Due to Torsion

General machine part:	0.001° to 0.01°/in of length
Moderate precision:	0.000 02° to 0.000 4°/in
High precision:	0.000 001° to 0.000 02°/in

See also Section 20–2 in Chapter 20 for additional suggestions for efficient design. The results of the deflection analysis may cause you to redesign the component. Typically, when high stiffness and precision are required, deflection, rather than strength, will govern the design.

19. Document the final design with drawings and specifications.

20. Maintain a careful record of the design analyses for future reference. Keep in mind that others may have to consult these records whether or not you are still involved in the project.

**5–11
DESIGN
EXAMPLES**

Example design problems are shown here to give you a feel for the application of the process outlined in Section 5–10. It is not practical to illustrate all possible situations, and you must develop the ability to adapt the design procedure to the specific characteristics of each problem. Also note that there are many possible solutions to any given design problem. The selection of a final solution is the responsibility of you, the designer.

In most design situations, a great deal more information will be available than is given in the problem statements in this book. But, often, you will have to seek out that information. We will make certain assumptions in the examples that allow the design to proceed. In your job, you must ensure that such assumptions are appropriate. The design examples focus on only one or a few of the components of the given systems. In

real situations, you must ensure that each design decision is compatible with the total-
ity of the design.

Design Example 5–1 A large electrical transformer is to be suspended from a roof truss of a building. The total
weight of the transformer is 32 000 lb. Design the means of support.

Solution   Objective   Design the means of supporting the transformer.

Given   The total load is 32 000 lb. The transformer will be suspended below a roof truss inside a
building. The load can be considered to be static. It is assumed that it will be protected from
the weather and that temperatures are not expected to be severely cold or hot in the vicinity of the transformer.

Basic Design Decisions   Two straight, cylindrical rods will be used to support the transformer, connecting the top of
its casing to the bottom chord of the roof truss. The ends of the rod will be threaded to allow them to be secured by nuts or by threading them into tapped holes. This design example will be concerned only with the rods. It is assumed that appropriate attachment points
are available to allow the two rods to share the load equally during service. However, it is
possible that only one rod will carry the entire load at some point during installation. Therefore, each rod will be designed to carry the full 32 000 lb.

We will use steel for the rods, and because neither weight nor physical size is critical
in this application, a plain, medium-carbon steel will be used. We specify AISI 1040 cold-
drawn steel. From Appendix 3, we find that it has a yield strength of 71 ksi and moderately
high ductility as represented by its 12% elongation. The rods should be protected from corrosion by appropriate coatings.

The objective of the design analysis that follows is to determine the size of the rod.

Analysis   The rods are to be subjected to direct normal tensile stress. Assuming that the threads at the
ends of the rods are cut or rolled into the nominal diameter of the rods, the critical place for
stress analysis is in the threaded portion.

Use the direct tensile stress formula, Equation (3–1): $\sigma = F/A$. We will first compute
the design stress and then compute the required cross-sectional area to maintain the stress
in service below that value. Finally, a standard thread will be specified from the data in
Chapter 18 on fasteners.

Case C1 from Section 5–9 applies for computing the design stress because the rod is
made from a ductile steel and it carries a static load. The design stress is

$$\sigma_d = s_y/N$$

We specify a design factor of $N = 3$, because it is typical for general machine design and
because there is some uncertainty about the actual installation procedures that may be used
(see Section 5–7). Then

$$\sigma_d = s_y/N = (71\ 000\ \text{psi})/3 = 23\ 667\ \text{psi}$$

Results   In the basic tensile stress equation, $\sigma = F/A$, we know F, and we will let $\sigma = \sigma_d$. Then the
required cross-sectional area is

$$A = F/\sigma_d = (32\ 000\ \text{lb})/(23\ 667\ \text{lb/in}^2) = 1.35\ \text{in}^2$$

A standard size thread will now be specified from the data in Chapter 18 on fasteners. You should be familiar with such data from earlier courses. Table A2–2(b) lists the tensile stress area for American Standard threads. A $1\frac{1}{2}$–6 UNC thread ($1\frac{1}{2}$-in-diameter rod with 6 threads per in) has a tensile stress area of 1.405 in^2 which should be satisfactory for this application.

Comments   The final design specifies a $1\frac{1}{2}$-in-diameter rod made from AISI 1040 cold-drawn steel with $1\frac{1}{2}$–6 UNC threads machined on each end to allow the attachment of the rods to the transformer and to the truss.

Design Example 5–2   A part of a conveyor system for a production operation is shown in Figure 5–18. Design the pin that connects the horizontal bar to the fixture. The empty fixture weighs 85 lb. A cast iron engine block weighing 225 lb is hung on the fixture to carry it from one process to another, where it is then removed. It is expected that the system will experience many thousands of cycles of loading and unloading of the engine blocks.

Solution   Objective   Design the pin for attaching the fixture to the conveyor system.

Given   The general arrangement is shown in Figure 5–18. The fixture places a shearing load that is alternately 85 lb and 310 lb (85 + 225) on the pin many thousands of times in the expected life of the system.

Basic Design Decisions   It is proposed to make the pin from AISI 1020 cold-drawn steel. Appendix 3 lists $s_y = 51$ ksi and $s_u = 61$ ksi. The steel is ductile with 15% elongation. This material is inexpensive, and it is not necessary to achieve a particularly small size for the pin.

The connection of the fixture to the bar is basically a clevis joint with two tabs at the top of the fixture, one on each side of the bar. There will be a close fit between the tabs and the bar to minimize bending action on the pin. Also, the pin will be a fairly close fit in the holes while still allowing rotation of the fixture relative to the bar.

Analysis   Case H from Section 5–9 applies for completing the design analysis because fluctuating shearing stresses are experienced by the pin. Therefore, we will have to determine relationships for the mean and alternating stresses (τ_m and τ_a) in terms of the applied loads and the cross-sectional area of the bar. Note that the pin is in double shear, so two cross sections resist the applied shearing force. In general, $\tau = F/2A$.

Now we will use the basic forms of Equations (5–1) and (5–2) to compute the values for the mean and alternating forces on the pin:

$$F_m = (F_{ma} + F_{min})/2 = (310 + 85)/2 = 198 \text{ lb}$$
$$F_a = (F_{max} - F_{min})/2 = (310 - 85)/2 = 113 \text{ lb}$$

The stresses will be found from $\tau_m = F_m/2A$ and $\tau_a = F_a/2A$.

The material strength values needed in Equation (5–28) for Case H are

$$s_{su} = 0.75 \, s_u = 0.75(51 \text{ ksi}) = 38.3 \text{ ksi} = 38\,300 \text{ psi}$$
$$s'_{sn} = 0.577 \, s'_n$$

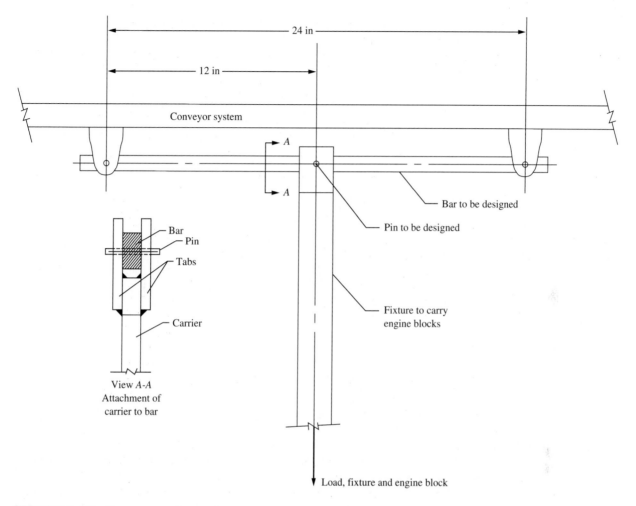

FIGURE 5–18 Conveyor system

We must find the value of s_n' using the method from Section 5–4. We find from Figure 5–8 that $s_n = 21$ ksi for the machined pin having a value of $s_u = 61$ ksi. It is expected that the pin will be fairly small, so we will use $C_s = 1.0$. The material is wrought steel rod, so $C_m = 1.0$. Let's use $C_{st} = 1.0$ to be conservative because there is little information about such factors for direct shearing stress. A high reliability is desired for this application, so let's use $C_R = 0.75$ to produce a reliability of 0.999 (see Table 5–1). Then

$$s_n' = C_R \, (s_n) = (0.75)(21 \text{ ksi}) = 15.75 \text{ ksi} = 15\,750 \text{ psi}$$

Finally,

$$s_{sn}' = 0.577 \, s_n' = 0.577 \, (15\,750 \text{ psi}) = 9088 \text{ psi}$$

We can now apply Equation (5–28) from Case H:

$$\frac{1}{N} = \frac{\tau_m}{s_{su}} + \frac{K_t \tau_a}{s_{sn}'}$$

Because the pins will be of uniform diameter, $K_t = 1.0$.

Substituting $\tau_m = F_m/2A$ and $\tau_a = F_a/2A$ found earlier gives

$$\frac{1}{N} = \frac{F_m}{2As_{su}} + \frac{F_a}{2As'_{sn}}$$

Let's use $N = 4$ because mild shock can be expected.

Note that we now know all factors in this equation except the cross-sectional area of the pin, A. We can solve for the required area:

$$A = \frac{N}{2}\left[\frac{F_m}{s_{su}} + \frac{F_a}{s'_{sn}}\right]$$

Finally, we can compute the minimum allowable pin diameter, D, from $A = \pi D^2/4$ and $D = \sqrt{4A/\pi}$.

Results   The required area is

$$A = \frac{4}{2}\left[\frac{198\ \text{lb}}{38\ 300\ \text{lb/in}^2} + \frac{113\ \text{lb}}{9\ 088\ \text{lb/in}^2}\right] = 0.0352\ \text{in}^2$$

Now the required diameter is

$$D = \sqrt{4A/\pi} = \sqrt{4(0.0352\ \text{in}^2/\pi} = 0.212\ \text{in}$$

Final Design Decisions   The computed value for the minimum required diameter for the pin, 0.212 in, is quite small.
and Comments   Other considerations such as bearing stress and wear at the surfaces that contact the tabs of the fixture and the bar indicate that a larger diameter would be preferred. Let's specify $D = 0.50$ in for the pin at this location. The pin will be of uniform diameter within the area of the bar and the tabs. It should extend beyond the tabs, and it could be secured with cotter pins or retaining rings.

This completes the design of the pin. But the next design example deals with the horizontal bar for this same system. There are pins at the conveyor hangers to support the bar. They would also have to be designed. However, note that each of these pins carries only half the load of the pin in the fixture connection. These pins would experience less relative motion as well, so wear should not be so severe. Thus, let's use pins with $D = 3/8$ in $= 0.375$ in at the ends of the horizontal bar.

Design Example 5–3   A part of a conveyor system for a production operation is shown in Figure 5–18. The complete system will include several hundred hanger assemblies like this one. Design the horizontal bar that extends between two adjacent conveyor hangers and that supports a fixture at its midpoint. The empty fixture weighs 85 lb. A cast iron engine block weighing 225 lb is hung on the fixture to carry it from one process to another, where it is then removed. It is expected that the bar will experience several thousand cycles of loading and unloading of the engine blocks. Design Example 5–2 considered this same system with the objective of specifying the diameter of the pins. The pin at the middle of the horizontal bar where the fixture is hung has been specified to have a diameter of 0.50 in.

Those at each end where the horizontal bar is connected to the conveyor hangers are each 0.375 in.

Solution   Objective   Design the horizontal bar for the conveyor system.

Given   The general arrangement is shown in Figure 5–18. The bar is simply supported at points 24 in apart. A vertical load that is alternately 85 lb and 310 lb (85 + 225) is applied at the middle of the bar through the pin connecting the fixture to the bar. The load will cycle between these two values many thousands of times in the expected life of the bar. The pin at the middle of the bar has a diameter of 0.50 in, while the pins at each end are 0.375 in.

Basic Design Decisions   It is proposed to make the bar from steel in the form of a rectangular bar with the long dimension of its cross section vertical. Cylindrical holes will be machined on the neutral axis of the bar at the support points and at its center to receive cylindrical pins that will attach the bar to the conveyor carriers and to the fixture. Figure 5–19 shows the basic design for the bar.

    The thickness of the bar, *t*, should be fairly large to provide a good bearing surface for the pins and to ensure lateral stability of the bar when subjected to the bending stress. A relatively thin bar would tend to buckle along its top surface where the stress is compressive. As a design decision, we will use a thickness of $t = 0.50$ in. The design analysis will determine the required height of the bar, *h*, assuming that the primary mode of failure is stress due to bending. Other possible modes of failure are discussed in the comments at the end of this example.

    An inexpensive steel is desirable because several hundred bars will be made. We specify AISI 1020 hot-rolled steel having a yield strength of $s_y = 30$ ksi and an ultimate strength of $s_u = 55$ ksi (Appendix 3).

Analysis   Case G from Section 5–9 applies for completing the design analysis because fluctuating normal stress due to bending is experienced by the bar. Equation (5–20) will be used:

$$\frac{1}{N} = \frac{\sigma_m}{s_u} + \frac{K_t \sigma_a}{s_n'}$$

In general, the bending stress in the bar will be computed from the flexure formula:

$$\sigma = M/S$$

where M = bending moment

S = section modulus of the cross section of the bar

    Our approach will be to first determine the values for both the mean and the alternating bending moments experienced by the bar at its middle. Then the yield and endurance strength values for the steel will be found. Furthermore, as shown in Figure A15–3, the stress concentration factor for this case can be taken as $K_t = 1.0$ if the ratio of the hole diameter, *d*, to the height of the bar, *h*, is less than 0.50. We will make that assumption and check it later. Finally, Equation (5–20) includes the design factor, *N*. Based on the application conditions, let's use $N = 4$ as advised in item 4 in Section 5–7 because the actual use pattern for this conveyor system in a factory environment is somewhat uncertain and shock loading is likely.

Bending Moments Figure 5–19 shows the shearing force and bending moment diagrams for the bar when carrying just the fixture and then both the fixture and the engine block. The

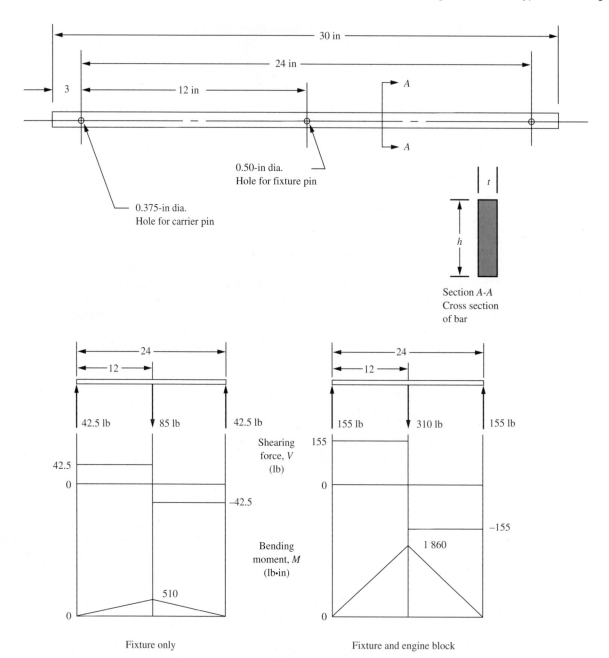

FIGURE 5–19   Basic design of the horizontal bar and the load, shearing force, and bending moment diagrams

maximum bending moment occurs at the middle of the bar where the load is applied. The values are $M_{max} = 1860$ lb·in with the engine block on the fixture and $M_{min} = 510$ lb·in for the fixture alone. Now the values for the mean and alternating bending moments are calculated using modified forms of Equations (5–1) and (5–2):

$$M_m = (M_{max} + M_{min})/2 = (1860 + 510)/2 = 1185 \text{ lb·in}$$
$$M_a = (M_{max} - M_{min})/2 = (1860 - 510)/2 = 675 \text{ lb·in}$$

The stresses will be found from $\sigma_m = M_m/S$ and $\sigma_a = M_a/S$.

Material Strength Values The material strength properties required are the ultimate strength s_u and the estimated actual endurance strength s_n'. We know that the ultimate strength $s_u = 55$ ksi. We now find s_n' using the method outlined in Section 5–4.

Size factor, Cs: From Section 5–4, Equation 5–8 defines an equivalent diameter, D_e, for the rectangular section as,

$$D_e = 0.808\sqrt{ht}$$

We have specified the thickness of the bar to be $t = 0.50$ in. The height is unknown at this time. As an estimate, let's assume $h \approx 2.0$ in. Then,

$$D_e = 0.808\sqrt{ht} = 0.808\sqrt{(2.0)(0.50)} = 0.808 \text{ in}$$

We can now use Figure 5–9 or the equations in Table 5–2 to find $C_s = 0.90$. This value should be checked later after a specific height dimension is proposed.

Material factor, C_m: Use $C_m = 1.0$ for the wrought, hot-rolled steel.

Stress-type factor, C_{st}: Use $C_{st} = 1.0$ for repeated bending stress.

Reliability factor, C_R: A high reliability is desired. Let's use $C_R = 0.75$ to achieve a reliability of 0.999 as indicated in Table 5–1.

The value of $s_n = 20$ ksi is found from Figure 5–8 for hot-rolled steel having an ultimate strength of 55 ksi.

Now, applying Equation (5–4) from Section 5–5, we have

$$s_n' = (C_m)(C_{st})(C_R)(C_s)\, s_n = (1.0)(1.0)(0.75)(0.90)(20 \text{ ksi}) = 13.5 \text{ ksi}$$

Solution for the Required Section Modulus At this point, we have specified all factors in Equation (5–20) except the section modulus of the cross section of the bar that is involved in each expression for stress as shown above. We will now solve the equation for the required value of S.

Recall that we showed earlier that $\sigma_m = M_m/S$ and $\sigma_a = M_a/S$. Then

$$\frac{1}{N} = \frac{\sigma_m}{s_u} + \frac{K_t\sigma_a}{s_n'} = \frac{M_m}{Ss_u} + \frac{K_t M_a}{Ss_n'} = \frac{1}{S}\left[\frac{M_m}{s_u} + \frac{K_t M_a}{s_n'}\right]$$

$$S = N\left[\frac{M_m}{s_u} + \frac{K_t M_a}{s_n'}\right] = 4\left[\frac{1185 \text{ lb}\cdot\text{in}}{55\,000 \text{ lb/in}^2} + \frac{1.0\,(675 \text{ lb}\cdot\text{in})}{13\,500 \text{ lb/in}^2}\right]$$

$$S = 0.286 \text{ in}^3$$

Results   The required section modulus has been found to be $S = 0.286$ in³. We observed earlier that $S = th^2/6$ for a solid rectangular cross section, and we decided to use this form to find an initial estimate for the required height of the section, h. We have specified $t = 0.50$ in. Then the estimated minimum acceptable value for the height h is

$$h = \sqrt{6S/t} = \sqrt{6(0.286 \text{ in}^3)/(0.50 \text{ in})} = 1.85 \text{ in}$$

The table of preferred basic sizes in the decimal-inch system (Table A2–1) recommends $h = 2.00$ in. We should first check the earlier assumption that the ratio $d/h < 0.50$ at the middle of the bar. The actual ratio is

$$d/h = (0.50 \text{ in})/(2.00 \text{ in}) = 0.25 \text{ (okay)}$$

This indicates that our earlier assumption that $K_t = 1.0$ is correct. Also, our assumed value of $C_s = 0.90$ is correct because the actual height, $h = 2.0$ in, is identical to our assumed value.

We will now compute the actual value for the section modulus of the cross section with the hole in it.

$$S = \frac{t(h^3 - d^3)}{6h} = \frac{(0.50 \text{ in})[(2.00 \text{ in})^3 - (0.50 \text{ in})^3]}{6(2.00 \text{ in})} = 0.328 \text{ in}^3$$

This value is larger than the minimum required value of 0.286 in^3. Therefore, the size of the cross section is satisfactory with regard to stress due to bending.

Final Design Decisions and Comments

In summary, the following are the design decisions for the horizontal bar of the conveyor hanger shown in Figure 5–19.

1. **Material:** AISI 1020 hot-rolled steel.

2. **Size:** Rectangular cross section. Thickness $t = 0.50$ in; height $h = 2.00$ in.

3. **Overall design:** Figure 5–19 shows the basic features of the bar.

4. **Other considerations:** Remaining to be specified are the tolerances on the dimensions for the bar and the finishing of its surfaces. The potential for corrosion should be considered and may call for paint, plating, or some other corrosion protection. The size of the cross section can likely be used with the as-received tolerances on thickness and height, but this is somewhat dependent on the design of the fixture that holds the engine block and the conveyor hangers. So the final tolerances will be left open pending later design decisions. The holes in the bar for the pins should be designed to produce a close sliding fit with the pins, and the details of specifying the tolerances on the hole diameters for such a fit are discussed in Chapter 13.

5. **Other possible modes of failure:** The analysis used in this problem assumed that failure would occur due to the bending stresses in the rectangular bar. The dimensions were specified to preclude this from happening. Other possible modes are discussed here:

   a. *Deflection of the bar as an indication of stiffness:* The type of conveyor system described in this problem should not be expected to have extreme rigidity because moderate deflection of members should not impair its operation. However, if the horizontal bar deflects so much that it appears to be rather flexible, it would be deemed unsuitable. This is a subjective judgment. We can use Case (a) in Table A14–2 to compute the deflection.

$$y = FL^3/48EI$$

In this design,

$$F = 310 \text{ lb} = \text{maximum load on the bar}$$
$$L = 24.0 \text{ in} = \text{distance between supports}$$
$$E = 30 \times 10^6 \text{ psi} = \text{modulus of elasticity of steel}$$
$$I = th^3/12 = \text{moment of inertia of the cross section}$$
$$I = (0.50 \text{ in})(2.00 \text{ in})^3/12 = 0.333 \text{ in}^4$$

Then

$$y = \frac{(310 \text{ lb})(24.0 \text{ in})^3}{48(30 \times 10^6 \text{ lb/in}^2)(0.333 \text{ in}^4)} = 0.0089 \text{ in}$$

This value seems satisfactory. In Section 5–10, some guidelines were given for deflection of machine elements. One stated that bending deflections for general machine parts should be limited to the range of 0.000 5 to 0.003 in/in of beam length. For the bar in this design, the ratio of y/L can be compared to this range:

$$y/L = (0.0089 \text{ in})/(24.0 \text{ in}) = 0.0004 \text{ in/in of beam length}$$

Therefore, this deflection is well within the recommended range.

b. *Buckling of the bar:* When a beam with a tall, thin, rectangular cross section is subjected to bending, it would be possible for the shape to distort due to buckling before the bending stresses would cause failure of the material. This is called *elastic instability,* and a complete discussion is beyond the scope of this book. However, Reference 16 shows a method of computing the critical buckling load for this kind of loading. The pertinent geometrical feature is the ratio of the thickness t of the bar to its height h. It can be shown that the bar as designed will not buckle.

c. *Bearing stress on the inside surfaces of the holes in the beam:* Pins transfer loads between the bar and the mating elements in the conveyor system. It is possible that the bearing stress at the pin–hole interface could be large, leading to excessive deformation or wear. Reference 3 in Chapter 3 indicates that the allowable bearing stress for a steel pin in a steel hole is $0.90s_y$.

$$\sigma_{bd} = 0.90s_y = 0.90(30\,000 \text{ psi}) = 27\,000 \text{ psi}$$

The actual bearing stress at the center hole is found using the projected area, $D_p t$.

$$\sigma_b = F/D_p t = (310 \text{ lb})/(0.50 \text{ in})(0.50 \text{ in}) = 1240 \text{ psi}$$

Thus the pin and hole are very safe for bearing.

Design Example 5–4 A bracket is made by welding a rectangular bar to a circular rod as shown in Figure 5–20. Design the bar and the rod to carry a static load of 250 lb.

Solution   Objective   The design process will be divided into two parts:

1. Design the rectangular bar for the bracket.
2. Design the circular rod for the bracket.

Rectangular Bar

Given   The bracket design is shown in Figure 5–20. The rectangular bar carries a load of 250 lb vertically downward at its end. Support is provided by the weld at its left end where the loads are transferred to the circular rod. The bar acts as a cantilever beam, 12 in long. The design task is to specify the material for the bar and the dimensions of its cross section.

FIGURE 5–20
Bracket design

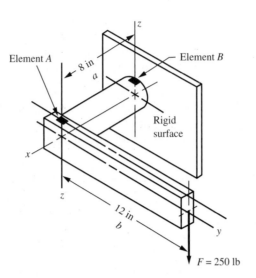

FIGURE 5–21 Free-body diagram of bar

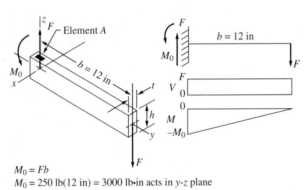

$M_0 = Fb$
$M_0 = 250$ lb$(12$ in$) = 3000$ lb·in acts in y-z plane

Basic Design Decisions

We will use steel for both parts of the bracket because of its relatively high stiffness, the ease of welding, and the wide range of strengths available. Let's specify AISI 1340 annealed steel having $s_y = 63$ ksi and $s_u = 102$ ksi (Appendix 3). The steel is highly ductile, with a 26% elongation.

The objective of the design analysis that follows is to determine the size of the cross section of the rectangular bar. Assuming that the loading and processing conditions are well known, we will use a design factor of $N = 2$ because of the static load.

Analysis and Results

The free-body diagram of the cantilever bar is shown in Figure 5–21, along with the shearing force and bending moment diagrams. This should be a familiar case, leading to the judgment that the maximum tensile stress occurs at the top of the bar near to where it is supported by the circular rod. This point is labeled element A in Figure 5–21. The maximum bending moment there is $M = 3000$ lb · in. The stress at A is

$$\sigma_A = M/S$$

where S = section modulus of the cross section of the bar.

We will first compute the minimum allowable value for S and then determine the dimensions for the cross section.

Case C1 from Section 5–9 applies because of the static loading. We will first compute the design stress from

$$\sigma_d = s_y/N$$
$$\sigma_d = s_y/N = (63\ 000\ \text{psi})/2 = 31\ 500\ \text{psi}$$

Now we must ensure that the expected maximum stress $\sigma_A = M/S$ does not exceed the design stress. We can substitute $\sigma_A = \sigma_d$ and solve for S.

$$S = M/\sigma_d = (3000\ \text{lb}\cdot\text{in})/(31\ 500\ \text{lb/in}^2) = 0.095\ \text{in}^3$$

The relationship for S is

$$S = th^2/6$$

As a design decision, let's specify the approximate proportion for the cross-sectional dimensions to be $h = 3t$. Then

$$S = th^2/6 = t(3t)^2/6 = 9t^3/6 = 1.5t^3$$

The required minimum thickness is then

$$t = \sqrt[3]{S/1.5} = \sqrt[3]{(0.095\ \text{in}^3)/1.5} = 0.399\ \text{in}$$

The nominal height of the cross section should be, approximately,

$$h = 3t = 3(0.399\ \text{in}) = 1.20\ \text{in}$$

Final Design Decisions and Comments

In the fractional-inch system, standard sizes are selected to be $t = 3/8\ \text{in} = 0.375\ \text{in}$ and $h = 1\frac{1}{4}\ \text{in} = 1.25\ \text{in}$ (see Table A2–1). Note that we chose a slightly smaller value for t but a slightly larger value for h. We must check to see that the resulting value for S is satisfactory.

$$S = th^2/6 = (0.375\ \text{in})(1.25\ \text{in})^2/6 = 0.0977\ \text{in}^3$$

This is larger than the required value of 0.095 in³, so the design is satisfactory.

Circular Rod

Given

The bracket design is shown in Figure 5–20. The design task is to specify the material for the rod and the diameter of its cross section.

Basic Design Decisions

Let's specify AISI 1340 annealed steel, the same as that used for the rectangular bar. Its properties are $s_y = 63$ ksi and $s_u = 102$ ksi.

Analysis and Results

Figure 5–22 is the free-body diagram for the rod. The rod is loaded at its left end by the reactions at the end of the rectangular bar, namely, a downward force of 250 lb and a moment of

FIGURE 5–22 Free-body diagram of rod

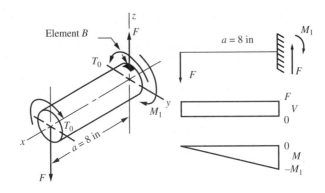

$x = F = 250$ lb

$M_1 = Fa = 250$ lb(8 in) $= 2000$ lb·in (acts in y-z plane)

$T_0 = Fb = 250$ lb(12 in) $= 3000$ lb·in (see figure 5–21. T_0-M_0)

3000 lb · in. The figure shows that the moment acts as a torque on the circular rod, and the 250-lb force causes bending with a maximum bending moment of 2000 lb · in at the right end. Reactions are provided by the weld at its right end where the loads are transferred to the support. The rod then is subjected to a combined stress due to torsion and bending. Element B on the top of the rod is subjected to the maximum combined stress.

The manner of loading on the circular rod is identical to that analyzed earlier in Section 4–6 in Chapter 4. It was shown that when only bending and torsional shear occur, a procedure called the *equivalent torque method* can be used to complete the analysis. First we define the equivalent torque, T_e:

$$T_e = \sqrt{M^2 + T^2} = \sqrt{(2000)^2 + (3000)^2} = 3606 \text{ lb} \cdot \text{in}$$

Then the shear stress in the bar is

$$\tau = T_e/Z_p$$

where Z_p = polar section modulus

For a solid circular rod,

$$Z_p = \pi D^3/16$$

Our approach is to determine the design shear stress and T_e and then solve for Z_p. Case C2 using the maximum shear stress theory of failure can be applied. The design shear stress is

$$\tau_d = 0.50 s_y/N = (0.5)(63\ 000 \text{ psi})/2 = 15\ 750 \text{ psi}$$

We let $\tau = \tau_d$ and solve for Z_p:

$$Z_p = T_e/\tau_d = (3606 \text{ lb} \cdot \text{in})/(15\ 750 \text{ lb/in}^2) = 0.229 \text{ in}^3$$

Now that we know Z_p, we can compute the required diameter from

$$D = \sqrt[3]{16Z_p/\pi} = \sqrt[3]{16(0.229 \text{ in}^3)/\pi} = 1.053 \text{ in}$$

This is the minimum acceptable diameter for the rod.

Final Design Decisions and Comments

The circular rod is to be welded to the side of the rectangular bar, and we have specified the height of the bar to be $1\frac{1}{4}$ in. Let's specify the diameter of the circular rod to be machined to 1.10 in. This will allow welding all around its periphery.

5–12 STATISTICAL APPROACHES TO DESIGN

The design approaches presented in this chapter are somewhat deterministic in the sense that data are taken to be discrete values, and analyses use the data to determine specific results. The method of accounting for uncertainty with regard to the data themselves lies with the selection of an acceptable value for the design factor represented by the final design decision. Obviously this is a subjective judgment. Often decisions that are made to ensure the safety of a design cause many designs to be quite conservative.

Competitive pressures call for ever more efficient, less conservative designs. Throughout this book, recommendations are made to seek more reliable data for loads, material properties, and environmental factors, providing more confidence in the results of design analyses and allowing lower values for the design factor as discussed in Section 5–7. A more robust and reliable product results from testing samples of the actual material to be used in the product; performing extensive measurements of loads to be experienced; investing in more detailed performance testing, experimental stress analysis, and finite element analysis; exerting more careful control of manufacturing processes; and life testing of prototype products in realistic conditions where possible. All of these measures typically come with significant additional costs, and difficult decisions must be made about whether or not to implement them.

In combination with the approaches listed previously, a greater use of statistical methods (also called *stochastic methods*) is emerging to account for the inevitable variability of data by determining the mean values of critical parameters from several sets of data and quantifying the variability using the concepts of distributions and standard deviations. References 13 and 14 provide guidance in these methods. Section 2–4 provided a modest discussion of this approach to account for the variability of materials property data.

Industries such as automotive, aerospace, construction equipment, and machine tools devote considerable resources to acquiring useful data for operating conditions that will aid designers in producing more efficient designs.

Samples of Statistical Terminology and Tools

- Statistical methods analyze data to present useful information about the source of the data.

- Stochastic methods apply probability theories to characterize variability in the data.

- Data sets can be analyzed to determine the mean (average), the range of variation, and the standard deviation.

- Inferences can be made about the nature of the distribution of the data, such as normal or lognormal.

FIGURE 5–23
Illustration of statistical
variation in failure
potential

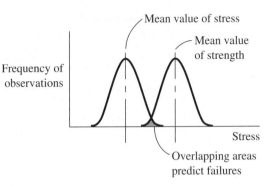

- Linear regression and other means of curve fitting can be employed to represent a set of data by mathematical functions.
- The distributions for applied loads and stresses can be compared with the distribution for the strength of the material to determine to what degree they overlap and the probability that a certain number of failures will occur. See Figure 5–23.
- The reliability of a component or a complete product can be quantified.
- The optimum assignment of tolerances can be made to reasonably ensure satisfactory performance of a product while allowing as broad a range of tolerances as practical.

5–13
FINITE LIFE AND
DAMAGE
ACCUMULATION
METHOD

The fatigue design methods described thus far in this chapter have the goal to design a component for infinite life by using the estimated actual endurance strength as the basis for design and assuming that this value is the *endurance limit*. Applied repeated stress levels below this value will provide infinite life. Furthermore, the analyses were based on the assumption that the loading pattern was uniform over the life of the component. That is, the mean and alternating stresses do not vary over time.

There are, however, many common examples for which a finite life is adequate for the application and where the loading pattern does vary with time. Consider the following.

Finite Life Examples

First, let's discuss the concept of finite life. Refer to the endurance strength curves in Figure 5–7 in Section 5–3. Data are plotted on a stress vs. number of cycles to failure graph (σ *vs. N*) with both axes using logarithmic scales. For the materials that exhibit an endurance limit, it can be seen that the limit occurs for a life of approximately 10^6 cycles. How long would it take to accumulate 1 million cycles of load applications? Here are some examples.

Bicycle brake lever: Assume that the brake is applied every 5.0 minutes while riding 4.0 hours per day every day of each year. It would take more than 57 years to apply the brake 1 million times.

Lawn mower height-adjustment mechanism: Consider a lawn mower used by a commercial lawn maintenance company. Assume that the cutting height for the mower is adjusted to accommodate varying terrain three times per mowing and that the mower is used

40 times per week for all 12 months per year. It would take 160 years to accumulate 1 million cycles of load on the height adjustment mechanism.

Automotive lift in a service station:　Assume that the service technician lifts four automobiles per hour, 10 hours per day, 6 days per week, each week of the year. It would take more than 80 years to accumulate 1 million cycles of load on the lift mechanism.

　　Each of these examples indicates that it may be appropriate to design the load-carrying members of the example systems for something less than infinite life. But many industrial examples do require design for infinite life. Here is an example.

Parts feeding device:　On an automated assembly system, a feeding device inserts 120 parts per minute. If the system operates 16 hours per day, 6 days per week, each week of the year, it would take only 8.7 days to accumulate 1 million cycles of loading. It would see 35.9 million cycles in one year.

　　When it can be justified to design for a finite life less than the number of cycles corresponding to the endurance limit, you will need data similar to that in Figure 5–7 for the actual material to be used in the component. Testing the material yourself is preferred, but it would be a time-consuming and costly exercise to acquire sufficient data to make statistically valid σ-*N* curves. References 1, 3, 6, and 14 may provide suitable data, or an additional literature search may be required. Once reliable data are identified, use the endurance strength at the specified number of cycles as the starting point for computing the estimated actual endurance strength as described in Section 5–5. Then use that value in subsequent analyses as described in Section 5–9.

Varying Stress Amplitude Examples

Here we are looking for examples where the component experiences cyclical loading for a large number of cycles but for which the amplitude of the stress varies over time.

Bicycle brake lever:　Let's reconsider the braking action on a bicycle. Sometimes you need to bring the bike to a stop very quickly from a high speed, requiring a rather high force on the brake lever. Other times you may apply a lighter force to simply slow down a bit to safely negotiate a curve.

Automotive suspension member:　Suspension parts such as a strut, spring, shock absorber, control arm, or fastener pass loads from the wheel to the frame of a car. The magnitude of the load depends on vehicle speed, the condition of the road, and driver action. Roads may be smoothly paved, potholed, or rough surfaced gravel. The vehicle may even be driven off-road where violent peaks of stress will be encountered.

Machine tool drive system:　Consider the lifetime of a milling machine. Its primary function is to cut metal, and it takes a certain amount of torque to drive the cutter depending on the machinability of the material, the depth of cut, and the feed rate. Surely the torque will vary significantly from job to job. During part of its operating time, there may be no cutting action at all as a new part is positioned or as the operation completes one cut and adjusts before making another. At times the cutter will encounter locally harder material, requiring higher torque for a short period of time.

Cranes, power shovels, bulldozers, and other construction equipment:　Here there are obviously varying loads as the equipment is used for numerous tasks, such as hoisting large steel beams or small bracing members, digging through hard clay or soft sandy soil, rapidly grading a hillside or performing the final smoothing of a driveway, or encountering a tree stump or large rock.

How would you determine the loads that these devices would experience over time? One method involves building a prototype and instrumenting critical elements with strain gages, load cells, or accelerometers. Then the system would be "put through its paces" over a wide range of tasks while loads and stresses are recorded as a function of time. Similar vehicles could be monitored to determine the frequency that different kinds of loading would be encountered over its expected life. Combining such data would produce a record from which the total number of cycles of stress at any given level can be estimated. Statistical techniques such as spectrum analysis, Fast Fourier Transform analysis, and time compression analysis produce charts that summarize stress amplitude and frequency data that are useful for fatigue and vibration analysis. See Reference 14 for extended discussion of these techniques.

Damage Accumulation Method

The basic principle of damage accumulation is based on the assumption that any given level of stress applied for one cycle of loading contributes a certain amount of damage to a component. Refer again to Figure 5–7 and observe curve A for a particular heat-treated alloy steel. If this material were subjected to a repeated and reversed stress with a constant amplitude of 120 ksi, its predicted life is 5×10^4 cycles. If the specimen experiences 100 cycles at this stress level, it would exhibit damage amounting to the ratio of $100/(5 \times 10^4)$. A stress amplitude of 100 ksi corresponds to a life of approximately 1.8×10^5. A total of 2000 cycles at this stress level would produce damage of $2000/(1.8 \times 10^5)$. A stress below 82 ksi would not be expected to produce any damage because it is below the endurance limit of the material.

This kind of logic can be used to predict the total life of a component subjected to a sequence of loading levels. Let n_i represent the number of cycles of a specific stress level experienced by a component. Let N_i be the number of cycles to failure for this stress level as indicated by an σ-N curve such as those shown in Figure 5–7. Then the damage contribution from this loading is

$$D_i = n_i/N_i$$

When several stress levels are experienced for different numbers of cycles, the cumulative damage can be represented as,

$$D_c = \sum_{i=l}^{i=k} (n_i/N_i) \tag{5–31}$$

Failure is predicted when $D_c = 1.0$. This process is called the *Miner linear cumulative-damage rule* or simply *Miner's rule* in honor of his work in 1945. An example problem will now demonstrate the application of Miner's rule.

Example Problem 5–5   Determine the cumulative damage experienced by a ground circular rod, 1.50 in diameter, subjected to the combination of the cycles of loading and varying levels of reversed, repeated bending stress shown in Table 5–3. The bar is made from AISI 6150 OQT1100 alloy steel. The σ-N curve for the steel is shown in Figure 5–7, curve A for the standard, polished R. R. Moore type specimen.

TABLE 5–3 Loading pattern for Example Problem 5–5

Stress level (ksi)	Cycles n_i
80	4000
70	6000
65	10 000
60	25 000
55	15 000
45	1500

Solution

Given

AISI 6150 OQT1100 alloy steel rod. $D = 1.50$ in. Ground surface.

Endurance strength data (σ-N) shown in Figure 5–7, curve A.

Loading is reversed, repeated bending. Load history shown in Table 5–3.

Analysis First adjust σ-N data for actual conditions using methods of Section 5–4. Use Miner's rule to estimate portion of life used by loading pattern.

Results

For AISI 6150 OQT1100, $s_u = 162$ ksi (Appendix A4–6)

From Figure 5–8, basic $s_n = 74$ ksi for ground surface

Material factor, $C_m = 1.00$ for wrought steel

Type-of-stress factor, $C_{st} = 1.0$ for reversed, rotating bending stress

Reliability factor, $C_R = 0.81$ (Table 5–1) for $R = 0.99$ (Design decision)

Size factor, $C_s = 0.84$ (Figure 5–9 and Table 5–2 for $D = 1.50$ in)

Estimated actual endurance strength, s_n' – Computed:

$$s_n' = s_n C_m C_{st} C_R C_s = (74 \text{ ksi})(1.0)(1.0)(0.81)(0.84) = 50.3 \text{ ksi}$$

This is the estimate for the endurance limit of the steel. In Figure 5–7, the endurance limit for the standard specimen is 82 ksi. The ratio of the actual to the standard data is, $50.3/82 = 0.61$. We can now adjust the entire σ-N curve by this factor. The result is shown in Figure 5–24.

FIGURE 5–24 σ-N Curve for Example Problem 5–5

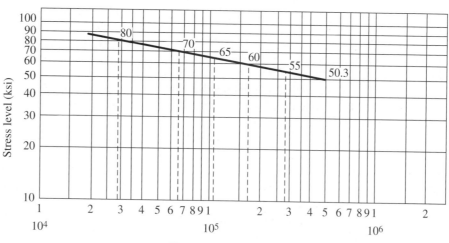

Number of cycles to failure (N)

Now we can read the number of cycles of life, N_i, corresponding to each of the given loading levels from Table 5–3. The combined data for the number of applied load cycles, n_i, and the life cycles, N_i, are now used in Miner's rule, Equation 5–31, to determine the cumulative damage, D_c.

Stress level (ksi)	Cycles n_i	Life cycles N_i	n_i/N_i
80	4000	2.80×10^4	0.143
70	6000	6.60×10^4	0.0909
65	10 000	1.05×10^5	0.0952
60	25 000	1.70×10^5	0.147
55	15 000	2.85×10^5	0.0526
45	1500	∞	0.00
		Total:	0.529

Comment   We can conclude from this number that approximately 53% of the life of the component has been accumulated by the given loading. For these data, the greatest damage occurs from the 60 ksi loading for 25 000 cycles. An almost equal amount of damage is caused by the 80 ksi loading for only 4000 cycles. Note that the cycles of loading at 45 ksi contributed nothing to the damage because they are below the endurance limit of the steel.

REFERENCES

1. Altshuler, Thomas. *S/N Fatigue Life Predictions for Materials Selection and Design (Software)*. Materials Park, OH: ASM International, 2000.

2. American Society of Mechanical Engineers. ANSI Standard B106.1M-1985. Design of Transmission Shafting. New York: American Society of Mechanical Engineers, 1985.

3. ASM International. *ASM Handbook Volume 19, Fatigue and Fracture*. Materials Park, OH: ASM International, 1996.

4. Balandin, D. V., N. N. Bolotnik, and W. D. Pilkey. *Optimal Protection from Impact, Shock, and Vibration*. London, UK: Taylor and Francis, 2001.

5. Bannantine, J. A., J. J. Comer, and J. L. Handrock. *Fundamentals of Metal Fatigue Analysis*. Upper Saddle River, NJ: Prentice Hall, 1997.

6. Boyer, H. E. *Atlas of Fatigue Curves*. Materials Park, OH: ASM International, 1986.

7. Frost, N. E., L. P. Pook, and K. J. Marsh. *Metal Fatigue*. Dover Publications, Mineola, NY: 1999.

8. Fuchs, H. O., R. I. Stephens, and R. R. Stephens. *Metal Fatigue in Engineering*. 2nd ed. New York: John Wiley & Sons, 2000.

9. Harris, C. M., and A. G. Piersol. *Harris' Shock and Vibration Handbook*. 5th ed. New York: McGraw-Hill, 2001.

10. Juvinall, R. C. *Engineering Considerations of Stress, Strain, and Strength*. New York: McGraw-Hill, 1967.

11. Juvinall, R. C., and K. M. Marshek. *Fundamentals of Machine Component Design*. 3rd ed. New York: John Wiley & Sons, 2000.

12. Marin, Joseph. *Mechanical Behavior of Engineering Materials*. Englewood Cliffs, NJ: Prentice Hall, 1962.

13. Shigley, J. E., and C. R. Mischke. *Mechanical Engineering Design*. 6th ed. New York: McGraw-Hill, 2001.

14. Society of Automotive Engineers, *SAE Fatigue Design Handbook*. 3rd ed. Warrendale, PA: SAE International, 1997.

15. Spotts, M. F., and T. E. Shoup. *Design of Machine Elements*. 7th ed. Upper Saddle River, NJ: Prentice Hall, 1998.

16. Young, W. C., and R. G. Budynas. *Roark's Formulas for Stress and Strain*. 7th ed. New York: McGraw-Hill, 2002.

PROBLEMS

Stress Ratio

For each of Problems 1–9, draw a sketch of the variation of stress versus time, and compute the maximum stress, minimum stress, mean stress, alternating stress, and stress ratio, R. For Problems 6–9, analyze the beam at the place where the maximum stress would occur at any time in the cycle.

1. A link in a mechanism is made from a round bar having a diameter of 10.0 mm. It is subjected to a tensile force that varies from 3500 to 500 N in a cyclical fashion as the mechanism runs.

2. A strut in a space frame has a rectangular cross section of 10.0 mm by 30.0 mm. It sees a load that varies from a tensile force of 20.0 kN to a compressive force of 8.0 kN.

3. A link in a packaging machine mechanism has a square cross section of 0.40 in on a side. It is subjected to a load that varies from a tensile force of 860 lb to a compressive force of 120 lb.

4. A circular rod with a diameter of 3/8 in supports part of a storage shelf in a warehouse. As products are loaded and unloaded, the rod is subjected to a tensile load that varies from 1800 to 150 lb.

5. A part of a latch for a car door is made from a circular rod having a diameter of 3.0 mm. With each actuation, it sees a tensile force that varies from 780 to 360 N.

6. A part of the structure for a factory automation system is a beam that spans 30.0 in as shown in Figure P5–6. Loads are applied at two points, each 8.0 in from a support. The left load $F_1 = 1800$ lb remains constantly applied, while the right load $F_2 = 1800$ lb is applied and removed frequently as the machine cycles.

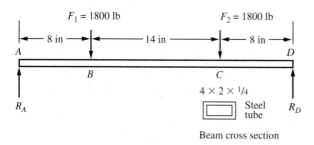

FIGURE P5–6 (Problems 6 and 23)

7. A cantilevered boom is part of an assembly machine and is made from an American Standard steel beam, S4 × 7.7. A tool with a weight of 500 lb moves continuously from the end of the 60-in beam to a point 10 in from the support.

8. A part of a bracket in the seat assembly of a bus is shown in Figure P5–8. The load varies from 1450 to 140 N as passengers enter and exit the bus.

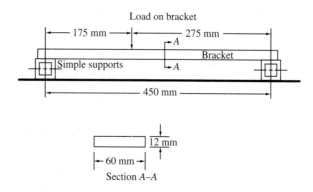

FIGURE P5–8 Seat bracket (Problems 8, 19, and 20)

9. A flat steel strip is used as a spring to maintain a force against part of a cabinet latch in a commercial printer as shown in Figure P5–9. When the cabinet door is open, the spring is deflected by the latch pin by an amount $y_1 = 0.25$ mm. The pin causes the deflection to increase to 0.40 mm when the door is closed.

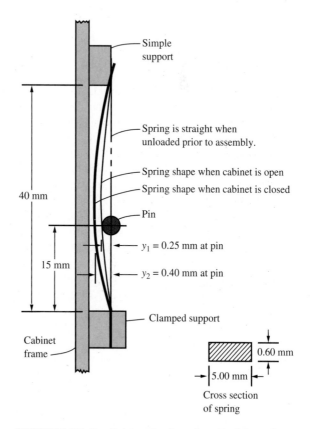

FIGURE P5–9 Cabinet latch spring (Problems 9 and 22)

Endurance Strength

For Problems 10–14, use the method outlined in Section 5–14 to determine the expected actual endurance strength for the material.

10. Compute the estimated actual endurance strength for a 0.75-in-diameter rod made from AISI 1040 cold-drawn steel. It is to be used in the as-rolled condition and subjected to repeated bending stress. A reliability of 99% is desired.

11. Compute the estimated actual endurance strength for AISI 5160 OQT 1300 steel rod with a diameter of 20.0 mm. It is to be machined and subjected to repeated bending stress. A reliability of 99% is desired.

12. Compute the estimated actual endurance strength for AISI 4130 WQT 1300 steel bar with a rectangular cross section of 20.0 mm by 60 mm. It is to be machined and subjected to repeated bending stress. A reliability of 99% is desired.

13. Compute the estimated actual endurance strength for AISI 301 stainless steel rod, 1/2 hard, with a diameter of 0.60 in. It is to be machined and subjected to repeated axial tensile stress. A reliability of 99.9% is desired.

14. Compute the estimated actual endurance strength for a machined rectangular steel bar (ASTM A242) 3.5 in high by 0.375 in thick, subjected to repeated bending stress. A reliability of 99% is desired.

Design and Analysis

15. A link in a mechanism is to be subjected to a tensile force that varies from 3500 to 500 N in a cyclical fashion as the mechanism runs. It has been decided to use AISI 1040 cold-drawn steel. Complete the design of the link, specifying a suitable cross-sectional shape and dimensions.

16. A circular rod is to support part of a storage shelf in a warehouse. As products are loaded and unloaded, the rod is subjected to a tensile load that varies from 1800 to 150 lb. Specify a suitable shape, material, and dimensions for the rod.

17. A strut in a space frame sees a load that varies from a tensile force of 20.0 kN to a compressive force of 8.0 kN. Specify a suitable shape, material, and dimensions for the strut.

18. A part of a latch for a car door is to be made from a circular rod. With each actuation, it sees a tensile force that varies from 780 to 360 N. Small size is important. Complete the design, and specify a suitable shape, material, and dimensions for the rod.

19. A part of a bracket in the seat assembly of a bus is shown in Figure P5–8. The load varies from 1450 to 140 N as passengers enter and exit the bus. The bracket is made from AISI 1020 hot-rolled steel. Determine the resulting design factor.

20. For the bus seat bracket described in Problem 19 and shown in Figure P5–8, propose an alternate design for the bracket, different from that shown in the figure, to achieve a lighter design with a design factor of approximately 4.0.

21. A cantilevered boom is part of an assembly machine. A tool with a weight of 500 lb moves continuously from the end of the 60-in beam to a point 10 in from the support. Specify a suitable design for the boom, giving the material, the cross-sectional shape, and the dimensions.

22. A flat steel strip is used as a spring to maintain a force against part of a cabinet latch in a commercial printer as shown in Figure P5–9. When the cabinet door is open, the spring is deflected by the latch pin by an amount y_1 = 0.25 mm. The pin causes the deflection to increase to 0.40 mm when the door is closed. Specify a suitable material for the spring if it is made to the dimensions shown in the figure.

23. A part of the structure for a factory automation system is a beam that spans 30.0 in as shown in Figure P5–6. Loads are applied at two points, each 8.0 in from a support. The left load F_1 = 1800 lb remains constantly applied, while the right load F_2 = 1800 lb is applied and removed frequently as the machine cycles. If the rectangular tube is made from ASTM A500 Grade B steel, is the proposed design satisfactory? Improve the design to achieve a lighter beam.

24. Figure P5–24 shows a hydraulic cylinder that pushes a heavy tool during the outward stroke, placing a compressive load of 400 lb in the piston rod. During the return stroke, the rod pulls on the tool with a force of 1500 lb. Compute the resulting design factor for the 0.60-in-diameter rod when subjected to this pattern of forces for many cycles. The material is AISI 4130 WQT 1300 steel. If the resulting design factor is much different from 4.0, determine the size of rod that would produce N = 4.0.

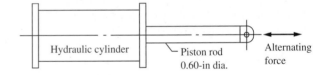

FIGURE P5–24   (Problem 24)

25. The cast iron cylinder shown in Figure P5–25 carries only an axial compressive load of 75 000 lb. (The torque T = 0.) Compute the design factor if it is made from gray cast iron, Grade 40, having a tensile ultimate strength of 40 ksi and a compressive ultimate strength of 140 ksi.

26. Repeat Problem 25, except using a tensile load with a magnitude of 12 000 lb.

27. Repeat Problem 25, except using a load that is a combination of an axial compressive load of 75 000 lb and a torsion of 20 000 lb·in.

28. The shaft shown in Figure P5–28 is supported by bearings at each end, which have bores of 20.0 mm. Design the shaft to carry the given load if it is steady and the shaft is stationary. Make the dimension a as large as possible while keeping the stress safe. Determine the required diameter in the middle portion. The maximum fillet per-

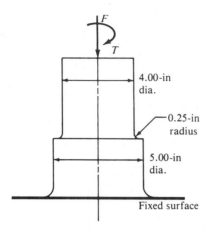

FIGURE P5–25 (Problems 25, 26, and 27)

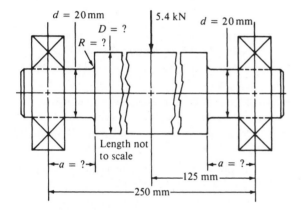

FIGURE P5–28 (Problems 28, 29, and 30)

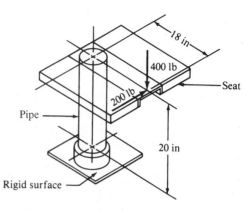

FIGURE P5–31 (Problem 31)

ial tensile load of 1500 lb and a bending load that varies from zero to a maximum of 800 lb at the center of the 48-in length of the bar. Use a design factor of 3.

34. Repeat Problem 33, but add a constant torsional moment of 1200 lb·in to the other loads.

In some of the following problems, you are asked to compute the design factor resulting from the design proposed for the given loading. Unless stated otherwise, assume that the element being analyzed has a machined surface. If the design factor is significantly different from $N = 3$, redesign the component to achieve approximately $N = 3$. (See the figures in Chapter 3.)

35. A tensile member in a machine structure is subjected to a steady load of 4.50 kN. It has a length of 750 mm and is made from a steel tube, AISI 1040 hot-rolled, having an outside diameter of 18 mm and an inside diameter of 12 mm. Compute the resulting design factor.

36. A steady tensile load of 5.00 kN is applied to a square bar, 12 mm on a side and having a length of 1.65 m. Compute the stress in the bar and the resulting design factor if it is made from (a) AISI 1020 hot-rolled steel; (b) AISI 8650 OQT 1000 steel; (c) ductile iron A536-84 (60-40-18); (d) aluminum alloy 6061-T6; (e) titanium alloy Ti-6Al-4V, annealed; (f) rigid PVC plastic; and (g) phenolic plastic.

37. An aluminum rod, made from alloy 6061-T6, is made in the form of a hollow square tube, 2.25 in outside with a wall thickness of 0.125 in. Its length is 16.0 in. It carries an axial compressive force of 12 600 lb. Compute the resulting design factor. Assume that the tube does not buckle.

38. Compute the design factor in the middle portion only of the rod AC in Figure P3–8 if the steady vertical force on the boom is 2500 lb. The rod is rectangular, 1.50 in by 3.50 in, and is made from AISI 1144 cold-drawn steel.

39. Compute the forces in the two angled rods in Figure P3–9 for a steady applied force, $F = 1500$ lb, if the angle θ is 45°. Then design the middle portion of each rod to be circular and made from AISI 1040 hot-rolled steel. Specify a suitable diameter.

missible is 2.0 mm. Use AISI 1137 cold-drawn steel. Use a design factor of 3.

29. Repeat Problem 28, except using a rotating shaft.

30. Repeat Problem 28, except using a shaft that is rotating and transmitting a torque of 150 N·m from the left bearing to the middle of the shaft. Also, there is a profile keyseat at the middle under the load.

31. Figure P5–31 shows a proposed design for a seat support. The vertical member is to be a standard pipe (see Table A16–6). Specify a suitable pipe to resist static loads simultaneously in the vertical and horizontal directions, as shown. The tube has properties similar to those of AISI 1020 hot-rolled steel. Use a design factor of 3.

32. A torsion bar is to have a solid circular cross section. It is to carry a fluctuating torque from 30 to 65 N·m. Use AISI 4140 OQT 1000 for the bar, and determine the required diameter for a design factor of 2. Attachments produce a stress concentration of 2.5 near the ends of the bar.

33. Determine the required size for a square bar to be made from AISI 1213 cold-drawn steel. It carries a constant ax-

40. Repeat Problem 39 if the angle θ is $15°$.

41. Figure 3–26 shows a portion of a circular bar that is subjected to a repeated and reversed force of 7500 N. If the bar is made from AISI 4140 OQT 1000, compute the resulting design factor.

42. Compute the torsional shear stress in a circular shaft having a diameter of 50 mm when subjected to a torque of 800 N·m. If the torque is completely reversed and repeated, compute the resulting design factor. The material is AISI 1040 WQT 1000.

43. If the torque in Problem 42 fluctuates from zero to the maximum of 800 N·m, compute the resulting design factor.

44. Compute the torsional shear stress in a circular shaft 0.40 inch in diameter that is due to a steady torque of 88.0 lb·in. Specify a suitable aluminum alloy for the rod.

45. Compute the required diameter for a solid circular shaft if it is transmitting a maximum of 110 hp at a speed of 560 rpm. The torque varies from zero to the maximum. There are no other significant loads on the shaft. Use AISI 4130 WQT 700.

46. Specify a suitable material for a hollow shaft with an outside diameter of 40 mm and an inside diameter of 30 mm when transmitting 28 kilowatts (kW) of steady power at a speed of 45 radians per second (rad/s).

47. Repeat Problem 46 if the power fluctuates from 15 to 28 kW.

48. Figure P5–48 shows part of a support bar for a heavy machine, suspended on springs to soften applied loads. The tensile load on the bar varies from 12 500 lb to a minimum of 7500 lb. Rapid cycling for many million cycles is expected. The bar is made from AISI 6150 OQT 1300 steel. Compute the design factor for the bar in the vicinity of the hole.

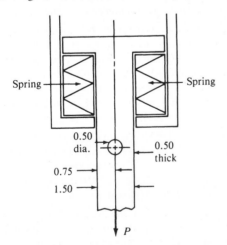

FIGURE P5–48   (Problem 48)

49. Figure P3–61 shows a valve stem from an engine subjected to an axial tensile load applied by the valve spring. The force varies from 0.80 to 1.25 kN. Compute the re-

sulting design factor at the fillet under the shoulder. The valve is made from AISI 8650 OQT 1300 steel.

50. A conveyor fixture shown in Figure P3–62 carries three heavy assemblies (1200 lb each). The fixture is machined from AISI 1144 OQT 900 steel. Compute the resulting design factor in the fixture, considering stress concentrations at the fillets and assuming that the load acts axially. The load will vary from zero to the maximum as the conveyor is loaded and unloaded.

51. For the flat plate in tension in Figure P3–63, compute the minimum resulting design factor, assuming that the holes are sufficiently far apart that their effects do not interact. The plate is machined from stainless steel, UNS S17400 in condition H1150. The load is repeated and varies from 4000 to 6200 lb.

For Problems 52–56, select a suitable material for the member, considering stress concentrations, for the given loading to produce a design factor of $N = 3$.

52. Use Figure P3–64. The load is steady. The material is to be some grade of gray cast iron, ASTM A48.

53. Use Figure P3–65. The load varies from 20.0 to 30.3 kN. The material is to be titanium.

54. Use Figure P3–66. The torque varies from zero to 2200 lb·in. The material is to be steel.

55. Use Figure P3–67. The bending moment is steady. The material is to be ductile iron, ASTM A536.

56. Use Figure P3–68. The bending moment is completely reversed. The material is to be stainless steel.

57. Figure P5–57 shows part of an automatic screwdriver designed to drive several million screws. The maximum torque required to drive a screw is 100 lb·in. Compute the design factor for the proposed design if the part is made from AISI 8740 OQT 1000.

58. The beam in Figure P5–58 carries two steady loads, $P = 750$ lb. Evaluate the design factor that would result if the beam were made from class 40 gray cast iron.

59. A tension link is subjected to a repeated, one-direction load of 3000 lb. Specify a suitable material if the link is to be steel and is to have a diameter of 0.50 in.

60. One member of an automatic transfer device in a factory must withstand a repeated tensile load of 800 lb and must not elongate more than 0.010 inch in its 25.0-in length. Specify a suitable steel material and the dimensions for the rod if it has a square cross section.

61. Figure P5–61 shows two designs for a beam to carry a repeated central load of 600 lb. Which design would have the highest design factor for a given material?

62. Refer to Figure P5–61. By reducing the 8.0-in dimension, redesign the beam in Part (b) of the figure so that it has a design factor equal to or higher than that for the design in Part (a).

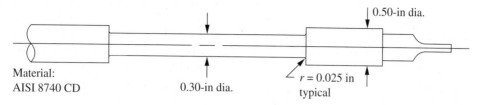

Material:
AISI 8740 CD

0.30-in dia.

0.50-in dia.

$r = 0.025$ in
typical

FIGURE P5–57   Screwdriver for Problem 57

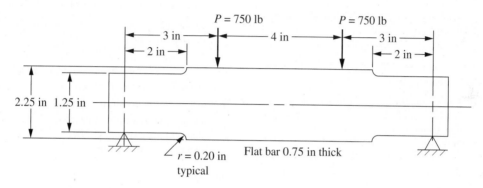

$P = 750$ lb $P = 750$ lb

3 in   4 in   3 in

2 in   2 in

2.25 in   1.25 in

$r = 0.20$ in
typical

Flat bar 0.75 in thick

FIGURE P5–58   Beam for Problem 58

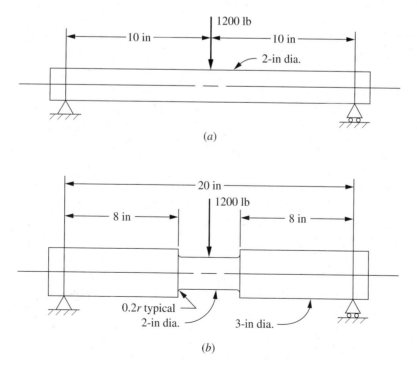

10 in   1200 lb   10 in

2-in dia.

(a)

20 in

1200 lb

8 in   8 in

0.2r typical
2-in dia.   3-in dia.

(b)

FIGURE P5–61   Beam for Problems 61, 62, and 63

63. Refer to Figure P5–61. Redesign the beam in Part (b) of the figure by first increasing the fillet radius to 0.40 in and then by reducing the 8.0-in dimension so that the new design has a design factor equal to or higher than that for the design in Part (a).

64. The part shown in Figure P5–64 is made from AISI 1040 HR steel. It is to be subjected to a repeated, one-direction force of 5000 lb applied through two 0.25-in-diameter pins in the holes at each end. Compute the resulting design factor.

65. For the part described in Problem 64, make at least three improvements in the design that will significantly reduce the stress without increasing the weight. The dimensions marked © are critical and cannot be changed. After the redesign, specify a suitable material to achieve a design factor of at least 3.

66. The link shown in Figure P5–66 is subjected to a tensile force that varies from 3.0 to 24.8 kN. Evaluate the design factor if the link is made from AISI 1040 CD steel.

67. The beam shown in Figure P5–67 carries a repeated, reversed load of 400 N applied at section *C*. Compute the resulting design factor if the beam is made from AISI 1340 OQT 1300.

68. For the beam described in Problem 67, change the steel's tempering temperature to achieve a design factor of at least 3.0.

69. The cantilever shown in Figure P5–69 carries a downward load that varies from 300 to 700 lb. Compute the resulting design factor if the bar is made from AISI 1050 HR steel.

70. For the cantilever described in Problem 69, increase the size of the fillet radius to improve the design factor to at least 3.0 if possible.

71. For the cantilever described in Problem 69, specify a suitable material to achieve a design factor of at least 3.0 without changing the geometry of the beam.

72. Figure P5–72 shows a rotating shaft carrying a steady downward load of 100 lb at *C*. Specify a suitable material.

73. The stepped rod shown in Figure P5–73 is subjected to a direct tensile force that varies from 8500 to 16 000 lb. If the rod is made from AISI 1340 OQT 700 steel, compute the resulting design factor.

74. For the rod described in Problem 73, complete a redesign that will achieve a design factor of at least 3.0. The two diameters cannot be changed.

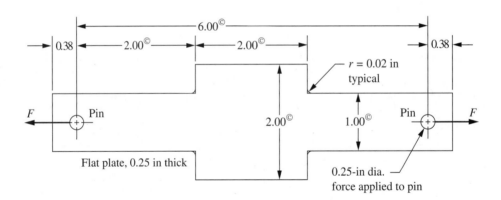

FIGURE P5–64   Beam for Problems 64 and 65

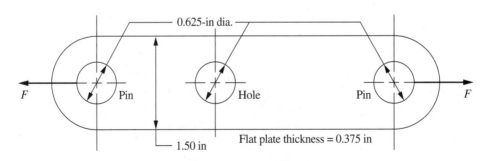

FIGURE P5–66   Link for Problem 66

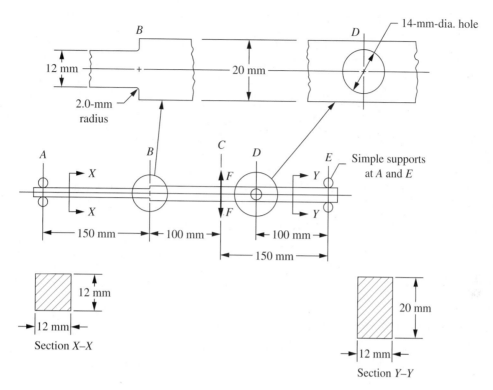

FIGURE P5–67   Beam for Problems 67 and 68

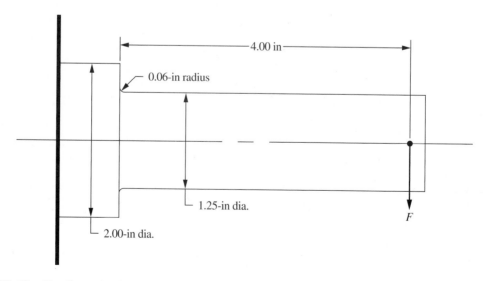

FIGURE P5–69   Cantilever for Problems 69, 70, and 71

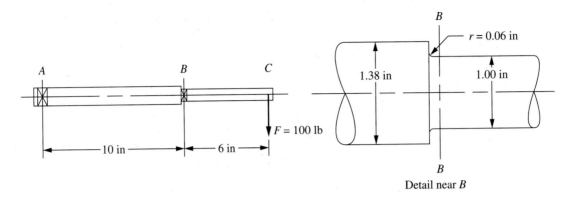

FIGURE P5–72   Shaft for Problem 72

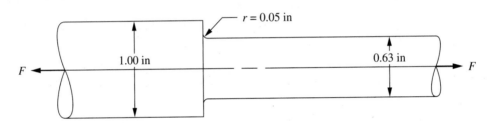

FIGURE P5–73   Rod for Problems 73 and 74

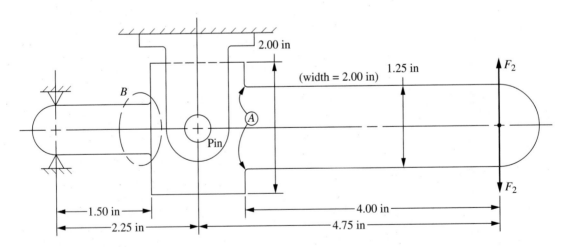

FIGURE P5–75   Beam for Problems 75 and 76

75. The beam shown in Figure P5–75 carries a repeated, re-versed load of 800 lb alternately applied upward and downward. If the beam is made from AISI 1144 OQT 1100, specify the smallest acceptable fillet radius at A to ensure a design factor of 3.0.

76. For the beam described in Problem 75, design the section at B to achieve a minimum design factor of 3.0. Specify the shape, dimensions, and fillet radius where the smaller part joins the 2.00-by-2.00-in section.

Design Problems

For each of the following problems, complete the requested design to achieve a minimum design factor of 3.0. Specify the shape, the dimensions, and the material for the part to be designed. Work toward an efficient design that will have a low weight.

77. The link shown in Figure P5–77 carries a load of 3000 N that is applied and released many times. The link is machined from a square bar, 12.0 mm on a side, from AISI 1144 OQT 1100 steel. The ends must remain 12.0 mm square to facilitate the connection to mating parts. It is desired to reduce the size of the middle part of the link to reduce weight. Complete the design.

78. Complete the design of the beam shown in Figure P5–78 to carry a large hydraulic motor. The beam is attached to the two side rails of the frame of a truck. Because of the vertical accelerations experienced by the truck, the load on the beam varies from 1200 lb upward to 5000 lb downward. One-half of the load is applied to the beam by each foot of the motor.

79. A tensile member in a truss frame is subjected to a load that varies from 0 to 6500 lb as a traveling crane moves across the frame. Design the tensile member.

80. A hanger for a conveyor system extends outward from two supports as shown in Figure P5–80. The load at the right end varies from 600 to 3800 lb. Design the hanger.

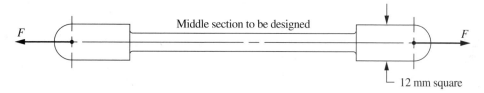

FIGURE P5–77 Link for Problem 77

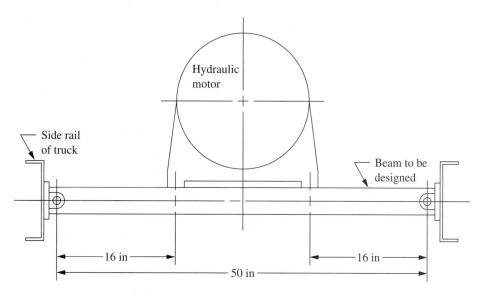

FIGURE P5–78 Beam for Problem 78

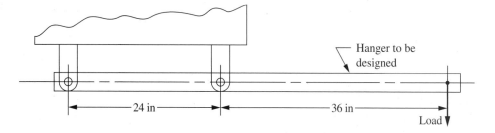

FIGURE P5–80 Hanger for the conveyor system for Problem 80

81. Figure P5–81 shows a yoke suspended beneath a crane beam by two rods. Design the yoke if the loads are applied and released many times.

82. For the system shown in Figure P5–81, design the two vertical rods if the loads are applied and released many times.

83. Design the connections between the rods and the yoke and the crane beam shown in Figure P5–81.

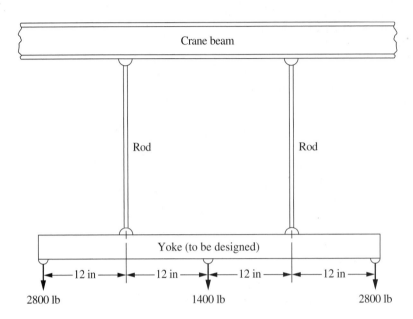

FIGURE P5–81   Yoke and rods for Problems 81, 82, and 83

6

Columns

The Big Picture

You Are the Designer

6–1 Objectives of This Chapter

6–2 Properties of the Cross Section of a Column

6–3 End Fixity and Effective Length

6–4 Slenderness Ratio

6–5 Transition Slenderness Ratio

6–6 Long Column Analysis: The Euler Formula

6–7 Short Column Analysis: The J. B. Johnson Formula

6–8 Column Analysis Spreadsheet

6–9 Efficient Shapes for Column Cross Sections

6–10 The Design of Columns

6–11 Crooked Columns

6–12 Eccentrically Loaded Columns

<table>
<tr><td>

The Big Picture

</td><td>

Columns

Discussion Map

</td></tr>
</table>

☐ A column is a long, slender member that carries an axial compressive load and that fails due to buckling rather than due to failure of the material of the column.

Discover

Find at least 10 examples of columns. Describe them and how they are loaded, and discuss them with your colleagues.

Try to find at least one column that you can load conveniently by hand, and observe the buckling phenomenon.

Discuss with your colleagues the variables that seem to affect how a column fails and how much load it can carry before failing.

This chapter will help you acquire some of the analytical tools necessary to design and analyze columns.

A *column* is a structural member that carries an axial compressive load and that tends to fail by elastic instability, or buckling, rather than by crushing the material. *Elastic instability* is the condition of failure in which the shape of the column is insufficiently rigid to hold it straight under load. At the point of buckling, a radical deflection of the axis of the column occurs suddenly. Then, if the load is not reduced, the column will collapse. Obviously this kind of catastrophic failure must be avoided in structures and machine elements.

Columns are ideally straight and relatively long and slender. If a compression member is so short that it does not tend to buckle, failure analysis must use the methods presented in Chapter 5. This chapter presents several methods of analyzing and designing columns to ensure safety under a variety of loading conditions.

Take a few minutes to visualize examples of column buckling. Find any object that appears to be long and slender, for example, a meter stick, a plastic ruler, a long wooden dowel with a small diameter, a drinking straw, or a thin metal or plastic rod. Carefully apply a downward load on your column while resting the bottom on a desk or the floor. Try to make sure that it does not slide. Gradually increase the load, and observe the behavior of the column until it begins to bend noticeably in the middle. Then hold that level of load. Don't increase it much beyond that level, or the column will likely break!

Now release the load; the column should return to its original shape. The material should not have broken or yielded. But wouldn't you have considered the column to have failed at the point of buckling? Wouldn't it be important to keep the applied load well below the load that caused the buckling to be initiated?

Now look around you. Think of things you are familiar with, or take time to go out and find other examples of columns. Remember, look for relatively long, slender, load-carrying members subjected to compressive loads. Consider parts of furniture, buildings, cars, trucks, toys, play structures, industrial machinery, and construction machinery. Try to find at least 10 examples. Describe their appearance: the material from which they are made, the way they are supported, and the way they are loaded. Do this activity in the classroom or with colleagues; bring the descriptions to class next session for discussion.

Notice that you were asked to find *relatively long*, *slender*, load-carrying members. How will you know when a member is long and slender? At this point, you should just use your judgment. If the column is available and you are strong enough to load it to buckling, go ahead and try it. Later in this chapter, we will quantify what the terms *long* and *slender* mean.

If the columns you have seen buckle did not actually collapse, what property of the material is highly related to the phenomenon of buckling failure? Remember that the failure was described as *elastic instability*. Then it should seem that the *modulus of elasticity* of the material is a key property, and it is. Review the definition of this property from Chapter 1, and look up representative values in the tables of material properties in Appendices 3–13.

Also note that we specified that the columns are to be initially straight and that loads are to be applied axially. What if these conditions are not met? What if the column is a little crooked before loading? Do you think that it would carry as much compressive loading as one that was straight? Why or why not? What if the column is loaded *eccentrically*, that is, the load is directed off-center, away from the centroidal axis of the column? How will that affect the load-carrying ability? How does the manner of supporting the ends of the column affect its load-carrying ability? What standards exist that guide designers when dealing with columns?

These and other questions will be addressed in this chapter. Any time you are involved in a design in which a compressive load is applied, you should think about analyzing it as a column. The following **You Are the Designer** situation is a good example of such a machine design problem.

You Are the Designer

You are a member of a team that is designing a commercial compactor to reduce the volume of cardboard and paper waste so that the waste can be transported easily to a processing plant. Figure 6–1 is a sketch of the compaction ram that is driven by a hydraulic cylinder under several

thousand pounds of force. The connecting rod between the hydraulic cylinder and the ram must be designed as a column because it is a relatively long, slender compression member. What shape should the cross section of the connecting rod be? From what material should it be made? How is it to be connected to the ram and to the hydraulic cylinder? What are the final dimensions of the rod to be? You, the designer, must specify all of these factors.

FIGURE 6–1 Waste paper compactor

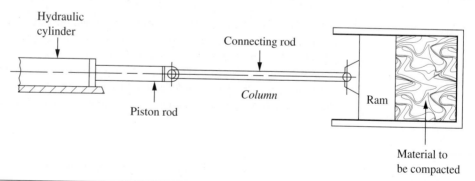

6–1
OBJECTIVES OF
THIS CHAPTER

After completing this chapter, you will be able to:

1. Recognize that any relatively long, slender compression member must be analyzed as a column to prevent buckling.

2. Specify efficient shapes for the cross section of columns.

3. Compute the *radius of gyration* of a column cross section.

4. Specify a suitable value for the *end-fixity factor*, *K*, and determine the *effective length* of a column.

5. Compute the *slenderness ratio* for columns.

6. Select the proper method of analysis or design for a column based on the manner of loading, the type of support, and the magnitude of the slenderness ratio.

7. Determine whether a column is *long* or *short* based on the value of the slenderness ratio in comparison with the *column constant*.

8. Use the *Euler formula* for the analysis and design of long columns.

9. Use the *J. B. Johnson formula* for the analysis and design of short columns.

10. Analyze crooked columns to determine the allowable load.

11. Analyze columns in which the load is applied with a modest amount of eccentricity to determine the maximum predicted stress and the maximum amount of deflection of the centerline of such columns under load.

**6–2
PROPERTIES OF
THE CROSS
SECTION OF A
COLUMN**

The tendency for a column to buckle is dependent on the shape and the dimensions of its cross section, along with its length and the manner of attachment to adjacent members or supports. Cross-sectional properties that are important are as follows:

1. The cross sectional area, *A*.

2. The moment of inertia of the cross section, *I*, with respect to the axis about which the value of *I* is minimum.

3. The least value of the radius of gyration of the cross section, *r*.

The radius of gyration is computed from

Radius of Gyration

$$r = \sqrt{I/A} \qquad\qquad (6\text{–}1)$$

A column tends to buckle about the axis for which the radius of gyration and the moment of inertia are minimum. Figure 6–2 shows a sketch of a column that has a rectangular cross section. The expected buckling axis is *Y-Y* because both *I* and *r* are much smaller for that axis than for the *X-X* axis. You can demonstrate this phenomenon by loading a common ruler or meter stick with an axial load of sufficient magnitude to cause buckling. See Appendix 1 for formulas for *I* and *r* for common shapes. See Appendix 16 for structural shapes.

**6–3
END FIXITY AND
EFFECTIVE
LENGTH**

The term *end fixity* refers to the manner in which the ends of a column are supported. The most important variable is the amount of restraint offered at the ends of the column to the tendency for rotation. Three forms of end restraint are *pinned*, *fixed*, and *free*.

A *pinned-end* column is guided so that the end cannot sway from side to side, but it offers no resistance to rotation of the end. The best approximation of the pinned end would be a frictionless ball-and-socket joint. A cylindrical pin joint offers little resistance about one axis, but it may restrain the axis perpendicular to the pin axis.

A *fixed end* is one that is held against rotation at the support. An example is a cylindrical column inserted into a tight-fitting sleeve that itself is rigidly supported. The sleeve

FIGURE 6–2
Buckling of a thin,
rectangular column.
(a) General appearance
of the buckled column.
(b) Radius of gyration
for *Y-Y* axis. (c) Radius
of gyration for *X-X* axis

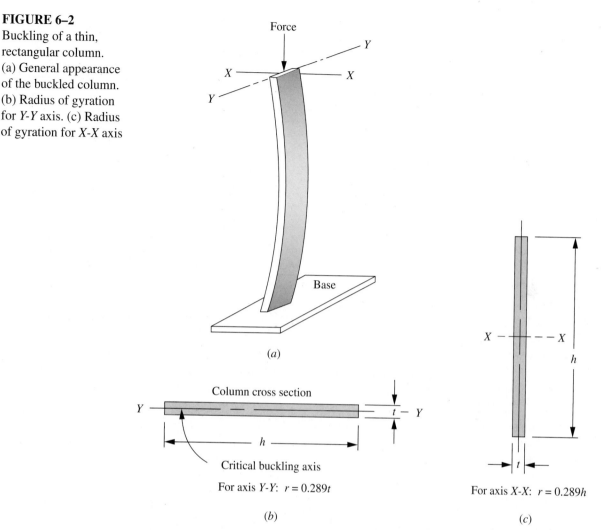

prohibits any tendency for the fixed end of the column to rotate. A column end securely welded to a rigid base plate is also a good approximation of a fixed-end column.

The *free end* can be illustrated by the example of a flagpole. The top end of a flagpole is unrestrained and unguided, the worst case for column loading.

The manner of support of both ends of the column affects the *effective length* of the column, defined as

 **Effective Length**

$$L_e = KL$$

(6–2)

where L = actual length of the column between supports
K = constant dependent on the end fixity, as illustrated in Figure 6–3

The first values given for K are theoretical values based on the shape of the deflected column. The second values take into account the expected fixity of the column ends in real, practical structures. It is particularly difficult to achieve a true fixed-end column because of the lack of complete rigidity of the support or the means of attachment. Therefore, the higher value of K is recommended.

FIGURE 6–3 Values of K for effective length, $L_e = KL$, for different end connections

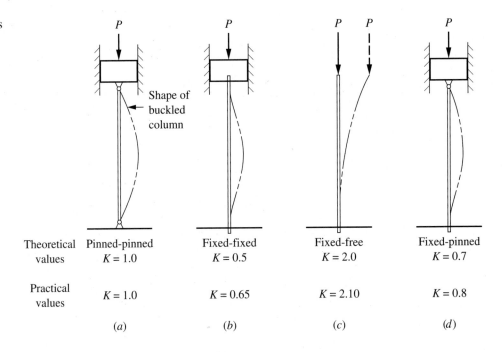

	Pinned-pinned	Fixed-fixed	Fixed-free	Fixed-pinned
Theoretical values	$K = 1.0$	$K = 0.5$	$K = 2.0$	$K = 0.7$
Practical values	$K = 1.0$	$K = 0.65$	$K = 2.10$	$K = 0.8$
	(*a*)	(*b*)	(*c*)	(*d*)

6–4 SLENDERNESS RATIO

⇨ Slenderness Ratio

The *slenderness ratio* is the ratio of the effective length of the column to its least radius of gyration. That is,

$$\text{Slenderness ratio} = L_e/r_{min} = KL/r_{min} \qquad (6\text{–}3)$$

We will use the slenderness ratio to aid in the selection of the method of performing the analysis of straight, centrally loaded columns.

6–5 TRANSITION SLENDERNESS RATIO

In the following sections, two methods for analyzing straight, centrally loaded columns are presented: (1) the Euler formula for long, slender columns and (2) the J. B. Johnson formula for short columns.

The choice of which method to use depends on the value of the actual slenderness ratio for the column being analyzed in relation to the *transition slenderness ratio*, or *column constant*, C_c, defined as

⇨ Column Constant

$$C_c = \sqrt{\frac{2\pi^2 E}{s_y}} \qquad (6\text{–}4)$$

where E = modulus of elasticity of the material of the column
   s_y = yield strength of the material

The use of the column constant is illustrated in the following procedure for analyzing straight, centrally loaded columns.

Procedure for Analyzing Straight, Centrally Loaded Columns

1. For the given column, compute its actual slenderness ratio.
2. Compute the value of C_c.

3. Compare C_c with KL/r. Because C_c represents the value of the slenderness ratio that separates a long column from a short one, the result of the comparison indicates which type of analysis should be used.

4. If the actual KL/r is greater than C_c, the column is *long*. Use Euler's equation, as described in Section 6–6.

5. If KL/r is less than C_c, the column is *short*. Use the J. B. Johnson formula, described in Section 6–7.

Figure 6–4 is a logical flowchart for this procedure.

The value of the column constant, or transition slenderness ratio, is dependent on the material properties of modulus of elasticity and yield strength. For any given class of material, for example, steel, the modulus of elasticity is nearly constant. Thus, the value of C_c varies inversely as the square root of the yield strength. Figures 6–5 and 6–6 show the resulting values for steel and aluminum, respectively, for the range of yield strengths expected for each material. The figures show that the value of C_c decreases as the yield strength increases. The importance of this observation is discussed in the following section.

6–6
LONG COLUMN ANALYSIS: THE EULER FORMULA

Analysis of a long column employs the Euler formula (see Reference 3):

Euler Formula for Long Columns

$$P_{cr} = \frac{\pi^2 EA}{(KL/r)^2} \tag{6–5}$$

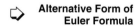

The equation gives the critical load, P_{cr}, at which the column would begin to buckle.

An alternative form of the Euler formula is often desirable. Note that, from Equation (6–5),

$$P_{cr} = \frac{\pi^2 EA}{(KL/r)^2} = \frac{\pi^2 EA}{(KL)^2/r^2} = \frac{\pi^2 EA\, r^2}{(KL)^2}$$

But, from the definition of the radius of gyration, r,

$$r = \sqrt{I/A}$$
$$r^2 = I/A$$

Then

Alternative Form of Euler Formula

$$P_{cr} = \frac{\pi^2 EA\, I}{(KL)^2 A} = \frac{\pi^2 EI}{(KL)^2} \tag{6–6}$$

This form of the Euler equation aids in a design problem in which the objective is to specify a size and a shape of a column cross section to carry a certain load. The moment of inertia for the required cross section can be easily determined from Equation (6–6).

Notice that the buckling load is dependent only on the geometry (length and cross section) of the column and the stiffness of the material represented by the modulus of elasticity. The strength of the material is not involved at all. For these reasons, it is often of no benefit to specify a high-strength material in a long column application. A lower-strength material having the same stiffness, E, would perform as well.

FIGURE 6–4
Analysis of a straight,
centrally loaded
column

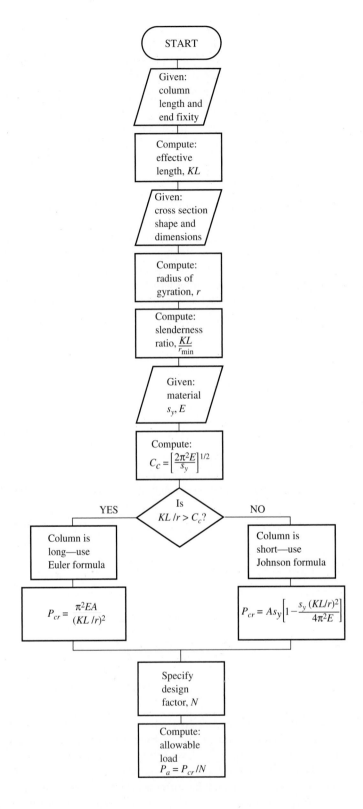

FIGURE 6–5
Transition slenderness ratio C_c vs. yield strength for steel

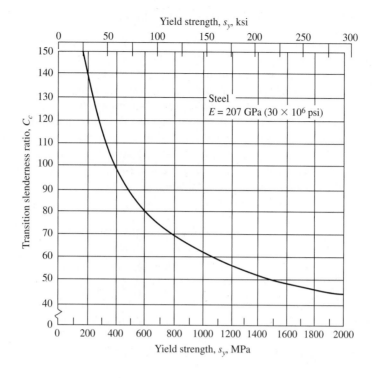

FIGURE 6–6
Transition slenderness ratio C_c vs. yield strength for aluminum

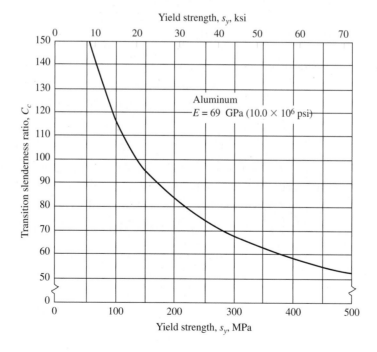

Design Factor and Allowable Load

Because failure is predicted to occur at a limiting load, rather than a stress, the concept of a design factor is applied differently than it is for most other load-carrying members. Rather than applying the design factor to the yield strength or the ultimate strength of the material, we apply it to the critical load, from Equation (6–5) or (6–6). For typical machine design applications, a design factor of 3 is used. For stationary columns with well-known loads and end fixity, a lower factor can be used, such as 2.0. A factor of 1.92 is used in some construction applications. Conversely, for very long columns, where there is some uncertainty about the loads or the end fixity, or where special dangers are presented, larger factors are advised. (See References 1 and 2.)

In summary, the objective of column analysis and design is to ensure that the load applied to a column is safe, well below the critical buckling load. The following definitions of terms must be understood:

$$P_{cr} = \text{critical buckling load}$$
$$P_a = \text{allowable load}$$
$$P = \text{actual applied load}$$
$$N = \text{design factor}$$

Then

➪ **Allowable Load**

$$P_a = P_{cr}/N$$

The actual applied load, P, must be less than P_a.

Example Problem 6–1   A column has a solid circular cross section, 1.25 inches in diameter; it has a length of 4.50 ft and is pinned at both ends. If it is made from AISI 1020 cold-drawn steel, what would be a safe column loading?

Solution   Objective   Specify a safe loading for the column.

Given   Solid circular cross section: diameter = d = 1.25 in; length = L = 4.50 ft.

Both ends of the column are pinned.

Material: AISI 1020 cold-drawn steel.

Analysis   Use the procedure in Figure 6–4.

Results   **Step 1.** For the pinned-end column, the end-fixity factor is $K = 1.0$. The effective length equals the actual length; $KL = 4.50$ ft = 54.0 in.

Step 2. From Appendix 1, for a solid round section,

$$r = D/4 = 1.25/4 = 0.3125 \text{ in}$$

Step 3. Compute the slenderness ratio:

$$\frac{KL}{r} = \frac{1.0(54)}{0.3125} = 173$$

Step 4. Compute the column constant from Equation (6–4). For AISI 1020 cold-drawn steel, the yield strength is 51 000 psi, and the modulus of elasticity is 30×10^6 psi. Then

$$C_c = \sqrt{\frac{2\pi^2 E}{s_y}} = \sqrt{\frac{2\pi^2 (30 \times 10^6)}{51\ 000}} = 108$$

Step 5. Because KL/r is greater than C_c, the column is long, and Euler's formula should be used. The area is

$$A = \frac{\pi D^2}{4} = \frac{\pi (1.25)^2}{4} = 1.23 \text{ in}^2$$

Then the critical load is

$$P_{cr} = \frac{\pi^2 EA}{(KL/r)^2} = \frac{\pi^2 (30 \times 10^6)(1.23)}{(173)^2} = 12\ 200 \text{ lb}$$

At this load, the column should just begin to buckle. A safe load would be a reduced value, found by applying the design factor to the critical load. Let's use $N = 3$ to compute the *allowable load*, $P_a = P_{cr}/N$:

$$P_a = (12\ 200)/3 = 4067 \text{ lb}$$

Comment   The safe load on the column is 4067 lb.

6–7
SHORT COLUMN ANALYSIS: THE J. B. JOHNSON FORMULA

When the actual slenderness ratio for a column, KL/r, is less than the transition value, C_c, then the column is short, and the J. B. Johnson formula should be used. Use of the Euler formula in this range would predict a critical load greater than it really is.

The J. B. Johnson formula is written as follows:

⇨ **J. B. Johnson Formula for Short Columns**

$$P_{cr} = As_y \left[1 - \frac{s_y (KL/r)^2}{4\pi^2 E} \right] \tag{6–7}$$

Figure 6–7 shows a plot of the results of this equation as a function of the slenderness ratio, KL/r. Notice that it becomes tangent to the result of the Euler formula at the transition

FIGURE 6–7
Johnson formula curves

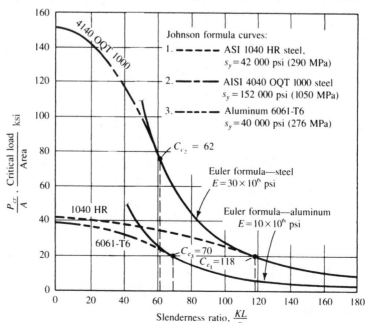

slenderness ratio, the limit of its application. Also, at very low values for the slenderness ratio, the second term of the equation approaches zero, and the critical load approaches the yield load. Curves for three different materials are included in the figure to illustrate the effect of E and s_y on the critical load and the transition slenderness ratio.

The critical load for a short column is affected by the strength of the material in addition to its stiffness, E. As shown in the preceding section, strength is not a factor for a long column when the Euler formula is used.

Example Problem 6–2

Determine the critical load on a steel column having a rectangular cross section, 12 mm by 18 mm, and a length of 280 mm. It is proposed to use AISI 1040 hot-rolled steel. The lower end of the column is inserted into a close-fitting socket and is welded securely. The upper end is pinned (see Figure 6–8).

Solution   Objective

Compute the critical load for the column.

Given

Solid rectangular cross section: $B = 12$ mm; $H = 18$ mm; $L = 280$ mm.

The bottom of column is fixed; the top is pinned (see Figure 6–8).

Material: AISI 1040 hot-rolled steel.

Analysis

Use the procedure in Figure 6–4.

Results

Step 1. Compute the slenderness ratio. The radius of gyration must be computed about the axis that gives the least value. This is the Y-Y axis, for which

$$r = \frac{B}{\sqrt{12}} = \frac{12 \text{ mm}}{\sqrt{12}} = 3.46 \text{ mm}$$

The column has a fixed-pinned end fixity for which $K = 0.8$. Then

$$KL/r = [(0.8)(280)]/3.46 = 64.7$$

FIGURE 6–8 Steel column

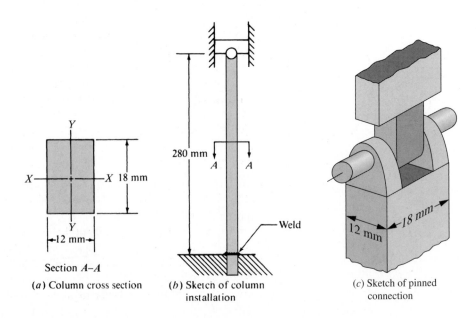

Section A–A

(a) Column cross section

(b) Sketch of column installation

(c) Sketch of pinned connection

Step 2. Compute the transition slenderness ratio. For the AISI 1040 hot-rolled steel, $E = 207$ GPa and $s_y = 290$ MPa. Then, from Equation (6–4),

$$C_c = \sqrt{\frac{2\pi^2 \left(207 \times 10^9 \text{ Pa}\right)}{290 \times 10^6 \text{ Pa}}} = 119$$

Step 3. Then $KL/r < C_c$; thus the column is short. Use the J. B. Johnson formula to compute the critical load:

$$P_{cr} = As_y\left[1 - \frac{s_y(KL/r)^2}{4\pi^2 E}\right]$$

$$P_{cr} = (216 \text{ mm}^2)(290 \text{ N/mm}^2)\left[1 - \frac{(290 \times 10^6 \text{ Pa})(64.7)^2}{4\pi^2(207 \times 10^9 \text{ Pa})}\right] \qquad \textbf{(6–7)}$$

$$P_{cr} = 53.3 \times 10^3 \text{ N} = 53.3 \text{ kN}$$

Comments   This is the critical buckling load. We would have to apply a design factor to determine the allowable load. Specifying $N = 3$ results in $P_a = 17.8$ kN.

**6–8
COLUMN
ANALYSIS
SPREADSHEET**

Completing the process described in Figure 6–4 using a calculator, pencil, and paper is tedious. A spreadsheet automates the calculations after you have entered the pertinent data for the particular column to be analyzed. Figure 6–9 shows the output of a spreadsheet used to solve Example Problem 6–1. The layout of the spreadsheet could be done in many ways, and you are encouraged to develop your own style. The following comments describe the features of the given spreadsheet:

1. At the top of the sheet, instructions to the user are given for entering data and for units. This sheet is for U.S. Customary units only. A different sheet would be used if SI metric data were to be used. (See Figure 6–10, which gives the solution for Example Problem 6–2.)

2. On the left side of the sheet are listed the various data that must be provided by the user to run the calculations. On the right are listed the output values. Formulas for computing L_e, C_c, KL/r, and allowable load are written directly into the cell where the computed values show. The output data for the message "Column is: *long*" and the critical buckling load are produced by *functions* set up within *macros* written in Visual Basic and placed on a separate sheet of the spreadsheet. Figure 6–11 shows the two macros used. The first (*LorS*) carries out the decision process to test whether the column is long or short as indicated by comparison of its slenderness ratio with the column constant. The second (*Pcr*) computes the critical buckling load using either the Euler formula or the J. B. Johnson formula, depending on the result of the *LorS* macro. These functions are called by statements in the cells where "long" and the computed value of the critical buckling load (12 197 lb) are located.

3. Having such a spreadsheet can enable you to analyze several design options quickly. For example, the given problem statement indicated that the ends were pinned, resulting in an end-fixity value of $K = 1$. What would happen if both ends were fixed? Simply changing the value of that one cell to $K = 0.65$ would cause the entire sheet to be recalculated, and the revised value of critical buckling load would be available almost instantly. The result is that $P_{cr} = 28\,868$ lb, an increase of 2.37 times the original value. With that kind of improvement, you, the designer, might be inclined to change the design to produce fixed ends.

COLUMN ANALYSIS PROGRAM		Data from:	Example Problem 6–1
Refer to Figure 6–4 for analysis logic.			
Enter data for variables in *italics in shaded boxes.*		Use consistent U.S. Customary units.	
Data to Be Entered:		**Computed Values:**	
Length and End Fixity:			
Column length, $L =$ 54 in End fixity, $K =$ 1	$\rightarrow$	Eq. length, $L_e = KL =$ 54.0 in	
Material Properties:			
Yield strength, $s_y =$ 51,000 psi Modulus of elasticity, $E =$ 3.00E + 07 psi	$\rightarrow$	Column constant, $C_c =$ 107.8	
Cross Section Properties:			
[Note: Enter r or compute $r = \text{sqrt}(I/A)$.] [Always enter Area.] [Enter zero for I or r if not used.]			
Area, $A =$ 1.23 in² Moment of inertia, $I =$ 0 in⁴ **Or** Radius of gyration, $r =$ 0.3125 in	$\rightarrow$	Slender ratio, $KL/r =$ 172.8	
		Column is: **long**	
		Critical buckling load = 12,197 lb	
Design Factor:			
Design factor on load, $N =$ 3	$\rightarrow$	**Allowable load = 4,066 lb**	

FIGURE 6–9 Spreadsheet for column analysis with data from Example Problem 6–1

COLUMN ANALYSIS PROGRAM	Data from:	Example Problem 6–2

Refer to Figure 6–4 for analysis logic.	Use consistent SI metric units.

Enter data for variables in *italics in shaded boxes.*	

Data to Be Entered:	**Computed Values:**

Length and End Fixity:	
Column length, L = 280 mm *End fixity, K = 0.8*	→ Eq. length, L_e = KL = 224.0 mm

Material Properties:	
Yield strength, s_y = 290 MPa *Modulus of elasticity, E = 207 GPa*	→ Column constant, C_c = 118.7

Cross Section Properties:

[Note: Enter *r* or compute $r = \sqrt{I/A}$.]
[Always enter Area.]
[Enter zero for *I* or *r* if not used.]

Area, A = 216 mm² *Moment of inertia, I = 0 mm⁴* ***Or*** *Radius of gyration, r = 3.5 mm*	→ Slender ratio, KL/r = 64.7

	Column is: ***short***

Design Factor:	**Critical buckling load =** ***53.32 kN***
Design factor on load, N = 3	→ **Allowable load =** ***17.77 kN***

FIGURE 6–10 Spreadsheet for column analysis with data from Example Problem 6–2

FIGURE 6–11
Macros used in the
column analysis
spreadsheet

```
' LorS Macro
' Determines if column is long or short.
Function LorS(SR, CC)
    If SR > CC Then
        LorS = "long"
    Else
        LorS = "short"
    End If
End Function


' Critical Load Macro
' Uses Euler formula for long columns
' Uses Johnson formula for short columns
Function Pcr(LorS, SR, E, A, Sy)
Const Pi = 3.1415926
    If LorS = "long" Then
        Pcr = Pi ^ 2 * E * A / SR ^ 2
        ' Euler Equation; Eq. (6-4)
    Else
        Pcr = A * Sy(1 - (Sy * SR ^ 2 / (4 * Pi ^ 2 * E)))
        ' Johnson Equation; Eq. (6-7)
    End If
End Function
```

6–9 EFFICIENT SHAPES FOR COLUMN CROSS SECTIONS

An *efficient shape* is one that provides good performance with a small amount of material. In the case of columns, the shape of the cross section and its dimensions determine the value of the radius of gyration, r. From the definition of the slenderness ratio, KL/r, we can see that as r gets larger, the slenderness ratio gets smaller. In the critical load equations, a smaller slenderness ratio results in a larger critical load, the most desirable situation. Therefore, it is desirable to maximize the radius of gyration to design an efficient column cross section.

Unless end fixity varies with respect to the axes of the cross section, the column will tend to buckle with respect to the axis with the *least* radius of gyration. So a column with equal values for the radius of gyration in any direction is desirable.

Review again the definition of the radius of gyration:

$$r = \sqrt{I/A}$$

This equation indicates that for a given area of material, we should try to maximize the moment of inertia to maximize the radius of gyration. A shape with a high moment of inertia has its area distributed far away from its centroidal axis.

Shapes that have the desirable characteristics described include circular hollow pipes and tubes, square hollow tubing, and fabricated column sections made from structural shapes placed at the outer boundaries of the section. Solid circular sections and solid square sections are also good, although not as efficient as the hollow sections. Figure 6–12(a–d) illustrates some of these shapes. The built-up section in (e) gives a rigid, boxlike section approximating the hollow square tube in larger sizes. In the case of the section in Figure 6–12(f), the angle sections at the corners provide the greatest contribution to the moment of inertia. The lacing bars merely hold the angles in position. The H-column in (g) has an equal depth and width and relatively heavy flanges and web. The moment of inertia with respect to the *y-y* axis is still smaller than for the *x-x* axis, but they are more nearly equal than

FIGURE 6–12
Column cross sections

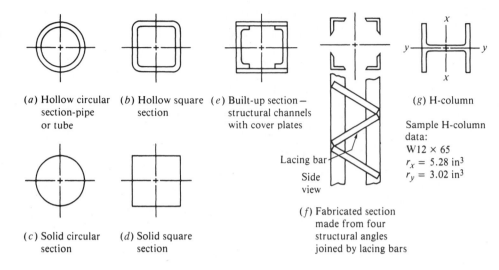

(*a*) Hollow circular
section-pipe
or tube

(*b*) Hollow square
section

(*e*) Built-up section —
structural channels
with cover plates

(*g*) H-column

Sample H-column
data:
W12 × 65
$r_x = 5.28$ in^3
$r_y = 3.02$ in^3

Lacing bar

Side
view

(*f*) Fabricated section
made from four
structural angles
joined by lacing bars

(*c*) Solid circular
section

(*d*) Solid square
section

for most other I-sections designed to be used as beams with bending in only one direction. Thus, this shape would be more desirable for columns.

6–10 THE DESIGN OF COLUMNS

In a design situation, the expected load on the column would be known, along with the length required by the application. The designer would then specify the following:

1. The manner of attaching the ends to the structure that affects the end fixity.
2. The general shape of the column cross section (for example, round, square, rectangular, and hollow tube).
3. The material for the column.
4. The design factor, considering the application.
5. The final dimensions for the column.

It may be desirable to propose and analyze several different designs to approach an optimum for the application. A computer program or spreadsheet facilitates the process.

It is assumed that items 1 through 4 are specified by the designer for any given trial. For some simple shapes, such as the solid round or square section, the final dimensions are computed from the appropriate formula: the Euler formula, Equation (6–5) or (6–6), or the J. B. Johnson formula, Equation (6–7). If an algebraic solution is not possible, iteration can be done.

In a design situation, the unknown cross-sectional dimensions make computing the radius of gyration and therefore the slenderness ratio, KL/r, impossible. Without the slenderness ratio, we cannot determine whether the column is long (Euler) or short (Johnson). Thus, the proper formula to use is not known.

We overcome this difficulty by making an assumption that the column is either long or short and proceeding with the corresponding formula. Then, after the dimensions are determined for the cross section, the actual value of KL/r will be computed and compared with C_c. This will show whether or not the correct formula has been used. If so, the computed answer is correct. If not, the alternate formula must be used and the computation repeated to determine new dimensions. Figure 6–13 shows a flowchart for the design logic described here.

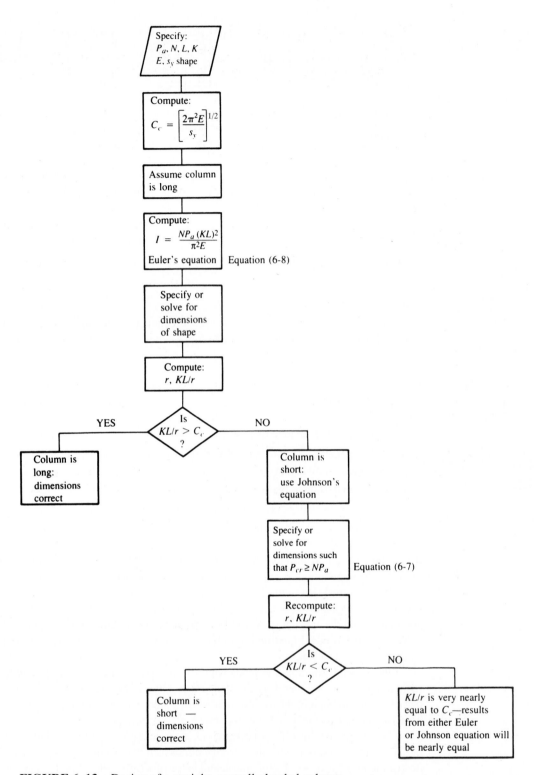

FIGURE 6–13 Design of a straight, centrally loaded column

Design: Assuming a Long Column

Euler's formula is used if the assumption is that the column is long. Equation (6–6) would be the most convenient form because it can be solved for the moment of inertia, I:

Euler's Formula Solved for Required Value of *I*

$$I = \frac{P_{cr}\,(KL)^2}{\pi^2\,E} = \frac{NP_a(KL)^2}{\pi^2\,E} \tag{6–8}$$

where P_a = allowable load, usually set equal to the actual maximum expected load

Having the required value for I, we can determine the dimensions for the shape by additional computations or by scanning tables of data of the properties of commercially available sections.

The solid circular section is one for which it is possible to derive a final equation for the characteristic dimension, the diameter. The moment of inertia is

$$I = \frac{\pi D^4}{64}$$

Substituting this into Equation (6–8) gives

$$I = \frac{\pi D^4}{64} = \frac{NP_a\,(KL)^2}{\pi^2\,E}$$

Solving for D yields

Required Diameter for a Long, Solid Circular Column

$$D = \left[\frac{64NP_a\,(KL)^2}{\pi^3\,E}\right]^{1/4} \tag{6–9}$$

Design: Assuming a Short Column

The J. B. Johnson formula is used to analyze a short column. It is difficult to derive a convenient form for use in design. In the general case, then, trial and error is used.

For some special cases, including the solid circular section, it is possible to solve the Johnson formula for the characteristic dimension, the diameter:

$$P_{cr} = As_y\left[1 - \frac{s_y\,(KL/r)^2}{4\pi^2\,E}\right] \tag{6–7}$$

But

$$A = \pi D^2/4$$
$$r = D/4\ (\text{from Appendix}\,1)$$
$$P_{cr} = NP_a$$

Then

$$NP_a = \frac{\pi D^2}{4}\,s_y\left[1 - \frac{s_y\,(KL)^2}{4\pi^2\,E(D/4)^2}\right]$$

$$\frac{4NP_a}{\pi s_y} = D^2\left[1 - \frac{s_y\,(KL)^2\,(16)}{4\pi^2\,ED^2}\right]$$

Solving for D gives

Required Diameter for a Short, Solid Circular Column

$$D = \left[\frac{4NP_a}{\pi s_y} + \frac{4s_y\,(KL)^2}{\pi^2\,E}\right]^{1/2} \tag{6–10}$$

Example Problem 6–3 Specify a suitable diameter of a solid, round cross section for a machine link if it is to carry 9800 lb of axial compressive load. The length will be 25 in, and the ends will be pinned. Use a design factor of 3. Use AISI 1020 hot-rolled steel.

Solution Objective Specify a suitable diameter for the column.

Given

Solid circular cross section: $L = 25$ in; use $N = 3$.

Both ends are pinned.

Material: AISI 1020 hot-rolled steel.

Analysis Use the procedure in Figure 6–13. Assume first that the column is long.

Results From Equation (6–9),

$$D = \left[\frac{64NP_a(KL)^2}{\pi^3 E}\right]^{1/4} = \left[\frac{64(3)(9800)(25)^2}{\pi^3 (30 \times 10^6)}\right]^{1/4}$$

$$D = 1.06 \text{ in}$$

The radius of gyration can now be found:

$$r = D/4 = 1.06/4 = 0.265 \text{ in}$$

The slenderness ratio is

$$KL/r = [(1.0)(25)]/0.265 = 94.3$$

For the AISI 1020 hot-rolled steel, $s_y = 30\,000$ psi. The graph in Figure 6–5 shows C_c to be approximately 138. Thus, the actual KL/r is less than the transition value, and the column must be redesigned as a short column, using Equation (6–10) derived from the Johnson formula:

$$D = \left[\frac{4NP_a}{\pi s_y} + \frac{4s_y (KL)^2}{\pi^2 E}\right]^{1/2}$$

$$D = \left[\frac{4(3)(9800)}{(\pi)(30\,000)} + \frac{4 (30\,000)(25)^2}{\pi^2 (30 \times 10^6)}\right]^{1/2} = 1.23 \text{ in} \qquad \textbf{(6–10)}$$

Checking the slenderness ratio again, we have

$$KL/r = [(1.0)(25)]/(1.23/4) = 81.3$$

Comments This is still less than the transition value, so our analysis is acceptable. A preferred size of $D = 1.25$ in could be specified.

An alternate method of using spreadsheets to design columns is to use an analysis approach similar to that shown in Figure 6–9 but to use it as a convenient "trial and error" tool. You could compute data by hand, or you could look them up in a table for A, I, and r for any desired cross-sectional shape and dimensions and insert the values into the spreadsheet. Then you could compare the computed allowable load with the required value and choose smaller or larger sections to bring the computed value close to the required value. Many iterations could be completed in a short amount of time. For shapes

CIRCULAR COLUMN ANALYSIS	*Data from:* Example Problem 6–3
Refer to Figure 6–4 for analysis logic.	
Enter data for variables in *italics in shaded boxes.*	Use consistent U.S. Customary units.
Data to Be Entered:	**Computed Values:**
Length and End Fixity:	
Column length, L = 25 in *End fixity, K =* 1 →	Eq. length, $L_e = KL =$ 25.0 in
Material Properties:	
Yield strength, s_y = 30 000 psi *Modulus of elasticity, E = 3.00E + 07 psi* →	Column constant, $C_c =$ 140.5
Cross Section Properties:	
[Note: *A* and *r* computed from] [dimensions for circular cross section] [in following section of this spreadsheet.]	
Area, A = 1.188 in²	
Radius of gyration, r = 0.3075 in →	Slender ratio, $KL/r =$ 81.3
Properties for round column:	
Diameter for round column = 1.23 in Area, A = 1.188 in² Radius of gyration, r = 0.3075 in	Column is: **short**
	Critical buckling load = 29,679 lb
Design Factor:	
Design factor on load, N = 3 →	**Allowable load = 9,893 lb**

FIGURE 6–14   Spreadsheet for column analysis used as a tool to design a column with a round cross section

that allow computing *r* and *A* fairly simply, you could add a new section to the spreadsheet to calculate these values. An example is shown in Figure 6–14, where the differently shaded box shows the calculations for the properties of a round cross section. The data are from Example Problem 6–3, and the result shown was arrived at with only four iterations.

FIGURE 6–15
Illustration of crooked
column

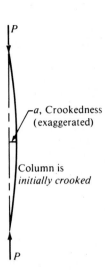

a, Crookedness
(exaggerated)

Column is
initially crooked

**6–11
CROOKED
COLUMNS**

The Euler and Johnson formulas assume that the column is straight and that the load acts in line with the centroid of the cross section of the column. If the column is somewhat crooked, bending occurs in addition to the column action (see Figure 6–15).

The crooked column formula allows an initial crookedness, *a*, to be considered (see References 6, 7, and 8):

Crooked Column
Formula

$$P_a^2 - \frac{1}{N}\left[s_y A + \left(1 + \frac{ac}{r^2}\right)P_{cr}\right]P_a + \frac{s_y A P_{cr}}{N^2} = 0 \qquad \textbf{(6–11)}$$

where c = distance from the neutral axis of the cross section about which bending occurs
to its outer edge
P_{cr} is defined to be the critical load found from the *Euler formula*.

Although this formula may become increasingly inaccurate for shorter columns, it is not appropriate to switch to the Johnson formula as it is for straight columns.

The crooked column formula is a quadratic with respect to the allowable load P_a. Evaluating all constant terms in Equation (6–11) produces an equation of the form

$$P_a^2 + C_1 P_a + C_2 = 0$$

Then, from the solution for a quadratic equation,

$$P_a = 0.5\left[-C_1 - \sqrt{C_1^2 - 4C_2}\right]$$

The smaller of the two possible solutions is selected.

Example Problem 6–4

A column has both ends pinned and has a length of 32 in. It has a circular cross section with a diameter of 0.75 in and an initial crookedness of 0.125 in. The material is AISI 1040 hot-rolled steel. Compute the allowable load for a design factor of 3.

Solution   Objective   Specify the allowable load for the column.

Given   Solid circular cross section: $D = 0.75$ in; $L = 32$ in; use $N = 3$.
Both are ends pinned. Initial crookedness $= a = 0.125$ in.
Material: AISI 1040 hot-rolled steel.

Analysis   Use Equation (6–11). First evaluate C_1 and C_2. Then solve the quadratic equation for P_a.

Results

$$s_y = 42\,000 \text{ psi}$$
$$A = \pi D^2/4 = (\pi)(0.75)^2/4 = 0.442 \text{ in}^2$$
$$r = D/4 = 0.75/4 = 0.188 \text{ in}$$
$$c = D/2 = 0.75/2 = 0.375 \text{ in}$$
$$KL/r = [(1.0)(32)]/0.188 = 171$$
$$P_{cr} = \frac{\pi^2 EA}{(KL/r)^2} = \frac{\pi^2 (30\,000\,000)(0.442)}{(171)^2} = 4476 \text{ lb}$$
$$C_1 = \frac{-1}{N}\left[s_y A + \left(1 + \frac{ac}{r^2}\right)P_{cr}\right] = -9649$$
$$C_2 = \frac{s_y A P_{cr}}{N^2} = 9.232 \times 10^6$$

The quadratic is therefore

$$P_a^2 - 9649\,P_a + 9.232 \times 10^6 = 0$$

Comment   From this, $P_a = 1077$ lb is the allowable load.

Figure 6–16 shows the solution of Example Problem 6–4 using a spreadsheet. Whereas its appearance is similar to that of the earlier column analysis spreadsheets, the details follow the calculations needed to solve Equation (6–11). On the lower left, two special data values are needed: (1) the crookedness a and (2) the distance c from the neutral axis for buckling to the outer surface of the cross section. In the middle of the right part are listed some intermediate values used in Equation (6–11): C_1 and C_2 as defined in the solution to Example Problem 6–4. The result, the allowable load, P_a, is at the lower right of the spreadsheet. Above that, for comparison, the computed value of the critical buckling load is given for a straight column of the same design. Note that this solution procedure is most accurate for long columns. If the analysis indicates that the column is *short* rather than *long*, the designer should take note of how short it is by comparing the slenderness ratio, KL/r, with the column constant, C_c. If the column is quite short, the designer should not rely on the accuracy of the result from Equation (6–11).

**6–12
ECCENTRICALLY
LOADED
COLUMNS**

An *eccentric load* is one that is applied away from the centroidal axis of the cross section of the column, as shown in Figure 6–17. Such a load exerts bending in addition to the column action that results in the deflected shape shown in the figure. The maximum stress in the deflected column occurs in the outermost fibers of the cross section at the midlength of

CROOKED COLUMN ANALYSIS		Data from:	Example Problem 6–4
Solves Equation 6–11 for allowable load.			
Enter data for variables in *italics in shaded boxes*.		Use consistent U.S. Customary units.	
Data to Be Entered:		**Computed Values:**	
Length and End Fixity:			
Column length, L = 32 in End fixity, K = 1	→	Eq. length, $L_e = KL$ =	32.0 in
Material Properties:			
Yield strength, s_y = 42,000 psi Modulus of elasticity, E = 3.00E+07 psi	→	Column constant, C_c =	18.7
Cross Section Properties:		Euler buckling load =	4,491 lb
[Note: Enter r or compute r = sqrt(I/A).] [Always enter Area.] [Enter zero for I or r if not used.]		C_1 in Eq. (6–11) = −9,678 C_2 in Eq. (6–11) = 9.259E+06	
Area, A = 0.442 in^2 Moment of inertia, I = 0 in^4			
Radius of gyration, r = 0.188 in	→	Slender ratio, KL/r =	170.7
Values for Eq. (6–11):		Column is: ***long***	
Initial crookedness = a = 0.125 in Neutral axis to outside = c = 0.375 in		***Straight Column***	
		Critical buckling load =	***4,491 lb***
Design Factor:		***Crooked Column***	
Design factor on load, N = 3	→	**Allowable load** =	***1,076 lb***

FIGURE 6–16 Spreadsheet for analysis of crooked columns

FIGURE 6–17
Illustration of
eccentrically loaded
columns

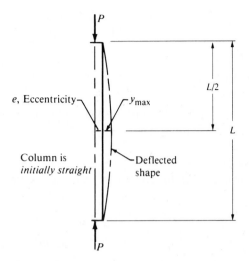

the column where the maximum deflection, y_{max}, occurs. Let's denote the stress at this point as $\sigma_{L/2}$. Then, for any applied load, P,

> **Secant Formula
> for Eccentrically
> Loaded Columns**

$$\sigma_{L/2} = \frac{P}{A}\left[1 + \frac{ec}{r^2}\sec\left(\frac{KL}{2r}\sqrt{\frac{P}{AE}}\right)\right] \tag{6–12}$$

(See References 4, 5, and 9.) Note that this stress is *not* directly proportional to the load. When evaluating the secant in this formula, note that its argument in the parentheses is in *radians*. Also, because most calculators do not have the secant function, recall that the secant is equal to 1/cosine.

For design purposes, we would like to specify a design factor, N, that can be applied to the *failure load* similar to that defined for straight, centrally loaded columns. However, in this case, failure is predicted when the maximum stress in the column exceeds the yield strength of the material. Let's now define a new term, P_y, to be the load applied to the eccentrically loaded column when the maximum stress is equal to the yield strength. Equation (6–12) then becomes

$$s_y = \frac{P_y}{A}\left[1 + \frac{ec}{r^2}\sec\left(\frac{KL}{2r}\sqrt{\frac{P_y}{AE}}\right)\right]$$

Now, if we define the *allowable load* to be

$$P_a = P_y/N$$

or

$$P_y = NP_a$$

this equation becomes

> **Design Equation
> for Eccentrically
> Loaded Columns**

$$\text{Required } s_y = \frac{NP_a}{A}\left[1 + \frac{ec}{r^2}\sec\left(\frac{KL}{2r}\sqrt{\frac{NP_a}{AE}}\right)\right] \tag{6–13}$$

This equation cannot be solved for either A or P_a. Therefore, an iterative solution is required, as will be demonstrated in Example Problem 6–6.

Another critical factor may be the amount of deflection of the axis of the column due to the eccentric load:

Maximum Deflection in an Eccentrically Loaded Column

$$y_{max} = e\left[\sec\left(\frac{KL}{2r}\sqrt{\frac{P}{AE}}\right) - 1\right] \qquad (6\text{--}14)$$

Note that the argument of the secant is the same as that used in Equation (6–12).

Example Problem 6–5   For the column of Example Problem 6–4, compute the maximum stress and deflection if a load of 1075 lb is applied with an eccentricity of 0.75 in. The column is initially straight.

Solution   **Objective**   Compute the stress and the deflection for the eccentrically loaded column.

Given

Data from Example Problem 6–4, but eccentricity = e = 0.75 in.

Solid circular cross section: D = 0.75 in; L = 32 in; Initially straight

Both ends are pinned; KL = 32 in; r = 0.188 in; $c = D/2$ = 0.375 in.

Material: AISI 1040 hot-rolled steel; $E = 30 \times 10^6$ psi.

Analysis   Use Equation (6–12) to compute maximum stress. Then use Equation (6–14) to compute maximum deflection.

Results   All terms have been evaluated before. Then the maximum stress is found from Equation (6–12):

$$\sigma_{L/2} = \frac{1075}{0.422}\left[1 + \frac{(0.75)(0.375)}{(0.188)^2}\sec\left(\frac{32}{2(0.188)}\sqrt{\frac{1075}{(0.442)(30 \times 10^6)}}\right)\right]$$

$$\sigma_{L/2} = 29\,300 \text{ psi}$$

The maximum deflection is found from Equation (6–14):

$$y_{max} = 0.75\left[\sec\left(\frac{32}{2(0.188)}\sqrt{\frac{1075}{(0.442)(30 \times 10^6)}}\right) - 1\right] = 0.293 \text{ in}$$

Comments   The maximum stress is 29 300 psi at the midlength of the column. The deflection there is 0.293 in from the original straight central axis of the column.

Example Problem 6–6   The stress in the column found in Example Problem 6–5 seems high for the AISI 1040 hot-rolled steel. Redesign the column to achieve a design factor of at least 3.

Solution   **Objective**   Redesign the eccentrically loaded column of Example Problem 6–5 to reduce the stress and achieve a design factor of at least 3.

Given   Data from Example Problems 6–4 and 6–5.

Analysis   Use a larger diameter. Use Equation (6–13) to compute the required strength. Then compare
             that with the strength of AISI 1040 hot-rolled steel. Iterate until the stress is satisfactory.

Results   Appendix 3 gives the value for the yield strength of AISI 1040 HR to be 42 000 psi. If we
             choose to retain the same material, the cross-sectional dimensions of the column must be in-
             creased to decrease the stress. Equation (6–13) can be used to evaluate a design alternative.
                 The objective is to find suitable values for A, c, and r for the cross section such that
             $P_a = 1075$ lb; $N = 3$; $L_e = 32$ in; $e = 0.75$ in; and the value of the entire right side of the
             equation is less than 42 000 psi. The original design had a circular cross section with a di-
             ameter of 0.75 in. Let's try increasing the diameter to $D = 1.00$ in. Then

$$A = \pi D^2/4 = \pi(1.00 \text{ in})^2/4 = 0.785 \text{ in}^2$$
$$r = D/4 = (1.00 \text{ in})/4 = 0.250 \text{ in}$$
$$r^2 = (0.250 \text{ in})^2 = 0.0625 \text{ in}^2$$
$$c = D/2 = (1.00 \text{ in})/2 = 0.50 \text{ in}$$

Now let's call the right side of Equation (6–13) s_y'. Then

$$s_y' = \frac{3(1075)}{0.785}\left[1 + \frac{(0.75)(0.50)}{(0.0625)}\sec\left(\frac{32}{2(0.250)}\sqrt{\frac{(3)(1075)}{(0.785)(30 \times 10^6)}}\right)\right]$$
$$s_y' = 37\ 740 \text{ psi} = \text{required value of } s_y$$

This is a satisfactory result because it is just slightly less than the value of s_y of 42 000 psi
for the steel.
    Now we can evaluate the expected maximum deflection with the new design using
Equation (6–14):

$$y_{max} = 0.75\left[\sec\left(\frac{32}{2(0.250)}\sqrt{\frac{1075}{(0.785)(30 \times 10^6)}}\right) - 1\right]$$
$$y_{max} = 0.076 \text{ in}$$

Comments   The diameter of 1.00 in is satisfactory. The maximum deflection for the column is 0.076 in.

Figure 6–18 shows the solution of the eccentric column problem of Example Problem 6–6 using a spreadsheet to evaluate Equations (6–13) and (6–14). It is a design aid
that facilitates the iteration required to determine an acceptable geometry for a column
to carry a specified load with a desired design factor. Note that the data are in U.S. Customary units. At the lower left of the spreadsheet, data required for Equations (6–13) and
(6–14) are entered by the designer, along with the other data discussed for earlier column analysis spreadsheets. The **"FINAL RESULTS"** at the lower right show the computed value of the required yield strength of the material for the column and compare it
with the given value entered by the designer near the upper left. The designer must ensure that the actual value is greater than the computed value. The last part of the right
side of the spreadsheet gives the computed maximum deflection of the column that occurs at its midlength.

ECCENTRIC COLUMN ANALYSIS		Data from:	Example Problem 6–6

Solves Equation (6–13) for design stress and Equation (6–14) for maximum deflection.

Enter data for variables in *italics in shaded boxes*.		Use consistent U.S. Customary units.
Data to Be Entered:		**Computed Values:**
Length and End Fixity:		
Column length, L = 32 in End fixity, K = 1	$\rightarrow$	Eq. length, $L_e = KL$ = 32.0 in
Material Properties:		
Yield strength, s_y = 42,000 psi Modulus of elasticity, E = 3.00E + 07 psi	$\rightarrow$	Column constant, C_c = 118.7
Cross Section Properties:		
[Note: Enter r or compute $r = \text{sqrt}(I/A)$.] [Always enter Area.] [Enter zero for I or r if not used.] Area, A = 0.785 in^2 Moment of inertia, I = 0 in^4 **OR**		Argument of sec = 0.749 for strength Value of secant = 1.3654 Argument of sec = 0.432 for deflection Value of secant = 1.1014
Radius of gyration, r = 0.250 in	$\rightarrow$	Slender ratio, KL/r = 128.0
Values for Eqs. (6–13) and (6–14): Eccentricity, e = 0.75 in Neutral axis to outside, c = 0.5 in Allowable load, P_a = 1,075 lb		Column is: **long**
		FINAL RESULTS
Design Factor: Design factor on load, N = 3		**Req'd yield strength = 37,764 psi** ***Must be less than actual yield strength:*** ***s_y = 42,000 psi***
		Max deflection, y_{max} = 0.076 in

FIGURE 6–18   Spreadsheet for analysis of eccentric columns

REFERENCES

1. Aluminum Association. *Aluminum Design Manual.* Washington, DC: Aluminum Association, 2000.

2. American Institute of Steel Construction. *Manual of Steel Construction*. LRFD 3rd ed. Chicago: American Institute of Steel Construction, 2001.

3. Hibbeler, R. C. *Mechanics of Materials.* 4th ed. Upper Saddle River, NJ: Prentice Hall, 2000.

4. Popov, E. P. *Engineering Mechanics of Solids.* 2d ed. Upper Saddle River, NJ: Prentice Hall, 1998.

5. Shigley, J. E., and C. R. Mischke. *Mechanical Engineering Design.* 6th ed. New York: McGraw-Hill, 2001.

6. Spotts, M. F., and T. E. Shoup. *Design of Machine Elements.* 7th ed. Upper Saddle River, NJ: Prentice Hall, 1998.

7. Timoshenko, S. *Strength of Materials.* Vol. 2. 2d ed. New York: Van Nostrand Reinhold, 1941.

8. Timoshenko, S., and J. M. Gere. *Theory of Elastic Stability.* 2d ed. New York: McGraw-Hill, 1961.

9. Young, W. C., and R. G. Budynas. *Roark's Formulas for Stress and Strain.* 7th ed. New York: McGraw-Hill, 2002.

PROBLEMS

1. A column has both ends pinned and has a length of 32 in. It is made of AISI 1040 HR steel and has a circular shape with a diameter of 0.75 in. Determine the critical load.

2. Repeat Problem 1 using a length of 15 in.

3. Repeat Problem 1 with the bar made of aluminum 6061-T4.

4. Repeat Problem 1 assuming both ends fixed.

5. Repeat Problem 1 using a square cross section, 0.65 in on a side, instead of the circular cross section.

6. Repeat Problem 1 with the bar made from high-impact acrylic plastic.

7. A rectangular steel bar has a cross section 0.50 by 1.00 in and is 8.5 in long. The bar has pinned ends and is made of AISI 4150 OQT 1000 steel. Compute the critical load.

8. A steel pipe has an outside diameter of 1.60 in, a wall thickness of 0.109 in, and a length of 6.25 ft. Compute the critical load for each of the end conditions shown in Figure 6–2. Use AISI 1020 HR steel.

9. Compute the required diameter of a circular bar to be used as a column carrying a load of 8500 lb with pinned ends. The length is 50 in. Use AISI 4140 OQT 1000 steel and a design factor of 3.0.

10. Repeat Problem 9 using AISI 1020 HR steel.

11. Repeat Problem 9 with aluminum 2014-T4.

12. In Section 6–10, equations were derived for the design of a solid circular column, either long or short. Perform the derivation for a solid square cross section.

13. Repeat the derivations called for in Problem 12 for a hollow circular tube for any ratio of inside to outside diameter. That is, let the ratio $R = ID/OD$, and solve for the required OD for a given load, material, design factor, and end fixity.

14. Determine the required dimensions of a column with a square cross section to carry an axial compressive load of 6500 lb if its length is 64 in and its ends are fixed. Use a design factor of 3.0. Use aluminum 6061-T6.

15. Repeat Problem 14 for a hollow aluminum tube (6061-T6) with the ratio of $ID/OD = 0.80$. Compare the weight of this column with that of Problem 14.

16. A toggle device is being used to compact scrap steel shavings, as illustrated in Figure P6–16. Design the two links of the toggle to be steel, AISI 5160 OQT 1000, with a circular cross section and pinned ends. The force P required to crush the shavings is 5000 lb. Use $N = 3.50$.

17. Repeat Problem 16, but propose a design that will be lighter than the solid circular cross section.

18. A sling, sketched in Figure P6–18, is to carry 18 000 lb. Design the spreader.

19. For the sling in Problem 18, design the spreader if the angle shown is changed from 30° to 15°.

20. A rod for a certain hydraulic cylinder behaves as a fixed-free column when used to actuate a compactor of industrial waste. Its maximum extended length will be 10.75 ft. If it is to be made of AISI 1144 OQT 1300 steel, determine the required diameter of the rod for a design factor of 2.5 for an axial load of 25 000 lb.

21. Design a column to carry 40 000 lb. One end is pinned, and the other is fixed. The length is 12.75 ft.

22. Repeat Problem 21 using a length of 4.25 ft.

23. Repeat Problem 1 if the column has an initial crookedness of 0.08 in. Determine the allowable load for a design factor of 3.

24. Repeat Problem 7 if the column has an initial crookedness of 0.04 in. Determine the allowable load for a design factor of 3.

25. Repeat Problem 8 if the column has an initial crookedness of 0.15 in. Determine the allowable load for a design factor of 3 and pinned ends only.

FIGURE P6–16 (Problems 16 and 17)

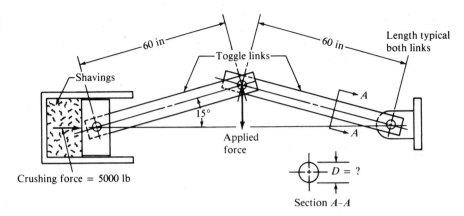

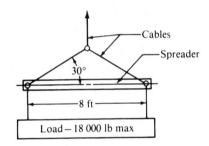

FIGURE P6–18 (Problems 18 and 19)

26. An aluminum (6063-T4) column is 42 in long and has a square cross section, 1.25 in on a side. If it carries a compressive load of 1250 lb, applied with an eccentricity of 0.60 in, compute the maximum stress in the column and the maximum deflection.

27. A steel (AISI 1020 hot-rolled) column is 3.2 m long and is made from a standard 3-in Schedule 40 steel pipe (see Table A16–6). If a compressive load of 30.5 kN is applied with an eccentricity of 150 mm, compute the maximum stress in the column and the maximum deflection.

28. A link in a mechanism is 14.75 in long and has a square cross section, 0.250 in on a side. It is made from annealed AISI 410 stainless steel. Use $E = 28 \times 10^6$ psi. If it car-

ries a compressive load of 45 lb with an eccentricity of 0.30 in, compute the maximum stress and the maximum deflection.

29. A hollow square steel tube, 40 in long, is proposed for use as a prop to hold up the ram of a punch press during installation of new dies. The ram weighs 75 000 lb. The prop is made from $4 \times 4 \times 1/4$ structural tubing. It is made from steel similar to structural steel, ASTM A500 Grade C. If the load applied by the ram could have an eccentricity of 0.50 in, would the prop be safe?

30. Determine the allowable load on a column 16.0 ft long made from a wide-flange beam shape, W5 × 19. The load will be centrally applied. The end conditions are somewhat between fixed and hinged, say, $K = 0.8$. Use a design factor of 3. Use ASTM A36 structural steel.

31. Determine the allowable load on a fixed-end column having a length of 66 in if it is made from a steel American Standard Beam, S4 × 7.7. The material is ASTM A36 structural steel. Use a design factor of 3.

32. Compute the maximum stress and deflection that can be expected in the steel machine member carrying an eccentric load as shown in Figure P6–32. The load P is 1000 lb. If a design factor of 3 is desired, specify a suitable steel.

33. Specify a suitable steel tube from Table A16–5 to support one side of a platform as shown in Figure P6–33. The ma-

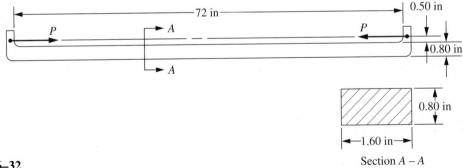

FIGURE P6–32

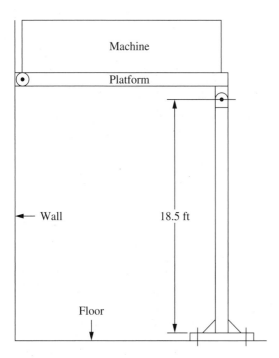

FIGURE P6–33

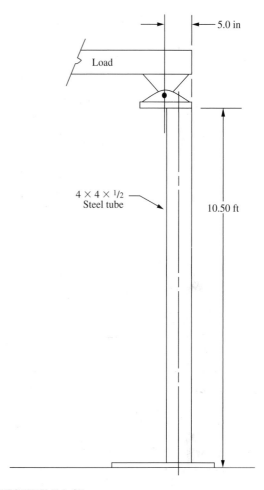

FIGURE P6–37

terial has a yield strength of 36 ksi. The total load on the platform is 55 000 lb, uniformly distributed.

34. Compute the allowable axial load on a steel channel, C5 × 9, made from ASTM A36 structural steel. The channel is 112 in long and can be considered to be pinned at its ends. Use a design factor of 3.

35. Repeat Problem 34 with the ends fixed rather than pinned.

36. Repeat Problem 34, except consider the load to be applied along the outside of the web of the channel instead of being axial.

37. Figure P6–37 shows a 4 × 4 × 1/2 steel column made from ASTM A500 Grade B structural steel. To accommodate a special mounting restriction, the load is applied eccentrically as shown. Determine the amount of load that the column can safely support. The column is supported laterally by the structure.

38. The device shown in Figure P6–38 is subjected to opposing forces F. Determine the maximum allowable load to achieve a design factor of 3. The device is made from aluminum 6061-T6.

39. A hydraulic cylinder is capable of exerting a force of 5200 N to move a heavy casting along a conveyor. The design of the pusher causes the load to be applied eccentrically to the piston rod as shown in Figure P6–39. Is the piston rod safe under this loading if it is made from AISI 416 stainless steel in the Q&T 1000 condition? Use $E = 200$ GPa.

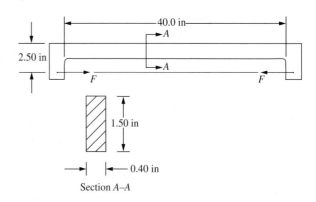

FIGURE P6–38

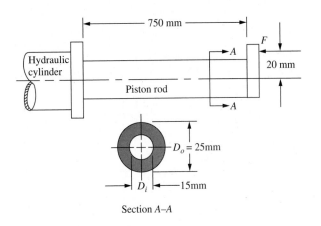

750 mm

Hydraulic cylinder

Piston rod

F

20 mm

A

A

$D_o = 25$mm

D_i 15mm

Section A–A

FIGURE P6–39

40. A standard 2-in schedule 40 steel pipe is proposed to be used to support the roof of a porch during renovation. Its length is 13.0 ft. The pipe is made from ASTM A501 structural steel.

(a) Determine the safe load on the pipe to achieve a design factor of 3 if the pipe is straight.

(b) Determine the safe load if the pipe has an initial crookedness of 1.25 in.

PART II

Design of a
Mechanical Drive

**OBJECTIVES
AND CONTENT
OF PART II**

Part II of this book contains nine chapters (Chapters 7–15) that help you gain experience in approaching the design of an important complete device—a *mechanical drive*. The drive, sometimes called a *power transmission*, serves the following functions:

- It receives power from some kind of rotating source such as an electric motor, an internal combustion engine, a gas turbine engine, a hydraulic or pneumatic motor, a steam or water turbine, or even hand rotation provided by the operator.

- The drive typically causes some change in the speed of rotation of the shafts that make up the drive so that the output shaft operates more slowly or faster than the input shaft. Speed reducers are more prevalent than speed increasers.

- The active elements of the drive transmit the power from the input to the output shafts.

- When there is a speed reduction, there is a corresponding increase in the torque transmitted. Conversely, a speed increase causes a reduction in torque at the output compared with the input of the drive.

The chapters of Part II provide the detailed descriptions of the various machine elements that are typically used in power transmissions: *belt drives*, *chain drives*, *gears*, *shafts*, *bearings*, *keys*, *couplings*, *seals*, and *housings to hold all the elements together*. You will learn the important features of these elements and the methods of analyzing and designing them.

Of equal importance is the information provided on how the various elements interact with each other. You must be sensitive, for example, to how gears are mounted on shafts, how the shafts are supported by bearings, and how the bearings must be mounted securely in a housing that holds the system together. The final completed design must function as an integrated unit.

**THE PROCESS
OF DESIGNING
A MECHANICAL
DRIVE**

In the design of a power transmission, you would typically know the following:

- *The nature of the driven machine:* It might be a machine tool in a factory that cuts metal parts for engines; an electric drill used by professional carpenters or home craft workers; the axle of a farm tractor; the propeller shaft of a turbojet for an airplane; the propeller shaft for a large ship; the wheels of a toy train; a mechanical timing mechanism; or any other of the numerous products that need a controlled-speed drive.

- *The level of power to be transmitted:* From the examples just listed, the power demanded may range from thousands of horsepower for a ship, hundreds of horsepower for a large farm tractor or airplane, or a few watts for a timer or a toy.

- *The rotational speed of the drive motor or other prime mover:* Typically the prime mover operates at a rather high speed of rotation. The shafts of standard electric motors rotate at about 1200, 1800, or 3600 revolutions per minute (rpm). Automotive engines operate from about 1000 to 6000 rpm. Universal motors in some hand tools (drills, saws, and routers) and household appliances (mixers, blenders, and vacuum cleaners) operate from 3500 to 20 000 rpm. Gas turbine engines for aircraft rotate many thousands of rpm.

- *The desired output speed of the transmission:* This is highly dependent on the application. Some gear motors for instruments rotate less than 1.0 rpm. Production machines in factories may run a few hundred rpm. Drives for assembly conveyors may run fewer than 100 rpm. Aircraft propellers may operate at several thousand rpm.

You, the designer, must then do the following:

- Choose the type of power transmission elements to be used: gears, belt drives, chain drives, or other types. In fact, some power transmission systems use two or more types in series to optimize the performance of each.

- Specify how the rotating elements are arranged and how the power transmission elements are mounted on shafts.

- Design the shafts to be safe under the expected torques and bending loads and properly locate the power transmission elements and the bearings. It is likely that the shafts will have several diameters and special features to accommodate keys, couplings, retaining rings, and other details. The dimensions of all features must be specified, along with the tolerances on the dimensions and surface finishes.

- Specify suitable bearings to support the shafts and determine how they will be mounted on the shafts and how they will be held in a housing.

- Specify keys to connect the shaft to the power transmission elements; couplings to connect the shaft from the driver to the input shaft of the transmission or to connect the output shaft to the driven machine; seals to effectively exclude contaminants from entering the transmission; and other accessories.

- Place all of the elements in a suitable housing that provides for the mounting of all elements and for their protection from the environment and their lubrication.

CHAPTERS THAT MAKE UP PART II

To guide you through this process of designing a mechanical drive, Part II includes the following chapters.

Chapter 7: Belt Drives and Chain Drives emphasizes recognizing the variety of commercially available belt and chain drives, the critical design parameters, and the methods used to specify reasonably optimum components of the drive systems.

Chapter 8: Kinematics of Gears describes and defines the important geometric features of gears. Methods of manufacturing gears are discussed, along with the importance of precision in the operation of the gears. The details of how a pair of gears operates are described, and the design and the operation of two or more gear pairs in a gear train are analyzed.

Chapter 9: Spur Gear Design illustrates how to compute forces exerted by one tooth of a gear on its mating teeth. Methods of computing the stresses in the gear teeth are presented, and design procedures for specifying gear-tooth geometry and material are given to produce a safe, long-life gear transmission system.

Chapter 10: Helical Gears, Bevel Gears, and Wormgearing contains approaches similar to those described for spur gears, with special attention to the unique geometry of these types of gears.

Chapter 11: Keys, Couplings, and Seals discusses how to design keys to be safe under the prevailing loads caused by the torque transmitted by them from the shaft to the gears or other elements. Couplings must be specified that accommodate the possible misalignment of connected shafts while transmitting the required torque at operating speeds. Seals must be specified for shafts that project through the sides of the housing and for bearings that must be kept free of contaminants. The maintenance of a reliable supply of clean lubricant for the active elements is essential.

Chapter 12: Shaft Design discusses the fact that in addition to being designed to safely transmit the required torque levels at given speeds, the shafts will probably have several diameters and special features to accommodate keys, couplings, retaining rings, and other details. The dimensions of all features must be specified, along with the tolerances on the dimensions and surface finishes. Completion of these tasks requires some of the skills developed in following chapters. So you will have to come back to this task later.

Chapter 13: Tolerances and Fits discusses the fit of elements that are assembled together and that may operate on one another; this fit is critical to the performance and life of the elements. In some cases, such as fitting the inner race of a ball or roller bearing onto a shaft, the bearing manufacturer specifies the allowable dimensional variation on the shaft so that it is compatible with the tolerances to which the bearing is produced. There is typically an interference fit between the bearing inner race and the shaft diameter where the bearing is to be mounted. But there is a close sliding fit between the outer race and the housing that holds the bearing in place. In general, it is important for you to take charge of specifying the tolerances for all dimensions to ensure proper operation while allowing economical manufacture.

Chapter 14: Rolling Contact Bearings focuses on commercially available rolling contact bearings such as ball bearings, roller bearings, tapered roller bearings, and others. You must be able to compute or specify the loads that the bearings will support, their speed of operation, and their expected life. From these data, standard bearings from manufacturers' catalogs will be specified. Then you must review the design process for the shafts as described for Chapter 12 to complete the specification of dimensions and tolerances. It is likely that iteration among the design processes for the power transmission elements, the shafts, and the bearings will be needed to achieve an optimum arrangement.

Chapter 15: Completion of the Design of a Power Transmission merges all of the preceding topics together. You will resolve the details of the design of each element and ensure the compatibility of mating elements. You will review all previous design decisions and assumptions and verify that the design meets specifications. After the individual elements have been analyzed and the iteration among them is complete, they must be packaged in a suitable housing to hold them securely, to protect them from contaminants, and to protect the people who may work around them. The housing must also be designed to be compatible with the driver and the driven machine. That often requires special fastening provisions and means of locating all connected devices relative to one another. Assembly and service must be considered. Then you will present a final set of specifications for the entire power transmission system and document your design with suitable drawings and a written report.

7

Belt Drives and Chain Drives

The Big Picture

You Are the Designer

7–1   Objectives of This Chapter

7–2   Types of Belt Drives

7–3   V-Belt Drives

7–4   V-Belt Drive Design

7–5   Chain Drives

7–6   Design of Chain Drives

<table>
<tr><td>

The Big Picture

</td><td>

Belt Drives and Chain Drives

Discussion Map

☐ Belts and chains are the major types of flexible power transmission elements. Belts operate on sheaves or pulleys, whereas chains operate on toothed wheels called *sprockets*.

</td></tr>
</table>

Discover

Look around and identify at least one mechanical device having a belt drive and one having a chain drive system.

Describe each system, and make a sketch showing how it receives power from some source and how it transfers power to a driven machine.

Describe the differences between belt drives and chain drives.

In this chapter, you will learn how to select suitable components for belt drives and chain drives from commercially available designs.

Belts and chains represent the major types of flexible power transmission elements. Figure 7–1 shows a typical industrial application of these elements combined with a gear-type speed reducer. This application illustrates where belts, gear drives, and chains are each used to best advantage.

Rotary power is developed by the electric motor, but motors typically operate at too high a speed and deliver too low a torque to be appropriate for the final drive application. Remember, for a given power transmission, the torque is increased in proportion to the amount that rotational speed is reduced. So some speed reduction is often desirable. The high speed of the motor makes belt drives somewhat ideal for that first stage of reduction. A smaller drive pulley is attached to the motor shaft, while a larger diameter pulley is attached to a parallel shaft that operates at a correspondingly lower speed. Pulleys for belt drives are also called *sheaves*.

However, if very large ratios of speed reduction are required in the drive, gear reducers are desirable because they can typically accomplish large reductions in a rather small package. The output shaft of the gear-type speed reducer is generally at low speed and high torque. If both speed and torque are satisfactory for the application, it could be directly coupled to the driven machine.

However, because gear reducers are available only at discrete reduction ratios, the output must often be reduced more before meeting the requirements of the machine. At the low-speed, high-torque condition, chain drives become desirable. The high torque causes high tensile forces to be developed in the chain. The elements of the chain are typically metal, and they are sized to withstand the high forces. The links of chains are engaged in toothed wheels called *sprockets* to provide positive mechanical drive, desirable at the low-speed, high-torque conditions.

In general, belt drives are applied where the rotational speeds are relatively high, as on the first stage of speed reduction from an electric motor or engine. The linear speed of a belt is usually 2500 to 6500 ft/min, which results in relatively low tensile forces in the belt. At lower speeds, the tension in the belt becomes too large for typical belt cross sections, and slipping may occur between the sides of the belt and the sheave or pulley that carries it. At higher speeds, dynamic effects such as centrifugal forces, belt whip, and vibration reduce the effectiveness of the drive and its life. A speed of 4000 ft/min is generally ideal. Some belt designs employ high-strength, reinforcing strands and a cogged design that engages matching grooves in the pulleys to enhance their ability to transmit the high forces at low speeds. These designs compete with chain drives in many applications.

FIGURE 7–1

Combination drive employing V-belts, a gear reducer, and a chain drive [Source for Part (*b*): Browning Mfg. Division, Emerson Electric Co., Maysville, KY]

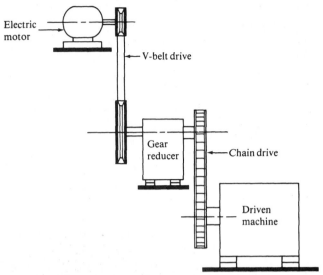

(a) Sketch of combination drive

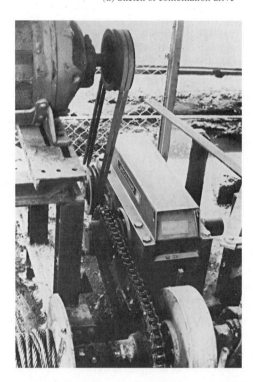

(b) Photograph of an actual drive installation. Note that guards have been removed from the belt and chain drives to show detail.

Where have you seen belt drives? Consider mechanical devices around your home or office; vehicles; construction equipment; heating, air conditioning, and ventilation systems; and industrial machinery. Describe their general appearance. To what was the input pulley attached? Was it operating at a fairly high speed? What was the size of the next pulley? Did it cause the second shaft to rotate at a slower speed? How much slower? Were there more stages of reduction accomplished by belts or by some other reducer? Make a sketch of the layout of the drive system. Make measurements if you can get access to the equipment safely.

Where have you seen chain drives? One obvious place is likely to be the chain on a bicycle where the sprocket attached to the pedal-crank assembly is fairly large and that attached to the rear wheel is smaller. The drive sprocket and/or the driven sprocket assemblies may have several sizes to allow the rider to select many different speed ratios to permit optimum operation under different conditions of speed and hill-climbing demands. Where else have you seen chain drives? Again consider vehicles, construction equipment, and industrial machinery. Describe and sketch at least one chain drive system.

This chapter will help you learn to identify the typical design features of commercially available belt and chain drives. You will be able to specify suitable types and sizes to transmit a given level of power at a certain speed and to accomplish a specified speed ratio between the input and the output of the drive. Installation considerations are also described so that you can put your designs into successful systems.

 You Are the Designer

A plant in Louisiana that produces sugar needs a drive system designed for a machine that chops long pieces of sugar cane into short lengths prior to processing. The machine's drive shaft is to rotate slowly at 30 rpm so that the cane is chopped smoothly and not beaten. The large machine requires a torque of 31 500 lb·in to drive the chopping blades.

Your company is asked to design the drive, and you are given the assignment. What kind of power source should be used? You might consider an electric motor, a gasoline engine, or a hydraulic motor. Most of these run at relatively high speeds, significantly higher than 30 rpm. Therefore, some type of speed reduction is needed. Perhaps you decide to use a drive similar to that shown in Figure 7–1.

Three stages of speed reduction are used. The input sheave of the belt drive rotates at the speed of the motor, while the larger driven sheave rotates at a slower speed and delivers the power to the input of the gear reducer. The larger part of the speed reduction is likely to be accomplished in the gear reducer, with the output shaft rotating slowly and providing a large torque. Remember, as the speed of rotation of a rotating shaft decreases, the torque delivered increases for a given power transmitted. But because there are only a limited number of reducer designs available, the output speed of the reducer will probably not be ideal for the cane chopper input shaft. The chain drive then provides the last stage of reduction.

As the designer, you must decide what type and size of belt drive to use and what the speed ratio between the driving and the driven sheave should be. How is the driving sheave attached to the motor shaft? How is the driven sheave attached to the input shaft of the gear reducer? Where should the motor be mounted in relation to the gear reducer, and what will be the resulting center distance between the two shafts? What speed reduction ratio will the gear reducer provide? What type of gear reducer should be used: helical gears, a worm and worm-gear drive, or bevel gears? How much additional speed reduction must the chain drive provide to deliver the proper speed to the cane-chopper shaft? What size and type of chain should be specified? What is the center distance between the output of the gear reducer and the input to the chopper? Then what length of chain is required? Finally, what motor power is required to drive the entire system at the stated conditions? The information in this chapter will help you answer questions about the design of power transmission systems incorporating belts and chains. Gear reducers are discussed in Chapter 8–10.

7–1
OBJECTIVES OF
THIS CHAPTER

After completing this chapter, you will be able to:

1. Describe the basic features of a belt drive system.
2. Describe several types of belt drives.
3. Specify suitable types and sizes of belts and sheaves to transmit a given level of power at specified speeds for the input and output sheaves.
4. Specify the primary installation variables for belt drives, including center distance and belt length.
5. Describe the basic features of a chain drive system.
6. Describe several types of chain drives.

FIGURE 7–2   Basic belt drive geometry

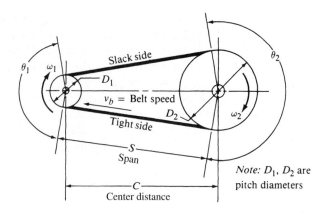

7. Specify suitable types and sizes of chains and sprockets to transmit a given level of power at specified speeds for the input and output sprockets.

8. Specify the primary installation variables for chain drives, including center distance between the sheaves, chain length, and lubrication requirements.

7–2
TYPES OF BELT
DRIVES

A belt is a flexible power transmission element that seats tightly on a set of pulleys or sheaves. Figure 7–2 shows the basic layout. When the belt is used for speed reduction, the typical case, the smaller sheave is mounted on the high-speed shaft, such as the shaft of an electric motor. The larger sheave is mounted on the driven machine. The belt is designed to ride around the two sheaves without slipping.

The belt is installed by placing it around the two sheaves while the center distance between them is reduced. Then the sheaves are moved apart, placing the belt in a rather high initial tension. When the belt is transmitting power, friction causes the belt to grip the driving sheave, increasing the tension in one side, called the "tight side," of the drive. The tensile force in the belt exerts a tangential force on the driven sheave, and thus a torque is applied to the driven shaft. The opposite side of the belt is still under tension, but at a smaller value. Thus, it is called the "slack side."

Many types of belts are available: flat belts, grooved or cogged belts, standard V-belts, double-angle V-belts, and others. See Figure 7–3 for examples. References 2–5 and 8–15 give more examples and technical data.

The *flat belt* is the simplest type, often made from leather or rubber-coated fabric. The sheave surface is also flat and smooth, and the driving force is therefore limited by the pure friction between the belt and the sheave. Some designers prefer flat belts for delicate machinery because the belt *will* slip if the torque tends to rise to a level high enough to damage the machine.

Synchronous belts, sometimes called *timing belts* [see Figure 7–3(c)], ride on sprockets having mating grooves into which the teeth on the belt seat. This is a positive drive, limited only by the tensile strength of the belt and the shear strength of the teeth.

Some cog belts, such as that shown in Figure 7–3(b), are applied to standard V-grooved sheaves. The cogs give the belt greater flexibility and higher efficiency compared with standard belts. They can operate on smaller sheave diameters.

A widely used type of belt, particularly in industrial drives and vehicular applications, is the *V-belt drive*, shown in Figures 7–3(a) and 7–4. The V-shape causes the belt to wedge tightly into the groove, increasing friction and allowing high torques to be transmitted before slipping occurs. Most belts have high-strength cords positioned at the pitch diameter of the belt cross section to increase the tensile strength of the belt. The cords, made from natural fibers, synthetic strands, or steel, are embedded in a firm rubber compound to pro-

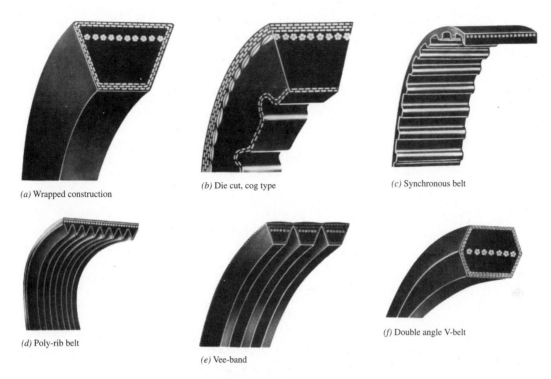

(a) Wrapped construction

(b) Die cut, cog type

(c) Synchronous belt

(d) Poly-rib belt

(e) Vee-band

(f) Double angle V-belt

FIGURE 7–3 Examples of belt construction (Dayco Corp., Dayton, OH)

FIGURE 7–4 Cross section of V-belt and sheave groove

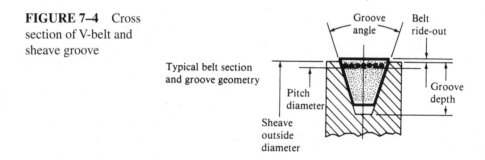

vide the flexibility needed to allow the belt to pass around the sheave. Often an outer fabric cover is added to give the belt good durability.

The selection of commercially available V-belt drives is discussed in the next section.

7–3
V-BELT DRIVES

The typical arrangement of the elements of a V-belt drive is shown in Figure 7–2. The important observations to be derived from this arrangement are summarized here:

1. The pulley, with a circumferential groove carrying the belt, is called a *sheave* (usually pronounced "shiv").

2. The size of a sheave is indicated by its pitch diameter, slightly smaller than the outside diameter of the sheave.

3. The speed ratio between the driving and the driven sheaves is inversely proportional to the ratio of the sheave pitch diameters. This follows from the observation

that there is no slipping (under normal loads). Thus, the linear speed of the pitch line of both sheaves is the same as and equal to the belt speed, v_b. Then

$$v_b = R_1\omega_1 = R_2\omega_2 \tag{7-1}$$

But $R_1 = D_1/2$ and $R_2 = D_2/2$. Then

$$v_b = \frac{D_1\omega_1}{2} = \frac{D_2\omega_2}{2} \tag{7-1A}$$

The angular velocity ratio is

$$\frac{\omega_1}{\omega_2} = \frac{D_2}{D_1} \tag{7-2}$$

4. The relationships between pitch length, L, center distance, C, and the sheave diameters are

$$L = 2C + 1.57\,(D_2 + D_1) + \frac{(D_2 - D_1)^2}{4C} \tag{7-3}$$

$$C = \frac{B + \sqrt{B^2 - 32\,(D_2 - D_1)^2}}{16} \tag{7-4}$$

where $B = 4L - 6.28(D_2 + D_1)$

5. The angle of contact of the belt on each sheave is

$$\theta_1 = 180° - 2\sin^{-1}\left[\frac{D_2 - D_1}{2C}\right] \tag{7-5}$$

$$\theta_2 = 180° + 2\sin^{-1}\left[\frac{D_2 - D_1}{2C}\right] \tag{7-6}$$

These angles are important because commercially available belts are rated with an assumed contact angle of 180°. This will occur only if the drive ratio is 1 (no speed change). The angle of contact on the smaller of the two sheaves will always be less than 180°, requiring a lower power rating.

6. The length of the span between the two sheaves, over which the belt is unsupported, is

$$S = \sqrt{C^2 - \left[\frac{D_2 - D_1}{2}\right]^2} \tag{7-7}$$

This is important for two reasons: You can check the proper belt tension by measuring the amount of force required to deflect the belt at the middle of the span by a given amount. Also, the tendency for the belt to vibrate or whip is dependent on this length.

7. The contributors to the stress in the belt are as follows:
   (a) The tensile force in the belt, maximum on the tight side of the belt.
   (b) The bending of the belt around the sheaves, maximum as the tight side of the belt bends around the smaller sheave.
   (c) Centrifugal forces created as the belt moves around the sheaves.

The maximum total stress occurs where the belt enters the smaller sheave, and the bending stress is a major part. Thus, there are recommended minimum sheave diameters for standard belts. Using smaller sheaves drastically reduces belt life.

8. The design value of the ratio of the tight side tension to the slack side tension is 5.0 for V-belt drives. The actual value may range as high as 10.0.

Standard Belt Cross Sections

Commercially available belts are made to one of the standards shown in Figures 7–5 through 7–8. The alignment between the inch sizes and the metric sizes indicates that the paired sizes are actually the same cross section. A "soft conversion" was used to rename the familiar inch sizes with the number for the metric sizes giving the nominal top width in millimeters.

The nominal value of the included angle between the sides of the V-groove ranges from 30° to 42°. The angle on the belt may be slightly different to achieve a tight fit in the groove. Some belts are designed to "ride out" of the groove somewhat.

Many automotive applications use synchronous belt drives similar to that called a *timing belt* in Figure 7–3(c) or V-ribbed belts similar to that called a *poly-rib belt* in Figure

FIGURE 7–5 Heavy-duty industrial V-belts

FIGURE 7–6 Industrial narrow-section V-belts

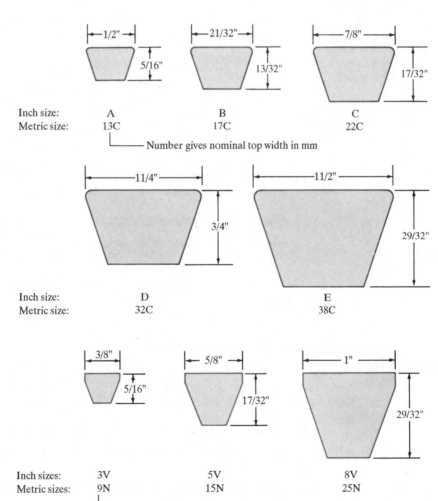

FIGURE 7–7 Light-duty, fractional horsepower (FHP) V-belts

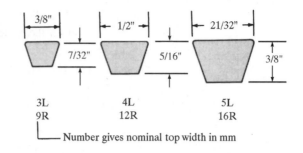

	3L	4L	5L
Inch sizes:	3L	4L	5L
Metric sizes:	9R	12R	16R

Number gives nominal top width in mm

FIGURE 7–8 Automotive V-belts

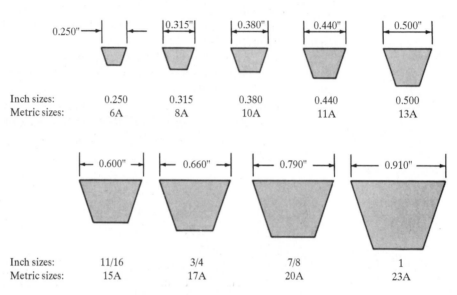

Inch sizes:	0.250	0.315	0.380	0.440	0.500
Metric sizes:	6A	8A	10A	11A	13A

Inch sizes:	11/16	3/4	7/8	1
Metric sizes:	15A	17A	20A	23A

7–3(d). The following standards of the Society of Automotive Engineers (SAE) give dimensions and performance standards for automotive belts.

SAE Standard J636: V-belts and pulleys

SAE Standard J637: Automotive V-belt drives

SAE Standard J1278: SI (metric) synchronous belts and pulleys

SAE Standard J1313: Automotive synchronous belt drives

SAE Standard J1459: V-ribbed belts and pulleys

**7–4
V-BELT DRIVE
DESIGN**

The factors involved in selection of a V-belt and the driving and driven sheaves and proper installation of the drive are summarized in this section. Abbreviated examples of the data available from suppliers are given for illustration. Catalogs contain extensive data, and step-by-step instructions are given for their use. The basic data required for drive selection are the following:

■ The rated power of the driving motor or other prime mover
■ The service factor based on the type of driver and driven load
■ The center distance
■ The power rating for one belt as a function of the size and speed of the smaller sheave
■ The belt length
■ The size of the driving and driven sheaves

- The correction factor for belt length
- The correction factor for the angle of wrap on the smaller sheave
- The number of belts
- The initial tension on the belt

Many design decisions depend on the application and on space limitations. A few guidelines are given here:

- Adjustment for the center distance must be provided in both directions from the nominal value. The center distance must be shortened at the time of installation to enable the belt to be placed in the grooves of the sheaves without force. Provision for increasing the center distance must be made to permit the initial tensioning of the drive and to take up for belt stretch. Manufacturers' catalogs give the data. One convenient way to accomplish the adjustment is the use of a take-up unit, as shown in Figure 14–10(b) and (c).
- If fixed centers are required, idler pulleys should be used. It is best to use a grooved idler on the inside of the belt, close to the large sheave. Adjustable tensioners are commercially available to carry the idler.
- The nominal range of center distances should be

$$D_2 < C < 3 \, (D_2 + D_1) \tag{7–8}$$

- The angle of wrap on the smaller sheave should be greater than 120°.
- Most commercially available sheaves are cast iron, which should be limited to 6500-ft/min belt speed.
- Consider an alternative type of drive, such as a gear type or chain, if the belt speed is less than 1000 ft/min.
- Avoid elevated temperatures around belts.
- Ensure that the shafts carrying mating sheaves are parallel and that the sheaves are in alignment so that the belts track smoothly into the grooves.
- In multibelt installations, matched belts are required. Match numbers are printed on industrial belts, with 50 indicating a belt length very close to nominal. Longer belts carry match numbers above 50; shorter belts below 50.
- Belts must be installed with the initial tension recommended by the manufacturer. Tension should be checked after the first few hours of operation because seating and initial stretch occur.

Design Data

Catalogs typically give several dozen pages of design data for the various sizes of belts and sheave combinations to ease the job of drive design. The data typically are given in tabular form (see Reference 2). Graphical form is also used here so that you can get a feel for the variation in performance with design choices. Any design made from the data in this book should be checked against a particular manufacturer's ratings before use.

The data given here are for the narrow-section belts: 3V, 5V, and 8V. These three sizes cover a wide range of power transmission capacities. Figure 7–9 can be used to choose the basic size for the belt cross section. Note that the power axis is *design power*, the rated power of the prime mover times the service factor from Table 7–1.

Figures 7–10, 7–11, and 7–12 give the rated power per belt for the three cross sections as a function of the pitch diameter of the smaller sheave and its speed of rotation. The labeled vertical lines in each figure give the standard sheave pitch diameters available.

FIGURE 7–9

Selection chart for
narrow-section
industrial V-belts
(Dayco Corp., Dayton,
OH)

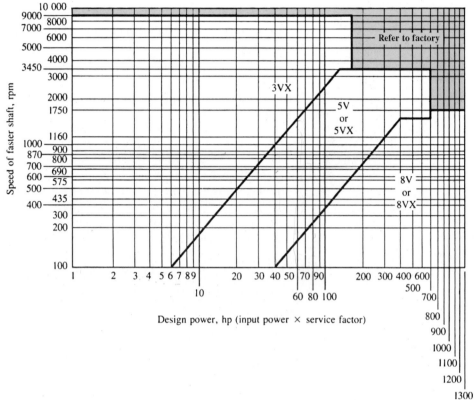

Design power, hp (input power × service factor)

TABLE 7–1   V-belt service factors

	Driver type					
	AC motors: Normal torque[a] DC motors: Shunt-wound Engines: Multiple-cylinder			AC motors: High torque[b] DC motors: Series-wound, compound-wound Engines: 4-cylinder or less		
Driven machine type	<6 h per day	6–15 h per day	>15 h per day	<6 h per day	6–15 h per day	>15 h per day
Agitators, blowers, fans, centrifugal pumps, light conveyors	1.0	1.1	1.2	1.1	1.2	1.3
Generators, machine tools, mixers, gravel conveyors	1.1	1.2	1.3	1.2	1.3	1.4
Bucket elevators, textile machines, hammer mills, heavy conveyors	1.2	1.3	1.4	1.4	1.5	1.6
Crushers, ball mills, hoists, rubber extruders	1.3	1.4	1.5	1.5	1.6	1.8
Any machine that can choke	2.0	2.0	2.0	2.0	2.0	2.0

[a]Synchronous, split-phase, three-phase with starting torque or breakdown torque less than 175% of full-load torque.
[b]Single-phase, three-phase with starting torque or breakdown torque greater than 175% of full-load torque.

The basic power rating for a speed ratio of 1.00 is given as the solid curve. A given belt can carry a greater power as the speed ratio increases, up to a ratio of approximately 3.38. Further increases have little effect and may also lead to trouble with the angle of wrap on the smaller sheave. Figure 7–13 is a plot of the data for power to be added to the basic rating as a function of speed ratio for the 5V belt size. The catalog data are given in a stepwise fashion.

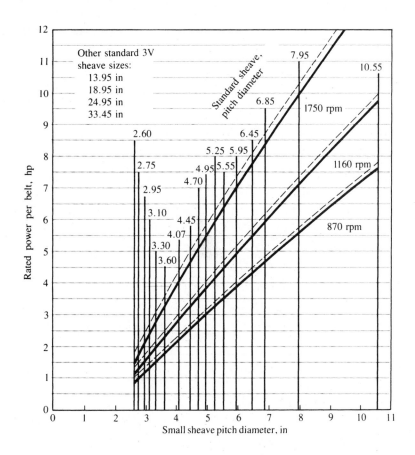

FIGURE 7–10
Power rating: 3V belts

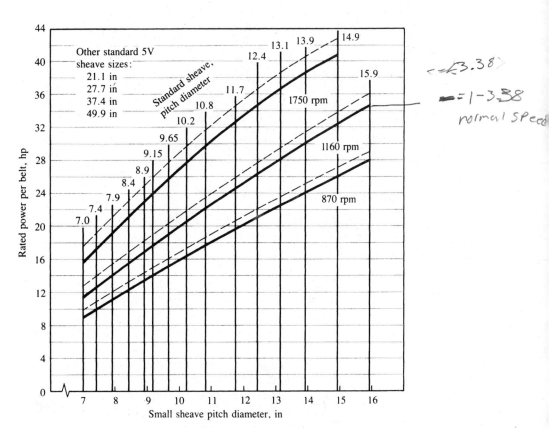

FIGURE 7–11
Power rating: 5V belts

FIGURE 7–12
Power rating: 8V belts

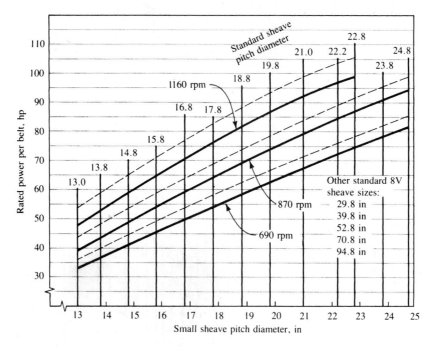

FIGURE 7–13
Power added versus
speed ratio: 5V belts

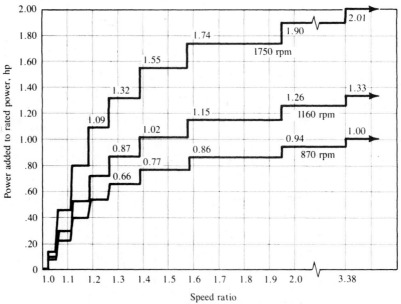

The maximum power added, for ratios of above 3.38, was used to draw the dashed curves in Figures 7–10, 7–11, and 7–12. In most cases, a rough interpolation between the two curves is satisfactory.

Figure 7–14 gives the value of a correction factor, C_θ, as a function of the angle of wrap of the belt on the small sheave.

Figure 7–15 gives the value of the correction factor, C_L, for belt length. A longer belt is desirable because it reduces the frequency with which a given part of the belt encounters the stress peak as it enters the small sheave. Only certain standard belt lengths are available (Table 7–2).

Example Problem 7–1 illustrates the use of the design data.

FIGURE 7–14 Angle of wrap correction factor, C_θ

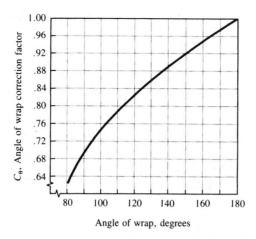

C_θ, Angle of wrap correction factor

Angle of wrap, degrees

FIGURE 7–15 Belt length correction factor, C_L

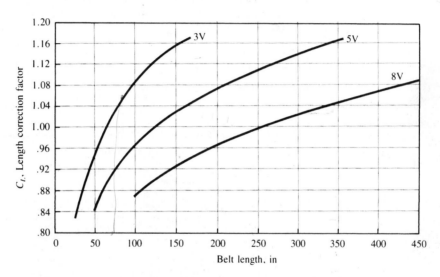

C_L, Length correction factor

Belt length, in

TABLE 7–2 Standard belt lengths for 3V, 5V, and 8V belts (in)

3V only	3V and 5V	3V, 5V, and 8V	5V and 8V	8V only
25	50	100	150	375
26.5	53	106	160	400
28	56	112	170	425
30	60	118	180	450
31.5	63	125	190	475
33.5	67	132	200	500
35.5	71	140	212	
37.5	75		224	
40	80		236	
42.5	85		250	
45	90		265	
47.5	95		280	
			300	
165			315	
			335	
			355	

Example Problem 7–1   Design a V-belt drive that has the input sheave on the shaft of an electric motor (normal torque) rated at 50.0 hp at 1160-rpm, full-load speed. The drive is to a bucket elevator in a potash plant that is to be used 12 hours (h) daily at approximately 675 rpm.

Solution   **Objective**   Design the V-belt drive.

Given
> Power transmitted = 50 hp to bucket elevator
> Speed of motor = 1160 rpm; output speed = 675 rpm

Analysis   Use the design data presented in this section. The solution procedure is developed within the Results section of the problem solution.

Results   *Step 1.* Compute the design power. From Table 7–1, for a normal torque electric motor running 12 h daily driving a bucket elevator, the service factor is 1.30. Then the design power is 1.30(50.0 hp) = 65.0 hp.

Step 2. Select the belt section. From Figure 7–9, a 5V belt is recommended for 70.0 hp at 1160-rpm input speed.

Step 3. Compute the nominal speed ratio:

$$\text{Ratio} = 1160/675 = 1.72$$

Step 4. Compute the driving sheave size that would produce a belt speed of 4000 ft/min, as a guide to selecting a standard sheave:

$$\text{Belt speed} = v_b = \frac{\pi D_1 n_1}{12} \text{ft/min}$$

Then the required diameter to give v_b = 4000 ft/min is

$$D_1 = \frac{12 \, v_b}{\pi n_1} = \frac{12(4000)}{\pi n_1} = \frac{15\,279}{n_1} = \frac{15\,279}{1160} = 13.17 \text{ in}$$

Step 5. Select trial sizes for the input sheave, and compute the desired size of the output sheave. Select a standard size for the output sheave, and compute the actual ratio and output speed.

For this problem, the trials are given in Table 7–3 (diameters are in inches).

The two trials in **boldface** in Table 7–3 give only about 1% variation from the desired output speed of 675 rpm, and the speed of a bucket elevator is not critical. Because no space limitations were given, let's choose the larger size.

Step 6. Determine the rated power from Figure 7–10, 7–11, or 7–12.

For the 5V belt that we have selected, Figure 7–11 is appropriate. For a 12.4-in sheave at 1160 rpm, the basic rated power is 26.4 hp. Multiple belts will be required. The ratio is relatively high, indicating that some added power rating can be used. This value can be estimated from Figure 7–11 or taken directly from Figure 7–13 for the 5V belt. Power added is 1.15 hp. Then the actual rated power is 26.4 + 1.15 = 27.55 hp.

Step 7. Specify a trial center distance.

We can use Equation (7–8) to determine a nominal acceptable range for *C*:

$$D_2 < C < 3(D_2 + D_1)$$
$$21.1 < C < 3(21.1 + 12.4)$$
$$21.1 < C < 100.5 \text{ in}$$

In the interest of conserving space, let's try *C* = 24.0 in.

TABLE 7–3 Trial sheave sizes for Example Problem 7–1

Standard driving sheave size, D_1	Approximate driven sheave size ($1.72D_1$)	Nearest standard sheave, D_2	Actual output speed (rpm)
13.10	22.5	21.1	720
12.4	**21.3**	**21.1**	**682**
11.7	20.1	21.1	643
10.8	18.6	21.1	594
10.2	17.5	15.9	744
9.65	16.6	15.9	704
9.15	**15.7**	**15.9**	**668**
8.9	15.3	14.9	693

Step 8. Compute the required belt length from Equation (7–3):

$$L = 2C + 1.57(D_2 + D_1) + \frac{(D_2 - D_1)^2}{4C}$$

$$L = 2(24.0) + 1.57(21.1 + 12.4) + \frac{(21.1 - 12.4)^2}{4(24.0)} = 101.4 \text{ in}$$

Step 9. Select a standard belt length from Table 7–2, and compute the resulting actual center distance from Equation (7–4).

In this problem, the nearest standard length is 100.0 in. Then, from Equation (7–4),

$$B = 4L - 6.28(D_2 + D_1) = 4(100) - 6.28(21.1 + 12.4) = 189.6$$

$$C = \frac{189.6 + \sqrt{(189.6)^2 - 32(21.1 - 12.4)^2}}{16} = 23.30 \text{ in}$$

Step 10. Compute the angle of wrap of the belt on the small sheave from Equation (7–5):

$$\theta_1 = 180° - 2\sin^{-1}\left[\frac{D_2 - D_1}{2C}\right] = 180° - 2\sin^{-1}\left[\frac{21.1 - 12.4}{2(23.30)}\right] = 158°$$

Step 11. Determine the correction factors from Figures 7–14 and 7–15. For $\theta = 158°$, $C_\theta = 0.94$. For $L = 100$ in, $C_L = 0.96$.

Step 12. Compute the corrected rated power per belt and the number of belts required to carry the design power:

$$\text{Corrected power} = C_\theta C_L P = (0.94)(0.96)(27.55 \text{ hp}) = 24.86 \text{ hp}$$

$$\text{Number of belts} = 65.0/24.86 = 2.61 \text{ belts (Use 3 belts.)}$$

Comments

Summary of Design

Input: Electric motor, 50.0 hp at 1160 rpm

Service factor: 1.4

Design power: 70.0 hp

Belt: 5V cross section, 100-in length, 3 belts

Sheaves: Driver, 12.4-in pitch diameter, 3 grooves, 5V. Driven, 21.1-in pitch diameter, 3 grooves, 5V

Actual output speed: 682 rpm

Center distance: 23.30 in

Belt Tension

The initial tension given to a belt is critical because it ensures that the belt will not slip under the design load. At rest, the two sides of the belt have the same tension. As power is being transmitted, the tension in the tight side increases while the tension in the slack side decreases. Without the initial tension, the slack side would go totally loose, and the belt would not seat in the groove; thus, it would slip. Manufacturers' catalogs give data for the proper belt-tensioning procedures.

Synchronous Belt Drives

Synchronous belts are constructed with ribs or teeth across the underside of the belt, as shown in Figure 7–3(c). The teeth mate with corresponding grooves in the driving and driven pulleys, called *sprockets*, providing a positive drive without slippage. Therefore, there is a fixed relationship between the speed of the driver and the speed of the driven sprocket. For this reason synchronous belts are often called *timing belts*. In contrast, V-belts can creep or slip with respect to their mating sheaves, especially under heavy loads and varying power demand. Synchronous action is critical to the successful operation of such systems as printing, material handling, packaging, and assembly. Synchronous belt drives are increasingly being considered for applications in which gear drives or chain drives had been used previously.

Figure 7–16 shows a synchronous belt mating with the toothed driving sprocket. Typical driving and driven sprockets are shown in Figure 7–17. At least one of the two sprockets will have side flanges to ensure that the belt does not move axially. Figure 7–18 shows

FIGURE 7–16
Synchronous belt on driving sprocket (Copyright Rockwell Automation, used by permission)

FIGURE 7–17
Driving and driven sprockets for synchronous belt drive (Copyright Rockwell Automation, used by permission)

FIGURE 7–18
Dimensions of standard synchronous belts

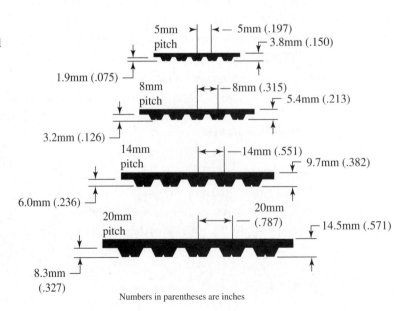

Numbers in parentheses are inches

the four common tooth pitches and sizes for commercially available synchronous belts. The pitch is the distance from the center of one tooth to the center of the next adjacent tooth. Standard pitches are 5 mm, 8 mm, 14 mm, and 20 mm.

Figure 7–3(c) shows detail of the construction of the cross section of a synchronous belt. The tensile strength is provided predominantly by high-strength cords made from fiberglass or similar materials. The cords are encased in a flexible rubber backing material, and the teeth are formed integrally with the backing. Often a fabric covering is used on those parts of the belt that contact the sprockets to provide additional wear resistance and higher net shear strength for the teeth. Various widths of the belts are available for each given pitch to provide a wide range of power transmission capacity.

Commercially available sprockets typically employ split-taper bushings in their hubs with a precise bore that provides a clearance of only 0.001 to 0.002 in (0.025 to 0.050 mm) relative to the shaft diameter on which it is to be mounted. Smooth, balanced, concentric operation results.

The process of selecting appropriate components for a synchronous belt drive is similar to that already discussed for V-belt drives. Manufacturers provide selection guides similar to those shown in Figure 7–19 that give the relationship between design power and the rotational speed of the smaller sprocket. These are used to determine the basic belt pitch required. Also, numerous pages of performance data are given showing the power transmission capacity for many combinations of belt width, driving and driven sprocket size, and center distances between the axes of the sprockets for specific belt lengths. In general the selection process involves the following steps. Refer to data and design procedures for specific manufacturers as listed in Internet sites 2–5.

General Selection Procedure for Synchronous Belt Drives

1. Specify the speed of the driving sprocket (typically a motor or engine) and the desired speed of the driven sprocket.

2. Specify the rated power for the driving motor or engine.

3. Determine a service factor, using manufacturers' recommendations, considering the type of driver and the nature of the driven machine.

4. Calculate the design power by multiplying the driver rated power by the service factor.

5. Determine the required pitch of the belt from a specific manufacturer's data.

6. Calculate the speed ratio between the driver and the driven sprocket.

7. Select several candidate combinations of the number of teeth in the driver sprocket to that in the driven sprocket that will produce the desired ratio.

8. Using the desired range of acceptable center distances, determine a standard belt length that will produce a suitable value.

9. A belt-length correction factor may be required. Catalog data will show factors less than 1.0 for shorter center distances and greater than 1.0 for longer center distances. This reflects the frequency with which a given part of the belt encounters the high-stress area as it enters the smaller sprocket. Apply the factor to the rated power capacity for the belt.

10. Specify the final design details for the sprockets such as flanges, type and size of bushings in the hub, and the bore size to match the mating shafts.

11. Summarize the design, check compatibility with other components of the system, and prepare purchasing documents.

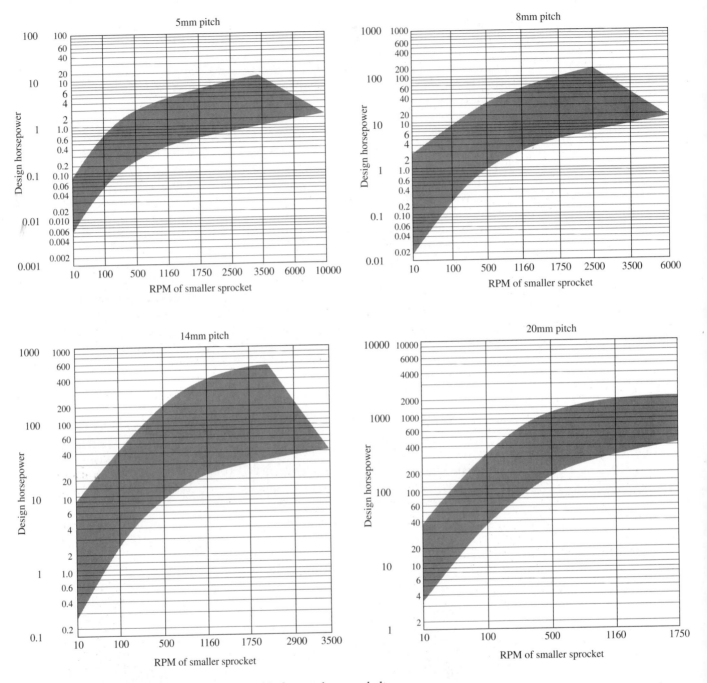

FIGURE 7–19 Belt pitch selection guide for synchronous belts

      Installation of the sprockets and the belt requires a nominal amount of center distance allowance to enable the belt teeth to slide into the sprocket grooves without force. Subsequently, the center distance will normally have to be adjusted outward to provide a suitable amount of initial tension as defined by the manufacturer. The initial tension is typically less than that required for a V-belt drive. Idlers can be used to take up slack if fixed centers are required between the driver and driven sprockets. However, they may decrease the life of the belt. Consult the manufacturer.

In operation, the final tension in the tight side of the belt is much less than that developed by a V-belt and the slack side tension is virtually zero. The results are lower net forces in the belt, lower side loads on the shafts carrying the sprockets, and reduced bearing loads.

7–5 CHAIN DRIVES

A chain is a power transmission element made as a series of pin-connected links. The design provides for flexibility while enabling the chain to transmit large tensile forces. See References 1, 6, and 7 for more technical information and manufacturers' data.

When transmitting power between rotating shafts, the chain engages mating toothed wheels, called sprockets. Figure 7–20 shows a typical chain drive.

The most common type of chain is the *roller chain,* in which the roller on each pin provides exceptionally low friction between the chain and the sprockets. Other types include a variety of extended link designs used mostly in conveyor applications (see Figure 7–21).

Roller chain is classified by its *pitch,* the distance between corresponding parts of adjacent links. The pitch is usually illustrated as the distance between the centers of adjacent pins. Standard roller chain carries a size designation from 40 to 240, as listed in Table 7–4. The digits (other than the final zero) indicate the pitch of the chain in eighths of an inch, as in the table. For example, the no. 100 chain has a pitch of 10/8 or $1\frac{1}{4}$ in. A series of heavy-duty sizes, with the suffix *H* on the designation (60H–240H), has the same basic dimensions as the standard chain of the same number except for thicker side plates. In addition, there are the smaller and lighter sizes: 25, 35, and 41.

The average tensile strengths of the various chain sizes are also listed in Table 7–4. These data can be used for very low speed drives or for applications in which the function of the chain is to apply a tensile force or to support a load. It is recommended that only 10% of the average tensile strength be used in such applications. For power transmission, the rating of a given chain size as a function of the speed of rotation must be determined, as explained later in this chapter.

A wide variety of attachments are available to facilitate the application of roller chain to conveying or other material handling uses. Usually in the form of extended plates

FIGURE 7–20
Roller chain drive (Rexnord, Inc., Milwaukee, WI)

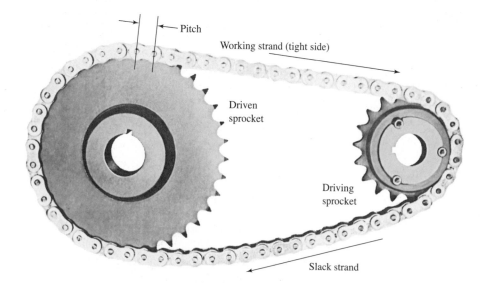

FIGURE 7–21 Some roller chain styles (Rexnord, Inc., Milwaukee, WI)

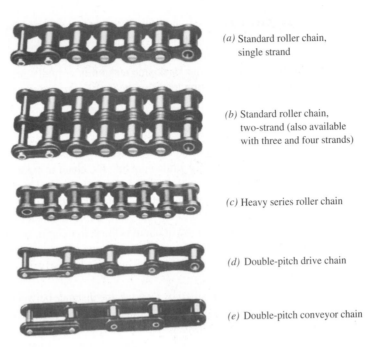

(a) Standard roller chain, single strand

(b) Standard roller chain, two-strand (also available with three and four strands)

(c) Heavy series roller chain

(d) Double-pitch drive chain

(e) Double-pitch conveyor chain

TABLE 7–4 Roller chain sizes

Chain number	Pitch (in)	Roller diameter	Roller width	Link plate thickness	Average tensile strength (lb)
25	1/4	None	–	0.030	925
35	3/8	None	–	0.050	2100
41	1/2	0.306	0.250	0.050	2000
40	1/2	0.312	0.312	0.060	3700
50	5/8	0.400	0.375	0.080	6100
60	3/4	0.469	0.500	0.094	8500
80	1	0.626	0.625	0.125	14 500
100	$1\frac{1}{4}$	0.750	0.750	0.156	24 000
120	$1\frac{1}{2}$	0.875	1.000	0.187	34 000
140	$1\frac{3}{4}$	1.000	1.000	0.219	46 000
160	2	1.125	1.250	0.250	58 000
180	$2\frac{1}{4}$	1.406	1.406	0.281	80 000
200	$2\frac{1}{2}$	1.562	1.500	0.312	95 000
240	3	1.875	1.875	0.375	130 000

or tabs with holes provided, the attachments make it easy to connect rods, buckets, parts pushers, part support devices, or conveyor slats to the chain. Figure 7–22 shows some attachment styles.

Figure 7–23 shows a variety of chain types used especially for conveying and similar applications. Such chain typically has a longer pitch than standard roller chain (usually twice the pitch), and the link plates are heavier. The larger sizes have cast link plates.

FIGURE 7–22 Chain attachments (Rexnord, Inc., Milwaukee, WI)

(*a*) Slats assembled to attachments to form a flat conveying surface

(*b*) *V* block assembled to attachments to convey round objects of varying diameters

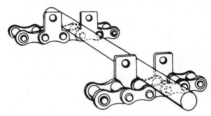

(*c*) Attachments used as spacers to convey and position long objects

7–6 DESIGN OF CHAIN DRIVES

The rating of chain for its power transmission capacity considers three modes of failure: (1) fatigue of the link plates due to the repeated application of the tension in the tight side of the chain, (2) impact of the rollers as they engage the sprocket teeth, and (3) galling between the pins of each link and the bushings on the pins.

The ratings are based on empirical data with a smooth driver and a smooth load (service factor = 1.0) and with a rated life of approximately 15 000 h. The important variables are the pitch of the chain and the size and rotational speed of the smaller sprocket. Lubrication is critical to the satisfactory operation of a chain drive. Manufacturers recommend the type of lubrication method for given combinations of chain size, sprocket size, and speed. Details are discussed later.

Tables 7–5, 7–6, and 7–7 list the rated power for three sizes of standard chain: no. 40 (1/2 in), no. 60 (3/4 in), and no. 80 (1.00 in). These are typical of the types of data available for all chain sizes in manufacturers' catalogs. Notice these features of the data:

1. The ratings are based on the speed of the smaller sprocket and an expected life of approximately 15 000 hours.

2. For a given speed, the power capacity increases with the number of teeth on the sprocket. Of course, the larger the number of teeth, the larger the diameter of the sprocket. Note that the use of a chain with a small pitch on a large sprocket produces the quieter drive.

3. For a given sprocket size (a given number of teeth), the power capacity increases with increasing speed up to a point; then it decreases. Fatigue due to the tension in the chain governs at the low to moderate speeds; impact on the sprockets governs at the higher speeds. Each sprocket size has an absolute upper-limit speed due to the onset of galling between the pins and the bushings

FIGURE 7–23
Conveyor chains
(Rexnord, Inc.,
Milwaukee, WI)

Mill, narrow series
(drive and conveyor sizes)
Offset cast-link chain used primarily in the lumber
industry for conveyor applications.

Combination mill
(wide conveyor sizes)
Cast block links and steel sidebar construction for
drag conveyor applications.

Heavy-duty drag chain
Cast steel offset block links. Used in ash and
clinker conveyors.

Pintle chain
Chain constructed of a series of cast offset links
coupled by steel pins or rivets. Suitable for slow-to
moderate-speed drive, conveyor and elevator
service.

Roller-top transfer
Cast links with top rollers used in several strands to
convey material transversely.

Roof-top
Cast root-shaped links used in several strands on
transfer conveyors.

Detachable
Consists of unit links, each with an open-type hook
that flexes on the end bar of the adjacent link. Used
for slow- to moderate-speed drive and conveyor
application.

Drop-forged
Drop-forged inner and outer links coupled by
headed pins. Used for trolley, scraper, flight and
similar conveyors.

of the chain. This explains the abrupt drop in power capacity to zero at the limiting speed.

4. The ratings are for a single strand of chain. Although multiple strands do increase the power capacity, they do not provide a direct multiple of the single-strand capacity. Multiply the capacity in the tables by the following factors.

     Two strands: Factor = 1.7

     Three strands: Factor = 2.5

     Four strands: Factor = 3.3

5. The ratings are for a service factor of 1.0. Specify a service factor for a given application according to Table 7–8.

TABLE 7–5   Horsepower ratings—single strand roller chain no. 40

	0.500 inch pitch							Rotational speed of small sprocket, rev/min																	
No. of teeth	10	25	50	100	180	200	300	500	700	900	1000	1200	1400	1600	1800	2100	2500	3000	3500	4000	5000	6000	7000	8000	9000
11	0.06	0.14	0.27	0.52	0.91	1.00	1.48	2.42	3.34	4.25	4.70	5.60	6.49	5.57	4.66	3.70	2.85	2.17	1.72	1.41	1.01	0.77	0.61	0.50	0.00
12	0.06	0.15	0.29	0.56	0.99	1.09	1.61	2.64	3.64	4.64	5.13	6.11	7.09	6.34	5.31	4.22	3.25	2.47	1.96	1.60	1.15	0.87	0.69	0.57	0.00
13	0.07	0.16	0.31	0.61	1.07	1.19	1.75	2.86	3.95	5.02	5.56	6.62	7.68	7.15	5.99	4.76	3.66	2.79	2.21	1.81	1.29	0.98	0.78	0.00	
14	0.07	0.17	0.34	0.66	1.15	1.28	1.88	3.08	4.25	5.41	5.98	7.13	8.27	7.99	6.70	5.31	4.09	3.11	2.47	2.02	1.45	1.10	0.87	0.00	
15	0.08	0.19	0.36	0.70	1.24	1.37	2.02	3.30	4.55	5.80	6.41	7.64	8.86	8.86	7.43	5.89	4.54	3.45	2.74	2.24	1.60	1.22	0.97	0.00	
16	0.08	0.20	0.39	0.75	1.32	1.46	2.15	3.52	4.86	6.18	6.84	8.15	9.45	9.76	8.18	6.49	5.00	3.80	3.02	2.47	1.77	1.34	0.00		
17	0.09	0.21	0.41	0.80	1.40	1.55	2.29	3.74	5.16	6.57	7.27	8.66	10.04	10.69	8.96	7.11	5.48	4.17	3.31	2.71	1.94	1.47	0.00		
18	0.09	0.22	0.43	0.84	1.48	1.64	2.42	3.96	5.46	6.95	7.69	9.17	10.63	11.65	9.76	7.75	5.97	4.54	3.60	2.95	2.11	1.60	0.00		
19	0.10	0.24	0.46	0.89	1.57	1.73	2.56	4.18	5.77	7.34	8.12	9.66	11.22	12.64	10.59	8.40	6.47	4.92	3.91	3.20	2.29	0.09	0.00		
20	0.10	0.25	0.48	0.94	1.65	1.82	2.69	4.39	6.07	7.73	8.55	10.18	11.81	13.42	11.44	9.07	6.99	5.31	4.22	3.45	2.47	0.00			
21	0.11	0.26	0.51	0.98	1.73	1.91	2.83	4.61	6.37	8.11	8.98	10.69	12.40	14.10	12.30	9.76	7.52	5.72	4.54	3.71	2.65	0.00			
22	0.11	0.27	0.53	1.03	1.81	2.01	2.96	4.83	6.68	8.50	9.40	11.20	12.99	14.77	13.19	10.47	8.06	6.13	4.87	3.98	2.85	0.00			
23	0.12	0.28	0.56	1.08	1.90	2.10	3.10	5.05	6.98	8.89	9.83	11.71	13.58	15.44	14.10	11.19	8.62	6.55	5.20	4.26	3.05	0.00			
24	0.12	0.30	0.58	1.12	1.98	2.19	3.23	5.27	7.28	9.27	10.26	12.22	14.17	16.11	15.03	11.93	9.18	6.99	5.54	4.54	0.87	0.00			
25	0.13	0.31	0.60	1.17	2.06	2.28	3.36	5.49	7.59	9.66	10.69	12.73	14.76	16.78	15.98	12.68	9.76	7.43	5.89	4.82	0.00				
26	0.13	0.32	0.63	1.22	2.14	2.37	3.50	5.71	7.89	10.04	11.11	13.24	15.35	17.45	16.95	13.45	10.36	7.88	6.25	5.12	0.00				
28	0.14	0.35	0.67	1.31	2.31	2.55	3.77	6.15	8.50	10.82	11.97	14.26	16.53	18.79	18.94	15.03	11.57	8.80	6.99	5.72	0.00				
30	0.15	0.37	0.72	1.41	2.47	2.74	4.04	6.59	9.11	11.59	12.82	15.28	17.71	20.14	21.01	16.67	12.84	9.76	7.75	6.34	0.00				
32	0.16	0.40	0.77	1.50	2.64	2.92	4.31	7.03	9.71	12.38	13.68	16.30	18.89	21.48	23.14	18.37	14.14	10.76	8.54	1.41					
35	0.18	0.43	0.84	1.64	2.88	3.19	4.71	7.69	10.62	13.52	14.96	17.82	20.67	23.49	26.30	21.01	16.17	12.30	9.76	0.00					
40	0.21	0.50	0.96	1.87	3.30	3.65	5.38	8.79	12.14	15.45	17.10	20.37	23.62	26.85	30.06	25.67	19.76	15.03	0.00						
45	0.23	0.56	1.08	2.11	3.71	4.10	6.08	9.89	13.66	17.39	19.24	22.92	26.57	30.20	33.82	30.63	23.58	5.53	0.00						

Type A   Type B   Type C

Type A: Manual or drip lubrication
Type B: Bath or disc lubrication
Type C: Oil stream lubrication

Source: American Chain Association, Naples, FL

TABLE 7–6 Horsepower ratings—single strand roller chain no. 60

No. of teeth	0.750 inch pitch					Rotational speed of small sprocket, rev/min																			
	10	25	50	100	120	200	300	400	500	600	800	1000	1200	1400	1600	1800	2000	2500	3000	3500	4000	4500	5000	5500	6000
11	0.19	0.46	0.89	1.72	2.05	3.35	4.95	6.52	8.08	9.63	12.69	15.58	11.85	9.41	7.70	6.45	5.51	3.94	3.00	2.38	1.95	1.63	1.39	1.21	0.00
12	0.21	0.50	0.97	1.88	2.24	3.66	5.40	7.12	8.82	10.51	13.85	17.15	13.51	10.72	8.77	7.35	6.28	4.49	3.42	2.71	2.22	1.86	1.59	1.38	0.00
13	0.22	0.54	1.05	2.04	2.43	3.96	5.85	7.71	9.55	11.38	15.00	18.58	15.23	12.08	9.89	8.29	7.08	5.06	3.85	3.06	2.50	2.10	1.79	0.00	
14	0.24	0.58	1.13	2.19	2.61	4.27	6.30	8.30	10.29	12.26	16.15	20.01	17.02	13.51	11.05	9.26	7.91	5.66	4.31	3.42	2.80	2.34	0.41	0.00	
15	0.26	0.62	1.21	2.35	2.80	4.57	6.75	8.90	11.02	13.13	17.31	21.44	18.87	14.98	12.26	10.27	8.77	6.28	4.77	3.79	3.10	2.60	0.00		
16	0.27	0.66	1.29	2.51	2.99	4.88	7.20	9.49	11.76	14.01	18.46	22.87	20.79	16.50	13.51	11.32	9.66	6.91	5.26	4.17	3.42	1.78	0.00		
17	0.29	0.70	1.37	2.66	3.17	5.18	7.65	10.08	12.49	14.88	19.62	24.30	22.77	18.07	14.79	12.40	10.58	7.57	5.76	4.57	3.74	0.00			
18	0.31	0.75	1.45	2.82	3.36	5.49	8.10	10.68	13.23	15.76	20.77	25.73	24.81	19.69	16.11	13.51	11.53	8.25	6.28	4.98	4.08	0.00			
19	0.33	0.79	1.53	2.98	3.55	5.79	8.55	11.27	13.96	16.63	21.92	27.16	26.91	21.35	17.48	14.65	12.50	8.95	6.81	5.40	0.20	0.00			
20	0.34	0.83	1.61	3.13	3.73	6.10	9.00	11.86	14.70	17.51	23.08	28.59	29.06	23.06	18.87	15.82	13.51	9.66	7.35	5.83	0.00				
21	0.36	0.87	1.69	3.29	3.92	6.40	9.45	12.46	15.43	18.38	24.23	30.02	31.26	24.81	20.31	17.02	14.53	10.40	7.91	6.28	0.00				
22	0.38	0.91	1.77	3.45	4.11	6.71	9.90	13.05	16.17	19.26	25.39	31.45	33.52	26.60	21.77	18.25	15.58	11.15	8.48	0.00					
23	0.40	0.95	1.85	3.61	4.29	7.01	10.35	13.64	16.90	20.13	26.54	32.88	35.84	28.44	23.28	19.51	16.66	11.92	9.07	0.00					
24	0.41	0.99	1.93	3.76	4.48	7.32	10.80	14.24	17.64	21.01	27.69	34.31	38.20	30.31	24.81	20.79	17.75	12.70	9.66	0.00					
25	0.43	1.04	2.01	3.92	4.67	7.62	11.25	14.83	18.37	21.89	28.85	35.74	40.61	32.23	26.38	22.11	18.87	13.51	10.27	0.00					
26	0.45	1.08	2.09	4.08	4.85	7.93	11.70	15.42	19.11	22.76	30.00	37.17	43.07	34.18	27.98	23.44	20.02	14.32	10.90	0.00					
28	0.48	1.16	2.26	4.39	5.23	8.54	12.60	16.61	20.58	24.51	32.31	40.03	47.68	38.20	31.26	26.20	22.37	16.01	0.00						
30	0.52	1.24	2.42	4.70	5.60	9.15	13.50	17.79	22.05	26.26	34.62	42.89	51.09	42.36	34.67	29.06	24.81	17.75	0.00						
32	0.55	1.33	2.58	5.02	5.98	9.76	14.40	18.98	23.52	28.01	36.92	45.75	54.50	46.67	38.20	32.01	27.33	19.56	0.00						
35	0.60	1.45	2.82	5.49	6.54	10.67	15.75	20.76	25.72	30.64	40.39	50.03	59.60	53.38	43.69	36.62	31.26	1.35	0.00						
40	0.69	1.66	3.22	6.27	7.47	12.20	18.00	23.73	29.39	35.02	46.16	57.18	68.12	65.22	53.38	44.74	38.20	0.00							
45	0.77	1.86	3.63	7.05	8.40	13.72	20.25	26.69	33.07	38.39	51.92	64.33	76.63	77.83	63.70	53.38	12.45	0.00							

Type A   Type B   Type C

Type A: Manual or drip lubrication
Type B: Bath of disc lubrication
Type C: Oil stream lubrication

Source: American Chain Association, Naples, FL.

TABLE 7-7 Horsepower ratings—single strand roller chain no. 80

No. of teeth	1.000 inch pitch — Rotational speed of small sprocket, rev/min																								
	10	25	50	75	88	100	200	300	400	500	600	700	800	900	1000	1200	1400	1600	1800	2000	2500	3000	3500	4000	4500
11	0.44	1.06	2.07	3.05	3.56	4.03	7.83	11.56	15.23	18.87	22.48	26.07	27.41	22.97	19.61	14.92	11.84	9.69	8.12	6.83	4.96	3.77	3.00	2.45	0.00
12	0.48	1.16	2.26	3.33	3.88	4.39	8.54	12.61	16.82	20.59	24.53	28.44	31.23	26.17	22.35	17.00	13.49	11.04	9.25	7.90	5.65	4.30	3.41	2.79	0.00
13	0.52	1.26	2.45	3.61	4.21	4.76	9.26	13.66	18.00	22.31	26.57	30.81	35.02	29.51	25.20	19.17	15.21	12.45	10.43	8.91	6.37	4.85	3.85	3.15	
14	0.56	1.35	2.63	3.89	4.53	5.12	9.97	14.71	19.39	24.02	28.62	33.18	37.72	32.98	28.16	21.42	17.00	13.91	11.66	9.96	7.12	5.42	4.30	3.52	
15	0.60	1.45	2.82	4.16	4.86	5.49	10.68	15.76	20.77	25.74	30.66	35.55	40.41	36.58	31.23	23.76	18.85	15.43	12.93	11.04	7.90	6.01	4.77	0.00	
16	0.64	1.55	3.01	4.44	5.18	5.86	11.39	16.81	22.16	27.45	32.70	37.92	43.11	40.30	34.41	26.17	20.77	17.00	14.25	12.16	8.70	6.62	5.25		
17	0.68	1.64	3.20	4.72	5.50	6.22	12.10	17.86	23.54	29.17	34.75	40.29	45.80	44.13	37.68	28.66	22.75	18.62	15.60	13.32	9.53	7.25	0.00		
18	0.72	1.74	3.39	5.00	5.83	6.59	12.81	18.91	24.93	30.88	36.79	42.66	48.49	48.08	41.05	31.23	24.78	20.29	17.00	14.51	10.39	7.90	0.00		
19	0.76	1.84	3.57	5.28	6.15	6.95	13.53	19.96	26.31	32.60	38.84	45.03	51.19	52.15	44.52	33.87	26.88	22.00	18.44	15.74	11.26	0.36	0.00		
20	0.80	1.93	3.76	5.55	6.47	7.32	14.24	21.01	27.70	34.32	40.88	47.40	53.88	56.32	48.08	36.58	29.03	23.76	19.91	17.00	12.16	0.00			
21	0.84	2.03	3.95	5.83	6.80	7.69	14.95	22.07	29.08	36.03	42.92	49.77	56.58	60.59	51.73	39.36	31.23	25.56	21.42	18.29	13.09	0.00			
22	0.88	2.13	4.14	6.11	7.12	8.05	15.66	23.12	30.47	37.75	44.97	52.14	59.27	64.97	55.47	42.20	33.49	27.41	22.97	19.61	14.03				
23	0.92	2.22	4.33	6.39	7.45	8.42	16.37	24.17	31.85	39.46	47.01	54.51	61.97	69.38	59.30	45.11	35.80	29.30	24.55	20.97	15.00				
24	0.96	2.32	4.52	6.66	7.77	8.78	17.09	25.22	33.24	41.18	49.06	56.88	64.66	72.40	63.21	48.08	38.16	31.23	26.17	22.35	15.99				
25	1.00	2.42	4.70	6.94	8.09	9.15	17.80	26.27	34.62	42.89	51.10	59.25	67.35	75.42	67.20	51.12	40.57	33.20	27.83	23.76	8.16				
26	1.04	2.51	4.89	7.22	8.42	9.52	18.51	27.32	36.01	44.61	53.14	61.62	70.05	78.43	71.27	54.22	43.02	36.22	29.51	25.20	0.00				
28	1.12	2.71	5.27	7.77	9.06	10.25	19.93	29.42	38.78	48.04	57.23	66.36	75.44	84.47	79.65	60.59	48.08	39.36	32.98	28.16	0.00				
30	1.20	2.90	5.64	8.33	9.71	10.98	21.36	31.52	41.55	51.47	61.32	71.10	80.82	90.50	88.33	67.20	53.33	43.65	36.58	31.23					
32	1.28	3.09	6.02	8.89	10.36	11.71	22.78	33.62	44.32	54.91	65.41	75.84	86.21	96.53	97.31	74.03	58.75	48.08	40.30	5.65					
35	1.40	3.38	6.58	9.72	11.33	12.81	24.92	36.78	48.47	60.05	71.54	82.95	94.29	105.58	111.31	84.68	67.20	55.00	28.15	0.00					
40	1.61	3.87	7.53	11.11	12.95	14.64	28.48	42.03	55.40	68.63	81.76	94.80	107.77	120.67	133.51	103.46	82.10	40.16	0.00						
45	1.81	4.35	8.47	12.49	14.57	16.47	32.04	47.28	62.32	77.21	91.98	106.65	121.24	135.75	150.20	123.45	72.28	0.00							

Type A   Type B   Type C

Type A: Manual or drip lubrication
Type B: Bath of disc lubrication
Type C: Oil stream lubrication

Source: American Chain Association, Naples, FL

TABLE 7–8 Service factors for chain drives

	Type of driver		
Load type	Hydraulic drive	Electric motor or turbine	Internal combustion engine with mechanical drive
Smooth (agitators; fans; light, uniformly loaded conveyors)	1.0	1.0	1.2
Moderate shock (machine tools, cranes, heavy conveyors, food mixers and grinders)	1.2	1.3	1.4
Heavy shock (punch presses, hammer mills, reciprocating conveyors, rolling mill drive)	1.4	1.5	1.7

Design Guidelines for Chain Drives

The following are general recommendations for designing chain drives:

1. The minimum number of teeth in a sprocket should be 17 unless the drive is operating at a very low speed, under 100 rpm.

2. The maximum speed ratio should be 7.0, although higher ratios are feasible. Two or more stages of reduction can be used to achieve higher ratios.

3. The center distance between the sprocket axes should be approximately 30 to 50 pitches (30 to 50 times the pitch of the chain).

4. The larger sprocket should normally have no more than 120 teeth.

5. The preferred arrangement for a chain drive is with the centerline of the sprockets horizontal and with the tight side on top.

6. The chain length must be an integral multiple of the pitch, and an even number of pitches is recommended. The center distance should be made adjustable to accommodate the chain length and to take up for tolerances and wear. Excessive sag on the slack side should be avoided, especially on drives that are not horizontal. A convenient relation between center distance (C), chain length (L), number of teeth in the small sprocket (N_1), and number of teeth in the large sprocket (N_2), expressed in pitches, is

$$L = 2C + \frac{N_2 + N_1}{2} + \frac{(N_2 - N_1)^2}{4\pi^2 C} \tag{7-9}$$

The center distance for a given chain length, again in pitches, is

$$C = \frac{1}{4}\left[L - \frac{N_2 + N_1}{2} + \sqrt{\left[L - \frac{N_2 + N_1}{2}\right]^2 - \frac{8\,(N_2 - N_1)^2}{4\pi^2}} \right] \tag{7-10}$$

The computed center distance assumes no sag in either the tight or the slack side of the chain, and thus it is a *maximum*. Negative tolerances or adjustment must be provided. Adjustment for wear must also be provided.

7. The pitch diameter of a sprocket with N teeth for a chain with a pitch of p is

$$D = \frac{p}{\sin(180°/N)} \qquad (7\text{--}11)$$

8. The minimum sprocket diameter and therefore the minimum number of teeth in a sprocket are often limited by the size of the shaft on which it is mounted. Check the sprocket catalog.

9. The arc of contact, θ_1, of the chain on the smaller sprocket should be greater than 120°.

$$\theta_1 = 180° - 2 \sin^{-1}\left[(D_2 - D_1)/2C\right] \qquad (7\text{--}12)$$

10. For reference, the arc of contact, θ_2, on the larger sprocket is,

$$\theta_2 = 180° + 2 \sin^{-1}\left[(D_2 - D_1)/2C\right] \qquad (7\text{--}13)$$

Lubrication

It is essential that adequate lubrication be provided for chain drives. There are numerous moving parts within the chain, along with the interaction between the chain and the sprocket teeth. The designer must define the lubricant properties and the method of lubrication.

Lubricant Properties. Petroleum-based lubricating oil similar to engine oil is recommended. Its viscosity must enable the oil to flow readily between chain surfaces that move relative to each other while providing adequate lubrication action. The oil should be kept clean and free of moisture. Table 7–9 gives the recommended lubricant for different ambient temperatures.

Method of Lubrication. The American Chain Association recommends three different types of lubrication depending on the speed of operation and the power being transmitted. See Tables 7–5 to 7–7 or manufacturer's catalogs for recommendations. Refer to the following descriptions of the methods and the illustrations in Figure 7–24.

Type A. Manual or drip lubrication: For manual lubrication, oil is applied copiously with a brush or a spout can, at least once every 8 h of operation. For drip feed lubrication, oil is fed directly onto the link plates of each chain strand.

TABLE 7–9 Recommended lubricant for chain drives

Ambient temperature		Recommended lubricant
°F	°C	
20 to 40	-7 to 5	SAE 20
40 to 100	5 to 38	SAE 30
100 to 120	38 to 49	SAE 40
120 to 140	49 to 60	SAE 50

FIGURE 7–24
Lubrication methods
(American Chain
Association,
Naples, FL)

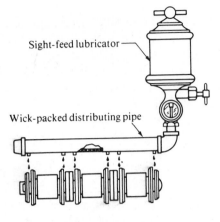

Sight-feed lubricator

Wick-packed distributing pipe

(a) Drip feed lubrication (Type A)

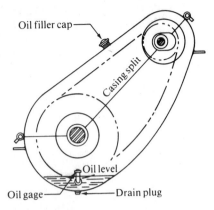

Oil filler cap

Casing split

Oil level

Oil gage   Drain plug

(b) Shallow bath lubrication (Type B)

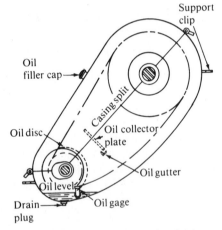

Support
clip

Oil
filler cap

Casing split

Oil disc

Oil collector
plate

Oil gutter

Oil level

Drain
plug   Oil gage

(c) Disc or slinger lubrication (Type B)

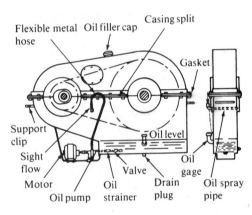

Flexible metal   Oil filler cap   Casing split
hose

Gasket

Support
clip

Oil level

Sight
flow   Oil
Motor   Oil   Valve   Drain   gage
Oil pump   strainer   plug   Oil spray
pipe

(d) Oil stream lubrication (Type C)

Type B.   *Bath or disc lubrication*: The chain cover provides a sump of oil into which the chain dips continuously. Alternatively, a disc or a slinger can be attached to one of the shafts to lift oil to a trough above the lower strand of chain. The trough then delivers a stream of oil to the chain. The chain itself, then, does not need to dip into the oil.

Type C.   *Oil stream lubrication*: An oil pump delivers a continuous stream of oil on the lower part of the chain.

Example Problem 7–2   Design a chain drive for a heavily loaded coal conveyor to be driven by a gasoline engine through a mechanical drive. The input speed will be 900 rpm, and the desired output speed is 230 to 240 rpm. The conveyor requires 15.0 hp.

Solution   Objective   Design the chain drive.

Given

Power transmitted = 15 hp to a coal conveyor

Speed of motor = 900 rpm; output speed range = 230 to 240 rpm

Analysis Use the design data presented in this section. The solution procedure is developed within the Results section of the problem solution.

Results

Step 1. Specify a service factor and compute the design power. From Table 7–8, for moderate shock and a gasoline engine drive through a mechanical drive, $SF = 1.4$.

$$\text{Design power} = 1.4(15.0) = 21.0 \text{ hp}$$

Step 2. Compute the desired ratio. Using the middle of the required range of output speeds, we have

$$\text{Ratio} = (900 \text{ rpm})/(235 \text{ rpm}) = 3.83$$

Step 3. Refer to the tables for power capacity (Tables 7–5, 7–6, and 7–7), and select the chain pitch. For a single strand, the no. 60 chain with $p = 3/4$ in seems best. A 17-tooth sprocket is rated at 21.96 hp at 900 rpm by interpolation. At this speed, type B lubrication (oil bath) is required.

Step 4. Compute the required number of teeth on the large sprocket:

$$N_2 = N_1 \times \text{ratio} = 17(3.83) = 65.11$$

Let's use the integer: 65 teeth.

Step 5. Compute the actual expected output speed:

$$n_2 = n_1(N_1/N_2) = 900 \text{ rpm}(17/65) = 235.3 \text{ rpm (Okay!)}$$

Step 6. Compute the pitch diameters of the sprockets using Equation (7–11):

$$D_1 = \frac{p}{\sin(180°/N_1)} = \frac{0.75 \text{ in}}{\sin(180°/17)} = 4.082 \text{ in}$$

$$D_2 = \frac{p}{\sin(180°/N_2)} = \frac{0.75 \text{ in}}{\sin(180°/65)} = 15.524 \text{ in}$$

Step 7. Specify the nominal center distance. Let's use the middle of the recommended range, 40 pitches.

Step 8. Compute the required chain length in pitches from Equation (7–9):

$$L = 2C + \frac{N_2 + N_1}{2} + \frac{(N_2 - N_1)^2}{4\pi^2 C}$$

$$L = 2(40) + \frac{65 + 17}{2} + \frac{(65 - 17)^2}{4\pi^2 (40)} = 122.5 \text{ pitches} \qquad \textbf{(7–9)}$$

Step 9. Specify an integral number of pitches for the chain length, and compute the actual theoretical center distance. Let's use 122 pitches, an even number. Then, from Equation (7–10),

$$C = \frac{1}{4}\left[L - \frac{N_2 + N_1}{2} + \sqrt{\left[L - \frac{N_2 + N_1}{2}\right]^2 - \frac{8\,(N_2 - N_1)^2}{4\pi^2}}\,\right]$$

$$C = \frac{1}{4}\left[122 - \frac{65 + 17}{2} + \sqrt{\left[122 - \frac{65 + 17}{2}\right]^2 - \frac{8(65 - 17)^2}{4\pi^2}}\,\right] \tag{7–10}$$

$C = 39.766$ pitches $= 39.766(0.75 \text{ in}) = 29.825 \text{ in}$

Step 10. Compute the angle of wrap of the chain for each sprocket using Equations (7–12) and (7–13). Note that the minimum angle of wrap should be 120 degrees.
For the small sprocket,

$$\theta_1 = 180° - 2 \sin^{-1}\left[(D_2 - D_1)/2C\right]$$
$$\theta_1 = 180° - 2 \sin^{-1}\left[(15.524 - 4.082)/(2(29.825))\right] = 158°$$

Because this is greater than 120°, it is acceptable.
For the larger sprocket,

$$\theta_2 = 180° + 2 \sin^{-1}\left(D_2 - D_1)/2C\right]$$
$$\theta_2 = 180° + 2 \sin^{-1}\left[(15.524 - 4.082)/(2(29.825))\right] = 202°$$

Comments   **Summary of Design**

Figure 7–25(a) shows a sketch of the design to scale.

  Pitch: No. 60 chain, 3/4-in pitch
  Length: 122 pitches = 122(0.75) = 91.50 in
  Center distance: $C = 29.825$ in (maximum)
  Sprockets: Single-strand, no. 60, 3/4-in pitch
  Small: 17 teeth, $D = 4.082$ in
  Large: 65 teeth, $D = 15.524$ in

Type B lubrication is required. The large sprocket can dip into an oil bath.

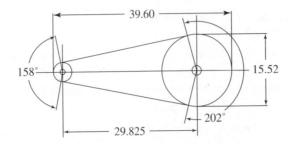 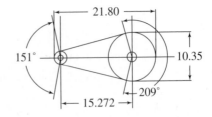

(a) Chain drive system for Example Problem 7–2   (b) Chain drive system for Example Problem 7–3

FIGURE 7–25   Scale drawings of layouts for chain drives for Example Problems 7–2 and 7–3

Example Problem 7–3 Create an alternate design for the conditions of Example Problem 7–2 to produce a smaller drive.

Solution Objective Design a smaller chain drive for the application in Example Problem 7–2.

Given Power transmitted = 15 hp to a conveyor
Speed of motor = 900 rpm; output speed range = 230 to 240 rpm

Analysis Use a multistrand design to permit a smaller-pitch chain to be used to transmit the same design power (21.0 hp) at the same speed (900 rpm). Use the design data presented in this section. The solution procedure is developed within the Results section of the problem solution.

Results Let's try a four-strand chain for which the power capacity factor is 3.3. Then the required power per strand is

$$P = 21.0/3.3 = 6.36 \text{ hp}$$

From Table 7–5, we find that a no. 40 chain (1/2-in pitch) with a 17-tooth sprocket will be satisfactory. Type B lubrication, oil bath, can be used.
For the required large sprocket,

$$N_2 = N_1 \times \text{ratio} = 17(3.83) = 65.11$$

Let's use $N_2 = 65$ teeth.
The sprocket diameters are

$$D_1 = \frac{p}{\sin(180°/N_1)} = \frac{0.500 \text{ in}}{\sin(180°/17)} = 2.721 \text{ in}$$

$$D_2 = \frac{p}{\sin(180°/N_2)} = \frac{0.500 \text{ in}}{\sin(180°/65)} = 10.349 \text{ in}$$

For the center distance, let's try the minimum recommended: $C = 30$ pitches.

$$30(0.50 \text{ in}) = 15.0 \text{ in}$$

The chain length is

$$L = 2(30) + \frac{65 + 17}{2} + \frac{(65 - 17)^2}{4\pi^2(30)} = 102.9 \text{ pitches}$$

Specify the integer length, $L = 104$ pitches $= 104(0.50) = 52.0$ in. The actual maximum center distance is

$$C = \frac{1}{4}\left[104 - \frac{65 + 17}{2} + \sqrt{\left[104 - \frac{65 + 17}{2}\right]^2 - \frac{8(65 - 17)^2}{4\pi^2}}\right]$$

$$C = 30.54 \text{ pitches} = 30.54(0.50) = 15.272 \text{ in}$$

Compute the angle of wrap of the chain for each sprocket using Equations (7–12) and (7–13). Note that the minimum angle of wrap should be 120 degrees.

For the small sprocket,

$$\theta_1 = 180° - 2 \sin^{-1}\left[(D_2 - D_1)/2C\right]$$
$$\theta_1 = 180° - 2 \sin^{-1}\left[(10.349 - 2.721)/(2(15.272))\right] = 151.1°$$

Because this is greater than 120°, it is acceptable.
For the larger sprocket,

$$\theta_1 = 180° + 2 \sin^{-1}\left[(D_2 - D_1)/2C\right]$$
$$\theta_2 = 180° + 2 \sin^{-1}\left[(10.349 - 2.721)/(2(15.272))\right] = 208.9°$$

Comments   **Summary**

Figure 7–25(b) shows the new design to the same scale as the first design. The space reduction is significant.

Chain: No. 40, 1/2-in pitch, four-strand, 104 pitches, 52.0 in length

Sprockets: No. 40–4 (four strands), $\frac{1}{2}$-in pitch

Small: 17 teeth, $D_1 = 2.721$ in

Large: 65 teeth, $D_2 = 10.349$ in

Maximum center distance: 15.272 in

Type B lubrication (oil bath)

Spreadsheet for Chain Design

Figure 7–26 shows a spreadsheet that assists in the design of chain drives using the procedure developed in this section. The user enters data shown in italics in the gray-shaded cells. Refer to Tables 7–4 to 7–8 for required data. Results for Example Problem 7–3 are shown in the figure.

REFERENCES

1. American Chain Association. *Chains for Power Transmission and Material Handling*. New York: Marcel Dekker, 1982.

2. Dayco CPT. *Industrial V-Belt Drives Design Guide*. Dayton, OH: Carlisle Power Transmission Products.

3. Dayco Products. *Engineering Handbook for Automotive V-Belt Drives*. Rochester Hills, MI: Mark IV Automotive Co.

4. Emerson Power Transmission Company. *Power Transmission Equipment Catalog*. Maysville, KY: Browning Manufacturing Division.

5. The Gates Rubber Company. *V-Belt Drive Design Manual*. Denver, CO: The Gates Rubber Company.

6. Putnam Precision Molding. *Plastic Chain Products*. Putnam, CT.

7. Rexnord, Incorporated. *Catalog of Power Transmission and Conveying Components*. Milwaukee, WI: Rexnord.

8. Rockwell Automation/Dodge. *Power Transmission Products*. Greenville, SC. Rockwell Automation.

9. Rubber Manufacturers Association. Power Transmission Belt Publication IP-3-10. *V-Belt Drives with Twist and Non-Alignment Including Quarter Turn*. 3rd ed. Washington, DC: Rubber Manufacturers Association. 1999.

10. Society of Automotive Engineers. *SAE Standard J636— V-Belts and Pulleys*. Warrendale, PA: Society of Automotive Engineers, 2001.

11. Society of Automotive Engineers. *SAE Standard J637— Automotive V-Belt Drives*. Warrendale, PA: Society of Automotive Engineers, 2001.

CHAIN DRIVE DESIGN					
Initial Input Data:		Example Problem 7–3—Multiple strands			
Application:	*Coal Conveyor*				
Drive/type:	*Engine-Mechanical drive*				
Driven machine:	*Heavily loaded conveyor*				
Power input:	*15* hp				
Service factor:	*1.4*	Table 7–8			
Input speed:	*900* rpm				
Desired output speed:	*235* rpm				
Computed Data:					
Design power:	21 hp				
Speed ratio:	3.83				
Design Decisions—Chain Type and Teeth Numbers:					
Number of strands:	*4*	1	2	3	4
Strand factor:	*3.3*	1.0	1.7	2.5	3.3
Required power per strand:	6.36 hp				
Chain number:	*40*	Tables 7–5, 7–6, or 7–7			
Chain pitch:	*0.5* in				
Number of teeth-Driver sprocket:	*17*				
Computed no. of teeth-Driven sprocket:	65.11				
Enter: Chosen number of teeth:	*65*				
Computed Data:					
Actual output speed:	235.4 rpm				
Pitch diameter-Driver sprocket:	2.721 in				
Pitch diameter-Driven sprocket:	10.349 in				
Center Distance, Chain Length and Angle of Wrap:					
Enter: Nominal center distance:	*30* pitches	30 to 50 pitches recommended			
Computed nominal chain length:	102.9 pitches				
Enter: Specified no. of pitches:	*104* pitches	Even number recommended			
Actual chain length:	52.00 in				
Computed actual center distance:	30.545 pitches				
Actual center distance:	15.272 in				
Angle of wrap–Driver sprocket:	151.1 degrees	Should be greater than 120 degrees			
Angle of wrap–Driven sprocket:	208.9 degrees				

FIGURE 7–26 Spread sheet for chain design

12. Society of Automotive Engineers. *SAE Standard J1278—SI (Metric) Synchronous Belts and Pulleys*. Warrendale, PA: Society of Automotive Engineers, 1993.

13. Society of Automotive Engineers. *SAE Standard J1313—Automotive Synchronous Belt Drives*. Warrendale, PA: Society of Automotive Engineers, 1993.

14. Society of Automotive Engineers. *SAE Standard J1459—V-Ribbed Belts and Pulleys*. Warrendale, PA: Society of Automotive Engineering, 2001.

15. T. B. Wood's Sons Company. *V-Belt Drive Manual*. Chambersburg, PA: T. B. Wood's Sons Company.

INTERNET SITES RELATED TO BELT DRIVES AND CHAIN DRIVES

1. **American Chain Association.** *www.americanchainassn.org* A national trade organization for companies providing products for the chain drive industry. Publishes standards and design aids for designing, applying, and maintaining chain drives and engineering chain conveyor systems.

2. **Dayco Belt Drives.** *www.dayco.com* and *www.markivauto.com* Manufacturer of Dayco industrial belt drive systems under Carlisle Power Transmission Products, Inc., and Dayco automotive belt drive systems under the MarkIV Automotive Company.

3. **Dodge Power Transmission.** *www.dodge-pt.com* Manufacturer of numerous power transmission components, including V-belt and synchronous belt drive systems. Part of Rockwell Automation, Inc., which includes Reliance Electric motors and drives and Allen-Bradley controls.

4. **Emerson Power Transmission.** *www.emerson-ept.com* Manufacturer of numerous power transmission components including V-belt drives, synchronous belt drives, and roller chain drives through their Browning and Morse divisions.

5. **Gates Rubber Company.** *www.gates.com* Rubber products for the automotive and industrial markets including V-belt drives and synchronous belt drives.

6. **Power Transmission** *www.powertransmission.com* A comprehensive website for companies providing products for the power transmission industry, many of which supply belt and chain drive systems.

7. **Putnam Precision Molding, Inc.** *www.putnamprecisionmolding.com* Producer of plastic injection molded mechanical drive components, including plastic chain, sprockets, and synchronous belt pulleys.

8. **Rexnord Corporation.** *www.rexnord.com* Manufacturer of power transmission and conveying components, including roller chain drives and engineered chain drive systems.

9. **Rubber Manufacturers Association.** *www.rma.org* National trade association for the finished rubber products industry. Provides many standards and technical publications for the application of rubber products, including V-belt drives.

10. **SAE International.** *www.sae.org* The Society of Automotive Engineers, the engineering society for advancing mobility on land or sea, in air or space. Offers standards on V-belts, synchronous belts, pulleys, and drives for automotive applications.

11. **T. B. Wood's Sons Company** *www.tbwoods.com* Manufacturer of many mechanical drives products, including V-belt drives, synchronous belt drives, and adjustable speed drives.

PROBLEMS

V-Belt Drives

1. Specify the standard 3V belt length (from Table 7–2) that would be applied to two sheaves with pitch diameters of 5.25 in and 13.95 in with a center distance of no more than 24.0 in.

2. For the standard belt specified in Problem 1, compute the actual center distance that would result.

3. For the standard belt specified in Problem 1, compute the angle of wrap on both of the sheaves.

4. Specify the standard 5V belt length (from Table 7–2) that would be applied to two sheaves with pitch diameters of 8.4 in and 27.7 in with a center distance of no more than 60.0 in.

5. For the standard belt specified in Problem 4, compute the actual center distance that would result.

6. For the standard belt specified in Problem 4, compute the angle of wrap on both of the sheaves.

7. Specify the standard 8V belt length (from Table 7–2) that would be applied to two sheaves with pitch diameters of 13.8 in and 94.8 in with a center distance of no more than 144 in.

8. For the standard belt specified in Problem 7, compute the actual center distance that would result.

9. For the standard belt specified in Problem 7, compute the angle of wrap on both of the sheaves.

10. If the small sheave of Problem 1 is rotating at 1750 rpm, compute the linear speed of the belt.

11. If the small sheave of Problem 4 is rotating at 1160 rpm, compute the linear speed of the belt.

12. If the small sheave of Problem 7 is rotating at 870 rpm, compute the linear speed of the belt.

13. For the belt drive from Problems 1 and 10, compute the rated power, considering corrections for speed ratio, belt length, and angle of wrap.

14. For the belt drive from Problems 4 and 11, compute the rated power, considering corrections for speed ratio, belt length, and angle of wrap.

15. For the belt drive from Problems 7 and 12, compute the rated power, considering corrections for speed ratio, belt length, and angle of wrap.

TABLE 7-10

Problem number	Driver type	Driven machine	Service (h/day)	Input speed (rpm)	Input power (hp)	Nominal output speed (rpm)
18.	AC motor (HT)	Hammer mill	8	870	25	310
19.	AC motor (NT)	Fan	22	1750	5	725
20.	6-cylinder engine	Heavy conveyor	16	1500	40	550
21.	DC motor (compound)	Milling machine	16	1250	20	695
22.	AC motor (HT)	Rock crusher	8	870	100	625

Note: NT indicates a normal-torque electric motor. *HT* indicates a high-torque electric motor.

16. Describe a standard 15N belt cross section. To what size belt (inches) would it be closest?

17. Describe a standard 17A belt cross section. To what size belt (inches) would it be closest?

    For Problems 18–22 (Table 7–10), design a V-belt drive. Specify the belt size, the sheave sizes, the number of belts, the actual output speed, and the center distance.

Roller Chain

23. Describe a standard roller chain, no. 140.

24. Describe a standard roller chain, no. 60.

25. Specify a suitable standard chain to exert a static pulling force of 1250 lb.

26. Roller chain is used in a hydraulic forklift truck to elevate the forks. If two strands support the load equally, which size would you specify for a design load of 5000 lb?

27. List three typical failure modes of roller chain.

28. Determine the power rating of a no. 60 chain, single-strand, operating on a 20-tooth sprocket at 750 rpm. Describe the preferred method of lubrication. The chain connects a hydraulic drive with a meat grinder.

29. For the data of Problem 28, what would be the rating for three strands?

30. Determine the power rating of a no. 40 chain, single-strand, operating on a 12-tooth sprocket at 860 rpm. De-scribe the preferred method of lubrication. The small sprocket is applied to the shaft of an electric motor. The output is to a coal conveyor.

31. For the data of Problem 30, what would be the rating for four strands?

32. Determine the power rating of a no. 80 chain, single-strand, operating on a 32-tooth sprocket at 1160 rpm. Describe the preferred method of lubrication. The input is an internal combustion engine, and the output is to a fluid agitator.

33. For the data of Problem 32, what would be the rating for two strands?

34. Specify the required length of no. 60 chain to mount on sprockets having 15 and 50 teeth with a center distance of no more than 36 in.

35. For the chain specified in Problem 34, compute the actual center distance.

36. Specify the required length of no. 40 chain to mount on sprockets having 11 and 45 teeth with a center distance of no more than 24 in.

37. For the chain specified in Problem 36, compute the actual center distance.

    For Problems 38–42 (Table 7–11), design a roller chain drive. Specify the chain size, the sizes and number of teeth in the sprockets, the number of chain pitches, and the center distance.

TABLE 7-11

Problem number	Driver type	Driven machine	Input speed (rpm)	Input power (hp)	Nominal output speed (rpm)
38.	AC motor	Hammer mill	310	25	160
39.	AC motor	Agitator	750	5	325
40.	6-cylinder engine	Heavy conveyor	500	40	250
41.	Steam turbine	Centrifugal pump	2200	20	775
42.	Hydraulic drive	Rock crusher	625	100	225

8

Kinematics of Gears

The Big Picture

You Are the Designer

8–1   Objectives of This Chapter

8–2   Spur Gear Styles

8–3   Spur Gear Geometry: Involute-Tooth Form

8–4   Spur Gear Nomenclature and Gear-Tooth Features

8–5   Interference between Mating Spur Gear Teeth

8–6   Velocity Ratio and Gear Trains

8–7   Helical Gear Geometry

8–8   Bevel Gear Geometry

8–9   Types of Wormgearing

8–10 Geometry of Worms and Wormgears

8–11 Typical Geometry of Wormgear Sets

8–12 Train Value for Complex Gear Trains

8–13 Devising Gear Trains

<table>
<tr><td rowspan="4">

The Big Picture

</td></tr>
</table>

The Big Picture

Kinematics of Gears

Discussion Map

☐ Gears are toothed, cylindrical wheels used for transmitting motion and power from one rotating shaft to another.

☐ Most gear drives cause a change in the speed of the output gear relative to the input gear.

☐ Some of the most common types of gears are *spur gears, helical gears, bevel gears,* and *worm/wormgear sets.*

Discover

Identify at least two machines or devices that employ gears. Describe the operation of the machines or devices and the appearance of the gears.

This chapter will help you learn about the features of different kinds of gears, the kinematics of a pair of gears operating together, and the operation of gear trains having more than two gears.

Gears are toothed, cylindrical wheels used for transmitting motion and power from one rotating shaft to another. The teeth of a driving gear mesh accurately in the spaces between teeth on the driven gear as shown in Figure 8–1. The driving teeth push on the driven teeth, exerting a force perpendicular to the radius of the gear. Thus, a torque is transmitted, and because the gear is rotating, power is also transmitted.

Speed Reduction Ratio. Often gears are employed to produce a change in the speed of rotation of the driven gear relative to the driving gear. In Figure 8–1, if the smaller top gear, called a *pinion,* is driving the larger lower gear, simply called the *gear,* the larger gear will rotate more slowly. The amount of speed reduction is dependent on the ratio of the number of teeth in the pinion to the number of teeth in the gear according to this relationship:

$$n_P/n_G = N_G/N_P \qquad \textbf{(8–1)}$$

The basis for this equation will be shown later in this chapter. But to show an example of its application here, consider that the pinion in Figure 8–1 is rotating at 1800 rpm. You can count the number of teeth in the pinion to be 11 and the number of teeth in the gear to be 18. Then we can compute the rotational speed of the gear by solving Equation (8–1) for n_G:

$$n_G = n_P(N_P/N_G) = (1800 \text{ rpm})(11/18) = 1100 \text{ rpm}$$

When there is a reduction in the speed of rotation of the gear, there is a simultaneous proportional *increase* in the torque transmitted to the shaft carrying the gear. More will be said about this later, also.

Kinds of Gears. Several kinds of gears having different tooth geometries are in common use. To acquaint you with the general appearance of some, their basic descriptions are given here. Later we will describe their geometry more completely.

Figure 8–2 shows a photograph of many kinds of gears. Labels indicate the major types of gears that are discussed in this chapter: *spur gears, helical gears, bevel gears,* and *worm/wormgear sets.* Obviously, the shafts that would carry the gears are not included in this photograph.

FIGURE 8–1 Pair of spur gears. The pinion drives the gear.

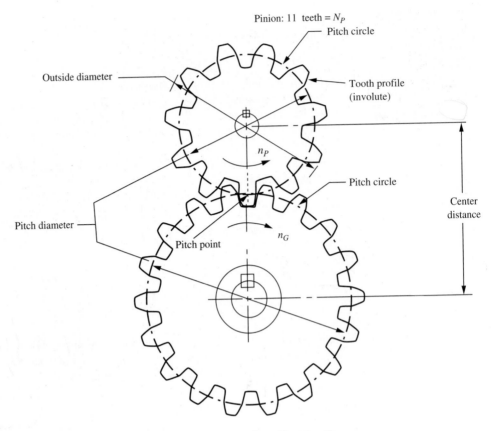

Pinion: 11 teeth = N_P

Pitch circle

Outside diameter

Tooth profile (involute)

n_P

Pitch circle

Center distance

Pitch diameter

Pitch point

n_G

Gear: 18 teeth = N_G

Spur gears have teeth that are straight and arranged parallel to the axis of the shaft that carries the gear. The curved shape of the faces of the spur gear teeth have a special geometry called an *involute curve*, described later in this chapter. This shape makes it possible for two gears to operate together with smooth, positive transmission of power. Figure 8–1 also shows the side view of spur gear teeth, and the involute curve shape is evident there. The shafts carrying the gears are parallel.

The teeth of *helical gears* are arranged so that they lie at an angle with respect to the axis of the shaft. The angle, called the *helix angle,* can be virtually any angle. Typical helix angles range from approximately 10° to 30°, but angles up to 45° are practical. The helical teeth operate more smoothly than equivalent spur gear teeth, and stresses are lower. Therefore, a smaller helical gear can be designed for a given power-transmitting capacity as compared with spur gears. One disadvantage of helical gears is that an axial force, called a *thrust force,* is generated in addition to the driving force that acts tangent to the basic cylinder on which the teeth are arranged. The designer must consider the thrust force when selecting bearings that will hold the shaft during operation. Shafts carrying helical gears are typically arranged parallel to each other. However, a special design, called *crossed helical* gears, has 45° helix angles, and their shafts operate 90° to each other.

Bevel gears have teeth that are arranged as elements on the surface of a cone. The teeth of straight bevel gears appear to be similar to spur gear teeth, but they are tapered, being wider at the outside and narrower at the top of the cone. Bevel gears typically operate on shafts that are 90° to each other. Indeed, this is often the reason for specifying bevel gears in a drive system. Specially designed bevel gears can operate on shafts that are at some angle other than 90°. When bevel gears are made with teeth that form a helix angle similar to

FIGURE 8–2 A variety of gear types (Boston Gear, Quincy, MA)

that in helical gears, they are called *spiral bevel gears*. They operate more smoothly than straight bevel gears and can be made smaller for a given power transmission capacity. When both bevel gears in a pair have the same number of teeth, they are called *miter gears* and are used only to change the axes of the shafts to 90 degrees. No speed change occurs.

A *rack* is a straight gear that moves linearly instead of rotating. When a circular gear is mated with a rack, as shown toward the right side of Figure 8–2, the combination is called *a rack and pinion drive*. You may have heard that term applied to the steering mechanism of a car or to a part of other machinery. See Section 8–6 for more discussion about a rack.

A *worm and its mating wormgear* operate on shafts that are at 90° to each other. They typically accomplish a rather large speed reduction ratio compared with other types of gears. The worm is the driver, and the wormgear is the driven gear. The teeth on the worm appear similar to screw threads, and, indeed, they are often called *threads* rather than *teeth*. The teeth of the wormgear can be straight like spur gear teeth, or they can be helical. Often the shape of the tip of the wormgear teeth is enlarged to partially wrap around the threads of the worm to improve the power transmission capacity of the set. One disadvantage of the worm/wormgear drive is that it has a somewhat lower mechanical efficiency than most other kinds of gears because there is extensive rubbing contact between the surfaces of the worm threads and the sides of the wormgear teeth.

Where Have You Observed Gears?   Think of examples where you have seen gears in actual equipment. Describe the operation of the equipment, particularly the power transmission system. Sometimes, of course, the gears and the shafts are enclosed in a housing, making it difficult for you to observe the actual gears. Perhaps you can find a manual for the equipment that shows the drive system. Or look elsewhere in this chapter and in Chapters 9 and 10 for some photographs of commercially available gear reducers. *(Note: If the equipment you are observing is operating, be very careful not to come in contact with any moving parts!)* Try to answer these questions:

- What was the source of the power? An electric motor, a gasoline engine, a steam turbine, a hydraulic motor? Or were the gears operated by hand?
- How were the gears arranged together, and how were they attached to the driving source and the driven machine?
- Was there a speed change? Can you determine how much of a change?
- Were there more than two gears in the drive system?
- What types of gears were used? (You should refer to Figure 8–2.)
- What materials were the gears made from?
- How were the gears attached to the shafts that supported them?
- Were the shafts for mating gears aligned parallel to each other, or were they perpendicular to one another?
- How were the shafts themselves supported?
- Was the gear transmission system enclosed in a housing? If so, describe it.

This chapter will help you learn the basic geometries and kinematics of gears and pairs of gears operating together. You will also learn how to analyze gear trains having more than two gears so that you can describe the motion of each gear. Then you will learn how to devise a gear train to produce a given speed reduction ratio. In later chapters, you will learn how to analyze gears for their power transmission capacity and to design gear trains to transmit a given amount of power at a specified ratio of the speed of the input shaft to the speed of the output shaft.

You Are the Designer

A gear-type speed reducer was described in Chapter 1, and a sketch of the layout of the gears within the reducer was shown in Figure 1–1. You are advised to review that discussion now because it will help you understand how the present chapter on *gear geometry* and *kinematics* fits into the design of the complete speed reducer.

Assume that you are responsible for the design of a speed reducer that will take the power from the shaft of an electric motor rotating at 1750 rpm and deliver it to a machine that is to operate at approximately 292 rpm. You have decided to use gears to transmit the power, and you are proposing a double-reduction speed reducer like that shown in Figure 8–3. This chapter will give you the information you need to define the general nature of the gears, including their arrangement and their relative sizes.

The input shaft (shaft 1) is coupled to the motor shaft. The first gear of the gear train is mounted on this shaft and rotates at the same speed as the motor, 1750 rpm. Gear 1 drives the mating gear 2, which is larger, causing the speed of rotation of shaft 2 to be slower than that of shaft 1. But the speed is not yet down to 292 rpm as desired.

The next step is to mount a third gear (gear 3) on shaft 2 and mate it with gear 4 mounted on the output shaft, shaft 3. With proper sizing of all four gears, you should be able to produce an output speed equal or quite close to the desired speed. This process requires knowledge of the concept of *velocity ratio* and the techniques of designing gear trains as presented in this chapter.

But you will also need to specify the appearance of the gears and the geometry of the several features that make up each gear. Whereas the final specification also requires the information from following chapters, you will learn how to recognize common forms of gears and to compute the dimensions of key features. This will be important when completing the design for strength and wear resistance in later chapters.

Let's say that you have chosen to use spur gears in your design. What design decisions must you make to complete the specification of all four gears? The following list gives some of the important parameters for each gear:

- The number of teeth
- The form of the teeth
- The size of the teeth as indicated by the *pitch*
- The width of the face of the teeth
- The style and dimensions of the gear blank into which the gear teeth are to be machined
- The design of the hub for the gear that facilitates its mounting to the shaft
- The degree of precision of the gear teeth and the corresponding method of manufacture that can produce that precision
- The means of attaching the gear to its shaft
- The means of locating the gear axially on the shaft

To make reliable decisions about these parameters, you must understand the special geometry of spur gears as presented first in this chapter. However, there are other forms of gears that you could choose. Later sections give the special geometry of helical gears, bevel gears, and worm/wormgear sets. The methods of analyzing the forces on these various kinds of gears are described in later chapters, including the stress analysis of the gear teeth and recommendations on material selection to ensure safe operation with long life.

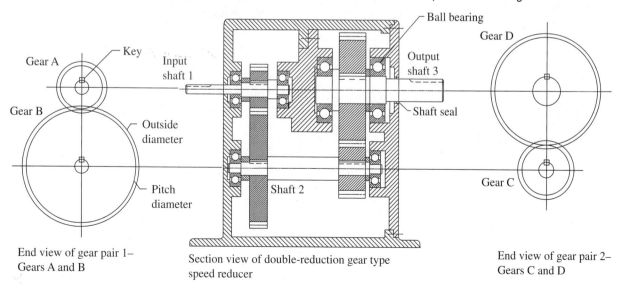

End view of gear pair 1– Gears A and B

Section view of double-reduction gear type speed reducer

End view of gear pair 2– Gears C and D

FIGURE 8–3 Conceptual design for a speed reducer

8–1
OBJECTIVES OF THIS CHAPTER

After completing this chapter, you will be able to:

1. Recognize and describe the main features of *spur gears, helical gears, bevel gears,* and *worm/wormgear sets.*

2. Describe the important operating characteristics of these various types of gears with regard to the similarities and differences among them and their general advantages and disadvantages.

3. Describe the *involute-tooth form* and discuss its relationship to the *law of gearing.*

4. Describe the basic functions of the American Gear Manufacturers Association (AGMA) and identify pertinent standards developed and published by this organization.

5. Define *velocity ratio* as it pertains to two gears operating together.

6. Specify appropriate numbers of teeth for a mating pair of gears to produce a given velocity ratio.

7. Define *train value* as it pertains to the overall speed ratio between the input and output shafts of a gear-type speed reducer (or speed increaser) that uses more than two gears.

8–2
SPUR GEAR STYLES

Figure 8–4 shows several different styles of commercially available spur gears. When gears are large, the spoked design in Part (a) is often used to save weight. The gear teeth are machined into a relatively thin rim that is held by a set of spokes connecting to the hub. The bore of the hub is typically designed to be a close sliding fit with the shaft that carries the

(a) Spur gear with spoked design

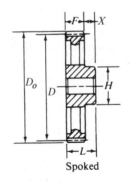

Spoked

(b) Spur gear with solid hub

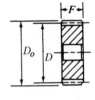

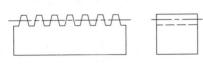

(c) Rack-straight spur gear

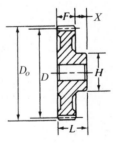

(d) Spur gear with thinned web

D_o = outside dia.

D = pitch dia.

F = face width

L = length of hub

X = extension of hub beyond face

H = hub dia.

FIGURE 8–4 Spur gears (Emerson Power Transmission Corporation, Browning Division, Maysville, KY)

gear. A keyway is usually machined into the bore to allow a key to be inserted for positive transmission of torque. The illustration does not include a keyway because this gear is sold as a stock item, and the ultimate user finishes the bore to match a given piece of equipment.

The solid hub design in Figure 8–4(b) is typical of smaller spur gears. Here the finished bore with a keyway is visible. The set screw over the keyway allows the locking of the key in place after assembly.

When spur gear teeth are machined into a straight, flat bar, the assembly is called a *rack,* as shown in Figure 8–4(c). The rack is essentially a spur gear with an infinite radius. In this form, the teeth become straight-sided, rather than the curved, involute form typical of smaller gears.

Gears with diameters between the small solid form [Part (b)] and the larger spoked form [Part (a)] are often produced with a thinned web as shown in Part (d), again to save weight.

You as a designer may create special designs for gears that you implement into a mechanical device or system. One useful approach is to machine the gear teeth of small pinions directly into the surface of the shaft that carries the gear. This is very often done for the input shaft of gear reducers.

8–3
SPUR GEAR GEOMETRY INVOLUTE-TOOTH FORM

The most widely used spur gear tooth form is the full-depth involute form. Its characteristic shape is shown in Figure 8–5.

The involute is one of a class of geometric curves called *conjugate curves.* When two such gear teeth are in mesh and rotating, there is a *constant angular velocity* ratio between them: From the moment of initial contact to the moment of disengagement, the speed of the driving gear is in a constant proportion to the speed of the driven gear. The resulting action of the two gears is very smooth. If this were not the case, there would be some speeding up and slowing down during the engagement, with the resulting accelerations causing vibration, noise, and dangerous torsional oscillations in the system.

You can easily visualize an involute curve by taking a cylinder and wrapping a string around its circumference. Tie a pencil to the end of the string. Then start with the pencil tight against the cylinder, and hold the string taut. Move the pencil away from the cylinder while keeping the string taut. The curve that you will draw is an involute. Figure 8–6 is a sketch of the process.

The circle represented by the cylinder is called the *base circle.* Notice that at any position on the curve, the string represents a line tangent to the base circle and, at the same time, perpendicular to the involute. Drawing another base circle along the same centerline in such a position that the resulting involute is tangent to the first one, as shown in Figure 8–7, demonstrates that at the point of contact, the two lines tangent to the base circles are coincident and will stay in the same position as the base circles rotate. This is what happens when two gear teeth are in mesh.

It is a fundamental principle of *kinematics,* the study of motion, that if the line drawn perpendicular to the surfaces of two rotating bodies at their point of contact always crosses the centerline between the two bodies at the same place, the angular velocity ratio of the two bodies will be constant. This is a statement of the *law of gearing.* As demonstrated here, the gear teeth made in the involute-tooth form obey the law.

Of course, only the part of the gear tooth that actually comes into contact with the mating tooth needs to be in the involute form.

FIGURE 8–5
Involute-tooth form

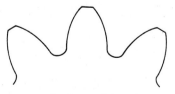

Involute tooth form

FIGURE 8–6
Graphical generation of
an involute curve

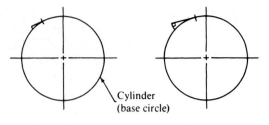

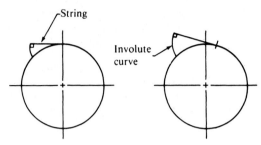

FIGURE 8–7   Mating
involutes

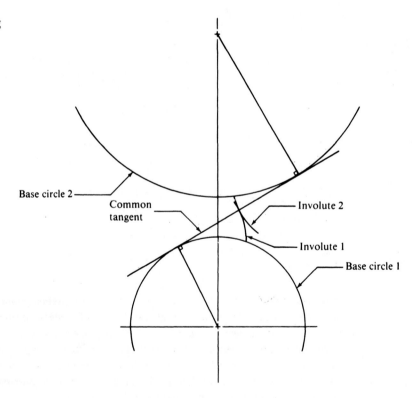

**8–4
SPUR GEAR
NOMENCLATURE
AND GEAR-
TOOTH
FEATURES**

This section describes several features of individual spur gear teeth and complete gears. Terms
and symbols used conform to American Gear Manufacturers Association (AGMA) standards.
(See Reference 1 for a more complete set of definitions.) Figure 8–8 shows drawings of spur
gear teeth, with the symbols for the various features indicated. These features are described next.

Pitch Diameter

Figure 8–9 shows teeth from two gears in mesh to demonstrate the relative positions of the
teeth at several stages of engagement. One of the most important observations that can be

FIGURE 8–8 Spur
gear teeth features

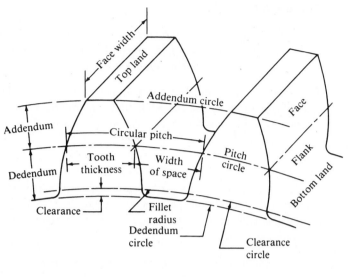

FIGURE 8–9 Cycle
of engagement of gear
teeth

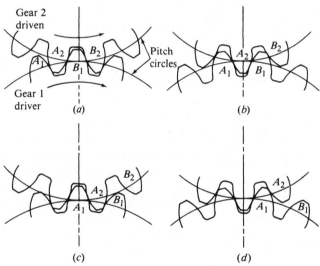

made from Figure 8–9 is that throughout the engagement cycle there are two circles, one from each gear, that remain tangent. These are called the *pitch circles.* The diameter of the pitch circle of a gear is its *pitch diameter;* the point of tangency is the *pitch point.*

When two gears mesh, the smaller gear is called the *pinion,* and the larger is the *gear.* We will use the symbol D_P to indicate the pitch diameter of the pinion, and the symbol D_G for the pitch diameter of the gear. When referring to the number of teeth, we will use N_P for the pinion and N_G for the gear.

Notice that the pitch diameter lies somewhere within the height of the gear tooth, and thus it is not possible to measure its diameter directly. It must be calculated from other known features of the gear; this calculation depends on understanding the concept of *pitch,* discussed in the following section.

Pitch

The spacing between adjacent teeth and the size of the teeth are controlled by the pitch of the teeth. Three types of pitch designation systems are in common use for gears: (1) circular pitch, (2) diametral pitch, and (3) the metric module.

Circular Pitch, *p.*

The distance from a point on a tooth of a gear at the pitch circle to a corresponding point on the next adjacent tooth, measured along the pitch circle, is the circular pitch (see Figure 8–8).

Note that it is an arc length, usually in inches. To compute the value of the circular pitch, take the circumference of the pitch circle and divide it into a number of equal parts corresponding to the number of teeth in the gear. Using N for the number of teeth, we have

⇨ Circular Pitch

$$p = \pi D/N \tag{8–2}$$

Notice that the tooth size increases as the value of the circular pitch increases because there is a larger pitch circle for the same number of teeth. Also note that the basic sizes of the mating gear teeth must be the same for them to mesh properly. This observation leads to a very important rule:

The pitch of two gears in mesh must be identical.

This must be true whether the pitch is indicated as the circular pitch, the diametral pitch, or the metric module. Then Equation (8–2) can be written in terms of either the pinion or the gear diameter:

⇨ Circular Pitch

$$p = \pi D_G/N_G = \pi D_P/N_P \tag{8–3}$$

Circular pitch is infrequently used now. It is sometimes an advantage to use this system when large gears are to be made by casting. To facilitate the layout of the pattern for the casting, lay off the chord of the arc length of the circular pitch. Also, some machines and product lines of machines have traditionally used circular pitch gears and continue to do so. Table 8–1 lists the recommended standard circular pitches for large gear teeth.

Diametral Pitch, P_d. The most common pitch system used today in the United States is the *diametral pitch* system, the number of teeth per inch of pitch diameter. Its basic definition is

⇨ Diametral Pitch

$$P_d = N_G/D_G = N_P/D_P \tag{8–4}$$

As such, it has units of in $^{-1}$. However, the units are rarely reported, and gears are referred to as 8-pitch or 20-pitch, for example. One of the advantages of the diametral pitch system is that there is a set list of standard pitches, and most of the pitches have integer values. Table 8–2 lists the recommended standard pitches, with those of 20 and above called *fine pitch* and those below 20 called *coarse pitch.*

TABLE 8–2 Standard diametral pitches (teeth/in)

Coarse pitch ($P_d < 20$)				Fine pitch ($P_d \geq 20$)	
1	2	5	12	20	72
1.25	2.5	6	14	24	80
1.5	3	8	16	32	96
1.75	4	10	18	48	120
				64	

TABLE 8–1 Standard circular pitches (in)

10.0	7.5	5.0
9.5	7.0	4.5
9.0	6.5	4.0
8.5	6.0	3.5
8.0	5.5	

FIGURE 8–10 Gear-tooth size as a function of diametral pitch (Barber-Colman Company, Loves Park, IL)

Other intermediate values are available, but most manufacturers produce gears from this list of pitches. In any case, it is advisable to check availability before finally specifying a pitch. In problem solutions in this book, it is expected that one of the pitches listed in Table 8–2 will be used if possible.

As stated before, the pitch of the gear teeth determines their size, and two mating gears must have the same pitch. Figure 8–10 shows the profiles of some of the standard diametral pitch gear teeth, drawn actual size. That is, you can lay a given gear down on the page and compare its size with the drawing to obtain a good estimate of the pitch of the teeth. Notice that as the numerical value of the diametral pitch increases, the physical size of the tooth decreases, and vice versa.

Sometimes it is necessary to convert from diametral pitch to circular pitch, or vice versa. Their definitions provide a simple means of doing this. Solving for the pitch diameter in both Equations (8–2) and (8–4) gives

$$D = Np/\pi$$
$$D = N/P_d$$

Equating these two gives

Relation between Circular and Diametral Pitches

$$N/P_d = Np/\pi \quad \text{or} \quad P_d p = \pi \qquad (8\text{–}5)$$

From this equation, the equivalent circular pitch for a gear having a diametral pitch of 1 is $p = \pi/1 = 3.1416$. Referring to Tables 8–1 and 8–2, notice that the circular pitches listed are for the larger gear teeth, being preferred when the diametral pitch is less than 1. Diametral pitch is preferred for sizes equivalent to 1 pitch or smaller.

Metric Module System.   In the SI, a common unit of length is the *millimeter.* The pitch of gears in the metric system is based on this unit and is designated the *module, m.* To find the module of a gear, divide the pitch diameter of the gear in millimeters by the number of teeth. That is,

  Metric Module

$$m = D_G/N_G = D_p/N_p \qquad (8\text{--}6)$$

There is rarely a need to convert from the module system to the diametral pitch system. However, it is important to have a feel for the physical size of gear teeth. Because at this time people are more familiar with the standard diametral pitches, as shown in Figure 8–10, we will develop the relationship between m and P_d. From their definitions, Equations (8–4) and (8–6), we can say

$$m = 1/P_d$$

But recall that diametral pitch uses the inch unit, and module uses the millimeter. Therefore, the conversion factor of 25.4 mm per inch must be applied.

$$m = \frac{1}{P_d\,\text{in}^{-1}} \cdot \frac{25.4 \text{ mm}}{\text{in}}$$

This reduces to

Relation between Module and Diametral Pitch

$$m = 25.4/P_d \qquad (8\text{--}7)$$

For example, if a gear has a diametral pitch of 10, the equivalent module is

$$m = 25.4/10 = 2.54 \text{ mm}$$

This is not a standard value for module, but it is close to the standard value of 2.5. So it can be concluded that a 10-pitch gear is of similar size to a gear with module 2.5. Table 8–3 gives selected standard modules with their equivalent diametral pitches.

Gear-Tooth Features

In design and inspection of gear teeth, several special features must be known. Figure 8–8, presented earlier, and Figure 8–11, which shows segments of two gears in mesh, identify these features. They are defined in the list that follows. Table 8–4 gives the relationships needed to compute their values. See References 1, 2, 4, 7, and 8 for the relevant AGMA standards. References 9, 10, and 12 provide much additional data. Note that many of the computations involve the diametral pitch, again illustrating that the physical size of a gear tooth is determined by its diametral pitch. The definitions are for external gears. Internal gears are discussed in Section 8–6.

- *Addendum (a):* The radial distance from the pitch circle to the outside of a tooth.
- *Dedendum (b):* The radial distance from the pitch circle to the bottom of the tooth space.
- *Clearance (c):* The radial distance from the top of a tooth to the bottom of the tooth space of the mating gear when the tooth is fully engaged. Note that

Clearance

$$c = b - a \qquad (8\text{--}8)$$

TABLE 8–3   Standard modules

Module (mm)	Equivalent P_d	Closest standard P_d (teeth/in)
0.3	84.667	80
0.4	63.500	64
0.5	50.800	48
0.8	31.750	32
1	25.400	24
1.25	20.320	20
1.5	16.933	16
2	12.700	12
2.5	10.160	10
3	8.466	8
4	6.350	6
5	5.080	5
6	4.233	4
8	3.175	3
10	2.540	2.5
12	2.117	2
16	1.587	1.5
20	1.270	1.25
25	1.016	1

FIGURE 8–11   Gear pair features

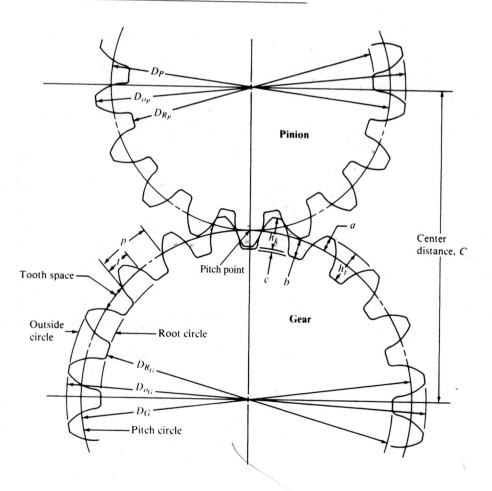

TABLE 8–4   Formulas for gear-tooth features for 20° pressure angle

Feature	Symbol	Full-depth involute system		Metric module system
		Coarse pitch $(P_d < 20)$	Fine pitch $(P_d \geq 20)$	
Addendum	a	$1/P_d$	$1/P_d$	$1.00m$
Dedendum	b	$1.25/P_d$	$1.200/P_d + 0.002$	$1.25m$
Clearance	c	$0.25/P_d$	$0.200/P_d + 0.002$	$0.25m$

■ *Outside diameter* (D_o): The diameter of the circle that encloses the outside of the gear teeth. Note that

⇨ **Outside Diameter Basic Definition**

$$D_o = D + 2a \tag{8–9}$$

Also note that both the pitch diameter, D, and the addendum, a, are defined in terms of the diametral pitch, P_d. Making these substitutions produces a very useful form of the equation for outside diameter:

⇨ **Outside Diameter in Terms of P$_d$ and N**

$$D_o = \frac{N}{P_d} + 2\frac{1}{P_d} = \frac{N+2}{P_d} \tag{8–10}$$

In the metric module system, a similar equation can be derived:

⇨ **Outside Diameter in Metric Module system**

$$D_o = mN + 2m = m(N + 2) \tag{8–11}$$

■ *Root diameter* (D_R): The diameter of the circle that contains the bottom of the tooth space; this circle is called the *root circle*. Note that

⇨ **Root Diameter**

$$D_R = D - 2b \tag{8–12}$$

■ *Whole depth* (h_t): The radial distance from the top of a tooth to the bottom of the tooth space. Note that

⇨ **Whole Depth**

$$h_t = a + b \tag{8–13}$$

■ *Working depth* (h_k): The radial distance that a gear tooth projects into the tooth space of the mating gear. Note that

⇨ **Working Depth**

$$h_k = a + a = 2a \tag{8–14}$$

and

⇨ **Whole Depth**

$$h_t = h_k + c \tag{8–15}$$

■ *Tooth thickness (t):* The arc length, measured on the pitch circle from one side of a tooth to the other side. This is sometimes called the *circular thickness* and has the theoretical value of one-half of the circular pitch. That is,

⇨ **Tooth Thickness**

$$t = p/2 = \pi/2P_d \tag{8–16}$$

■ *Tooth space:* The arc length, measured on the pitch circle, from the right side of one tooth to the left side of the next tooth. Theoretically, the tooth space equals the tooth thickness. But for practical reasons, the tooth space is made larger (see "Backlash").

- *Backlash:* If the tooth thickness were made identical in value to the tooth space, as it theoretically is, the tooth geometry would have to be absolutely precise for the gears to operate, and there would be no space available for lubrication of the tooth surfaces. To alleviate these problems, practical gears are made with the tooth space slightly larger than the tooth thickness, the difference being called the *backlash.* To provide backlash, the cutter generating the gear teeth can be fed more deeply into the gear blank than the theoretical value on either or both of the mating gears. Alternatively, backlash can be created by adjusting the center distance to a larger value than the theoretical value.

  The magnitude of backlash depends on the desired precision of the gear pair and on the size and the pitch of the gears. It is actually a design decision, balancing cost of production with desired performance. The American Gear Manufacturers Association (AGMA) provides recommendations for backlash in their standards. (See Reference 2.) Table 8–5 lists recommended ranges for several values of pitch.

- *Face width (F):* The width of the tooth measured parallel to the axis of the gear.

- *Fillet:* The arc joining the involute-tooth profile to the root of the tooth space.

- *Face:* The surface of a gear tooth from the pitch circle to the outside circle of the gear.

- *Flank:* The surface of a gear tooth from the pitch circle to the root of the tooth space, including the fillet.

TABLE 8–5   Recommended minimum backlash for coarse pitch gears

A. Diametral pitch system (backlash in inches)

P_d	\multicolumn{5}{c}{Center distance, C (in)}
	2	4	8	16	32
18	0.005	0.006			
12	0.006	0.007	0.009		
8	0.007	0.008	0.010	0.014	
5		0.010	0.012	0.016	
3		0.014	0.016	0.020	0.028
2			0.021	0.025	0.033
1.25				0.034	0.042

B. Metric module system (backlash in millimeters)

Module, m	\multicolumn{5}{c}{Center distance, C (mm)}
	50	100	200	400	800
1.5	0.13	0.16			
2	0.14	0.17	0.22		
3	0.18	0.20	0.25	0.35	
5		0.26	0.31	0.41	
8		0.35	0.40	0.50	0.70
12			0.52	0.62	0.82
18				0.80	1.00

Source: Extracted from AGMA 2002-B88 Standard, *Tooth Thickness Specification and Measurement*, with permission of the publisher, American Gear Manufacturers Association, 1500 King Street, Suite 201, Alexandria, VA 22314.

- ■ *Center distance (C):* The distance from the center of the pinion to the center of the gear; the sum of the pitch radii of two gears in mesh. That is, because radius = diameter/2,

⇨ **Center Distance**

$$C = D_G/2 + D_P/2 = (D_G + D_P)/2 \qquad (8\text{–}17)$$

Also note that both pitch diameters can be expressed in terms of the diametral pitch:

⇨ **Center Distance in terms of N_G, N_P, and P_d**

$$C = \frac{1}{2}\left[\frac{N_G}{P_d} + \frac{N_P}{P_d}\right] = \frac{(N_G + N_P)}{2\,P_d} \qquad (8\text{–}18)$$

It is recommended that Equation (8–18) be used for center distance because all of the terms are usually integers, giving greater accuracy in the computation. In the metric module system, a similar equation can be derived:

$$C = (D_G + D_P)/2 = (mN_G + mN_P)/2 = [(N_G + N_P)m]/2 \qquad (8\text{–}19)$$

Pressure Angle

*The **pressure angle** is the angle between the tangent to the pitch circles and the line drawn normal (perpendicular) to the surface of the gear tooth (see Figure 8–12).*

The normal line is sometimes referred to as the *line of action.* When two gear teeth are in mesh and are transmitting power, the force transferred from the driver to the driven gear tooth acts in a direction along the line of action. Also, the actual shape of the gear tooth depends on the pressure angle, as illustrated in Figure 8–13. The teeth in this figure were drawn according to the proportions for a 20-tooth, 5-pitch gear having a pitch diameter of 4.000 in.

All three teeth have the same tooth thickness because, as stated in Equation (8–16), the thickness at the pitch line depends only on the pitch. The difference between the three teeth shown is due to the different pressure angles because the pressure angle determines the size of the base circle. Remember that the base circle is the circle from which the involute is generated. The line of action is always tangent to the base circle. Therefore, the size of the base circle can be found from

⇨ **Base Circle Diameter**

$$D_b = D \cos \phi \qquad (8\text{–}20)$$

FIGURE 8–12
Pressure angle

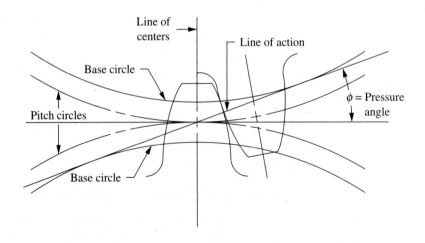

FIGURE 8–13 Full-depth, involute-tooth form for varying pressure angles

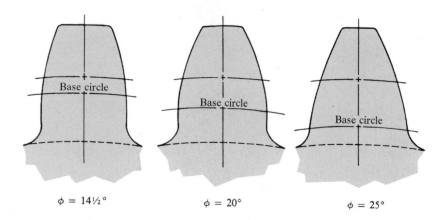

$\phi = 14\frac{1}{2}°$ $\phi = 20°$ $\phi = 25°$

Standard values of the pressure angle are established by gear manufacturers, and the pressure angles of two gears in mesh must be the same. Current standard pressure angles are $14\frac{1}{2}°$, $20°$, and $25°$ as illustrated in Figure 8–13. Actually, the $14\frac{1}{2}°$, tooth form is considered to be obsolete. Although it is still available, it should be avoided for new designs. The $20°$ tooth form is the most readily available at this time. The advantages and disadvantages of the different values of pressure angle relate to the strength of the teeth, the occurrence of interference, and the magnitude of forces exerted on the shaft. Interference is discussed in Section 8–5. The other points are discussed in a later chapter.

Contact Ratio

When two gears mesh, it is essential for smooth operation that a second tooth begin to make contact before a given tooth disengages. The term *contact ratio* is used to indicate the average number of teeth in contact during the transmission of power. A recommended minimum contact ratio is 1.2 and typical spur gear combinations often have values of 1.5 or higher.

The contact ratio is defined as the ratio of the length of the line-of–action to the base pitch for the gear. The line-of-action is the straight-line path of a tooth from where it encounters the outside diameter of the mating gear to the point where it leaves engagement. The base pitch is the diameter of the base circle divided by the number of teeth in the gear. A convenient formula for computing the contact ratio, m_f, is,

$$m_f = \frac{\sqrt{R_{oP}^2 - R_{bP}^2} + \sqrt{R_{oG}^2 - R_{bG}^2} - C \sin \phi}{p \cos \phi}$$

where,

ϕ = Pressure angle

R_{oP} = Outside radius of the pinion = $D_{oP}/2 = (N_P + 2)/(2P_d)$

R_{bP} = Radius of the base circle for the pinion = $D_{bP}/2 = (D_P/2) \cos \phi = (N_P/2P_d) \cos \phi$

R_{oG} = Outside radius of the gear = $D_{oG}/2 = (N_G + 2)/(2P_d)$

R_{bG} = Radius of the base circle for the gear = $D_{bG}/2 = (D_G/2) \cos \phi = (N_G/2P_d) \cos \phi$

C = Center distance = $(N_P + N_G)/(2P_d)$

p = Circular pitch = $(\pi D_P/N_P) = \pi/P_d$

For example, consider a pair of gears with the following data:

$$N_P = 18, N_G = 64, P_d = 8, \phi = 20°$$

Then,

$$R_{oP} = (N_P + 2)/(2P_d) = (18 + 2)/[2(8)] = 1.250 \text{ in}$$
$$R_{bP} = (N_P/2P_d) \cos \phi = 18/[2(8)] \cos 20° = 1.05715 \text{ in}$$
$$R_{oG} = (N_G + 2)/(2P_d) = (64 + 2)/[2(8)] = 4.125 \text{ in}$$
$$R_{bG} = (N_G/2P_d) \cos \phi = 64/[2(8)] \cos 20° = 3.75877 \text{ in}$$
$$C = (N_P + N_G)/(2P_d) = (18 + 64)/[2(8)] = 5.125 \text{ in}$$
$$p = \pi/P_d = \pi/8 = 0.392699 \text{ in}$$

Finally, the contact ratio is,

$$m_f = \frac{\sqrt{(1.250)^2 - (1.05715)^2} + \sqrt{(4.125)^2 - (3.75877)^2} - (5.125)\sin 20°}{(0.392699)\cos 20°}$$

$$m_f = 1.66$$

This value is comfortably above the recommended minimum value of 1.20.

Example Problem 8–1   For the pair of gears shown in Figure 8–1, compute all of the features of the gear teeth described in this section. The gears conform to the standard AGMA form and have a diametral pitch of 12 and a 20° pressure angle.

Solution

Given $P_d = 12$; $N_P = 11$; $N_G = 18$; $\phi = 20°$.

Analysis   We use Equations (8–2) through (8–20) and Table 8–4 to compute the features. Note that gears are precision mechanical components. Dimensions are typically produced to at least the nearest thousandth of an inch (0.001 in). Often, for more accurate gears, controlling to the nearest 0.0001 in is important. Also, in the inspection of gear features using metrology techniques, it is important to know the standard dimension to a high degree of precision.

The results for this problem are presented to a minimum of three decimal places or, for small dimensions, to four decimal places. A similar level of precision is expected in practice problems using this book.

Results   *Pitch Diameters*
For the pinion,

$$D_P = N_P/P_d = 11/12 = 0.9167 \text{ in}$$

For the gear,

$$D_G = N_G/P_d = 18/12 = 1.500 \text{ in}$$

Circular Pitch
Three different approaches could be used. First, using Equation (8–5) is preferred.

$$p = \pi/P_d = \pi/12 = 0.2618 \text{ in}$$

Now, we can also use Equation (8–5): Note that either the pinion or the gear data may be used. For the pinion,

$$p = \pi D_P/N_P = \pi(0.9167 \text{ in})/11 = 0.2618 \text{ in}$$

For the gear,

$$p = \pi D_G/N_G = \pi(1.500 \text{ in})/18 = 0.2618 \text{ in}$$

Addendum
From Table 8–4,

$$a = 1/P_d = 1/12 = 0.833 \text{ in}$$

Dedendum
From Table 8–4, note that the 12-pitch gear is considered to be coarse. Thus,

$$b = 1.25/P_d = 1.25/12 = 0.1042 \text{ in}$$

Clearance
From Table 8–4,

$$c = 0.25/P_d = 0.25/12 = 0.0208 \text{ in}$$

Outside Diameters
Use of Equation (8–10) is preferred for accuracy. For the pinion,

$$D_{oP} = (N_P + 2)/P_d = (11 + 2)/12 = 1.0833 \text{ in}$$

For the gear,

$$D_{oG} = (N_G + 2)/P_d = (18 + 2)/12 = 1.6667 \text{ in}$$

Root Diameters
We use Equation (8–12). First, for the pinion,

$$D_{RP} = D_P - 2b = 0.9167 \text{ in} - 2(0.1042 \text{ in}) = 0.7083 \text{ in}$$

For the gear,

$$D_{RG} = D_G - 2b = 1.500 \text{ in} - 2(0.1042 \text{ in}) = 1.2917 \text{ in}$$

Whole Depth
Using Equation (8–13), we have

$$h_t = a + b = 0.0833 \text{ in} + 0.104 \text{ in} = 0.1875 \text{ in}$$

Working Depth
Using Equation (8–14), we have

$$h_k = 2a = 2(0.0833 \text{ in}) = 0.1667 \text{ in}$$

Tooth Thickness
Using Equation (8–16), we have

$$t = \pi/2P_d = \pi/2(12) = 0.1309 \text{ in}$$

Center Distance
Use of Equation (8–18) is preferred:

$$C = (N_G + N_P)/(2P_d) = (18 + 11)/[2(12)] = 1.2083 \text{ in}$$

Base Circle Diameter
Using Equation (8–20), we have

$$D_{bP} = D_P \cos \phi = (0.9167 \text{ in}) \cos (20°) = 0.8614 \text{ in}$$
$$D_{bG} = D_G \cos \phi = (1.500 \text{ in}) \cos (20°) = 1.4095 \text{ in}$$

8–5 INTERFERENCE BETWEEN MATING SPUR GEAR TEETH

For certain combinations of numbers of teeth in a gear pair, there is interference between the tip of the teeth on the pinion and the fillet or root of the teeth on the gear. Obviously this cannot be tolerated because the gears simply will not mesh. The probability that interference will occur is greatest when a small pinion drives a large gear, with the worst case being a small pinion driving a rack. A *rack* is a gear with a straight pitch line; it can be thought of as a gear with an infinite pitch diameter [see Figure 8–4(c)].

It is the designer's responsibility to ensure that interference does not occur in a given application. The surest way to do this is to control the minimum number of teeth in the pinion to the limiting values shown on the left side of Table 8–6. With this number of teeth or a greater number, there will be no interference with a rack or with any other gear. A designer who desires to use fewer than the listed number of teeth can use a graphical layout to test the combination of pinion and gear for interference. Texts on kinematics provide the necessary procedure. The right side of Table 8–6 indicates the maximum number of gear teeth that you can use for a given number of pinion teeth to avoid interference. (See References 9 and 11.)

Using the information in Table 8–6, we can draw the following conclusions:

1. If a designer wants to be sure that there will not be interference between any two gears when using the $14\frac{1}{2}°$, full-depth, involute system, the pinion of the gear pair must have no fewer than 32 teeth.

TABLE 8–6 Number of pinion teeth to ensure no interference

For a pinion meshing with a rack		For a 20°, full-depth pinion meshing with a gear	
Tooth form	Minimum number of teeth	Number of pinion teeth	Maximum number of gear teeth
$14\frac{1}{2}°$, involute, full-depth	32	17	1309
20°, involute, full-depth	18	16	101
25°, involute, full-depth	12	15	45
		14	26
		13	16

2. For the 20°, full-depth, involute system, using no fewer than 18 teeth will ensure that no interference occurs.

3. For the 25°, full-depth, involute system, using no fewer than 12 teeth will ensure that no interference occurs.

4. If a designer desires to use fewer than 18 teeth in a pinion having 20°, full-depth teeth, there is an upper limit to the number of teeth that can be used on the mating gear without interference. For 17 teeth in the pinion, any number of teeth on the gear can be used up to 1309, a very high number. Most gear drive systems use no more than about 200 teeth in any gear. But a 17-tooth pinion *would* have interference with a *rack* which is effectively a gear with an infinite number of teeth or an infinite pitch diameter. Similarly, the following requirements apply for 20° full-depth teeth:

A 16-tooth pinion requires a gear having 101 or fewer teeth, producing a maximum velocity ratio of $N_G/N_P = 101/16 = 6.31$.

A 15-tooth pinion requires a gear having 45 or fewer teeth, producing a maximum velocity ratio of 45/15 = 3.00.

A 14-tooth pinion requires a gear having 26 or fewer teeth, producing a maximum velocity ratio of 26/14 = 1.85.

A 13-tooth pinion requires a gear having 16 or fewer teeth, producing a maximum velocity ratio of 16/13 = 1.23.

As noted earlier, the $14\frac{1}{2}°$ system is considered to be obsolete. The data in Table 8–6 indicate one of the main disadvantages with that system: its potential for causing interference.

Overcoming Interference

If a proposed design encounters interference, there are ways to make it work. But caution should be exercised because the tooth form or the alignment of the mating gears is changed, causing the stress and wear analysis to be inaccurate. With this in mind, the designer can provide for undercutting, modification of the addendum on the pinion or the gear, or modification of the center distance:

> *Undercutting is the process of cutting away the material at the fillet or root of the gear teeth, thus relieving the interference.*

Figure 8–14 shows the result of undercutting. It should be obvious that this process weakens the tooth; this point is discussed further in the section on stresses in gear teeth.

FIGURE 8–14
Undercutting of a gear tooth

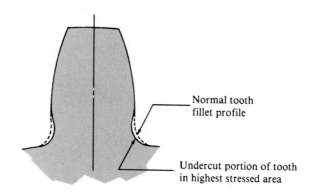

Normal tooth
fillet profile

Undercut portion of tooth
in highest stressed area

To alleviate the problem of interference, increase the addendum of the pinion while decreasing the addendum of the gear. The center distance can remain the same as its theoretical value for the number of teeth in the pair. But the resulting gears are, of course, nonstandard. (See Reference 10.) It is possible to make the pinion of a gear pair larger than standard while keeping the gear standard if the center distance for the pair is enlarged. (See Reference 9.)

8–6
VELOCITY RATIO
AND GEAR
TRAINS

A gear train is one or more pairs of gears operating together to transmit power.

Normally there is a speed change from one gear to the next due to the different sizes of the gears in mesh. The fundamental building block of the total speed change ratio in a gear train is the *velocity ratio* between two gears in a single pair.

Velocity Ratio

The velocity ratio (VR) is defined as the ratio of the rotational speed of the input gear to that of the output gear for a single pair of gears.

To develop the equation for computing the velocity ratio, it is helpful to view the action of two gears in mesh, as shown in Figure 8–15. The action is equivalent to the action of two smooth wheels rolling on each other without slipping, with the diameters of the two wheels equal to the pitch diameters of the two gears. Remember that when two gears are in mesh, their pitch circles are tangent, obviously, the gear teeth prohibit any slipping.

As shown in Figure 8–15, without slipping there is no relative motion between the two pitch circles at the pitch point, and therefore the linear velocity of a point on either pitch circle is the same. We will use the symbol v_t for this velocity. The linear velocity of a point that is in rotation at a distance R from its center of rotation and rotating with an angular velocity, ω, is found from

⇨ **Pitch Line**
Speed of a Gear

$$v_t = R\omega \qquad (8\text{–}21)$$

Using the subscript P for the pinion and G for the gear for two gears in mesh, we have

$$v_t = R_P\omega_P \quad \text{and} \quad v_t = R_G\omega_G$$

FIGURE 8–15 Two gears in mesh

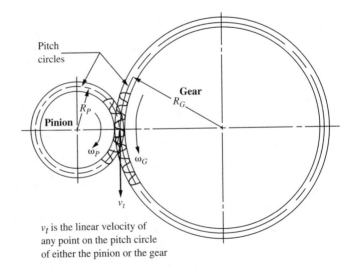

Pitch circles

Gear R_G

R_P

Pinion

ω_P

ω_G

v_t

v_t is the linear velocity of any point on the pitch circle of either the pinion or the gear

This set of equations says that the pitch line speeds of the pinion and the gear are the same. Equating these two and solving for ω_P/ω_G gives our definition for the velocity ratio, *VR:*

$$VR = \omega_P/\omega_G = R_G/R_P$$

In general, it is convenient to express the velocity ratio in terms of the pitch diameters, the rotational speeds, or the numbers of teeth of the two gears in mesh. Remember that

$$R_G = D_G/2$$
$$R_P = D_P/2$$
$$D_G = N_G/P_d = 4.71$$
$$D_P = N_P/P_d$$
$$n_P = \text{rotational speed of the pinion} \quad \text{(in rpm)}$$
$$n_G = \text{rotational speed of the gear} \quad \text{(in rpm)}$$

The velocity ratio can then be defined in any of the following ways:

**Velocity Ratio
for Gear Pair**

$$VR = \frac{\omega_P}{\omega_G} = \frac{n_P}{n_G} = \frac{R_G}{R_P} = \frac{D_G}{D_P} = \frac{N_G}{N_P} = \frac{\text{speed}_P}{\text{speed}_G} = \frac{\text{size}_G}{\text{size}_P} \qquad \text{(8–22)}$$

Most gear drives are *speed reducers;* that is, their output speed is lower than their input speed. This results in a velocity ratio greater than 1. If a *speed increaser* is desired, then *VR* is less than 1. Note that not all books and articles use the same definition for velocity ratio. Some define it as the ratio of the output speed to the input speed, the inverse of our definition. It is thought that the use of *VR* greater than 1 for the reducer—that is, the majority of the time—is more convenient.

Train Value

*When more than two gears are in mesh, the term **train value (TV)** refers to the ratio of the input speed (for the first gear in the train) to the output speed (for the last gear in the train). By definition the train value is the product of the values of VR for each **gear pair** in the train. In this definition, a gear pair is any set of two gears with a driver and a follower (driven) gear.*

Again, *TV* will be greater than 1 for a reducer and less than 1 for an increaser. For example, consider the gear train sketched in Figure 8–16. The input is through the shaft carrying gear *A*. Gear *A* drives gear *B*. Gear *C* is on the same shaft with gear *B* and rotates at the same speed. Gear *C* drives gear *D,* which is connected to the output shaft. Then gears *A* and *B* constitute the first gear pair, and gears *C* and *D* constitute the second pair. The velocity ratios are

$$VR_1 = n_A/n_B$$
$$VR_2 = n_C/n_D$$

The train value is

$$TV = (VR_1)(VR_2) = \frac{n_A}{n_B}\frac{n_C}{n_D}$$

But, because they are on the same shaft, $n_B = n_C$, and the preceding equation reduces to

$$TV = n_A/n_D$$

FIGURE 8–16
Double-reduction gear
train

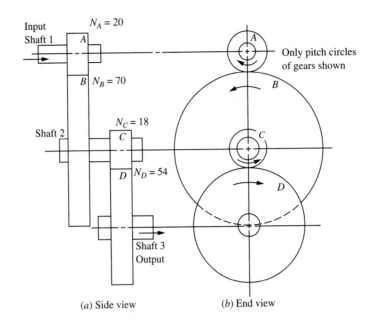

(a) Side view   (b) End view

This is the input speed divided by the output speed, the basic definition of the train value. This process can be expanded to any number of stages of reduction in a gear train.

Remember that any of the forms for velocity ratio shown in Equation (8–22) can be used for computing the train value. In design, it is often most convenient to express the velocity ratio in terms of the numbers of teeth in each gear because they must be integers. Then, once the diametral pitch or module is defined, the values of the diameters or radii can be determined.

The train value of the double-reduction gear train in Figure 8–16 can be expressed in terms of the numbers of teeth in the four gears as follows:

$$VR_1 = N_B/N_A$$

Note that this is the number of teeth in the *driven gear B* divided by the number of teeth in the *driving gear A*. This is the typical format for velocity ratio. Then VR_2 can be found similarly:

$$VR_2 = N_D/N_C$$

Thus, the train value is

$$TV = (VR_1)(VR_2) = (N_B/N_A)(N_D/N_C)$$

This is usually shown in the form

⇨   **Train Value** $$TV = \frac{N_B}{N_A}\frac{N_D}{N_C} = \frac{\text{product of number of teeth in the driven gears}}{\text{product of number of teeth in the driving gears}} \qquad \textbf{(8–23)}$$

This is the form for train value that we will use most often.

The direction of rotation can be determined by observation, noting that there is a direction reversal for each pair of external gears.

We will use the term **positive train value** *to refer to one in which the input and output gears rotate in the same direction. Conversely, if they rotate in the opposite direction, the train value will be negative.*

Example Problem 8–2   For the gear train shown in Figure 8–16, if the input shaft rotates at 1750 rpm clockwise, compute the speed of the output shaft and its direction of rotation.

Solution   We can find the output speed if we can determine the train value:

$$TV = n_A/n_D = \text{input speed/output speed}$$

Then

$$n_D = n_A/TV$$

But

$$TV = (VR_1)(VR_2) = \frac{N_B}{N_A}\frac{N_D}{N_C} = \frac{70}{20}\frac{54}{18} = \frac{3.5}{1}\frac{3.0}{1} = \frac{10.5}{1} = 10.5$$

Now

$$n_D = n_A/TV = (1750 \text{ rpm})/10.5 = 166.7 \text{ rpm}$$

Gear A rotates clockwise; gear B rotates counterclockwise.
Gear C rotates counterclockwise; gear D rotates clockwise.

Thus, the train in Figure 8–16 is a positive train.

Example Problem 8–3   Determine the train value for the train shown in Figure 8–17. If the shaft carrying gear A rotates at 1750 rpm clockwise, compute the speed and the direction of the shaft carrying gear E.

Solution   Look first at the direction of rotation. Remember that a gear pair is defined as any two gears in mesh (a driver and a follower). There are actually three gear pairs:

Gear A drives gear B: A rotates clockwise; B, counterclockwise.
Gear C drives gear D: C rotates counterclockwise; D, clockwise.
Gear D drives gear E: D rotates clockwise; E, counterclockwise.

Because gears A and E rotate in opposite directions, the train value is negative. Now

$$TV = -(VR_1)(VR_2)(VR_3)$$

In terms of the number of teeth,

$$TV = -\frac{N_B}{N_A}\frac{N_D}{N_C}\frac{N_E}{N_D}$$

Note that the number of teeth in gear D appears in both the numerator and the denominator and thus can be canceled. The train value then becomes

$$TV = -\frac{N_B}{N_A}\cdot\frac{N_E}{N_C} = -\frac{70}{20}\cdot\frac{50}{18} = -\frac{3.5}{1}\frac{3.0}{1} = -10.5$$

Gear D is called an *idler*. As demonstrated here, it has no effect on the magnitude of the train value, but it does cause a direction reversal. The output speed is then found from

$$TV = n_A/n_E$$

$$n_E = n_A/TV = (1750 \text{ rpm})/(-10.5) = -166.7 \text{ rpm} \quad (\text{counterclockwise})$$

FIGURE 8–17
Double-reduction gear train with an idler. Gear D is an idler.

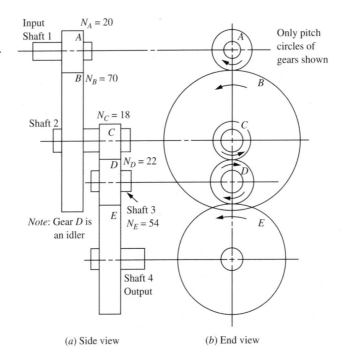

(*a*) Side view   (*b*) End view

Idler Gear

Example Problem 8–2 introduced the concept of an *idler gear,* defined as follows:

> *Any gear in a gear train that performs as both a driving gear and a driven gear is called an* **idler gear** *or simply an* **idler.**

The main features of an idler are as follows:

1. An idler does not affect the train value of a gear train because, since it is both a driver and a driven gear, its number of teeth appears in both the numerator and the denominator of the train value equation, Equation (8–23). Thus, any pitch diameter size and any number of teeth may be used for the idler.

2. Placing an idler in a gear train causes a direction reversal of the output gear.

3. An idler gear may be used to fill a space between two gears in a gear train when the desired distance between their centers is greater than the center distance for the two gears alone.

Internal Gear

> *An* **internal gear** *is one for which the teeth are machined on the inside of a ring instead of on the outside of a gear blank.*

An internal gear mating with a standard, external pinion is illustrated at the lower left in Figure 8–2, along with a variety of other kinds of gears.

Figure 8–18 is a sketch of an external pinion driving an internal gear. Note the following:

1. The gear rotates in the *same direction* as the pinion. This is different from the case when an external pinion drives an external gear.

FIGURE 8–18
Internal gear driven by
an external pinion

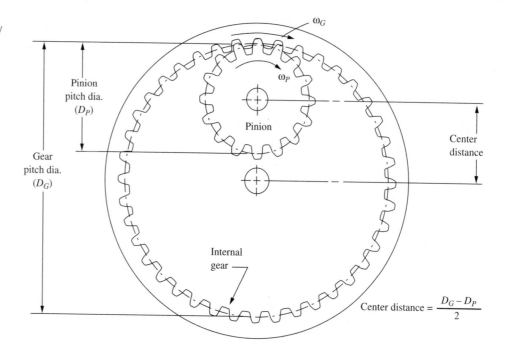

Center distance $= \dfrac{D_G - D_P}{2}$

2. The center distance is

Center Distance
Internal Gear

$$C = D_G/2 - D_P/2 = (D_G - D_P)/2 = (N_G/P_d - N_P/P_d)/2 = (N_G - N_P)/(2P_d) \quad \textbf{(8–24)}$$

The last form is preferred because its factors are all integers for typical gear trains.

3. The descriptions of most other features of internal gears are the same as those for external gears presented earlier. Exceptions for an internal gear are as follows:

The addendum, *a,* is the radial distance from the pitch circle to the inside of a tooth.

The inside diameter, D_i, is

$$D_i = D - 2a$$

The root diameter, D_R, is

$$D_R = D + 2b$$

where b = dedendum

Internal gears are used when it is desired to have the same direction of rotation for the input and the output. Also note that less space is taken for an internal gear mating with an external pinion compared with two external gears in mesh.

Velocity of a Rack

Figure 8–19 shows the basic configuration of a *rack-and-pinion* drive. The function of such a drive is to produce a linear motion of the rack from the rotational motion of the driving pinion. The opposite is also true: If the driver produces the linear motion of the rack, it produces a rotational motion of the pinion.

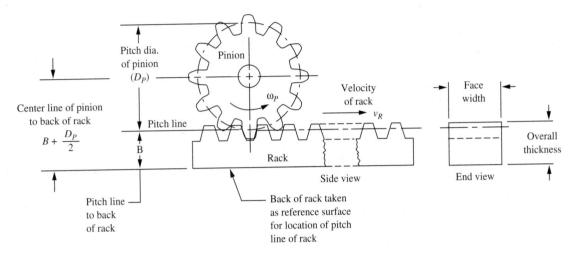

FIGURE 8–19 Rack driven by a pinion

The linear velocity of the rack, v_R, must be the same as the pitch line velocity of the pinion, v_t, as defined by Equation (8–21), repeated here. Recall that w_P is the angular velocity of the pinion:

$$v_R = v_t = R_P \omega_P = (D_P/2)\omega_P$$

You must carefully consider units when using this equation. The angular velocity ω_P should be expressed in rad/s. Then the units for linear velocity will be in/s if the pitch diameter is in inches. If the pitch diameter is in mm, as in the metric module system, then the units for velocity will be mm/s. These units could be converted to m/s if that is more convenient.

The concept of center distance does not apply directly for a rack-and-pinion set because the center of the rack is at infinity. But it is critical that the pitch circle of the pinion be tangent to the pitch line of the rack as shown in Figure 8–19. The rack will be machined so that there is a specified dimension between the pitch line and a reference surface, typically the back of the rack. This is dimension B in Figure 8–19. Then the location of the center of the pinion can be computed using the relationships shown in the figure.

Example Problem 8–4   Determine the linear velocity of the rack in Figure 8–19 if the driving pinion rotates at 125 rpm. The pinion has 24 teeth and a diametral pitch of 6.

Solution   We will use Equation (8–21). First the pitch diameter of the pinion is computed using Equation (8–4):

$$D_P = N_P/P_d = 24/6 = 4.000 \text{ in}$$

Now the rotational speed is converted to rad/s:

$$\omega_P = (125 \text{ rev/min}(2\pi \text{ rad/rev})(1 \text{ min/60 s}) = 13.09 \text{ rad/s}$$

Then the pitch line speed of the pinion and the linear velocity of the rack are both equal to

$$v_R = v_t = (D_P/2)\omega_P = (4.000 \text{ in}/2)(13.09 \text{ rad/s}) = 26.2 \text{ in/s}$$

**8–7
HELICAL GEAR
GEOMETRY**

Helical and spur gears are distinguished by the orientation of their teeth. On spur gears, the teeth are straight and are aligned with the axis of the gear. On helical gears, the teeth are inclined at an angle with the axis, that angle being called the *helix angle*. If the gear were very wide, it would appear that the teeth wind around the gear blank in a continuous, helical path. However, practical considerations limit the width of the gears so that the teeth normally appear to be merely inclined with respect to the axis. Figure 8–20 shows two examples of commercially available helical gears.

The forms of helical gear teeth are very similar to those discussed for spur gears. The basic task is to account for the effect of the helix angle.

Helix Angle

The helix for a given gear can be either *left-hand* or *right-hand.* The teeth of a right-hand helical gear would appear to lean to the right when the gear is lying on a flat surface. Conversely, the teeth of a left-hand helical gear would lean to the left. In normal installation, helical gears would be mounted on parallel shafts as shown in Figure 8–20(a). To achieve this arrangement, it is required that one gear be of the right-hand design and that the other be left-hand with an equal helix angle. If both gears in mesh are of the same hand, as shown in Figure 8–20(b), the shafts will be at 90 to each other. Such gears are called *crossed helical gears.*

The parallel shaft arrangement for helical gears is preferred because it results in a much higher power-transmitting capacity for a given size of gear than the crossed helical arrangement. In this book, we will assume that the parallel shaft arrangement is being used unless otherwise stated.

Figure 8–21(a) shows the pertinent geometry of helical gear teeth. To simplify the drawing, only the pitch surface of the gear is shown. The pitch surface is the cylinder that passes through the gear teeth at the pitch line. Thus, the diameter of the cylinder is equal to the pitch diameter of the gear. The lines drawn on the pitch surface represent elements of each tooth where the surface would cut into the face of the tooth. These elements are inclined with respect to a line parallel to the axis of the cylinder, and the angle of inclination is the *helix angle,* ψ (the Greek letter *psi*).

FIGURE 8–20
Helical gears. These gears have a 45° helix angle. (Emerson Power Transmission Corporation, Browning Division, Maysville, KY)

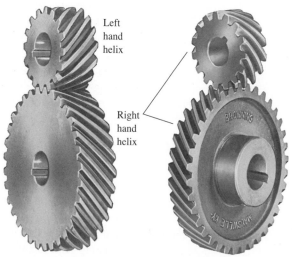

Left hand helix

Right hand helix

(a) Helical gears with parallel shafts   *(b)* Crossed helical gears, shafts at right angle

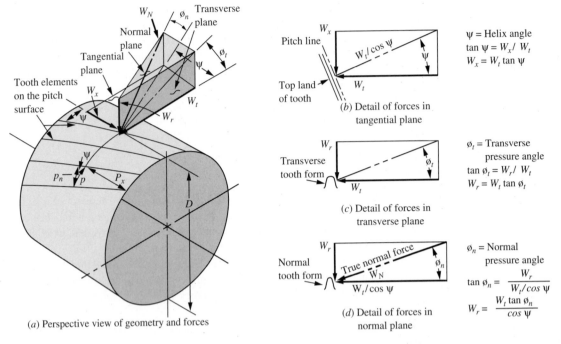

(a) Perspective view of geometry and forces

(b) Detail of forces in tangential plane

ψ = Helix angle
$\tan \psi = W_x / W_t$
$W_x = W_t \tan \psi$

(c) Detail of forces in transverse plane

ϕ_t = Transverse pressure angle
$\tan \phi_t = W_r / W_t$
$W_r = W_t \tan \phi_t$

(d) Detail of forces in normal plane

ϕ_n = Normal pressure angle
$\tan \phi_n = \dfrac{W_r}{W_t / \cos \psi}$
$W_r = \dfrac{W_t \tan \phi_n}{\cos \psi}$

FIGURE 8–21   Helical gear geometry and forces

The main advantage of helical gears over spur gears is smoother engagement because a given tooth assumes its load gradually instead of suddenly. Contact starts at one end of a tooth near the tip and progresses across the face in a path downward across the pitch line to the lower flank of the tooth, where it leaves engagement. Simultaneously, other teeth are coming into engagement before a given tooth leaves engagement, with the result that a larger average number of teeth are engaged and are sharing the applied loads compared with a spur gear. The lower average load per tooth allows a greater power transmission capacity for a given size of gear, or a smaller gear can be designed to carry the same power.

The main disadvantage of helical gears is that an *axial thrust load* is produced as a natural result of the inclined arrangement of the teeth. The bearings that hold the shaft carrying the helical gear must be capable of reacting against the thrust load.

The helix angle is specified for each given gear design. A balance should be sought to take advantage of the smoother engagement of the gear teeth when the helix angle is high while maintaining a reasonable value of the axial thrust load that increases with increasing helix angle. A typical range of values of helix angles is from 15° to 45°.

Pressure Angles, Primary Planes, and Forces for Helical Gears

To completely describe the geometry of helical gear teeth, we need to define two different pressure angles in addition to the helix angle. The two pressure angles are related to the three primary planes that are illustrated in Figure 8–21: (1) the *tangential plane*, (2) the *transverse plane*, and (3) the *normal plane*. Note that these planes contain the three orthogonal components of the true total normal force that is exerted by a tooth of one gear on a tooth of the mating gear. It may help you to understand the geometry of the teeth, and the importance of that geometry, to see how it affects the forces.

We refer first to the *true normal force* as W_N. It acts normal (perpendicular) to the curved surface of the tooth. In reality, we typically do not use the normal force itself in analyzing the performance of the gear. Instead we use its three orthogonal components:

■ The *tangential force* (also called the *transmitted force*), W_t, acts tangential to the pitch surface of the gear and perpendicular to the axis of the shaft carrying the gear. This is the force that actually drives the gear. Stress analysis and pitting resistance are both related to the magnitude of the tangential force. It is similar to W_t used in spur gear design and analysis.

■ The *radial force*, W_r, acts toward the center of the gear along a radius and tends to separate the two gears in mesh. It is similar to W_r used in spur gear design and analysis.

■ The *axial force*, W_x, acts in the tangential plane parallel to the axis of the shaft carrying the gear. Another name for the axial force is the *thrust*. It tends to push the gear along the shaft. The thrust must be reacted by one of the bearings that carry the shaft, so this force is generally undesirable. Spur gears generate no such force because the teeth are straight and are parallel to the axis of the gear.

The plane containing the tangential force, W_t, and the axial force, W_x, is the *tangential plane* [see Figure 8–21(b)]. It is tangential to the pitch surface of the gear and acts through the pitch point at the middle of the face of the tooth being analyzed.

The plane containing the tangential force, W_t, and the radial force, W_r, is the *transverse plane* [see Figure 8–21(c)]. It is perpendicular to the axis of the gear and acts through the pitch point at the middle of the face of the tooth being analyzed. The *transverse pressure angle*, ϕ_t, is defined in this plane as shown in the figure.

The plane containing the true normal force, W_N, and the radial force, W_r, is the *normal plane* [see Figure 8–21(d)]. The angle between the normal plane and the transverse plane is the helix angle, ψ. Within the normal plane we can see that the angle between the tangential plane and the true normal force, W_N, is the *normal pressure angle*, ϕ_n.

In design of a helical gear, there are three angles of interest: (1) the helix angle, ψ; (2) the *normal pressure angle*, ϕ_n; and (3) the *transverse pressure angle*, ϕ_t. Designers must specify the helix angle and one of the two pressure angles. The other pressure angle can be computed from the following relationship:

$$\tan \phi_n = \tan \phi_t \cos \psi \qquad\qquad (8\text{–}25)$$

For example, one manufacturer's catalog offers standard helical gears with a normal pressure angle of $14\frac{1}{2}°$ and a $45°$ helix angle. Then the transverse pressure angle is found from

$$\tan \phi_n = \tan \phi_t \cos \psi$$

$$\tan \phi_t = \tan \phi_n / \cos \psi = \tan(14.5°)/\cos(45°) = 0.3657$$

$$\phi_t = \tan^{-1}(0.3657) = 20.09°$$

Pitches for Helical Gears

To obtain a clear picture of the geometry of helical gears, you must understand the following five different pitches.

Circular Pitch, *p*.   *Circular pitch* is the distance from a point on one tooth to the corresponding point on the next adjacent tooth, measured at the pitch line in the transverse plane. This is the same definition used for spur gears. Then

⇨ **Circular Pitch**

$$p = \pi D/N \qquad \qquad \text{(8–26)}$$

Normal Circular Pitch, p_n.   *Normal circular pitch* is the distance between corresponding points on adjacent teeth measured on the pitch surface in the normal direction. Pitches p and p_n are related by the following equation:

⇨ **Normal Circular Pitch**

$$p_n = p \cos \psi \qquad \qquad \text{(8–27)}$$

Diametral Pitch, P_d.   *Diametral pitch* is the ratio of the number of teeth in the gear to the pitch diameter. This is the same definition as the one for spur gears; it applies in considerations of the form of the teeth in the diametral or transverse plane. Thus, this pitch is sometimes called the *transverse diametral pitch:*

⇨ **Diametral Pitch**

$$P_d = N/D \qquad \qquad \text{(8–28)}$$

Normal Diametral Pitch, P_{nd}.   Normal diametral pitch is the equivalent diametral pitch in the plane normal to the teeth:

⇨ **Normal Diametral Piltch**

$$P_{nd} = P_d/\cos \psi \qquad \qquad \text{(8–29)}$$

It is helpful to remember these relationships:

$$P_d p = \pi \qquad \qquad \text{(8–30)}$$
$$P_{nd} p_n = \pi \qquad \qquad \text{(8–31)}$$

Axial Pitch, P_x.   *Axial pitch* is the distance between corresponding points on adjacent teeth, measured on the pitch surface in the axial direction:

⇨ **Axial Pitch**

$$P_x = p/\tan \psi = \pi/(P_d \tan \psi) \qquad \qquad \text{(8–32)}$$

It is necessary to have at least two axial pitches in the face width to have the benefit of full helical action and its smooth transfer of the load from tooth to tooth.

The use of equations (8–25) through (8–29) and Equation (8–32) is now illustrated in the following example problem.

Example Problem 8–5   A helical gear has a transverse diametral pitch of 12, a transverse pressure angle of $14\frac{1}{2}°$, 28 teeth, a face width of 1.25 in, and a helix angle of 30°. Compute circular pitch, normal circular pitch, normal diametral pitch, axial pitch, pitch diameter, and the normal pressure angle. Compute the number of axial pitches in the face width.

Solution   *Circular Pitch*
Use Equation (8–30):

$$p = \pi/P_d = \pi/12 = 0.262 \text{ in}$$

Normal Circular Pitch
Use Equation (8–27):

$$p_n = p \cos \psi = (0.262)\cos(30) = 0.227 \text{ in}$$

Normal Diametral Pitch
Use Equation (8–29):

$$P_{nd} = P_d/\cos \psi = 12/\cos(30) = 13.856$$

Axial Pitch
Use Equation (8–32):

$$P_x = p/\tan \psi = 0.262/\tan(30) = 0.453 \text{ in}$$

Pitch Diameter
Use Equation (8–28):

$$D = N/P_d = 28/12 = 2.333 \text{ in}$$

Normal Pressure Angle
Use Equation (8–25):

$$\phi_n = \tan^{-1}(\tan \phi_t \cos \psi)$$
$$\phi_n = \tan^{-1}[\tan(14\tfrac{1}{2})\cos(30)] = 12.62°$$

Number of Axial Pitches in the Face Width

$$F/P_x = 1.25/0.453 = 2.76 \text{ pitches}$$

Since this is greater than 2.0, there will be full helical action.

**8–8
BEVEL GEAR
GEOMETRY**

Bevel gears are used to transfer motion between nonparallel shafts, usually at 90° to one another. The four primary styles of bevel gears are straight bevel, spiral bevel, zero spiral bevel, and hypoid. Figure 8–22 shows the general appearance of these four types of bevel gear sets. The surface on which bevel gear teeth are machined is inherently a part of a cone. The differences occur in the specific shape of the teeth and in the orientation of the pinion relative to the gear. (See References 3, 5, 13, and 14.)

Straight Bevel Gears

The teeth of a straight bevel gear are straight and lie along an element of the conical surface. Lines along the face of the teeth through the pitch circle meet at the apex of the pitch cone. As shown in Figure 8–22(f), the centerlines of both the pinion and the gear also meet at this apex. In the standard configuration, the teeth are tapered toward the center of the cone.

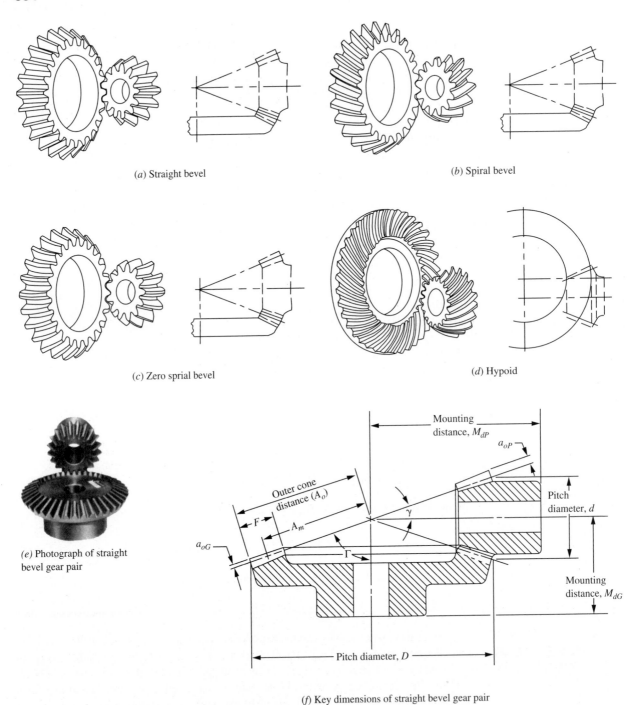

(a) Straight bevel

(b) Spiral bevel

(c) Zero sprial bevel

(d) Hypoid

(e) Photograph of straight bevel gear pair

(f) Key dimensions of straight bevel gear pair

FIGURE 8–22 Types of bevel gears [Parts (a) through (d) extracted from ANSI/AGMA 2005-C96, *Design Manual for Bevel Gears,* with the permission of the publisher, the American Gear Manufacturers Association, 1500 King Street, Suite 201, Alexandria, VA 22314. Photograph in (e) courtesy of Emerson Power Transmission Corporation, Browning Division, Maysville, KY

Key dimensions are specified either at the outer end of the teeth or at the mean, mid-face position. The relationships that control some of these dimensions are listed in Table 8–7 for the case when the shafts are at the 90° angle. The pitch cone angles for the pinion and the gear are determined by the ratio of the number of teeth, as shown in the table. Note that their sum is 90°. Also, for a pair of bevel gears having a ratio of unity, each has a pitch cone angle of 45°. Such gears, called *miter gears,* are used simply to change the direction of the shafts in a machine drive without affecting the speed of rotation.

You should understand that many more features need to be specified before the gears can be produced. Furthermore, many successful, commercially available gears are made in some nonstandard form. For example, the addendum of the pinion is often made longer than that of the gear. Some manufacturers modify the slope of the root of the teeth to produce a uniform depth, rather than using the standard, tapered form. Reference 5 gives many more data.

TABLE 8–7   Geometrical features of straight bevel gears

Given   Diametral pitch $= P_d = N_P/d = N_G/D$

where $N_P =$ number of teeth in pinion
$N_G =$ number of teeth in gear

Dimension	Formula
Gear ratio	$m_G = N_G/N_P$
Pitch diameters:	
Pinion	$d = N_P/P_d$
Gear	$D = N_G/P_d$
Pitch cone angles:	
Pinion	$\gamma = \tan^{-1}(N_P/N_G)$ (lowercase Greek *gamma*)
Gear	$\Gamma = \tan^{-1}(N_G/N_P)$ (uppercase Greek *gamma*)
Outer cone distance	$A_o = 0.5D/\sin(\Gamma)$
Face width must be specified.	$F =$
Nominal face width	$F_{\text{nom}} = 0.30A_o$
Maximum face width	$F_{\max} = A_o/3$ or $F_{\max} = 10/P_d$ (whichever is less)
Mean cone distance	$A_m = A_o - 0.5F$
	(*Note:* A_m is defined for the gear, also called A_{mG}.)
Mean circular pitch	$p_m = (\pi/P_d)(A_m/A_o)$
Mean working depth	$h = (2.00/P_d)(A_m/A_o)$
Clearance	$c = 0.125h$
Mean whole depth	$h_m = h + c$
Mean addendum factor	$c_1 = 0.210 + 0.290/(m_G)^2$
Gear mean addendum	$a_G = c_1 h$
Pinion mean addendum	$a_P = h - a_G$
Gear mean dedendum	$b_G = h_m - a_G$
Pinion mean dedendum	$b_P = h_m - a_P$
Gear dedendum angle	$\delta_G = \tan^{-1}(b_G/A_{mG})$
Pinion dedendum angle	$\delta_P = \tan^{-1}(b_P/A_{mG})$
Gear outer addendum	$a_{oG} = a_G + 0.5F \tan \delta_P$
Pinion outer addendum	$a_{oP} = a_P + 0.5F \tan \delta_G$
Gear outside diameter	$D_o = D + 2a_{oG} \cos \Gamma$
Pinion outside diameter	$d_o = d + 2a_{oP} \cos \gamma$

Source: Extracted from ANSI/AGMA 2005-C96, *Design Manual for Bevel Gears,* with the permission of the publisher, the American Gear Manufacturers Association, 1500 King Street, Suite 201, Alexandria, VA 22314.

The pressure angle, ϕ, is typically 20°, but 22.5° and 25° are often used to avoid interference. The minimum number of teeth for straight bevel gears is typically 12. More is said about the design of straight bevel gears in Chapter 10.

The mounting of bevel gears is critical if satisfactory performance is to be achieved. Most commercial gears have a defined mounting distance similar to that shown in Figure 8–22(f). It is the distance from some reference surface, typically the back of the hub of the gear, to the apex of the pitch cone. Because the pitch cones of the mating gears have coincident apexes, the mounting distance also locates the axis of the mating gear. If the gear is mounted at a distance smaller than the recommended mounting distance, the teeth will likely bind. If it is mounted at a greater distance, there will be excessive backlash, causing noisy and rough operation.

Spiral Bevel Gears

The teeth of a spiral bevel gear are curved and sloped with respect to the surface of the pitch cone. Spiral angles, ψ, of 20° to 45° are used, with 35° being typical. Contact starts at one end of the teeth and moves along the tooth to its end. For a given tooth form and number of teeth, more teeth are in contact for spiral bevel gears than for straight bevel gears. The gradual transfer of loads and the greater average number of teeth in contact make spiral bevel gears smoother and allow smaller designs than for typical straight bevel gears. Recall that similar advantages were described for a helical gear relative to a spur gear.

The pressure angle, ϕ, is typically 20° for spiral bevel gears, and the minimum number of teeth is typically 12 to avoid interference. But nonstandard spiral gears allow as few as five teeth in the pinion of high-ratio sets if the tips of the teeth are trimmed to avoid interference. The rather high average number of teeth in contact (high contact ratio) for spiral gears makes this approach acceptable and can result in a very compact design. Reference 5 gives the relationships for computing the geometric features of spiral bevel gears that are extensions of those given in Table 8–7.

Zero Spiral Bevel Gears

The teeth of a zero spiral bevel gear are curved somewhat as in a spiral bevel gear, but the spiral angle is zero. These gears can be used in the same mountings as straight bevel gears, but they operate more smoothly. They are sometimes called ZEROL® bevel gears

Hypoid Gears

The major difference between hypoid gears and the others just described is that the centerline of the pinion for a set of hypoid gears is offset either above or below the centerline of the gear. The teeth are designed specially for each combination of offset distance and spiral angle of the teeth. A major advantage is the more compact design that results, particularly when applied to vehicle drive trains and machine tools. (See References 5, 13, and 14 for more data.)

The hypoid gear geometry is the most general form, and the others are special cases. The hypoid gear has an offset axis for the pinion, and its curved teeth are cut at a spiral angle. Then the spiral bevel gear is a hypoid gear with a zero offset distance. A ZEROL® bevel gear is a hypoid gear with a zero offset and a zero spiral angle. A straight bevel gear is a hypoid gear with a zero offset, a zero spiral angle, and straight teeth.

Example Problem 8–6   Compute the values for the geometrical features listed in Table 8–7 for a pair of straight bevel gears having a diametral pitch of 8, a 20° pressure angle, 16 teeth in the pinion, and 48 teeth in the gear. The shafts are at 90°.

Solution   Given $P_d = 8$; $N_p = 16$; $N_G = 48$.

Computed Values   *Gear Ratio*

$$m_G = N_G/N_P = 48/16 = 3.000$$

Pitch Diameter
For the pinion,

$$d = N_P/P_d = 16/8 = 2.000 \text{ in}$$

For the gear,

$$D = N_G/P_d = 48/8 = 6.000 \text{ in}$$

Pitch Cone Angles
For the pinion,

$$\gamma = \tan^{-1}(N_P/N_G) = \tan^{-1}(16/48) = 18.43°$$

For the gear,

$$\Gamma = \tan^{-1}(N_G/N_P) = \tan^{-1}(48/16) = 71.57°$$

Outer Cone Distance

$$A_o = 0.5\, D/\sin(\Gamma) = 0.5(6.00 \text{ in})/\sin(71.57°) = 3.162 \text{ in}$$

Face Width
The face width must be specified:

$$F = 1.000 \text{ in}$$

Based on the following guidelines:
    Nominal face width:

$$F_{nom} = 0.30A_o = 0.30(3.162 \text{ in}) = 0.949 \text{ in}$$

Maximum face width:

$$F_{max} = A_o/3 = (3.162 \text{ in})/3 = 1.054 \text{ in}$$

or

$$F_{max} = 10/P_d = 10/8 = 1.25 \text{ in}$$

Mean Cone Distance

$$A_m = A_{mG} = A_o - 0.5\, F = 3.162 \text{ in} - 0.5(1.00 \text{ in}) = 2.662 \text{ in}$$

Ratio $A_m/A_o = 2.662/3.162 = 0.842$ (This ratio occurs in several following calculations.)

Mean Circular Pitch

$$p_m = (\pi/P_d)(A_m/A_o) = (\pi/8)(0.842) = 0.331 \text{ in}$$

Mean Working Depth

$$h = (2.00/P_d)(A_m/A_o) = (2.00/8)(0.842) = 0.210 \text{ in}$$

Clearance

$$c = 0.125h = 0.125(0.210 \text{ in}) = 0.026 \text{ in}$$

Mean Whole Depth

$$h_m = h + c = 0.210 \text{ in} + 0.026 \text{ in} = 0.236 \text{ in}$$

Mean Addendum Factor

$$c_1 = 0.210 + 0.290/(m_G)^2 = 0.210 + 0.290/(3.00)^2 = 0.242$$

Gear Mean Addendum

$$a_G = c_1 h = (0.242)(0.210 \text{ in}) = 0.051 \text{ in}$$

Pinion Mean Addendum

$$a_p = h - a_G = 0.210 \text{ in} - 0.051 \text{ in} = 0.159 \text{ in}$$

Gear Mean Dedendum

$$b_G = h_m - a_G = 0.236 \text{ in} - 0.051 \text{ in} = 0.185 \text{ in}$$

Pinion Mean Dedendum

$$b_P = h_m - a_P = 0.236 \text{ in} - 0.159 \text{ in} = 0.077 \text{ in}$$

Gear Dedendum Angle

$$\delta_G = \tan^{-1}(b_G/A_{mG}) = \tan^{-1}(0.185/2.662) = 3.98°$$

Pinion Dedendum Angle

$$\delta_P = \tan^{-1}(b_P/A_{mG}) = \tan^{-1}(0.077/2.662) = 1.66°$$

Gear Outer Addendum

$$a_oG = a_G + 0.5\,F \tan \delta_P$$
$$a_{oG} = (0.051 \text{ in}) + (0.5)(1.00 \text{ in}) \tan(1.657°) = 0.0655 \text{ in}$$

Pinion Outer Addendum

$$a_oP = a_P + 0.5\,F \tan \delta_G$$
$$a_oP = (0.159 \text{ in}) + (0.5)(1.00 \text{ in}) \tan(3.975°) = 0.1937 \text{ in}$$

Gear Outside Diameter

$$D_o = D + 2a_{oG} \cos \Gamma$$
$$D_o = 6.000 \text{ in} + 2(0.0655 \text{ in}) \cos(71.57°) = 6.041 \text{ in}$$

Pinion Outside Diameter

$$d_o = d + 2a_{oP} \cos \gamma$$
$$d_o = 2.000 \text{ in} + 2(0.1937 \text{ in}) \cos(18.43°) = 2.368 \text{ in}$$

8–9 TYPES OF WORMGEARING

Wormgearing is used to transmit motion and power between nonintersecting shafts, usually at 90° to each other. The drive consists of a worm on the high-speed shaft which has the general appearance of a power screw thread: a cylindrical, helical thread. The worm drives a wormgear, which has an appearance similar to that of a helical gear. Figures 8–23 shows a typical worm and wormgear set. Sometimes the wormgear is referred to as a *worm wheel* or simply a *wheel* or *gear*. (See Reference 6.) Worms and wormgears can be provided with either right hand or left hand threads on the worm and correspondingly designed teeth on the wormgear affecting the rotational direction of the wormgear.

Several variations of the geometry of wormgear drives are available. The most common one, shown in Figures 8–23 and 8–24, employs a cylindrical worm mating with a wormgear having teeth that are throated, wrapping partially around the worm. This is called a *single-enveloping type* of wormgear drive. The contact between the threads of the worm and wormgear teeth is along a line, and the power transmission capacity is quite good. Many manufacturers offer this type of wormgear set as a stocked item. Installation of the worm is relatively easy because axial alignment is not very critical. However, the wormgear must be carefully aligned radially in order to achieve the benefit of the enveloping action. Figure 8–25 shows a cutaway of a commercial worm-gear reducer.

FIGURE 8–23 Worms and wormgears (Emerson Power Transmission Corporation, Browning Division, Maysville, KY)

FIGURE 8–24
Single-enveloping
wormgear set

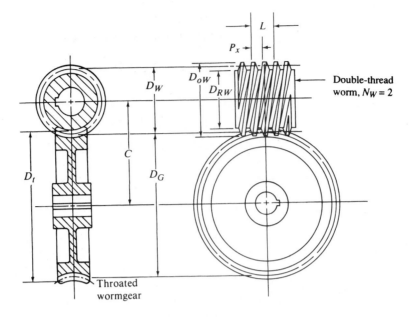

FIGURE 8–25
Wormgear reducer
(Rockwell
Automation/Dodge
Greenville, SC)

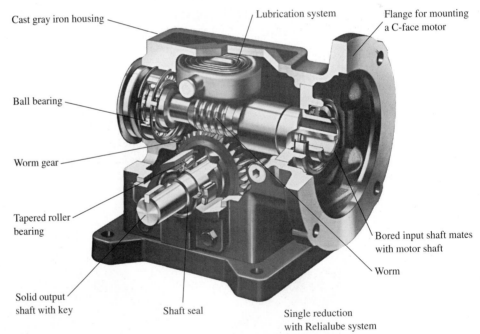

A simpler form of wormgear drive allows a special cylindrical worm to be used with a standard spur gear or helical gear. Neither the worm nor the gear must be aligned with great accuracy, and the center distance is not critical. However, the contact between the worm threads and the wormgear teeth is theoretically a point, drastically reducing the power transmission capacity of the set. Thus, this type is used mostly for nonprecision positioning applications at low speeds and low power levels.

A third type of wormgear set is the *double-enveloping type* in which the worm is made in an hourglass shape and mates with an enveloping type of wormgear. This results in area contact rather than line or point contact and allows a much smaller system to transmit a given power at a given reduction ratio. However, the worm is more difficult to manufacture, and the alignment of both the worm and the wormgear is very critical.

**8–10
GEOMETRY OF
WORMS AND
WORMGEARS**

Pitches, p and P_d

A basic requirement of the worm and wormgear set is that the *axial pitch* of the worm must be equal to the *circular pitch* of the wormgear in order for them to mesh. Figure 8–24 shows the basic geometric features of a single-enveloping worm and wormgear set. *Axial pitch, P_x,* is defined as the distance from a point on the worm thread to the corresponding point on the next adjacent thread, measured axially on the pitch cylinder. As before, the circular pitch is defined for the wormgear as the distance from a point on a tooth on the pitch circle of the gear to the corresponding point on the next adjacent tooth, measured along the pitch circle. Thus, the circular pitch is an arc distance that can be calculated from

▷ **Circular Pitch**

$$p = \pi D_G/N_G \qquad (8\text{–}33)$$

where D_G = pitch diameter of the gear
N_G = number of teeth in the gear

Some wormgears are made according to the circular pitch convention. But, as noted with spur gears, commercially available wormgear sets are usually made to a diametral pitch convention with the following pitches readily available: 48, 32, 24, 16, 12, 10, 8, 6, 5, 4, and 3. The diametral pitch is defined for the gear as

▷ **Diametral Pitch**

$$P_d = N_G/D_G \qquad (8\text{–}34)$$

The conversion from diametral pitch to circular pitch can be made from the following equation:

$$P_d p = \pi \qquad (8\text{–}35)$$

Number of Worm Threads, N_W

Worms can have a single thread, as in a typical screw, or multiple threads, usually 2 or 4, but sometimes 3, 5, 6, 8, or more. It is common to refer to the number of threads as N_W and then to treat that number as if it were the number of teeth in the worm. The number of threads in the worm is frequently referred to as the number of *starts;* this is convenient because if you look at the end of a worm, you can count the number of threads that start at the end and wind down the cylindrical worm.

Lead, L

The *lead* of a worm is the axial distance that a point on the worm would move as the worm is rotated one revolution. Lead is related to the axial pitch by

▷ **Lead**

$$L = N_W P_x \qquad (8\text{–}36)$$

Lead Angle, λ

The *lead angle* is the angle between the tangent to the worm thread and the line perpendicular to the axis of the worm. To visualize the method of calculating the lead angle, refer to Figure 8–26, which shows a simple triangle that would be formed if one thread of the worm

FIGURE 8–26 Lead angle

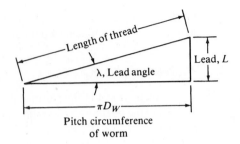

were unwrapped from the pitch cylinder and laid flat on the paper. The length of the hypotenuse is the length of the thread itself. The vertical side is the lead, L. The horizontal side is the circumference of the pitch cylinder, πD_W, where D_W is the pitch diameter of the worm. Then

 **Lead Angle**

$$\tan \lambda = L/\pi D_W \qquad (8\text{–}37)$$

Pitch Line Speed, v_t

As before, the pitch line speed is the linear velocity of a point on the pitch line for the worm or the wormgear. For the worm having a pitch diameter D_W in, rotating at n_W rpm,

Pitch Line Speed for worm

$$v_{tW} = \frac{\pi D_W n_W}{12} \text{ ft/min}$$

For the wormgear having a pitch diameter D_G in, rotating at n_G rpm,

Pitch Line Speed for Gear

$$V_{tG} = \frac{\pi D_G n_G}{12} \text{ ft/min}$$

Note that these two values for pitch line speed are *not* equal.

Velocity Ratio, *VR*

It is most convenient to calculate the velocity ratio of a worm and wormgear set from the ratio of the input rotational speed to the output rotational speed:

Velocity Ratio for Worm/Wormgear Set

$$VR = \frac{\text{speed of worm}}{\text{speed of gear}} = \frac{n_W}{n_G} = \frac{N_G}{N_W} \qquad (8\text{–}38)$$

Example Problem 8–7   A wormgear has 52 teeth and a diametral pitch of 6. It mates with a triple-threaded worm that rotates at 1750 rpm. The pitch diameter of the worm is 2.000 in. Compute the circular pitch, the axial pitch, the lead, the lead angle, the pitch diameter of the wormgear, the center distance, the velocity ratio, and the rotational speed of the wormgear.

Solution   *Circular Pitch*

$$p = \pi/P_d = \pi/6 = 0.5236 \text{ in}$$

Axial Pitch

$$P_x = p = 0.5236 \text{ in}$$

Lead

$$L = N_W P_x = (3)(0.5236) = 1.5708 \text{ in}$$

Lead Angle

$$\lambda = \tan^{-1}(L/\pi D_W) = \tan^{-1}(1.5708/\pi 2.000)$$
$$\lambda = 14.04°$$

Pitch Diameter

$$D_G = N_G/P_d = 52/6 = 8.667 \text{ in}$$

Center Distance

$$C = (D_W + D_G)/2 = (2.000 + 8.667)/2 = 5.333 \text{ in}$$

Velocity Ratio

$$VR = N_G/N_W = 52/3 = 17.333$$

Gear rpm

$$n_G = n_W/VR = 1750/17.333 = 101 \text{ rpm}$$

Pressure Angle

Most commercially available wormgears are made with pressure angles of $14\frac{1}{2}°$, 20°, 25°, or 30°. The low pressure angles are used with worms having a low lead angle and/or a low diametral pitch. For example, a $14\frac{1}{2}°$ pressure angle may be used for lead angles up to about 17°. For higher lead angles and with higher diametral pitches (smaller teeth), the 20° or 25° pressure angle is used to eliminate interference without excessive undercutting. The 20° pressure angle is the preferred value for lead angles up to 30°. From 30° to 45° of lead angle, the 25° pressure angle is recommended. Either the normal pressure angle, ϕ_n, or the transverse pressure angle, ϕ_t, may be specified. These are related by

☞ **Pressure Angle**

$$\tan\phi_n = \tan\phi_t \cos\lambda \tag{8–39}$$

Self-Locking Wormgear Sets

Self-locking is the condition in which the worm drives the wormgear, but if torque is applied to the gear shaft, the worm does not turn. It is locked! The locking action is produced by the friction force between the worm threads and the wormgear teeth, and this is highly dependent on the lead angle. It is recommended that a lead angle no higher than about 5.0°

be used in order to ensure that self-locking will occur. This lead angle usually requires the use of a single-threaded worm. Note that the triple-threaded worm in Example Problem 8–7 has a lead angle of 14.04°. It is *not* likely to be self-locking.

**8–11
TYPICAL
GEOMETRY OF
WORMGEAR
SETS**

Considerable latitude is permissible in the design of wormgear sets because the worm and wormgear combination is designed as a unit. However, there are some guidelines.

**General
Guidelines for
Worm and
Wormgear
Dimensions**

Typical Tooth Dimensions

Table 8–8 shows typical values used for the dimensions of worm threads and gear teeth.

Worm Diameter

The diameter of the worm affects the lead angle, which in turn affects the efficiency of the set. For this reason, small diameters are desirable. But for practical reasons and proper proportion with respect to the wormgear, it is recommended that the worm diameter be approximately $C^{0.875}/2.2$, where C is the center distance between the worm and the wormgear. Variation of about 30% is allowed. (See Reference 6.) Thus, the worm diameter should fall in the range

$$1.6 < \frac{C^{0.875}}{D_W} < 3.0 \qquad\qquad \textbf{(8–40)}$$

But some commercially available wormgear sets fall outside this range, especially in the smaller sizes. Also, those worms designed to have a through-hole bored in them for installation on a shaft are typically larger than you would find from Equation (8–40). Proper proportion and efficient use of material should be the guide. The worm shaft must also be checked for deflection under operating loads. For worms machined integral with the shaft, the root of the worm threads determines the minimum shaft diameter. For worms having bored holes, sometimes called *shell worms,* care must be exercised to leave sufficient material between the thread root and the keyway in the bore. Figure 8–27 shows the recommended thickness above the keyway to be a minimum of one-half the whole depth of the threads.

TABLE 8–8 Typical tooth dimensions for worms and wormgears

Dimension	Formula
Addendum	$a = 0.3183P_x = 1/P_d$
Whole depth	$h_t = 0.6866P_x = 2.157/P_d$
Working depth	$h_k = 2a = 0.6366P_a = 2/P_d$
Dedendum	$b = h_t - a = 0.3683P_x = 1.157/P_d$
Root diameter of worm	$D_{rW} = D_W - 2b$
Outside diameter of worm	$D_{oW} = D_W + 2a = D_W + h_k$
Root diameter of gear	$D_{rG} = D_G - 2b$
Throat diameter of gear	$D_t = D_G + 2a$

Source: Standard AGMA *Design Manual Cylindrical Wormgearing,* with the permission of the publisher, American Gear Manufacturers Association, 1500 King Street, Suite 201, Alexandria, VA 22314.

FIGURE 8–27   Shell worm

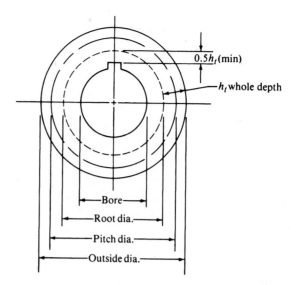

FIGURE 8–28
Wormgear details

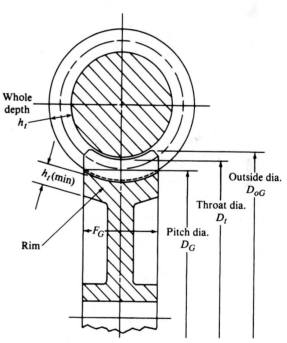

Wormgear Dimensions

We are concerned here with the single-enveloping type of wormgear, as shown in Figure 8–28. Its addendum, dedendum, and depth dimensions are assumed to be the same as those listed in Table 8–8, measured at the throat of the wormgear teeth. The throat is in line with the vertical centerline of the worm. The recommended face width for the wormgear is

▷ **Face Width of Wormgear**

$$F_G = (D_{oW}^2 - D_W^2)^{1/2} \qquad (8–41)$$

This corresponds to the length of the line tangent to the pitch circle of the worm and limited by the outside diameter of the worm. Any face width beyond this value would not be effective in resisting stress or wear, but a convenient value slightly greater than the minimum should be used. The outer edges of the wormgear teeth should be chamfered approximately as shown in Figure 8–28.

Another recommendation, which is convenient for initial design, is that the face width of the gear should be approximately 2.0 times the circular pitch. Because we are working in the diametral pitch system, we will use

$$F_G = 2p = 2\pi/P_d \qquad (8\text{--}42)$$

However, since this is only approximate and 2π is approximately 6, we will use

$$F_G = 6/P_d \qquad (8\text{--}43)$$

If the gear web is thinned, a rim thickness at least equal to the whole depth of the teeth should be left.

Face Length of the Worm

For maximum load sharing, the worm face length should extend to at least the point where the outside diameter of the worm intersects the throat diameter of the wormgear. This length is

⇨ **Face Length of Worm**

$$F_W = 2[D_t/2)^2 - (D_G/2 - a)^2]^{1/2} \qquad (8\text{--}44)$$

Example Problem 8–8

A worm and wormgear set is to be designed to produce a velocity ratio of 40. It has been proposed that the diametral pitch of the wormgear be 8, based on the torque that must be transmitted. (This will be discussed in Chapter 10.) Using the relationships presented in this section, specify the following:

Worm diameter, D_W

Number of threads in the worm, N_W

Number of teeth in the gear, N_G

Actual center distance, C

Face width of the gear, F_G

Face length of the worm, F_W

Minimum thickness of the rim of the gear

Solution

Many design decisions need to be made, and multiple solutions could satisfy the requirements. Presented here is one solution, along with comparisons with the various guidelines discussed in this section. This type of analysis precedes the stress analysis and the determination of the power-transmitting capacity of the worm and wormgear drive which is discussed in Chapter 10.

Trial Design: Let's specify a double-threaded worm: $N_W = 2$. Then there must be 80 teeth in the wormgear to achieve a velocity ratio of 40. That is,

$$VR = N_G/N_W = 80/2 = 40$$

With the known diametral pitch, $P_d = 8$, the pitch diameter of the wormgear is

$$D_G = N_G/P_d = 80/8 = 10.000 \text{ in}$$

An initial estimate for the magnitude of the center distance is approximately $C = 6.50$ in. We know that it will be greater than 5.00 in, the radius of the wormgear. Using Equation (8–40), the recommended minimum size of the worm is

$$D_W = C^{0.875}/3.0 = 1.71 \text{ in}$$

Similarly, the maximum diameter should be

$$D_W = C^{0.875}/1.6 = 3.21 \text{ in}$$

A small worm diameter is desirable. Let's specify $D_W = 2.25$ in. The actual center distance is

$$C = (D_W + D_G)/2 = 6.125 \text{ in}$$

Worm Outside Diameter

$$D_{oW} = D_W + 2a = 2.25 + 2(1/P_d) = 2.25 + 2(1/8) = 2.50 \text{ in}$$

Whole Depth

$$h_t = 2.157/P_d = 2.157/8 = 0.270 \text{ in}$$

Face Width for Gear
Let's use Equation (8–41):

$$F_G = (D_{oW}^2 - D_W^2)^{1/2} = (2.50^2 - 2.25^2) = 1.090 \text{ in}$$

Let's specify $F_G = 1.25$ in.

Addendum

$$a = 1/P_d = 1/8 = 0.125 \text{ in}$$

Throat Diameter of Wormgear

$$D_t = D_G + 2a = 10.000 + 2(0.125) = 10.250 \text{ in}$$

Recommended Minimum Face Length of Worm

$$F_W = 2[(D_t/2)^2 - (D_G/2 - a)^2]^{1/2} = 3.16 \text{ in}$$

Let's specify $F_W = 3.25$ in.

Minimum Thickness of the Rim of the Gear
The rim thickness should be greater than the whole depth:

$$h_t = 0.270 \text{ in}$$

**8–12
TRAIN VALUE
FOR COMPLEX
GEAR TRAINS**

The concept of train value was introduced in Section 8–6 and was applied to gear trains having all spur gears and a modest reduction ratio. Trains of two to five gears having a single speed reduction or a double reduction were included. This section expands on the concept of train value to include a wider variety of gear types, higher reduction ratios, and the opportunity to use several different arrangements of gears.

The basic definition of *train value* as shown in Equation (8–23) will continue to be used. Its general form is repeated here:

$$TV = \frac{\text{product of number of teeth in the driven gears}}{\text{product of number of teeth in the driving gears}} = \frac{\text{input speed}}{\text{output speed}}$$

This is equivalent to saying that the train value is the product of the velocity ratios of the individual gear pairs in the train.

When spur, helical, and bevel gears are used, the specific geometry of the gears and their teeth does not affect the train value. Also, when a worm/wormgear set is a part of the train, its data can be entered into the equation if we recall that the number of *threads* in the worm can be considered equivalent to the number of *teeth* in the driver of that set.

Sketches of the gear trains are valuable to illustrate the arrangement of gears and to enable you to track the flow of power through the train, that is, how the motion is transferred from the input shaft through the entire train to the output shaft. The sketches can be somewhat schematic, but they should show the relative positions of the gears and the shafts that carry the gears. Although sizes need not be drawn to scale, it is helpful to suggest the nominal size of the gears to aid in judging whether a given pair will produce a speed reduction or a speed increase. Showing the pitch diameters of the gears will suffice. Recall that the motion of one gear on its mating gear is kinematically similar to the rolling of smooth cylinders on one another without slipping where the diameters for the cylinders are equal to the pitch diameters of the gears. But be sure to understand that any two gears in mesh must have compatible tooth forms. Only the kinematics of the train are being considered here. Later, when the actual gears, shafts, bearings, and the housing are designed, many other geometric features and operating parameters must be considered.

The sketching technique shown in Figure 8–29 will be used in this section and in the Problems section of this chapter. Gears are given letter designations, and the shafts carrying the gears are numbered. The data for the train shown in the figure should be interpreted as follows:

■ Gears *A, B, C,* and *D* are external type, either spur or helical. Their shafts (shafts 1, 2, and 3) are parallel and are indicated only by their centerlines. Shaft 1 is the input to the train. It would typically be connected directly to a drive such as an electric motor.

■ Gear *A* drives gear *B* with a speed reduction because gear *A* is smaller than gear *B*. A change in direction of rotation occurs. Gear *C* is on the same shaft as gear *B* and rotates at the same speed and in the same direction. Gear *C* drives gear *D* with a speed reduction and a direction reversal.

■ Gears *E* and *F* are bevel gears, straight, or spiral, or some other tooth form. Their shafts (3 and 4) are perpendicular.

■ Gear *G* is a worm, and it is driving wormgear *H*. Their shafts (4 and 5) are perpendicular, with shaft 5 directed out of the page. Note that we are seeing the side view of the pitch cylinder of the worm rotating about a vertical axis (shaft 4). We see the end view of wormgear *H*.

■ The small pinion *I* is also mounted on shaft 5 and rotates at the same speed and in the same direction as wormgear *H*.

■ Pinion *I* drives the internal gear *J* mounted on shaft 6 which is the output gear of the train. Note that shaft 6 rotates in the same direction as shaft 5.

Data for the gears in the train may be provided in a number of ways. If only speed of rotation is to be considered, the number of teeth in each gear will normally be given. If physical size of the train, center distances, bearing placement, housing design, and detailed layout are to be considered, then other geometrical features such as diametral pitch, face width, gear blank style, hub style, and gear diameters will have to be specified. Most of these features are discussed in later chapters. We will give only the numbers of teeth in most cases in this chapter. But notice that if pitch diameter and diametral pitch of a pinion and gear are given, their numbers of teeth can be determined from

$$P_d = N_P/D_P = N_G/D_G$$
$$N_P = P_d D_P$$
$$N_G = P_d D_G$$

Also, we can use the definition of diametral pitch to show the relationship between the ratio of the numbers of teeth in a pair of gears and the ratio of their diameters:

$$P_d = N_P/D_P = N_G/D_G$$
$$N_G/N_P = D_G/D_P$$

This shows that the diameter ratio could replace the teeth ratio in the train value equation.

Example Problem 8–9

Refer to Figure 8–29. Shaft 1 is a motor shaft and rotates at 1160 rpm. Compute the rotational speed of the output shaft, shaft 6. Data for the gears are as follows:

$N_A = 18$	$N_B = 34$	$N_C = 20$	$N_D = 62$
$N_E = 30$	$N_F = 60$	$N_G = 2$ (worm threads)	$N_H = 40$
		$N_I = 16$	$N_J = 88$

Solution

We will compute the train value first:

$$TV = \frac{(N_B)(N_D)(N_F)(N_H)(N_J)}{(N_A)(N_C)(N_E)(N_G)(N_I)} = \frac{(34)(62)(60)(40)(88)}{(18)(20)(30)(2)(16)} = 1288.2$$

FIGURE 8–29 Gear train for Example Problem 8–9

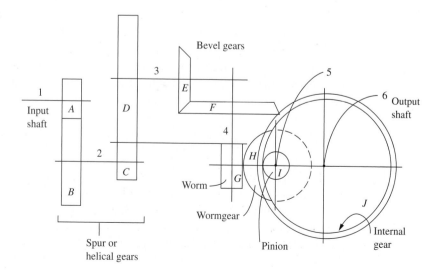

But $TV = n_1/n_6$. Then

$$n_6 = n_1/TV = (1160 \text{ rpm})/1288.2 = 0.900 \text{ rpm}$$

Shaft 6 rotates 0.900 rpm.

8–13 DEVISING GEAR TRAINS

Now we will show several methods for devising gear trains to produce a desired train value. The result will typically be the specification of the number of teeth in each gear and the general arrangement of the gears relative to each other. The determination of the types of gears will generally not be considered except for how they may affect the direction of rotation or the general alignment of the shafts. Additional details can be specified after completion of the study of the design procedures in later chapters.

A few general principles that were discussed earlier in this chapter are reviewed first.

General Principles for Devising Gear Trains

1. The velocity ratio for any pair of gears can be computed in a variety of ways as indicated in Equation (8–22).
2. The number of teeth in any gear must be an integer.
3. Gears in mesh must have the same tooth form and the same pitch.
4. When external gears mesh, there is a direction reversal of their shafts.
5. When an external pinion meshes with an internal gear, their shafts rotate in the same direction.
6. An idler is a gear that performs as both a driver and a driven gear in the same train. Its size and number of teeth have no effect on the magnitude of the train value, but the direction of rotation is changed.
7. Spur and helical gears operate on parallel shafts.
8. Bevel gears and a worm/wormgear set operate on shafts perpendicular to each other.
9. The number of teeth in the pinion of a gear pair should not be such that it causes interference with the teeth of its mating gear. Refer to Table 8–6.
10. In general, the number of teeth in the gear should not be larger than about 150. This is somewhat arbitrary, but it is typically more desirable to use a double-reduction gear train rather than a very large, single-reduction gear pair.

Hunting Tooth

Some designers recommend that integer velocity ratios be avoided, if possible, because the same two teeth would come into contact frequently and produce uneven wear patterns on the teeth. For example when using a velocity ratio of exactly 2.0, a given tooth on the pinion would contact the same two teeth on the gear with every two revolutions. In Chapter 9 you will learn that the pinion teeth are often made harder than the gear because the pinion experiences higher stresses. As the gears rotate, the pinion teeth tend to smooth any inherent roughness of the gear teeth, a process sometimes called *wearing in*. Each tooth on the pinion has a slightly different geometry causing unique wear patterns on the few teeth with which it mates.

A more uniform wear pattern will result if the velocity ratio in not an integer. Adding or subtracting one tooth from the number of teeth in the gear produces the result that each pinion tooth would contact a different gear tooth with each revolution and the wear pattern would be more uniform. The added or subtracted tooth is called the *hunting tooth*. Obviously the velocity ratio for the gear pair will be slightly different, but that is often not a concern unless precise timing between the driver and driven gears is required. Consider the following example.

An initial design for a gear pair calls for the pinion to be mounted to the shaft of an electric motor having a nominal speed of 1750 rpm. The pinion has 18 teeth and the gear has 36 teeth, resulting in a velocity ratio of 36/18 or 2.000. The output speed would then be,

Initial design: $\qquad n_2 = n_1 \ (N_P/N_G) = 1750 \text{ rpm } (18/36) = 875 \text{ rpm}$

Now consider adding or subtracting one tooth from the gear. The output speeds would be,

Modified design: $\qquad n_2 = n_1 \ (N_P/N_G) = 1750 \text{ rpm } (18/35) = 900 \text{ rpm}$

Modified design: $\qquad n_2 = n_1 \ (N_P/N_G) = 1750 \text{ rpm } (18/37) = 851 \text{ rpm}$

The output speeds for the modified designs are less than 3.0 percent different from the original design. You would have to decide if that is acceptable in a given design project. However, be aware that the motor speed is typically not exactly 1750 rpm. As discussed in Chapter 21, 1750 rpm is the *full load speed* of a four-pole alternating current electric motor. When operating at a torque less than the full load torque the speed would be greater than 1750 rpm. Conversely, a greater torque would result in a slower speed. When precise speeds are required, a variable speed drive that can be adjusting according to actual loads is recommended.

Several design procedures will now be demonstrated through example problems. It is not practical to outline a completely general procedure because of the many variables in any given design situation. You are advised to study the examples for their general approach, which you can adapt to future problems as needed.

Design of a Single Pair of Gears to Produce a Desired Velocity Ratio

---

Example Problem 8–10   Devise a gear train to reduce the speed of rotation of a drive from an electric motor shaft operating at 3450 rpm to approximately 650 rpm.

Solution   First we will compute the nominal train value:

$$TV = (\text{input speed})/(\text{output speed}) = 3450/650 = 5.308$$

If a single pair of gears is used then the train value is equal to the velocity ratio for that pair. That is, $TV = VR = N_G/N_P$.

Let's decide that spur gears having 20°, full-depth, involute teeth are to be used. Then we can refer to Table 8–6 and determine that no fewer than 16 teeth should be used for the pinion in order to avoid interference. We can specify the number of teeth in the pinion and use the velocity ratio to compute the number of teeth in the gear:

$$N_G = (VR)(N_P) = (5.308)(N_P)$$

Some examples are given in Table 8–9.

Conclusion and   The combination of $N_P = 26$ and $N_G = 138$ gives the most ideal result for the output speed.
Comments   But all of the trial values give output speeds reasonably close to the desired value. Only two

TABLE 8–9

N_P	Computed $N_G = (5.308)(N_P)$	Nearest integer N_G	Actual VR: $VR = N_G/N_P$	Actual output speed (rpm): $n_G = n_P/VR = n_P(N_P/N_G)$
16	84.92	85	85/16 = 5.31	649.4
17	90.23	90	90/17 = 5.29	651.7
18	95.54	96	96/18 = 5.33	646.9
19	100.85	101	101/19 = 5.32	649.0
20	106.15	106	106/20 = 5.30	650.9
21	111.46	111	111/21 = 5.29	652.7
22	116.77	117	117/22 = 5.32	648.7
23	122.08	122	122/23 = 5.30	650.4
24	127.38	127	127/24 = 5.29	652.0
25	132.69	133	133/25 = 5.32	648.5
26	138.00	138	138/26 = 5.308	650.0 Exact
27	143.31	143	143/27 = 5.30	651.4
28	148.61	149	149/28 = 5.32	648.3
29	153.92	154	**Too large**	

are more than 2.0 rpm off the desired value. It remains a design decision as to how close the output speed must be to the stated value of 650 rpm. Note that the input speed is given as 3450 rpm, the full load speed of an electric motor. But how accurate is that? The actual speed of the input will vary depending on the load on the motor. Therefore, it is not likely that the ratio must be precise.

Equal Reduction Ratios for Compound Gear Trains

Example Problem 8–11   Devise a gear train for a machine tool drive. The input is a shaft that rotates at exactly 1800 rpm. The output speed must be within the range of 31.5 and 32.5 rpm. Use 20°, full-depth, involute teeth and no more than 150 teeth in any gear.

Solution   *Permissible Train Values*
First let's compute the nominal train value that will produce an output speed of 32.0 rpm at the middle of the allowable range:

$$TV_{nom} = \text{(input speed)/(nominal output speed)} = 1800/32 = 56.25$$

Similarly, we can compute the minimum and maximum allowable ratio:

$$TV_{min} = \text{(input speed)/(maximum output speed)} = 1800/32.5 = 55.38$$
$$TV_{max} = \text{(input speed)/(minimum output speed)} = 1800/31.5 = 57.14$$

Possible Ratio for Single Pair
The maximum ratio that any one pair of gears can produce occurs when the gear has 150 teeth and the pinion has 17 teeth (see Table 8–6). Then

$$VR_{max} = N_G/N_P = 150/17 = 8.82$$

This is too low.

FIGURE 8–30
General layout of
proposed gear train

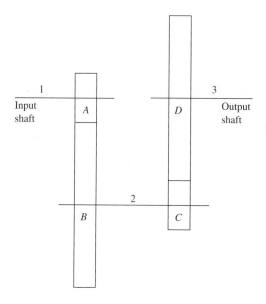

Possible Train Value for Double-Reduction Train
If a double-reduction train is proposed, the train value will be

$$TV = (VR_1)(VR_2)$$

But the maximum value for either *VR* is 8.82. Then the maximum train value for the double-reduction train is

$$TV_{max} = (8.82)(8.82) = (8.82)^2 = 77.9$$

It can be concluded that a double-reduction train should be practical.

Optional Designs
The general layout of the proposed train is shown in Figure 8–30. Its train value is

$$TV = (VR_1)(VR_2) = (N_B/N_A)(N_D/N_C)$$

We need to specify the number of teeth in each of the four gears to achieve a train value within the range just computed. Our approach is to specify two ratios, VR_1 and VR_2, such that their product is within the desired range. One possibility is to let the two ratios be equal. To produce the middle train value, each velocity ratio must be the square root of the target ratio, 56.25. That is,

$$VR_1 = VR_2 = \sqrt{56.25} = 7.50$$

As shown in Table 8–10, we will use a process similar to that used in the previous example problem to select possible numbers of teeth.

Any of the possible designs shown in Table 8–10 would produce acceptable results. For example, we could specify

$$N_A = 18 \qquad N_B = 135 \qquad N_C = 18 \qquad N_D = 135$$

This combination would produce an output speed of exactly 32.0 rpm when the input speed was exactly 1800 rpm.

TABLE 8–10

N_P	Computed $N_G = (7.5)(N_P)$	Nearest integer N_G	Actual VR: $VR = N_G/N_P$	Actual output speed (rpm): $n_G = n_P(VR)^2$
17	127.5	128	128/17 = 7.529	31.75
17	127.5	127	127/17 = 7.470	32.25
18	135	135	135/18 = 7.500	32.00
19	142.5	143	143/19 = 7.526	31.78
19	142.5	142	142/19 = 7.474	32.23
20	150	150	150/20 = 7.500	32.00

Factoring Approach for Compound Gear Trains

Example Problem 8–12   Devise a gear train for a recorder for a precision measuring instrument. The input is a shaft that rotates at exactly 3600 rpm. The output speed must be within the range of 11.0 and 11.5 rpm. Use 20°, full-depth, involute teeth; no fewer than 18 teeth; and no more than 150 teeth in any gear.

Solution   *Nominal Target TV*

$$TV_{nom} = 3600/11.25 = 320$$

Maximum TV

$$TV_{max} = 3600/11.0 = 327.3$$

Minimum TV

$$TV_{min} = 3600/11.5 = 313.0$$

Maximum Single VR

$$VR_{max} = 150/18 = 8.33$$

Maximum TV for Double Reduction

$$TV_{max} = (8.333)^2 = 69.4 \text{ (too low)}$$

Maximum TV for Triple Reduction

$$TV_{max} = (8.333)^3 = 578 \text{ (okay)}$$

Design a triple-reduction gear train such as that shown in Figure 8–31. The train value is the product of the three individual velocity ratios:

$$TV = (VR_1)(VR_2)(VR_3)$$

If we can find three factors of 320 that are within the range of the possible ratio for a single pair of gears, they can be specified for each velocity ratio.

FIGURE 8–31
Triple-reduction gear train

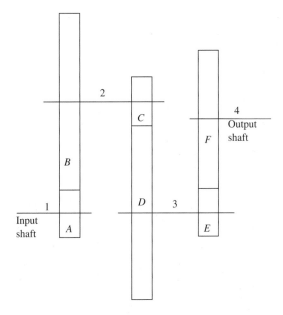

Factors of 320
One method is to divide by the smallest prime numbers that will divide evenly into the given number, typically 2, 3, 5, or 7. For example,

$$320/2 = 160$$
$$160/2 = 80$$
$$80/2 = 40$$
$$40/2 = 20$$
$$20/2 = 10$$
$$10/2 = 5$$

Then the prime factors of 320 are 2, 2, 2, 2, 2, 2, and 5. We desire a set of three factors, which we can find by combining each set of three "2" factors into their product. That is,

$$(2)(2)(2) = 8$$

Then the three factors of 320 are

$$(8)(8)(5) = 320$$

Now let the number of teeth in the pinion of each pair be 18. The number of teeth in the gears will then be $(8)(18) = 144$ or $(5)(18) = 90$. Finally, we can specify

$N_A = 18$	$N_C = 18$	$N_E = 18$
$N_B = 144$	$N_D = 144$	$N_F = 90$

Residual Ratio

Example Problem 8–13   Devise a gear train for a conveyor drive. The drive motor rotates at 1150 rpm, and it is desired that the output speed for the shaft that drives the conveyor be in the range of 24 to 28 rpm. Use a double-reduction gear train. Power transmission analysis indicates that it would be desirable for the reduction ratio for the first pair of gears to be somewhat greater than that for the second pair.

Solution   The start of this problem is similar to that for Example Problem 8–12.

Permissible Train Values

First let's compute the nominal train value that will produce an output speed of 26.0 rpm at the middle of the allowable range:

$$TV_{\text{nom}} = (\text{input speed})/(\text{nominal output speed}) = 1150/26 = 44.23$$

Now we can compute the minimum and maximum allowable speed ratio:

$$TV_{\text{min}} = (\text{input speed})/(\text{maximum output speed}) = 1150/28 = 41.07$$
$$TV_{\text{max}} = (\text{input speed})/(\text{minimum output speed}) = 1150/24 = 47.92$$

Possible Ratio for Single Pair

The maximum ratio that any one pair of gears can produce occurs when the gear has 150 teeth and the pinion has 17 teeth (see Table 8–6). Then

$$VR_{\text{max}} = N_G/N_P = 150/17 = 8.82 \text{ (too low)}$$

Possible Train Value for Double-Reduction Train

$$TV = (VR_1)(VR_2)$$

But the maximum value for either *VR* is 8.82. Then the maximum train value is

$$TV_{\text{max}} = (8.82)(8.82) = (8.82)^2 = 77.9$$

A double-reduction train is practical.

Optional Designs

The general layout of the proposed train is shown in Figure 8–30. Its train value is

$$TV = (VR_1)(VR_2) = (N_B/N_A)(N_D/N_C)$$

We need to specify the number of teeth in each of the four gears to achieve a train value within the range just computed. Our approach is to specify two ratios, VR_1 and VR_2, such that their product is within the desired range. If the two ratios were equal as before, each would be the square root of the target ratio, 44.23. That is,

$$VR_1 = VR_2 = \sqrt{44.23} = 6.65$$

But we want the first ratio to be somewhat larger than the second. Let's specify

$$VR_1 = 8.0 = (N_B/N_A)$$

If we let pinion *A* have 17 teeth, the number of teeth in gear *B* must be

$$N_B = (N_A)(8) = (17)(8) = 136$$

Then the second ratio should be approximately

$$(VR_2) = TV(VR_1) = 44.23/8.0 = 5.53$$

This is the *residual ratio* left after the first ratio has been specified. Now if we specify 17 teeth for pinion *C*, gear *D* must be

$$VR_2 = 5.53 = N_D/N_C = N_D/17$$
$$N_D = (5.53)(17) = 94.01$$

Rounding this off to 94 is likely to produce an acceptable result. Finally,

$$N_A = 17 \qquad N_B = 136 \qquad N_C = 17 \qquad N_D = 94$$

We should check the final design:

$$TV = (136/17)(94/17) = 44.235 = n_A/n_D$$

The actual output speed is

$$n_D = n_A/TV = (1150 \text{ rpm})/44.235 = 26.0 \text{ rpm}$$

This is right in the middle of the desired range.

REFERENCES

1. American Gear Manufacturers Association. Standard 1012-F90. *Gear Nomenclature, Definitions of Terms with Symbols.* Alexandria, VA: American Gear Manufacturers Association, 1990.

2. American Gear Manufacturers Association. Standard 2002-B88 (R1996). *Tooth Thickness Specification and Measurement.* Alexandria, VA: American Gear Manufacturers Association, 1996.

3. American Gear Manufacturers Association. Standard 2008. *Standard for Assembling Bevel Gears.* Alexandria, VA: American Gear Manufacturers Association, 2001.

4. American Gear Manufacturers Association. Standard 917-B97. *Design Manual for Parallel Shaft Fine-Pitch Gearing.* Alexandria, VA: American Gear Manufacturers Association, 1997.

5. American Gear Manufacturers Association. Standard 2005-C96. *Design Manual for Bevel Gears.* Alexandria, VA: American Gear Manufacturers Association, 1996.

6. American Gear Manufacturers Association. Standard 6022-C93. *Design Manual for Cylindrical Wormgearing.* Alexandria, VA: American Gear Manufacturers Association, 1993.

7. American Gear Manufacturers Association. Standard 6001 D97. *Design and Selection of Components for Enclosed Gear Drives.* Alexandria, VA: American Gear Manufacturers Association, 1997.

8. American Gear Manufacturers Association. Standard 2000-A88. *Gear Classification and Inspection Handbook— Tolerances and Measuring Methods for Unassembled Spur and Helical Gears (Including Metric Equivalents).* Alexandria, VA: American Gear Manufacturers Association, 1988.

9. Drago, Raymond J. *Fundamentals of Gear Design.* Boston: Butterworths, 1988.

10. Dudley, Darle W. *Dudley's Gear Handbook.* New York: McGraw-Hill, 1991.

11. Lipp, Robert. "Avoiding Tooth Interference in Gears." *Machine Design* 54, no. 1 (January 7, 1982).

12. Oberg, Erik, et al. *Machinery's Handbook.* 26th ed. New York: Industrial Press, 2000.

13. Shtipelman, Boris. *Design and Manufacture of Hypoid Gears.* New York : John Wiley & Sons, 1978.

14. Wildhaber, Ernst. *Basic Relationship of Hypoid Gears.* Cleveland American Machinist, 1946.

INTERNET SITES RELATED TO KINEMATICS OF GEARS

1. **American Gear Manufacturers Association (AGMA).** *www.agma.org* Develops and publishes voluntary, consensus standards for gears and gear drives. Some standards are jointly published with the American National Standards Institute (ANSI).

2. **Boston Gear Company.** *www.bostongear.com* A manufacturer of gears and complete gear drives. Part of the Colfax Power Transmission Group. Data provided for spur, helical, bevel, and worm gearing.

3. **Emerson Power Transmission Corporation.** *www.emerson-ept.com* The Browning Division produces spur, helical, bevel, and worm gearing and complete gear drives.

4. **Gear Industry Home Page.** *www.geartechnology.com* Information source for many companies that manufacture or use gears or gearing systems. Includes gear machinery, gear cutting tools, gear materials, gear drives, open gearing, tooling

& supplies, software, training and education. Publishes *Gear Technology Magazine, The Journal of Gear Manufacturing.*

5. Power Transmission Home Page.
www.powertransmission.com   Clearinghouse on the Internet for buyers, users, and sellers of power transmission-related products and services. Included are gears, gear drives, and gearmotors.

6. Rockwell Automation/Dodge.   *www,dodge-pt.com* Manufacturer of many power transmission components, including complete gear-type speed reducers, bearings, and components such as belt drives, chain drives, clutches, brakes, and couplings.

PROBLEMS

Gear Geometry

1. A gear has 44 teeth of the 20°, full-depth, involute form and a diametral pitch of 12. Compute the following:
   (a) Pitch diameter
   (b) Circular pitch
   (c) Equivalent module
   (d) Nearest standard module
   (e) Addendum
   (f) Dedendum
   (g) Clearance
   (h) Whole depth
   (i) Working depth
   (j) Tooth thickness
   (k) Outside diameter

   Repeat Problem 1 for the following gears:

2. $N = 34; P_d = 24$
3. $N = 45; P_d = 2$
4. $N = 18; P_d = 8$
5. $N = 22; P_d = 1.75$
6. $N = 20; P_d = 64$
7. $N = 180; P_d = 80$
8. $N = 28; P_d = 18$
9. $N = 28; P_d = 20$

   For Problems 10–17, repeat Problem 1 for the following gears in the metric module system. Replace Part (c) with equivalent P_d and Part (d) with nearest standard P_d.

10. $N = 34; m = 3$
11. $N = 45; m = 1.25$
12. $N = 18; m = 12$
13. $N = 22; m = 20$
14. $N = 20; m = 1$
15. $N = 180; m = 0.4$
16. $N = 28; m = 1.5$
17. $N = 28; m = 0.8$

18. Define *backlash,* and discuss the methods used to produce it.

19. For the gears of Problems 1 and 12, recommend the amount of backlash.

Velocity Ratio

20. An 8-pitch pinion with 18 teeth mates with a gear having 64 teeth. The pinion rotates at 2450 rpm. Compute the following:
    (a) Center distance
    (b) Velocity ratio
    (c) Speed of gear
    (d) Pitch line speed

    Repeat Problem 20 for the following data:

21. $P_d = 4; N_P = 20; N_G = 92; n_P = 225$ rpm
22. $P_d = 20; N_P = 30; N_G = 68; n_P = 850$ rpm
23. $P_d = 64; N_P = 40; N_G = 250; n_P = 3450$ rpm
24. $P_d = 12; N_P = 24; N_G = 88; n_P = 1750$ rpm
25. $m = 2; N_P = 22; N_G = 68; n_P = 1750$ rpm
26. $m = 0.8; N_P = 18; N_G = 48; n_P = 1150$ rpm
27. $m = 4; N_P = 36; N_G = 45; n_P = 150$ rpm
28. $m = 12; N_P = 15; N_G = 36; n_P = 480$ rpm

    For Problems 29–32, all gears are made in standard 20°, full-depth, involute form. Tell what is wrong with the following statements:

29. An 8-pitch pinion having 24 teeth mates with a 10-pitch gear having 88 teeth. The pinion rotates at 1750 rpm, and the gear at approximately 477 rpm. The center distance is 5.900 in.

30. A 6-pitch pinion having 18 teeth mates with a 6-pitch gear having 82 teeth. The pinion rotates at 1750 rpm, and the gear at approximately 384 rpm. The center distance is 8.3 in.

31. A 20-pitch pinion having 12 teeth mates with a 20-pitch gear having 62 teeth. The pinion rotates at 825 rpm, and the gear at approximately 160 rpm. The center distance is 1.850 in.

32. A 16-pitch pinion having 24 teeth mates with a 16-pitch gear having 45 teeth. The outside diameter of the pinion is 1.625 in. The outside diameter of the gear is 2.938 in. The center distance is 2.281 in.

Housing Dimensions

33. The gear pair described in Problem 20 is to be installed in a rectangular housing. Specify the dimensions X and Y as sketched in Figure P8–33 that would provide a minimum clearance of 0.10 in.

34. Repeat Problem 33 for the data of Problem 23.

35. Repeat Problem 33 for the data of Problem 26, but make the clearance 2.0 mm.

36. Repeat Problem 33 for the data of Problem 27, but make the clearance 2.0 mm.

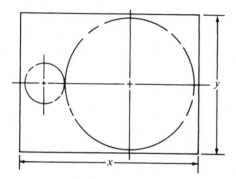

FIGURE P8–33   (Problems 33, 34, 35, and 36)

Analysis of Simple Gear Trains

For the gear trains sketched in the given figures, compute the output speed and the direction of rotation of the output shaft if the input shaft rotates at 1750 rpm clockwise.

37. Use Figure P8–37.

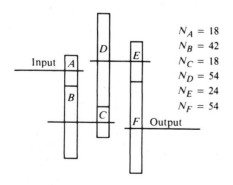

$N_A = 18$
$N_B = 42$
$N_C = 18$
$N_D = 54$
$N_E = 24$
$N_F = 54$

FIGURE P8–37

38. Use Figure P8–38.

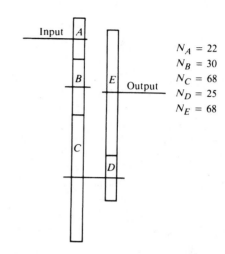

$N_A = 22$
$N_B = 30$
$N_C = 68$
$N_D = 25$
$N_E = 68$

FIGURE P8–38

39. Use Figure P8–39.

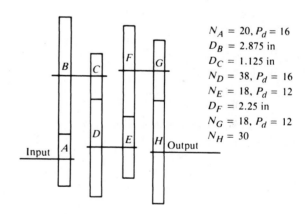

$N_A = 20, P_d = 16$
$D_B = 2.875$ in
$D_C = 1.125$ in
$N_D = 38, P_d = 16$
$N_E = 18, P_d = 12$
$D_F = 2.25$ in
$N_G = 18, P_d = 12$
$N_H = 30$

FIGURE P8–39

40. Use Figure P8–40.

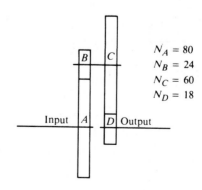

$N_A = 80$
$N_B = 24$
$N_C = 60$
$N_D = 18$

FIGURE P8–40

Helical Gearing

41. A helical gear has a transverse diametral pitch of 8, a transverse pressure angle of $14\frac{1}{2}°$, 45 teeth, a face width of 2.00 in, and a helix angle of 30°. Compute the circular pitch, normal circular pitch, normal diametral pitch, axial pitch, pitch diameter, and normal pressure angle. Then compute the number of axial pitches in the face width.

42. A helical gear has a normal diametral pitch of 12, a normal pressure angle of 20°, 48 teeth, a face width of 1.50 in, and a helix angle of 45°. Compute the circular pitch, normal circular pitch, transverse diametral pitch, axial pitch, pitch diameter, and transverse pressure angle. Then compute the number of axial pitches in the face width.

43. A helical gear has a transverse diametral pitch of 6, a transverse pressure angle of $14\frac{1}{2}°$, 36 teeth, a face width of 1.00 in, and a helix angle of 45°. Compute the circular pitch, normal circular pitch, normal diametral pitch, axial pitch, pitch diameter, and normal pressure angle. Then compute the number of axial pitches in the face width.

44. A helical gear has a normal diametral pitch of 24, a normal pressure angle of $14\frac{1}{2}°$, 72 teeth, a face width of 0.25 in, and a helix angle of 45°. Compute the circular pitch, normal circular pitch, transverse diametral pitch, axial pitch, pitch diameter, and transverse pressure angle. Then compute the number of axial pitches in the face width.

Bevel Gears

45. A straight bevel gear pair has the following data: $N_P = 15$; $N_G = 45$; $P_d = 6$; 20° pressure angle. Compute all of the geometric features from Table 8–7.

46. Draw the gear pair of Problem 45 to scale. The following additional dimensions are given (refer to Figure 8–22). Mounting distance (M_{dP}) for the pinion = 5.250 in; M_{dG} for the gear = 3.000 in; face width = 1.250 in. Supply any other needed dimensions.

47. A straight bevel gear pair has the following data: $N_P = 25$; $N_G = 50$; $P_d = 10$; 20° pressure angle. Compute all of the geometric features from Table 8–7.

48. Draw the gear pair of Problem 47 to scale. The following additional dimensions are given (refer to Figure 8–22). Mounting distance (M_{dP}) for the pinion = 3.375 in; M_{dG} for the gear = 2.625 in; face width = 0.700 in. Supply any other needed dimensions.

49. A straight bevel gear pair has the following data: $N_P = 18$; $N_G = 72$; $P_d = 12$; 20° pressure angle. Compute all of the geometric features from Table 8–7.

50. A straight bevel gear pair has the following data: $N_P = 16$; $N_G = 64$; $P_d = 32$; 20° pressure angle. Compute all of the geometric features from Table 8–7.

51. A straight bevel gear pair has the following data: $N_P = 12$; $N_G = 36$; $P_d = 48$; 20° pressure angle. Compute all of the geometric features from Table 8–7.

Wormgearing

52. A wormgear set has a single-thread worm with a pitch diameter of 1.250 in, a diametral pitch of 10, and a normal pressure angle of 14.5°. If the worm meshes with a wormgear having 40 teeth and a face width of 0.625 in, compute the lead, axial pitch, circular pitch, lead angle, addendum, dedendum, worm outside diameter, worm root diameter, gear pitch diameter, center distance, and velocity ratio.

53. Three designs are being considered for a wormgear set to produce a velocity ratio of 20 when the wormgear rotates at 90 rpm. All three have a diametral pitch of 12, a worm pitch diameter of 1.000 in, a gear face width of 0.500 in, and a normal pressure angle of 14.5°. One has a single-thread worm and 20 wormgear teeth; the second has a double-thread worm and 40 wormgear teeth; the third has a four-thread worm and 80 wormgear teeth. For each design, compute the lead, axial pitch, circular pitch, lead angle, gear pitch diameter, and center distance.

54. A wormgear set has a double-threaded worm with a normal pressure angle of 20°, a pitch diameter of 0.625 in, and a diametral pitch of 16. Its mating wormgear has 100 teeth and a face width of 0.3125 in. Compute the lead, axial pitch, circular pitch, lead angle, addendum, dedendum, worm outside diameter, center distance, and velocity ratio.

55. A wormgear set has a four-threaded worm with a normal pressure angle of $14\frac{1}{2}°$, a pitch diameter of 2.000 in, and a diametral pitch of 6. Its mating wormgear has 72 teeth and a face width of 1.000 in. Compute the lead, axial pitch, circular pitch, lead angle, addendum, dedendum, worm outside diameter, center distance, and velocity ratio.

56. A wormgear set has a single-threaded worm with a normal pressure angle of $14\frac{1}{2}°$, a pitch diameter of 4.000 in, and a diametral pitch of 3. Its mating wormgear has 54 teeth and a face width of 2.000 in. Compute the lead, axial pitch, circular pitch, lead angle, addendum, dedendum, worm outside diameter, center distance, and velocity ratio.

57. A wormgear set has a four-threaded worm with a normal pressure angle of 25°, a pitch diameter of 0.333 in, and a diametral pitch of 48. Its mating wormgear has 80 teeth and a face width of 0.156 in. Compute the lead, axial pitch, circular pitch, lead angle, addendum, dedendum, worm outside diameter, center distance, and velocity ratio.

Analysis of Complex Gear Trains

58. The input shaft for the gear train shown in Figure P8–58 rotates at 3450 rpm. Compute the rotational speed of the output shaft.

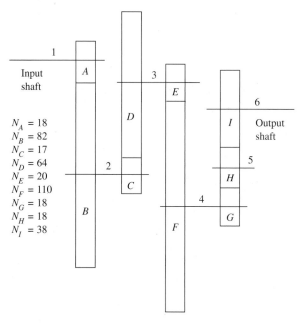

$N_A = 18$
$N_B = 82$
$N_C = 17$
$N_D = 64$
$N_E = 20$
$N_F = 110$
$N_G = 18$
$N_H = 18$
$N_I = 38$

FIGURE P8–58   Gear train for Problem 58

59. The input shaft for the gear train shown in Figure P8–59 rotates at 12 200 rpm. Compute the rotational speed of the output shaft.

60. The input shaft for the gear train shown in Figure P8–60 rotates at 6840 rpm. Compute the rotational speed of the output shaft.

61. The input shaft for the gear train shown in Figure P8–61 rotates at 2875 rpm. Compute the rotational speed of the output shaft.

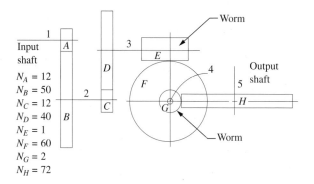

$N_A = 12$
$N_B = 50$
$N_C = 12$
$N_D = 40$
$N_E = 1$
$N_F = 60$
$N_G = 2$
$N_H = 72$

FIGURE P8–59   Gear train for Problem 59

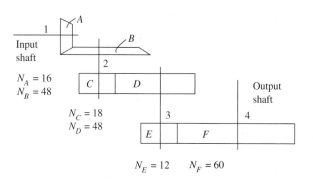

$N_A = 16$
$N_B = 48$

$N_C = 18$
$N_D = 48$

$N_E = 12$ $N_F = 60$

FIGURE P8–60   Gear train for Problem 60

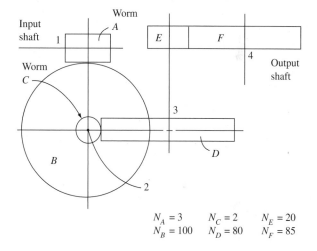

$N_A = 3$	$N_C = 2$	$N_E = 20$
$N_B = 100$	$N_D = 80$	$N_F = 85$

FIGURE P8–61   Gear train for Problem 61

Kinematic Design of a Single Gear Pair

62. Specify the numbers of teeth for the pinion and gear of a single gear pair to produce a velocity ratio of π as closely as possible. Use no fewer than 16 teeth nor more than 24 teeth in the pinion.

63. Specify the numbers of teeth for the pinion and gear of a single gear pair to produce a velocity ratio of $\sqrt{3}$ as closely as possible. Use no fewer than 16 teeth nor more than 24 teeth in the pinion.

64. Specify the numbers of teeth for the pinion and gear of a single gear pair to produce a velocity ratio of $\sqrt{38}$ as closely as possible. Use no fewer than 18 teeth nor more than 24 teeth in the pinion.

65. Specify the numbers of teeth for the pinion and gear of a single gear pair to produce a velocity ratio of 7.42 as closely as possible. Use no fewer than 18 teeth nor more than 24 teeth in the pinion.

Kinematic Design of Gear Trains

For Problems 66–75, devise a gear train using all external gears on parallel shafts. Use 20° full-depth involute teeth and no more than 150 teeth in any gear. Ensure that there is no interference. Sketch the layout for your design.

Problem no.	Input speed (rpm)	Output speed range (rpm)
66.	1800	2.0 Exactly
67.	1800	21.0 to 22.0
68.	3360	12.0 Exactly
69.	4200	13.0 to 13.5
70.	5500	221 to 225
71.	5500	13.0 to 14.0
72.	1750	146 to 150
73.	850	40.0 to 44.0
74.	3000	548 to 552 Use two pairs
75.	3600	3.0 to 5.0

For Problems 76–80, devise a gear train using any type of gears. Try for the minimum number of gears while avoiding interference and having no more that 150 teeth in any gear. Sketch your design.

Problem no.	Input speed (rpm)	Output speed (rpm)
76.	3600	3.0 to 5.0
77.	1800	8.0 Exactly
78.	3360	12.0 Exactly
79.	4200	13.0 to 13.5
80.	5500	13.0 to 14.0

9

Spur Gear Design

The Big Picture

You Are the Designer

9–1   Objectives of This Chapter

9–2   Concepts from Previous Chapters

9–3   Forces, Torque, and Power in Gearing

9–4   Gear Manufacture

9–5   Gear Quality

9–6   Allowable Stress Numbers

9–7   Metallic Gear Materials

9–8   Stresses in Gear Teeth

9–9   Selection of Gear Material Based on Bending Stress

9–10   Pitting Resistance of Gear Teeth

9–11   Selection of Gear Material Based on Contact Stress

9–12   Design of Spur Gears

9–13   Gear Design for the Metric Module System

9–14   Computer-Aided Spur Gear Design and Analysis

9–15   Use of the Spur Gear Design Spreadsheet

9–16   Power-Transmitting Capacity

9–17   Practical Considerations for Gears and Interfaces with Other Elements

9–18   Plastics Gearing

Spur Gear Design

Discussion Map

☐ A spur gear has involute
teeth that are straight and
parallel to the axis of the
shaft that carries the gear.

Discover

*Describe the action of the teeth of the driving gear on
those of the driven gear. What kinds of stresses are
produced?*

*How do the geometry of the gear teeth, the materials
from which they are made, and the operating conditions
affect the stresses and the life of the drive system?*

This chapter will help you acquire the skills to perform
the necessary analyses and to design safe spur gear
drive systems that demonstrate long life.

A *spur gear* is one of the most fundamental types of gears. Its teeth are straight and parallel to the axis of the shaft that carries the gear. The teeth have the involute form described in Chapter 8. So, in general, the action of one tooth on a mating tooth is like that of two convex, curved members in contact: As the driving gear rotates, its teeth exert a force on the mating gear that is tangential to the pitch circles of the two gears. Because this force acts at a distance equal to the pitch radius of the gear, a torque is developed in the shaft that carries the gear. When the two gears rotate, they transmit power that is proportional to the torque. Indeed, that is the primary purpose of the spur gear drive system.

Consider the action described in the preceding paragraph:

- How does that action relate to the design of the gear teeth? Look back at Figure 8–1 in Chapter 8 as you consider this question and those that follow.
- As the force is exerted by the driving tooth on the driven tooth, what kinds of stresses are produced in the teeth? Consider both the point of contact of one tooth on the other and the whole tooth. Where are stresses a maximum?
- How could the teeth fail under the influence of these stresses?
- What material properties are critical to allow the gears to carry such loads safely and with a reasonable life span?
- What important geometric features affect the level of stress produced in the teeth?
- How does the precision of the tooth geometry affect its operation?
- How does the nature of the application affect the gears? What if the machine that the gears drive is a rock crusher that takes large boulders and reduces them to gravel made up of small stones? How would that loading compare with that of a gear system that drives a fan providing ventilation air to a building?
- What is the influence of the driving machine? Would the design be different if an electric motor were the driver or if a gasoline engine were used?
- The gears are typically mounted on shafts that deliver power from the driver to the input gear of a gear train and that take power from the output gear and transmit it to the driven machine. Describe various ways that the gears can be attached to the shafts and located with respect to each other. How can the shafts be supported?

This chapter contains the kinds of information that you can use to answer such questions and to complete the analysis and design of spur gear power transmission systems.

Later chapters cover similar topics for helical gears, bevel gears, and wormgearing, along with the design and specification of keys, couplings, seals, shafts, and bearings—all of which are needed to design a complete mechanical drive.

You Are the Designer

You have already made the design decision that a spur gear type of speed reducer is to be used for a particular application. How do you complete the design of the gears themselves?

This is a continuation of a design scenario that was started in Chapter 1 of this book when the original goals were stated and when an overview of the entire book was given. The introduction to Part II continued this theme by indicating that the arrangement of the chapters is aligned with the design process that you could use to complete the design of the speed reducer.

Then in Chapter 8, you, as the designer, dealt with the kinematics of a gear reducer that would take power from the shaft of an electric motor rotating at 1750 rpm and deliver it to a machine that was to operate at approximately 292 rpm. There you limited your interest to the decisions that affected motion and the basic geometry of the gears. It was decided that you would use a double-reduction gear train to reduce the speed of rotation of the drive system in two stages using two pairs of gears in series. You also learned how to specify the layout of the gear train, along with key design decisions such as the numbers of

teeth in all of the gears and the relationships among the diametral pitch, the number of teeth in the gears, the pitch diameters, and the distance between the centers of the shafts that carry those gears. For a chosen diametral pitch, you learned how to compute the dimensions of key features of the gear teeth such as the addendum, dedendum, and tooth width.

But the design is not complete until you have specified the material from which the gears are to be made and until you have verified that the gears will withstand the forces exerted on the gears as they transmit power and the corresponding torque. The teeth must not break, and they must have a sufficiently long life to meet the needs of the customer who uses the reducer.

To complete the design, you need more data: How much power is to be transmitted? To what kind of machine is the power from the output of the reducer being delivered? How does that affect the design of the gears? What is the anticipated duty cycle for the reducer in terms of the number of hours per day, days per week, and years of life expected? What options do you have for materials that are suitable for gears? Which material will you specify, and what will be its heat treatment?

You are the designer. The information in this chapter will help you complete the design.

9–1 OBJECTIVES OF THIS CHAPTER

After completing this chapter, you will be able to demonstrate the competencies listed below. They are presented in the order that they are covered in this chapter. The primary objectives are numbers 6, 7, and 8, which involve (a) the calculation of the bending strength and the ability of the gear teeth to resist pitting and (b) the design of gears to be safe with regard to both strength and pitting resistance. The competencies are as follows:

1. Compute the forces exerted on gear teeth as they rotate and transmit power.
2. Describe various methods for manufacturing gears and the levels of precision and quality to which they can be produced.
3. Specify a suitable level of quality for gears according to the use to which they are to be put.
4. Describe suitable metallic materials from which to make the gears, in order to provide adequate performance for both strength and pitting resistance.
5. Use the standards of the American Gear Manufacturers Association (AGMA) as the basis for completing the design of the gears.
6. Use appropriate stress analyses to determine the relationships among the applied forces, the geometry of the gear teeth, the precision of the gear teeth, and other factors specific to a given application, in order to make final decisions about those variables.

7. Perform the analysis of the tendency for the contact stresses exerted on the surfaces of the teeth to cause pitting of the teeth, in order to determine an adequate hardness of the gear material that will provide an acceptable level of pitting resistance for the reducer.

8. Complete the design of the gears, taking into consideration both the stress analysis and the analysis of pitting resistance. The result will be a complete specification of the gear geometry, the material for the gear, and the heat treatment of the material.

9–2
CONCEPTS
FROM PREVIOUS
CHAPTERS

As you study this chapter, it is assumed that you are familiar with the geometry of gear features and the kinematics of one gear driving another as presented in Chapter 8 and illustrated in Figures 8–1, 8–8, 8–11, 8–12, 8–13, and 8–15. (See also References 4 and 23.) Key relationships that you should be able to use include the following:

$$\text{Pitch line speed} = v_t = R\omega = (D/2)\omega$$

where R = radius of the pitch circle
        D = pitch diameter
        ω = angular velocity of the gear

Because the pitch line speed is the same for both the pinion and the gear, values for R, D, and can be for either. In the computation of stresses in gear teeth, it is usual to express the pitch line speed in the units of ft/min, while the size of the gear is given as its pitch diameter expressed in inches. Speed of rotation is typically given as n rpm—that is, n rev/min. Let's compute the unit-specific equation that gives pitch line speed in ft/min:

☞ **Pitch Line Speed**
$$v_t = (D/2)\omega = \frac{D \text{ in}}{2} \cdot \frac{n \text{ rev}}{\text{min}} \cdot \frac{2\pi \text{ rad}}{\text{rev}} \cdot \frac{1 \text{ ft}}{12 \text{ in}} = (\pi D n/12) \text{ ft/min} \qquad \textbf{(9–1)}$$

The velocity ratio can be expressed in many ways. For the particular case of a pinion driving a larger gear,

☞ **Velocity Ratio**
$$\text{Velocity ratio} = VR = \frac{\omega_P}{\omega_G} = \frac{n_P}{n_G} = \frac{R_G}{R_P} = \frac{D_G}{D_P} = \frac{N_G}{N_P} \qquad \textbf{(9–2)}$$

A related ratio, m_G, called the *gear ratio*, is often used in analysis of the performance of gears. It is always defined as the ratio of the number of teeth in the larger gear to the number of teeth in the pinion, regardless of which is the driver. Thus, m_G is always greater than or equal to 1.0. When the pinion is the driver, as it is for a speed reducer, m_G is equal to VR. That is,

☞ **Gear Ratio**
$$\text{Gear ratio} = m_G = N_G/N_P \geq 1.0 \qquad \textbf{(9–3)}$$

The diametral pitch, P_d, characterizes the physical size of the teeth of a gear. It is related to the pitch diameter and the number of teeth as follows:

☞ **Diametral Pitch**
$$P_d = N_G/D_G = N_P/D_P \qquad \textbf{(9–4)}$$

The pressure angle, ϕ, is an important feature that characterizes the form of the involute curve that makes up the active face of the teeth of standard gears. See Figure 8–13. Also notice in Figure 8–12 that the angle between a normal to the involute curve and the tangent to the pitch circle of a gear is equal to the pressure angle.

9–3
FORCES,
TORQUE, AND
POWER IN
GEARING

Torque

To understand the method of computing stresses in gear teeth, consider the way power is transmitted by a gear system. For the simple single-reduction gear pair shown in Figure 9–1, power is received from the motor by the input shaft rotating at motor speed. Thus, torque in the shaft can be computed from the following equation:

$$\text{Torque} = \text{power/rotational speed} = P/n \qquad (9\text{--}5)$$

The input shaft transmits the power from the coupling to the point where the pinion is mounted. The power is transmitted from the shaft to the pinion through the key. The teeth of the pinion drive the teeth of the gear and thus transmit the power to the gear. But again, power transmission actually involves the application of a torque during rotation at a given speed. The torque is the product of the force acting tangent to the pitch circle of the pinion times the pitch radius of the pinion. We will use the symbol W_t to indicate the *tangential force*. As described,

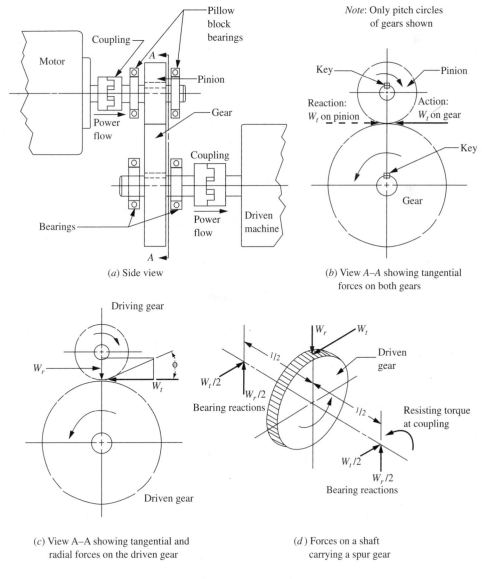

(a) Side view

(b) View A–A showing tangential forces on both gears

(c) View A–A showing tangential and radial forces on the driven gear

(d) Forces on a shaft carrying a spur gear

FIGURE 9–1   Power flow through a gear pair

W_t is the force exerted *by the pinion teeth on the gear teeth*. But if the gears are rotating at a constant speed and are transmitting a uniform level of power, the system is in equilibrium. Therefore, there must be an equal and opposite tangential force exerted by the gear teeth back on the pinion teeth. This is an application of the principle of action and reaction.

To complete the description of the power flow, the tangential force on the gear teeth produces a torque on the gear equal to the product of W_t times the pitch radius of the gear. Because W_t is the same on the pinion and the gear, but the pitch radius of the gear is larger than that of the pinion, the torque on the gear (the output torque) is greater than the input torque. However, note that the power transmitted is the same or slightly less because of mechanical inefficiencies. The power then flows from the gear through the key to the output shaft and finally to the driven machine.

From this description of power flow, we can see that gears transmit power by exerting a force by the driving teeth on the driven teeth while the reaction force acts back on the teeth of the driving gear. Figure 9–2 shows a single gear tooth with the tangential force W_t acting on it. But this is not the total force on the tooth. Because of the involute form of the tooth, the total force transferred from one tooth to the mating tooth acts normal to the involute profile. This action is shown as W_n. The tangential force W_t is actually the horizontal component of the total force. To complete the picture, note that there is a vertical component of the total force acting radially on the gear tooth, indicated by W_r.

We will start the computation of forces with the transmitted force, W_t, because its value is based on the given data for power and speed. It is convenient to develop unit-specific equations for W_t because standard practice typically calls for the following units for key quantities pertinent to the analysis of gear sets:

Forces in pounds (lb)

Power in horsepower (hp) (Note that 1.0 hp = 550 lb·ft/s.)

Rotational speed in rpm, that is, rev/min

Pitch line speed in ft/min

Torque in lb·in

FIGURE 9–2 Forces on gear teeth

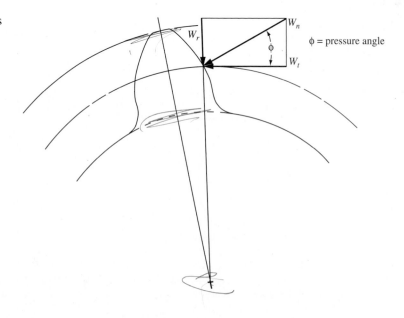

ϕ = pressure angle

The torque exerted on a gear is the product of the transmitted load, W_t, and the pitch radius of the gear. The torque is also equal to the power transmitted divided by the rotational speed. Then

$$T = W_t(R) = W_t(D/2) = P/n$$

Then we can solve for the force, and the units can be adjusted as follows:

$$W_t = \frac{2P}{Dn} = \frac{2P(\text{hp})}{D(\text{in}) \cdot n\,(\text{rev/min})} \cdot \frac{550\ \text{lb} \cdot \text{ft/s}}{(\text{hp})} \cdot \frac{1.0\ \text{rev}}{2\pi\ \text{rad}} \cdot \frac{60\ \text{s/min}}{} \cdot \frac{12\ \text{in}}{\text{ft}}$$

Tangential Force
$$W_t = (126\,000)(P)/(n\,D)\ \text{lb} \tag{9–6}$$

Data for either the pinion or the gear can be used in this equation. Other relationships are now developed because they are needed in other parts of the process of analyzing the gears or the shafts that carry them.

Power is also the product of transmitted force, W_t, and the pitch line velocity:

$$P = W_t \cdot v_t$$

Then, solving for the force and adjusting units, we have

Tangential Force
$$W_t = \frac{P}{v_t} = \frac{P\,(\text{hp})}{v_t\,(\text{ft/min})} \cdot \frac{550\ \text{lb} \cdot \text{ft/s}}{1.0\ \text{hp}} \cdot \frac{60\text{s/min}}{} = 33\,000\,(P)/(v_t)\ \text{lb} \tag{9–7}$$

We may also need to compute torque in lb · in:

$$T = \frac{P}{\omega} = \frac{P\,(\text{hp})}{n\,(\text{rev/min})} \cdot \frac{550\ \text{lb} \cdot \text{ft/s}}{1.0\ \text{hp}} \cdot \frac{1.0\ \text{rev}}{2\pi\ \text{rad}} \cdot \frac{60\ \text{s/min}}{} \cdot \frac{12\ \text{in}}{\text{ft}}$$

Torque
$$T = 63\,000\,(P)/n\ \text{lb} \cdot \text{in} \tag{9–8}$$

These values can be computed for either the pinion or the gear by appropriate substitutions. Remember that the pitch line speed is the same for the pinion and the gear and that the transmitted loads on the pinion and the gear are the same, except that they act in opposite directions.

The normal force, W_n, and the radial force, W_r, can be computed from the known W_t by using the right triangle relations evident in Figure 9–2:

Radial Force
$$W_r = W_t \tan \phi \tag{9–9}$$

Normal Force
$$W_n = W_t/\cos \phi \tag{9–10}$$

where ϕ = pressure angle of the tooth form

In addition to causing the stresses in the gear teeth, these forces act on the shaft. In order to maintain equilibrium, the bearings that support the shaft must provide the reactions. Figure 9–1(d) shows the free-body diagram of the output shaft of the reducer.

Power Flow and Efficiency

The discussion thus far has focused on power, torque, and forces for a single pair of gears. For compound gearing having two or more pairs of gears, the flow of power and the overall efficiency become increasingly important.

Power losses in gear drives made from spur, helical, and bevel gears depend on the action of each tooth on its mating tooth, a combination of rolling and sliding. For accurate, well-lubricated gears, the power loss ranges from 0.5% to 2.0% and is typically taken to be approximately 1.0% (See Reference 26). *Because this is quite small, it is customary to neglect it in sizing individual gear pairs; we do that in this book.*

Compound gear drives employ several pairs of gears in series to produce large speed reduction ratios. With 1.0% power loss in each pair, the accumulated power loss for the system can become significant, and it can affect the size of motor to drive the system or the ultimate power and torque available for use at the output. Furthermore, the power loss is transferred to the environment or into the gear lubricant and, for large power transmissions, the management of the heat generated is critical to the overall performance of the unit. The viscosity and load-carrying ability of lubricants is degraded with increasing temperature.

Tracking power flow in a simple or compund gear train is simple, the power is transferred from one gear pair to the next with only a small power loss at each mesh. More complex designs may split the power flow at some point to two or more paths. This is typical of planetary gear trains. In such cases, you should consider the basic relationship among power, torque, and rotational speed shown in Equation (9–5), $P = T \times n$. We can present this in another form. Let the rotational speed, n, that is typically taken to be in the units of rpm, be the more general term *angular velocity*, ω, in the units of rad/s. Now express the torque in terms of the transmitted forces, W_t, and the pitch radius of the gear, R. That is, $T = W_t R$. Equation (9–5) then becomes,

$$P = T \times n = W_t R \omega$$

But $R \omega$ is the pitch line velocity for the gears, v_t. Then,

$$P = W_t R \omega = W_t v_t$$

Knowing how the power splits enables the determination of the transmitted load at each mesh.

9–4
GEAR
MANUFACTURE

The discussion of gear manufacture will begin with the method of producing the gear blank. Small gears are frequently made from wrought plate or bar, with the hub, web, spokes, and rim machined to final or near-final dimensions before the gear teeth are produced. The face width and the outside diameter of the gear teeth are also produced at this stage. Other gear blanks may be forged, sand cast, or die cast to achieve the basic form prior to machining. A few gears in which only moderate precision is required may be die cast with the teeth in virtually final form.

Large gears are frequently fabricated from components. The rim and the portion into which the teeth are machined may be rolled into a ring shape from a flat bar and then welded. The web or spokes and the hub are then welded inside the ring. Very large gears may be made in segments with the final assembly of the segments by welding or by mechanical fasteners.

The popular methods of machining the gear teeth are form milling, shaping, and hobbing. (See References 23 and 25.)

FIGURE 9–3 A
variety of gear cutting
tools (Gleason Cutting
Tools Corporation,
Loves Park, IL)

(a) Form milling cutter

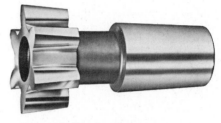

(b) Spur gear shaper cutter

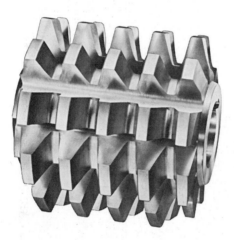

(c) Hob for small pitch gears having large teeth

(d) Hob for high pitch gears having small teeth

In *form milling* [Figure 9–3(a)], a milling cutter that has the shape of the tooth space is used, and each space is cut completely before the gear blank is indexed to the position of the next adjacent space. This method is used mostly for large gears, and great care is required to achieve accurate results.

Shaping [Figures 9–3(b) and 9–4] is a process in which the cutter reciprocates, usually on a vertical spindle. The shaping cutter rotates as it reciprocates and is fed into the gear blank. Thus, the involute-tooth form is generated gradually. This process is frequently used for internal gears.

Hobbing [Figures 9–3(c) and (d) and 9–5] is a process similar to milling except that the workpiece (the gear blank) and the cutter (the hob) rotate in a coordinated fashion. Here also, the tooth form is generated gradually as the hob is fed into the blank.

The gear teeth are finished to greater precision after form milling, shaping, or hobbing by the processes of grinding, shaving, and honing. Being products of secondary processes, they are expensive and should be used only where the operation requires high accuracy in the tooth form and spacing. Figure 9–6 shows a gear grinding machine.

(a) Shaping small external gear   *(b)* Shaping large internal gear

FIGURE 9–4   Gear shaping operations (Bourn & Koch, Inc., Rockford, IL)

(a) Gear hobbing machine   *(b)* Close-up view of hobbing process

FIGURE 9–5   CNC 4-axis fear hobber close-up of gear hobbing process (Bourn & Koch, Inc., Rockford, IL)

**9–5
GEAR QUALITY**

Quality in gearing is the precision of specific features of a single gear or the composite error of a gear rotating in mesh with a precise master gear. The factors typically measured to determine quality are:

> *Index variation:* The difference between the actual location of a point on the face of a gear tooth at the pitch circle and the corresponding point of a reference tooth, measured on the pitch circle. The variation causes inaccuracy in the action of mating gear teeth.

(a) Gear grinding machine

(b) Close-up view of grinding process

FIGURE 9–6   CNC gear grinder and close-up of gear grinding setup (Bourn & Koch, Inc., Rockford, IL)

Tooth alignment: The deviation of the actual line on the gear tooth surface at the pitch circle from the theoretical line. Measurements are made across the face from one end to the other. For a spur gear the theoretical line is straight. For a helical gear it is a part of a helix. Measurement of tooth alignment is sometimes called the *helix* measurement. It is important because excessive misalignment causes nonuniform loading on the gear teeth.

Tooth profile: The measurement of the actual profile of the surface of a gear tooth from the point of the start of the active profile to the tip of the tooth. The theoretical profile is a true involute curve. Variations of the actual profile from the theoretical profile cause variations in the instantaneous velocity ratio between the two gears in mesh, affecting the smoothness of the motion.

Root radius: The radius of the fillet at the base of the tooth. Variations from the theoretical value can affect the meshing of mating gears, creating possible interference, and the stress concentration factors related to bending stress in the tooth.

Runout: A measure of the eccentricity and out-of-roundness of a gear. Excessive runout causes the contact point on mating gear teeth to move radially during each revolution.

Total composite variation: A measure of the variation in the center distance between a precise master gear and the test gear for a full revolution. The shaft of one gear is fixed and the shaft of the mating gear is permitted to move while the teeth are kept in tight mesh. Figure 9–7(a) shows a sketch of one arrangement.

Standards for Gear Quality

The allowable amounts of variations of the actual tooth form from the theoretical form, or the composite variation, are specified by the AGMA as a *quality number.* Detailed charts giving the tolerances for many features are included in AGMA Standard 2000-A88 *Gear Classification and Inspection Handbook, Tolerances and Measuring Methods for Unassembled Spur and Helical Gears.* The quality numbers range from 5 to 15 with increasing precision.

FIGURE 9–7

Recording of errors in gear geometry (Extracted from AGMA Standard 2000-A88, *Gear Classification and Inspection Handbook, Tolerances and Measuring Methods for Unassembled Spur and Helical Gears (Including Metric Equivalents),* with the permission of the publisher, American Gear Manufacturers Association, 1500 King Street, Suite 201, Alexandria, VA 22314)

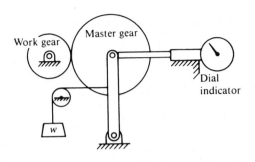

(*a*) Schematic diagram of a typical gear rolling fixture

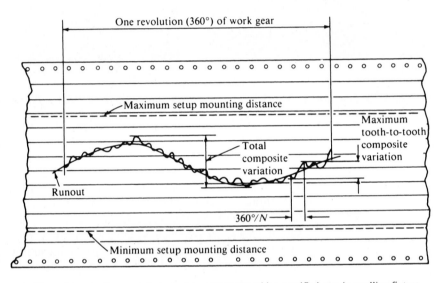

(*b*) Chart of gear-tooth errors of a typical gear when run with a specified gear in a rolling fixture

The actual tolerances are a function of the quality number, diametral pitch of the gear teeth, and the number of teeth of the gear. Table 9–1 shows representative data for the total composite tolerance for several quality numbers.

The International Organization for Standardization (ISO) defines a different set of quality numbers in its Standard 1328-1-1995, *Cylindrical gears–ISO system of accuracy–Part 1: Definitions and allowable values of deviations relevant to corresponding flanks of gear teeth* and Standard 1328-2-1997 *Cylindrical gears–ISO system of accuracy–Part 2: Definitions and allowable values of deviations relevant to radical composite deviations and runout information.* These standards differ greatly from the AGMA Standard 2000-A88. One major difference is that the quality numbering system is reversed. Whereas in the AGMA Standard 2008, higher numbers indicate greater precision, in the ISO standard lower numbers indicate greater precision.

The AGMA released two new standards just prior to the time this book was prepared. AGMA 2015-1-A01 *Accuracy Classification System–Tangential Measurements for Cylindrical Gears* applies a system in which lower grade numbers indicate lower tolerance values, similar but not identical to the ISO method. Comparisons are shown in this standard among the new AGMA 2015, the former AGMA 2008, and ISO 1328. AGMA 915-1-A02 *Inspection Practices–Part 1: Cylindrical Gears–Tangential Measurements,* deals with im-

TABLE 9–1   Selected values for total composite tolerance

AGMA quality number	Diametral pitch, P_d	Number of gear teeth				
		20	40	60	100	200
Q5	2	0.0260	0.0290	0.0320	0.0350	0.0410
	8	0.0120	0.0130	0.0140	0.0150	0.0170
	20	0.0074	0.0080	0.0085	0.0092	0.0100
	32	0.0060	0.0064	0.0068	0.0073	0.0080
Q8	2	0.0094	0.0110	0.0120	0.0130	0.0150
	8	0.0043	0.0047	0.0050	0.0055	0.0062
	20	0.0027	0.0029	0.0031	0.0034	0.0037
	32	0.0022	0.0023	0.0025	0.0027	0.0029
Q10	2	0.0048	0.0054	0.0059	0.0066	0.0076
	8	0.0022	0.0024	0.0026	0.0028	0.0032
	20	0.0014	0.0015	0.0016	0.0017	0.0019
	32	0.0011	0.0012	0.0013	0.0014	0.0015
Q12	2	0.0025	0.0028	0.0030	0.0034	0.0039
	8	0.0011	0.0012	0.0013	0.0014	0.0016
	20	0.000 71	0.000 77	0.000 81	0.000 87	0.000 97
	32	0.000 57	0.000 60	0.000 64	0.000 69	0.000 76
Q14	2	0.0013	0.0014	0.0015	0.0017	0.0020
	8	0.000 57	0.000 62	0.000 67	0.000 73	0.000 82
	20	0.000 36	0.000 39	0.000 41	0.000 45	0.000 50
	32	0.000 29	0.000 31	0.000 33	0.000 35	0.000 39

Source: Extracted from AGMA Standard 2000-A88, *Gear Classification and Inspection Handbook, Tolerances and Measuring Methods for Unassembled Spur and Helical Gears (Including Metric Equivalents),* with the permission of the publisher, American Gear Manufacturers Association, 1500 King Street, Suite 201, Alexandria, VA 22314.

plementing the new grades. While the quality grades are not precisely the same, a rough set of equivalents follows.

AGMA 2008	AGMA 2015	ISO 1328		AGMA 2008	AGMA 2015	ISO 1328
Q5	—	12		Q11	A6	6
Q6	A11	11		Q12	A5	5
Q7	A10	10		Q13	A4	4
Q8	A9	9		Q14	A3	3
Q9	A8	8		Q15	A2	2
Q10	A7	7			(Most precise)	

Note that the sum of the quality number from AGMA 2008 and the corresponding accuracy classification number from AGMA 2015 or ISO 1328 is always 17.

Some European manufacturers employ standards of the German DIN (Deutsche Industrie Normen) system whose quality numbers are similar to those of the ISO, although the detailed specifications of tolerances and measurement methods are not identical.

In this book, because of the recent introduction of AGMA 2015, we use the quality numbers from AGMA 2000-A88 unless stated otherwise. Also, AGMA 2008 is integrated into the gear design methodology that follows in terms of the dynamic factor, K_v.

(a) Overall view

(b) Closeup view of probe
and test gear

FIGURE 9–8   Analytical measurement system for gear quality
(Process Equipment Company, Tipp City, Ohio)

Methods of Gear Measurement

Two different approaches to determining gear quality are in use, functional measurement
and analytical measurement.

Functional measurement typically uses a system such as that sketched in Figure
9–7(a) to measure total composite error. The variation in center distance is recorded for a
complete revolution as shown in Figure 9–7(b). The total composite variation is the maximum spread between the highest and lowest points on the chart. In addition, the maximum
spread on the chart for any two adjacent teeth is determined as a measure of the tooth-to-
tooth composite variation. The runout can also be determined from the total excursion of
the mean line through the plot as shown. These data allow the determination of the AGMA
quality number based primarily on the total composite variation and are often considered
adequate for general-purpose gears in industrial machinery.

Analytical measurement measures individual errors of *index, alignment (helix), involute profile,* and other features. The equipment is a specially designed coordinate measurement system (CMM) with a highly accurate probe that scans the various important surfaces
of the test gear and produces electronic and printed records of the variations. Figure 9–8
shows one commercially available model of an analytical measurement system. Part (a) is an
overall view while part (b) shows the probe engaging the teeth of the test gear. Figure 9–9
shows two different types of output charts from an analytical measurement system. The *index
variation* chart (a) shows the amount of index variation for each tooth relative to a specified
datum tooth. The *profile variation* chart (b) shows a plot of the deviation of the actual tooth
profile from the true involute. The tooth alignment chart, measuring the helix accuracy, is
similar. Tabulated data are also given along with the corresponding quality number related
to each measurement.

Comparisons are automatically made with the theoretical tooth forms and with tolerance values to report the resulting quality number in the AGMA, ISO, DIN, or a user-

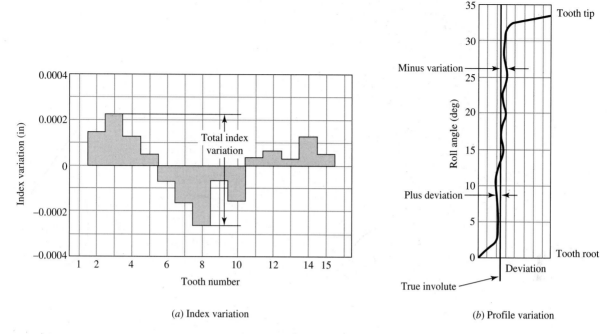

(a) Index variation   (b) Profile variation

FIGURE 9–9   Typical output charts from an analytical measurement system (Process Equipment Company, Tipp City, Ohio)

defined standard. Besides giving the quality numbers, the detailed data from analytical measurement systems are useful to the manufacturing staff in making adjustments to cutters or equipment settings to improve the accuracy of the total process.

Using the general capabilities of the analytical measurement system, dimensions of features other than those of the gear teeth may also be determined while the gear is in its fixture. For example, when a gear is machined onto a shaft, key shaft diameters and geometric features may be checked for dimensions, perpendicularity, parallelism, and concentricity. Gear segments, composite gears having two or more gears on the same shaft, splines, cam surfaces, and other special features can be inspected along with the gear teeth.

Recommended Quality Numbers

The data in Table 9–1 should impress you with the precision that is normally exercised in the manufacture and installation of gears. The design of the entire gear system, including the shafts, bearings, and housing, must be consistent with this precision. Of course, the system should not be made more precise than necessary because of cost. For this reason, manufacturers have recommended quality numbers that will give satisfactory performance at a reasonable cost for a variety of applications. Table 9–2 lists several of these recommendations.

Also shown in Table 9–2 are recommendations for quality numbers for machine tool drives. Because this is such a wide range of specific applications, the recommended quality numbers are related to the *pitch line speed*, defined as the linear velocity of a point on the pitch circle of the gear. Use Equation (9–1). We recommend using these values for any high-accuracy machinery.

TABLE 9–2   Recommended AGMA quality numbers

Application	Quality number	Application	Quality number
Cement mixer drum drive	3–5	Small power drill	7–9
Cement kiln	5–6	Clothes washing machine	8–10
Steel mill drives	5–6	Printing press	9–11
Grain harvester	5–7	Computing mechanism	10–11
Cranes	5–7	Automotive transmission	10–11
Punch press	5–7	Radar antenna drive	10–12
Mining conveyor	5–7	Marine propulsion drive	10–12
Paper-box-making machine	6–8	Aircraft engine drive	10–13
Gas meter mechanism	7–9	Gyroscope	12–14

Machine tool drives and drives for other high-quality mechanical systems

Pitch line speed (fpm)	Quality number	Pitch line speed (m/s)
0–800	6–8	0–4
800–2000	8–10	4–11
2000–4000	10–12	11–22
Over 4000	12–14	Over 22

9–6
ALLOWABLE
STRESS
NUMBERS

Later in this chapter, design procedures are presented in which two forms of gear-tooth failure are considered.

A gear tooth acts like a cantilever beam in resisting the force exerted on it by the mating tooth. The point of highest tensile bending stress is at the root of the tooth where the involute curve blends with the fillet. The AGMA has developed a set of *allowable bending stress numbers*, called s_{at}, which are compared to computed bending stress levels in the tooth to rate the acceptability of a design.

A second, independent form of failure is the pitting of the surface of the teeth, usually near the pitch line, where high contact stresses occur. The transfer of force from the driving to the driven tooth theoretically occurs across a line contact because of the action of two convex curves on each other. Repeated application of these high contact stresses can cause a type of fatigue failure of the surface, resulting in local fractures and an actual loss of material. This is called *pitting*. The AGMA has developed a set of *allowable contact stress numbers*, called s_{ac}, which are compared to computed contact stress levels in the tooth to rate the acceptability of a design. (See References 10–12.)

Representative data for s_{at} and s_{ac} are given in the following section for general information and use in problems in this book. More extensive data are given in the AGMA standards listed at the end of the chapter. (See References 6, 8, and 9.)

Many of the data given in this book for the design of spur and helical gears are taken from the AGMA Standard 2001-C95, *Fundamental Rating Factors and Calculation Methods for Involute Spur and Helical Gear Teeth*, with permission of the publisher, American Gear Manufacturers Association, 1500 King Street, Suite 201, Alexandria, VA 22314. That document should be consulted for details beyond the discussion in this book. Data in the U.S. Customary System only are given here. A separate document, AGMA Standard 2101-C95, has been published as a Metric Edition of AGMA 2001-C95. A brief summary of the differences between the terminology in these two standards is given later in this chapter.

9–7
METALLIC
GEAR
MATERIALS

Gears can be made from a wide variety of materials to achieve properties appropriate to the application. From a mechanical design standpoint, strength and pitting resistance are the most important properties. But, in general, the designer should consider the producibility of the gear, taking into account all of the manufacturing processes involved, from the preparation of the gear blank, through the forming of the gear teeth, to the final assembly of the gear into a machine. Other considerations are weight, appearance, corrosion resistance, noise, and, of course, cost. This section discusses several types of metals used for gears. Plastics are covered in a later section.

Steel Gear Materials

Through-Hardened Steels. Gears for machine tool drives and many kinds of medium- to heavy-duty speed reducers and transmissions are typically made from medium-carbon steels. Among a wide range of carbon and alloy steels used are

AISI 1020	AISI 1040	AISI 1050	AISI 3140
AISI 4140	AISI 4340	AISI 4620	AISI 5120
AISI 6150	AISI 8620	AISI 8650	AISI 9310

(See Reference 17.) AGMA Standard 2001-C95 gives data for the allowable bending stress number, s_{at}, and the allowable contact stress number, s_{ac}, for steels in the through-hardened condition. Figures 9–10 and 9–11 are graphs relating the stress numbers to the Brinell hardness number for the teeth. Notice that only knowledge of the hardness is required because of the direct relationship between hardness and the tensile strength of steels. See Appendix 19

FIGURE 9–10
Allowable bending stress number for through-hardened steel gears, s_{at} (Extracted from AGMA 2001-C95 Standard, *Fundamental Rating Factors and Calculation Methods for Involute Spur and Helical Gear Teeth*, with permission of the publisher, American Gear Manufacturers Association, 1500 King Street, Suite 201, Alexandria, VA 22314)

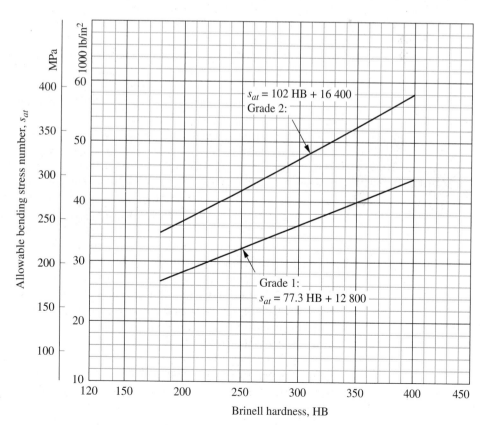

FIGURE 9–11

Allowable contact stress number for through-hardened steel gears, s_{ac} (Extracted from AGMA 2001-C95 Standard, *Fundamental Rating Factors and Calculation Methods for Involute Spur and Helical Gear Teeth*, with permission of the publisher, American Gear Manufacturers Association, 1500 King Street, Suite 201, Alexandria, VA 22314)

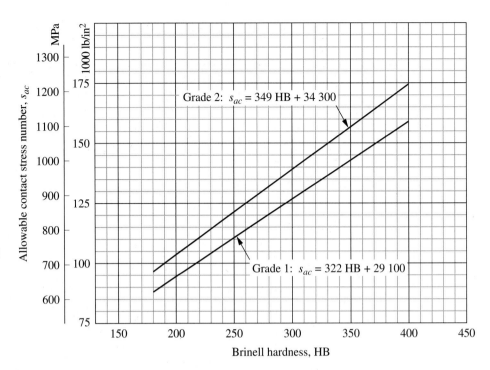

for data that correlate the Brinell hardness number, HB, with the tensile strength of steel in ksi. The range of hardnesses covered by the AGMA data is from 180 to 400 HB, corresponding to a tensile strength of approximately 87 to 200 ksi. It is not recommended to use through-hardening above 400 HB because of inconsistent performance of the gears in service. Typically, case hardening is used when there is a desire to achieve a surface hardness above 400 HB. This is described later in this section.

The hardness measurement for the allowable bending stress number is to be taken at the root of the teeth because that is where the highest bending stress occurs. The allowable contact stress number is related to the surface hardness on the face of the gear teeth where the mating teeth experience high contact stresses.

When selecting a material for gears, the designer must specify one that can be hardened to the desired hardness. Review Chapter 2 for discussions about heat treatment techniques. Consult Appendices 3 and 4 for representative data. For the higher hardnesses, say, above 250 HB, a medium-carbon-alloy steel with good hardenability is desirable. Examples are AISI 3140, 4140, 4340, 6150, and 8650. Ductility is also rather important because of the numerous cycles of stress experienced by gear teeth and the likelihood of occasional overloads, impact, or shock loading. A percent elongation value of 12% or higher is desired.

The curves in Figures 9–10 and 9–11 include two grades of steel: Grade 1 and Grade 2. *Grade 1 is considered to be the basic standard and will be used for problem solutions in this book.* Grade 2 requires a higher degree of control of the microstructure, alloy composition, greater cleanliness, prior heat treatment, nondestructive testing performed, core hardness values, and other factors. See AGMA Standard 2001-C95 (Reference 6) for details.

Case-Hardened Steels.   Flame hardening, induction hardening, carburizing, and nitriding are processes used to produce a high hardness in the surface layer of gear teeth. See Figure 2–13 and the related discussion in Section 2–6. These processes provide surface

TABLE 9–3 Allowable stress numbers for case-hardened steel gear materials

Hardness at surface	Allowable bending stress number, s_{at} (ksi)			Allowable contact stress number, s_{ac} (ksi)		
	Grade 1	Grade 2	Grade 3	Grade 1	Grade 2	Grade 3
Flame- or induction-hardened:						
50 HRC	45	55		170	190	
54 HRC	45	55		175	195	
Carburized and case-hardened:						
55–64 HRC	55			180		
58–64 HRC	55	65	75	180	225	275
Nitrided, through-hardened steels:						
83.5 HR15N	See Figure 9–14.			150	163	175
84.5 HR15N	See Figure 9–14.			155	168	180
Nitrided, nitralloy 135M:[a]						
87.5 HR15N	See Figure 9–15.					
90.0 HR15N	See Figure 9–15.			170	183	195
Nitrided, nitralloy N:[a]						
87.5 HR15N	See Figure 9–15.					
90.0 HR15N	See Figure 9–15.			172	188	205
Nitrided, 2.5% chrome (no aluminum):						
87.5 HR15N	See Figure 9–15.			155	172	189
90.0 HR15N	See Figure 9–15.			176	196	216

Source: Extracted from AGMA Standard 2001-C95. *Fundamental Rating Factors and Calculation Methods for Involute Spur and Helical Gear Teeth,* with the permission of the publisher, American Gear Manufacturers Association, 1500 King Street, Suite 201, Alexandria, VA 22314.

[a]Nitralloy is a proprietary family of steels containing approximately 1.0% aluminum which enhances the formation of hard nitrides.

hardness values from 50 to 64 HRC (Rockwell C) and correspondingly high values of s_{at} and s_{ac}, as shown in Table 9–3. Special discussions are given below for each of the types of case-hardening processes.

In addition to Grade 1 and Grade 2 as described earlier, case-hardened steel gears can be produced to Grade 3 which requires an even higher standard of control of the metallurgy and processing of the material. See AGMA Standard 2001-C95 (References 6 and 20) for details.

Flame- and Induction-Hardened Gear Teeth. Recall that these processes involve the local heating of the surface of the gear teeth by high-temperature gas flames or electrical induction coils. By controlling the time and energy input, the manufacturer can control the depth of heating and the depth of the resulting case. It is essential that the heating occur around the entire tooth, producing the hard case on the face of the teeth *and in the fillet and root areas*, in order to use the stress values listed in Table 9–3. This may require a special design for the flame shape or the induction heater.

The specifications for flame- or induction-hardened steel gear teeth call for a resulting hardness of HRC 50 to 54. Because these processes rely on the inherent hardenability of the steels, you must specify a material that can be hardened to these levels. Normally, medium-carbon-alloy steels (approximately 0.40% to 0.60% carbon) are specified. Appendices 3 and 4 list some suitable materials.

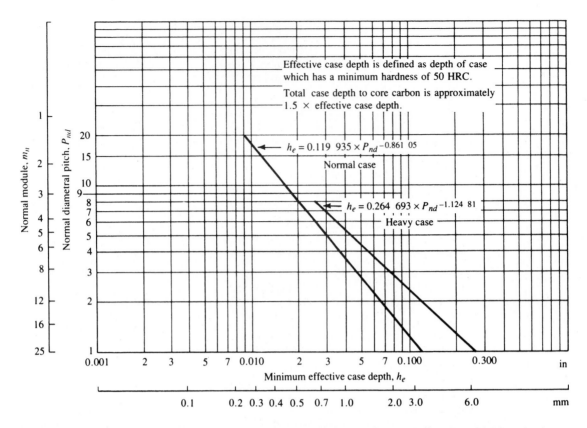

FIGURE 9–12 Effective case depth for carburized gears, h_e (Extracted from AGMA 2001-C95 Standard, *Fundamental Rating Factors and Calculation Methods for Involute Spur and Helical Gear Teeth,* with permission of the publisher, American Gear Manufacturers Association, 1500 King Street, Suite 201, Alexandria, VA 22314)

Carburizing.   Carburizing produces surface hardnesses in the range of 55 to 64 HRC. It results in some of the highest strengths in common use for gears. Special carburizing steels are listed in Appendix 5. Figure 9–12 shows the AGMA recommendation for the thickness of the case for carburized gear teeth. The effective case depth is defined as the depth from the surface to the point where the hardness has reached 50 HRC.

Nitriding.   Nitriding produces a very hard *but very thin* case. It is specified for applications in which loads are smooth and well known. Nitriding should be avoided when overloading or shock can be experienced, because the case is not sufficiently strong or well supported to resist such loads. Because of the thin case, the Rockwell 15N scale is used to specify hardness.

Figure 9–13 shows the AGMA recommendation for the case depth of nitrided gears, defined as the depth below the surface at which the hardness has dropped to 110% of that at the core of the teeth. The values for the allowable bending stress number, s_{at}, are dependent on the conditions of the material in the core of the teeth because of the thin case for nitrided gears. Figure 9–14 gives the values for the general group of alloy steels used for gears that are through-hardened and then nitrided. Examples are AISI 4140 and AISI 4340 and similar alloys. As with other through-hardened materials, the primary variable is

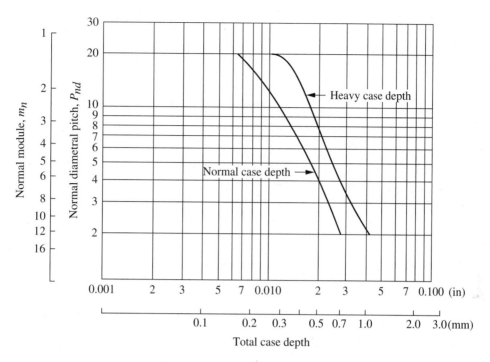

Equations for Minimum Total Case Depth for Nitrided, Case-Hardened Gears

Normal case depth (in inches):
$$h_{c\,min} = 0.0432896 - 0.00968115*P_d + 0.00120185*P_d{}^2 - 6.79721 \times 10^{-5}*P_d{}^3 + 1.37117 \times 10^{-6}*P_d{}^4$$

Heavy case depth (in inches):
$$h_{c\,min} = 0.0660090 - 0.0162224 P_d + 0.00209361 P_d{}^2 - 1.17755 \times 10^{-4} P_d{}^3 + 2.33160 \times 10^{-6} P_d{}^4$$

Note: $P_d = P_{nd}$ for helical gear teeth

FIGURE 9–13   Recommended case depth for nitrided gears, h_c (Extracted from AGMA 2001-C95 Standard, *Fundamental Rating Factors and Calculation Methods for Involute Spur and Helical Gear Teeth,* with permission of the publisher, American Gear Manufacturers Association, 1500 King Street, Suite 201, Alexandria, VA 22314)

the Brinell hardness number HB. Also, special alloys have been developed for use in gears with the nitriding process. Data are shown in Table 9–3 and Figure 9–15 for *nitralloy* and an alloy called *2.5% chrome.*

Iron and Bronze Gear Materials

Cast Irons.   Two types of iron used for gears are *gray cast iron* and *ductile* (sometimes called *nodular*) iron. Table 9–4 gives the common ASTM grades used, with their corresponding allowable bending stress numbers and contact stress numbers. Remember that gray cast iron is brittle, so care should be exercised when shock loading is possible. Also, the higher-strength forms of the other irons have low ductility. Austempered ductile iron (ADI) is being used in some important automotive applications. However, standardized allowable stress numbers have not yet been specified.

FIGURE 9–14
Allowable bending stress numbers for nitrided, through-hardened steel gears (that is, AISI 4140, AISI 4340), s_{at} (Extracted from AGMA 2001-C95 Standard, *Fundamental Rating Factors and Calculation Methods for Involute Spur and Helical Gear Teeth*, with permission of the publisher, American Gear Manufacturers Association,1500 King Street, Suite 201, Alexandria, VA 22314)

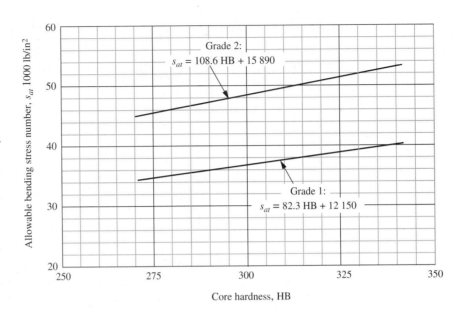

FIGURE 9–15
Allowable bending stress numbers for nitriding steel gears, s_{at} (Extracted from AGMA 2001-C95 Standard, *Fundamental Rating Factors and Calculation Methods for Involute Spur and Helical Gear Teeth*, with permission of the publisher, American Gear Manufacturers Association, 1500 King Street, Suite 201. Alexandira, VA 22314

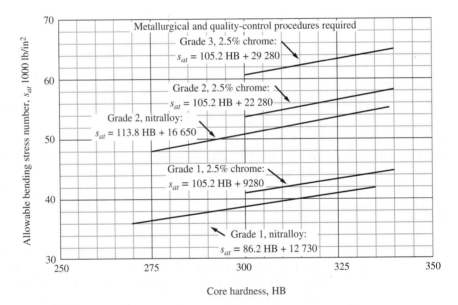

Bronzes. Four families of bronzes are typically used for gears: (1) phosphor or tin bronze, (2) manganese bronze, (3) aluminum bronze, and (4) silicon bronze. Yellow brass is also used. Most bronzes are cast, but some are available in wrought form. Corrosion resistance, good wear properties, and low friction coefficients are some reasons for choosing bronzes for gears. Table 9–4 shows allowable stress numbers for one bronze alloy in two common forms.

TABLE 9–4 Allowable stress numbers for iron and bronze gears

Material designation	Minimum hardness at surface (HB)	Allowable bending stress number		Allowable contact stress number	
		(ksi)	(MPa)	(ksi)	(MPa)
Gray cast iron, ASTM A48, as cast					
Class 20		5	35	50	345
Class 30	174	8.5	59	65	448
Class 40	201	13	90	75	517
Ductile (nodular) iron, ASTM A536					
60-40-18 annealed	140	22	152	77	530
80-55-06 quenched and tempered	179	22	152	77	530
100-70-03 quenched and tempered	229	27	186	92	634
120-90-02 quenched and tempered	269	31	214	103	710
Bronze, sand-cast, $s_{u\ min}$ = 40 ksi (275 MPa)					
		5.7	39	30	207
Bronze, heat-treated, $s_{u\ min}$ = 90 ksi (620 MPa)					
		23.6	163	65	448

Source: Extracted from AGMA Standard 2001-C95 *Fundamental Rating Factors and Calculation Methods for Involute Spur and Helical Gear Teeth,* with the permission of the publisher, American Gear Manufacturers Association, 1500 King Street, Suite 201, Alexandria, VA 22314.

9–8 STRESSES IN GEAR TEETH

⇨ **Lewis Equation for Bending Stress In Gear Teeth**

The stress analysis of gear teeth is facilitated by consideration of the orthogonal force components, W_t and W_r, as shown in Figure 9–2.

The tangential force, W_t, produces a bending moment on the gear tooth similar to that on a cantilever beam. The resulting bending stress is maximum at the base of the tooth in the fillet that joins the involute profile to the bottom of the tooth space. Taking the detailed geometry of the tooth into account, Wilfred Lewis developed the equation for the stress at the base of the involute profile, which is now called the *Lewis equation*:

$$\sigma_t = \frac{W_t P_d}{FY} \qquad (9\text{--}12)$$

where W_t = tangential force
P_d = diametral pitch of the tooth
F = face width of the tooth
Y = *Lewis form factor*, which depends on the tooth form, the pressure angle, the diametral pitch, the number of teeth in the gear, and the place where W_t acts

While the theoretical basis for the stress analysis of gear teeth is presented, the Lewis equation must be modified for practical design and analysis. One important limitation is that it does not take into account the stress concentration that exists in the fillet of the tooth. Figure 9–16 is a photograph of a photoelastic stress analysis of a model of a gear tooth. It indicates a stress concentration in the fillet at the root of the tooth as well as high contact stresses at the mating surface (the contact stress is discussed in the following section). Comparing the actual stress at the root with that predicted by the Lewis equation enables

FIGURE 9–16
Photoelastic study of
gear teeth under load
(Measurements Group,
Inc., Raleigh, NC)

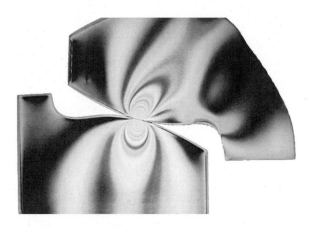

us to determine the stress concentration factor, K_t, for the fillet area. Placing this into equation (9–12) gives

$$\sigma_t = \frac{W_t P_d K_t}{FY} \tag{9–13}$$

The value of the stress concentration factor is dependent on the form of the tooth, the shape and size of the fillet at the root of the tooth, and the point of application of the force on the tooth. Note that the value of the Lewis form factor, Y, also depends on the tooth geometry. Therefore, the two factors are combined into one term, the *geometry factor, J*, where $J = Y/K_t$. The value of J also, of course, varies with the location of the point of application of the force on the tooth because Y and K_t vary.

Figure 9–17 shows graphs giving the values for the geometry factor for 20° and 25°, full-depth, involute teeth. The safest value to use is the one for the load applied at the tip of the tooth. However, this value is overly conservative because there is some load sharing by another tooth at the time that the load is initially applied at the tip of a tooth. The critical load on a given tooth occurs when the load is at the highest point of single-tooth contact, when the tooth carries the entire load. The upper curves in Figure 9–17 give the values for J for this condition.

Using the geometry factor, J, in the stress equation gives

$$\sigma_t = \frac{W_t P_d}{FJ} \tag{9–14}$$

The graphs in Figure 9–17 are taken from the former AGMA Standard 218.01 which has been superseded by the two new standards: AGMA 2001-C95, *Fundamental Rating Factors and Calculation Methods for Involute Spur and Helical Gear Teeth*, 1995, and AGMA 908-B89 (R1995), *Geometry Factors for Determining the Pitting Resistance and Bending Strength of Spur, Helical and Herringbone Gear Teeth*, 1995. Standard 908-B89 includes an analytical method for calculating the geometry factor, J. But the values for J are unchanged from those in the former standard. Rather than graphs, the new standard reports values for J for a variety of tooth forms in tables. The graphs from the former standard are shown in Figure 9–17 so that you can visualize the variation of J with the number of teeth in the pinion and the gear.

Note also that J factors for only two tooth forms are included in Figure 9–17 and that the values are valid only for those forms. Designers must ensure that J factors for

FIGURE 9–17
Geometry factor, *J*
(Extract from AGMA
218.01 Standard,
*Rating the Pitting
Resistance and
Bending Strength of
Spur and Helical
Involute Gear Teeth*,
with the permission of
the publisher, American
Gear Manufacturers
Association, 1500 King
Street, Suite 201,
Alexandria, VA 22314)

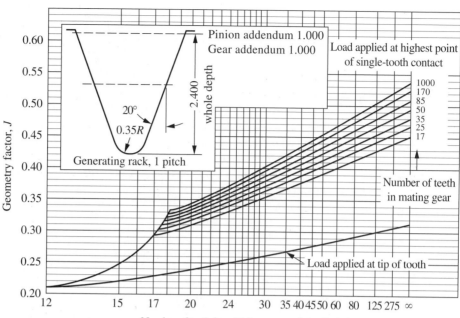

(*a*) 20° spur gear: standard addendum

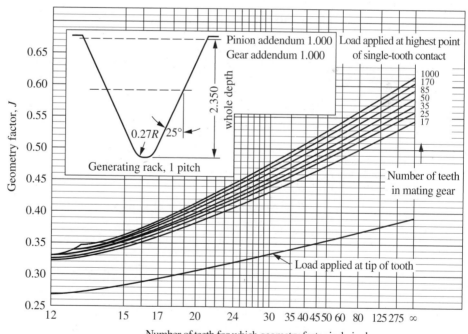

(*b*) 25° spur gear: standard addendum

the tooth form actually used, including the form of the fillet, are included in the stress analysis.

Equation (9–14) can be called the *modified Lewis equation*. Other modifications to the equation are recommended by the AGMA in Standard 2001-C95 for practical design to account for the variety of conditions that can be encountered in service.

The approach used by the AGMA is to apply a series of additional modifying factors to the bending stress from the modified Lewis equation to compute a value called the *bending stress number, s_t*. These factors represent the degree to which the actual loading case differs from the theoretical basis of the Lewis equation. The result is a better estimate of the real level of bending stress that is produced in the teeth of the gear and the pinion.

Then, separately, the allowable bending stress number, s_{at}, is modified by a series of factors that affect that value when the environment is different from the nominal situation assumed when the values for s_{at} are set. The result here is a better estimate of the real level of the bending strength of the material from which the gear or the pinion is made.

The design is completed in a manner that ensures that the bending stress number is less than the modified allowable bending stress number. This process should be completed for both the pinion and the gear of a given pair because materials may be different; the geometry factor, J, is different; and other operating conditions may be different. This is demonstrated in example problems later in this chapter.

Often the major decision to be made is the specification of suitable materials from which to make the pinion and the gear. In such cases, the required basic allowable bending stress number, s_{at}, will be computed. When steel is used, the required hardness of the material is found from the data described in Section 9–7. Finally, the material and its heat treatment are specified to ensure that it will have at least the required hardness.

We proceed now with the discussion of the bending stress number, s_t.

Bending Stress Number, s_t

The design analysis method used here is based primarily on AGMA Standard 2001-C95. However, because values for some of the factors are not included in the standard, data from other sources are added. These data illustrate the kinds of conditions that affect the final design. The designer ultimately has the responsibility for making appropriate design decisions.

The following equation will be used in this book:

Bending Stress Number, s_t

$$s_t = \frac{W_t P_d}{FJ} K_o K_s K_m K_B K_v \qquad (9\text{–}15)$$

where K_o = overload factor for bending strength
K_s = size factor for bending strength
K_m = load distribution factor for bending strength
K_B = rim thickness factor
K_v = dynamic factor for bending strength

Methods for specifying values for these factors are discussed below.

Overload Factor, K_o

Overload factors consider the probability that load variations, vibrations, shock, speed changes, and other application-specific conditions may result in peak loads greater than W_t being applied to the gear teeth during operation. A careful analysis of actual conditions should be made, and the AGMA Standard 2001-C95 gives no specific values for K_o. Reference 15 gives some recommended values, and many industries have established suitable values based on experience.

For problem solutions in this book, we will use the values shown in Table 9–5. The primary considerations are the nature of *both* the driving power source and the driven machine. An overload factor of 1.00 would be applied for a perfectly smooth electric motor driving a perfectly smooth generator through a gear-type speed reducer. Any rougher conditions call for a value of K_o greater than 1.00. For power sources we will use the following:

Uniform: Electric motor or constant-speed gas turbine

Light shock: Water turbine, variable-speed drive

Moderate shock: Multicylinder engine

Examples of the roughness of driven machines include the following:

Uniform: Continuous-duty generator

Light shock: Fans and low-speed centrifugal pumps, liquid agitators, variable-duty generators, uniformly loaded conveyors, rotary positive displacement pumps

Moderate shock: High-speed centrifugal pumps, reciprocating pumps and compressors, heavy-duty conveyors, machine tool drives, concrete mixers, textile machinery, meat grinders, saws

Heavy shock: Rock crushers, punch press drives, pulverizers, processing mills, tumbling barrels, wood chippers, vibrating screens, railroad car dumpers

Size Factor, K_s

The AGMA indicates that the size factor can be taken to be 1.00 for most gears. But for gears with large-size teeth or large face widths, a value greater than 1.00 is recommended. Reference 15 recommends a value of 1.00 for diametral pitches of 5 or greater or for a metric module of 5 or smaller. For larger teeth, the values shown in Table 9–6 can be used.

Load-Distribution Factor, K_m

The determination of the load-distribution factor is based on many variables in the design of the gears themselves as well as in the shafts, bearings, housings, and the structure in which the gear drive is installed. Therefore, it is one of the most difficult factors to specify. Much analytical and experimental work is continuing on the determination of values for K_m.

TABLE 9–5   Suggested overload factors, K_o

	Driven Machine			
Power source	Uniform	Light shock	Moderate shock	Heavy shock
Uniform	1.00	1.25	1.50	1.75
Light shock	1.20	1.40	1.75	2.25
Moderate shock	1.30	1.70	2.00	2.75

TABLE 9–6   Suggested size factors, K_s

Diametral pitch, P_d	Metric module, m	Size factor, K_s
≥5	≤5	1.00
4	6	1.05
3	8	1.15
2	12	1.25
1.25	20	1.40

If the intensity of loading on all parts of all teeth in contact at any given time were uniform, the value of K_m would be 1.00. However, this is seldom the case. Any of the following factors can cause misalignment of the teeth on the pinion relative to those on the gear:

1. Inaccurate gear teeth
2. Misalignment of the axes of shafts carrying gears
3. Elastic deformations of the gears, shafts, bearings, housings, and support structures
4. Clearances between the shafts and the gears, the shafts and the bearings, or the bearings and the housing
5. Thermal distortions during operation
6. Crowning or end relief of gear teeth

AGMA Standard 2001-C95 presents extensive discussions of methods of determining values for K_m. One is empirical and considers gears up to 40 in (1000 mm) wide. The other method is analytical and considers the stiffness and mass of individual gears and gear teeth and the total mismatch between mating teeth. We will not provide so much detail. However, rough guidelines are given below.

The designer can minimize the load-distribution factor by specifying the following:

1. Accurate teeth (a high quality number)
2. Narrow face widths
3. Gears centered between bearings (straddle mounting)
4. Short shaft spans between bearings
5. Large shaft diameters (high stiffness)
6. Rigid, stiff housings
7. High precision and small clearances on all drive components

You are advised to study the details of AGMA Standard 2001-C95 which covers a wide range of physical sizes for gear systems. But the gear designs discussed in this book are of moderate size, typical of power transmissions in light industrial and vehicular applications. A more limited set of data are reported here to illustrate the concepts that must be considered in gear design.

We will use the following equation for computing the value of the load-distribution factor:

$$K_m = 1.0 + C_{pf} + C_{ma} \qquad (9\text{--}16)$$

where C_{pf} = pinion proportion factor (see Figure 9–18)
C_{ma} = mesh alignment factor (see Figure 9–19)

In this book, we are limiting designs to those with face widths of 15 in or less. Wider face widths call for additional factors. Also, some commercially successful designs employ modifications to the basic tooth form to achieve a more uniform meshing of the teeth. Such methods are not discussed in this book.

Figure 9–18 shows that the pinion proportion factor is dependent on the actual face width of the pinion and on the ratio of the face width to the pinion pitch diameter. Figure 9–19 relates the mesh alignment factor to expected accuracy of different methods of applying gears. *Open gearing* refers to drive systems in which the shafts are supported in bearings that are mounted on structural elements of the machine with the expectation that relatively large misalignments will result. In *commercial-quality enclosed gear units*, the bearings are mounted in a specially designed housing that provides more rigidity than for

FIGURE 9–18 Pinion proportion factor, C_{pf} (Extracted from AGMA 2001-C95 *Standard, Fundamental Rating Factors and Calculation Methods for Involute Spur and Helical Gear Teeth*, with permission of the publisher, American Gear Manufacturers Association, 1500 King Street, Suite 201, Alexandria, VA 22314)

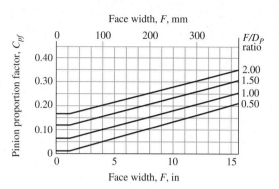

D_P = Pinion diameter

For $F/D_P < 0.50$, use curve for $F/D_P = 0.50$

When $F \leq 1.0$ in. ($F \leq 25$ mm)

$$C_{pf} = \frac{F}{10D_P} - 0.025$$

When $1.0 \leq F < 15$,

$$C_{pf} = \frac{F}{10D_P} - 0.0375 + 0.0125F$$

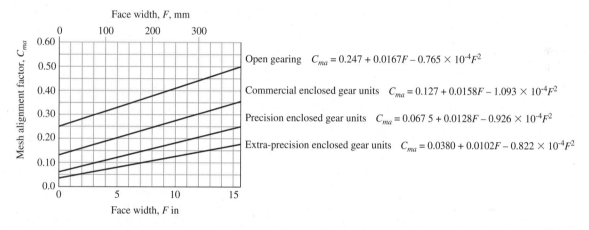

Open gearing $C_{ma} = 0.247 + 0.0167F - 0.765 \times 10^{-4}F^2$

Commercial enclosed gear units $C_{ma} = 0.127 + 0.0158F - 1.093 \times 10^{-4}F^2$

Precision enclosed gear units $C_{ma} = 0.067\,5 + 0.0128F - 0.926 \times 10^{-4}F^2$

Extra-precision enclosed gear units $C_{ma} = 0.0380 + 0.0102F - 0.822 \times 10^{-4}F^2$

FIGURE 9–19 Mesh alignment factor, C_{ma} (Extracted from AGMA 2001-C95 Standard, *Fundamental Rating Factors and Calculation Methods for Involute Spur and Helical Gear Teeth*, with permission of the publisher, American Gear Manufacturers Association, 1500 King Street, Suite 201, Alexandria, VA 22314)

open gearing, but for which the tolerances on individual dimensions are fairly loose. The *precision enclosed gear units* are made to tighter tolerances. *Extra-precision enclosed gear* units are made to exacting precision and are often adjusted at assembly to achieve excellent alignment of the gear teeth. Experience with similar units in the field will help you gain better understanding among the different types of designs.

Rim Thickness Factor, K_B

The basic analysis used to develop the Lewis equation assumes that the gear tooth behaves as a cantilever attached to a perfectly rigid support structure at its base. If the rim of the gear is too thin, it can deform and cause the point of maximum stress to shift from the area of the gear-tooth fillet to a point within the rim.

Figure 9–20 can be used to estimate the influence of rim thickness. The key geometry parameter is called the *backup ratio, m_B*, where

$$m_B = t_R/h_t$$

t_R = rim thickness

h_t = whole depth of the gear tooth

FIGURE 9–20 Rim thickness factor, K_B (Extracted from AGMA 2001-C95 Standard, *Fundamental Rating Factors and Calculation Methods for Involute Spur and Helical Gear Teeth*, with permission of the publisher, American Gear Manufacturers Association, 1500 King Street, Suite 201, Alexandria, VA 22314)

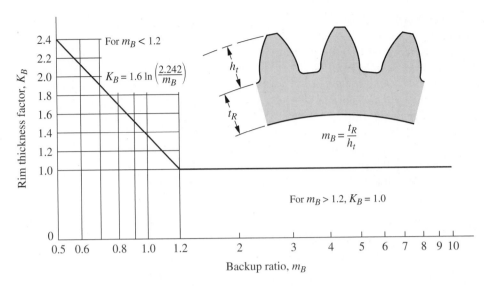

For $m_B > 1.2$, the rim is sufficiently strong and stiff to support the tooth, and $K_B = 1.0$. The K_B factor can also be used in the vicinity of a keyseat where a small thickness of metal occurs between the top of the keyseat and the bottom of the tooth space.

Dynamic Factor, K_v

The dynamic factor accounts for the fact that the load is assumed by a tooth with some degree of impact and that the actual load subjected to the tooth is higher than the transmitted load alone. The value of K_v depends on the accuracy of the tooth profile, the elastic properties of the tooth, and the speed with which the teeth come into contact.

Figure 9–21 shows a graph of the AGMA-recommended values for K_v, where the Q_v numbers are the AGMA-quality numbers referred to earlier in Section 9–5. Gears in typical machine design would fall into the classes represented by curves 5, 6, or 7, which are for gears made by hobbing or shaping with average to good tooling. If the teeth are finish-ground or shaved to improve the accuracy of the tooth profile and spacing, curve 8, 9, 10, or 11 should be used. Under special conditions where teeth of high precision are used in applications where there is little chance of developing external dynamic loads, the shaded area can be used. If the teeth are cut by form milling, factors lower than those found from curve 5 should be used. Note that the quality 5 gears should not be used at pitch line speeds above 2500 ft/min. Note that the dynamic factors are approximate. For severe applications, especially those operating above 4000 ft/min, approaches taking into account the material properties, the mass and inertia of the gears, and the actual error in the tooth form should be used to predict the dynamic load. (See References 12, 15, and 18.)

Example Problem 9–1   Compute the bending stress numbers for the pinion and the gear of the pair of gears in Figure 9–1. The pinion rotates at 1750 rpm, driven directly by an electric motor. The driven machine is an industrial saw requiring 25 hp. The gear unit is enclosed and is made to commercial standards. Gears are straddle-mounted between bearings. The following gear data apply:

$$N_P = 20 \quad N_G = 70 \quad P_d = 8 \quad F = 1.50 \text{ in} \quad Q_v = 6$$

The gear teeth are 20°, full-depth, involute teeth, and the gear blanks are solid.

FIGURE 9–21
Dynamic factor, K_V
(Extracted from
AGMA 2001-C95
Standard
*Fundamental, Rating
Factors and
Calculation Methods
for Involute Spur and
Helical Gear, Teeth.*
with the permission
of the publisher,
American Gear
Manufacturers
Association, 1500
King Street, Suite
201, Alexandria, VA
22314)

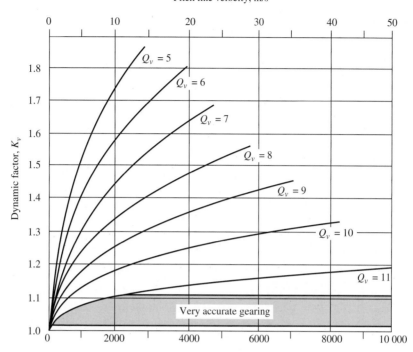

$v_{t\,max} = [A + (Q_v - 3)]^2$ (U.S. units)

$v_{t\,max} = \dfrac{[A + (Q_v - 3)]^2}{200}$ (SI units)

where $v_{t\,max}$ = end-point of K_v curves (ft/min or m/s)

Curves 5–11

$K_v = \left(\dfrac{A + \sqrt{v_t}}{A}\right)^B$ (U.S. units)

$K_v = \left(\dfrac{A + \sqrt{200v_t}}{A}\right)^B$ (SI units)

where $A = 50 + 56(1.0 - B)$

$B = \dfrac{(12 - Q_v)^{0.667}}{4}$

Q_v = transmission accuracy grade number

Solution   We will use Equation (9–15) to compute the expected stress:

$$s_t = \frac{W_t P_d}{FJ} K_o K_s K_m K_B K_v$$

We can first use the principles from Section 9–3 to compute the transmitted load on the gear teeth:

$$D_P = N_P/P_d = 20/8 = 2.500 \text{ in}$$
$$v_t = \pi D_P n_P/12 = \pi(2.5)(1750)/12 = 1145 \text{ ft/min}$$
$$W_t = 33\,000(P)/v_t = (33\,000)(25)/(1145) = 720 \text{ lb}$$

From Figure 9–17, we find that $J_P = 0.335$ and $J_G = 0.420$.

The overload factor is found from Table 9–5. For a smooth, uniform electric motor driving an industrial saw generating moderate shock, $K_o = 1.50$ is a reasonable value.

The size factor $K_s = 1.00$ because the gear teeth with $P_d = 8$ are relatively small. See Table 9–6.

The load distribution factor, K_m, can be found from Equation (9–16) for commercial enclosed gear drives. For this design, $F = 1.50$ in, and

$$F/D_P = 1.50/2.50 = 0.60$$
$$C_{pf} = 0.04 \text{ (Figure 9–18)}$$
$$C_{ma} = 0.15 \text{ (Figure 9–19)}$$
$$K_m = 1.0 + C_{pf} + C_{ma} = 1.0 + 0.04 + 0.15 = 1.19$$

The rim thickness factor, K_B, can be taken as 1.00 because the gears are to be made from solid blanks.

The dynamic factor can be read from Figure 9–21. For $v_t = 1145$ ft/min and $Q_v = 6$, $K_v = 1.45$.

The stress can now be computed from Equation (9–15). We will compute the stress in the pinion first:

$$s_{tP} = \frac{(720)(8)}{(1.50)(0.335)}(1.50)(1.0)(1.19)(1.0)(1.45) = 29\ 700 \text{ psi}$$

Notice that all factors in the stress equation are the same for the gear except the value of the geometry factor, J. Then the stress in the gear can be computed from

$$s_{tG} = \sigma_{tP}(J_P/J_G) = (29\ 700)(0.335/0.420) = 23\ 700 \text{ psi}$$

The stress in the pinion teeth will always be higher than the stress in the gear teeth because the value of J increases as the number of teeth increases.

9–9
SELECTION OF GEAR MATERIAL BASED ON BENDING STRESS

For safe operation, it is the designer's responsibility to specify a material that has an allowable bending stress greater than the computed stress due to bending from Equation (9–15). Recall that in Section 9–6, allowable stress numbers, s_{at}, were given for a variety of commonly used gear materials. Then it is necessary that

$$s_t < s_{at}$$

These data are valid for the following conditions:

Temperature less than 250°F

10^7 cycles of tooth loading

Reliability of 99%: Less than one failure in 100

Safety factor of 1.00

In this book we will assume that the operating temperature of the gears is less than 250°F. For higher temperatures, testing is recommended to determine the degree of reduction in the strength of the gear material.

Adjusted Allowable Bending Stress Numbers, s'_{at}

Data have been generated for different levels of expected life and reliability as discussed below. Designers may also choose to apply a factor of safety to the allowable bending stress number to account for uncertainties in the design analysis, material characteristics, or manufacturing tolerances, or to provide an extra measure of safety in critical applications. These factors are applied to the value of s_{at} to produce an *adjusted allowable bending stress number* which we will refer to as s'_{at}:

$$s'_{at} = s_{at} Y_N/(SF \cdot K_R) \qquad \text{(9–17)}$$

Stress Cycle Factor, Y_N

Figure 9–22 allows the determination of the life adjustment factor, Y_N, if the teeth of the gear being analyzed are expected to experience a number of cycles of loading much different from 10^7. Note that the general type of material is a factor in this chart for the lower number of cycles. For the higher number of cycles, a range is indicated by a shaded area. General design practice would use the upper line of this range. Critical applications where pitting and tooth wear must be minimal may use the lower part of the range.

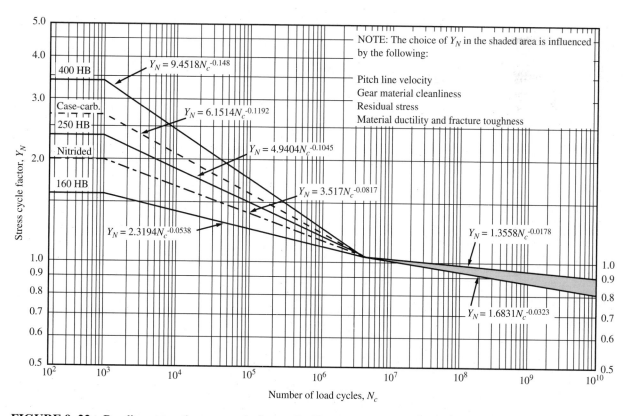

FIGURE 9–22 Bending strength stress cycle factor, Y_N (Extracted from AGMA Standard 2001-C95, *Fundamental Rating Factors and Calculation Methods for Involute Spur and Helical Gear Teeth,* with permission of the publisher, American Gear Manufacturers Association, 1500 King Street, Suite 201, Alexandria, VA 22314)

Calculation of the expected number of cycles of loading can be done using

$$N_c = (60)(L)(n)(q) \tag{9–18}$$

where N_c = expected number of cycles of loading
L = design life in hours
n = rotational speed of the gear in rpm
q = number of load applications per revolution

Design life is, indeed, a design decision based on the application. As a guideline, we will use a set of data created for use in bearing design and reported here as Table 9–7. Unless stated otherwise, we will use a design life of L = 20 000 h as listed for general industrial machines. The normal number of load applications per revolution for any given tooth is typically, of course, one. But consider the case of an idler gear that serves as both a driven and a driving gear in a gear train. It receives two cycles of load per revolution: one as it receives power from and one as it delivers power to its mating gears. Also, in certain types of gear trains, one gear may deliver power to two or more gears mating with it. Gears in a planetary gear train often have this characteristic.

As an example of the application of Equation (9–18), consider that the pinion in Example Problem 9–1 is designed to have a life of 20 000 h. Then

$$N_c = (60)(L)(n)(q) = (60)(20\,000)(1750)(1) = 2.1 \times 10^9 \text{ cycles}$$

Because this is higher than 10^7, an adjustment in the allowable bending stress number must be made.

Reliability Factor, K_R

Table 9–8 gives data that adjust for the design reliability desired. These data are based on statistical analyses of failure data.

Factor of Safety, SF

The factor of safety may be used to account for the following:

TABLE 9–7 Recommended design life

Application	Design life (h)
Domestic appliances	1000–2000
Aircraft engines	1000–4000
Automotive	1500–5000
Agricultural equipment	3000–6000
Elevators, industrial fans, multipurpose gearing	8000–15 000
Electric motors, industrial blowers, general industrial machines	20 000–30 000
Pumps and compressors	40 000–60 000
Critical equipment in continuous 24-h operation	100 000–200 000

Source: Eugene A. Avallone and Theodore Baumeister III, eds. *Marks' Standard Handbook for Mechanical Engineers.* 9th ed. New York: McGraw-Hill, 1986.

TABLE 9–8 Reliability factor, K_R

Reliability	K_R
0.90, one failure in 10	0.85
0.99, one failure in 100	1.00
0.999, one failure in 1000	1.25
0.9999, one failure in 10 000	1.50

- Uncertainties in the design analysis
- Uncertainties in material characteristics
- Uncertainties in manufacturing tolerances

It may also be used to provide an extra measure of safety in critical applications.

No general guidelines are published, and designers must evaluate the conditions of each application. Note, however, that many of the factors often considered to be a part of a factor of safety in general design practice have already been included in the calculations for s_t and s_{at}. Therefore, a modest value for factor of safety should suffice, say, between 1.00 and 1.50.

Procedure for Selecting Gear Materials for Bending Stress

The logic of the material selection process can be summarized as follows: The bending stress number from Equation (9–15) must be less than the adjusted allowable bending stress number from Equation (9–17). That is,

$$s_t < s'_{at}$$

Let's equate the expressions for these two values:

$$\frac{W_t P_d}{FJ} K_o K_s K_m K_B K_v = s_t < s_{at} \frac{Y_N}{SF \cdot K_R} \tag{9–19}$$

To use this relationship for material selection, it is convenient to solve for s_{at}:

$$\frac{K_R\,(SF)}{Y_N} s_t < s_{at} \tag{9–20}$$

We will use this equation for selecting gear materials based on bending stress. The list below summarizes the terms included in Equations (9–19) and (9–20) for your reference. You should review the more complete discussion of each before doing practice designs:

W_t = tangential force on gear teeth = (63 000)P/n

P_d = diametral pitch of the gear

F = face width of the gear

J = geometry factor for bending stress (see Figure 9–17)

K_o = overload factor (see Table 9–5)

K_s = size factor (see Table 9–6)

K_m = mesh alignment factor = $1.0 + C_{pf} + C_{ma}$ (see Figures 9–18 and 9–19)

K_B = rim thickness factor (see Figure 9–20)

K_v = velocity factor (see Figure 9–21)

K_R = reliability factor (see Table 9–8)

SF = factor of safety (design decision)

Y_N = bending strength stress cycle number (see Figure 9–22)

Completing the calculation of the value of the left side of Equation (9–20) gives the required value for the allowable bending stress number, s_{at}. You should then go to the data in Section 9–7, "Gear Materials," to select a suitable material. Consider first whether the material should be steel, cast iron, or bronze. Then consult the related tables of data.

For steel materials, the following review should aid your selection:

1. Start by checking Figure 9–10 to see whether a through-hardened steel will give the needed s_{at}. If so, determine the required hardness. Then specify a steel material and its heat treatment by referring to Appendices 3 and 4.

2. If a higher s_{at} is needed, see Table 9–3 and Figures 9–14 and 9–15 for properties of case-hardened steels.

3. Appendix 5 will aid in the selection of carburized steels.

4. If flame or induction hardening is planned, specify a material with a good hardenability, such as AISI 4140 or 4340 or similar medium-carbon-alloy steels. See Appendix 4.

5. Refer to Figure 9–12 or 9–13 for recommended case depths for surface-hardened steels.

For cast iron or bronze, refer to Table 9–4.

Example Problem 9–2 Specify suitable materials for the pinion and the gear from Example Problem 9–1. Design for a reliability of fewer than one failure in 10 000. The application is an industrial saw that will be fully utilized on a normal, one-shift, five-day-per-week operation.

Solution The results of Example Problem 9–1 include the expected bending stress number for both the pinion and the gear as follows:

$$s_{tP} = 29\,700 \text{ psi} \quad s_{tG} = 23\,700 \text{ psi}$$

We should consider the number of stress cycles, the reliability, and the safety factor to complete the calculation indicated in Equation (9–20).

Stress Cycle Factor, Y_N: From the problem statement, $n_P = 1750$ rpm, $N_P = 20$ teeth, and $N_G = 70$ teeth. Let's use these data to determine the expected number of cycles of stress that the pinion and gear teeth will experience. The application conforms to common industry practice, calling for a design life of approximately 20 000 h as suggested in Table 9–7. The number of stress cycles for the pinion is

$$N_{cP} = (60)(L)(n_p)(q) = (60)(20\,000)(1750)(1) = 2.10 \times 10^9 \text{ cycles}$$

The gear rotates more slowly because of the speed reduction. Then

$$n_G = n_P(N_P/N_G) = (1750 \text{ rpm})(20/70) = 500 \text{ rpm}$$

We can now compute the number of stress cycles for each gear tooth:

$$N_{cG} = (60)(L)(n_G)(q) = (60)(20\,000)(500)(1) = 6.00 \times 10^8 \text{ cycles}$$

Because both values are above the nominal value of 10^7 cycles, a value of Y_N must be determined from Figure 9–22 for both the pinion and the gear:

$$Y_{NP} = 0.92 \quad Y_{NG} = 0.96$$

Reliability Factor, K_R: For the design goal of fewer than one failure in 10 000, Table 9–8 recommends $K_R = 1.50$.

Factor of Safety: This is a design decision. Reviewing the discussion of factors throughout Example Problem 9–1 and this problem, we see that virtually all factors typically considered in adjusting stress on the teeth and strength of the material have been taken into account. Furthermore, when we select a material, it will likely have strength and hardness values somewhat above the minimum acceptable values. Therefore, as a design decision, let's use $SF = 1.00$.

Adjusted Value of s_{at}: We can now complete Equation (9–20) and use it for material selection. For the pinion,

$$\frac{K_R\,(SF)}{Y_{NP}} s_t = \frac{(1.50)(1.00)}{0.92} (29\ 700\ \text{psi}) = 48\ 450\ \text{psi} < s_{at}$$

For the gear,

$$\frac{K_R\,(SF)}{Y_{NG}} s_t = \frac{(1.50)(1.00)}{0.96} (23\ 700\ \text{psi}) = 37\ 050\ \text{psi} < s_{at}$$

Now, referring to Figure 9–10, and deciding to use Grade 1 steel, we find that the required allowable bending stress number for the pinion is greater than permitted for a through-hardened steel. But Table 9–3 indicates that a carburized, case-hardened steel with a case hardness of 55 to 64 HRC would be satisfactory, having a value of $s_{at} = 55$ ksi = 55 000 psi. Referring to Appendix 5, we see that virtually any of the listed carburized materials could be used. Let's specify AISI 4320 SOQT 300, having a core tensile strength of 218 ksi, 13% elongation, and a case hardness of 62 HRC.

For the gear, Figure 9–10 indicates that a through-hardened steel with a hardness of 320 HB would be satisfactory. From Appendix 3, let's specify AISI 4340 OQT 1000 having a hardness of 363 HB, a tensile strength of 171 ksi, and 16% elongation.

Comments   These materials should provide satisfactory service, considering bending strength. The following section considers the other major failure mode: pitting resistance. It is possible, perhaps likely, that the requirements to meet that condition will be more severe than for bending.

**9–10
PITTING
RESISTANCE OF
GEAR TEETH**

In addition to being safe from bending, gear teeth must also be capable of operating for the desired life without significant pitting of the tooth form. *Pitting* is the phenomenon in which small particles are removed from the surface of the tooth faces because of the high contact stresses, causing fatigue. Refer again to Figure 9–16 showing the high, localized contact stresses. Prolonged operation after pitting begins causes the teeth to roughen, and eventually the form is deteriorated. Rapid failure follows. Note that both the driving and driven teeth are subjected to these high contact stresses.

The action at the contact point on gear teeth is that of two externally curved surfaces. If the gear materials were infinitely rigid, the contact would be a simple line. Actually, because of the elasticity of the materials, the tooth shape deforms slightly, resulting in the transmitted force acting on a small rectangular area. The resulting stress is called a *contact stress* or *Hertz stress*. Reference 24 gives the following form of the equation for the Hertz stress,

⇨ **Hertz Contact Stress on Gear Teeth**

$$\sigma_c = \sqrt{\frac{W_c}{F} \frac{1}{\pi\{[(1 - v_1^2)/E_1] + [(1 - v_2^2)/E_2]\}} \left(\frac{1}{r_1} + \frac{1}{r_2}\right)} \qquad (9\text{-}21)$$

where the subscripts 1 and 2 refer to the materials of the two bodies in contact. The tensile modulus of elasticity is E and the Poisson's ratio is v. W_c is the contact force exerted between the two bodies, and F is the length of the contacting surfaces. The radii of curvature of the two surfaces are called r_1 and r_2.

When Equation 9–21 is applied to gears, F is the face width of the gear teeth and W_c is the normal force delivered by the driving tooth on the driven tooth, found from Equation (9–10) to be,

$$W_N = W_t/\cos \phi$$

The second term is Equation 9–21 (including the square root) can be computed if the elastic properties of the materials for the pinion and gear are known. It is given the name *elastic coefficient, C_P.* That is,

⇨ **Elastic Coefficient**

$$C_P = \sqrt{\frac{1}{\pi\{[(1 - v_P^2)/E_P] + [(1 - v_G^2)/E_G]\}}} \qquad (9\text{-}22)$$

Table 9–9 gives values for the most common combinations of materials for pinions and gears.

TABLE 9–9 Elastic coefficient, C_p

Pinion material	Modulus of elasticity, E_P, lb/in^2 (MPa)	Gear material and modulus of elasticity, E_G, lb/in^2 (MPa)					
		Steel 30×10^6 (2×10^5)	Malleable iron 25×10^6 (1.7×10^5)	Nodular iron 24×10^6 (1.7×10^5)	Cast iron 22×10^6 (1.5×10^5)	Aluminum bronze 17.5×10^6 (1.2×10^5)	Tin bronze 16×10^6 (1.1×10^5)
Steel	30×10^6 (2×10^5)	2300 (191)	2180 (181)	2160 (179)	2100 (174)	1950 (162)	1900 (158)
Mall. iron	25×10^6 (1.7×10^5)	2180 (181)	2090 (174)	2070 (172)	2020 (168)	1900 (158)	1850 (154)
Nod. iron	24×10^6 (1.7×10^5)	2160 (179)	2070 (172)	2050 (170)	2000 (166)	1880 (156)	1830 (152)
Cast iron	22×10^6 (1.5×10^5)	2100 (174)	2020 (168)	2000 (166)	1960 (163)	1850 (154)	1800 (149)
Al. bronze	17.5×10^6 (1.2×10^5)	1950 (162)	1900 (158)	1880 (156)	1850 (154)	1750 (145)	1700 (141)
Tin bronze	16×10^6 (1.1×10^5)	1900 (158)	1850 (154)	1830 (152)	1800 (149)	1700 (141)	1650 (137)

Source: Extracted from AGMA Standard 2001-C95, *Fundamental Rating Factors and Calculation Methods for Involute Spur and Helical Gear Teeth,* with the permission of the publisher, American Gear Manufacturers Association, 1500 King Street, Suite 201, Alexandria, VA 22314.
Note: Poisson's ratio = 0.30; units for C_p are (lb/in^2)$^{0.5}$ or (MPa)$^{0.5}$.

The terms r_1 and r_2 are the radii of curvature of the involute tooth forms of the two mating teeth. These radii change continuously during the meshing cycle as the contact point moves from the top of the tooth through the pitch circle, and onto the lower flank of the tooth before leaving engagement. We can write the following equations for the radius of curvature when contact is at the pitch point,

$$r_1 = (D_P/2) \sin \phi \quad \text{and} \quad r_2 = (D_G/2) \sin \phi \qquad \textbf{(9–23)}$$

However, the AGMA calls for the computation of the contact stress to be made at the lowest point of single tooth contact (LPSTC) because above that point, the load is being shared with other teeth. Computation of the radii of curvature for the LPSTC is somewhat complex. A geometry factor for pitting, I, is defined by the AGMA to include the radii of curvature terms and the cos φ term in Equation (9–21) because they all involve the specific geometry of the tooth. The variables required to compute I are the pressure angle ϕ, the gear ratio $m_G = N_G/N_P$, and the number of teeth in the pinion N_p. Another factor is the pinion diameter that is not included in I. The contact stress equation then becomes,

$$\sigma_c = C_p \sqrt{\frac{W_t}{FD_PI}} \qquad \textbf{(9–24)}$$

Values for the elastic coefficient, I, for a few common cases are graphed in Figure 9–23 and should be used for problem solving in this book. Appendix 19 provides an approach to computing the value for I for spur gears as given in Reference 3.

As with the equation for bending stress in gear teeth, several factors are added to the equation for contact stress as shown below. The resulting quantity is called the *contact stress number*, s_c:

⇨ **Contact Stress Number**

$$s_c = C_p \sqrt{\frac{W_t K_o K_s K_m K_v}{FD_PI}} \qquad \textbf{(9–25)}$$

This is the form of the contact stress equation that we will use in problem solutions.

The values for the overload factor, K_o; the size factor, K_s; the load-distribution factor, K_m; and the dynamic factor, K_v, can be taken to be the same as the corresponding values for the bending stress analysis in the preceding sections.

Example Problem 9–3 Compute the contact stress number for the gear pair described in Example Problem 9–1.

Solution Data from Example Problem 9–1 are summarized as follows:

$N_P = 20$	$N_G = 70,$	$F = 1.50$ in	$W_t = 720$ lb	$D_P = 2.500$ in
$K_o = 1.50$	$K_s = 1.00$	$K_m = 1.19$	$K_v = 1.45$	

The gear teeth are 20°, full-depth, involute teeth. We also need the geometry factor for pitting resistance, I. From Figure 9–23(a), at a gear ratio of $m_G = N_G/N_P = 70/20 = 3.50$ and for $N_P = 20$, we read $I = 0.108$, approximately.

The design analysis for bending strength indicated that two steel gears should be used. Then, from Table 9–9, we find that $C_p = 2300$. Then the contact stress number is

$$s_c = C_p \sqrt{\frac{W_t K_o K_s K_m K_v}{FD_PI}} = 2300 \sqrt{\frac{(720)(1.50)(1.0)(1.19)(1.45)}{(1.50)(2.50)(0.108)}}$$

$$s_c = 156\,000 \text{ psi}$$

FIGURE 9–23
External spur pinion geometry factor, *I*, for standard center distances. All curves area for the lowest point of single-tooth contact on the pinion. (Extracted from AGMA Standard 218.01, *Rating the Pitting Resistance and Bending Strength of Spur and Helical Involute Gear Teeth*, with the permission of the publisher, American Gear Manufacturers Association, 1500 King Street, Suite 201, Alexandria, VA 22314)

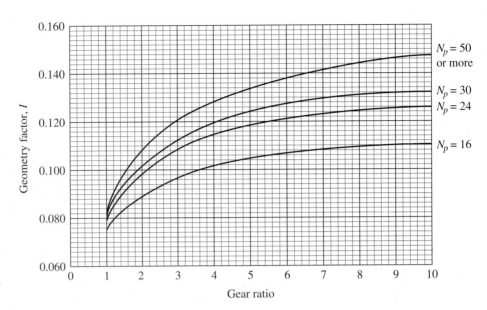

(*a*) 20° pressure angle, full-depth teeth (standard addendum = $1/P_d$)

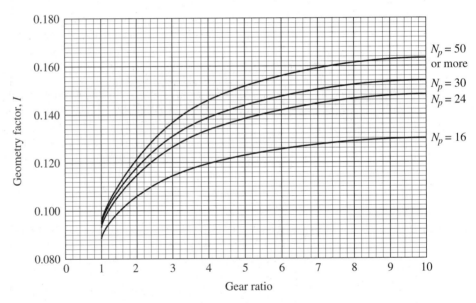

(*b*) 25° pressure angle, full-depth teeth (standard addendum = $1/P_d$)

9–11 SELECTION OF GEAR MATERIAL BASED ON CONTACT STRESS

Because the pitting resulting from the contact stress is a different failure phenomenon from tooth failure due to bending, an independent specification for suitable materials for the pinion and the gear must now be made. In general, the designer must specify a material having an allowable contact stress number, s_{ac}, greater than the computed contact stress number, s_c. That is,

$$s_c < s_{ac}$$

In Section 9–6, values for s_{ac} were given for several materials that are valid for 10^7 cycles of loading at a reliability of 99% if the material temperature is under 250°F. For different life expectancy and reliability, other factors are added:

$$s_c < s_{ac} \frac{Z_N C_H}{(SF)K_R} \tag{9-26}$$

Designs in this book are limited to applications where the operating temperature is less than 250°F and so no temperature factor is applied. Data for the reduction in hardness and strength as a function of temperature should be sought if higher temperatures are experienced.

The reliability factor, K_R, is the same as that for bending stress; it is given in Table 9–8. The other factors in Equation (9–26) are discussed below.

Pitting Resistance Stress Cycle Factor, Z_N

The term Z_N is the *pitting resistance stress cycle factor* and accounts for an expected number of contacts different from 10^7 as was assumed when the data were produced for the allowable contact stress number. Figure 9–24 shows values for Z_N where the solid-line curve is for most steels and the dashed-line curve is for nitrided steels. The number of cycles of contact is computed from Equation (9–18) and is the same as that used for bending. For the higher number of cycles, a range is indicated by the shaded area. General design practice would use the upper line of this range. Critical applications where pitting and tooth wear must be minimal may use the lower part of the range.

Factor of Safety, *SF*

The factor of safety is based on the same conditions as described for bending, and often the same value would be used for both bending and pitting resistance. Review that discussion

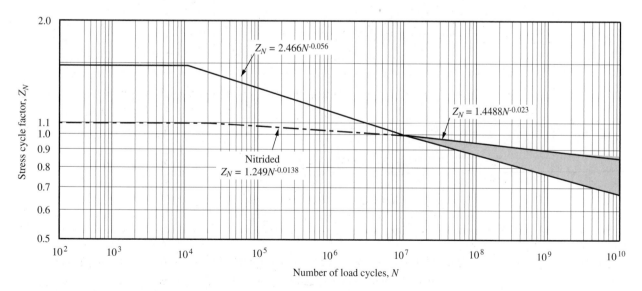

FIGURE 9–24 Pitting resistance stress cycle factor, Z_N (Extracted from AGMA Standard 2001-C95, *Fundamental Rating Factors and Calculation Methods for Involute Spur and Helical Gear Teeth*, with permission of the publisher, American Gear Manufacturers Association, 1500 King Street, Suite 201, Alexandria, VA 22314)

in Section 9–9. However, if there are different levels of uncertainty, a different value should be chosen. No general guidelines are published. Because many factors are already considered in the pitting resistance calculations, a modest value for factor of safety should suffice, say, between 1.00 and 1.50.

Hardness Ratio Factor, C_H

Good gear design practice calls for making the pinion teeth harder than the gear teeth so that the gear teeth are smoothed and work-hardened during operation. This increases the gear capacity with regard to pitting resistance and is accounted for by the factor C_H. Figure 9–25 shows data for C_H for through-hardened gears that depend on the ratio of the hardness of the pinion and the hardness of the gear, expressed as the Brinell hardness number, and on the gear ratio where $m_G = N_G/N_P$. Use the given curves for hardness ratios between 1.2 and 1.7. For hardness ratios under 1.2, use $C_H = 1.00$. For hardness ratios over 1.7, use the value of C_H for 1.7, as no substantial additional improvement is gained.

Figure 9–26 shows data for C_H when pinions are surface-hardened to 48 HRC or higher and the gear is through-hardened up to 400 HB. The parameters are the Brinell hardness number for the gear and the surface finish of the pinion teeth, expressed as f_p and measured as the average roughness, R_a. Smoother teeth give a higher value of hardness factor and generally increase the pitting resistance of the gear teeth.

Note that C_H is applied for the calculations for only the gear, not the pinion.

When designing gears, the specification of the materials for the pinion and the gear is the final step. Therefore, the hardness for the two gears is unknown and a specific value for C_H cannot be determined. It is recommended that an initial value of $C_H = 1.00$ be used. Then, after specifying the materials, a final value for C_H can be determined and used in Equation 9–27 to determine the final value of S_{ac}.

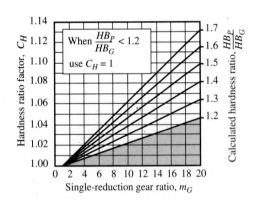

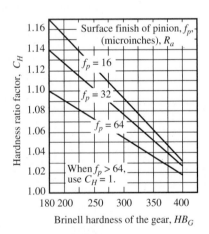

FIGURE 9–25 Hardness ratio facator, C_H (through-hardenced gears) (Extract from AGMA Standard 2001-C95, *Fundamental Rating Factors and Calculation Methods for Involute Spur and Helical Gear Teeth,* with the permission of the publisher, American Gear Manufacturers Association, 1500 King Street, Suite 201, Alexandria, VA 22314)

FIGURE 9–26 Hardness ratio factor, C_H (surface-hardened pinions) (Extracted from AGMA Standard 2001-C95, *Fundamental Rating Factors and Calculation Methods for Involute Spur and Helical Gear Teeth,* with permission of the publisher, American Gear Manufacturers Association, 1500 King Street, Suite 201, Alexandria, VA 22314)

Procedure for Selecting Gear Materials for Pitting Resistance

We can refer to the value on the right side of Equation (9–26) as the *modified allowable contact stress number*, as it accounts for nonstandard conditions under which the gears operate that are different from those assumed when the data for s_{ac} were determined as reported in Section 9–6.

Transposing the modifying factors from Equation (9–26) to the left side of the equation gives

**Required
Allowable Contact
Stress Number**

$$\frac{K_R (SF)}{Z_N C_H} s_c < s_{ac} \tag{9–27}$$

This is the equation we will use to determine the required properties of most metallic materials used for gears. We can summarize the procedure as follows.

**Procedure for
Determining the
Required
Properties of
Most Metallic
Materials**

1. Solve for the contact stress number, s_c, from Equation (9–25) using the same factors as those used for bending stress number.
2. Use the value for K_R from the bending stress analysis, or evaluate it from Table 9–8.
3. Use Figure 9–22 to find Z_N.
4. Assume an initial value for the hardness ratio factor $C_H = 1.00$.
5. Specify a safety factor, typically between 1.00 and 1.50, considering the degree of uncertainty in material property data, the gear precision, the severity of the application, or the danger presented by the application.
6. Compute s_{ac} from Equation (9–27).
7. Refer to data in Section 9–7, "Metallic Gear Materials," to select a suitable material. You should consider first whether the material should be steel, cast iron, or bronze. Then consult the related tables of data.

For cast iron or bronze, refer to Table 9–4. For steel materials, the following review should aid your selection:

1. Start by checking Figure 9–11 to see whether a through-hardened steel will give the needed s_{ac}. If so, determine the required hardness. Then specify a steel material and its heat treatment by referring to Appendices 3 and 4.
2. If a higher s_{ac} is needed, see Table 9–3 for properties of case-hardened steels.
3. Appendix 5 will aid in the selection of carburized steels.
4. If flame or induction hardening is planned, specify a material with a good hardenability, such as AISI 4140 or 4340 or similar medium-carbon-alloy steels. See Appendix 4.
5. Refer to Figure 9–12 or 9–13 for recommended case depths for surface-hardened steels.
6. If the specified pinion material has a significantly harder surface than the gear, refer to Figures 9–25 or 9–26 to determine a value for the hardness ratio factor C_H. Use Equation 9–27 to recompute the required s_{ac} and adjust the material selection if the data warrant a change.

Example Problem 9–4 Specify suitable materials for the pinion and the gear from Example Problem 9–3 based on contact stress. Application conditions are described in Example Problems 9–1 and 9–2.

Solution In Example Problem 9–3, we found that the expected contact stress number is $s_c = 156\,000$ psi. This should be modified as indicated in Equation (9–27).

In Example Problem 9–2, we determined that a carburized, case-hardened steel would be used for the pinion, and a through-hardened steel should be used for the gear. We must complete the material selection for pitting resistance independently from the bending stress analysis. However, we could not specify a material with lower properties than those specified in Example Problem 9–2 because then the bending strength would be inadequate.

In Example Problem 9–2, we used $K_R = 1.50$ for the desired reliability of fewer than one failure in 10 000. We decided to use $S_F = 1.00$ because we anticipated no unusual factors in the application that have not already been taken into account by other factors.

We can find Z_N from Figure 9–24 using 2.10×10^9 cycles of loading on the pinion and 6.00×10^8 cycles for the gear as computed in Example Problem 9–2. We find, then, that

$$Z_{NP} = 0.88 \quad Z_{NG} = 0.91$$

Let's complete the analysis for the pinion first. The hardness ratio factor does not apply to the pinion. Then Equation (9–27) gives

$$\frac{K_R(SF)}{Z_N} s_c = \frac{(1.50)(1.00)}{(0.88)} (156\,000 \text{ psi}) = 265\,900 \text{ psi} < s_{ac}$$

This value is quite high. Referring to Table 9–3, note that the only suitable listed material is a Grade 3 carburized and case-hardened steel having an allowable contact stress number of 275 ksi. Let's complete the calculations for the gear material and then discuss the results.

Assume an initial value for the hardness ratio factor $C_H = 1.00$. Then

Equation (9–27) gives

$$\frac{K_R(SF)}{Z_N C_H} s_c = \frac{(1.50)(1.00)}{(0.91)(1.00)} (156\,000 \text{ psi}) = 257\,100 \text{ psi} < s_{ac}$$

This value is also quite high, requiring the same Grade 3 carburized and case-hardened steel to provide adequate pitting resistance.

Comments and Design Decisions Specifying the Grade 3 carburized and case-hardened steel would be expected to provide adequate strength and pitting resistance for this pair of gears. However, the design is marginal, and it would be expensive because of the special requirements of cleanliness for the material and other guarantees related to material composition and microstructure. Most designs are executed using Grade 1 steel. It is recommended that the gears be redesigned to produce a lower bending stress and contact stress. In general, that can be achieved by using larger teeth (a smaller value for diametral pitch, P_d), a larger diameter for each gear, and a larger face width. Greater precision in the manufacture of the gears, producing a higher quality number, Q_v, would lower the dynamic factor and therefore reduce the bending stress number and the contact stress number. See section 9–15.

The next section outlines a design methodology for gears and provides guidelines for iterating on a design to produce alternatives from which we can select an optimum design for given conditions. In Section 9–14, a spreadsheet is developed that is useful for producing multiple iterations.

In Section 9–15, we use the same design requirements for the gear pair from Example Problems 9–1 through 9–4 as an example problem, making the necessary adjustments to our design decisions to ensure a satisfactory, economical gear design.

Then, in Chapter 15, this same design is used as the basis for a comprehensive discussion of the completion of the design of a power transmission. Considered there are the final selection of the gear design parameters, the design of the shafts for both the pinion and the gear (Chapter 12), the selection of two rolling contact bearings for each shaft (Chapter 14), and the enclosure of the transmission components in a suitable housing. Consideration is given for the inclusion of belt or chain drives for the input or output shafts of the gear reducer to provide more flexibility of its use. Chapter 15, then, is the culmination of virtually all of the design procedures presented in Part II of the book from Chapter 7 through Chapter 14.

9–12 DESIGN OF SPUR GEARS

In designs involving gear drives, normally the required speeds of rotation of the pinion and the gear and the amount of power that the drive must transmit are known. These factors are determined from the application. Also, the environment and operating conditions to which the drive will be subjected must be understood. It is particularly important to know the type of driving device and the driven machine, in order to judge the proper value for the overload factor.

The designer must decide the type of gears to use; the arrangement of the gears on their shafts; the materials of the gears, including their heat treatment; and the geometry of the gears: numbers of teeth, diametral pitch, pitch diameters, tooth form, face width, and quality numbers.

This section presents a design procedure that accounts for the bending fatigue strength of the gear teeth and the pitting resistance, called *surface durability*. This procedure makes extensive use of the design equations presented in the preceding sections of the chapter and of the tables of material properties in Appendices 3 through 5, 8, and 12.

You should understand that there is no one best solution to a gear design problem; several good designs are possible. Your judgment and creativity and the specific requirements of the application will greatly affect the final design selected. The purpose here is to provide a means of approaching the problem to create a reasonable design.

Design Objectives

Some overall objectives of a design are listed below. The resulting drive should

> Be compact and small
>
> Operate smoothly and quietly
>
> Have long life
>
> Be low in cost
>
> Be easy to manufacture
>
> Be compatible with the other elements in the machine, such as bearings, shafts, the housing, the driver, and the driven machine

The major objective of the design procedure is to define a safe, long-lasting gear drive. General steps and guidelines are outlined here to produce a reasonable initial design. However, because of the numerous variables involved, several iterations are typically made to work toward an optimum design. Details of the procedure are presented in Example Problem 9–5.

Procedure for Designing a Safe and Long-Lasting Gear Drive

1. From the design requirements, identify the input speed of the pinion, n_P, the desired output speed of the gear, n_G, and the power to be transmitted, P.

2. Choose the type of material for the gears, such as steel, cast iron, or bronze.

3. Considering the type of driver and the driven machine, specify the overload factor, K_o, using Table 9–5. The primary concern is the expected level of shock or impact loading.

4. Specify a trial value for the diametral pitch. When steel gears are used, Figure 9–27 provides initial guidance. The graph of design power transmitted versus the pinion rotational speed was derived for selected pitches and pinion diameters. Design power, $P_{des} = K_o P$. Steel that is through hardened to HB 300 is used. Because of the numerous variables involved, the value of P_d read from the figure is only an initial target value. Subsequent iterations may require considering a different value.

5. Specify the face width within the following recommended range for general machine drive gears:

$$8/P_d < F < 16/P_d$$

Nominal Face Width ⇨

$$\text{Nominal value of } F = 12/P_d \qquad (9\text{–}28)$$

The upper limit given tends to minimize alignment problems and ensure reasonably uniform loading across the face. When the face width is less than the lower limit, it is probable that a more compact design can be achieved with a different pitch. Also, the face width normally is less than twice the pitch diameter of the pinion.

6. Compute or specify the transmitted load, pitch line speed, quality number, geometry factor, and other factors required in the equations for bending stress and contact stress.

7. Compute the bending stress and the contact stress on the pinion and gear teeth. Judge whether the stresses are reasonable (nether too low nor too high) in terms of being able to specify a suitable material. If not, select a new pitch or revise the number of teeth, pitch diameter, or face width. Typically the contact stress on the pinion is the limiting value for gears designed for a long life.

8. Iterate the design process to seek more optimum designs. It is not unusual to make several trials before settling on a particular design. Using computer aids such as the spreadsheets described in Section 9–12 can make successive trials quickly.

Guidelines for Adjustments in Successive Iterations

The following relationships should help you determine what changes in your design assumptions you should make after the first set of calculations to achieve a more optimum design:

- Decreasing the numerical value of the diametral pitch results in larger teeth and generally lower stresses. Also, the lower value of the pitch usually means a larger face width, which decreases stress and increases surface durability.

- Increasing the diameter of the pinion decreases the transmitted load, generally lowers the stresses, and improves the surface durability.

- Increasing the face width lowers the stress and improves the surface durability, but to a generally lesser extent than either the pitch or the pitch diameter changes discussed previously.

- Gears with more and smaller teeth tend to run more smoothly and quietly than gears with fewer and larger teeth.

- Standard values of diametral pitch should be used for ease of manufacture and lower cost (see Table 8–2).

- Using high-alloy steels with high surface hardness results in the most compact system, but the cost in higher.

- Using very accurate gears (with ground or shaved teeth) results in a higher quality number, lower dynamic loads, and consequently lower stresses and improved surface durability, but the cost is higher.

- The number of teeth in the pinion should generally be as small as possible to make the system compact. But the possibility of interference is greater with fewer teeth. Check Table 8–6 to ensure that no interference will occur. (See Reference 22.)

FIGURE 9–27

Design power transmitted vs. pinion speed for spur gears with different pitches and diameters

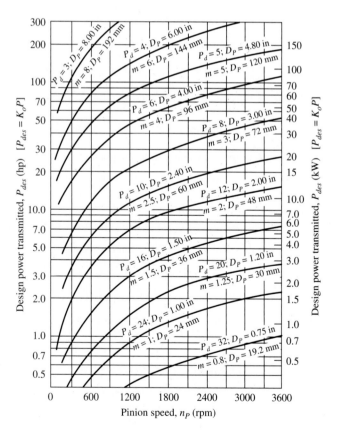

For all curves: 20° full depth teeth;
$N_P = 24$; $N_G = 96$; $m_G = 4.00$; $F = 12/P_d$; $Q_v = 6$
Steel gears, HB 300; $s_{at} = 36\,000$ psi; $s_{ac} = 126\,000$ psi

Example Problem 9–5

Design a pair of spur gears to be used as a part of the drive for a chipper to prepare pulp-wood for use in a paper mill. Intermittent use is expected. An electric motor transmits 3.0 horsepower to the pinion at 1750 rpm and the gear must rotate between 460 and 465 rpm. A compact design is desired.

Solution and General Design Procedure

Step 1. Considering the transmitted power, P, the pinion speed, n_P, and the application, refer to Figure 9–27 to determine a trial value for the diametral pitch, P_d. The overload factor, K_o, can be determined from Table 9–5, considering both the power source and the driven machine.

For this problem, $P = 3.0$ hp and $n_P = 1750$ rpm, $K_o = 1.75$ (uniform driver, heavy shock driven machine). Then $P_{des} = (1.75) (3.0 \text{ hp}) = 5.25$ hp. Try $P_d = 12$ for the initial design.

Step 2. Specify the number of teeth in the pinion. For small size, use 17 to 20 teeth as a start.

For this problem, let's specify $N_P = 18$.

Step 3. Compute the nominal velocity ratio from $VR = n_P/n_G$.

For this problem, use $n_G = 462.5$ rpm at the middle of the acceptable range.

$$VR = n_P/n_G = 1750/462.5 = 3.78$$

Step 4. Compute the approximate number of teeth in the gear from $N_G = N_P(VR)$.

For this problem, $N_G = N_P(VR) = 18(3.78) = 68.04$. Specify $N_G = 68$.

Step 5. Compute the actual velocity ratio from $VR = N_G/N_P$

For this problem, $VR = N_G/N_P = 68/18 = 3.778$.

Step 6. Compute the actual output speed from $n_G = n_P (N_P/N_G)$.

For this problem, $n_G = n_P (N_P/N_G) = (1750 \text{ rpm})(18/68) = 463.2$ rpm. OK.

Step 7. Compute the pitch diameters, center distance, pitch line speed, and transmitted load and judge the general acceptability of the results.

For this problem, the pitch diameters are:

$$D_P = N_P/P_d = 18/12 = 1.500 \text{ in}$$
$$D_G = N_G/P_d = 68/12 = 5.667 \text{ in}$$

Center distance:

$$C = (N_P + N_G)/(2P_d) = (18 + 68)/(24) = 3.583 \text{ in}$$

Pitch line speed: $v_t = \pi D_P n_P/12 = [\pi(1.500)(1.750)]/12 = 687$ ft/min

Transmitted load: $W_t = 33\,000(P)/v_t = 33\,000(3.0)/687 = 144$ lb

These values seem to be acceptable.

Step 8. Specify the face width of the pinion and the gear using Equation (9–28) as a guide.

For this problem: Lower limit $= 8/P_d = 8/12 = 0.667$ in.

Upper limit $= 16/P_d = 16/12 = 1.333$ in

Nominal value $= 12/P_d = 12/12 = 1.00$ in. Use this value.

Step 9. Specify the type of material for the gears and determine C_p from Table 9–9.

For this problem, specify two steel gears. $C_p = 2300$.

Step 10. Specify the quality number, Q_v, using Table 9–2 as a guide. Determine the dynamic factor from Figure 9–21.

For this problem, specify $Q_v = 6$ for a wood chipper. $K_v = 1.35$.

Step 11. Specify the tooth form, the bending geometry factors for the pinion and the gear from Figure 9–17 and the pitting geometry factor from Figure 9–23.

For this problem, specify $20°$ full depth teeth. $J_P = 0.325$, $J_G = 0.410$, $I = 0.104$.

Step 12. Determine the load distribution factor, K_m, from Equation (9–16) and Figures 9–18 and 9–19. The precision class of the gear system design must be specified. Values may be computed from equations in the figures or read from the graphs.

For this problem: $F = 1.00$ in, $D_P = 1.500$. $F/D_P = 0.667$. Then $C_{pf} = 0.042$.

Specify open gearing for the wood chipper, mounted to the frame. $C_{ma} = 0.264$.

Compute: $K_m = 1.0 + C_{pf} + C_{ma} + 0.042 + 0.264 = 1.31$

Step 13. Specify the size factor, K_s from Table 9–6.

For this problem, $K_s = 1.00$ for $P_d = 12$

Step 14. Specify the rim thickness factor, K_B, from Figure 9–20.

For this problem, specify a solid gear blank. $K_B = 1.00$.

Step 15. Specify a service factor, SF, typically from 1.00 to 1.50, based on uncertainty of data.

For this problem, there is no unusual uncertainty. Let $SF = 1.00$.

Step 16. Specify a hardness ratio factor, C_H, for the gear, if any. Use $C_H = 1.00$ for the early trials until materials have been specified. Then adjust C_H if significant differences exist in the hardness of the pinion and the gear.

Step 17. Specify a reliability factor using Table 9–8 as a guideline.

For this problem, specify a reliability of 0.99. $K_R = 1.00$.

Step 18. Specify a design life. Compute the number of loading cycles for the pinion and the gear. Determine the stress cycle factors for bending (Y_N) and pitting (Z_N) for the pinion and the gear.

For this problem, intermittent use is expected. Specify the design life to be 3000 hours, similar to agricultural machinery. The numbers of loading cycles are:

$$N_{cP} = (60)(3000 \text{ hr})(1750 \text{ rpm})(1) = 3.15 \times 10^8 \text{ cycles}$$
$$N_{cG} = (60)(3000 \text{ hr})(462.5 \text{ rpm})(1) = 8.33 \times 10^7 \text{ cycles}$$

Then, from Figure 9–22, $Y_{NP} = 0.96$, $Y_{NG} = 0.98$. From Figure 9–24, $Z_{NP} = 0.92$, $Z_{NG} = 0.95$.

Step 19. Compute the expected bending stresses in the pinion and the gear using Equation (9–15).

$$s_{tP} = \frac{W_t P_d}{F J_P} K_o K_s K_m K_B K_v = \frac{(144)(12)}{(1.00)(0.325)}(1.75)(1.0)(1.31)(1.0)(1.35) = 16\,400 \text{ psi}$$

$$s_{tG} = s_{tP}(J_P/J_G) = (16\,400)(0.325/0.410) = 13\,000 \text{ psi}$$

Step 20. Adjust the bending stresses using Equation 9–20.

For this problem, for the pinion:

$$S_{atP} > S_{tP}\frac{K_R(SF)}{Y_{NP}} = (16\,400)\frac{(1.00)(1.00)}{0.96} = 17\,100 \text{ psi}$$

For the gear:

$$S_{atG} > S_{tG}\frac{K_R(SF)}{Y_{NG}} = (13\,000)\frac{(1.00)(1.00)}{0.98} = 13\,300 \text{ psi}$$

Step 21. Compute the expected contact stress in the pinion and the gear from Equation (9–25). Note that this value will be the same for both the pinion and the gear.

$$s_c = C_p\sqrt{\frac{W_t K_o K_s K_m K_v}{F D_P I}} = 2300\sqrt{\frac{(144)(1.75)(1.0)(1.31)(1.35)}{(1.00)(1.50)(0.104)}} = 122\,900 \text{ psi}$$

Step 22. Adjust the contact stresses for the pinion and the gear using Equation (9–27).

$$S_{acP} > S_{cP}\frac{K_R(SF)}{Z_{NP}} = (122\,900)\frac{(1.00)(1.00)}{(0.92)} = 133\,500 \text{ psi}$$

For the gear:

$$S_{acG} > S_{cG}\frac{K_R(SF)}{Z_{NG}C_H} = (122\,900)\frac{(1.00)(1.00)}{(0.95)(1.00)} = 129\,300 \text{ psi}$$

Step 23. Specify materials for the pinion and the gear that will have suitable through hardening or case hardening to provide allowable bending and contact stresses greater than those required from Steps 20 and 22. Typically the contact stress is the controlling factor. Refer to Figures 9–10 and 9–11 and Tables 9–3 and 9–4 for data on required hardness. Refer to Appendices 3 to 5 for properties of steel to specify a particular alloy and heat treatment.

For this problem contact stress is the controlling factor. Figure 9–11 shows that through hardening of steel with a hardness of HB 320 is required for both the pinion and the gear. From Figure A4–4, we can specify AISI 4140 OQT 1000 steel that has a hardness of HB 341, giving a value of s_{ac} = 140 000 psi. Ductility is adequate as indicated by the 18% elongation value. Other materials could be specified.

**9–13
GEAR DESIGN
FOR THE METRIC
MODULE SYSTEM**

Section 8-4, "Gear Nomenclature and Gear-Tooth Features," described the metric module system of gearing and its relation to the diametral pitch system. As the design process was being developed in Sections 9–6 through 9–11, the data for stress analysis and surface durability analysis were taken from charts using U.S. Customary units (in, lb, hp, ft/min, and ksi). Data for the metric module system were also available in the charts in units of millimeters (mm), newtons (N), kilowatts (kW), meters per second (m/s), and megapascals (MPa). But to use the SI data, we must modify some of the formulas.

The following example problem uses SI units. The procedure will be virtually the same as that used to design with U.S. Customary units. Those formulas that are converted to SI units are identified.

Example Problem 9–6

A gear pair is to be designed to transmit 15.0 kilowatts (kW) of power to a large meat grinder in a commercial meat processing plant. The pinion is attached to the shaft of an electric motor rotating at 575 rpm. The gear must operate at 270 to 280 rpm. The gear unit will be enclosed and of commercial quality. Commercially hobbed (quality number 5), 20°, full-depth, involute gears are to be used in the metric module system. The maximum center distance is to be 200 mm. Specify the design of the gears. Use $K_R = C_H = SF = Z_N = 1.00$.

Solution

The nominal velocity ratio is

$$VR = 575/275 = 2.09$$

Specify an overload factor of $K_o = 1.50$ from Table 9–5 for a uniform power source and moderate shock for the meat grinder. Then computer design power,

$$P_{des} = K_o P = (1.50)(15 \text{ kW}) = 22.5 \text{ kW}$$

From Figure 9–27, $m = 4$ is a reasonable trial module. Then

$$N_P = 18 \quad \text{(design decision)}$$
$$D_P = N_P m = (18)(4) = 72 \text{ mm}$$
$$N_G = N_p(VR) = (18)(2.09) = 37.6 \quad \text{(Use 38.)}$$
$$D_G = N_G m = (38)(4) = 152 \text{ mm}$$
$$\text{Final output speed} = n_G = n_P(N_P/N_G)$$
$$n_G = 575 \text{ rpm} \times (18/38) = 272 \text{ rpm} \quad \text{(okay)}$$
$$\text{Center distance} = C = (N_P + N_G)m/2 \quad [\text{Equation (8–18)}]$$
$$C = (18 + 38)(4)/2 = 112 \text{ mm} \quad \text{(okay)}$$

In SI units, the pitch line speed in meters per second (m/s) is

$$v_t = \pi D_P n_P/(60\ 000)$$

where D_P is in mm and n_P is in revolutions per minute (rpm). Then

$$v_t = [(\pi)(72)(575)]/(60\ 000) = 2.17 \text{ m/s}$$

In SI units, the transmitted load, W_t, is in newtons (N). If the power, P, is in kW, and v_t is in m/s,

$$W_t = 1000(P)/v_t = (1000)(15)/(2.17) = 6920 \text{ N}$$

In the U.S. Customary Unit System, it was recommended that the face width be approximately $F = 12/P_d$ in. The equivalent SI value is $F = 12(m)$ mm. For this problem, $F = 12(4) = 48$ mm. Let's use $F = 50$ mm.

Other factors are found as before.

$$K_s = K_B = 1.00$$
$$K_v = 1.34 \quad \text{(Figure 9–21)}$$
$$K_m = 1.21 \quad \text{(Figures 9–19 and 9–19)} \quad (F/D_P = 50/72 = 0.69)$$
$$J_P = 0.315 \quad J_G = 0.380 \quad \left[\text{Figure 9–17(a)}\right]$$

Then the stress in the pinion is found from Equation (9–15), modified by letting $P_d = 1/m$:

$$S_{tP} = \frac{W_t K_o K_s K_B K_m K_v}{Fm J_P} = \frac{(6920)(1.50)(1)(1.21)(1.34)}{(50)(4)(0.315)} = 269 \text{ MPa}$$

This is a reasonable stress level. The required hardness of grade 1 material is HB 360, as found in Figure 9–10. Proceed with design for pitting resistance.

For two steel gears,

$$K_s = 1.0$$
$$C_p = 191 \quad \text{(Table 9–10)}$$
$$I = 0.092 \quad \text{(Figure 9–23)}$$
$$K_v = 1.34$$
$$K_o = 1.50$$
$$K_m = 1.21$$

The contact stress [Equation (9–25)] gives

$$s_c = C_p\sqrt{\frac{W_t K_o K_s K_m K_v}{FD_P I}} = 191\sqrt{\frac{(6920)(1.50)(1.0)(1.21)(1.34)}{(50)(72)(0.092)}} = 1367 \text{ MPa}$$

Converting to ksi gives

$$s_c = 1367 \text{ MPa} \times 1 \text{ ksi}/6.895 \text{ MPa} = 198 \text{ ksi}$$

From Table 9–3, the required surface hardness is HRC 58-64, case-carburized, Grade 2. Material selection from Appendix 5, for carburized steels, is as follows:

AISI 4320 SOQT 300; $s_u = 1500$ MPa; 13% elongation, Grade 2

Case-harden by carburizing to HRC 58 minimum

Case depth: 0.6 mm minimum (Figure 9–12)

Comment: Redesigning the gears to permit using Grade 1 material is recommended.

**9–14
COMPUTER-
AIDED SPUR
GEAR DESIGN
AND ANALYSIS**

This section presents one approach to assisting the gear designer with the many calculations and judgments that must be made to produce an acceptable design. The spreadsheet shown in Figure 9–28 facilitates the completion of a prospective design for a pair of gears in a few minutes by an experienced designer. You must have studied all of the material in Chapters 8 and 9 in order to understand the data needed in the spreadsheet and to use it effectively.

The recommended use of the spreadsheet is to create a series of design iterations that allow you to progress toward an optimum design in a short amount of time. It follows the process outlined in Section 9–12 up to the point of computing the required allowable bending stress number and the allowable contact stress number for both the pinion and the gear. The designer must use those data to specify suitable materials for the gears and their heat treatments.

Given below is a discussion of the essential features of the spreadsheet. In general, it first calls for the input of basic performance data, allowing a proposed geometry to be specified. The final result is the completion of the stress analyses for bending and pitting resistance for both the pinion and the gear. Equations (9–19) and (9–20) are combined for the bending analysis. The analysis of pitting resistance uses Equations (9–25), and (9–27). The designer must provide data for the several factors in those equations taken from appropriate figures and charts or based on design decisions. Virtually all computations are performed by the spreadsheet, allowing the designer to exercise judgment based on the intermediate results.

The format used for the spreadsheet helps the designer follow the process. After defining the problem at the top of the sheet, the first column at the left calls for several pieces of input data. Any value in italics within a gray-shaded area must be entered by the designer. White areas offer the results of calculations and provide guidance. The upper part of the second column also guides the designer in determining values for the several factors needed to complete the stress analyses for bending and pitting resistance. The area at the lower right of the spreadsheet gives the primary output data on which the design decisions for materials and heat treatments are based.

The data in Figure 9–28 are taken from Example Problem 9–5 which was completed in the traditional manner in Section 9–12.

Discussion of the Use of the Spur Gear Design Spreadsheet

1. *Describing the application:* In the heading of the sheet, the designer is asked to describe the application for identification purposes and to focus on the basic uses for the gears. Of particular interest is the nature of the prime mover and the driven machine.

2. *Initial input data:* It is assumed that designers begin with a knowledge of the power transmission requirement, the rotational speed of the pinion of the gear pair, and the desired output speed. Use Figure 9–27 to determine a trial value of the diametral pitch based on the design power to be transmitted and the rotational speed of the pinion. The number of teeth in the pinion is a critical design decision because the size of the system depends on this value. Ensure against interference.

3. *Number of gear teeth:* The spreadsheet computes the approximate number of gear teeth to produce the desired output speed from $N_G = N_P(n_G/n_P)$. But, of course, the number of teeth in any gear must be an integer, and the actual value of N_G is entered by the designer.

DESIGN SPUR OF GEARS

Initial Input Data:

Input power:	$P =$	3	hp
Input speed:	$n_P =$	1,750.	rpm
Diametral pitch:	$P_d =$	12	
Number of pinion teeth:	$N_P =$	18	
Desired output speed:	$n_G =$	462.5	rpm

Computed number of gear teeth: 68.1

Enter: Chosen no. of gear teeth: $N_G=$ 68

Computed Data:

Actual output speed:	$n_G =$	463.2	rpm
Gear ratio:	$m_G =$	3.78	
Pitch diameter, pinion:	$D_P =$	1.500	in
Pitch diameter, gear:	$D_G =$	5.667	in
Center distance:	$C =$	3.583	in
Pitch line speed:	$v_t =$	687	ft/min
Transmitted load:	$W_t =$	144	lb

Secondary Input Data:

	Min.	Nom.	Max.
Face width guidelines (in):	0.667	1.000	1.333

Enter: Face width: $F = 1.000$ in

Ratio: Face width/pinion diameter: $F/D_P = 0.67$
Recommended range of ratio: $F/D_P < 2.00$

Enter: Elastic coefficient: $C_p = 2300$ Table 9–9

Enter: Quality number: $Q_v = 6$ Table 9–2

Enter: Bending geometry factors:
Pinion: $J_P = 0.325$ Fig. 9–17
Gear: $J_G = 0.410$ Fig. 9–17
Enter: Pitting geometry factor: $I = 0.104$ Fig. 9–23

FIGURE 9–28   Spreadsheet solution for Example Problem 9–5

Application:	Wood chipper driven by an electric motor Example Problem 9–5			

<table>
<tr><td colspan="5" align="center">Factors in Design Analysis:</td></tr>
<tr><td>Alignment factor, $K_m = 1.0 + C_{pf} + C_{ma}$</td><td>If $F < 1.0$</td><td>If $F > 1.0$</td><td colspan="2">$F/D_P = 0.67$</td></tr>
<tr><td>Pinion proportion factor, $C_{pf} =$</td><td>0.042</td><td>0.042</td><td colspan="2">$[0.50 < F/D_P < 2.00]$</td></tr>
<tr><td align="right">Enter: $C_{pf} =$</td><td>0.042</td><td colspan="3">Fig. 9–18</td></tr>
<tr><td align="right">Type of gearing:
Mesh alignment factor, $C_{ma} =$</td><td>Open
0.264</td><td>Commer.
0.143</td><td>Precision
0.080</td><td>Ex. Prec.
0.048</td></tr>
<tr><td align="right">Enter: $C_{ma} =$</td><td>0.264</td><td colspan="3">Fig. 9–19</td></tr>
<tr><td align="right">Alignment factor: $K_m =$</td><td>1.31</td><td colspan="3">[Computed]</td></tr>
<tr><td align="right">Overload factor: $K_o =$
Size factor: $K_s =$
Pinion rim thickness factor: $K_{BP} =$
Gear rim thickness factor: $K_{BG} =$</td><td>1.75
1.00
1.00
1.00</td><td colspan="3">Table 9–5
Table 9–6: Use 1.00 if $P_d \geq 5$.
Fig. 9–20: Use 1.00 if solid blank.
Fig. 9–20: Use 1.00 if solid blank.</td></tr>
<tr><td align="right">Dynamic factor: $K_v =$</td><td>1.35</td><td colspan="3">[Computed: See Fig. 9–21.]</td></tr>
<tr><td align="right">Service factor: $SF =$
Hardness ratio factor: $C_H =$
Reliability factor: $K_R =$</td><td>1.00
1.00
1.00</td><td colspan="3">Use 1.00 if no unusual conditions.
Fig. 9–25 or 9–26; gear only
Table 9–8: Use 1.00 for R = 0.99.</td></tr>
<tr><td align="right">Enter: Design life: =</td><td>3000 hours</td><td colspan="3">See Table 9–7.</td></tr>
</table>

Pinion—Number of load cycles: $N_P =$	3.2E+08		Guidelines: Y_N, Z_N	
Gear—Number of load cycles: $N_G =$	8.3E+07	10^7 cycles	$>10^7$	$<10^7$
Bending stress cycle factor: $Y_{NP} =$	0.96	1.00	0.96	Fig. 9–22
Bending stress cycle factor: $Y_{NG} =$	0.98	1.00	0.98	Fig. 9–22
Pitting stress cycle factor: $Z_{NP} =$	0.92	1.00	0.92	Fig. 9–24
Pitting stress cycle factor: $Z_{NG} =$	0.95	1.00	0.95	Fig. 9–24

Stress Analysis: Bending

Pinion: Required $s_{at} =$ 17,102 psi   See Fig. 9–10 or
Gear: Required $s_{at} =$ 13,280 psi   Table 9–3 or 9–4.

Stress Analysis: Pitting

Pinion: Required $s_{ac} =$ 133,471 psi   See Fig. 9–11 or
Gear: Required $s_{ac} =$ 129,256 psi   Table 9–3 or 9–4.

Specify materials, alloy and heat treatment, for most severe requirement.

One possible material specification:

Pinion: Requires HB > 320; AISI 4140 OQT 1000, HB = 341, $S_{ac} =$ 140,000 psi

Gear: Requires HB > 320; AISI 4140 OQT 1000, HB = 341, $S_{ac} =$ 140,000 psi

FIGURE 9–28 *Continued*

4. *Computed data:* The seven values reported in the middle of the first column are all determined from the input data, and they allow the designer to evaluate the suitability of the geometry of the proposed design at this point. Changes to the input data can be made at this time if any value is out of the desired range in the judgment of the designer.

5. *Secondary input data:* When a suitable geometry for the gears is obtained, the designer enters the data called for at the lower part of the first column of the spreadsheet. The locations of data in pertinent tables and figures are listed.

6. *Factors in design analysis:* The stress analysis requires many factors to account for the unique situation of the design being pursued. Again, guidance is offered, but the designer must enter the values of the required factors. Many of the factors can have a value of 1.00 for normal conditions or to produce a conservative result.

7. *Alignment factor:* The alignment factor depends on two other factors: the pinion proportion factor and the mesh alignment factor as shown in Figures 9–18 and 9–19. The suggested values in the white areas are computed from the equations given in the figures. Note the listed value of F/D_P. If $F/D_P < 0.50$, use $F/D_P = 0.50$ to find C_{pf}. The designer must decide on the type of gearing to be used (open or closed) and the degree of precision to be designed into the system. The final result is computed from the input data.

8. *Overload, size, and rim thickness factors:* Consult Tables 9–5 and 9–6 along with Figure 9–20. Note that the rim thickness factor can be different for the pinion and the gear. Sometimes the smaller pinion is made from a solid blank while the larger gear can use a rim-and-spoke design.

9. *Dynamic factor:* The spreadsheet uses the equations included in Figure 9–21 to compute the dynamic factor using the quality number and pitch line speed found from data in the first column.

10. *Service factor:* This is a design decision as discussed in Section 9–12. Often a value of 1.00 is used if no unusual conditions are expected that are not already accounted for in other factors. Larger service factors allow for a higher degree of safety or to account for uncertainties.

11. *Hardness ratio factor:* This value depends on the ratio of the hardness of the pinion to the hardness of the gear teeth as shown in Figures 9–25 and 9–26. At the start of the design, these data are not known, and it is suggested that a value of $C_H = 1.00$ be used initially. Then, after one or more iterations are completed with tentative specifications for the pinion and gear materials, the value can be adjusted to refine the design. The factor C_H is applied only in the pitting resistance stress analysis for the gear, and it may allow the use of a less expensive or more ductile material with a lower hardness.

12. *Reliability factor:* The designer must select a value from Table 9–8 according to the desired level of reliability.

13. *Stress cycle factors:* Here the designer must specify the design life in hours of operation for the gear pair being designed. Table 9–7 provides suggestions according to the use of the system. The number of cycles of stress is then computed for both the pinion and the gear, assuming the normal case of one cycle of one-direction stress per revolution. If the gears operate in a reversing mode, as idlers, or in planetary gear trains, this calculation must be adjusted to account for the multiple cycles of stress experienced in each revolution. Guidelines recommend factors of 1.00 for 10^7 cycles

for which the allowable stress numbers are computed. For a larger number of cycles, equations given in Figures 9–22 and 9–24 are used to compute the recommended factors. Because a variety of data are given for the case of fewer than 10^7 cycles, the designer is referred to the figures to determine the factors. In any case, the user of the spreadsheet must enter the selected values.

14. ***Stress analyses for bending and pitting resistance:*** Finally, the required allowable bending stress number and the required allowable contact stress number are computed using Equations (9–20) and (9–27), adjusted for the special values of factors for the pinion and the gear.

15. ***Specification of the materials and their heat treatment:*** The final step is left to the designer to use the computed values from the stress analyses and to specify materials that will provide an adequate strength and surface hardness of the gear teeth. Pertinent data are listed in Figures 9–10 and 9–11 and Tables 9–3 and 9–4. The Appendices tables for material properties may also be consulted once the required hardnesses of the materials are determined.

9–15
USE OF THE
SPUR GEAR
DESIGN
SPREADSHEET

The spreadsheet developed in Section 9–14 is a useful tool that aids the designer in the process of completing a design for a pair of gears to be safe with regard to bending stresses in the teeth of the gears and for pitting resistance. The use of the spreadsheet was demonstrated for the data in Example Problem 9–5 as shown in Figure 9–28.

A more important use for the spreadsheet is to propose and analyze several design alternatives and to work toward a goal of optimizing the design with regard to size, cost, or other parameters important to a particular design objective.

In this section, we use the data for Example Problems 9–1 through 9–4 to work toward an improved solution. The basic requirements were to produce a satisfactory design to transmit 25 hp with a pinion speed of 1750 rpm and a gear speed of 500 rpm. The initial trial started in Example Problem 9–1 and carried through the other problems called for gears with a diametral pitch of 8 and 20 teeth in the pinion and 70 teeth in the gear. A quality number of 6 was chosen. Although those sound like reasonable choices, it was shown that the resulting stresses were significantly higher than desirable, particularly the required allowable contact stress number. It was shown in Example Problem 9–4 that a Grade 3 steel, case-hardened by carburizing, was required. This is a very expensive design because of the extreme controls on the material composition and cleanliness and because of the time-consuming heat treatment process.

Designers for typical machine and vehicle drives would plan to use Grade 1 steels and standard quenching and tempering heat treatments. Where small size is critical or where cost is not a major concern, case hardening by carburizing, induction or flame hardening, or nitriding can be used. Therefore, it is usually desirable to produce several design alternatives that can be analyzed for cost and manufacturability. Then the final selection can be made with assurance that a reasonably optimum design has been identified.

The spreadsheet in Figure 9–29 shows the combined results of Example Problems 9–1 through 9–4. The resulting stresses are slightly different because of small differences in the factors computed by the spreadsheet formulas and those read from charts. This summary can lead you to alternatives that meet the design goal of using only Grade 1 steels and achieving a cost-effective design.

Successive Iterations. We now continue the design process by making carefully selected changes in design decisions using the *Guidelines for Adjustments in Successive Iterations* from Section 9–12, just before Example Problem 9–5. The goal in this case is to reduce the required contact stress number to permit the use of lower cost Grade 1 steels.

Figures 9–30, 9–31, and 9–32 show three additional trial designs for the system described in Example Problem 9–1. Each of these designs reaches the goal of producing a practical design that allows the use of Grade 1 steels. You should study these designs to ascertain the differences among the design decisions and the resulting required allowable stress numbers for both bending and pitting resistance. Figure 9–33 summarizes the major results from all of the trials.

DESIGN SPUR OF GEARS

Initial Input Data:

Input power:	$P =$ 25	hp
Input speed:	$n_P =$ 1750	rpm
Diametral pitch:	$P_d =$ 8	
Number of pinion teeth:	$N_P =$ 20	
Desired output speed:	$n_G =$ 500	rpm

Computed number of gear teeth: 70.0

Enter: Chosen no. of gear teeth: $N_G =$ 70

Computed Data:

Actual output speed:	$n_G =$ 500.0	rpm
Gear ratio:	$m_G =$ 3.50	
Pitch diameter, pinion:	$D_P =$ 2.500	in
Pitch diameter, gear:	$D_G =$ 8.750	in
Center distance:	$C =$ 5.625	in
Pitch line speed:	$v_t =$ 1145	ft/min
Transmitted load:	$W_t =$ 720	lb

Secondary Input Data:

	Min.	Nom.	Max.
Face width guidelines (in):	1.000	1.500	2.000

Enter: Face width: $F =$ 1.500 in

Ratio: Face width/pinion diameter:	$F/D_P =$ 0.60
Recommended range of ratio:	$F/D_P <$ 2.00

Enter: Elastic coefficient: $C_p =$ 2300   Table 9–9

Enter: Quality number: $Q_v =$ 6   Table 9–2

Enter: Bending geometry factors:
Pinion:	$J_P =$ 0.335	Fig. 9–17
Gear:	$J_G =$ 0.420	Fig. 9–17
Enter: Pitting geometry factor:	$I =$ 0.108	Fig. 9–23

FIGURE 9–29 Spreadsheet solution for Example Problem 9–1 through 9–4

Factors in Design Analysis:				
Application:	Industrial saw driven by an electric motor Example Problems 9–1 through 9–4			

Factors in Design Analysis:				
Alignment factor, $K_m = 1.0 + C_{pf} + C_{ma}$	If $F < 1.0$	If $F > 1.0$	$F/D_P = 0.60$	
Pinion proportion factor, $C_{pf} =$	0.035	0.041	[$0.50 < F/D_P < 2.00$]	
Enter: $C_{pf} =$	0.041		Fig. 9–18	
Type of gearing: Mesh alignment factor, $C_{ma} =$	Open 0.272	Commer. 0.150	Precision 0.086	Ex. Prec. 0.053
Enter: $C_{ma} =$	0.15		Fig. 9–19	
Alignment factor: $K_m =$	1.19		[Computed]	
Overload factor: $K_o =$	1.50		Table 9–5	
Size factor: $K_s =$	1.00		Table 9–6: Use 1.00 if $P_d \geq 5$.	
Pinion rim thickness factor: $K_{BP} =$	1.00		Fig. 9–20: Use 1.00 if solid blank.	
Gear rim thickness factor: $K_{BG} =$	1.00		Fig. 9–20: Use 1.00 if solid blank.	
Dynamic factor: $K_v =$	1.45		[Computed: See Fig. 9–21.]	
Service factor: $SF =$	1.00		Use 1.00 if no unusual conditions.	
Hardness ratio factor: $C_H =$	1.00		Fig. 9–25 or 9–26; gear only	
Reliability factor: $K_R =$	1.50		Table 9–8: Use 1.00 for R = 0.99.	
Enter: Design life: =	20,000 hours		See Table 9–7.	

Pinion—Number of load cycles: $N_P =$	2.1E+09	Guidelines: Y_N, Z_N			
Gear—Number of load cycles: $N_G =$	6.0E+08	10^7 cycles	$>10^7$	$<10^7$	
Bending stress cycle factor: $Y_{NP} =$	0.93	1.00	0.93	Fig. 9–22	
Bending stress cycle factor: $Y_{NG} =$	0.95	1.00	0.95	Fig. 9–22	
Pitting stress cycle factor: $Z_{NP} =$	0.88	1.00	0.88	Fig. 9–24	
Pitting stress cycle factor: $Z_{NG} =$	0.91	1.00	0.91	Fig. 9–24	

Stress Analysis: Bending

Pinion: Required $s_{at} =$ 47,871 psi   See Fig. 9–10 or
Gear: Required $s_{at} =$ 37,379 psi   Table 9–3 or 9–4.

Stress Analysis: Pitting

Pinion: Required $s_{ac} =$ 265,989 psi   See Fig. 9–11 or
Gear: Required $s_{ac} =$ 257,170 psi   Table 9–3 or 9–4.

Specify materials, alloy and heat treatment, for most severe requirement.

One possible material specification:

Pinion: Requires grade 3 carburized, case hardened steel

Gear: Requires grade 3 carburized, case hardened steel

FIGURE 9–29 *Continued*

Note that in all of the trials, the contact stress is critical. That is, the required hardness or case-hardening treatment for the gear teeth is most severe with regard to achieving adequate pitting resistance. The bending stress numbers are quite moderate. This situation is typical for steel gears, and designers often work first toward a satisfactory contact stress and then check to ensure that the bending stress is acceptable.

The trials shown in Figure 9–33 are arranged in order of increasing pinion diameter and increasing center distance, with the consequent decreasing of the bending and contact

---

DESIGN SPUR OF GEARS

Initial Input Data:

Input power:	$P =$	25 hp
Input speed:	$n_P =$	1750 rpm
Diametral pitch:	$P_d =$	6
Number of pinion teeth:	$N_P =$	18
Desired output speed:	$n_G =$	500 rpm

Computed number of gear teeth: 63.0

Enter: Chosen no. of gear teeth: $N_G =$ 63

Computed Data:

Actual output speed:	$n_G =$	500.0 rpm
Gear ratio:	$m_G =$	3.50
Pitch diameter, pinion:	$D_P =$	3.000 in
Pitch diameter, gear:	$D_G =$	10.500 in
Center distance:	$C =$	6.750 in
Pitch line speed:	$v_t =$	1374 ft/min
Transmitted load:	$W_t =$	600 lb

Secondary Input Data:

	Min.	Nom.	Max.
Face width guidelines (in):	1.333	2.000	2.667

Enter: Face width: $F = 2.750$ in

Ratio: Face width/pinion diameter:	$F/D_P =$	0.92
Recommended range of ratio:	$F/D_P <$	2.00

Enter: Elastic coefficient: $C_p = 2300$ Table 9–9

Enter: Quality number: $Q_v =$ 8   Table 9–2

Enter: Bending geometry factors:
Pinion: $J_P = 0.325$ Fig. 9–17
Gear: $J_G = 0.410$ Fig. 9–17
Enter: Pitting geometry factor: $I = 0.105$ Fig. 9–23

FIGURE 9–30 Redesign for data of Example Problem 9–1, Trial 1

Application:	Industrial saw driven by an electric motor Redesign for data of Example Problem 9–1, Trial 1			

Factors in Design Analysis:

Alignment factor, $K_m = 1.0 + C_{pf} + C_{ma}$	If $F < 1.0$	If $F > 1.0$	$F/D_P = 0.92$	
Pinion proportion factor, $C_{pf} =$	0.067	0.089	[$0.50 < F/D_P < 0.92$]	
Enter: $C_{pf} =$	0.089	Fig. 9–18		

Type of gearing: Mesh alignment factor, $C_{ma} =$	Open 0.292	Commer. 0.170	Precision 0.102	Ex. Prec. 0.065
Enter: $C_{ma} =$	0.17	Fig. 9–19		
Alignment factor: $K_m =$	1.26	[Computed]		

Overload factor: $K_o =$	1.50	Table 9–5
Size factor: $K_s =$	1.00	Table 9–6: Use 1.00 if $P_d \geq 5$.
Pinion rim thickness factor: $K_{BP} =$	1.00	Fig. 9–20: Use 1.00 if solid blank.
Gear rim thickness factor: $K_{BG} =$	1.00	Fig. 9–20: Use 1.00 if solid blank.
Dynamic factor: $K_v =$	1.30	[Computed: See Fig. 9–21.]
Service factor: $SF =$	1.00	Use 1.00 if no unusual conditions.
Hardness ratio factor: $C_H =$	1.00	Fig. 9–25 or 9–26; gear only
Reliability factor: $K_R =$	1.50	Table 9–8: Use 1.00 for R = 0.99.
Enter: Design life:	20 000 hours	See Table 9–7.

Pinion—Number of load cycles: $N_P =$	2.1E+09	Guidelines: Y_N, Z_N		
Gear—Number of load cycles: $N_G =$	6.0E+08	10^7 cycles	$>10^7$	$<10^7$
Bending stress cycle factor: $Y_{NP} =$	0.93	1.00	0.93	Fig. 9–22
Bending stress cycle factor: $Y_{NG} =$	0.95	1.00	0.95	Fig. 9–22
Pitting stress cycle factor: $Z_{NP} =$	0.88	1.00	0.88	Fig. 9–24
Pitting stress cycle factor: $Z_{NG} =$	0.91	1.00	0.91	Fig. 9–24

Stress Analysis: Bending

Pinion: Required $s_{at} =$ 16,009 psi	See Fig. 9–10 or
Gear: Required $s_{at} =$ 12,423 psi	Table 9–3 or 9–4.

Stress Analysis: Pitting

Pinion: Required $s_{ac} =$ 161,968 psi	See Fig. 9–11 or
Gear: Required $s_{ac} =$ 156,629 psi	Table 9–3 or 9–4.

Specify materials, alloy and heat treatment, for most severe requirement.

One possible material specification:

Pinion: AISI 4140 Induction Hardened to 50 HRC, Grade 1

Gear: AISI 4140 OQT 800, HB 429, Grade 1

FIGURE 9–30 *Continued*

stresses. These are the most critical factors. Diametral pitch, face width, and quality numbers are also varied, but their effects are secondary.

The designs represented by Figures 9–30, 9–31, and 9–32 could all use the same material, provided that it could be hardened to the required levels. For example, specifying AISI 4340 Grade 1 steel or some similar medium-carbon-alloy steel would ensure that it could be flame or induction-hardened to a minimum of 50 HRC or through-hardened by quenching and tempering to the required 350 to 400 HB hardness on the Brinell scale.

DESIGN SPUR OF GEARS

Initial Input Data:

Input power:	$P =$	25 hp
Input speed:	$n_P =$	1750 rpm
Diametral pitch:	$P_d =$	8
Number of pinion teeth:	$N_P =$	28
Desired output speed:	$n_G =$	500 rpm

Computed number of gear teeth:	98.0

Enter: Chosen no. of gear teeth:	$N_G =$	98

Computed Data:

Actual output speed:	$n_G =$	500.0 rpm
Gear ratio:	$m_G =$	3.50
Pitch diameter, pinion:	$D_P =$	3.500 in
Pitch diameter, gear:	$D_G =$	12.250 in
Center distance:	$C =$	7.875 in
Pitch line speed:	$v_t =$	1604 ft/min
Transmitted load:	$W_t =$	514 lb

Secondary Input Data:

	Min.	Nom.	Max.
Face width guidelines (in):	1.000	1.500	2.000

Enter: Face width:	$F =$	2.000 in

Ratio: Face width/pinion diameter:	$F/D_P =$	0.57
Recommended range of ratio:	$F/D_P <$	2.00

Enter: Elastic coefficient:	$C_p =$	2300	Table 9–9

Enter: Quality number:	$Q_v =$	8	Table 9–2

Enter: Bending geometry factors:			
Pinion:	$J_P =$	0.380	Fig. 9–17
Gear:	$J_G =$	0.440	Fig. 9–17
Enter: Pitting geometry factor:	$I =$	0.115	Fig. 9–23

FIGURE 9–31   Redesign for data for Example Problem 9–1, Trial 2

Application:	Industrial saw driven by an electric motor				
	Redesign for data of *Example Problem 9–1, Trial 2*				

Factors in Design Analysis:

Alignment factor, $K_m = 1.0 + C_{pf} + C_{ma}$	If $F < 1.0$	If $F > 1.0$	$F/D_P = 0.57$		
Pinion proportion factor, C_{pf} =	0.032	0.045	[0.50 < F/D_P < 2.00]		

Enter: C_{pf} =	*0.045*		Fig. 9–18		

Type of gearing: Mesh alignment factor, C_{ma} =	Open 0.280	Commer. 0.158	Precision 0.093	Ex. Prec. 0.058	

Enter: C_{ma} =	*0.158*		Fig. 9–19		

Alignment factor: K_m =	1.20		[Computed]		

Overload factor: K_o =	*1.50*	Table 9–5	
Size factor: K_s =	*1.00*	Table 9–6: Use 1.00 if $P_d \geq 5$.	
Pinion rim thickness factor: K_{BP} =	*1.00*	Fig. 9–20: Use 1.00 if solid blank.	
Gear rim thickness factor: K_{BG} =	*1.00*	Fig. 9–20: Use 1.00 if solid blank.	

Dynamic factor: K_v =	1.30	[Computed: See Fig. 9–21.]

Service factor: SF =	*1.00*	Use 1.00 if no unusual conditions.
Hardness ratio factor: C_H =	*1.00*	Fig. 9–25 or 9–26; gear only
Reliability factor: K_R =	*1.50*	Table 9–8: Use 1.00 for R = 0.99.

Enter: Design life: =	*20,000 hours*	See Table 9–7.

Pinion—Number of load cycles N_P =	2.1E+09	Guidelines: Y_N, Z_N		
Gear—Number of load cycles: N_G =	6.0E+08	10^7 cycles	$>10^7$	$<10^7$
Bending stress cycle factor: Y_{NP} =	*0.93*	1.00	0.93	Fig. 9–22
Bending stress cycle factor: Y_{NG} =	*0.95*	1.00	0.95	Fig. 9–22
Pitting stress cycle factor: Z_{NP} =	*0.88*	1.00	0.88	Fig. 9–24
Pitting stress cycle factor: Z_{NG} =	*0.91*	1.00	0.91	Fig. 9–24

Stress Analysis: Bending

Pinion: Required s_{at} = 20,915 psi	See Fig. 9–10 or
Gear: Required s_{at} = 17,682 psi	Table 9–3 or 9–4.

Stress Analysis: Pitting

Pinion: Required s_{ac} = 153,363 psi	See Fig. 9–11 or
Gear: Required s_{ac} = 148,307 psi	Table 9–3 or 9–4.

Specify materials, alloy and heat treatment, for most severe requirement.

One possible material specification:

Pinion: AISI 4140 OQT 800, HB 429, Grade 1

Gear: AISI 4140 OQT 900, HB 388, Grade 1

FIGURE 9–31 *Continued*

DESIGN SPUR OF GEARS

Initial Input Data:

Input power:	$P =$	25	hp
Input speed:	$n_P =$	1,750	rpm
Diametral pitch:	$P_d =$	6	
Number of pinion teeth:	$N_P =$	24	
Desired output speed:	$n_G =$	500	rpm

Computed number of gear teeth:	84.0

Enter: Chosen no. of gear teeth:	$N_G =$	84

Computed Data:

Actual output speed:	$n_G =$	500.0	rpm
Gear ratio:	$m_G =$	3.50	
Pitch diameter, pinion:	$D_P =$	4.000	in
Pitch diameter, gear:	$D_G =$	14.000	in
Center distance:	$C =$	9.000	in
Pitch line speed:	$v_t =$	1833	ft/min
Transmitted load:	$W_t =$	450	lb

Secondary Input Data:

	Min.	Nom.	Max.
Face width guidelines (in):	1.333	2.000	2.667

Enter: Face width:	$F =$	2.000	in

Ratio: Face width/pinion diameter:	$F/D_P =$	0.50
Recommended range of ratio:	$F/D_P <$	2.00

Enter: Elastic coefficient:	$C_p =$	2300	Table 9–9

Enter: Quality number:	$Q_v =$	6	Table 9–2

Enter: Bending geometry factors:			
Pinion:	$J_P =$	0.360	Fig. 9–17
Gear:	$J_G =$	0.430	Fig. 9–17
Enter: Pitting geometry factor:	$I =$	0.112	Fig. 9–23

FIGURE 9–32   Redesign for data for Example Problem 9–1, Trial 3

Application:	Industrial saw driven by an electric motor
	Redesign for data of Example Problem 9–1, Trial 3

Factors in Design Analysis:

Alignment factor, $K_m = 1.0 + C_{pf} + C_{ma}$	If $F < 1.0$	If $F > 1.0$	$F/D_P = 0.50$
Pinion proportion factor, C_{pf} =	0.025	0.038	$[0.50 < F/D_P < 2.00]$

Enter: C_{pf} =	*0.038*	Fig. 9–18		

Type of gearing:	Open	Commer.	Precision	Ex. Prec.
Mesh alignment factor, C_{ma} =	0.280	0.158	0.093	0.058

Enter: C_{ma} =	*0.158*	Fig. 9–19

Alignment factor: K_m =	1.20	[Computed]

Overload factor: K_o =	*1.50*	Table 9–5
Size factor: K_s =	1.00	Table 9–6: Use 1.00 if $P_d \geq 5$.
Pinion rim thickness factor: K_{BP} =	1.00	Fig. 9–20: Use 1.00 if solid blank.
Gear rim thickness factor: K_{BG} =	1.00	Fig. 9–20: Use 1.00 if solid blank.

Dynamic factor: K_v =	1.56	[Computed: See Fig. 9–21.]

Service factor: SF =	*1.00*	Use 1.00 if no unusual conditions.
Hardness ratio factor: C_H =	*1.00*	Fig. 9–25 or 9–26; gear only
Reliability factor: K_R =	*1.50*	Table 9–8: Use 1.00 for R=0.99.

Enter: Design life: =	*20,000 hours*	See Table 9–7.

Pinion—Number of load cycles: N_P =	2.1E+09	Guidelines: Y_N, Z_N		
Gear—Number of load cycles: N_G =	6.0E+08	10^7 cycles	$>10^7$	$<10^7$
Bending stress cycle factor: Y_{NP} =	*0.93*	1.00	0.93	Fig. 9–22
Bending stress cycle factor: Y_{NG} =	*0.95*	1.00	0.95	Fig. 9–22
Pitting stress cycle factor: Z_{NP} =	*0.88*	1.00	0.88	Fig. 9–24
Pitting stress cycle factor: Z_{NG} =	*0.91*	1.00	0.91	Fig. 9–24

Stress Analysis: Bending

Pinion: Required s_{at} = 16,961 psi See Fig. 9–10 or
Gear: Required s_{at} = 13,901 psi Table 9–3 or 9–4.

Stress Analysis: Pitting

Pinion: Required s_{ac} = 147,128 psi See Fig. 9–11 or
Gear: Required s_{ac} = 142,277 psi Table 9–3 or 9–4.

Specify materials, alloy and heat treatment, for most severe requirement.

One possible material specification:

Pinion: AISI 4140 OQT 800, HB 429, Grade 1

Gear: AISI 4140 OQT 900, HB 388, Grade 1

FIGURE 9–32 *Continued*

COMPARISON OF DESIGN ALTERNATIVES FOR DATA OF EXAMPLE PROBLEMS 9–1 THROUGH 9–4

P = 25 hp   Pinion speed = 1750 rpm   Gear speed = 500 rpm

	Problems 9–1 through 9–4	Alternative Trial 1	Alternative Trial 2	Alternative Trial 3	Comments:
Geometry, Quality, and Transmitted Load:					
N_P	20	18	28	24	All designs produce same gear ratio.
N_G	70	63	98	84	
P_d	8	6	8	6	Note change in diametral pitch.
D_P (in)	2.500	3.000	3.500	4.000	Increasing pinion diameter.
D_G (in)	8.750	10.500	12.250	14.000	Increasing gear diameter.
C (in)	5.625	6.750	7.875	9.000	Increasing center distance.
W_t (lb)	720	600	514	450	Decreasing transmitted load.
F (in)	1.50	2.75	2.00	2.00	Face width varies.
Q_v	6	8	8	6	Note change in quality number.
K_v	1.45	1.30	1.33	1.56	Note change in dynamic factor.
Stresses:					
s_{atP} (psi)	47,900	16,000	20,900	17,000	Moderate bending stress.
s_{atG} (psi)	37,400	12,450	17,700	13,900	Moderate bending stress.
s_{acP} (psi)	266,000	162,000	153,400	147,150	Contact stress critical.
s_{acG} (psi)	257,100	156,700	148,300	142,300	Contact stress critical.
Materials:					
Pinion	Grade 3 steel, carburized, case-hardened	Grade 1 steel, induction-hardened, 50 HRC	Grade 1 steel, through-hardened, 400 HB	Grade 1 steel, through-hardened, 400 HB	Design goal is to use only Grade 1 steels.
Gear	Grade 3 steel, carburized, case-hardened	Grade 1 steel, through-hardened, 400 HB	Grade 1 steel, through-hardened, 370 HB	Grade 1 steel, through-hardened, 350 HB	

FIGURE 9–33 Comparison of design alternative trials for data of Example Problem 9–1

9–16 POWER-TRANSMITTING CAPACITY

It is sometimes desirable to compute the amount of power that a gear pair can safely transmit after it has been completely defined. The *power-transmitting capacity* is the capacity when the tangential load causes the expected stress to equal the allowable stress number with all of the modifying factors considered. The capacity should be computed for both bending and pitting resistance and for both the pinion and the gear.

When similar materials are used for both the pinion and the gear, it is likely that the pinion will be critical for bending stress. But the most critical condition is usually pitting resistance. The following relationships can be used to compute the power-transmitting capacity. In this analysis, it is assumed that the operating temperature of the gears and their lubricants is 250°F and that gears are produced with the appropriate surface finish.

Bending

We start with Equation (9–19) in which the computed bending stress number is compared with the modified allowable bending stress number for the gear:

$$\frac{W_t P_d}{F J} K_o K_s K_m K_B K_v = s_t < s_{at} \frac{Y_N}{(SF) K_R}$$

But solving for W_t gives

$$W_t = \frac{s_{at} Y_N F J}{(SF) K_R K_o K_s K_m K_B K_v P_d} \tag{9–29}$$

It was shown in Equation (9–6) that

$$W_t = (126\ 000)(P)/(n_P D_P)$$

Then substituting into Equation (9–29) gives

$$\frac{(126\ 000)(P)}{n_P D_P} = \frac{s_{at} Y_N F J}{(SF) K_R K_o K_s K_m K_B K_v P_d}$$

Solving for P, we have

$$P = \frac{s_{at}\ Y_N\ F J\ n_P D_P}{(126\ 000)(P_d)(SF)\ K_R\ K_o\ K_s\ K_m\ K_B\ K_v} \tag{9–30}$$

This equation should be solved for both the pinion and the gear. Most variables will be the same except for s_{at}, Y_N, J, and possibly K_B.

Pitting Resistance

Here we start with Equations (9–25), (9–26), and (9–27) in which the computed contact stress number is compared with the modified allowable contact stress number for the gear. Equation (9–26) can be expressed in the form

$$s_c = C_p \sqrt{\frac{W_t K_o K_s K_m K_v}{D_P F I}} = \frac{s_{ac} Z_N C_H}{(SF) K_R}$$

Squaring both sides of this equation and solving for W_t gives

$$\frac{W_t K_o K_s K_m K_v}{D_P F I} = \left[\frac{s_{ac} Z_N C_H}{(SF) K_R C_P}\right]^2$$

$$W_t = \frac{D_P F I}{K_o K_s K_m K_v}\left[\frac{s_{ac} Z_N C_H}{(SF) K_R C_P}\right]^2$$

Now substituting this into Equation (9–6) and solving for the power P gives

$$P = \frac{W_t D_P n_P}{126\,000} = \frac{D_P n_P D_P FI}{126\,000\ K_o K_s K_m K_v} \left[\frac{s_{ac} Z_N C_H}{(SF) K_R C_p} \right]^2$$

$$P = \frac{n_P FI}{126\,000\ K_o K_s K_m K_v} \left[\frac{s_{ac} D_P Z_N C_H}{(SF)\ K_R C_P} \right]^2 \qquad\qquad (9\text{–}31)$$

Equations (9–30) and (9–31) should be used to compute the power-transmitting capacity for a pair of gears of known design with particular materials.

9–17 PRACTICAL CONSIDERATIONS FOR GEARS AND INTERFACES WITH OTHER ELEMENTS

It is important to consider the design of the entire gear system when designing the gears because they must work in harmony with the other elements in the system. This section will briefly discuss some of these practical considerations and will show commercially available speed reducers.

Our discussion so far has been concerned primarily with the gear teeth, including the tooth form, pitch, face width, material selection, and heat treatment. Also to be considered is the type of gear blank. Figures 8–2 and 8–4 show several styles of blanks. Smaller gears and lightly loaded gears are typically made in the plain style. Gears with pitch diameters of approximately 5.0 in through 8.0 in are frequently made with thinned webs between the rim and the hub for lightening, with some having holes bored in the webs for additional lightening. Larger gears, typically with pitch diameters greater than 8.0 in, are made from cast blanks with spokes between the rim and the hub.

In many precision special machines and gear systems produced in large quantities, the gears are machined integral with the shaft carrying the gears. This, of course, eliminates some of the problems associated with mounting and location of the gears, but it may complicate the machining operations.

In general machine design, gears are usually mounted on separate shafts, with the torque transmitted from the shaft to the gear through a key. This setup provides a positive means of transmitting the torque while permitting easy assembly and disassembly. The axial location of the gear must be provided by another means, such as a shoulder on the shaft, a retaining ring, or a spacer (see Chapters 11 and 12).

Other considerations include the forces exerted on the shaft and the bearings that are due to the action of the gears. These subjects are discussed in Section 9–3. The housing design must provide adequate support for the bearings and protection of the interior components. Normally, it must also provide a means of lubricating the gears.

See References 12–15 and 18–19 for additional practical considerations.

Lubrication

The action of spur gear teeth is a combination of rolling and sliding. Because of the relative motion, and because of the high local forces exerted at the gear faces, adequate lubrication is critical to smoothness of operation and gear life. A continuous supply of oil at the pitch line is desirable for most gears unless they are lightly loaded or operate only intermittently.

In splash-type lubrication, one of the gears in a pair dips into an oil supply sump and carries the oil to the pitch line. At higher speeds, the oil may be thrown onto the inside surfaces of the case; then it flows down, in a controlled fashion, onto the pitch line. Simultaneously, the oil can be directed to the bearings that support the shafts. One difficulty with the splash type of lubrication is that the oil is churned; at high gear speeds, excessive heat can be generated, and foaming can occur.

A positive oil circulation system is used for high-speed and high-capacity systems. A separate pump draws the oil from the sump and delivers it at a controlled rate to the meshing teeth.

The primary functions of gear lubricants are to reduce friction at the mesh and to keep operating temperatures at acceptable levels. It is essential that a continuous film of lubricant be maintained between the mating tooth surfaces of highly loaded gears and that there be a sufficient flow rate and total quantity of oil to maintain cool temperatures. Heat is generated by the meshing gear teeth, by the bearings, and by the churning of the oil. This heat must be dissipated from the oil to the case or to some other external heat-exchange device in order to keep the oil itself below 160°F (approximately 70°C). Above this temperature, the lubricating ability of the oil, as indicated by its viscosity, is severely decreased. Also, chemical changes can be produced in the oil, decreasing its lubricity. Because of the wide variety of lubricants available and the many different conditions under which they must operate, it is recommended that suppliers of lubricants be consulted for proper selection. (See also Reference 10.)

The AGMA, in Reference 10, defines several types of lubricants for use in gear drives.

- ■ *Rust and oxidation inhibited gear oils* (called R & O) are petroleum based with chemical additives.
- ■ *Compounded gear lubricants* (Comp) blend 3% to 10% of fatty oils with petroleum oils.
- ■ *Extreme pressure lubricants* (EP) include chemical additives that inhibit scuffing of gear tooth faces.
- ■ *Synthetic gear lubricants* (S) are special chemical formulations applied mostly in severe operating conditions.

R & O lubricants are supplied in 14 viscosity grades (0 to 13) where the lower numbers refer to the lower viscosities. Similar numbers are used for the other types with modified grade designations carrying suffixes *Comp*, *EP*, or *S*. The recommended lubricant grade depends on the ambient temperature around the drive and the pitch line velocity of the lowest speed pair of gears in a reducer. See the Table 9–10. Wormgear drives call for higher viscosity grades.

Commercially Available Gear-Type Speed Reducers

By studying the design of commercially available gear-type speed reducers, you should get a better feel for design details and the relationships among the component parts: the gears, the shafts, the bearings, the housing, the means of providing lubrication, and the coupling to the driving and driven machines.

Figure 9–34 shows a double-reduction spur gear speed reducer with an electric motor rigidly attached. Such a unit is often called a *gear motor*. Figure 9–35 is similar, except that one of the stages of reduction uses helical gears (discussed in the next chapter). The cross-sectional drawing shown with Figure 9–36 gives a clear picture of the several components of a reducer.

The planetary reducer in Figure 9–37 has quite a different design to accommodate the placement of the sun, planet, and ring gears. Figure 9–38 shows the eight-speed transmission from a large farm tractor and illustrates the high degree of complexity that may be involved in the design of transmissions.

TABLE 9–10   Recommended lubricant grade for spur, helical, herringbone, and bevel gear drives.

Pitch line velocity	Ambient temperature			
	−40°F to 14°F −40°C to −10°C	14°F to 50°F −10°C to 10°C	50°F to 95°F 10°C to 35°C	95°F to 131°F 35°C to 55°C
	Lubricant grade			
Less than 1000 ft/min Less than 5m/s	3S	4	6	8
1000 to 3000 ft/min 5 to 15 m/s	3S	3	5	7
3000 to 5000 ft/min 15 to 25 m/s	2S	2	4	6
Over 5000 ft/min Over 25 m/s	0S	0	2	3

Extracted from AGMA Standard 9005-D94, *Industrial Gear Lubrication,* with permission of the publisher, American Gear Manufactures Association, 1500 King Street, Suite 201, Alexandria, VA 22314.

FIGURE 9–34   Double-reduction spur gear reducer (Bison Gear & Engineering Corporation, Downers Grove, IL)

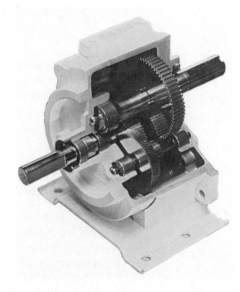

FIGURE 9–35   Double-reduction gear reducer. First stage, helical gears; second stage, spur gears, (Bison Gear & Engineering Corporation, Downers Grove, IL)

FIGURE 9–36

Concentric helical gear reducer (Peerless-Winsmith Subsidiary HBD Industries, Springville, NY)

(a) Cutaway of a concentric helical gear reducer   *(b)* Complete reducer

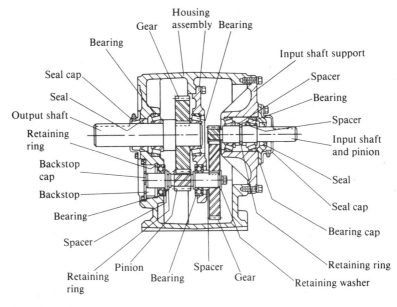

(c) **Parts index**

FIGURE 9–37   Planetary gear reducer (Rexnord, Milwaukee, WI)

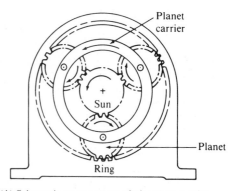

(b) Schematic arrangement of planetary gearing

FIGURE 9–38
Eight-speed tractor
transmission
(Case IH, Racine, WI)

**9–18
PLASTICS
GEARING**

Plastics are satisfying an important and growing part of the applications for gearing. Some of the numerous advantages of plastics in gearing systems compared with steels and other metals are:

- Lighter weight
- Lower inertia
- Possibility of running with little or no external lubrication
- Quieter operation
- Low sliding friction, which results in efficient gear meshing
- Chemical resistance and ability to operate in corrosive environments
- Ability to operate well under conditions of moderate vibration, shock, and impact
- Relatively low cost when made in large quantities
- Ability to combine several features into one part
- Accommodation of larger tolerances because of resiliency
- Material properties that can be tailored to meet the needs of the application
- Less wear among some plastics compared to metals in certain applications

The advantages must be weighed against disadvantages such as:

- Relatively lower strength of plastics as compared with metals
- Lower modulus of elasticity
- Higher coefficients of thermal expansion
- Difficulty operating at high temperatures
- Initial high cost for design, development, and mold manufacture
- Dimensional change with moisture absorption that varies with conditions
- Wide range of possible material formulations, which makes design more difficult

Some plastic gears are cut using hobbing or shaping processes similar to those used to cut metallic gears. However, most plastic gears are produced with the injection molding process because of its ability to make large quantities rapidly with low unit cost. Mold design is critical because it must accommodate the shrinking that occurs as the molten plastic solidifies. The typical successful approach accounts for predicted shrinkage by making the die larger than the required finished gear size. However, the allowance is not uniform throughout the gear, and significant amounts of data are required about the material molding properties and the molding process itself to produce plastic gears with high dimensional accuracy. Computer-assisted mold design software that simulates the flow of molten plastic through the mold cavities and the curing process is often used. The gear mold or the gear cutting tools are designed to produce dimensionally accurate gear teeth with tooth thickness controlled to produce a proper amount of backlash during operation. The electrical discharge machining process (EDM) is typically used to produce accurate gear tooth forms in molds made from high-hardness, wear-resistant steels to ensure that large production runs can be made without replacing tooling.

Plastic Materials for Gears

The great variety of plastics available makes material selection difficult, and it is recommended that gear system designers consult with material suppliers, mold designers, and manufacturing staff during the design process. While simulation can aid in reaching a suitable design, it is recommended that testing be done in realistic conditions before committing the design to production. Some of the more popular types of materials used for gears are

Nylon	Acetal	ABS (acrylonitrile-butadiene-styrene)
Polycarbonate	Polyurethane	Polyester thermoplastic
Polyimide	Phenolic	Polyphenylene sulfide
Polysufones	Phenylene oxides	Styrene-acrylonitrile (SAN)

Designers must seek a balance of material characteristics appropriate to the application, considering, for example:

- Strength in flexure under fatigue conditions
- High modulus of elasticity for stiffness
- Impact strength and toughness
- Wear and abrasion resistance
- Dimensional stability under expected temperatures
- Dimensional stability due to moisture absorption from liquids and humidity
- Frictional performance and need for lubrication, if any
- Operation in vibration environments
- Chemical resistance and compatibility with the operating environment
- Sensitivity to ultraviolet radiation
- Creep resistance if operated under load for long periods of time
- Flame retarding ability
- Cost
- Ease of processing and molding
- Assembly and disassembly considerations
- Compatibility with mating parts
- Environmental impact during processing, use, recycling, and disposal

The basic plastic materials listed previously are typically modified with fillers and additives to produce optimum as-molded properties. Some of these are:

Fillers for strength reinforcement, toughness, moldability, long-term stability, thermal conductivity, and dimensional stability: Long glass fibers, chopped glass fibers, milled glass, woven glass fibers, carbon fibers, glass beads, aluminum flake, mineral, cellulose, rubber modifiers, wood flour, cotton, fabric, mica, talc, and calcium carbonate.

Fillers to improve lubricity and overall frictional performance: PTFE (polytetrafluoroethylene), silicone, carbon fibers, graphite powders, and molybdenum disulfide (MoS_2).

Refer also to Section 2–17 in Chapter 2 for additional discussion about plastic materials, their properties, and special considerations for selecting plastics.

Design Strength for Plastic Gear Materials

Data are provided here for typical plastic materials used for gears. They can be applied to problem solving in this book. However, verification of properties for materials to be actually used in a commercial application, with due regard for the operating conditions, should be acquired from the material supplier. The effects of temperature on strength, modulus, toughness, chemical stability, and dimensional precision are particularly important. Manufacturing processes must be controlled to ensure that final properties are consistent with prescribed values.

Table 9–11 lists some selected data for allowable tooth bending stress in plastic gears. Much additional data for other materials can be found in References 2, 15, and 16. Note the significant increase in allowable strength provided by the glass reinforcement. The combination of glass fibers and the basic plastic matrix performs like a composite material with the amount of filler typically ranging from 20% to 50%.

Material suppliers may be able to provide fatigue data for plastics in charts such as those shown in Figure 9–39, showing allowable bending stress versus number of cycles to failure for DuPont Zytel© nylon resin and Delrin© acetal resin. These data are for molded gears operating at room temperature with diametral pitches shown, pitch line velocity below 4000 ft/min, and continuous lubrication. Reductions should be applied for cut gears, higher temperatures, different pitches, and different lubrication conditions. See Reference 16.

TABLE 9–11   Approximate allowable tooth bending stress in plastic gears

Material	Approximate allowable bending stress, ksi (MPa)	
	Unfilled	Glass-filled
ABS	3000 (21)	6000 (41)
Acetal	5000 (34)	7000 (48)
Nylon	6000 (41)	12 000 (83)
Polycarbonate	6000 (41)	9000 (62)
Polyester	3500 (24)	8000 (55)
Polyurethane	2500 (17)	

Source: Plastics Gearing. Manchester, CT: ABA/PGT Publishing, 1994.

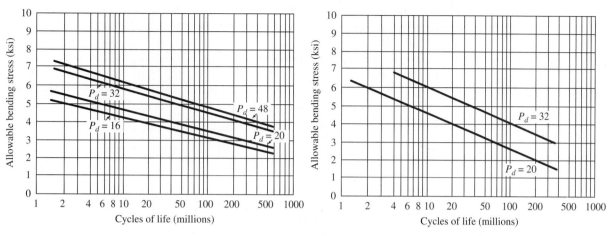

(a) Fatigue life data for DuPont© Zytel nylon resin   (b) Fatigue life data for DuPont© Delrin acetal resin

FIGURE 9–39   Fatigue life data for two types of plastic materials used for gears
(DuPont Polymers, Wilmington, DE)

Tooth Geometry

In general, standard tooth geometry for plastic gears conforms to the configurations described in Section 8–4 in Chapter 8. Standard diametral pitches from Table 8–2 and standard metric modules from Table 8–3 should be used unless there are major advantages to
using other values. Suppliers' ability to provide nonstandard pitches should be investigated.
Pressure angles of 14½°, 20°, and 25° are used, with 20° usually preferred. Standard formulas for addendum, dedendum, and clearance for full-depth involute teeth are listed in
Table 8–4. Gear quality values are set similarly to those for metallic gears as discussed in
Section 9–5. The typical AGMA quality number produced by injection molding is in range
of 6 to 10.

Designers sometimes use special tooth forms to tailor the strength of plastic gear
teeth to the demands of particular applications. The 20° stub tooth system provides a
shorter, broader tooth than the standard 20° full-depth tooth system, decreasing tooth-
bending stress. The Plastics Gearing Technology unit of the ABA-PGT company has
developed another system that is finding favor with some designers. See References 1
and 2.

Many designers of plastic gears prefer to use a longer addendum on the pinion
and a shorter addendum on the mating gear to produce more favorable operation because of the grater flexibility of plastics as compared with metals. Tooth thickness is
typically thinned on either or both of the pinion and gear to provide acceptable backlash and to ensure that mating gears do not bind. Binding may result from deflection of
the teeth under load or from expansions due to increased temperature or moisture absorption from exposure to water or high humidity. Enlarging the center distance is another method employed to adjust for backlash. Designers must specify these feature
sizes on drawings and in specifications. Consult AGMA Standard 1106-A97 *Tooth
Proportions for Plastic Gears* for details. Reference 2 provides useful tables of formulas and data for adjustments to tooth form and center distance. Reference 16 recommends the range of backlash values shown in Figure 9–40.

FIGURE 9–40
Recommended
backlash for plastic
gears

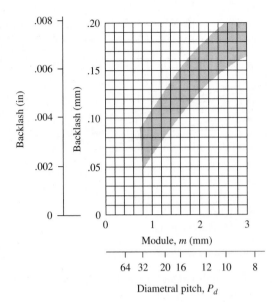

Shrinkage

During manufacture of plastic gears using injection molding, enlarging the effective diametral pitch and the pitch diameter of the gear teeth cut into the mold accommodates shrinkage. The pressure is also adjusted. The nominal corrections are computed as follows:

$$P_{dc} = \frac{P_d}{(1 + S)} \tag{9–32}$$

$$\cos \phi_1 = \frac{\cos \phi}{(1 + S)} \tag{9–33}$$

$$D_c = N/P_{dc} \tag{9–34}$$

where
S = Shrinkage of material
P_d = Standard diametral pitch for the gear
P_{dc} = Modified diametral pitch of the teeth in the mold
ϕ = Standard pressure angle for the gear
ϕ_1 = Modified pressure angle of the teeth in the mold
N = Number of teeth
D_c = Modified pitch diameter of teeth in the mold

After molding, the teeth should very nearly conform to standard geometry. Additional adjustments are sometimes made, relieving the tips of the teeth for smoother engagement and increasing the tooth width at the base near the point of highest bending stress.

Stress Analysis

Bending stress analysis for plastic gears relies on the basic Lewis formula introduced in Section 9–8, Equation (9–12). The modifying factors called for by the AGMA standards for

TABLE 9–12   Lewis tooth form factor, Y, for load near the pitch point

Number Teeth	Tooth Form		
	14½° Full Depth	20° Full Depth	20° Stub
14	–	–	0.540
15	–	–	0.566
16	–	–	0.578
17	–	0.512	0.587
18	–	0.521	0.603
19	–	0.534	0.616
20	–	0.544	0.628
22	–	0.559	0.648
24	0.509	0.572	0.664
26	0.522	0.588	0.678
28	0.535	0.597	0.688
30	0.540	0.606	0.698
34	0.553	0.628	0.714
38	0.566	0.651	0.729
43	0.575	0.672	0.739
50	0.588	0.694	0.758
60	0.604	0.713	0.774
75	0.613	0.735	0.792
100	0.622	0.757	0.808
150	0.635	0.779	0.830
300	0.650	0.801	0.855
Rack	0.660	0.823	0.881

Source: DuPont Polymers, Wilmington, DE

steel gears are not specified for plastic gears at this time. We can account for uncertainty or shock loading by inserting a safety factor. The overload factor from Table 9–5 can be used a guide. Testing of the proposed design in realistic conditions should be completed. The bending stress equation then becomes

$$\sigma_t = \frac{W_t \, P_d \, (SF)}{FY} \tag{9–35}$$

Values for the Lewis form factor, Y, shown in Table 9–12, describe the geometry of the involute gear teeth acting as a cantilever beam with the load applied near the pitch point. Thus Equation (9–35) gives the bending stress at the root of the tooth. Most plastic gear designs call for a generous fillet radius between the start of the active involute profile on the flank of the tooth and the root, resulting in little if any stress concentration.

Wear Considerations

Wear of tooth surfaces in plastic gear teeth is a function of the contact stress between mating teeth as it is with metal teeth. Equation (9–21) can be used compute the contact stress. However, published data are lacking for allowable contact stress values.

In reality, lubrication and the *combination of materials in mating gears* play major roles in the wear life of the pair. Presented here are some general guidelines from References 2 and 16. Communication with material suppliers and testing of proposed designs is recommended.

- Continuously lubricated gearing promotes the longest life.

- With continuous lubrication and light loads, fatigue resistance, not wear, typically determines life.

- Unlubricated gears tend to fail by wear, not fatigue, provided proper design bending stresses are used.

- When continuous lubrication is not practical, initially lubricating the gearing can aid in the run-in process and add life compared with gears that are never lubricated.

- When continuous lubrication is not practical, the combination of a nylon pinion and an acetal gear exhibits low friction and wear.

- Excellent wear performance for relatively high loads and pitch line speeds can be obtained by using a lubricated pair of a hardened steel pinion (HRC > 50) mating with a plastic gear made from nylon, acetal, or polyurethane.

- Wear accelerates when operating temperatures rise. Cooling to promote heat dissipation can increase life.

Gear Shapes and Assembly

References 2 and 16 include many recommendations for the geometric design of gears considering strength, inertia, and molding conditions. Many smaller gears are simply made with uniform thickness equal to the face width of the gear teeth. Larger gears often have a rim to support the teeth, a thinned web for lightening and material savings, and a hub to facilitate mounting on a shaft. Figure 9–41 shows recommended proportions. Symmetrical cross sections are preferred, along with balanced section thicknesses to promote good flow of material and to minimize distortion during molding.

Fastening gears to shafts requires careful design. Keys placed in shaft key seats and keyways in the hub of the gear provide reliable transmission of torque. For light torques, setscrews can be used, but slippage and damage of the shaft surface are possible. The bore of the gear hub can be lightly press fit onto the shaft with care to ensure that a sufficient torque can be transmitted while not overstressing the plastic hub. Knurling the shaft before pressing the gear on increases the torque capability. Some designers prefer to use metal hubs to facilitate the use of keys. Plastic is then molded onto the hub to form the rim and gear teeth.

Design Procedure

Design of plastic gearing should consider a variety of possibilities, and it is likely to be an iterative process. The following procedure outlines the steps for a given trial using U.S. Customary units.

FIGURE 9–41
Suggested plastic gear proportions
(DuPont Polymers, Wilmington, DE)

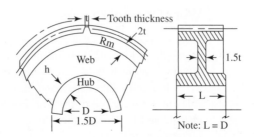

Procedure for Designing Plastic Gears

1. Determine the required horsepower, P, to be transmitted and the speed of rotation, n_P, of the pinion in rpm.

2. Specify the number of teeth, N, and select a trial diametral pitch diameter for the pinion.

3. Compute the pinion diameter from $D_P = N_P/P_d$.

4. Compute the transmitted load, W_t (in lb) from Equation (9–6), repeated here.

$$W_t = (126\,000)(P)/(n_P D_P)$$

5. Specify the tooth form and determine the Lewis form factor Y from Table 9–12.

6. Specify a safety factor, SF. Refer to Table 9–5 for guidance.

7. Specify the material to be used and determine the allowable stress from Table 9–10 or Figure 9–39.

8. Solve Equation (9–35) for the face width, F, and compute its value from,

$$F = \frac{W_t P_d (SF)}{s_{at} Y} \tag{9-36}$$

9. Judge the suitability of the computed face width as it relates to the application. Consider its mounting on a shaft, space available in the diametral and axial directions, and whether the general proportions are acceptable for injection molding. See References 2 and 16. No general recommendations are published for the face width of plastic gears and often they are narrower than similar metallic gears.

10. Repeat steps 2 to 9 until a satisfactory design for the pinion is achieved. Specify convenient dimensions for the final value of the face width and other features of the pinion.

11. Considering the desired velocity ratio between the pinion and the gear, compute the required number of teeth in the gear and repeat steps 3 to 9 using the same diametral pitch as the pinion. Using the same face width as for the pinion, the stress in the gear teeth will always be lower than in the pinion because the form factor Y will increase and all other factors will be the same. When the same material is to be used for the gear, it will always be safe. Alternatively, you could compute the bending stress directly from Equation 9–35 and specify a different material for the gear that has a suitable allowable bending stress.

Example Problem 9–6

Design a pair of plastic gears for a paper shredder to transmit 0.25 horsepower at a pinion speed of 1160 rpm. The pinion will be mounted on the shaft of an electric motor that has a diameter of 0.625 in with a keyway for a 3/16 × 3/16 in key. The gear is to rotate approximately 300 rpm.

Given Data

$P = 0.25$ hp, $n_P = 1160$ rpm,

Shaft diameter = $D_s = 0.625$ in, Keyway for a 3/16 × 3/16 in key.

Approximate gear speed = $n_G = 300$ rpm

Solution   Use the design procedure outlined in this section.

Step 1: Consider the given data.

Step 2: Specify $N_P = 18$ and $P_d = 16$

Step 3: $D_P = N_P/P_d = 18/16 = 1.125$ in. This seems reasonable for mounting on the 0.625 in motor shaft.

Step 4: Compute the transmitted load,

$$W_t = (126\,000)(P)/(n_P\,D_P) = (126\,000)(0.25)/[(1160)(1.125)] = 24.1 \text{ lb}$$

Step 5: Specify 20° full depth teeth. Then $Y = 0.521$ for 18 teeth from Table 9–12.

Step 6: Specify a safety factor, *SF*. The shredder will likely experience light shock; the preference is to operate the gears without lubrication. Specify $SF = 1.50$ from Table 9–5.

Step 7: Specify unfilled nylon. From Table 9–11, $s_{at} = 6000$ psi.

Step 8: Compute the required face width using Equation (9–36).

$$F = \frac{W_t\,P_d\,(SF)}{s_{at}\,Y} = \frac{(24.1)(16)(1.50)}{(6000)(0.521)} = 0.185 \text{ in}$$

Step 9: The dimensions seem reasonable.

Step 10: Appendix 2 lists a preferred size for the face width of 0.200 in.

Comment   In summary, the proposed pinion has the following features:

$$P_d = 16, N_P = 18 \text{ teeth}, D_P = 1.125 \text{ in}, F = 0.200 \text{ in}, \text{Bore} = 0.625 \text{ in},$$

Keyway for a 3/16 × 3/16 in key. Unfilled nylon material.

Step 11: Gear design: Specify $F = 0.200$ in, $P_d = 16$. Compute the number of teeth in the gear.

$$N_G = N_P\,(n_p/n_G) = 18(1160/300) = 69.6 \text{ teeth}$$

Specify $N_G = 70$ teeth

Pitch diameter of gear $= D_G = N_G/P_d = 70/16 = 4.375$ in

From Table 9–12, $Y_G = 0.728$ by interpolation.
Stress in gear teeth using Equation 9–35:

$$\sigma_t = \frac{W_t\,P_d\,(SF)}{FY} = \frac{(24.1)(16)(1.50)}{(0.200)(0.728)} = 3973 \text{ psi}$$

Comment   This stress level is safe for nylon. The gear could also be made from acetal to achieve better wear performance.

REFERENCES

1. ABA-PGT, Inc. *Plastics Gearing. Manchester*, CT:ABA-PGT Publishing, 1994.

2. Adams, Clifford E. *Plastics Gearing, Selection and Application*. New York: Marcel Dekker, 1986.

3. American Gear Manufacturers Association. Standard 908-B89 (R1995). *Geometry Factors for Determining the Pitting Resistance and Bending Strength of Spur, Helical, and Herringbone Gear Teeth*. Alexandria, VA.: American Gear Manufacturers Association, 1995.

4. American Gear Manufacturers Association. Standard 1012-F90. *Gear Nomenclature, Definitions of Terms with Symbols*. Alexandria, VA: American Gear Manufacturers Association, 1990.

5. American Gear Manufacturers Association, Standard 1106—A97. *Tooth Proportions for Plastic Gears*. Washington, DC: American Gear Manufacturers Association.

6. American Gear Manufacturers Association. Standard 2001-C95. *Fundamental Rating Factors and Calculation Methods for Involute Spur and Helical Gear Teeth*. Alexandria, VA: American Gear Manufacturers Association, 1995.

7. American Gear Manufacturers Association. Standard 2002-B88 (R1996). *Tooth Thickness Specification and Measurement*. Alexandria, VA: American Gear Manufacturers Association, 1996.

8. American Gear Manufacturers Association. Standard 2004-B89 (R2000). *Gear Materials and Heat Treatment Manual*. Alexandria, VA: American Gear Manufacturers Association, 1995.

9. American Gear Manufacturers Association. Standard 6010-F97 *Standard for Spur, Helical, Herringbone, and Bevel Enclosed Drives*. Alexandria, VA: American Gear Manufacturers Association, 1997.

10. American Gear Manufacturers Association. Standard 9005-D94 (R2000). *Industrial Gear Lubrication*. Alexandria, VA: American Gear Manufacturers Association, 1994.

11. American Gear Manufacturers Association. Standard 1010-E95. *Appearance of Gear Teeth-Terminology of Wear and Failure*. Alexandria, VA: American Gear Manufacturers Association, 1995.

12. American Gear Manufacturers Association. AGMA 427.01. *Information Sheet—Systems Considerations for Critical Service Gear Drives*. Alexandria, VA: American Gear Manufacturers Association, 1994.

13. American Society for Metals. *Source Book on Gear Design, Technology and Performance*. Metals Park, OH: American Society for Metals, 1980.

14. Drago, Raymond J. *Fundamentals of Gear Design*. Boston: Butterworths, 1988.

15. Dudley, Darle W. *Dudley's Gear Handbook*. New York: McGraw-Hill, 1991.

16. DuPont Polymers. *Design Handbook for DuPont Engineering Polymers, Module I—General Design Principles*. Wilmington, DE: DuPont Polymers, 1992.

17. Ewert, Richard H. *Gears and Gear Manufacture*. New York: Chapman & Hall, 1997.

18. Hosel, Theodor. *Comparison of Load Capacity Ratings for Involute Gears Due to ANSI/AGMA, ISO, DIN and Comecon Standards*. AGMA Technical Paper 89FTM4. Alexandria, VA: American Gear Manufacturers Association, 1989.

19. Lynwander, Peter. *Gear Drive Systems, Design and Application*. New York: Marcel Dekker, 1983.

20. Kern, Roy F. *Achievable Carburizing Specifications*. AGMA Technical Paper 88FTM1. Alexandria, VA: American Gear Manufacturers Association, 1988.

21. Kern, Roy F., and M. E. Suess. *Steel Selection*. New York: John Wiley & Sons, 1979.

22. Lipp, Robert. "Avoiding Tooth Interference in Gears." *Machine Design 54,* no. 1 (January 7, 1982).

23. Oberg, Erik, et al. *Machinery's Handbook*. 26th ed. New York: Industrial Press, 2000.

24. Shigley, Joseph E., and C. R. Mischke. *Mechanical Engineering Design*. 6th ed. New York: McGraw-Hill, 2001.

25. Society of Automotive Engineers. *Gear Design, Manufacturing and Inspection Manual*. Warrendale, PA: Society of Automotive Engineers, 1990.

26. Stock Drive Products–Sterling Instruments. *Handbook of Design Components*. New Hyde Park, NY: Designatronics Corp., 1992.

INTERNET SITES RELATED TO SPUR GEAR DESIGN

1. **ABA-PGT, Inc.** *www.abapgt.com* The ABA division produces molds for making plastic gears using injection molding; the PGT division is dedicated to plastic gearing technology.

2. **American Gear Manfuactures Association (AGMA)** *www.agma.org* Develops and publishes voluntary, consensus standards for gears and gear drives.

3. **Bison Gear, Inc.** *www.bisongear.com* Manufacturer of fractional horsepower gear reducers and gear motors.

4. **Boston Gear, Company.** *www.bostongear.com* Manufacturer of gears and complete gear drives. Part of the Colfax Power Transmission Group. Data provided for spur, helical, and worm gearing.

5. **DuPont Polymers** *www.plastics.dupont.com* Information and data on plastics and their properties. Searchable database by type of plastic or application.

6. **Emerson Power Transmission Corporation.** *www.emerson-ept.com* The Browning Division produces spur, helical, bevel, and worm gearing and complete gear drives.

7. **Gear Industry Home Page.** *www.geartechnology.com* Information source for many companies that manufacture or use gears or gearing systems. Includes gear machinery, gear cutting tools, gear materials, gear drives, open gearing, tooling & supplies, software, training and education. Publishes

Gear Technology Magazine: The Journal of Gear Manufacturing.

8. **Power Transmission Home Page.**
www.powertransmission.com   Clearinghouse on the Internet for buyers, users and sellers of power transmission-related products and services. Included are gears, gear drives, and gear motors.

9. **Rockwell Automation/Dodge.**   *www.dodge-pt.com*
Manufacturer of many power transmission components, including complete gear-type speed reducers, bearings, and components such as belt drives, chain drives, clutches, brakes, and couplings.

10. **Stock Drive Products–Sterling Instruments.**
www.sdp-si.com   Manufacturer and distributor of commercial and precision mechanical components, including gear reducers. Site includes an extensive handbook of design and information on metallic and plastic gears.

11. **Peerless-Winsmith, Inc.**   *www.winsmith.com*
Manufacturer of a wide variety of gear reducers and power transmission products, including worm gearing,

planetary gearing, and combined helical/worm gearing. Subsidiary of HBD Industries, Inc.

12. **Drivetrain Technology Center.**
www.arl.psu.edu/areas/drivetrain/drivetrain.html
Research center for gear-type drivetrain technolgy. Part of the Applied Research Laboratory of Penn State University.

13. **Gleason Corporation.**   *www.gleason.com*
Manufacturer of many types of gear cutting machines for hobbing, milling, and grinding. The Gleason Cutting Tools Corporation manufactures a wide variety of milling cutters, hobs, shaper cutters, shaving cutters, and grinding wheels for gear production equipment.

14. **Bourn & Koch, Inc.**   *www.bourn-koch.com*
Manufacturer of hobbing, grinding, and other types of machines to produce gears, including the Barber-Colman line. Also provides remanufacturing services for a wide variety of existing macahine tools.

15. **Star-SU, Inc.**   *www.star-su.com*   Manufacturer of a wide variety of cutting tools for the gear industry, including hobs, shaping cutters, shaving cutters, bevel gear cutting tools, and grinding tools.

PROBLEMS

Forces on Spur Gear Teeth

1. A pair of spur gears with 20°, full-depth, involute teeth transmits 7.5 hp. The pinion is mounted on the shaft of an electric motor operating at 1750 rpm. The pinion has 20 teeth and a diametral pitch of 12. The gear has 72 teeth. Compute the following:
   a. The rotational speed of the gear
   b. The velocity ratio and the gear ratio for the gear pair
   c. The pitch diameter of the pinion and the gear
   d. The center distance between the shafts carrying the pinion and the gear
   e. The pitch line speed for both the pinion and the gear
   f. The torque on the pinion shaft and on the gear shaft
   g. The tangential force acting on the teeth of each gear
   h. The radial force acting on the teeth of each gear
   i. The normal force acting on the teeth of each gear

2. A pair of spur gears with 20°, full-depth, involute teeth transmits 50 hp. The pinion is mounted on the shaft of an electric motor operating at 1150 rpm. The pinion has 18 teeth and a diametral pitch of 5. The gear has 68 teeth. Compute the following:
   a. The rotational speed of the gear
   b. The velocity ratio and the gear ratio for the gear pair
   c. The pitch diameter of the pinion and the gear

   d. The center distance between the shafts carrying the pinion and the gear
   e. The pitch line speed for both the pinion and the gear
   f. The torque on the pinion shaft and on the gear shaft
   g. The tangential force acting on the teeth of each gear
   h. The radial force acting on the teeth of each gear
   i. The normal force acting on the teeth of each gear

3. A pair of spur gears with 20°, full-depth, involute teeth transmits 0.75 hp. The pinion is mounted on the shaft of an electric motor operating at 3450 rpm. The pinion has 24 teeth and diametral pitch of 24. The gear has 110 teeth. Compute the following:
   a. The rotational speed of the gear
   b. The velocity ratio and the gear ratio for the gear pair
   c. The pitch diameter of the pinion and the gear
   d. The center distance between the shafts carrying the pinion and the gear
   e. The pitch line speed for both the pinion and the gear
   f. The torque on the pinion shaft and on the gear shaft
   g. The tangential force acting on the teeth of each gear
   h. The radial force acting on the teeth of each gear
   i. The normal force acting on the teeth of each gear

4. For the data of Problem 1, repeat Parts (g), (h), and (i) if the teeth have 25° full depth instead of 20°.

5. For the data of Problem 2, repeat Parts (g), (h), and (i) if the teeth have 25° full depth instead of 20°.

6. For the data of Problem 3, repeat Parts (g), (h), and (i) if the teeth have 25° full depth instead of 20°.

Gear Manufacture and Quality

7. List three methods for producing gear teeth, and describe each method. Include a description of the cutter for each method along with its motion relative to the gear blank.

8. Specify a suitable quality number for the gears in the drive for a grain harvester. List the total composite tolerance for a pinion in the drive having a diametral pitch of 8 and 40 teeth and for its mating gear having 100 teeth.

9. Specify a suitable quality number for the gears in the drive for a high-speed printing press. List the total composite tolerance for a pinion in the drive having a diametral pitch of 20 and 40 teeth and for its mating gear having 100 teeth.

10. Specify a suitable quality number for the gears in the drive for an automotive transmission. List the total composite tolerance for a pinion in the drive having a diametral pitch of 8 and 40 teeth and its mating gear having 100 teeth.

11. Specify a suitable quality number for the gears in the drive for a gyroscope used in the guidance system for a spacecraft. List the total composite tolerance for a pinion in the drive having a diametral pitch of 32 and 40 teeth and its mating gear having 100 teeth.

12. Compare the values of total composite tolerance for the gears of Problems 8 and 10.

13. Compare the values of total composite tolerance for the gears of Problems 8, 9, and 11.

14. Specify a suitable quality number for the gears of Problem 1 if the drive is part of a precision machine tool.

15. Specify a suitable quality number for the gears of Problem 2 if the drive is part of a precision machine tool.

16. Specify a suitable quality number for the gears of Problem 3 if the drive is part of a precision machine tool.

Gear Materials

17. Identify the two major types of stresses that are created in gear teeth as they transmit power. Describe how the stresses are produced and where the maximum values of such stresses are expected to occur.

18. Describe the nature of the data contained in AGMA standards that relate to the ability of a given gear tooth to withstand the major types of stresses that it sees in operation.

19. Describe the general nature of steels that are typically used for gears, and list at least five examples of suitable alloys.

20. Describe the range of hardness that can typically be produced by through-hardening techniques and used successfully in steel gears.

21. Describe the general nature of the differences among steels produced as Grade 1, Grade 2, and Grade 3.

22. Suggest at least three applications in which Grade 2 or Grade 3 steel might be appropriate.

23. Describe three methods of producing gear teeth with strengths greater than can be achieved with through-hardening.

24. What AGMA standard should be consulted for data on the allowable stresses for steels used for gears?

25. In the AGMA standard identified in Problem 24, for what other materials besides steels are strength data provided?

26. Determine the allowable bending stress number and the allowable contact stress number for the following materials:

    a. Through-hardened, Grade 1 steel with a hardness of 200 HB

    b. Through-hardened, Grade 1 steel with a hardness of 300 HB

    c. Through-hardened, Grade 1 steel with a hardness of 400 HB

    d. Through-hardened, Grade 1 steel with a hardness of 450 HB

    e. Through-hardened, Grade 2 steel with a hardness of 200 HB

    f. Through-hardened, Grade 2 steel with a hardness of 300 HB

    g. Through-hardened, Grade 2 steel with a hardness of 400 HB

27. If the design of a steel gear indicates that an allowable bending stress number of 36 000 psi is needed, specify a suitable hardness level for Grade 1 steel. What hardness level would be required for Grade 2 steel?

28. What level of hardness can be expected for gear teeth that are case-hardened by carburizing?

29. Name three typical steels that are used in carburizing.

30. What is the level of hardness that can be expected for gear teeth that are case-hardened by flame or induction hardening?

31. Name three typical steels that are used for flame or induction hardening. What is an important property of such steels?

32. What level of hardness can be expected for gear teeth that are nitrided?

33. Determine the allowable bending stress number and the allowable contact stress number for the following materials:

  a. Flame-hardened AISI 4140 steel, Grade 1, with a surface hardness of 50 HRC

  b. Flame-hardened AISI 4140 steel, Grade 1, with a surface hardness of 54 HRC

  c. Carburized and case-hardened AISI 4620 Grade 1 steel, DOQT 300

  d. Carburized and case-hardened AISI 4620 Grade 2 steel, DOQT 300

  e. Carburized and case-hardened AISI 1118 Grade 1 steel, SWQT 350

  f. Nitrided, through-hardened, Grade 1 steel with a surface hardness of 84.5 HRN and a core hardness of 325 HB

  g. Nitrided, through-hardened, Grade 2 steel with a surface hardness of 84.5 HRN and a core hardness of 325 HB

  h. Nitrided, 2.5% chrome, Grade 3 steel with a surface hardness of 90.0 HRN and a core hardness of 325

  i. Gray cast iron, class 20

  j. Gray cast iron, class 40

  k. Ductile iron, 100-70-03

  l. Sand-cast bronze with a minimum tensile strength of 40 ksi (275 MPa)

  m. Heat-treated bronze with a minimum tensile strength of 90 ksi (620 MPa)

  n. Glass-filled nylon

  o. Glass-filled polycarbonate

34. What depth should be specified for the case for a carburized gear tooth having a diametral pitch of 6?

35. What depth should be specified for the case for a carburized gear tooth having a metric module of 6?

Bending Stresses in Gear Teeth

For Problems 36–41, compute the bending stress number, s_t, using Equation (9–15). Assume that the gear blank is solid unless otherwise stated. *(Note that the data in these problems are used in later problems through Problem 59. You are advised to keep solutions to earlier problems accessible so that you can use data and results in later problems. The four problems that are keyed to the same set of data require the analysis of bending stress and contact stress and the corresponding specification of suitable materials based on those stresses. Later design problems, 60–70, use the complete analysis within each problem.)*

36. A pair of gears with 20°, full-depth, involute teeth transmits 10.0 hp while the pinion rotates at 1750 rpm. The diametral pitch is 12, and the quality number is 6. The pinion has 18 teeth, and the gear has 85 teeth. The face width is 1.25 in. The input power is from an electric mo-

tor, and the drive is for an industrial conveyor. The drive is a commercial enclosed gear unit.

37. A pair of gears with 20°, full-depth, involute teeth transmits 40 hp while the pinion rotates at 1150 rpm. The diametral pitch is 6, and the quality number is 6. The pinion has 20 teeth, and the gear has 48 teeth. The face width is 2.25 in. The input power is from an electric motor, and the drive is for a cement kiln. The drive is a commercial enclosed gear unit.

38. A pair of gears with 20°, full-depth, involute teeth transmits 0.50 hp while the pinion rotates at 3450 rpm. The diametral pitch is 32, and the quality number is 10. The pinion has 24 teeth, and the gear has 120 teeth. The face width is 0.50 in. The input power is from an electric motor, and the drive is for a small machine tool. The drive is a precision enclosed gear unit.

39. A pair of gears with 25° full-depth, involute teeth transmits 15.0 hp while the pinion rotates at 6500 rpm. The diametral pitch is 10, and the quality number is 12. The pinion has 30 teeth, and the gear has 88 teeth. The face width is 1.50 in. The input power is from a universal electric motor, and the drive is for an actuator on an aircraft. The drive is an extra-precision, enclosed gear unit.

40. A pair of gears with 25°, full-depth, involute teeth transmits 125 hp while the pinion rotates at 2500 rpm. The diametral pitch is 4, and the quality number is 8. The pinion has 32 teeth, and the gear has 76 teeth. The face width is 1.50 in. The input power is from a gasoline engine, and the drive is for a portable industrial water pump. The drive is a commercial enclosed gear unit.

41. A pair of gears with 25°, full-depth, involute teeth transmits 2.50 hp while the pinion rotates at 680 rpm. The diametral pitch is 10, and the quality number is 6. The pinion has 24 teeth, and the gear has 62 teeth. The face width is 1.25 in. The input power is from a vane-type fluid motor, and the drive is for a small lawn and garden tractor. The drive is a commercial enclosed gear unit.

Required Allowable Bending Stress Number

For Problems 42–47, compute the required allowable bending stress number, s_{at}, using Equation (9–20). Assume that no unusual conditions exist unless stated otherwise. That is, use a service factor, *SF*, of 1.00. Then specify a suitable steel and its heat treatment for both the pinion and the gear based on bending stress.

42. Use the data and results from Problem 36. Design for a reliability of 0.99 and a design life of 20 000 h.

43. Use the data and results from Problem 37. Design for a reliability of 0.99 and a design life of 8000 h.

44. Use the data and results from Problem 38. Design for a reliability of 0.9999 and a design life of 12 000 h. Consider that the machine tool is a critical part of a production system calling for a service factor of 1.25 to avoid unexpected down time.

45. Use the data and results from Problem 39. Design for a reliability of 0.9999 and a design life of 4000 h.

46. Use the data and results from Problem 40. Design for a reliability of 0.99 and a design life of 8000 h.

47. Use the data and results from Problem 41. Design for a reliability of 0.90 and a design life of 2000 h. The uncertainty of the actual use of the tractor calls for a service factor of 1.25. Consider using cast iron or bronze if the conditions permit.

Pitting Resistance

For Problems 48–53, compute the expected contact stress number, s_c, using Equation (9–25). Assume that both gears are to be steel unless stated otherwise.

48. Use the data and results from Problems 36 and 42.

49. Use the data and results from Problems 37 and 43.

50. Use the data and results from Problems 38 and 44.

51. Use the data and results from Problems 39 and 45.

52. Use the data and results from Problems 40 and 46.

53. Use the data and results from Problems 41 and 47.

Required Allowable Contact Stress Number

For Problems 54–59, compute the required allowable contact stress number, s_{ac}, using Equation (9–27). Use a service factor, *SF,* of 1.00 unless stated otherwise. Then specify suitable material for the pinion and the gear based on pitting resistance. Use steel unless an earlier decision has been made to use another material. Then evaluate whether the earlier decision is still valid. If not, specify a different material according to the most severe requirement. If no suitable material can be found, consider redesigning the original gears to enable reasonable materials to be used.

54. Use the data and results from Problems 36, 42, and 48.

55. Use the data and results from Problems 37, 43, and 49.

56. Use the data and results from Problems 38, 44, and 50.

57. Use the data and results from Problems 39, 45, and 51.

58. Use the data and results from Problems 40, 46, and 52.

59. Use the data and results from Problems 41, 47, and 53.

Design Problems

Problems 60–70, describe design situations. For each, design a pair of spur gears, specifying (at least) the diametral pitch, the number of teeth in each gear, the pitch diameters of each gear, the center distance, the face width, and the material from which the gears are to be made. Design for recommended life with regard to both strength and pitting resistance. Work toward designs that are compact. Use standard values of diametral pitch, and avoid designs for which interference could

occur. See Example Problem 9–5. Assume that the input to the gear pair is from an electric motor unless otherwise stated.

If the data are given in SI units, complete the design in the metric module system with dimensions in millimeters, forces in newtons, and stresses in megapascals. See Example Problem 9–6.

60. A pair of spur gears is to be designed to transmit 5.0 hp while the pinion rotates at 1200 rpm. The gear must rotate between 385 and 390 rpm. The gear drives a reciprocating compressor.

61. A gear pair is to be a part of the drive for a milling machine requiring 20.0 hp with the pinion speed at 550 rpm and the gear speed to be between 180 and 190 rpm.

62. A drive for a punch press requires 50.0 hp with the pinion speed of 900 rpm and the gear speed of 225 to 230 rpm.

63. A single-cylinder gasoline engine has the pinion of a gear pair on its output shaft. The gear is attached to the shaft of a small cement mixer. The mixer requires 2.5 hp while rotating at approximately 75 rpm. The engine is governed to run at approximately 900 rpm.

64. A four-cylinder industrial engine runs at 2200 rpm and delivers 75 hp to the input gear of a drive for a large wood chipper used to prepare pulpwood chips for paper making. The output gear must run between 4500 and 4600 rpm.

65. A small commercial tractor is being designed for chores such as lawn mowing and snow removal. The wheel drive system is to be through a gear pair in which the pinion runs at 600 rpm while the gear, mounted on the hub of the wheel, runs at 170 to 180 rpm. The wheel is 300 mm in diameter. The gasoline engine delivers 3.0 kW of power to the gear pair.

66. A water turbine transmits 75 kW of power to a pair of gears at 4500 rpm. The output of the gear pair must drive an electric power generator at 3600 rpm. The center distance for the gear pair must not exceed 150 mm.

67. A drive system for a large commercial band saw is to be designed to transmit 12.0 hp. The saw will be used to cut steel tubing for automotive exhaust pipes. The pinion rotates at 3450 rpm, while the gear must rotate between 725 and 735 rpm. It has been specified that the gears are to be made from AISI 4340 steel, oil quenched and tempered. Case hardening is *not* to be used.

68. Repeat Problem 67, but consider a case-hardened carburized steel from Appendix 5. Try to achieve the smallest practical design. Compare the result with the design from Problem 67.

69. A gear drive for a special-purpose, dedicated machine tool is being designed to mill a surface on a rough steel casting. The drive must transmit 20 hp with a pinion speed of 650 rpm and an output speed between 110 and 115 rpm. The mill is to be used continuously, two shifts per day, six days per week, for at least five years. Design the drive to be as small as practical to permit its being mounted close to the milling head.

70. A cable drum for a crane is to rotate between 160 and 166 rpm. Design a gear drive for 25 hp in which the input pinion rotates at 925 rpm and the output rotates with the drum. The crane is expected to operate with a 50% duty cycle for 120 hours per week for at least 10 years. The pinion and the gear of the drive must fit within the 24-in inside diameter of the drum, with the gear mounted on the drum shaft.

Power-Transmitting Capacity

71. Determine the power-transmitting capacity for a pair of spur gears having 20°, full-depth teeth, a diametral pitch of 10, a face width of 1.25 in, 25 teeth in the pinion, 60 teeth in the gear, and an AGMA quality class of 8. The pinion is made from AISI 4140 OQT 1000, and the gear is made from AISI 4140 OQT 1100. The pinion will rotate at 1725 rpm on the shaft of an electric motor. The gear will drive a centrifugal pump.

72. Determine the power-transmitting capacity for a pair of spur gears having 20°, full-depth teeth, a diametral pitch of 6, 35 teeth in the pinion, 100 teeth in the gear, a face width of 2.00 in, and an AGMA quality class of 6. A gasoline engine drives the pinion at 1500 rpm. The gear drives a conveyor for crushed rock in a quarry. The pinion is made from AISI 1040 WQT 800. The gear is made from gray cast iron, ASTM A48-83, class 30. Design for 15 000 hr life.

73. It was found that the gear pair described in Problem 72 wore out when driven by a 25-hp engine. Propose a redesign that would be expected to give indefinite life under the conditions described. Design for 15 000 hr life.

Design of Double-Reduction Drives

74. Design a double-reduction gear train that will transmit 10.0 hp from an electric motor running at 1750 rpm to an assembly conveyor whose drive shaft must rotate between 146 and 150 rpm. Note that this will require the design of two pairs of gears. Sketch the arrangement of the train, and compute the actual output speed.

75. A commercial food waste grinder in which the final shaft rotates at between 40 and 44 rpm is to be designed. The input is from an electric motor running at 850 rpm and delivering 0.50 hp. Design a double-reduction spur gear train for the grinder.

76. A small, powered hand drill is driven by an electric motor running at 3000 rpm. The drill speed is to be approx-

imately 550 rpm. Design the speed reduction for the drill. The power transmitted is 0.25 hp.

77. The output from the drill described in Problem 76 provides the drive for a small bench-scale band saw similar to the one in Figure 1–1. The saw blade is to move with a linear velocity of 375 ft/min. The saw blade rides on 9.0-in-diameter wheels. Design a spur gear reduction to drive the band saw. Consider using plastic gears.

78. Design a rack-and-pinion drive to lift a heavy access panel on a furnace. A fluid power motor rotating 1500 rpm will provide 5.0 hp at the input to the drive. The linear speed of the rack is to be at least 2.0 ft/s. The rack moves 6.0 ft each way during the opening and closing of the furnace doors. More than one stage of reduction may be used, but attempt to design with the fewest number of gears. The drive is expected to operate at least six times per hour for three shifts per day, seven days per week, for at least 15 years.

79. Design the gear drive for the wheels of an industrial lift truck. Its top speed is to be 20 mph. It has been decided that the wheels will have a diameter of 12.0 in. A DC motor supplies 20 hp at a speed of 3000 rpm. The design life is 16 hours per day, six days per week, for 20 years.

Plastics Gearing

80. Design a pair of plastic gears to drive a small band saw. The input is from a 0.50 hp electric motor rotating at 860 rpm, and the pinion will be mounted on its 0.75-inch diameter shaft with a keyway for a $3/16 \times 3/16$ in key. The gear is to rotate between 265 and 267 rpm.

81. Design a pair of plastic gears to drive a paper feed roll for an office printer. The pinion rotates at 88 rpm and the gear must rotate between 20 and 22 rpm. The power required is 0.06 hp. Work toward the smallest practical size.

82. Design a pair of plastic gears to drive the wheels of a small remote control car. The gear is mounted on the axle of the wheel and must rotate between 120 and 122 rpm. The pinion rotates at 430 rpm. The power required is 0.025 hp. Work toward the smallest practical size using unfilled nylon.

83. Design a pair of plastic gears to drive a commercial food-chopping machine. The input is from a 0.65 hp electric motor rotating at 1560 rpm, and the pinion will be mounted on its 0.875-inch diameter shaft with a keyway for a $1/4 \times 1/4$ in key. The gear is to rotate between 468 and 470 rpm.

10

Helical Gears, Bevel Gears, and Wormgearing

The Big Picture

You Are the Designer

10–1 Objectives of This Chapter

10–2 Forces on Helical Gear Teeth

10–3 Stresses in Helical Gear Teeth

10–4 Pitting Resistance for Helical Gear Teeth

10–5 Design of Helical Gears

10–6 Forces on Straight Bevel Gears

10–7 Bearing Forces on Shafts Carrying Bevel Gears

10–8 Bending Moments on Shafts Carrying Bevel Gears

10–9 Stresses in Straight Bevel Gear Teeth

10–10 Design of Bevel Gears for Pitting Resistance

10–11 Forces, Friction, and Efficiency in Wormgear Sets

10–12 Stress in Wormgear Teeth

10–13 Surface Durability of Wormgear Drives

Helical Gears, Bevel Gears, and Wormgearing

Discussion Map

☐ The geometry of helical gears, bevel gears, and wormgearing was described in Chapter 8.

☐ The principles of stress analysis of gears were discussed in Chapter 9 for spur gears. Much of that information is applicable to the types of gears discussed in this chapter.

Discover

Review Chapters 8 and 9 now.

Recall some of the discussion at the beginning of Chapter 8 about uses for gears that you see in your world. Review that information now, and focus your discussion on helical gears, bevel gears, and wormgearing.

In this chapter, you acquire the skills to perform the necessary analyses to design safe gear drives that use helical gears, bevel gears, and wormgearing and that demonstrate long life.

Much was said in Chapters 8 and 9 about the kinematics of gears and about the stress analysis and design of spur gears. That information is important to the objectives of this chapter, in which we extend the application of those concepts to the analysis and design of helical gears, bevel gears, and wormgearing.

The basic geometry of helical gears was described in Section 8–7. The force system on helical gear teeth was also described, and that system is important to your understanding of the stresses and modes of potential failure for helical gears that we discuss in this chapter.

In Chapter 9, you learned how to analyze spur gears for bending strength and resistance to pitting of the surface of the teeth. This chapter modifies that same approach for application to the special geometry of helical gears. In fact, AGMA Standard 2001-C95, which we often referred to in Chapter 9 for spur gears, is the same reference that we use for helical gears. So you will need to refer to Chapter 9 from time to time.

Similarly, the geometry of bevel gears was described in Section 8–8, and wormgearing was described in Sections 8–9 and 8–10. This chapter includes information about stresses in bevel gears and wormgearing.

Figure 10–1 shows an example of a large, commercially available, double-reduction, parallel-shaft reducer employing helical gears. Notice that the shafts are supported in tapered roller bearings that have the ability to withstand the thrust loads created by the helical gears. Chapter 14 elaborates on the selection of such bearings.

Figure 10–2 shows another form of helical gear reducer in which the drive motor is mounted above the reducer, and the shaft of the driven machine is inserted directly through the hollow output shaft of the reducer. This allows the reducer to be supported by the frame of the driven machine.

Refer to Figure 8–22 for drawings and photographs of bevel gear systems. The three-dimensional force system that acts between the teeth of bevel gears requires great care in the installation and alignment of the gears and requires the use of bearings that can withstand forces in all directions. The stress analysis is adapted from the approach described in Chapter 9, modified for the geometry of the bevel gear teeth.

Refer to Figure 8–25 for a photograph of a commercially available wormgear reducer. Notice the use of tapered roller bearings here also to withstand the thrust forces created by

FIGURE 10–1
Parallel shaft reducer
(Emerson Power
Transmission
Corporation, Drive and
Component Division,
Ithaca, NY)

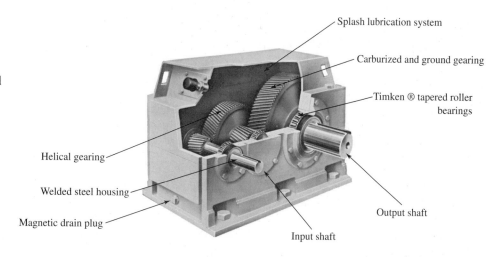

Splash lubrication system

Carburized and ground gearing

Timken ® tapered roller
bearings

Helical gearing

Welded steel housing

Magnetic drain plug

Output shaft

Input shaft

FIGURE 10–2
Helical shaft mount
reducer (Emerson
Power Transmission
Corporation, Drive and
Component Division,
Ithaca, NY)

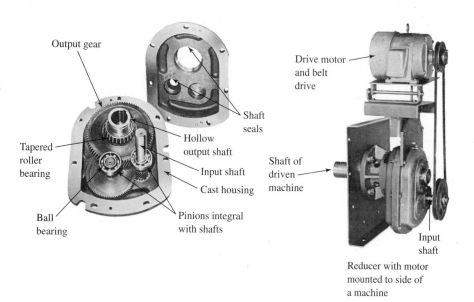

Output gear

Shaft
seals

Hollow
output shaft

Tapered
roller
bearing

Input shaft

Cast housing

Ball
bearing

Pinions integral
with shafts

Drive motor
and belt
drive

Shaft of
driven
machine

Input
shaft

Reducer with motor
mounted to side of
a machine

wormgearing. The lubrication of the gears is highlighted in the photograph. Lubrication is
important because sliding action between the worm threads and the wormgear teeth inherently occurs, causing frictional heat to be generated. Consistent lubrication of the gear mesh
is critical to the performance, efficiency, and life of the system.

This chapter presents more information about these factors in wormgearing system
design. The design for strength and pitting resistance is presented for each type of gear, with
an analysis of the geometry and the forces exerted on the gears.

You Are the Designer

The gear drives that were designed in Chapter 9 all assumed that spur gears would be used to accomplish the speed reduction or speed increase between the input and the output of the drive. But many other types of gears could have been used. Assume that you are the designer of the drive for the wood chipper described in Example Problem 9–5. How would the design be different if helical gears were used instead of spur gears? What forces would be created and transferred to the shafts carrying the gears and to the bearings carrying the shafts? Would you be able to use smaller gears? How is the geometry of helical gears different from that of spur gears?

Rather than having the input and output shafts parallel as they were in designs up to this time, how can we design drives that deliver power to an output shaft at right angles to the input shaft? What special analysis techniques are applied to bevel gears and wormgearing?

The information in this chapter will help you answer these and other questions.

10–1 OBJECTIVES OF THIS CHAPTER

After completing this chapter, you will be able to:

1. Describe the geometry of helical gears and compute the dimensions of key features.
2. Compute the forces exerted by one helical gear on its mating gear.
3. Compute the stress due to bending in helical gear teeth and specify suitable materials to withstand such stresses.
4. Design helical gears for surface durability.
5. Describe the geometry of bevel gears and compute the dimensions of key features.
6. Analyze the forces exerted by one bevel gear on another and show how those forces are transferred to the shafts carrying the gears.
7. Design and analyze bevel gear teeth for strength and surface durability.
8. Describe the geometry of worms and wormgears.
9. Compute the forces created by a wormgear drive system and analyze their effect on the shafts carrying the worm and the wormgear.
10. Compute the efficiency of wormgear drives.
11. Design and analyze wormgear drives to be safe for bending strength and wear. References 3, 4, 7, 8, and 17–22 are recommended for general guidelines for design and application of helical gears, bevel gears, and wormgearing.

10–2 FORCES ON HELICAL GEAR TEETH

Figure 10–3 shows a photograph of two helical gears in mesh and designed to be mounted on parallel shafts. This is the basic configuration that we analyze in this chapter. Refer to Figure 10–4 for a representation of the force system that acts between the teeth of two helical gears in mesh. In Chapter 8, using the same figure, we defined the following forces:

- W_N is the *true normal force* that acts perpendicular to the face of the tooth in the plane normal to the surface of the tooth. The normal plane is shown in Part (d) of Figure 10–4. We seldom need to use the value of W_N because its three orthogonal components, defined next, are used in the analyses performed for helical gears. The values for the orthogonal components depend on the following three angles that help define the geometry of the helical gear teeth:

   Normal pressure angle: ϕ_n
   Transverse pressure angle: ϕ_t
   Helix angle: ψ

FIGURE 10–3
Helical gears. These
gears have a 45° helix
angle. (Emerson Power
Transmission
Corporation, Drive and
Component Division,
Ithaca, NY)

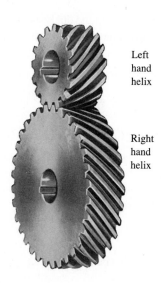

Left
hand
helix

Right
hand
helix

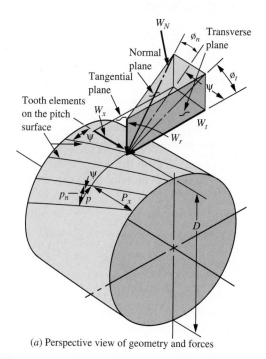

(a) Perspective view of geometry and forces

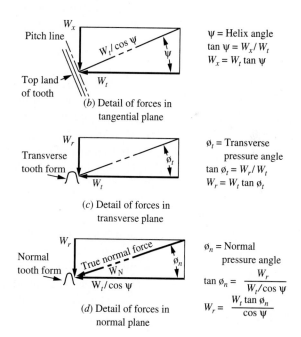

ψ = Helix angle
$\tan \psi = W_x / W_t$
$W_x = W_t \tan \psi$

(b) Detail of forces in
tangential plane

ϕ_t = Transverse
pressure angle
$\tan \phi_t = W_r / W_t$
$W_r = W_t \tan \phi_t$

(c) Detail of forces in
transverse plane

ϕ_n = Normal
pressure angle
$\tan \phi_n = \dfrac{W_r}{W_t / \cos \psi}$
$W_r = \dfrac{W_t \tan \phi_n}{\cos \psi}$

(d) Detail of forces in
normal plane

FIGURE 10–4   Helical gear geometry and forces

For helical gears, the helix angle and one of the other two are specified. The third
angle can be computed from

$$\tan \phi_n = \tan \phi_t \cos \psi \tag{10–1}$$

■ W_t is the *tangential force* that acts in the transverse plane and tangent to the pitch
circle of the helical gear and that causes the torque to be transmitted from the
driver to the driven gear. Therefore, this force is often called the *transmitted force*.

454 Chapter 10 ■ Helical Gears, Bevel Gears, and Wormgearing

It is functionally similar to W_t used in the analysis of spur gears in Chapter 8. We can compute its value from the same equations, as follows:

If the torque being transmitted (T) and the size of the gear (D) are known,

$$W_t = T/(D/2) \tag{10-2}$$

If the power being transmitted (P) and the rotational speed (n) are known,

$$T = (P/n) \tag{10-3}$$

For the unit-specific situation where power is expressed in horsepower and the rotational speed is in rpm, the torque in lb·in is

$$T = 63\,000(P)/n \tag{10-4}$$

Then the tangential force can also be expressed as

$$W_t = 63\,000(P)/[(n)(D/2)] = 126\,000(P)/[(n)(D)] \tag{10-5}$$

If the pitch line speed, v_t (ft/min) of the gear is known, along with the power being transmitted (hp), the tangential load is

$$W_t = 33\,000(P)/v_t \tag{10-6}$$

The value of the tangential load is the most fundamental of the three orthogonal components of the true normal force. The calculation of the bending stress number and the pitting resistance of the gear teeth depends on W_t.

■ W_r is the *radial force* that acts toward the center of the gear perpendicular to the pitch circle and to the tangential force. It tends to push the two gears apart. As can be seen in Figure 10–4(c),

$$W_r = W_t \tan \phi_t \tag{10-7}$$

where ϕ_t = transverse pressure angle for the helical teeth

■ W_x is the *axial force* that acts parallel to the axis of the gear and causes a thrust load that must be resisted by the bearings carrying the shaft. With the tangential force known, the axial force is computed from

$$W_x = W_t \tan \psi \tag{10-8}$$

Example Problem 10–1 A helical gear has a normal diametral pitch of 8, a normal pressure angle of 20°, 32 teeth, a face width of 3.00 in, and a helix angle of 15°. Compute the diametral pitch, the transverse pressure angle, and the pitch diameter. If the gear is rotating at 650 rpm while transmitting 7.50 hp, compute the pitch line speed, the tangential force, the axial force, and the radial force.

Solution *Diametral Pitch [Equation (8–28)]*

$$P_d = P_{nd} \cos \psi = 8 \cos (15) = 7.727$$

Transverse Pressure Angle [Equation (10–1)]

$$\phi_t = \tan^{-1}(\tan \phi_n / \cos \psi)$$

$$\phi_t = \tan^{-1}[\tan(20)/\cos(15)] = 20.65°$$

Pitch Diameter [Equation (8–27)]

$$D = N/P_d = 32/7.727 = 4.141 \text{ in}$$

Pitch Line Speed, v_t [Equation (9–1)]

$$v_t = \pi Dn/12 = \pi(4.141)(650)/12 = 704.7 \text{ ft/min}$$

Tangential Force, W_t [Equation (10–6)]

$$W_t = 33\,000(P)/v_t = 33\,000(7.5)/704.7 = 351 \text{ lb}$$

Axial Force, W_x [Equation (10–8)]

$$W_x = W_t \tan \psi = 351 \tan(15) = 94 \text{ lb}$$

Radial Force, W_r [Equation (10–7)]

$$W_r = W_t \tan \phi_t = 351 \tan(20.65) = 132 \text{ lb}$$

10–3 STRESSES IN HELICAL GEAR TEETH

We will use the same basic equation for computing the bending stress number for helical gear teeth as we did for spur gear teeth in Chapter 9, given in Equation (9–15) and repeated here:

$$s_t = \frac{W_t P_d}{FJ} K_o K_s K_m K_B K_v$$

Figures 10–5, 10–6, and 10–7 show the values for the geometry factor, *J*, for helical gear teeth with 15°, 20°, and 22° normal pressure angles, respectively.[1] The *K* factors are the same as those used for spur gears. See References 9 and 18 and the following locations for values:

K_o = overload factor (Table 9–5)

K_s = size factor (Table 9–6)

K_m = load-distribution factor [Figures 9–18 and 9–19 and Equation (9–16)]

K_B = rim thickness factor (Figure 9–20)

K_v = dynamic factor (Figure 9–21)

[1] Figures 10–5, 10–6, and 10–7:
Graphs for the geometry factor, *J*, for helical gears have been taken from AGMA Standard 218.01-1982, *Standard for Rating the Pitting Resistance and Bending Strength of Spur and Helical Involute Gear Teeth,* with the permission of the publisher, American Gear Manufacturers Association, 1500 King Street, Suite 201, Alexandria, VA 22314. This standard has been superseded by two standards: (1) Standard 908-B89 (R1995), *Geometry Factors for Determining the Pitting Resistance and Bending Strength of Spur, Helical and Herringbone Gear Teeth,* 1989; (2) Standard 2001-C95, *Fundamental Rating Factors and Calculation Methods for Involute Spur and Helical Gear Teeth,* 1995. The method of calculating the value for *J* has not been changed. However, the new standards do not contain the graphs. Users are cautioned to ensure that geometry factors for a given design conform to the specific cutter geometry used to manufacture the gears. Standards 908-B89 (R 1995) and 2001-C95 should be consulted for the details of computing the values for *J* and for rating the performance of the gear teeth.

FIGURE 10–5

Geometry factor (J) for 15° normal pressure angle

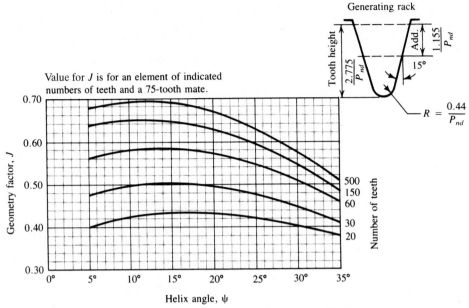

Value for J is for an element of indicated numbers of teeth and a 75-tooth mate.

(a) Geometry factor (J) for 15° normal pressure angle and indicated addendum

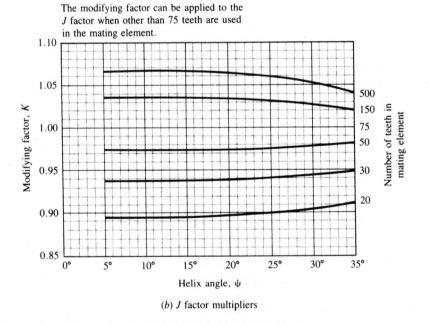

The modifying factor can be applied to the J factor when other than 75 teeth are used in the mating element.

(b) J factor multipliers

For design, a material must be specified that has an allowable bending stress number, s_{at}, greater than the computed bending stress number, s_t. Design values of s_{at} can be found:

> Figure 9–10: Steel, through-hardened, Grades 1 and 2
>
> Table 9–3: Case-hardened steels
>
> Figures 9–14 and 9–15: Nitrided gears
>
> Table 9–4: Cast iron and bronze

FIGURE 10–6
Geometry factor (J) for
20° normal pressure
angle

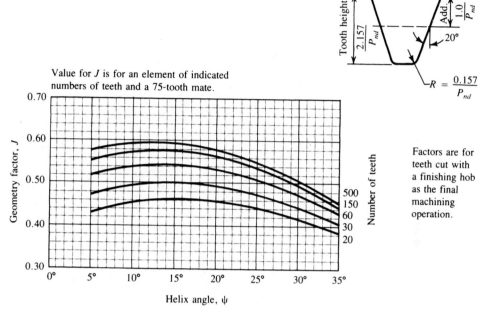

Value for J is for an element of indicated
numbers of teeth and a 75-tooth mate.

Factors are for
teeth cut with
a finishing hob
as the final
machining
operation.

(a) Geometry factor (J) for 20° normal pressure angle, standard addendum, and finishing hob

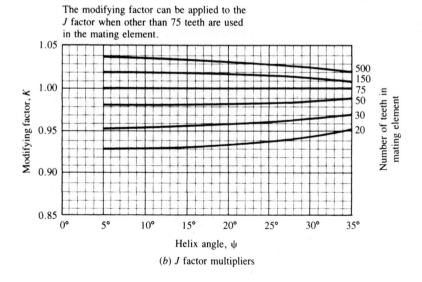

The modifying factor can be applied to the
J factor when other than 75 teeth are used
in the mating element.

(b) J factor multipliers

(See also References 11, 16, and 21.) The data for steel, iron, and bronze apply to a design life of 10^7 cycles at a reliability of 99% (fewer than one failure in 100). If other values for design life or reliability are desired, the allowable stress can be modified using the procedure described in Section 9–9.

FIGURE 10–7

Geometry factor (J) for 22° normal pressure angle

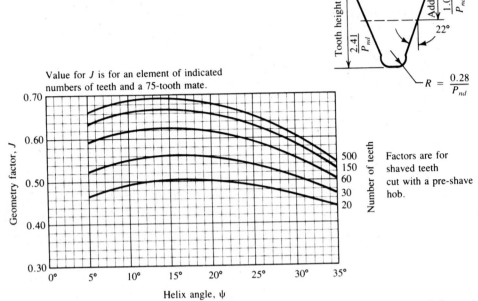

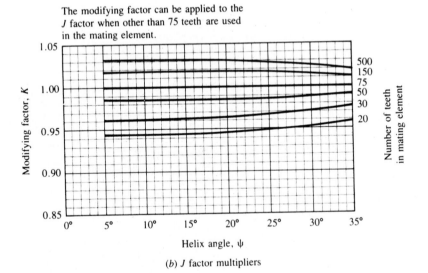

Value for J is for an element of indicated numbers of teeth and a 75-tooth mate.

(a) Geometry factor (J), for 22° normal pressure angle, standard addendum, and pre-shave hob

The modifying factor can be applied to the J factor when other than 75 teeth are used in the mating element.

(b) J factor multipliers

**10–4
PITTING
RESISTANCE
FOR HELICAL
GEAR TEETH**

Pitting resistance for helical gear teeth is evaluated using the same procedure as that discussed in Chapter 9 for spur gears. Equation (9–25) is repeated here:

$$s_c = C_p \sqrt{\frac{W_t K_o K_s K_m K_v}{F D_p I}} \qquad (9\text{–}25)$$

All of the factors are the same for helical gears except the geometry factor for pitting resistance, I. The values for C_p are found in Table 9–9. Note that the other K factors have the same values as the K factors discussed and identified in Section 10–3.

Because of the larger variety of geometric features needed to define the form of helical gears, it is not reasonable to reproduce all of the necessary tables of values or the complete formulas for computing I. Values change with the gear ratio, the number of teeth in the pinion, the tooth form, the helix angle, and the specific values for addendum, whole depth, and fillet radius. See References 6 and 13 for extensive discussions about the procedures. To facilitate problem solving in this book, Tables 10–1 and 10–2 give a few values for I.

TABLE 10–1 Geometry factors for pitting resistance, I, for helical gears with 20° normal pressure angle and standard addendum

A. Helix angle $\psi = 15.0°$

Gear teeth	Pinion teeth				
	17	21	26	35	55
17	0.124				
21	0.139	0.128			
26	0.154	0.143	0.132		
35	0.175	0.165	0.154	0.137	
55	0.204	0.196	0.187	0.171	0.143
135	0.244	0.241	0.237	0.229	0.209

B. Helix angle $\psi = 25.0°$

Gear teeth	Pinion teeth					
	14	17	21	26	35	55
14	0.123					
17	0.137	0.126				
21	0.152	0.142	0.130			
26	0.167	0.157	0.146	0.134		
35	0.187	0.178	0.168	0.156	0.138	
55	0.213	0.207	0.199	0.189	0.173	0.144
135	0.248	0.247	0.244	0.239	0.230	0.210

Source: Extracted from AGMA Standard 908-B89 (R 1995), *Geometry Factors for Determining the Pitting Resistance and Bending Strength of Spur, Helical and Herringbone Gear Teeth,* with the permission of the publisher, American Gear Manufacturers Association, 1500 King Street, Suite 201, Alexandria, VA 22314.

TABLE 10–2   Geometry factors for pitting resistance, *I*, for helical gears with 25° normal pressure angle and standard addendum

A. Helix angle $\psi = 15.0°$

Gear teeth	Pinion teeth					
	14	17	21	26	35	55
14	0.130					
17	0.144	0.133				
21	0.160	0.149	0.137			
26	0.175	0.165	0.153	0.140		
35	0.195	0.186	0.175	0.163	0.143	
55	0.222	0.215	0.206	0.195	0.178	0.148
135	0.257	0.255	0.251	0.246	0.236	0.214

B. Helix angle $\psi = 25.0°$

Gear teeth	Pinion teeth						
	12	14	17	21	26	35	55
12	0.129						
14	0.141	0.132					
17	0.155	0.146	0.135				
21	0.170	0.162	0.151	0.138			
26	0.185	0.177	0.166	0.154	0.141		
35	0.203	0.197	0.188	0.176	0.163	0.144	
55	0.227	0.223	0.216	0.207	0.196	0.178	0.148
135	0.259	0.258	0.255	0.251	0.246	0.235	0.213

Source: Extracted from AGMA Standard 908-B89, *Geometry Factors for Determining the Pitting Resistance and Bending Strength of Spur, Helical and Herringbone Gear Teeth,* with the permission of the publisher, American Gear Manufacturers Association, 1500 King Street, Suite 201, Alexandria, VA 22314.

For design, when the computed contact stress number is known, a material must be specified that has an allowable contact stress number, s_{ac}, greater than s_c. Design values for s_{ac} can be found from the following:

Figure 9–11: Steel, through-hardened, Grades 1 and 2

Table 9–3: Steel, case-hardened, Grades 1, 2, and 3; flame- or induction-hardened, carburized, or nitrided

Table 9–4: Cast iron and bronze

The data from these sources apply to a design life of 10^7 cycles at a reliability of 99% (fewer than one failure in 100). If other values for design life or reliability are desired, or if a service factor is to be applied, the allowable contact stress number can be modified using the procedure described in Section 9–11.

10–5
DESIGN OF
HELICAL GEARS

The example problem that follows illustrates the procedure to design helical gears.

Example Problem 10–2 A pair of helical gears for a milling machine drive is to transmit 65 hp with a pinion speed of 3450 rpm and a gear speed of 1100 rpm. The power is from an electric motor. Design the gears.

Solution Of course, there are several possible solutions. Here is one. Let's try a normal diametral pitch of 12, 24 teeth in the pinion, a helix angle of 15°, a normal pressure angle of 20°, and a quality number of 8.

Now compute the transverse diametral pitch, the axial pitch, the transverse pressure angle, and the pitch diameter. Then we will choose a face width that will give at least two axial pitches to ensure true helical action.

$$P_d = P_{dn} \cos \psi = 12 \cos(15°) = 11.59$$

$$P_x = \frac{\pi}{P_d \tan \psi} = \frac{\pi}{11.59 \tan(15°)} = 1.012 \text{ in}$$

$$\phi_t = \tan^{-1}(\tan \phi_n / \cos \psi) = \tan^{-1}[\tan(20°)/\cos(15°)] = 20.65°$$

$$d = D_P/P_d = 24/11.59 = 2.071 \text{ in}$$

$$F = 2P_x = 2(1.012) = 2.024 \text{ in} \quad (\text{nominal face width})$$

Let's use 2.25 in, a more convenient value. The pitch line speed and the transmitted load are

$$v_t = \pi D_P n/12 = \pi(2.071)(3450)/12 = 1871 \text{ ft/min}$$

$$W_t = 33\,000 \, (\text{hp})/v_t = 33\,000(65)/1871 = 1146 \text{ lb}$$

Now we can calculate the number of teeth in the gear:

$$VR = N_G/N_P = n_P/n_G = 3450/1100 = 3.14$$

$$N_G = N_P \,(VR) = 24(3.14) = 75 \text{ teeth} \quad (\text{integer value})$$

The values for the factors in Equation (9–15) must now be determined to enable the calculation of the bending stress. The geometry factor for the pinion is found in Figure 10–6 for 24 teeth in the pinion and 75 teeth in the gear: $J_P = 0.48$. The value of J_G will be greater than the value of J_P, resulting in a lower stress in the gear.

The K factors are

K_o = overload factor = 1.5 (moderate shock)

K_s = size factor = 1.0

K_m = load-distribution factor = 1.26 for F/D_P = 1.09 and commercial-quality, enclosed gearing

K_B = rim thickness factor = 1.0 (solid gears)

K_v = dynamic factor = 1.35 for Q_v = 8 and v_t = 1871 ft/min

The bending stress in the pinion can now be computed:

$$s_{tP} = \frac{W_t P_d}{F J_P} K_o K_s K_m K_B K_v$$

$$s_{tP} = \frac{(1146)(11.59)}{(2.25)(0.48)} (1.50)(1.0)(1.26)(1.0)(1.35) = 31\,400 \text{ psi}$$

From Figure 9–10, a Grade 1 steel with a hardness of approximately 250 HB would be required. Let's proceed to the design for pitting resistance.

Use Equation (9–25):

$$s_c = C_p \sqrt{\frac{W_t K_o K_s K_m K_v}{F D_P I}}$$

For two steel gears, $C_p = 2300$. Rough interpolation from the data in Table 10–1 for $N_P = 24$ and $N_G = 75$ gives $I = 0.202$. It is recommended that the computational procedure described in the AGMA standards be used to compute a more precise value for critical work. The contact stress is then

$$s_c = 2300 \sqrt{\frac{(1146)(1.50)(1.0)(1.26)(1.35)}{(2.25)(2.071)(0.202)}} = 128\ 200 \text{ psi}$$

Figure 9–11 indicates that a Grade 1 steel with a hardness of 310 HB would be recommended. Assuming that standard life and reliability factors are acceptable, we could specify AISI 5150 OQT 1000, which has a hardness of 321 HB as listed in Appendix 3.

Comments   It is obvious that the contact stress governs this design. Let's adjust the solution for a higher reliability and to account for the expected number of cycles of operation. Certain design decisions must be made. For example, consider the following:

Design for a reliability of 0.999 (less than one failure in 1000): $K_R = 1.25$ (Table 9–8).

Design life: Let's design for 10 000 h of life as suggested in Table 9–7 for multipurpose gearing. Then, using Equation (9–18), we can compute the number of cycles of loading. For the pinion rotating at 3450 rpm with one cycle of loading per revolution,

$$N_c = (60)(L)(n)(q) = (60)(10\ 000)(3450)(1.0) = 2.1 \times 10^9 \text{ cycles}$$

From Figure 9–24, we find that $Z_N = 0.89$.

No unusual conditions seem to exist in this application beyond those already considered in the various K factors. Therefore, we use a service factor, SF, of 1.00.

We can use Equation (9–27) to apply these factors. For the pinion, we use $C_H = 1.00$:

$$\frac{K_R(SF)}{Z_N C_H} s_c = s_{ac} = \frac{(1.25)(1.00)}{(0.89)(1.00)} (128\ 200 \text{ psi}) = 180\ 000 \text{ psi}$$

Table 9–3 indicates that Grade 1 steel, case-hardened by carburizing, would be suitable. From Appendix 5, let's specify AISI 4320 SOQT 450, having a case hardness of HRC 59 and a core hardness of 415 HB. This should be satisfactory for both bending and pitting resistance. Both the pinion and the gear should be of this material. There is a modest difference in the Z_N factor, but it should not lower the required allowable contact stress number below that requiring case hardening. Also, when both the pinion and the gear are case-hardened, the hardness ratio factor, C_H, is 1.00.

10–6
FORCES ON
STRAIGHT BEVEL
GEARS

Review Section 8–8 and Figure 8–22 for the geometry of bevel gears. Also see References 1, 5, and 12.

Because of the conical shape of bevel gears and because of the involute-tooth form, a three-component set of forces acts on bevel gear teeth. Using notation similar to that for helical gears, we will compute the tangential force, W_t; the radial force, W_r; and the axial force, W_x. It is assumed that the three forces act concurrently at the midface of the teeth and on the pitch cone (see Figure 10–8). Although the actual point of application of the resultant force is a little displaced from the middle, no serious error results.

The tangential force acts tangential to the pitch cone and is the force that produces the torque on the pinion and the gear. The torque can be computed from the known power transmitted and the rotational speed:

$$T = 63\,000\,P/n$$

Then, using the pinion, for example, the transmitted load is

$$W_{tP} = T/r_m \tag{10–9}$$

where r_m = mean radius of the pinion

FIGURE 10–8
Forces on bevel gears

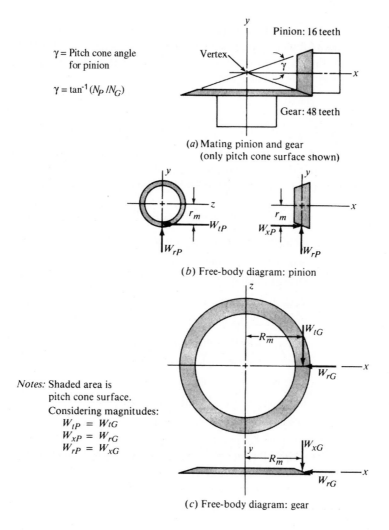

The value of r_m can be computed from

$$r_m = d/2 - (F/2)\sin\gamma \qquad\qquad (10\text{--}10)$$

Remember that the pitch diameter, d, is measured to the pitch line of the tooth at its large end. The angle, γ, is the pitch cone angle for the pinion as shown in Figure 10–8(a). The radial load acts toward the center of the pinion, perpendicular to its axis, causing bending of the pinion shaft. Thus,

$$W_{rP} = W_t\tan\phi\cos\gamma \qquad\qquad (10\text{--}11)$$

The angle, ϕ, is the pressure angle for the teeth.

The axial load acts parallel to the axis of the pinion, tending to push it away from the mating gear. It causes a thrust load on the shaft bearings. It also produces a bending moment on the shaft because it acts at the distance from the axis equal to the mean radius of the gear. Thus,

$$W_{xP} = W_t\tan\phi\sin\gamma \qquad\qquad (10\text{--}12)$$

The values for the forces on the gear can be calculated by the same equations shown here for the pinion, if the geometry for the gear is substituted for that of the pinion. Refer to Figure 10–8 for the relationships between the forces on the pinion and the gear in both magnitude and direction.

Example Problem 10–3   For the gear pair described in Example Problem 8–6, calculate the forces on the pinion and the gear if they are transmitting 2.50 hp with a pinion speed of 600 rpm. The geometry factors computed in Example Problem 8–6 apply. The data are summarized here.

Summary of Pertinent Results from Example Problem 8–6 and Given Data

Number of teeth in the pinion: $N_P = 16$

Number of teeth in the gear: $N_G = 48$

Diametral pitch: $P_d = 8$

Pitch diameter of pinion: $d = 2.000$ in

Pressure angle: $\phi = 20°$

Pinion pitch cone angle: $\gamma = 18.43°$

Gear pitch cone angle: $\Gamma = 71.57°$

Face width: $F = 1.00$ in

Rotational speed of pinion: $n_P = 600$ rpm

Power transmitted: $P = 2.50$ hp

Solution   Forces on the pinion are described by the following equations:

$$W_t = T/r_m$$

But

$$T_P = 63\,000(P)/n_P = [63\,000(2.50)]/600 = 263 \text{ lb} \cdot \text{in}$$
$$r_m = d/2 - (F/2)\sin \gamma$$
$$r_m = (2.000/2) - (1.00/2)\sin(18.43°) = 0.84 \text{ in}$$

Then

$$W_t = T_P/r_m = 263 \text{ lb} \cdot \text{in}/0.84 \text{ in} = 313 \text{ lb}$$
$$W_r = W_t \tan \phi \cos \gamma = 313 \text{ lb} \tan(20°)\cos(18.43°) = 108 \text{ lb}$$
$$W_x = W_t \tan \phi \sin \gamma = 313 \text{ lb} \tan(20°)\sin(18.43°) = 36 \text{ lb}$$

To determine the forces on the gear, first let's calculate the rotational speed of the gear:

$$n_G = n_P(N_P/N_G) = 600 \text{ rpm}(16/48) = 200 \text{ rpm}$$

Then

$$T_G = 63\,000(2.50)/200 = 788 \text{ lb} \cdot \text{in}$$
$$R_m = D/2 - (F/2)\sin \Gamma$$
$$R_m = 6.000/2 - (1.00/2)\sin(71.57°) = 2.53 \text{ in}$$
$$W_t = T_G/R_m = (788 \text{ lb} \cdot \text{in})/(2.53 \text{ in}) = 313 \text{ lb}$$
$$W_r = W_t \tan \phi \cos \Gamma = 313 \text{ lb} \tan(20°)\cos(71.57°) = 36 \text{ lb}$$
$$W_x = W_t \tan \phi \sin \Gamma = 313 \text{ lb} \tan(20°)\sin(71.57°) = 108 \text{ lb}$$

Note from Figure 10–8 that the forces on the pinion and the gear form an *action-reaction pair*. That is, the forces on the gear are equal to those on the pinion, but they act in the opposite direction. Also, because of the 90° orientation of the shafts, the radial force on the pinion becomes the axial thrust load on the gear, and the axial thrust load on the pinion becomes the radial load on the gear.

**10–7
BEARING
FORCES ON
SHAFTS
CARRYING
BEVEL GEARS**

Because of the three-dimensional force system that acts on bevel gears, the calculation of the forces on shaft bearings can be cumbersome. An example is worked out here to show the procedure. In order to obtain numerical data, the arrangement shown in Figure 10–9 is proposed for the bevel gear pair that was the subject of Example Problems 8–6 and 10–3. The locations for the bearings are given with respect to the vertex of the two pitch cones where the shaft axes intersect.

Note that both the pinion and the gear are *straddle-mounted;* that is, each gear is positioned between the supporting bearings. This is the most preferred arrangement because it usually provides the greatest rigidity and maintains the alignment of the teeth during power transmission. Care should be exercised to provide rigid mountings and stiff shafts when using bevel gears.

The arrangement of Figure 10–9 is designed so that the bearing on the right resists the axial thrust load on the pinion, and the lower bearing resists the axial thrust load on the gear.

FIGURE 10–9
Layout of bevel gear
pair for Example
Problem 10–4

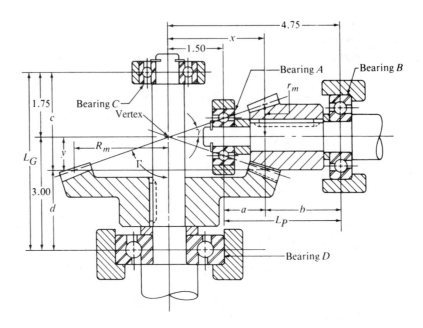

Example Problem 10–4   Compute the reaction forces on the bearings that support the shafts carrying the bevel gear
pair shown in Figure 10–9. The values of Example Problems 8–6 and 10–3 apply.

Solution   Referring to the results of Example Problem 10–3 and Figure 10–8, we have listed the
forces acting on the gears:

Force	Pinion	Gear
Tangential	$W_{tP} = 313$ lb	$W_{tG} = 313$ lb
Radial	$W_{rP} = 108$ lb	$W_{rG} = 36$ lb
Axial	$W_{xP} = 36$ lb	$W_{xG} = 108$ lb

It is critical to be able to visualize the directions in which these forces are acting
because of the three-dimensional force system. Notice in Figure 10–8 that a rectangular coordinate system has been set up. Figure 10–10 is an isometric sketch of the free-
body diagrams of the pinion and the gear, simplified to represent the concurrent forces
acting at the pinion/gear interface and at the bearing locations. Although the two free-
body diagrams are separated for clarity, notice that you can bring them together by moving the point called *vertex* on each sketch together. This is the point in the actual gear
system where the vertices of the two pitch cones lie at the same point. The two pitch
points also coincide.

For setting up the equations of static equilibrium needed to solve for the bearing reactions, the distances a, b, c, d, L_P, and L_G are needed, as shown in Figure 10–9. These require the two dimensions labeled x and y. Note from Example Problem 10–3 that

$$x = R_m = 2.53 \text{ in}$$
$$y = r_m = 0.84 \text{ in}$$

FIGURE 10–10
Free-body diagrams for
pinion and gear shafts

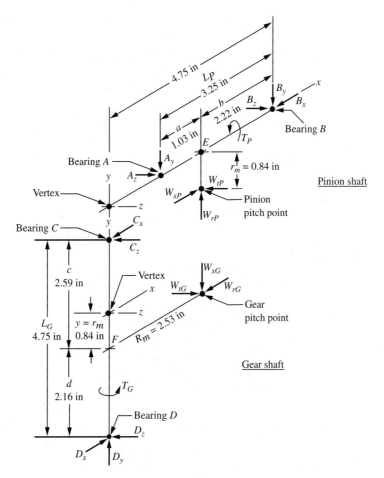

Then

$$a = x - 1.50 = 2.53 - 1.50 = 1.03 \text{ in}$$
$$b = 4.75 - x = 4.75 - 2.53 = 2.22 \text{ in}$$
$$c = 1.75 + y = 1.75 + 0.84 = 2.59 \text{ in}$$
$$d = 3.00 - y = 3.00 - 0.84 = 2.16 \text{ in}$$
$$L_P = 4.75 - 1.50 = 3.25 \text{ in}$$
$$L_G = 1.75 + 3.00 = 4.75 \text{ in}$$

These values are shown on Figure 10–10.

To solve for the reactions, we need to consider the horizontal (x-z) and the vertical (x-y) planes separately. It may help you to look also at Figure 10–11, which breaks out the forces on the pinion shaft in these two planes. Then we can analyze each plane using the fundamental equations of equilibrium.

Bearing Reactions, Pinion Shaft: Bearings A and B

Step 1. To find B_z and A_z: In the x-z plane, only W_{tP} acts. Summing moments about A yields

$$0 = W_{tP}(a) - B_z(L_P) = 313(1.03) - B_z(3.25)$$
$$B_z = 99.2 \text{ lb}$$

FIGURE 10–11
Pinion shaft bending
moments

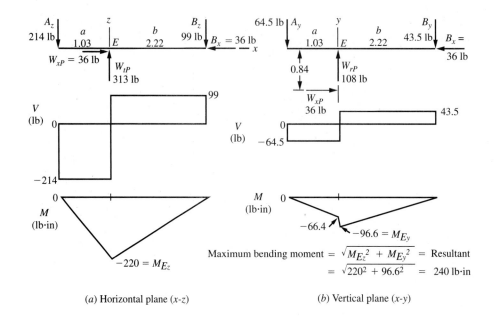

(a) Horizontal plane (x-z)

(b) Vertical plane (x-y)

Summing moments about B yields

$$0 = W_{tP}(b) - A_z(L_P) = 313(2.22) - A_z(3.25)$$
$$A_z = 214 \text{ lb}$$

Step 2. To find B_y and A_y: In the x-y plane, both W_{rP} and W_{xP} act. Summing moments about A yields

$$0 = w_{rP}(a) + W_{xP}(r_m) - B_y(L_P)$$
$$0 = 108(1.03) + 36(0.84) - B_y(3.25)$$
$$B_y = 43.5 \text{ lb}$$

Summing moments about B yields

$$0 = W_{rP}(b) + W_{xP}(r_m) - A_y(L_P)$$
$$0 = 108(2.22) + 36(0.84) - A_y(3.25)$$
$$A_y = 64.5 \text{ lb}$$

Step 3. To find B_x: Summing forces in the x-direction yields

$$B_x = W_{xP} = 36 \text{ lb}$$

This is the thrust force on bearing B.

Step 4. To find the total radial force on each bearing: Compute the resultant of the y- and z-components.

$$A = \sqrt{A_y^2 + A_z^2} = \sqrt{64.5^2 + 214^2} = 224 \text{ lb}$$
$$B = \sqrt{B_y^2 + B_z^2} = \sqrt{43.5^2 + 99.2^2} = 108 \text{ lb}$$

FIGURE 10–12 Gear shaft bending moments

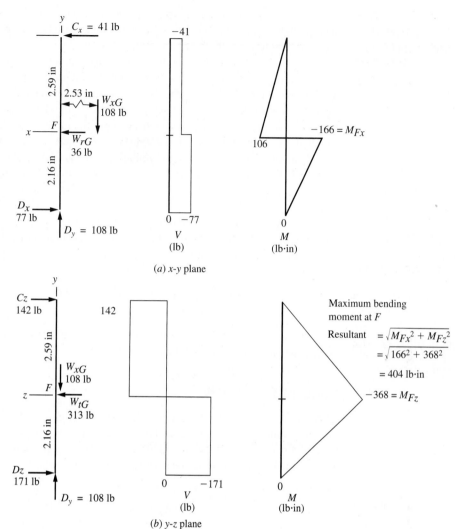

(a) x-y plane

(b) y-z plane

Bearing Reactions, Gear Shaft: Bearings C and D
Using similar methods, we can find the forces in Figure 10–12.

$$\left.\begin{array}{l} C_z = 142 \text{ lb} \\ C_x = 41.1 \text{ lb} \end{array}\right\} C = 148 \text{ lb (radial force on } C)$$

$$\left.\begin{array}{l} D_z = 171 \text{ lb} \\ D_x = 77.1 \text{ lb} \end{array}\right\} D = 188 \text{ lb (radial force on } D)$$

$$D_y = W_{xG} = 108 \text{ lb (thrust force on } D)$$

Summary   In selection of the bearings for these shafts, the following capacities are required:

Bearing *A:* 224-lb radial

Bearing *B:* 108-lb radial; 36-lb thrust

Bearing *C:* 148-lb radial

Bearing *D:* 188-lb radial; 108-lb thrust

**10–8
BENDING
MOMENTS ON
SHAFTS
CARRYING
BEVEL GEARS**

Because there are forces acting in two planes on bevel gears, as discussed in the preceding section, there is also bending in two planes. The analysis of the shearing force and bending moment diagrams for the shafts must take this into account.

Figures 10–11 and 10–12 show the resulting diagrams for the pinion and the gear shafts, respectively, for the gear pair used for Example Problems 8–6, 10–3, and 10–4. Notice that the axial thrust load on each gear provides a concentrated moment to the shaft equal to the axial load times the distance that it is offset from the axis of the shaft. Also notice that the maximum bending moment for each shaft is the resultant of the moments in the two planes. On the pinion shaft, the maximum moment is 240 lb·in at E, where the lines of action for the radial and tangential forces intersect the shaft. Similarly, on the gear shaft, the maximum moment is 404 lb·in at F. These data are used in the shaft design (as discussed in Chapter 12).

**10–9
STRESSES IN
STRAIGHT BEVEL
GEAR TEETH**

The stress analysis for bevel gear teeth is similar to that already presented for spur and helical gear teeth. The maximum bending stress occurs at the root of the tooth in the fillet. This stress can be computed from

**Bending Stress
Number**

$$s_t = \frac{W_t P_d}{FJ} \frac{K_o K_s K_m}{K_v} \tag{10–13}$$

The terms have all been used before. But there are minor differences in the manner of evaluating the factors, so they will be reviewed here.

Tangential Force, W_t

Contrary to the way W_t was computed in the preceding section, we compute it here by using the diameter of the gear at its large end, rather than the diameter at the middle of the tooth. This is more convenient, and the adjustment for the actual force distribution on the teeth is made in the value of the geometry factor, J. Then

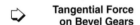
**Tangential Force
on Bevel Gears**

$$W_t = \frac{T}{r} = \frac{63\,000(P)}{n_P} \frac{1}{d/2} \tag{10–14}$$

where T = torque transmitted (lb·in)
  r = pitch radius of the pinion (in)
  P = power transmitted (hp)
  n_P = rotational speed of the pinion (rpm)
  d = pitch diameter of the pinion at its large end (in)

Dynamic Factor, K_v

The values for the dynamic factor for bevel gears are different from those for spur or helical gears. Factors affecting the dynamic factor include the accuracy of manufacture of the gear teeth (quality number Q); the pitch line speed, v_t; the tooth load; and the stiffness of the teeth. AGMA Standard 2003-A86 recommends the following procedure for computing K_v for the bending strength calculation and C_v for pitting resistance:

**Dynamic Factor for
Bevel Gears**

$$C_v = K_v = \left[\frac{K_z}{K_z + \sqrt{v_t}} \right]^u \tag{10–15}$$

where

$$u = \frac{8}{(2)^{0.5Q}} - s_{at} \left[\frac{125}{E_P + E_G} \right]$$

$$K_z = 85 - 10(u)$$

If the equation for u results in a negative value, use $u = 0$. As a check on the specification of an appropriate value for the quality number, a minimum value for C_v should be computed from

$$C_{v\,min} = \frac{2}{\pi} \tan^{-1}(v_t/333)$$

The value of the result of the inverse tangent calculation must be in radians. If the actual value of C_v is less than $C_{v\,min}$, a higher quality number should be specified.

Size Factor, K_s

Use the values from Table 9–6.

Load-Distribution Factor, K_m

Values are highly dependent on the manner of mounting both the pinion and the gear. The preferred mounting is called *straddle mounting,* in which the gear is between its supporting bearings. Figure 10–9 shows straddle mounting of both the pinion and the gear. In addition, stiff, short shafts are recommended to minimize shaft deflections that cause misalignment of the gear teeth.

Refer to AGMA Standard 2003-A86 (see Reference 10) for general methods of evaluating K_m. For enclosed gearing with special care in mounting the gears and in controlling the form of the gear teeth, AGMA Standard 6010-F97 (Reference 13) recommends the values in Table 10–3. The gears should be tested under load to ensure an optimum tooth contact pattern. We will use these values in problem solutions.

Geometry Factor, J

Use Figure 10–13 if the pressure angle is 20° and the shaft angle is 90°.

TABLE 10–3 Load-distribution factors for bevel gears, K_m

Type of gearing	Both gears straddle-mounted	One gear straddle-mounted	Neither gear straddle-mounted
General commercial-quality	1.44	1.58	1.80
High-quality, commercial gearing	1.20	1.32	1.50

Source: Extracted from AGMA 6010–E88, *Standard for Spur, Helical, Herringbone and Bevel Enclosed Drives,* with the permission of the publisher, American Gear Manufacturers Association, 1500 King Street, Suite 201, Alexandria, VA 22314.

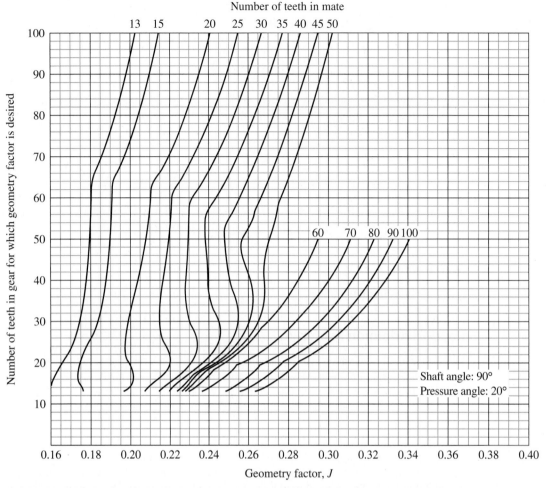

FIGURE 10–13   Geometry factor, *J*, for straight bevel gears with 20° pressure angle and $0.120/P_d$ tool edge radius (Extracted from AGMA 6010-F97, *Standard for Spur, Helical, Herringbone and Bevel Enclosed Drives,* with permission of the publisher, American Gear Manufacturers Association, 1500 King Street, Suite 201, Alexandria, VA 22314)

Allowable Bending Stress Number

The computed value for stress from Equation (10–13) can be compared with the allowable bending stress number from Tables 9–3 and 9–4 and Figure 9–10. A life factor, Y_N, or a reliability factor, K_R, can be applied as discussed in Chapter 9 if the design life is different from 10^7 cycles or if the desired reliability is different from 0.99.

Example Problem 10–5   Compute the bending stress in the teeth of the bevel pinion shown in Figure 10–9. The data from Example Problem 10–3 apply: $N_P = 16$; $N_G = 48$; $n_P = 600$ rpm; $P = 2.50$ hp; $P_d = 8$; $d = 2.000$ in; $F = 1.00$ in. Assume that the pinion is driven by an electric motor and that the load provides moderate shock. The quality number, Q_v, is to be 6.

Solution

$$W_t = \frac{T}{r} = \frac{63\,000(P)}{n_p}\,\frac{1}{d/2} = \frac{63\,000(2.50)}{600}\,\frac{1}{2.000/2} = 263 \text{ lb}$$

$$v_t = \pi dn_P/12 = \pi(2.000)(600)/12 = 314 \text{ ft/min}$$

$$K_o = 1.50 \text{ (from Table 9–5)}$$

$$K_s = 1.00$$

$$K_m = 1.44 \text{ (both gears straddle-mounted, general commercial-quality)}$$

$$J = 0.230 \text{ (from Figure 10–13)}$$

The value for the dynamic factor, K_v, must be computed from Equation (10–15), for $Q = 6$ and $v_t = 314$ ft/min. As a design decision, let's use two Grade 1 steel gears that are through-hardened at 300 HB with $s_{at} = 36\,000$ psi (Figure 9–10). The modulus of elasticity for both gears is 30×10^6 psi. Then

$$u = \frac{8}{(2)^{0.5(6)}} - (36\,000)\left[\frac{125}{60 \times 10^6}\right] = 0.925$$

$$K_z = 85 - 10(u) = 85 - 10(0.85) = 75.8$$

$$C_v = K_v = \left[\frac{K_z}{K_z + \sqrt{v_t}}\right]^u = \left[\frac{76.5}{75.8 + \sqrt{314}}\right]^{0.925} = 0.823$$

Checking $C_{v\,\min} = (2/\pi)\tan^{-1}(314/333) = 0.481$. The value of C_v is acceptable.
Then, from Equation (10-13),

$$s_t = \frac{W_t\,P_d}{FJ}\,\frac{K_oK_sK_m}{K_v} = \frac{(263)(8)}{(1.00)(0.230)}\,\frac{(1.50)(1.00)(1.44)}{(0.823)} = 24\,000 \text{ psi}$$

Referring to Figure 9–10, note that this is a very modest stress level for steel gears, requiring only about HB 180 for the hardness. If stress were the only consideration, a redesign might be attempted to achieve a more compact system. However, normally the pitting resistance or surface durability of the teeth requires a harder material. The next section discusses pitting.

**10–10
DESIGN OF
BEVEL GEARS
FOR PITTING
RESISTANCE**

The approach to the design of bevel gears for pitting resistance is similar to that for spur gears. The failure mode is fatigue of the surface of the teeth under the influence of the contact stress between the mating gears.

The contact stress, called the *Hertz stress,* s_c, can be computed from

⇨ **Contact Stress
Number**

$$s_c = C_pC_b\sqrt{\frac{W_t}{FdI}\,\frac{C_oC_m}{C_v}} \qquad \textbf{(10–16)}$$

The factors C_o, C_v, and C_m are the same as K_o, K_v, and K_m, used for computing the bending stress in the preceding section. The terms W_t, F, and d have the same meanings as well. The factor C_p is the elastic coefficient and is the same as shown in Equation (9–23) and Table 9–9. For steel or cast iron gears,

FIGURE 10–14

Geometry factors for straight and ZEROL® bevel gears (Extracted from AGMA 2003-A86, *Rating the Pitting Resistance and Bending Strength of Generated Straight Bevel, ZEROL® Bevel, and Spiral Bevel Gear Teeth,* with the permission of the publisher, American Gear Manufacturers Association, 1500 King Street, Suite 201, Alexandria, VA 22314)

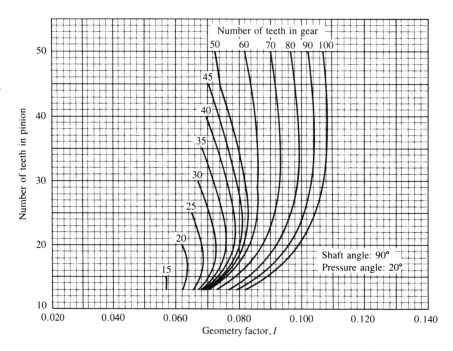

$C_p = 2300$ for two steel gears $(E = 30 \times 10^6 \text{ psi})$

$C_p = 1960$ for two cast iron gears $(E = 19 \times 10^6 \text{ psi})$

$C_p = 2100$ for a steel pinion and a cast iron gear

Using $C_b = 0.634$ allows the use of the same allowable contact stress as for spur and helical gears.

The factor I is the geometry factor for surface durability and can be found from Figure 10–14.

The Hertz contact stress, computed from Equation (10–16), should be compared with the allowable contact stress number, s_{ac}, from Figure 9–11 or Table 9–3 if the material is steel. For cast iron, use the values in Table 9–4.

Example Problem 10–6

Compute the Hertz stress for the gear pair in Figure 10–9 for the conditions used in Example Problem 10–5: $N_P = 16$; $N_G = 48$; $n_P = 600$ rpm; $P_d = 8$; $F = 1.00$ in; $d = 2.000$ in. Both gears are to be steel. Specify a suitable steel for the gears, along with its heat treatment.

Solution

From Example Problem 10–5: $W_t = 263$ lb, $C_o = 1.50$, $C_v = 0.83$, and $C_m = 1.44$. For two steel gears, $C_p = 2300$. From Figure 10–14, $I = 0.077$. Then

$$s_c = C_p C_b \sqrt{\frac{W_t}{FdI} \frac{C_o C_m}{C_v}} = (2300)(0.634)\sqrt{\frac{263}{(1.00)(2.000)(0.077)} \frac{(1.50)(1.44)}{0.823}}$$

$$s_c = 97\,500 \text{ psi}$$

Comparing this value with the allowable contact stress number from Figure 9–11 shows that a through-hardened, Grade 1 steel with a hardness of HB 220 is capable of withstanding this level of stress. Because this value is higher than that required for bending strength, it controls the design.

Practical Considerations for Bevel Gearing

Factors similar to those discussed for spur and helical gears should be considered in the design of systems using bevel gears. The accuracy of alignment and the accommodation of thrust loads discussed in the example problems are critical. Figures 10–15 and 10–16 show commercial applications.

**10–11
FORCES,
FRICTION, AND
EFFICIENCY IN
WORMGEAR
SETS**

See Chapter 8 for the geometry of wormgear sets. Also see References 2, 14, 15, and 17.

The force system acting on the worm/wormgear set is usually considered to be made of three perpendicular components as was done for helical and bevel gears. There are a tangential force, a radial force, and an axial force acting on the worm and the wormgear. We will use the same notation here as in the bevel gear system.

Figure 10–17 shows two orthogonal views (front and side) of a worm/wormgear pair, showing only the pitch diameters of the gears. The figure shows the separate worm

Spiral bevel
gearing, first stage

Input shaft

Fabricated steel housing

Helical gearing
second and third stages

Fabricated steel housing

FIGURE 10–15   Spiral bevel, right-angle reducer (Sumitomo Machinery Corporation of America, Teterboro, NJ)

FIGURE 10–16
Final drive for a tractor
(Case IH, Racine, WI)

FIGURE 10–17

Forces on a worm and a wormgear

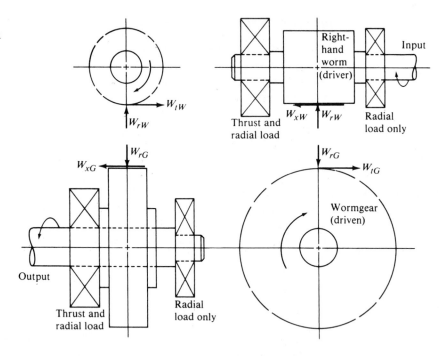

and wormgear with the forces acting on each. Note that because of the 90° orientation of the two shafts,

⇨ **Forces on Worms and Wormgears**

$$\left.\begin{array}{l} W_{tG} = W_{xW} \\ W_{xG} = W_{tW} \\ W_{rG} = W_{rW} \end{array}\right\} \qquad (10\text{–}17)$$

Of course, the directions of the paired forces are opposite because of the action/reaction principle.

The tangential force on the wormgear is computed first and is based on the required operating conditions of torque, power, and speed at the output shaft.

Coefficient of Friction, μ

Friction plays a major part in the operation of a wormgear set because there is inherently sliding contact between the worm threads and the wormgear teeth. The coefficient of friction is dependent on the materials used, the lubricant, and the sliding velocity. Based on the pitch line speed of the gear, the sliding velocity is

⇨ **Sliding Velocity for Gear**

$$v_s = v_{tG}/\sin\lambda \qquad (10\text{–}18)$$

Based on the pitch line speed of the worm,

⇨ **Sliding Velocity for Worm**

$$v_s = v_{tW}/\cos\lambda \qquad (10\text{–}19)$$

The term λ is the lead angle for the worm thread as defined in equation (8–37).

The AGMA (see Reference 15) recommends the following formulas to estimate the coefficient of friction for a hardened steel worm (58 HRC minimum), smoothly ground, or polished, or rolled, or with an equivalent finish, operating on a bronze wormgear. The

FIGURE 10–18
Coefficient of friction vs. sliding velocity for steel worm and bronze wormgear

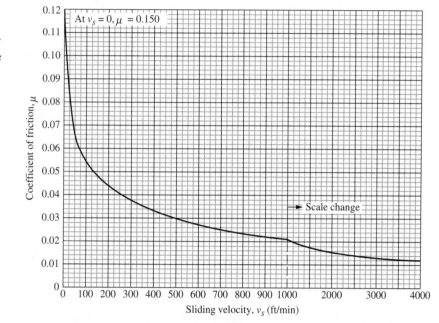

choice of formula depends on the sliding velocity. *Note:* v_s must be in ft/min in the formulas; 1.0 ft/min = 0.0051 m/s.

Static Condition: $v_s = 0$

$$\mu = 0.150$$

Low Speed: $v_s < 10$ ft/min (0.051 m/s)

$$\mu = 0.124e^{(-0.074v_s^{0.645})} \qquad \textbf{(10–20)}$$

Higher Speed: $v_s > 10$ ft/min

$$\mu = 0.103e^{(-0.110v_s^{0.450})} + 0.012 \qquad \textbf{(10–21)}$$

Figure 10–18 is a plot of the coefficient μ versus the sliding velocity v_s.

Output Torque from Wormgear Drive, T_o

In most design problems for wormgear drives, the output torque and the rotating speed of the output shaft will be known from the requirements of the driven machine. Torque and speed are related to the output power by

Output Torque from Wormgear

$$T_o = \frac{63\,000(P_o)}{n_G} \qquad \textbf{(10–22)}$$

By referring to the end view of the wormgear in Figure 10–17, you can see that the output torque is

$$T_o = W_{tG} \cdot r_G = W_{tG}\,(D_G/2)$$

Then the following procedure can be used to compute the forces acting in a worm/wormgear drive system.

Procedure for Calculating the Forces on a Worm/Wormgear Set

Given:

> Output torque, T_o, in lb·in
> Output speed, n_G, in rpm
> Pitch diameter of the wormgear, D_G, in inches
> Lead angle, λ
> Normal pressure angle, ϕ_n

Compute:

$$W_{tG} = 2\,T_o/D_G \tag{10-23}$$

$$W_{xG} = W_{tG}\,\frac{\cos\phi_n\,\sin\lambda\,+\,\mu\,\cos\lambda}{[\cos\phi_n\,\cos\lambda\,-\,\mu\,\sin\lambda]} \tag{10-24}$$

$$W_{rG} = \frac{W_{tG}\,\sin\phi_n}{\cos\phi_n\,\cos\lambda\,-\,\mu\,\sin\lambda} \tag{10-25}$$

The forces on the worm can be obtained by observation, using Equation (10–17). Equations (10–24) and (10–25) were derived using the components of both the tangential driving force on the wormgear and the friction force at the location of the meshing worm threads and wormgear teeth. The complete development of the equations is shown in Reference 20.

Friction Force, W_f

The friction force, W_f, acts parallel to the face of the worm threads and the gear teeth and depends on the tangential force on the gear, the coefficient of friction, and the geometry of the teeth:

$$W_f = \frac{\mu W_{tG}}{(\cos\lambda)(\cos\phi_n)\,-\,\mu\,\sin\lambda} \tag{10-26}$$

Power Loss Due to Friction, P_L

Power loss is the product of the friction force and the sliding velocity at the mesh. That is,

$$P_L = \frac{v_s W_f}{33\,000} \tag{10-27}$$

In this equation, the power loss is in hp, v_s is in ft/min, and W_f is in lb.

Input Power, P_i

The input power is the sum of the output power and the power loss due to friction:

$$P_i = P_o + P_L \tag{10-28}$$

Efficiency, η

Efficiency is defined as the ratio of the output power to the input power:

$$\eta = P_o/P_i \tag{10-29}$$

FIGURE 10–19
Efficiency of wormgear
drive vs. lead angle

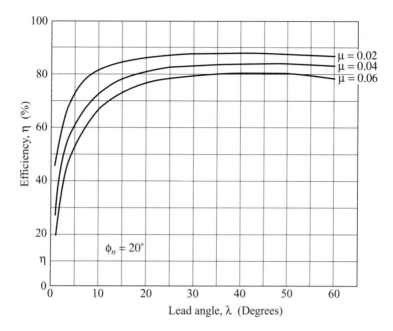

Efficiency for a wormgear drive with the usual case of the input coming through the worm can also be computed directly from the following equation.

$$\eta = \frac{\cos \phi_n - \mu \tan \lambda}{\cos \phi_n + \mu/\tan \lambda} \qquad \textbf{(10–30)}$$

Factors Affecting Efficiency

As can be seen in Equation (10–26), the lead angle, the normal pressure angle, and the co-efficient of friction all affect the efficiency. The one that has the largest effect, and the one over which the designer has the most control, is the lead angle, λ. The larger the lead angle, the higher the efficiency, up to approximately $\lambda = 45°$. (See figure 10–19.)

Now, looking back to the definition of the lead angle, note that the number of threads in the worm has a major effect on the lead angle. Therefore, to obtain a high efficiency, use multiple-threaded worms. But there is a disadvantage to this conclusion. More worm threads require more gear teeth to achieve the same ratio, resulting in a larger system over-all. The designer is often forced to compromise.

Example Problem: Forces and Efficiency in Wormgearing

Review now the results of Example Problem 8–7, in which the geometry factors for a particular worm and wormgear set were computed. The following example problem extends the analysis to include the forces acting on the system for a given output torque.

Example Problem 10–7   The wormgear drive described in Example Problem 8–7 is transmitting an output torque of 4168 lb·in. The transverse pressure angle is 20°. The worm is made from hardened and ground steel, and the wormgear is bronze. Compute the forces on the worm and the worm-gear, the output power, the input power, and the efficiency.

Solution   Recall from Example Problem 8–7 that

$$\lambda = 14.04° \qquad D_G = 8.667 \text{ in} \qquad n_G = 101 \text{ rpm}$$
$$n_W = 1750 \text{ rpm} \qquad D_W = 2.000 \text{ in}$$

The normal pressure angle is required. From Equation (8–41),

$$\phi_n = \tan^{-1}(\tan \phi_t \cos \lambda) = \tan^{-1}(\tan 20° \cos 14.04°) = 19.45°$$

Because they recur in several formulas, let's compute the following:

$$\sin \phi_n = \sin 19.45° = 0.333$$
$$\cos \phi_n = \cos 19.45° = 0.943$$
$$\cos \lambda = \cos 14.04° = 0.970$$
$$\sin \lambda = \sin 14.04° = 0.243$$
$$\tan \lambda = \tan 14.04° = 0.250$$

We can now compute the tangential force on the wormgear using Equation (10–23)

$$W_{tG} = \frac{2T_o}{D_G} = \frac{(2)(4168 \text{ lb} \cdot \text{in})}{8.667 \text{ in}} = 962 \text{ lb}$$

The calculations of the axial and radial forces require a value for the coefficient of friction that, in turn, depends on the pitch line speed and the sliding velocity.

Pitch Line Speed of the Gear

$$v_{tG} = \pi D_G n_G / 12 = \pi(8.667)(101)/12 = 229 \text{ ft/min}$$

Sliding Velocity [Equation (10–18)]

$$v_s = v_s/\sin \lambda = 229/\sin 14.04° = 944 \text{ ft/min}$$

Coefficient of Friction: From Figure 10–18, at a sliding velocity of 944 ft/min, we can read $\mu = 0.022$.

Now the axial and radial forces on the wormgear can be computed.

Axial Force on the Wormgear [Equation 10–24]

$$W_{xG} = 962 \text{ lb} \left[\frac{(0.943)(0.243) + (0.022)(0.970)}{(0.943)(0.970) - (0.022)(0.243)} \right] = 265 \text{ lb}$$

Radial Force on the Wormgear [Equation 10–25]

$$W_{rG} = \left[\frac{(962)(0.333)}{(0.943)(0.970) - (0.022)(0.243)} \right] = 352 \text{ lb}$$

Now the output power, input power, and efficiency can be computed.

Output Power [Equation 10–22]

$$P_o = \frac{T_o n_G}{63\,000} = \frac{(4168 \text{ lb} \cdot \text{in})(101 \text{ rpm})}{63\,000} = 6.68 \text{ hp}$$

The input power depends on the friction force and the consequent power loss due to friction.

Friction Force [Equation 10–26]

$$W_f = \frac{\mu W_{tG}}{(\cos \lambda)(\cos \phi_n) - \mu \sin \lambda} = \frac{(0.022)(962 \text{ lb})}{(0.970)(0.943) - (0.022)(0.243)} = 23.3 \text{ lb}$$

Power Loss Due to Friction [Equation (10–27)]

$$P_L = \frac{v_s W_f}{33\ 000} = \frac{(944 \text{ ft/min})(23.3 \text{ lb})}{33\ 000} = 0.666 \text{ lb}$$

Input Power [Equation (10–28)]

$$P_i = P_o + P_L = 6.68 + 0.66 = 7.35 \text{ hp}$$

Efficiency [Equation (10–29)]

$$\eta = \frac{P_o}{P_i} (100\%) = \frac{6.68 \text{ hp}}{7.35 \text{ hp}} (100\%) = 90.9\%$$

Equation (10–30) could also be used to compute efficiency directly without computing friction power loss.

Self-Locking Wormgear Sets

Self-locking is the condition in which the worm drives the wormgear, but, if torque is applied to the gear shaft, the worm does not turn. It is locked! The locking action is produced by the friction force between the worm threads and the wormgear teeth, and this is highly dependent on the lead angle. It is recommended that a lead angle no higher than about 5.0° be used in order to ensure that self-locking will occur. This lead angle usually requires the use of a single-threaded worm; the low lead angle results in a low efficiency, possibly as low as 60% or 70%.

**10–12
STRESS IN
WORMGEAR
TEETH**

We present here an approximate method of computing the bending stress in the teeth of the wormgear. Because the geometry of the teeth is not uniform across the face width, it is not possible to generate an exact solution. However, the method given here should predict the bending stress with sufficient accuracy to check a design because most worm/wormgear systems are limited by pitting, wear, or thermal considerations rather than strength.

The AGMA, in its Standard 6034-B92, does not include a method of analyzing wormgears for strength. The method shown here was adapted from Reference 20. Only the wormgear teeth are analyzed because the worm threads are inherently stronger and are typically made from a stronger material.

The stress in the gear teeth can be computed from

$$\sigma = \frac{W_d}{yFp_n} \qquad \qquad \textbf{(10–31)}$$

TABLE 10–4 Approximate Lewis form factor for wormgear teeth

ϕ_n	y
$14\frac{1}{2}°$	0.100
$20°$	0.125
$25°$	0.150
$30°$	0.175

where W_d = dynamic load on the gear teeth
  y = Lewis form factor (see Table 10–4)
  F = face width of the gear
  p_n = normal circular pitch = $p \cos \lambda = \pi \cos \lambda / P_d$ **(10–32)**

The dynamic load can be estimated from

$$W_d = W_{tG}/K_v \tag{10–33}$$

and

$$K_v = 1200/(1200 + v_{tG}) \tag{10–34}$$

$$v_{tG} = \pi D_G n_G/12 = \text{pitch line speed of the gear} \tag{10–35}$$

Only one value is given for the Lewis form factor for a given pressure angle because the actual value is very difficult to calculate precisely and does not vary much with the number of teeth. The actual face width should be used, up to the limit of two-thirds of the pitch diameter of the worm.

The computed value of tooth bending stress from Equation (10–31) can be compared with the fatigue strength of the material of the gear. For manganese gear bronze, use a fatigue strength of 17 000 psi; for phosphor gear bronze, use 24 000 psi. For cast iron, use approximately 0.35 times the ultimate strength, unless specific data are available for fatigue strength.

**10–13
SURFACE
DURABILITY OF
WORMGEAR
DRIVES**

AGMA Standard 6034-B92 (see Reference 15) gives a method for rating the surface durability of hardened steel worms operating with bronze gears. The ratings are based on the ability of the gears to operate without significant damage from pitting or wear.

The procedure calls for the calculation of a *rated tangential load*, W_{tR}, from

⇨ **Rated Tangential
Load on Wormgears**

$$W_{tR} = C_s D_G^{0.8} F_e C_m C_v \tag{10–36}$$

MDESIGN

where C_s = materials factor (from Figure 10–20)
  D_G = pitch diameter of the wormgear, in inches
  F_e = effective face width, in inches. Use the actual face width of the wormgear up to a maximum of $0.67\,D_W$.
  C_m = ratio correction factor (from Figure 10–21)
  C_v = velocity factor (from Figure 10–22)

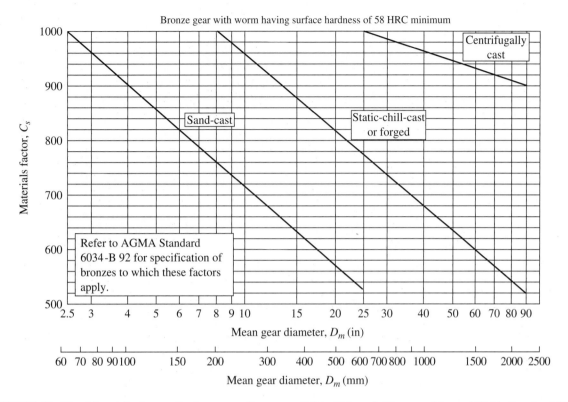

Bronze gear with worm having surface hardness of 58 HRC minimum

FIGURE 10–20   Materials factor, C_s, for center distance > 3.0 in (76 mm) (Extracted from AGMA Standard 6034-B92, *Practice for Enclosed Cylindrical Wormgear Speed Reducers and Gearmotors,* with the permission of the publisher, American Gear Manufacturers Association, 1500 King Street, Suite 201, Alexandria, VA 22314)

Conditions on the Use of Equation (10–36)

1. The analysis is valid only for a hardened steel worm (58 HRC minimum) operating with gear bronzes specified in AGMA Standard 6034-B92. The classes of bronzes typically used are tin bronze, phosphor bronze, manganese bronze, and aluminum bronze. The materials factor, C_s, is dependent on the method of casting the bronze, as indicated in Figure 10–20. The values for C_s can be computed from the following formulas.

Sand-Cast Bronzes:

For $D_G > 2.5$ in,

$$C_s = 1189.636 - 476.545 \log_{10}(D_G) \qquad \textbf{(10–37)}$$

For $D_G < 2.5$ in,

$$C_s = 1000$$

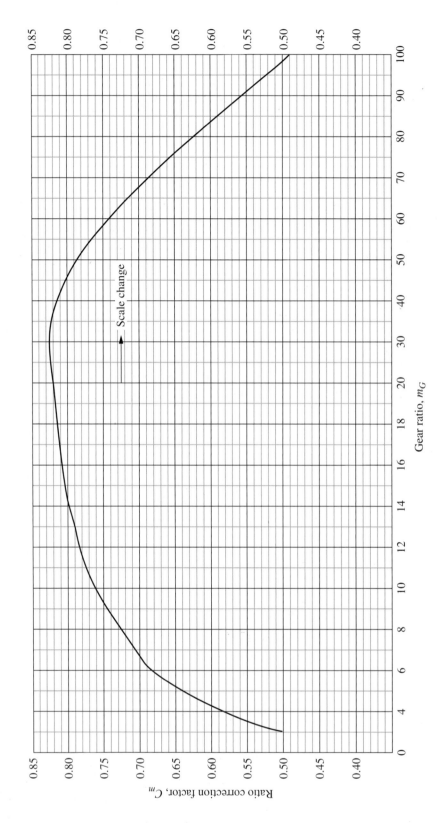

FIGURE 10–21 Ratio correction factor, C_m, vs. gear ratio, m_G

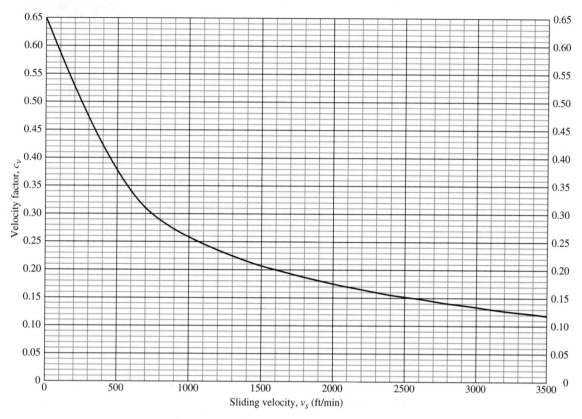

FIGURE 10–22 Velocity factor, C_v, vs. sliding velocity.

Static-Chill-Cast or Forged Bronzes:

For $D_G > 8.0$ in,

$$C_s = 1411.651 - 455.825 \log_{10}(D_G) \qquad \textbf{(10–38)}$$

For $D_G < 8.0$ in,

$$C_s = 1000$$

Centrifugally Cast Bronzes:

For $D_G > 25$ in,

$$C_s = 1251.291 - 179.750 \log_{10}(D_G) \qquad \textbf{(10–39)}$$

For $D_G < 25$ in,

$$C_s = 1000$$

2. The wormgear diameter is the second factor in determining C_s. The *mean diameter* at the midpoint of the working depth of the gear teeth should be used. If standard addendum gears are used, the mean diameter is equal to the pitch diameter.

3. Use the actual face width, F, of the wormgear as F_e if $F < 0.667(D_W)$. For larger face widths, use $F_e = 0.67\,(D_W)$, because the excess width is not effective.

4. The ratio correction factor, C_m, can be computed from the following formulas.

For Gear Ratios, m_G, from 6 to 20

$$C_m = 0.0200\,(-m_G^2 + 40m_G - 76)^{0.5} + 0.46 \qquad \text{(10–40)}$$

For Gear Ratios, m_G, from 20 to 76

$$C_m = 0.0107\,(-m_G^2 + 56m_G + 5145)^{0.5} \qquad \text{(10–41)}$$

For $m_G > 76$

$$C_m = 1.1483 - 0.00658m_G \qquad \text{(10–42)}$$

5. The velocity factor depends on the sliding velocity, v_s, computed from Equation (10–18) or (10–19). Values for C_v can be computed from the following formulas.

For v_s from 0 to 700 ft/min

$$C_v = 0.659e^{(-0.0011\,v_s)} \qquad \text{(10–43)}$$

For v_s from 700 to 3000 ft/min

$$C_v = 13.31v_s^{(-0.571)} \qquad \text{(10–44)}$$

For $v_s > 3000$ ft/min

$$C_v = 65.52v_s^{(-0.774)} \qquad \text{(10–45)}$$

6. The proportions of the worm and the wormgear must conform to the following limits defining the maximum and minimum pitch diameters of the worm in relation to the center distance, C, for the gear set. All dimensions are in inches.

$$\text{Maximum } D_W = C^{0.875}/1.6 \qquad \text{(10–46)}$$
$$\text{Minimum } D_W = C^{0.875}/3.0 \qquad \text{(10–47)}$$

7. The shaft carrying the worm must be sufficiently rigid to limit the deflection of the worm at the pitch point to the maximum value of $0.005\sqrt{P_x}$, where P_x is the axial pitch of the worm, which is numerically equal to the circular pitch, p, of the wormgear.

8. When you are analyzing a given wormgear set, the value of the rated tangential load, W_{tR}, must be greater than the actual tangential load, W_t, for satisfactory life.

9. Ratings given in this section are valid only for smooth systems such as fans or centrifugal pumps driven by an electric or hydraulic motor operating under 10 hours per day. More severe conditions, such as shock loading, internal combustion engine drives, or longer hours of operation, require the application of a service factor. Reference 15 lists several such factors based on field experience with specific types of equipment. For problems in this book, factors from Table 9–5 can be used.

Example Problem 10–8   Is the wormgear set described in Example Problem 8–7 satisfactory with regard to strength and wear when operating under the conditions of Example Problem 10–7? The wormgear has a face width of 1.25 in.

Solution   From previous problems and solutions,

$$W_{tG} = 962 \text{ lb} \qquad VR = m_G = 17.33$$
$$v_{tG} = 229 \text{ ft/min} \qquad v_s = 944 \text{ ft/min}$$
$$D_G = 8.667 \text{ in} \qquad D_W = 2.000 \text{ in}$$

Assume 58 HRC minimum for the steel worm. Assume that the bronze gear is sand-cast.

Stress

$$K_v = 1200/(1200 + v_{tG}) = 1200/(1200 + 229) = 0.84$$
$$W_d = W_{tG}/K_v = 962/0.84 = 1145 \text{ lb}$$
$$F = 1.25 \text{ in}$$
$$y = 0.125 \text{ (from Table 10–4)}$$
$$p_n = p \cos \lambda = (0.5236)\cos 14.04° = 0.508 \text{ in}$$

Then

$$\sigma = \frac{W_d}{yFp_n} = \frac{1145}{(0.125)(1.25)(0.508)} = 14\,430 \text{ psi}$$

The guidelines in Section 10–12 indicate that this stress level would be adequate for either manganese or phosphor gear bronze.

Surface Durability: Use Equation (10–36):

$$W_{tR} = C_s D_G^{0.8} F_e C_m C_v \qquad\qquad\qquad \textbf{(10–36)}$$

C Factors: The values for the *C* factors can be found from Figures 10–20, 10–21, and 10–22. We find

$$C_s = 740 \text{ for sand-cast bronze} \quad \text{and} \quad D_G = 8.667 \text{ in}$$
$$C_m = 0.184 \quad \text{for } m_G = 17.33$$
$$C_v = 0.265 \quad \text{for } v_s = 944 \text{ ft/min}$$

We can use $F_e = F = 1.25$ in if this value is not greater than 0.67 times the worm diameter. For $D_W = 2.000$ in,

$$0.67 D_W = (0.67)(2.00 \text{ in}) = 1.333 \text{ in}$$

Therefore, use $F_e = 1.25$ in. Then the rated tangential load is

$$W_{tR} = (740)(8.667)^{0.8}(1.25)(0.814)(0.265) = 1123 \text{ lb}$$

Because this value is greater than the actual tangential load of 962 lb, the design should be satisfactory, provided that the conditions defined for the application of Equation (10–36) are met.

REFERENCES

1. American Gear Manufacturers Association. Standard 2008-C01. *Standard for Assembling Bevel Gears.* Alexandria, VA: American Gear Manufacturers Association, 2001.

2. American Gear Manufacturers Association. Standard 6022-C93. *Design Manual for Cylindrical Wormgearing.* Alexandria, VA: American Gear Manufacturers Association, 1993.

3. American Gear Manufacturers Association. Standard AGMA 917-B97. *Design Manual for Parallel Shaft Fine-Pitch Gearing.* Alexandria, VA: American Gear Manufacturers Association, 1997.

4. American Gear Manufacturers Association. Standard 6001-D97. *Design and Selection of Components for Enclosed Gear Drives.* Alexandria, VA: American Gear Manufacturers Association, 1997.

5. American Gear Manufacturers Association. Standard AGMA 390.03a (R1980). *Gear Handbook: Gear Classification, Materials and Measuring Methods for Bevel, Hypoid, Fine Pitch Wormgearing and Racks Only as Unassembled Gears.* Alexandria, VA: American Gear Manufacturers Association, 1988. Those portions of 390.03 not included in AGMA 2000-A88, June 1988.

6. American Gear Manufacturers Association. Standard AGMA 908-B89 (R1995). *Geometry Factors for Determining the Pitting Resistance and Bending Strength of Spur, Helical, and Herringbone Gear Teeth.* Alexandria, VA: American Gear Manufacturers Association, 1995.

7. American Gear Manufacturers Association. Standard AGMA 1012-F90. *Gear Nomenclature, Definitions of Terms with Symbols.* Alexandria, VA: American Gear Manufacturers Association, 1990.

8. American Gear Manufacturers Association. Standard AGMA 2000-A88. *Gear Classification and Inspection Handbook—Tolerances and Measuring Methods for Unassembled Spur and Helical Gears (Including Metric Equivalents).* Alexandria, VA: American Gear Manufacturers Association, 1988. Partial replacement of AGMA 390.03.

9. American Gear Manufacturers Association. Standard AGMA 2001-C95. *Fundamental Rating Factors and Calculation Methods for Involute Spur and Helical Gear Teeth.* Alexandria, VA: American Gear Manufacturers Association, 1994.

10. American Gear Manufacturers Association. Standard AGMA 2003-B97. *Rating the Pitting Resistance and Bending Strength of Generated Straight Bevel, ZEROL® Bevel, and Spiral Bevel Gear Teeth.* Alexandria, VA: American Gear Manufacturers Association, 1997.

11. American Gear Manufacturers Association. Standard AGMA 2004-B89 (R1995). *Gear Materials and Heat Treatment Manual.* Alexandria, VA: American Gear Manufacturers Association, 1995.

12. American Gear Manufacturers Association. Standard AGMA 2005-C96. *Design Manual for Bevel Gears.* Alexandria, VA: American Gear Manufacturers Association, 1996.

13. American Gear Manufacturers Association. Standard AGMA 6010-F97. *Standard for Spur, Helical, Herringbone, and Bevel Enclosed Drives.* Alexandria, VA: American Gear Manufacturers Association, 1997.

14. American Gear Manufacturers Association. Standard AGMA 6030-C87 (R1994). *Design of Industrial Double-Enveloping Wormgears.* Alexandria, VA: American Gear Manufacturers Association, 1994.

15. American Gear Manufacturers Association. Standard AGMA 6034-B92. *Practice for Enclosed Cylindrical Wormgear Speed Reducers and Gearmotors.* Alexandria, VA: American Gear Manufacturers Association, 1992.

16. American Society for Metals. *Source Book on Gear Design, Technology and Performance.* Metals Park, OH: American Society for Metals, 1980.

17. Drago, Raymond J. *Fundamentals of Gear Design.* Boston: Butterworths, 1988.

18. Dudley, Darle W. *Dudley's Gear Handbook.* New York: McGraw-Hill, 1991.

19. Lynwander, Peter. *Gear Drive Systems: Design* and *Applications.* New York: Marcel Dekker, 1983.

20. Shigley, Joseph E., and Charles R. Mischke. *Mechanical Engineering Design:* 6th ed. New York: McGraw-Hill, 2001.

21. Shigley, Joseph E., and Charles R. Mischke. *Gearing: A Mechanical Designers' Workbook.* New York: McGraw-Hill, 1990.

22. Society of Automotive Engineers. *Gear Design, Manufacturing and Inspection Manual AE-15.* Warrendale, PA: Society of Automotive Engineers, 1990.

INTERNET SITES RELATED TO HELICAL GEARS, BEVEL GEARS, AND WORMGEARING

Refer to the list of Internet sites at the end of Chapter 9, Spur Gears. Virtually all sites listed there are also relevant to the design of helical gears, bevel gears, and wormgearing.

PROBLEMS

Helical Gearing

1. A helical gear has a transverse diametral pitch of 8, a transverse pressure angle of 14½°, 45 teeth, a face width of 2.00 in, and a helix angle of 30°.

   (a) If the gear transmits 5.0 hp at a speed of 1250 rpm, compute the tangential force, the axial force, and the radial force.

   (b) If the gear operates with a pinion having 15 teeth, compute the bending stress in the pinion teeth. The power comes from an electric motor, and the drive is to a reciprocating pump. Specify a quality number for the teeth.

   (c) Specify a suitable material for the pinion and the gear considering both strength and pitting resistance.

2. A helical gear has a normal diametral pitch of 12, a normal pressure angle of 20°, 48 teeth, a face width of 1.50 in, and a helix angle of 45°.

   (a) If the gear transmits 2.50 hp at a speed of 1750 rpm, compute the tangential force, the axial force, and the radial force.

   (b) If the gear operates with a pinion having 16 teeth, compute the bending stress in the pinion teeth. The power comes from an electric motor, and the drive is to a centrifugal blower. Specify a quality number for the teeth.

   (c) Specify a suitable material for the pinion and the gear considering both strength and pitting resistance.

3. A helical gear has a transverse diametral pitch of 6, a transverse pressure angle of 14½°, 36 teeth, a face width of 1.00 in, and a helix angle of 45°.

   (a) If the gear transmits 15.0 hp at a speed of 2200 rpm, compute the tangential force, the axial force, and the radial force.

   (b) If the gear operates with a pinion having 12 teeth, compute the bending stress in the pinion teeth. The power comes from a six-cylinder gasoline engine, and the drive is to a concrete mixer. Specify a quality number for the teeth.

   (c) Specify a suitable material for the pinion and the gear considering both strength and pitting resistance.

4. A helical gear has a normal diametral pitch of 24, a normal pressure angle of 14½°, 72 teeth, a face width of 0.25 in, and a helix angle of 45°.

   (a) If the gear transmits 0.50 hp at a speed of 3450 rpm, compute the tangential force, the axial force, and the radial force.

   (b) If the gear operates with a pinion having 16 teeth, compute the bending stress in the pinion teeth. The

power comes from an electric motor, and the drive is to a winch that will experience moderate shock. Specify a quality number for the teeth.

   (c) Specify a suitable material for the pinion and the gear considering both strength and pitting resistance.

For Problems 5–11, complete the design of a pair of helical gears to operate under the stated conditions. Specify the geometry of the gears and the material and its heat treatment. Assume that the drive is from an electric motor unless otherwise specified. Consider both strength and pitting resistance.

5. A pair of helical gears is to be designed to transmit 5.0 hp while the pinion rotates at 1200 rpm. The gear drives a reciprocating compressor and must rotate between 385 and 390 rpm.

6. A helical gear pair is to be a part of the drive for a milling machine requiring 20.0 hp with the pinion speed at 550 rpm and the gear speed to be between 180 and 190 rpm.

7. A helical gear drive for a punch press requires 50.0 hp with the pinion rotating at 900 rpm and the gear speed at 225 to 230 rpm.

8. A single-cylinder gasoline engine has the pinion of a helical gear pair on its output shaft. The gear is attached to the shaft of a small cement mixer. The mixer requires 2.5 hp while rotating at approximately 75 rpm. The engine is governed to run at approximately 900 rpm.

9. A four-cylinder industrial engine runs at 2200 rpm and delivers 75 hp to the input gear of a helical gear drive for a large wood chipper used to prepare pulpwood chips for paper making. The output gear must run between 4500 and 4600 rpm.

10. A small commercial tractor is being designed for chores such as lawn mowing and snow removal. The wheel drive system is to be through a helical gear pair in which the pinion runs at 450 rpm while the gear, mounted on the hub of the wheel, runs at 75 to 80 rpm. The wheel has an 18-in diameter. The two-cylinder gasoline engine delivers 20.0 hp to the wheels.

11. A water turbine transmits 15.0 hp to a pair of helical gears at 4500 rpm. The output of the gear pair must drive an electric power generator at 3600 rpm. The center distance for the gear pair must not exceed 4.00 in.

12. Determine the power-transmitting capacity of a pair of helical gears having a normal pressure angle of 20°, a helix angle of 15°, a normal diametral pitch of 10, 20 teeth in the pinion, 75 teeth in the gear, and a face width of 2.50 in, if they are made from AISI 4140 OQT 1000 steel. They are of typical commercial quality. The pinion will rotate at 1725 rpm on the shaft of an electric motor. The gear will drive a centrifugal pump.

13. Repeat Problem 12 with the gears made from AISI 4620 DOQT 300 carburized, case-hardened steel. Then compute the axial and radial forces on the gears.

Bevel Gears

14. A straight bevel gear pair has the following data: N_P =15; N_G = 45; P_d = 6; 20° pressure angle. If the gear pair is transmitting 3.0 hp, compute the forces on both the pinion and the gear. The pinion speed is 300 rpm. The face width is 1.25 in. Compute the bending stress and the contact stress for the teeth, and specify a suitable material and heat treatment. The gears are driven by a gasoline engine, and the load is a concrete mixer providing moderate shock. Assume that neither gear is straddle-mounted.

15. A straight bevel gear pair has the following data: N_P = 25; N_G = 50; P_d = 10; 20° pressure angle. If the gear pair is transmitting 3.5 hp, compute the forces on both the pinion and the gear. The pinion speed is 1250 rpm. The face width is 0.70 in. Compute the bending stress and the contact stress for the teeth, and specify a suitable material and heat treatment. The gears are driven by a gasoline engine, and the load is a conveyor providing moderate shock. Assume that neither gear is straddle-mounted.

16. Design a pair of straight bevel gears to transmit 5.0 hp at a pinion speed of 850 rpm. The gear speed should be approximately 300 rpm. Consider both strength and pitting resistance. The driver is a gasoline engine, and the driven machine is a heavy-duty conveyor.

17. Design a pair of straight bevel gears to transmit 0.75 hp at a pinion speed of 1800 rpm. The gear speed should be approximately 475 rpm. Consider both strength and pitting resistance. The driver is an electric motor, and the driven machine is a reciprocating saw.

Wormgearing

18. A wormgear set has a single-thread worm with a pitch diameter of 1.250 in, a diametral pitch of 10, and a normal pressure angle of 14.5°. If the worm meshes with a wormgear having 40 teeth and a face width of 0.625 in, compute the gear pitch diameter, the center distance, and the velocity ratio. If the wormgear set is transmitting 924 lb·in of torque at its output shaft, which is rotating at 30 rpm, compute forces on the gear, efficiency, input speed, input power, and stress on the gear teeth. If the worm is hardened steel and the gear is chilled bronze, evaluate the rated load, and determine whether the design is satisfactory for pitting resistance.

19. Three designs are being considered for a wormgear set to produce a velocity ratio of 20 when the worm gear rotates at 90 rpm. All three have a diametral pitch of 12, a worm pitch diameter of 1.000 in, a gear face width of 0.500 in, and a normal pressure angle of 14.5°. One has a single-thread worm and 20 wormgear teeth; the second has a double-thread worm and 40 wormgear teeth; the third has a four-thread worm and 80 wormgear teeth. For each design, compute the rated output torque, considering both strength and pitting resistance. The worms are hardened steel, and the wormgears are chilled bronze.

20. For each of the three designs proposed in Problem 19, compute the efficiency.

    The data for Problems 21, 22, and 23 are given in Table 10–5. Design a wormgear set to produce the desired velocity ratio when transmitting the given torque at the output shaft for the given output rotational speed.

TABLE 10–5 Data for Problems 21–23

Problem	VR	Torque (lb·in)	Output speed (rpm)
21.	7.5	984	80
22.	3	52.5	600
23.	40	4200	45

24. Compare the two designs described in Table 10–6 when each is transmitting 1200 lb·in of torque at its output shaft, which rotates at 20 rpm. Compute the forces on the worm and the wormgear, the efficiency, and the input power required.

TABLE 10–6 Designs for Problem 24

Design	P_d	N_t	N_G	D_w	F_G	Pressure angle
A	6	1	30	2.000	1.000	14.5°
B	10	2	60	1.250	0.625	14.5°

11

Keys, Couplings, and Seals

The Big Picture

You Are the Designer

11–1 Objectives of This Chapter

11–2 Keys

11–3 Materials for Keys

11–4 Stress Analysis to Determine Key Length

11–5 Splines

11–6 Other Methods of Fastening Elements to Shafts

11–7 Couplings

11–8 Universal Joints

11–9 Retaining Rings and Other Means of Axial Location

11–10 Types of Seals

11–11 Seal Materials

<table>
<tr>
<td>

The Big Picture

</td>
<td>

Keys, Couplings, and Seals

Discussion Map

☐ Keys and couplings connect functional parts of mechanisms and machines, allowing moving parts to transmit power or to locate parts relative to each other.

☐ Retaining rings hold assemblies together or hold parts on shafts, such as keeping a sprocket in position or holding a wheel on an axle.

☐ Seals protect critical components by excluding contaminants or by retaining fluids inside the housing of a machine.

Discover

Look around you and identify several examples of the use of keys, couplings, retaining rings, and seals in automobiles, trucks, home appliances, shop tools, gardening equipment, or bicycles.

This chapter will help you understand the functions and design requirements of such devices. In addition, you will learn to recognize commercially available designs and apply them properly.

</td>
</tr>
</table>

Think of how two or more parts of a machine can be connected for the purpose of locating one part with respect to another. Now think about how that connection must be designed if the parts are moving and if power must be transmitted between them.

This chapter presents information on commercially available products that accomplish these functions. The generic categories of keys, couplings, and seals actually encompass numerous different designs.

A *key* is used to connect a drive member such as a belt pulley, chain sprocket, or gear to the shaft that carries it. See Figure 11–1. Torque and power are transmitted across the key to or from the shaft. But how does the power get into or out of the shaft? One way might be that the output from the shaft of a motor or engine is connected to the input shaft of a transmission through a *flexible coupling* that reliably transmits power but allows for some misalignment between the shafts during operation because of the flexing of frame members or through progressive misalignment due to wear.

Seals may be difficult to see because they are typically encased in a housing or are covered in some way. Their function is to protect critical elements of a machine from contamination due to dust, dirt, or water or other fluids while allowing rotating or translating machine elements to move to accomplish their desired functions. Seals exclude undesirable materials from the inside of a mechanism, or they hold critical lubrication or cooling fluids inside the housing of the mechanism.

Look at machines that you interact with each day, and identify parts that fit the descriptions given here. Look in the engine compartment of a car or a truck. How are drive pulleys, linkages, latches, the hinges for the hood, the fan, the windshield wipers, and any other moving part connected to something else—the frame of the car, a rotating shaft, or some other moving part? If you are familiar with the inner workings of the engine and transmission, describe how those parts are connected. Look at the steering and suspension systems, the water pump, the fuel pump, the brake fluid reservoir, and the suspension struts or shock absorbers. Try to see where seals are used. Can you see the universal joints, sometimes called the *constant-velocity* (CV) joints, in the drive train? They should be connecting the output shaft from the transmission to the final parts of the drive train as the power is delivered to the wheels.

Find a small lawn or garden tractor at home or at a local home store. Typically their mechanisms are accessible, although protected from casual contact for safety reasons. Trace how power is transmitted from the engine, through a transmission, through a drive chain or belt, and all the way to the wheels or to the blade of a mower. How are functional parts connected together?

Look at home appliances, power tools in a home shop, and gardening equipment. Can you see parts that are held in place with *retaining rings?* These are typically thin, flat rings that are pressed onto shafts or inserted into grooves to hold a wheel on a shaft, to hold a gear or a pulley in position along the length of a shaft, or to simply hold some part of the device in place.

How are the keys, couplings, seals, retaining rings, and other connecting devices made? What materials are used? How are they installed? Can they be removed? What kinds of forces must they resist? How does their special geometry accomplish the desired function? How could they fail?

This chapter will help you become familiar with such mechanical components, with some of the manufacturers that offer commercial versions, and with correct application methods.

 You Are the Designer

In the first part of Chapter 8, you were the designer of a gear-type speed reducer whose conceptual design is shown in Figure 8–3. It has four gears, three shafts, and six bearings, all contained within a housing. How are the gears attached to the shafts? One way is to use keys at the interface between the hub of the gears and the shaft. You should be able to design the keys. How is the input shaft connected to the motor or engine that delivers the power? How is the output shaft connected to the driven machine? One way is to use flexible couplings. You should be able to specify commercially available couplings and apply them properly, considering the amount of torque they must transmit and how much misalignment they should permit.

How are the gears located axially along the shafts? Part of this function may be provided by shoulders machined on the shaft. But that works only on one side. On the other side, one type of locating means is a retaining ring that is installed into a groove in the shaft after the gear is in place. Rings or spacers may be used at the left of gears *A* and *B* and at the right of gears *C* and *D*. Notice that the input and output shafts extend outside the housing. How can you keep contaminants from the outside from getting inside? How can you keep lubricating oil inside? Shaft seals can accomplish that function. Seals may also be provided on the bearings to retain lubricant inside and in full contact with the rotating balls or rollers of the bearing.

You should be familiar with the kinds of materials used for seals and with their special geometries. These concepts are discussed in this chapter.

**11–1
OBJECTIVES OF
THIS CHAPTER**

After completing this chapter, you will be able to:

1. Describe several kinds of *keys.*

2. Specify a suitable size key for a given size shaft.

3. Specify suitable materials for keys.

4. Complete the design of keys and the corresponding keyways and keyseats, giving the complete geometries.

5. Describe *splines* and determine their torque capacity.

6. Describe several alternate methods of fastening machine elements to shafts.

7. Describe *rigid couplings* and *flexible couplings.*

8. Describe several types of flexible couplings.

9. Describe *universal joints.*

10. Describe *retaining rings* and other means of locating elements on shafts.

11. Specify suitable seals for shafts and other types of machine elements.

11–2
KEYS

A *key* is a machinery component placed at the interface between a shaft and the hub of a power-transmitting element for the purpose of transmitting torque [see Figure 11–1(a)]. The key is demountable to facilitate assembly and disassembly of the shaft system. It is installed in an axial groove machined into the shaft, called a *keyseat.* A similar groove in the hub of the power-transmitting element is usually called a *keyway,* but it is more properly also a keyseat. The key is typically installed into the shaft keyseat first; then the hub keyseat is aligned with the key, and the hub is slid into position.

Square and Rectangular Parallel Keys

The most common type of key for shafts up to $6\frac{1}{2}$ inches in diameter is the square key, as illustrated in Figure 11–1(b). The rectangular key, Figure 11–1(c), is recommended for larger shafts and is used for smaller shafts where the shorter height can be tolerated. Both the square and the rectangular keys are referred to as *parallel keys* because the top and bottom and the sides of the key are parallel. (See Internet site 1.)

Table 11–1 gives the preferred dimensions for parallel keys as a function of shaft diameter, as specified in ANSI Standard B17.1-1967. The width is nominally one-quarter of the diameter of the shaft. (See Reference 6.)

The keyseats in the shaft and the hub are designed so that exactly one-half of the height of the key is bearing on the side of the shaft keyseat and the other half on the side of the hub keyseat. Figure 11–2 shows the resulting geometry. The distance Y is the radial distance from the theoretical top of the shaft, before the keyseat is machined, to the top edge of the finished keyseat to produce a keyseat depth of exactly $H/2$. To assist in machining

FIGURE 11–1
Parallel keys (Source
for (d): Driv-Lok, Inc.,
Sycamore, IL)

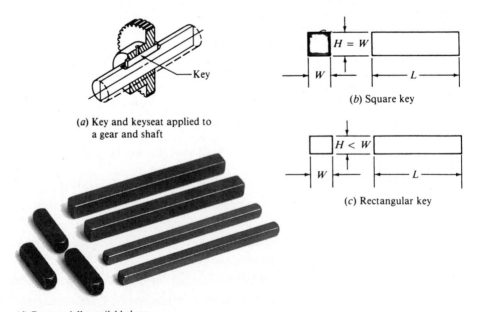

(a) Key and keyseat applied to
a gear and shaft

(b) Square key

(c) Rectangular key

(d) Commercially available keys

TABLE 11–1 Key size vs. shaft diameter

Nominal shaft diameter		Nominal key size		
			Height, H	
Over	To (incl.)	Width, W	Square	Rectangular
5/16	7/16	3/32	3/32	
7/16	9/16	1/8	1/8	3/32
9/16	7/8	3/16	3/16	1/8
7/8	$1\frac{1}{4}$	1/4	1/4	3/16
$1\frac{1}{4}$	$1\frac{3}{8}$	5/16	5/16	1/4
$1\frac{3}{8}$	$1\frac{3}{4}$	3/8	3/8	1/4
$1\frac{3}{4}$	$2\frac{1}{4}$	1/2	1/2	3/8
$2\frac{1}{4}$	$2\frac{3}{4}$	5/8	5/8	7/16
$2\frac{3}{4}$	$3\frac{1}{4}$	3/4	3/4	1/2
$3\frac{1}{4}$	$3\frac{3}{4}$	7/8	7/8	5/8
$3\frac{3}{4}$	$4\frac{1}{2}$	1	1	3/4
$4\frac{1}{2}$	$5\frac{1}{2}$	$1\frac{1}{4}$	$1\frac{1}{4}$	7/8
$5\frac{1}{2}$	$6\frac{1}{2}$	$1\frac{1}{2}$	$1\frac{1}{2}$	1
$6\frac{1}{2}$	$7\frac{1}{2}$	$1\frac{3}{4}$	$1\frac{3}{4}$	$1\frac{1}{2}$
$7\frac{1}{2}$	9	2	2	$1\frac{1}{2}$
9	11	$2\frac{1}{2}$	$2\frac{1}{2}$	$1\frac{3}{4}$
11	13	3	3	2
13	15	$3\frac{1}{2}$	$3\frac{1}{2}$	$2\frac{1}{2}$
15	18	4		3
18	22	5		$3\frac{1}{2}$
22	26	6		4
26	30	7		5

Source: Reprinted from ANSI Standard B17.1-1967(R98), by permission of the American Society of Mechanical Engineers. All rights reserved.

Note: Values in nonshaded areas are preferred. Dimensions are in inches.

and inspecting the shaft or the hub, the dimensions S and T can be computed and shown on the part drawings. The equations are given in Figure 11–2. Tabulated values of Y, S, and T are available in References 6 and 9.

As discussed later in Chapter 12, keyseats in shafts are usually machined with either an end mill or a circular milling cutter, producing the profile or sled runner keyseat, respectively (refer to Figure 12–6). In general practice, the keyseats and keys are left with essentially square corners. But radiused keyseats and chamfered keys can be used to reduce the stress concentrations. Table 11–2 shows suggested values from ANSI Standard B17.1. When using chamfers on a key, be sure to take that factor into account when computing the bearing stress on the side of the key [see Equation (11–3)].

As alternates to the use of parallel keys, taper keys, gib head keys, pin keys, and Woodruff keys can be used to provide special features of installation or operation. Figure 11–3 shows the general geometry of these types of keys.

Taper Keys and Gib Head Keys

Taper keys are designed to be inserted from the end of the shaft after the hub is in position rather than installing the key first and then sliding the hub over the key as with parallel keys.

FIGURE 11–2

Dimensions for parallel keyseats

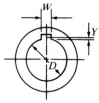

(*a*) Chordal height

$$Y = \frac{D - \sqrt{D^2 - W^2}}{2}$$

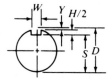

(*b*) Depth of shaft keyseat

$$S = D - Y - \frac{H}{2} = \frac{D - H + \sqrt{D^2 - W^2}}{2}$$

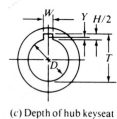

(*c*) Depth of hub keyseat

$$T = D - Y + \frac{H}{2} + C = \frac{D + H + \sqrt{D^2 - W^2}}{2} + C$$

Symbols

C = Allowance
+ 0.005-in clearance for parallel keys
− 0.020-in interference for taper keys
D = Nominal shaft or bore dia., in
H = Nominal key height, in
W = Nominal key width, in
Y = Chordal height, in

TABLE 11–2 Suggested fillet radii and key chamfers

$H/2$, keyseat depth			
Over	To (incl.)	Fillet radius	45° chamfer
1/8	1/4	1/32	3/64
1/4	1/2	1/16	5/64
1/2	7/8	1/8	5/32
7/8	$1\frac{1}{4}$	3/16	7/32
$1\frac{1}{4}$	$1\frac{3}{4}$	1/4	9/32
$1\frac{3}{4}$	$2\frac{1}{2}$	3/8	13/32

Source: Reprinted from ASME B17.1-1967 by permission of the American Society of Mechanical Engineers. All rights reserved.

Note: All dimensions are given in inches.

The taper extends over at least the length of the hub, and the height, *H,* measured at the end of the hub, is the same as for the parallel key. The taper is typically 1/8 in per foot. Note that this design gives a smaller bearing area on the sides of the key, and the bearing stress must be checked.

The *gib head key* [Figure 11–3(c)] has a tapered geometry inside the hub that is the same as that of the plain taper key. But the extended head provides the means of extracting the key from the same end at which it was installed. This is very desirable if the opposite end is not accessible to drive the key out.

FIGURE 11–3 Key types

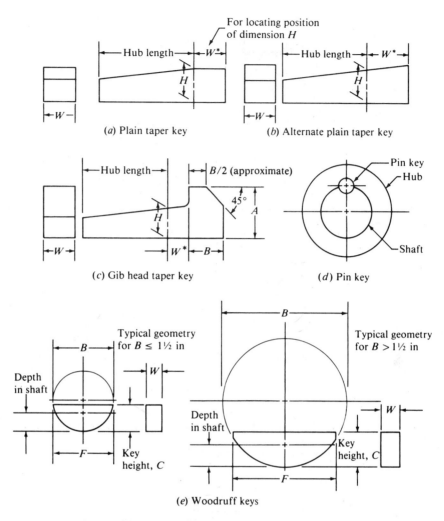

(a) Plain taper key

(b) Alternate plain taper key

(c) Gib head taper key

(d) Pin key

(e) Woodruff keys

Note: Plain and gib head taper keys have a 1/8″ taper in 12″.

Pin Keys

The *pin key,* shown in Figure 11–3(d), is a cylindrical pin placed in a cylindrical groove in the shaft and hub. Lower stress concentration factors result from this design as compared with parallel or taper keys. A close fit between the pin and the groove is required to ensure that the pin does not move and that the bearing is uniform along the length of the pin.

Woodruff Keys

Where light loading and relatively easy assembly and disassembly are desired, the *Woodruff key* should be considered. Figure 11–3(e) shows the standard configuration. The circular groove in the shaft holds the key in position while the mating part is slid over the key. The stress analysis for this type of key proceeds in the manner discussed below for the parallel key, taking into consideration the particular geometry of the Woodruff key. ANSI Standard B17.2-1967 lists the dimensions for a large number of standard Woodruff keys and their

TABLE 11–3   Woodruff key dimensions

Key number	Nominal key size, $W \times B$	Actual length, F	Height of key, C	Shaft keyseat depth	Hub keyseat depth
202	1/16 × 1/4	0.248	0.104	0.0728	0.0372
204	1/16 ×1/2	0.491	0.200	0.1668	0.0372
406	1/8 × 3/4	0.740	0.310	0.2455	0.0685
608	3/16 × 1	0.992	0.435	0.3393	0.0997
810	1/4 × $1\frac{1}{4}$	1.240	0.544	0.4170	0.1310
1210	3/8 × $1\frac{1}{4}$	1.240	0.544	0.3545	0.1935
1628	1/2 × $3\frac{1}{2}$	2.880	0.935	0.6830	0.2560
2428	3/4 × $3\frac{1}{2}$	2.880	0.935	0.5580	0.3810

Source: Reprinted from ASME B17.2-1967, by permission of the American Society of Mechanical Engineers. All rights reserved.

Note: All dimensions are given in inches.

mating keyseats. (See Reference 7.) Table 11–3 provides a sampling. Note that the *key number* indicates the nominal key dimensions. The last two digits give the nominal diameter, *B,* in eighths of an inch, and the digits preceding the last two give the nominal width, *W,* in thirty-seconds of an inch. For example, key number 1210 has a diameter of 10/8 in ($1\frac{1}{4}$ in), and a width of 12/32 in (3/8 in). The actual size of the key is slightly smaller than half of the full circle, as shown in dimensions *C* and *F* in Table 11–3.

Selection and Installation of Keys and Keyseats

The key and the keyseat for a particular application are usually designed after the shaft diameter is specified by the methods of Chapter 12. Then, with the shaft diameter as a guide, the size of the key is selected from Table 11–1. The only remaining variables are the length of the key and its material. One of these can be specified, and the requirements for the other can then be computed.

Typically the length of a key is specified to be a substantial portion of the hub length of the element in which it is installed to provide for good alignment and stable operation. But if the keyseat in the shaft is to be in the vicinity of other geometric changes, such as shoulder fillets and ring grooves, it is important to provide some axial clearance between them so that the effects of the stress concentrations are not compounded.

The key can be cut off square at its ends or provided with a radius at each end when installed in a profile keyseat to improve location. Square cut keys are usually used with the sled-runner-type keyseat.

The key is sometimes held in position with a set screw in the hub over the key. However, the reliability of this approach is questionable because of the possibility of the set screw's backing out with vibration of the assembly. Axial location of the assembly should be provided by more positive means, such as shoulders, retaining rings, or spacers.

11–3 MATERIALS FOR KEYS

Keys are most often made from low-carbon, cold-drawn steel. For example, AISI 1020 CD is listed in Appendix 3 as having an ultimate tensile strength of 61 ksi (420 MPa), a yield strength of 51 ksi (352 MPa), and a 15% elongation. This is adequate strength and ductility for most applications. Standard key stock conforming to the dimensions of ANSI Standard B17.1 (Table 11–1) is available from industrial supply centers in similar materials.

You should check the actual material and guaranteed strength of key stock used in critical applications.

If the low-carbon steel is not sufficiently strong, higher-carbon steel, such as AISI 1040 or 1045, could be used, also in the cold-drawn condition. Heat-treated steels could be used to gain even higher strength. However, the material should retain good ductility as indicated by a percent elongation value greater than about 10%, particularly when shock or impact loads could be encountered.

**11–4
STRESS
ANALYSIS TO
DETERMINE KEY
LENGTH**

There are two basic modes of potential failure for keys transmitting power: (1) shear across the shaft/hub interface and (2) compression failure due to the bearing action between the sides of the key and the shaft or hub material. The analysis for either failure mode requires an understanding of the forces that act on the key. Figure 11–4 shows the idealized case in which the torque on the shaft creates a force on the left side of the key. The key in turn exerts a force on the right side of the hub keyseat. The reaction force of the hub back on the key then produces a set of opposing forces that place the key in direct shear over its cross section, $W = L$. The magnitude of the shearing force can be found from

$$F = T/(D/2)$$

The shearing stress is then

$$\tau = \frac{F}{A_s} = \frac{T}{(D/2)(WL)} = \frac{2T}{DWL} \tag{11–1}$$

In design, we can set the shearing stress equal to a design stress in shear for the maximum shear stress theory of failure:

$$\tau_d = 0.5s_y/N$$

Then the required length of the key is

$$L = \frac{2T}{\tau_d DW} \tag{11–2}$$

FIGURE 11–4
Forces on a key

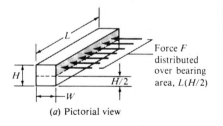

(a) Pictorial view

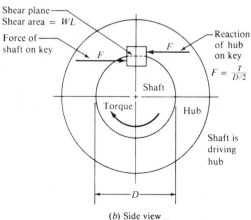

(b) Side view

The failure in bearing is related to the compressive stress on the side of the key, the side of the shaft keyseat, or the side of the hub keyseat. The area in compression is the same for either of these zones, $L \times (H/2)$. Thus, the failure occurs on the surface with the lowest compressive yield strength. Let's define a *design stress for compression* as

$$\sigma_d = s_y/N$$

Then the compressive stress is

$$\sigma = \frac{F}{A_c} = \frac{T}{(D/2)(L)(H/2)} = \frac{4T}{DLH} \tag{11–3}$$

Letting this stress equal the design compressive stress allows the computation of the required length of the key for this mode of failure:

$$L = \frac{4T}{\sigma_d DH} \tag{11–4}$$

In typical industrial applications, $N = 3$ is adequate.

For the design of a square key in which the strength of the key material is lower than that of the shaft or the hub, Equations (11–2) and (11–4) produce the same result. Substituting the design stress into either equation would give

$$L = \frac{4TN}{DWs_y} \tag{11–5}$$

But be sure to evaluate the length from Equation (11–4) if either the shaft or the hub has a lower yield strength than the key.

Design Procedure for Parallel Keys

1. Complete the design of the shaft into which the key will be installed, and specify the actual diameter at the location of the keyseat.

2. Select the size of the key from Table 11–1. Use a square key with $W = H$ if the shaft size is 6.50 in or smaller. Use a rectangular key if the diameter is greater than 6.50 in. Then the key width W will be greater than the height H, requiring that both Equation (11–2) and (11–4) be used to determine the length.

3. Specify the material for the key, usually AISI 1020 CD steel. A higher-strength material can be used.

4. Determine the yield strength of the materials for the key, the shaft, and the hub.

5. If a square key is used and the key material has the lowest strength, use Equation (11–5) to compute the minimum required length of the key. This length will be satisfactory for both shear and bearing stress.

6. If a rectangular key is used, or if either the shaft or the hub has a lower strength than the key, use Equation (11–4) to compute the minimum required length of the key based on bearing stress. Also, use Equation (11–2) or (11–5) to compute the minimum required length based on shear of the key. The larger of the two computed lengths governs the design. Check to be sure that the computed length is shorter than the hub length. If not, a higher-strength material must be selected and

the design process repeated. Alternatively, two keys or a spline can be used instead of a single key.

7. Specify the actual length of the key to be equal to or longer than the computed minimum length. A convenient standard size should be specified using the preferred basic sizes shown in Appendix A2–1. The key should extend over all or a substantial part of the length of the hub. But the keyseat should not run into other stress raisers such as shoulders or grooves.

8. Complete the design of the keyseat in the shaft and the keyway in the hub using the equations in Figure 11–2. ANSI Standard B17.1 should be consulted for standard tolerances on dimensions for the key and the keyseats.

Example Problem 11–1 A portion of a shaft where a gear is to be mounted has a diameter of 2.00 in. The gear transmits 2965 lb·in of torque. The shaft is to be made of AISI 1040 cold-drawn steel. The gear is made from AISI 8650 OQT 1000 steel. The width of the hub of the gear mounted at this location is 1.75 in. Design the key.

Solution From Table 11–1, the standard key dimension for a 2.00-in-diameter shaft would be 1/2 in square. See Figure 11–5 for the proposed design.

Material selection is a design decision. Let's choose AISI 1020 CD steel with $s_y = 51\,000$ psi.

A check of the yield strengths of the three materials in the key, the shaft, and the hub indicates that the key is the weakest material. Then Equation (11–5) can be used to compute the minimum required length of the key:

$$L = \frac{4TN}{DWs_y} = \frac{4(2965)(3)}{(2.00)(0.50)(51\,000)} = 0.698 \text{ in}$$

This length is well below the width of the hub of the gear. Notice that the design of the shaft includes retaining rings on both sides of the gear. It is desirable to keep the keyseat well clear of the ring grooves. Therefore, let's specify the length of the key to be 1.50 in.

Summary In summary, the key has the following characteristics:

Material: AISI 1020 CD steel

Width: 0.500 in

Height: 0.500 in

Length: 1.50 in.

Figure 11–5 shows some details of the completed design. A profile keyseat in the shaft is shown.

Shear and Bearing Areas for Woodruff Keys

The geometry of Woodruff keys makes it more difficult to determine the shear area and the bearing area for use in stress analyses. Figure 11–3(e) shows that the bearing area on the side of the key in the keyseat is a segment of a circle. The shear area is the product of the

FIGURE 11–5
Details for proposed
design of key and
keyseats

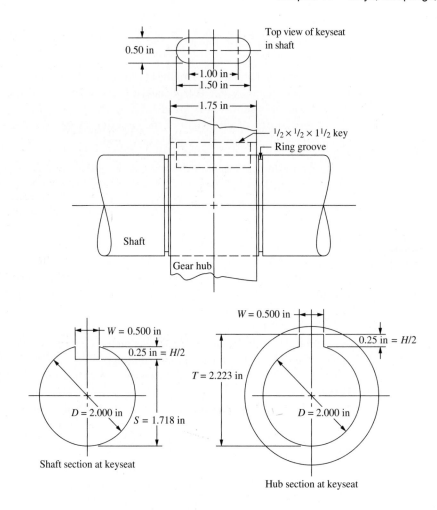

chord of that segment times the thickness of the key. The following equations describe the
geometry:

Given

B = nominal diameter of the cylinder of which the key is a part

W = width (thickness) of the key

C = full height of the key

d_s = depth of the keyseat in the shaft

Results

$$\text{Shear area} = A_s = 2W\sqrt{d_s(B - d_s)} \tag{11–6}$$

To define the equations for the bearing areas on the side of the key in the shaft and in
the hub, we first define three geometric variables, G, L, and J, as follows:

$$G = (\pi/180)B \cos^{-1}\{2[(B/2) - d_s]/B\}$$

$$L = 2\sqrt{d_s(B - d_s)}$$

$$J = (\pi/180)B \cos^{-1}\{2[(B/2) - C]/B\}$$

Bearing Area in Shaft

$$A_{c\,shaft} = 0.5\{G(B/2) - L[(B/2) - d_s]\} \qquad \textbf{(11–7)}$$

$$A_{c\,hub} = 0.5\{J(B/2) - F[(B/2) - C]\} - A_{c\,shaft} \qquad \textbf{(11–8)}$$

11–5
SPLINES

A *spline* can be described as a series of axial keys machined into a shaft, with corresponding grooves machined into the bore of the mating part (gear, sheave, sprocket, and so on; see Figure 11–6). The splines perform the same function as a key in transmitting torque from the shaft to the mating element. The advantages of splines over keys are many. Because usually four or more splines are used, as compared with one or two keys, a more uniform transfer of the torque and a lower loading on a given part of the shaft/hub interface result. The splines are integral with the shaft, so no relative motion can occur as between a key and the shaft. Splines are accurately machined to provide a controlled fit between the mating internal and external splines. The surface of the spline is often hardened to resist wear and to facilitate its use in applications in which axial motion of the mating element is desired. Sliding motion between a standard parallel key and the mating element should not be permitted. Because of the multiple splines on the shaft, the mating element can be indexed to various positions.

Splines can be either straight-sided or involute. The involute form is preferred because it provides for self-centering of the mating element and because it can be machined with standard hobs used to cut gear teeth.

Straight-Sided Splines

Straight splines are made according to the specifications of the Society of Automotive Engineers (SAE) and usually contain 4, 6, 10, or 16 splines. Figure 11–6 shows the six-spline

FIGURE 11–6
Straight-sided spline

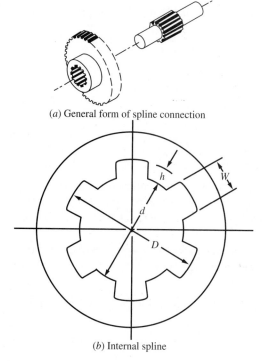

(*a*) General form of spline connection

(*b*) Internal spline

TABLE 11–4   Formulas for SAE straight splines

No. of splines	W, for all fits	A: Permanent fit		B: To slide without load		C: To slide under load	
		h	d	h	d	h	d
Four	0.241D	0.075D	0.850D	0.125D	0.750D		
Six	0.250D	0.050D	0.900D	0.075D	0.850D	0.100D	0.800D
Ten	0.156D	0.045D	0.910D	0.070D	0.860D	0.095D	0.810D
Sixteen	0.098D	0.045D	0.910D	0.070D	0.860D	0.095D	0.810D

Note: These formulas give the maximum dimensions for *W, h,* and *d.*

version, from which you can see the basic design parameters of D (major diameter), d (minor diameter), W (spline width), and h (spline depth). The dimensions for d, W, and h are related to the nominal major diameter D by the formulas given in Table 11–4. Note that the values of h and d differ according to the use of the spline. The permanent fit, A, is used when the mating part is not to be moved after installation. The B fit is used if the mating part will be moved along the shaft without a torque load. When the mating part must be moved under load, the C fit is used.

The torque capacity for SAE splines is based on the limit of 1000-psi bearing stress on the sides of the splines, from which the following formula is derived:

$$T = 1000NRh \tag{11–9}$$

where N = number of splines
  R = mean radius of the splines
  h = depth of the splines (from Table 11–4)

The torque capacity is per inch of length of the spline. But note that

$$R = \frac{1}{2}\left[\frac{D}{2} + \frac{d}{2}\right] = \frac{D + d}{4}$$

$$h = \frac{1}{2}(D - d)$$

Then

$$T = 1000N\frac{(D + d)}{4}\frac{(D - d)}{2} = 1000N\frac{(D^2 - d^2)}{8} \tag{11–10}$$

This equation can be further refined for each of the types of splines in Table 11–4 by substitution of the appropriate relationships for N and d. For example, for the six-spline version and the B fit, $N = 6$, $d = 0.850D$, and $d^2 = 0.7225D^2$. Then

$$T = 1000(6)\frac{[D^2 - 0.7225D^2]}{8} = 208D^2$$

Thus, the required diameter to transmit a given torque would be

$$D = \sqrt{T/208}$$

In these formulas, dimensions are in inches and the torque is in pound-inches. We use this same approach to find the torque capacities and required diameters for the other versions of straight splines (Table 11–5).

The graphs in Figure 11–7 enable you to choose an acceptable diameter for a spline to carry a given torque, depending on the desired fit, *A, B,* or *C.* The data were taken from Table 11–5.

Involute Splines

Involute splines are typically made with pressure angles of 30°, 37.5°, or 45°. The 30° form is illustrated in Figure 11–8, showing the two types of fit that can be specified. The *major diameter fit* produces accurate concentricity between the shaft and the mating element. In

TABLE 11–5 Torque capacity for straight splines per inch of spline length

Number of splines	Fit	Torque capacity	Required diameter
4	A	$139D^2$	$\sqrt{T/139}$
4	B	$219D^2$	$\sqrt{T/219}$
6	A	$143D^2$	$\sqrt{T/143}$
6	B	$208D^2$	$\sqrt{T/208}$
6	C	$270D^2$	$\sqrt{T/270}$
10	A	$215D^2$	$\sqrt{T/215}$
10	B	$326D^2$	$\sqrt{T/326}$
10	C	$430D^2$	$\sqrt{T/430}$
16	A	$344D^2$	$\sqrt{T/344}$
16	B	$521D^2$	$\sqrt{T/521}$
16	C	$688D^2$	$\sqrt{T/688}$

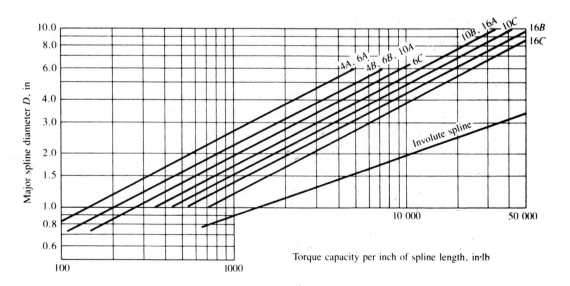

FIGURE 11–7 Torque capacity per inch of spline length, lb·in

FIGURE 11–8
30° involute spline

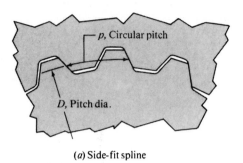

(a) Side-fit spline

N = Number of spline teeth
P = Diametral pitch
$D = N/P$ = Pitch dia.
$p = \pi/P$ = Circular pitch

Minor dia.:

Internal: $\dfrac{N-1}{P}$

External: $\dfrac{N-1.35}{P}$

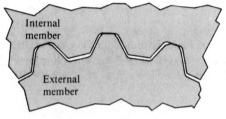

Note chamfer on tips of
external spline teeth

(b) Major dia. fit spline

Major dia.:

Internal: $\dfrac{N+1.35}{P}$ side fit

$\dfrac{N+1}{P}$ major dia. fit

External: $\dfrac{N+1}{P}$

the *side fit,* contact occurs only on the sides of the teeth, but the involute form tends to center the shaft in the mating splined hub.

Figure 11–8 also gives some of the basic formulas for key features of involute splines in the U.S. Customary Unit System with dimensions in inches. (See Reference 4.) The terms are similar to those for involute spur gears which are discussed more completely in Chapter 8. The basic spline size is governed by its *diametral pitch, P:*

$$P = N/D \qquad\qquad (11\text{–}11)$$

where N = number of spline teeth
D = pitch diameter

The diametral pitch, then, is *the number of teeth per inch of pitch diameter.* Only even numbers of teeth from 6 to 60 are typically used. Up to 100 teeth are used on some 45° splines. Note that the pitch diameter lies *within* the tooth and is related to the major and minor diameters by the relationships shown in Figure 11–8.

The *circular pitch, p,* is the distance from one point on a tooth to the corresponding point on the next adjacent tooth, measured along the pitch circle. To find the nominal value of p, divide the circumference of the pitch circle by the number of spline teeth. That is,

$$p = \pi D/N \qquad\qquad (11\text{–}12)$$

But because $P = N/D$, we can also say

$$p = \pi/P \qquad\qquad (11\text{–}13)$$

The *tooth thickness, t,* is the thickness of the tooth measured along the pitch circle. Then the theoretical value is

$$t = p/2 = \pi/2P$$

The nominal value of the width of the tooth space is equal to t.

Standard Diametral Pitches. The following are the 17 standard diametral pitches in common use:

2.5	3	4	5	6	8	10	12	16
20	24	32	40	48	64	80	128	

The common designation for an involute spline is given as a fraction, P/P_s, where P_s is called the *stub pitch* and is always equal to $2P$. Thus, if a spline had a diametral pitch of 4, it would be called a 4/8 pitch spline. For convenience, we will use only the diametral pitch.

Length of Splines. Common designs use spline lengths from $0.75D$ to $1.25D$, where D is the pitch diameter of the spline. If these standards are used, the shear strength of the splines will exceed that of the shaft on which they are machined.

Metric Module Splines. The dimensions of splines made to metric standards are related to the *module, m,* where

$$m = D/N \qquad\qquad (11\text{--}14)$$

and both D and m are in millimeters. (See Reference 5.) Note that the symbol Z is used in place of N for the number of teeth in standards describing metric splines. Other features of metric splines can be found from the following formulas:

$$\text{Pitch diameter} = D = mN \qquad\qquad (11\text{--}15)$$
$$\text{Circular pitch} = p = \pi m \qquad\qquad (11\text{--}16)$$
$$\text{Basic tooth thickness} = t = \pi m/2 \qquad\qquad (11\text{--}17)$$

Standard Modules. There are 15 standard modules:

0.25	0.50	0.75	1.00	1.25	1.50	1.75	2.00
2.50	3	4	5	6	8	10	

Design Aids

Refer to References 9, 11, and 12 for additional design guidelines. Reference 9 includes extensive tables of data for splines and offers application and design information. Reference 11 gives information on the application, operation, dimensioning, and manufacture of involute splines as applied to automotive applications. Data are included on allowable shear stress, allowable compressive stress, wear life factor, spline overload factor, and fatigue life factor. Reference 12 is an SAE aerospace standard that defines an involute spline with a 30° pressure angle and a full fillet radius that reduces the stress concentration in the area of the root.

**11–6
OTHER
METHODS OF
FASTENING
ELEMENTS TO
SHAFTS**

The following discussion will acquaint you with some of the ways in which power-transmitting elements can be attached to shafts without keys or splines. (See Reference 10.) In most cases the designs have not been standardized, and analysis of individual cases, considering the forces exerted on the elements and the manner of loading of the fastening means, is necessary. In several of the designs, the analysis of shear and bearing will follow a procedure similar to that shown for keys. If a satisfactory analysis is not possible, testing of the assembly is recommended.

Pinning

With the element in position on the shaft, a hole can be drilled through both the hub and the shaft, and a pin can be inserted in the hole. Figure 11–9 shows three examples of this approach. The straight, solid, cylindrical pin is subjected to shear over two cross sections. If there is a force, F, on each end of the pin at the shaft/hub interface, and if the shaft diameter is D, then

$$T = 2F(D/2) = FD$$

or $F = T/D$. With the symbol d representing the pin diameter, the shear stress in the pin is

$$\tau = \frac{F}{A_s} = \frac{T}{D(\pi d^2/4)} = \frac{4T}{D(\pi d^2)} \tag{11–18}$$

Letting the shear stress equal the design stress in shear as before, solving for d gives the required pin diameter:

$$d = \sqrt{\frac{4T}{D(\pi)(\tau_d)}} \tag{11–19}$$

Sometimes the diameter of the pin is purposely made small to ensure that the pin will break if a moderate overload is encountered, in order to protect critical parts of a mechanism. Such a pin is called a *shear pin*.

One problem with a cylindrical pin is that fitting it adequately to provide precise location of the hub and to prevent the pin from falling out is difficult. The *taper pin* overcomes some of these problems, as does the *split spring pin* shown in Figure 11–9(c). For the split spring pin, the hole is made slightly smaller than the pin diameter so that a light force is required to assemble the pin in the hole. The spring force retains the pin in the hole and holds

FIGURE 11–9
Pinning

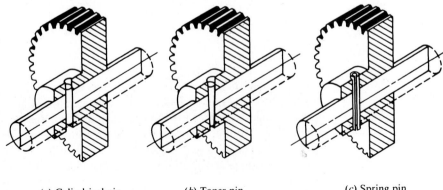

(*a*) Cylindrical pin   (*b*) Taper pin   (*c*) Spring pin

the assembly in position. But, of course, the presence of any of the pin-type connections produces stress concentrations in the shaft. (See Internet site 1.)

Keyless Hub to Shaft Connections

Using a steel ring compressed tightly around a smooth shaft allows torque to be transmitted between the hub of a power-transmitting element and a shaft without having a key between the two elements. Figure 11–10 shows a commercially available product called a *Locking Assembly*™ from Ringfeder® Corporation that employs this principle.

The Locking Assembly™ employs steel rings with opposing mating tapers held together with a series of fasteners. With the Locking Assembly™ placed completely within a counter bore of the hub of virtually any kind of power-transmitting element such as a gear, sprocket, fan wheel, cam, coupling, or turbine rotor, the fasteners can then be tightened. Initially there is a small clearance between the inside diameter of the locking device and the shaft as well as the hub bore. This clearance facilitates easy assembly and positioning of the hub. After the hub is positioned in the desired location on the shaft, the fasteners are tightened to a specified torque in a specific sequence. As the bolts are tightened, they draw the opposing tapered rings together, generating a radial movement of the inner ring toward the shaft and a simultaneous outward movement of the outer ring toward the ID of the hub. Once the initial clearances are eliminated, further tightening of the bolts results in a high pressure against the shaft and the hub. When the bolts are properly torqued using a torque wrench, the final contact pressure combined with friction allows for the transmission of a predetermined and predictable amount of torque between the hub and the shaft.

The connection can transmit axial forces in the form of thrust loads as well as torque as with helical gears, for example. Rated torque values range from a few hundred lb-ft for smaller units to more than 1 million lb-ft for shafts with diameters of 20 inches. Unlike a thermal or pressure fit connection, the locking device can be easily removed since it is a *mechanical shrink fit*. The pressures generated within the locking device itself allow the stresses to remain within the elastic limits of the materials. Removal is simply done by carefully loosening the screws, thus allowing the rings to slide apart and return the Locking Assembly™ to its original relaxed condition. The element can then be repositioned or removed at any time. Some locking devices have self-releasing taper angles, which allow them to self-release when the fasteners are loosened. Other locking devices have self-locking tapers that require the gentle pressing apart of the locking device parts. Different applications dictate which type of device is better suited.

Advantages of the keyless connection are the elimination of keys, keyways, or splines, and the cost of machining them; tight fit of the driving element around the shaft; the ability to transmit reversing or dynamically changing loads; and easy assembly, disassembly, and adjustment of the elements. General engineering information is provided on the Internet site for Ringfeder® (Corporation listed at the end of this chapter for installation dimensions, tolerances, lubrication, and surface finishes required. The rated torque values are for use on

FIGURE 11–10
Ringfeder® Locking Assemblies (Ringfeder® Corporation, Westwood, NJ)

(a) A variety of styles

(b) Locking assembly applied to a gear

Three-sided P3 external profile

Four-sided PC4 external profile

Polygon profile produced in a variety of products

FIGURE 11–11   Polygon hub to shaft connections (General Polygon Systems, Inc., Millville, NJ)

solid shafts; hollow shafts require additional analysis. Special considerations are necessary for the hub design to ensure that stresses remain below the yield strength of the hub material.

Polygon Hub to Shaft Connection

Figure 11–11 shows a shaft to hub connection that employs special mating polygon shapes to transmit torque without keys or splines. See Internet site 17 for available sizes and application information. German standards DIN 32711 and 32712 describe the forms. They can be produced on shaft sizes from 0.188 in (4.76 mm) to 8.00 in (203 mm). The three-sided configuration is called the P3 profile, and the four-sided design is called the PC4 profile. CNC turning and grinding can be used to produce the external form while broaching typically produces the internal form. Torque is transmitted by distributing the load on each side of the polygon, eliminating the shearing action inherent with keys or splines. Dimensions can be controlled for tight or press fits for precision, backlash-free location or with a sliding fit for ease of assembly.

Split Taper Bushing

A *split taper bushing* (see Figure 11–12) uses a key to transmit torque. Axial location on the shaft is provided by the clamping action of a split bushing having a small taper on its outer surface. When the bushing is pulled into a mating hub with a set of capscrews, the bushing is brought into tight contact with the shaft to hold the assembly in the proper axial position. The small taper locks the assembly together. Removal of the bushing is accomplished by removing the capscrews and using them in push-off holes to force the hub off the taper. The assembly can then be easily disassembled.

Set Screws

A *set screw* is a threaded fastener driven radially through a hub to bear on the outer surface of a shaft (see Figure 11–13). The point of the set screw is flat, oval, cone-shaped, cupped,

FIGURE 11–12 Split taper bushing (Emerson Power Transmission Corporation, Drive and Component Division, Ithaca, NY)

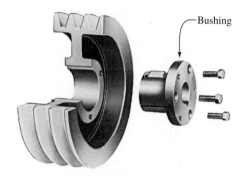
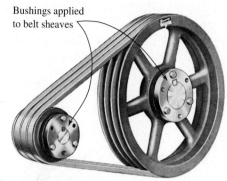

FIGURE 11–13 Set screws

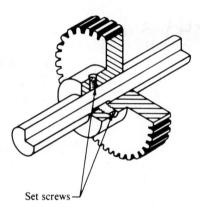

or any of several proprietary forms. The point bears on the shaft or digs slightly into its surface. Thus, the set screw transmits torque by the friction between the point and the shaft or by the resistance of the material in shear. The capacity for torque transmission is somewhat variable, depending on the hardness of the shaft material and the clamping force created when the screw is installed. Furthermore, the screw may loosen during operation because of vibration. For these reasons, set screws should be used with care. Some manufacturers provide set screws with plastic inserts in the side among the threads. When the set screw is screwed into a tapped hole, the plastic is deformed by the threads and holds the screw securely, resisting vibration. Using a liquid adhesive also helps resist loosening.

Another problem with using set screws is that the shaft surface is damaged by the point; this damage may make disassembly difficult. Machining a flat on the surface of the shaft may help reduce the problem and also produce a more consistent assembly.

When set screws are properly assembled on typical industrial shafting, their force capability is approximately as follows (see Reference 9):

Screw Diameter	Holding Force
1/4 in	100 lb
3/8 in	250 lb
1/2 in	500 lb
3/4 in	1300 lb
1 in	2500 lb

FIGURE 11–14
Tapered shaft for
fastening machine
elements to shafts

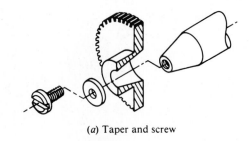

(*a*) Taper and screw

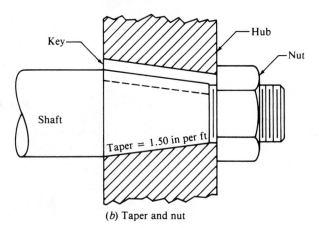

(*b*) Taper and nut

Taper and Screw

The power-transmitting element (gear, sheave, sprocket, or other) that is to be mounted at the end of a shaft can be secured with a screw and a washer in the manner shown in Figure 11–14(a). The taper provides good concentricity and moderate torque transmission capacity. Because of the machining required, the connection is fairly costly. A modified form uses the tapered shaft with a threaded end for the application of a nut, as shown in Figure 11–14(b). The inclusion of a key lying in a keyseat machined parallel to the taper increases the torque-transmitting capacity greatly and ensures positive alignment.

Press Fit

Making the diameter of the shaft greater than the bore diameter of the mating element results in an interference fit. The resulting pressure between the shaft and the hub permits the transmission of torque at fairly high levels, depending on the degree of interference. This is discussed more fully in Chapter 13. Sometimes the press fit is combined with a key, with the key providing the positive drive and the press fit ensuring concentricity.

Molding

Plastic and die cast gears can be molded directly to their shafts. Often the gear is applied to a location that is knurled to improve the ability to transmit torque. A modification of this procedure is to take a separate gear blank with a prepared hub, locate it over the proper position on a shaft, and then cast zinc into the space between the shaft and the hub to lock them together.

**11–7
COUPLINGS**

The term *coupling* refers to a device used to connect two shafts together at their ends for the purpose of transmitting power. There are two general types of couplings: rigid and flexible.

Rigid Couplings

Rigid couplings are designed to draw two shafts together tightly so that no relative motion can occur between them. This design is desirable for certain kinds of equipment in which precise alignment of two shafts is required and can be provided. In such cases, the coupling must be designed to be capable of transmitting the torque in the shafts.

A typical rigid coupling is shown in Figure 11–15, in which flanges are mounted on the ends of each shaft and are drawn together by a series of bolts. The load path is then from the driving shaft to its flange, through the bolts, into the mating flange, and out to the driven shaft. The torque places the bolts in shear. The total shear force on the bolts depends on the radius of the bolt circle, $D_{bc}/2$, and the torque, T. That is,

$$F = T/(D_{bc}/2) = 2T/D_{bc}$$

Letting N be the number of bolts, the shear stress in each bolt is

$$\tau = \frac{F}{A_s} = \frac{F}{N(\pi d^2/4)} = \frac{2T}{D_{bc}N(\pi d^2/4)} \qquad \textbf{(11–20)}$$

Letting the stress equal the design stress in shear and solving for the bolt diameter, we have

$$d = \sqrt{\frac{8T}{D_{bc}N\pi\tau_d}} \qquad \textbf{(11–21)}$$

Notice that this analysis is similar to that for pinned connections in Section 11–6. The analysis assumes that the bolts are the weakest part of the coupling.

Rigid couplings should be used only when the alignment of the two shafts can be maintained very accurately, not only at the time of installation but also during operation of the machines. If significant angular, radial, or axial misalignment occurs, stresses that are difficult to predict and that may lead to early failure due to fatigue will be induced in the shafts. These difficulties can be overcome by the use of flexible couplings.

Flexible Couplings

Flexible couplings are designed to transmit torque smoothly while permitting some axial, radial, and angular misalignment. The flexibility is such that when misalignment does

FIGURE 11–15
Rigid coupling
(Rockwell
Automation/Dodge,
Greenville, SC)

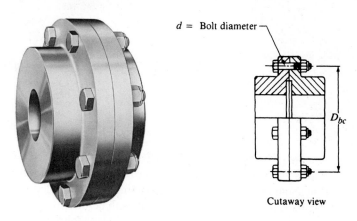

Cutaway view

occur, parts of the coupling move with little or no resistance. Thus, no significant axial or bending stresses are developed in the shaft.

Many types of flexible couplings are available commercially, as shown in Figures 11–16 through 11–24. Each is designed to transmit a given limiting torque. The manufacturer's catalog lists the design data from which you can choose a suitable coupling. Re-

Elastometric member bonded
to steel flanges and hubs

FIGURE 11–16   Chain coupling. Torque is transmitted through a double roller chain. Clearances between the chain and the sprocket teeth on the two coupling halves accommodate misalignment. (Emerson Power Transmission Corporation, Ithaca, NY)

FIGURE 11–17   Ever-Flex coupling. The features of this coupling are that it (1) generally minimizes torsional vibration; (2) cushions shock loads; (3) compensates for parallel misalignment up to 1/32 in; (4) accommodates angular misalignment of ±3°; and (5) provides adequate end float, ±1/32 in. (Emerson Power Transmission Corporation, Ithaca, NY)

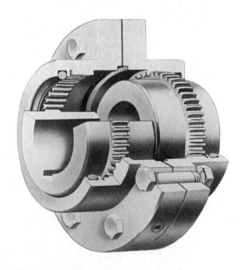

FIGURE 11–18   Grid-Flex coupling. Torque is transmitted through a flexible spring steel grid. Flexing of the grid permits misalignment and makes it torsionally resilient to resist shock loads. (Emerson Power Transmission Corporation, Ithaca, NY)

FIGURE 11–19   Gear coupling. Torque is transmitted between crown-hobbed teeth from the coupling half to the sleeve. The crown shape on the gear teeth permits misalignment. (Emerson Power Transmission Corporation, Ithaca, NY)

FIGURE 11–20
Bellows coupling. The inherent flexibility of the bellows accommodates the misalignment. (Stock Drive Products, New Hyde Park, NY)

FIGURE 11–21
PARA-FLEX® coupling. Using an elastomeric element permits misalignment and cushions shocks. (Rockwell Automation/Dodge, Greenville, SC)

 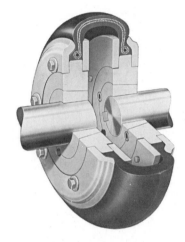

FIGURE 11–22
Dynaflex® coupling. Torque is transmitted through elastomeric material that flexes to permit misalignment and to attenuate shock loads. (Lord Corporation, Erie, PA)

member that torque equals power divided by rotational speed. So for a given size of coupling, as the speed of rotation increases, the amount of power that the coupling can transmit also increases, although not always in direct proportion. Of course, centrifugal effects determine the upper limit of speed.

The degree of misalignment that can be accommodated by a given coupling should be obtained from the manufacturer's catalog data, with values varying with the size and

FIGURE 11–23
Jaw-Type coupling
(Emerson Power
Transmission
Corporation, Ithaca,
NY)

Neoprene
(normal duty
applications)

Bronze,
oil impregnated
(low-speed,
high-torque
applications)

Polyurethane
(extra capacity
at medium to
high speed)

Inserts

(a) Assembled coupling

(b) Types of inserts

FIGURE 11–24
FORM-FLEX®
coupling. Torque is
transmitted from hubs
through laminated
flexible elements to the
spacer. (T. B. Wood's
Incorporated,
Chambersberg, PA)

design of the coupling. Small couplings may be limited to parallel misalignment of 0.005 in, although larger couplings may allow 0.030 in or more. Typical allowable angular misalignment is ±3°. Axial movement allowed, sometimes called *end float,* is up to 0.030 in for many types of couplings. (See References 1–3 and the Internet sites for the manufacturers.)

**11–8
UNIVERSAL
JOINTS**

When an application calls for accommodating misalignment between mating shafts that is greater than the three degrees typically provided by flexible couplings, a *universal joint* is often employed. Figures 11–25 to 11–29 show some of the styles that are available. Angular misalignments of up to 45 degrees are possible at low rotational speeds with single universal joints like that shown in Figure 11–25, consisting of two yokes, a center bearing block, and two pins that pass through the block at right angles. Approximately 20 to 30 degrees is more reasonable for speeds above 10 rpm. Single universal joints have the disadvantage that the rotational speed of the output shaft is nonuniform in relation to the input shaft.

A double universal joint as shown in Figure 11–26 allows the connected shafts to be parallel and offset by large amounts. Furthermore, the second joint cancels the nonuniform oscillation of the first joint so the input and output rotate at the same uniform speed.

Figure 11–27 shows a vehicular universal joint connecting an engine or transmission to the drive wheels used in some rear-wheel drive cars, light and heavy trucks, agricultural equipment, and construction vehicles. The spider assembly contains needle-bearing rollers on each arm. The right end shows a ball stud yoke, a flange yoke, and a center coupling yoke that make up a *double Cardan universal joint.* Another style, called a *constant velocity joint,* or simply a *CV joint,* is often used as a key component of front wheel drive and all wheel drive vehicle drivelines.

FIGURE 11–25   Universal joint components
(Curtis Universal Joint Co., Inc., Springfield, MA)

FIGURE 11–26   Double universal joint
(Curtis Universal Joints Co., Inc., Springfield, MA)

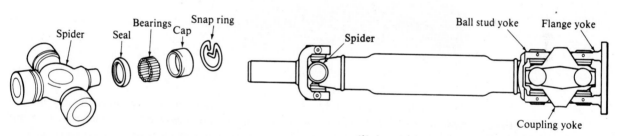

(*a*) Automotive universal joint components

(*b*) Automotive propeller shaft with
double Cardan universal joint

FIGURE 11–27   Automotive universal joints (Pontiac Motor Division, General Motors Corp., Pontiac, MI)

Figure 11–28 shows a heavy-duty industrial type double universal joint. Some of this type will have a two-part connecting tube that is splined to allow for sizable changes in axial position as well as accommodating the angular or parallel misalignment.

Figure 11–29 shows a novel design called the Cornay™ universal joint that produces true constant velocity of the output shaft through all drive angles up to 90 degrees. Compared with standard universal joint designs, the Cornay™ joint can operate at higher speeds, carry higher torque levels, and produce less vibration.

Manufacturers' literature should be consulted for specifying a suitable size of universal joint for a given application. The primary variables involved are the torque to be transmitted, the rotational speed, and the angle at which the joint will be operating. A *speed/angle factor* is computed as the product of the rotational speed and the operating angle. From this value, an *operating use factor* is determined that is applied to the basic torque load to compute the required rated torque for the joint. All manufacturers provide such data in their catalogs.

FIGURE 11–28
Industrial universal
joint (Spicer
Driveshaft, Inc., Dana
Corp., Toledo, OH)

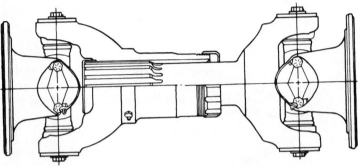

FIGURE 11–29
Cornay™ universal
joint (Drive
Technologies, Inc.,
Longmont, CO)

11–9
RETAINING
RINGS AND
OTHER MEANS
OF AXIAL
LOCATION

The preceding sections of this chapter have focused on means of connecting machine elements to shafts for the purpose of transmitting power. Therefore, they emphasized the ability of the elements to withstand a given torque at a given speed of rotation. It must be recognized that the axial location of the machine elements must also be ensured by the designer.

The choice of the means for axial location depends heavily on whether or not axial thrust is transmitted by the element. Spur gears, V-belt sheaves, and chain sprockets produce no significant thrust loads. Therefore, the need for axial location affects only incidental forces due to vibration, handling, and shipping. Although not severe, these forces should not be taken lightly. Movement of an element in relation to its mating element in an axial direction can cause noise, excessive wear, vibration, or complete disconnection of the drive. Any bicycle rider who has experienced the loss of a chain can appreciate the consequences of misalignment. Recall that for spur gears, the strength of the gear teeth and the wear resistance are both directly proportional to the face width of the gear. Axial misalignment decreases the effective face width.

Some of the methods discussed in Section 11–6 for fastening elements to shafts for the purpose of transmitting power also provide some degree of axial location. Refer to the discussions of pinning, keyless hub to shaft connections, split taper bushings, set screws, the taper and screw, the taper and nut, the press fit, molding, and the several methods of mechanically locking the elements to the shaft.

Among the wide variety of other means available for axial location, we will discuss the following:

- Retaining rings
- Collars
- Shoulders
- Spacers
- Locknuts

Retaining Rings

Retaining rings are placed on a shaft, in grooves in the shaft, or in a housing to prevent the axial movement of a machine element. Figure 11–30 shows the wide variety of styles of rings available from one manufacturer. The different designs allow for either internal or external mounting of the ring. The amount of axial thrust load capacity and the shoulder height provided by the different ring styles also vary. The manufacturer's catalog provides data on the capacity and installation data for the rings. Figure 11–31 shows examples of retaining rings in use.

Collars

A *collar* is a ring slid over the shaft and positioned adjacent to a machine element for the purpose of axial location. It is held in position, typically, by set screws. Its advantage is that axial location can be set virtually anywhere along the shaft to allow adjustment of the position at the time of assembly. The disadvantages are chiefly those related to the use of set screws themselves (Section 11–6).

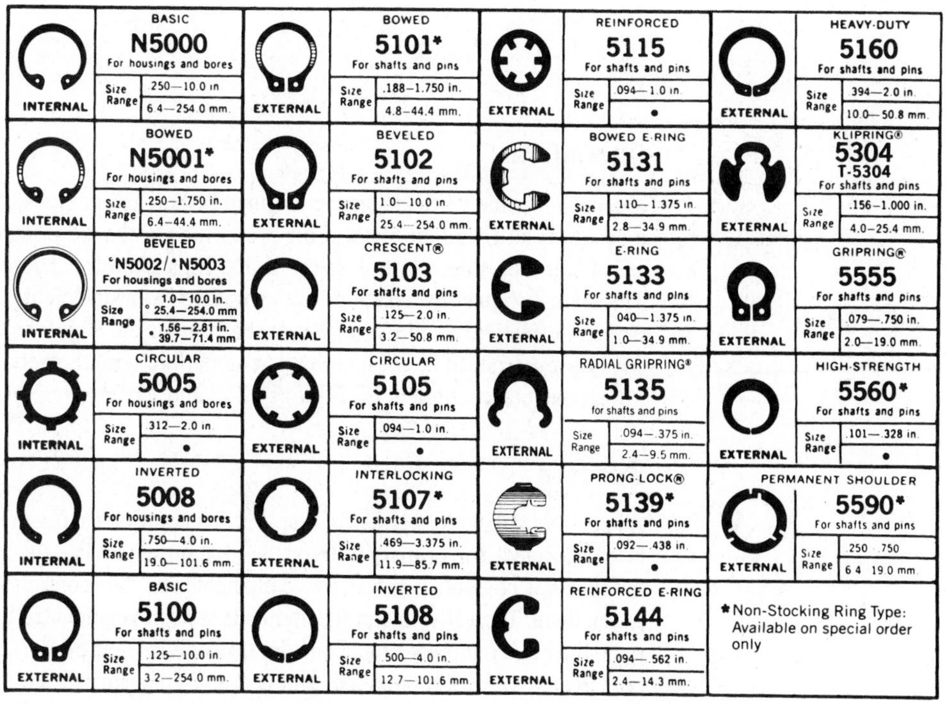

FIGURE 11–30
Standard Truarc®
retaining ring series
(Courtesy Truarc
Company LLC,
Millburn, NJ)

FIGURE 11–31
Retaining rings applied
to a water conditioner
mechanism (Courtesy
Truarc Company LLC,
Millburn, NJ)

FIGURE 11–32
Application of spacers

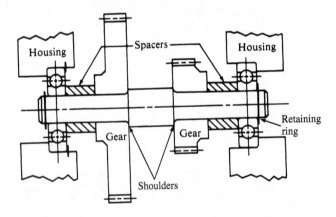

Shoulders

A *shoulder* is the vertical surface produced when a diameter change occurs on a shaft. Such a design is an excellent method for providing for the axial location of a machine element, at least on one side. Several of the shafts illustrated in Chapter 12 incorporate shoulders. The main design considerations are providing (1) a sufficiently large shoulder to locate the element effectively and (2) a fillet at the base of the shoulder that produces an acceptable stress concentration factor and that is compatible with the geometry of the bore of the mating element (see Chapter 12, Figures 12–2 and 12–7).

Spacers

A *spacer* is similar to a collar in that it is slid over the shaft against the machine element that is to be located. The primary difference is that set screws and the like are not necessary because the spacer is positioned *between* two elements and thus controls only the relative position between them. Typically, one of the elements is positively located by some other means, such as a shoulder or a retaining ring. Consider the shaft in Figure 11–32, on which two spacers locate each gear with respect to its adjacent bearing. The bearings are in turn located on one side by

FIGURE 11–33
Locknut and
lockwasher for
retaining bearings.
Dimensions are given
in manufacturer's
catalog; part sizes are
compatible with
standard bearing sizes.
(SKF Industries, Inc.,
Norristown, PA)

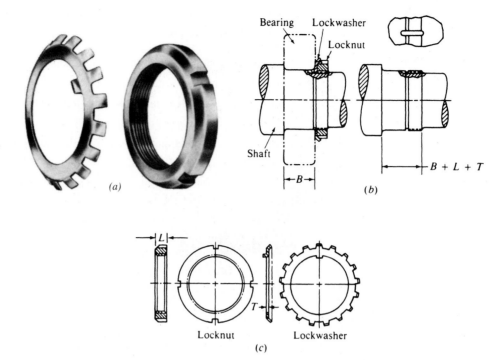

the housing supporting their outer races. The gears also seat against shoulders on one side. The retaining ring on the shaft secures the subassembly prior to installing it into the housing.

Locknuts

When an element is located at the end of a shaft, a *locknut* can be used to retain it on one side. Figure 11–33 shows a bearing retainer type of locknut. These are available as stock items from bearing suppliers.

Caution against Overconstraint

One practical consideration in the matter of axial location of machine elements is exercising care that elements are not *overconstrained*. Under certain conditions of differential thermal expansion or with an unfavorable tolerance stack, the elements may be forced together so tightly as to cause dangerous axial stresses. At times it may be desirable to locate only one bearing positively on a shaft and to permit the other to float slightly in the axial direction. The floating element may be held lightly with an axial spring force accommodating the thermal expansion without creating dangerous forces.

**11–10
TYPES OF SEALS**

Seals are an important part of machine design in situations where the following conditions apply:

1. Contaminants must be excluded from critical areas of a machine.
2. Lubricants must be contained within a space.
3. Pressurized fluids must be contained within a component such as a valve or a hydraulic cylinder.

Some of the parameters affecting the choice of the type of sealing system, the materials used, and the details of its design are as follows:

1. The nature of the fluids to be contained or excluded.
2. Pressures on both sides of the seal.
3. The nature of any relative motion between the seal and the mating components.
4. Temperatures on all parts of the sealing system.
5. The degree of sealing required: Is some small amount of leakage permissible?
6. The life expectancy of the system.
7. The nature of the solid materials against which the seal must act: corrosion potential, smoothness, hardness, wear resistance.
8. Ease of service for replacement of worn sealing elements.

The number of designs for sealing systems is virtually limitless, and only a brief overview will be presented here. More comprehensive coverage can be found in Reference 8. Often, designers rely on technical information provided by manufacturers of complete sealing systems or specific sealing elements. Also, in critical or unusual situations, testing of a proposed design is advised. See Internet sites 12–15.

The choice of a type of sealing system depends on the duty that it must perform. Common conditions in which seals must operate are listed here, along with some of the types of seals used.

1. Static conditions such as sealing a closure on a pressurized container: elastomeric O-rings; T-rings; hollow metal O-rings; and sealants such as epoxies, silicones, and butyl caulking (Figure 11–34).

FIGURE 11–34
O-rings and T-rings
used as static seals

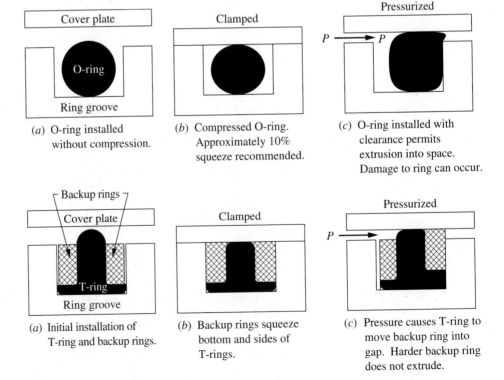

(a) O-ring installed without compression.

(b) Compressed O-ring. Approximately 10% squeeze recommended.

(c) O-ring installed with clearance permits extrusion into space. Damage to ring can occur.

(a) Initial installation of T-ring and backup rings.

(b) Backup rings squeeze bottom and sides of T-rings.

(c) Pressure causes T-ring to move backup ring into gap. Harder backup ring does not extrude.

2. Sealing a closed container while allowing relative movement of some part, such as diaphragms, bellows, and boots (Figure 11–35).

3. Sealing around a continuously reciprocating rod or piston, such as in a hydraulic cylinder or a spool valve in a hydraulic system: lip seal; U-cup seal; V-packing; and split ring seals, sometimes called *piston rings* (Figure 11–36).

FIGURE 11–35
Application of a
diaphragm seal

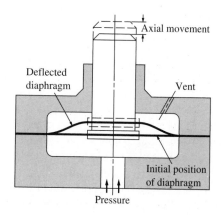

FIGURE 11–36 Lip
seals, U-cup seal,
wiper, and O-rings
applied to a hydraulic
actuator

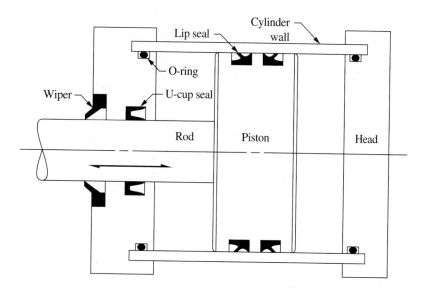

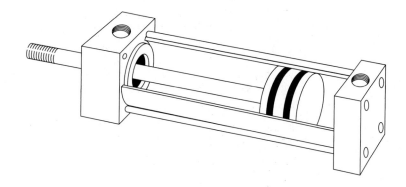

4. Sealing around a rotating shaft such as the input or output shafts of a speed reducer, transmission, or engine: lip seal; wipers and scrapers; and face seals (Figure 11–37).

5. Protection of rolling-element bearings supporting shafts to keep contaminants from the balls and rollers (Figure 11–38).

6. Sealing the active elements of a pump to retain the pumped fluid: face seals and V-packing.

FIGURE 11–37   Face seals

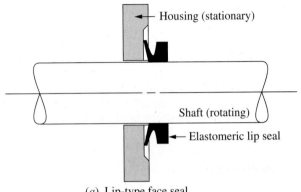

(a) Lip-type face seal

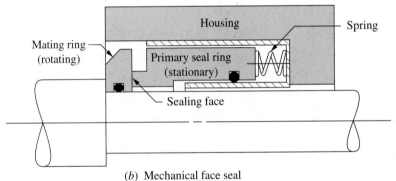

(b) Mechanical face seal

FIGURE 11–38   Seal for ball bearing

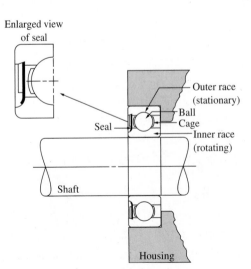

7. Sealing infrequently moved elements such as a valve stem of a fluid-flow control valve: compression packings and V-packings.

8. Sealing between hard, rigid surfaces such as between a cylinder head and the block of an engine: resilient gaskets.

9. Circumferential seals, such as at the tips of turbine blades, and on large, high-speed rotating elements: labyrinth seals; abradable seals; and hydrostatic seals.

11–11
SEAL MATERIALS

Most seal materials are resilient to permit the sealing points to follow minor variations in the geometry of mating surfaces. Flexing of parts of the seal cross section also occurs in some designs, calling for resiliency in the materials. Alternatively, as in the case of hollow metal O-rings, the shape of the seal allows the flexing of hard materials to occur. Face seals require rigid, hard materials that can withstand constant sliding motion and that can be produced with fine accuracy, flatness, and smoothness.

Elastomers

Resilient seals such as O-rings, T-rings, and lip seals are often made from synthetic elastomers such as the following:

Neoprene	Butyl	Nitrile (Buna-*N*)
Fluorocarbon	Silicone	Fluorosilicone
Butadiene	Polyester	Ethylene propylene
Polysulfide	Polyurethane	Epichlorohydrin
Polyacrylate	PNF (Phosphonitrilic fluoroelastomer)	

Many proprietary formulations within these general classifications are available under trade names from seal producers and plastics manufacturers.

Properties needed in a given installation will limit the selection of candidate materials. The following list gives some of the more prevalent requirements for seals and some of the materials that meet those requirements:

Weather resistance: Silicone, fluorosilicone, fluorocarbon, ethylene propylene, polyurethane, polysulfide, polyester, neoprene, epichlorohydrin, and PNF.

Petroleum fluid resistance: Polyacrylate, polyester, PNF, nitrile, polysulfide, polyurethane, fluorocarbon, and epichlorohydrin.

Acid resistance: Fluorocarbon.

High-temperature operation: Ethylene propylene, fluorocarbon, polyacrylate, silicone, and PNF.

Cold-temperature operation: Silicone, fluorosilicone, ethylene propylene, and PNF.

Tensile strength: Butadiene, polyester, and polyurethane.

Abrasion resistance: Butadiene, polyester, and polyurethane.

Impermeability: Butyl, polyacrylate, polysulfide, and polyurethane.

Rigid Materials

Face seals, and the parts of other types of sealing systems against which elastomers seal, require rigid materials that can withstand the sliding action and that are compatible with the environment around the seal. Some typical rigid materials used in sealing systems are described in the following list:

Metals: Carbon steel, stainless steel, cast iron, nickel alloys, bronze, and tool steels.

Plastics: Nylon, filled polytetrafluoroethylene (PTFE), and polyimide.

Carbon, ceramics, tungsten-carbide.

Plating: Chromium, cadmium, tin, nickel, and silver.

Flame-sprayed compounds.

Packings

Packings for sealing shafts, rods, valve stems, and similar applications are made from a variety of materials, including leather, cotton, flax, several types of plastics, braided or twisted wire made from copper or aluminum, laminated cloth and elastomeric materials, and flexible graphite.

Gaskets

Common gasket materials are cork, cork and rubber compounds, filled rubber, paper, resilient plastics, and foams.

Shafts

When radial lip seals are required around shafts, the shafts are typically steel. They should be hardened to HRC 30 to resist scoring of the surface. Tolerance on the diameter of the shaft on which the seal bears should conform to the following recommendations to ensure that the seal lip can follow the variations:

Shaft Diameter (in)	*Tolerance (in)*
$D \leq 4.000$	± 0.003
$4.000 < D \leq 6.000$	± 0.004
$D > 6.000$	± 0.005

The surface of the shaft and any areas over which the seal must pass during installation should be free of burrs to protect against tearing the seal. A surface finish of 10 to 20 μin is recommended, with adequate lubrication to ensure full contact and to reduce friction between the seal and the shaft surface.

REFERENCES

1. American Gear Manufacturers Association. AGMA 9009-D02. *Nomenclature for Flexible Couplings.* Alexandria, VA: American Gear Manufacturers Association, 2002.

2. American Gear Manufacturers Association. ANSI/AGMA 9002-A86. *Bores and Keyways for Flexible Couplings.* Alexandria, VA: American Gear Manufacturers Association, 1986.

3. American Gear Manufacturers Association. ANSI/AGMA 9001-B97. *Lubrication of Flexible Couplings.* Alexandria, VA: American Gear Manufacturers Association, 1997.

4. American National Standards Institute. ANSI B92.1-1986. *Involute Splines.* New York: American National Standards Institute, 1996.

5. American National Standards Institute. ANSI B92.2-1980 R1989. *Metric Module Involute Splines.* New York: American National Standards Institute, 1989.

6. American Society of Mechanical Engineers. ANSI B17.1-67.R98. *Keys and Keyseats.* New York: American Society of Mechanical Engineers, 1998.

7. American Society of Mechanical Engineers. ANSI B17.2-67.R98. *Woodruff Keys and Keyseats.* New York: American Society of Mechanical Engineers, 1998.

8. Lebeck, Alan O. *Principles and Design of Mechanical Face Seals.* New York: John Wiley & Sons, 1991.

9. Oberg, Erik, et al. *Machinery's Handbook.* 26th ed. New York: Industrial Press, 2000.

10. Penton Publishing Co. *Machine Design Magazine Power and Motion Control Volume.* Vol. 61, no. 12. Cleveland, OH: Penton Publishing, June 1989.

11. Society of Automotive Engineers. *Design Guide for Involute Splines, Product Code M-117.* Warrendale, PA: Society of Automotive Engineers, 1994.

12. Society of Automotive Engineers. *Standard AS-84, Splines, Involute (Full Fillet).* Warrendale, PA: Society of Automotive Engineers, 2000.

INTERNET SITES FOR KEYS, COUPLINGS, AND SEALS

1. **Driv-Lok, Inc.** *www.driv-lok.com* Manufacturer of a wide variety of parallel keys and press-fit fastening devices such as grooved pins, dowel pins, spring pins, and studs.

2. **Ringfeder Corporation.** *www.ringfeder-usa.com* Manufacturer of keyless hub to shaft locking devices.

3. **Rockwell Automation/Dodge.** *www.dodge-pt.com* Manufacturer of a wide variety of power transmission components, including flexible couplings, gearing, belt drives, clutches and brakes, and bearings.

4. **Emerson Power Transmission, Inc.** *www.emerson-ept.com* Manufacturer of a wide variety of power transmission equipment including flexible couplings and universal joints under the Kop-Flex, Browning, and Morse brands. Other brands include McGill, Rollway, Sealmaster, and Van Gorp for gearing, belt drives, chain drives, bearings, conveyor pulleys, and clutches.

5. **Dana Corporation.** *www.dana.com* and *www.spicerdriveshaft.com* Manufacturer of flexible couplings, universal joints, and driveshafts for vehicular and industrial applications using the Dana and Spicer brand names.

6. **GKN Drivetech, Inc.** *www.gkndrivetech.com* Manufacturer of constant velocity (CV) joints and driveline components for vehicular applications.

7. **Stock Drive Products/Sterling Instrument.** *www.sdp-si.com* Manufacturer and distributor of precision machine components and assemblies, including flexible couplings, gearing, clutches and brakes, fasteners, and many others.

8. **T. B. Wood's Incorporated.** *www.tbwoods.com* Manufacturer of mechanical, electrical, and electronic industrial power transmission products, including flexible couplings, synchronous belt drives, V-belt drives, gearmotors, and gearing.

9. **Curtis Universal Joint Company.** *www.curtisuniversal.com* Manufacturer of universal joints for the industrial and aerospace markets.

10. **Cooper Power Tools/Apex Operation.** *www.cooperindustries.com* Manufacturer of universal joints for the military, aerospace, performance racing, and industrial power transmission markets.

11. **Truarc Company LLC.** *www.truarc.com* Manufacturer of a wide variety of retaining rings for industrial, commercial, military, and consumer products.

12. **Federal Mogul Corporation/National Seals.** *www.federal-mogul.com/national* Manufacturer of seals for engines, transmissions, wheels, differentials, and industrial applications. Site includes an electronic catalog for vehicular seals.

13. **American Seal Company.** *www.americanseal.com* Manufacturer of O-rings, backups, square rings, and custom molded parts.

14. **Industrial Gasket & Shim Company.** *www.igscorp.com* Manufacturer of gaskets, shims, and custom fabricated seals, expansion joints, and industrial sealants.

15. **American High Performance Seals, Inc.** *www.ahps.net* Manufacturer of a wide variety of wipers and seals for rods, pistons, and rotary applications. Site includes numerous graphic representations of seal cross sections and tables of materials and their properties.

16. **Lord Corporation.** *www.lordmpd.com* Manufacturer of flexible couplings, vibration mounts, and shock isolation mounts using elastomeric materials bonded to metals. Site includes catalog data for selecting couplings and other products.

17. **General Polygon Systems, Inc.** *www.generalpolygon.com* Providers of mechanical shaft to hub connections using the polygon system.

18. **Drive Technologies, LLC** *www.drivetechnologies.com* Developer and manufacturer of the Cornay™ universal joint, drive shafts, power take-offs, CV joints, and complete drive systems.

PROBLEMS

For Problems 1–4 and 7, determine the required key geometry: length, width, and height. Use AISI 1020 cold-drawn steel for the keys if a satisfactory design can be achieved. If not, use a higher-strength material. Unless otherwise stated, assume that the key material is weakest when compared with the shaft material or the mating elements.

1. Specify a key for a gear to be mounted on a shaft with a 2.00-in diameter. The gear transmits 21 000 lb·in of torque and has a hub length of 4.00 in.

2. Specify a key for a gear carrying 21 000 lb·in of torque if it is mounted on a 3.60-in-diameter shaft. The hub length of the gear is 4.00 in.

3. A V-belt sheave transmits 1112 lb·in of torque to a 1.75-in-diameter shaft. The sheave is made from ASTM class 20 cast iron and has a hub length of 1.75 in.

4. A chain sprocket delivers 110 hp to a shaft at a rotational speed of 1700 rpm. The bore of the sprocket is 2.50 inches in diameter. The hub length is 3.25 in.

5. Specify a suitable spline having a *B* fit for each of the applications in Problems 1–4.

6. Design a cylindrical pin to transmit the power, as in Problem 4. But design it so that it will fail in shear if the power exceeds 220 hp.

7. Specify a key for both the sprocket and the wormgear from Example Problem 12–4. Note the specifications for the final shaft diameters at the end of the problem.

8. Describe a Woodruff key no. 204.

9. Describe a Woodruff key no. 1628.

10. Make a detailed drawing of a Woodruff key connection between a shaft and the hub of a gear. The shaft has a diameter of 1.500 in. Use a no. 1210 Woodruff key. Dimension the keyseat in the shaft and the hub.

11. Repeat Problem 10, using a no. 406 Woodruff key in a shaft having a 0.500-in diameter.

12. Repeat Problem 10, using a no. 2428 Woodruff key in a shaft having a 3.250-in diameter.

13. Compute the torque that could be transmitted by the key of Problem 10 on the basis of shear and bearing if the key is made from AISI 1020 cold-drawn steel with a design factor of $N = 3$.

14. Repeat Problem 13 for the key of Problem 11.

15. Repeat Problem 13 for the key of Problem 12.

16. Make a drawing of a four-spline connection having a major diameter of 1.500 in and an *A* fit. Show critical dimensions.

17. Make a drawing of a 10-spline connection having a major diameter of 3.500 in and a *B* fit. Show critical dimensions.

18. Make a drawing of a 16-spline connection having a major diameter of 2.500 in and a *C* fit. Show critical dimensions.

19. Determine the torque capacity of the splines in Problems 16–18.

20. Describe the manner in which a set screw transmits torque if it is used in place of a key. Discuss the disadvantages of such an arrangement.

21. Describe a press fit as it would be used to secure a power transmission element to a shaft.

22. Describe the main differences between rigid and flexible couplings as they affect the stresses in the shafts that they connect.

23. Discuss a major disadvantage of using a single universal joint to connect two shafts with angular misalignment.

24. Describe five ways to locate power transmission elements positively on a shaft axially.

25. Describe three situations in which seals are applied in machine design.

26. List eight parameters that should be considered when selecting a type of seal and specifying a particular design.

27. Name three means of sealing a pressurized container under static conditions.

28. Name three methods of sealing a closed container while allowing relative movement of some part.

29. Name four types of seals used on reciprocating rods or pistons.

30. Name three types of seals applied to rotating shafts.

31. Describe the method of sealing a ball bearing from contaminants.

32. Describe an O-ring seal, and sketch its installation.

33. Describe a T-ring seal, and sketch its installation.

34. Describe some advantages of T-ring seals over O-rings.

35. Describe a diaphragm seal and the type of situation in which it is used.

36. Describe suitable methods of sealing the sides of a piston against the inner walls of the cylinder of a hydraulic actuator.

37. Describe the function of a scraper or wiper on a cylinder rod.

38. Describe the essential elements of a mechanical face seal.

39. Name at least six kinds of elastomers commonly used for seals.

40. Name at least three kinds of elastomers that are recommended for use when exposed to weather.

41. Name at least three kinds of elastomers that are recommended for use when exposed to petroleum-based fluids.

42. Name at least three kinds of elastomers that are recommended for use when exposed to cold-temperature operation.

43. Name at least three kinds of elastomers that are recommended for use when exposed to high-temperature operation.

44. A sealing application involves the following conditions: exposure to high-temperature petroleum fluids; impermeability. Specify a suitable elastomer for the seal.

45. A sealing application involves the following conditions: exposure to high temperature and weather; impermeability; and high strength and abrasion resistance. Specify a suitable elastomer for the seal.

46. Describe suitable shaft design details where elastomeric seals contact the shaft.

12

Shaft Design

The Big Picture

You Are the Designer

12–1   Objectives of This Chapter

12–2   Shaft Design Procedure

12–3   Forces Exerted on Shafts by Machine Elements

12–4   Stress Concentrations in Shafts

12–5   Design Stresses for Shafts

12–6   Shafts in Bending and Torsion Only

12–7   Shaft Design Example

12–8   Recommended Basic Sizes for Shafts

12–9   Additional Design Examples

12–10 Spreadsheet Aid for Shaft Design

12–11 Shaft Rigidity and Dynamic Considerations

12–12 Flexible Shafts

Shaft Design

Discussion Map

☐ A shaft is a rotating machine component that transmits power.

Discover

Identify examples of mechanical systems that incorporate power-transmitting shafts. Describe their geometry and the forces and torques that are exerted on them.

What kinds of stresses are produced in the shaft?

How are other elements mounted on the shaft? How does the shaft geometry accommodate them? How is the shaft supported? What kinds of bearings are used?

This chapter provides approaches that you can use to design shafts that are safe for their intended use. But you have the final responsibility for the design.

A *shaft* is the component of a mechanical device that transmits rotational motion and power. It is integral to any mechanical system in which power is transmitted from a prime mover, such as an electric motor or an engine, to other rotating parts of the system. Can you identify some kinds of mechanical systems incorporating rotating elements that transmit power?

Here are some examples: gear-type speed reducers, belt or chain drives, conveyors, pumps, fans, agitators, and many types of automation equipment. What others can you think of? Consider household appliances, lawn maintenance equipment, parts of a car, power tools, and machines around an office or in your workplace. Describe them and discuss how shafts are used. From what source is power delivered into the shaft? What kind of power-transmitting element, if any, is on the shaft itself? Or is the shaft simply transmitting the rotational motion and torque to some other element? If so, how is the shaft connected to that element?

Visualize the forces, torques, and bending moments that are created in the shaft during operation. In the process of transmitting power at a given rotational speed, the shaft is inherently subjected to a torsional moment, or *torque*. Thus, torsional shear stress is developed in the shaft. Also, a shaft usually carries power-transmitting components, such as gears, belt sheaves, or chain sprockets, which exert forces on the shaft in the transverse direction (perpendicular to its axis). These transverse forces cause bending moments to be developed in the shaft, requiring analysis of the stress due to bending. In fact, most shafts must be analyzed for combined stress.

Describe the specific geometry of shafts from some types of equipment that you can examine. Make a sketch of any variations in geometry that may occur, such as changes in diameter, to produce shoulders, grooves, keyseats, or holes. How are any power-transmitting elements held in position along the length of the shaft? How are the shafts supported? Typically, bearings are used to support the shaft while permitting rotation relative to the housing of the machine. What kinds of bearings are used? Do they have rolling elements such as ball bearings? Or are they smooth-surfaced bearings? What materials are used?

It is likely that you will find much diversity in the design of the shafts in different kinds of equipment. You should see that the functions of a shaft have a large influence on its design. Shaft geometry is greatly affected by the mating elements such as bearings, couplings, gears, chain sprockets, or other kinds of power-transmitting elements.

This chapter provides approaches that you can use to design shafts that are safe for their intended use. But the final responsibility for the design is yours because it is impractical to predict in a book all conditions to which a given shaft will be subjected.

You Are the Designer

Consider the gear-type speed reducer shown in Figures 1–12 and 8–3. As you continue the design, three shafts must be designed. The input shaft carries the first gear in the gear train and rotates at the speed of the prime mover, typically an electric motor or an engine. The middle shaft carries two gears and rotates more slowly than the input shaft because of the first stage of speed reduction. The final gear in the train is carried by the third shaft, which also transmits the power to the driven machine. From what material should each shaft be made? What

torque is being transmitted by each shaft, and over what part of the shaft is it acting? How are the gears to be located on the shafts? How is the power to be transmitted from the gears to the shafts, or vice versa? What forces are exerted on the shaft by the gears, and what bending moments result? What forces must the bearings that support each shaft resist? What are the minimum acceptable diameters for the shafts at all sections to ensure safe operation? What should be the final dimensional specifications for the many features of the shafts, and what should be the tolerances on those dimensions? The material in this chapter will help you make these and other shaft design decisions.

12–1
OBJECTIVES OF THIS CHAPTER

After completing this chapter, you will be able to:

1. Propose reasonable geometries for shafts to carry a variety of types of power-transmitting elements, providing for the secure location of each element and the reliable transmission of power.

2. Compute the forces exerted on shafts by gears, belt sheaves, and chain sprockets.

3. Determine the torque distribution on shafts.

4. Prepare shearing force and bending moment diagrams for shafts in two planes.

5. Account for stress concentration factors commonly encountered in shaft design.

6. Specify appropriate design stresses for shafts.

7. Apply the shaft design procedure recommended by the standard, ANSI B106.1M-1985, *Design of Transmission Shafting*, to determine the required diameter of shafts at any section to resist the combination of torsional shear stress and bending stress.

8. Specify reasonable final dimensions for shafts that satisfy strength requirements and installation considerations and that are compatible with the elements mounted on the shafts.

9. Consider the influence of shaft rigidity on its dynamic performance.

12–2
SHAFT DESIGN PROCEDURE

Because of the simultaneous occurrence of torsional shear stresses and normal stresses due to bending, the stress analysis of a shaft virtually always involves the use of a combined stress approach. The recommended approach for shaft design and analysis is the *distortion energy theory of failure*. This theory was introduced in Chapter 5 and will be discussed more fully in Section 12–5. Vertical shear stresses and direct normal stresses due to axial loads may also occur. On very short shafts or on portions of shafts where no bending or torsion occurs, such stresses may be dominant. The discussions in Chapters 3, 4, and 5 explain the appropriate analysis.

The specific tasks to be performed in the design and analysis of a shaft depend on the shaft's proposed design in addition to the manner of loading and support. With this in mind, the following is a recommended procedure for the design of a shaft.

Procedure for Design of a Shaft

1. Determine the rotational speed of the shaft.

2. Determine the power or the torque to be transmitted by the shaft.

3. Determine the design of the power-transmitting components or other devices that will be mounted on the shaft, and specify the required location of each device.

4. Specify the location of bearings to support the shaft. Normally two and only two bearings are used to support a shaft. The reactions on bearings supporting radial loads are assumed to act at the midpoint of the bearings. For example, if a single-row ball bearing is used, the load path is assumed to pass directly through the balls. If thrust (axial) loads exist in the shaft, you must specify which bearing is to be designed to react against the thrust load. Then the bearing that does not resist the thrust should be permitted to move slightly in the axial direction to ensure that no unexpected and undesirable thrust load is exerted on that bearing.

   Bearings should be placed on either side of the power-transmitting elements if possible to provide stable support for the shaft and to produce reasonably well-balanced loading of the bearings. The bearings should be placed close to the power-transmitting elements to minimize bending moments. Also, the overall length of the shaft should be kept small to keep deflections at reasonable levels.

5. Propose the general form of the geometry for the shaft, considering how each element on the shaft will be held in position axially and how power transmission from each element to the shaft is to take place. For example, consider the shaft in Figure 12–1, which is to carry two gears as the intermediate shaft in a double-reduction, spur gear-type speed reducer. Gear A accepts power from gear P by way of the input shaft. The power is transmitted from gear A to the shaft through the key at the interface between

FIGURE 12–1

Intermediate shaft for a double-reduction, spur gear-type speed reducer

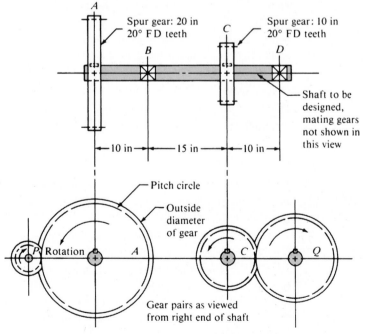

Input: Gear *P* drives gear *A*   *Output*: Gear *C* drives gear *Q*

FIGURE 12–2

Proposed geometry for the shaft in Figure 12–1. Sharp fillets at r_3, r_5; well-rounded fillets at r_1, r_2, r_4; profile keyseats at A, C.

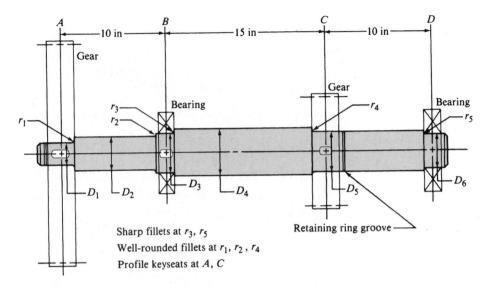

Sharp fillets at r_3, r_5
Well-rounded fillets at r_1, r_2, r_4
Profile keyseats at A, C

the gear hub and the shaft. The power is then transmitted down the shaft to point C, where it passes through another key into gear C. Gear C then transmits the power to gear Q and thus to the output shaft. The locations of the gears and bearings are dictated by the overall configuration of the reducer.

It is now decided that the bearings will be placed at points B and D on the shaft to be designed. But how will the bearings and the gears be located so as to ensure that they stay in position during operation, handling, shipping, and so forth? Of course, there are many ways to do this. One way is proposed in Figure 12–2. Shoulders are to be machined in the shaft to provide surfaces against which to seat the bearings and the gears on one side of each element. The gears are restrained on the other side by retaining rings snapped into grooves in the shaft. The bearings will be held in position by the housing acting on the outer races of the bearings. Keyseats will be machined in the shaft at the location of each gear. This proposed geometry provides for positive location of each element.

6. Determine the magnitude of torque that the shaft sees at all points. It is recommended that a torque diagram be prepared, as will be shown later.

7. Determine the forces that are exerted on the shaft, both radially and axially.

8. Resolve the radial forces into components in perpendicular directions, usually vertically and horizontally.

9. Solve for the reactions on all support bearings in each plane.

10. Produce the complete shearing force and bending moment diagrams to determine the distribution of bending moments in the shaft.

11. Select the material from which the shaft will be made, and specify its condition: cold-drawn, heat-treated, and so on. Refer to Table 2–4 in Chapter 2 for suggestions for steel materials for shafts. Plain carbon or alloy steels with medium carbon content are typical, such as AISI 1040, 4140, 4340, 4640, 5150, 6150, and 8650. Good ductility with percent elongation above about 12% is recommended. Determine the ultimate strength, yield strength, and percent elongation of the selected material.

12. Determine an appropriate design stress, considering the manner of loading (smooth, shock, repeated and reversed, or other).

13. Analyze each critical point of the shaft to determine the minimum acceptable diameter of the shaft at that point in order to ensure safety under the loading at that point. In general, the critical points are several and include those where a change of diameter takes place, where the higher values of torque and bending moment occur, and where stress concentrations occur.

14. Specify the final dimensions for each point on the shaft. Normally, the results from Step 13 are used as a guide, and convenient values are then chosen. Design details such as tolerances, fillet radii, shoulder heights, and keyseat dimensions must also be specified. Sometimes the size and the tolerance for a shaft diameter are dictated by the element to be mounted there. For example, ball bearing manufacturers' catalogs specify limits for bearing seat diameters on shafts.

This process will be demonstrated after the concepts of force and stress analysis are presented.

12–3 FORCES EXERTED ON SHAFTS BY MACHINE ELEMENTS

Gears, belt sheaves, chain sprockets, and other elements typically carried by shafts exert forces on the shaft that cause bending moments. The following is a discussion of the methods for computing these forces for some cases. In general, you will have to use the principles of statics and dynamics to determine the forces for any particular element.

Spur Gears

The force exerted on a gear tooth during power transmission acts normal (perpendicular) to the involute-tooth profile, as discussed in Chapter 9 and shown in Figure 12–3. It is convenient for the analysis of shafts to consider the rectangular components of this force acting in the radial and tangential directions. It is most convenient to compute the tangential force, W_t, directly from the known torque being transmitted by the gear. For U.S. Customary units,

$$\text{units } \frac{lbf \cdot in \cdot rpm}{hp}$$

➡ **Torque**

$$T = 63\ 000\ (P)/n \qquad (12\text{–}1)$$

FIGURE 12–3
Forces on teeth of a driven gear

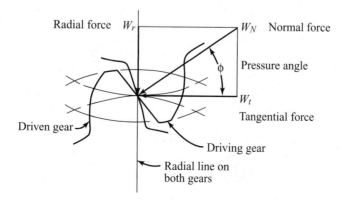

⇨ **Tangential Force**

$$W_t = T/(D/2) = \frac{T}{\frac{HP}{D \cdot RPm}} \qquad (12\text{–}2)$$

where P = power being transmitted in hp
   n = rotational speed in rpm
   T = torque on the gear in lb·in
   D = pitch diameter of the gear in inches

The angle between the total force and the tangential component is equal to the pressure angle, ϕ, of the tooth form. Thus, if the tangential force is known, the radial force can be computed directly from

⇨ **Radial Forces**

$$W_r = W_t \tan \phi = 20° \text{ always} \qquad (12\text{–}3)$$

and there is no need to compute the normal force at all. For gears, the pressure angle is typically $14\frac{1}{2}°$, $20°$, or $25°$.

Directions for Forces on Mating Spur Gears

Representing the forces on gears in their correct directions is essential to an accurate analysis of forces and stresses in the shafts that carry the gears. The force system shown in Figure 12–4(a) represents the action of the driving gear A on the driven gear B. The tangential force, W_t, pushes perpendicular to the radial line causing the driven gear to rotate. The radial force, W_r, exerted by the driving gear A, acts along the radial line tending to push the driven gear B away.

An important principle of mechanics states that for each action force, there is an equal and opposite reaction force. Therefore, as shown in Figure 12–4(b), the driven gear pushes back on the driving gear with a tangential force opposing that of the driving gear and a radial force that tends to push the driving gear away. For the orientation of the gears shown in Figure 12–4, note the following directions for forces:

FIGURE 12–4
Directions for forces on mating spur gears.

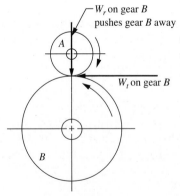

(a) Forces exerted on
   gear B by gear A.
   Action forces—gear A
   drives gear B.

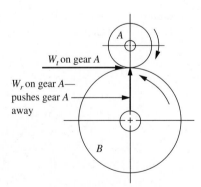

(b) Forces exerted on
   gear A by gear B.
   Reaction forces.

Action: Driver pushes on driven gear   *Reaction:* Driven gear pushes back on driver

W_t: Acts to the left W_t: Acts to the right

W_r: Acts downward W_r: Acts upward

In summary, whenever you need to determine the direction of forces acting on a given gear, first determine whether it is a driver or a driven gear. Then visualize the action forces of the driver. If the gear of interest is the driven gear, these are the forces on it. If the gear of interest is the driver gear, the forces on it act in the opposite directions to the action forces.

Helical Gears

In addition to the tangential and radial forces encountered with spur gears, helical gears produce an axial force (as discussed in Chapter 10). First compute the tangential force from Equations (12–1) and (12–2). Then, if the helix angle of the gear is ψ, and if the normal pressure angle is ϕ_n, the radial load can be computed from

⇨ **Radial Force**

$$W_r = W_t \tan \phi_n / \cos \psi \qquad \text{(12–4)}$$

The axial load is

⇨ **Axial Force**

$$W_x = W_t \tan \psi \qquad \text{(12–5)}$$

Bevel Gears

Refer to Chapter 10 to review the formulas for the three components of the total force on bevel gear teeth in the tangential, radial, and axial directions. Example Problem 10–4 gives a comprehensive analysis of the forces, torques, and bending moments on shafts carrying bevel gears.

Worms and Wormgears

Chapter 10 also gives the formulas for computing the forces on worms and wormgears in the tangential, radial, and axial directions. See Example Problem 10–10.

Chain Sprockets

Figure 12–5 shows a pair of chain sprockets transmitting power. The upper part of the is in tension and produces the torque on either sprocket. The lower part of the ferred to as the *slack side*, exerts no force on either sprocket. Therefore, the tot force on the shaft carrying the sprocket is equal to the tension in the tight side If the torque on a certain sprocket is known,

⇨ **Force in Chain**

$$F_c = T/(D/2)$$

where D = pitch diameter of that sprocket

FIGURE 12–5

Forces on chain sprockets

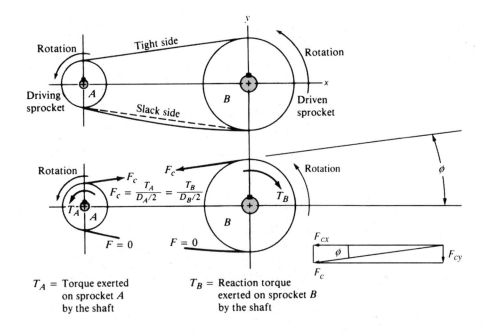

T_A = Torque exerted on sprocket A by the shaft

T_B = Reaction torque exerted on sprocket B by the shaft

Notice that the force, F_c, acts along the direction of the tight side of the belt. Because of the size difference between the two sprockets, that direction is at some angle from the centerline between the shaft centers. A precise analysis would call for the force, F_c, to be resolved into components parallel to the centerline and perpendicular to it. That is,

$$F_{cx} = F_c \cos \theta \quad \text{and} \quad F_{cy} = F_c \sin \theta$$

where the x-direction is parallel to the centerline

the y-direction is perpendicular to it

the angle θ is the angle of inclination of the tight side of the chain with respect to the x-direction

These two components of the force would cause bending in both the x-direction and the y-direction. Alternatively, the analysis could be carried out in the direction of the force, F_c, in which single plane bending occurs.

If the angle θ is small, little error will result from the assumption that the entire force, ̄cts along the x-direction. *Unless stated otherwise, this book will use this assumption.*

̄es

̄of the V-belt drive system looks similar to the chain drive system. ̄difference: Both sides of the V-belt are in tension, as indicated ̄t side tension, F_1, is greater than the "slack side" tension, F_2, and ̄ng force on the sheaves equal to

$$F_N = F_1 - F_2 \tag{12–7}$$

of the net driving force can be computed from the torque transmitted:

$$F_N = T/(D/2) \tag{12–8}$$

(12–6)

FIGURE 12–6

Forces on belt sheaves or pulleys

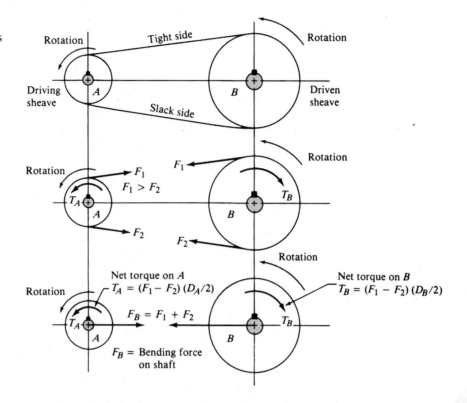

But notice that the bending force on the shaft carrying the sheave is dependent on the *sum*, $F_1 + F_2 = F_B$. To be more precise, the components of F_1 and F_2 parallel to the line of centers of the two sprockets should be used. But unless the two sprockets are radically different in diameter, little error will result from $F_B = F_1 + F_2$.

To determine the bending force, F_B, a second equation involving the two forces F_1 and F_2 is needed. This is provided by assuming a ratio of the tight side tension to the slack side tension. For V-belt drives, the ratio is normally taken to be

$$F_1/F_2 = 5 \qquad \text{(12–9)}$$

It is convenient to derive a relationship between F_N and F_B of the form

$$F_B = CF_N \qquad \text{(12–10)}$$

where C = constant to be determined

$$C = \frac{F_B}{F_N} = \frac{F_1 + F_2}{F_1 - F_2} \qquad \text{(12–11)}$$

But from Equation (12–9), $F_1 = 5F_2$. Then

$$C = \frac{F_1 + F_2}{F_1 - F_2} = \frac{5F_2 + F_2}{5F_2 - F_2} = \frac{6F_2}{4F_2} = 1.5$$

Equation (12–10) then becomes, for V-belt drives,

⇨ **Bending Force on Shaft for V-Belt Drive**

$$F_B = 1.5 \, F_N = 1.5T/(D/2) \qquad \text{(12–12)}$$

It is customary to consider the bending force, F_B, to act as a single force in the direction along the line of centers of the two sheaves as shown in Figure 12–6.

Flat-Belt Pulleys

The analysis of the bending force exerted on shafts by flat-belt pulleys is identical to that for V-belt sheaves except that the ratio of the tight side to the slack side tension is typically taken to be 3 instead of 5. Using the same logic as with V-belt sheaves, we can compute the constant C to be 2.0. Then, for flat-belt drives,

⇨ **Bending Force on Shaft for Flat-Belt Drive**

$$F_B = 2.0\,F_N = 2.0T/(D/2) \qquad\qquad (12\text{–}13)$$

Flexible Couplings

More detailed discussion of flexible couplings was presented in Chapter 11, but it is important to observe here how the use of a flexible coupling affects the design of a shaft.

A flexible coupling is used to transmit power between shafts while accommodating minor misalignments in the radial, angular, or axial directions. Thus, the shafts adjacent to the couplings are subjected to torsion, but the misalignments cause no axial or bending loads.

**12–4
STRESS
CONCENTRATIONS
IN SHAFTS**

In order to mount and locate the several types of machine elements on shafts properly, a final design typically contains several diameters, keyseats, ring grooves, and other geometric discontinuities that create stress concentrations. The shaft design proposed in Figure 12–2 is an example of this observation.

These stress concentrations must be taken into account during the design analysis. But a problem exists because the true design values of the stress concentration factors, K_t, are unknown at the start of the design process. Most of the values are dependent on the diameters of the shaft and on the fillet and groove geometries, and these are the objectives of the design.

You can overcome this dilemma by establishing a set of preliminary design values for commonly encountered stress concentration factors, which can be used to produce initial estimates for the minimum acceptable shaft diameters. Then, after the refined dimensions are selected, you can analyze the final geometry to determine the real values for stress concentration factors. Comparing the final values with the preliminary values will enable you to judge the acceptability of the design. The charts from which the final values of K_t can be determined are found in Figures A15–1 and A15–4.

Preliminary Design Values for K_t

Considered here are the types of geometric discontinuities most often found in power-transmitting shafts: keyseats, shoulder fillets, and retaining ring grooves. In each case, a suggested design value is relatively high in order to produce a conservative result for the first approximation to the design. Again it is emphasized that the final design should be checked for safety. That is, if the final value is less than the original design value, the design is still safe. Conversely, if the final value is higher, the stress analysis for the design must be rechecked.

Keyseats. A *keyseat* is a longitudinal groove cut into a shaft for the mounting of a key, permitting the transfer of torque from the shaft to a power-transmitting element, or vice versa. The detail design of keys was covered in Chapter 11.

FIGURE 12–7
Keyseats

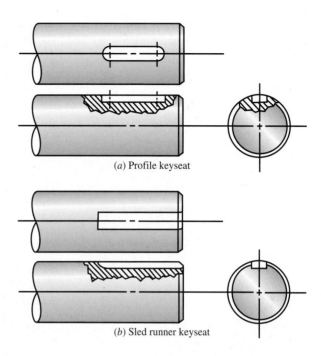

(*a*) Profile keyseat

(*b*) Sled runner keyseat

Two types of keyseats are most frequently used: profile and sled runner (see Figure 12–7). The profile keyseat is milled into the shaft, using an end mill having a diameter equal to the width of the key. The resulting groove is flat-bottomed and has a sharp, square corner at its end. The sled runner keyseat is produced by a circular milling cutter having a width equal to the width of the key. As the cutter begins or ends the keyseat, it produces a smooth radius. For this reason, the stress concentration factor for the sled runner keyseat is lower than that for the profile keyseat. Normally used design values are

$$K_t = 2.0 \quad (\text{profile})$$
$$K_t = 1.6 \quad (\text{sled runner})$$

Each of these is to be applied to the bending stress calculation for the shaft, using the full diameter of the shaft. The factors take into account both the reduction in cross section and the effect of the discontinuity. Consult the references listed for more detail about stress concentration factors for keyseats. (See Reference 6.) If the torsional shear stress is fluctuating rather than steady, the stress concentration factor is also applied to that.

Shoulder Fillets. When a change in diameter occurs in a shaft to create a shoulder against which to locate a machine element, a stress concentration dependent on the ratio of the two diameters and on the radius in the fillet is produced (see Figure 12–8). It is recommended that the fillet radius be as large as possible to minimize the stress concentration, but at times the design of the gear, bearing, or other element affects the radius that can be used. For the purpose of design, we will classify fillets into two categories: sharp and well-rounded.

The term *sharp* here does not mean truly sharp, without any fillet radius at all. Such a shoulder configuration would have a very high stress concentration factor and should be avoided. Instead, *sharp* describes a shoulder with a relatively small fillet radius. One situation in which this is likely to occur is where a ball or roller bearing is to be located. The inner race of the bearing has a factory-produced radius, but it is small. The fillet radius on the

FIGURE 12–8
Fillets on shafts

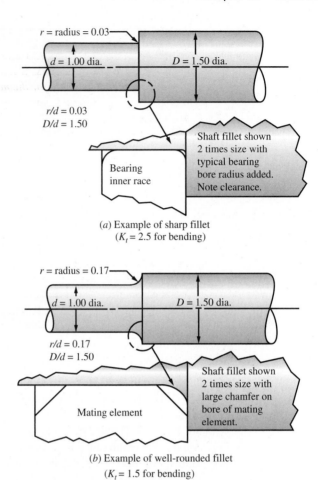

(a) Example of sharp fillet
($K_t = 2.5$ for bending)

(b) Example of well-rounded fillet
($K_t = 1.5$ for bending)

shaft must be smaller yet in order for the bearing to be seated properly against the shoulder. When an element with a large chamfer on its bore is located against the shoulder, or when nothing at all seats against the shoulder, the fillet radius can be much larger (*well-rounded*), and the corresponding stress concentration factor is smaller. We will use the following values for design for bending:

$$K_t = 2.5 \quad \text{(sharp fillet)}$$
$$K_t = 1.5 \quad \text{(well-rounded fillet)}$$

Referring to the chart for stress concentration factors in Figure A15–1, you can see that these values correspond to ratios of r/d of approximately 0.03 for the sharp fillet case and 0.17 for the well-rounded fillet for a D/d ratio of 1.50.

Retaining Ring Grooves.   Retaining rings are used for many types of locating tasks in shaft applications. The rings are installed in grooves in the shaft after the element to be retained is in place. The geometry of the groove is dictated by the ring manufacturer. Its usual configuration is a shallow groove with straight side walls and bottom and a small fillet at the base of the groove. The behavior of the shaft in the vicinity of the groove can be approximated by considering two sharp-filleted shoulders positioned close together. Thus, the stress concentration factor for a groove is fairly high.

For preliminary design, we will apply $K_t = 3.0$ to the bending stress at a retaining ring groove to account for the rather sharp fillet radii. The stress concentration factor is not applied to the torsional shear stress if it is steady in one direction.

The computed estimate for the minimum required diameter at a ring groove is at the base of the groove. You should increase this value by approximately 6% to account for the typical groove depth to determine the nominal size for the shaft. Apply a ring groove factor of 1.06 to the computed required diameter.

12–5 DESIGN STRESSES FOR SHAFTS

In a given shaft, several different stress conditions can exist at the same time. For any part of the shaft that transmits power, there will be a torsional shear stress, while bending stress is usually present on the same parts. Only bending stresses may occur on other parts. Some points may not be subjected to either bending or torsion but will experience vertical shearing stress. Axial tensile or compressive stresses may be superimposed on the other stresses. Then there may be some points where no significant stresses at all are created.

Thus, the decision of what design stress to use depends on the particular situation at the point of interest. In many shaft design and analysis projects, computations must be done at several points to account completely for the variety of loading and geometry conditions that exist.

Several cases discussed in Chapter 5 for computing design factors, N, are useful for determining design stresses for shaft design. The bending stresses will be assumed to be completely reversed and repeated because of the rotation of the shaft. Because ductile materials perform better under such loads, it will be assumed that the material for the shaft is ductile. It will also be assumed that the torsional loading is relatively constant and acting in one direction. If other situations exist, consult the appropriate case from Chapter 5.

The symbol τ_d will be used for the design stress when a shear stress is the basis for the design. The symbol σ_d will be used when a normal stress is the basis.

Design Shear Stress—Steady Torque

It was stated in Chapter 5 that the best predictor of failure in ductile materials due to a steady shear stress was the distortion energy theory in which the design shear stress is computed from

$$\tau_d = s_y/(N\sqrt{3}) = (0.577s_y)/N \qquad \textbf{(12–14)}$$

We will use this value for steady torsional shear stress, vertical shear stress, or direct shear stress in a shaft.

Design Shear Stress—Reversed Vertical Shear

Points on a shaft where no torque is applied and where the bending moments are zero or very low are often subjected to significant vertical shearing forces which then govern the design analysis. This typically occurs where a bearing supports an end of a shaft and where no torque is transmitted in that part of the shaft.

Figure 12–9(a) shows the distribution of vertical shearing stresses on such a circular cross section. Note that the maximum shearing stress is at the neutral axis of the shaft, that is, at the diameter. The stress decreases in a roughly parabolic manner to zero at the outer surface of the shaft. Recall from strength of materials that the maximum vertical shearing stress for the special case of a solid circular cross section can be computed from

$$\tau_{\max} = 4V/3A$$

FIGURE 12–9
Shearing stress in a
rotating shaft due to
vertical shearing
force, V

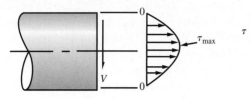
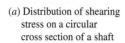
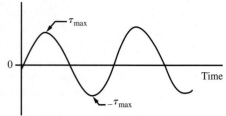

(a) Distribution of shearing
    stress on a circular
    cross section of a shaft

(b) Variation of shearing stress
    on a given element at the
    surface of a circular rotating shaft

where V = vertical shearing force
      A = area of the cross section

Where stress concentration factors are to be considered,

$$\tau_{max} = K_t(4V/3A)$$

Also note, as shown in Figure 12–9, that the rotation of the shaft causes any point at the outer part of the cross section to experience a reversing shearing stress that varies from $+\tau_{max}$ to zero to $-\tau_{max}$ to zero in each revolution. Then the stress analysis should be completed using Equation (5–15) in Case E in Section 5–9 of Chapter 5 dealing with design stresses:

$$N = s'_{sn}/\tau_{max}$$

We recommend using the distortion energy theory. Then the endurance strength in shear is

$$s'_{sn} = 0.577s'_n$$

Thus, Equation (5–15) can be written in the form

$$N = 0.577s'_n/\tau_{max}$$

Expressed as a design stress, this is

$$\tau_d = 0.577s'_n/N$$

Now letting $\tau_{max} = \tau_d = K_t(4V)/3A$ gives

$$\frac{K_t(4V)}{3A} = \frac{0.577s'_n}{N}$$

Solving for N gives

$$N = \frac{0.577s'_n(3A)}{K_t(4V)} = \frac{0.433s'_n(A)}{K_t(V)} \tag{12–15}$$

Equation (12–15) is useful if the goal is to evaluate the design factor for a giv
of loading, a given geometry for the shaft, and given material properties.

Now, solving for the required area gives

$$A = \frac{K_t(V)N}{0.433s'_n} = \frac{2.31K_t(V)N}{s'_n}$$

But our usual objective is the design of the shaft to determine the required diameter. By
substituting

$$A = \pi D^2/4$$

we can solve for D:

Required Shaft Diameter

$$D = \sqrt{2.94\ K_t(V)N/s'_n} \qquad \text{(12–16)}$$

This equation should be used to compute the required diameter for a shaft where a vertical shearing force V is the only significant loading present. In most shafts, the resulting
diameter will be much smaller than that required at other parts of the shaft where significant values of torque and bending moment occur. Also, practical considerations may require that the shaft be somewhat larger than the computed minimum to accommodate a
reasonable bearing at the place where the shearing force is equal to the radial load on the
bearing.

Implementation of Equations (12–15) and (12–16) has the complication that values
for the stress concentration factor under conditions of vertical shearing stress are not well
known. Published data, such as that in the Appendices of this book and in Reference 5, report values for stress concentration factors for axial normal stress, bending normal stress,
and torsional shear stress. But values for vertical shearing stress are rarely reported. As an
approximation, we will use the values for K_t for torsional shear stress when using these
equations.

Design Normal Stress—Fatigue Loading

For the repeated, reversed bending in a shaft caused by transverse loads applied to the rotating shaft, the design stress is related to the endurance strength of the shaft material. The
actual conditions under which the shaft is *manufactured* and *operated* should be considered
when specifying the design stress.

Refer to the discussion in Section 5–4 in Chapter 5 for the method of computing the
estimated actual endurance strength, s'_n, for use in shaft design. The process starts with using the graph in Figure 5–8 to determine the endurance strength as a function of the ultimate tensile strength of the material, adjusted for the surface finish. Equation (5–4) adjusts
this value by applying four factors for the type of material, the type of stress, reliability, and
the size of the cross section. When rotating steel shafts are being designed, the values for
the material factor and the type of stress factor are both equal to 1.0. Use Table 5–1 for the
reliability factor. Use Figure 5–9 or the equations in Table 5–2 to determine the size factor.

Note that any stress concentration factor will be accounted for in the design equation
developed later. Other factors, not considered here, that could have an adverse effect on the
endurance strength of the shaft material, and therefore on the design stress, are temperatures above approximately 400°F (200°C); variation in peak stress levels above the nominal endurance strength for some periods of time; vibration; residual stresses; case

hardening; interference fits; corrosion; thermal cycling; plating or surface coatings; and stresses not accounted for in the basic stress analysis. Testing on actual components is recommended for such conditions.

For parts of the shaft subjected to only reversed bending, let the design stress be

$$\sigma_d = s_n'/N \qquad (12\text{--}17)$$

Design Factor, N

Refer to the discussion in Section 5–7 in Chapter 5 for recommended values of the design factor. We will use $N = 2.0$ for typical shaft designs where there is average confidence in the data for material strength and loads. Higher values should be used for shock and impact loading and where uncertainty in the data exists.

12–6 SHAFTS IN BENDING AND TORSION ONLY

Examples of shafts subjected to bending and torsion only are those carrying spur gears, V-belt sheaves, or chain sprockets. The power being transmitted causes the torsion, and the transverse forces on the elements cause bending. In the general case, the transverse forces do not all act in the same plane. In such cases, the bending moment diagrams for two perpendicular planes are prepared first. Then the resultant bending moment at each point of interest is determined. The process will be illustrated in Example Problem 12–1.

A design equation is now developed based on the assumption that the bending stress in the shaft is repeated and reversed as the shaft rotates, but that the torsional shear stress is nearly uniform. The design equation is based on the principle shown graphically in Figure 12–10 in which the vertical axis is the ratio of the reversed bending stress to the endurance strength of the material. (See Reference 8.) The horizontal axis is the ratio of the torsional shear stress to the yield strength of the material in shear. The points having the value of 1.0 on these axes indicate impending failure in pure bending or pure torsion, respectively. Experimental data show that failure under combinations of bend-

FIGURE 12–10
Basis for shaft design equation for repeated reversed bending stress and steady torsional shear stress

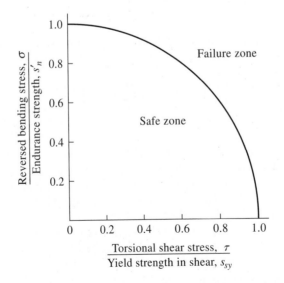

ing and torsion roughly follows the curve connecting these two points, which obeys the following equation:

$$(\sigma/s_n')^2 + (\tau/s_{ys})^2 = 1 \qquad (12\text{--}18)$$

We will use $s_{ys} = s_y/\sqrt{3}$ for the distortion energy theory. Also, a design factor can be introduced to each term on the left side of the equation to yield an expression based on *design stresses:*

$$(N\sigma/s_n')^2 + (N\tau\sqrt{3}/s_y)^2 = 1$$

Now we can introduce a stress concentration factor for bending in the first term only, because this stress is repeated. No factor is needed for the torsional shear stress term because it is assumed to be steady, and stress concentrations have little or no effect on the failure potential. Then

$$(K_t N\sigma/s_n')^2 + (N\tau\sqrt{3}/s_y)^2 = 1 \qquad (12\text{--}19)$$

For rotating, solid, circular shafts, the bending stress due to a bending moment, M, is

$$\sigma = M/S \qquad (12\text{--}20)$$

where $S = \pi D^3/32$ is the rectangular section modulus. The torsional shear stress is

$$\tau = T/Z_p \qquad (12\text{--}21)$$

where $Z_p = \pi D^3/16$ is the polar section modulus.

Note that $Z_p = 2S$ and that, therefore,

$$\tau = T/(2S)$$

Substituting these relationships into Equation (12–19) gives

$$\left[\frac{K_t N M}{S s_n'}\right]^2 + \left[\frac{N T \sqrt{3}}{2 S s_y}\right]^2 = 1 \qquad (12\text{--}22)$$

Now the terms N and S can be factored out, and the terms $\sqrt{3}$ and 2 can be brought outside the bracket in the torsion term:

$$\left[\frac{N}{S}\right]^2 \left[\left[\frac{K_t M}{s_n'}\right]^2 + \frac{3}{4}\left[\frac{T}{s_y}\right]^2\right] = 1$$

We now take the square root of the entire equation:

$$\frac{N}{S}\sqrt{\left[\frac{K_t M}{s_n'}\right]^2 + \frac{3}{4}\left[\frac{T}{s_y}\right]^2} = 1$$

Let $S = \pi D^3/32$ for a solid circular shaft.

$$\frac{32N}{\pi D^3}\sqrt{\left[\frac{K_t M}{s_n'}\right]^2 + \frac{3}{4}\left[\frac{T}{s_y}\right]^2} = 1 \qquad (12\text{--}23)$$

$$\frac{32N}{\pi D^3}\sqrt{\left[\frac{K_t M}{s_n'}\right]^2 + \frac{3}{4}\left[\frac{T}{s_y}\right]^2} = 1 \qquad \textbf{(12–23)}$$

Design
Equation for
Shaft Design

Now we can solve for the diameter D:

$$D = \left[\frac{32N}{\pi}\sqrt{\left[\frac{K_t M}{s_n'}\right]^2 + \frac{3}{4}\left[\frac{T}{s_y}\right]^2}\right]^{1/3} \qquad \textbf{(12–24)}$$

Equation (12–24) is used for shaft design in this book. It is compatible with the standard ANSI B106.1M-1985. (See Reference 1.) Note that Equation (12–24) can also be used for pure bending or pure torsion.

12–7
SHAFT DESIGN
EXAMPLE

Design Example 12–1   Design the shaft shown in Figures 12–1 and 12–2. It is to be machined from AISI 1144 OQT 1000 steel. The shaft is part of the drive for a large blower system supplying air to a furnace. Gear A receives 200 hp from gear P. Gear C delivers the power to gear Q. The shaft rotates at 600 rpm.

Solution   First determine the properties of the steel for the shaft. From Figure A4–2, $s_y = 83\,000$ psi, $s_u = 118\,000$ psi, and the percent elongation is 19%. Thus, the material has good ductility. Using Figure 5–8, we can estimate $s_n = 42\,000$ psi.

A size factor should be applied to the endurance strength because the shaft will be quite large to be able to carry 200 hp. Although we do not know the actual size at this time, we might select $C_s = 0.75$ from Figure 5–9 as an estimate.

A reliability factor should also be specified. This is a design decision. For this problem, let's design for a reliability of 0.99 and use $C_R = 0.81$. Now we can compute the estimated actual endurance strength:

$$s_n' = s_n C_s C_R = (42\,000)(0.75)(0.81) = 25\,500 \text{ psi}$$

The design factor is taken to be $N = 2$. The blower is not expected to present any unusual shock or impact.

Now we can compute the torque in the shaft from Equation (12–1):

$$T = 63\,000(P)/n = 63\,000(200)/600 = 21\,000 \text{ lb} \cdot \text{in}$$

Note that only that part of the shaft from A to C is subjected to this torque. There is zero torque from the right of gear C over to bearing D.

Forces on the Gears: Figure 12–11 shows the two pairs of gears with the forces acting *on gears* A *and* C *shown.* Observe that gear A is driven by gear P, and gear C drives gear Q. It is very important for the directions of these forces to be correct. The values of the forces are found from Equations (12–2) and (12–3).

$$W_{tA} = T_A/(D_A/2) = 21\,000/(20/2) = 2100 \text{ lb} \downarrow$$
$$W_{rA} = W_{tA}\tan(\phi) = 2100\tan(20°) = 764 \text{ lb} \rightarrow$$
$$W_{tC} = T_C/(D_C/2) = 21\,000/(10/2) = 4200 \text{ lb} \downarrow$$
$$W_{rC} = W_{tC}\tan(\phi) = 4200\tan(20°) = 1529 \text{ lb} \leftarrow$$

FIGURE 12–11

Forces on gears A and C

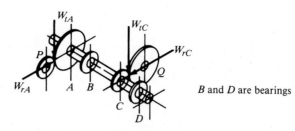

B and D are bearings

(*a*) Pictorial view of forces on gears A and C

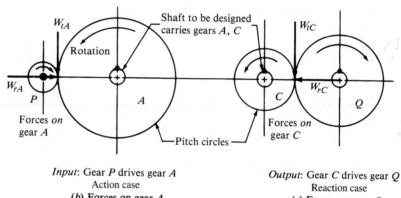

Input: Gear P drives gear A
Action case
(*b*) Forces on gear A

Output: Gear C drives gear Q
Reaction case
(*c*) Forces on gear C

Forces on the Shaft: The next step is to show these forces on the shaft in their proper planes of action and in the proper direction. The reactions at the bearings are computed, and the shearing force and bending moment diagrams are prepared. The results are shown in Figure 12–12.

We continue the design by computing the minimum acceptable diameter of the shaft at several points along the shaft. At each point, we will observe the magnitude of torque and the bending moment that exist at the point, and we will estimate the value of any stress concentration factors. If more than one stress concentration exist in the vicinity of the point of interest, the larger value is used for design. This assumes that the geometric discontinuities themselves do not interact, which is good practice. For example, at point A, the keyseat should end well before the shoulder fillet begins.

1. ***Point A:*** Gear A produces torsion in the shaft from A and to the right. To the left of A, where there is a retaining ring, there are no forces, moments, or torques.

   The moment at A is zero because it is a free end of the shaft. Now we can use Equation (12–24) to compute the required diameter for the shaft at A, using only the torsion term.

$$D_1 = \left[\frac{32 N}{\pi} \sqrt{\frac{3}{4}\left(\frac{T}{s_y}\right)^2} \right]^{1/3}$$

$$D_1 = \left[\frac{32(2)}{\pi} \sqrt{\frac{3}{4}\left(\frac{21\,000}{83\,000}\right)^2} \right]^{1/3} = 1.65 \text{ in}$$

2. ***Point B:*** Point B is the location of a bearing with a sharp fillet to the right of B and a well-rounded fillet to the left. It is desirable to make D_2 at least slightly smaller than D_3 at the bearing seat to permit the bearing to be slid easily onto the shaft up

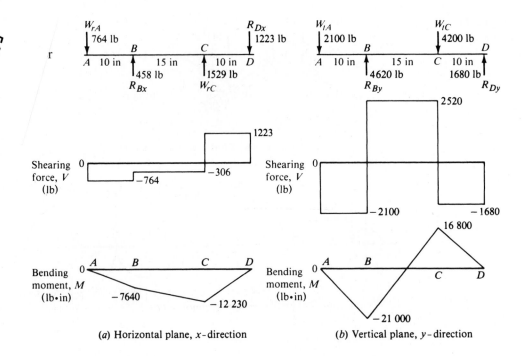

(a) Horizontal plane, x-direction   (b) Vertical plane, y-direction

to the place where it is pressed to its final position. There is usually a light press fit between the bearing bore and the shaft seat.

To the left of B (diameter D_2),

$$T = 21\,000 \text{ lb} \cdot \text{in}$$

The bending moment at B is the resultant of the moment in the x- and y-planes from Figure 12–12:

$$M_B = \sqrt{M_{Bx}^2 + M_{By}^2} = \sqrt{(7640)^2 + (21\,000)^2} = 22\,350 \text{ lb} \cdot \text{in}$$

$$K_t = 1.5 \text{ (well-rounded fillet)}$$

Using Equation (12–24) because of the combined stress condition,

$$D_2 = \left[\left(\frac{32N}{\pi} \right) \sqrt{\left(\frac{K_t M}{s_n'} \right)^2 + \frac{3}{4} \left(\frac{T}{s_y} \right)^2} \right]^{1/3}$$

$$D_2 = \left[\frac{32(2)}{\pi} \sqrt{\left[\frac{1.5(22\,350)}{25\,500} \right]^2 + \frac{3}{4} \left[\frac{21\,000}{83\,000} \right]^2} \right]^{1/3} = 3.30 \text{ in}$$

(12–24a)

At B and to the right of B (diameter D_3), everything is the same, except the value of $K_t = 2.5$ for the sharp fillet. Then

$$D_3 = \left[\frac{32(2)}{\pi} \sqrt{\left[\frac{2.5(22\,350)}{25\,500} \right]^2 + \frac{3}{4} \left[\frac{21\,000}{83\,000} \right]^2} \right]^{1/3} = 3.55 \text{ in}$$

Notice that D_4 will be larger than D_3 in order to provide a shoulder for the bearing. Therefore, it will be safe. Its actual diameter will be specified after we have completed the stress analysis and selected the bearing at B. The bearing manufacturer's catalog will specify the minimum acceptable diameter to the right of the bearing to provide a suitable shoulder against which to seat the bearing.

3. **Point C:** Point C is the location of gear C with a well-rounded fillet to the left, a profile keyseat at the gear, and a retaining ring groove to the right. The use of a well-rounded fillet at this point is actually a design decision that requires that the design of the bore of the gear accommodate a large fillet. Usually this means that a chamfer is produced at the ends of the bore. The bending moment at C is

$$M_c = \sqrt{M_{Cx}^2 + M_{Cy}^2} = \sqrt{(12\,230)^2 + (16\,800)^2} = 20\,780\ \text{lb}\cdot\text{in}$$

To the left of C the torque of 21 000 lb·in exists with the profile keyseat giving $K_t = 2.0$. Then

$$D_5 = \left[\frac{32(2)}{\pi}\sqrt{\left[\frac{2.0(20\,780)}{25\,500}\right]^2 + \frac{3}{4}\left[\frac{21\,000}{83\,000}\right]^2}\right]^{1/3} = 3.22\ \text{in}$$

To the right of C there is no torque, but the ring groove suggests $K_t = 3.0$ for design, and there is reversed bending. We can use Equation (12–24) with $K_t = 3.0$, $M = 20\,780$ lb·in and $T = 0$.

$$D_5 = \left[\frac{32(2)}{\pi}\sqrt{\left(\frac{(3.0)(20\,780)}{25\,500}\right)^2}\right]^{1/3} = 3.68\ \text{in}$$

Applying the ring groove factor of 1.06 raises the diameter to 3.90 in.

This value is higher than that computed for the left of C, so it governs the design at point C.

4. **Point D:** Point D is the seat for bearing D, and there is no torque or bending moment here. However, there is a vertical shearing force equal to the reaction at the bearing. Using the resultant of the x- and y-plane reactions, the shearing force is

$$V_D = \sqrt{(1223)^2 + (1680)^2} = 2078\ \text{lb}$$

We can use Equation (12–16) to compute the required diameter for the shaft at this point:

$$D = \sqrt{2.94\ K_t(V)N/s_n'} \tag{12–16a}$$

Referring to Figure 12–2, we see a sharp fillet near this point on the shaft. Then a stress concentration factor of 2.5 should be used:

$$D_6 = \sqrt{\frac{2.94(2.5)(2078)(2)}{25\,500}} = 1.094\ \text{in}$$

This is very small compared to the other computed diameters, and it will usually be so. In reality, the diameter at D will probably be made much larger than this computed value because of the size of a reasonable bearing to carry the radial load of 2078 lb.

Summary   The computed minimum required diameters for the various parts of the shaft in Figure 12–2 are as follows:

$$D_1 = 1.65 \text{ in}$$
$$D_2 = 3.30 \text{ in}$$
$$D_3 = 3.55 \text{ in}$$
$$D_5 = 3.90 \text{ in}$$
$$D_6 = 1.094 \text{ in}$$

Also, D_4 must be somewhat greater than 3.90 in in order to provide adequate shoulders for gear C and bearing B.

12–8 RECOMMENDED BASIC SIZES FOR SHAFTS

When mounting a commercially available element, of course, follow the manufacturer's recommendation for the basic size of the shaft and the tolerance.

In the U.S. Customary Unit System, diameters are usually specified to be common fractions or their decimal equivalents. Appendix 2 lists the preferred basic sizes that you can use for dimensions over which you have control in decimal-inch, fractional-inch, and metric units. (See Reference 2.)

When commercially available, unmounted bearings are to be used on a shaft, it is likely that their bores will be in metric dimensions. Typical sizes available and their decimal equivalents are listed in Table 14–3.

Design Example 12–2   Specify convenient decimal-inch dimensions for the six diameters from Design Example 12–1. Choose the bearing seat dimensions from Table 14–3. Choose all other dimensions from Appendix 2.

Solution   Table 12–1 shows one possible set of recommended diameters.

Diameters D_3 and D_6 are the decimal equivalents of the metric diameters of the inner races of bearings from Table 14–3. The procedures in Chapter 14 would have to be used to determine whether bearings having those diameters are suitable to carry the given radial loads. Also, D_4 would have to be checked to see whether it provides a sufficiently high shoulder against which to seat the bearing mounted at point B on the shaft. Then detailed specifications for fillet radii, lengths, keyseats, and retaining ring grooves would have to be defined. The actual values for stress concentration factors and the size factor should then be determined. Finally, the stress analysis should be repeated to ensure that the resulting design factor is acceptable. Equation (12–23) can be solved for N and evaluated for actual conditions.

TABLE 12–1   Recommended diameters

Mating part	Diameter number (from Design Example 12–1 and Fig 12–2)	Minimum diameter	Specified diameter (basic size)
Gear	D_1	1.65 in	1.800 in
Nothing	D_2	3.30 in	3.400 in
Bearing	D_3	3.55 in	3.7402 in (95 mm)
Nothing	D_4	>D_3 or D_5	4.400 in
Gear	D_5	3.90 in	4.000 in
Bearing	D_6	1.094 in	3.1496 in (80 mm)

**12–9
ADDITIONAL
DESIGN
EXAMPLES**

Two additional design examples are given in this section. The first is for a shaft that contains three different types of power transmission devices: a V-belt sheave, a chain sprocket, and a spur gear. Some of the forces are acting at angles other than vertical and horizontal, requiring the resolution of the bending forces on the shaft into components before the shearing force and bending moment diagrams are created. Design Example 12–4 involves a shaft carrying a wormgear and a chain sprocket. The axial force on the wormgear presents a slight modification of the design procedure. Except for these differences, the design procedure is the same as that for Design Example 12–1. Thus, much of the detailed manipulation of formulas is omitted.

Design Example 12–3

The shaft shown in Figure 12–13 receives 110 hp from a water turbine through a chain sprocket at point C. The gear pair at E delivers 80 hp to an electrical generator. The V-belt sheave at A delivers 30 hp to a bucket elevator that carries grain to an elevated hopper. The shaft rotates at 1700 rpm. The sprocket, sheave, and gear are located axially by retaining rings. The sheave and gear are keyed with sled runner keyseats, and there is a profile keyseat at the sprocket. Use AISI 1040 cold-drawn steel for the shaft. Compute the minimum acceptable diameters D_1 through D_7 as defined in Figure 12–13.

Solution

First, the material properties for the AISI 1040 cold-drawn steel are found from Appendix 3:

$$s_y = 71\ 000 \text{ psi} \qquad s_u = 80\ 000 \text{ psi}$$

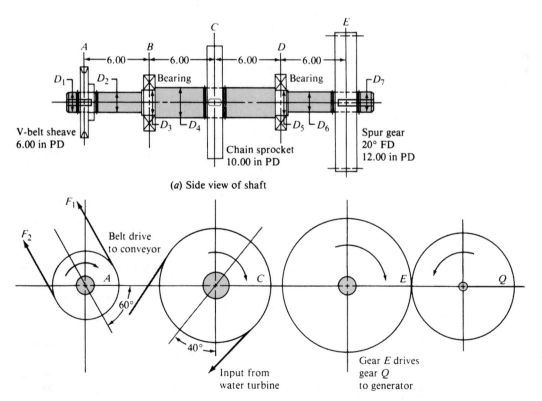

(a) Side view of shaft

(b) Orientation of elements A, C, and E as viewed from right end of shaft

FIGURE 12–13   Shaft design

Then from Figure 5–8, $s_n = 30\ 000$ psi. Let's design for a reliability of 0.99 and use $C_R = 0.81$. The shaft size should be moderately large, so we can assume $C_s = 0.85$ as a reasonable estimate. Then the modified endurance strength is

$$s'_n = s_n C_s C_R = (30\ 000)(0.85)(0.81) = 20\ 650 \text{ psi}$$

This application is fairly smooth: a turbine drive and a generator and a conveyor at the output points. A design factor of $N = 2$ should be satisfactory.

Torque Distribution in the Shaft: Recalling that all of the power comes into the shaft at C, we can then observe that 30 hp is delivered down the shaft from C to the sheave at A. Also, 80 hp is delivered down the shaft from C to the gear at E. From these observations, the torque in the shaft can be computed:

$$T = 63\ 000(30)/1700 = 1112 \text{ lb·in} \quad \text{from } A \text{ to } C$$
$$T = 63\ 000(80)/1700 = 2965 \text{ lb·in} \quad \text{from } C \text{ to } E$$

Figure 12–14 shows a plot of the torque distribution *in the shaft* superimposed on the sketch of the shaft. When designing the shaft at C, we will use 2965 lb·in *at C and to the right*, but we can use 1112 lb·in *to the left of C*. Notice that no part of the shaft is subjected to the full 110 hp that comes into the sprocket at C. The power splits into two parts as it enters the shaft. When analyzing the sprocket itself, we must use the full 110 hp and the corresponding torque:

$$T = 63\ 000(110)/1700 = 4076 \text{ lb·in} \quad \text{(torque in the sprocket)}$$

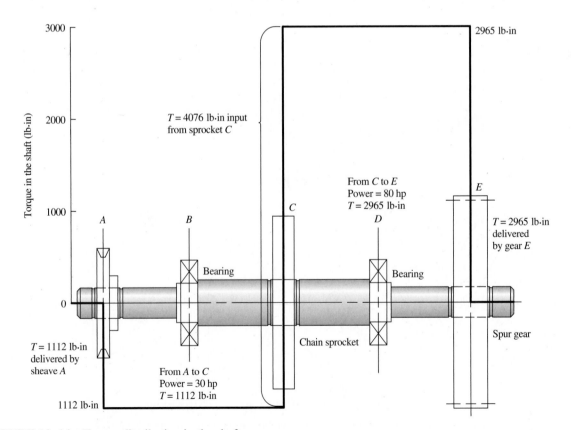

FIGURE 12–14 Torque distribution in the shaft

Forces: We will compute the forces at each element separately and show the component forces that act in the vertical and horizontal planes, as in Design Example 12–1. Figure 12–15 shows the directions of the applied forces and their components for each element.

1. Forces on sheave A: Use Equations (12–7), (12–8), and (12–12):

$$F_N = F_1 - F_2 = T_A/(D_A/2) = 1112/3.0 = 371 \text{ lb} \quad \text{(net driving force)}$$
$$F_A = 1.5\, F_N = 1.5(371) = 556 \text{ lb} \quad \text{(bending force)}$$

The bending force acts upward and to the left at an angle of 60° from the horizontal. As shown in Figure 12–15, the components of the bending force are

$$F_{Ax} = F_A \cos(60°) = (556)\cos(60°) = 278 \text{ lb} \leftarrow \text{(toward the left)}$$
$$F_{Ay} = F_A \sin(60°) = (556)\sin(60°) = 482 \text{ lb} \uparrow \text{(upward)}$$

2. Forces on sprocket C: Use Equation (12–6):

$$F_C = T_C/(D_C/2) = 4076/5.0 = 815 \text{ lb}$$

This is the bending load on the shaft. The components are

$$F_{Cx} = F_C \sin(40°) = (815)\sin(40°) = 524 \text{ lb} \leftarrow \text{(to the left)}$$
$$F_{Cy} = F_C \cos(40°) = (815)\cos(40°) = 624 \text{ lb} \downarrow \text{(downward)}$$

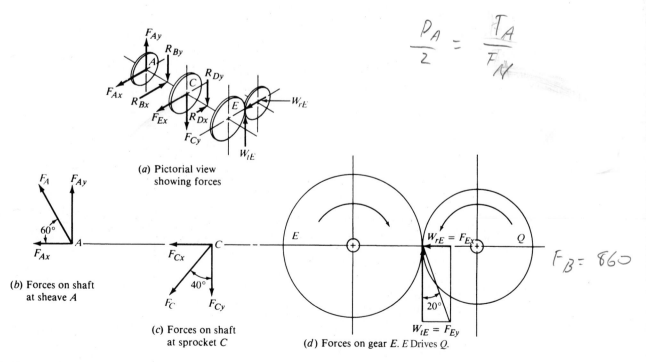

(a) Pictorial view showing forces

$$\frac{P_A}{2} = \frac{T_A}{F_N}$$

(b) Forces on shaft at sheave A

(c) Forces on shaft at sprocket C

(d) Forces on gear E. E Drives Q.

$F_B = 860$

FIGURE 12–15 Forces resolved into *x*- and *y*-components

3. **Forces on gear E:** The transmitted load is found from Equation (12–2), and the radial load from Equation (12–3). The directions are shown in Figure 12–15.

$$F_{Ey} = W_{tE} = T_E/(D_E/2) = 2965/6.0 = 494 \text{ lb} \uparrow \text{ (upward)}$$

$$F_{Ex} = W_{rE} = W_{tE}\tan(\phi) = (494)\tan(20°) = 180 \text{ lb} \leftarrow \text{ (to the left)}$$

Load, Shear, and Moment Diagrams: Figure 12–16 shows the forces acting on the shaft at each element, the reactions at the bearings, and the shearing force and bending moment diagrams for both the horizontal (*x*-) and vertical (*y*-) planes. In the figure, the computations of the resultant bending moment at points *B*, *C*, and *D* are also shown.

Design of the Shaft: We will use Equation (12–24) to determine the minimum acceptable diameter of the shaft at each point of interest. Because the equation requires a fairly large number of individual operations, and because we will be using it at least seven times, it may be desirable to write a computer program just for that operation. Or the use of a spreadsheet would be nearly ideal. See Section 12–10. Note that we can use Equation (12–24) even though there is only torsion or only bending by entering zero for the missing value.

    Equation (12–24) is repeated here for reference. In the solution below, the data used for each design point are listed. You may want to verify the calculations for the required minimum diameters. The design factor of $N = 2$ has been used.

$$D = \left[\frac{32N}{\pi}\sqrt{\left(\frac{K_t M}{s_n'}\right)^2 + \frac{3}{4}\left(\frac{T}{s_y}\right)^2}\right]^{1/3}$$

1. **Point A:** Torque = 1112 lb·in; moment = 0. The sheave is located with retaining rings. Because the torque is steady, we will not use a stress concentration factor in this calculation, as discussed in Section 12–4. But then we will find the nominal diameter at the groove by increasing the computed result by about 6%. The result should be conservative for typical groove geometries.

    Using Equation (12–24), $D_1 = 0.65$ in. Increasing this by 6% gives $D_1 = 0.69$ in.

FIGURE 12–16
Load, shear, and moment diagrams

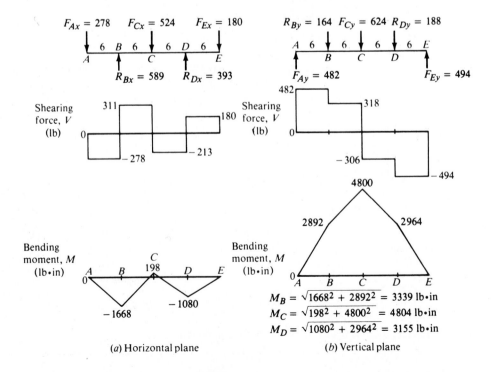

$$M_B = \sqrt{1668^2 + 2892^2} = 3339 \text{ lb·in}$$
$$M_C = \sqrt{198^2 + 4800^2} = 4804 \text{ lb·in}$$
$$M_D = \sqrt{1080^2 + 2964^2} = 3155 \text{ lb·in}$$

(*a*) Horizontal plane   (*b*) Vertical plane

2. *To the left of point B:* This is the relief diameter leading up to the bearing seat. A well-rounded fillet radius will be specified for the place where D_2 joins D_3. Thus,

$$\text{Torque} = 1112 \text{ lb·in} \qquad \text{Moment} = 3339 \text{ lb·in} \qquad K_t = 1.5$$

Then $D_2 = 1.70$ in.

3. *At point B and to the right:* This is the bearing seat with a shoulder fillet at the right, requiring a fairly sharp fillet:

$$\text{Torque} = 1112 \text{ lb·in} \qquad \text{Moment} = 3339 \text{ lb·in} \qquad K_t = 2.5$$

Then $D_3 = 2.02$ in.

4. *At point C:* It is planned that the diameter be the same all the way from the right of bearing B to the left of bearing D. The worst condition is at the right of C, where there is a ring groove and the larger torque value is

$$\text{Torque} = 2965 \text{ lb·in} \qquad \text{Moment} = 4804 \text{ lb·in} \qquad K_t = 3.0$$

Then $D_4 = 2.57$ in after applying the ring groove factor of 1.06.

5. *At point D and to the left:* This is a bearing seat similar to that at B:

$$\text{Torque} = 2965 \text{ lb·in} \qquad \text{Moment} = 3155 \text{ lb·in} \qquad K_t = 2.5$$

Then $D_5 = 1.98$ in.

6. *To the right of point D:* This is a relief diameter similar to D_2:

$$\text{Torque} = 2965 \text{ lb·in} \qquad \text{Moment} = 3155 \text{ lb·in} \qquad K_t = 1.5$$

Then $D_6 = 1.68$ in.

7. *At point E:* The gear is mounted with retaining rings on each side:

$$\text{Torque} = 2965 \text{ lb·in} \qquad \text{Moment} = 0 \qquad K_t = 3.0$$
$$\text{Then } D_7 = 0.96 \text{ in after applying the ring groove factor of 1.06.}$$

Summary with Convenient Values Specified

Using Appendix 2, we specify convenient fractions at all places, including bearing seats (see Table 12–2). It is assumed that inch bearings of the pillow block type will be used.

We decided to make the diameters D_1, D_2, D_6, and D_7 the same to minimize machining and to provide a little extra safety factor at the ring grooves. Again the bearing bore sizes would have to be checked against the load rating of the bearings. The size of D_4 would have to be checked to see that it provides a sufficient shoulder for the bearings at B and D.

The size factor and the stress concentration factors must also be checked.

TABLE 12–2 Specification of values

Mating part	Diameter number	Minimum diameter	Specified diameter	
			Fraction	Decimal
Sheave	D_1	0.69	$1\frac{3}{4}$	1.750
Nothing	D_2	1.70	$1\frac{3}{4}$	1.750
Bearing	D_3	2.02	$2\frac{1}{4}$	2.250
Sprocket	D_4	2.57	$2\frac{3}{4}$	2.750
Bearing	D_5	1.98	2	2.000
Nothing	D_6	1.68	$1\frac{3}{4}$	1.750
Gear	D_7	0.96	$1\frac{3}{4}$	1.750

Design Example 12–4   A wormgear is mounted at the end of the shaft as shown in Figure 12–17. The gear has the same design as that discussed in Example Problem 10–7 and delivers 6.68 hp to the shaft at a speed of 101 rpm. The magnitudes and directions of the forces on the gear are given in the figure. Notice that there is a system of three orthogonal forces acting on the gear. The power is transmitted by a chain sprocket at B to drive a conveyor removing cast iron chips from a machining system. Design the shaft.

Solution   The torque on the shaft from the wormgear at point D to the chain sprocket at B is

$$T = W_{tG}(D_G/2) = 962(4.333) = 4168 \text{ lb·in}$$

The force on the chain sprocket is

$$F_c = T/(D_s/2) = 4168/(6.71/2) = 1242 \text{ lb}$$

This force acts horizontally toward the right as viewed from the end of the shaft.

Bending Moment Diagrams: Figure 12–18 shows the forces acting on the shaft in both the vertical and the horizontal planes and the corresponding shearing force and bending moment diagrams. You should review these diagrams, especially that for the vertical plane, to grasp the effect of the axial force of 265 lb. Notice that because it acts above the shaft, it creates a bending moment at the end of the shaft of 1148 lb·in. It also affects the reactions at the bearings. The resultant bending moments at B, C, and D are also shown in the figure.

In the design of the entire system, we must decide which bearing will resist the axial force. For this problem, let's specify that the bearing at C will transfer the axial thrust force to the housing. This decision places a compressive stress in the shaft from C to D and requires that means be provided to transmit the axial force from the wormgear to the bearing. The geometry proposed in Figure 12–17 accomplishes this, and it will be adopted for the following stress analysis. The procedures are the same as those used in Design Examples 12–1 through 12–3, and only summary results will be shown. The consideration of the axial compressive stress is discussed, along with the computations at point C on the shaft.

Material Selection and Design Strengths: A medium-carbon steel with good ductility and a fairly high strength is desired for this demanding application. We will use AISI 1340 OQT 1000 (Appendix 3), having an ultimate strength of 144 000 psi, a yield strength of 132 000

FIGURE 12–17
Shaft design

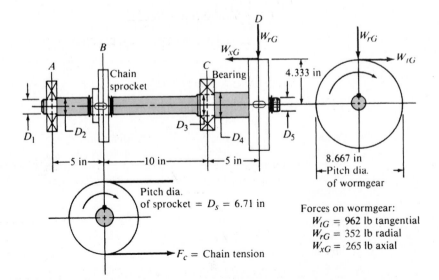

Forces on wormgear:
$W_{tG} = 962$ lb tangential
$W_{rG} = 352$ lb radial
$W_{xG} = 265$ lb axial

FIGURE 12–18
Load, shear, and
moment diagrams for
the shaft in Figure
12–16

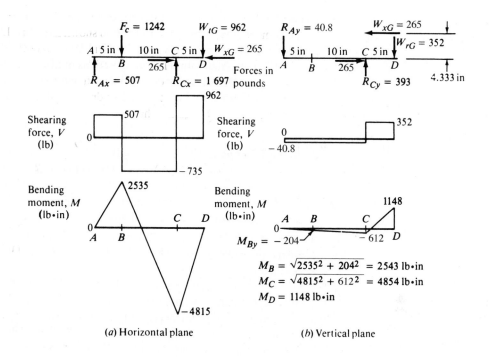

(a) Horizontal plane

(b) Vertical plane

$$M_B = \sqrt{2535^2 + 204^2} = 2543 \text{ lb} \cdot \text{in}$$
$$M_C = \sqrt{4815^2 + 612^2} = 4854 \text{ lb} \cdot \text{in}$$
$$M_D = 1148 \text{ lb} \cdot \text{in}$$

psi, and a 17% elongation. From Figure 5–8 we estimate $s_n = 50\,000$ psi. Let's use an initial size factor of 0.80 and a reliability factor of 0.81. Then

$$s'_n = (50\,000 \text{ psi})(0.80)(0.81) = 32\,400 \text{ psi}$$

Because the use of the conveyor is expected to be rough, we will use a design factor of $N = 3$, higher than average.

Except at point A, where only a vertical shear stress exists, the computation of the minimum required diameter is done using Equation (12–24).

1. **Point A:** The left bearing mounts at point A, carrying the radial reaction force only, which acts as a vertical shearing force in the shaft. There is no torque or bending moment here.

   The vertical shearing force is

$$V = \sqrt{R_{Ax}^2 + R_{Ay}^2} = \sqrt{(507)^2 + (40.8)^2} = 509 \text{ lb}$$

We can use Equation (12–16) to compute the required diameter for the shaft at this point:

$$D = \sqrt{2.94\, K_t(V)N/s'_n} \qquad \text{(12–16)}$$

Referring to Figure 12–17, we see a sharp fillet near this point on the shaft. Then a stress concentration factor of 2.5 should be used:

$$D = \sqrt{\frac{2.94(2.5)(509)(3)}{32\,400}} = 0.588 \text{ in}$$

As seen before, this is quite small, and the final specified diameter will probably be larger, depending on the bearing selected.

2. ***Point B:*** The chain sprocket mounts at point *B*, and it is located axially by retaining rings on both sides. The critical point is at the right of the sprocket at the ring groove, where $T = 4168$ lb·in, $M = 2543$ lb·in, and $K_t = 3.0$ for bending.

   The computed minimum diameter required is $D_2 = 1.93$ in at the base of the groove. We should increase this by approximately 6%, as discussed in Section 12–3. Then

$$D_2 = 1.06(1.93 \text{ in}) = 2.05 \text{ in}$$

3. ***To the left of point C:*** This is the relief diameter for the bearing seat. The diameter here will be specified to be the same as that at *B*, but different conditions occur: Torque $= 4168$ lb·in, $M = 4854$ lb·in, and $K_t = 1.5$ for the well-rounded fillet for bending only. The required diameter is 1.91 in. Because this is smaller than that at *B,* the previous calculation will govern.

4. ***At point C and to the right:*** The bearing will seat here, and it is assumed that the fillet will be rather sharp. Thus, $T = 4168$ lb·in, $M = 4854$ lb·in, and $K_t = 2.5$ for bending only. The required diameter is $D_3 = 2.26$ in.

   The axial thrust load acts between points *C* and *D*. The inclusion of this load in the computations would greatly complicate the solution for the required diameters. In most cases, the axial normal stress is relatively small compared with the bending stress. Also, the fact that the stress is compressive improves the fatigue performance of the shaft, because fatigue failures normally initiate at points of tensile stress. For these reasons, the axial stress is ignored in these calculations. The computed diameters are also interpreted as nominal minimum diameters, and the final selected diameter is larger than the minimum. This, too, tends to make the shaft safe even when there is an added axial load. When in doubt, or when a relatively high axial tensile stress is encountered, the methods of Chapter 5 should be applied. Long shafts in compression should also be checked for buckling.

5. ***Point D:*** The wormgear mounts at point *D*. We will specify that a well-rounded fillet will be placed to the left of *D* and that there will be a sled runner keyseat. Thus, $T = 4168$ lb·in, $M = 1148$ lb·in, and $K_t = 1.6$ for bending only. The computed required diameter is $D_5 = 1.24$ in. Notice that D_4 must be greater than either D_3 or D_5 because it provides the means to transfer the thrust load from the wormgear to the inner race of the bearing at *C*.

Summary and Selection of Convenient Diameters

Table 12–3 presents a summary of the required diameters and the specified diameters for all parts of the shaft in this design example. See Figure 12–17 for the locations of the five diameters.

For this application we have chosen to use fractional-inch dimensions from Appendix 2 except at the bearing seats, where the use of metric bearing bores from Table 14–3 are selected.

TABLE 12–3   Summary of shaft diameters

Mating part	Diameter number	Minimum diameter	Specified diameter Fraction (metric)	Decimal
Bearing *A*	D_1	0.59 in	(35 mm)	1.3780 in
Sprocket *B*	D_2	2.05 in	$2\frac{1}{4}$ in	2.250 in
Bearing *C*	D_3	2.26 in	(65 mm)	2.5591 in
(Shoulder)	D_4	$>D_3$	3 in	3.000 in
Wormgear *D*	D_5	1.24 in	$1\frac{1}{2}$ in	1.500 in

12–10 SPREADSHEET AID FOR SHAFT DESIGN

A spreadsheet is useful to organize the data required to compute the minimum required shaft diameter at various points along a shaft and to complete the calculation using Equations (12–16) and (12–24). Note that Equation (12–24) can be used for bending only, torsion only, or the combination of bending and torsion.

Figure 12–19 shows a typical example, using U.S. Customary units, for data from Design Example 12–1. Describe the application in the upper panel for future reference. Then complete the following steps:

- Enter the shaft material specification along with its ultimate and yield strength properties found from tables in the Appendices.

DESIGN OF SHAFTS			

Application: Design Example 12–1, Drive for a Blower System
Diameter D_3—To right of point B—Bending and torsion

This design aid computes the minimum acceptable diameter for shafts using Equation (12–24) for shafts subjected to steady torsion and/or rotating bending.

Equation (12–16) is used when only vertical shear stress is present.

Input Data:		(Insert values in italics.)	
Shaft material specification:	AISI 1144 OQT 1000 Steel		
Tensile strength:	$s_u =$	118 000 psi	
Yield strength:	$s_y =$	83 000 psi	
Basic endurance strength:	$s_n =$	42 000 psi	From Figure 5–8
Size factor:	$C_s =$	0.75	From Figure 5–9
Reliability factor:	$C_R =$	0.81	From Table 5–1
Modified endurance strength:	$s_n' =$	25 515 psi	**Computed**
Stress concentration factor:	$K_t =$	2.5 Sharp fillet	
Design factor:	$N =$	2 Nominal $N=2$	

Shaft Loading Data: Bending and Torsion			
Bending moment components:	$M_x =$	21 000 lb·in	$M_y = 7640$ lb·in
Combined bending moment:	$M =$	22 347 lb·in	**Computed**
Torque:	$T =$	21 000 lb·in	
Minimum shaft diameter:	**D =**	3.55 in	**Computed from Eq. (12–24)**

Shaft Loading Data: Vertical Shearing Force Only			
Shearing force components:	$V_x =$	764 lb	$V_y = 2520$ lb
Combined shearing force:	$V =$	2633 lb	**Computed**
Minimum shaft diameter:	**D =**	1.232 in	**Computed from Eq. (12–16)**

FIGURE 12–19 Spreadsheet aid for shaft design

- Find the basic endurance strength from Figure 5–8 considering the ultimate tensile strength and the manner of production (ground, machined, etc.).

- Enter values for the size factor and the reliability factor. The spreadsheet then computes the modified endurance strength, s_n'.

- Enter the stress concentration factor for the point of interest.

- Enter the design factor.

- After analysis such as that shown in Design Example 12–1, enter the torque and the components of the bending moment in the x- and y-planes that exist at the point of interest along the shaft. The spreadsheet computes the combined bending moment.

- Enter the components of the vertical shearing force in the x- and y-planes. The spreadsheet computes the combined shearing force.

The minimum acceptable shaft diameters from both Equation (12–16) (vertical shear only) and Equation (12–24) (torsion and/or bending) are computed. You must observe which required diameter is larger.

**12–11
SHAFT RIGIDITY
AND DYNAMIC
CONSIDERATIONS**

The design processes outlined in this chapter thus far have concentrated on stress analysis and ensuring that the shaft is safe with regard to the bending and torsional shear stresses imposed on the shaft by power-transmitting elements. Rigidity of the shaft is also a major concern for several reasons:

1. Excessive radial deflection of the shaft may cause active elements to become misaligned, resulting in poor performance or accelerated wear. For example, the center distance between shafts carrying precision gears should not vary more than approximately 0.005 in (0.13 mm) from the theoretical dimension. Improper meshing of the gear teeth would occur and actual bending and contact stresses could be significantly higher than those predicted in the design for the gears.

2. Guidelines are reported in Section 20–2 in Chapter 20 for the recommended limits for bending and torsional deflection of a shaft according to its intended precision.

3. Shaft deflection is also an important contributor to a tendency for shafts to vibrate as they rotate. A flexible shaft will oscillate in both bending and torsional modes, causing movements that are greater than the static deflections due just to gravity and applied loads and torques. A long slender shaft tends to whip and whirl with relatively large deformations from its theoretical straight axis.

4. The shaft itself and the elements mounted on it should be balanced. Any amount of unbalance causes centrifugal forces to be created that rotate with the shaft. Large unbalances and high rotating speeds may create unacceptable force levels and shaking of the rotating system. An example with which you may be familiar is an automobile wheel that is "out of balance." You can actually feel the vibration through the steering wheel as you drive. Having the tire and wheel balanced reduces the vibration to acceptable levels.

5. The dynamic behavior of the shaft may become dangerously destructive if it is operated near its *critical speed*. At critical speed, the system is in resonance, the deflection of the shaft continues to increase virtually without bound, and it will eventually self-destruct.

Critical speeds of typical machinery shaft designs are several thousand revolutions per minute. However, many variables are involved. Long shafts used for vehicular drive shafts, power screws, or agitators may have much lower critical speeds and they must be checked. It is typical to keep operating speeds to at least 20% below or above the critical speed. When operating above the critical speed, it is necessary to accelerate the shaft rapidly through the critical speed because it takes time for dangerous oscillations to develop.

Analysis to determine the critical speed is complex, and computer programs are available to assist in the calculations. See Internet site 1. The objective is to determine the natural frequency of the shaft while carrying the static weight of elements such as gears, sprockets, and pulleys. The stiffness of bearings is also a factor. One fundamental expression for natural frequency, ω_n, is,

$$\omega_n = \sqrt{k/m}$$

where k is the stiffness of the shaft and m is its mass. It is desirable to have a high critical speed, well above operating speed, so stiffness should be high and mass low. The primary variables over which a designer has control are the material and its modulus of elasticity, E, its density, ρ, the shaft diameter, D, and the shaft length, L. The following functional relationship can help to understand the influence of each of these variables:

$$\omega_n \propto (D/L^2)\sqrt{E/\rho}$$

where the symbol $\propto$ indicates proportionality among the variables. Using this function as a guide, the following actions can reduce potential problems with deflections or critical speeds.

1. Making the shaft more rigid can prevent undesirable dynamic behavior.
2. Larger shaft diameters add rigidity.
3. Shorter shaft lengths reduce deflections and lower critical speeds.
4. Placing active elements on the shaft close to support bearings is recommended.
5. Reducing the weight of elements carried by the shaft reduces static deflection and lowers critical speed.
6. Selection of a material for the shaft with a high ratio of E/ρ is desirable. While most metals have similar ratios, that for composite materials is typically high. For example, long driveshafts for vehicles that must operate at high speeds are frequently made from hollow tubular composite materials using carbon fibers.
7. Bearings should have a high stiffness in terms of radial deflection as a function of load.
8. Mountings for bearings and housings should be designed with high rigidity.
9. References 3, 4, 5, and 7 provide additional information and analytical methods for estimating critical speeds. References 9–11 discuss machinery vibration.

**12–12
FLEXIBLE
SHAFTS**

At times it is desirable to transmit rotational motion and power between two points that are not aligned with each other. Flexible shafts can be used in such situations to couple a driver, such as a motor, to a driven device along a curved or dynamically moving path. The flexibility allows the driven point to be offset from the driving point in either parallel or angular fashion. Unidirectional shafts are used for power transmission in applications such as automation systems, industrial machinery, agricultural equipment, aircraft actuators, seat adjusters, medical devices, dentistry, speedometers, woodworking, and jeweler's tools. Bidirectional flexible shafts are used in remote control, valve actuation, and safety devices. They can often provide more design flexibility than other design options such as right-angle gear drives or universal joints. The offset between the driver and driven machine can be as much as 180° provided that the bend radius is above a specified minimum. Torque capacity increases with increasing bend radius. Efficiencies are in the range from 90% to 95%.

The construction consists of a flexible cable-like core connected to fittings on each end. The fittings facilitate connection with the driving and driven machines. The casing protects nearby equipment or people from contact with the rotating core. Designs are available for transmitting power in either direction, clockwise or counterclockwise. Some will operate bidirectionally.

The capacity of a flexible shaft is reported as the torque that can be transmitted reliably versus the minimum bend radius of the shaft. Internet site 7 includes selection data for a large range of sizes with capacities from as low as 0.20 to almost 2500 lb-in (0.03 to 282 N-m).

REFERENCES

1. American Society of Mechanical Engineers. ANSI Standard B106.1M-1985. *Design of Transmission Shafting*. New York: American Society of Mechanical Engineers, 1985.

2. American Society of Mechanical Engineers. ANSI Standard B4.1-67. *Preferred Limits and Fits for Cylindrical Parts*. (R87/94). New York: American Society of Mechanical Engineers, 1994.

3. Avallone, E., and T. Baumeister, et al. *Marks' Standard Handbook for Mechanical Engineers*. 10th ed. New York: McGraw-Hill, 1996.

4. Harris, Cyril M., and Allan G. Piersol (Editors). *Harris' Shock and Vibration Handbook*. 5th ed. New York: McGraw-Hill, 2001.

5. Oberg, Erik, et al. *Machinery's Handbook*. 26th ed. New York: Industrial Press, 2000.

6. Pilkey, Walter D. *Peterson's Stress Concentration Factors*. 2d ed. New York: John Wiley & Sons, 1997.

7. Shigley, J. E., and C. R. Mischke. *Mechanical Engineering Design*. 6th ed. New York: McGraw-Hill, 2001.

8. Soderberg, C. R. "Working Stresses." *Journal of Applied Mechanics* 57 (1935): A–106.

9. Wowk, Victor. *Machinery Vibration: Alignment*. New York: McGraw-Hill, 2000.

10. Wowk, Victor. *Machinery Vibration: Balancing*. New York: McGraw-Hill, 1998.

11. Wowk, Victor. *Machinery Vibration: Measurement and Analysis*. New York: McGraw-Hill, 1991.

INTERNET SITES FOR SHAFT DESIGN

1. **Hexagon Software.** *www.hexagon.de* Computer software called WL1+ for shaft calculations. Can handle up to 100 cylindrical or conical shaft segments and up to 50 individual forces. Computes shearing force and bending moment diagrams, slope, deflection, bending stresses, and shearing stresses. Critical speeds for flexure and torsional vibration are calculated.

2. **Spinning Composites.** *www.spinning-composites.com* Manufacturer of composite shafts for vehicular and industrial applications. Includes discussion of critical speeds.

3. **Advanced Composite Products and Technology, Inc.** *www.acpt.com* Manufacturer of composite products for aerospace, military, commercial and industrial markets, including driveshafts and high-speed rotors.

4. **Kerk Motion Products, Inc.** *www.kerkmotion.com* Manufacturer of lead screws for industrial applications. Site includes data for critical speeds and discussion about the factors involved.

5. **Orand Systems, Inc.** *www.orandsystems.com* Developer and distributor of *Beam 2D Stress Analysis* software. Analyzes beams with numerous loads, cross sections, and support conditions. Load, shearing force, bending moment, slope, and deflection outputs in graphical and numerical data formats. Can be used for the bending calculations for shafts.

6. **MDSolids.** *www.mdsolids.com* Educational software for stress analysis that includes several modules including beams, torsion, trusses, and columns.

7. **S. S. White Company.** *www.sswt.com* Manufacturer of flexible shafts. Site includes rating charts for torque

capacity versus the minimum radius of the shaft during operation.

8. **Elliott Manufacturing Co.** *www.elliottmfg.com*
   Manufacturer of flexible shafts.

9. **Suhner Transmission Co.** *www.suhner.com*
   Manufacturer of flexible shafts, spiral bevel gears, and small, high speed air motors and electric motors. Site includes rating charts for power and torque capacity of flexible shafts.

PROBLEMS

Forces and Torque on Shafts—Spur Gears

1. See Figure P12–1. The shaft rotating at 550 rpm carries a spur gear *B* having 96 teeth with a diametral pitch of 6. The teeth are of the 20°, full-depth, involute form. The gear receives 30 hp from a pinion directly above it. Compute the torque delivered to the shaft and the tangential and radial forces exerted on the shaft by the gear.

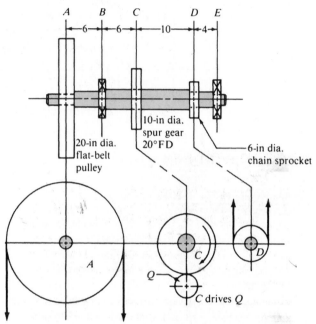

FIGURE P12–2 (Problems 2, 12, 13, and 25)

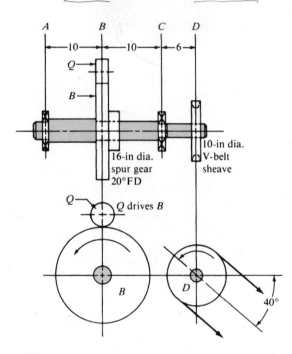

FIGURE P12–1 (Problems 1, 14, and 24)

2. See Figure P12–2. The shaft rotating at 200 rpm carries a spur gear *C* having 80 teeth with a diametral pitch of 8. The teeth are of the 20°, full-depth, involute form. The gear delivers 6 hp to a pinion directly below it. Compute the torque delivered by the shaft to gear *C* and the tangential and radial forces exerted on the shaft by the gear.

3. See Figure P12–3. The shaft rotating at 480 rpm carries a spur pinion *B* having 24 teeth with a diametral pitch of 8. The teeth are of the 20°, full-depth, involute form. The pinion delivers 5 hp to a gear directly below it. Compute the torque delivered by the shaft to pinion *B* and the tangential and radial forces exerted on the shaft by the pinion.

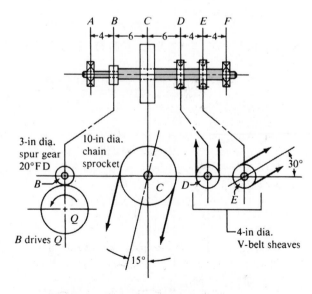

FIGURE P12–3 (Problems 3, 15, 16, and 26)

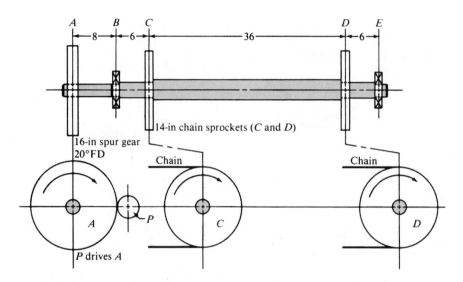

14-in chain sprockets (*C* and *D*)

16-in spur gear
20°FD

Chain

Chain

P drives *A*

FIGURE P12–4   (Problems 4, 19, and 27)

4. See Figure P12–4. The shaft rotating at 120 rpm carries a spur gear *A* having 80 teeth with a diametral pitch of 5. The teeth are of the 20°, full-depth, involute form. The gear receives 40 hp from a pinion to the right as shown. Compute the torque delivered to the shaft and the tangential and radial forces exerted on the shaft by the gear.

5. See Figure P12–5. The shaft rotating at 240 rpm carries a spur gear *D* having 48 teeth with a diametral pitch of 6. The teeth are of the 20°, full-depth, involute form. The gear receives 15 hp from pinion *Q* located as shown. Compute the torque delivered to the shaft and the tangential and radial forces exerted on the shaft by the gear.

Resolve the tangential and radial forces into their horizontal and vertical components, and determine the net forces acting on the shaft at *D* in the horizontal and vertical directions.

6. See Figure P12–6. The shaft rotating at 310 rpm carries a spur pinion *E* having 36 teeth with a diametral pitch of 6. The teeth are of the 20°, full-depth, involute form. The pinion delivers 20 hp to a gear to the left as shown. Compute the torque delivered by the shaft to pinion *E* and the tangential and radial forces exerted on the shaft by the pinion. Include the weight of the pinion.

7. See Figure P12–7. The shaft rotating at 480 rpm carries a spur gear *C* having 50 teeth with a diametral

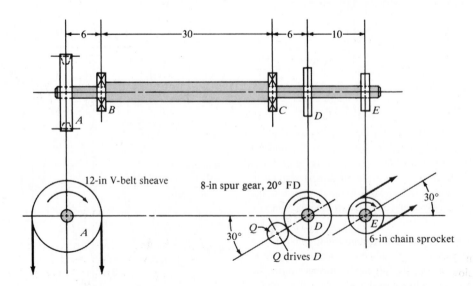

12-in V-belt sheave

8-in spur gear, 20° FD

30°

6-in chain sprocket

Q drives *D*

FIGURE P12–5   (Problems 5, 20, 21, and 28)

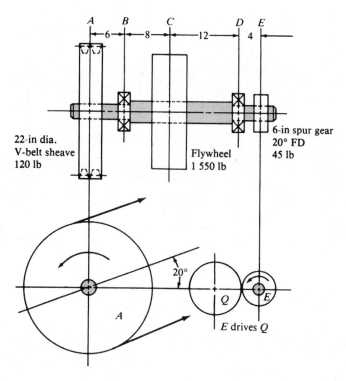

FIGURE P12–6 (Problems 6 and 29)

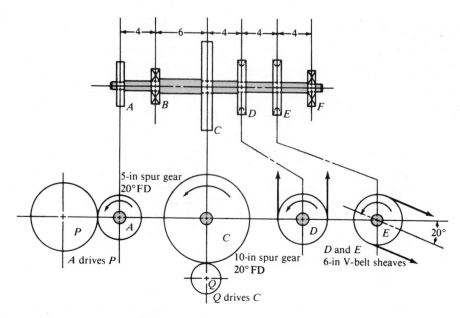

FIGURE P12–7 (Problems 7, 8, and 30)

pitch of 5. The teeth are of the 20°, full-depth, involute form. The gear receives 50 hp from a pinion directly below it. Compute the torque delivered to the shaft and the tangential and radial forces exerted on the shaft by the gear.

8. See Figure P12–7. The shaft rotating at 480 rpm carries a spur pinion *A* having 30 teeth with a diametral pitch of 6. The teeth are of the 20°, full-depth, involute form. The pinion delivers 30 hp to a gear to the left as shown. Compute the torque delivered by the shaft to pinion *A* and the

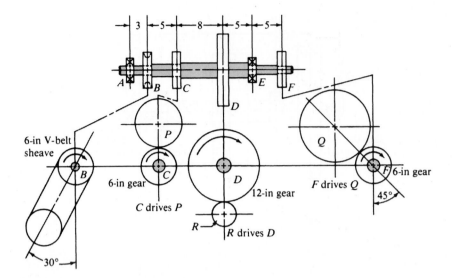

FIGURE P12–9   (Problems 9, 10, 11, and 31)

tangential and radial forces exerted on the shaft by the pinion.

9. See Figure P12–9. The shaft rotating at 220 rpm carries a spur pinion *C* having 60 teeth with a diametral pitch of 10. The teeth are of the 20°, full-depth, involute form. The pinion delivers 5 hp to a gear directly above it. Compute the torque delivered by the shaft to pinion *C* and the tangential and radial forces exerted on the shaft by the pinion.

10. See Figure P12–9. The shaft rotating at 220 rpm carries a spur gear *D* having 96 teeth with a diametral pitch of 8. The teeth are of the 20°, full-depth, involute form. The gear receives 12.5 hp from a pinion directly below it. Compute the torque delivered to the shaft and the tangential and radial forces exerted on the shaft by the gear.

11. See Figure P12–9. The shaft rotating at 220 rpm carries a spur pinion *F* having 60 teeth with a diametral pitch of 10. The teeth are of the 20°, full-depth, involute form. The pinion delivers 5 hp to a gear located as shown. Compute the torque delivered by the shaft to pinion *F* and the tangential and radial forces exerted on the shaft by the pinion. Resolve the forces into their horizontal and vertical components, and determine the net forces acting on the shaft at *F* in the horizontal and vertical directions.

Forces and Torque on Shafts—Sprockets and Belt Sheaves

12. See Figure P12–2. The shaft rotating at 200 rpm carries a 6-in-diameter chain sprocket *D* that delivers 4 hp to a mating sprocket above. Compute the torque delivered by the shaft to sprocket *D* and the force exerted on the shaft by the sprocket.

13. See Figure P12–2. The shaft rotating at 200 rpm carries a 20-in-diameter flat-belt pulley at *A* that receives 10 hp from below. Compute the torque delivered by the pulley to the shaft and the force exerted on the shaft by the pulley.

14. See Figure P12–1. The shaft rotating at 550 rpm carries a 10-in-diameter V-belt sheave at *D* that delivers 30 hp to a mating sheave as shown. Compute the torque delivered by the shaft to the sheave and the total force exerted on the shaft by the sheave. Resolve the force into its horizontal and vertical components, and show the net forces acting on the shaft at *D* in the horizontal and vertical directions.

15. See Figure 12–3. The shaft rotating at 480 rpm carries a 10-in-diameter chain sprocket at *C* that receives 11 hp from a mating sprocket below and to the left as shown. Compute the torque delivered to the shaft by the sprocket and the total force exerted on the shaft by the sprocket. Resolve the force into its horizontal and vertical components, and show the net forces acting on the shaft at *C* in the horizontal and vertical directions.

16. See Figure 12–3. The shaft rotating at 480 rpm carries two 4-in-diameter sheaves at *D* and *E* that each deliver 3 hp to mating sheaves as shown. Compute the torque delivered by the shaft to each sheave and the total force exerted on the shaft by each sheave. Resolve the force at *E* into its horizontal and vertical components, and show the net forces acting on the shaft at *E* in the horizontal and vertical directions.

17. See Figure P12–17. The shaft rotating at 475 rpm carries a 10-in V-belt sheave at *C* that receives 15.5 hp from a mating sheave to the left as shown. Compute the torque delivered to the shaft at *C* by the sheave and the total force exerted on the shaft at *C* by the sheave.

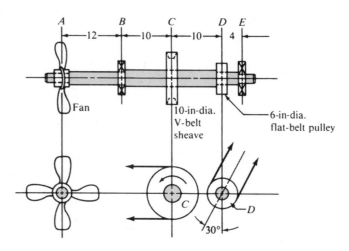

FIGURE P12–17   (Problems 17, 18, and 32)

18. See Figure P12–17. The shaft rotating at 475 rpm carries a 6-in-diameter flat-belt pulley at D that delivers 3.5 hp to a mating pulley above and to the right as shown. Compute the torque delivered to the shaft by the pulley and the total force exerted on the shaft by the pulley. Resolve the force into its horizontal and vertical components, and show the net forces acting on the shaft at D in the horizontal and vertical directions.

19. See Figure P12–4. The shaft rotating at 120 rpm carries two identical 14-in-diameter chain sprockets at C and D. Each sprocket delivers 20 hp to mating sprockets to the left as shown. Compute the torque delivered by the shaft to each sprocket and the total force exerted on the shaft by each sprocket.

20. See Figure P12–5. The shaft rotating at 240 rpm carries a 12-in-diameter V-belt sheave at A that delivers 10 hp to a mating sheave directly below. Compute the torque delivered by the shaft to the sheave and the total force exerted on the shaft at A by the sheave.

21. See Figure P12–5. The shaft rotating at 240 rpm carries a 6-in-diameter chain sprocket at E that delivers 5.0 hp to a mating sprocket to the right and above as shown. Compute the torque delivered by the shaft to the sprocket and the total force exerted on the shaft by the sprocket. Resolve the force into its horizontal and vertical components, and show the net forces acting on the shaft at E in the horizontal and vertical directions.

Forces and Torque on Shafts—Helical Gears

22. See Figure P12–22. The shaft is rotating at 650 rpm, and it receives 7.5 hp through a flexible coupling. The

power is delivered to an adjacent shaft through a single helical gear B having a normal pressure angle of 20° and a helix angle of 15°. The pitch diameter for the gear is 4.141 in. Review the discussion of forces on helical gears in Chapter 10, and use Equations (12–1), (12–2), (12–4), and (12–5) in this chapter to verify the values for the forces given in the figure. Draw the complete free-body diagrams for the shaft in both the vertical plane and the horizontal plane. Then draw the complete shearing force and bending moment diagrams for the shaft in both planes.

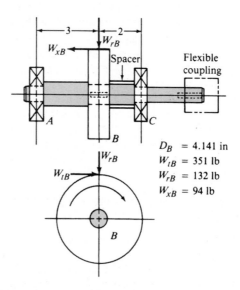

D_B = 4.141 in
W_{tB} = 351 lb
W_{rB} = 132 lb
W_{xB} = 94 lb

FIGURE P12–22   (Problems 22 and 33)

Forces and Torque on Shafts—Wormgears

23. See Figure P12–23. The shaft is rotating at 1750 rpm, and it receives 7.5 hp through a 5.00-in V-belt drive from a mating belt sheave directly below. The power is delivered by a worm with a pitch diameter of 2.00 in. The forces on the worm were computed in Design Example 12–4 and are shown in the figure. Review the description of these forces and the method of computing them. Draw the complete free-body diagrams for the shaft in both the vertical plane and the horizontal plane. Then draw the complete shearing force and bending moment diagrams for the shaft in both planes.

Comprehensive Shaft Design Problems

For each of the following problems, it will be necessary to do the following:

(a) Determine the magnitude of the torque in the shaft at all points

(b) Compute the forces acting on the shaft at all power-transmitting elements

(c) Compute the reactions at the bearings

(d) Draw the complete load, shear, and bending moment diagrams

Neglect the weight of the elements on the shafts, unless otherwise noted.

The objective of any problem, at the discretion of the instructor, could be any of the following:

■ Design the complete shaft, including the specification of the overall geometry and the consideration of stress concentration factors. The analysis would show the minimum acceptable diameter at each point on the shaft to be safe from the standpoint of strength.

■ Given a suggested geometry of one part of the shaft, specify the minimum acceptable diameter for the shaft at that point.

■ Specify the required geometry at any selected element on the shaft: a gear, sheave, bearing, or other.

■ Make a working drawing of the design for the shaft, following the appropriate stress analysis, and specify the final dimensions.

■ Suggest how the given shaft can be redesigned by moving or reorienting the elements on the shaft to improve the design to produce lower stresses, smaller shaft size, more convenient assembly, and so on.

■ Incorporate the given shaft into a more comprehensive machine, and complete the design of the entire machine. In most problems, the type of machine for which the shaft is being designed is suggested.

24. The shaft in Figure P12–1 is part of a drive for an automated transfer system in a metal stamping plant. Gear Q delivers 30 hp to gear B. Sheave D delivers the power to its mating sheave as shown. The shaft carrying B and D rotates at 550 rpm. Use AISI 1040 cold-drawn steel.

25. The shaft in Figure P12–2 rotates at 200 rpm. Pulley A receives 10 hp from below. Gear C delivers 6 hp to the mating gear below it. Chain sprocket D delivers 4 hp to a shaft above. Use AISI 1117 cold-drawn steel.

26. The shaft in Figure P12–3 is part of a special machine designed to reclaim scrap aluminum cans. The gear at B delivers 5 hp to a chopper that cuts the cans into small pieces. The V-belt sheave at D delivers 3 hp to a blower that draws air through the chopper. V-belt sheave E delivers 3 hp to a conveyor that raises the shredded aluminum to an elevated hopper. The shaft rotates at 480 rpm. All power comes into the shaft through the chain sprocket at C. Use AISI 1137 OQT 1300 steel for the shaft. The elements at B, C, D, and E are held in position with retaining rings and keys in profile keyseats. The shaft is to be of uniform diameter, except at its ends, where the bearings are to be mounted. Determine the required diameter.

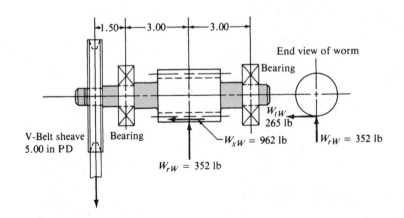

FIGURE P12–23   (Problems 23 and 34)

27. The shaft in Figure P12–4 is the drive shaft for a large, bulk material conveyor. The gear receives 40 hp and rotates at 120 rpm. Each chain sprocket delivers 20 hp to one side of the conveyor. Use AISI 1020 cold-drawn steel.

28. The shaft in Figure P12–5 is part of a conveyor drive system that feeds crushed rock into a railroad car. The shaft rotates at 240 rpm and is subjected to moderate shock during operation. All power is input to the gear at D. The V-belt sheave at A delivers 10.0 hp vertically downward. Chain sprocket E delivers 5.0 hp. Note the position of the gear Q, which drives gear D.

29. Figure P12–6 illustrates an intermediate shaft of a punch press that rotates at 310 rpm while transmitting 20 hp from the V-belt sheave to the gear. The flywheel is not absorbing or giving up any energy at this time. Consider the weight of all elements in the analysis.

30. The shaft in Figure P12–7 is part of a material handling system aboard a ship. All power comes into the shaft through gear C, which rotates at 480 rpm. Gear A delivers 30 hp to a hoist. V-belt sheaves D and E each deliver 10 hp to hydraulic pumps. Use AISI 3140 OQT 1000 steel.

31. The shaft in Figure P12–9 is part of an automatic machining system. All power is input through gear D. Gears C and F drive two separate tool feed devices requiring 5.0 hp each. The V-belt sheave at B requires 2.5 hp to drive a coolant pump. The shaft rotates at 220 rpm. All gears are spur gears with 20°, full-depth teeth. Use AISI 1020 cold-drawn steel for the shaft.

32. The shaft in Figure P12–17 is part of a grain-drying system. At A is a propeller-type fan that requires 12 hp when rotating at 475 rpm. The fan weighs 34 lb, and its weight should be included in the analysis. The flat-belt pulley at D delivers 3.5 hp to a screw conveyor handling the grain. All power comes into the shaft through the V-belt sheave at C. Use AISI 1144 cold-drawn steel.

33. Figure P12–22 shows a helical gear mounted on a shaft that rotates at 650 rpm while transmitting 7.5 hp. The gear is also analyzed in Example Problem 10–1, and the tangential, radial, and axial forces on it are shown in the figure. The pitch diameter of the gear is 4.141 in. The power is delivered from the shaft through a flexible coupling at its right end. A spacer is used to position the gear relative to bearing C. The thrust load is taken at bearing A.

34. The shaft shown in Figure P12–23 is the input shaft of a wormgear drive. The V-belt sheave receives 7.5 hp from directly downward. The worm rotates at 1750 rpm and has a pitch diameter of 2.000 in. This is the driving worm for the wormgear described in Design Example 12–4. The tangential, radial, and axial forces are shown in the figure. The worm is to be machined integrally with the

shaft, and it has a root diameter of 1.614 in. Assume that the geometry of the root area presents a stress concentration factor of 1.5 for bending. Analyze the stress in the area of the worm thread root, and specify a suitable material for the shaft.

35. The double-reduction, helical gear reducer shown in Figure P12–35 transmits 5.0 hp. Shaft 1 is the input, rotating at 1800 rpm and receiving power directly from an electric motor through a flexible coupling. Shaft 2 rotates at 900 rpm. Shaft 3 is the output, rotating at 300 rpm. A chain sprocket is mounted on the output shaft as shown and delivers the power upward. The data for the gears are given in Table 12–4.

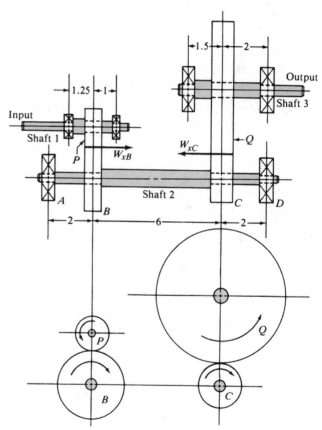

FIGURE P12–35   (Problem 35)

TABLE 12–4

Gear	Diametral pitch	Pitch diameter	Number of teeth	Face width
P	8	1.500 in	12	0.75 in
B	8	3.000 in	24	0.75 in
C	6	2.000 in	12	1.00 in
Q	6	6.000 in	36	1.00 in

Each gear has a 14½° normal pressure angle and a 45° helix angle. The combinations of left- and right-hand helixes are arranged so that the axial forces oppose each other on shaft 2 as shown. Use AISI 4140 OQT 1200 for the shafts.

36. Complete the design of the shafts carrying the bevel gear and pinion shown in Figure 10–8. The forces on the gears, the bearing reactions, and the bending moment diagrams are developed in Example Problems 10–3 and 10–4 and are shown in Figures 10–8 through 10–12. Assume that the 2.50 hp enters the pinion shaft from the right through a flexible coupling. The power is delivered by way of the lower extension of the gear shaft through another flexible coupling. Use AISI 1040 OQT 1200 for the shafts.

37. The vertical shaft shown in Figure P12–37 is driven at a speed of 600 rpm with 4.0 hp entering through the bevel gear. Each of the two chain sprockets delivers 2.0 hp to the side to drive mixer blades in a chemical reactor vessel. The bevel gear has a diametral pitch of 5, a pitch diameter of 9.000 in, a face width of 1.31 in, and a pressure angle of 20°. Use AISI 4140 OQT 1000 steel for the shaft.

See Chapter 10 for the methods for computing the forces on the bevel gear.

38. Figure P12–38 shows a sketch for a double-reduction spur gear train. Shaft 1 rotates at 1725 rpm, driven directly by an electric motor that transmits 15.0 hp to the reducer. The gears in the train all have 20°, full-depth teeth and the following numbers of teeth and pitch diameters:

Gear A	Gear B
18 teeth	54 teeth
1.80-in dia.	5.40-in dia.

Gear C	Gear D
24 teeth	48 teeth
4.00-in dia.	8.00-in dia.

Note that the speed reduction for each gear pair is proportional to the ratio of the number of teeth. Therefore, shaft 2 rotates at 575 rpm, and shaft 3 rotates at 287.5 rpm. Assume that all shafts carry 15.0 hp. The distance from the middle of each bearing to the middle of the face of the nearest gear is 3.00 in. Shaft 2 has a

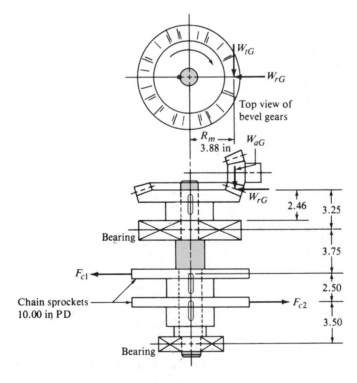

FIGURE P12–37   Bevel gears shown in section for Problem 37. See Chapter 10.

length of 5.00 in between the two gears, making the total distance between the middle of the two bearings equal to 11.00 in. The extensions of the input and the output shafts carry torque, but no bending loads are exerted on them. Complete the design of all three shafts. Provide for profile keyseats for each gear and sled-runner keyseats at the outboard ends of the input and output shafts. Provide for the location of each gear and bearing on the shaft.

39. Figure P12–39 shows a speed reducer with a V-belt drive delivering power to the input shaft and a chain drive tak-

ing power from the output shaft and delivering it to a conveyor. The drive motor supplies 12.0 hp and rotates 1150 rpm. The speed reductions of the V-belt drive and the chain drive are proportional to the ratio of the diameter of the driving and driven sheaves or sprockets. The arrangement of the gears in the reducer is the same as that described in Figure P12–38 and Problem 38. Determine the forces applied to the motor shaft, to each of the three shafts of the reducer, and to the conveyor drive shaft. Then complete the design of the three shafts of the reducer, assuming that all shafts transmit 12.0 hp.

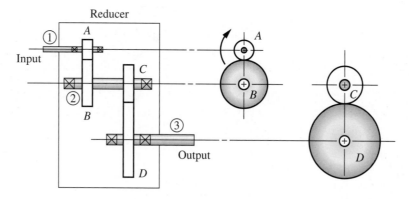

FIGURE P12–38   Gear reducer for Problems 38 and 39

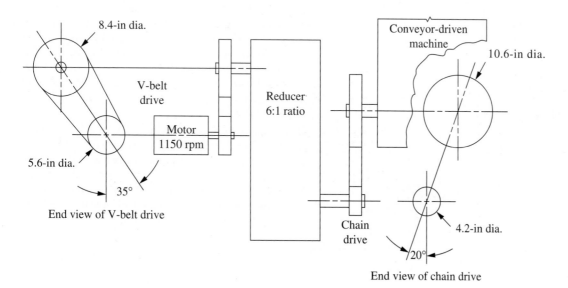

FIGURE P12–39   Drive system for Problem 39

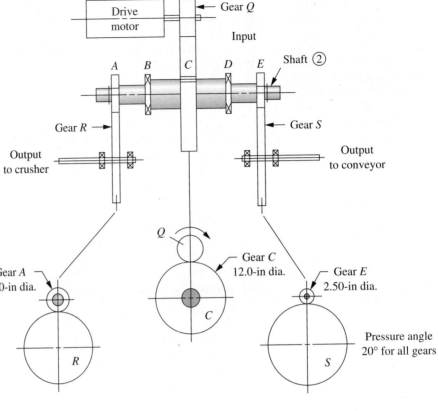

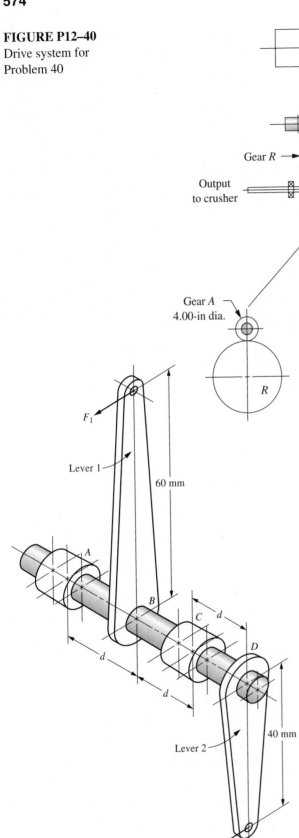

40. Figure P12–40 shows a drive for a system to crush coal
and deliver it by conveyor to a railroad car. Gear A delivers 15 kW to the crusher, and gear E delivers 7.5 kW
to the conveyor. All power enters the shaft through gear
C. The shaft carrying gears A, C, and E rotates at 480
rpm. Design that shaft. The distance from the middle of
each bearing to the middle of the face of the nearest gear
is 100 mm.

41. A shaft is to be designed for a linkage of a windshield
wiper mechanism for a truck (see Figure P12–41). A 20-N
force is applied to lever 1 by an adjacent link. The reaction
force, F_2, on lever 2 is transmitted to another link. The distance, d, between elements is 20.0 mm. Bearings A and C
are straight, cylindrical, bronze bushings, 10.0 mm long.
Design the shaft and levers 1 and 2.

FIGURE P12–41   Shaft and levers for windshield
wiper system for Problem 41

13

Tolerances and Fits

The Big Picture

You Are the Designer

13–1 Objectives of This Chapter

13–2 Factors Affecting Tolerances and Fits

13–3 Tolerances, Production Processes, and Cost

13–4 Preferred Basic Sizes

13–5 Clearance Fits

13–6 Interference Fits

13–7 Transition Fits

13–8 Stresses for Force Fits

13–9 General Tolerancing Methods

13–10 Robust Product Design

Tolerances and Fits

Discussion Map

- A *tolerance* is the permissible variation on key dimensions of mechanical parts. You, as a designer, must specify the tolerance for each dimension, considering how it will perform and how it will be manufactured.

- Two or more parts may be assembled together either with a *clearance fit*, allowing free relative motion, or with an *interference fit*, where the two parts must be pressed together and therefore will not move during operation of the device.

Discover

Identify examples of products for which precise dimensional tolerances would be appropriate and some for which looser tolerances are permitted.

Look for examples of mechanical parts that have clearance fits and some that have interference fits.

Describe the reasons why the designer may have made the design in this way.

This chapter will help you acquire the competency to specify suitable fits for mating parts and the dimensional tolerances that will produce the desired fit.

The objective of most of the analysis procedures discussed in this book is to determine the minimum acceptable geometric size at which a component is safe and performs properly under specified conditions. As a designer, you must also then specify the final dimensions for the components, including the tolerances on those dimensions.

The term *tolerance* refers to the permissible deviation of a dimension from the specified basic size. The proper performance of a machine can depend on the tolerances specified for its parts, particularly those that must fit together for location or for suitable relative motion.

The term *fit* usually refers to the *clearances* that are permissible between mating parts in a mechanical device that must assemble easily and that must often move relative to each other during normal operation of the device. Such fits are usually called *running* or *sliding fits*. *Fit* can also refer to the amount of *interference* that exists when the inside part should be larger than the outside part. Interference fits are used to ensure that mating parts do not move relative to each other.

Look for examples of parts for mechanical devices that have clearance fits. Any machine having shafts that rotate in plane surface bearings should have such parts. This type of bearing is called a *journal bearing*, and there must be a small but reliable clearance between the shaft and the bearing to permit smooth rotation of the shaft. But the clearance cannot be too large, or the operation of the machine will appear to be too crude and rough.

Consider also any assembly having hinges that allow one part to rotate relative to the other, such as an access door or panel or the cover of a container. The hinge parts will have a clearance fit. Many types of measuring equipment, such as calipers, dial indicators, and electronic probes, that have moving parts must be designed carefully to maintain the expected precision of the measurements while allowing reliable motion. On the other end of the spectrum is the rather loose fit that is typical of toys and some other types of recreation equipment. The fit of the wheel of a child's wagon onto its axle is typically quite large to allow free rotation and to permit easy assembly. We will show in this chapter that the wide range of fits that are encountered in practice are designed by specifying a *class of fit* and then determining the allowable range of key dimensions of mating parts.

Look also for examples where two parts are assembled so that they cannot move relative to one another. They are held tightly together because the inside part is larger than the outside part. Perhaps some parts of your car are like that.

This chapter will help you design parts that must have clearance or interference fits. When an interference fit occurs, it is often desirable to predict the level of stress to which the mating parts are subjected. That topic will also be covered in this chapter.

You Are the Designer

Assume that you are responsible for the design of the gear-type speed reducer discussed in Chapter 1 and sketched in Figure 1–12. The sketch is repeated here as Figure 13–1. The input and output shafts each carry one gear while being supported on two bearings mounted in the housing. The gears are keyed to the shafts to permit the transmission of torque from the shaft to the gear, or vice versa. The material in Chapter 12 will help you to design the shafts themselves. Chapter 14 discusses bearing selection and application. The results of these design decisions include the calculation of the minimum acceptable diameter for the shaft at any section and the specification of suitable bearings to carry the loads applied with a reasonable life. The bearing manufacturer will specify the tolerances on the various dimensions of the bearing. Your job is to specify the final dimensions for the shaft at all points, including the tolerances on those dimensions.

Consider the part of the input shaft where the first gear in the train is mounted. What is a convenient nominal size to specify for the shaft diameter at the location of the gear? Do you want to be able to slide the gear easily over the left end of the shaft and up against the shoulder that locates it? If so, how much clearance should be allowed between the shaft and the bore of the gear to ensure ease of assembly while still providing accurate location of the gear and smooth operation? When the shaft is machined, what range of dimensions will you allow the manufacturing staff to produce? What surface finish should be specified for the shaft, and what manufacturing process is

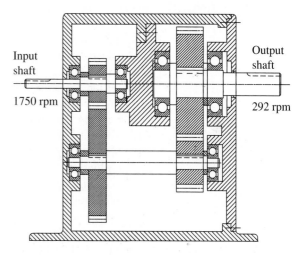

FIGURE 13–1 Conceptual design for a speed reducer

required to produce that finish? What is the relative cost of the manufacturing operation? Similar questions must be answered for the bore of the gear.

Rolling contact bearings, such as the familiar ball bearing, are designed to be installed on the shaft with an *interference fit*. That is, the inside diameter of the bearing is smaller than the outside diameter of the shaft where the bearing is to be seated. A significant force is required to press the bearing onto the shaft. What dimensions do you specify for the shaft at the bearing seat? How much interference should be specified? How much stress is produced in the shaft because of the interference fit? As the designer you must answer these questions.

13–1
OBJECTIVES OF
THIS CHAPTER

After completing this chapter, you will be able to:

1. Define the terms *tolerance, allowance, unilateral tolerance,* and *bilateral tolerance.*

2. Describe the relationships among tolerances, production processes, and cost.

3. Specify basic sizes for dimensions according to a set of preferred sizes.

4. Use the ANSI Standard B4.1, *Preferred Limits and Fits for Cylindrical Parts,* to specify tolerances, fits, and clearances.

5. Specify transitional, interference, and force fits.

6. Compute the pressure created between parts subjected to interference fits and the resulting stresses in the mating members.

7. Use a spreadsheet to assist in the calculation of stresses for interference fits.

8. Specify appropriate geometric dimensions and tolerancing controls for mating parts.

13–2
FACTORS
AFFECTING
TOLERANCES
AND FITS

Consider the plain surface bearings designed in Chapter 16. A critical part of the design is the specification of the diametral clearance between the journal and the bearing. The typical value is just a few thousandths of an inch. But some variation must be allowed on both the journal outside diameter and the bearing inside diameter for reasons of economy of manufacture. Thus, there will be a variation of the actual clearance in production devices, depending on where the individual mating components fall within their own tolerance bands. Such variations must be accounted for in the analysis of the bearing performance. Too small a clearance could cause seizing. Conversely, too large a clearance would reduce the precision of the machine and adversely affect the lubrication.

The mounting of power transmission elements onto shafts is another situation in which tolerances and fits must be considered. A chain sprocket for general-purpose mechanical drives is usually produced with a bore that will easily slide into position on the shaft during assembly. But once in place, it will transmit power smoothly and quietly only if it is not excessively loose. A high-speed turbine rotor must be installed on its shaft with an interference fit, eliminating any looseness that would cause vibrations at the high rotational speeds.

When relative motion between two parts is required, a clearance fit is necessary. But here, too, there are differences. Some measuring instruments have parts that must move with no perceptible looseness (sometimes called *play*) between mating parts which would adversely affect the accuracy of measurement. An idler sheave in a belt drive system must rotate on its shaft reliably, without the tendency to seize, but with only a small amount of play. The requirement for the mounting of a wheel of a child's wagon on its axle is much different. A loose clearance fit is satisfactory for the expected use of the wheel, and it permits wide tolerances on the wheel bore and on the axle diameter for economy.

13–3
TOLERANCES,
PRODUCTION
PROCESSES,
AND COST

A *unilateral tolerance* deviates in only one direction from the basic size. A *bilateral tolerance* deviates both above and below the basic size. The *total tolerance* is the difference between the maximum and minimum permissible dimensions.

The term *allowance* refers to an intentional difference between the maximum material limits of mating parts. For example, a positive allowance for a hole/shaft pair would define the minimum *clearance* between the mating parts from the largest shaft mating with the smallest hole. A negative allowance would result in the shaft being larger than the hole (*interference*).

The term *fit* refers to the relative looseness (clearance fit) or tightness (interference fit) of mating parts, especially as it affects the motion of the parts or the force between them after assembly. Specifying the degree of clearance or interference is one of the tasks of the designer.

It is costly to produce components with very small tolerances on dimensions. It is the designer's responsibility to set the tolerances at the highest possible level that results in satisfactory operation of the machine. Certainly judgment and experience must be exercised in such a process. In quantity production situations, it may be cost-effective to test prototypes with a range of tolerances to observe the limits of acceptable performance.

In general, the production of parts with small tolerances on their dimensions requires multiple processing steps. A shaft may first have to be turned on a lathe and then ground to produce the final dimensions and surface finish. In extreme cases, lapping may be required. Each subsequent step in the manufacture adds cost. Even if different operations are not required, the holding of small tolerances on a single machine, such as a lathe, may require

FIGURE 13–2

Machining costs versus surface finish specified [Reprinted with permission from the Association for Integrated Manufacturing Technology (formerly the Numerical Control Society)]

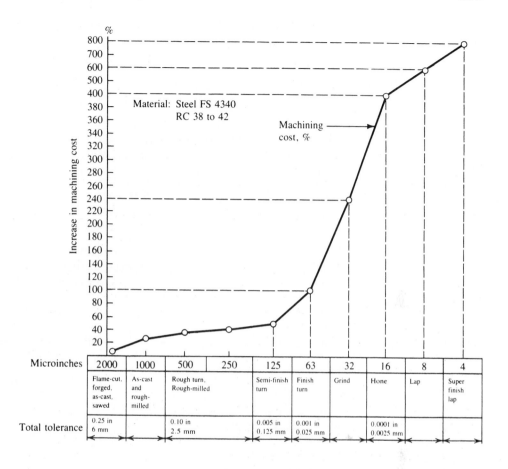

several passes, ending with a fine finish cut. Cutting tool changes must be more frequent, also, as wear of the tool takes the part out of tolerance.

The production of part features with small tolerances usually involves finer surface finishes as well. Figure 13–2 shows the general relationship between the surface finish and the relative cost of producing a part. The typical total tolerance produced by the processes described is included in the figure. The increase in cost is dramatic for the small tolerances and fine finishes.

Figure 13–3 presents the relationship between the surface finish and the machining operations available to produce it.

The basic reference for tolerances and fits in the United States is the ANSI Standard B4.1-1967, *Preferred Limits and Fits for Cylindrical Parts*. Metric dimensions should conform to ANSI B4.2-1978, *Preferred Metric Limits and Fits*. The most recent version should be consulted. (See References 1 and 2.)

An international body, the International Organization for Standardization (ISO), gives metric data for limits and fits in Recommendation 286 (ISO R286), used in Europe and many other countries.

The term *tolerance grade* refers to a set of tolerances that can be produced with an approximately equal production capability. The actual total tolerance allowed within each grade depends on the nominal size of the dimension. Smaller tolerances can be achieved for smaller dimensions, and vice versa. Standards ISO R286 and ANSI B4.1 include complete data for tolerance grades from 01 through 16 as shown in Table 13–1. Tolerances are smaller for the smaller grade number.

A sampling of tolerance data for machined parts for a few grades and ranges of size is given in Table 13–2. Figure 13–2 shows the capability of selected manufacturing processes to produce work within the given tolerance grades.

FIGURE 13–3
Finishes produced by
various techniques
(roughness average, R_a)

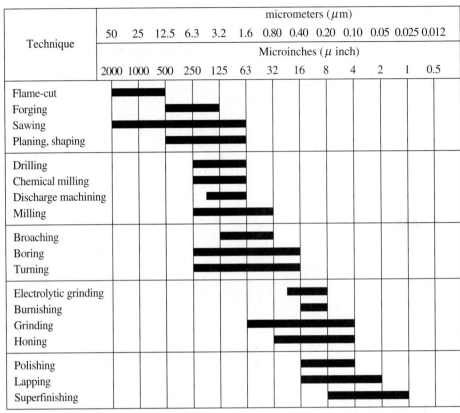

■ = Average industrial application (values may vary under special conditions)

TABLE 13–1   Tolerance grades

Application	Tolerance grades								
Measuring tools	01	0	1	2	3	4	5	6	7
Fits of machined parts	4	5	6	7	8	9	10	11	
Material, as supplied	8	9	10	11	12	13	14		
Rough forms (casting, sawing, forging, etc.)	12	13	14	15	16				

TABLE 13–2   Tolerances for some tolerance grades

Nominal size (in)	Tolerance grade							
	4	5	6	7	8	9	10	11
	Tolerances in thousandths of an inch							
0.24–0.40	0.15	0.25	0.4	0.6	0.9	1.4	2.2	3.5
0.40–0.71	0.20	0.3	0.4	0.7	1.0	1.6	2.8	4.0
0.71–1.19	0.25	0.4	0.5	0.8	1.2	2.0	3.5	5.0
1.19–1.97	0.3	0.4	0.6	1.0	1.6	2.5	4.0	6.0
1.97–3.15	0.3	0.5	0.7	1.2	1.8	3.0	4.5	7.0
3.15–4.73	0.4	0.6	0.9	1.4	2.2	3.5	5.0	9.0
4.73–7.09	0.5	0.7	1.0	1.6	2.5	4.0	6.0	10.0

**13–4
PREFERRED
BASIC SIZES**

The first step in specifying a dimension for a part is to decide on the basic size, that dimension to which the tolerances are applied. The analysis for strength, deflection, or performance of the part determines the nominal or minimum size required. Unless special conditions exist, the basic size should be chosen from the lists of preferred basic sizes in Table A2–1 for fractional-inch sizes, decimal-inch sizes, and metric sizes from the SI. If possible, select from the first-choice column. If a size between two first-choice sizes is required, the second-choice column should be used. This will limit the number of sizes typically encountered in the manufacture of products and will lead to cost-effective standardization. The choice of system depends on the policies of the company and the market for the product.

**13–5
CLEARANCE FITS**

When there must always be a clearance between mating parts, a clearance fit is specified. The designation for standard clearance fits from ANSI Standard B4.1 for members that must move together is the *running* or *sliding clearance fit (RC)*. Within this standard, there are nine classes, RC1 through RC9, with RC1 providing the smallest clearance, and RC9 the largest. The following descriptions for the individual members of this class should help you to decide which is most appropriate for a given application.

> *RC1 (close sliding fit):* Accurate location of parts that must assemble without perceptible play.
>
> *RC2 (sliding fit):* Parts that will move and turn easily but are not intended to run freely. Parts may seize with small temperature changes, especially in the larger sizes.
>
> *RC3 (precision running fit):* Precision parts operating at slow speeds and light loads that must run freely. Changes in temperature may cause difficulties.
>
> *RC4 (close running fit):* Accurate location with minimum play for use under moderate loads and speeds. A good choice for accurate machinery.
>
> *RC5 (medium running fit):* Accurate machine parts for higher speeds and/or loads than for RC4.
>
> *RC6 (medium running fit):* Similar to RC5 for applications in which larger clearance is desired.
>
> *RC7 (free running fit):* Reliable relative motion under wide temperature variations in applications where accuracy is not critical.
>
> *RC8 (loose running fit):* Permits large clearances, allowing the use of parts with commercial, "as-received" tolerances.
>
> *RC9 (loose running fit):* Similar to RC8, with approximately 50% larger clearances.

The complete standard ANSI B4.1 lists the tolerances on the mating parts and the resulting limits of clearances for all nine classes and for sizes from zero to 200 in. (See also References 4, 5, and 9.) Table 13–3 is abstracted from the standard. Let RC2 represent the precision fits (RC1, RC2, RC3); let RC5 represent the accurate, reliable running fits (RC4 to RC7); and let RC8 represent the loose fits (RC8, RC9).

The numbers in Table 13–3 are in thousandths of an inch. Thus, a clearance of 2.8 from the table means a difference in size between the inside and outside parts of 0.0028 in. The tolerances on the hole and the shaft are to be applied to the basic size to determine the limits of size for that dimension.

In applying tolerances and the allowance to the dimensions of mating parts, designers use the basic hole system and the basic shaft system. In the *basic hole system*, the

TABLE 13–3   Clearance fits (RC)

Nominal size range (in) Over To	Class RC2			Class RC5			Class RC8			Nominal size range (in) Over To
	Limits of clearance	Standard limits		Limits of clearance	Standard limits		Limits of clearance	Standard limits		
		Hole	Shaft		Hole	Shaft		Hole	Shaft	
0–0.12	0.1 0.55	+0.25 0	−0.1 −0.3	0.6 1.6	+0.6 −0	−0.6 −1.0	2.5 5.1	+1.6 0	−2.5 −3.5	0–0.12
0.12–0.24	0.15 0.65	+0.3 0	−0.15 −0.35	0.8 2.0	+0.7 −0	−0.8 −1.3	2.8 5.8	+1.8 0	−2.8 −4.0	0.12–0.24
0.24–0.40	0.2 0.85	+0.4 0	−0.2 −0.45	1.0 2.5	+0.9 −0	−1.0 −1.6	3.0 6.6	+2.2 0	−3.0 −4.4	0.24–0.40
0.40–0.71	0.25 0.95	+0.4 0	−0.25 −0.55	1.2 2.9	+1.0 −0	−1.2 −1.9	3.5 7.9	+2.8 0	−3.5 −5.1	0.40–0.71
0.71–1.19	0.3 1.2	+0.5 0	−0.3 −0.7	1.6 3.6	+1.2 −0	−1.6 −2.4	4.5 10.0	+3.5 0	−4.5 −6.5	0.71–1.19
1.19–1.97	0.4 1.4	+0.6 0	−0.4 −0.8	2.0 4.6	+1.6 −0	−2.0 −3.0	5.0 11.5	+4.0 0	−5.0 −7.5	1.19–1.97
1.97–3.15	0.4 1.6	+0.7 0	−0.4 −0.9	2.5 5.5	+1.8 −0	−2.5 −3.7	6.0 13.5	+4.5 0	−6.0 −9.0	1.97–3.15
3.15–4.73	0.5 2.0	+0.9 0	−0.5 −1.1	3.0 6.6	+2.2 −0	−3.0 −4.4	7.0 15.5	+5.0 0	−7.0 −10.5	3.15–4.73
4.73–7.09	0.6 2.3	+1.0 0	−0.6 −1.3	3.5 7.6	+2.5 −0	−3.5 −5.1	8.0 18.0	+6.0 0	−8.0 −12.0	4.73–7.09
7.09–9.85	0.6 2.6	+1.2 0	−0.6 −1.4	4.0 8.6	+2.8 −0	−4.0 −5.8	10.0 21.5	+7.0 0	−10.0 −14.5	7.09–9.85
9.85–12.41	0.7 2.8	+1.2 0	−0.7 −1.6	5.0 10.0	+3.0 −0	−5.0 −7.0	12.0 25.0	+8.0 0	−12.0 −17.0	9.85–12.41

Source: Reprinted from ANSI Standard B4. 1-1967 by permission of The American Society of Mechanical Engineers. All rights reserved.

Note: Limits are in thousandths of an inch.

design size of the hole is the basic size, and the allowance is applied to the shaft; the basic size is the minimum size of the hole. In the *basic shaft system*, the design size of the shaft is the basic size, and the allowance is applied to the hole; the basic size is the maximum size of the shaft. The basic hole system is preferred.

Figure 13–4 shows a graphical display of the tolerances and fits for all nine RC classes when applied to a shaft/hole combination in which the basic size is 2.000 in and the basic hole system is used. Note that such a diagram shows the total tolerance on both the shaft and the hole. The tolerance for the hole always starts at the basic size, while the shaft tolerance is offset below the basic size to provide for the minimum clearance (smallest hole combined with the largest shaft). The maximum clearance combines the largest hole with the smallest shaft. This figure also shows the dramatic range of clearances provided by the nine classes within the RC system.

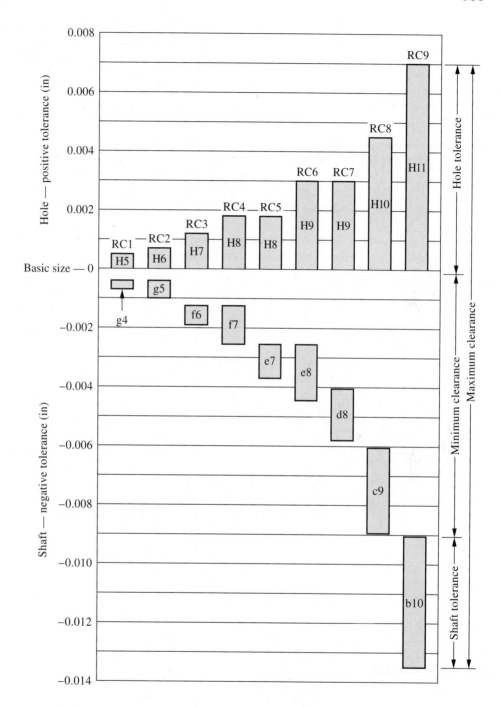

FIGURE 13–4 RC fits for basic hole size of 2.00 in showing tolerances for hole and shaft, minimum clearance and maximum clearance

The codes within the tolerance bars in Figure 13–4 refer to the tolerance grades mentioned earlier. The capital H combined with a tolerance grade number is used for the hole in the basic hole system for which there is no fundamental deviation from the basic size. The lowercase letters in the shaft tolerance bars indicate some fundamental deviation of the shaft size from the basic size. Then the tolerance is added to the fundamental deviation. The size of the tolerance is indicated by the number.

ISO standard R286 for limits and fits also uses the letter/number codes. For example, a specification for a shaft/hole combination with a basic size of 50 mm that is to provide a free running fit (similar to RC7) could be defined on a drawing as

Hole: Ø50 H9   Shaft: Ø50 d8

No tolerance values need to be given on the drawing.

Example Problem 13–1

A shaft carrying an idler sheave for a belt drive system is to have a nominal size of 2.00 in. The sheave must rotate reliably on the shaft, but with the smoothness characteristic of accurate machinery. Specify the limits of size for the shaft and the sheave bore, and list the limits of clearance that will result. Use the basic hole system.

Solution

An RC5 fit should be satisfactory in this application. From Table 13–3, the hole tolerance limits are +1.8 and −0. The sheave hole then should be within the following limits:

Sheave Hole

$$2.0000 + 0.0018 = 2.0018 \text{ in} \quad \text{(largest)}$$
$$2.0000 - 0.0000 = 2.0000 \text{ in} \quad \text{(smallest)}$$

Notice that the smallest hole is the basic size.

The shaft tolerance limits are −2.5 and −3.7. The resulting size limits are as follows:

Shaft Diameter

$$2.0000 - 0.0025 = 1.9975 \text{ in} \quad \text{(largest)}$$
$$2.0000 - 0.0037 = 1.9963 \text{ in} \quad \text{(smallest)}$$

Figure 13–5 illustrates these results.

Combining the smallest shaft with the largest hole gives the largest clearance. Conversely, combining the largest shaft with the smallest hole gives the smallest clearance. Therefore, the limits of clearance are

$$2.0018 - 1.9963 = 0.0055 \text{ in} \quad \text{(largest)}$$
$$2.0000 - 1.9975 = 0.0025 \text{ in} \quad \text{(smallest)}$$

These values check with the limits of clearance in Table 13–3. Notice that the total tolerance for the shaft is 0.0012 in, and for the hole 0.0018 in, both relatively small values.

FIGURE 13–5 An RC5 fit using the basic hole system

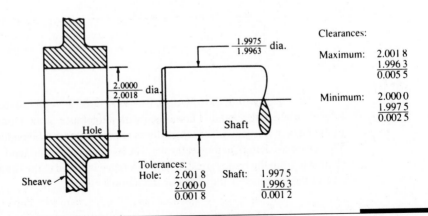

Locational Clearance Fits

Another clearance fit system is available for parts for which control of the location is desired although the parts will normally not move in relation to each other in operation. Called *locational clearance (LC)* fits, they include eleven classes. The first four, LC1 to LC4, have a zero clearance (size to size) as the lower limit of the fit, regardless of size or class. The upper limit of fit increases with both the size of the parts and the class number. Classes LC5 through LC11 provide some positive clearance for all sizes, increasing with the size of the parts and with the class. Numerical values for the tolerances and fits for these classes have been published. (See References 1, 4, 5, and 9.)

13–6 INTERFERENCE FITS

Interference fits are those in which the inside member is larger than the outside member, requiring the application of force during assembly. There is some deformation of the parts after assembly, and a pressure exists at the mating surfaces.

Force fits are designed to provide a controlled pressure between mating parts throughout the range of sizes for a given class. They are used where forces or torques are transmitted through the joint. Instead of being assembled through the application of force, similar fits are obtained by *shrink fitting*, in which one member is heated to expand it while the other remains cool. The parts are then assembled with little or no force. After cooling, the same dimensional interference exists as for the force fit. *Locational interference fits* are used for location only. There is no movement between parts after assembly, but there is no special requirement for the resulting pressure between mating parts.

Force Fits (FN)

As shown in Table 13–4, five classes of force fits are defined in ANSI Standard B4.1. (See References 1, 4, 5, and 9.)

FN1 (light drive fit): Only light pressure required to assemble mating parts. Used for fragile parts and where no large forces must be transmitted across the joint.

FN2 (medium drive fit): General-purpose class used frequently for steel parts of moderate cross section.

FN3 (heavy drive fit): Used for heavy steel parts.

FN4 (force fit): Used for high-strength assemblies where high resulting pressures are required.

FN5 (force fit): Similar to FN4 for higher pressures.

The use of shrink fit methods is desirable in most cases of interference fits and is virtually required in the heavier classes and larger-size parts. The temperature increase required to produce a given expansion for assembly can be computed from the basic definition of the coefficient of thermal expansion:

$$\delta = \alpha L(\Delta t) \tag{13–1}$$

where δ = total deformation desired (in or mm)
α = coefficient of thermal expansion (in/in·°F or mm/mm·°C)
L = nominal length of member being heated (in or mm)
Δt = temperature difference (°F or °C)

For cylindrical parts, L is the diameter and δ is the change in diameter required. Table 13–5 gives the values for α for several materials. (See Reference 9 for more data.)

TABLE 13–4   Force and shrink fits (FN)

Nominal size range (in) Over To	Class FN1 Limits of interference	Standard limits Hole	Shaft	Class FN2 Limits of interference	Standard limits Hole	Shaft	Class FN3 Limits of interference	Standard limits Hole	Shaft	Class FN4 Limits of interference	Standard limits Hole	Shaft	Class FN5 Limits of interference	Standard limits Hole	Shaft
0–0.12	0.05 / 0.5	+0.25 / −0	+0.5 / +0.3	0.2 / 0.85	+0.4 / −0	+0.85 / +0.6				0.3 / 0.95	+0.4 / −0	+0.95 / +0.7	0.3 / 1.3	+0.6 / −0	+1.3 / +0.9
0.12–0.24	0.1 / 0.6	+0.3 / −0	+0.6 / +0.4	0.2 / 1.0	+0.5 / −0	+1.0 / +0.7				0.4 / 1.2	+0.5 / −0	+1.2 / +0.9	0.5 / 1.7	+0.7 / −0	+1.7 / +1.2
0.24–0.40	0.1 / 0.75	+0.4 / −0	+0.75 / +0.5	0.4 / 1.4	+0.6 / −0	+1.4 / +1.0				0.6 / 1.6	+0.6 / −0	+1.6 / +1.2	0.5 / 2.0	+0.9 / −0	+2.0 / +1.4
0.40–0.56	0.1 / 0.8	+0.4 / −0	+0.8 / +0.5	0.5 / 1.6	+0.7 / −0	+1.6 / +1.2				0.7 / 1.8	+0.7 / −0	+1.8 / +1.4	0.6 / 2.3	+1.0 / −0	+2.3 / +1.6
0.56–0.71	0.2 / 0.9	+0.4 / −0	+0.9 / +0.6	0.5 / 1.6	+0.7 / −0	+1.6 / +1.2				0.7 / 1.8	+0.7 / −0	+1.8 / +1.4	0.8 / 2.5	+1.0 / −0	+2.5 / +1.8
0.71–0.95	0.2 / 1.1	+0.5 / −0	+1.1 / +0.7	0.6 / 1.9	+0.8 / −0	+1.9 / +1.4				0.8 / 2.1	+0.8 / −0	+2.1 / +1.6	1.0 / 3.0	+1.2 / −0	+3.0 / +2.2
0.95–1.19	0.3 / 1.2	+0.5 / −0	+1.2 / +0.8	0.6 / 1.9	+0.8 / −0	+1.9 / +1.4	0.8 / 2.1	+0.8 / −0	+2.1 / +1.6	1.0 / 2.3	+0.8 / −0	+2.3 / +1.8	1.3 / 3.3	+1.2 / −0	+3.3 / +2.5
1.19–1.58	0.3 / 1.3	+0.6 / −0	+1.3 / +0.9	0.8 / 2.4	+1.0 / −0	+2.4 / +1.8	1.0 / 2.6	+1.0 / −0	+2.6 / +2.0	1.5 / 3.1	+1.0 / −0	+3.1 / +2.5	1.4 / 4.0	+1.6 / −0	+4.0 / +3.0
1.58–1.97	0.4 / 1.4	+0.6 / −0	+1.4 / +1.0	0.8 / 2.4	+1.0 / −0	+2.4 / +1.8	1.2 / 2.8	+1.0 / −0	+2.8 / +2.2	1.8 / 3.4	+1.0 / −0	+3.4 / +2.8	2.4 / 5.0	+1.6 / −0	+5.0 / +4.0
1.97–2.56	0.6 / 1.8	+0.7 / −0	+1.8 / +1.3	0.8 / 2.7	+1.2 / −0	+2.7 / +2.0	1.3 / 3.2	+1.2 / −0	+3.2 / +2.5	2.3 / 4.2	+1.2 / −0	+4.2 / +3.5	3.2 / 6.2	+1.8 / −0	+6.2 / +5.0
2.56–3.15	0.7 / 1.9	+0.7 / −0	+1.9 / +1.4	1.0 / 2.9	+1.2 / −0	+2.9 / +2.2	1.8 / 3.7	+1.2 / −0	+3.7 / +3.0	2.8 / 4.7	+1.2 / −0	+4.7 / +4.0	4.2 / 7.2	+1.8 / −0	+7.2 / +6.0
3.15–3.94	0.9 / 2.4	+0.9 / −0	+2.4 / +1.8	1.4 / 3.7	+1.4 / −0	+3.7 / +2.8	2.1 / 4.4	+1.4 / −0	+4.4 / +3.5	3.6 / 5.9	+1.4 / −0	+5.9 / +5.0	4.8 / 8.4	+2.2 / −0	+8.4 / +7.0
3.94–4.73	1.1 / 2.6	+0.9 / −0	+2.6 / +2.0	1.6 / 3.9	+1.4 / −0	+3.9 / +3.0	2.6 / 4.9	+1.4 / −0	+4.9 / +4.0	4.6 / 6.9	+1.4 / −0	+6.9 / +6.0	5.8 / 9.4	+2.2 / −0	+9.4 / +8.0
4.73–5.52	1.2 / 2.9	+1.0 / −0	+2.9 / +2.2	1.9 / 4.5	+1.6 / −0	+4.5 / +3.5	3.4 / 6.0	+1.6 / −0	+6.0 / +5.0	5.4 / 8.0	+1.6 / −0	+8.0 / +7.0	7.5 / 11.6	+2.5 / −0	+11.6 / +10.0
5.52–6.30	1.5 / 3.2	+1.0 / −0	+3.2 / +2.5	2.4 / 5.0	+1.6 / −0	+5.0 / +4.0	3.4 / 6.0	+1.6 / −0	+6.0 / +5.0	5.4 / 8.0	+1.6 / −0	+8.0 / +7.0	9.5 / 13.6	+2.5 / −0	+13.6 / +12.0

Source: Reprinted from ASME B4.1-1967, by permission of The American Society of Mechanical Engineers. All rights reserved.

Note: Limits are in thousandths of an inch.

13–7 TRANSITION FITS

The *location transition (LT)* fit is used where accuracy of location is important, but where a small amount of clearance or a small amount of interference is acceptable. There are six classes, LT1 to LT6. In any class there is an overlap in the tolerance limits for both the hole and the shaft so that possible combinations produce a small clearance, a small interference, or even a size-to-size fit. Complete tables of data for these fits are published in References 1, 4, 5, and 9.

TABLE 13–5 Coefficient of thermal expansion

Material	Coefficient of thermal expansion, α	
	in/in·°F	mm/mm·°C
Steel:		
AISI 1020	6.5×10^{-6}	11.7×10^{-6}
AISI 1050	6.1×10^{-6}	11.0×10^{-6}
AISI 4140	6.2×10^{-6}	11.2×10^{-6}
Stainless steel:		
AISI 301	9.4×10^{-6}	16.9×10^{-6}
AISI 430	5.8×10^{-6}	10.4×10^{-6}
Aluminum:		
2014	12.8×10^{-6}	23.0×10^{-6}
6061	13.0×10^{-6}	23.4×10^{-6}
Bronze:	10.0×10^{-6}	18.0×10^{-6}

13–8 STRESSES FOR FORCE FITS

When force fits are used to secure mechanical parts, the interference creates a pressure acting at the mating surfaces. The pressure causes stresses in each part. Under heavy force fits or even with lighter fits in fragile parts, the stresses developed can be great enough to yield ductile materials. A permanent set results, which normally destroys the usefulness of the assembly. With brittle materials such as cast iron, actual fracture may result.

The stress analysis applicable to force fits is related to the analysis of thick-walled cylinders. The outer member expands under the influence of the pressure at the mating surface, with the tangential tensile stress developed being a maximum at the mating surface. There is a radial stress equal to the pressure itself. Also, the inner member contracts because of the pressure and is subjected to a tangential compressive stress along with the radial compressive stress equal to the pressure. (See Reference 10.)

The usual objective of the analysis is to determine the magnitude of the pressure due to a given interference fit that would be developed at the mating surfaces. Then the stresses due to this pressure in the mating members are computed. The following procedure can be used:

Procedure for Computing Stresses for Force Fits

1. Determine the amount of interference from the design of the parts. For standard force fits, Table 13–4 can be used. The maximum limit of interference would, of course, give the maximum stresses for the parts. Note that the interference values are based on the total interference on the diameter, which is the sum of the expansion of the outer ring plus the contraction of the inner member (see Figure 13–6).

2. Compute the pressure at the mating surface from Equation (13–2) if both members are of the same material:

Pressure Created by Force Fit

$$p = \frac{E\delta}{2b}\left[\frac{(c^2 - b^2)(b^2 - a^2)}{2b^2(c^2 - a^2)}\right] \tag{13–2}$$

Use Equation (13–3) if they are of different materials:

Pressure Created by Force Fit, Two Different Materials

$$p = \frac{\delta}{2b\left[\dfrac{1}{E_o}\left(\dfrac{c^2 + b^2}{c^2 - b^2} + v_o\right) + \dfrac{1}{E_i}\left(\dfrac{b^2 + a^2}{b^2 - a^2} - v_i\right)\right]} \tag{13–3}$$

FIGURE 13–6
Terminology for
interference fit

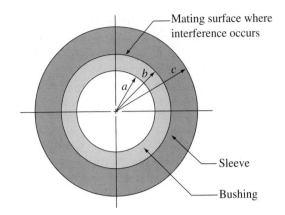

where p = pressure at the mating surface
   δ = total diametral interference
   E = modulus of elasticity of each member if they are the same
   E_o = modulus of elasticity of outer member
   E_i = modulus of elasticity of inner member
   v_o = Poisson's ratio for outer member
   v_i = Poisson's ratio for inner member

3. Compute the tensile stress in the outer member from

$$\sigma_o = p\left(\frac{c^2 + b^2}{c^2 - b^2}\right) \quad \begin{pmatrix}\text{tensile at the inner surface} \\ \text{in the tangential direction}\end{pmatrix} \tag{13–4}$$

4. Compute the compressive stress in the inner member from

$$\sigma_i = -p\left(\frac{b^2 + a^2}{b^2 - a^2}\right) \quad \begin{pmatrix}\text{compressive at the outer surface} \\ \text{in the tangential direction}\end{pmatrix} \tag{13–5}$$

5. If desired, the increase in the diameter of the outer member due to the tensile stress
   can be computed from

$$\delta_o = \frac{2bp}{E_o}\left[\frac{c^2 + b^2}{c^2 - b^2} + v_o\right] \tag{13–6}$$

6. If desired, the decrease in the diameter of the inner member due to the compressive stress can be computed from

$$\delta_i = -\frac{2bp}{E_i}\left[\frac{b^2 + a^2}{b^2 - a^2} - v_i\right] \tag{13–7}$$

The stresses computed from Equations (13–4) and (13–5) were derived assuming that
the two cylinders are of equal length. If the outer member is shorter than the inner member,
the stresses are higher at its ends by as much as a factor of 2.0. Such a factor should be applied as a stress concentration factor.

In the absence of applied shear stresses, the tensile stress in the tangential direction in the outer member is the maximum principal stress and can be compared with the yield strength of the material to determine the resulting design factor.

Example Problem 13–2 shows the application of these relationships. Then a spreadsheet is listed to solve the same equations, with sample output showing the solution to Example Problem 13–2.

Example Problem 13–2   A bronze bushing is to be installed into a steel sleeve as indicated in Figure 13–6. The bushing has an inside diameter of 2.000 in and a nominal outside diameter of 2.500 in. The steel sleeve has a nominal inside diameter of 2.500 in and an outside diameter of 3.500 in.

1. Specify the limits of size for the outside diameter of the bushing and the inside diameter of the sleeve in order to obtain a heavy drive fit, FN3. Determine the limits of interference that would result.

2. For the maximum interference from 1, compute the pressure that would be developed between the bushing and the sleeve, the stress in the bushing and the sleeve, and the deformation of the bushing and the sleeve. Use $E = 30 \times 10^6$ psi for the steel and $E = 17 \times 10^6$ for the bronze. Use $v = 0.27$ for both materials.

Solution   For 1, from Table 13–4, for a part size of 2.50 in at the mating surface, the tolerance limits on the hole in the outer member are +1.2 and −0. Applying these limits to the basic size gives the dimension limits for the hole in the steel sleeve:

$$2.5012 \text{ in}$$
$$2.5000 \text{ in}$$

For the bronze insert, the tolerance limits are +3.2 and +2.5. Then the size limits for the outside diameter of the bushing are

$$2.5032 \text{ in}$$
$$2.5025 \text{ in}$$

The limits of interference would be 0.0013 to 0.0032 in.

For 2, the maximum pressure would be produced by the maximum interference, 0.0032 in. Then, using $a = 1.00$ in, $b = 1.25$ in, $c = 1.75$ in, $E_o = 30 \times 10^6$ psi, $E_i = 17 \times 10^6$ psi, and $v_o = v_i = 0.27$ from Equation (13–3),

$$p = \frac{\delta}{2b\left[\dfrac{1}{E_o}\left(\dfrac{c^2 + b^2}{c^2 - b^2} + v_o\right) + \dfrac{1}{E_i}\left(\dfrac{b^2 + a^2}{b^2 - a^2} - v_i\right)\right]}$$

$$p = \frac{0.0032}{(2)(1.25)\left[\dfrac{1}{30 \times 10^6}\left(\dfrac{1.75^2 + 1.25^2}{1.75^2 - 1.25^2} + 0.27\right) + \dfrac{1}{17 \times 10^6}\left(\dfrac{1.25^2 + 1.00^2}{1.25^2 - 1.00^2} - 0.27\right)\right]}$$

$$p = 3518 \text{ psi}$$

The tensile stress in the steel sleeve is

$$\sigma_o = p\left(\frac{c^2 + b^2}{c^2 - b^2}\right) = 3518\left(\frac{1.75^2 + 1.25^2}{1.75^2 - 1.25^2}\right) = 10\,846 \text{ psi}$$

The compressive stress in the bronze bushing is

$$\sigma_i = -p\left(\frac{b^2 + a^2}{b^2 - a^2}\right) = -3518\left(\frac{1.25^2 + 1.00^2}{1.25^2 - 1.00^2}\right) = -16\,025 \text{ psi}$$

The increase in the diameter of the sleeve is

$$\delta_o = \frac{2bp}{E_o}\left[\frac{c^2 + b^2}{c^2 - b^2} + v_o\right]$$

$$\delta_o = \frac{2(1.25)(3518)}{30 \times 10^6}\left[\frac{1.75^2 + 1.25^2}{1.75^2 - 1.25^2} + 0.27\right] = 0.00\,098 \text{ in}$$

The decrease in the diameter of the bushing is

$$\delta_i = -\frac{2bp}{E_i}\left[\frac{b^2 + a^2}{b^2 - a^2} - v_i\right]$$

$$\delta_i = \frac{2(1.25)(3518)}{17 \times 10^6}\left[\frac{1.25^2 + 1.00^2}{1.25^2 - 1.00^2} + 0.27\right] = 0.00\,222 \text{ in}$$

Note that the sum of δ_o and δ_i equals 0.0032, the total interference, δ.

The spreadsheet to analyze force fits is shown in Figure 13–7. The data shown are from Example Problem 13–2.

STRESSES FOR FORCE FITS Refer to Figure 13–6 for geometry.	*Data from:* Example Problem 13–2	
Input Data:	**Numerical values in *italics* must be inserted for each problem.**	
Inside radius of inner member: $a =$	*1.0000 in*	
Outside radius of inner member: $b =$	*1.2500 in*	
Outside radius of outer member: $c =$	*1.7500 in*	
Total interference: $d =$	*0.0032 in*	
Modulus of outer member: $E_o =$	*3.00E+07 psi*	
Modulus of inner member: $E_i =$	*1.70E+07 psi*	
Poisson's ratio for outer member: $v_o =$	*0.27*	
Poisson's ratio for inner member: $v_i =$	*0.27*	
Computed Results		
Pressure at mating surface: $p =$	**3,518 psi**	Using Equation (13–3)
Tensile stress in outer member: $\sigma_o =$	**10 846 psi**	Using Equation (13–4)
Compressive stress in inner member: $\sigma_i =$	**−16 025 psi**	Using Equation (13–5)
Increase in diameter of outer member: $\delta_o =$	**0.00 098 in**	Using Equation (13–6)
Decrease in diameter of inner member: $\delta_i =$	**0.00 222 in**	Using Equation (13–7)

FIGURE 13–7   Spreadsheet solution for pressure, stresses, and deformations in cylindrical members mating with a force fit

13–9
GENERAL
TOLERANCING
METHODS

It is the designer's responsibility to establish tolerances on each dimension of each component in a mechanical device. The tolerances must ensure that the component fulfills its function. But it should also be as large as practical to permit economical manufacture. This pair of conflicting principles must be dealt with. See comprehensive texts on technical drawing and interpretation of engineering drawings for general principles (References 4, 5, and 8).

Special attention should be paid to the features of a component that mate with other components and with which they must operate reliably or with which they must be accurately located. The fit of the inner races of the bearings on the shafts is an example of such features. Others in this reducer are the clearances between parts that must be assembled together easily but that must not have large relative motion during operation.

Where no other component mates with certain features of a given component, the tolerances should be as large as practical so that they can be produced with basic machining, molding, or casting processes without the need for subsequent finishing. It is often recommended that blanket tolerances be given for such dimensions and that the precision with which the basic size is stated on the drawing implies a certain tolerance. For decimal dimensions in U.S. Customary units, a note similar to the following is often given:

DIMENSIONS IN inches. TOLERANCES ARE AS FOLLOWS UNLESS OTHERWISE STATED.

$$XX.X = \pm 0.050$$
$$XX.XX = \pm 0.010$$
$$XX.XXX = \pm 0.005$$
$$XX.XXXX = \pm 0.0005$$
$$\text{ANGLES: } \pm 0.50°$$

where X represents a specified digit

For example, if a given dimension has a basic size of 2.5 inches, the dimension can be stated on the drawing in any of four ways with different interpretations:

2.5	means	2.5 ± 0.050 or limits of 2.550 to 2.450 in
2.50	means	2.50 ± 0.010 or limits of 2.510 to 2.490 in
2.500	means	2.500 ± 0.005 or limits of 2.505 to 2.495 in
2.5000	means	2.5000 ± 0.0005 or limits of 2.5005 to 2.4995 in

Any other desired tolerance must be specified on the dimension. Of course, you may select different standard tolerances according to the needs of the system being designed.

Similar data for metric drawings would appear as follows:

DIMENSIONS IN mm. TOLERANCES ARE AS FOLLOWS UNLESS OTHERWISE STATED.

$$XX = \pm 1.0$$
$$XX.X = \pm 0.25$$
$$XX.XX = \pm 0.15$$
$$XX.XXX = \pm 0.012$$
$$\text{ANGLES: } \pm 0.50°$$

Some tolerance notes also relate the degree of precision to the nominal size of the feature, with tighter tolerances on smaller dimensions and looser tolerances on larger dimensions. The international tolerance (IT) grades, discussed in Section 13–3, use this approach.

Geometric Dimensioning and Tolerancing (GD&T).   The geometric dimensioning and tolerancing (GD&T) method is used to control location, form, profile, orientation, and runout on a dimensioned feature. Its purpose is to ensure proper assembly and/or operation of parts, and it is especially useful in quantity production of interchangeable parts. The complete definition of the method is given in the ANSI Standard Y14.5M-1994 (Reference 3). Reference 8 shows some applications and demonstrates the interpretation of the numerous symbols.

Figure 13–8(a) shows some of the more common geometric symbols. Part (b) illustrates their use in a *feature control frame*, containing the symbol for the feature geometric characteristic being controlled, the tolerance on the dimension or form, and the *datum* to which the given feature is to be related. For example, in part (c), the smaller diameter is to be concentric with the larger diameter (datum –A–) within a tolerance of 0.010 in.

Surface Finish.   The designer should also control the surface finish of all features critical to the performance of the device being designed. This includes the mating surfaces which have already been discussed. But also any surface that experiences relatively high stresses, particularly reversed bending, should have a smooth surface.

See Figure 5–8 for a rough indication of the effect of surface finishes between, for example, *machined* and *ground* surfaces on the basic endurance strength of steels. In general *ground* surfaces have an average roughness, R_a, of 16 μin (0.4 μm). Figure 13–3 shows the expected range of surface finish for many kinds of machining processes. Note that turning is reported to be capable of producing that level of surface finish, but it is at the limit of its capability and will likely require very fine finish cuts with sharp-edged tooling having a broad radius nose. The more nominal surface finish from turning, milling, broaching, and boring is 63 μin (1.6 μm). This would correspond to the *machined* category in Figure 5–8.

Bearing seats on shafts for accurate machinery are typically ground, particularly in the smaller sizes under 3.0 in (80 mm), with a maximum allowable average roughness of 16 μin (0.4 μm). Above that size and up to 20 in (500 mm), 32 μin (0.8 μm) is allowed. See manufacturers' catalogs.

13–10 ROBUST PRODUCT DESIGN

Careful control of the tolerances on the dimensions of machine parts is the responsibility of both the product designer and the manufacturing staff. The goal is to ensure the functionality of the product while permitting economical manufacture. Often these goals seem incompatible.

Robust product design is a tool that can help. (See Reference 11 and Internet sites 3 and 6.) It is a technique in which a sequence of experiments is undertaken to determine the variables in a product's design that most affect its performance. Then the optimum limits on those variables are defined. The concepts can be used for dimensions, material properties, control settings, and many other factors.

The design of the experiments used to implement a robust product is critical to the success of the design. An initial screening experiment is used to select the most significant variables. Then additional experiments are run to determine the response of the system to combinations of these variables. Computer-generated plots of results establish limits for the significant variables.

Tolerance	Characteristic	Symbol
Form	Straightness	—
	Flatness	▱
	Circularity	○
	Cylindricity	⌭
Profile	Profile of a line	⌒
	Profile of a surface	⌓
Orientation	Angularity	∠
	Perpendicularity	⊥
	Parallelism	//
Location	Position/symmetry	⊕
	Concentricity	◎
Runout	Circular runout	↗
	Total runout	↗↗

(a) Geometric symbols

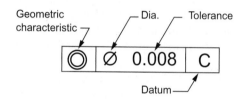

(b) Feature control frame

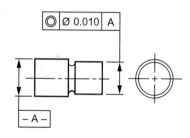

(c) Concentricity

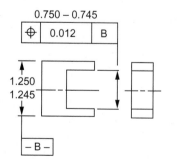

(d) Symmetry

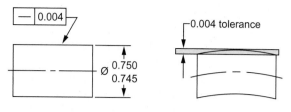

(e) Straightness

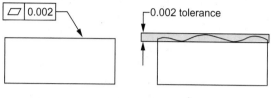

(f) Flatness

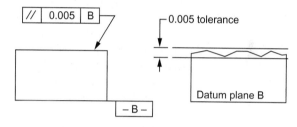

(g) Parallelism: planes

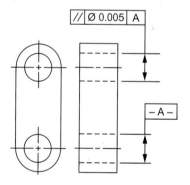

(h) Parallelism: cylinders

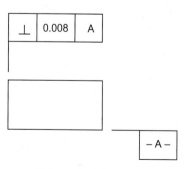

(i) Perpendicularity

FIGURE 13–8 Examples of geometric tolerancing

Robust product design is based on the *Taguchi method*, an important element in the process of improving manufacturing quality developed by Dr. Genichi Taguchi. He showed how design of experiments can be used to produce products of consistently high quality that are insensitive to the operating environment. (See Reference 11.)

Controlling dimensions and tolerances to minimize variation can be accomplished using statistical analysis tools. The designer models the type and magnitude of the expected size variation for a set of mating components. Simulation software predicts the final assembly configuration considering the "tolerance stack" of the individual components. The analysis can be done in either two or three dimensions. See Internet site 5.

The major goal of these methods is the allocation of tolerances on critical dimensions of mating parts that ensure satisfactory operation under all foreseeable conditions of component manufacture, assembly, environment, and use. Maintaining reasonable manufacturing costs is also an important consideration. Many manufacturers practice these methods, particularly in the automotive, aerospace, military, and high-production consumer products markets. Computer-assisted techniques are available to facilitate the process. Most comprehensive, solid modeling, computer-aided design systems include statistically based tolerance analysis as a standard feature. Internet site 7 describes the activities of the Association for the Development of Computer-Aided Tolerancing Systems (ADCATS), a consortium of several industries and Brigham Young University.

REFERENCES

1. The American Society of Mechanical Engineers. ANSI Standard B4.1-1967 (R 1974, Reaffirmed 1999). *Preferred Limits and Fits for Cylindrical Parts*. New York: American Society of Mechanical Engineers, 1999.

2. The American Society of Mechanical Engineers. ANSI Standard B4.2-1978 (R84). *Preferred Metric Limits and Fits*. New York: American Society of Mechanical Engineers, 1984.

3. The American Society of Mechanical Engineers. ANSI Standard Y14.5M (R1999). *Dimensioning and Tolerancing*. New York: American Society of Mechanical Engineers, 1999.

4. Earle, James H. *Graphics for Engineers with AutoCAD 2002*. 6th ed. Upper Saddle River, NJ: Prentice Hall, 2003.

5. Giesecke, F. E., et al. *Technical Drawing*. 12th ed. Upper Saddle River, NJ: Prentice Hall, 2003.

6. Gooldy, Gary. *Geometric Dimensioning & Tolerancing*. Rev. ed. Upper Saddle River, NJ: Prentice Hall, 1995.

7. Griffith, Gary K. *Geometric Dimensioning & Tolerancing: Application and Inspection*. 2d ed. Upper Saddle River, NJ: Prentice Hall, 2002.

8. Jensen, C., and R. Hines. *Interpreting Engineering Drawings*. 6th ed. Albany, NY: Delmar Publishers, 2001.

9. Oberg, E., et al. *Machinery's Handbook*. 26th ed. New York: Industrial Press, 2000.

10. Shigley, J. E., and C. R. Mischke. *Mechanical Engineering Design*. 6th ed. New York: McGraw-Hill, 2001.

11. Taguchi, Genichi, and Shin Taguchi. *Robust Engineering*. New York: McGraw-Hill, 1999.

INTERNET SITES RELATED TO TOLERANCES AND FITS

1. **Engineering Fundamentals.** *www.efunda.com* A comprehensive site offering design information on numerous topics, including geometric dimensioning and tolerancing.

2. **Engineers Edge.** *www.engineersedge.com* A comprehensive site offering design information on numerous topics, including geometric dimensioning and tolerancing.

3. **iSixSigma.** *www.isixsigma.com/me/taguchi/* A free information resource about quality management, including robust design, Taguchi methods, and six sigma quality management.

4. **Engineering Bookstore.** *www.engineeringbookstore.com* A site offering technical books in many fields, including geometric dimensioning and tolerancing.

5. **Dimensional Control Systems, Inc.** *www.3dcs.com* A developer and vendor of software for managing and controlling the dimensions and tolerances of components and assemblies in three dimensions. Variations can be modeled and simulated to ensure that quality objectives for fit, finish, and functionality are met.

6. **DRM Associates.** *www.npd-solutions.com/robust.html* An extensive site on the subject of new product development (NPD), including robust design, design of experiments, Taguchi methods, geometric dimensioning and tolerancing, variability reduction, and tolerance design and analysis for components and assemblies. Select the navigation button for *NPD Body of Knowledge* for a comprehensive index of topics.

7. **ADCATS.** *http://adcats.et.byu.edu* The Association for the Development of Computer-Aided Tolerancing Systems (ADCATS) at Brigham Young University.

8. **Drafting Zone.** *www.draftingzone.com* Information on mechanical drafting standards and practices. Genium Publishing Corporation.

PROBLEMS

Clearance Fits

1. Specify the class of fit, the limits of size, and the limits of clearance for the bore of a slow-moving but heavily loaded conveyor roller that must rotate freely on a stationary shaft under heavy load. The nominal diameter of the shaft is 3.500 in. Use the basic hole system.

2. A sine plate is a measuring device that pivots at its base, allowing the plate to assume different angles that are set with gage blocks for precision. The pin in the pivot has a nominal diameter of 0.5000 in. Specify the class of fit, the limits of size, and the limits of clearance for the pivot. Use the basic hole system.

3. A child's toy wagon has an axle with a nominal diameter of 5/8 in. For the wheel bore and axle, specify the class of fit, the limits of size, and the limits of clearance. Use the basic shaft system.

4. The planet gear of an epicyclic gear train must rotate reliably on its shaft while maintaining accurate position with respect to the mating gears. For the planet gear bore and its shaft, specify the class of fit, the limits of size, and the limits of clearance. Use the basic hole system. The nominal shaft diameter is 0.800 in.

5. The base of a hydraulic cylinder is mounted to the frame of a machine by means of a clevis joint, which allows the cylinder to oscillate during operation. The clevis must provide reliable motion, but some play is acceptable. For the clevis holes and the pin, which have a nominal diameter of 1.25 in, specify the class of fit, the limits of size, and the limits of clearance. Use the basic hole system.

6. A heavy door on a furnace swings upward to permit access to the interior of the furnace. During various modes of operation, the door and its hinge assembly see temperatures of 50°F to 500°F. The nominal diameter of each hinge pin is 4.00 in. For the hinge and its pin, specify the class of fit, the limits of size, and the limits of clearance. Use the basic shaft system.

7. The stage of an industrial microscope pivots to permit the mounting of a variety of shapes. The stage must move with precision and reliability under widely varying temperatures. For the pin mount for the stage, having a nominal diameter of 0.750 in, specify the class of fit, the limits of size, and the limits of clearance. Use the basic hole system.

8. An advertising sign is suspended from a horizontal rod and is allowed to swing under wind forces. The rod is to be commercial bar stock, nominally 1.50 inches in diameter. For the mating hinges on the sign, specify the class of fit, the limits of size, and the limits of clearance. Use the basic shaft system.

9. For any of Problems 1–8, make a diagram of the tolerances, fits, and clearances for your specifications, using a method similar to that in Figure 13–4.

Force Fits

10. A spacer made of AISI 1020 hot-rolled steel is in the form of a hollow cylinder with a nominal inside diameter of 3.25 in and an outside diameter of 4.000 in. It is to be mounted on a solid steel shaft with a heavy force fit. Specify the dimensions for the shaft and sleeve, and compute the stress in the sleeve after installation. Use the basic hole system.

11. A bronze bushing, with an inside diameter of 3.50 in and a nominal outside diameter of 4.00 in, is pressed into a steel sleeve having an outside diameter of 4.50 in. For an FN3 class of fit, specify the dimensions of the parts, the limits of interference, and the stresses created in the bushing and the sleeve. Use the basic hole system.

12. It is proposed to install a steel rod with a nominal diameter of 3.00 inches in a hole of an aluminum cylinder with an outside diameter of 5.00 in, with an FN5 force fit. Would this be satisfactory?

13. The allowable compressive stress in the wall of an aluminum tube is 8500 psi. Its outside diameter is 2.000 in, and the wall thickness is 0.065 in. What is the maximum amount of interference between this tube and a steel sleeve that can be tolerated? The sleeve has an outside diameter of 3.00 in.

14. To what temperature would the spacer of Problem 10 have to be heated so that it would slide over the shaft with a clearance of 0.002 in? The ambient temperature is 75°F.

15. For the bronze bushing and the steel sleeve of Problem 11, the ambient temperature is 75°F. How much would the bronze bushing shrink if placed in a freezer at −20°F? Then to what temperature would the steel sleeve have to be heated to provide a clearance of 0.004 in for assembly on the cold bushing?

16. For the bronze bushing of Problem 11, what would be the final inside diameter after assembly into the sleeve if it started with an inside diameter of 3.5000 in?

14

Rolling Contact Bearings

The Big Picture

You Are the Designer

14–1 Objectives of This Chapter

14–2 Types of Rolling Contact Bearings

14–3 Thrust Bearings

14–4 Mounted Bearings

14–5 Bearing Materials

14–6 Load/Life Relationship

14–7 Bearing Manufacturers' Data

14–8 Design Life

14–9 Bearing Selection: Radial Loads Only

14–10 Bearing Selection: Radial and Thrust Loads Combined

14–11 Mounting of Bearings

14–12 Tapered Roller Bearings

14–13 Practical Considerations in the Application of Bearings

14–14 Importance of Oil Film Thickness in Bearings

14–15 Life Prediction under Varying Loads

Rolling Contact Bearings

Discussion Map

☐ *Bearings* are used to support a load while permitting relative motion between two elements of a machine.

☐ Some bearings use rolling elements, such as spherical balls or cylindrical or tapered rollers. This results in a very low coefficient of friction.

Discover

Look for examples of bearings on machines, cars, trucks, bicycles, and consumer products.

Describe the bearings, including how they are installed and what kinds of forces are exerted on them.

This chapter presents information about such bearings and gives methods for analyzing bearings and for selecting commercially available bearings.

The purpose of a bearing is to support a load while permitting relative motion between two elements of a machine. The term *rolling contact bearings* refers to the wide variety of bearings that use spherical balls or some other type of roller between the stationary and the moving elements. The most common type of bearing supports a rotating shaft, resisting purely radial loads or a combination of radial and axial (thrust) loads. Some bearings are designed to carry only thrust loads. Most bearings are used in applications involving rotation, but some are used in linear motion applications.

The components of a typical rolling contact bearing are the inner race, the outer race, and the rolling elements. Figure 14–1 shows the common single-row, deep-groove ball bearing. Usually the outer race is stationary and is held by the housing of the machine. The inner race is pressed onto the rotating shaft and thus rotates with it. Then the balls roll between the outer and inner races. The load path is from the shaft, to the inner race, to the balls, to the outer race, and finally to the housing. The presence of the balls allows a very smooth, low-friction rotation of the shaft. The typical coefficient of friction for a rolling contact bearing is approximately 0.001 to 0.005. These values reflect only the rolling elements themselves and the means of retaining them in the bearing. The presence of seals, excessive lubricant, or unusual loading increases these values.

Look for consumer products, industrial machinery, or transportation equipment (cars, trucks, bicycles, and so on), and identify any uses of rolling contact bearings. Be sure that the machine is turned off, and then try to gain access to the mechanical drive shafts used to transmit power from a motor or an engine to some moving parts of the machine. Are these shafts supported in ball or roller bearings? Or are they supported in plain surface bearings in which the shaft passes through cylindrical members called *bushings* or *bearings*, typically with lubricants present between the rotating shaft and the stationary bearing? Plain surface bearings are discussed in Chapter 16.

For the shafts that are supported in ball or roller bearings, describe the bearing. How is it mounted on the shaft? How is it mounted in the housing of the machine? Can you identify what kinds of forces are acting on the bearing and in what directions they act? Are the forces directed radially toward the centerline of the shaft? Is there any force that acts parallel to the axis of the shaft? Compare the bearings that you find with the photographs that are in this chapter. Which varieties of bearings have you found? Measure or estimate the physical size of the bearings, particularly the diameter of the bore that is in contact with the shaft, the outside diameter, and the width. Can you see the rolling elements—balls or

FIGURE 14–1
Single-row, deep-
groove ball bearing
(NSK Corporation,
Ann Arbor, MI)

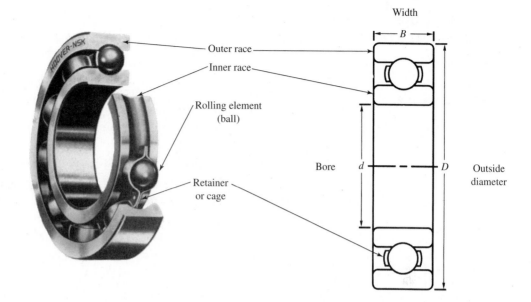

rollers? If so, sketch them and estimate their diameters and/or lengths. Are any of the rolling elements tapered rollers like those shown later in this chapter in Figure 14–7?

When you complete this chapter, you will be able to identify several kinds of rolling contact bearings and specify suitable bearings to carry specified loads. You will also be able to apply such bearings properly, planning for their installation on shafts and in housings.

You Are the Designer

In Chapter 12, you were the designer of a shaft that was rotating at 600 rpm carrying two gears as a part of a power transmission system. Figures 12–1 and 12–2 showed the basic layout that was proposed. The shaft was designed to be supported on two bearings at points B and D. Then, in Example Problem 12–1, we completed the force analysis by computing the forces applied to the shaft by the gears and then computing the *reactions at the bearings*. Figure 12–11 showed the results, summarized here:

$$R_{Bx} = 458 \text{ lb} \quad R_{By} = 4620 \text{ lb}$$
$$R_{Dx} = 1223 \text{ lb} \quad R_{Dy} = 1680 \text{ lb}$$

where *x* refers to the horizontal direction and *y* refers to the vertical direction. All forces on the bearings are in the

radial direction. Your task now is to specify suitable rolling contact bearings for the shaft to withstand those forces and transfer them from the shaft to the housing of the speed reducer.

What kind of bearing should be selected? How do the forces just identified affect the choice? What life expectancy is reasonable for the bearings, and how does that affect the selection of the bearings? What size should be specified? How are the bearings to be installed on the shaft, and how does that affect the detailed design of the shaft? What limit dimensions should be defined for the bearing seats on the shaft? How is the bearing to be located axially on the shaft? How is it to be installed in the housing and located there? How is lubrication to be provided for the bearings? Is there a need for shields and seals to keep contaminants out of the bearings? The information in this chapter will help you make these and other design decisions.

14–1
OBJECTIVES OF
THIS CHAPTER

After completing this chapter, you will be able to:

1. Identify the types of rolling contact bearings that are commercially available, and select the appropriate type for a given application, considering the manner of loading and installation conditions.

2. Use the relationship between forces on bearings and the life expectancy for the bearings to determine critical bearing selection factors.

3. Use manufacturers' data for the performance of ball bearings to specify suitable bearings for a given application.

4. Recommend appropriate values for the design life of bearings.

5. Compute the *equivalent load* on a bearing corresponding to combinations of radial and thrust loads applied to it.

6. Specify mounting details for bearings that affect the design of the shaft onto which the bearing is to be seated and the housing into which it is to be installed.

7. Compute the equivalent loads on tapered roller bearings.

8. Describe the special design of thrust bearings.

9. Describe several types of commercially available mounted bearings and their application to machine design.

10. Understand certain practical considerations involved in the application of bearings, including lubrication, sealing, limiting speeds, bearing tolerance classes, and standards related to the manufacture and application of bearings.

11. Consider the effects of varying loads on the life expectancy and specification of bearings.

14–2
TYPES OF
ROLLING
CONTACT
BEARINGS

Here we will discuss seven different types of rolling contact bearings and the applications in which each is typically used. Many variations on the designs shown are available. As each is discussed, refer to Table 14–1 for a comparison of the performance relative to the others.

Radial loads act toward the center of the bearing along a radius. Such loads are typical of those created by power transmission elements on shafts such as spur gears, V-belt drives, and chain drives. *Thrust loads* are those that act parallel to the axis of the shaft. The axial components of the forces on helical gears, worms and wormgears, and bevel gears are thrust loads. Also, bearings supporting shafts with vertical axes are subjected to thrust loads due to the weight of the shaft and the elements on the shaft as well as from axial operating forces. *Misalignment* refers to the angular deviation of the axis of the shaft at the bearing from the true axis of the bearing itself. An excellent rating for misalignment in Table 14–1 indicates that the bearing can accommodate up to 4.0° of angular deviation. A bearing with

TABLE 14–1 Comparison of bearing types

Bearing type	Radial load capacity	Thrust load capacity	Misalignment capability
Single-row, deep-groove ball	Good	Fair	Fair
Double-row, deep-groove ball	Excellent	Good	Fair
Angular contact	Good	Excellent	Poor
Cylindrical roller	Excellent	Poor	Fair
Needle	Excellent	Poor	Poor
Spherical roller	Excellent	Fair/good	Excellent
Tapered roller	Excellent	Excellent	Poor

a fair rating can withstand up to 0.15°, while a poor rating indicates that rigid shafts with less than 0.05° of misalignment are required. Manufacturers' catalogs should be consulted for specific data. Check Internet sites 1–8.

Single-Row, Deep-Groove Ball Bearing

Sometimes called *Conrad bearings*, the single-row, deep-groove ball bearing (Figure 14–1) is what most people think of when the term *ball bearing* is used. The inner race is typically pressed on the shaft at the bearing seat with a slight interference fit to ensure that it rotates with the shaft. The spherical rolling elements, or balls, roll in a deep groove in both the inner and the outer races. The spacing of the balls is maintained by retainers or "cages." While designed primarily for radial load-carrying capacity, the deep groove allows a fairly sizable thrust load to be carried. The thrust load would be applied to one side of the inner race by a shoulder on the shaft. The load would pass across the side of the groove, through the ball, to the opposite side of the outer race, and then to the housing. The radius of the ball is slightly smaller than the radius of the groove to allow free rolling of the balls. The contact between a ball and the race is theoretically at a point, but it is actually a small circular area because of the deformation of the elements. Because the load is carried on a small area, very high local contact stresses occur. To increase the capacity of a single-row bearing, a bearing with a greater number of balls, or larger balls operating in larger-diameter races, should be used.

Double-Row, Deep-Groove Ball Bearing

Adding a second row of balls (Figure 14–2) increases the radial load-carrying capacity of the deep-groove type of bearing compared with the single-row design because more balls share the load. Thus, a greater load can be carried in the same space, or a given load can be carried in a smaller space. The greater width of double-row bearings often adversely affects the misalignment capability.

Angular Contact Ball Bearing

One side of each race in an angular contact bearing is higher to allow the accommodation of greater thrust loads compared with the standard single-row, deep-groove bearing. The sketch in Figure 14–3 shows the preferred angle of the resultant force (radial and thrust loads combined), with commercially available bearings having angles of 15° to 40°.

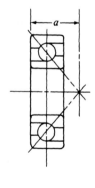

FIGURE14–2   Double-row, deep-groove ball bearing (NSK Corporation, Ann Arbor, MI)

FIGURE 14–3   Angular contact ball bearing (NSK Corporation, Ann Arbor, MI)

Cylindrical Roller Bearing

Replacing the spherical balls with cylindrical rollers (Figure 14–4), with corresponding changes in the design of the races, gives a greater radial load capacity. The pattern of contact between a roller and its race is theoretically a line, and it becomes a rectangular shape as the members deform under load. The resulting contact stress levels are lower than for equivalent-sized ball bearings, allowing smaller bearings to carry a given load or a given-size bearing to carry a higher load. Thrust load capacity is poor because any thrust load would be applied to the side of the rollers, causing rubbing, not true rolling motion. It is recommended that *no* thrust load be applied. Roller bearings are often fairly wide, giving them only fair ability to accommodate angular misalignment.

Needle Bearing

Needle bearings (Figure 14–5) are actually roller bearings, but they have much smaller-diameter rollers, as you can see by comparing Figures 14–4 and 14–5. A smaller radial space is typically required for needle bearings to carry a given load than for any other type of rolling contact bearing. This makes it easier to design them into many types of equipment and components such as pumps, universal joints, precision instruments, and household appliances. The cam follower shown in Figure 14–5(b) is another example in which the antifriction operation of needle bearings can be built-in with little radial space required. As with other roller bearings, thrust and misalignment capabilities are poor.

Spherical Roller Bearing

The spherical roller bearing (Figure 14–6) is one form of *self-aligning bearing,* so called because there is actual relative rotation of the outer race relative to the rollers and the inner race when angular misalignments occur. This gives the excellent rating for misalignment capability while retaining virtually the same ratings on radial load capacity.

FIGURE 14–4
Cylindrical roller bearing (NSK Corporation, Ann Arbor, MI)

FIGURE 14–5
Needle bearings
(McGill Manufacturing
Co., Inc., Bearing
Division,
Valparaiso, IN)

(a) Single- and double-row needle bearings

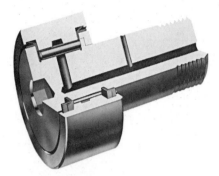

(b) Needle bearings adapted to cam followers

FIGURE 14–6
Spherical roller bearing
(NSK Corporation,
Ann Arbor, MI)

FIGURE 14–7
Tapered roller bearing
(NSK Corporation,
Ann Arbor, MI)

Tapered Roller Bearing

Tapered roller bearings (Figure 14–7) are designed to take substantial thrust loads along with high radial loads, resulting in excellent ratings on both. They are often used in wheel bearings for vehicles and mobile equipment and in heavy-duty machinery having inherently high thrust loads. Section 14–12 gives additional information about their application. Figures 8–25, 9–36, 10–1, and 10–2 show tapered roller bearings applied in gear-type speed reducers.

FIGURE 14–8
Thrust bearings
(Andrews Bearing
Corp., Spartanburg,
SC)

(a) Example of ball thrust bearings

(c) Examples of roller thrust bearings

(b) Typical cross section of ball thrust bearing

Standard cylindrical roller-thrust bearing

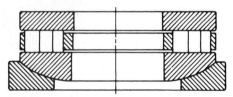

Self-aligning thrust bearing

**14–3
THRUST
BEARINGS**

The bearings discussed so far in this chapter have been designed to carry radial loads or a combination of radial and thrust loads. Many machine design projects demand a bearing that resists only thrust loads, and several types of standard thrust bearings are commercially available. The same types of rolling elements are used: spherical balls, cylindrical rollers, and tapered rollers (see Figure 14–8).

Most thrust bearings can take little or no radial load. Then the design and the selection of such bearings are dependent only on the magnitude of the thrust load and the design life. The data for basic dynamic load rating and basic static load rating are reported in manufacturers' catalogs in the same way as they are for radial bearings.

**14–4
MOUNTED
BEARINGS**

In many types of heavy machines and special machines produced in small quantities, mounted bearings rather than unmounted bearings are selected. The mounted bearings provide a means of attaching the bearing unit directly to the frame of the machine with bolts rather than inserting it into a machined recess in a housing as is required in unmounted bearings.

Figure 14–9 shows the most common configuration for a mounted bearing: the *pillow block*. The housing is made from formed steel, cast iron, or cast steel, with holes or slots provided for attachment during assembly of the machine, at which time alignment of the bearing unit is adjusted. The bearings themselves can be of virtually any of the types discussed in the preceding sections; ball, tapered roller, or spherical roller is preferred. Misalignment capability is an important application consideration because of the conditions of use of such bearings. This capability is provided either in the bearing construction itself or in the housing.

FIGURE 14–9 Ball bearing pillow block (Rockwell Automation/Dodge)

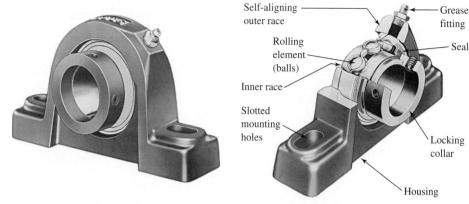

Self-aligning outer race

Rolling element (balls)

Inner race

Slotted mounting holes

Grease fitting

Seal

Locking collar

Housing

FIGURE 14–10 Forms of mounted bearings (Rockwell Automation/Dodge)

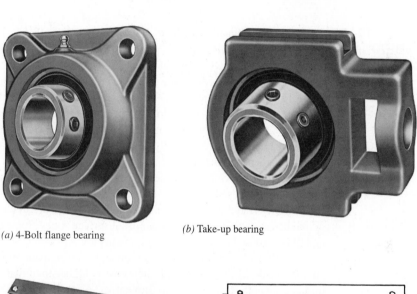

(a) 4-Bolt flange bearing

(b) Take-up bearing

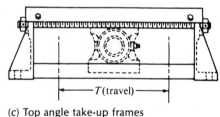

T (travel)

(c) Top angle take-up frames

Because the bearing itself is similar to those already discussed, the selection process is also similar. Most catalogs provide extensive charts of data listing the load-carrying capacity at specified rated life values. Internet site 7 includes an example.

Other forms of mounted bearings are shown in Figure 14–10. The *flange units* are designed to be mounted on the vertical side frames of machines, holding horizontal shafts. Again, several bearing types and sizes are available. The term *take-up unit* refers to a bearing mounted in a housing, which in turn is mounted in a frame that allows movement of the bearing with the shaft in place. Used on conveyors, chain drives, belt drives, and similar applications, the take-up unit permits adjustment of the center distance of the drive components at the time of installation and during operation to accommodate wear or stretch of parts of the assembly.

TABLE 14–2   Comparison of bearing materials

	Material			
	Silicon nitride	52100 steel	440C stainless steel	M50 steel
Room-temperature hardness, HRC	78	62	60	64
Room-temperature elastic modulus	45×10^6 psi 310 GPa	30×10^6 psi 207 GPa	29×10^6 psi 200 GPa	28×10^6 psi 193 GPa
Maximum operating temperature	2200°F 1200°C	360°F 180°C	500°F 260°C	600°F 320°C
Density, kg/m^3	3200	7800	7800	7600

**14–5
BEARING
MATERIALS**

The load on a rolling contact bearing is exerted on a small area. The resulting contact stresses are quite high, regardless of the type of bearing. Contact stresses of approximately 300 000 psi are not uncommon in commercially available bearings. To withstand such high stresses, the balls, rollers, and races are made from a very hard, high-strength steel or ceramic.

The most widely used bearing material is AISI 52100 steel, which has a very high carbon content, 0.95% to 1.10%, along with 1.30% to 1.60% chromium, 0.25% to 0.45% manganese, 0.20% to 0.35% silicon, and other alloying elements with low, but controlled, amounts. Impurities are carefully minimized to obtain a very clean steel. The material is through-hardened to the range of 58–65 on the Rockwell C scale to give it the ability to resist high contact stress. Some tool steels, particularly M1 and M50, are also used. Case hardening by carburizing is used with such steels as AISI 3310, 4620, and 8620 to achieve the high surface hardness required while retaining a tough core. Careful control of the case depth is required because critical stresses occur in subsurface zones. Some more lightly loaded bearings and those exposed to corrosive environments use AISI 440C stainless steel elements.

Rolling elements and other components can be made from ceramic materials such as silicon nitride (Si_3N_4). Although their cost is higher than that of steel, ceramics offer significant advantages as shown in Table 14–2. (See Reference 6.) Their light weight, high strength, and high temperature capability make them desirable for aerospace, engine, military, and other demanding applications.

**14–6
LOAD/LIFE
RELATIONSHIP**

Despite using steels with very high strength, all bearings have a finite life and will eventually fail due to fatigue because of the high contact stresses. But, obviously, the lighter the load, the longer the life, and vice versa. The relationship between load, P, and life, L, for rolling contact bearings can be stated as

**Relationship
between Bearing
Load and Life**

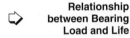

$$\frac{L_2}{L_1} = \left(\frac{P_1}{P_2}\right)^k \qquad (14\text{–}1)$$

where $k = 3.00$ for ball bearings
$k = 3.33$ for roller bearings

**14–7
BEARING
MANUFACTURERS'
DATA**

The selection of a rolling contact bearing from a manufacturer's catalog involves considerations of load-carrying capacity and the geometry of the bearing. Table 14–3 shows a portion of the data from a catalog for two sizes of single-row, deep-groove ball bearings.

TABLE 14–3 Bearing selection data for single-row, deep-groove, Conrad-type ball bearings

A. Series 6200

Bearing number	Nominal bearing dimensions						r^*	Preferred shoulder diameter		Bearing weight	Basic static load rating, C_o	Basic dynamic load rating, C
	d		D		B			Shaft	Housing			
	mm	in	mm	in	mm	in	in	in	in	lb	lb	lb
6200	10	0.3937	30	1.1811	9	0.3543	0.024	0.500	0.984	0.07	520	885
6201	12	0.4724	32	1.2598	10	0.3937	0.024	0.578	1.063	0.08	675	1180
6202	15	0.5906	35	1.3780	11	0.4331	0.024	0.703	1.181	0.10	790	1320
6203	17	0.6693	40	1.5748	12	0.4724	0.024	0.787	1.380	0.14	1010	1660
6204	20	0.7874	47	1.8504	14	0.5512	0.039	0.969	1.614	0.23	1400	2210
6205	25	0.9843	52	2.0472	15	0.5906	0.039	1.172	1.811	0.29	1610	2430
6206	30	1.1811	62	2.4409	16	0.6299	0.039	1.406	2.205	0.44	2320	3350
6207	35	1.3780	72	2.8346	17	0.6693	0.039	1.614	2.559	0.64	3150	4450
6208	40	1.5748	80	3.1496	18	0.7087	0.039	1.811	2.874	0.82	3650	5050
6209	45	1.7717	85	3.3465	19	0.7480	0.039	2.008	3.071	0.89	4150	5650
6210	50	1.9685	90	3.5433	20	0.7874	0.039	2.205	3.268	1.02	4650	6050
6211	55	2.1654	100	3.9370	21	0.8268	0.059	2.441	3.602	1.36	5850	7500
6212	60	2.3622	110	4.3307	22	0.8661	0.059	2.717	3.996	1.73	7250	9050
6213	65	2.5591	120	4.7244	23	0.9055	0.059	2.913	4.390	2.18	8000	9900
6214	70	2.7559	125	4.9213	24	0.9449	0.059	3.110	4.587	2.31	8800	10 800
6215	75	2.9528	130	5.1181	25	0.9843	0.059	3.307	4.783	2.64	9700	11 400
6216	80	3.1496	140	5.5118	26	1.0236	0.079	3.504	5.118	3.09	10 500	12 600
6217	85	3.3465	150	5.9055	28	1.1024	0.079	3.740	5.512	3.97	12 300	14 600
6218	90	3.5433	160	6.2992	30	1.1811	0.079	3.937	5.906	4.74	14 200	16 600
6219	95	3.7402	170	6.6929	32	1.2598	0.079	4.213	6.220	5.73	16 300	18 800
6220	100	3.9370	180	7.0866	34	1.3386	0.079	4.409	6.614	6.94	18 600	21 100
6221	105	4.1339	190	7.4803	36	1.4173	0.079	4.606	7.008	8.15	20 900	23 000
6222	110	4.3307	200	7.8740	38	1.4961	0.079	4.803	7.402	9.59	23 400	24 900
6224	120	4.7244	215	8.4646	40	1.5748	0.079	5.197	7.992	11.4	26 200	26 900

TABLE 14–3 (continued)

A. *Series 6200, continued*

Bearing number	Nominal bearing dimensions								Preferred shoulder diameter		Bearing weight	Basic static load rating, C_o	Basic dynamic load rating, C
	d		D		B		r*		Shaft	Housing			
	mm	in	mm	in	mm	in	in		in	in	lb	lb	lb
6226	130	5.1181	230	9.0551	40	1.5748	0.098		5.669	8.504	12.7	29 100	28 700
6228	140	5.5118	250	9.8425	42	1.6535	0.098		6.063	9.291	19.6	29 300	28 700
6230	150	5.9055	270	10.6299	45	1.7717	0.098		6.457	10.079	25.3	32 500	30 000
6232	160	6.2992	290	11.4173	48	1.8898	0.098		6.850	10.886	32.0	35 500	32 000
6234	170	6.6929	310	12.2047	52	2.0472	0.118		7.362	11.535	38.5	43 000	36 500
6236	180	7.0866	320	12.5984	52	2.0472	0.118		7.758	11.929	41.0	46 500	39 000
6238	190	7.4803	340	13.3858	55	2.1654	0.118		8.150	12.717	50.5	54 500	44 000
6240	200	7.8740	360	14.1732	58	2.2835	0.118		8.543	13.504	61.5	60 000	46 500

B. *Series 6300*

Bearing number	Nominal bearing dimensions								Preferred shoulder diameter		Bearing weight	Basic static load rating, C_o	Basic dynamic load rating, C
	d		D		B		r*		Shaft	Housing			
	mm	in	mm	in	mm	in	in		in	in	lb	lb	lb
6300	10	0.3937	35	1.3780	11	0.4331	0.024		0.563	1.181	0.12	805	1400
6301	12	0.4724	37	1.4567	12	0.4724	0.039		0.656	1.220	0.13	990	1680
6302	15	0.5906	42	1.6535	13	0.5118	0.039		0.781	1.417	0.18	1200	1980
6303	17	0.6693	47	1.8504	14	0.5512	0.039		0.875	1.614	0.25	1460	2360
6304	20	0.7874	52	2.0472	15	0.5906	0.039		1.016	1.772	0.32	1730	2760
6305	25	0.9843	62	2.4409	17	0.6693	0.039		1.220	2.165	0.52	2370	3550
6306	30	1.1811	72	2.8346	19	0.7480	0.039		1.469	2.559	0.76	3150	4600
6307	35	1.3780	80	3.1496	21	0.8268	0.059		1.688	2.795	1.01	4050	5800
6308	40	1.5748	90	3.5433	23	0.9055	0.059		1.929	3.189	1.40	5050	7050
6309	45	1.7717	100	3.9370	25	0.9843	0.059		2.126	3.583	1.84	6800	9150
6310	50	1.9685	110	4.3307	27	1.0630	0.079		2.362	3.937	2.42	8100	10 700
6311	55	2.1654	120	4.7244	29	1.1417	0.079		2.559	4.331	2.98	9450	12 300
6312	60	2.3622	130	5.1181	31	1.2205	0.079		2.835	4.646	3.75	11 000	14 100
6313	65	2.5591	140	5.5118	33	1.2992	0.079		3.031	5.039	4.63	12 600	16 000
6314	70	2.7559	150	5.9055	35	1.3780	0.079		3.228	5.433	5.51	14 400	18 000
6315	75	2.9528	160	6.2992	37	1.4567	0.079		3.425	5.827	6.61	16 300	19 600

TABLE 14–3 *(continued)*

B. *Series 6300,* continued

Bearing number	Nominal bearing dimensions								Preferred shoulder diameter		Bearing weight	Basic static load rating, C_o	Basic dynamic load rating, C
	d		D		B		r^*		Shaft	Housing			
	mm	in	mm	in	mm	in	in		in	in	lb	lb	lb
6316	80	3.1496	170	6.6929	39	1.5354	0.079		3.622	6.220	7.93	18 300	21 300
6317	85	3.3465	180	7.0866	41	1.6142	0.098		3.898	6.535	9.37	20 400	22 900
6318	90	3.5433	190	7.4803	43	1.6929	0.098		4.094	6.929	10.8	22 500	24 700
6319	95	3.7402	200	7.8740	45	1.7717	0.098		4.291	7.323	12.5	24 900	26 400
6320	100	3.9370	215	8.4646	47	1.8504	0.098		4.488	7.913	15.3	29 800	30 000
6321	105	4.1339	225	8.8583	49	1.9291	0.098		4.685	8.307	17.9	32 500	31 700
6322	110	4.3307	240	9.4488	50	1.9685	0.098		4.882	8.898	21.0	38 000	35 500
6324	120	4.7244	260	10.2362	55	2.1654	0.098		5.276	9.685	27.6	38 500	36 000
6326	130	5.1181	280	11.0236	58	2.2835	0.118		5.827	10.315	40.8	44 500	39 500
6328	140	5.5118	300	11.8110	62	2.4409	0.118		6.220	11.102	48.5	51 000	43 500
6330	150	5.9055	320	12.5984	65	2.5591	0.118		6.614	11.890	57.3	58 000	47 500
6332	160	6.2992	340	13.3858	68	2.6772	0.118		7.008	12.677	58	58 500	48 000
6334	170	6.6929	360	14.1732	72	2.8346	0.118		7.402	13.465	84	73 500	56 500
6336	180	7.0866	380	14.9606	75	2.9528	0.118		7.795	14.252	98	84 000	61 500
6338	190	7.4803	400	15.7480	78	3.0709	0.157		8.346	14.882	112	84 000	61 500
6340	200	7.8740	420	16.5354	80	3.1496	0.157		8.740	15.669	127	91 500	65 500

Source: NSK Corporation, Ann Arbor, MI.

*Maximum fillet that corner radius will clear.

FIGURE 14–11
Relative sizes of
bearing series

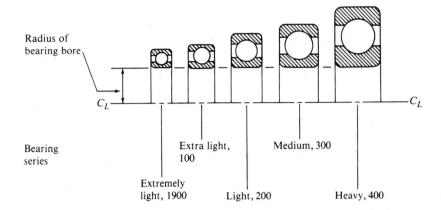

Standard bearings are available in several classes, typically extra-light, light, medium, and heavy classes. The designs differ in the size and number of load-carrying elements (balls or rollers) in the bearing. The bearing number usually indicates the class and the size of the bore of the bearing. Most bearings are manufactured with nominal dimensions in metric units, and the last two digits of the bearing number indicate the nominal bore size. The bore-size convention can be seen from the data in Table 14–3. Note that for bore sizes 04 and above, the nominal bore dimension in millimeters is five times the last two digits in the bearing number.

The number preceding the last two digits indicates the class. For example, several manufacturers use series 100 to indicate extra-light, 200 for light, 300 for medium, and 400 for heavy-duty classes. The three digits may be preceded by others to indicate a special design code of the manufacturer, as is the case in Table 14–3. Figure 14–11 shows the relative size of the different classes of bearings.

Inch-type bearings are available with bores ranging from 0.125 through 15.000 in.

Considering load-carrying capacity first, the data reported for each bearing design will include a basic dynamic load rating, C, and a basic static load rating, C_o.

The *basic static load rating* is the load that the bearing can withstand without permanent deformation of any component. If this load is exceeded, the most probable result would be the indentation of one of the bearing races by the rolling elements. The deformation would be similar to that produced in the Brinell hardness test, and the failure is sometimes referred to as *brinelling*. Operation of the bearing after brinelling would be very noisy, and the impact loads on the indented area would produce rapid wear and progressive failure of the bearing.

To understand the basic dynamic load rating, it is necessary first to discuss the concept of the rated life of a bearing. Fatigue occurs over a large number of cycles of loading; for a bearing, that would be a large number of revolutions. Also, fatigue is a statistical phenomenon with considerable spread of the actual life of a group of bearings of a given design. The rated life is the standard means of reporting the results of many tests of bearings of a given design. It represents the life that 90% of the bearings would achieve successfully at a rated load. Note that it also represents the life that 10% of the bearings would not achieve. The rated life is thus typically referred to as the L_{10} *life* at the rated load.

Now the *basic dynamic load* rating can be defined as that load to which the bearings can be subjected while achieving a rated life (L_{10}) of 1 million revolutions (rev). Thus, the manufacturer supplies you with one set of data relating load and life. Equation (14–1) can be used to compute the expected life at any other load.

You should be aware that different manufacturers use other bases for the rated life. For example, some use 90 million cycles as the rated life and determine the rated load for that life. Also, some will report *average life,* for which 50% of the bearings will not survive. Thus, average life can be called the L_{50} life, not the L_{10}. Note that the average life is approximately five times as long as the L_{10} life. (See Internet site 2.) Be sure that you understand the basis for the ratings in any given catalog. (See References 2, 5, 7, 8, and 11 for additional analysis of roller bearing performance.)

Example Problem 14–1   A catalog lists the basic dynamic load rating for a ball bearing to be 7050 lb for a rated life of 1 million rev. What would be the expected L_{10} life of the bearing if it were subjected to a load of 3500 lb?

Solution   In Equation (14–1),

$$P_1 = C = 7050 \text{ lb} \qquad \text{(basic dynamic load rating)}$$
$$P_2 = P_d = 3500 \text{ lb} \qquad \text{(design load)}$$
$$L_1 = 10^6 \text{ rev} \qquad (L_{10} \text{ life at load } C)$$
$$k = 3 \qquad \text{(ball bearing)}$$

Then letting the life, L_2, be called the *design life*, L_d, at the design load,

$$L_2 = L_d = L_1 \left(\frac{P_1}{P_2}\right)^k = 10^6 \left(\frac{7050}{3500}\right)^{3.00} = 8.17 \times 10^6 \text{ rev}$$

This must be interpreted as the L_{10} life at a load of 3500 lb.

**14–8
DESIGN LIFE**   We will use the method developed in Example Problem 14–1 to refine the procedure for computing the required basic dynamic load rating C for a given design load P_d and a given design life L_d. If the reported load data in the manufacturer's literature is for 10^6 revolutions, Equation (14–1) can be written as

 **Design Life** $$L_d = (C/P_d)^k \,(10^6) \qquad\qquad \textbf{(14–2)}$$

The required C for a given design load and life would be

**Basic Dynamic
Load Rating** $$C = P_d \,(L_d/10^6)^{1/k} \qquad\qquad \textbf{(14–3)}$$

Most people do not think in terms of the number of revolutions that a shaft makes. Rather, they consider the speed of rotation of the shaft, usually in rpm, and the design life of the machine, usually in hours of operation. The design life is specified by the designer, considering the application. As a guide, Table 14–4 can be used. Now, for a specified design life in hours, and a known speed of rotation in rpm, the number of design revolutions for the bearing would be

$$L_d = (\text{h})(\text{rpm})(60 \text{ min/h})$$

Example Problem 14–2   Compute the required basic dynamic load rating, C, for a ball bearing to carry a radial load of 650 lb from a shaft rotating at 600 rpm that is part of an assembly conveyor in a manufacturing plant.

TABLE 14–4   Recommended design life for bearings

Application	Design life L_{10}, h
Domestic appliances	1000–2000
Aircraft engines	1000–4000
Automotive	1500–5000
Agricultural equipment	3000–6000
Elevators, industrial fans, multipurpose gearing	8000–15 000
Electric motors, industrial blowers, general industrial machines	20 000–30 000
Pumps and compressors	40 000–60 000
Critical equipment in continuous, 24-h operation	100 000–200 000

Source: Eugene A. Avallone and Theodore Baumeister III, eds., *Marks' Standard Handbook for Mechanical Engineers,* 9th ed. New York: McGraw-Hill, 1986.

Solution   From Table 14–4, let's select a design life of 30 000 h. Then L_d is

$$L_d = (30\ 000\ \text{h})(600\ \text{rpm})(60\ \text{min/h}) = 1.08 \times 10^9\ \text{rev}$$

From Equation (14–3),

$$C = 650(1.08 \times 10^9/10^6)^{1/3} = 6670\ \text{lb}$$

To facilitate calculations, some manufacturers provide charts or tables of life factors and speed factors that make computing the number of revolutions unnecessary. Note that the rated life of 1 million rev would be achieved by a shaft rotating at $33\frac{1}{3}$ rpm for 500 h. If the actual speed or desired life is different from these two values, a speed factor, f_N, and a life factor, f_L, can be determined from charts such as those shown in Figure 14–12. The factors account for the load/life relationship of Equation (14–1). The required basic dynamic load rating, C, for a bearing to carry a design load, P_d, would then be

 Required Basic Dynamic Load Rating

$$C = P_d f_L / f_N \tag{14–4}$$

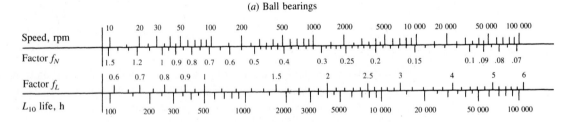

(a) Ball bearings

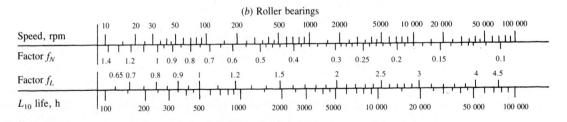

(b) Roller bearings

FIGURE 14–12   Life and speed factors for ball and roller bearings

Other catalogs use different approaches, but they are all based on the load/life relationship of Equation (14–1).

If we solved Example Problem 14–2 by using the charts of Figure 14–12, the following would result:

$$f_N = 0.381 \quad \text{(for 600 rpm)}$$
$$f_L = 3.90 \quad \text{(for 30 000-h life)}$$
$$C = 650(3.90)/(0.381) = 6654 \text{ lb}$$

This compares closely with the value of 6670 lb found previously.

**14–9
BEARING
SELECTION:
RADIAL LOADS
ONLY**

The selection of a bearing takes into consideration the load capacity, as discussed, and the geometry of the bearing that will ensure that it can be installed conveniently in the machine. We will first consider unmounted bearings carrying radial loads only. Then we will consider unmounted bearings carrying a combination of radial and thrust loads. The term *unmounted* refers to the case in which the designer must provide for the proper application of the bearing onto the shaft and into a housing.

The bearing is normally selected after the shaft design has progressed to the point where the required minimum diameter of the shaft has been determined, using the techniques presented in Chapter 12. The radial loads are also known, along with the orientation of the bearings with respect to other elements in the machine.

**Procedure for
Selecting a
Bearing—Radial
Load Only**

⇨ **Equivalent Load,
Radial Load Only**

1. Specify the design load on the bearing, usually called *equivalent load*. The method of determining the equivalent load when only a radial load, *R*, is applied takes into account whether the inner or the outer race rotates.

$$P = VR \qquad\qquad \text{(14–5)}$$

   The factor *V* is called a *rotation factor* and takes the value of 1.0 if the inner race of the bearing rotates, the usual case. Use $V = 1.2$ if the outer race rotates.

2. Determine the minimum acceptable diameter of the shaft that will limit the bore size of the bearing.

3. Select the type of bearing, using Table 14–1 as a guide.

4. Specify the design life of the bearing, using Table 14–4.

5. Determine the speed factor and the life factor if such tables are available for the selected type of bearing. We will use Figure 14–12.

6. Compute the required basic dynamic load rating, *C*, from Equation (14–1), (14–3), or (14–4).

7. Identify a set of candidate bearings that have the required basic dynamic load rating.

8. Select the bearing having the most convenient geometry, also considering its cost and availability.

9. Determine mounting conditions, such as shaft seat diameter and tolerance, housing bore diameter and tolerance, means of locating the bearing axially, and special needs such as seals or shields.

Example Problem 14–3   Select a single-row, deep-groove ball bearing to carry 650 lb of pure radial load from a shaft that rotates at 600 rpm. The design life is to be 30 000 h. The bearing is to be mounted on a shaft with a minimum acceptable diameter of 1.48 in.

Solution   Note that this is a pure radial load and the inner race is to be pressed onto the shaft and rotate with it. Therefore, the factor $V = 1.0$ in Equation (14–5), and the design load is equal to the radial load. These are the same data used in Example Problem 14–2, where we found the required basic dynamic load rating, C, to be 6670 lb. From Table 14–3, giving design data for two classes of bearings, we find that we could use a bearing 6211 or a bearing 6308. Either has a rated C of just over 6670 lb. But note that the 6211 has a bore of 55 mm (2.1654 in), and the 6308 has a bore of 40 mm (1.5748 in). The 6308 is more nearly in line with the desired shaft size.

Summary of Data for the Selected Bearing

       Bearing number: 6308, single-row, deep-groove ball bearing

       Bore: $d = 40$ mm (1.5748 in)

       Outside diameter: $D = 90$ mm (3.5433 in)

       Width: $B = 23$ mm (0.9055 in)

       Maximum fillet radius: $r = 0.059$ in

       Basic dynamic load rating: $C = 7050$ lb

14–10 BEARING SELECTION: RADIAL AND THRUST LOADS COMBINED

When both radial and thrust loads are exerted on a bearing, the equivalent load is the constant radial load that would produce the same rated life for the bearing as the combined loading. The method of computing the equivalent load, P, for such cases is presented in the manufacturer's catalog and takes the form

$$P = VXR + YT \tag{14–6}$$

where P = equivalent load
    V = rotation factor (as defined)
    R = applied radial load
    T = applied thrust load
    X = radial factor
    Y = thrust factor

 Equivalent Load with Radial and Thrust Loads

The values of X and Y vary with the specific design of the bearing and with the magnitude of the thrust load relative to the radial load. For relatively small thrust loads, $X = 1$ and $Y = 0$, so the equivalent load equation reverts to the form of Equation (14–5) for pure radial loads. To indicate the limiting thrust load for which this is the case, manufacturers list a factor called e. If the ratio $T/R > e$, Equation (14–6) must be used to compute P. If $T/R < e$, Equation (14–5) is used. Table 14–5 shows one set of data for a single-row, deep-groove ball bearing. Note that

TABLE 14–5   Radial and thrust factors for single-row, deep-groove ball bearings

e	T/C_o	Y	e	T/C_o	Y
0.19	0.014	2.30	0.34	0.170	1.31
0.22	0.028	1.99	0.38	0.280	1.15
0.26	0.056	1.71	0.42	0.420	1.04
0.28	0.084	1.55	0.44	0.560	1.00
0.30	0.110	1.45			

Note: $X = 0.56$ for all values of Y.

both e and Y depend on the ratio T/C_o, where C_o is the static load rating of a particular bearing. This presents a difficulty in bearing selection because the value of C_o is not known until the bearing has been selected. Therefore, a simple trial-and-error method is applied. If a significant thrust load is applied to a bearing along with a radial load, perform the following steps:

Procedure for Selecting a Bearing—Radial and Thrust Load

1. Assume a value of Y from Table 14–5. The value $Y = 1.50$ is reasonable, being at about the middle of the range of possible values.

2. Compute $P = VXR + YT$.

3. Compute the required basic dynamic load rating C from Equation (14–1), (14–3), or (14–4).

4. Select a candidate bearing having a value of C at least equal to the required value.

5. For the selected bearing, determine C_o.

6. Compute T/C_o.

7. From Table 14–5, determine e.

8. If $T/R > e$, then determine Y from Table 14–5.

9. If the new value of Y is different from that assumed in Step 1, repeat the process.

10. If $T/R < e$, use Equation (14–5) to compute P, and proceed as for a pure radial load.

Example Problem 14–4

Select a single-row, deep-groove ball bearing from Table 14–3 to carry a radial load of 1850 lb and a thrust load of 675 lb. The shaft is to rotate at 1150 rpm, and a design life of 20 000 h is desired. The minimum acceptable diameter for the shaft is 3.10 in.

Solution

Use the procedure outlined above.

Step 1. Assume $Y = 1.50$.

Step 2. $P = VXR + YT = (1.0)(0.56)(1850) + (1.50)(675) = 2049$ lb.

Step 3. From Figure 14–12, the speed factor $f_N = 0.30$, and the life factor $f_L = 3.41$. Then the required basic dynamic load rating C is
$$C = Pf_L/f_N = 2049(3.41)/(0.30) = 23\ 300\ \text{lb}$$

Step 4. From Table 14–3, we could use either bearing number 6222 or 6318. The 6318 has a bore of 3.5433 in and is well suited to this application.

Step 5. For bearing number 6318, $C_o = 22\ 500$ lb.

Step 6. $T/C_o = 675/22\ 500 = 0.03$.

Step 7. From Table 14–5, $e = 0.22$ (approximately).

Step 8. $T/R = 675/1850 = 0.36$. Because $T/R > e$, we can find $Y = 1.97$ from Table 14–5 by interpolation based on $T/C_o = 0.03$.

Step 9. Recompute $P = (1.0)(0.56)(1850) + (1.97)(675) = 2366$ lb:
$$C = 2366(3.41)/(0.30) = 26\ 900\ \text{lb}$$

The bearing number 6318 is not satisfactory at this load. Let's choose bearing number 6320 and repeat the process from Step 5.

Step 5. $C_o = 29\ 800$ lb.

Step 6. $T/C_o = 675/29\ 800 = 0.023$.

Step 7. $e = 0.20$.

Step 8. $T/R > e$. Then $Y = 2.10$ using $T/C_o = 0.023$.

Step 9. $P = (1.0)(0.56)(1850) + (2.10)(675) = 2454$ lb. Thus,
$$C = 2454(3.41)/(0.30) = 27\ 900 \text{ lb}$$

Because bearing number 6320 has a value of $C = 30\ 000$ lb, it is satisfactory.

Adjustment of Life Rating for Reliability

Thus far we have used the basic L_{10} life for selecting rolling contact bearings. This is the general industrial practice and the basis for data published by most bearing manufacturers. Recall that L_{10} life indicates a 90% probability that the selected bearing would carry its rated dynamic load for the specified number of design hours. That leaves a 10% probability that any given bearing would have a lower life.

Certain applications call for greater reliability. Examples can be found in the aerospace, military, instrumentation, and medical fields. It is then desirable to be able to adjust the expected life of a bearing for higher reliability. The following equation provides a method.

$$L_{aR} = C_R L_{10} \tag{14–7}$$

where

L_{10} = Life in millions of revolutions for 90% reliability
L_{aR} = Life adjusted for reliability
C_R = Adjustment factor for reliability

Table 14–6 gives values of C_R for reliabilities between 90% and 99%.
It should be noted that one result of designing for higher reliability is that the bearings would be larger and more expensive.

14–11
MOUNTING OF
BEARINGS

Up to this point, we have considered the load-carrying capacity of the bearings and the bore size in selecting a bearing for a given application. Although these are the most critical parameters, the successful application of a bearing must consider its proper mounting. Bearings are precision machine elements. Great care must be exercised in their handling, mounting, installation, and lubrication.

The primary considerations in the mounting of a bearing are as follows:

- The shaft seat diameter and tolerances
- The housing internal bore and tolerances
- The shaft shoulder diameter against which the inner race of the bearing will be located

Table 14–6   Life adjustment factors for reliability, C_R

Reliability (%)	C_R	Life designation
90	1.0	L_{10}
95	0.62	L_5
96	0.53	L_4
97	0.44	L_3
98	0.33	L_2
99	0.21	L_1

- The housing shoulder diameter provided for locating the outer race
- The radius of the fillets at the base of the shaft and housing shoulders
- The means of retaining the bearing in position

In a typical installation, the bore of the bearing makes a light interference fit on the shaft, and the outside diameter of the outer race makes a close clearance fit in the housing bore. To ensure proper operation and life, the mounting dimensions must be controlled to a total tolerance of only *a few ten-thousandths of an inch*. Most catalogs specify the limit dimensions for both the shaft seat diameter and the housing bore diameter.

Likewise, the catalog will specify the desirable shoulder diameters for the shaft and the housing that will provide a secure surface against which to locate the bearing while ensuring that the shaft shoulder contacts only the inner race and the housing shoulder contacts only the outer race. Table 14–3 includes these values.

The fillet radius specified in the catalog (see *r* in Table 14–3) is the maximum permissible radius *on the shaft and in the housing* that will clear the external radius on the bearing races. Using too large a radius would not permit the bearing to seat tightly against the shoulder. Of course, the actual fillet radius should be made as large as possible up to the maximum to minimize the stress concentration at the shoulder.

Bearings can be retained in the axial direction by many of the means described in Chapter 11. Three popular methods are retaining rings, end caps, and locknuts. Figure 14–13 shows one possible arrangement. Note that for the left bearing, the shaft diameter is slightly smaller to the left of the bearing seat. This allows the bearing to be slid easily over the shaft up to the place where it must be pressed on. The retaining ring for the outer race could be supplied as a part of the outer race instead of as a separate piece.

The right bearing is held on the shaft with a locknut threaded on the end of the shaft. See Figure 14–14 for the design of standard locknuts. The internal tab on the lockwasher engages a groove in the shaft, and one of the external tabs is bent into a groove on the nut after it is seated to keep the nut from backing off. The external cap not only protects the bearing but also retains the outer race in position.

Care must be exercised to ensure that the bearings are not overly constrained. If both bearings are held tightly, any changes in dimensions due to thermal expansion or unfavorable tolerance stackup would cause binding and could lead to dangerous unexpected loads

FIGURE 14–13
Bearing mounting
illustration

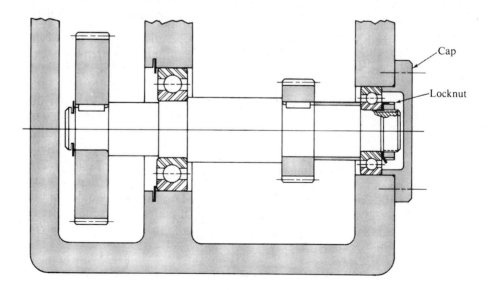

FIGURE 14–14
Locknut and
lockwasher for
retaining bearings
(SKF USA, Inc.,
Norristown, PA)

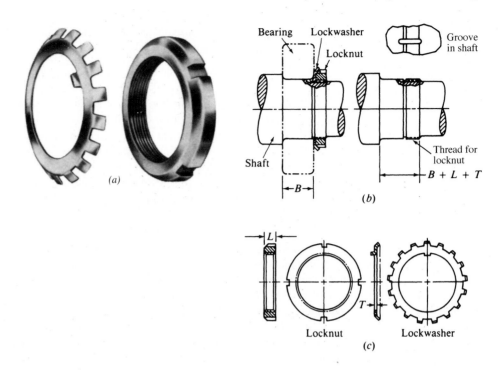

on the bearings. It is desirable to give one bearing complete location while allowing the other bearing to float axially.

**14–12
TAPERED
ROLLER
BEARINGS**

The taper on the rollers of tapered roller bearings, evident in Figure 14–7, results in a load path different from that for the bearings discussed thus far. Figure 14–15 shows two tapered roller bearings supporting a shaft with a combination of a radial load and a thrust load. The design of the shaft is such that the thrust load is resisted by the left bearing. But a peculiar feature of this type of bearing is that a radial load on one of the bearings creates a thrust on the opposing bearing, also; this feature must be considered in analysis of the bearing.

The location of the radial reaction must also be determined with care. Part (b) of Figure 14–15 shows a dimension a that is found by the intersection of a line perpendicular to the axis of the roller and the centerline of the shaft. The radial reaction at the bearing acts through this point. The distance a is reported in the tables of data for the bearings.

The American Bearings Manufacturers' Association (ABMA) recommends the following approach in computing the equivalent loads on a tapered roller bearing:

⇨ **Equivalent Load
for Tapered
Roller Bearing**

$$P_A = 0.4F_{rA} + 0.5\frac{Y_A}{Y_B}F_{rB} + Y_A T_A \qquad (14\text{–}8)$$

$$P_B = F_{rB} \qquad (14\text{–}9)$$

where P_A = equivalent radial load on bearing A
P_B = equivalent radial load on bearing B
F_{rA} = applied radial load on bearing A
F_{rB} = applied radial load on bearing B
T_A = thrust load on bearing A
Y_A = thrust factor for bearing A from tables
Y_B = thrust factor for bearing B from tables

FIGURE 14–15
Example of tapered
roller bearing
installation

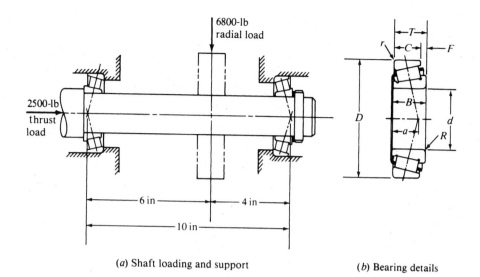

(*a*) Shaft loading and support

(*b*) Bearing details

TABLE 14–7   Tapered roller bearing data

Bore	Outside diameter	Width	a	Thrust factor, Y	Basic dynamic load rating, C
1.0000	2.5000	0.8125	0.583	1.71	8370
1.5000	3.0000	0.9375	0.690	1.98	12 800
1.7500	4.0000	1.2500	0.970	1.50	21 400
2.0000	4.3750	1.5000	0.975	2.02	26 200
2.5000	5.0000	1.4375	1.100	1.65	29 300
3.0000	6.0000	1.6250	1.320	1.47	39 700
3.5000	6.3750	1.8750	1.430	1.76	47 700

Note: Dimensions are in inches. Load C is in pounds for an L_{10} life of 1 million rev.

Table 14–7 shows an abbreviated set of data from a catalog to illustrate the method of computing equivalent loads.

For the several hundred designs of standard tapered roller bearings available commercially, the value of the thrust factor varies from as small as 1.07 to as high as 2.26. In design problems, a trial-and-error procedure is usually necessary. Example Problem 14–5 illustrates one approach.

Example Problem 14–5   The shaft shown in Figure 14–15 carries a transverse load of 6800 lb and a thrust load of 2500 lb. The thrust is resisted by bearing *A*. The shaft rotates at 350 rpm and is to be used in a piece of agricultural equipment. Specify suitable tapered roller bearings for the shaft.

Solution   The radial loads on the bearings are

$$F_{rA} = 6800(4 \text{ in}/10 \text{ in}) = 2720 \text{ lb}$$
$$F_{rB} = 6800(6 \text{ in}/10 \text{ in}) = 4080 \text{ lb}$$
$$T_A = 2500 \text{ lb}$$

To use Equation (14–8), we must assume values of Y_A and Y_B. Let's use $Y_A = Y_B = 1.75$. Then

$$P_A = 0.40(2720) + 0.5 \frac{1.75}{1.75} 4080 + 1.75(2500) = 7503 \text{ lb}$$

$$P_B = F_{rB} = 4080 \text{ lb}$$

Using Table 14–4 as a guide, let's select 4000 h as a design life. Then the number of revolutions would be

$$L_d = (4000 \text{ h})(350 \text{ rpm})(60 \text{ min/h}) = 8.4 \times 10^7 \text{ rev}$$

The required basic dynamic load rating can now be calculated from Equation (14–3), using $k = 3.33$:

$$C_A = P_A(L_d/10^6)^{1/k}$$
$$C_A = 7503(8.4 \times 10^7/10^6)^{0.30} = 28\ 400 \text{ lb}$$

Similarly,

$$C_B = 4080(8.4 \times 10^7/10^6)^{0.30} = 15\ 400 \text{ lb}$$

From Table 14–7, we can choose the following bearings.

Bearing A

$$d = 2.5000 \text{ in} \qquad D = 5.0000 \text{ in}$$
$$C = 29\ 300 \text{ lb} \qquad Y_A = 1.65$$

Bearing B

$$d = 1.7500 \text{ in} \qquad D = 4.0000 \text{ in}$$
$$C = 21\ 400 \text{ lb} \qquad Y_B = 1.50$$

We can now recompute the equivalent loads:

$$P_A = 0.40(2720) + 0.5 \frac{1.65}{1.50} 4080 + 1.65(2500) = 7457 \text{ lb}$$

$$P_B = F_{rB} = 4080 \text{ lb}$$

From these, the new values of $C_A = 28\ 200$ lb and $C_B = 15\ 400$ lb are still satisfactory for the selected bearings.

One caution must be observed in using the equations for equivalent loads for tapered roller bearings. If, from Equation (14–8), the equivalent load on bearing A is less than the applied radial load, the following equation should be used.

If $P_A < F_{rA}$, then let $P_A = F_{rA}$ and compute P_B.

$$P_B = 0.4F_{rB} + 0.5 \frac{Y_B}{Y_A} F_{rA} - Y_B T_A \qquad \text{(14–10)}$$

A similar analysis is used for angular contact ball bearings in which the design of the races results in a load path similar to that for tapered roller bearings. Figure 14–3 shows an angular contact bearing and the angle through the pressure center. This is equivalent to the line perpendicular to the axis of the tapered roller bearing. The radial reaction on the bearing acts through the intersection of this line and the axis of the shaft. Also, a radial load on one bearing induces a thrust load on the opposing bearing, requiring the application of the equivalent load formulas of the type used in Equations (14–8) and (14–10). The angle of the load line in commercially available angular contact bearings ranges from 15° to 40°.

14–13 PRACTICAL CONSIDERATIONS IN THE APPLICATION OF BEARINGS

This section will discuss lubrication of bearings, installation, preloading, stiffness, operation under varying loads, sealing, limiting speeds, and standards and bearing tolerance classes related to the manufacture and application of bearings.

Lubrication

The functions of lubrication in a bearing unit are as follows:

1. To provide a low-friction film between the rolling elements and the races of the bearing and at points of contact with cages, guiding surfaces, retainers, and so on.
2. To protect the bearing components from corrosion.
3. To help dissipate heat from the bearing unit.
4. To carry heat away from the bearing unit.
5. To help dispel contaminants and moisture from the bearing.

Rolling contact bearings are usually lubricated with either grease or oil. Under normal ambient temperatures (approximately 70°F) and relatively slow speeds (under 500 rpm), grease is satisfactory. At higher speeds or higher ambient temperatures, oil lubrication applied in a continuous flow is required, possibly with external cooling of the oil.

Oils used in bearing lubrication are usually clean, stable mineral oils. Under lighter loads and lower speeds, light oil is used. Heavier loads and/or higher speeds require heavier oils up to SAE 30. A recommended upper limit for lubricant temperature is 160°F. The choice of the correct oil or grease depends on many factors, so each application should be discussed with the bearing manufacturer. In general, a kinematic viscosity of 13 to 21 centistokes should be maintained at the operating temperature of the lubricant in the bearing. Manufacturer's recommendations should be sought.

In some critical applications such as bearings in jet engines and very-high-speed devices, lubricating oil is pumped under pressure to an enclosed housing for the bearing where the oil is directed at the rolling elements themselves. A controlled return path is also provided. The temperature of the oil in the sump is monitored and controlled with heat exchangers or refrigeration to maintain oil viscosity within acceptable limits. Such systems provide reliable lubrication and ensure the removal of heat from the bearing.

See Section 14–14 for additional discussion about the importance of oil film thickness in bearings. Reference 7 contains an extensive amount of information on this subject.

Greases used in bearings are mixtures of lubricating oils and thickening agents, usually soaps such as lithium or barium. The soaps act as carriers for the oil which is drawn out at the point of need within the bearing. Additives to resist corrosion or oxidation of the oil itself are sometimes added. Classifications of greases specify the operating temperatures to which the greases will be exposed, as defined by the American Bearing Manufacturers' Association (ABMA), and outlined in Table 14–8.

TABLE 14–8 Types of greases used for bearing lubrication

Group	Type of grease	Operation temperature range (°F)
I	General-purpose	−40–250
II	High-temperature	0–300
III	Medium-temperature	32–200
IV	Low-temperature	−67–225
V	Extremely high-temperature	Up to 450

Installation

It has already been stated that most bearings should be installed with a light interference fit between the bore of the bearing and the shaft to preclude the possibility of rotation of the inner race of the bearing with respect to the shaft. Such a condition would result in uneven wear of the bearing elements and early failure. To install the bearing then requires rather heavy forces applied axially. Care must be exercised so that the bearing is not damaged during installation. The installation force should be applied directly to the inner race of the bearing.

If the force were applied through the outer race, the load would be transferred through the rolling elements to the inner race. Because of the small contact area, it is likely that such transfer of forces would overstress some element, exceeding the static load capacity. Brinelling would result, along with the noise and rapid wear that accompany this condition. For large bearings it may be necessary to heat the bearing to expand its diameter in order to keep the installation forces within reason. Removal of bearings intended for reuse requires similar precautions. Bearing pullers are available to facilitate this task.

Preloading

Some bearings are made with internal clearances that must be taken up in a particular direction to ensure satisfactory operation. In such cases, preloading must be provided, usually in the axial direction. On horizontal shafts, springs are typically used, with axial adjustment of the spring deflection sometimes provided to adjust the amount of preload. When space is limited, the use of Belleville washers is desirable because they provide high forces with small deflections. Shims can be used to adjust the actual deflection and preload obtained (see Chapter 19). On vertical shafts the weight of the shaft assembly itself may be sufficient to provide the required preload.

Bearing Stiffness

Stiffness is the deflection that a given bearing undergoes when carrying a given load. Usually the radial stiffness is most important because the dynamic behavior of the rotating shaft system is affected. The critical speed and the mode of vibration are both functions of the bearing stiffness. Generally speaking, the softer the bearing (low stiffness), the lower the critical speed of the shaft assembly. Stiffness is measured in the units used for springs, such as pounds per inch or newtons per millimeter. Of course, the stiffness values are quite high, with values of 500 000 to 1 000 000 lb/in reasonable. The manufacturer should be consulted when such information is needed, because it is rarely included in standard catalogs.

Operation under Varying Loads

The load/life relationships used thus far assume that the load is reasonably constant in magnitude and direction. If the load varies considerably, an effective mean load must be used for determining the expected life of the bearing. (See References 4 and 7.) Oscillating loads also require special analysis because only a few of the rolling elements share the load. See Section 14–15 for additional information about life prediction under varying loads.

Sealing

When the bearing is to operate in dirty or moist environments, special shields and seals are usually specified. They can be provided on either or both sides of the rolling elements. Shields are typically metal and are fixed to the stationary race, but they remain clear of the rotating race. Seals are made of elastomeric materials and do contact the rotating race. Bearings fitted with both seals and shields and precharged at the factory with grease are sometimes called *permanently lubricated.* Although such bearings are likely to give many years of satisfactory service, extreme conditions can produce a degradation of the lubricating properties of the grease. The presence of seals also increases the friction in a bearing. Sealing can be provided outside the bearing in the housing or at the shaft/housing interface. On high-speed shafts, a *labyrinth seal,* consisting of a noncontacting ring around the shaft with a few thousandths of an inch radial clearance, is frequently used. Grooves, sometimes in the form of a thread, are machined in the ring; the relative motion of the shaft with respect to the ring creates the sealing action.

Limiting Speeds

Most catalogs list limiting speeds for each bearing. Exceeding these limits may result in excessively high operating temperatures due to friction between the cages supporting the rolling elements. Generally, the limiting speed is lower for larger bearings than for smaller bearings. Also, a given bearing will have a lower limiting speed as loads increase. With special care, either in the fabrication of the bearing cage or in the lubrication of the bearing, bearings can be operated at higher speeds than those listed in the catalog. The manufacturer should be consulted in such applications. The use of ceramic rolling elements with their lower mass can result in higher limiting speeds.

Standards

Several groups are involved in standard setting for the bearing industry. A partial list is given here:

American Bearing Manufacturers Association	(ABMA)
Annular Bearing Engineers Committee	(ABEC)
Roller Bearing Engineers Committee	(RBEC)
Ball Manufacturers Engineers Committee	(BMEC)
American National Standards Institute	(ANSI)
International Standards Organization	(ISO)

Many standards are listed by the ANSI and ABMA organizations. Two of them are listed here:

> *Load Ratings and Fatigue Life for Ball Bearings,* ABMA 9
> *Load Ratings and Fatigue Life for Roller Bearings,* ABMA 11

Tolerances

Several different tolerance classes are recognized in the bearing industry to accommodate the needs of the wide variety of equipment using rolling contact bearings. In general, of course, all bearings are precision machine elements and should be treated as such. As noted before, the general range of tolerances is of the order of a few ten-thousandths of an inch. The standard tolerance classes are defined by ABEC, as identified below.

ABEC 1:   Standard radial ball and roller bearings
ABEC 3:   Semiprecision instrument ball bearings
ABEC 5:   Precision radial ball and roller bearings
ABEC 5P: Precision instrument ball bearings
ABEC 7:   High-precision radial ball bearings
ABEC 7P: High-precision instrument ball bearings

Most machine applications would use ABEC 1 tolerances, the data for which are usually provided in the catalogs. Machine tool spindles requiring extra smooth and accurate running would use ABEC 5 or ABEC 7 classes.

14–14 IMPORTANCE OF OIL FILM THICKNESS IN BEARINGS

In bearings operated at heavy loads or high speeds, it is critical to maintain a film of lubricating oil at the surface of the rolling elements. A steady supply of clean lubricant with an adequate viscosity is required. With careful analysis of the geometry of the bearing, the speed of rotation, and the properties of the lubricant, it is possible to estimate the thickness of the oil film between the rolling elements and the races. Although the film thickness may be only a few microinches, it has been shown that oil starvation of the contact area is a chief cause of premature failure of rolling contact bearings. Conversely, if a film thickness significantly greater than the height of surface roughness features can be maintained, the expected life will be several times greater than published data from catalogs. (See References 4 and 7.)

The nature of the lubrication at the interface between rolling elements and the races is called *elastohydrodynamic lubrication* because it depends on the specific elastic deformation of the mating surfaces under the influence of high contact stresses and the creation of a pressurized film of lubricant by the dynamic action of the rolling elements.

The data needed to evaluate film thickness for ball bearings are given in the following lists:

Bearing Geometry Factors

Ball diameter

Number of balls

Radius of curvature of the inner race groove in both the circumferential and the axial directions

Pitch diameter of the bearing; the average of the bore diameter and the outside diameter

Surface roughness of the balls and the races

Contact angle for angular contact bearings

Bearing Material Factors

Modulus of elasticity of balls and races

Poisson's ratio

Lubricant Factors

Dynamic viscosity at the operating temperature *within* the bearing

Pressure coefficient of viscosity; the change of viscosity with pressure

Operational Factors

Rotational speed of both the inner and the outer races

Radial load

Thrust load

The details of the analysis are given in Reference 7.
The results of the analysis include the following:

The minimum film thickness of the lubricant, h_o

The composite roughness of the balls and the races, S

The ratio $\Lambda = h_o/S$

The service life of the bearing is dependent on the value of Λ:

- If $\Lambda < 0.90$, a life less than the manufacturer's rated life is to be expected because of surface damage due to inadequate lubricant film.
- If Λ is in the range of 0.90 to 1.50, rated service life can be expected.
- If Λ is in the range of 1.50 to 3.0, an increase of life of up to three times the rated life is possible.
- If Λ is greater than 3.0, a life of up to six times the rated life is possible.

General Recommendations for Achieving Long Life for Bearings

1. Choose a bearing with an adequate rated life using the procedures outlined in this chapter.

2. Ensure that the bearing has a fine surface finish and is not damaged by rough handling, poor installation practices, corrosion, vibration, or exposure to electrical current flow.

3. Ensure that operating loads are within the design values.

4. Supply the bearing with a copious flow of clean lubricant at an adequate viscosity at the operating temperature within the bearing according to the bearing manufacturer's recommendations. Provide external cooling for the lubricant if necessary. For an existing system, this is the factor over which you have the most control without a significant redesign of the system itself.

5. If redesign is possible, design the system to operate at as low a speed as possible.

**14–15
LIFE PREDICTION
UNDER VARYING
LOADS**

The design and analysis procedures used so far in this book assumed that the bearing would operate at a single design load throughout its life. It is possible to predict the life of the bearing under such conditions fairly accurately using manufacturers' data published in catalogs. If loads vary with time, a modified procedure is required.

One procedure that is recommended by bearing manufacturers is called the *Palmgren-Miner rule,* or sometimes simply *Miner's rule*. References 10 and 11 describe their work, and Reference 9 discusses a modified approach more tailored to bearings.

The basis of Miner's rule is that if a given bearing is subjected to a series of loads of different magnitudes for known lengths of time, each load contributes to the eventual failure of the bearing in proportion to the ratio of the load to the expected life that the bearing would have under that load. Then the cumulative effect of the series of loads must account for all such contributions to failure.

A similar approach, described in Reference 7, introduces the concept of *mean effective load, F_m:*

$$F_m = \left(\frac{\Sigma_i (F_i)^p N_i}{N} \right)^{1/p} \tag{14-11}$$

where F_i = individual load among a series of i loads
N_i = number of revolutions at which F_i operates
N = total number of revolutions in a complete cycle
p = exponent on the load/life relationship; $p = 3$ for ball bearings, and $p = 10/3$ for rollers

Alternatively, if the bearing is rotating at a constant speed, and because the number of revolutions is proportional to the time of operation, N_i can be the number of minutes of operation at F_i, and N is the sum of the number of minutes in the total cycle. That is,

$$N = N_1 + N_2 + \cdots + N_i$$

Then the total expected life, in millions of revolutions of the bearing, would be

$$L = \left(\frac{C}{F_m} \right)^p \tag{14-12}$$

Example Problem 14-6   A single-row, deep-groove ball bearing number 6308 is subjected to the following set of loads for the given times:

Condition	F_i	Time
1	650 lb	30 min
2	750 lb	10 min
3	250 lb	20 min

This cycle of 60 min is repeated continuously throughout the life of the bearing. The shaft carried by the bearing rotates at 600 rpm. Estimate the total life of the bearing.

Solution   Using Equation (14-11), we have

$$F_m = \left(\frac{\Sigma_i (F_i)^p N_i}{N} \right)^{1/p} \tag{14-11a}$$

$$F_m = \left(\frac{30(650)^3 + 10(750)^3 + 20(250)^3}{30 + 10 + 20} \right)^{1/3} = 597 \text{ lb}$$

Now use Equation (14–12):

$$L = \left(\frac{C}{F_m}\right)^p \qquad\qquad \textbf{(14–12a)}$$

From Table 14–3, for the 6308 bearing, we find that $C = 7050$ lb. Then

$$L = \left(\frac{7050}{597}\right)^3 = 1647 \text{ million rev}$$

At a rotational speed of 600 rpm, the number of hours of life would be

$$L = \frac{1647 \times 10^6 \text{ rev}}{1} \cdot \frac{\min}{600 \text{ rev}} \cdot \frac{h}{60 \min} = 45\,745 \text{ h}$$

Note that this is the same bearing used in Example Problem 14–3 which was selected to operate at least 30 000 h when carrying a steady load of 650 lb.

REFERENCES

1. Association of Iron and Steel Engineers. *Lubrication Engineers Manual*. 2d ed. Pittsburgh, PA: AISE, 1996.

2. Avallone, Eugene A., and Theodore Baumeister III, eds. *Marks' Standard Handbook for Mechanical Engineers*. 10th ed. New York: McGraw-Hill, 1996.

3. Bloch, H. P. *Practical Lubrication for Industrial Facilities*. New York: Marcel Dekker, 2000.

4. Brandlein, J., and Karl Weigand. *Ball and Roller Bearings*. New York: John Wiley & Sons, 1999.

5. Eschmann, Paul, Ludwig Hasbargen, Karl Weisgand, and Johannes Brandlein. *Ball and Roller Bearings: Theory, Design, and Application*. 2d ed. New York: John Wiley & Sons, 1985.

6. Hannoosh, J. G. "Ceramic Bearings Enter the Mainstream." *Design News* (November 23, 1988).

7. Harris, Tedric A. *Rolling Bearing Analysis*. 4th ed. New York: John Wiley & Sons, 2000.

8. Juvinall, Robert C., and Kurt M. Marshek. *Fundamentals of Machine Component Design*. 3rd ed. New York: John Wiley & Sons, 2000.

9. Kauzlarich, James J. "The Palmgren-Miner Rule Derived." *Proceedings of the 15th Leeds/Lyon Symposium of Tribology*. (Leeds, United Kingdom, September 6–9, 1988).

10. Miner, M. A. "Cumulative Damage in Fatigue." *Journal of Applied Mechanics* 67 (1945): A159–A164.

11. Palmgren, A. *Ball and Roller Bearing Engineering*. 3rd ed. Philadelphia, PA: Burbank, 1959.

INTERNET SITES RELATED TO ROLLING CONTACT BEARINGS

1. **Torrington Company.** *www.torrington.com* Manufacturer of rolling contact bearings and motion control components and assemblies under the Torrington, Fafnir, and Kilian brands. Online catalog.

2. **SKF USA, Inc.** *www.skfusa.com* or *http://products. skf.com* Manufacturer of rolling contact bearings under the SKF brand. Online catalog.

3. **FAG Bearings.** *http://fag.com* or *www.fagauto.com* Manufacturer of rolling contact bearings under the FAG brand. Online catalog. A subsidiary of the INA Corporation.

4. **INA USA Corporation.** *www.ina.com/us* Manufacturer of rolling contact bearings, cam followers, linear motion bearings, plain bearings, thrust bearings, and turntable bearings under the INA brand.

5. **NSK Corporation.** *www.nsk.com* or *www.tec.nsk.com* Manufacturer of rolling contact bearings under the NSK brand. Online catalog.

6. Timken Corporation. *www.timken.com*
Manufacturer of rolling contact bearings under the Timken brand. Online catalog.

7. Rockwell Automation/Dodge. *www.dodge-pt.com*
Manufacturer of numerous power transmission products, including mounted rolling contact bearings along with many other types of bearings. Online catalog.

8. Emerson Power Transmission. *www.emerson-ept. com* Manufacturer of numerous power transmission products, including bearings from their Browning, McGill, Rollway, and Sealmaster units. Online catalog.

9. PowerTransmission.com. *www.powertransmission. com* A site that lists numerous manufacturers of

power transmission components including rolling contact bearings, gears, gear drives, clutches, couplings, motors, and more. Descriptive information about the listed companies is provided along with links to their websites.

10. Machinery Lubrication. *www.machinerylubrication. com* On online service of *Machinery Lubrication Magazine* providing technical articles and information on lubrication of industrial machinery, including bearings.

11. American Bearing Manufacturers Association. *www.abma-dc.org* A non-profit association of manufacturers of anti-friction bearings. ABMA defines national standards for bearings.

PROBLEMS

1. A radial ball bearing has a basic dynamic load rating of 2350 lb for a rated (L_{10}) life of 1 million rev. What would its L_{10} life be when operating at a load of 1675 lb?

2. Determine the required basic dynamic load rating for a bearing to carry 1250 lb from a shaft rotating at 880 rpm if the design life is to be 20 000 h.

3. A catalog lists the basic dynamic load rating for a ball bearing to be 3150 lb for a rated life of 1 million rev. What would be the L_{10} life of the bearing if it were subjected to a load of (a) 2200 lb and (b) 4500 lb?

4. Compute the required basic dynamic load rating, *C,* for a ball bearing to carry a radial load of 1450 lb at a shaft speed of 1150 rpm for an industrial fan.

5. Specify suitable bearings for the shaft of Example Problem 12–1. Note the data contained in Figures 12–1, 12–2, 12–10, and 12–11.

6. Specify suitable bearings for the shaft of Example Problem 12–3. Note the data contained in Figures 12–12, 12–14, and 12–15.

7. Specify suitable bearings for the shaft of Example Problem 12–4. Note the data contained in Figures 12–17 and 12–18.

8. For any of the bearings specified in Problem 2–7, make a scale drawing of the shaft, the bearings, and the part of the housing to support the outer races of the bearings. Be sure to consider fillet radii and axial location of the bearings.

9. A bearing is to carry a radial load of 455 lb and no thrust load. Specify a suitable bearing from Table 14–3 if the shaft rotates at 1150 rpm and the design life is 20 000 h.

For each of the problems in Table 14–9, repeat Problem 9 with the new data.

TABLE 14–9

Problem number	Radial load	Thrust load	rpm	Design life, h
10.	857 lb	0	450	30 000
11.	1265 lb	645 lb	210	5000
12.	235 lb	88 lb	1750	20 000
13.	2875 lb	1350 lb	600	15 000
14.	3.8 kN	0	3450	15 000
15.	5.6 kN	2.8 kN	450	2000
16.	10.5 kN	0	1150	20 000
17.	1.2 kN	0.85 kN	860	20 000

18. In Chapter 12, Figures P12–1 through P12–40 showed shaft design exercises related to the problems at the end of the chapter. For each bearing of each shaft, specify a suitable bearing from Table 14–3. If the shaft design has already been completed to the point that the minimum acceptable diameter of the shaft at the bearing seat is known, consider that diameter when specifying the bearing. Note the Chapter 12 problem statements for the speed of the shaft and the loading data.

19. Bearing number 6332 from Table 14–3 is carrying the set of loads in Table P14–19 while rotating at 600 rpm. Com-

Compute the expected L_{10} life for the bearing under these conditions if the cycle repeats continuously.

TABLE P14–19

Condition	Load, F_i	Time, N_i
1	4500 lb	25 min
2	2500 lb	15 min

20. Bearing number 6318 from Table 14–3 is carrying the set of loads in Table P14–20 while rotating at 600 rpm. Compute the expected L_{10} life for the bearing under these conditions if the cycle repeats continuously.

TABLE P14–20

Condition	Load, F_i	Time, N_i
1	2500 lb	25 min
2	1500 lb	15 min

21. Bearing number 6211 from Table 14–3 is carrying the set of loads in Table P14–21 while rotating at 1700 rpm. Compute the expected L_{10} life for the bearing under these conditions if the cycle repeats continuously.

TABLE P14–21

Condition	Load, F_i	Time, N_i
1	600 lb	480 min
2	200 lb	115 min
3	100 lb	45 min

22. Bearing number 6211 from Table 14–3 is carrying the set of loads in Table P14–22 while rotating at 1700 rpm. Compute the expected L_{10} life for the bearing under these conditions if the cycle repeats continuously.

TABLE P14–22

Condition	Load, F_i	Time, N_i
1	450 lb	480 min
2	180 lb	115 min
3	50 lb	45 min

23. Bearing number 6206 from Table 14–3 is carrying the set of loads in Table P14–23 while rotating at 101 rpm for one 8-h shift. Compute the expected L_{10} life for the bearing under these conditions if the cycle repeats continuously. If the machine operates two shifts per day, six days per week, in how many weeks would you expect to have to replace the bearing?

TABLE P14–23

Condition	Load, F_i	Time, N_i
1	500 lb	6.75 h
2	800 lb	0.40 h
3	100 lb	0.85 h

24. Bearing number 6212 from Table 14–3 is carrying the set of loads in Table P14–24 while rotating at 101 rpm for one 8-h shift. Compute the expected L_{10} life for the bearing under these conditions if the cycle repeats continuously. If the machine operates two shifts per day, six days per week, in how many weeks would you expect to have to replace the bearing?

TABLE P14–24

Condition	Radial load	Thrust load	Time, N_i
1	1750 lb	350 lb	6.75 h
2	600 lb	250 lb	0.40 h
3	280 lb	110 lb	0.85 h

25. Compute the basic dynamic load rating, C, for a ball bearing to carry a radial load of 1450 lb at a shaft speed of 1150 rpm for 15 000 hours. Use a reliability of 95%.

26. Compute the basic dynamic load rating, C, for a ball bearing to carry a radial load of 509 lb at a shaft speed of 101 rpm for 20 000 hours. Use a reliability of 99%.

27. Compute the basic dynamic load rating, C, for a ball bearing to carry a radial load of 436 lb at a shaft speed of 1700 rpm for 5000 hours. Use a reliability of 97%.

28. Compute the basic dynamic load rating, C, for a ball bearing to carry a radial load of 1250 lb at a shaft speed of 880 rpm for a design life of 20 000 hours. Use a reliability of 95%.

15

Completion of the Design of a Power Transmission

The Big Picture

15–1 Objectives of This Chapter

15–2 Description of the Power Transmission to Be Designed

15–3 Design Alternatives and Selection of the Design Approach

15–4 Design Alternatives for the Gear-Type Reducer

15–5 General Layout and Design Details of the Reducer

15–6 Final Design Details for the Shafts

15–7 Assembly Drawing

Completion of the Design of a Power Transmission

Discussion Map

☐ Now we bring together the concepts and design procedures from the last eight chapters to complete the design of the power transmission.

Discover

Consider how all of the machine elements that you have studied in Chapters 7–14 fit together.

Think also about the entire life cycle of the transmission, from its design to its disposal.

This chapter presents a summary of the steps that you should take to complete the design. Some procedures are quite detailed. You should be able to apply this experience to any design you are responsible for in the future.

This is where all of the work of Part II of this book comes together. In Chapters 7–14, you learned important concepts and design procedures for many kinds of machine elements that could all be part of a given power transmission. In each case, we mentioned how the elements need to work together. Now we show the approach to completing the design of a power transmission which demonstrates an integrated approach and illustrates the title of this book, *Machine Elements in Mechanical Design.* The emphasis is on the whole design.

The lesson of this chapter is that you as a designer must constantly keep in mind how the part you are currently working on fits with other parts and how its design can affect the design of other parts. You must also consider how the part is to be manufactured, how it may be serviced and repaired as necessary, and how it will eventually be taken out of service. What will happen to the materials in the product when they have served their useful life as part of your current project?

Although we are using a power transmission in this example, the skills and insights that you will acquire should be transferable to the design of almost any other mechanical device or system.

**15–1
OBJECTIVES OF
THIS CHAPTER**

After completing this chapter, you will be able to:

1. Bring together the individual components of a mechanical, gear-type power transmission into a unified, complete system.

2. Resolve the interface questions where two components fit together.

3. Establish reasonable tolerances and limit dimensions on key dimensions of components, especially where assembly and operation of the components are critical.

4. Verify that the final design is safe and suitable for its intended purpose.

5. Add details to some of the components that were not considered in earlier analyses.

**15–2
DESCRIPTION OF
THE POWER
TRANSMISSION
TO BE DESIGNED**

The project to be completed in this chapter is the design of a single-reduction speed reducer that uses spur gears. We will use the data from Example Problem 9–1 in which the gears for the drive for an industrial saw were designed. You should review that problem now and note that it was carried forward through Example Problems 9–1 through 9–4. It was considered again in Section 9–15 where we refined the design using the spreadsheet developed in Section 9–14.

We will also use elements of the mechanical design process that were first outlined in Chapter 1 in Sections 1–4 and 1–5. We will state the functions and design requirements for the power transmission, establish a set of criteria for evaluating design decisions, and implement the design tasks that were outlined in Section 1–5. References 4, 7, 8, and 9 provide additional approaches that you may find useful for other design projects.

Basic Statement of the Problem

We will design a power transmission for an industrial saw that will be used to cut tubing for vehicle exhaust pipes to length prior to the forming processes. The saw will receive 25 hp from the shaft of an electric motor rotating at 1750 rpm. The drive shaft for the saw should rotate at approximately 500 rpm.

Functions, Design Requirements, and Selection Criteria for the Power Transmission

Functions.   The functions of the power transmission are as follows:

1. To receive power from an electric motor through a rotating shaft.
2. To transmit the power through machine elements that reduce the rotational speed to a desired value.
3. To deliver the power at the lower speed to an output shaft which ultimately drives the saw.

Design Requirements.   Additional information is presented here for the specific case of the industrial saw. You would normally be responsible for acquiring the necessary information and for making design decisions at this point in the design process. You would be involved in the design of the saw and would be able to discuss its desirable features with colleagues in marketing, sales, manufacturing planning, and production management and field service, and, perhaps, with customers. The kinds of information that you should seek are illustrated in the following list:

1. The reducer must transmit 25 hp.
2. The input is from an electric motor whose shaft rotates at a full-load speed of 1750 rpm. It has been proposed to use a NEMA frame 284T motor having a shaft diameter of 1.875 in and a keyway to accommodate a $1/2 \times 1/2$ in key. See Chapter 21, Figure 21–18 and Table 21–3, for more data on the dimensions of the motor.
3. The output of the reducer delivers power to the saw through a shaft that rotates between 495 and 505 rpm. The speed reduction ratio should then be in the range of 3.46 to 3.53.
4. A mechanical efficiency of greater than 95% is desirable.
5. The minimum torque delivered to the saw should be 2950 lb·in.
6. The saw is a band saw. The cutting operation is generally smooth, but moderate shock may be encountered as the saw blade engages the tubes and if there is any binding of the blade in the cut.
7. The speed reducer will be mounted on a rigid plate that is part of the base of the saw. The means of mounting the reducer should be specified.
8. It has been decided that flexible couplings may be used to connect the motor shaft to the input shaft of the reducer and to connect the output shaft directly to the shaft

of the main drive wheel for the band saw. The design for the shaft for the band-saw drive has not yet been completed. It is likely that its diameter will be the same as that of the output shaft of the reducer.

9. Whereas a small, compact size for the reducer is desirable, space in the machine base should be able to accommodate most reasonable designs.

10. The saw is expected to operate 16 hours per day, 5 days per week, with a design life of 5 years. This is approximately 20 000 hours of operation.

11. The machine base will be enclosed and will prevent any casual contact with the reducer. However, the functional components of the reducer should be enclosed in their own rigid housing to protect them from contaminants and to provide for the safety of those who work with the equipment.

12. The saw will operate in a factory environment and should be capable of operation in the temperature range of 50°F to 100°F.

13. The saw is expected to be produced in quantities of 5000 units per year.

14. A moderate cost is critical to the marketing success of the saw.

Selection Criteria. The list of criteria should be produced by an interdisciplinary team composed of people having broad experience with the market for and use of such equipment. The details will vary according to the specific design. As an illustration of the process, the following criteria are suggested for the present design:

1. *Safety:* The speed reducer should operate safely and provide a safe environment for people near the machine.

2. *Cost:* Low cost is desirable so that the saw appeals to a large set of customers.

3. *Small size.*

4. *High reliability.*

5. *Low maintenance.*

6. *Smooth operation; low noise; low vibration.*

15–3 DESIGN ALTERNATIVES AND SELECTION OF THE DESIGN APPROACH

There are many ways that the speed reduction for the saw can be accomplished. Figure 15–1 shows four possibilities: (a) belt drive, (b) chain drive, (c) gear-type drive connected through flexible couplings, and (d) gear-type drive with a belt drive on the input side and connected to the saw with a flexible coupling.

Reference 6 includes a more extensive set of alternatives and a more detailed analysis.

Selection of the Basic Design Approach

Table 15–1 shows an example of the rating that could be done to select the type of design to be produced for the speed reducer for the saw. A 10-point scale is used, with 10 being the highest rating. Of course, with more information about the actual application, a different design approach could be selected. Also, it may be desirable to proceed with more than one design to determine more of the details, thus allowing a more rational decision. A modification of the design decision matrix calls for weighting factors to be assigned to each criterion to reflect its relative importance. See References 3 and 7 and Internet site 1 for extensive discussions of rational decision analysis techniques.

On the basis of this decision analysis, let's proceed with (c), the design of a *gear-type speed reducer using flexible couplings* to connect with the drive motor and the driven shaft

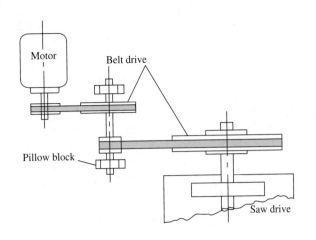

(a) Belt drive option

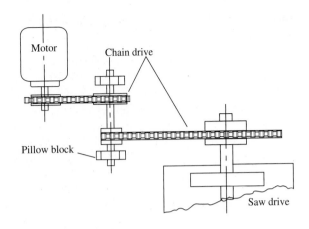

(b) Chain drive option

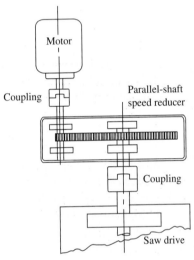

(c) Single-stage gear-reducer option

(d) Belt drive with single-stage gear-reducer option

FIGURE 15–1   Options for speed reduction for saw drive design project

TABLE 15–1   Decision analysis chart

| | | Alternatives | | |
| | (a) Belt | (b) Chain | (c) Gear with flexible couplings | (d) Gear with input belt drive |
Criteria				
1. Safety	6	6	9	7
2. Cost	9	8	7	6
3. Size	5	6	9	6
4. Reliability	7	6	10	7
5. Maintenance	6	5	9	6
6. Smoothness	8	6	9	8
Totals:	41	31	53	40

of the saw. It is considered to have a higher level of safety for operators and maintenance people because its rotating components are enclosed. The input and output shafts and the couplings can be covered at the time of installation. Reliability is expected to be higher because precision metallic parts are used and the drive is enclosed in a sealed housing. The flexing of belts and the significant number of moving parts in a chain drive are judged to provide lower reliability. Initial cost may be higher than for belt or chain drives. However, it is expected that maintenance will be somewhat less, leading to lower cost overall. The space taken by the design should be small, simplifying the design of other parts of the saw. Design alternative (d) is attractive if there is some interest in providing a variable speed operation in the future. By using different belt drive ratios, we can achieve different cutting speeds for the saw. A further alternative would be to consider a variable-speed electric drive motor, either to replace the need for a reducer at all or to be used in conjunction with the gear-type reducer.

15–4 DESIGN ALTERNATIVES FOR THE GEAR-TYPE REDUCER

Now that we have selected the gear-type reducer, we need to decide which type to use. Here are some alternatives:

1. *Single-reduction spur gears:* The nominal ratio of 3.50 : 1 is reasonable for a single pair of gears. Spur gears produce only radial loads which simplify selection of the bearings that support the shafts. Efficiency should be greater than 95% with reasonable precision of the gears, bearings, and seals. Spur gears are relatively inexpensive to produce. Shafts would be parallel and should be fairly easy to align with the motor and the drive shaft for the saw.

2. *Single-reduction helical gears:* These gears are equally practical as spur gears. Shaft alignment is similar. A smaller size should be possible because of the greater capacity of helical gears. However, axial thrust loads would be created which must be accommodated by the bearings and the housing. Cost is likely to be somewhat higher.

3. *Bevel gears:* These gears produce a right-angle drive which may be desirable, but not necessary in the present design. They are also somewhat more difficult to design and assemble to achieve adequate precision.

4. *Worm and wormgear drive:* This drive also produces a right-angle drive. It is typically used to achieve a higher reduction ratio than 3.50 : 1. Efficiency is usually much lower than the 95% called for in the design requirements. Heat generation could be a problem with 25 hp and the lower efficiency. A larger motor could be required to overcome the loss of power and still provide the required torque at the output shaft.

Design Decision for the Gear Type

For the present design, we choose the *single-reduction spur gear reducer*. Its simplicity is desirable, and the final cost is likely to be lower than that of the other proposed designs. The smaller size of the helical reducer is not considered to be of high priority.

15–5 GENERAL LAYOUT AND DESIGN DETAILS OF THE REDUCER

Figure 15–2 shows the proposed arrangement of the components for the single-reduction spur gear–type speed reducer. Note that the illustration in Part (b) is the top view. The design involves the following tasks:

1. Design a pinion and gear to transmit 25 hp with a pinion speed of 1750 rpm and a gear speed in the range from 495 to 505 rpm. The nominal ratio is 3.50 : 1. Design

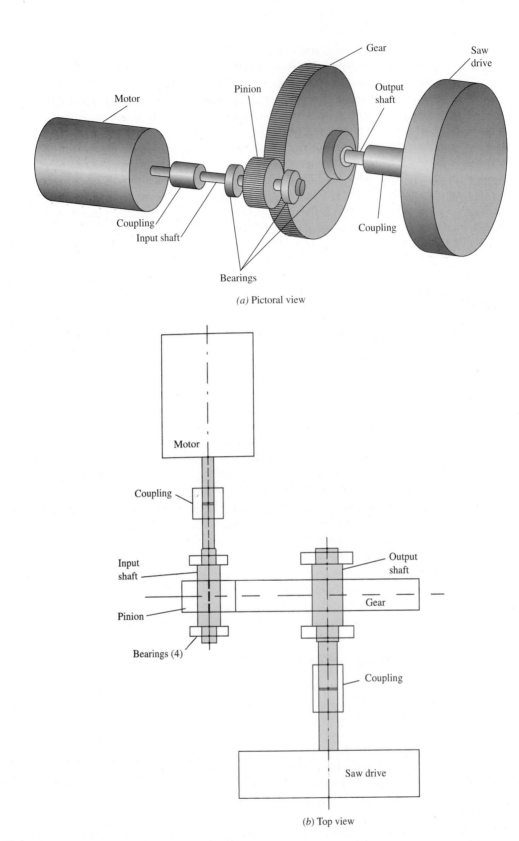

(a) Pictoral view

(b) Top view

FIGURE 15–2 General layout of saw drive with single-stage gear reducer

for both strength and pitting resistance to achieve approximately 20 000 hours of life and a reliability of at least 0.999.

2. Design two shafts, one for the pinion and one for the gear. Provide positive axial location for the gears on the shaft. The input shaft must be designed to extend beyond the housing to enable the motor shaft to be coupled to it. The output shaft must accommodate a coupling that mates with the drive shaft of the saw. Use a design reliability of 0.999.

3. Design six keys: one for each gear; one for the motor; one for the input shaft at the coupling; one for the output shaft at the coupling; and one for the drive shaft for the saw.

4. Specify two flexible couplings: one for the input shaft and one for the output shaft.

5. Specify four commercially available rolling contact bearings, two for each shaft. The L_{10} design life should be 20 000 hours.

6. Design a housing to enclose the gears and the bearings and to support them rigidly.

7. Provide a means of lubricating the gears within the housing.

8. Provide seals for the input and output shafts at the place where they pass through the housing. We will not specify the particular seals because of lack of data in this book. However, refer to Chapter 11 for suggestions for the types of seals that may be suitable.

Gear Design

The conditions for this design are the same as those considered in Example Problems 9–1 through 9–4. Several design iterations were produced in Section 9–15 using the aid of the gear design spreadsheet. The design alternatives all used a reliability factor of 1.50 to gain an expected reliability of 0.9999, fewer than one failure in 10 000. This is more conservative than the suggested reliability of 0.999. An overload factor of 1.50 was used to account for the moderate shock expected from the operation of the saw. We will use the design documented in Figure 9–31 having the following major features:

- Diametral pitch: $P_d = 8$; 20°, full-depth, involute teeth
- Number of teeth in the pinion: $N_P = 28$
- Number of teeth in the gear: $N_G = 98$
- Diameter of the pinion: $D_P = 3.500$ in
- Diameter of the gear: $D_G = 12.250$ in
- Center distance: $C = 7.875$ in
- Face width: $F = 2.00$ in
- Quality number: $Q_v = 8$
- Tangential force: $W_t = 514$ lb
- Required bending stress number for pinion: $s_{at} = 20\ 900$ psi
- Required contact stress number for pinion: $s_{ac} = 153\ 000$ psi; requires 390 HB steel
- Material specified: AISI 4140 OQT 800; 429 HB; $s_u = 210$ ksi; 16% elongation

Shaft Design

1. *Forces:* Figure 15–3(a) shows the proposed configuration for the input shaft that carries the pinion and connects to the motor shaft through a flexible coupling. Figure 15–3(b) shows the output shaft, similarly configured. The only active

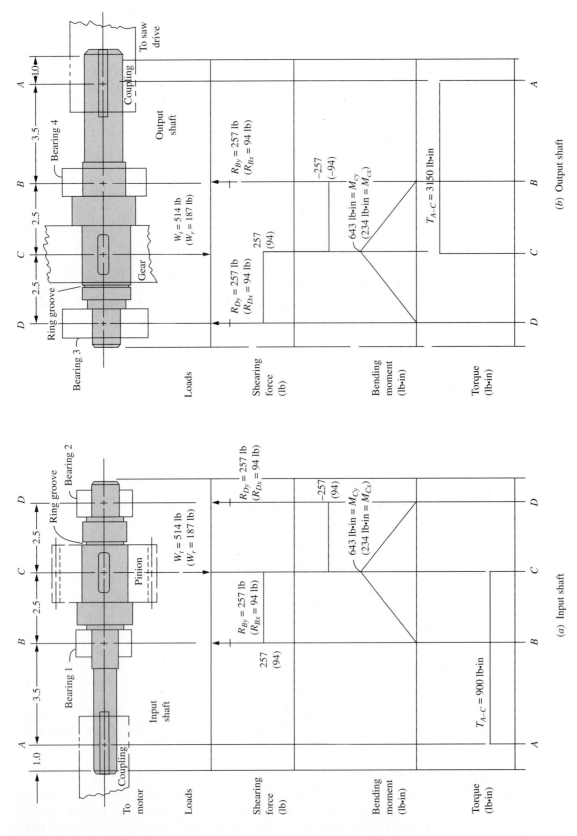

FIGURE 15–3 Diagrams of shaft configurations, shearing force, bending moment, and torque.

forces on the shafts are the tangential force and the radial force from the gear teeth. The flexible couplings at the ends of the shafts allow torque transmission, but no radial or axial forces are transmitted when the alignment of the shafts is within the recommended limits for the coupling. See Chapter 11. Without the flexible couplings, it is very likely that significant radial loads would be produced, requiring somewhat larger diameters for the shaft and larger bearings. See Chapter 12.

The spreadsheet analysis for the gears gives the tangential force as $W_t = 514$ lb. It acts downward in the vertical plane on the pinion and upward on the gear. The radial force is

$$W_r = W_t \tan \phi = (514 \text{ lb})\tan(20°) = 187 \text{ lb}$$

The radial force acts horizontally toward the left on the pinion, tending to separate the pinion from the gear. It acts toward the right on the gear.

2. *Torque values:* The torque on the input shaft is

$$T_1 = (63\ 000)(P)/n_P = (63\ 000)(25)/1750 = 900 \text{ lb} \cdot \text{in}$$

This value acts from the coupling at the left end of the shaft to the pinion where the power is delivered through the key to the pinion and thus to the mating gear.

The torque on the output shaft is computed next, assuming that no power is lost. The resulting value of torque is conservative for use in the shaft design:

$$T_2 = (63\ 000)(P)/n_G = (63\ 000)(25)/500 = 3150 \text{ lb} \cdot \text{in}$$

The torque acts in the output shaft from the gear to the coupling at the right end of the shaft. Assuming that the system is 95% efficient, the actual output torque is approximately

$$T_o = T_2 (0.95) = 2992 \text{ lb} \cdot \text{in}$$

This value is within the required range, as indicated in the design requirements, item 5.

3. *Shearing force and bending moment diagrams:* Figure 15–3 also shows the shearing force and bending moment diagrams for the two shafts. Because the active loading occurs only at the gears, the form of each diagram is the same in the vertical and horizontal directions. The first number given is the value for load, the shearing force, or the bending moment in the vertical plane. The second number in parentheses is the value in the horizontal plane. The maximum bending moment in each shaft occurs where the gears are mounted. The values are

$$M_y = 643 \text{ lb} \cdot \text{in} \qquad M_x = 234 \text{ lb} \cdot \text{in}$$

The resultant moment is $M_{max} = 684 \text{ lb} \cdot \text{in}$.

The bending moment is zero at the bearings and in the extensions for the input and output shafts.

4. *Support reactions—bearing forces:* The reactions at all bearings are the same for this example because of the simplicity of the loading pattern and the symmetry of the design. The horizontal and vertical components are

$$F_y = 257 \text{ lb} \qquad F_x = 93.5 \text{ lb}$$

The resultant force is the radial force that must be carried by the bearings: $F_r = 274$ lb. This value also produces vertical shearing stress in the shaft at the bearings.

5. **Material selection for shafts:** Each shaft will have a series of different diameters, shoulders with fillets, keyseats, and a ring groove as shown in Figure 15–3. Thus, much machining will be required. The shafts will be subjected to a combination of steady torque and reversed, repeated bending during normal use as the saw cuts steel tubing for vehicular exhaust systems. Occasional moderate shock loading is expected as the saw engages the tubing and when binding occurs in the cut due to dullness of the blade or unusually hard steel in the tubing.

   These conditions call for a steel that has moderately high strength, good fatigue resistance, good ductility, and good machinability. Such shafts are typically made from a medium-carbon-alloy steel (0.30% to 0.60% carbon) in either the cold-drawn or the oil-quenched and tempered condition. Good machinability is obtained from a steel having moderately high sulfur content, a characteristic of the 1100 series. Where good hardenability is also desired, higher manganese content is used.

   An example of such an alloy is AISI 1144 which has 0.40% to 0.48% carbon, 1.35% to 1.65% manganese, and 0.24% to 0.33% sulfur. It is called a *resulfurized, free-machining* grade of steel. Figure A4–2 shows the range of properties available for this material when it is oil-quenched and tempered. We select a tempering temperature of 1000°F, which produces good balance between strength and ductility.

   In summary, the material specified is

   AISI 1144 OQT 1000 steel: $s_u = 118\ 000$ psi; $s_y = 83\ 000$ psi; 20% elongation

   The endurance strength of the material can be estimated using the method outlined in Chapters 5 and 12:

   Basic endurance strength: $s_n = 43\ 000$ psi (from Figure 5–8 for a machined surface)

   Size factor: $C_s = 0.81$ (from Figure 5–9 with an estimate of a 2.0-in diameter)

   Reliability factor: $C_R = 0.75$ (desired reliability of 0.999)

   Modified endurance strength: $s_n' = s_n(C_s)(C_R) = (43\ 000\ \text{psi})(0.81)(0.75) = 26\ 100$ psi

6. **Design factor N:** The choice of a design factor N should consider the many factors discussed in Chapter 5 where a nominal value of $N = 2$ was suggested for general machine design. With the expectation of moderate shock and impact loading, let's specify $N = 4$ for extra safety.

7. **Minimum allowable shaft diameters:** The minimum allowable shaft diameters are now computed at several sections along the shaft using Equation (12–24) if there is any combination of torsion or bending loads at the section of interest. For those sections subjected only to vertical shearing loads, such as at the bearings labeled D in Figure 15–3, Equation (12–16) is used. Table 15–2 summarizes the data used in these equations for each section and reports the computed minimum diameter. The spreadsheet from Section 12–10 was used to complete the analysis.

TABLE 15–2 Summary of the results of shaft diameter calculations for the preliminary design sizing for the shaft

A. Input shaft

Section	Diameter (and related component)	Torque (lb·in)	M_x (lb·in)	M_y (lb·in)	V_x (lb)	V_y (lb)	K_t	Feature	Minimum	Design
			Bending moments		Shearing forces				Diameter (in)	
A	D_1 (coupling)	900	0	0	0	0	1.60	S.R. keyseat	0.73	1.000
B (to right)	D_2 (bearing)	900	0	0	0	0	2.50	Sharp fillet	0.73	*
Note: D_3 must be greater than D_2 or D_4 to provide shoulders for bearing and gear.										2.000
C	D_4 (gear)	900	234	643	94	257	2.00	Prof. keyseat	1.29	1.750
C (to right)	D_4 (gear)	0	234	643	94	257	3.00	Ring groove	1.47	1.750
D	D_5 (bearing)	0	0	0	94	257	2.50	Sharp fillet	0.56	*

B. Output shaft

Section	Diameter (and related component)	Torque (lb·in)	M_x (lb·in)	M_y (lb·in)	V_x (lb)	V_y (lb)	K_t	Feature	Minimum	Design
			Bending moments		Shearing forces				Diameter (in)	
A	D_1 (coupling)	3150	0	0	0	0	1.60	S.R. keyseat	1.10	1.250
B (to left)	D_2 (bearing)	3150	0	0	0	0	2.50	Sharp fillet	1.10	*
Note: D_3 must be greater than D_2 or D_4 to provide shoulders for bearing and gear.										2.000
C	D_4 (gear)	3150	234	643	94	257	2.00	Prof. keyseat	1.36	1.750
C (to left)	D_4 (gear)	0	234	643	94	257	3.00	Ring groove	1.47	1.750
D	D_5 (bearing)	0	0	0	94	257	2.50	Sharp fillet	0.56	*

Note: Bearing seat diameters denoted by * are to be specified. S.R. = sled runner; Prof. = profile.

The last column in Table 15–2 also lists some *preliminary* design decisions for convenient diameters at the given locations. These will be reevaluated and refined as the design is completed. The diameters suggested for the shaft extensions at A on both the input and the output shafts have been set to standard values available for the bores of flexible couplings. The actual flexible couplings are discussed later.

Note that the diameters for the bearing seats at sections B and D have not been given. The reason is that the next task in the design project is to specify commercially available rolling contact bearings to carry the radial loads with suitable life. The diameters of the shafts must be specified according to the limit dimensions recommended by the bearing manufacturer. Therefore, we will leave Table 15–2 as it is for now and revisit it after completing the bearing selection process.

Bearing Selection

We will use the method outlined in Section 14–9 to select commercially available, single-row, deep-groove ball bearings from the data given in Table 14–3. The design load is equal to the radial load, and the value can be found from the shaft analysis shown in Figure 15–3. The reactions at the supports for each shaft are, in fact, the radial loads to which the bearings are subjected.

Because of the symmetry of the design of this system, and because there are no radial loads on the shaft except those produced by the action of the gear teeth, the radial loads on each of the four bearings in this design are the same. Earlier in this chapter, in the shaft design section, we determined that the bearing load is 274 lb.

Recall that design life for the bearings, L_d, is the total number of revolutions expected in service. Therefore, it is dependent on both the rotational speed of the shaft and the design life in hours. We are using a design life of 20 000 hours for all bearings. Shaft 1, the input shaft, rotates at 1750 rpm, resulting in a total number of revolutions of

$$L_d = (20\ 000\ \text{h})(1750\ \text{rev/min})(60\ \text{min/h}) = 2.10 \times 10^9\ \text{rev}$$

Shaft 2, the output shaft, rotates at 500 rpm. Then its design life is

$$L_d = (20\ 000\ \text{h})(500\ \text{rev/min})(60\ \text{min/h}) = 6.0 \times 10^8\ \text{rev}$$

Data for the bearings in Table 14–3 are for a life of 1.0 million rev (10^6 rev).

Now we will use Equation (14–3) with $k = 3$ to compute the required basic dynamic load rating C for each ball bearing. For the bearings on shaft 1,

$$C = P_d(L_d/10^6)^{1/k} = (274\ \text{lb})(2.10 \times 10^9/10^6)^{1/3} = 3510\ \text{lb}$$

Similarly, for the bearings on shaft 2,

$$C = P_d(L_d/10^6)^{1/k} = (274\ \text{lb})(6.0 \times 10^8/10^6)^{1/3} = 2310\ \text{lb}$$

Listed in Table 15–3 are candidate bearings for each shaft having basic dynamic load ratings at least as high as those just computed. We also need to refer to Table 15–2 to determine the minimum acceptable diameters for the shafts at each bearing seat to ensure that the inner race diameter for the bearing is compatible.

We have selected the smallest bearing for each location on shaft 1 that had a suitable value for the basic dynamic load rating. For shaft 2, we decided that the diameter of the shaft extension should be 1.25 in and that the bearing bore must be larger. Bearing 6207 provides a suitable bore and an additional safety factor on the load rating.

Note that the dimensions for the bearings are actually listed in mm by the manufacturer as indicated in Table 14–3. The decimal-inch equivalents are somewhat inconvenient, but they must be used. The dimensions in mm are listed below:

Inch Dimension	mm	Inch Dimension	mm
0.5512	14	1.8504	47
0.5906	15	2.0472	52
0.6693	17	2.4409	62
0.9843	25	2.8346	72
1.3780	35	0.039	1.00-mm (fillet radius)

Bearing Mounting on the Shafts and in the Housing

With the specifications for the bearings, we can finalize the basic dimensions for the shaft diameters. Table 15–4 is an update of the data in Table 15–2 with the bearing bore dimensions given. A few other changes are included as well, which is typical of the iterative

TABLE 15–3 Candidate bearings for shafts 1 and 2

A. For shaft 1: Minimum bore = 0.60 in at section B; 0.54 in at section D

Bearing no.	C (lb)	d (in)	D (in)	B (in)	r_{max} (in)	Comment
6207	4450	1.3780	2.8346	0.6693	0.039	Large for both B and D
6305	3550	0.9843	2.4409	0.6693	0.039	Specify for both B and D

B. For shaft 2: Minimum bore = 1.10 in at section B; 0.54 in at section D

Bearing no.	C (lb)	d (in)	D (in)	B (in)	r_{max} (in)	Comment
6205	2430	0.9843	2.0472	0.5906	0.039	Small for B; specify for D
6207	4450	1.3780	2.8346	0.6693	0.039	Specify for B

TABLE 15–4 Summary of the results of the shaft diameter calculations for the revised design sizing for the shaft

A. Input shaft

Section	Diameter (and related component)	Torque (lb·in)	Bending moments M_x (lb·in)	M_y (lb·in)	Shearing forces V_x (lb)	V_y (lb)	K_t	Feature	Diameter (in) Minimum	Design
A	D_1 (coupling)	900	0	0	0	0	1.60	S.R. keyseat	0.73	0.875
B (to right)	D_2 (bearing)	900	0	0	0	0	2.50	Sharp fillet	0.73	0.984
Note: D_3 must be greater than D_2 or D_4 to provide shoulders for bearing and gear.										2.000
C	D_4 (gear)	900	234	643	94	257	2.00	Prof. keyseat	1.29	1.750
C (to right)	D_4 (gear)	0	234	643	94	257	3.00	Ring groove	1.47	1.750
D	D_5 (bearing)	0	0	0	94	257	2.50	Sharp fillet	0.56	0.984

B. Output shaft

Section	Diameter (and related component)	Torque (lb·in)	Bending moments M_x (lb·in)	M_y (lb·in)	Shearing forces V_x (lb)	V_y (lb)	K_t	Feature	Diameter (in) Minimum	Design
A	D_1 (coupling)	3150	0	0	0	0	1.60	S.R. keyseat	1.10	1.250
B (to left)	D_2 (bearing)	3150	0	0	0	0	2.50	Sharp fillet	1.10	1.378
Note: D_3 must be greater than D_2 or D_4 to provide shoulders for bearing and gear.										2.000
C	D_4 (gear)	3150	234	643	94	257	2.00	Prof. keyseat	1.36	1.750
C (to left)	D_4 (gear)	0	234	643	94	257	3.00	Ring groove	1.47	1.750
D	D_5 (bearing)	0	0	0	94	257	2.50	Sharp fillet	0.56	0.984

Note: S.R. = sled runner; Prof. = profile.

nature of design. For example, the diameter of the input shaft at the coupling (section A) was made slightly smaller than the bearing seat diameter. This permits the bearing to be slid onto the shaft easily to the point where it is then pressed into position on its seat at section B and against the shoulder. Another check needs to be made at section D for both shafts where a 1.750-in diameter steps down to the bearing seat diameter of 0.984 in. There is the possibility that the step is too large and that it may interfere with the outer race of the bearing. That will be checked as we complete the details of mounting the bearings. If interference does occur, it should be a simple matter to provide another small step to make the bearing shoulder an acceptable height.

Mounting of ball and roller bearings on shafts and into housings requires very careful consideration of limit dimensions on all mating parts to ensure proper fits as defined by the bearing manufacturer. The total tolerances on shaft diameters are only a few ten-thousandths of an inch in sizes up to about 6.00 in. Total tolerances on housing bore diameters range from about 0.001 to 0.004 in for sizes from about 1.00 in to over 16.0 in. Violation of the recommended fits will likely cause unsatisfactory performance and possibly early failure of the bearing.

The bore of a bearing is typically pressed onto the shaft seat with a light interference fit to ensure that the inner race rotates with the shaft. The *OD* of the bearing is typically a close sliding fit in the housing, with the minimum clearance being zero. This facilitates installation and allows some slight movement of the bearing as thermal deformation occurs during operation. Tighter fits than those recommended by the manufacturer may cause the rolling elements to bind between the inner and outer races, resulting in higher loads and higher friction. Looser fits may permit the outer race to rotate relative to the housing, a very undesirable situation.

Only one of the two bearings on a shaft should be located and held fixed axially in the housing to provide proper alignment of functional components such as the gears in this design. The second bearing should be installed in a way that allows some small axial movement during operation. If the second bearing is held fixed also, it is likely that extra axial loads will be developed for which the bearing has not been designed.

We first discuss the specification of shaft limit dimensions at the bearing seats.

Bearing Seat Diameters.   Because most commercially available bearings are produced to metric dimensions, the fits are specified according to the tolerance system of the International Standards Organization (ISO). Only a sampling of the data are listed here to illustrate the process of specifying limit dimensions for shafts and housings to accommodate bearings. Manufacturers' catalogs include much more extensive data.

For bearings carrying moderate to heavy loads such as those in this example design, the following tolerance grades are recommended for the bearing seats on shafts and housing bore fits with the outer race:

Bearing Bore Diameter Range	Tolerance Grade
10–18 mm	j5
20–100 mm	k5
105–140 mm	m5
150–200 mm	m6
Housing bore (any)	H8

Table 15–5 shows representative data for the actual limit dimensions for these grades over the size ranges included for the bearings listed in Table 14–3. Note that the bearing bore and the bearing *OD* dimensions are those expected from the bearing manufacturer. You must control the shaft diameter and the housing bore to the specified minimum and maximum dimensions. The table also lists the minimum and maximum fits that result. The symbol L indicates that there is a net clearance (loose) fit; T indicates an interference (tight) fit. So bearings must be pressed onto the shaft seat. Sometimes heating of the bearing and cool-

ing of the shaft are used to produce a clearance to facilitate assembly. When the parts return to normal temperatures, the final fit is produced.

We now show the determination of the limit dimensions for the shaft at each bearing seat.

Shaft 1: Input Shaft. Both bearings 1 and 2 are number 6305.

Nominal bore = 25 mm (0.9843 in)

From Table 15–5: k5 ISO tolerance grade on the shaft seat; limits of 0.9847–0.9844 in

Resulting fit between bearing bore and shaft seat: 0.0001 in tight to 0.0008 in tight

OD of the outer race = 62 mm (2.4409 in)

TABLE 15–5 Shaft and housing fits for bearings

A. Shaft fits

Bearing bore			ISO tolerance grade	Shaft diameter		Limits of fit	
Nominal (mm)	Maximum (in)	Minimum (in)		Maximum (in)	Minimum (in)	Minimum (in)	Maximum (in)
10	0.3937	0.3934	j5	0.3939	0.3936	0.0001L	0.0005T
12	0.4724	0.4721	j5	0.4726	0.4723	0.0001L	0.0005T
15	0.5906	0.5903	j5	0.5908	0.5905	0.0001L	0.0005T
17	0.6693	0.6690	j5	0.6695	0.6692	0.0001L	0.0005T
20	0.7874	0.7870	k5	0.7878	0.7875	0.0001T	0.0008T
25	0.9843	0.9839	k5	0.9847	0.9844	0.0001T	0.0008T
30	1.1811	1.1807	k5	1.1815	1.1812	0.0001T	0.0008T
35	1.3780	1.3775	k5	1.3785	1.3781	0.0001T	0.0010T
40	1.5748	1.5743	k5	1.5753	1.5749	0.0001T	0.0010T
45	1.7717	1.7712	k5	1.7722	1.7718	0.0001T	0.0010T
50	1.9685	1.9680	k5	1.9690	1.9686	0.0001T	0.0010T
55	2.1654	2.1648	k5	2.1660	2.1655	0.0001T	0.0012T
60	2.3622	2.3616	k5	2.3628	2.3623	0.0001T	0.0012T
65	2.5591	2.5585	k5	2.5597	2.5592	0.0001T	0.0012T
70	2.7559	2.7553	k5	2.7565	2.7560	0.0001T	0.0012T
75	2.9528	2.9522	k5	2.9534	2.9529	0.0001T	0.0012T
80	3.1496	3.1490	k5	3.1502	3.1497	0.0001T	0.0012T
85	3.3465	3.3457	k5	3.3472	3.3466	0.0001T	0.0015T
90	3.5433	3.5425	k5	3.5440	3.5434	0.0001T	0.0015T
95	3.7402	3.7394	k5	3.7409	3.7403	0.0001T	0.0015T
100	3.9370	3.9362	k5	3.9377	3.9371	0.0001T	0.0015T
105	4.1339	4.1331	m5	4.1350	4.1344	0.0005T	0.0019T
110	4.3307	4.3299	m5	4.3318	4.3312	0.0005T	0.0019T
115	4.5276	4.5268	m5	4.5287	4.5281	0.0005T	0.0019T
120	4.7244	4.7236	m5	4.7255	4.7249	0.0005T	0.0019T
125	4.9213	4.9203	m5	4.9226	4.9219	0.0006T	0.0023T
130	5.1181	5.1171	m5	5.1194	5.1187	0.0006T	0.0023T
140	5.5118	5.5108	m5	5.5131	5.5124	0.0006T	0.0023T
150	5.9055	5.9045	m6	5.9071	5.9061	0.0006T	0.0026T
160	6.2992	6.2982	m6	6.3008	6.2998	0.0006T	0.0026T
170	6.6929	6.6919	m6	6.6945	6.6935	0.0006T	0.0026T
180	7.0866	7.0856	m6	7.0882	7.0872	0.0006T	0.0026T
190	7.4803	7.4791	m6	7.4821	7.4810	0.0007T	0.0030T
200	7.8740	7.8728	m6	7.8758	7.8747	0.0007T	0.0030T

(*continued*)

TABLE 15–5   (*continued*)

B. Housing Fits

Bearing *OD*			ISO tolerance grade	Housing bore		Limits of fit	
Nominal (mm)	Maximum (in)	Minimum (in)		Minimum (in)	Maximum (in)	Minimum (in)	Maximum (in)
30	1.1811	1.1807	H8	1.1811	1.1824	0	0.0017L
32	1.2598	1.2594	H8	1.2598	1.2613	0	0.0019L
35	1.3780	1.3776	H8	1.3780	1.3795	0	0.0019L
37	1.4567	1.4563	H8	1.4567	1.4582	0	0.0019L
40	1.5748	1.5744	H8	1.5748	1.5763	0	0.0019L
42	1.6535	1.6531	H8	1.6535	1.6550	0	0.0019L
47	1.8504	1.8500	H8	1.8504	1.8519	0	0.0019L
52	2.0472	2.0467	H8	2.0472	2.0490	0	0.0023L
62	2.4409	2.4404	H8	2.4409	2.4427	0	0.0023L
72	2.8346	2.8341	H8	2.8346	2.8364	0	0.0023L
80	3.1496	3.1491	H8	3.1496	3.1514	0	0.0023L
85	3.3465	3.3459	H8	3.3465	3.3486	0	0.0027L
90	3.5433	3.5427	H8	3.5433	3.5454	0	0.0027L
100	3.9370	3.9364	H8	3.9370	3.9391	0	0.0027L
110	4.3307	4.3301	H8	4.3307	4.3328	0	0.0027L
120	4.7244	4.7238	H8	4.7244	4.7265	0	0.0027L
125	4.9213	4.9206	H8	4.9213	4.9238	0	0.0032L
130	5.1181	5.1174	H8	5.1181	5.1206	0	0.0032L
140	5.5118	5.5111	H8	5.5118	5.5143	0	0.0032L
150	5.9055	5.9048	H8	5.9055	5.9080	0	0.0032L
160	6.2992	6.2982	H8	6.2992	6.3017	0	0.0035L
170	6.6929	6.6919	H8	6.6929	6.6954	0	0.0035L
180	7.0866	7.0856	H8	7.0866	7.0891	0	0.0035L
190	7.4803	7.4791	H8	7.4803	7.4831	0	0.0040L
200	7.8740	7.8728	H8	7.8740	7.8768	0	0.0040L
215	8.4646	8.4634	H8	8.4646	8.4674	0	0.0040L
225	8.8583	8.8571	H8	8.8583	8.8611	0	0.0040L
230	9.0551	9.0539	H8	9.0551	9.0579	0	0.0040L
240	9.4488	9.4476	H8	9.4488	9.4516	0	0.0040L
250	9.8425	9.8413	H8	9.8425	9.8453	0	0.0040L
260	10.2362	10.2348	H8	10.2362	10.2394	0	0.0046L
270	10.6299	10.6285	H8	10.6299	10.6331	0	0.0046L
280	11.0236	11.0222	H8	11.0236	11.0268	0	0.0046L
290	11.4173	11.4159	H8	11.4173	11.4205	0	0.0046L
300	11.8110	11.8096	H8	11.8110	11.8142	0	0.0046L
310	12.2047	12.2033	H8	12.2047	12.2079	0	0.0046L
320	12.5984	12.5968	H8	12.5984	12.6019	0	0.0051L
340	13.3858	13.3842	H8	13.3858	13.3893	0	0.0051L
360	14.1732	14.1716	H8	14.1732	14.1767	0	0.0051L
380	14.9606	14.9590	H8	14.9606	14.9641	0	0.0051L
400	15.7480	15.7464	H8	15.7480	15.7515	0	0.0051L
420	16.5354	16.5336	H8	16.5354	16.5392	0	0.0056L

Note: L = loose; T = tight.

From Table 15–5: H8 ISO tolerance grade on the housing bore; limits of 2.4409–2.4427 in

Resulting fit between outer race and housing bore: 0.0 to 0.0023 in loose

Shaft 2: Output Shaft.   Bearing 3 at D is number 6205.

Nominal bore = 25 mm (0.9843 in)

From Table 15–5: k5 ISO tolerance grade on the shaft seat; limits of 0.9847–0.9844 in

Resulting fit between bearing bore and shaft seat: 0.0001 in tight to 0.0008 in tight

OD of the outer race = 52 mm (2.0472 in)

From Table 15–5: H8 ISO tolerance grade on the housing bore; limits of 2.0472–2.0490 in

Resulting fit between outer race and housing bore: 0.0 to 0.0023 in loose

Shaft 2: Output Shaft.   Bearing 4 at B is number 6207.

Nominal bore = 35 mm (1.3780 in)

From Table 15–5: k5 ISO tolerance grade on the shaft seat; limits of 1.3785–1.3781 in

Resulting fit between bearing bore and shaft seat: 0.0001 in tight to 0.0010 in tight

OD of the outer race = 72 mm (2.8346 in)

From Table 15–5: H8 ISO tolerance grade on the housing bore; limits of 2.8346–2.8364 in

Resulting fit between outer race and housing bore: 0.0 to 0.0023 in loose

Shaft and Housing Shoulder Diameters.   Each of the bearings in this design is to be seated against a shoulder on one side of the bearing. The shaft shoulder must be sufficiently large to provide a solid, flat surface against which to seat the side of the inner race. But the shoulder must not be so high that it contacts the outer race because the inner race rotates at shaft speed and the outer race is stationary.

Similarly, a shoulder in the housing must provide for the solid location of the outer race but not be such that it contacts the inner race.

Bearing manufacturers' catalogs provide data such as those shown in Table 15–6 to guide you in specifying suitable shoulder heights. The value of S is the minimum shaft shoulder diameter. The nominal maximum diameter is the mean diameter for the bearing at the middle of the balls. The value of H is the maximum housing shoulder diameter, with the nominal minimum diameter being the mean diameter for the bearing.

For example, in the present design, the minimum shoulder diameter at each bearing on shaft 1 should be 1.14 in as indicated for the bearing number 305 in Table 15–6. (Note that the bearing number 6305 specified for the shaft is of the same *series* as the number 305, indicating that it would have similar dimensions.) The maximum housing shoulder diameter for the number 305 bearing is 2.17 in, where the outer race is to seat against a shoulder.

On shaft 2, the shoulder for bearing 6205 should also be at least 1.14 in, and the shoulder for bearing 6207 should be 1.53 in minimum. The maximum housing shoulder diameter for the 6205 bearing on shaft 2 is 1.81. For the 6207 bearing, the maximum housing shoulder diameter is 2.56 in.

TABLE 15–6   Shaft and housing shoulder diameters

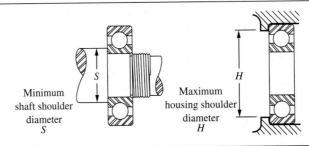

Minimum shaft shoulder diameter S

Maximum housing shoulder diameter H

200 series bearings						300 series bearings					
No.	S	H	No.	S	H	No.	S	H	No.	S	H
200	0.50	0.98	216	3.55	5.12	300	0.50	1.18	316	3.62	6.22
201	0.58	1.06	217	3.75	5.51	301	0.63	1.22	317	3.90	6.54
202	0.69	1.18	218	3.94	5.91	302	0.75	1.42	318	4.09	6.93
203	0.77	1.34	219	4.21	6.22	303	0.83	1.61	319	4.29	7.32
204	0.94	1.61	220	4.41	6.61	304	0.94	1.77	320	4.49	7.91
205	1.14	1.81	221	4.61	7.01	305	1.14	2.17	321	4.69	8.31
206	1.34	2.21	222	4.80	7.40	306	1.34	2.56	322	4.88	8.90
207	1.53	2.56	224	5.20	7.99	307	1.69	2.80	324	5.28	9.69
208	1.73	2.87	226	5.67	8.50	308	1.93	3.19	326	5.83	10.32
209	1.94	3.07	228	6.06	9.29	309	2.13	3.58	328	6.22	11.10
210	2.13	3.27	230	6.46	10.08	310	2.36	3.94	330	6.61	11.89
211	2.41	3.68	232	6.85	10.87	311	2.56	4.33	332	7.01	12.68
212	2.67	3.98	234	7.40	11.50	312	2.84	4.65	334	7.40	13.47
213	2.86	4.37	236	7.80	11.89	313	3.03	5.04	336	7.80	14.25
214	3.06	4.57	238	8.19	12.68	314	3.23	5.43	338	8.35	14.88
215	3.25	4.76	240	8.58	13.47	315	3.43	5.83	340	8.74	15.67

Notes:

S = minimum shaft shoulder diameter
    Maximum diameter should not exceed the mean diameter of the bearing at the middle of the balls.

H = maximum housing shoulder diameter
    Minimum diameter should not be less than the mean diameter of the bearing at the middle of the balls.

Table 15–7 shows the pertinent data used to decide on the values for the shoulder diameters and, in the last two columns, gives the specified values. Where the specified shoulder diameter is less than the preliminary value shown in Table 15–4, another step in the shaft will be used to provide the proper shoulder for the bearing and for the gear. This can be seen in the drawings of the shafts given at the end of this chapter.

The use of a 1.75-in diameter for the shoulder at D on shaft 1 was specified because that is the diameter of the shaft chosen earlier. It is a bit higher than the mean diameter of the bearing, but it should still be lower than that of the outer race. More complete data in a manufacturer's catalog indicates the diameter of the inner surface of the outer race to be 2.00 in, so the 1.75-in diameter is acceptable.

Fillet Radii.   Each of the bearings specified for the reducer calls for the maximum fillet radius at the shoulder that locates the bearing to be 0.039 in. See Table 14–3. Let's specify the limits on the radius to be 0.039 to 0.035. Before committing to the design, we will check the stress concentration factor at each shoulder.

TABLE 15–7   Shaft and housing shoulder diameters for bearings in the present design

A. Shaft 1

Bearing no.	Bore	OD	Mean diameter	S Minimum	H Maximum	Specified shaft shoulder	Specified housing shoulder
6305 at B	0.9843	2.4409	1.713	1.14	2.17	1.50	2.00
6305 at D	0.9843	2.4409	1.713	1.14	2.17	1.75	2.00

B. Shaft 2

Bearing no.	Bore	OD	Mean diameter	S Minimum	H Maximum	Specified shaft shoulder	Specified housing shoulder
6207 at B	1.3780	2.8346	2.106	1.53	2.56	2.00	2.25
6205 at D	0.9843	2.0472	1.516	1.14	1.81	1.50	1.75

Flexible Couplings.   The use of flexible couplings on both the input and the output shafts has been taken into account in the shaft design and analysis. They allow the transmission of torque between two shafts but do not exert significant radial or axial forces on the shaft. In the present design, the use of flexible couplings made the shaft design simpler and decreased the loads on bearings compared with having a device such as a belt sheave or a chain sprocket on the shaft.

Now we specify suitable couplings for the input and output shafts. Chapter 11 showed many examples for such couplings, and you should review them now. It is impractical to reproduce data for all couplings in this book. As you read this section, it would be good for you to seek a copy of the catalog of one of the manufacturers of couplings and study their recommended selection procedures. See the Internet sites for Chapter 11.

We have selected couplings of the type pictured in Figure 11–16, called the *Browning Ever-flex Coupling* from Emerson Power Transmission, a division of the Emerson Electric Company. Rubber flex members are permanently bonded to steel hubs, and the flexing of the rubber accommodates parallel misalignment of the mating shafts up to 0.032 in, angular misalignment of $\pm 3°$, and axial end float of the shafts of up to ± 0.032 in. It is important for you to design the drive system for the saw to provide this alignment of the input shaft to the drive motor and of the output shaft to the drive shaft of the saw.

The selection of a suitable coupling relies on the power transmission rating of the various sizes available. But the power rating must be correlated to the speed of rotation because the real variable is the torque to which the coupling is subjected. Both the input and the output couplings transmit nominally 25 hp in this design for the drive for the saw. But the input shaft rotates at 1750 rpm, and the output shaft rotates at 500 rpm. Because torque is inversely proportional to speed of rotation, the torque experienced by the coupling on the output shaft is approximately 3.5 times higher than that on the input shaft. The coupling catalog data also call for the use of a *service factor* based on the kind of machine being driven, and some suggested values are included in the catalog. We judge that a service factor of 1.5 is suitable for the saw which will see mostly smooth power transmission with occasional moderate shock loading.

The service factor is applied to the nominal power being transmitted to compute a value for the *normal rating* for the couplings. Then

$$\text{Normal rating} = \text{power input} \times \text{service factor} = 25 \text{ hp}(1.5) = 37.5 \text{ hp}$$

The catalog tables list coupling number CFR6 with a suitable normal rating at 1750 rpm for the input shaft and number CFR9 for the output shaft at 500 rpm.

We specify hubs for the couplings that have machined bores and keyways with a range of bores allowed. Each coupling half can have a different bore according to the shaft size on which it is to be mounted. For the input shaft, we have specified the diameter to be 0.875 in (7/8 in), and this will be the specification for the bore of that half for the CFR6 coupling. The keyway is 3/16 × 3/32 to accept a 3/16-in square key. The nominal maximum length of shaft inside each half of the coupling is 2.56 in.

The other half of the CFR6 coupling mounts on the motor shaft. Recall that the design requirements listed at the beginning of this design process specified a 25-hp motor with a NEMA Frame 284T. Table 21–3 gives the shaft diameter for this motor to be 1.875 in (1 7/8 in) with a 1/2 × 1/4 in keyway to accept a 1/2-in square key. This will be the bore specified for the motor half of the CFR6 coupling. The reason for the large difference in the sizes of the shafts for the motor and for our reducer is that the general-purpose motor must be designed to carry a significant side load, and our shaft does not.

The output shaft of the reducer at the coupling has a diameter of 1.250 in, and the input shaft for the saw will have the same size. Therefore, both halves of the CFR9 coupling will have that bore with a 1/4 × 1/8 in keyway to accept a 1/4-in square key. The nominal maximum shaft length inside each half of the coupling is 3.125 in.

Keys and Keyseats. A total of six keys need to be specified: two for each half of the flexible couplings on the input and output shafts, and one for each gear in the reducer. The methods of Chapter 11 are used to verify the suitability of the keys and to specify the required length using Equation (11–5). We will use standard key sizes made from AISI 1020 CD steel having a yield strength of 51 000 psi.

1. *Keys for CFR6 coupling on the input shaft:* First let's check the keys inside the couplings because their sizes have already been specified by the coupling manufacturer. The coupling half that mounts on the input shaft is critical because its bore diameter of 0.875 in is smaller, resulting in larger forces on the key when transmitting the torque of 900 lb·in which was computed earlier during the shaft design. The key is 3/16 in square (0.188 in). We use a design factor N of 4 as we did in the shaft design. Then, from Equation (11–5),

$$L = \frac{4TN}{DWs_y} = \frac{4(900 \text{ lb·in})(4)}{(0.875 \text{ in})(0.188 \text{ in})(51\,000 \text{ psi})} = 1.72 \text{ in}$$

We can specify a key length of 2.50 in for extra safety and to match the length of the hub of the CFR6 coupling. The 1/2-in key for the motor shaft should be made to match the length of the coupling hub also, and it should be very safe because of the larger key size and the larger shaft size carrying the same torque.

2. *Keys for the CFR9 coupling on the output shaft:* For the output shaft and the drive shaft for the saw,

$$T = 3150 \text{ lb·in}$$
$$D = 1.25 \text{ in}$$
$$W = 0.250 \text{ in} \quad \text{(key width)}$$
$$L = \frac{4TN}{DWs_y} = \frac{4(3150 \text{ lb·in})(4)}{(1.25 \text{ in})(0.250 \text{ in})(51\,000 \text{ psi})} = 3.16 \text{ in}$$

We will make the key length 3.125 in (3 1/8 in), the full length of the hub of the CFR9 coupling. The conservative design factor of 4 should make this length acceptable.

3. *Key for the pinion on shaft 1:* The bore of the pinion is to be nominally 1.75 in as determined in the shaft design and shown in Table 15–4. The key size for this diameter should be 3/8 in square according to Table 11–1. The torque being transmitted is 900 lb·in. Then Equation (11–5) gives

$$L = \frac{4TN}{DWs_y} = \frac{4(900 \text{ lb·in})(4)}{(1.75 \text{ in})(0.375 \text{ in})(51\ 000 \text{ psi})} = 0.430 \text{ in}$$

The face width of the gear is 2.00 in. Let's use a key length of 1.50 in and center the profile keyseat at section C on the shaft so that the keyseat does not significantly interact with the ring groove to the right or with the shoulder fillet to the left.

4. *Key for the gear on shaft 2:* The bore of the gear is to be nominally 1.75 in as determined in the shaft design and shown in Table 15–4. The key size for this diameter should be 3/8 in square according to Table 11–1. The torque being transmitted is 3150 lb·in. Then Equation (11–5) gives

$$L = \frac{4TN}{DWs_y} = \frac{4(3150 \text{ lb·in})(4)}{(1.75 \text{ in})(0.375 \text{ in})(51\ 000 \text{ psi})} = 1.50 \text{ in}$$

The face width of the gear is 2.00 in. Let's use a key length of 1.50 in for this key also.

The key designs are summarized in the following list:

Summary of Key Designs

Motor shaft: 1/2-in square key × 2.50 in long

Input shaft of reducer at coupling: 3/16-in square key × 2.50 in long; sled runner keyseat

Input shaft at pinion: 3/8-in square key × 1.50 in long; profile shaft keyseat

Output shaft at gear: 3/8-in square key × 1.50 in long; profile shaft keyseat

Output shaft at coupling: 1/4-in square key × 3.125 in long; sled runner keyseat

Drive shaft for saw at coupling: 1/4-in square key × 3.125 in long; sled runner keyseat

The tolerances for the keys and keyseats are summarized as follows: Standard square bar stock in AISI 1020 or 1030 steel is available to be used for keys. Typical tolerances are given in Table 15–8. Also given are recommended tolerances on the keyseat width dimension and the resulting fit between the key and the keyseat. A small clearance fit is desirable to permit easy assembly while not allowing the key to rock noticeably when installed.

Tolerances on Other Shaft Dimensions. You should review the discussion in Chapter 13, Sections 13–5 and 13–9, on tolerances and fits. See also References 1, 2, and 5 in Chapter 15, along with other comprehensive texts on technical drawing and interpretation of engineering drawings.

TABLE 15–8   Tolerances and fits for keys and keyseats

Key size (width in inches)	Tolerance on key (all +0.000)	Tolerance on keyseat (all −0.000)	Fit range
Up to 1/2	−0.002	+0.002	0.000–0.004
Over 1/2 to 3/4	−0.002	+0.003	0.000–0.005
Over 3/4 to 1	−0.003	+0.003	0.000–0.006
Over 1 to $1\frac{1}{2}$	−0.003	+0.004	0.000–0.007
Over $1\frac{1}{2}$ to $2\frac{1}{2}$	−0.004	+0.004	0.000–0.008

The fit of the bore of the gears on the shafts or the fit of the *ID* of the couplings on the ends of the shafts must be specified. Either a close sliding fit or a close locational fit is recommended for such components. Data are given in Table 13–3 for the RC2, RC5, and RC8 fits. More complete data are available in References 1, 2, 3, 4, 5, and 9 in Chapter 13. We will apply the RC5 fit to accurate mating parts that must assemble easily but where little perceptible play between the parts is desired. The RC fits use the *basic hole system*, as illustrated in Chapter 13.

15–6 FINAL DESIGN DETAILS FOR THE SHAFTS

Figures 15–4 and 15–5 show the final design for the input and output shafts. Data from throughout this chapter have been used to specify pertinent dimensions. Where fillets are specified, a final check on the stress condition has been made to ensure that the estimated stress concentration factors used in the earlier design analysis are satisfactory and that the final stress levels are safe.

Keyseat details have been shown in sections below the main dimensioned shaft drawings. See Chapter 11 for computing the vertical dimension from the bottom of the shaft to the bottom of the keyseat.

The retaining ring grooves are drawn to the dimensions specified for a basic external ring (Type 5100) for a 1.75-in-diameter shaft from the Truarc Company as pictured in Figure 11–30.

Four shaft diameters are sized to the RC5 fit in Figures 15–4 and 15–5. On shaft 1 they are at the extension where the coupling mounts and at the pinion. On shaft 2 they are at the gear and at the coupling location on the output extension. Table 15–9 summarizes the data for the fits. You should verify these using the procedure shown in Chapter 13. Note that the limit dimensions for both the shaft diameter and the bore of the mating element are given and that the total tolerance on each dimension is small, less than 0.002 in for any dimension. Note also the small variation of the clearance on mating parts as indicated in the last column called "Fit."

For the shafts in this project, we have specified geometric tolerances for concentricity of four critical diameters for each shaft. Figure 15–5(b) documents the approach for the output shaft. Because of the similarity of the two shafts, the nature of the callouts would be the same for the input shaft. The datum or reference diameter is specified at the gear. Then the diameters at the two bearing seats and at the end of the shaft where the coupling mounts are controlled with concentricity feature control blocks. Shoulders for locating bearings and gears are controlled for perpendicularity to the axis of the shaft as represented by the gear diameter. The keyseat is controlled for parallelism to the axis of the shaft.

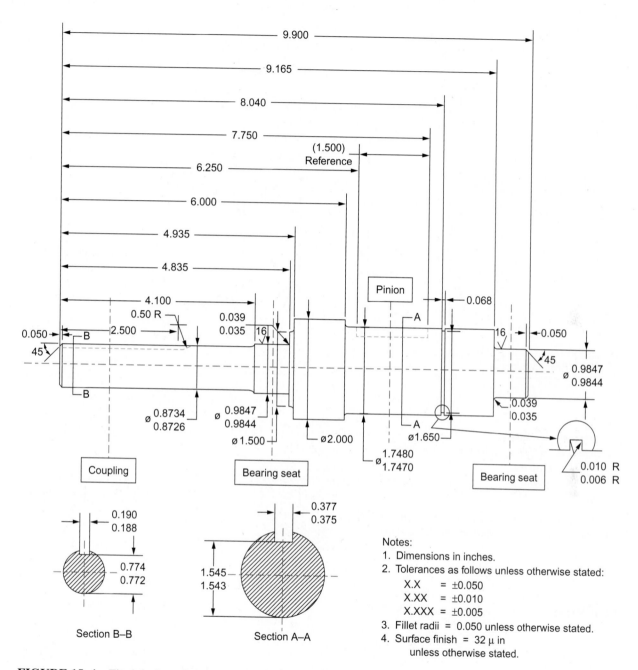

FIGURE 15–4 Final design of the input shaft

653

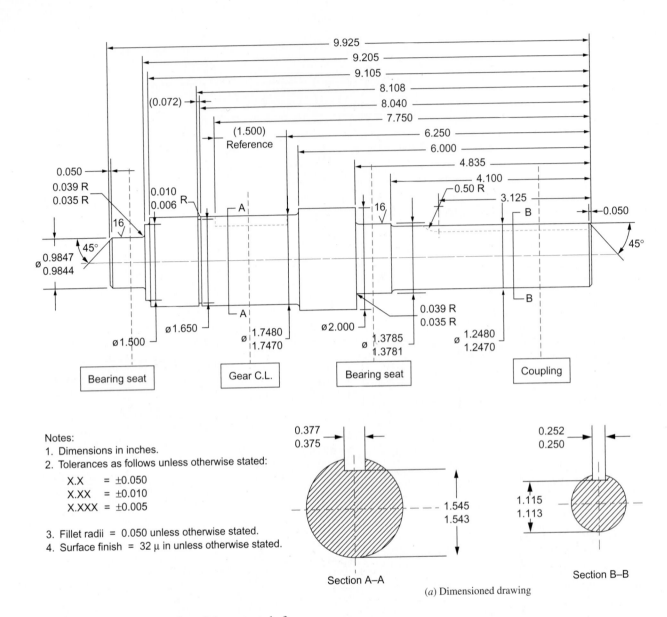

Notes:
1. Dimensions in inches.
2. Tolerances as follows unless otherwise stated:

   X.X   = ±0.050
   X.XX   = ±0.010
   X.XXX = ±0.005

3. Fillet radii = 0.050 unless otherwise stated.
4. Surface finish = 32 μ in unless otherwise stated.

Section A–A

Section B–B

(a) Dimensioned drawing

FIGURE 15–5 Final design of the output shaft

TABLE 15–9 Calculations and specifications for fits of elements on the shafts of the transmission

Location	Nominal diameter	Bore limit dimensions for external element	OD limit dimensions for shaft	Fit
Shaft 1:				
Input at coupling	0.8750	0.8762/0.8750	0.8734/0.8726	+0.0026/+0.0036
Pinion	1.7500	1.7516/1.7500	1.7480/1.7470	+0.0020/+0.0046
Shaft 2:				
Gear	1.7500	1.7516/1.7500	1.7480/1.7470	+0.0020/+0.0046
Output at coupling	1.2500	1.2516/1.2500	1.2480/1.2470	+0.0020/+0.0046

Note: RC5 fit is used for all locations; dimensions are in inches.

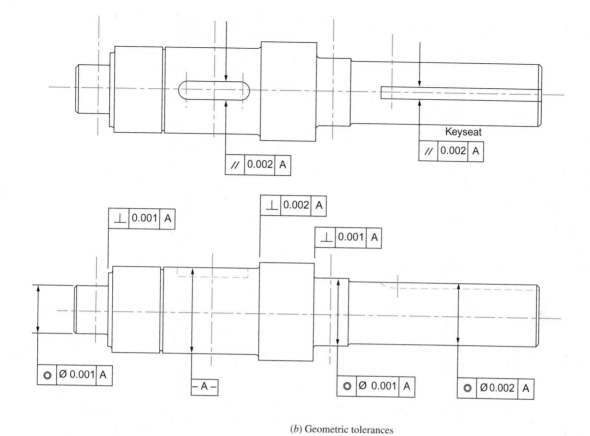

(b) Geometric tolerances

FIGURE 15–5 (continued)

15–7 ASSEMBLY DRAWING

Figure 15–6 is an assembly drawing for the reducer with all features drawn to scale. The housing has been shown as a rectangular box shape for simplicity. Bearings are held in bearing retainers which are then fastened to the housing walls. Special attention to the alignment of the retainers is required and would be an important task for the detailing of the housing, not shown here.

The assembly of all components into the housing is facilitated by having the right side removable. Again, alignment of the cover piece with the main housing is critical.

Seals have been shown in the bearing retainers where the shafts penetrate the side walls of the housing. See Chapter 11 for more information on seals.

Critique of the Design

Figures 15–4, 15–5, and 15–6 present a design that meets the basic design requirements established at the beginning of this chapter. It is likely that refinements could be made if more detail were available about the saw for which the reducer is being designed.

It appears that the length of the shafts could be made somewhat shorter. The distances between the center of the gears and the bearings was arbitrarily set at 2.50 in at the start of the design process when dimensions for any components were unknown. Now that the nominal size of the gears, bearings, and couplings are known, further iterations on the design could result in a smaller package.

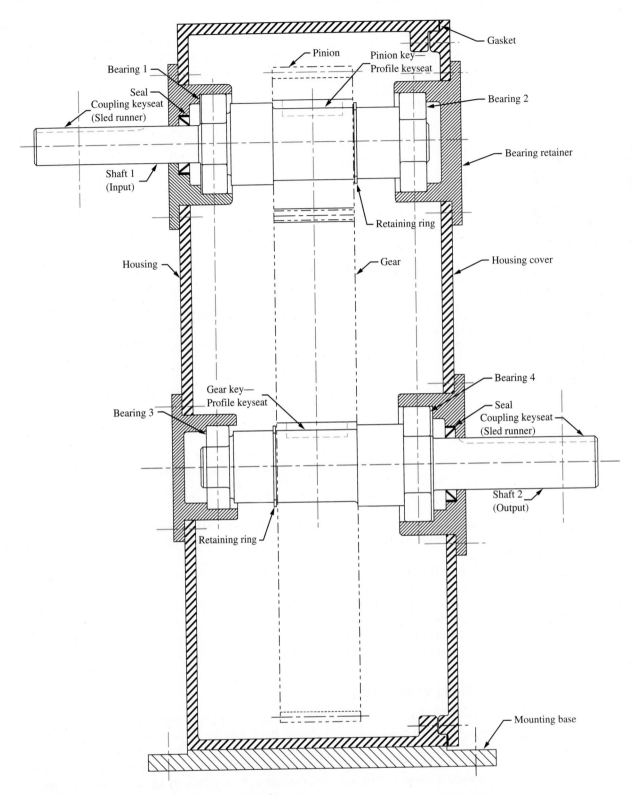

FIGURE 15–6 Assembly drawing for the reducer

You should look at other commercially available gear-type speed reducers for other features that might be built into this design. Note particularly Figures 9–34, 9–35, and 9–36 in Chapter 9 and Figures 10–1, 10–2, and 10–15 in Chapter 10.

REFERENCES

1. American Society of Mechanical Engineers. ASME Standard Y14.5M. *Dimensioning and Tolerancing*. New York: American Society of Mechanical Engineers, 1995.

2. Earle, James H. *Graphics for Engineers with AutoCAD 2002*. 6th ed. Upper Saddle River, NJ: Prentice Hall, 2003.

3. Kepner, Charles H., and Benjamin B. Tregoe. *The New Rational Manager: An Updated Edition for a New World*. Princeton, NJ: Princeton Research Press, 1997.

4. National Research Council. *Improving Engineering Design: Designing for Competitive Advantage*. Washington, DC: National Academy Press, 1991.

5. Oberg, Erik, et al. *Machinery's Handbook*. 26th ed. New York: Industrial Press, 2000.

6. Peerless-Winsmith, Inc. *The Speed Reducer Book: A Practical Guide to Enclosed Gear Drives*. Springville, NY: Peerless-Winsmith, 1980.

7. Pugh, Stuart. *Total Design: Integrated Methods for Successful Product Engineering*. Reading, MA: Addison-Wesley, 1991.

8. Pugh, Stuart, and Don Clausing. *Creating Innovative Products Using Total Design: The Living Legacy of Stuart Pugh*. Reading, MA: Addison-Wesley, 1996.

9. Ullman, David G. *The Mechanical Design Process*. 2nd ed. New York: McGraw-Hill, 1997.

INTERNET SITES RELATED TO TRANSMISSION DESIGN

1. **Kepner-Tregoe, Inc.** *www.kepner-tregoe.com* A management consulting and training firm specializing in strategic and operational decision making. The firm's work evolves from the popular book listed as Reference 3. The Kepner-Tregoe approach is applicable to problem solving and decision making in product development and manufacturing operations.

2. **Peerless-Winsmith, Inc.** *www.winsmith.com* Manufacturer of a broad line of speed reducers. This company produced the book listed in Reference 6.

PART III

Design Details and Other Machine Elements

OBJECTIVES AND CONTENT OF PART III
Chapters 16–23 present methods of analysis and design of several important machine elements that were not particularly pertinent to the design of a power transmission as presented in Part II of this book. These chapters can be covered in any order or can be used as reference material for general design projects.

Chapter 16: Plain Surface Bearings discusses plain surface bearings, sometimes called *journal bearings*. These bearings employ smooth-surfaced elements to support loads from shafts or other devices in which relative motion occurs. You will learn how to specify materials for the parts of the bearing, the geometry of the bearing, and the lubricant. Full-film hydrodynamic lubrication and boundary lubrication are discussed.

Chapter 17: Linear Motion Elements discusses devices that convert rotary motion to linear motion, or vice versa. They are often used in machine tools, automation equipment, and parts of construction machinery. You will learn about their geometry and how to analyze their performance.

Chapter 18: Fasteners describes machine screws, bolts, nuts, and set screws. You will learn about the kinds of materials used for fasteners and how to design them for safe, reliable performance. Also discussed briefly are rivets, quick-operating fasteners, welding, brazing, soldering, and adhesive bonding.

Chapter 19: Springs discusses how to design and analyze helical compression springs, helical extension springs, and torsional springs.

Chapter 20: Machine Frames, Bolted Connections, and Welded Joints presents the important skill of designing a frame for rigidity as well as strength. You will gain experience with analyzing the forces and stresses in bolted and welded connections that hold load-carrying members together. You will also learn how to analyze eccentric loads on connections.

Chapter 21: Electric Motors and Contrals discusses the many types of AC and DC motors that are commercially available. Because a great many mechanical design projects involve the use of an electric motor as the prime mover, you will learn how to match their performance characteristics to the needs of the machine being designed and to specify motor controls.

Chapter 22: Motion Control: Clutches and Brakes discusses the many types of clutches and brakes that can be applied. Clutches provide the path to connect power from a prime mover to a driven machine. Brakes bring the moving equipment to a stop or slow it down. You will learn how to analyze their performance and to either design them or specify commercially available units.

Chapter 23: Design Projects presents several projects that you can complete yourself.

The Big Picture

You Are the Designer

16–1   Objectives of This Chapter

16–2   The Bearing Design Task

16–3   Bearing Parameter, $\mu n/p$

16–4   Bearing Materials

16–5   Design of Boundary-Lubricated Bearings

16–6   Full-Film Hydrodynamic Bearings

16–7   Design of Full-Film Hydrodynamically Lubricated Bearings

16–8   Practical Considerations for Plain Surface Bearings

16–9   Hydrostatic Bearings

16–10 Tribology: Friction, Lubrication, and Wear

Plain Surface Bearings

☐ The purpose of a bearing is to support a load while permitting relative motion between two elements of a machine. This chapter discusses *plain surface* bearings where the two parts that move relative to each other do not have rolling elements between them.

Discover

Look around your home and in your car for products that have plain surface bearings. Find some that experience rotational motion and some that have linear sliding contact. Consider simple items such as hinges, door locks, latches, or the wheels of a lawn mower. Most of these experience boundary lubrication.

Now see if you can find information about the crankshaft bearings in the engine of your car. Such bearings usually employ full-film hydrodynamic *lubrication. What can you discover about them?*

How do you think a large telescope or an astronomical radio antenna is supported to permit it to be moved easily and positioned precisely? One method is called hydrostatic bearings. *What can you discover about such bearings?*

This chapter will help you explore all of these kinds of bearings and complete the basic design analyses required to ensure satisfactory operation.

The purpose of a bearing is to support a load while permitting relative motion between two elements of a machine. This chapter discusses *plain surface* bearings where the two parts that move relative to each other do not have rolling elements between them. Note that rolling contact bearings were discussed in Chapter 14.

Plain surface bearings for rotating parts are, of course, cylindrical with a typical arrangement as shown in Figure 16–1. The inner member, called the *journal*, is usually that part of a shaft where any radial reaction forces are transferred to the base of the machine.

FIGURE 16–1
Bearing geometry

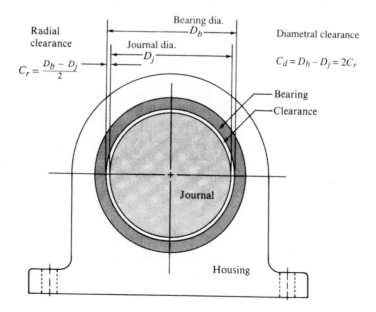

Radial clearance
$$C_r = \frac{D_b - D_j}{2}$$

Bearing dia.
D_b

Journal dia.
D_j

Diametral clearance
$$C_d = D_b - D_j = 2C_r$$

Bearing

Clearance

Journal

Housing

The stationary member that mates with the journal is called simply the *bearing*. Other names used for plain surface bearings are *journal bearings* and *sleeve bearings*.

Where have you seen such plain surface bearings in operation? As you did in Chapter 14 on rolling contact bearings, look for consumer products, industrial machinery, or transportation equipment (cars, trucks, bicycles, and so on—any device with rotating shafts). If the bearings are not visible from the outside (and that is often the case), you may need to partially disassemble the product to see inside.

But there are some obvious examples that should be accessible to you. The wheels of a household lawn mower, wheelbarrow, or simple cart are typically mounted directly on axles to form the plain surface bearing. Hand tools such as grass and hedge clippers, pruning shears, pliers, and adjustable or ratchet wrenches use plain surface bearings at points where one part must rotate relative to another.

Look at almost any hinge for a door. It incorporates a solid pin riding inside a cylindrical outer member. That is an example of a plain surface bearing. Garage doors often have rollers that ride in channels, but the shafts that carry the rollers are operating in plain cylindrical bearings, often with very large clearances. The cable system that connects to the counterbalance springs runs over pulleys having plain surface bearings rotating on stationary shafts. If the door has an electric door opener, several of its components probably use plain surface bearings. Get a ladder and climb up to inspect them. *But be careful of the moving parts!*

Besides the rotational motion of the kinds of applications described previously, some plain surface bearings have linear sliding motion. Look inside a printer for a computer of the ink-jet or dot-matrix-pin type. Identify the part that delivers the actual image to the paper as it traverses across the page. It typically slides on a polished rod with great precision. Many linkages contain parts that slide. Look at locking mechanisms, latches, staplers, switches, car seat adjusters, gear shifters for cars, parts of window-lift devices in cars, and coin-operated machines. Can you identify parts that have linear sliding motion?

If you visit a factory, you will likely see numerous examples of both rotational and linear sliding action. Transfer devices, reciprocating machinery, machine tool slides, and packaging equipment employ many plain surface bearings.

Most of the examples just mentioned have intermittent and relatively slow motion between the mating parts, whether the motion is rotational or linear. In such applications, there is often a lubricant between the moving parts. Or the materials are carefully chosen to produce low friction and reliable, smooth motion. This kind of bearing experiences *boundary lubrication* as discussed in this chapter.

You may be familiar with the bearings on the crankshaft of an internal combustion engine. They are often called *journal bearings* because they have the classic configuration shown in Figure 16–1. But consider their general operation. After the engine has been started, the crank rotates at several thousand revolutions per minute, typically from about 1500 to 6000 rpm. This is one of the most important applications of plain surface bearings using *full-film hydrodynamic lubrication*. In such bearings, there is a continuous film of lubricant, usually oil, that actually lifts the rotating shaft journal off the stationary bearing. Thus, there is no metal-to-metal contact between the members. Much of this chapter is devoted to this kind of bearing and the analysis of the conditions required to maintain the supporting oil film. Engine bearings are some of the most highly engineered mechanical devices.

Can you identify other examples of bearings that use full-film hydrodynamic lubrication?

Another example of a class of plain surface bearings is called *hydrostatic*. Pumps create high pressure in a fluid such as oil that is delivered by carefully shaped pads where the pressure lifts the part to be moved. Then it can be moved easily, either slowly or rapidly. Think about a huge telescope or radio antenna that must be positioned accurately and moved with ease. Such applications are rare but important, justifying the large engineering and technical effort of designing, installing, and maintaining them.

This chapter will help you understand boundary lubrication, full-film hydrodynamic lubrication, and hydrostatic lubrication. Also discussed are more general aspects of friction, lubrication, wear, and the lubricants that are used to minimize friction.

 You Are the Designer

Your company is designing a conveyor system to transport products in a shipping and receiving system of a major manufacturer. The design has progressed to the point where it has been decided to use a flexible belt conveyor driven by flat pulleys at either end. The shafts that carry the pulleys must be supported in bearings in the side frames of the conveyor. Your task is to design the bearings.

First you must decide what type of bearing to use: plain surface or journal bearings as discussed in this chapter, or rolling contact bearings, which are discussed in Chapter 14. For the plain surface bearings, you must determine whether full-film hydrodynamic lubrication can be achieved with its advantages of low friction and long life. Or will the shaft operate in the bearing with boundary lubrication? What materials will the bearing and the journal be made from? What dimensions will be specified for all components? What lubricant should be used? These and other questions are discussed in this chapter.

16–1 OBJECTIVES OF THIS CHAPTER

After completing this chapter, you will be able to:

1. Describe the three modes of operation of a plain surface bearing (boundary, mixed-film, and full-film hydrodynamic lubrication), and discuss the conditions under which each will normally occur.

2. Discuss the significance of the bearing parameter, $\mu n/p$.

3. List the decisions that a bearing designer must make to completely define a plain surface bearing system.

4. List the materials often used for journals and bearings, and describe their important properties.

5. Define the *pV factor* and use it in the design of boundary-lubricated bearings.

6. Describe the operation of full-film hydrodynamically lubricated bearings.

7. Complete the design of full-film bearings, defining the size of the journal and bearing, the diametral clearance, the bearing length, the minimum film thickness, the surface finish, the lubricant, and the resulting frictional performance of the bearing system.

8. Describe a hydrostatic bearing system and complete the basic design of such bearings.

9. Define *tribology* and discuss the essential characteristics of friction, lubrication, and wear as applied to machinery.

10. Describe the general nature of oils and greases and their effects on lubrication and wear.

16–2 THE BEARING DESIGN TASK

The term *plain surface bearing* refers to the kind of bearing in which two surfaces move relative to each other without the benefit of rolling contact. Thus, there is sliding contact. The actual shape of the surfaces can be anything that permits the relative motion. The most common shapes are flat surfaces and concentric cylinders. Figure 16–1 shows the basic geometry of a cylindrical plain surface bearing.

A given bearing system can operate with any of three types of lubrication:

Boundary lubrication: There is actual contact between the solid surfaces of the moving and the stationary parts of the bearing system, although a film of lubricant is present.

Mixed-film lubrication: There is a transition zone between boundary and full-film lubrication.

Full-film lubrication: The moving and stationary parts of the bearing system are separated by a complete film of lubricant that supports the load. The term *hydrodynamic lubrication* is often used to describe this type.

All of these types of lubrication can be encountered in a bearing without external pressurization of the bearing. If lubricant under pressure is supplied to the bearing, it is called a *hydrostatic bearing*, which is discussed separately. Running dry surfaces together is not recommended unless there is inherently good lubricity between the mating materials. Some plastics are used dry, as discussed in Sections 16–4 and 16–5.

Bearing design involves so many design decisions that it is not possible to develop a procedure that produces the single best design. Thus, several feasible designs could be proposed, and the designer must make judgments based on knowledge of the application and on the principles of bearing operation to define the final design. The following lists identify the information needed to design a bearing system and the types of design decisions that must be made. (See Reference 18.) In this discussion, it is assumed that the bearing will be cylindrical, like that used to support a rotating shaft. Modified lists could be made for linear sliding surfaces or some other geometry.

Bearing Requirements

Magnitude, direction, and degree of variation of the radial load

Magnitude and direction of the thrust load, if any

Rotational speed of the journal (shaft)

Frequency of starts and stops, and duration of idle periods

Magnitude of the load when the system is stopped and when it is started

Life expectancy of the bearing system

Environment in which the bearing will operate

Design Decisions

Materials for the journal and the bearing

Diameters, including tolerances, of the journal and the bearing

Nominal value and range of journal clearance in the bearing

Surface finish for the journal and the bearing

Length of the bearing

Method of manufacturing the bearing system

Type of lubricant to be used and the means of supplying it

Operating temperature of the bearing system and of the lubricant

Method of maintaining the lubricant cleanliness and temperature

Analyses Required

Type of lubrication: boundary, mixed-film, full-film

Coefficient of friction

Frictional power loss

Minimum film thickness

Thermal expansion

Heat dissipation required and the means of accomplishing it

Shaft stiffness and slope of the shaft in the bearing

16–3 BEARING PARAMETER, $\mu n/p$

The performance of a bearing differs radically, depending on which type of lubrication occurs. There is a marked decrease in the coefficient of friction when the operation changes from boundary to full-film lubrication. Wear also decreases with full-film lubrication. Thus, it is desirable to understand the conditions under which one or the other type of lubrication occurs.

The creation of full-film lubrication, the most desirable type, is encouraged by light loads, high relative speed between the moving and stationary parts, and presence of a high-viscosity lubricant in the bearing in copious supply. For a rotating journal bearing, the combined effect of these three factors, as it relates to the friction in the bearing, can be evaluated by computing the *bearing parameter*, $\mu n/p$. The viscosity of the lubricant is indicated by μ, the rotational speed by n, and the bearing load by the pressure, p. To compute the pressure, divide the applied radial load on the bearing by the *projected area* of the bearing, that is, the product of the length times the diameter.

The bearing parameter $\mu n/p$ is dimensionless when each term is expressed in consistent units. Some unit systems that can be used are listed here:

Unit System	Viscosity, μ	Rotational Speed, n	Pressure, p
SI metric	N·s/m² or Pa· s	rev/s	N/m² or Pa
English	lb·s/in² or reyn	rev/s	lb/in²
Old metric (obsolete)	dyne·s/cm² or poise	rev/s	dynes/cm²

The effect of the bearing parameter is shown in Figure 16–2, sometimes called the *Stribeck curve*, which is a plot of the coefficient of friction, f, for the bearing versus the value of $\mu n/p$. At low values of $\mu n/p$, boundary lubrication occurs, and the coefficient of friction is high. For example, with a steel shaft sliding slowly in a lubricated bronze bearing (boundary lubrication), the value of f would be approximately 0.08 to 0.14. At high values of $\mu n/p$, the full hydrodynamic film is created, and the value of f is normally in the range of 0.001 to 0.005. Note that this compares favorably with precision rolling contact bearings. Between boundary and full-film lubrication, the *mixed-film* type, which is some combination of the other two, occurs. The dashed curve shows the general nature of the variation in film thickness in the bearing.

It is recommended that designers avoid the mixed-film zone because it is virtually impossible to predict the performance of the bearing system. Also notice that the curve is very steep in this zone. Thus, a small change in any of the three factors, μ, n, or p, produces a large change in f, resulting in erratic performance of the machine.

The value of $\mu n/p$ at which full-film lubrication is produced is difficult to predict. Besides the individual factors of speed, pressure (load), and viscosity (a function of the type

FIGURE 16–2
Bearing performance
and types of lubrication
related to the bearing
parameter, $\mu n/p$. The
Stribeck curve.

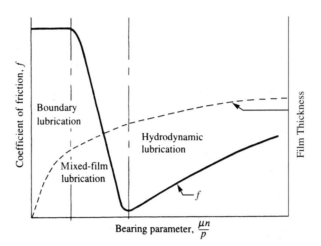

of lubricant and its temperature), the variables affecting the production of the film include the quantity of lubricant supplied, the adhesion of the lubricant to the surfaces, the materials of the journal and the bearing, the structural rigidity of the journal and the bearing, and the surface roughness of the journal and the bearing. After completing the design process presented later in this chapter, you are advised to test the design.

In general, boundary lubrication should be expected for slow-speed operation with a surface speed less than about 10 ft/min (0.05 m/s). Reciprocating or oscillating motion or a combination of a light lubricant and high pressure would also produce boundary lubrication.

The design of bearings to produce full-film lubrication is described in Section 16–6. In general, it requires a surface speed of greater than 25 ft/min (0.13 m/s) continuously in one direction with an adequate supply of oil at a proper viscosity.

**16–4
BEARING
MATERIALS**

In rotational applications, the journal on the shaft is frequently steel. The stationary bearing may be made of any of a wide variety of materials, including the following:

> Bronze
>
> Babbitt
>
> Aluminum
>
> Zinc
>
> Porous metals
>
> Plastics (nylon, TFE, PTFE, phenolic, acetal, polycarbonate, filled polyimide)

The properties desirable for materials used for plain bearings are unique, and compromises must frequently be made. The following list discusses these properties.

1. *Strength:* The function of the bearing is to carry the applied load and deliver it to the supporting structure. At times, the loads are varied, requiring fatigue strength as well as static strength.

2. *Embeddability:* The property of embeddability relates to the ability of the material to hold contaminants in the bearing without causing damage to the rotating journal. Thus, a relatively soft material is desirable.

3. *Corrosion resistance:* The total environment of the bearing must be considered, including journal material, lubricant, temperature, air-borne particulates, and corrosive gases or vapors.

4. *Cost:* Always an important factor, cost includes not only the material cost but also processing and installation costs.

A brief discussion of the performance of some of the bearing materials follows.

Cast Bronze

The name *bronze* refers to several alloys of copper with tin, lead, zinc, or aluminum, either singly or in combination. The leaded bronzes contain 25% to 35% lead, which gives them good embeddability and resistance to seizing under conditions of boundary lubrication. However, their strength is relatively low. The cast bearing bronze, SAE CA932, has 83% copper, 7% tin, 7% lead, and 3% zinc. It possesses a good combination of properties for such uses as pumps, machinery, and appliances. Tin and aluminum bronzes have higher strength and hardness and can carry greater loads, particularly in impact situations. But they rate lower on embeddability. See Internet sites 1, 6, 7, 9, and 10.

Babbitt

Babbitts may be lead-based or tin-based, having nominally 80% of the parent metal. Various alloy compositions of copper and antimony (as well as lead and tin) can tailor the properties to suit a particular application. Because of their softness, babbitts have outstanding embeddability and resistance to seizure, important properties in applications in which boundary lubrication occurs. They have rather low strength, however, and are often applied as liners in steel or cast iron housings. See Internet site 9.

Aluminum

With the highest strength of the commonly used bearing materials, aluminum is used in severe applications in engines, pumps, and aircraft. The high hardness of aluminum bearings results in poor embeddability, requiring clean lubricants.

Zinc

Bearings made from zinc alloys offer good protection to running without a continuous supply of lubricant, although they perform best if lubricated. Standard bearing greases are often used. When operating on steel journals, a thin film of the softer zinc material is transferred to the steel to protect it from wear and damage. It performs well in most atmospheric conditions except for continuously wet environments and exposure to seawater. See Internet site 8.

Porous Metals

Products of the powder metal industry, porous metals are sintered from powders of bronze, iron, and aluminum; some are mixed with lead or copper. The sintering leaves a large number of voids in the bearing material into which lubricating oil is forced. Then, during operation, the oil migrates out of the pores, supplying the bearing. Such bearings are particularly good for slow-speed, reciprocating, or oscillating motions. See Internet sites 6 and 7.

Plastics

Generally referred to as *self-lubricating materials*, plastics used in bearing applications have inherently low friction characteristics. They can be operated dry, but most improve in performance with a lubricant present. Embeddability is usually good, as is resistance to seizure. But many have low strength, limiting their load-carrying capacity. Backing with metal sleeves is frequently done to improve load-carrying capacity. Major advantages are corrosion resistance and, when operated dry, freedom from contamination. These properties are particularly important in the processing of foodstuffs and chemical products. See Internet sites 2, 3, 4, 5, and 7.

Because of the complex chemical names for plastic materials and the virtually infinite combinations of base materials, reinforcements, and fillers used, it is difficult to characterize bearing plastics. Most are composites of several components. The group known as *fluoropolymers* are popular because of the very low coefficient of friction (0.05 to 0.15) and good wear resistance. Phenolics, polycarbonates, acetals, nylons, and many other plastics are also used for bearings. Among the chemical names and abbreviations found in this field are the following:

> *PTFE:* Polytetrafluoroethylene
>
> *PA:* Polyamide
>
> *PPS:* Polyphenylene sulfide
>
> *PVDF:* Polyvinylidene fluoride
>
> *PEEK:* Polyetheretherketone
>
> *PEI:* Polyetherimide
>
> *PES:* Polyethersulfone
>
> *PFA:* Perfluoroalkoxy-modified tetrafluoroethylene

Reinforcements and fillers used with plastic bearing materials include glass fibers, milled glass, carbon fibers, bronze powders, PTFE, PPS, and some solid lubricants, such as graphite and molybdenum disulfide.

16–5 DESIGN OF BOUNDARY-LUBRICATED BEARINGS

The factors to be considered when selecting materials for bearings and specifying the design details include the following:

> *Coefficient of friction:* Both static and dynamic conditions should be considered.
>
> *Load capacity, p:* Radial load divided by the projected area of the bearing (lb/in^2 or Pa).
>
> *Speed of operation, V:* The relative speed between the moving and stationary components, usually in ft/min or m/s.
>
> *Temperature at operating conditions.*
>
> *Wear limitations.*
>
> *Producibility:* Machining, molding, fastening, assembly, and service.

pV Factor

In addition to the individual consideration of load capacity, p, and speed of operation, V, the product pV is an important performance parameter for bearing design when boundary lubrication occurs. The pV value is a measure of the ability of the bearing material to accom-

TABLE 16–1 Typical performance parameters for bearing materials in boundary lubrication at room temperature

Material	pV		
	psi-fpm	kPa-m/s	
Vespel® SP-21 polyimide	300 000	10 500	Trademark of DuPont Co.
Manganese bronze (C86200)	150 000	5250	Also called SAE 430A
Aluminum bronze (C95200)	125 000	4375	Also called SAE 68A
Leaded tin bronze (C93200)	75 000	2625	Also called SAE 660
KU dry lubricant bearing	51 000	1785	See note 1
Porous bronze/oil impregnated	50 000	1750	
Babbitt: high tin content (89%)	30 000	1050	
Rulon® PTFE: M-liner	25 000	875	Metal backed
Rulon® PTFE: FCJ	20 000	700	Oscillatory and linear motion
Babbitt: low tin content (10%)	18 000	630	
Graphite/Metallized	15 000	525	Graphite Metallizing Corp.
Rulon® PTFE: 641	10 000	350	Food and drug applications (see note 2)
Rulon® PTFE: J	7500	263	Filled PTFE
Polyurethane: UHMW	4000	140	Ultra high molecular weight
Nylon® 101	3000	105	Trademark of DuPont Co.

Source: Bunting Bearings Corp., Holland, OH

[1]KU Bearings consist of bonded layers of a steel backing and a porous bronze matrix overlaid with PTFE/lead bearing material. A film from the bearing material is transferred to the journal during operation.

[2]Rulon® is a registered trademark of Saint-Gobain Performance Plastics Company. Bearings are made from Rulon® PTFE (polytetrafluoroethylene) material in a variety of formulations and physical constructions.

modate the frictional energy generated in the bearing. At the limiting pV value, the bearing will not achieve a stable temperature limit, and rapid failure will occur. A practical design value for pV is one-half of the limiting pV value, given in Table 16–1.

Units for pV. The nominal units for pV are simply the product of the units for pressure and the units for velocity. In the U.S. Customary System, this is, considering only units,

$$p = F/LD = \text{lb/in}^2 = \text{psi}$$
$$V = \pi Dn/12 = \text{ft/min} = \text{fpm}$$
$$pV = (\text{lb/in}^2)(\text{ft/min}) = \text{psi-fpm}$$

Another way of looking at these units is to rearrange them into the form

$$pV = (\text{ft·lb/min})/\text{in}^2$$

The numerator represents one unit for power or energy transfer per unit time. The denominator represents area. Therefore, pV can be thought of as the rate of energy input to the bearing per unit of projected area of the bearing if the coefficient of friction is 1.0. Of course, the actual coefficient of friction is normally much less than one. Then you can think of pV as a comparative measure of the bearing's ability to absorb energy without overheating.

In SI units, force, F, is in newtons (N) and bearing dimensions are in mm. The pressure is then

⇨ Bearing Pressure

$$p = F/LD = \text{N/mm}^2$$

For typically encountered values, it is convenient to convert this to kPa or 10^3 N/m²:

$$p = \frac{N}{mm^2} \cdot \frac{(10^3 \text{ mm})^2}{m^2} = \frac{10^6 \text{ N}}{m^2} \cdot \frac{1 \text{ kN}}{10^3 \text{ N}} = \frac{10^3 \text{ kN}}{m^2} = 10^3 \text{ kPa}$$

In summary, pressure, p, in N/mm² is the same as 10^3 kPa.

The linear velocity of the surface of the journal is typically computed from

$$v = \pi Dn/(60\ 000) \text{ m/s}$$

with D in mm and n in rpm. Then the units for pV are

$$pV = (\text{kPa})(\text{m/s})$$

A useful conversion to the U.S. Customary System is

$$1.0 \text{ psi-fpm} = 0.035 \text{ kPa·m/s}$$

Again, the units for pV can be reformatted to reflect the rate of energy transfer per unit area:

$$pV = \text{kPa} \cdot \text{m/s} = \frac{\text{kN}}{m^2} \cdot \frac{\text{m}}{\text{s}} = \frac{\text{kW}}{m^2}$$

where 1 kW = 1 kN · m/s

Operating Temperature

Most plastics are limited to approximately 200°F (93°C). However, PTFE can operate at 500°F (260°C). Babbitt is limited to 300°F (150°C), while tin-bronze and aluminum can operate at 500°F (260°C). A major advantage of carbon-graphite bearings is their ability to operate at up to 750°F (400°C).

Design Procedure

The following is one method of completing the preliminary design of boundary-lubricated plain surface bearings.

Procedure for Designing Boundary-lubricated Plain Surface Bearings

> **Given information:** Radial load on the bearing, F (lb or N); speed of rotation, n (rpm); nominal minimum shaft diameter, D_{min} (in or mm) (based on stress or deflection analysis).
>
> **Objectives of the design process:** To specify the nominal diameter and length of the bearing and a material that will have a safe value of pV.
>
> **1.** Specify a trial diameter, D, for the journal and the bearing.
>
> **2.** Specify a ratio of bearing length to diameter, L/D, typically in the range of 0.5 to 2.0. For nonlubricated (dry-rubbing) or oil-impregnated porous bearings, $L/D = 1$

is recommended. (See Reference 18.) For carbon-graphite bearings, $L/D = 1.5$ is recommended.

3. Compute $L = D(L/D)$ = nominal length of the bearing.

4. Specify a convenient value for L.

5. Compute the bearing pressure (lb/in² or Pa):

$$p = F/LD$$

6. Compute the linear speed of the journal surface:

U.S. Customary units: $V = \pi D n / 12$ ft/min or fpm

SI metric units: $V = \pi D n / (60\ 000)$ m/s

Also note: 1.0 ft/min = 0.005 08 m/s

                   1.0 m/s = 197 ft/min

7. Compute pV (psi-fpm or Pa·m/s or kW/m²).

8. Multiply $2(pV)$ to obtain a design value for pV.

9. Specify a material from Table 16–1 with a rated value of pV equal to or greater than the design value.

10. Complete the design of the bearing system considering diametral clearance, lubricant selection, lubricant supply, surface finish specification, thermal control, and mounting considerations. Often the supplier of the bearing material provides recommendations for many of these design decisions.

11. Nominal diametral clearance: Many factors affect the final specification for clearance, such as need for precision, thermal expansion of all parts of the bearing system, load variations, shaft deflection expected, means of providing lubricant, and manufacturing capability. One long-used rule of thumb is to provide 0.001 in of clearance per inch of journal diameter. Figure 16–3 shows recommended minimum values for clearance based on journal diameter and rotational speed under steady loads. These values apply to the smallest clearance under any combination of tolerances on the dimensions of the bearing system to avoid problems of bearing heating and eventual seizure. The operating clearance will then be higher than these values because of manufacturing tolerances. Performance over the entire range of clearances should be evaluated, preferably by testing.

Example Problem 16–1 A bearing is to be designed to carry a radial load of 150 lb from a shaft having a minimum acceptable diameter of 1.50 in and rotating at 500 rpm. Design the bearing to operate under boundary-lubrication conditions.

Solution We will use the design procedure previously outlined.

Step 1. Trial diameter: Try $D = D_{min} = 1.50$ in.

Steps 2–4. Try $L/D = 1.0$. Then $L = D = 1.50$ in.

Step 5. Bearing pressure:

$p = F/LD = (150\ \text{lb})/(1.50\ \text{in})(1.50\ \text{in}) = 66.7$ psi

FIGURE 16–3
Minimum
recommended
diametral clearance for
bearings considering
journal diameter and
rotational speed
(Taken from
R. J. Welsh. *Plain
Bearing Design
Handbook*. London:
Butterworths, 1983)

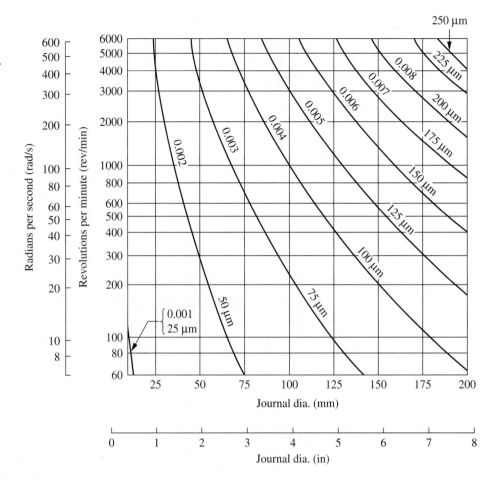

Step 6. Journal speed:
$$V = \pi D n/12 = \pi(1.50)(500)/12 = 196 \text{ ft/min}$$

Step 7. pV factor:
$$pV = (66.7 \text{ psi})(196 \text{ fpm}) = 13\ 100 \text{ psi-fpm}$$

Step 8. Design value of $pV = 2(13\ 100) = 26\ 200$ psi-fpm.

Step 9. From Table 16–1, we could use a bearing made from high tin babbit having a rated value of pV of 30 000 psi-fpm.

Steps 10–11. Nominal diametral clearance: From Figure 16–3, we can recommend a minimum $C_d = 0.002$ in based on $D = 1.50$ in and $n = 500$ rpm. Other design details are dependent on the details of the system into which the bearing will be placed.

Example Problem 16–2   Design a boundary-lubricated plain surface bearing to carry a radial load of 2.50 kN from a shaft rotating at 1150 rpm. The nominal minimum diameter of the journal is 65 mm.

Solution   We will use the design procedure previously outlined.

Step 1. Trial diameter: Try $D = 75$ mm.

Steps 2–4. Try $L/D = 1.0$. Then $L = D = 75$ mm.

Step 5. Bearing pressure:

$$p = F/LD = (2500 \text{ N})/(75 \text{ mm})(75 \text{ mm}) = 0.444 \text{ N/mm}^2$$

Converting to kPa, we have

$$p = 0.444 \text{ N/mm}^2 (10^3 \text{ kPa})/(\text{N/mm}^2) = 444 \text{ kPa}$$

Step 6. Journal speed:

$$V = \pi Dn/(60\,000) = \pi(75)(1150)/(60\,000) = 4.52 \text{ m/s}$$

Step 7. pV factor:

$$pV = (444 \text{ kPa})(4.52 \text{ m/s}) = 2008 \text{ kPa} \cdot \text{m/s}$$

Step 8. Design value for $pV = 2(2008) = 4016 \text{ kPa} \cdot \text{m/s}$.

Step 9. From Table 16–1, we can specify aluminum bronze (C95200) having a pV rating of 4375 kPa-m/s.

Steps 10–11. From Figure 16–3, we can recommend a minimum $C_d = 100 \text{ μm}$ (0.100 mm or 0.004 in) based on $D = 75$ mm and 1150 rpm.

Alternate design: The pV factor for the initial design, while satisfactory, is somewhat high and may require careful lubrication. Consider the following alternate design having a larger bearing diameter.

Step 1. Try $D = 150$ mm.

Step 2. Let $L/D = 1.25$.

Step 3. Then

$$L = D(L/D) = (150 \text{ mm})(1.25) = 187.5 \text{ mm}$$

Step 4. Let's use the more convenient value of 175 mm for L.

Step 5. Bearing pressure:

$$p = F/LD = (2500 \text{ N})/(175 \text{ mm})(150 \text{ mm}) = 0.095 \text{ N/mm}^2 = 95 \text{ kPa}$$

Step 6. Journal speed:

$$V = \pi Dn/(60\,000) = \pi(150)(1150)/(60\,000) = 9.03 \text{ m/s}$$

Step 7. pV factor:

$$pV = (95 \text{ kPa})(9.03 \text{ m/s}) = 860 \text{ kPa·m/s} = 860 \text{ kW/m}^2$$

Step 8. Design value of $pV = 2(860) = 1720 \text{ kW/m}^2$.

Step 9. From Table 16–1, we can specify an oil impregnated porous bronze bearing having a pV rating of 1750 kPa-m/s or a KU dry lubricated bearing having a pV rating of 1785 kPa-m/s.

Steps 10–11. From Figure 16–3, we can recommend a minimum $C_d = 150\,\mu m$ (0.150 mm or 0.006 in) based on $D = 150$ mm and 1150 rpm. Other design details are dependent on the system into which the bearing will be placed.

16–6 FULL-FILM HYDRODYNAMIC BEARINGS

In the *full-film hydrodynamic bearing*, the load on the bearing is supported on a continuous film of lubricant, usually oil, so that no contact between the bearing and the rotating journal occurs. A pressure must be developed in the oil in order to carry the load. With proper design, the motion of the journal inside the bearing creates the necessary pressure.

Figure 16–4 shows the progressive action in a plain surface bearing from start-up to the steady-state hydrodynamic operation. Observe that boundary lubrication and mixed-film lubrication precede the establishment of full-film hydrodynamic lubrication. At start-up, the radial load applied through the journal to the bearing forces the journal off center in the direction of the load, taking up all clearance [Figure 16–4(a)]. At the initial slow rotational speeds, the friction between the journal and the bearing causes the journal to climb up the bearing wall, somewhat as shown in Figure 16–4(b). Because of the viscous shear stresses developed in the oil, the moving journal draws oil into the converging, wedge-shaped area above the region of contact. The resulting pumping action produces a pressure in the oil film; when the pressure is sufficiently high, the journal is lifted from the bearing. The frictional forces are greatly reduced under this operating condition, and the journal moves eventually to the steady-state position shown in Figure 16–4(c). Note that the journal is offset from the direction of the load; that there is a certain eccentricity, e, between the geometric center of the bearing and the center of the journal; and that there is a point of minimum film thickness, h_o, at the nose of the wedge-shaped pressurized zone.

Figure 16–5 illustrates the general form of the pressure distribution within a full-film hydrodynamically lubricated bearing. The clearance between the bearing and the journal is highly exaggerated. Part (a) of the figure shows the rise of pressure as the rotating shaft draws oil into the converging wedge, approaching the point of minimum film thickness. The maximum pressure occurs there and then falls rapidly to nearly zero as the space between the journal and the bearing diverges again. The integrated effect of the pressure distribution is a force sufficient to support the shaft on a film of oil without metal-to-metal contact.

FIGURE 16–4

Position of the journal relative to the bearing as a function of mode of operation

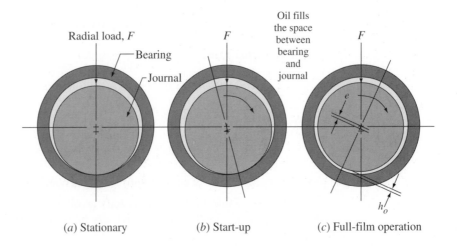

(*a*) Stationary   (*b*) Start-up   (*c*) Full-film operation

FIGURE 16–5
Pressure distribution in
the oil film for
hydrodynamic
lubrication

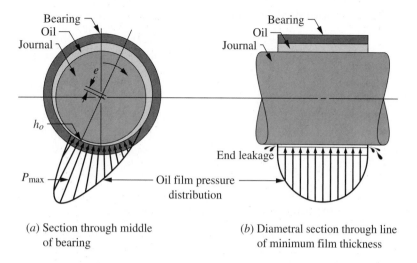

(a) Section through middle
of bearing

(b) Diametral section through line
of minimum film thickness

Figure 16–5(b) shows the pressure distribution axially along the shaft through the line of minimum film thickness or maximum pressure. The highest pressure value occurs in the middle of the bearing length, and it falls rapidly as the ends are approached because the pressure outside the bearing is ambient pressure, typically atmospheric pressure. There is continual leakage flow from both ends of the bearing. This illustrates the importance of providing a means of continually supplying oil to the bearing to maintain the full-film operation. Without a steady, adequate supply of oil, the system would not be able to create the pressurized film to carry the shaft, and boundary lubrication would result. The significantly higher frictional forces thus created would cause rapid heating of the bearing/journal interface, and seizure would likely occur very rapidly.

16–7
DESIGN OF
FULL-FILM
HYDRO-
DYNAMICALLY
LUBRICATED
BEARINGS

The following discussion presents several guidelines for bearing design for typical industrial applications. The design procedure is based primarily on information in References 1–3.

Surface Roughness

A ground journal with a surface roughness average of 16 to 32 microinches (μin), or 0.40 to 0.80 μm, is recommended for bearings of good quality. The bearing should be equally smooth or made from one of the softer materials so that "wearing in" can smooth the high spots, creating a good fit between the journal and the bearing. In high-precision equipment, polishing or lapping can be used to produce a surface finish of the order of 8 to 16 μin (0.20 to 0.40 μm).

Minimum Film Thickness

The limiting acceptable value of the minimum film thickness depends on the surface roughness of the journal and the bearing because the film must be thick enough to eliminate any solid contact during expected operating conditions. The suggested design value also depends on the size of the journal. For ground journals, the following relationship can be used to estimate the design value:

$$h_o = 0.00025D \qquad \qquad \textbf{(16–1)}$$

where D = diameter of the bearing

Diametral Clearance

The clearance between the journal and the bearing depends on the nominal diameter of the bearing, the precision of the machine for which the bearing is being designed, the speed of rotation, and the surface roughness of the journal. The coefficient of thermal expansion for both the journal and the bearing must also be considered to ensure a satisfactory clearance under all expected operating conditions. An overall guideline of making the clearance in the range of 0.001 to 0.002 times the bearing diameter can be used. Figure 16–3 shows a graph of the recommended minimum diametral clearance as a function of the bearing diameter and the rotational speed. Some variation above the values from the curve is permissible.

Ratio of Bearing Length to Diameter

Because the journal is a part of the shaft itself, its minimum diameter is usually limited by stress and deflection considerations of the type discussed in Chapter 12. Then the length of the bearing is specified to provide for a suitable level of bearing pressure. General-purpose industrial machinery bearings typically operate at approximately 200 to 500 psi (1.4 to 3.4 MPa) bearing pressure, based on the projected area of the bearing [$p = $ load/(LD)]. The pressure may range from as low as 50 psi (0.34 MPa) for light-duty equipment to 2000 psi (13.4 MPa) for heavy machinery under varying loads, such as in internal combustion engines. The leakage of oil from the bearing is also affected by the bearing length. The typical range of length to diameter ratio (L/D) for full-film hydrodynamic bearings is from 0.35 to 1.5. But many successful bearings operate outside this range. Reference 18 recommends $L/D = 0.60$ for most industrial applications.

Lubricant Temperature

The viscosity of the oil is a critical parameter in the performance of a bearing. Figure 16–6 shows the great variation of viscosity with temperature, which indicates that temperature control is advisable. Also, most petroleum lubricating oils should be limited to approximately 160°F (70°C) in order to retard oxidation. The temperature of interest is, of course, that inside the bearing. Frictional energy or thermal energy from the equipment itself can increase the oil temperature above that in the supply reservoir. In design examples we will select the lubricant that will ensure a satisfactory viscosity at 160°F, unless otherwise noted. If the actual operating temperature is lower, the resulting film thickness will be greater than the design value—a conservative result. It is incumbent on the designer to ensure that the limiting temperature is not exceeded, using forced cooling if necessary.

Lubricant Viscosity

The specification of the lubricant for the bearing is one of the final decisions to be made in the design procedure that follows. In the calculations, it is the dynamic viscosity, μ, which is used. In the U.S. Customary Unit (lb-in-s) System, the dynamic viscosity is expressed in lb·s/in², which is given the name *reyn* in honor of Osbourne Reynolds, who performed much significant work in fluid flow. In SI units, the standard unit is N·s/m² or Pa·s. Some prefer the unit of poise or centipoise, derived from the dyne-cm-s metric system. Some useful conversions are

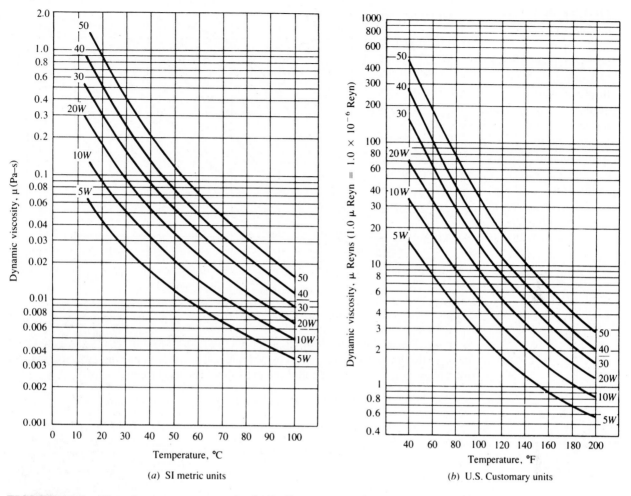

FIGURE 16–6   Viscosity vs. temperature for SAE oils

$$1.0 \text{ reyn} = 6895 \text{ Pa·s}$$
$$1.0 \text{ Pa·s} = 1000 \text{ centipoise}$$

Other viscosity conversions may also be useful. (See Reference 12.) Figure 16–6 shows graphs of dynamic viscosity versus temperature for both SI and U.S. Customary units. Note in Figure 16–6(b) that common values in U.S. Customary units are very small. Values from the scale must be multiplied by 10^{-6}.

Sommerfeld Number

The combined effect of many of the variables involved in the operation of a bearing under hydrodynamic lubrication can be characterized by the dimensionless number, S, called the *Sommerfeld number*. Some refer to this as the *bearing characteristic number*, defined as follows:

Bearing Characteristic Number

$$S = \frac{\mu n_s (R/C_r)^2}{p} \tag{16–2}$$

FIGURE 16–7 Film thickness variable, h_o/C_r, versus Sommerfeld number, S (Adapted from John Boyd and Albert A. Raimondi, "A Solution for the Finite Journal Bearing and Its Application to Analysis and Design," Parts I and II. *Transactions of the American Society of Lubrication Engineers*, Vol. 1, No. 1, 1958)

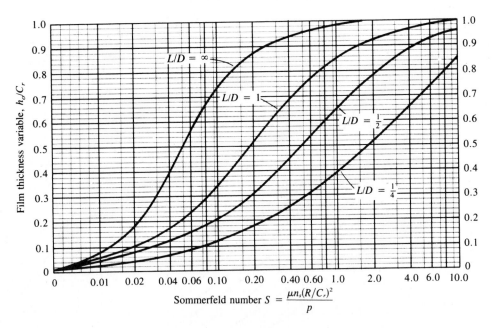

Note that S is similar to the bearing parameter, $\mu n/p$, discussed in Section 16–3, in that it involves the combined effect of viscosity, rotational speed, and bearing pressure. In order for S to be dimensionless, the following units should be used for the factors:

	U.S. Customary Units	SI Units
μ	lb·s/in² (reyns)	Pa·s (N·s/m²)
n_s	rev/s	rev/s
p	lb/in² (psi)	Pa (N/m²)
R, C_r	in	m or mm

Any consistent units can be used. Figure 16–7, adapted from Reference 3, shows the relationship between the Sommerfeld number and the film thickness ratio, h_o/C_r. Figure 16–8 shows the relationship between S and the coefficient of friction variable, $f(R/C_r)$. These values are used in the design procedure that follows. Because many design decisions are required, several acceptable solutions are possible.

Design Procedure

Procedure for Designing Full-Film Hydrodynamically Lubricated Bearings

Because the bearing design is usually done after the stress analysis of the shaft is completed, the following items are typically known:

Radial load on the bearing, F, usually in lb or N

Rotational speed, n, usually in rpm

Nominal shaft diameter at the journal, sometimes specified as a minimum acceptable diameter based on strength and stiffness

The results of the design procedure produce values for the actual journal diameter, the bearing length, the diametral clearance, the minimum film thickness of the lubricant during operation, the surface finish for the journal, the lubricant and its

FIGURE 16–8
Coefficient of friction
variable, $f(R/C_r)$, versus
Sommerfeld number, S

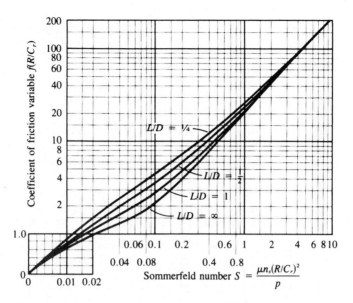

maximum operating temperature, the coefficient of friction, the friction torque, and the power dissipated because of friction.

1. Specify a trial value for the journal diameter, D, and the radius, $R = D/2$.

2. Specify a nominal operating bearing pressure, usually 200 to 500 psi (1.4 to 3.4 MPa), where $p = F/LD$. Solve for L:

$$L = F/pD$$

   Then compute L/D. It might be desirable to redefine L/D to be a convenient value from 0.25 to 1.5 in order to use the design charts available. Finally, specify the actual design value of L/D and L, and compute the actual $p = F/LD$.

3. Referring to Figure 16–3, specify the diametral clearance, C_d, based on the values for D and n. Then compute $C_r = C_d/2$ and the ratio R/C_r.

4. Specify the desired surface finish for the journal and bearing, also based on the application. A typical value is 16–32 μin average (0.40 to 0.80 μm).

5. Compute the nominal minimum film thickness from Equation (16–1),

$$h_o = 0.00025D$$

6. Compute h_o/C_r, the film thickness ratio.

7. From Figure 16–7, determine the value of the Sommerfeld number for the selected film thickness ratio and the L/D ratio. Exercise care in interpolating in this chart because of the logarithmic axis and the nonlinear spread between curves. For $L/D > 1$, only approximate data can be obtained. For $L/D \doteq 1.5$, interpolate approximately one-fourth of the way between the curves for $L/D = 1$ and $L/D =$ infinity. For $L/D = 2$, go about halfway.

8. Compute the rotational speed n_s in revolutions per second:

$$n_s = n/60$$

   where n is in rpm

9. Because each factor of the Sommerfeld number is known except the lubricant viscosity, μ, solve for the required minimum viscosity that will produce the desired minimum film thickness:

⇨ **Required Minimum Lubricant Viscosity**

$$\mu = \frac{Sp}{n_s(R/C_r)^2} \tag{16–3}$$

10. Specify a maximum acceptable lubricant temperature, usually 160°F or 70°C. Select a lubricant from Figure 16–6 that will have at least the required viscosity at the operating temperature. If the selected lubricant has a viscosity greater than that computed in Step 9, recompute S for the new value of viscosity. The resulting value for the minimum film thickness will now be somewhat greater than the design value, a generally desirable result. You can consult Figure 16–7 again to determine the new value of the minimum film thickness, if desired.

11. From Figure 16–8, obtain the coefficient of friction variable, $f(R/C_r)$.

12. Compute $f = f(R/C_r)/(R/C_r)$ = coefficient of friction.

13. Compute the friction torque. The product of the coefficient of friction and the load F gives the frictional force at the surface of the journal. That value multiplied by the radius gives the torque:

⇨ **Friction Torque**

$$T_f = F_f R = fFR \tag{16–4}$$

14. Compute the power dissipated in the bearing from the relationship between power, torque, and speed that we have used many times:

⇨ **Frictional Power Loss**

$$P_f = T_f n/63\,000 \text{ hp} \tag{16–5}$$

This frictional power loss represents the rate of energy input to the lubricant within the bearing that can cause the temperature to rise. It is a part of the energy that must be removed from the bearing to maintain satisfactory lubricant viscosity.

Example Problem 16–3 will illustrate the procedure.

Example Problem 16–3   Design a plain surface journal bearing to carry a steady radial load of 1500 lb while the shaft rotates at 850 rpm. The shaft stress analysis determines that the minimum acceptable diameter at the journal is 2.10 in. The shaft is part of a machine requiring good precision.

Solution   *Step 1.* Let's select $D = 2.50$ in. Then $R = 1.25$ in.

Step 2. For $p = 200$ psi, L must be

$$L = F/pD = 1500/(200)(2.50) = 3.00 \text{ in}$$

For this value of L, $L/D = 3.00/2.50 = 1.20$. In order to use one of the standard design charts, let's change L to 2.50 in so that $L/D = 1.0$. This is not essential, but it eliminates interpolation. The actual pressure is then

$$p = F/LD = 1500/(2.50)(2.50) = 240 \text{ psi}$$

This is an acceptable pressure.

Step 3. From Figure 16–3, $C_d = 0.003$ in is suitable for the diametral clearance, based on $D = 2.50$ in and $n = 850$ rpm, and $C_r = C_d/2 = 0.0015$ in. Also,

$$R/C_r = 1.25/0.0015 = 833$$

This value is used in later calculations.

Step 4. For the precision desired in this machine, use a surface finish of 16 to 32 μin, requiring a ground journal.

Step 5. Minimum film thickness (design value):

$$h_o = 0.00025D = 0.00025(2.50) = 0.0006 \text{ in (approx.)}$$

Step 6. Film thickness variable:

$$h_o/C_r = 0.0006/0.0015 = 0.40$$

Step 7. From Figure 16–7, for $h_o/C_r = 0.40$ and $L/D = 1$, we can read $S = 0.13$.

Step 8. Rotational speed in revolutions per second:

$$n_s = n/60 = 850/60 = 14.2 \text{ rev/s}$$

Step 9. Solve for the viscosity from the Sommerfeld number, S:

$$\mu = \frac{Sp}{n_s(R/C_r)^2} = \frac{(0.13)(240)}{(14.2)(833)^2} = 3.17 \times 10^{-6} \text{ reyn}$$

Step 10. From the viscosity chart, Figure 16–6, SAE 30 oil is required to ensure a sufficient viscosity at 160°F. The actual expected viscosity of SAE 30 at 160°F is approximately 3.3×10^{-6} reyn.

Step 11. For the actual viscosity, the Sommerfeld number would be

$$S = \frac{\mu n_s(R/C_r)^2}{p} = \frac{(3.3 \times 10^{-6})(14.2)(833)^2}{240} = 0.135$$

Step 12. Coefficient of friction (from Figure 16–8): $f(R/C_r) = 3.5$ for $S = 0.135$ and $L/D = 1$. Now, because $R/C_r = 833$,

$$f = 3.5/833 = 0.0042$$

Step 13. Friction torque:

$$T_f = fFR = (0.0042)(1500)(1.25) = 7.88 \text{ lb·in}$$

Step 14. Frictional power:

$$P_f = T_f n/63\,000 = (7.88)(850)/63\,000 = 0.106 \text{ hp}$$

Comment   A qualitative evaluation of the result would require more knowledge about the application. But note that a coefficient of friction of 0.0042 is quite low. It is likely that a machine requiring such a large shaft and with such high bearing forces also requires a large power to drive it. Then the 0.106-hp friction power would appear small.

Thermal considerations are also important to determine how much energy must be dissipated from the bearing. Converting the frictional power loss to thermal power gives

$$P_f = 0.106 \text{ hp} \frac{745.7 \text{ W}}{\text{hp}} = 79.0 \text{ W}$$

Expressing this in U.S. Customary units gives

$$P_f = 79.0 \text{ W} \frac{1 \text{ Btu/hr}}{0.293 \text{ W}} = 270 \text{ Btu/hr}$$

16–8
PRACTICAL CONSIDERATIONS FOR PLAIN SURFACE BEARINGS

The design of the bearing system must consider the method of delivery of the lubricant to the bearing, the distribution of the lubricant within the bearing, the quantity of lubricant required, the amount of heat generated in the bearing and its effect on the temperature of the lubricant, the dissipation of heat from the bearing, the maintenance of the lubricant in a clean condition, and the performance of the bearing over the complete range of operating conditions that the bearing is likely to experience.

Many of these factors are simply design details that must be worked out along with the other aspects of the machine design. But some guidelines and general recommendations will be presented here.

The lubricant can be delivered to the bearing by a pump, perhaps driven from the same source that drives the entire machine. In some gear drives, one of the gears is designed to dip into an oil sump and carry oil up to the gear mesh and to the bearings. An external oil cup can be used to supply oil by gravity if the lubricant quantity required is small.

Methods of estimating the quantity of oil required, considering the leakage of oil from the ends of the bearing, are available. (See References 1, 3, 5, 14–18.)

The delivery of oil to the bearing should always be in an area opposite the location of the hydrodynamic pressure that supports the load. Otherwise, the oil delivery hole would destroy the pressurization of the film.

Grooving is frequently used to distribute the oil along the length of the bearing. Oil would be delivered through a radial hole in the bearing at the midlength point. The groove would extend axially in both directions from the hole, but it would terminate somewhat before the end of the bearing in order to keep the oil from leaking to the side. The rotation of the journal then carries the oil around to the area where the hydrodynamic film is generated. Figure 16–9 shows several styles of grooving in use.

Cooling of the bearing itself, or of the oil in the sump that supplies the oil, must always be considered. Natural convection may be sufficient to transfer heat away and maintain an acceptable bearing temperature. If not, forced convection can be used. In severe cases of heat generation, especially where the bearing system operates in a hot area such as a furnace, liquid coolant can be pumped through a jacket around the bearing. Some commercially available bearings provide this feature. Placing a heat exchanger in the oil sump or pumping the oil through an external heat exchanger can also be done. Figure 16–10 shows one commercially available style of a plain bearing with coolant chambers to allow use of water, air, or oil as the internal coolant.

FIGURE 16-9 Examples of styles of grooving for plain bearings (Bunting Bearings Corp., Holland, OH)

FIGURE 16-10 Sleevoil RTL pillow block hydrodynamically lubricated bearing with coolant chambers (Copyright Rockwell Automation/Dodge, Milwaukee, WI. Used by permission.)

The lubricant can be cleaned by passing it through filters as it is pumped to the bearing. Magnetic plugs within the sump can be effective in attracting and holding metallic particles that can score the bearing if allowed to enter the clearance space between the journal and the bearing. Of course, frequent changing of the oil is also desirable.

The design procedure used in the preceding section was completed for one set of conditions: a given temperature, diametral clearance, load, and rotational speed. If any of these factors varies during the operation of the machine, the performance of the bearing must be evaluated under the new conditions. The testing of a prototype under a variety of conditions is also desirable. References 1, 2, 8, and 13–19 discuss other practical considerations for the design of plain surface bearings.

**16-9
HYDROSTATIC
BEARINGS**

Recall that hydrodynamic lubrication results from the creation of a pressurized film of oil sufficient to carry the load on the bearing, with the film being generated by the motion of the journal itself within the bearing. It was noted that a steady relative motion between the journal and the bearing is required to generate and maintain the film.

In some types of equipment, the conditions are such that a hydrodynamic film cannot be developed. Reciprocating or oscillating devices or very-slow-moving machines are examples. If the load on the bearing is very high, it may not be possible to generate a sufficiently high pressure in the film to support the load. Even in cases in which hydrodynamic lubrication can be developed during normal operation of the machine, there is still mixed-film or boundary lubrication during the start-up and shut-down cycles. This may be unacceptable.

Consider the design of the mount for a telescope or an antenna system in which rotation of the base is required to be at very slow speed and to have smooth motion. Also, low friction is desired in order to keep the drive system small and to provide fast response and accurate positioning. The mount is essentially a thrust bearing supporting the weight of the system.

In this type of application, *hydrostatic lubrication* is desirable. Lubricant is supplied to the bearing under high pressure, several hundred psi or higher, and the pressure acting on the area of the bearing literally lifts the load off the bearing, even with the equipment stationary.

FIGURE 16–11

Main elements of a hydrostatic bearing system

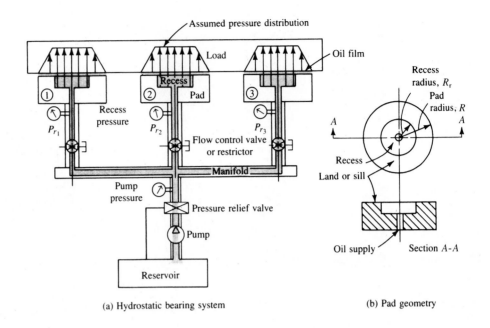

(a) Hydrostatic bearing system   (b) Pad geometry

Figure 16–11 shows the main elements of a hydrostatic bearing system. A positive displacement pump draws oil from a reservoir and delivers it under pressure to a supply manifold from which several bearing pads can be supplied. At each bearing pad, the oil passes through a control element that permits the balancing of the system. The control element may be a flow-control valve, a length of small-diameter tubing, or an orifice, any of which offers a resistance to the flow of oil and permits the several bearing pads to operate at a sufficiently high pressure to lift the load on that pad. When the system is in operation, the oil enters a recess within the bearing pad. For example, Figure 16–11(b) shows a circular pad with a circular recess in its center supplied with oil through a central hole. The load initially rests on the land area, sealing the recess. As the pressure in the recess reaches the level where the product of the pressure times the recess area equals the applied load, the load is lifted from the pad. Immediately there is a flow of oil across the land area under the lifted load, and the pressure decreases to atmospheric pressure at the outside of the pad. The flow of oil must be maintained at a level that matches the outflow from the pad. When equilibrium occurs, the integrated product of the local pressure times the area lifts the load a certain distance, h, usually in the range of 0.001 to 0.010 in (0.025 to 0.25 mm). The film thickness, h, must be large enough to ensure no solid contact over the range of operating conditions, but it should be kept as small as possible to minimize the flow of oil through each bearing and the pump power required to drive the system.

Hydrostatic Bearing Performance

Three factors characterizing the performance of a hydrostatic bearing are its load-carrying capacity, the flow of oil required, and the pumping power required, as indicated by the dimensionless coefficients a_f, q_f, and H_f. The magnitudes of the coefficients depend on the design of the pad:

⇨ **Load-Carrying Capacity**

$$F = a_f A_p p_r \tag{16–6}$$

⇨ **Oil Flow Required**

$$Q = q_f \frac{F}{A_p} \frac{h^3}{\mu} \tag{16–7}$$

⇨ **Pumping Power Required**

$$P = p_r Q = H_f \left(\frac{F}{A_p} \right)^2 \frac{h^3}{\mu} \qquad (16\text{–}8)$$

where F = load on the bearing, lb or N
  Q = volume flow rate of oil, in³/s or m³/s
  P = pumping power, lb·in/s or N·m/s (watts)
  a_f = pad load coefficient, dimensionless
  q_f = pad flow coefficient, dimensionless
  H_f = pad power coefficient, dimensionless (*Note: $H_f = q_f/a_f$.*)
  A_p = pad area, in² or m²
  p_r = oil pressure in the recess of the pad, psi or Pa
  h = film thickness, in or m
  μ = dynamic viscosity of the oil, lb·s/in² (reyn) or Pa·s

Figure 16–12 shows the typical variation of the dimensionless coefficients as a function of the pad geometry for a circular pad with a circular recess. As the size of the recess, R_r/R, increases, the load-carrying capacity increases, as indicated by a_f. But at the same time, the flow through the bearing increases, as indicated by q_f. The increase is gradual up to a value of R_r/R of approximately 0.7, and then rapid for higher ratios. This higher flow rate requires much higher pumping power, as indicated by the rapidly increasing power coefficient. At very low ratios of R_r/R, the load coefficient decreases rapidly. The pressure in the recess would have to increase to compensate in order to lift the load. The higher pressure requires more pumping power. Therefore, the power coefficient is high either at very small ratios of R_r/R or at high ratios. The minimum power is required for ratios between 0.4 and 0.6.

These general characteristics of hydrostatic bearing pad performance are typical for many different geometries of pads. Extensive data for the performance of different pad shapes are published. (See Reference 6.)

Example Problem 16–4 illustrates the basic design procedure for hydrostatic bearings.

FIGURE 16–12
Dimensionless performance coefficients for circular pad hydrostatic bearing (Cast Bronze Institute. *Cast Bronze Hydrostatic Bearing Design Manual.* New York: Copper Development Association, 1975)

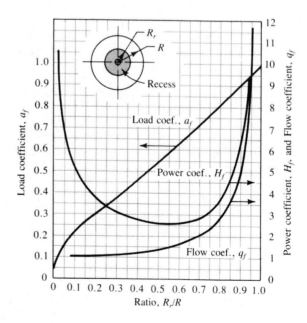

Example Problem 16–4   A large antenna mount weighing 12 000 lb is to be supported on three hydrostatic bearings such that each bearing pad carries 4000 lb. A positive displacement pump will be used to deliver oil at a pressure up to 500 psi. Design the hydrostatic bearings.

Solution   We will choose the circular pad design for which the performance coefficients are available in Figure 16–12. The results of the design will specify the dimensions of the pads, the required oil pressure in the recess of each pad, the type of oil required and its temperature, the thickness of the film of oil when the bearings are supporting the load, the flow rate of oil required, and the pumping power required.

Step 1. From Figure 16–12, the minimum power required for a circular pad bearing would occur with the ratio R_r/R of approximately 0.50. For that ratio, the value of the load coefficient is $a_f = 0.55$. The pressure at the bearing recess will be somewhat below the maximum available of 500 psi because of the pressure drop in the restrictor placed between the supply manifold and the pad. Let's design for a recess pressure of approximately 400 psi. Then, from Equation (16–6),

$$A_p = \frac{F}{a_f p_r} = \frac{4000 \text{ lb}}{0.55(400 \text{ lb/in}^2)} = 18.2 \text{ in}^2$$

But $A_p = \pi D^2/4$. Then the required pad diameter is

$$D = \sqrt{4A_p/\pi} = \sqrt{4(18.2)/\pi} = 4.81 \text{ in}$$

For convenience, let's specify $D = 5.00$ in. Then the actual pad area will be

$$A_p = \pi D^2/4 = (\pi)(5.00 \text{ in})^2/4 = 19.6 \text{ in}^2$$

The required recess pressure is then

$$p_r = \frac{F}{a_f A_p} = \frac{4000 \text{ lb}}{0.55(19.6 \text{ in}^2)} = 370 \text{ lb/in}^2$$

Also,

$$R = D/2 = 5.00 \text{ in}/2 = 2.50 \text{ in}$$
$$R_r = 0.50R = 0.50(2.50 \text{ in}) = 1.25 \text{ in}$$

Step 2. Specify the design value of the film thickness, h. It is recommended that h be between 0.001 and 0.010 in. Let's use $h = 0.005$ in.

Step 3. Specify the lubricant and the operating temperature. Let's select SAE 30 oil and assume that the maximum oil temperature in the oil film will be 120°F. A method of estimating the actual film temperature during operation may be consulted. (See Reference 6.) From the viscosity/temperature curves, Figure 16–6, the viscosity is approximately 8.3×10^{-6} reyn (lb·s/in²).

Step 4. Compute the oil flow through the bearing from Equation (16–7). The value of $q_f = 1.4$ can be found from Figure 16–12:

$$Q = q_f \frac{F}{A_p} \frac{h^3}{\mu} = (1.4)\frac{4000 \text{ lb}}{19.6 \text{ in}^2} \frac{(0.005 \text{ in})^3}{8.3 \times 10^{-6} \text{ lb·s/in}^2}$$
$$Q = 4.30 \text{ in}^3/\text{s}$$

Step 5. Compute the pumping power required from Equation (16–8). The value of $H_f = 2.6$ can be found from Figure 16–12:

$$P = p_r Q = H_f \left(\frac{F}{A_p} \right)^2 \frac{h^3}{\mu} = 2.6 \left(\frac{4000}{19.6} \right)^2 \frac{(0.005 \text{ in})^3}{8.3 \times 10^{-6}} = 1631 \text{ lb} \cdot \text{in/s}$$

For convenience, we can convert this to horsepower:

$$P = \frac{1631 \text{ lb} \cdot \text{in}}{\text{s}} \frac{1.0 \text{ ft}}{12 \text{ in}} \frac{1.0 \text{ hp}}{550 \text{ lb·ft/s}} = 0.247 \text{ hp}$$

16–10 TRIBOLOGY: FRICTION, LUBRICATION, AND WEAR

The study of friction, lubrication, and wear, called *tribology*, involves many disciplines, such as solid mechanics, fluid mechanics, material science, and chemistry. It is appropriate to engage several specialists on design teams where critical bearing or general lubrication designs are completed. For study beyond the scope of this book, see References 1, 2, 8–10, 11, 13–15, and 17.

This section presents some general principles of lubrication that can apply to a variety of design situations where relative motion occurs between mating machine elements. The goal is to help you recognize the many parameters that must be considered in designing machinery and in analyzing failures or unsatisfactory operation of existing machines. Much of the discussion relates to minimizing or controlling *friction*, defined generally as the resistance to parallel motion of mating surfaces. *Lubrication* is used to minimize friction by introducing a film of material that inherently reduces the force required to move one component relative to its mating component. Some materials have inherently low friction coefficients and can operate satisfactorily without external lubrication. When the relative motion results in physical contact between the surfaces of the mating components, some of the surface material can be removed, resulting in *wear*.

Friction

Not all friction is undesirable. Consider the need for drive wheels to use friction to develop propulsive force against floors, rails, or roads. Clutches and brakes employ friction to engage machinery, accelerate it, decelerate it, bring it to a stop, or hold it in position. See Chapter 22. Clamps and collets use friction to hold work pieces during manufacturing operations. In such applications, large, consistent frictional forces are desirable.

Most other applications where sliding contact occurs between mating components call for minimizing friction in order to minimize the forces, torques, and power required to drive the system. We are most concerned with these applications in this section.

The primary phenomena involved in creating friction are adhesion, elastic effects such as rolling resistance, viscoelastic effects, and hydrodynamic resistance. *Adhesion* is the bonding between dissimilar materials. The strength of adhesion depends on the structure and chemistry of the mating materials. Surface characteristics also contribute, such as the height of roughness peaks and valleys, called *asperities*. Sometimes asperities on mating parts are deformed or fractured during relative motion, while for other conditions, the motion is resisted as the asperities ride up and down on one another. *Rolling resistance* is caused by the elastic deformation of the moving body or the surface on which it moves. The geometry of the members in rolling contact, the magnitude of applied forces, and the elasticity of the contacting materials all play a part in determining the amount of resistance. *Viscoelastic effects* relate to the forces caused by deformation of flexible materials, such as elastomers, during contact. *Hydrodynamic resistance*, also called the *viscous effect*, is caused by the relative motion among the molecules of fluid lubricants between the moving mating components. This is the primary form of resistance in full-film hydrodynamically lubricated bearings. All or many of these forms of friction exist simultaneously in most practical machines.

Lubricants

Some of the important functions of lubricants are to reduce friction, remove heat from bearings and other machine elements where friction occurs, and to suspend contaminants. Several properties contribute to satisfactory lubricant performance:

- Good lubricity or oiliness to promote low friction
- Adequate viscosity for the application
- Low volatility under operating conditions
- Satisfactory flow characteristics at the temperatures encountered in use
- Appropriate thermal conductivity and specific heat to perform the heat transfer function
- Good chemical and thermal stability and the ability to maintain desirable characteristics for a reasonable period of use
- Compatibility with other materials in the system such as bearings, seals, and machine parts, particularly with regard to corrosion protection and degradation
- Environmental friendliness

Here we discuss the basic nature of oils, greases, and solid lubricants.

Oils.   Vendors of lubricating oils offer a huge variety of grades. One general classification is between refined natural petroleum oil and synthetic lubricants. Typically, natural oils are lower cost and can provide satisfactory service for general-purpose lubrication. Additives are frequently compounded with the natural oil to enhance viscosity, improve viscosity index (decrease the variation of viscosity with temperature), reduce corrosion potential, retard oxidation or other forms of chemical breakdown, or increase the ability to withstand localized high pressure. Synthetic lubricants are specially designed chemical formulations that can be tailored to specific applications. While performance is better than natural oils, cost is typically higher.

Lubricants are offered in categories such as engine oils, gear oils, compressor oils, turbine oils, general-purpose lubricating oils, chain lubricants, bearing oils, food machinery lubricants, automatic transmission fluids, hydraulic oils, and metalworking fluids. Internet sites 11–13 and 16 provide good examples of the types of lubricants available. Properties of lubricants that can affect selection are viscosity grade, viscosity index, and corrosion protection.

Viscosity grades are typically reported using the ISO (International Standards Organization) rating system. The grade number is the kinematic viscosity of the oil in centistokes measured at 40°C (104°F). Common ISO grades for lubricants are 32, 46, 68, 100, 150, 220, 320, 460, 680, and 1000. Gear oils are typically grades 150 to 680, depending on environmental temperature and degree of loading. Often the viscosity will also be reported at 100°C (212°F) as an indication of the variation of viscosity with temperature. This is not a part of the ISO classification and is influenced primarily by the property of viscosity index, discussed later in this section.

The American Gear Manufacturers Association (AGMA) defines lubricant numbers from 0 through 15, with numbers 3 to 8 being the most common for general power transmission uses. The basic viscosities for AGMA grades 0 to 8 correlate with ISO viscosity grades as follows.

AGMA	ISO	AGMA	ISO	AGMA	ISO
0	32	3	100	6	320
1	46	4	150	7	460
2	68	5	220	8	680

Suffixes are added to the AGMA grade designations; EP for extreme pressure, and S for synthetic oils. See References 10 and 26 for Chapter 9. See also Section 9–17.

SAE viscosity grades are also used to report oil viscosity. Similar to the grades used in automotive engines, SAE oils are also good for general lubrication and for gear-type power transmissions. Common grades are SAE 20, 30, 40, 50, 60, 85, 90, 140, and 250. These grades must conform to limits of kinematic viscosity in centistokes measured at 100°C (212°F). The W-grades of SAE oils, such as SAE 20W, must have viscosities less than specified limits at low temperatures ranging from −5°C to −55°C (23°F to −67°F). See Reference 12. This ensures that the lubricant can flow to critical surfaces in cold environments, particularly during equipment startup. Gear oils are typically SAE grades 80 to 250.

Viscosity index (VI) is a measure of how greatly the viscosity of a fluid changes with temperature. VI is determined by measuring the viscosity of the sample fluid at 40°C (104°F) and 100°C (212°F) and comparing these values with those of certain reference fluids that were assigned VI values of 0 and 100.

A fluid with a high VI exhibits a small change in viscosity with temperature.

A fluid with a low VI exhibits a large change in viscosity with temperature.

For most lubricants, a high VI is desirable because it would provide more reliable protection and exhibit a more uniform performance as the temperature varies. Commercially available lubricants report VI values from approximately 90 to 250. Gear lubricants typically have a VI of approximately 150. Additives, typically organic polymers, are used to improve viscosity and adjust viscosity index.

Corrosion protection can be tailored for specific applications by blending various additives with the base oil. Rust inhibitors for ferrous materials or copper and bronze corrosion protection are common. Oxidation inhibitors are used to prolong the life of oils. Extreme pressure additives help prevent scuffing in heavily loaded applications. Foam control additives prevent foaming where gears or other machine elements churn in an oil bath.

Greases.   Grease is a two-phase lubricant composed of a thickener dispersed in a base fluid, typically an oil. When applied to the interface between moving components, the grease tends to remain in place and adhere to the surfaces. The oil provides lubrication in a manner similar to that already discussed. As long as there is a sufficient amount of grease at the interface, continuous lubrication is provided. Some mechanisms are said to be lubricated for life. However, the designer must take great care to ensure that grease is not displaced from the critical areas that need it or to design the system for periodic reapplication of the grease. Some components such as bearings are provided with grease fittings for replenishing the supply and for discharging contaminated or oxidized grease.

Several kinds of thickeners are used in combination with natural or synthetic oils. See Internet sites 14 and 15 for examples. Thickeners are soaps formed by the reaction of animal or vegetable fats with alkaline substances such as lithium, calcium, an aluminum complex, clay, polyurea, and others. Lithium 12-hydroxy stearate is the most often used form. Soaps have a smooth buttery consistency and hold the oil in suspension until it is drawn out in the zone to be lubricated. Additives provide extreme pressure (EP) capability, rust protection, oxidation stability, and improved pumpability of the grease.

The National Lubricating Grease Institute (NLGI) defines nine ranges of consistency, labeled 000 through 6, from semifluid to soft, medium hard, hard, and rigid block. Grade #2 is good for general industrial applications.

Solid Lubricants.   Some applications cannot employ oils or greases because of contamination of other parts of the system, exposure to foods, excessively high or low temperatures, operation in a vacuum, or other environmental considerations. In such cases the designer may specify solid materials that have inherently good lubricating properties or add solid lubricants on critical surfaces. The section on boundary lubrication has already discussed various formulations of PTFE (polytetrafluoroethylene) as examples of materials with good lubricity.

A solid lubricant is a thin solid film that reduces friction and wear. Some are applied in powder form by brushing, spraying, or dipping, and they then adhere to mating surfaces. Binders are often blended with the base material to facilitate application and to promote adhesion. Curing in air or by baking is required.

Molybdenum disulfide (MoS_2) and graphite are two often-used types of solid lubricants. Others are lead iodide (PbI_2), silver sulfate ($AgSO_4$), tungsten disulfide, and stearic acid. An example of their effectiveness is the reduction of the coefficient of sliding friction for steel on steel from approximately 0.50 for dry clean surfaces to the range from 0.03 to 0.06.

Wear

Wear is the gradual removal of surface material from sliding surfaces. It is a complex process with numerous variables. Only testing under real conditions of service can predict actual wear in a given system. Several types of wear occur:

- Pitting, galling, scuffing, or scoring that typically result from high contact stresses and fatigue of surface material during rolling or sliding contact

- Abrasive wear, mechanical scraping, cutting, or scratching such as by hard contaminants in the interface between mating parts

- Fretting, the cyclic sliding of very small amplitude that displaces surface material. The subsequent accumulation of the debris tends to accelerate the process. Continued operation produces a surface appearance similar to corrosion and can generate small cracks from which ultimate fatigue failure can occur. It often occurs when tight-fitting parts are subjected to oscillating loads or vibration.

- Impingement wear caused by the eroding of material due to hard moving materials hitting the surface, perhaps carried by air or fluids. High velocity fluids, such as the discharge from high-pressure washers, can themselves cause such wear.

While it is not possible to prescribe specific approaches to reduce wear, the following are the things a designer may try. Again, testing is the only way to ensure satisfactory operation.

1. Keep contact force low between sliding surfaces.

2. Maintain low temperature at the mating surfaces.

3. Use hard contacting surfaces.

4. Produce smooth mating surfaces.

5. Maintain continuous lubrication to reduce friction.

6. Keep the relative velocity between mating surfaces low.

7. Specify materials that have inherently good wear properties.

Many material suppliers will report wear properties of their materials when operating against a similar or dissimilar material. These data are acquired by testing under carefully controlled laboratory conditions. Typically one member of a pair of materials is moved at a known velocity such as by rotation. The mating material is held stationary and under a known load. Careful measurements are made of the original weight and dimensions of the specimens of mating materials. After a sizable time of operation, the specimens are again weighed and measured to determine how much material has been removed. The results are reported as wear, computed from an equation such as,

$$K = W/FVT \qquad\qquad (16\text{--}9)$$

where K = wear factor for the materials
W = wear measured as loss of weight or volume
F = applied load
V = linear relative velocity between the sliding members
T = time of operation

Comparing the K factors for a variety of materials being considered can aid the designer in material selection.

REFERENCES

1. Avraham, Hanroy. *Bearing Design in Machinery.* New York: Marcel Dekker, 2002.

2. Bloch, Heinz P. *Practical Lubrication for Industrial Facilities.* New York: Marcel Dekker, 2000.

3. Boyd, John, and Albert A. Raimondi. "A Solution for the Finite Journal Bearing and Its Application to Analysis and Design." Parts I, II, and III. *Transactions of the American Society of Lubrication Engineers.* Vol. 1, No. 1, pp. 159–209, 1958.

4. Boyd, John, and Albert A. Raimondi. "Applying Bearing Theory to the Analysis and Design of Journal Bearings." Parts I and II. *Journal of Applied Mechanics* 73 (1951): 298–316.

5. Cast Bronze Institute. *Cast Bronze Bearing Design Manual.* New York: Copper Development Association, 1979.

6. Cast Bronze Institute. *Cast Bronze Hydrostatic Bearing Design Manual.* New York: Copper Development Association, 1975.

7. Juvinall, R. C., and Kurt M. Marshek. *Fundamentals of Machine Component Design.* 3rd ed. New York: John Wiley & Sons, 2000.

8. Khonsan, Michael, and E. R. Booser. *Applied Tribology: Bearing Design and Lubrication.* New York: John Wiley & Sons, 2001.

9. Ludema, Kenneth C. *Friction, Wear, Lubrication: A Textbook in Tribology.* New York: CRC Press, 1996.

10. Mang, Theo, and Wilfried Dresel. *Lubricants and Lubrications.* New York: Wiley-VCH, 2001.

11. Miuoshi, Kazuhisa. *Solid Lubrication Fundamentals and Applications.* New York: Marcel Dekker, 2001.

12. Mott, R. L. *Applied Fluid Mechanics.* 5th ed. Upper Saddle River, New Jersey: Prentice Hall, 2000.

13. Neale, John J. *Lubrication and Reliability Handbook.* London: Butterworth-Heinmann, 2000.

14. Neale, Michael J. *Bearings: A Tribology Handbook.* West Conshohocken, PA: Society of Automotive Engineers, 1993.

15. Pirro, D. M., A. A. Wessol, and J. G. Willis. *Lubrication Fundamentals.* 2d ed. New York: Marcel Dekker, 2001.

16. Someya, Tsuneo, ed. *Journal Bearing Data Book.* New York: Springer-Verlag, 1989.

17. Stolarski, T. A. *Tribology in Machine Design.* London, Butterworth-Heinemann, 2000.

18. Welsh, R. J. *Plain Bearing Design Handbook.* London: Butterworth-Heinemann, 1983.

19. Wolverton, M. P., et al. "How Plastic Composites Wear at High Temperatures." *Machine Design Magazine* (February 10, 1983).

INTERNET SITES RELATED TO PLAIN BEARINGS AND LUBRICATION

1. **Copper Development Association (CDA).** *www.copper.org* Industry association of companies and organizations involved in the production and use of copper, including cast bronze bearings and porous bronze bushings. Site includes some technical information on the design of plain bronze bearings. Also publishes the useful *Cast Bronze Bearing Design Manual.*

2. **Thomson Industries, Inc.** *www.thomsonindustries.com* Manufacturer of plain plastic bearings under the brand names Nyliner® and Nyliner Plus®.

3. **Saint-Gobain Performance Plastics.** *www.saint-gobain.com* Manufacturer of plain plastic bearings under the brand name Rulon®, available in 15 formulations.

4. **Glacier Garlock Bearings.** *www.garlockbearings.com* Manufacturer of a variety of plain bearings made from a composite of PTFE plastic, porous bronze, and steel or fiberglass under the brand names DU®, DX®, Gar-Fil®, and Gar-Max®.

5. **Graphite Metallizing Corporation.** *www.graphalloy.com* Manufacturer of plain bearings under the Graphalloy® brand name. Graphalloy® is a graphite-metal alloy formed from molten metal and graphite to make a uniform, solid, self-lubricating bushing material, available in over 100 grades.

6. **Beemer Precision, Inc.** *www.beemerprecision.com* Manufacturer of plain bearings made from cast bronze and porous sintered bronze impregnated with oil under the Oilite® brand.

7. **Bunting Bearings Corporation.** *www.buntingbearings.com* Manufacturer of plain bearings made from cast bronze, porous sintered bronze impregnated with oil, and plastic bearings made from Nylon, Rulon®, and Vespel®. Also makes the KU bearing consisting of a steel backing strip, a porous bronze matrix impregnated and overlaid with a PTFE/lead lining for low friction coefficient.

8. **Zincaloy, Inc.** *www.teamtube.com* Manufacturer of continuous cast ZA-12 zinc-aluminum alloy (Zincaloy™) used for bearings in mining, construction, forestry, and off-road vehicle industries. Team Tube Ltd. is an authorized distributor of Zincaloy bearing stock.

9. **Rockwell Automation/Dodge.** *www.dodge-pt.com* Manufacturer of mounted plain self-lubricating sleeve bearings under the Solidlube® brand used in general industrial, material handling, and bulk materials industries. Also available are Bronzoil® bearings with oil impregnated porous bronze, babbitted and bronze bushed bearings, and flange-mounted M Series Hydrodynamic bearings specifically designed for large motors and generators.

10. **Waukesha Bearings Corporation.** *www.waukbearing.com* Manufacturer of fluid film hydrodynamic sleeve bearings, tilting pad thrust and journal bearings, and housings for the bearings. These bearings are used for turbines, compressors, generators, gearboxes, and other types of rotating equipment.

11. **Exxon-Mobol, Inc.** *www.mobil.com* Producer of a wide variety of lubricants for general industry, automotive, turbine, compressor, and engine applications. Data sheets provided under the *Products and Applications* page.

12. **BP-Amoco, Inc.** *www.bplubricants.co.uk* Producer of a wide variety of lubricants for general industry, automotive, turbine, compressor, and engine applications. Data sheets provided under the *Products and Services* page.

13. **Shell Oil Company.** *www.shell-lubricants.com* Producer of a wide variety of lubricants for general industry, automotive, turbine, compressor, and engine applications. Data sheets provided under the *Industrial Lubricants* page found through the "Browse by application" feature.

14. **National Lubricating Grease Institute.** *www.nlgi.com* An association of companies and research organizations that develop, manufacture, distribute, and use greases. Establishes standards and provides publications and technical articles on grease.

15. **Lubrizol Corporation.** *www.lubrizol.com* Manufacturer of a large variety of greases and additives for the lubrication industry. The *Knowledge* section of the site provides much information about the science and technology of lubrication, a listing of numerous product formulations, a guide for the choice of additives, and recommendations for use.

16. **LubeLink.com** *www.lubelink.com* "The Portal for Information on the Worldwide Lubricant Market." Lists live links to companies providing lubrication products and services, testing laboratories, original equipment manufacturers that are heavy users of lubricants for vehicular and industrial applications, and industry associations related to the lubrication field.

PROBLEMS

For Problems 1–8 and the data listed in Table 16–2, design a plain surface bearing, using the boundary-lubricated approach of Section 16–5. Use an *L/D* ratio for the bearing in the range of 0.50 to 1.50. Compute the *pV* factor, and specify a material from Table 16–1.

TABLE 16–2

Problem number	Radial load (lb)	Shaft diameter (in)	Shaft speed (rpm)
1.	225	3.00	1750
2.	100	1.50	1150
3.	200	1.25	850
4.	75	0.50	600
5.	850	4.50	625
6.	500	3.75	450
7.	800	3.00	350
8.	60	0.75	750

For Problems 9–18 and the data listed in Table 16–3, design a hydrodynamically lubricated bearing, using the method outlined in Section 16–7. Specify the journal nominal diameter, the bearing length, the diametral clearance, the minimum film thickness of the lubricant during operation, the surface finish of the journal and bearing, the lubricant, and its maximum operating temperature. For your design, compute the coefficient of friction, the friction torque, and the power dissipated as the result of friction.

For Problems 19–28 and the data listed in Table 16–4, design a hydrostatic bearing of circular shape. Specify the pad diameter, the recess diameter, the recess pressure, the film thickness, the lubricant and its temperature, the oil flow rate, and the pumping power. The load specified is for a single bearing. You may choose to use multiple bearings. (The supply pressure is the maximum available at the pump.)

TABLE 16–3

Problem number	Radial load	Min. shaft diameter	Shaft speed (rpm)	Application
9.	1250 lb	2.60 in	1750	Electric motor
10.	2250 lb	3.50 in	850	Conveyor
11.	1875 lb	2.25 in	1150	Air compressor
12.	1250 lb	1.75 in	600	Precision spindle
13.	500 lb	1.15 in	2500	Precision spindle
14.	850 lb	1.45 in	1200	Idler sheave
15.	4200 lb	4.30 in	450	Shaft for chain drive
16.	18.7 kN	100 mm	500	Conveyor
17.	2.25 kN	25 mm	2200	Machine tool
18.	5.75 kN	65 mm	1750	Printer

TABLE 16–4

Problem number	Load	Supply pressure
19.	1250 lb	300 psi
20.	5000 lb	300 psi
21.	3500 lb	500 psi
22.	750 lb	500 psi
23.	250 lb	150 psi
24.	500 lb	150 psi
25.	22.5 kN	2.0 MPa
26.	1.20 kN	750 kPa
27.	8.25 kN	1.5 MPa
28.	12.5 kN	1.5 MPa

17

Linear Motion Elements

The Big Picture

You Are the Designer

17–1 Objectives of This Chapter

17–2 Power Screws

17–3 Ball Screws

17–4 Application Considerations for Power Screws and Ball Screws

Linear Motion Elements

Discussion Map

☐ Many kinds of mechanical devices produce linear motion for machines such as automation equipment, packaging systems, and machine tools.

☐ *Power screws*, *jacks* and *ball screws* are designed to convert rotary motion to linear motion and to exert the necessary force to move a machine element along a desired path. They use the principle of a screw thread and its mating nut.

Discover

Visit a machine shop and see if you can identify power screws, ball screws, or other linear motion devices. Look in particular at the lathes and the milling machines. Manual machines likely use power screws. Those using computer numerical control should have ball screws. Describe the shape of the threads. How are the screws powered? How are they fastened to other machine parts?

Can you find other equipment that uses linear motion devices? Look in laboratories where materials are tested or where large forces must be generated.

This chapter will help you learn to analyze power screws and ball screw drives and to specify suitable sizes for a given application.

A common requirement in mechanical design is to move components in a straight line. Elevators move vertically up or down. Machine tools move cutting tools or parts to be machined in straight lines, either horizontally or vertically, to shape metal into desired forms. A precision measuring device moves a probe in straight lines to determine electronically the dimensions of a part. Assembly machines require many straight-line motions to insert components and to fasten them together. A packaging machine moves products into cartons, closes flaps, and seals the cartons.

Some examples of components and systems that facilitate linear motion are:

Power screws	Ball screws	Jacks	Fluid power cylinders
Linear actuators	Linear slides	Ball bushings	Rack and pinion sets
Linear solenoids	Positioning stages	X-Y-Z tables	Gantry tables

Figure 17–1(a) shows a cutaway view of a jack that employs a power screw to produce linear motion. Power is delivered to the input shaft by an electric motor. The worm, machined integral with the input shaft, drives the wormgear, resulting in a reduction in the speed of rotation. The inside of the wormgear has machined threads that engage the external threads of the power screw, driving it vertically. Figure 17–1(b) shows the screw jack mated with an external gear reducer and motor to provide a complete linear motion system. Limit switches, position sensors, and programmable logic controllers can be used to control the motion cycle. These and other types of linear actuators can be seen on Internet sites 1–10.

Rack and pinion sets are discussed in Chapter 8. Fluid power actuators employ oil hydraulic or pneumatic fluid pressure to extend or retract a piston rod within a cylinder as discussed in textbooks on fluid power. Positioning stages, X-Y-Z tables, and gantry tables typically are driven by precision stepper motors or servomechanisms that allow precision location of components anywhere within their control volume. Linear solenoids are devices that cause a rodlike core to be extended or retracted as power is supplied to an electrical coil, producing rapid movement over small distances. Applications are seen in office equipment, automation devices, and packaging systems. See Internet site 11.

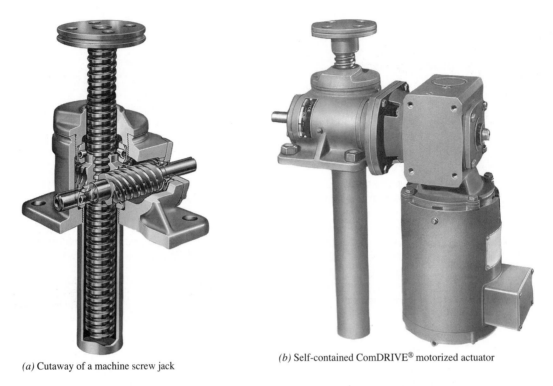

(a) Cutaway of a machine screw jack

(b) Self-contained ComDRIVE® motorized actuator

FIGURE 17–1   Examples of linear motion machine elements (Joyce/Dayton Corporation, Dayton, OH)

Linear slides and ball bushings are designed to guide mechanical components along a precise linear track. Low friction materials or rolling contact elements are used to produce smooth motion with low power required. See Internet sites 1, 3, 5, and 7–9.

Power screws and ball screws are designed to convert rotary motion to linear motion and to exert the necessary force to move a machine element along a desired path. Power screws operate on the classic principle of the screw thread and its mating nut. If the screw is supported in bearings and rotated while the nut is restrained from rotating, the nut will translate along the screw. If the nut is made an integral part of a machine, for example, the tool holder for a lathe, the screw will drive the tool holder along the bed of the machine to take a cut. Conversely, if the nut is supported while it is rotating, the screw can be made to translate. The screw jack uses this approach.

A ball screw is similar in function to a power screw, but the configuration is different. The nut contains many small, spherical balls that make rolling contact with the threads of the screw, giving low friction and high efficiencies when compared with power screws. Modern machine tools, automation equipment, vehicle steering systems, and actuators on aircraft use ball screws for high precision, fast response, and smooth operation.

Visit a machine shop where there are metal-cutting machine tools. Look for examples of power screws that convert rotary motion to linear motion. They are likely to be on manual lathes moving the tool holder. Or look at the table drive for a milling machine. Inspect the form of the threads of the power screw. Are they of a form similar to that of a screw thread with sloped sides? Or are the sides of the threads straight? Compare the screw threads with those shown in Figure 17–2 for square, Acme, and buttress forms.

While in the shop, do you see any type of material-testing equipment or a device called an *arbor press* that exerts large axial forces? Such machines often employ square-thread power screws to produce the axial force and motion from rotational input, through

FIGURE 17–2
Forms of power screw
threads [(b), ANSI
Standard B1.5–1973;
(c), ANSI Standard
B1.9–1973]

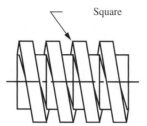

Square

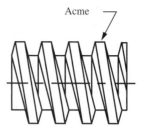

Acme

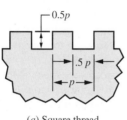

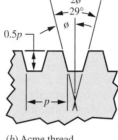

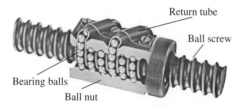

(*a*) Square thread

(*b*) Acme thread
(Ref.: ANSI B1.5-1973)

(*c*) Buttress thread
(Ref.: ANSI B1.9-1973)

FIGURE 17–3 Ball
bearing screw
(Thomson Industries,
Inc., Port Washington,
NY)

Return tube

Ball screw

Bearing balls

Ball nut

either a hand crank or an electric motor drive. If they are not in the machine shop, look for them in the metallurgy lab or another room where materials testing is done.

Now look further in the machine shop. Are there machines that use digital readouts to indicate position of the table or the tool? Are there computer numerical control machine tools? Any of these types of machines should have ball screws rather than the traditional power screws because ball screws require significantly less power and torque to drive them against a given load. They can also be moved faster and positioned more accurately than power screws. You may or may not be able to see the recirculating balls in the nut of the power screw, as illustrated in Figure 17–3. But you should be able to see the different-shaped threads looking like grooves with circular bottoms in which the spherical balls roll.

Have you seen such power screws or ball screws anywhere outside a machine shop? Some garage door openers employ a screw drive, but others use chain drives. Perhaps your home has a screw jack or a scissors jack for raising the car to change a tire. Both use power screws. Have you ever sat in a seat on an airplane where you can see the mechanisms that actuate the flaps on the rear edge of the wings? Try it sometime, and observe the actuators during takeoff or landing. It is likely that you will see a ball screw in action.

This chapter will help you learn the methods of analyzing the performance of power screws and ball screws and to specify the proper size for a given application.

You Are the Designer

You are a member of a plant engineering team for a large steel processing plant. One of the furnaces in the plant in which steel is heated prior to final heat treatment is installed beneath the floor, and the large ingots are lowered vertically into it. While the ingots are soaking in the furnace, a large, heavy hatch is placed over the opening to minimize the escape of heat and provide more uniform temperature. The hatch weighs 25 000 lb.

You are asked to design a system that will permit the hatch to be raised at least 15 in above the floor within 12.0 s and to lower it again within 12.0 s.

What design concept would you propose? Of course, there are many feasible concepts, but suppose that you proposed a system like that sketched in Figure 17–4. An overhead support structure is suggested on which a worm/wormgear drive system would be mounted. One shaft would be driven directly by the gear drive while a second shaft would be driven simultaneously by a chain drive. The shafts are power

screws, supported in bearings at the top and bottom. A yoke is connected to the hatch and is mounted on the screws with the nuts that mate with the screw integral with the yoke. Therefore, as the screw rotates, the nuts carry the yoke and the hatch vertically upward or downward.

As the designer of the hatch lift system, you must make several decisions. What size screw is required to ensure that it can safely raise the 25 000-lb hatch? Notice that the screws are placed in tension as they are supported on the collars on the upper support system. What diameter, thread type, and thread size should be used? The sketch suggests the Acme thread form. What other styles are available? At what speed must the screws be rotated to raise the hatch in 12.0 s or less? How much power is required to drive the screws? What safety concerns exist while the system is handling this heavy load? What advantage would there be to using a ball screw rather than a power screw?

The material in this chapter will help you make these decisions, along with providing methods of computing stresses, torques, and efficiencies.

FIGURE 17–4 An Acme screw-driven system for raising a hatch

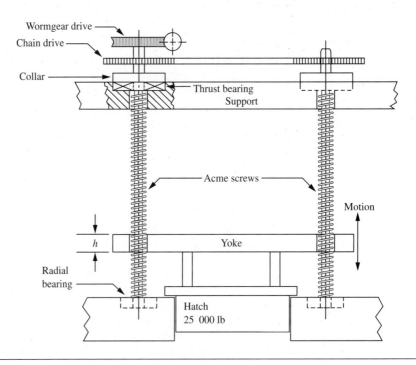

17–1
OBJECTIVES OF THIS CHAPTER

After completing this chapter, you will be able to:

1. Describe the operation of a power screw and the general form of *square threads*, *Acme threads*, and *buttress threads* as they are applied to power screws.

2. Compute the torque that must be applied to a power screw to raise or lower a load.

3. Compute the efficiency of power screws.

4. Compute the power required to drive a power screw.
5. Describe the design of a ball screw and its mating nut.
6. Specify suitable ball screws for a given set of requirements of load, speed, and life.
7. Compute the torque required to drive a ball screw, and compute its efficiency.

17–2
POWER SCREWS

Figure 17–2 shows three types of power screw threads: the square thread, the Acme thread, and the buttress thread. Of these, the square and buttress threads are the most efficient. That is, they require the least torque to move a given load along the screw. However, the Acme thread is not greatly less efficient, and it is easier to machine. The buttress thread is desirable when force is to be transmitted in only one direction.

Table 17–1 gives the preferred combinations of basic major diameter, D, and number of threads per inch, n, for Acme screw threads. The pitch, p, is the distance from a point on one thread to the corresponding point on the adjacent thread, and $p = 1/n$.

Other pertinent dimensions listed in Table 17–1 include the minimum minor diameter and the minimum pitch diameter of a screw with an external thread. When you are performing stress analyses on the screw, the safest approach is to compute the area corresponding to the minor diameter for tensile or compressive stresses. However, a more accurate stress computation results from using the *tensile stress area* (listed in Table 17–1), found from

➪ **Tensile Stress Area for Screw Threads**

$$A_t = \frac{\pi}{4}\left[\frac{D_r + D_p}{2}\right]^2 \qquad\qquad (17\text{–}1)$$

TABLE 17–1 Preferred Acme screw threads

Nominal major diameter, D (in)	Threads per in, n	Pitch, $p = 1/n$ (in)	Minimum minor diameter, D_r (in)	Minimum pitch diameter, D_p (in)	Tensile stress area, A_t (in²)	Shear stress area, A_s (in²)[a]
1/4	16	0.0625	0.1618	0.2043	0.026 32	0.3355
5/16	14	0.0714	0.2140	0.2614	0.044 38	0.4344
3/8	12	0.0833	0.2632	0.3161	0.065 89	0.5276
7/16	12	0.0833	0.3253	0.3783	0.097 20	0.6396
1/2	10	0.1000	0.3594	0.4306	0.1225	0.7278
5/8	8	0.1250	0.4570	0.5408	0.1955	0.9180
3/4	6	0.1667	0.5371	0.6424	0.2732	1.084
7/8	6	0.1667	0.6615	0.7663	0.4003	1.313
1	5	0.2000	0.7509	0.8726	0.5175	1.493
$1\frac{1}{8}$	5	0.2000	0.8753	0.9967	0.6881	1.722
$1\frac{1}{4}$	5	0.2000	0.9998	1.1210	0.8831	1.952
$1\frac{3}{8}$	4	0.2500	1.0719	1.2188	1.030	2.110
$1\frac{1}{2}$	4	0.2500	1.1965	1.3429	1.266	2.341
$1\frac{3}{4}$	4	0.2500	1.4456	1.5916	1.811	2.803
2	4	0.2500	1.6948	1.8402	2.454	3.262
$2\frac{1}{4}$	3	0.3333	1.8572	2.0450	2.982	3.610
$2\frac{1}{2}$	3	0.3333	2.1065	2.2939	3.802	4.075
$2\frac{3}{4}$	3	0.3333	2.3558	2.5427	4.711	4.538
3	2	0.5000	2.4326	2.7044	5.181	4.757
$3\frac{1}{2}$	2	0.5000	2.9314	3.2026	7.388	5.700
4	2	0.5000	3.4302	3.7008	9.985	6.640
$4\frac{1}{2}$	2	0.5000	3.9291	4.1991	12.972	7.577
5	2	0.5000	4.4281	4.6973	16.351	8.511

[a]Per inch of length of engagement.

FIGURE 17–5
Screw thread force
analysis

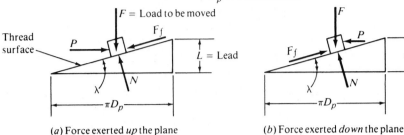

P = Force required to move the load
F_f = Friction force
N = Normal force
λ = Lead angle
D_p = Pitch diameter

(a) Force exerted *up* the plane　　　　　　　　(b) Force exerted *down* the plane

This is the area corresponding to the average of the minor (or root) diameter, D_r, and the pitch diameter, D_p. The data reflect the minimums for commercially available screws according to recommended tolerances.

Another failure mode for a power screw is the shearing of the threads in the axial direction, cutting them away from the main shaft near the pitch diameter. The shear stress is computed from the direct stress formula,

$$\tau = F/A_s$$

The shear stress area, A_s, listed in Table 17–1 is also found in published data and represents the area in shear approximately at the pitch line of the threads for a 1.0-in length of engagement. Other lengths would require that the area be modified by the ratio of the actual length to 1.0 in.

Torque Required to Move a Load

When using a power screw to exert a force, as with a jack raising a load, you need to know how much torque must be applied to the nut of the screw to move the load. The parameters involved are the force to be moved, F; the size of the screw, as indicated by its pitch diameter, D_p; the lead of the screw, L; and the coefficient of friction, f. Note that the *lead* is defined as the axial distance that the screw would move in one complete revolution. For the usual case of a single-threaded screw, the lead is equal to the pitch and can be read from Table 17–1 or computed from $L = p = 1/n$.

In the development of Equation (17–2) for the torque required to turn the screw, Figure 17–5(a), which depicts a load being pushed up an inclined plane against a friction force, is used. This is a reasonable representation for a square thread if you think of the thread as being unwrapped from the screw and laid flat. The torque for an Acme thread is slightly different from this due to the thread angle. The revised equation for the Acme thread will be shown later.

The torque computed from Equation (17–2) is called T_u, implying that the force is applied to move a load up the plane, that is, to raise the load. This observation is completely appropriate if the load is raised vertically, as with a jack. If, however, the load is horizontal or at some angle, Equation (17–2) is still valid if the load is to be advanced along the screw "up the thread." Equation (17–4) shows the required torque, T_d, to lower a load or move a load "down the thread."

The torque to move a load up the thread is

**Torque Required to
Move a Load up a
Power Screw with a
Square Thread**

$$T_u = \frac{FD_p}{2}\left[\frac{L + \pi f D_p}{\pi D_p - fL}\right] \tag{17–2}$$

This equation accounts for the force required to overcome friction between the screw and the nut in addition to the force required just to move the load. If the screw or the nut bears against a stationary surface while rotating, there will be an additional friction torque developed at that surface. For this reason, many jacks and similar devices incorporate antifriction bearings at such points.

The coefficient of friction for use in Equation (17–2) depends on the materials used and the manner of lubricating the screw. For well-lubricated steel screws acting in steel nuts, $f = 0.15$ should be conservative.

An important factor in the analysis for torque is the angle of inclination of the plane. In a screw thread, the angle of inclination is referred to as the *lead angle*, λ. It is the angle between the tangent to the helix of the thread and the plane transverse to the axis of the screw. It can be seen from Figure 17–5 that

$$\tan \lambda = L/(\pi D_p) \tag{17–3}$$

where πD_p = circumference of the pitch line of the screw

Then if the rotation of the screw tends to raise the load (move it up the incline), the friction force opposes the motion and acts down the plane.

Conversely, if the rotation of the screw tends to lower the load, the friction force will act up the plane, as shown in Figure 17–5(b). The torque analysis changes, producing Equation (17–4):

Torque Required to Lower a Load Down a Power Screw with a Square Thread

$$T_d = \frac{FD_p}{2} \left[\frac{\pi f D_p - L}{\pi D_p + fL} \right] \tag{17–4}$$

If the screw thread is steep (that is, if it has a high lead angle), the friction force may not be able to overcome the tendency for the load to "slide" down the plane, and the load will fall due to gravity. In most cases for power screws with single threads, however, the lead angle is rather small, and the friction force is large enough to oppose the load and keep it from sliding down the plane. Such a screw is called *self-locking*, a desirable characteristic for jacks and similar devices. Quantitatively, the condition that must be met for self-locking is

$$f > \tan \lambda \tag{17–5}$$

The coefficient of friction must be greater than the tangent of the lead angle. For $f = 0.15$, the corresponding value of the lead angle is 8.5°. For $f = 0.1$, for very smooth, well-lubricated surfaces, the lead angle for self-locking is 5.7°. The lead angles for the screw designs listed in Table 17–1 range from 1.94° to 5.57°. Thus, it is expected that all would be self-locking. However, operation under vibration should be avoided, as this may still cause movement of the screw.

Efficiency of a Power Screw

Efficiency for the transmission of a force by a power screw can be expressed as the ratio of the torque required to move the load without friction to that with friction. Equation (17–2) gives the torque required with friction, T_u. Letting $f = 0$, the torque required without friction, T', is

$$T' = \frac{FD_p}{2} \frac{L}{\pi D_p} = \frac{FL}{2\pi} \tag{17–6}$$

Then the efficiency, e, is

Efficiency of a Power Screw

$$e = \frac{T'}{T_u} = \frac{FL}{2\pi T_u} \tag{17–7}$$

Alternate Forms of the Torque Equations

Equations (17–2) and (17–4) can be expressed in terms of the lead angle, rather than the lead and the pitch diameter, by noting the relationship in Equation (17–3). With this substitution, the torque required to move the load would be

⇨ **Torque to Raise a Load with a Square Thread**

$$T_u = \frac{FD_p}{2}\left[\frac{(\tan\lambda + f)}{(1 - f\tan\lambda)}\right]$$ (17–8)

and the torque required to lower the load is

⇨ **Torque to Lower a Load with a Square Thread**

$$T_d = \frac{FD_p}{2}\left[\frac{(f - \tan\lambda)}{(1 + f\tan\lambda)}\right]$$ (17–9)

Adjustment for Acme Threads

The difference between Acme threads and square threads is the presence of the thread angle, ϕ. Note from Figure 17–1 that $2\phi = 29°$, and therefore $\phi = 14.5°$. This changes the direction of action of the forces on the thread from that depicted in Figure 17–5. Figure 17–6 shows that F would have to be replaced by $F/\cos\phi$. Carrying this through, the analysis for torque would give the following modified forms of Equations (17–8) and (17–9). The torque to move the load up the thread is

⇨ **Torque to Raise a Load with an Acme Thread**

$$T_u = \frac{FD_p}{2}\left[\frac{(\cos\phi\tan\lambda + f)}{(\cos\phi - f\tan\lambda)}\right]$$ (17–10)

and the torque to move the load down the thread is

⇨ **Torque to Lower a Load with an Acme Thread**

$$T_d = \frac{FD_p}{2}\left[\frac{(f - \cos\phi\tan\lambda)}{(\cos\phi + f\tan\lambda)}\right]$$ (17–11)

Power Required to Drive a Power Screw

If the torque required to rotate the screw is applied at a constant rotational speed, n, then the power in horsepower to drive the screw is

$$P = \frac{Tn}{63\,000}$$

References 1 and 2 include more detail about the formulas that characterize the performance of power screws.

FIGURE 17–6 Force on an Acme thread

(a) Force normal to a square thread

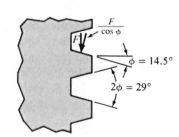

(b) Force normal to an Acme thread

Example Problem 17–1   Two Acme-threaded power screws are to be used to raise a heavy access hatch, as sketched in Figure 17–4. The total weight of the hatch is 25 000 lb, divided equally between the two screws. Select a satisfactory screw from Table 17–1 on the basis of tensile strength, limiting the tensile stress to 10 000 psi. Then determine the required thickness of the yoke that acts as the nut on the screw to limit the shear stress in the threads to 5000 psi. For the screw thus designed, compute the lead angle, the torque required to raise the load, the efficiency of the screw, and the torque required to lower the load. Use a coefficient of friction of 0.15.

Solution   The load to be lifted places each screw in direct tension. Therefore, the required tensile stress area is

$$A_t = \frac{F}{\sigma_d} = \frac{12\ 500\ \text{lb}}{10\ 000\ \text{lb/in}^2} = 1.25\ \text{in}^2$$

From Table 17–1, a $1\frac{1}{2}$-in-diameter Acme thread screw with four threads per inch would provide a tensile stress area of 1.266 in².

For this screw, each inch of length of a nut would provide 2.341 in² of shear stress area in the threads. The required shear area is then

$$A_s = \frac{F}{\tau_d} = \frac{12\ 500\ \text{lb}}{5000\ \text{lb/in}^2} = 2.50\ \text{in}^2$$

Then the required length of the yoke would be

$$h = 2.5\ \text{in}^2 \left[\frac{1.0\ \text{in}}{2.341\ \text{in}^2} \right] = 1.07\ \text{in}$$

For convenience, let's specify $h = 1.25$ in.

The lead angle is (remember that $L = p = 1/n = 1/4 = 0.250$ in)

$$\lambda = \tan^{-1} \frac{L}{\pi D_p} = \tan^{-1} \frac{0.250}{\pi(1.3429)} = 3.39°$$

The torque required to raise the load can be computed from Equation (17–10):

$$T_u = \frac{FD_p}{2} \left[\frac{(\cos\phi\tan\lambda + f)}{(\cos\phi - f\tan\lambda)} \right] \tag{17–10}$$

Using $\cos\phi = \cos(14.5°) = 0.968$, and $\tan\lambda = \tan(3.39°) = 0.0592$, we have

$$T_u = \frac{(12\ 500\ \text{lb})(1.3429\ \text{in})}{2} \frac{[(0.968)(0.0592) + 0.15]}{[0.968 - (0.15)(0.0592)]} = 1809\ \text{lb·in}$$

The efficiency can be computed from Equation (17–7):

$$e = \frac{FL}{2\pi T_u} = \frac{(12\ 500\ \text{lb})(0.250\ \text{in})}{2(\pi)(1809\ \text{lb·in})} = 0.275\ \text{or}\ 27.5\%$$

The torque required to lower the load can be computed from Equation (17–11):

$$T_d = \frac{FD_p}{2} \left[\frac{(f - \cos\phi\tan\lambda)}{(\cos\phi + f\tan\lambda)} \right] \tag{17–11}$$

$$T_d = \frac{(12\ 500\ \text{lb})(1.3429\ \text{in})}{2} \frac{[0.15 - (0.968)(0.0592)]}{[0.968 + (0.15)(0.0592)]} = 796\ \text{lb·in}$$

Example Problem 17–2   It is desired to raise the hatch in Figure 17–4 a total of 15.0 inches in no more than 12.0 s. Compute the required rotational speed for the screws and the power required.

Solution   The screw selected in the solution for Example Problem 17–1 was a $1\frac{1}{2}$-in Acme-threaded screw with four threads per inch. Thus, the load would be moved 1/4 in with each revolution. The linear speed required is

$$V = \frac{15.0 \text{ in}}{12.0 \text{ s}} = 1.25 \text{ in/s}$$

The required rotational speed would be

$$n = \frac{1.25 \text{ in}}{\text{s}} \; \frac{1 \text{ rev}}{0.25 \text{ in}} \; \frac{60 \text{ s}}{\text{min}} = 300 \text{ rpm}$$

Then the power required to drive each screw would be

$$P = \frac{Tn}{63\,000} = \frac{(1809 \text{ lb} \cdot \text{in})(300 \text{ rpm})}{63\,000} = 8.61 \text{ hp}$$

Alternative Thread Forms for Power Screws

While the standard Acme thread is probably the most widely used, others are available. The *stub Acme* thread has a similar form with a 29° angle between the sides, the depth of the thread is shorter, providing a stronger, more rigid thread. Metric power screws are typically made according to the ISO trapezoidal form that has a 30° included angle.

The relatively low efficiency of standard single-thread Acme screws (approximately 30% or less) can be a strong disadvantage. Higher efficiencies can be achieved using high lead, multiple thread designs. The higher lead angle produces efficiencies in the 30% to 70% range. It should be understood that some mechanical advantage is lost so that higher torques are required to move a particular load as compared with single thread screws. See Internet site 10.

17–3
BALL SCREWS
The basic action of using screws to produce linear motion from rotation was described in Section 17–2 on power screws. A special adaptation of this action that minimizes the friction between the screw threads and the mating nut is the *ball screw*.

Figure 17–3 shows a cutaway view of a commercially available ball screw. It replaces the sliding friction of the conventional power screw with the rolling friction of bearing balls. The bearing balls circulate in hardened steel races formed by concave helical grooves in the screw and the nut. All reactive loads between the screw and the nut are carried by the bearing balls which provide the only physical contact between these members. As the screw and the nut rotate relative to each other, the bearing balls are diverted from one end and are carried by the ball-guide return tubes to the opposite end of the ball nut. This recirculation permits unrestricted travel of the nut in relation to the screw. (See Internet site 3.)

Applications of ball screws occur in automotive steering systems, machine tool tables, linear actuators, jacking and positioning mechanisms, aircraft controls such as flap actuating devices, packaging equipment, and instruments. Figure 17–7 shows a machine with a ball screw installed to move a component along the ways of the bed.

FIGURE 17–7
Application of a ball screw (Thomson Industries, Inc., Port Washington, NY)

The application parameters to be considered in selecting a ball screw include the following:

The axial load to be exerted by the screw during rotation

The rotational speed of the screw

The maximum static load on the screw

The direction of the load

The manner of supporting the ends of the screw

The length of the screw

The expected life

The environmental conditions

Load-Life Relationship

When transmitting a load, a ball screw experiences stresses similar to those on a ball bearing, as discussed in Chapter 14. The load is transferred from the screw to the balls, from the balls to the nut, and from the nut to the driven device. The contact stress between the balls and the races in which they roll eventually causes fatigue failure, indicated by pitting of the balls or the races.

Thus, the rating of ball screws gives the load capacity of the screw for a given life which 90% of the screws of a given design will survive. This is similar to the L_{10} life of ball bearings. Because ball screws are typically used as linear actuators, the most pertinent life parameter is the distance traveled by the nut relative to the screw.

Manufacturers usually report the rated load that a given screw can exert for 1 million inches (25.4 km) of cumulative travel. The relationship between load, P, and life, L, is also similar to that for a ball bearing:

> **Relationship between Bearing Load and Life**

$$\frac{L_2}{L_1} = \left(\frac{P_1}{P_2}\right)^3$$

(17–12)

Thus, if the load on a ball screw is doubled, the life is reduced to one-eighth of the original life. If the load is cut in half, the life is increased by eight times. Figure 17–8 shows the nominal performance of ball screws of a small variety of sizes. Many more sizes and styles are available.

Torque and Efficiency

The efficiency of a ball bearing screw is typically taken to be 90%. This far exceeds the efficiency for power screws without rolling contact that are typically in the range of 20% to 30%. Thus, far less torque is required to exert a given load with a given size of screw. Power is correspondingly reduced. The computation of the torque to turn is adapted from Equation (17–7), relating efficiency to torque:

> **Efficiency of a Ball Screw**

$$e = \frac{FL}{2\pi T_u}$$

(17–7)

Then, using $e = 0.90$,

> **Torque to Drive a Ball Screw**

$$T_u = \frac{FL}{2\pi e} = 0.177FL$$

(17–13)

FIGURE 17–8 Ball screw performance

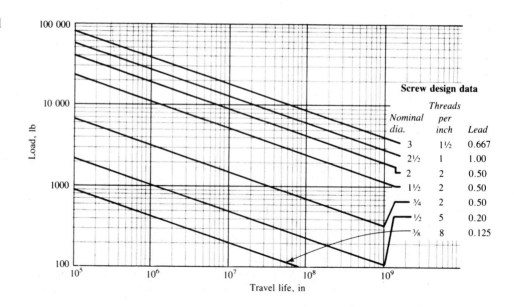

Screw design data		
Nominal dia.	Threads per inch	Lead
3	1½	0.667
2½	1	1.00
2	2	0.50
1½	2	0.50
¾	2	0.50
½	5	0.20
⅜	8	0.125

Because of the low friction, ball screws are virtually never self-locking. In fact, designers also use this property to advantage by purposely using the applied load on the nut to rotate the screw. This is called *backdriving*; backdriving torque can be computed from

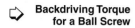

**Backdriving Torque
for a Ball Screw**

$$T_b = \frac{FLe}{2\pi} = 0.143FL \tag{17–14}$$

Example Problem 17–3

Select suitable ball screws for the application described in Example Problem 17–1 and illustrated in Figure 17–4. The hatch must be lifted 15.0 in to open it eight times per day, and then it must be closed. The design life is 10 years. The lifting or lowering is to be completed in no more than 12.0 s.

For the screw selected, compute the torque to turn the screw, the power required, and the actual expected life.

Solution

The data required to select a screw from Figure 17–8 are the load and the travel of the nut on the screw over the desired life. The load is 12 500 lb on each screw:

$$\text{Travel} = \frac{15.0 \text{ in}}{\text{stroke}} \frac{2 \text{ strokes}}{\text{cycle}} \frac{8 \text{ cycles}}{\text{day}} \frac{365 \text{ days}}{\text{year}} \frac{10 \text{ years}}{} = 8.76 \times 10^5 \text{ in}$$

From Figure 17–8, the 2-in screw with two threads per inch and a lead of 0.50 in is satisfactory.

The torque required to turn the screw is

$$T_u = 0.177FL = 0.177(12\ 500)(0.50) = 1106 \text{ lb} \cdot \text{in}$$

The rotational speed required is

$$n = \frac{1 \text{ rev}}{0.50 \text{ in}} \frac{15.0 \text{ in}}{12.0 \text{ s}} \frac{60 \text{ s}}{\text{min}} = 150 \text{ rpm}$$

The power required for each screw is

$$P = \frac{Tn}{63\ 000} = \frac{(1106)(150)}{63\ 000} = 2.63 \text{ hp}$$

Compare this with the 8.61 hp required for the Acme screw in Example Problem 17–1.

The actual travel life expected for this screw at a 12 500-lb load would be approximately 3.2×10^6 in, using Figure 17–8. This is 3.65 times longer than required.

**17–4
APPLICATION
CONSIDERATIONS
FOR POWER
SCREWS AND
BALL SCREWS**

This section discusses additional application considerations that apply to both power screws and ball screws. Details may change according to the specific geometry and manufacturing processes. Supplier data should be consulted.

Critical Speed

The proper application of ball screws must take into account their vibration tendencies, particularly when operating at relatively high speeds. Long slender screws may exhibit the phenomenon of *critical speed* at which the screw would tend to vibrate or whirl about its axis, possibly reaching dangerous amplitudes. Therefore, it is recommended that the operating speed of the screw be below 0.80 times the critical speed. An estimate for the critical speed, offered by Roton Products, Inc. (Internet site 10), is:

$$n_c = \frac{4.76 \times 10^6 \, dK_s}{(SF)L^2} \qquad\qquad\qquad \textbf{(17–15)}$$

where,

d = Minor diameter of the screw (in)
K_s = End fixity factor
L = Length between supports (in)
SF = Safety factor

The end fixity factor, K_s, depends on the manner of supporting the ends of the screw with the possibilities being:

1. Simply supported at each end by one bearing: $K_s = 1.00$
2. Fixed at each end by two bearings that prevent rotation at the support: $K_s = 2.24$
3. Fixed at one end and simply supported at the other: $K_s = 1.55$
4. Fixed at one end and free at the other: $K_s = 0.32$

The value of the safety factor is a design decision, often taken to be in the range from 1.25 to 3.0. Note that the screw length is squared in the denominator, indicating that a relatively long screw would have a low critical speed. The best designs would employ a short length, rigid fixed supports, and large diameter.

Column Buckling

Ball screws that carry axial compressive loads must be checked for column buckling. The parameters, similar to those discussed in Chapter 6, are the material from which the screw is made, the end fixity, the diameter, and the length. Long screws should be analyzed using the Euler formula, Equation (6–5) or (6–6), while the J. B. Johnson formula, Equation (6–7), is used for shorter screws. End fixity depends on the rigidity of supports similar to that described above for critical speed. However, the factors are different for column loading.

1. Simply supported at each end by one bearing: $K_s = 1.00$
2. Fixed at each end by two bearings that prevent rotation at the support: $K_s = 4.00$
3. Fixed at one end and simply supported at the other: $K_s = 2.00$
4. Fixed at one end and free at the other: $K_s = 0.25$

Suppliers of commercially available ball screws include data for allowable compressive load in their catalogs. See Internet sites 3, 5, and 10.

Material for Screws

Ball screws are typically made from carbon or alloy steels using thread-rolling technology. After the threads are formed, induction heating improves the hardness and strength of the surfaces on which the circulating balls roll for wear resistance and long life. Ball screw nuts are made from alloy steel that is case hardened by carburizing.

Power screws are typically produced from carbon or alloy steels such as AISI 1018, 1045, 1060, 4130, 4140, 4340, 4620, 6150, 8620, and others. For corrosive environments or where high temperatures are experienced, stainless steels are used, such as AISI 304, 305, 316, 384, 430, 431, or 440. Some are made from aluminum alloys 1100, 2014, or 3003.

Power screw nuts are made from steels for moderate loads and when operating at relatively low speeds. Grease lubrication is recommended. Higher speeds and loads call for lubricated bronze nuts that have superior wear performance. Applications requiring lighter loads can use plastic nuts that have inherently good lubricity without external lubrication. Examples of such applications are food processing equipment, medical devices, and clean manufacturing operations.

REFERENCES

1. Faires, V. M. *Design of Machine Elements*. 5th ed. New York: Macmillan, 1965.

2. Shigley, J. E., and C. R. Mischke. *Mechanical Engineering Design*. 6th ed. New York: McGraw-Hill, 2001.

INTERNET SITES FOR LINEAR MOTION ELEMENTS

1. **PowerTransmission.com.** *www.powertransmission.com* Online listing of numerous manufacturers and suppliers of linear motion devices, including ball screws, lead (power) screws, linear actuators, screw jacks, and linear slides.

2. **Ball-screws.net.** *www.ball-screws.net* Site lists numerous manufacturers of ball screws, some with online catalogs.

3. **Thomson Industries, Inc.** *www.thomsonindustries.com* Manufacturer of ball screws, ball bushings, linear motion guides, and a variety of other linear motion elements. Site includes catalog information and design data for load, life, and travel rate. Thomson is a part of the Linear Motion Systems Division of Danaher Motion Group.

4. **Danaher Motion Group.** *www.danahermcg.com* Part of Danaher Corporation. Manufactures and markets motion control components under the Thomson, BS&A (Ball Screws & Actuators), Deltran, Warner linear, and other brands.

5. **THK Linear Motion Systems.** *www.thk.com* Manufacturer of ball screws, ball splines, linear motion guides, linear bushings, linear actuators, and other motion control products.

6. **Joyce/Dayton Company.** *www.joycejacks.com* Manufacturer of a wide variety of jacks for commercial and industrial applications. Included are power screw and ball screw types with integral gear drives and complete motorized actuator systems.

7. **SKF Linear Motion.** *www.linearmotion.skf.com* Manufacturer of high efficiency screws, linear guiding systems, and actuators.

8. **Specialty Motions, Inc. (SMI).** *www.smi4motion.com* Manufacturer and supplier of linear motion systems and components from miniature to large sizes. Ball screws, positioning stages, linear bearings and bushings, ball slides and others.

9. **Techno, Inc.** *www.techno-isel.com* Manufacturer of a wide variety of linear motion products including slides, X-Y and X-Y-Z tables, gantry tables, and accessories.

10. **Roton Products, Inc.** *www.roton.com* Manufacturer of a large variety of power screws, ball screws, and worms.

11. **Ledex.** *www.ledex.com* Manufacturer of small linear and rotary solenoid actuators and related products under the Ledex and Dormeyer brands. A unit of Saia-Burgess Company.

PROBLEMS

1. Name three types of threads used for power screws.

2. Make a scale drawing of an Acme thread having a major diameter of $1\frac{1}{2}$ in and four threads per inch. Draw a section 2.0 in long.

3. Repeat Problem 2 for a buttress thread.

4. Repeat Problem 2 for a square thread.

5. If an Acme-thread power screw is loaded in tension with a force of 30 000 lb, what size screw from Table 17–1 should be used to maintain a tensile stress below 10 000 psi?

6. For the screw chosen in Problem 5, what would be the required axial length of the nut on the screw that transfers the load to the frame of the machine if the shear stress in the threads must be less than 6000 psi?

7. Compute the torque required to raise the load of 30 000 lb with the Acme screw selected in Problem 5. Use a coefficient of friction of 0.15.

8. Compute the torque required to lower the load with the screw from Problem 5.

9. If a square-thread screw having a major diameter of 3/4 in and six threads per inch is used to lift a load of 4000 lb, compute the torque required to rotate the screw. Use a coefficient of friction of 0.15.

10. For the screw of Problem 9, compute the torque required to rotate the screw when lowering the load.

11. Compute the lead angle for the screw of Problem 9. Is it self-locking?

12. Compute the efficiency of the screw of Problem 9.

13. If the load of 4000 lb is lifted by the screw described in Problem 9 at the rate of 0.5 in/s, compute the required rotational speed of the screw and the power required to drive it.

14. A ball screw for a machine table drive is to be selected. The axial force to be transmitted by the screw is 600 lb. The table moves 24 in per cycle, and it is expected to cycle 10 times per hour for a design life of 10 years. Select an appropriate screw.

15. For the screw selected in Problem 14, compute the torque required to drive the screw.

16. For the screw selected in Problem 14, the normal travel speed of the table is 10.0 in/min. Compute the power required to drive the screw.

17. If the cycle time for the machine in Problem 14 were decreased to obtain 20 cycles/h instead of 10, what would be the expected life in years of the screw originally selected?

18

Fasteners

The Big Picture

You Are the Designer

18–1 Objectives of This Chapter

18–2 Bolt Materials and Strength

18–3 Thread Designations and Stress Area

18–4 Clamping Load and Tightening of Bolted Joints

18–5 Externally Applied Force on a Bolted Joint

18–6 Thread Stripping Strength

18–7 Other Types of Fasteners and Accessories

18–8 Other Means of Fastening and Joining

Fasteners

Discussion Map

☐ *Fasteners* connect or join two or more components. Common types are *bolts* and *screws* such as those illustrated in Figures 18–1 through 18–4.

Discover

Look for examples of bolts and screws. List how many types you have found. For what functions were they being used? What kinds of forces are the fasteners subjected to? What materials are used for the fasteners?

In this chapter you will learn to analyze the performance of fasteners and to select suitable types and sizes.

A *fastener* is any device used to connect or join two or more components. Literally hundreds of fastener types and variations are available. The most common are threaded fasteners referred to by many names, among them bolts, screws, nuts, studs, lag screws, and set screws.

A *bolt* is a threaded fastener designed to pass through holes in the mating members and to be secured by tightening a nut from the end opposite the head of the bolt. See Figure 18–1(a), called a *hex head bolt*. Several other types of bolts are shown in Figure 18–2.

A *screw* is a threaded fastener designed to be inserted through a hole in one member to be joined and into a threaded hole in the mating member. See Figure 18–1(b). The threaded hole may have been preformed, for example, by tapping, or it may be formed by the screw itself as it is forced into the material. *Machine screws*, also called *cap screws*, are precision fasteners with straight-threaded bodies that are turned into tapped holes (see Figure 18–3). A popular type of machine screw is the socket head cap screw. The usual configuration, shown in Figure 18–3(f), has a cylindrical head with a recessed hex socket. Also readily available are flat head styles for countersinking to produce a flush surface, button head styles for a low profile appearance, and shoulder screws providing a precision bearing surface for location or pivoting. See Internet sites 9 and 11. *Sheet-metal screws, lag screws, self-tapping screws*, and *wood screws* usually form their own threads. Figure 18–4 shows a few styles.

Search for examples where the kinds of fasteners illustrated in Figures 18–1 through 18–4 are used. How many can you find? Make a list using the names for the fasteners in the figures. Describe the application. What function is the fastener performing? What kinds of forces are exerted on each fastener during service? How large is the fastener? Measure as many dimensions as you can. What material is each fastener made from?

Look in your car, particularly under the hood in the engine compartment. If you can, also look under the chassis to see where fasteners are used to hold different components onto the frame or some other structural member.

FIGURE 18–1

Comparison of a bolt with a screw
(R. P. Hoelscher et al., *Graphics for Engineers*, New York: John Wiley & Sons, 1968)

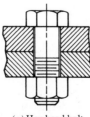 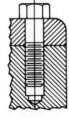

(*a*) Hex head bolt   (*b*) Hex head cap screw

FIGURE 18–2 Bolt styles. See also the hex head bolt in Figure 18–1. (R. P Hoelscher et al., *Graphics for Engineers,* New York: John Wiley & Sons, 1968)

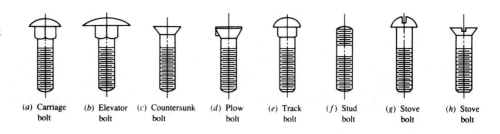

(a) Carriage bolt (b) Elevator bolt (c) Countersunk bolt (d) Plow bolt (e) Track bolt (f) Stud bolt (g) Stove bolt (h) Stove bolt

FIGURE 18–3 Cap screws or machine screws. See also the hex head cap screw in Figure 18–1. (R. P. Hoelscher et al., *Graphics for Engineers,* New York: John Wiley & Sons, 1968)

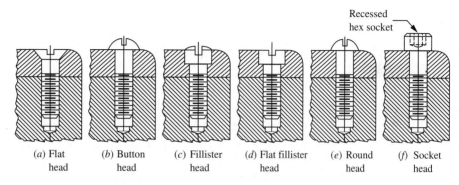

Recessed hex socket

(a) Flat head (b) Button head (c) Fillister head (d) Flat fillister head (e) Round head (f) Socket head

FIGURE 18–4 Sheet-metal and lag screws (R. P. Hoelscher et al., *Graphics for Engineers,* New York: John Wiley & Sons, 1968)

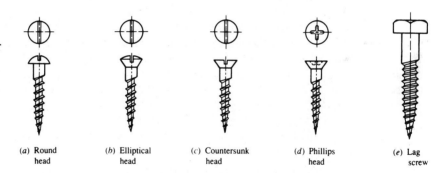

(a) Round head (b) Elliptical head (c) Countersunk head (d) Phillips head (e) Lag screw

Look also at bicycles, lawn and garden equipment, grocery carts, display units in a store, hand tools, kitchen appliances, toys, exercise equipment, and furniture. If you have access to a factory, you should be able to identify hundreds or thousands of examples. Try to get some insight about where certain types of fasteners are used and for what purposes.

In this chapter, you will learn about many of the types of fasteners that you will encounter, including how to analyze their performance.

You Are the Designer

Review Figure 15–6 which shows the assembly of the gear-type power transmission that was designed in that chapter. Fasteners are called for in several places on the housing for the transmission, but they were not specified in that chapter. The four *bearing retainers* are to be fastened to the housing and the cover by threaded fasteners. The cover itself is to be attached to the housing by fasteners. Finally, the mounting base has provisions for using fasteners to hold the entire transmission to a support structure.

You are the designer. What kinds of fasteners would you consider for these applications? What material should be used to make them? What strength should the material have? If threaded fasteners are used, what size should the threads be, and how long must they be? What head style would you specify? How much torque should be applied to the fastener to ensure that there is sufficient clamping force between the joined members? How does the design of the gasket between the cover and the housing affect the choice of the fasteners and the specification of the tightening torque for them? What alternatives are there to the use of threaded fasteners to hold the components together and to still allow disassembly?

This chapter presents information that you can use to make such design decisions. The references at the end of the chapter give other valuable sources of information from the large body of knowledge about fasteners.

18–1
OBJECTIVES OF THIS CHAPTER

After completing this chapter, you will be able to:

1. Describe a bolt in comparison with a machine screw.

2. Name and describe nine styles of heads for bolts.

3. Name and describe six styles of heads for machine screws.

4. Describe sheet-metal screws and lag screws.

5. Describe six styles of set screws and their application.

6. Describe nine types of locking devices that restrain a nut from becoming loose on a bolt.

7. Use tables of data for various grades of steel materials used for bolts as published by the Society of Automotive Engineers (SAE) and the American Society for Testing and Materials (ASTM), and for standard metric grades.

8. List at least 10 materials other than steel that are used for fasteners.

9. Use tables of data for standard screw threads in the American Standard and metric systems for dimensions and stress analysis.

10. Define *proof load*, *clamping load*, and *tightening torque* as applied to bolts and screws, and compute design values.

11. Compute the effect of adding an externally applied force on a bolted joint, including the final force on the bolts and the clamped members.

12. List and describe 16 different coating and finishing techniques that are used for metal fasteners.

13. Describe rivets, quick-operating fasteners, welding, brazing, and adhesives, and contrast them with bolts and screws for fastening applications.

18–2
BOLT MATERIALS AND STRENGTH

In machine design, most fasteners are made from steel because of its high strength, high stiffness, good ductility, and good machinability and formability. But varying compositions and conditions of steel are used. The strength of steels used for bolts and screws is used to determine its *grade*, according to one of several standards. Three strength ratings are frequently available: the familiar tensile strength and yield strength plus the proof strength. The *proof strength*, similar to the elastic limit, is defined as the stress at which the bolt or the screw would undergo permanent deformation. It usually ranges between 0.90 and 0.95 times the yield strength.

The SAE uses grade numbers ranging from 1 to 8, with increasing numbers indicating greater strength. Table 18–1 lists some aspects of this grading system taken from SAE Standard J429 (Reference 12). The markings shown are embossed into the head of the bolt.

The ASTM publishes five standards relating to bolt steel strength, as listed in Table 18–2 (Reference 2). These are often used in construction work.

Metric bolts and screws use a numerical code system ranging from 4.6 to 12.9, with higher numbers indicating higher strengths. The numbers before the decimal point are approximately 0.01 times the tensile strength of the material in MPa. The last digit with the decimal point is the approximate ratio of the yield strength of the material to the tensile strength. Table 18–3 shows pertinent data from SAE Standard J1199 (Reference 7).

Approximate Equivalencies among SAE, ASTM, and Metric Grades of Bolt Steels. The following list shows approximate equivalents that may be useful when comparing designs for which specifications include combinations of SAE, ASTM, and metric grades of bolt steels. The individual standards should be consulted for specific strength data.

SAE Grade	ASTM Grade	Metric Grade
J429 Grade 1	A307 Grade A	Grade 4.6
J429 Grade 2	——————	Grade 5.8
J429 Grade 5	A449	Grade 8.8
J429 Grade 8	A354 Grade BD	Grade 10.9

Socket head cap screws of the 1960 Series are made from a heat-treated alloy steel having the following strengths:

Size Range	Tensile Strength (ksi)	Yield Strength (ksi)
0–5/8	190	170
3/4–3	180	155

TABLE 18–1 SAE grades of steels for fasteners

Grade number	Bolt size (in)	Tensile strength (ksi)	Yield strength (ksi)	Proof strength (ksi)	Head marking
1	1/4–1$\frac{1}{2}$	60	36	33	None
2	1/4–3/4	74	57	55	None
	>3/4–1$\frac{1}{2}$	60	36	33	
4	1/4–1$\frac{1}{2}$	115	100	65	None
5	1/4–1	120	92	85	⬡
	>1–1$\frac{1}{2}$	105	81	74	
7	1/4–1$\frac{1}{2}$	133	115	105	⬡
8	1/4–1$\frac{1}{2}$	150	130	120	⬡

TABLE 18–2 ASTM standards for bolt steels

ASTM grade	Bolt size (in)	Tensile strength (ksi)	Yield strength (ksi)	Proof strength (ksi)	Head marking
A307	1/4–4	60	(Not reported)		None
A325	1/2–1	120	92	85	(A 325)
	>1–1½	105	81	74	
A354-BC	1/4–2½	125	109	105	(BC)
A354-BD	1/4–2½	150	130	120	
A449	1/4–1	120	92	85	
	>1–1½	105	81	74	
	>1½–3	90	58	55	
A574	0.060–1/2	180		140	(Socket head
	5/8–4	170		135	cap screws)

TABLE 18–3 Metric grades of steels for bolts

Grade	Bolt size	Tensile strength (MPa)	Yield strength (MPa)	Proof strength (MPa)
4.6	M5–M36	400	240	225
4.8	M1.6–M16	420	340[a]	310
5.8	M5–M24	520	415[a]	380
8.8	M17–M36	830	660	600
9.8	Ml.6–M16	900	720[a]	650
10.9	M6–M36	1040	940	830
12.9	M1.6–M36	1220	1100	970

[a]Yield strengths are approximate and are not included in the standard.

Roughly equivalent performance is obtained from metric socket head cap screws made to the metric strength grade 12.9. The same geometry is available in corrosion-resistant stainless steel, typically 18–8, at somewhat lower strength levels. Consult the manufacturers.

Aluminum is used for its corrosion resistance, light weight, and fair strength level. Its good thermal and electrical conductivity may also be desirable. The most widely used alloys are 2024-T4, 2011-T3, and 6061-T6. Properties of these materials are listed in Appendix 9.

Brass, copper, and *bronze* are also used for their corrosion resistance. Ease of machining and an attractive appearance are also advantages. Certain alloys are particularly good for resistance to corrosion in marine applications.

Nickel and its alloys, such as *Monel* and *Inconel* (from the International Nickel Company), provide good performance at elevated temperatures while also having good corrosion resistance, toughness at low temperatures, and an attractive appearance.

Stainless steels are used primarily for their corrosion resistance. Alloys used for fasteners include 18–8, 410, 416, 430, and 431. In addition, stainless steels in the 300 series are nonmagnetic. See Appendix 6 for properties.

A high strength-to-weight ratio is the chief advantage of *titanium* alloys used for fasteners in aerospace applications. Appendix 11 gives a list of properties of several alloys.

Plastics are used widely because of their light weight, corrosion resistance, insulating ability, and ease of manufacture. Nylon 6/6 is the most frequently used material, but others include ABS, acetal, TFE fluorocarbons, polycarbonate, polyethylene, polypropylene, and polyvinylchloride. Appendix 13 lists several plastics and their properties. In addition to being used in screws and bolts, plastics are used extensively where the fastener is designed specially for the particular application.

Coatings and *finishes* are provided for metallic fasteners to improve appearance or corrosion resistance. Some also lower the coefficient of friction for more consistent results relating tightening torque to clamping force. Steel fasteners can be finished with black oxide, bluing, bright nickel, phosphate, and hot-dip zinc. Plating can be used to deposit cadmium, copper, chromium, nickel, silver, tin, and zinc. Various paints, lacquers, and chromate finishes are also used. Aluminum is usually anodized. Check environmental hazards for coatings and finishes.

18–3 THREAD DESIGNATIONS AND STRESS AREA

Table 18–4 shows pertinent dimensions for threads in the American Standard styles, and Table 18–5 gives SI metric styles. For consideration of strength and size, the designer must know the basic major diameter, the pitch of the threads, and the area available to resist tensile loads. Note that the pitch is equal to $1/n$, where n is the number of threads per inch in

TABLE 18–4   American Standard thread dimensions

A. Numbered sizes

Size	Basic major diameter (in)	Coarse threads: UNC		Fine threads: UNF	
		Threads per in	Tensile stress area (in²)	Threads per in	Tensile stress area (in²)
0	0.0600			80	0.001 80
1	0.0730	64	0.00263	72	0.002 78
2	0.0860	56	0.00370	64	0.003 94
3	0.0990	48	0.00487	56	0.005 23
4	0.1120	40	0.00604	48	0.006 61
5	0.1250	40	0.00796	44	0.008 30
6	0.1380	32	0.00909	40	0.010 15
8	0.1640	32	0.0140	36	0.014 74
10	0.1900	24	0.0175	32	0.0200
12	0.2160	24	0.0242	28	0.0258

B. Fractional sizes

Size	Basic major diameter (in)	Coarse threads: UNC		Fine threads: UNF	
1/4	0.2500	20	0.0318	28	0.0364
5/16	0.3125	18	0.0524	24	0.0580
3/8	0.3750	16	0.0775	24	0.0878
7/16	0.4375	14	0.1063	20	0.1187
1/2	0.5000	13	0.1419	20	0.1599
9/16	0.5625	12	0.182	18	0.203
5/8	0.6250	11	0.226	18	0.256
3/4	0.7500	10	0.334	16	0.373
7/8	0.8750	9	0.462	14	0.509
1	1.000	8	0.606	12	0.663
$1\frac{1}{8}$	1.125	7	0.763	12	0.856
$1\frac{1}{4}$	1.250	7	0.969	12	1.073
$1\frac{3}{8}$	1.375	6	1.155	12	1.315
$1\frac{1}{2}$	1.500	6	1.405	12	1.581
$1\frac{3}{4}$	1.750	5	1.90		
2	2.000	$4\frac{1}{2}$	2.50		

TABLE 18–5   Metric thread dimensions

Basic major diameter (mm)	Coarse threads		Fine threads	
	Pitch (mm)	Tensile stress area (mm²)	Pitch (mm)	Tensile stress area (mm²)
1	0.25	0.460		
1.6	0.35	1.27	0.20	1.57
2	0.4	2.07	0.25	2.45
2.5	0.45	3.39	0.35	3.70
3	0.5	5.03	0.35	5.61
4	0.7	8.78	0.5	9.79
5	0.8	14.2	0.5	16.1
6	1	20.1	0.75	22.0
8	1.25	36.6	1	39.2
10	1.5	58.0	1.25	61.2
12	1.75	84.3	1.25	92.1
16	2	157	1.5	167
20	2.5	245	1.5	272
24	3	353	2	384
30	3.5	561	2	621
36	4	817	3	865
42	4.5	1121		
48	5	1473		

the American Standard system. In the SI, the pitch in millimeters is designated directly. The tensile stress area listed in Tables 18–4 and 18–5 takes into account the actual area cut by a transverse plane. Because of the helical path of the thread on the screw, such a plane will cut near the root on one side of the screw but will cut near the major diameter on the other. The equation for the tensile stress area for American Standard threads is

⇨ Tensile Stress Area for UNC or UNF Threads

$$A_t = (0.7854)[D - (0.9743)p]^2 \qquad \textbf{(18–1)}$$

where D = major diameter
p = pitch of the thread

For metric threads, the tensile stress area is

⇨ Tensile Stress Area for Metric Threads

$$A_t = (0.7854)[D - (0.9382)p]^2 \qquad \textbf{(18–2)}$$

For most standard screw thread sizes, at least two pitches are available: the *coarse* series and the *fine thread* series. Both are included in Tables 18–4 and 18–5.

The smaller American Standard threads use a number designation from 0 to 12. The corresponding major diameter is listed in Table 18–4(A). The larger sizes use fractional-inch designations. The decimal equivalent for the major diameter is shown in Table 18–4(B). Metric threads list the major diameter and the pitch in millimeters, as shown in Table 18–5. Samples of the standard designations for a thread are given next.

> *American Standard:* Basic size followed by number of threads per inch and the
> thread series designation.

10–24 UNC	10–32 UNF
1/2–13 UNC	1/2–20 UNF
$1\frac{1}{2}$–6 UNC	$1\frac{1}{2}$–12 UNF

Metric: M (for "metric"), followed by the basic major diameter and then the pitch in millimeters.

$$\text{M3} \times 0.5 \qquad \text{M3} \times 0.35 \qquad \text{M10} \times 1.5$$

**18–4
CLAMPING LOAD
AND TIGHTENING
OF BOLTED
JOINTS**

Clamping Load

When a bolt or a screw is used to clamp two parts, the force exerted between the parts is the *clamping load*. The designer is responsible for specifying the clamping load and for ensuring that the fastener is capable of withstanding the load. The maximum clamping load is often taken to be 0.75 times the proof load, where the proof load is the product of the proof stress times the tensile stress area of the bolt or screw.

Tightening Torque

The clamping load is created in the bolt or the screw by exerting a tightening torque on the nut or on the head of the screw. An approximate relationship between the torque and the axial tensile force in the bolt or screw (the clamping force) is

 **Tightening Torque**

$$T = KDP \qquad\qquad (18\text{–}3)$$

where T = torque, lb·in
  D = nominal outside diameter of threads, in
  P = clamping load, lb
  K = constant dependent on the lubrication present

For average commercial conditions, use $K = 0.15$ if any lubrication at all is present. Even cutting fluids or other residual deposits on the threads will produce conditions consistent with $K = 0.15$. If the threads are well cleaned and dried, $K = 0.20$ is better. Of course, these values are approximate, and variations among seemingly identical assemblies should be expected. Testing and statistical analysis of the results are recommended.

Example Problem 18–1

A set of three bolts is to be used to provide a clamping force of 12 000 lb between two components of a machine. The load is shared equally among the three bolts. Specify suitable bolts, including the grade of the material, if each is to be stressed to 75% of its proof strength. Then compute the required tightening torque.

Solution

The load on each screw is to be 4000 lb. Let's specify a bolt made from SAE Grade 5 steel, having a proof strength of 85 000 psi. Then the allowable stress is

$$\sigma_a = 0.75(85\ 000\ \text{psi}) = 63\ 750\ \text{psi}$$

The required tensile stress area for the bolt is then

$$A_t = \frac{\text{load}}{\sigma_a} = \frac{4000\ \text{lb}}{63\ 750\ \text{lb/in}^2} = 0.0627\ \text{in}^2$$

From Table 18–4(B), we find that the 3/8–16 UNC thread has the required tensile stress area. The required tightening torque will be

$$T = KDP = 0.15(0.375\ \text{in})(4000\ \text{lb}) = 225\ \text{lb}\cdot\text{in}$$

Equation (18–3) is adequate for general mechanical design. A more complete analysis of the torque to create a given clamping force requires more information about the joint design. There are three contributors to the torque. One, which we will refer to as T_1, is the torque required to develop the tensile load in the bolt, P_t, using the inclined plane nature of the thread.

$$T_1 = \frac{P_t l}{2\pi} = \frac{P_t}{2\pi n} \tag{18–4}$$

where l is the lead of the bolt thread and $l = p = 1/n$.

The second component of the torque, T_2, is that required to overcome friction between the mating threads, computed from,

$$T_2 = \frac{d_p \, \mu_1 \, P_t}{2 \cos \alpha} \tag{18–5}$$

where

d_p = pitch diameter of the thread
μ_1 = coefficient of friction between the thread surfaces
α = 1/2 of the thread angle, typically 30°

The third component of the torque, T_3, is the friction between the underside of the head of the bolt or nut and the clamped surface. This friction force is assumed to act at the middle of the friction surface and is computed from,

$$T_3 = \frac{(d + b) \, \mu_2 \, P_t}{4} \tag{18–6}$$

where

d = major diameter of the bolt
b = outside diameter of the friction surface on the underside of the bolt
μ_2 = coefficient of friction between the bolt head and the clamped surface

The total torque is then,

$$T_{tot} = T_1 + T_2 + T_3 \tag{18–7}$$

See References 3, 4, 9, and 10 for additional discussion on torque for bolts. It is important to note that many variables are involved in the contributors to the relationship between the applied torque and the tensile preload given to the bolt. Accurate prediction of the coefficients of friction is difficult. The accuracy with which the specified torque is applied is affected by the precision of the measurement device used, such as a torque wrench, pneumatic nut runner, or hydraulic nut driver, as well as the operator's skill. Reference 3 has an extensive discussion of the large variety of torque wrenches available.

Computer Assisted Bolt Analysis. Because of the many variables and the numerous calculations required to analyze a bolted joint, commercial computer software packages are available to perform the necessary analysis. See Internet site 12 for an example.

Other Methods of Bolt Tightening

Measuring the torque applied to the bolt, screw, or nut during installation is convenient. However, because of the many variables involved, the actual clamping force created may vary significantly. Fastening of critical connections often uses other methods of bolt tightening that more directly relate to the clamping force. Situations where these methods may be used are structural steel connections, flanges for high-pressure systems, nuclear power plant components, cylinder head and connecting rod bolts for engines, aerospace structures, turbine engine components, propulsion systems, and military equipment.

Turn of the Nut Method.   The bolt is first tightened to a snug fit to bring all of the parts of the joint into intimate contact. Then the nut is given an additional turn with a wrench of between one-third and one full turn, depending on the size of the bolt. One full turn would produce a stretch in the bolt equal to the lead of the thread, where $l = p = 1/n$. The elastic behavior of the bolt determines the amount of the resulting clamping force. References 1 and 3 give more details.

Tension Control Bolting Products.   Special bolts are available that include a carefully sized neck on one end connected to a splined section. The spline is held fixed as the nut is turned. When a predetermined torque is applied to the nut, the neck section breaks and the tightening stops. Consistent connection performance results.

Another form of tension control bolt employs a tool that exerts direct axial tension on the bolt, swages a collar into annular grooves or the threads of the fastener, and then breaks a small-diameter part of the bolt at a predetermined force. The result is a predictable amount of clamping force in the joint.

Wavy Flanged Bolt.   The underside of the head of this bolt is formed in a wavy pattern during manufacture. When torque is applied to the joint, the wavy surface is deformed to become flat against the clamped surface when the proper amount of tension has been created in the body of the bolt.

Direct Tension Indicator (DTI) Washers.   The DTI washer has several raised areas on its upper surface. A regular washer is then placed over the DTI washer and a nut tightens the assembly until the raised areas are flattened to a specified degree, creating a predictable tension in the bolt.

Ultrasonic Tension Measurement and Control.   Recent developments have resulted in the availability of equipment that imparts ultrasonic acoustic waves to bolts as they are tightened with the timing of the reflected waves being correlated to the amount of stretch and tension in the bolt. See Reference 3.

Tighten to Yield Method.   Most fasteners are provided with guaranteed yield strength; therefore, it takes a predictable amount of tensile force to cause the bolt to yield. Some automatic systems use this principle by sensing the relationship between applied torque and the rotation of the nut and stopping the process when the bolt begins to yield. During the elastic part of the stress-strain curve for the bolt, a linear change in torque versus rotation occurs. At yield, a dramatic increase in rotation with little or no increase in torque signifies yielding. A variation on this method, called the *logarithmic rate method (LRM)*, determines the peak of the curve of the logarithm of the rate of torque versus turn data and then applies a preset amount of additional turn to the nut. See Internet site 9.

18–5
EXTERNALLY
APPLIED FORCE
ON A BOLTED
JOINT

The analysis shown in Example Problem 18–1 considers the stress in the bolt only under static conditions and only for the clamping load. It was recommended that the tension on the bolt be very high, approximately 75% of the proof load for the bolt. Such a load will use the available strength of the bolt efficiently and will prevent the separation of the connected members.

When a load is applied to a bolted joint over and above the clamping load, special consideration must be given to the behavior of the joint. Initially, the force on the bolt (in tension) is equal to the force on the clamped members (in compression). Then some of the additional load will act to stretch the bolt beyond its length assumed after the clamping load was applied. Another portion will result in a *decrease* in the compressive force in the clamped member. Thus, only part of the applied force is carried by the bolt. The amount is dependent on the relative stiffness of the bolt and the clamped members.

If a stiff bolt is clamping a flexible member, such as a resilient gasket, most of the added force will be taken by the bolt because it takes little force to change the compression in the gasket. In this case, the bolt design must take into account not only the initial clamping force but also the added force.

Conversely, if the bolt is relatively flexible compared with the clamped members, virtually all of the externally applied load will initially go to decreasing the clamping force until the members actually separate, a condition usually interpreted as failure of the joint. Then the bolt will carry the full amount of the external load.

In practical joint design, a situation between the extremes previously described would normally occur. In typical "hard" joints (without a soft gasket), the stiffness of the clamped members is approximately three times that of the bolt. The externally applied load is then shared by the bolt and the clamped members according to their relative stiffnesses as follows:

$$F_b = P + \frac{k_b}{k_b + k_c} F_e \qquad (18\text{–}8)$$

$$F_c = P - \frac{k_c}{k_b + k_c} F_e \qquad (18\text{–}9)$$

where F_e = externally applied load
$\quad P$ = initial clamping load [as used in Equation (18–3)]
$\quad F_b$ = final force in bolt
$\quad F_c$ = final force on clamped members
$\quad k_b$ = stiffness of bolt
$\quad k_c$ = stiffness of clamped members

Example Problem 18–2

Assume that the joint described in Example Problem 18–1 was subjected to an additional external load of 3000 lb after the initial clamping load of 4000 lb was applied. Also assume that the stiffness of the clamped members is three times that of the bolt. Compute the force in the bolt, the force in the clamped members, and the final stress in the bolt after the external load is applied.

Solution

We will first use Equations (18–8) and (18–9) with $P = 4000$ lb, $F_e = 3000$ lb, and $k_c = 3k_b$:

$$F_b = P + \frac{k_b}{k_b + k_c} F_e = P + \frac{k_b}{k_b + 3k_b} F_e = P + \frac{k_b}{4k_b} F_e$$

$$F_b = P + F_e/4 = 4000 + 3000/4 = 4750 \text{ lb}$$

$$F_c = P - \frac{k_c}{k_b + k_c} F_e = P - \frac{3k_b}{k_b + 3k_b} F_e = P - \frac{3k_b}{4k_b} F_e$$

$$F_c = P - 3F_e/4 = 4000 - 3(3000)/4 = 1750 \text{ lb}$$

Because F_c is still greater than zero, the joint is still tight. Now the stress in the bolt can be found. For the 3/8–16 bolt, the tensile stress area is 0.0775 in². Thus,

$$\sigma = \frac{P}{A_t} = \frac{4750 \text{ lb}}{0.0775 \text{ in}^2} = 61\ 300 \text{ psi}$$

The proof strength of the Grade 5 material is 85 000 psi, and this stress is approximately 72% of the proof strength. Therefore, the selected bolt is still safe. But consider what would happen with a relatively "soft" joint, discussed in Example Problem 18–3.

Example Problem 18–3   Solve Example Problem 18–2 again, but assume that the joint has a flexible elastomeric gasket separating the clamping members and that the stiffness of the bolt is then 10 times that of the joint.

Solution   The procedure will be the same as that used previously, but now $k_b = 10k_c$. Thus,

$$F_b = P + \frac{k_b}{k_b + k_c}F_e = P + \frac{10k_c}{10k_c + k_c}F_e = P + \frac{10k_c}{11k_c}F_e$$

$$F_b = P + 10F_e/11 = 4000 + 10(3000)/11 = 6727 \text{ lb}$$

$$F_c = P - \frac{k_c}{k_b + k_c}F_e = P - \frac{k_c}{10k_c + k_c}F_e = P - \frac{k_c}{11k_c}F_e$$

$$F_c = P - F_e/11 = 4000 - 3000/11 = 3727 \text{ lb}$$

The stress in the bolt would be

$$\sigma = \frac{6727 \text{ lb}}{0.0775 \text{ in}^2} = 86\ 800 \text{ psi}$$

This exceeds the proof strength of the Grade 5 material and is dangerously close to the yield strength.

18–6
THREAD
STRIPPING
STRENGTH

In addition to sizing a bolt on the basis of axial tensile stress, the threads must be checked to ensure that they will not be stripped off by shearing. The variables involved in the shear strength of the threads are the materials of the bolt, the nut, or the internal threads of a tapped hole, the length of engagement, L_e, and the size of the threads. The details of analysis depend on the relative strength of the materials.

Internal Thread Material Stronger than Bolt Material.   For this case, the strength of the threads of the bolt will control the design. Here we present an equation for the required length of engagement, L_e, of the bolt threads that will have at least the same strength in shear as the bolt itself does in tension.

$$L_e = \frac{2\,A_{tB}}{\pi\,(ID_{Nmax})\,[0.5 + 0.57735\,n(PD_{Bmin} - ID_{Nmax})]} \qquad (18\text{–}10)$$

where

A_{tB} = tensile stress area of bolt
ID_{Nmax} = maximum inside (root) diameter of nut threads
n = number of threads per inch
PD_{Bmin} = minimum pitch diameter of bolt threads

The subscripts B and N refer to the bolt and nut, respectively. The subscripts *min* and *max* refer to the minimum and maximum values, respectively, considering the tolerances on thread dimensions. Reference 9 gives data for tolerances as a function of the class of thread specified.

For a given length of engagement, the resulting shear area for the bolt threads is

$$A_{sB} = \pi \, L_e \, ID_{Nmax} \, [0.5 + 0.57735 \, n(PD_{Bmin} - ID_{Nmax})] \qquad (18\text{--}11)$$

Nut Material Weaker than Bolt Material. This is particularly applicable when the bolt is inserted into a tapped hole in cast iron, aluminum, or some other material with relatively low strength. The required length of engagement to develop at least the full strength of the bolt is

$$L_e = \frac{S_{utB} \, (2 \, A_{tB})}{S_{utN} \, \pi \, OD_{Bmin} \, [0.5 + 0.57735 \, n \, (OD_{Bmin} - PD_{Nmax})]} \qquad (18\text{--}12)$$

where

S_{utB} = ultimate tensile strength of the bolt material
S_{utN} = ultimate tensile strength of the nut material
OD_{Bmin} = minimum outside diameter of the bolt threads
PD_{Nmax} = maximum pitch diameter of the nut threads

The shear area of the root of the threads of the nut is

$$A_{sN} = \pi \, L_e \, OD_{Bmin} \, [0.5 + 0.57735 \, n \, (OD_{Bmin} - PD_{Nmax})] \qquad (18\text{--}13)$$

Equal Strength for Nut and Bolt Material. For this case failure is predicted as shear of either part at the nominal pitch diameter, PD_{nom}. The required length of engagement to develop at least the full strength of the bolt is

$$L_e = \frac{4 \, A_{tB}}{\pi \, PD_{nom}} \qquad (18\text{--}14)$$

The shear stress area for the nut or bolt threads is

$$A_s = \pi \, PD_{nom} \, L_e/2 \qquad (18\text{--}15)$$

**18–7
OTHER TYPES OF
FASTENERS AND
ACCESSORIES**

Most bolts and screws have enlarged heads that bear down on the part to be clamped and thus exert the clamping force. *Set screws* are headless, are inserted into tapped holes, and are designed to bear directly on the mating part, locking it into place. Figure 18–5 shows several styles of points and drive means for set screws. Caution must be used with set screws, as with any threaded fastener, so that vibration does not loosen the screw.

FIGURE 18–5 Set screws with different head and point styles applied to hold a collar on a shaft (R. P Hoelscher et al., *Graphics for Engineers*, New York: John Wiley & Sons, 1968)

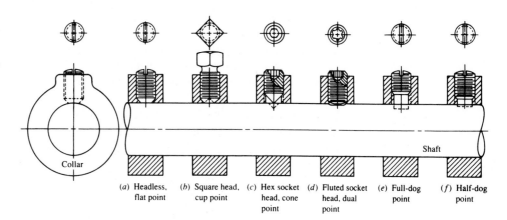

(*a*) Headless, flat point   (*b*) Square head, cup point   (*c*) Hex socket head, cone point   (*d*) Fluted socket head, dual point   (*e*) Full-dog point   (*f*) Half-dog point

FIGURE 18–6 Locking devices (R. P. Hoelscher et al., *Graphics for Engineers*, New York: John Wiley & Sons, 1968)

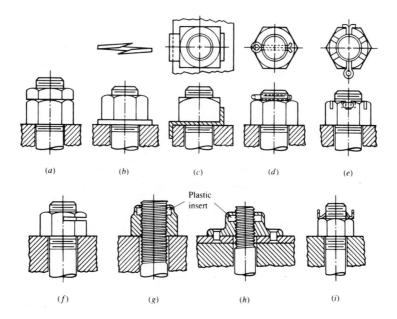

(*a*)   (*b*)   (*c*)   (*d*)   (*e*)

(*f*)   (*g*)   (*h*)   (*i*)

A *washer* may be used under either or both the bolt head and the nut to distribute the clamping load over a wide area and to provide a bearing surface for the relative rotation of the nut. The basic type of washer is the plain flat washer, a flat disc with a hole in it through which the bolt or screw passes. Other styles, called *lockwashers*, have axial deformations or projections that produce axial forces on the fastener when compressed. These forces keep the threads of the mating parts in intimate contact and decrease the probability that the fastener will loosen in service.

Figure 18–6 shows several means of using washers and other types of locking devices. Part (a) is a jam nut tightened against the regular nut. Part (b) is the standard lockwasher. Part (c) is a locking tab that keeps the nut from turning. Part (d) is a cotter inserted through a hole drilled through the bolt. Part (e) uses a cotter, but it also passes through slots in the nut. Part (f) is one of several types of thread-deformation techniques used. Part (g), an *elastic stop nut*, uses a plastic insert to keep the threads of the nut in tight contact with the bolt. This may be used on machine screws as well. In part (h), the elastic stop nut is riveted to a thin plate, allowing a mating part to be bolted from the opposite side. The thin metal device in (i) bears against the top of the nut and grips the threads, preventing axial motion of the nut.

A *stud* is like a stationary bolt attached permanently to a part of one member to be joined. The mating member is then placed over the stud, and a nut is tightened to clamp the parts together.

Additional variations occur when these types of fasteners are combined with different head styles. Several of these are shown in the figures already discussed. Others are listed next:

Square	Hex	Heavy hex	Hex jam
Hex castle	Hex flat	Hex slotted	12-point
High crown	Low crown	Round	T-head
Pan	Truss	Hex washer	Flat countersunk
Plow	Cross recess	Fillister	Oval countersunk
Hex socket	Spline socket	Button	Binding

Additional combinations are created by consideration of the American National Standards or British Standard (metric); material grades; finishes; thread sizes; lengths; class (tolerance grade); manner of forming heads (machining, forging, cold heading); and the manner of forming threads (machining, die cutting, tapping, rolling, and plastic molding).

Thus, you can see that comprehensive treatment of threaded fasteners encompasses extensive data. See the References and Internet sites listed at the end of the chapter.

18–8 OTHER MEANS OF FASTENING AND JOINING

Thus far, this chapter has focused on screws and bolts because of their wide applications. Other types of fastening means will now be discussed.

Rivets are nonthreaded fasteners, usually made of steel or aluminum. They are originally made with one head, and the opposite end is formed after the rivet is inserted through holes in the parts to be joined. Steel rivets are formed hot, whereas aluminum can be formed at room temperatures. Of course, riveted joints are not designed to be assembled more than once. (See Internet sites 3, 4, and 11.)

A large variety of *quick-operating fasteners* is available. Many are of the quarter-turn type, requiring just a 90° rotation to connect or disconnect the fastener. Access panels, hatches, covers, and brackets for removable equipment are attached with such fasteners. Similarly, many varieties of *latches* are available to provide quick action with, perhaps, added holding power. (See Internet sites 3 and 4.)

Welding involves the metallurgical bonding of metals, usually by the application of heat with an electric arc, a gas flame, or electrical resistance heating under heavy pressure. Welding is discussed in Chapter 20.

Brazing and *soldering* use heat to melt a bonding agent that flows into the space between parts to be joined, adhering to both parts and then solidifying as it cools. *Brazing* uses relatively high temperatures, above 840°F (450°C), using alloys of copper, silver, aluminum, silicon, or zinc. Of course, the metals to be joined must have a significantly higher melting temperature. Metals successfully brazed include plain carbon and alloy steels, stainless steels, nickel alloys, copper, aluminum, and magnesium. *Soldering* is similar to brazing, except that it is performed at lower temperatures, less than 840°F (450°C). Several soldering alloys of lead-tin, tin-zinc, tin-silver, lead-silver, zinc-cadmium, zinc-aluminum, and others are used. Brazed joints are generally stronger than soldered joints due to the inherently higher strength of the brazing alloys. Most soldered joints are fabricated with interlocking lap joints to provide mechanical strength, and then the solder is used to hold the assembly together and possibly to provide sealing. Joints in piping and tubing are frequently soldered.

Adhesives are seeing wide use. Versatility and ease of application are strong advantages of adhesives used in an array of products from toys and household appliances to automotive and aerospace structures. (See Internet site 3.) Some types include the following:

Acrylics: Used for many metals and plastics.

Cyanoacrylates: Very fast curing; flow easily between well-mated surfaces.

Epoxies: Good structural strength; joint is usually rigid. Some require two-part formulations. A large variety of formulations and properties are available.

Anaerobics: Used for securing nuts and bolts and other joints with small clearances; cures in the absence of oxygen.

Silicones: Flexible adhesive with good high-temperature performance (400°F, 200°C).

Polyester hot melt: Good structural adhesive; easy to apply with special equipment.

Polyurethane: Good bonding; provides a flexible joint.

REFERENCES

1. American Institute of Steel Construction. *Allowable Stress Design Specification for Structural Joints Using ASTM A325 or A490 Bolts*. Chicago: American Institute of Steel Construction, 2001.

2. American Society for Testing and Materials. *Fasteners, Volume 8*. Philadelphia: American Society for Testing and Materials, 2001.

3. Bickford, John H. *An Introduction to the Design and Behavior of Bolted Joints*. 3rd ed. New York: Marcel Dekker, 1995.

4. Bickford, John H. (Ed.), and Sayed Nassar (Ed.). *Handbook of Bolts and Bolted Joints*. New York: Marcel Dekker, 1998.

5. Bickford, John H. (Ed.). *Gaskets and Gasketed Joints*. New York: Marcel Dekker, 1997.

6. Industrial Fasteners Institute. *Fastener Standards*. 6th ed. Cleveland: Industrial Fasteners Institute, 1988.

7. Industrial Fasteners Institute. *Metric Fastener Standards*. 3rd ed. Cleveland: Industrial Fasteners Institute, 1999.

8. Kulak, G. L., J. W. Fisher, and J. H. A. Struik. *Guide to Design Criteria for Bolted and Riveted Joints*. 2d ed. New York: John Wiley & Sons, 1987.

9. Oberg, E., F. D. Jones, and H. L. Horton. *Machinery's Handbook*. 26th ed. New York: Industrial Press, 2000.

10. Parmley, Robert O. *Standard Handbook of Fastening and Joining*. 3rd ed. New York: McGraw-Hill, 1997.

11. Society of Automotive Engineers. *SAE Fastener Standards Manual*. Warrendale, PA: Society of Automotive Engineers, 1999.

12. Society of Automotive Engineers. *SAE Standard J429: Mechanical and Material Requirements for Externally Threaded Fasteners, SAE Handbook*. Warrendale, PA: Society of Automotive Engineers, 2001.

13. Hoelscher, R. P., et al. *Graphics for Engineers*. New York: John Wiley & Sons, 1968.

INTERNET SITES RELATED TO FASTENERS

1. **Industrial Fasteners Institute (IFI).** *www.industrial-fasteners.org* An association of manufacturers and suppliers of bolts, nuts, screws, rivets, and special formed parts and the materials and equipment to make them. IFI develops standards, organizes research, and conducts education programs related to the fasteners industry.

2. **Research Council on Structural Connections (RCSC).** *www.boltcouncil.org.* An organization that stimulates and supports research on structural connections, prepares and publishes standards, and conducts educational programs.

3. **Accurate Fasteners, Inc.** *www.actfast.com* A supplier of bolts, cap screws, nuts, rivets, and numerous other types of fasteners for general industry uses.

4. **The Fastener Group** *www.fastenergroup.com* A supplier of bolts, cap screws, nuts, rivets, and numerous other types of fasteners for general industry uses.

5. **Haydon Bolts, Inc.** *www.haydonbolts.com* Manufacturer of bolts, nuts, and numerous other types of fasteners for the construction industry.

6. **Nucor Fastener Division.** *www.nucor-fastener.com* Manufacturer of hex head cap screws in SAE, ASTM, and metric grades, hex nuts, and structural bolts, nuts and washers.

7. **Nylok Fastener Corporation.** *www.nylok.com* Manufacturer of Nylok® self-locking fasteners for automotive, aerospace, consumer products, agricultural, industrial, furniture, and many other applications.

8. **Phillips Screw Company.** *www.phillips-screw.com*
   Developer of the Phillips® screwdriver. Manufacturer of
   related fasteners for the aerospace, automotive,
   construction, and industrial markets.

9. **SPS Technologies, Inc.** *www.spstech.com/unbrako*
   Manufacturer of engineered fasteners under the Unbrako®,
   Flexloc®, and Durlok® brands, including socket head cap
   screws, locknuts, and vibration resistant nuts and bolts for
   industrial machinery, automotive, and aerospace
   applications. Site includes catalogs and engineering data.

10. **St. Louis Screw & Bolt Company.** *www.*
    stlouisscrewbolt.com   Manufacturer of bolts, nuts, and
    washers to ASTM standards for the construction
    industry.

11. **Textron Fastening Systems.**
    www.textronfasteningsystems.com   Manufacturer of a
    wide variety of threaded fasteners for the automotive,
    aerospace, commercial, electronics, and construction
    industries under the Camcar®, Elco®, Fabco®, Avdel®,
    and Cherry® brands. Site includes product descriptions,
    catalogs, and technical data.

12. **Sensor Products, Inc.** *www.sensorprod.com*
    Developer of the BoltFAST® software for bolted joint
    failure analysis and strength testing. Includes joint
    analysis, thread analysis, and tightening torque analysis
    programs. This company also provides consulting on
    bolted joint analysis and design.

PROBLEMS

1. Describe the difference between a screw and a bolt.

2. Define the term *proof strength*.

3. Define the term *clamping load*.

4. Specify suitable machine screws to be installed in a pattern of four, equally spaced around a flange, if the clamping force between the flange and the mating structure is
   to be 6000 lb. Then recommend a suitable tightening
   torque for each screw.

5. What would be the tensile force in a machine screw having an 8–32 thread if it is made from SAE Grade 5 steel
   and is stressed to its proof strength?

6. What would be the tensile proof force in newtons (N) in
   a machine screw having a major diameter of 4 mm with
   standard fine threads if it is made from steel having a metric strength grade of 8.6?

7. What would be the nearest standard metric screw thread
   size to the American Standard 7/8–14 thread? By how
   much do their major diameters differ?

8. A machine screw is found with no information given as to
   its size. The following data are found by using a standard
   micrometer caliper: The major diameter is 0.196 in; the axial length for 20 full threads is 0.630 in. Identify the thread.

9. A threaded fastener is made from nylon 6/6 with an
   M10 ×1.5 thread. Compute the maximum tensile force
   that can be permitted in the fastener if it is to be stressed
   to 75% of the tensile strength of the nylon 66 dry. See
   Appendix 13.

10. Compare the tensile force that can be carried by a 1/4–20
    screw if it is to be stressed to 50% of its tensile strength
    and if it is made from each of the following materials:

(a) Steel, SAE Grade 2

(b) Steel, SAE Grade 5

(c) Steel, SAE Grade 8

(d) Steel, ASTM Grade A307

(e) Steel, ASTM Grade A574

(f) Steel, metric Grade 8.8

(g) Aluminum 2024-T4

(h) AISI 430 annealed

(i) Ti–6A1–4V annealed

(j) Nylon 66 dry

(k) Polycarbonate

(l) High-impact ABS

11. Describe the differences among welding, brazing, and
    soldering.

12. What types of metals are typically brazed?

13. What are some common brazing alloys?

14. What materials make up commonly used solders?

15. Name five common adhesives, and give the typical prop-
    erties of each.

16. The label of a common household adhesive describes it
    as a *cyanoacrylate*. What would you expect its proper-
    ties to be?

17. Find three commercially available adhesives from your
    home, a laboratory, a machine shop, or your workplace.
    Try to identify the generic nature of the adhesive, and
    compare it with the list presented in this chapter.

19

Springs

The Big Picture

You Are the Designer

19–1 Objectives of This Chapter

19–2 Kinds of Springs

19–3 Helical Compression Springs

19–4 Stresses and Deflection for Helical Compression Springs

19–5 Analysis of Spring Characteristics

19–6 Design of Helical Compression Springs

19–7 Extension Springs

19–8 Helical Torsion Springs

19–9 Improving Spring Performance by Shot Peening

19–10 Spring Manufacturing

Springs

Discussion Map

Discover

Look around you and see if you can find one or more springs. Describe them, giving their basic geometry, the type of force or torque produced, the way in which they are used, and other features.

Share your observations about springs with your colleagues, and learn from their observations.

Write a brief report about at least two different kinds of springs. Include sketches that show their basic size, geometry, and appearance. Describe their functions, including how they work and how they affect the operation of the device of which they are a part.

This chapter will help you gain skill in designing and analyzing springs of the helical compression, helical tension, and torsional types.

A *spring* is a flexible element used to exert a force or a torque and, at the same time, to store energy. The force can be a linear push or pull, or it can be radial, acting similarly to a rubber band around a roll of drawings. The torque can be used to cause a rotation, for example, to close a door on a cabinet or to provide a counterbalance force for a machine element pivoting on a hinge.

Springs inherently store energy when they are deflected and return the energy when the force that causes the deflection is removed. Consider the child's toy called a *Jack-in-the-Box*. When you push Jack into the box, you are exerting a force on a spring and delivering energy to it. Then when you close the lid of the box, the spring is captured and is held in the compressed state. What happens when you trip the latch on the lid? Jack leaps out of the box! More precisely, the spring force causes Jack to push the lid open, and then the stored energy in the spring is released, causing the spring to expand to its initial, unloaded length. Some devices contain *power springs* or *motor springs* that are wound up, and then they deliver the energy at a measured pace to produce a long-term action. Examples are mechanical animated toys, toy race cars, some watches, timers, and clocks.

Look around you and see if you can find one or more springs. Or think about where you might have recently encountered a device that used springs. Consider various appliances, an automobile, a truck, a bicycle, an office machine, a door latch, a toy, a machine in a production operation, or some other device that has moving parts.

Describe the springs. Did they exert a push or a pull? Or did they exert a torque, tending to cause rotation? What were the springs made from? How big were they? Was the force or torque very large, or was it light enough for you to actuate the spring easily? How were the springs mounted in the device of which they were a part? Was there a load on the spring at all times? Or did the spring relax completely at one point in its total possible operating cycle? Was the spring designed to be actuated often so that it experienced a very large number of cycles of load (and stress) during its expected life? What was the environment in which the spring operated? Hot or cold? Wet or dry? Exposed to corrosives? How did the environment affect the kind of material used for the spring or the kind of coating on it?

Share your observations with others in your group and with your instructor. Listen to the observations of others, and compare them with your examples of springs. Take at least two springs that are quite different from each other, and prepare a brief report about them, including sketches that show the basic size, geometry, and appearance. Describe their functions, including how they work and how they affect the operation of the device of which they are a part. Refer to the preceding paragraph for some factors that you might describe. Also include a listing with a brief description of each different kind of spring that you or your colleagues found.

This chapter will present basic information about a variety of spring types. Design procedures will be developed for helical compression springs, helical tension springs, and torsion springs. We will consider loads and stresses, deflection characteristics, material selection, life expectancy, attachment, and installation.

You Are the Designer

One design for an automotive engine valve train is shown in Figure 19–1. As the cam rotates, it causes the push rod to move upward. The rocker arm then rotates and pushes the valve stem downward, opening the valve. Concurrently, the spring that surrounds the valve stem is compressed, and energy is stored. As the cam continues to rotate, it allows the movement of the train to return to its original position. The valve is aided in its upward motion and seating action by the spring exerting a force that closes the valve at the end of the cycle.

You are the designer of the spring for the valve train. What type of spring do you specify? What should its dimensions be, including the length, outside diameter, inside diameter, and diameter of the wire for the coils? How many coils should be used? What should the ends of the spring look like? How much force is exerted on the valve, and how does this force change as the valve train goes through a complete cycle? What material should be used? What stress levels are developed in the spring wire, and how can the spring be designed to be safe under the load, life, and environmental conditions in which it must operate? You must specify or calculate all of these factors to ensure a successful spring design.

FIGURE 19–1
Engine valve train showing the application of a helical compression spring

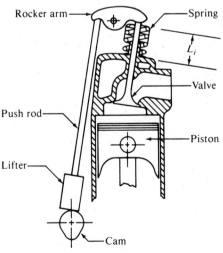

(a) Valve closed: spring length, L_i

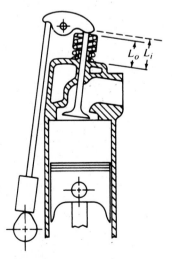

(b) Valve open: spring length, L_o

19–1
OBJECTIVES OF THIS CHAPTER

After completing this chapter, you will be able to:

1. Identify and describe several types of springs, including the helical compression spring, helical extension spring, torsion spring, Belleville spring, flat spring, draw-bar spring, garter spring, constant-force spring, and power spring.

2. Design and analyze helical compression springs to conform to design requirements such as force/deflection characteristics, life, physical size, and environmental conditions.

3. Compute the dimensions of various geometric features of helical compression springs.

4. Specify suitable materials for springs based on strength, life, and deflection parameters.

5. Design and analyze helical extension springs.

6. Design and analyze torsion springs.

7. Use computer programs to assist in the design and analysis of springs.

19–2
KINDS OF SPRINGS

Springs can be classified according to the direction and the nature of the force exerted by the spring when it is deflected. Table 19–1 lists several kinds of springs classified as *push*, *pull*, *radial*, and *torsion*. Figure 19–2 shows several typical designs.

Helical compression springs are typically made from round wire, wrapped into a straight, cylindrical form with a constant pitch between adjacent coils. Square or rectangular wire may also be used. Four practical end configurations are shown in Figure 19–3. Without an applied load, the spring's length is called the *free length*. When a compression force is applied, the coils are pressed more closely together until they all touch, at which time the length is the minimum possible called the *solid length*. A linearly increasing amount of force is required to compress the spring as its deflection is increased. Straight, cylindrical helical compression springs are among the most widely used types. Also shown in Figure 19–2 are the conical, barrel, hourglass, and variable-pitch types.

Helical extension springs appear to be similar to compression springs, having a series of coils wrapped into a cylindrical form. However, in extension springs, the coils either touch or are closely spaced under the no-load condition. Then as the external tensile load is applied, the coils separate. Figure 19–4 shows several end configurations for extension springs.

TABLE 19–1   Types of springs

Uses	Types of springs
Push	Helical compression spring
	Belleville spring
	Torsion spring: force acting at the end of the torque arm
	Flat spring, such as a cantilever or leaf spring
Pull	Helical extension spring
	Torsion spring: force acting at the end of the torque arm
	Flat spring, such as a cantilever or leaf spring
	Drawbar spring (special case of the compression spring)
	Constant-force spring
Radial	Garter spring, elastomeric band, spring clamp
Torque	Torsion spring, power spring

FIGURE 19–2
Several types of springs

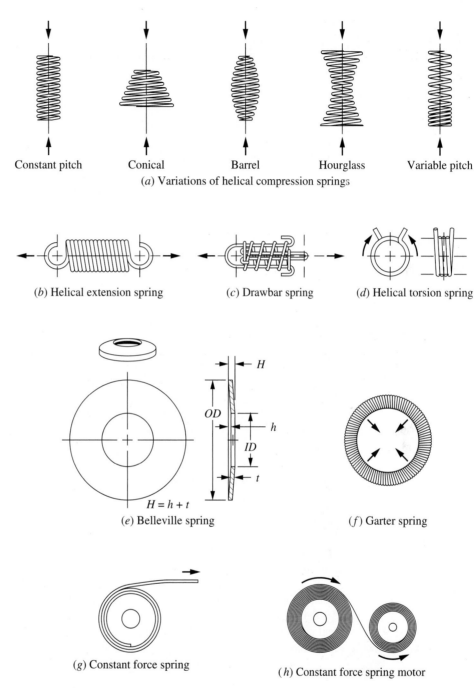

Constant pitch   Conical   Barrel   Hourglass   Variable pitch

(*a*) Variations of helical compression springs

(*b*) Helical extension spring   (*c*) Drawbar spring   (*d*) Helical torsion spring

$H = h + t$

(*e*) Belleville spring   (*f*) Garter spring

(*g*) Constant force spring   (*h*) Constant force spring motor

The *drawbar spring* incorporates a standard helical compression spring with two looped wire devices inserted through the inside of the spring. With such a design, a tensile force can be exerted by pulling on the loops while still placing the spring in compression. It also provides a definite stop as the compression spring is compressed to its solid height.

A *torsion spring*, as the name implies, is used to exert a torque as the spring is deflected by rotation about its axis. The common spring-action clothespin uses a torsion spring to provide the gripping action. Torsion springs are also used to rotate a door to its open or closed position or to counterbalance the lid of a container. Some timers and other controls use torsion

FIGURE 19–3
Appearance of helical
compression springs
showing end treatments

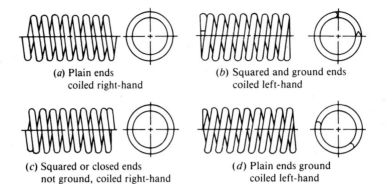

(*a*) Plain ends
coiled right-hand

(*b*) Squared and ground ends
coiled left-hand

(*c*) Squared or closed ends
not ground, coiled right-hand

(*d*) Plain ends ground
coiled left-hand

FIGURE 19–4 End
configurations for
extension springs

Type	End configurations
Twist loop or hook	
Cross center loop or hook	
Side loop or hook	
Extended hook	
Special ends	

springs to actuate switch contacts or to produce similar actions. Push or pull forces can be exerted by torsion springs if one end of the spring is attached to the member to be actuated.

Leaf springs are made from one or more flat strips of brass, bronze, steel, or other materials loaded as cantilevers or simple beams. They can provide a push or a pull force as they are deflected from their free condition. Large forces can be exerted within a small space by leaf springs. By tailoring the geometry of the leaves and by nesting leaves of different dimensions, the designer can achieve special force-deflection characteristics. The design of leaf springs uses the principles of stress and deflection analysis as presented in courses in strength of materials and as reviewed in Chapter 3.

A *Belleville spring* has the shape of a shallow, conical disk with a central hole. It is sometimes called a *Belleville washer* because its appearance is similar to that of a flat washer. A very high spring rate or spring force can be developed in a small axial space with such springs. By varying the height of the cone relative to the thickness of the disk, the designer can obtain a variety of load-deflection characteristics. Also, nesting several springs face-to-face or back-to-back provides numerous spring rates.

Garter springs are coiled wire formed into a continuous ring shape so that they exert a radial force around the periphery of the object to which they are applied. Either inward or outward forces can be obtained with different designs. The action of a garter spring with an inward force is similar to that of a rubber band, and the spring action is similar to that of an extension spring.

Constant-force springs take the form of a coiled strip. The force required to pull the strip off the coil is virtually constant over a long length of pull. The magnitude of the force is dependent on the width, thickness, and radius of curvature of the coil and on the elastic modulus of the spring material. Basically, the force is related to the deformation of the strip from its originally curved shape to the straight form.

Power springs, sometimes called *motor* or *clock springs*, are made from flat spring steel stock, wound into a spiral shape. A torque is exerted by the spring as it tends to un-wrap the spiral. Figure 19–2 shows a spring motor made from a constant-force spring.

A *torsion bar*, as its name implies, is a bar loaded in torsion. When a round bar is used, the analyses for torsional stress and deflection are similar to the analysis presented for circular shafts in Chapters 3 and 12. Other cross-sectional shapes can be used, and special care must be exercised at the points of attachment.

19–3 HELICAL COMPRESSION SPRINGS

In the most common form of helical compression spring, round wire is wrapped into a cylindrical form with a constant pitch between adjacent coils. This basic form is completed by a variety of end treatments, as shown in Figure 19–3.

For medium- to large-size springs used in machinery, the squared and ground-end treatment provides a flat surface on which to seat the spring. The end coil is collapsed against the adjacent coil (squared), and the surface is ground until at least 270° of the last coil is in contact with the bearing surface. Springs made from smaller wire (less than approximately 0.020 in, or 0.50 mm) are usually squared only, without grinding. In unusual cases the ends may be ground without squaring, or they may be left with plain ends, simply cut to length after coiling.

You are probably familiar with many uses of helical compression springs. The retractable ballpoint pen depends on the helical compression spring, usually installed around the ink supply barrel. Suspension systems for cars, trucks, and motorcycles frequently incorporate these springs. Other automotive applications include the valve springs in engines, hood linkage counterbalancing, and the clutch pressure-plate springs. In manufacturing, springs are used in dies to actuate stripper plates; in hydraulic control valves; as pneumatic cylinder return springs; and in the mounting of heavy equipment for shock isolation. Many small devices such as electrical switches and ball check valves incorporate helical compression springs. Desk chairs have stout springs to return the chair seat to its upright position. And don't forget the venerable pogo stick!

The following paragraphs define the many variables used to describe and analyze the performance of helical compression springs.

Diameters

Figure 19–5 shows the notation used in referring to the characteristic diameters of helical compression springs. The outside diameter (OD), the inside diameter (ID), and the wire diameter (D_w) are obvious and can be measured with standard measuring instruments. In calculating the stress and deflection of a spring, we use the mean diameter, D_m. Notice that

⇨ **Spring Diameters**

$$OD = D_m + D_w$$
$$ID = D_m - D_w$$

Standard Wire Diameters. The specification of the required wire diameter is one of the most important outcomes of the design of springs. Several types of materials are typically used for spring wire, and the wire is produced in sets of standard diameters covering a broad range. Table 19–2 lists the most common standard wire gages. Notice that except for music wire, the wire size gets smaller as the gage number gets larger. Also see the notes to the table.

FIGURE 19–5
Notation for diameters

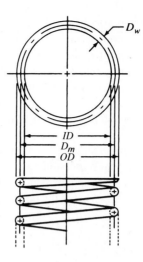

TABLE 19–2   Wire gages and diameters for springs

Gage no.	U.S. Steel Wire Gage $(in)^a$	Music Wire Gage $(in)^b$	Brown & Sharpe Gage $(in)^c$	Preferred Metric Diameters $(mm)^d$
7/0	0.4900			13.0
6/0	0.4615	0.004	0.5800	12.0
5/0	0.4305	0.005	0.5165	11.0
4/0	0.3938	0.006	0.4600	10.0
3/0	0.3625	0.007	0.4096	9.0
2/0	0.3310	0.008	0.3648	8.5
0	0.3065	0.009	0.3249	8.0
1	0.2830	0.010	0.2893	7.0
2	0.2625	0.011	0.2576	6.5
3	0.2437	0.012	0.2294	6.0
4	0.2253	0.013	0.2043	5.5
5	0.2070	0.014	0.1819	5.0
6	0.1920	0.016	0.1620	4.8
7	0.1770	0.018	0.1443	4.5
8	0.1620	0.020	0.1285	4.0
9	0.1483	0.022	0.1144	3.8
10	0.1350	0.024	0.1019	3.5
11	0.1205	0.026	0.0907	3.0
12	0.1055	0.029	0.0808	2.8
13	0.0915	0.031	0.0720	2.5
14	0.0800	0.033	0.0641	2.0
15	0.0720	0.035	0.0571	1.8
16	0.0625	0.037	0.0508	1.6
17	0.0540	0.039	0.0453	1.4
18	0.0475	0.041	0.0403	1.2
19	0.0410	0.043	0.0359	1.0
20	0.0348	0.045	0.0320	0.90
21	0.0317	0.047	0.0285	0.80
22	0.0286	0.049	0.0253	0.70

TABLE 19–2 (*continued*)

Gage no.	U.S. Steel Wire Gage (in)[a]	Music Wire Gage (in)[b]	Brown & Sharpe Gage (in)[c]	Preferred metric diameters (mm)[d]
23	0.0258	0.051	0.0226	0.65
24	0.0230	0.055	0.0201	0.60 or 0.55
25	0.0204	0.059	0.0179	0.50 or 0.55
26	0.0181	0.063	0.0159	0.45
27	0.0173	0.067	0.0142	0.45
28	0.0162	0.071	0.0126	0.40
29	0.0150	0.075	0.0113	0.40
30	0.0140	0.080	0.0100	0.35
31	0.0132	0.085	0.00893	0.35
32	0.0128	0.090	0.00795	0.30 or 0.35
33	0.0118	0.095	0.00708	0.30
34	0.0104	0.100	0.00630	0.28
35	0.0095	0.106	0.00501	0.25
36	0.0090	0.112	0.00500	0.22
37	0.0085	0.118	0.00445	0.22
38	0.0080	0.124	0.00396	0.20
39	0.0075	0.130	0.00353	0.20
40	0.0070	0.138	0.00314	0.18

Source: Associated Spring, Barnes Group, Inc. *Engineering Guide to Spring Design.* Bristol, CT, 1987. Carlson, Harold. *Spring Designer's Handbook.* New York: Marcel Dekker, 1978. Oberg, E., et al. *Machinery's Handbook.* 26th ed. New York: Industrial Press, 2000.

[a]Use the U.S. Steel Wire Gage for steel wire, except music wire. This gage has also been called the *Washburn and Moen Gage (W&M),* the *American Steel Wire Co. Gage,* and the *Roebling Wire Gage.*

[b]Use the Music Wire Gage only for music wire (ASTM A228).

[c]Use the Brown & Sharpe Gage for nonferrous wires such as brass and phosphor bronze.

[d]The preferred metric sizes are from Associated Spring, Barnes Group, Inc., and are listed as the nearest preferred metric size to the U.S. Steel Wire Gage. The gage numbers do not apply.

Lengths

It is important to understand the relationship between the length of the spring and the force exerted by it (see Figure 19–6). The *free length,* L_f, is the length that the spring assumes when it is exerting no force, as if it were simply sitting on a table. The *solid length,* L_s, is found when the spring is collapsed to the point where all coils are touching. This is obviously the shortest possible length that the spring can have. The spring is usually not compressed to the solid length during operation.

The shortest length for the spring during normal operation is the *operating length,* L_o. At times, a spring will be designed to operate between two limits of deflection. Consider the valve spring for an engine, for example, as shown in Figure 19–1. When the valve is open, the spring assumes its shortest length, which is L_o. Then when the valve is closed, the spring gets longer but still exerts a force to keep the valve securely on its seat. The length at this condition is called the *installed length,* L_i. So the valve spring length changes from L_o to L_i during normal operation as the valve itself reciprocates.

FIGURE 19–6
Notation for lengths
and forces

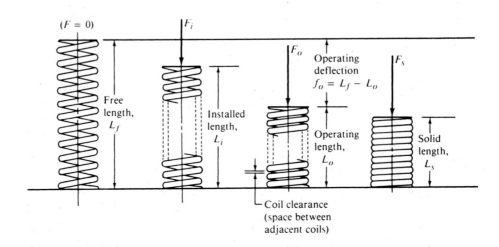

Forces

We will use the symbol F to indicate forces exerted by a spring, with various subscripts to specify which level of force is being considered. The subscripts correspond to those used for the lengths. Thus,

F_s = force at solid length, L_s; the maximum force that the spring ever sees.

F_o = force at operating length, L_o; the maximum force the spring sees in *normal operation*.

F_i = force at installed length, L_i; the force varies between F_o and F_i for a reciprocating spring.

F_f = force at free length, L_f; this force is zero.

Spring Rate

The relationship between the force exerted by a spring and its deflection is called its *spring rate*, k. Any change in force divided by the corresponding change in deflection can be used to compute the spring rate:

> **Spring Rate**

$$k = \Delta F / \Delta L \qquad (19\text{–}1)$$

For example,

$$k = \frac{F_o - F_i}{L_i - L_o} \qquad (19\text{–}1a)$$

or

$$k = \frac{F_o}{L_f - L_o} \qquad (19\text{–}1b)$$

or

$$k = \frac{F_i}{L_f - L_i} \qquad (19\text{–}1c)$$

In addition, if the spring rate is known, the force at any deflection can be computed. For example, if a spring had a rate of 42.0 lb/in, the force exerted at a deflection from free length of 2.25 in would be

$$F = k(L_f - L) = (42.0 \text{ lb/in})(2.25 \text{ in}) = 94.5 \text{ lb}$$

Spring Index

The ratio of the mean diameter of the spring to the wire diameter is called the *spring index, C*:

⇨ **Spring Index**

$$C = D_m/D_w$$

It is recommended that C be greater than 5.0, with typical machinery springs having C values ranging from 5 to 12. For C less than 5, the forming of the spring will be very difficult, and the severe deformation required may create cracks in the wire. The stresses and the deflections in springs are dependent on C, and a larger C will help to eliminate the tendency for a spring to buckle.

Number of Coils

The total number of coils in a spring will be called N. But in the calculation of stress and deflections for a spring, some of the coils are inactive and are neglected. For example, in a spring with squared and ground ends or simply squared ends, each end coil is inactive, and the number of *active coils*, N_a, is $N - 2$. For plain ends, all coils are active: $N_a = N$. For plain coils with ground ends, $N_a = N - 1$.

Pitch

Pitch, p, refers to the axial distance from a point on one coil to the corresponding point on the next adjacent coil. The relationships among the pitch, free length, wire diameter, and number of active coils are given next:

Squared and ground ends:	$L_f = pN_a + 2D_w$
Squared ends only:	$L_f = pN_a + 3D_w$
Plain and ground ends:	$L_f = p(N_a + 1)$
Plain ends:	$L_f = pN_a + D_w$

Pitch Angle

Figure 19–7 shows the pitch angle, λ; note that the larger the pitch angle, the steeper the coils appear to be. Most practical spring designs produce a pitch angle less than about 12°. If the angle is greater than 12°, undesirable compressive stresses develop in the wire, and the formulas presented later are inaccurate. The pitch angle can be computed by the formula

$$\lambda = \tan^{-1}\left[\frac{p}{\pi D_m}\right] \tag{19-2}$$

FIGURE 19–7 Pitch angle

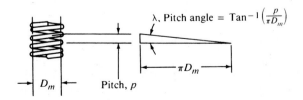

λ, Pitch angle $= \text{Tan}^{-1}\left(\frac{p}{\pi D_m}\right)$

πD_m

D_m Pitch, p

You can see the logic of this formula by taking one coil of a spring and unwrapping it onto a flat surface, as illustrated in Figure 19–7. The horizontal line is the mean circumference of the spring, and the vertical line is the pitch, p.

Installation Considerations

Frequently, a spring is installed in a cylindrical hole or around a rod. When it is, adequate clearances must be provided. When a compression spring is compressed, its diameter gets larger. Thus, the inside diameter of a hole enclosing the spring must be greater than the outside diameter of the spring to eliminate rubbing. An initial diametral clearance of one-tenth of the wire diameter is recommended for springs having a diameter of 0.50 in (12 mm) or greater. If a more precise estimate of the actual outside diameter of the spring is required, the following formula can be used for the OD at the solid length condition:

$$OD_s = \sqrt{D_m^2 + \frac{p^2 - D_w^2}{\pi^2}} + D_w \qquad (19\text{--}3)$$

Even though the spring ID gets larger, it is also recommended that the clearance at the ID be approximately $0.1D_w$.

Springs with squared ends or squared and ground ends are frequently mounted on a button-type seat or in a socket with a depth equal to the height of just a few coils for the purpose of locating the spring.

Coil Clearance. The term *coil clearance* refers to the space between adjacent coils when the spring is compressed to its operating length, L_o. The actual coil clearance, cc, can be estimated from

Coil Clearance

$$cc = (L_o - L_s)/N_a$$

One guideline is that the coil clearance should be greater than $D_w/10$, especially in springs loaded cyclically. Another recommendation relates to the overall deflection of the spring:

$$(L_o - L_s) > 0.15(L_f - L_s)$$

Materials Used for Springs

Virtually any elastic material can be used for a spring. However, most mechanical applications use metallic wire—high-carbon steel (most common), alloy steel, stainless steel, brass, bronze, beryllium copper, or nickel-base alloys. Most spring materials are made according to specifications of the ASTM. Table 19-3 lists some common types. See Internet sites 8–11.

Types of Loading and Allowable Stresses

The allowable stress to be used for a spring depends on the type of loading, the material, and size of the wire. Loading is usually classified into three types:

- *Light service:* Static loads or up to 10 000 cycles of loading with a low rate of loading (nonimpact).
- *Average service:* Typical machine design situations; moderate rate of loading and up to 1 million cycles.
- *Severe service:* Rapid cycling for above 1 million cycles; possibility of shock or impact loading; engine valve springs are a good example.

TABLE 19–3   Spring materials

Material type	ASTM no.	Relative cost	Temperature limits, °F
A. *High-carbon steels*			
Hard-drawn	A227	1.0	0–250
General-purpose steel with 0.60%–0.70% carbon; low cost			
Music wire	A228	2.6	0–250
High-quality steel with 0.80%–0.95% carbon; very high strength; excellent surface finish; hard-drawn; good fatigue performance; used mostly in smaller sizes up to 0.125 in			
Oil-tempered	A229	1.3	0–350
General-purpose steel with 0.60%–0.70% carbon; used mostly in larger sizes above 0.125 in; not good for shock or impact			
B. *Alloy steels*			
Chromium-vanadium	A231	3.1	0–425
Good strength, fatigue resistance, impact strength, high-temperature performance; valve-spring quality			
Chromium-silicon	A401	4.0	0–475
Very high strength and good fatigue and shock resistance			
C. *Stainless steels*			
Type 302	A313(302)	7.6	<0–550
Very good corrosion resistance and high-temperature performance; nearly nonmagnetic; cold-drawn; types 304 and 316 also fall under this ASTM class and have improved workability but lower strength			
Type 17-7 PH	A313(631)	11.0	0–600
Good high-temperature performance			
D. *Copper alloys:* All have good corrosion resistance and electrical conductivity.			
Spring brass	B134	High	0–150
Phosphor bronze	B159	8.0	<0–212
Beryllium copper	B197	27.0	0–300
E. *Nickel-base alloys:* All are corrosion-resistant, have good high- and low-temperature properties, and are nonmagnetic or nearly nonmagnetic (trade names of the International Nickel Company).			
Monel™			
K-Monel™			−100–425
Inconel™			−100–450
Inconel-X™		44.0	Up to 700
			Up to 850

Source: Associated Spring, Barnes Group, Inc. *Engineering Guide to Spring Design.* Bristol, CT, 1987. Carlson, Harold. *Spring Designer's Handbook.* New York: Marcel Dekker, 1978. Oberg, E., et al. *Machinery's Handbook.* 26th ed. New York: Industrial Press, 2000.

The strength of a given material is greater for the smaller sizes. Figures 19–8 through 19–13 show the design stresses for six different materials. Note that some curves can be used for more than one material by the application of a factor. As a conservative approach to design, we will use the average service curve for most design examples, unless true high cycling conditions exist. We will use the light service curve as the upper limit on stress when the spring is compressed to its solid height. If the stress exceeds the light service value by a small amount, the spring will undergo permanent set because of yielding.

FIGURE 19–8

Design shear stresses
for ASTM A227 steel
wire, hard-drawn
(Reprinted from Harold
Carlson, *Spring
Designer's Handbook*,
p. 144, by courtesy of
Marcel Dekker, Inc.)

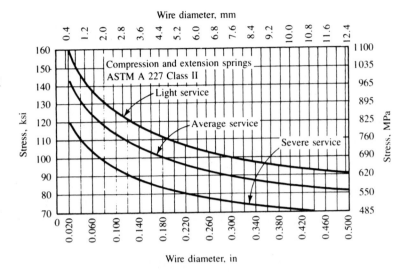

FIGURE 19–9

Design shear stresses
for ASTM A228 steel
wire (music wire)
(Reprinted from Harold
Carlson, *Spring
Designer's Handbook*,
p. 143, by courtesy of
Marcel Dekker, Inc.)

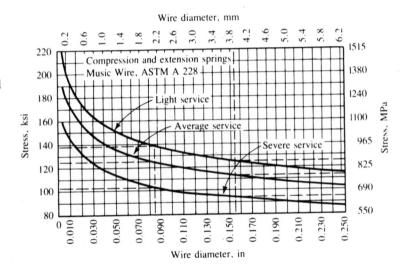

FIGURE 19–10

Design shear stresses
for ASTM A229 steel
wire, oil-tempered
(Reprinted from Harold
Carlson, *Spring
Designer's Handbook*,
p. 146, by courtesy of
Marcel Dekker, Inc.)

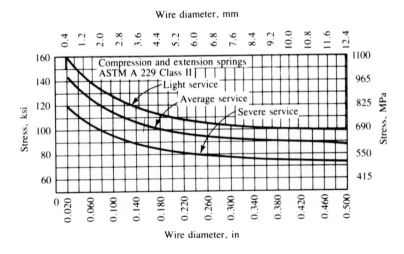

FIGURE 19–11

Design shear stresses
for ASTM A231 steel
wire, chromium-
vanadium alloy, valve-
spring quality
(Reprinted from Harold
Carlson, *Spring
Designer's Handbook*,
p. 147, by courtesy of
Marcel Dekker, Inc.)

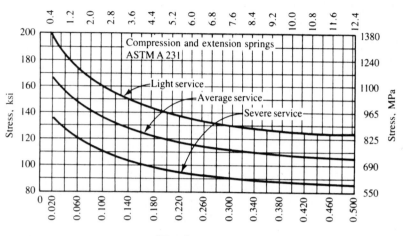

FIGURE 19–12

Design shear stresses
for ASTM A401 steel
wire, chromium-silicon
alloy, oil-tempered
(Reprinted from Harold
Carlson, *Spring
Designer's Handbook*,
p. 148, by courtesy of
Marcel Dekker, Inc.)

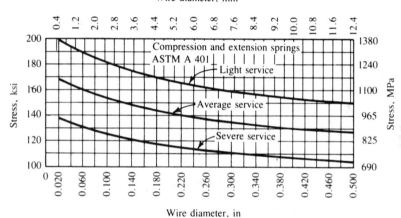

FIGURE 19–13

Design shear stresses
for ASTM A313
corrosion-resistant
stainless steel wire
(Reprinted from Harold
Carlson, *Spring
Designer's Handbook*,
p. 150, by courtesy of
Marcel Dekker, Inc.)

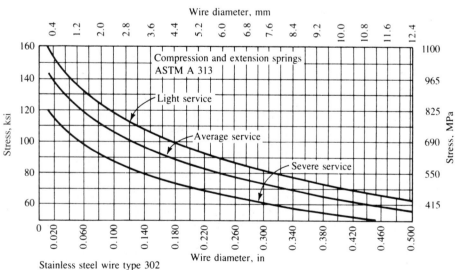

Stainless steel wire type 302
For type 304 multiply by 0.95
For type 316 multiply by 0.85

19–4
STRESSES AND
DEFLECTION
FOR HELICAL
COMPRESSION
SPRINGS

As a compression spring is compressed under an axial load, the wire is twisted. Therefore, the stress developed in the wire is *torsional shear stress*, and it can be derived from the classical equation $\tau = Tc/J$.

When the equation is applied specifically to a helical compression spring, some modifying factors are needed to account for the curvature of the spring wire and for the direct shear stress created as the coils resist the vertical load. Also, it is convenient to express the shear stress in terms of the design variables encountered in springs. The resulting equation for stress is attributed to Wahl. (See Reference 9.) The maximum shear stress, which occurs at the inner surface of the wire, is

⇨ **Shear Stress**
in a Spring

$$\tau = \frac{8KFD_m}{\pi D_w^3} = \frac{8KFC}{\pi D_w^2} \qquad (19\text{–}4)$$

These are two forms of the same equation as the definition of $C = D_m/D_w$ demonstrates. The shear stress for any applied force, F, can be computed. Normally we will be concerned about the stress when the spring is compressed to solid length under the influence of F_s and when the spring is operating at its normal maximum load, F_o. Notice that the stress is inversely proportional to the *cube* of the wire diameter. This illustrates the great effect that variation in the wire size has on the performance of the spring.

The Wahl factor, K, in Equation (19–4) is the term that accounts for the curvature of the wire and the direct shear stress. Analytically, K is related to C:

⇨ **Wahl Factor**

$$K = \frac{4C - 1}{4C - 4} + \frac{0.615}{C} \qquad (19\text{–}5)$$

Figure 19–14 shows a plot of K versus C for round wire. Recall that $C = 5$ is the recommended minimum value of C. The value of K rises rapidly for $C < 5$.

Deflection

Because the primary manner of loading on the wire of a helical compression spring is torsion, the deflection is computed from the angle of twist formula:

$$\theta = TL/GJ$$

FIGURE 19–14
Wahl factor vs. spring
index for round wire

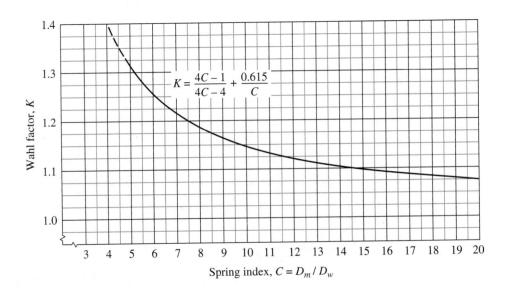

where θ = angle of twist in radians

T = applied torque

L = length of the wire

G = modulus of elasticity of the material in shear

J = polar moment of inertia of the wire

Again, for convenience, we will use a different form of the equation in order to calculate the linear deflection, f, of the spring from the typical design variables of the spring. The resulting equation is

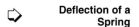

Deflection of a Spring

$$f = \frac{8FD_m^3 N_a}{GD_w^4} = \frac{8FC^3 N_a}{GD_w} \qquad (19\text{–}6)$$

Recall that N_a is the number of *active* coils, as discussed in Section 19–3. Table 19–4 lists the values for G for typical spring materials. Note again, in Equation (19–6), that the wire diameter has a strong effect on the performance of the spring.

Buckling

The tendency for a spring to buckle increases as the spring becomes tall and slender, much as for a column. Figure 19–15 shows plots of the critical ratio of deflection to the free length versus the ratio of free length to the mean diameter for the spring. Three different end conditions are described in the figure. As an example of the use of this figure, consider a spring having squared and ground ends, a free length of 6.0 in, and a mean diameter of 0.75 in. We want to know what deflection would cause the spring to buckle. First compute

$$\frac{L_f}{D_m} = \frac{6.0}{0.75} = 8.0$$

TABLE 19–4   Spring wire modulus of elasticity in shear (G) and tension (E)

Material and ASTM no.	Shear modulus, G		Tension modulus, E	
	(psi)	(GPa)	(psi)	(GPa)
Hard-drawn steel: A227	11.5×10^6	79.3	28.6×10^6	197
Music wire: A228	11.85×10^6	81.7	29.0×10^6	200
Oil-tempered: A229	11.2×10^6	77.2	28.5×10^6	196
Chromium-vanadium: A231	11.2×10^6	77.2	28.5×10^6	196
Chromium-silicon: A401	11.2×10^6	77.2	29.5×10^6	203
Stainless steels: A313				
Types 302, 304, 316	10.0×10^6	69.0	28.0×10^6	193
Type 17-7 PH	10.5×10^6	72.4	29.5×10^6	203
Spring brass: B134	5.0×10^6	34.5	15.0×10^6	103
Phosphor bronze: B159	6.0×10^6	41.4	15.0×10^6	103
Beryllium copper: B197	7.0×10^6	48.3	17.0×10^6	117
Monel and K-Monel	9.5×10^6	65.5	26.0×10^6	179
Inconel and Inconel-X	10.5×10^6	72.4	31.0×10^6	214

Note: Data are average values. Slight variations with wire size and treatment may occur.

FIGURE 19–15
Spring buckling
criteria. If the actual
ratio of f_o/L_f is greater
than the critical ratio,
the spring will buckle
at operating deflection.

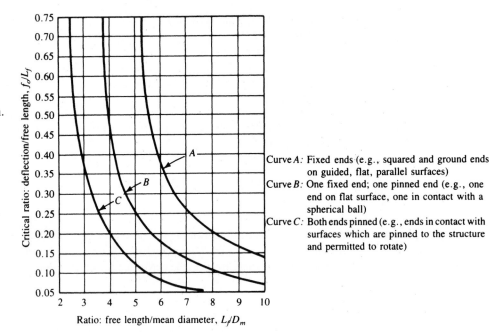

Curve A: Fixed ends (e.g., squared and ground ends
on guided, flat, parallel surfaces)
Curve B: One fixed end; one pinned end (e.g., one
end on flat surface, one in contact with a
spherical ball)
Curve C: Both ends pinned (e.g., ends in contact with
surfaces which are pinned to the structure
and permitted to rotate)

Then, from Figure 19–15, the critical deflection ratio is 0.20. From this we can compute the
critical deflection:

$$\frac{f_o}{L_f} = 0.20 \quad \text{or} \quad f_o = 0.20(L_f) = 0.20(6.0 \text{ in}) = 1.20 \text{ in}$$

That is, if the spring is deflected more than 1.20 in, the spring should buckle.

Additional development and discussion of formulas for stresses and deflection of helical compression springs can be found in References 4 and 6. Reference 3 provides valuable information on the analysis of spring failures.

19–5
ANALYSIS OF
SPRING
CHARACTERISTICS

This section demonstrates the use of the concepts developed in previous sections to analyze
the geometry and the performance characteristics of a given spring. Assume that you have
found a spring but there are no performance data available for it. By making a few measurements and computations, you should be able to determine those characteristics. One
piece of information you *would* have to know is the material from which the spring is made
so that you could evaluate the acceptability of calculated stress levels.

The method of analysis is presented in Example Problem 19–1.

Example Problem 19–1

A spring is known to be made from music wire, ASTM A228 steel, but no other data are
known. You are able to measure the following features using simple measurement tools:

Free length = L_f = 1.75 in

Outside diameter = OD = 0.561 in

Wire diameter = D_w = 0.055 in

The ends are squared and ground.

The *total* number of coils is 10.0.

This spring will be used in an application where the normal operating load is to be 14.0 lb. Approximately 300 000 cycles of loading are expected.

For this spring, compute and/or do the following:

1. The music wire gage number, mean diameter, inside diameter, spring index, and Wahl factor.
2. The expected stress at the operating load of 14.0 lb.
3. The deflection of the spring under the 14.0-lb load.
4. The operating length, solid length, and spring rate.
5. The force on the spring when it is at its solid length and the corresponding stress at solid length.
6. The design stress for the material; then compare it with the actual operating stress.
7. The maximum permissible stress; then compare it with the stress at solid length.
8. Check the spring for buckling and coil clearance.
9. Specify a suitable diameter for a hole in which to install the spring.

Solution   The solution is presented in the same order as the requested items just listed. The formulas used are found in the preceding sections of this chapter.

Step 1. The wire is 24-gage music wire (Table 19–2). Thus,

$$D_m = OD - D_w = 0.561 - 0.055 = 0.506 \text{ in}$$
$$ID = D_m - D_w = 0.506 - 0.055 = 0.451 \text{ in}$$
$$\text{Spring index} = C = D_m/D_w = 0.506/0.055 = 9.20$$
$$\text{Wahl factor} = K = (4C - 1)/(4C - 4) + 0.615/C$$
$$K = [4(9.20) - 1]/[4(9.20) - 4] + 0.615/9.20$$
$$K = 1.158$$

10 4.5

Step 2. Stress in spring at $F = F_o = 14.0$ lb [Equation (19–4)]:

load

$$\tau_o = \frac{8KF_oC}{\pi D_w^2} = \frac{8(1.158)(14.0)(9.20)}{\pi(0.055)^2} = 125\,560 \text{ psi}$$

wire Diameter

Step 3. Deflection at operating force [Equation (19–6)]:

$$f_o = \frac{8F_oC^3N_a}{GD_w} = \frac{8(14.0)(9.20)^3(8.0)}{(11.85 \times 10^6)(0.055)} = 1.071 \text{ in}$$

Note that the number of active coils for a spring with squared and ground ends is $N_a = N - 2 = 10.0 - 2 = 8.0$. Also, the spring wire modulus, G, was found in Table 19–4. The value of f_o is the deflection *from free length* to the operating length.

Step 4. Operating length: We compute operating length as

$$L_o = L_f - f_o = 1.75 - 1.071 = 0.679 \text{ in}$$
$$\text{Solid length} = L_s = D_w(N) = 0.055(10.0) = 0.550 \text{ in}$$

Spring Index: We use Equation (19–1).

$$k = \frac{\Delta F}{\Delta L} = \frac{F_o}{L_f - L_o} = \frac{F_o}{f_o} = \frac{14.0 \text{ lb}}{1.071 \text{ in}} = 13.07 \text{ lb/in}$$

Step 5. We can find the force at solid length by multiplying the spring rate times the deflection from the free length to the solid length. Then

$$F_s = k(L_f - L_s) = (13.07 \text{ lb/in})(1.75 \text{ in} - 0.550 \text{ in}) = 15.69 \text{ lb}$$

The stress at solid length, τ_s, could be found from Equation (19–4), using $F = F_s$. However, an easier method is to recognize that the stress is directly proportional to the force on the spring and that all of the other data in the formula are the same as those used to compute the stress under the operating force, F_o. We can then use the simple proportion

$$\tau_s = \tau_o \, (F_s/F_o) = (125\,560 \text{ psi})(15.69/14.0) = 140\,700 \text{ psi}$$

Step 6. Design stress, τ_d: From Figure 19–9, in the graph of design stress versus spring wire diameter for ASTM A228 steel, we can use the *average service* curve based on the expected number of cycles of loading. We read $\tau_d = 135\,000$ psi for the 0.055-in wire. Because the actual operating stress, τ_o, is less than this value, it is satisfactory.

Step 7. Maximum allowable stress, τ_{max}: It is recommended that the *light service* curve be used to determine this value. For $D_w = 0.055$, $\tau_{max} = 150\,000$ psi. The actual expected maximum stress that occurs at solid length ($\tau_s = 140\,700$ psi) is less than this value, and therefore the design is satisfactory with regard to stresses.

Step 8. Buckling: To evaluate buckling, we must compute

$$L_f/D_m = (1.75 \text{ in})/(0.506 \text{ in}) = 3.46$$

Referring to Figure 19–15 and using curve *A* for squared and ground ends, we see that the critical deflection ratio is very high and that buckling should not occur. In fact, for any value of $L_f/D_m < 5.2$, we can conclude that buckling will not occur.

Coil clearance, *cc*: We evaluate *cc* as follows:

$$cc = (L_o - L_s)/N_a = (0.679 - 0.550)/(8.0) = 0.016 \text{ in}$$

Comparing this to the recommended minimum clearance of

$$D_w/10 = (0.055 \text{ in})/10 = 0.0055 \text{ in}$$

we can judge this clearance to be acceptable.

Step 9. Hole diameter: It is recommended that the hole into which the spring is to be installed should be greater in diameter than the *OD* of the spring by the amount of $D_w/10$. Then

$$D_{hole} > OD + D_w/10 = 0.561 \text{ in} + (0.055 \text{ in})/10 = 0.567 \text{ in}$$

A diameter of 5/8 in (0.625 in) would be a satisfactory standard size.

This completes the example problem.

19–6 **DESIGN OF** **HELICAL** **COMPRESSION** **SPRINGS**	The objective of the design of helical compression springs is to specify the geometry for the spring to operate under specified limits of load and deflection, possibly with space limitations, also. We will specify the material and the type of service by considering the environment and the application.

A typical problem statement follows. Then two solution procedures are shown, and each is implemented with the aid of a spreadsheet.

Example Problem 19–2

A helical compression spring is to exert a force of 8.0 lb when compressed to a length of 1.75 in. At a length of 1.25 in, the force must be 12.0 lb. The spring will be installed in a machine that cycles slowly, and approximately 200 000 cycles total are expected. The temperature will not exceed 200°F. The spring will be installed in a hole having a diameter of 0.75 in.

For this application, specify a suitable material, wire diameter, mean diameter, *OD*, *ID*, free length, solid length, number of coils, and type of end condition. Check the stress at the maximum operating load and at the solid length condition.

The first of two solution procedures will be shown. The numbered steps can be used as a guide for future problems and as a kind of algorithm for the spreadsheet approach that follows the manual solution.

Solution Method 1

The procedure works directly toward the overall geometry of the spring by specifying the mean diameter to meet the space limitations. The process requires that the designer have tables of data available for wire diameters (such as Table 19–2) and graphs of design stresses for the material from which the spring will be made (such as Figures 19–8 through 19–13). We must make an initial estimate for the design stress for the material by consulting the charts of design stress versus wire diameter to make a reasonable choice. In general, more than one trial must be made, but the results of early trials will help you decide the values to use for later trials.

Step 1. Specify a material and its shear modulus of elasticity, *G*.

For this problem, several standard spring materials can be used. Let's select ASTM A231 chromium-vanadium steel wire, having a value of $G = 11\ 200\ 000$ psi (see Table 19–4).

Step 2. From the problem statement, identify the operating force, F_o; the operating length at which that force must be exerted, L_o; the force at some other length, called the *installed force*, F_i; and the installed length, L_i.

Remember, F_o is the maximum force that the spring experiences under normal operating conditions. Many times, the second force level is not specified. In that case, let $F_i = 0$, and specify a design value for the free length, L_f, in place of L_i.

For this problem, $F_o = 12.0$ lb; $L_o = 1.25$ in; $F_i = 8.0$ lb; and $L_i = 1.75$ in.

Step 3. Compute the spring rate, *k*, using Equation (19–1a):

$$k = \frac{F_o - F_i}{L_i - L_o} = \frac{12.0 - 8.0}{1.75 - 1.25} = 8.00 \text{ lb/in}$$

Step 4. Compute the free length, L_f:

$$L_f = L_i + F_i/k = 1.75 \text{ in} + [(8.00 \text{ lb})/(8.00 \text{ lb/in})] = 2.75 \text{ in}$$

The second term in the preceding equation is the amount of deflection from free length to the installed length in order to develop the installed force, F_i. Of course, this step becomes unnecessary if the free length is specified in the original data.

Step 5. Specify an initial estimate for the mean diameter, D_m.

Keep in mind that the mean diameter will be smaller than the *OD* and larger than the *ID*. Judgment is necessary to get started. For this problem, let's specify $D_m = 0.60$ in. This should permit the installation into the 0.75-in-diameter hole.

Step 6. Specify an initial design stress.

The charts for the design stresses for the selected material can be consulted, considering also the service. In this problem, we should use average service. Then for the ASTM A231 steel, as shown in Figure 19–11, a nominal design stress would be 130 000 psi. This is strictly an estimate based on the strength of the material. The process includes a check on stress later.

Step 7. Compute the trial wire diameter by solving Equation (19–4) for D_w. Notice that everything else in the equation is known except the Wahl factor, K, because it depends on the wire diameter itself. But K varies only little over the normal range of spring indexes, C. From Figure 19–14, note that $K = 1.2$ is a nominal value. This, too, will be checked later. With the assumed value of K, some simplification can be done:

$$D_w = \left[\frac{8KF_oD_m}{\pi\tau_d}\right]^{1/3} = \left[\frac{(8)(1.2)(F_o)(D_m)}{(\pi)(\tau_d)}\right]^{1/3}$$

Combining constants gives

  Trial Wire Diameter

$$D_w = \left[\frac{8KF_oD_m}{\pi\tau_d}\right]^{1/3} = \left[\frac{(3.06)(F_o)(D_m)}{(\tau_d)}\right]^{1/3} \qquad \textbf{(19–7)}$$

For this problem,

$$D_w = \left[\frac{(3.06)(F_o)(D_m)}{\tau_d}\right]^{1/3} = \left[\frac{(3.06)(12)(0.6)}{130\,000}\right]^{0.333}$$

$$D_w = 0.0553 \text{ in}$$

Step 8. Select a standard wire diameter from the tables, and then determine the design stress and the maximum allowable stress for the material at that diameter. The design stress will normally be for average service, unless high cycling rates or shock indicate that severe service is warranted. The light service curve should be used with care because it is very near to the yield strength. In fact, we will use the light service curve as an estimate of the maximum allowable stress.

For this problem, the next larger standard wire size is 0.0625 in, no. 16 on the U.S. Steel Wire Gage chart. For this size, the curves in Figure 19–11 for ASTM A231 steel wire show the design stress to be approximately 145 000 psi for average service, and the maximum allowable stress to be 170 000 psi from the light service curve.

Step 9. Compute the actual values of C and K, the spring index and the Wahl factor:

$$C = \frac{D_m}{D_w} = \frac{0.60}{0.0625} = 9.60$$

$$K = \frac{4C - 1}{4C - 4} + \frac{0.615}{C} = \frac{4(9.60) - 1}{4(9.60) - 4} + \frac{0.615}{9.60} = 1.15$$

Step 10. Compute the actual expected stress due to the operating force, F_o, from Equation (19–4):

$$\tau_o = \frac{8KF_oD_m}{\pi D_w^3} = \frac{(8)(1.15)(12.0)(0.60)}{(\pi)(0.0625)^3} = 86\,450 \text{ psi}$$

Comparing this with the design stress of 145 000 psi, we see that it is safe.

Step 11. Compute the number of active coils required to give the proper deflection characteristics for the spring. Using Equation (19–6) and solving for N_a, we have

$$f = \frac{8FC^3 N_a}{GD_w}$$

⇨ **Number of Active Coils**

$$N_a = \frac{fGD_w}{8FC^3} = \frac{GD_w}{8kC^3} \quad (\text{Note: } F/f = k, \text{ the spring rate.}) \tag{19–8}$$

Then, for this problem,

$$N_a = \frac{GD_w}{8kC^3} = \frac{(11\ 200\ 000)(0.0625)}{(8)(8.0)(9.60)^3} = 12.36 \text{ coils}$$

Notice that $k = 8.0$ lb/in is the spring rate. Do not confuse this with K, the Wahl factor.

Step 12. Compute the solid length, L_s; the force on the spring at solid length, F_s; and the stress in the spring at solid length, τ_s. This computation will give the maximum stress that the spring will receive.

The solid length occurs when all of the coils are touching, but recall that there are two inactive coils for springs with squared and ground ends. Thus,

$$L_s = D_w(N_a + 2) = 0.0625(14.36) = 0.898 \text{ in}$$

The force at solid length is the product of the spring rate times the deflection to solid length $(L_f - L_s)$:

$$F_s = k(L_f - L_s) = (8.0 \text{ lb/in})(2.75 - 0.898) \text{ in} = 14.8 \text{ lb}$$

Because the stress in the spring is directly proportional to the force, a simple method of computing the solid length stress is

$$\tau_s = (\tau_o)(F_s/F_o) = (86\ 450 \text{ psi})(14.8/12.0) = 106\ 750 \text{ psi}$$

When this value is compared with the maximum allowable stress of 170 000 psi, we see that it is safe, and the spring will not yield when compressed to solid length.

Step 13. Complete the computations of geometric features, and compare them with space and operational limitations:

$$OD = D_m + D_w = 0.60 + 0.0625 = 0.663 \text{ in}$$
$$ID = D_m - D_w = 0.60 - 0.0625 = 0.538 \text{ in}$$

These dimensions are satisfactory for installation in a hole having a diameter of 0.75 in.

Step 14. The tendency to buckle is checked, along with the coil clearance.

This procedure completes the design of one satisfactory spring for this application. It may be desirable to make other trials to attempt to find a more nearly optimum spring.

Spreadsheet for Spring Design Method 1

The 14 steps required to complete one trial design using Method 1 which was demonstrated in Example Problem 19–2 are fairly involved, tedious, and time-consuming. Furthermore, it is highly likely that several iterations are required to produce an optimum solution that meets application considerations of the physical size of the spring, acceptable levels of stress under all loads, cost, and other factors. You might want to investigate the use of different materials for the same basic design goals for the forces, lengths, and spring rate.

For these and many other reasons, it is recommended that computer-assisted design approaches be developed to perform most of the calculations and to guide you through the solution procedure. This could be done with a computer program, a spreadsheet, mathematical analysis software, or a programmable calculator. Once written, the program or spreadsheet can be used for any similar design problem in the future by yourself or others.

Figure 19–16 shows one approach using a spreadsheet with the data from Example Problem 19–2 used for illustration. This approach is attractive because the entire solution is presented on one page, and the user is guided through the solution procedure. Let's summarize the use of this spreadsheet. As you read this, you should compare the spreadsheet entries with the details of the solution of Example Problem 19–2. The various formulas used there are programmed into the appropriate cells of the spreadsheet.

1. The process begins, as with virtually any spring design procedure, with the specification of the relationship between force and length for two separate conditions, typically called the *operating force-length* and the *installed force-length*. From these data, the spring rate is effectively specified as well. Sometimes the free length is known, which then equals the installed length, and the installed force equals zero. Also, the designer would know roughly the space into which the spring is to be installed.

2. The heading for the spreadsheet gives a brief overview of the design approach built into Method 1. The designer specifies a target mean diameter for the spring to fit a particular application. The spring material is specified, and the graph of the strength of that material is used as a guide for estimating the design stress. This requires the user to make a rough estimate also of the wire diameter, but a specific size is not yet chosen. The service (light, average, or severe) must also be specified at this time.

3. The spreadsheet then computes the resulting spring rate, the free length, and a trial wire diameter to produce an acceptable stress. Equation (19–7) is used to compute wire size.

4. The designer then enters a standard wire size, typically greater than the computed value. Table 19–2 is a listing of standard wire sizes. At this time, also, the designer must look again at the design stress graph for the chosen material and must determine a revised value for the design stress corresponding with the new wire diameter. The maximum allowable stress is also entered, found from the light service curve for the material at the specified wire size.

5. With the data entered, the spreadsheet completes the entire set of remaining calculations. The spring must be manufactured with exactly the specified diameters and number of active coils. The end condition for the spring, from the possibilities shown in Figure 19–3, should also be specified.

6. The designer's task is to evaluate the suitability of the results for basic geometry, stresses, potential for buckling, coil clearance, and installation of the spring into a hole. The **Comments** section to the right includes several prompts. But this is the designer's responsibility: to make judgments and design decisions.

DESIGN OF HELICAL COMPRESSION SPRING—METHOD 1
Specify mean diameter and design stresses. Compute wire diameter and number of coils.

1. Enter data for forces and lengths.
2. Specify material; shear modulus, G; and an estimate for design stress.
3. Enter trial mean diameter for spring, considering space available.
4. Check computed values for spring rate, free length, and new trial wire diameter.
5. Enter your choice for wire diameter of a standard size.
6. Enter design stress and maximum allowable stress from Figures 19–8 through 19–13 for new D_w.

Numerical values in italics in shaded areas must be inserted for each problem.	Problem ID:	Example Problem 19–2

Initial Input Data:			Comments
Maximum operating force =	$F_o =$	12.0 lb	
Operating length =	$L_o =$	1.25 in	
Installed force =	$F_i =$	8.0 lb	Note: $F_i = 0$ if:
Installed length =	$L_i =$	1.75 in	$L_i =$ free length
Trial mean diameter =	$D_m =$	0.60 in	
Type of spring wire:	ASTM A231 steel		See Figures 19–8–19–13.
Shear modulus of elasticity of spring wire =	$G =$	1.12E + 07 psi	From Table 19–4
Initial estimate of design stress =	$\tau_{di} =$	130 000 psi	From design stress graph

Computed Values:		
Computed spring rate =	$k =$	8.00 lb/in
Computed free length =	$L_f =$	2.75 in
Computed trial wire diameter =	$D_{wt} =$	0.055 in

Secondary Input Data:			
Standard wire diameter =	$D_w =$	0.0625 in	From Table 19–2
Design stress =	$\tau_d =$	145 000 psi	From design stress graph
Maximum allowable stress =	$\tau_{max} =$	170 000 psi	Use light service curve.

Computed Values:			
Outside diameter =	$D_o =$	0.663 in	
Inside diameter =	$D_i =$	0.538 in	
Number of active coils =	$N_a =$	12.36	
Spring index =	$C =$	9.60	**Should not be <5.0**
Wahl factor =	$K =$	1.15	
Stress at operating force =	$\tau_o =$	86,459 psi	**Cannot be >145,000**
Solid length =	$L_s =$	0.898 in	**Cannot be >1.25**
Force at solid length =	$F_s =$	14.82 lb	
Stress at solid length =	$\tau_s =$	106 768 psi	**Cannot be >170,000**

Check Buckling, Coil Clearance, and Hole Size:			
Buckling: Ratio =	$L_f/D_m =$	4.58	Check Figure 19–15 if >5.2.
Coil clearance =	$cc =$	0.029 in	Should be >0.00625
If installed in hole, minimum hole diameter =	$D_{hole} >$	0.669 in	For side clearance

FIGURE 19–16 Spreadsheet for Spring Design Method 1 of Example Problem 19–2

Notice the advantage of using a spreadsheet: The designer does the thinking, and the spreadsheet does the computing. For subsequent design iterations, only the values that change need to be entered. For example, if the designer wants to try a different wire diameter for the same spring material, only the three data values in the section called **Secondary Input Data** need to be changed, and a new result is produced immediately. Many design iterations can be completed in a short time with this approach.

You might see ways to enhance the utility of the spreadsheet, and you are encouraged to do that.

Spreadsheet for Spring Design Method 2

An alternative spring design method is shown in Figure 19–17. This method allows the designer more freedom in manipulating the parameters. The data used are basically the same as those of Example Problem 19–2 except that there is no requirement for installing the spring in a certain size hole. We refer to this modified problem statement as Example Problem 19–3.

Example Problem 19–3

The process is similar to that of Method 1, so only the major differences are discussed below. Follow along with the spreadsheet in Figure 19–17 as we describe Method 2, Trials 1 and 2.

Solution Method 2, Trial 1

1. The general procedure is outlined at the top of the spreadsheet. The designer selects a material, estimates the wire diameter as an initial trial, and inputs a corresponding estimate of the design stress.

2. The spreadsheet then computes a new trial wire diameter using a formula derived from the fundamental equation for shear stress in a helical compression spring, Equation (19–4). The development is described here.

$$\tau = \frac{8KFC}{\pi D_w^2} \tag{19–4}$$

Let $F = F_o$ and $\tau = \tau_d$ (the design stress). Solving for the wire diameter gives

$$D_w = \sqrt{\frac{8KF_oC}{\pi \tau_d}} \tag{19–9}$$

The values of K and C are not yet known, but a good estimate for the wire diameter can be computed if the spring index is assumed to be approximately 7.0, a reasonable value. The corresponding value for the Wahl factor is $K = 1.2$, from Equation (19–5). Combining these assumed values with the other constants in the preceding equation gives

$$D_w = \sqrt{21.4(F_o)/(\tau_d)} \tag{19–10}$$

This formula is programmed into the spreadsheet in the cell to the right of the one labeled D_{wt}, the computed trial wire diameter.

3. The designer then enters a standard wire size and determines revised values for the design stress and the maximum allowable stress from the graphs of material properties, Figures 19–8 through 19–13.

DESIGN OF HELICAL COMPRESSION SPRING—METHOD 2
Specify wire diameter, design stresses, and number of coils. Compute mean diameter.

1. Enter data for forces and lengths.
2. Specify material; shear modulus, G; and an estimate for design stress.
3. Enter trial wire diameter.
4. Check computed values for spring rate, free length, and new trial wire diameter.
5. Enter your choice for wire diameter of a standard size.
6. Enter design stress and maximum allowable stress from Figures 19–8 through 19–13 for new D_w.
7. Check computed maximum number of coils. Enter selected actual number of coils.

Numerical values in italics in shaded areas must be inserted for each problem.	Problem ID:	Example Problem 19–3, Trial 1:
Initial Input Data:		**Comments**
Maximum operating force = F_o =	12.0 lb	
Operating length = L_o =	1.25 in	
Installed force = F_i =	8.0 lb	Note: $F_i = 0$ if:
Installed length = L_i =	1.75 in	L_i = free length
Type of spring wire: ASTM A231 steel		See Figures 19–8–19–13.
Shear modulus of elasticity of spring wire = G =	1.12E + 07 psi	From Table 19–4
Initial estimate of design stress = τ_{di} =	144 000 psi	From design stress graph
Initial trial wire diameter = D_w =	0.06 in	
Computed Values:		
Computed spring rate = k =	8.00 lb/in	
Computed free length = L_f =	2.75 in	
Computed trial wire diameter = D_{wt} =	0.042 in	
Secondary Input Data:		
Standard wire diameter = D_w =	0.0475 in	From Table 19–2
Design stress = τ_d =	149 000 psi	From design stress graph
Maximum allowable stress = τ_{max} =	174 000 psi	Use light service curve.
Computed: Maximum number of coils = N_{max} =	24.32	
Input: Number of active coils = N_a =	22	Suggest using an integer.
Computed Values:		
Spring index = C =	7.23	**Should not be <5.0**
Wahl factor = K =	1.21	
Mean diameter = D_m =	0.343 in	
Outside diameter = D_o =	0.391 in	
Inside diameter = D_i =	0.296 in	
Solid length = L_s =	1.140 in	**Cannot be >1.25**
Stress at operating force = τ_o =	118 030 psi	**Cannot be >149 000**
Force at solid length = F_s =	12.88 lb	
Strength at solid length = τ_s =	126 685 psi	**Cannot be >174 000**
Check Buckling, Coil Clearance, and Hole Size:		
Buckling: Ratio = L_f/D_m =	8.01	Check Figure 19–15 if >5.2.
Coil clearance = cc =	0.005 0 in	Should be >0.00475
If installed in hole, minimum hole diameter = D_{hole} >	0.396 in	For side clearance

FIGURE 19–17   Spreadsheet for Spring Design Method 2 of Example Problem 19–3, Trial 1

4. The spreadsheet then computes the maximum permissible number of active coils for the spring. The logic here is that *the solid length must be less than the operating length*. The solid length is the product of the wire diameter and the total number of coils. For squared and ground ends, this is

$$L_s = D_w(N_a + 2)$$

Note that different relationships are used for the total number of coils for springs with other end conditions. See the discussion "Number of Coils" in Section 19–3. Now letting $L_s = L_o$ as a limit and solving for the number of coils gives

$$(N_a)_{max} = (L_o - 2D_w)/D_w \qquad\qquad \textbf{(19–11)}$$

This is the formula programmed into the spreadsheet cell to the right of the one labeled N_{max}.

5. The designer now has the freedom to choose any number of active coils less than the computed maximum value. Note the effects of that decision. Choosing a small number of coils will provide more clearance between adjacent coils and will use less wire per spring. However, the stresses produced for a given load will be higher, so there is a practical limit. One approach is to try progressively fewer coils until the stress approaches the design stress. Whereas any number of coils can be produced, even fractional numbers, we suggest trying integer values for the convenience of the manufacturer.

6. After the selected number of coils has been entered, the spreadsheet can complete the remaining calculations. One additional new formula is used in this spreadsheet to compute the value of the spring index, C. It is developed from the second form of Equation (19–6) relating the deflection for the spring, f, to a corresponding applied force, F, the value of C, and other parameters that are already known. First we solve for C^3:

Deflection of a Spring

$$f = \frac{8FD_m^3 N_a}{GD_w^4} = \frac{8FC^3 N_a}{GD_w}$$

$$C^3 = \frac{fGD_w}{8FN_a}$$

Now notice that we have the force F in the denominator and the corresponding deflection f in the numerator. But the spring rate k is defined as the ratio of F/f. Then we can substitute k into the denominator and solve for C:

$$C = \left[\frac{GD_w}{8kN_a}\right]^{1/3} \qquad\qquad \textbf{(19–12)}$$

This formula is programmed into the spreadsheet cell to the right of the one labeled $C =$.

7. Recall that C is defined as the ratio, D_m/D_w. We can now solve for the mean diameter:

$$D_m = CD_w$$

This is used to compute the mean diameter in the cell to the right of $D_m =$.

8. The remaining calculations use equations already developed and used earlier. Again, the designer is responsible for evaluating the suitability of the results and for performing any additional iterations to search for an optimum result.

Solution Method 2, Trial 2

Now notice that the solution obtained in Trial 1 and shown in Figure 19–17 is far from optimum. Its free length of 2.75 in is quite long compared with the mean diameter of 0.343 in. The buckling ratio of $L_f/D_m = 8.01$ indicates that the spring is long and slender. Checking Figure 19–15, we can see that buckling is predicted.

One way to work toward a more suitable geometry is to increase the wire diameter and reduce the number of coils. The net result will be a larger mean diameter, improving the buckling ratio.

Figure 19–18 shows the result of several iterations, finally using $D_w = 0.0625$ in (larger than the former value of 0.0475 in) and 16 active coils (down from 22 in the first trial). The buckling ratio is down to 4.99, indicating that buckling is unlikely. The stress at operating force is comfortably lower than the design stress. The other geometrical features also appear to be satisfactory.

This example should demonstrate the value of using spreadsheets or other computer-based computational aids. You should also have a better feel for the kinds of design decisions that can work toward a more optimum design. See Internet sites 1 and 2 for commercially available spring design software.

19–7 EXTENSION SPRINGS

Extension springs are designed to exert a pulling force and to store energy. They are made from closely coiled helical coils similar in appearance to helical compression springs. Most extension springs are made with adjacent coils touching in such a manner that an initial force must be applied to separate the coils. Once the coils are separated, the force is linearly proportional to the deflection, as it is for helical compression springs. Figure 19–19 shows a typical extension spring, and Figure 19–20 shows the characteristic type of load-deflection curve. By convention, the initial force is found by projecting the straight line portion of the curve back to zero deflection.

The stresses and deflections for an extension spring can be computed by using the formulas used for compression springs. Equation (19–4) is used for the torsional shear stress, Equation (19–5) for the Wahl factor to account for the curvature of the wire and the direct shear stress, and Equation (19–6) for the deflection characteristics. All coils in an extension spring are active. In addition, since the end loops or hooks deflect, their deflection may affect the actual spring rate.

The initial tension in an extension spring is typically 10% to 25% of the maximum design force. Figure 19–21 shows one manufacturer's recommendation of the preferred torsional stress due to initial tension as a function of the spring index.

End Configurations for Extension Springs

A wide variety of end configurations may be obtained for attaching the spring to mating machine elements, some of which are shown in Figure 19–4. The cost of the spring can be greatly affected by its end type, and it is recommended that the manufacturer be consulted before ends are specified.

DESIGN OF HELICAL COMPRESSION SPRING—METHOD 2
Specify wire diameter, design stresses, and number of coils. Compute mean diameter.

1. Enter data for forces and lengths.
2. Specify material; shear modulus, G; and an estimate for design stress.
3. Enter trial wire diameter.
4. Check computed values for spring rate, free length, and new trial wire diameter.
5. Enter your choice for wire diameter of a standard size.
6. Enter design stress and maximum allowable stress from Figures 19–8 through 19–13 for new D_w.
7. Check computed maximum number of coils. Enter selected actual number of coils.

Numerical values in italics in shaded areas must be inserted for each problem.	Problem ID:	*Example Problem 19–3, Trial 2: $D_w = 0.0625$*

Initial Input Data:			Comments
Maximum operating force =	$F_o =$	*12.0 lb*	
Operating length =	$L_o =$	*1.25 in*	
Installed force =	$F_i =$	*8.0 lb*	*Note: $F_i = 0$ if:*
Installed length =	$L_i =$	*1.75 in*	L_i = free length
Type of spring wire:		*ASTM A231 steel*	See Figures 19–8–19–13.
Shear modulus of elasticity of spring wire =	$G =$	*1.12E+07 psi*	From Table 19–4
Initial estimate of design stress =	$\tau_{di} =$	*144 000 psi*	From design stress graph
Initial trial wire diameter =	$D_w =$	*0.06 in*	

Computed Values:		
Computed spring rate =	$k =$	8.00 lb/in
Computed free length =	$L_f =$	2.75 in
Computed trial wire diameter =	$D_{wt} =$	0.042 in

Secondary Input Data:			
Standard wire diameter =	$D_w =$	*0.0625 in*	From Table 19–2
Design stress =	$\tau_d =$	*142 600 psi*	From design stress graph
Maximum allowable stress =	$\tau_{max} =$	*167 400 psi*	Use light service curve.

Computed: Maximum number of coils =	$N_{max} =$	18.00	

Input: Number of active coils =	$N_a =$	*16*	Suggest using an integer.

Computed Values:			
Spring index =	$C =$	8.81	**Should not be <5.0**
Wahl factor =	$K =$	1.17	
Mean diameter =	$D_m =$	0.551 in	
Outside diameter =	$D_o =$	0.613 in	
Inside diameter =	$D_i =$	0.488 in	
Solid length =	$L_s =$	1.125 in	**Cannot be >1.25**
Stress at operating force =	$\tau_o =$	80 341 psi	**Cannot be >142 600**
Force at solid length =	$F_s =$	13.00 lb	
Stress at solid length =	$\tau_s =$	87 036 psi	**Cannot be >167 400**

Check Buckling, Coil Clearance, and Hole Size:			
Buckling: Ratio =	$L_f/D_m =$	4.99	Check Figure 19–15 if >5.2.
Coil clearance =	$cc =$	0.0078 in	Should be >0.00625
If installed in hole, minimum hole diameter =	$D_{hole} >$	0.619 in	For side clearance

FIGURE 19–18 Spreadsheet for Spring Design Method 2 of Example Problem 19–3, Trial 2

FIGURE 19–19
Extension spring

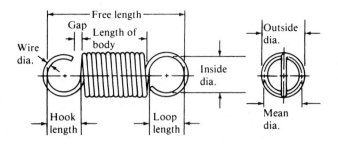

FIGURE 19–20
Load-deflection curve
for an extension spring

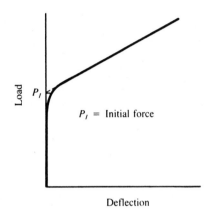

FIGURE 19–21
Recommended
torsional shear stress in
an extension spring due
to initial tension (Data
from Associated
Spring, Barnes Group,
Inc.)

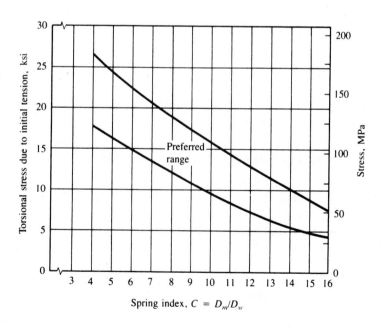

Frequently, the weakest part of an extension spring is its end, especially in fatigue loading cases. The loop end shown in Figure 19–22, for example, has a high bending stress at point A and a torsional shear stress at point B. Approximations for the stresses at these points can be computed as follows:

FIGURE 19–22
Stresses in ends of
extension spring

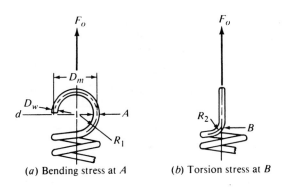

(a) Bending stress at A   (b) Torsion stress at B

Bending Stress at A

$$\sigma_A = \frac{16D_mF_oK_1}{\pi D_w^3} + \frac{4F_o}{\pi D_w^2} \qquad (19\text{--}13)$$

$$K_1 = \frac{4C_1^2 - C_1 - 1}{4C_1(C_1 - 1)} \qquad (19\text{--}14)$$

$$C_1 = 2R_1/D_w$$

Torsional Stress at B

$$\tau_B = \frac{8D_mF_oK_2}{\pi D_w^3} \qquad (19\text{--}15)$$

$$K_2 = \frac{4C_2 - 1}{4C_2 - 4} \qquad (19\text{--}16)$$

$$C_2 = 2R_2/D_w$$

The ratios C_1 and C_2 relate to the curvature of the wire and should be large, typically greater than 4, to avoid high stresses.

Allowable Stresses for Extension Springs

The torsional shear stress in the coils of the spring and in the end loops can be compared to the curves in Figures 19–8 through 19–13. Some designers reduce these allowable stresses by about 10%. The bending stress in the end loops, such as that from Equation (19–13), should be compared to the allowable bending stresses for torsion springs, as presented in the next section.

Example Problem 19–4   A helical extension spring is to be installed in a latch for a large, commercial laundry machine. When the latch is closed, the spring must exert a force of 16.25 lb at a length between attachment points of 3.50 in. As the latch is opened, the spring is pulled to a length of 4.25 in with a maximum force of 26.75 lb. An outside diameter of 5/8 in (0.625 in) is desired. The latch will be cycled only about 10 times per day, and thus the design stress will be based on average service. Use ASTM A227 steel wire. Design the spring.

Solution   As before, the suggested design procedure will be given as numbered steps, followed by the calculations for this set of data.

Step 1. Assume a trial mean diameter and a trial design stress for the spring.

Let the mean diameter be 0.500 in. For ASTM A227 wire under average service, a design stress of 110 000 psi is reasonable (from Figure 19–8).

Step 2. Compute a trial wire diameter from Equation (19–4) for the maximum operating force, the assumed mean diameter and design stress, and an assumed value for K of about 1.20.

$$D_w = \left[\frac{8KF_oD_m}{\pi\tau_d} \right]^{1/3} = \left[\frac{(8)(1.20)(26.75)(0.50)}{(\pi)(110\,000)} \right]^{1/3} = 0.072 \text{ in}$$

A standard wire size in the U.S. Steel Wire Gage system is available in this size. Use 15-gage.

Step 3. Determine the actual design stress for the selected wire size.

From Figure 19–8, at a wire size of 0.072 in, the design stress is 120 000 psi.

Step 4. Compute the actual values for outside diameter, mean diameter, inside diameter, spring index, and the Wahl factor, K. These factors are the same as those defined for helical compression springs.

Let the outside diameter be as specified, 0.625 in. Then

$$D_m = OD - D_w = 0.625 - 0.072 = 0.553 \text{ in}$$
$$ID = OD - 2D_w = 0.625 - 2(0.072) = 0.481 \text{ in}$$
$$C = D_m/D_w = 0.553/0.072 = 7.68$$
$$K = \frac{4C - 1}{4C - 4} + \frac{0.615}{C} = \frac{4(7.68) - 1}{4(7.68) - 4} + \frac{0.615}{7.68} = 1.19$$

Step 5. Compute the actual expected stress in the spring wire under the operating load from Equation (19–4):

$$\tau_o = \frac{8KF_oD_m}{\pi D_w^3} = \frac{8(1.19)(26.75)(0.553)}{\pi(0.072)^3} = 120\,000 \text{ psi} \qquad \text{(okay)}$$

Step 6. Compute the required number of coils to produce the desired deflection characteristics. Solve Equation (19–6) for the number of coils, and substitute $k = $ force/deflection $= F/f$:

$$k = \frac{26.75 - 16.25}{4.25 - 3.50} = 14.0 \text{ lb/in}$$

$$N_a = \frac{GD_w}{8C^3k} = \frac{(11.5 \times 10^6)(0.072)}{8(7.68)^3(14.0)} = 16.3 \text{ coils}$$

Step 7. Compute the body length for the spring, and propose a trial design for the ends.

$$\text{Body length} = D_w(N_a + 1) = (0.072)(16.3 + 1) = 1.25 \text{ in}$$

Let's propose to use a full loop at each end of the spring, adding a length equal to the *ID* of the spring at each end. Then the total free length is

$$L_f = \text{body length} + 2(ID) = 1.25 + 2(0.481) = 2.21 \text{ in}$$

Step 8. Compute the deflection from free length to operating length:

$$f_o = L_o - L_f = 4.25 - 2.21 = 2.04 \text{ in}$$

Step 9. Compute the initial force in the spring at which the coils just begin to separate. This is done by subtracting the amount of force due to the deflection, f_o:

$$F_I = F_o - kf_o = 26.75 - (14.0)(2.04) = -1.81 \text{ lb}$$

The negative force resulting from a free length that is too small for the specified conditions is clearly impossible.

Let's try $L_f = 2.50$ in, which will require a redesign of the end loops. Then

$$f_o = 4.25 - 2.50 = 1.75 \text{ in}$$
$$F_I = 26.75 - (14.0)(1.75) = 2.25 \text{ lb} \quad \text{(reasonable)}$$

Step 10. Compute the stress in the spring under the initial tension, and compare it with the recommended levels in Figure 19–21.

Because the stress is proportional to the load,

$$\tau_I = \tau_o(F_I/F_o) = (120\ 000)(2.25/26.75) = 10\ 100 \text{ psi}$$

For $C = 7.68$, this stress is in the preferred range from Figure 19–21.

At this point, the coiled portion of the spring is satisfactory. The final configuration of the end loops must be completed and analyzed for stress.

19–8
HELICAL
TORSION
SPRINGS

Many machine elements require a spring that exerts a rotational moment, or torque, instead of a push or a pull force. The helical torsion spring is designed to satisfy this requirement. The spring has the same general appearance as either the helical compression or the extension spring, with round wire wrapped into a cylindrical shape. Usually the coils are close together but with a small clearance, allowing no initial tension in the spring as there is for extension springs. Figure 19–23 shows a few examples of torsion springs with a variety of end treatments.

The familiar clip-type clothespin uses a torsion spring to provide the gripping force. Many cabinet doors are designed to close automatically under the influence of a torsion spring. Some timers and switches use torsion springs to actuate mechanisms or close contacts. Torsion springs often provide counterbalancing of machine elements that are mounted to a hinged plate.

FIGURE 19–23
Torsion springs
showing a variety
of end types

The following are some of the special features and design guides for torsion springs:

1. The moment applied to a torsion spring should always act in a direction that causes the coils to wind more tightly, rather than to open the spring up. This takes advantage of favorable residual stresses in the wire after forming.

2. In the free (no-load) condition, the definition of mean diameter, outside diameter, inside diameter, wire diameter, and spring index are the same as those used for compression springs.

3. As the load on a torsion spring increases, its mean diameter, D_m, decreases, and its length, L, increases, according to the following relations:

$$D_m = D_{mI} N_a/(N_a + \theta) \tag{19-17}$$

where D_{mI} = initial mean diameter at the free condition
N_a = number of active coils in the spring (to be defined later)
θ = angular deflection of the spring from the free condition, expressed in revolutions or a fraction of a revolution

$$L = D_w (N_a + 1 + \theta) \tag{19-18}$$

This equation assumes that all coils are touching. If any clearance is provided, as is often desirable to reduce friction, a length of N_a times the clearance must be added.

4. Torsion springs must be supported at three or more points. They are usually installed around a rod to provide location and to transfer reaction forces to the structure. The rod diameter should be approximately 90% of the *ID* of the spring at the maximum load.

Stress Calculations

The stress in the coils of a helical torsion spring is *bending stress* created because the applied moment tends to bend each coil into a smaller diameter. Thus, the stress is computed from a form of the flexure formula, $\sigma = Mc/I$, modified to account for the curved wire. Also, because most torsion springs are made from round wire, the section modulus I/c is $S = \pi D_w^3/32$. Then

$$\sigma = \frac{McK_b}{I} = \frac{MK_b}{S} = \frac{MK_b}{\pi D_w^3/32} = \frac{32MK_b}{\pi D_w^3} \tag{19-19}$$

K_b is the curvature correction factor and is reported by Wahl (see Reference 9) to be

$$K_b = \frac{4C^2 - C - 1}{4C(C - 1)} \tag{19-20}$$

where C = spring index

Deflection, Spring Rate, and Number of Coils

The basic equation governing deflection is

$$\theta' = ML_w/EI$$

where θ' = angular deformation of the spring in radians (rad)

M = applied moment, or torque

L_w = length of wire in the spring

E = tensile modulus of elasticity

I = moment of inertia of the spring wire

We can substitute equations for L_w and I and convert θ' (radians) to θ (revolutions) to produce a more convenient form for application to torsion springs:

$$\theta = \frac{ML_w}{EI} = \frac{M(\pi D_m N_a)}{E(\pi D_w^4/64)} \frac{1 \text{ rev}}{2\pi \text{ rad}} = \frac{10.2 M D_m N_a}{E D_w^4} \tag{19-21}$$

To compute the spring rate, k_θ (moment per revolution), solve for M/θ:

$$k_\theta = \frac{M}{\theta} = \frac{E D_w^4}{10.2 D_m N_a} \tag{19-22}$$

Friction between coils and between the ID of the spring and the guide rod may decrease the rate slightly from this value.

The number of coils, N_a, is composed of a combination of the number of coils in the body of the spring, called N_b, and the contribution of the ends as they are subjected to bending, also. Calling N_e the contribution of the ends having lengths L_1 and L_2, we have

$$N_e = (L_1 + L_2)/(3\pi D_m) \tag{19-23}$$

Then compute $N_a = N_b + N_e$.

Design Stresses

Because the stress in a torsion spring is bending and not torsional shear, the design stresses are different from those used for compression and extension springs. Figures 19–24 through 19–29 include six graphs of the design stress versus wire diameter for the same alloys used earlier for compression springs.

Design Procedure for Torsion Springs

A general procedure for designing torsion springs is presented and illustrated with Example Problem 19–5.

Example Problem 19–5   A timer incorporates a mechanism to close a switch after the timer rotates one complete revolution. The switch contacts are actuated by a torsion spring that is to be designed. A cam on the timer shaft slowly moves a lever attached to one end of the spring to a point where the maximum torque on the spring is 3.00 lb·in. At the end of the revolution, the cam permits the lever to rotate 60° suddenly with the motion produced by the energy stored in the spring. At this new position, the torque on the spring is 1.60 lb·in. Because of space limitations, the OD of the spring should not be greater than 0.50 in, and the length should not be greater than 0.75 in. Use music wire, ASTM A228 steel wire. The number of cycles of the spring will be moderate, so use design stresses for average service.

FIGURE 19–24
Design bending
stresses for torsion
springs ASTM A227
steel wire, hard-drawn
(Reprinted from Harold
Carlson, *Spring
Designer's Handbook*,
p. 144, by courtesy of
Marcel Dekker, Inc.)

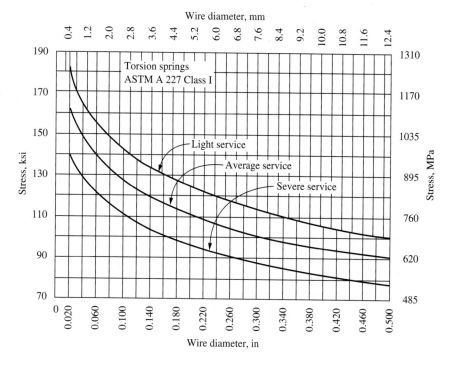

FIGURE 19–25
Design bending
stresses for torsion
springs ASTM A228
music wire (Reprinted
from Harold Carlson,
*Spring Designer's
Handbook*, p. 143, by
courtesy of Marcel
Dekker, Inc.)

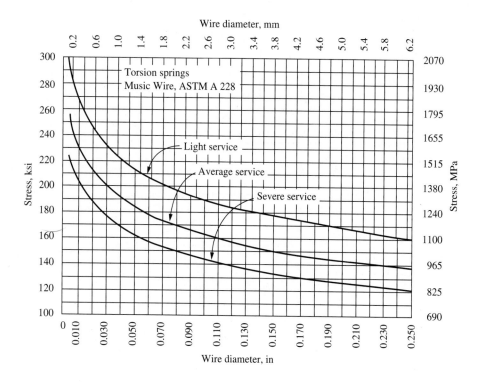

FIGURE 19–26

Design bending stresses for torsion springs ASTM A229 steel wire, oil-tempered, MB grade (Reprinted from Harold Carlson, *Spring Designer's Handbook*, p. 146, by courtesy of Marcel Dekker, Inc.)

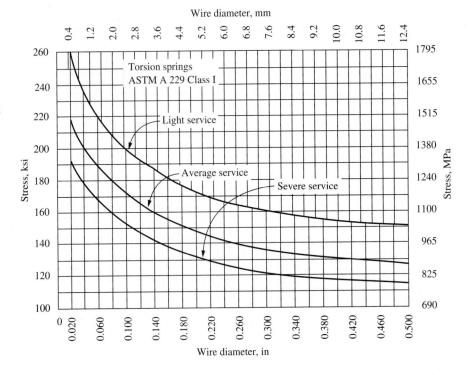

FIGURE 19–27

Design bending stresses for torsion springs ASTM A231 steel wire, chromium-vanadium alloy, valve–spring quality (Reprinted from Harold Carlson, *Spring Designer's Handbook*, p. 147, by courtesy of Marcel Dekker, Inc.)

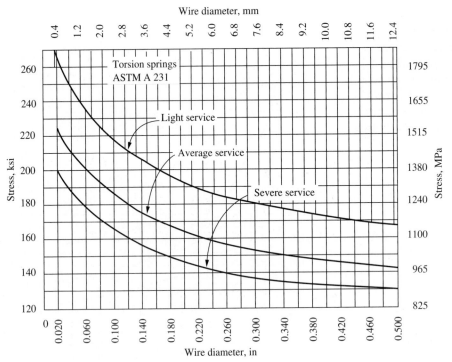

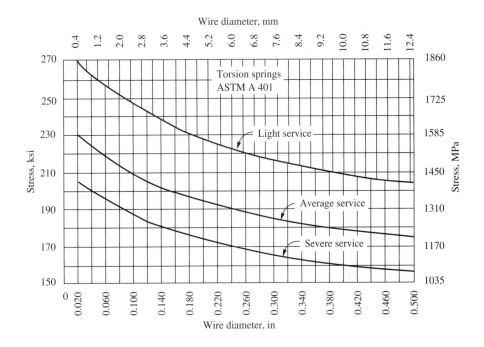

FIGURE 19–28
Design bending stresses for torsion springs ASTM A401 steel wire, chromium-silicon alloy, oil-tempered (Reprinted from Harold Carlson, *Spring Designer's Handbook*, p. 148, by courtesy of Marcel Dekker, Inc.)

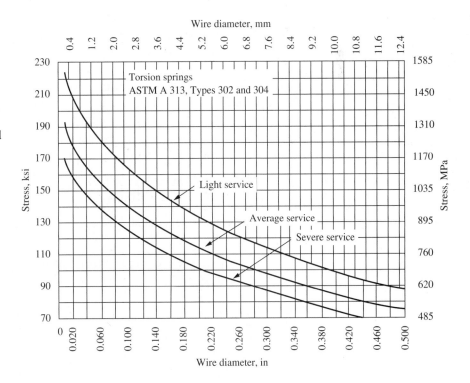

FIGURE 19–29
Design bending stresses for torsion springs ASTM A313 stainless steel wire, types 302 and 304, corrosion-resistant (Reprinted from Harold Carlson, *Spring Designer's Handbook*, p. 150, by courtesy of Marcel Dekker, Inc.)

Solution   *Step 1.* Assume a trial value for the mean diameter and an estimate for the design stress. Let's use a mean diameter of 0.400 in and estimate the design stress for A228 music wire, average service, to be 180 000 psi (Figure 19–25).

Step 2. Solve Equation (19–19) for the wire diameter, compute a trial size, and select a standard wire size. Let $K_b = 1.15$ as an estimate. Also use the largest applied torque:

$$D_w = \left[\frac{32MK_b}{\pi\sigma_d}\right]^{1/3} = \left[\frac{32(3.0)(1.15)}{\pi(180\ 000)}\right]^{1/3} = 0.058 \text{ in}$$

From Table 19–2, we can choose 25-gage music wire with a diameter of 0.059 in. For this size wire, the actual design stress for average service is 178 000 psi.

Step 3. Compute the *OD*, *ID*, spring index, and the new K_b:

$$OD = D_m + D_w = 0.400 + 0.059 = 0.459 \text{ in}\quad\text{(okay)}$$
$$ID = D_m - D_w = 0.400 - 0.059 = 0.341 \text{ in}$$
$$C = D_m/D_w = 0.400/0.059 = 6.78$$
$$K_b = \frac{4C^2 - C - 1}{4C(C - 1)} = \frac{4(6.78)^2 - 6.78 - 1}{4(6.78)(6.78 - 1)} = 1.123$$

Step 4. Compute the actual expected stress from Equation (19–19):

$$\sigma = \frac{32MK_b}{\pi D_w^3} = \frac{32(3.0)(1.123)}{(\pi)(0.059)^3} = 167\ 000 \text{ psi}\quad\text{(okay)}$$

Step 5. Compute the spring rate from the given data.

The torque exerted by the spring decreases from 3.00 to 1.60 lb·in as the spring rotates 60°. Convert 60° to a fraction of a revolution (rev):

$$\theta = \frac{60}{360} = 0.167 \text{ rev}$$
$$k_\theta = \frac{M}{\theta} = \frac{3.00 - 1.60}{0.167} = 8.40 \text{ lb·in/rev}$$

Step 6. Compute the required number of coils by solving for N_a from Equation (19–22):

$$N_a = \frac{ED_w^4}{10.2D_m k_\theta} = \frac{(29 \times 10^6)(0.059)^4}{(10.2)(0.400)(8.40)} = 10.3 \text{ coils}$$

Step 7. Compute the equivalent number of coils due to the ends of the spring from Equation (19–23).

This requires some design decisions. Let's use straight ends, 2.0 in long on one side and 1.0 in long on the other. These ends will be attached to the structure of the timer during operation. Then

$$N_e = (L_1 + L_2)/(3\pi\ D_m) = (2.0 + 1.0)/[3\pi\ (0.400)] = 0.80 \text{ coil}$$

Step 8. Compute the required number of coils in the body of the spring:

$$N_b = N_a - N_e = 10.3 - 0.8 = 9.5 \text{ coils}$$

Step 9. Complete the geometric design of the spring, including the size of the rod on which it will be mounted.

We first need the total angular deflection for the spring from the free condition to the maximum load. In this case, we know that the spring rotates 60° during operation. To this we must add the rotation from the free condition to the initial torque, 1.60 lb·in. Thus,

$$\theta_I = M_I/k_\theta = 1.60 \text{ lb·in}/(8.4 \text{ lb·in/rev}) = 0.19 \text{ rev}$$

Then the total rotation is

$$\theta_t = \theta_I + \theta_o = 0.19 + 0.167 = 0.357 \text{ rev}$$

From Equation (19–15), the mean diameter at the maximum operating torque is

$$D_m = D_{ml} N_a/(N_a + \theta_t) = [(0.400)(10.3)]/[(10.3 + 0.357)] = 0.387 \text{ in}$$

The minimum inside diameter is

$$ID_{\min} = 0.387 - D_w = 0.387 - 0.059 = 0.328 \text{ in}$$

The rod diameter on which the spring is mounted should be approximately 0.90 times this value. Then

$$D_r = 0.9(0.328) = 0.295 \text{ in} \qquad (\text{say, } 0.30 \text{ in})$$

The length of the spring, assuming that all coils are originally touching, is computed from Equation (19–18):

$$L_{\max} = D_w (N_a + 1 + \theta_t) = (0.059)(10.3 + 1 + 0.356) = 0.688 \text{ in} \qquad (\text{okay})$$

This value for length is the maximum space required in the direction along the axis of the coil when the spring is fully actuated. The specifications allow an axial length of 0.75 in, so this design is acceptable.

19–9 IMPROVING SPRING PERFORMANCE BY SHOT PEENING

The data presented in this chapter for spring wire materials are for commercially available spring wire with good surface finish. Care should be exercised to avoid nicks and scratches on the spring wire that may present sites for initiating fatigue cracks. When forming ends of extension springs or when creating other special geometry, bend radii should be as large as practical to avoid regions of high residual stress after the bending process.

Critical applications may call for the use of *shot peening* to enhance the fatigue performance of springs and other machine elements. Examples are valve springs for engines, rapid cycling springs in automation equipment, clutch springs, aerospace applications, medical equipment, gears, pump impellers, and military systems. Shot peening is a process in which small hard pellets, called shot, are directed with high velocity at the surface to be treated. The shot causes small local plastic deformation to occur near the surface. The material below the deformed area is subsequently placed in compression as it tries to return the surface to its original shape. High residual compressive stresses are produced at the surface, a favorable situation. Fatigue cracks typically initiate at points of high tensile stress. Therefore, the residual compressive stress tends to preclude such cracks from starting, and the fatigue strength of the material is significantly increased. See Internet site 12.

**19–10
SPRING
MANUFACTURING**

Spring forming machines are fascinating examples of high speed, flexible, programmable, multiaxis, multifunction devices. Internet site 6 shows images of several types that include wire feeding systems, coiling wheels, and multiple forming heads. The systems use CNC (computer numerical control) to permit rapid setup and adjustment to produce automatically the varied geometries for the body and ends of springs such as those shown in Figures 19–2 and 19–4. Furthermore, other complex shapes can be made for wire forms used for special clips, clamps, and brackets.

Accessories to the spring forming machines are:

- Systems to handle the large coils of raw wire, uncoiling it smoothly as needed
- Straighteners that remove the curved shape of the coiled wire prior to forming the spring
- Grinders to form the squared and ground ends of compression springs
- Heat treating ovens that perform stress relieving on springs immediately after forming so they will maintain the specified geometry
- Shot peening equipment as described in Section 19–9

REFERENCES

1. Associated Spring, Barnes Group, Inc. *Engineering Guide to Spring Design.* Bristol, CT: Associated Spring, 1987.

2. Carlson, Harold. *Spring Designer's Handbook.* New York: Marcel Dekker, 1978.

3. Carlson, Harold. *Springs-Troubleshooting and Failure Analysis.* New York: Marcel Dekker, 1980.

4. Faires, V. M. *Design of Machine Elements.* 4th ed. New York: Macmillan, 1965.

5. Oberg, E., et al. *Machinery's Handbook.* 26th ed. New York: Industrial Press, 2000.

6. Shigley, J. E., and C. R. Mischke. *Mechanical Engineering Design.* 6th ed. New York: McGraw-Hill, 2001.

7. Society of Automotive Engineers. *Spring Design Manual.* 2d ed. Warrendale, PA: Society of Automotive Engineers, 1995.

8. Spring Manufacturers Institute. *Handbook of Spring Design.* Oak Brook, IL: Spring Manufacturers Institute, 2002.

9. Wahl, A. M. *Mechanical Springs.* New York: McGraw-Hill, 1963.

INTERNET SITES RELEVANT TO SPRING DESIGN

1. **Spring Manufacturers Institute.** *www.smihq.org* Industrial association serving spring manufacturers and providers of technical publications, spring design software, education, and a variety of services. The software called *Advance Spring Design* facilitates the design of numerous types of springs, including compression, extension, torsion, and many others.

2. **Institute of Spring Technology.** *www.ist.org.uk* Provides a wide range of research, development, problem solving, failure analysis, training, CAD software, materials and mechanical testing for the spring industry.

3. **Associated Spring–Barnes Group, Inc.** *www.asbg.com* A large manufacturer of precision springs for industries such as transportation, telecommunications, electronics, home appliances, and farm equipment.

4. **Associated Spring Raymond.** *www.asraymond.com* Producer of a wide variety of springs and related components. Online catalog of springs from the SPEC line of stock compression, extension, and torsion springs, spring washers, gas springs, constant force springs, and urethane springs.

5. **Century Spring Corporation.** *www.centuryspring.com* Producer of a wide variety of springs. Online catalog of stock compression, extension, torsion, and die springs, urethane springs, and drawbar springs.

6. **Unidex Machinery Company.** *www.unidexmachinery. com* Manufacturer of CNC spring forming machinery, coiling machines, and related equipment.

7. **Oriimec Corporation of America.** *www.oriimec.com* Manufacturer of multiaxis CNC spring forming machinery, CNC coiling machines, and tension spring machines.

8. **American Spring Wire Corporation.**
   www.americanspringwire.com   Manufacturer of valve grade and commercial quality spring wire in carbon and alloy steels. Also produces 7-strand wire for high tensile load applications such as pre-tensioned or post-tensioned concrete structures.

9. **Mapes Piano String Company.** *www.mapeswire.com* Manufacturer of spring wire, piano wire, and speciality wire for guitar strings. Spring wires include music wire (plain and coated), high tensile missile wire, stainless steel, and others.

10. **Little Falls Alloys, Inc.** *www.littlefallsalloys.com* Manufacturer of nonferrous wire such as beryllium copper, brass, phosphor bronze, nickel, and cupro

nickel. Product listings include mechanical and physical properties data.

11. **Alloy Wire International.** *www.alloywire.com* Manufacturer of round and shaped wire in superalloys such as Inconel®, Incoloy®, Monel® (trademarks of Special Metals Group of Companies), hastelloy, and titanium.

12. **Metal Improvement Company.**
    www.metalimprovement.com   Provider of shot peening services to aerospace, automotive, chemical, marine, agriculture, mining, and medical industries. Shot peening improves the fatigue performance of springs, gears, and many other products. The website includes discussions of the technical aspects of shot peening.

PROBLEMS

Compression Springs

1. A spring has an overall length of 2.75 in when it is not loaded and a length of 1.85 in when carrying a load of 12.0 lb. Compute its spring rate.

2. A spring is loaded initially with a load of 4.65 lb and has a length of 1.25 in. The spring rate is given to be 18.8 lb/in. What is the free length of the spring?

3. A spring has a spring rate of 76.7 lb/in. At a load of 32.2 lb, it has a length of 0.830 in. Its solid length is 0.626 in. Compute the force required to compress the spring to solid height. Also compute the free length of the spring.

4. A spring has an overall length of 63.5 mm when it is not loaded and a length of 37.1 mm when carrying a load of 99.2 N. Compute its spring rate.

5. A spring is loaded initially with a load of 54.05 N, and it has a length of 39.47 mm. The spring rate is given to be 1.47 N/mm. What is the free length of the spring?

6. A spring has a spring rate of 8.95 N/mm. At a load of 134 N, it has a length of 29.4 mm. Its solid length is 21.4 mm. Compute the force required to compress the spring to solid height. Also compute the free length of the spring.

7. A helical compression spring with squared and ground ends has an outside diameter of 1.100 in, a wire diameter of 0.085 in, and a solid height of 0.563 in. Compute the *ID*, the mean diameter, the spring index, and the approximate number of coils.

8. The following data are known for a spring:

   Total number of coils = 19

   Squared and ground ends

   Outside diameter = 0.560 in

   Wire diameter = 0.059 in (25-gage music wire)

   Free length = 4.22 in

For this spring, compute the spring index, the pitch, the pitch angle, and the solid length.

9. For the spring of Problem 8, compute the force required to reduce its length to 3.00 in. At that force, compute the stress in the spring. Would that stress be satisfactory for average service?

10. Would the spring of Problems 8 and 9 tend to buckle when compressed to the length of 3.00 in?

11. For the spring of Problem 8, compute the estimate for the outside diameter when it is compressed to solid length.

12. The spring of Problem 8 must be compressed to solid length to install it. What force is required to do this? Compute the stress at solid height. Is that stress satisfactory?

13. A support bar for a machine component is suspended to soften applied loads. During operation, the load on each spring varies from 180 to 220 lb. The position of the rod is to move no more than 0.500 in as the load varies. Design a compression spring for this application. Several million cycles of load application are expected. Use ASTM A229 steel wire.

14. Design a helical compression spring to exert a force of 22.0 lb when compressed to a length of 1.75 in. When its length is 3.00 in, it must exert a force of 5.0 lb. The spring will be cycled rapidly, with severe service required. Use ASTM A401 steel wire.

15. Design a helical compression spring for a pressure relief valve. When the valve is closed, the spring length is 2.0 in, and the spring force is to be 1.50 lb. As the pressure on the valve increases, a force of 14.0 lb causes the valve to open and compress the spring to a length of 1.25 in. Use corrosion-resistant ASTM A313, type 302 steel wire, and design for average service.

16. Design a helical compression spring to be used to return a pneumatic cylinder to its original position after being actuated. At a length of 10.50 in, the spring must exert a force of 60 lb. At a length of 4.00 in, it must exert a force of 250 lb. Severe service is expected. Use ASTM A231 steel wire.

17. Design a helical compression spring using music wire that will exert a force of 14.0 lb when its length is 0.68 in. The free length is to be 1.75 in. Use average service.

18. Design a helical compression spring using stainless steel wire, ASTM A313, type 316, for average service, which will exert a force of 8.00 lb after deflecting 1.75 in from a free length of 2.75 in.

19. Repeat Problem 18 with the additional requirement that the spring must operate around a rod having a diameter of 0.625 in.

20. Repeat Problem 17 with the additional requirement that the spring is to be installed in a hole having a diameter of 0.750 in.

21. Design a helical compression spring using ASTM A231 steel wire for severe service. The spring will exert a force of 45.0 lb at a length of 3.05 in and a force of 22.0 lb at a length of 3.50 in.

22. Design a helical compression spring using round steel wire, ASTM A227. The spring will actuate a clutch and must withstand several million cycles of operation. When the clutch discs are in contact, the spring will have a length of 2.50 in and must exert a force of 20 lb. When the clutch is disengaged, the spring will be 2.10 in long and must exert a force of 35 lb. The spring will be installed around a round shaft having a diameter of 1.50 in.

23. Design a helical compression spring using round steel wire, ASTM A227. The spring will actuate a clutch and must withstand several million cycles of operation. When the clutch discs are in contact, the spring will have a length of 60 mm and must exert a force of 90 N. When the clutch is disengaged, the spring will be 50 mm long and must exert a force of 155 N. The spring will be installed around a round shaft having a diameter of 38 mm.

24. Evaluate the performance of a helical compression spring made from 17-gage ASTM A229 steel wire that has an outside diameter of 0.531 in. It has a free length of 1.25 in, squared and ground ends, and a total of 7.0 coils. Compute the spring rate and the deflection and stress when carrying 10.0 lb. At this stress level, for what service (light, medium, or severe) would the spring be suitable?

Extension Springs

For Problems 25–31, be sure that the stress in the spring under initial tension is within the range suggested in Figure 19–21.

25. Design a helical extension spring using music wire to exert a force of 7.75 lb when the length between attachment points is 2.75 in, and a force of 5.25 lb at a length of 2.25 in. The outside diameter must be less than 0.300 in. Use severe service.

26. Design a helical extension spring for average service using music wire to exert a force of 15.0 lb when the length between attachment points is 5.00 in, and a force of 5.20 lb at a length of 3.75 in. The outside diameter must be less than 0.75 in.

27. Design a helical extension spring for severe service using music wire to exert a maximum force of 10.0 lb at a length of 3.00 in. The spring rate should be 6.80 lb/in. The outside diameter must be less than 0.75 in.

28. Design a helical extension spring for severe service using music wire to exert a maximum force of 10.0 lb at a length of 6.00 in. The spring rate should be 2.60 lb/in. The outside diameter must be less than 0.75 in.

29. Design a helical extension spring for average service using music wire to exert a maximum force of 10.0 lb at a length of 9.61 in. The spring rate should be 1.50 lb/in. The outside diameter must be less than 0.75 in.

30. Design a helical extension spring for average service using stainless steel wire, ASTM A313, type 302, to exert a maximum force of 162 lb at a length of 10.80 in. The spring rate should be 38.0 lb/in. The outside diameter should be approximately 1.75 in.

31. An extension spring has an end similar to that shown in Figure 19–22. The pertinent data are as follows: U.S. Wire Gage no. 19; mean diameter = 0.28 in; $R_1 = 0.25$ in; $R_2 = 0.094$ in. Compute the expected stresses at points A and B in the figure for a force of 5.0 lb. Would those stresses be satisfactory for ASTM A227 steel wire for average service?

Torsion Springs

32. Design a helical torsion spring for average service using stainless steel wire, ASTM A313, type 302, to exert a torque of 1.00 lb·in after a deflection of 180° from the free condition. The outside diameter of the coil should be no more than 0.500 in. Specify the diameter of a rod on which to mount the spring.

33. Design a helical torsion spring for severe service using stainless steel wire, ASTM A313, type 302, to exert a torque of 12.0 lb·in after a deflection of 270° from the free condition. The outside diameter of the coil should be no more than 1.250 in. Specify the diameter of a rod on which to mount the spring.

34. Design a helical torsion spring for severe service using music wire to exert a maximum torque of 2.50 lb·in after a deflection of 360° from the free condition. The outside diameter of the coil should be no more than 0.750 in. Specify the diameter of a rod on which to mount the spring.

35. A helical torsion spring has a wire diameter of 0.038 in; an outside diameter of 0.368 in; 9.5 coils in the body; one end 0.50 in long; the other end 1.125 in long; and a material of ASTM A401 steel. What torque would cause the spring to rotate 180°? What would the stress be then? Would it be safe?

20

Machine Frames, Bolted Connections, and Welded Joints

The Big Picture

You Are the Designer

20–1 Objectives of This Chapter

20–2 Machine Frames and Structures

20–3 Eccentrically Loaded Bolted Joints

20–4 Welded Joints

Machine Frames, Bolted Connections, and Welded Joints

☐ As you develop the design of machine elements (as you have throughout this book), you must also design the housing, frame, or structure that supports them and protects them from the elements.

Discover

Select a variety of products, machines, vehicles, and even toys. Observe how they are built. What is the basic shape of the structure that holds everything together? Why was that shape chosen by the designer? What functions are performed by the frame?

What kinds of forces, bending moments, and torsional moments (torques) are produced when the product is operating? How are they managed and controlled? What is the load path that delivers them to the ultimate structure?

This chapter will help you identify some efficient approaches to the design of structures and frames and to analyze the performance of fasteners and welded joints loaded in many different ways.

Throughout this book, you have studied individual machine elements while continuously considering how those elements must work together in a more comprehensive machine. As the design progresses, there comes a time when *you must put it all together.* But then you are faced with the questions, "What do I put it in? How do I hold all of the functional components safely, allowing assembly and service while providing a secure, rigid structure?"

It is impractical to generate a completely general approach to the design of a structure or a frame for a machine, a vehicle, a consumer product, or even a toy. Each is different with regard to its functions; the number, size, and type of components in the product; the intended use; and the expected demand for an aesthetic design. For example, toys often exhibit clever design approaches because the manufacturer wants to provide a safe, functional toy while minimizing the material used and the amount of personnel time required to produce the toy.

In this chapter you will explore some basic concepts for creating a satisfactory frame design, considering the shape of structural components, material properties, the use of fasteners such as bolts, and the fabrication of welded assemblies. You will learn some of the techniques for analyzing and designing bolted assemblies to consider loads on the bolts in several directions. The design of welded joints to be safe and rigid is discussed.

Chapter 18 covered part of the story, that dealing with bolts loaded in pure tension, as in a clamping function. This chapter extends that chapter to consider eccentrically loaded joints: those that must resist a combination of direct shear and a bending moment on a bolt pattern.

The ability of a welded joint to carry a variety of loads is discussed with the objective of designing the weld. Here both uniformly loaded and eccentrically loaded joints are treated.

To appreciate the value of such study, select a variety of products, machines, and vehicles, and observe how they are built. What is the basic shape of the structure that holds everything together? Where are forces, bending moments, and torsional moments (torques) generated? What kinds of stress do they create? Consider the *load path* by following a force from the place where it is generated through all of the means by which that force or its effects are passed to a series of members and to the point where it is supported by a basic frame of the machine or where it is delivered out of the system of interest. Considering the load paths for all forces that exist in a mechanical device should give you an understanding of the desirable characteristics of the structure and should help you know how to produce a design that optimizes the management of the forces exerted on it.

Take apart a variety of mechanical devices to observe their structure. How did shape contribute to the rigidity and safety of the device? Is it rather stiff? Or is it more flexible? Are there ribs or thickened sections to provide special strength or rigidity for certain parts?

This chapter will help you identify some efficient approaches to the design of structures and frames and to analyze the performance of fasteners and welded joints loaded in many ways.

The subject of machine frames and structures is quite complex. It is discussed from the standpoint of general principles and guidelines, rather than specific design techniques. Critical frames are typically designed with computerized finite-element analysis. Also, experimental stress analysis techniques are often used to verify designs.

You Are the Designer

In Chapter 16, you were the designer of bearings for conveyor systems for a large product distribution center. Now, how do you design the frame and the structure of the conveyor system? What general form is desirable? What materials and shape should be used for the structural elements? Are the elements loaded in tension, compression, bending, shear, torsion, or some combination of these types of stress? How do the manner of loading and the nature of the stress in the structure affect these decisions? Should the frame be fabricated from standard structural shapes and bolted together? Or should it be made from steel plate and welded? How about using aluminum? Or should it be cast from cast iron or cast steel? Can it be molded from plastic? Can composites be used? How would the weight of the structure be affected? How much rigidity is desirable for this kind of structure? What shapes of load-carrying elements contribute to a stiff, rigid structure, as well as one that is safe in resisting applied stresses?

If the structure involves bolting or welding, how are the joints to be designed? What forces, both magnitude and direction, must be carried by the fasteners or the welds?

The material in this chapter will help you make some of these design decisions. Much of the information is general in nature rather than giving you specific design procedures. You must exercise judgment and creativity not only in the design of the conveyor frame but also in the analysis of its components. Because the frame design could evolve into a form too complex for analysis using traditional techniques of stress analysis, you may have to employ finite-element modeling to determine whether the design is adequate or, perhaps, overdesigned. Perhaps one or more prototypes should be built for testing.

20–1
OBJECTIVES OF THIS CHAPTER

After completing this chapter, you will be able to:

1. Apply the principles of stress and deflection analysis to propose a reasonable and efficient shape for a structure or frame and for the components involved.

2. Specify materials that are well suited to the demands of a given design, given certain conditions of load, environment, fabrication requirements, safety, and aesthetics.

3. Analyze eccentrically loaded bolted joints.

4. Design welded joints to carry many types of loading patterns.

**20–2
MACHINE
FRAMES AND
STRUCTURES**

The design of machine frames and structures is largely art in that the components of the machine must be accommodated. The designer is often restricted in where supports can be placed in order not to interfere with the operation of the machine or in order to provide access for assembly or service.

But, of course, technical requirements must be met as well for the structure itself. Some of the more important design parameters include the following:

Strength	Stiffness
Appearance	Cost to manufacture
Corrosion resistance	Weight
Size	Noise reduction
Vibration limitation	Life

Because of the virtually infinite possibilities for design details for frames and structures, this section will concentrate on general guidelines. The implementation of the guidelines would depend on the specific application. Factors to consider in starting a design project for a frame are now summarized:

- Forces exerted by the components of the machine through mounting points such as bearings, pivots, brackets, and feet of other machine elements
- Manner of support of the frame itself
- Precision of the system: allowable deflection of components
- Environment in which the unit will operate
- Quantity of production and facilities available
- Availability of analytical tools such as computerized stress analysis, past experience with similar products, and experimental stress analysis
- Relationship to other machines, walls, and so on

Again, many of these factors require judgment by the designer. The parameters over which the designer has the most control are material selection, the geometry of load-carrying parts of the frame, and manufacturing processes. A review of some possibilities follows.

Materials

As with machine elements discussed throughout this book, the material properties of strength and stiffness are of prime importance. Chapter 2 presented an extensive amount of information about materials, and the Appendices contain much useful information. In general, steel ranks high in strength compared with competing materials for frames. But it is often better to consider more than just yield strength, ultimate tensile strength, or endurance strength alone. The complete design can be executed in several candidate materials to evaluate the overall performance. Considering the *ratio of strength to density,* sometimes referred to as the *strength-to-weight ratio* or *specific strength,* may lead to a different material selection. Indeed, this is one reason for the use of aluminum, titanium, and composite materials in aircraft, aerospace vehicles, and transportation equipment.

Rigidity of a structure or a frame is frequently the determining factor in the design, rather than strength. In these cases, the stiffness of the material, indicated by its modulus of elasticity, is the most important factor. Again, the *ratio of stiffness to density,* called *specific stiffness,* may need to be evaluated. See Table 2–10 and Figures 2–21 and 2–22 for data.

Recommended Deflection Limits

Actually, only intimate knowledge of the application of a machine member or a frame can give a value for an acceptable deflection. But some guidelines are available to give you a place to start. (See Reference 3.)

Deflection Due to Bending

General machine part: 0.0005 to 0.003 in/in of beam length

Moderate precision: 0.00001 to 0.0005 in/in

High precision: 0.000 001 to 0.000 01 in/in

Deflection (Rotation) Due to Torsion

General machine part: 0.001° to 0.01°/in of length

Moderate precision: 0.000 02° to 0.0004°/in

High precision: 0.000 001° to 0.000 02°/in

Suggestions for Design to Resist Bending

Scrutiny of a table of deflection formulas for beams in bending such as those in Appendix 14 would yield the following form for the deflection:

$$\Delta = \frac{PL^3}{KEI}$$

(20–1)

where P = load

L = length between supports

E = modulus of elasticity of the material in the beam

I = moment of inertia of the cross section of the beam

K = a factor depending on the manner of loading and support

Some obvious conclusions from Equation (20–1) are that the load and the length should be kept small, and the values of E and I should be large. Note the cubic function of the length. This means, for example, that reducing the length by a factor of 2.0 would reduce the deflection by a factor of 8.0, obviously a desirable effect.

Figure 20–1 shows the comparison of four types of beam systems to carry a load, P, at a distance, a, from a rigid support. A beam simply supported at each end is taken as the "basic case." Using standard beam formulas, we computed the value of the bending moment and the deflection in terms of P and a, and these values were arbitrarily normalized to be 1.0. Then the values for the three other cases were computed and ratios determined relative to the basic case. The data show that a fixed-end beam gives both the lowest bending moment and the lowest deflection, while the cantilever gives the highest values for both.

In summary, the following suggestions are made for designing to resist bending:

1. Keep the length of the beam as short as possible, and place loads close to the supports.

2. Maximize the moment of inertia of the cross section in the direction of bending. In general, you can do so by placing as much of the material as far away from the neutral axis of bending as possible, as in a wide-flange beam or a hollow rectangular section.

FIGURE 20–1
Comparison of methods of supporting a load on a beam (Robert L. Mott, *Applied Strength of Materials*, 4th ed. Upper Saddle River, NJ: Prentice-Hall, 2001)

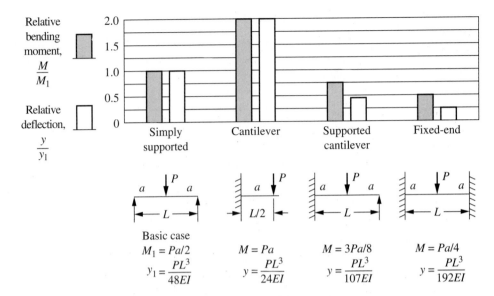

3. Use a material with a high modulus of elasticity.

4. Use fixed ends for the beam where possible.

5. Consider lateral deflection in addition to deflection in the primary load direction. Such loads could be encountered during fabrication, handling, shipping, careless use, or casual bumping.

6. Be sure to evaluate the final design with regard to both strength and rigidity. Some approaches to improving rigidity (increasing I) can actually increase the stress in the beam because the section modulus is decreased.

7. Provide rigid corner bracing in open frames.

8. Cover an open frame section with a sheet material to resist distortion. This process is sometimes called *panel stiffening.*

9. Consider a truss-type construction to obtain structural stiffness with lightweight members.

10. When designing an open space frame, use diagonal bracing to break sections into triangular parts, an inherently rigid shape.

11. Consider stiffeners for large panels to reduce vibration and noise.

12. Add bracing and gussets to areas where loads are applied or at supports to help transfer the forces into adjoining members.

13. Beware of load-carrying members with thin, extended flanges that may be placed in compression. Local buckling, sometimes called *crippling* or *wrinkling,* could occur.

14. Place connections at points of low stress if possible.

See also References 7, 8 and 10 for additional design and analysis techniques.

Suggestions for Design of Members to Resist Torsion

Torsion can be created in a machine frame member in a variety of ways: A support surface may be uneven; a machine or a motor may transmit a reaction torque to the frame; a load acting to the side of the axis of the beam (or any place away from the flexural center of the beam) would produce twisting.

FIGURE 20–2
Comparison of
torsional deformation
as a function of shape.

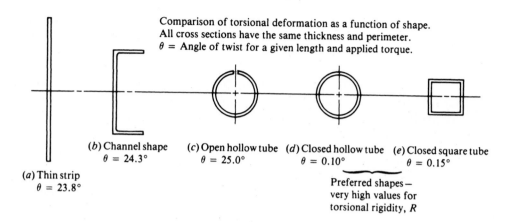

Comparison of torsional deformation as a function of shape.
All cross sections have the same thickness and perimeter.
θ = Angle of twist for a given length and applied torque.

(*a*) Thin strip
$\theta = 23.8°$

(*b*) Channel shape
$\theta = 24.3°$

(*c*) Open hollow tube
$\theta = 25.0°$

(*d*) Closed hollow tube
$\theta = 0.10°$

(*e*) Closed square tube
$\theta = 0.15°$

Preferred shapes —
very high values for
torsional rigidity, R

FIGURE 20–3
Comparison of
torsional angle of twist,
θ, for boxlike frames.
Each has the same
basic dimensions and
applied torque.

(*a*) Conventional cross bracing
$\theta = 10.8°$

(*b*) Single diagonal bracing
$\theta = 0.30°$

(*c*) Double diagonal bracing
$\theta = 0.10°$

In general, the torsional deflection of a member is computed from

$$\theta = \frac{TL}{GR} \tag{20–2}$$

where T = applied torque or twisting moment
L = length over which torque acts
G = shear modulus of elasticity of the material
R = torsional rigidity constant

The designer must choose the shape of the torsion member carefully to obtain a rigid structure. The following suggestions are made:

1. Use closed sections wherever possible. Examples are solid bars with large cross section, hollow pipe and tubing, closed rectangular or square tubing, and special closed shapes that approximate a tube.

2. Conversely, avoid open sections made from thin materials. Figure 20–2 shows a dramatic illustration.

3. For wide frames, brackets, tables, bases, and so on, use diagonal braces placed at 45° to the sides of the frame (see Figure 20–3).

4. Use rigid connections, such as by welding members together.

Most of the suggestions made in this section can be implemented regardless of the specific type of frame designed: castings made from cast iron, cast steel, aluminum, zinc, or magnesium; weldments made from steel or aluminum plate; formed housings from sheet metal or plate; or plastic moldings. The references in this chapter provide valuable additional guidance as you complete the design of frames, structures, and housings.

**20–3
ECCENTRICALLY
LOADED BOLTED
JOINTS**

Figure 20–4 shows an example of an eccentrically loaded bolted joint. The motor on the extended bracket places the bolts in shear because its weight acts directly downward. But there also exists a moment equal to $P \times a$ that must be resisted. The moment tends to rotate the bracket and thus to shear the bolts.

The basic approach to the analysis and design of eccentrically loaded joints is to determine the forces that act on each bolt because of all the applied loads. Then, by a process of superposition, the loads are combined vectorially to determine which bolt carries the greatest load. That bolt is then sized. The method will be illustrated in Example Problem 20–1.

The American Institute of Steel Construction (AISC) lists allowable stresses for bolts made from ASTM grade steels, as shown in Table 20–1. These data are for bolts used in standard-sized holes, 1/16 in larger than the bolt. Also, a *friction-type connection,* in which the clamping force is sufficiently large that the friction between the mating parts helps carry some of the shear load, is assumed. (See Reference 1.)

In the design of bolted joints, you should ensure that there are no threads in the plane where shear occurs. The body of the bolt will then have a diameter equal to the major diameter of the thread. You can use the tables in Chapter 18 to select the standard size for a bolt.

FIGURE 20–4
Eccentrically loaded
bolted joint

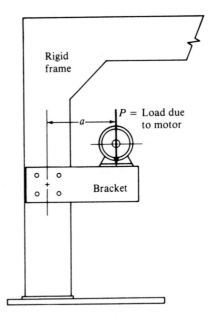

TABLE 20–1   Allowable stresses for bolts

ASTM grade	Allowable shear stress	Allowable tensile stress
A307	10 ksi (69 MPa)	20 ksi (138 MPa)
A325 and A449	17.5 ksi (121 MPa)	44 ksi (303 MPa)
A490	22 ksi (152 MPa)	54 ksi (372 MPa)

Example Problem 20–1

For the bracket in Figure 20–4, assume that the total force P is 3500 lb and the distance a is 12 in. Design the bolted joint, including the location and number of bolts, the material, and the diameter.

Solution

The solution shown is an outline of a procedure that can be used to analyze similar joints. The data of this problem illustrate the procedure.

Step 1. Propose the number of bolts and the pattern. This is a design decision, based on your judgment and the geometry of the connected parts. In this problem, let's try a pattern of four bolts placed as shown in Figure 20–5.

Step 2. Determine the direct shear force on the bolt pattern and on each individual bolt, assuming that all bolts share the shear load equally:

$$\text{Shear load} = P = 3500 \text{ lb}$$
$$\text{Load per bolt} = F_s = P/4 = 3500 \text{ lb}/4 = 875 \text{ lb/bolt}$$

The shear force acts directly downward on each bolt.

Step 3. Compute the **moment** to be resisted by the bolt pattern: the product of the overhanging load and the distance to the **centroid** of the bolt pattern. In this problem, $M = P \times a = (3500 \text{ lb})(12 \text{ in}) = 42\ 000 \text{ lb} \cdot \text{in}$.

FIGURE 20–5
Geometry of bolted joint and forces on bolt 1

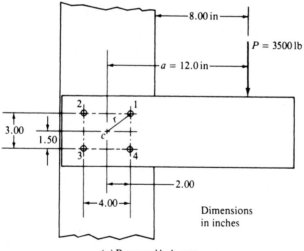

(a) Proposed bolt pattern

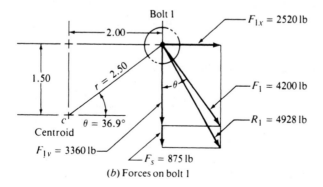

(b) Forces on bolt 1

Step 4. Compute the radial distance from the centroid of the bolt pattern to the center of each bolt. In this problem, each bolt has a radial distance of

$$r = \sqrt{(1.50\ \text{in})^2 + (2.00\ \text{in})^2} = 2.50\ \text{in}$$

Step 5. Compute the sum of the *squares* of all radial distances to all bolts. In this problem, all four bolts have the same r. Then

$$\Sigma r^2 = 4(2.50\ \text{in})^2 = 25.0\ \text{in}^2$$

Step 6. Compute the force on each bolt required to resist the bending moment from the relation

$$F_i = \frac{Mr_i}{\Sigma r^2} \tag{20–3}$$

where r_i = radial distance from the centroid of the bolt pattern to the ith bolt
F_i = force on the ith bolt due to the moment. The force acts perpendicular to the radius.

In this problem, all such forces are equal. For example, for bolt 1,

$$F_1 = \frac{Mr_1}{\Sigma r_2} = \frac{(42\ 000\ \text{lb}\cdot\text{in})(2.50\ \text{in})}{25.0\ \text{in}^2} = 4200\ \text{lb}$$

Step 7. Determine the resultant of all forces acting on each bolt. A vector summation can be performed either analytically or graphically, or each force can be resolved into horizontal and vertical components. The components can be summed and then the resultant can be computed.
Let's use the latter approach for this problem. The shear force acts directly downward, in the y-direction. The x- and y-components of F_1 are

$$F_{1x} = F_1 \sin\theta = (4200\ \text{lb})\sin(36.9°) = 2520\ \text{lb}$$
$$F_{1y} = F_1 \cos\theta = (4200)\cos(36.9°) = 3360\ \text{lb}$$

The total force in the y-direction is then

$$F_{1y} + F_s = 3360 + 875 = 4235\ \text{lb}$$

Then the resultant force on bolt 1 is

$$R_1 = \sqrt{(2520)^2 + (4235)^2} = 4928\ \text{lb}$$

Step 8. Specify the bolt material; compute the required area for the bolt; and select an appropriate size. For this problem, let's specify ASTM A325 steel for the bolts having an allowable shear stress of 17 500 psi from Table 20–1. Then the required area for the bolt is

$$A_s = \frac{R_1}{\tau_a} = \frac{4928\ \text{lb}}{17\ 500\ \text{lb/in}^2} = 0.282\ \text{in}^2$$

The required diameter would be

$$D = \sqrt{\frac{4A_s}{\pi}} = \sqrt{\frac{4(0.282\ \text{in}^2)}{\pi}} = 0.599\ \text{in}$$

Let's specify a 5/8-in bolt having a diameter of 0.625 in.

**20–4
WELDED JOINTS**

The design of welded joints requires consideration of the manner of loading on the joint, the types of materials in the weld and in the members to be joined, and the geometry of the joint itself. The load may be either uniformly distributed over the weld such that all parts of the weld are stressed to the same level, or the load may be eccentrically applied. Both are discussed in this section.

The materials of the weld and the parent members determine the allowable stresses. Table 20–2 lists several examples for steel and aluminum. The allowables listed are for shear on fillet welds. For steel, welded by the electric arc method, the type of electrode is an indication of the tensile strength of the filler metal. For example, the E70 electrode has a minimum tensile strength of 70 ksi (483 MPa). Additional data are available in publications of the American Welding Society (AWS), the American Institute for Steel Construction (AISC), and the Aluminum Association (AA). See Reference 1 and Internet sites 3 and 7.

Types of Joints

Joint type refers to the relationship between mating parts, as illustrated in Figure 20–6. The butt weld allows a joint to be the same nominal thickness as the mating parts and is usually loaded in tension. If the joint is properly made with the appropriate weld metal, the joint will be stronger than the parent metal. Thus, no special analysis of the joint is required if the joined members themselves are shown to be safe. Caution is advised, however, when the materials to be joined are adversely affected by the heat of the welding process. Heat-treated steels and many aluminum alloys are examples. The other types of joints in Figure 20–6 are assumed to place the weld in shear.

TABLE 20–2   Allowable shear stresses on fillet welds

A. Steel

Electrode type	Typical metals joined (ASTM grade)	Allowable shear stress
E60	A36, A500	18 ksi (124 MPa)
E70	A242, A441	21 ksi (145 MPa)
E80	A572, Grade 65	24 ksi (165 MPa)
E90		27 ksi (186 MPa)
E100		30 ksi (207 MPa)
E110		33 ksi (228 MPa)

B. Aluminum

	Filler Alloy							
	1100		4043		5356		5556	
	Allowable Shear Stress							
Metal joined	ksi	MPa	ksi	MPa	ksi	MPa	ksi	MPa
1100	3.2	22	4.8	33				
3003	3.2	22	5.0	34				
6061			5.0	34	7.0	48	8.5	59
6063			5.0	34	6.5	45	6.5	45

FIGURE 20–6 Types of weld joints

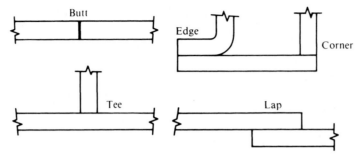

FIGURE 20–7 Some types of welds showing edge preparation

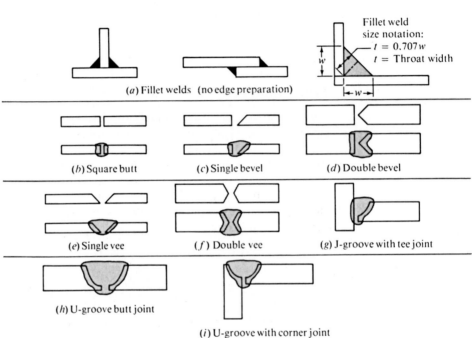

(a) Fillet welds (no edge preparation)

(b) Square butt   (c) Single bevel   (d) Double bevel

(e) Single vee   (f) Double vee   (g) J-groove with tee joint

(h) U-groove butt joint

(i) U-groove with corner joint

Types of Welds

Figure 20–7 shows several types of welds named for the geometry of the edges of the parts to be joined. Note the special edge preparation required, especially for thick plates, to permit the welding rod to enter the joint and build a continuous weld bead.

Size of Weld

The five types of groove-type welds in Figure 20–7 are made as complete penetration welds. Then, as indicated before for butt welds, the weld is stronger than the parent metals, and no further analysis is required.

Fillet welds are typically made as equal-leg right triangles, with the size of the weld indicated by the length of the leg. A fillet weld loaded in shear would tend to fail along the shortest dimension of the weld that is the line from the root of the weld to the theoretical face of the weld and normal to the face. The length of this line is found from simple trigonometry to be $0.707w$, where w is the leg dimension.

The objectives of the design of a fillet welded joint are to specify the length of the legs of the fillet; the pattern of the weld; and the length of the weld. Presented here is the

TABLE 20–3 Allowable shear stresses and forces on welds

Base metal ASTM grade	Electrode	Allowable shear stress	Allowable force per inch of leg
Building-type structures:			
A36, A441	E60	13 600 psi	9600 lb/in
A36, A441	E70	15 800 psi	11 200 lb/in
Bridge-type structures:			
A36	E60	12 400 psi	8800 lb/in
A441, A242	E70	14 700 psi	10 400 lb/in

method that treats the weld as a line having no thickness. The method involves determining the *maximum force per inch* of weld leg length. Comparing the actual force with an allowable force allows the calculation of the required leg length.

Table 20–3 gives data for the allowable shear stress and the allowable force per inch for some combinations of base metal and welding electrode. In general, the allowables for building-type structures are for steady loads. The values for bridge-type loading accounts for the cyclic effects. For true fatigue-type repeated loading, refer to the literature. (See References 2, 3, 4, and 6.)

Method of Treating Weld as a Line

Four different types of loading are considered here: (1) direct tension or compression, (2) direct vertical shear, (3) bending, and (4) twisting. The method allows the designer to perform calculations in a manner very similar to that used to design the load-carrying members themselves. In general, the weld is analyzed separately for each type of loading to determine the force per inch of weld size due to each load. The loads are then combined vectorially to determine the maximum force. This maximum force is compared with the allowables from Table 20–3 to determine the size of the weld required. See Reference 2.

The relationships used are summarized next:

Type of Loading	*Formula (and Equation Number) for Force per Inch of Weld*	
Direct tension or compression	$f = P/A_w$	**(20-4)**
Direct vertical shear	$f = V/A_w$	**(20-5)**
Bending	$f = M/S_w$	**(20-6)**
Twisting	$f = Tc/J_w$	**(20-7)**

In these formulas, the geometry of the weld is used to evaluate the terms A_w, S_w, and J_w, using the relationships shown in Figure 20–8. Note the similarity between these formulas and those used to perform the stress analysis. Note, also, the similarity between the geometry factors for welds and the properties of areas used for the stress analysis. Because the weld is treated as a line having no thickness, the units for the geometry factors are different from those of the area properties, as indicated in Figure 20–8.

The use of this method of weld analysis will be demonstrated with example problems. In general, the method requires the steps on page 787:

FIGURE 20–8
Geometry factors for weld analysis

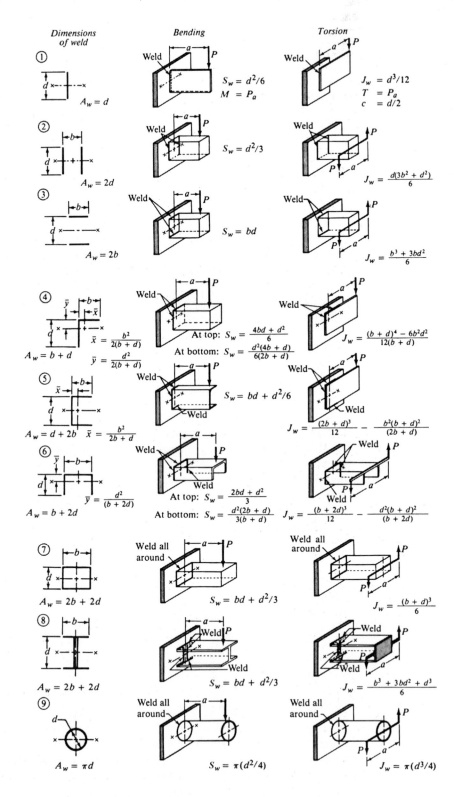

General Procedure for Designing Welded Joints

1. Propose the geometry of the joint and the design of the members to be joined.

2. Identify the types of stresses to which the joint is subjected (bending, twisting, vertical shear, direct tension, or compression).

3. Analyze the joint to determine the magnitude and the direction of the force on the weld due to each type of load.

4. Combine the forces vectorially at the point or points of the weld where the forces appear to be maximum.

5. Divide the maximum force on the weld by the allowable force from Table 20–3 to determine the required leg size for the weld. Note that when thick plates are welded, there are minimum acceptable sizes for the welds as listed in Table 20–4.

TABLE 20–4 Minimum weld sizes for thick plates

Plate thickness (in)	Minimum leg size for fillet weld (in)
$\leq 1/2$	3/16
$>1/2$–3/4	1/4
$>3/4$–$1\frac{1}{2}$	5/16
$>1\frac{1}{2}$–$2\frac{1}{4}$	3/8
$>2\frac{1}{4}$–6	1/2
>6	5/8

Example Problem 20–2 Design a bracket similar to that in Figure 20-4, but use welding to attach the bracket to the column. The bracket is 6.00 in high and is made from ASTM A36 steel having a thickness of 1/2 in. The column is also made from A36 steel and is 8.00 in wide.

Solution *Step 1.* The proposed geometry is a design decision and may have to be subjected to some iteration to achieve an optimum design. For a first trial, let's use the C-shaped weld pattern shown in Figure 20–9.

Step 2. The weld will be subjected to direct vertical shear and twisting caused by the 3500-lb load on the bracket.

Step 3. To compute the forces on the weld, we must know the geometry factors A_w and J_w. Also, the location of the centroid of the weld pattern must be computed [see Figure 20–9(b)]. Use Case 5 in Figure 20–8.

$$A_w = 2b + d = 2(4) + 6 = 14 \text{ in}$$
$$J_w = \frac{(2b+d)^3}{12} - \frac{b^2(b+d)^2}{(2b+d)} = \frac{(14)^3}{12} - \frac{16(10)^2}{14} = 114.4 \text{ in}^3$$
$$\bar{x} = \frac{b^2}{2b+d} = \frac{16}{14} = 1.14 \text{ in}$$

FIGURE 20–9

C-shaped weld bracket

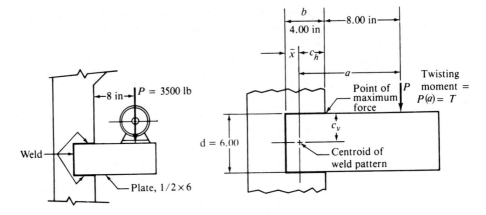

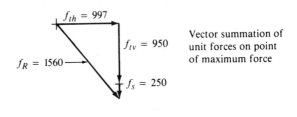

(a) Basic design of bracket

(b) Dimensions of bracket

(c) Analysis of forces

Force Due to Vertical Shear

$$V = P = 3500 \text{ lb}$$

$$f_s = P/A_w = (3500 \text{ lb})/14 \text{ in} = 250 \text{ lb/in}$$

This force acts vertically downward on all parts of the weld.

Forces Due to the Twisting Moment

$$T = P[8.00 + (b - \bar{x})] = 3500[8.00 + (4.00 - 1.14)]$$

$$T = 3500(10.86) = 38\,010 \text{ lb} \cdot \text{in}$$

The twisting moment causes a force to be exerted on the weld that is perpendicular to a radial line from the centroid of the weld pattern to the point of interest. In this case, the end of the weld to the upper right experiences the greatest force. It is most convenient to break the force down into horizontal and vertical components and then subsequently re-combine all such components to compute the resultant force:

$$f_{th} = \frac{Tc_v}{J_w} = \frac{(38\,010)(3.00)}{114.4} = 997 \text{ lb/in}$$

$$f_{tv} = \frac{Tc_h}{J_w} = \frac{(38\,010)(2.86)}{114.4} = 950 \text{ lb/in}$$

Step 4. The vectorial combination of the forces on the weld is shown in Figure 20–9(c). Thus, the maximum force is 1560 lb/in.

Step 5. Selecting an E60 electrode for the welding, we find that the allowable force per inch of weld leg size is 9600 lb/in (Table 20–3). Then the required weld leg size is

$$w = \frac{1560 \text{ lb/in}}{9600 \text{ lb/in per in of leg}} = 0.163 \text{ in}$$

Table 20–4 shows that the minimum size weld for a 1/2-in plate is 3/16 in (0.188 in). That size should be specified.

Example Problem 20–3   A steel strap, 1/4 in thick, is to be welded to a rigid frame to carry a dead load of 12 500 lb, as shown in Figure 20–10. Design the strap and its weld.

Solution   The basic objectives of the design are to specify a suitable material for the strap, the welding electrode, the size of the weld, and the dimensions W and h, as shown in Figure 20–10.
    Let's specify that the strap is to be made from ASTM A441 structural steel and that it is to be welded with an E70 electrode, using the minimum size weld, 3/16 in. Appendix 7 gives the yield strength of the A441 steel as 42 000 psi. Using a design factor of 2, we can compute an allowable stress of

$$\sigma_a = 42\ 000/2 = 21\ 000 \text{ psi}$$

Then the required area of the strap is

$$A = \frac{P}{\sigma_a} = \frac{12\ 500 \text{ lb}}{21\ 000 \text{ lb/in}^2} = 0.595 \text{ in}^2$$

But the area is $W \times t$, where $t = 0.25$ in. Then the required width W is

$$W = A/t = 0.595/0.25 = 2.38 \text{ in}$$

Let's specify that $W = 2.50$ in.
    To compute the required length of the weld h, we need the allowable force on the 3/16-in weld. Table 20–3 indicates the allowable force on the A441 steel welded with an E70 electrode to be 11 200 lb/in per in of leg size. Then

$$f_a = \frac{11\ 200 \text{ lb/in}}{1.0 \text{ in leg}} \times 0.188 \text{ in leg} = 2100 \text{ lb/in}$$

FIGURE 20–10
Steel strap

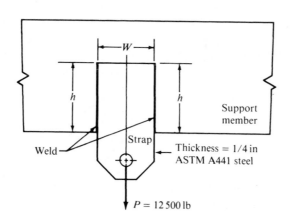

$P = 12\ 500$ lb

The actual force on the weld is

$$f_a = P/A_w = P/2h$$

Then solving for h gives

$$h = \frac{P}{2(f_a)} = \frac{12\,500\ \text{lb}}{2(2100\ \text{lb/in})} = 2.98\ \text{in}$$

Let's specify $h = 3.00$ in.

Example Problem 20–4   Evaluate the design shown in Figure 20–11 with regard to stress in the welds. All parts of the assembly are made of ASTM A36 structural steel and are welded with an E60 electrode. The 2500-lb load is a dead load.

Solution   The critical point would be the weld at the top of the tube where it is joined to the vertical surface. At this point, there is a three-dimensional force system acting on the weld as shown in Figure 20–12. The offset location of the load causes a twisting on the weld that produces a force f_t on the weld toward the left in the y-direction. The bending produces a force f_b acting outward along the x-axis. The vertical shear force f_s acts downward along the z-axis.

FIGURE 20–11
Bracket assembly

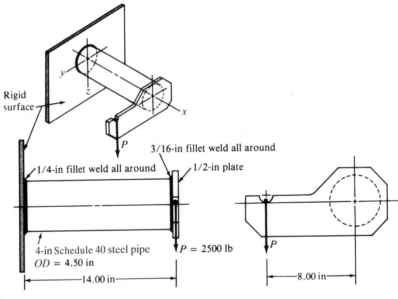

Rigid surface

P

3/16-in fillet weld all around

1/4-in fillet weld all around

1/2-in plate

4-in Schedule 40 steel pipe
OD = 4.50 in

P = 2500 lb

—14.00 in—

—8.00 in—

FIGURE 20–12
Force vectors

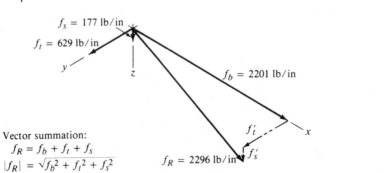

$f_s = 177$ lb/in

$f_t = 629$ lb/in

$f_b = 2201$ lb/in

Vector summation:
$f_R = f_b + f_t + f_s$
$|f_R| = \sqrt{f_b^2 + f_t^2 + f_s^2}$

$f_R = 2296$ lb/in

From statics, the resultant of the three force components would be

$$f_R = \sqrt{f_t^2 + f_b^2 + f_s^2}$$

Now each component force on the weld will be computed.

Twisting Force, f_t

$$f_t = \frac{Tc}{J_w}$$

$$T = (2500 \text{ lb})(8.00 \text{ in}) = 20\,000 \text{ lb} \cdot \text{in}$$

$$c = OD/2 = 4.500/2 = 2.25 \text{ in}$$

$$J_w = (\pi)(OD)^3/4 = (\pi)(4.500)^3/4 = 71.57 \text{ in}^3$$

Then

$$f_t = \frac{Tc}{J_w} = \frac{(20\,000)(2.25)}{71.57} = 629 \text{ lb/in}$$

Bending Force, f_b

$$f_b = \frac{M}{S_w}$$

$$M = (2500 \text{ lb})(14.00 \text{ in}) = 35\,000 \text{ lb} \cdot \text{in}$$

$$S_w = (\pi)(OD)^2/4 = (\pi)(4.500)^2/4 = 15.90 \text{ in}^2$$

Then

$$f_b = \frac{M}{S_w} = \frac{35\,000}{15.90} = 2201 \text{ lb/in}$$

Vertical Shear Force, f_s

$$f_s = \frac{P}{A_w}$$

$$A_w = (\pi)(OD) = (\pi)(4.500 \text{ in}) = 14.14 \text{ in}$$

$$f_s = \frac{P}{A_w} = \frac{2500}{14.14} = 177 \text{ lb/in}$$

Now the resultant can be computed:

$$f_R = \sqrt{f_1^2 + f_b^2 + f_s^2}$$

$$f_R = \sqrt{629^2 + 2201^2 + 177^2} = 2296 \text{ lb/in}$$

Comparing this with the allowable force on a 1.0-in weld gives

$$w = \frac{2296 \text{ lb/in}}{9600 \text{ lb/in per in of leg size}} = 0.239 \text{ in}$$

The 1/4-in fillet specified in Figure 20–11 is satisfactory.

REFERENCES

1. American Institute of Steel Construction. *Manual of Steel Construction.* 9th ed. New York: American Institute of Steel Construction, 1989.

2. Blodgett, Omer W. *Design of Welded Structures.* Cleveland, OH: James F. Lincoln Arc Welding Foundation, 1966.

3. Blodgett, Omer W. *Design of Weldments.* Cleveland, OH: James F. Lincoln Arc Welding Foundation, 1963.

4. Brandon, D. G., and W. D. Kaplan. *Joining Processes.* New York: John Wiley & Sons, 1997.

5. Fry, Gary T. *Weldments in Steel Frame Structures.* Reston, VA: American Society of Civil Engineers, 2002.

6. Hicks, John. *Welded Joint Design.* 3rd ed. New York: Industrial Press, 1999.

7. Marshek, Kurt M. *Design of Machine and Structural Parts.* New York: John Wiley & Sons, 1987.

8. Mott, Robert L. *Applied Strength of Materials.* 4th ed. Upper Saddle River, NJ: Prentice Hall, 2001.

9. North American Die Casting Association. *Product Design for Die Casting.* Rosemont, IL: North American Die Casting Association, 1998.

10. Slocum, Alexander H. *Precision Machine Design.* Dearborn, MI: Society of Manufacturing Engineers, 1998.

11. Society of Automotive Engineers. *Spot Welding and Weld Joint Failure Processes.* Warrendale, PA: Society of Automotive Engineers, 2001.

12. Weiser, Peter F., ed. *Steel Casting Handbook.* 6th ed. Rocky River, OH: Steel Founder's Society of America, 1980.

INTERNET SITES FOR MACHINE FRAMES, BOLTED CONNECTIONS, AND WELDED JOINTS

Also refer to the Internet sites from Chapter 18 on Fasteners that includes many related to bolted connections.

1. **Steel Founders' Society of America.** *www.sfsa.org* An association of companies providing foundry services.

2. **American Foundry Society.** *www.afsinc.org* A professional society that promotes research and technology for the foundry industry.

3. **American Welding Society.** *www.aws.org* A professional society that develops standards for the welding industry, including AWS D1.1 Structural Welding Code-Steel (2002), AWS D1.2 Structural Welding Code-Aluminum (1997), and many others.

4. **James F. Lincoln Foundation.** *www.jflf.org* An organization that promotes education and training in welding technology. The site includes technical papers and general information about welding processes, joint design, and guides to welded steel construction.

5. **Miller Electric Company.** *www.millerwelds.com* A manufacturer of a wide variety of welding equipment and accessories for the professional welding community and occasional welders. The site includes a Training/Education section that can be searched for information about numerous welding processes.

6. **Lincoln Electric Company.** *www.lincolnelectric.com* A manufacturer of a wide variety of welding equipment and accessories for the welding industry. The site includes a Knowledge section that provides technical articles about welding technology.

7. **Hobart Brothers Company.** *www.hobartbrothers.com* A manufacturer of numerous types of filler metal products, electrodes, and steel wire for the welding industry.

8. **Hobart Institute of Welding Technology.** *www.welding.org* An educational organization that provides instruction in the performance of welding techniques. The site includes welding tips for a variety of processes and a glossary of welding terms.

PROBLEMS

For Problems 1–6, design a bolted joint to join the two members shown in the appropriate figure. Specify the number of bolts, the pattern, the bolt grade, and the bolt size.

1. Figure P20–1

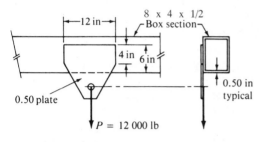

FIGURE P20–1 (Problems 1, 7, and 13)

2. Figure P20–2

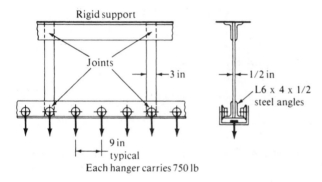

FIGURE P20–2

3. Figure P20–3

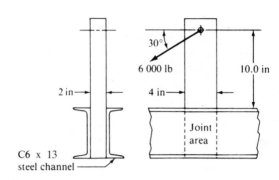

FIGURE P20–3

4. Figure P20–4

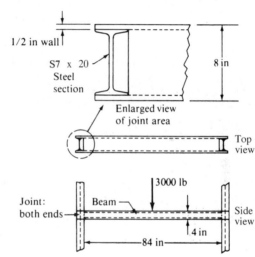

FIGURE P20–4 (Problems 4 and 8)

5. Figure P20–5

FIGURE P20–5
(Problems 5 and 9)

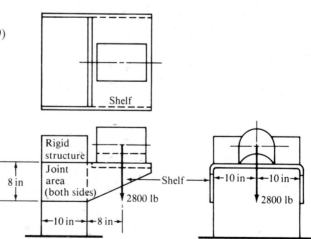

6. Figure P20–6

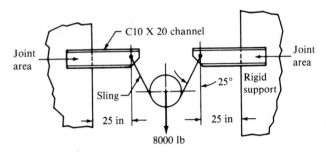

Load shared equally by *four* brackets (only two shown)

FIGURE P20–6   (Problems 6 and 10)

For Problems 7–12, design a welded joint to join the two members shown in the appropriate figure. Specify the weld pattern, the type of electrode to be used, and the size of weld. In Problems 7–9, the members are ASTM A36 steel. In Problems 10–12, the members are ASTM A441 steel. Use the method of treating the joint as a line, and use the allowable forces per inch of leg for building-type structures from Table 20–3.

7. Figure P20–1

8. Figure P20–4

9. Figure P20–5

10. Figure P20–6

11. Figure P20–11

12. Figure P20–11 (but $P_2 = 0$)

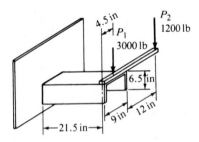

FIGURE P20–11   (Problems 11 and 12)

For Problems 13–16, design a welded joint to join the two aluminum members shown in the appropriate figure. Specify the weld pattern, the type of filler alloy, and the size of weld. The types of materials joined are listed in the problems.

13. Figure P20–1: 6061: alloy (but $P = 4000$ lb)

14. Figure P20–14: 6061 alloy

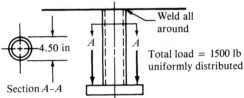

FIGURE P20–14

15. Figure P20–15: 6063 alloy

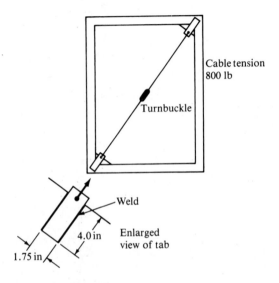

FIGURE P20–15

16. Figure P20–16: 3003 alloy

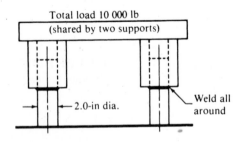

FIGURE P20–16

17. Compare the weight of a tensile rod carrying a dead load of 4800 lb if it is made from (a) AISI 1020 HR steel; (b) AISI 5160 OQT 1300 steel; (c) aluminum 2014-T6; (d) aluminum 7075-T6; (e) titanium 6A1-4V, annealed; and (f) titanium 3A1-13V-11Cr, aged. Use $N = 2$ based on yield strength.

21

Electric Motors and Controls

The Big Picture

You Are the Designer

21–1 Objectives of This Chapter

21–2 Motor Selection Factors

21–3 AC Power and General Information about AC Motors

21–4 Principles of Operation of AC Induction Motors

21–5 AC Motor Performance

21–6 Three-Phase, Squirrel-Cage Induction Motors

21–7 Single-Phase Motors

21–8 AC Motor Frame Types and Enclosures

21–9 Controls for AC Motors

21–10 DC Power

21–11 DC Motors

21–12 DC Motor Control

21–13 Other Types of Motors

Electric Motors

Discussion Map

- Electric motors provide the power for a huge array of products in homes, factories, schools, commercial facilities, transportation equipment, and many portable devices.

- The two major classifications of motors are *alternating current (AC)* and *direct current (DC)*. Some can operate on either type of power.

Discover

Look for a variety of machines and products that are powered by electric motors. Choose both large and small devices; some portable and some that plug into standard electrical outlets. Look around your home, at work, and in a factory.

Try to find the nameplate for each motor, and copy down as much information as you can. How are the nameplate data related to the performance characteristics of the motor? Is it an AC motor or a DC motor? How fast does it run? What is its electrical rating in terms of voltage and current?

Some motors that will operate well in the United States will run differently or will not run at all in other countries. Why? What are the electric voltage and frequency standards in various parts of the world?

This chapter will help you identify many kinds of motors, understand the general performance characteristics of each, and apply them properly.

The electric motor is widely used for providing the prime power to industrial machinery, consumer products, and business equipment. This chapter will describe the various types of motors and will discuss their performance characteristics. The objective is to provide you with the background needed to specify motors and to communicate with vendors to acquire the proper motor for a given application.

Types of motors discussed in this chapter are direct current (DC), alternating current (AC, both single-phase and three-phase), universal motors, stepper motors, and variable-speed AC motors. Motor controls are also discussed.

Look for a variety of machines and products that are powered by electric motors. Choose both large and small devices; some portable and some that plug into standard electrical outlets. Look around your home, at work, and in a factory if possible.

Try to find the nameplate for each motor and copy down as much information as you can. What does each piece of information mean? Are some electrical wiring diagrams included? How are the nameplate data related to the performance characteristics of the motor? Is it an AC motor or a DC motor? How fast does it run? What is its electrical rating in terms of voltage and current? For the motors in portable devices, what kind of power supply is used? What kinds of batteries are required? How many? How are they connected? In series? In parallel? How do those factors relate to the voltage rating of the motor? What is the relationship between the kind of work done by the motor and the time between charging sessions?

Compare some of the larger motors with the photographs in Section 21–8 of this chapter. Can you identify the frame type? What company manufactured the motor?

See if you can find more information about the motors or their manufacturers from the Internet. Look up some of the company names in this chapter to learn more about the lines of motors that they offer. Do a broader search to find as many different types as you can.

What is the rated power for each motor? What units are used for the power rating? Convert all of the power ratings to watts. (See Appendix 18.) Then convert them all to horsepower to gain an appreciation for the physical size of various motors and to visualize the comparison between the SI unit of watts and the older U.S. Customary unit of horsepower.

Separate the motors you find into those that are AC and those that are DC. Do you see any significant differences between the two classes of motors? Within each class, what variations do you see? Do any of the nameplates describe the type of motor, such as *synchronous*, *universal*, *split-phase*, *NEMA design C*, or other designations?

What kinds of controls are connected to the motors? Switches? Starters? Speed controls? Protection devices?

This chapter will help you identify many kinds of motors, understand the general performance characteristics of each, and apply them properly.

You Are the Designer

Consider a conveyor system that you are to design. One possibility for driving the conveyor system is to use an electric motor. What type should be used? At what speed will it operate? What type of electric power is available to supply the motor? What power is required? What type of housing and mounting style should be specified? What are the dimensions of the motor? How is the motor connected to the drive pulley of the conveyor system? The information in this chapter will help you to answer these and other questions.

21–1
OBJECTIVES OF
THIS CHAPTER

After completing this chapter, you will be able to:

1. Describe the factors that must be specified to select a suitable motor.
2. Describe the principles of operation of AC motors.
3. Identify the typical classifications of AC electric motors according to rated power.
4. Identify the common voltages and frequencies of AC power and the speed of operation of AC motors operating on those systems.
5. Describe single-phase and three-phase AC power.
6. Describe the typical frame designs, sizes, and enclosure styles of AC motors.
7. Describe the general form of a motor performance curve.
8. Describe the comparative performance of *shaded-pole*, *permanent-split capacitor*, *split-phase*, and *capacitor-start*, single-phase AC motors.
9. Describe three-phase, squirrel-cage AC motors.
10. Describe the comparative performance of *NEMA design B*, *NEMA design C*, *NEMA design D*, and *wound-rotor* three-phase AC motors.
11. Describe *synchronous motors*.
12. Describe *universal motors*.
13. Describe three means of producing DC power and the common voltages produced.
14. Describe the advantages and disadvantages of DC motors compared with AC motors.
15. Describe four basic designs of DC motors—*shunt-wound*, *series-wound*, *compound-wound*, and *permanent magnet*—and describe their performance curves.
16. Describe torque motors, servomotors, stepper motors, brushless DC motors, and printed circuit motors.

17. Describe motor control systems for system protection, speed control, starting, stopping, and overload protection.

18. Describe speed control of AC motors.

19. Describe the control of DC motors.

21–2
MOTOR
SELECTION
FACTORS

As a minimum, the following items must be specified for motors:

- Motor type: DC, AC, single-phase, three-phase, and so on
- Power rating and speed
- Operating voltage and frequency
- Type of enclosure
- Frame size
- Mounting details

In addition, there may be many special requirements that must be communicated to the vendor. The primary factors to be considered in selecting a motor include the following:

- Operating torque, operating speed, and power rating. Note that these are related by the equation

$$\text{Power} = \text{torque} \times \text{speed}$$

- Starting torque.
- Load variations expected and corresponding speed variations that can be tolerated.
- Current limitations during the running and starting phases of operation.
- Duty cycle: how frequently the motor is to be started and stopped.
- Environmental factors: temperature, presence of corrosive or explosive atmospheres, exposure to weather or to liquids, availability of cooling air, and so on.
- Voltage variations expected: Most motors will tolerate up to $\pm$ 10% variation from the rated voltage. Beyond this, special designs are required.
- Shaft loading, particularly side loads and thrust loads that can affect the life of shaft bearings.

Motor Size

A rough classification of motors by size is used to group motors of similar design. Horsepower (hp) is currently used most frequently, with the metric unit of watts or kilowatts also used at times. The conversion is

$$1.0 \text{ hp} = 0.746 \text{ kW} = 746 \text{ W}$$

The classifications are as follows:

- *Subfractional horsepower:* 1 to 40 millihorsepower (mhp), where 1 mhp = 0.001 hp. Thus, this range includes 0.001 to 0.040 hp (0.75 to 30 W, approximately).
- *Fractional horsepower:* 1/20 to 1.0 hp (37 to 746 W, approximately).
- *Integral horsepower:* 1.0 hp (0.75 kW) and larger.

References 1–4 provide additional information about the selection and application of electric motors. See also Internet sites 1–4.

**21–3
AC POWER AND
GENERAL
INFORMATION
ABOUT AC
MOTORS**

Alternating current (AC) power is produced by the electric utility and delivered to the industrial, commercial, or residential consumer in a variety of forms. In the United States, AC power has a frequency of 60 hertz (Hz) or 60 cycles/s. In many other countries, 50 Hz is used. Some aircraft use 400-Hz power from an on-board generator.

AC power is also classified as single-phase or three-phase. Most residential units and light commercial installations have only single-phase power, carried by two conductors plus ground. The waveform of the power would appear like that in Figure 21–1, a single continuous sine wave at the system frequency whose amplitude is the rated voltage of the power. Three-phase power is carried on a three-wire system and is composed of three distinct waves of the same amplitude and frequency, with each phase offset from the next by 120°, as illustrated in Figure 21–2. Industrial and large commercial installations use three-phase power for the larger electrical loads because smaller motors are possible and there are economies of operation.

FIGURE 21–1
Single-phase AC power

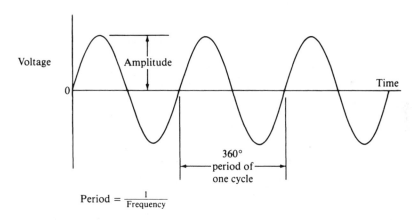

FIGURE 21–2
Three-phase AC power

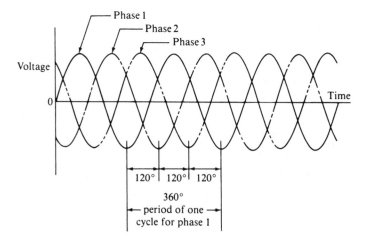

TABLE 21–1   AC motor voltages

	Motor voltage ratings	
System voltage	Single-phase	Three-phase
120	115	115
120/208	115	200
240	230	230
480		460
600		575

TABLE 21–2   AC motor speeds for 60 hertz power

Number of poles	Synchronous speed (rpm)	Full-load[a] speed (rpm)
2	3600	3450
4	1800	1725
6	1200	1140
8	900	850
10	720	690
12	600	575

[a]Approximately 95% of synchronous speed (normal slip).

AC Voltages

Some of the more popular voltage ratings available in AC power are listed in Table 21–1. Given are the nominal system voltage and the typical motor voltage rating for that system in both single-phase and three-phase. In most cases, the highest voltage available should be used because the current flow for a given power is smaller. This allows smaller conductors to be used.

Speeds of AC Motors

An AC motor at zero load would tend to operate at or near its *synchronous speed*, n_s, which is related to the frequency, f, of the AC power and to the number of electrical poles, p, wound into the motor, according to the equation

⇨ **Synchronous Speed**

$$n_s = \frac{120f}{p} \text{ rev/min} \qquad\qquad (21\text{–}1)$$

Motors have an even number of poles, usually from 2 to 12, resulting in the synchronous speeds listed in Table 21–2 for 60-Hz power. But the induction motor, the most widely used type, operates at a speed progressively slower than its synchronous speed as the load (torque) demand increases. When the motor is delivering its rated torque, it will be operating near its rated or full-load speed, also listed in Table 21–2. Note that the full-load speed is not precise and that those listed are for motors with normal slip of approximately 5%. Some motors described later are "high-slip" motors having lower full-load speeds. Some 4-pole motors are rated at 1750 rpm at full load, indicating only about 3% slip. *Synchronous motors* operate precisely at the synchronous speed with no slip.

21–4
PRINCIPLES OF
OPERATION OF
AC INDUCTION
MOTORS

Later in this chapter we will discuss the particular details of several different types of AC motors, but the most common of these is the *induction motor*. The two active parts of an induction motor are the *stator*, or stationary element, and the *rotor*, or rotating element. Figure 21–3 shows a longitudinal cross section of an induction motor showing the stator, in the form of a hollow cylinder, fixed in the housing. The rotor is positioned inside the stator and is carried on the shaft. The shaft, in turn, is supported by bearings in the housing.

The stator is constructed of many thin, flat discs of steel, called *laminations*, stacked together and insulated from one another. Figure 21–4 shows the shape of the laminations, including a series of slots around the inside. These slots are typically aligned as the stator laminations are stacked, thus forming channels along the length of the stator core. Several layers of copper wire are passed through the channels and are looped around to form a set

FIGURE 21–3
Longitudinal section through an induction motor

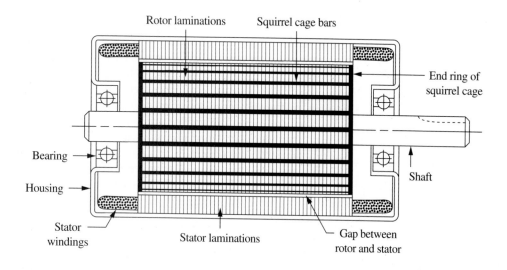

FIGURE 21–4   Induction motor laminations

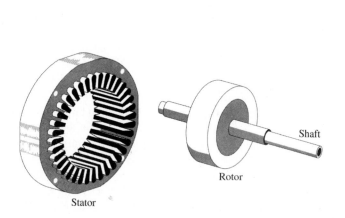

FIGURE 21–5   Squirrel cage

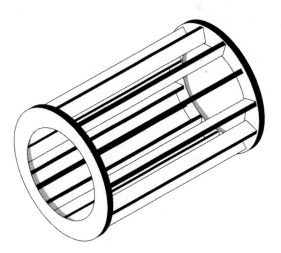

of continuous coils, called *windings*. The pattern of the coils in the stator determines the number of poles for the motor, usually 2, 4, 6, 8, 10, or 12. Table 21–2 shows that the rotational speed of the motor depends on the number of poles.

The rotor also has a lamination stack with channels along the length. The channels are typically filled with solid bars made from a good electrical conductor such as copper or aluminum, with the ends of all bars connected to continuous rings at each end. In some smaller rotors, the complete set of bars and end rings are cast from aluminum as a unit. As shown in Figure 21–5, if this casting were viewed without the laminations, it would appear similar to a squirrel cage. Thus, induction motors are often called squirrel-cage motors. The combination of the squirrel cage and the laminations is fixed onto the motor shaft with good precision to assure concentric alignment with the stator and good dynamic balance while rotating. When the rotor is installed in the supporting bearings and is inserted inside the stator, there is a small gap of approximately 0.020 in (0.50 mm) between the outer surface of the rotor and the inner surface of the stator.

Three-Phase Motors

The principles of operation of AC motors is first discussed for three-phase induction motors. Single-phase motor designs are discussed later. The three-phase electrical power, shown schematically in Figure 21–2, is connected to the stator windings. As the current flows in the windings, electromagnetic fields are created that are exposed to the conductors in the rotor. Because the three phases of power are displaced from each other in time, the effect created is a set of fields rotating around the stator. A conductor placed in a moving magnetic field has a current induced in it, and a force is exerted perpendicular to the conductor. The force acts near the periphery of the rotor, thus creating a torque to rotate the rotor.

It is the production of the induced current in the rotor that leads to calling such motors *induction motors*. Note also that there are no direct electrical connections to the rotor, thus greatly simplifying the design and construction of the motor and contributing to its high reliability.

21–5
AC MOTOR
PERFORMANCE

The performance of electric motors is usually displayed on a graph of speed versus torque, as shown in Figure 21–6. The *vertical axis* is the rotational speed of the motor as a percentage of synchronous speed. The *horizontal axis* is the torque developed by the motor as a percentage of the full-load or rated torque. When exerting its full-load torque, the motor operates at its full-load speed and delivers the rated power. See Table 21–2 for a listing of synchronous speeds and full-load speeds.

The torque at the bottom of the curve where the speed is zero is called the *starting torque* or *locked-rotor torque*. It is the torque available to initially get the load moving and begin its acceleration. This is one of the most important selection parameters for motors, as will be discussed in the descriptions of the individual types of motors.

The "knee" of the curve, called the *breakdown torque*, is the maximum torque developed by the motor during acceleration. The slope of the speed/torque curve in the vicinity of the full-load operating point is an indication of *speed regulation*. A flat curve (a low slope) indicates good speed regulation with little variation in speed as load varies. Conversely, a steep curve (a high slope) indicates poor speed regulation, and the motor will exhibit wide swings in speed as load varies. Such motors produce a "soft" acceleration of a load which may be an advantage in some applications. But where fairly constant speed is desired, a motor with good speed regulation should be selected.

FIGURE 21–6

General form of motor performance curve

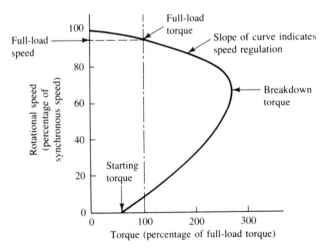

21–6
THREE-PHASE, SQUIRREL-CAGE INDUCTION MOTORS

Three of the most commonly used three-phase AC motors are simply designated as designs B, C, and D by the National Electrical Manufacturers Association (NEMA). They differ primarily in the value of starting torque and in the speed regulation near full load. Figure 21–7 shows the performance curves for these three designs for comparison. Each of these designs employs the solid squirrel-cage type of rotor, and thus there is no electrical connection to the rotor.

The 4-pole design with a synchronous speed of 1800 rpm is the most common and is available in virtually all power ratings from 1/4 hp to 500 hp. Certain sizes are available in 2-pole (3600 rpm), 6-pole (1200 rpm), 8-pole (900 rpm), 10-pole (720 rpm), and 12-pole (600 rpm) designs.

NEMA Design B

The performance of the three-phase design B motor is similar to that of the single-phase split-phase motor described later. It has a moderate starting torque (about 150% of full-load torque) and good speed regulation. The breakdown torque is high, usually 200% of full-load torque or more. Starting current is fairly high, at approximately six times full-load current. The starting circuit must be selected to be able to handle this current for the short time required to bring the motor up to speed.

Typical uses for the design B motor are centrifugal pumps, fans, blowers, and machine tools such as grinders and lathes.

NEMA Design C

High starting torque is the main advantage of the design C motor. Loads requiring 200% to 300% of full-load torque to start can be driven. Starting current is typically lower than for the design B motor for the same starting torque. Speed regulation is good and is about the same as for the design B motor. Reciprocating compressors, refrigeration systems, heavily loaded conveyors, and ball-and-rod mills are typical uses.

NEMA Design D

The design D motor has a high starting torque, about 300% of full-load torque. But it also has poor speed regulation, which results in large speed changes with varying loads. Sometimes called a *high-slip motor*, it operates at 5% to 13% slip at full load, whereas the designs B and C operate at 3% to 5% slip. Thus, the full-load speed will be lower for the design D motor.

FIGURE 21–7
Performance curves for three-phase motors: designs B, C, and D

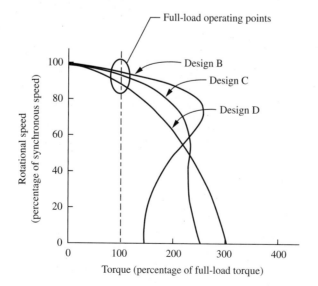

The poor speed regulation is considered an advantage in some applications and is the main reason for selecting the design D motor for such uses as punch presses, shears, sheet-metal press brakes, cranes, elevators, and oil well pumps. Allowing the motor to slow down significantly when loads increase gives the system a "soft" response, reducing shock and jerk felt by the drive system and the driven machine. Consider an elevator: When a heavily loaded elevator cab is started, the acceleration should be smooth and soft, and the cruising speed should be approached without excessive jerk. This comment also applies to a crane. If a large jerk occurs when the crane hook is heavily loaded, the peak acceleration will be high. The resulting high inertia force may break the cable.

Wound-Rotor Motors

As the name implies, the rotor of the *wound-rotor motor* has electrical windings that are connected through slip rings to the external power circuit. The selective insertion of resistance in the rotor circuit allows the performance of the motor to be tailored to the needs of the system and to be changed with relative ease to accommodate system changes or to actually vary the speed of the motor.

Figure 21–8 shows the results obtained by changing the resistance in the rotor circuit. Note that the four curves are all for the same motor, with curve 0 giving the performance with zero external resistance. This is similar to design B. Curves 1, 2, and 3 show the performance with progressively higher levels of resistance in the rotor circuit. Thus, the starting torque and the speed regulation (softness) can be tuned to the load. Speed adjustment under a given load up to approximately 50% of full-load speed can be obtained.

FIGURE 21–8

(a) Performance curves for a three-phase, wound-rotor motor with varying external resistance in the rotor circuit. (b) Schematic of wound-rotor motor with external resistance control.

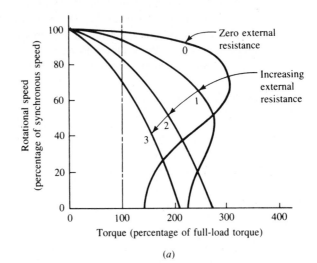

(a)

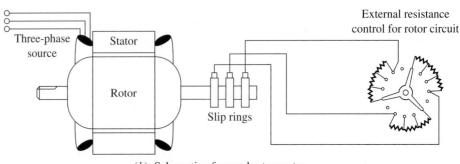

(b) Schematic of wound rotor motor with external resistance control

The wound-rotor design is used in such applications as printing presses, crushing equipment, conveyors, and hoists.

Synchronous Motors

Entirely different from the squirrel-cage induction motor or the wound-rotor motor, the *synchronous motor* operates precisely at the synchronous speed with no slip. Such motors are available in sizes from subfractional, used for timers and instruments, to several hundred horsepower to drive large air compressors, pumps, or blowers.

The synchronous motor must be started and accelerated by a means separate from the synchronous motor components themselves because they provide very little torque at zero speed. Typically, there will be a separate squirrel-cage type of winding within the normal rotor which initially accelerates the motor shaft. When the speed of the rotor is within a few percent of the synchronous speed, the field poles of the motor are excited, and the rotor is pulled into synchronism. At that point the squirrel cage becomes ineffective, and the motor continues to run at speed regardless of load variations, up to a limit called the *pull-out torque*. A load above the pull-out torque will pull the motor out of synchronism and cause it to stop.

Universal Motors

Universal motors operate on either AC or direct current (DC) power. Their construction is similar to that of a series-wound DC motor, which is described later. The rotor has electrical coils in it that are connected to the external circuit through a commutator on the shaft, a kind of slip ring assembly made of several copper segments on which stationary carbon brushes ride. Contact is maintained with light spring pressure.

Universal motors usually run at high speeds, 3500 to 20 000 rpm. This results in a high power-to-weight and high power-to-size ratio for this type of motor, making it desirable for hand-held tools such as drills, saws, and food mixers. Vacuum cleaners and sewing machines also frequently use universal motors. Figure 21–9 shows a typical set of speed/torque curves for a high-speed version of the universal motor, showing the performance for 60-Hz and 25-Hz AC power and DC power. Note that performance near the rated

FIGURE 21–9
Performance curves for a universal motor

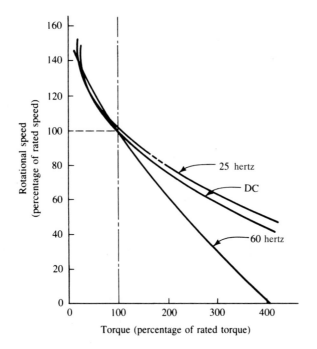

load is similar, regardless of the nature of the input power. Note also that these motors have poor speed regulation; that is, the speed varies greatly with load.

21–7 SINGLE-PHASE MOTORS

The four most common types of single-phase motors are the *split-phase, capacitor-start, permanent-split capacitor*, and *shaded-pole*. Each is unique in its physical construction and in the manner in which the electrical components are connected to provide for starting and running of the motor. The emphasis here is not on how to design the motors, but rather on the performance, so that a suitable motor can be selected.

Figure 21–10 shows the performance characteristics of these four types of motors so that they can be compared. The special features of the performance curves for the four types of motors are discussed in later sections.

In general, the construction of single-phase motors is similar to that for three-phase motors, consisting of a fixed stator, a solid rotor, and a shaft carried on bearings. The induction principle discussed earlier applies also to single-phase motors. Differences occur because single-phase power does not inherently rotate around the stator to create a moving field. Each type uses a different scheme for initially starting the motor. See Figure 21–11.

Single-phase motors are usually in the subfractional or fractional horsepower range from 1/50 hp (15 W) to 1.0 hp (750 W), although some are available up to 10 hp (7.5 kW).

Split-Phase Motors

The stator of the split-phase motor [Figure 21–11(b)] has two windings: the *main winding*, which is continuously connected to the power line, and the *starting winding*, which is connected only during the starting of the motor. The starting winding creates a slight phase shift that creates the initial torque to start and accelerate the rotor. After the rotor reaches approximately 75% of its synchronous speed, the starting winding is cut out by a centrifugal switch, and the rotor continues to run on the main winding.

The performance curve for the split-phase motor is shown in Figure 21–10. It has moderate starting torque, approximately 150% of full-load torque. It has good efficiency and is designed for continuous operation. Speed regulation is good. One of the disadvantages is that it requires a centrifugal switch to cut out the starting winding. The step in the speed/torque curve indicates this cutout.

FIGURE 21–10
Performance curves of four types of single-phase electric motors

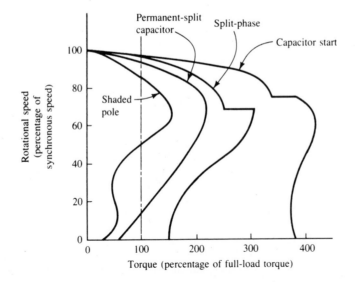

FIGURE 21–11
Schematic diagrams of
single-phase motors

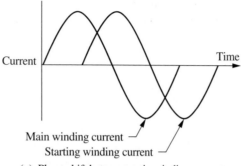

(a) Phase shift between main winding current
and starting winding current

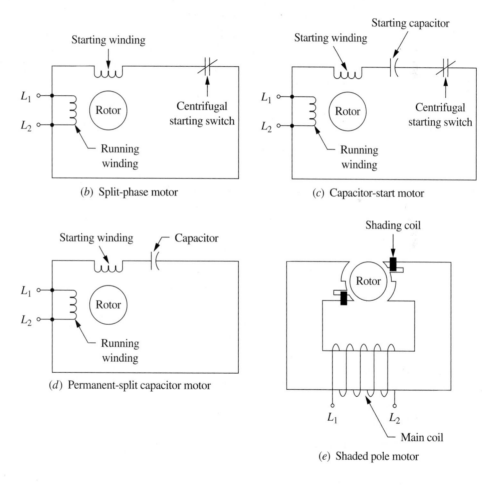

(b) Split-phase motor

(c) Capacitor-start motor

(d) Permanent-split capacitor motor

(e) Shaded pole motor

These characteristics make the split-phase motor one of the most popular types, used in business machines, machine tools, centrifugal pumps, electric lawn mowers, and similar applications.

Capacitor-Start Motors

Like the split-phase motor, the capacitor-start motor [Figure 21–11(c)] also has two windings: a *main* or *running winding* and a *starting winding*. But in it, a capacitor is connected in series with the starting winding, giving a much higher starting torque than with the split-phase motor. A starting torque of 250% of full load or higher is common. Again a centrifugal switch is used to cut out the starting winding and the capacitor. The running characteristics of the motor are then very similar to those of the split-phase motor: good speed regulation and good efficiency for continuous operation.

Disadvantages include the switch and the relatively bulky capacitor. Frequently the capacitor is mounted conspicuously right on top of the motor (see Figure 21–17). It may also be integrated into a package containing the starting switch, a relay, or other control elements.

Uses for the capacitor-start motor include the many types of machines that need the high starting torque. Examples include heavily loaded conveyors, refrigeration compressors, and pumps and agitators for heavy fluids.

Permanent-Split Capacitor Motors

A capacitor is connected in series with the starting winding at all times. The starting torque of the permanent-split capacitor motor [Figure 21–11(d)] is typically quite low, approximately 40% or less of full-load torque. Thus, only low-inertia loads such as fans and blowers are usually used. An advantage is that you can tailor the running performance and the speed regulation to match the load by selecting the appropriate capacitor value. Also, no centrifugal switch is required.

Shaded-Pole Motors

The shaded-pole motor [Figure 21–11(e)] has only one winding, the *main* or *running winding*. The starting reaction is created by the presence of a copper band around one side of each pole. The low-resistance band "shades" the pole to produce a rotating magnetic field to start the motor.

The shaded-pole motor is simple and inexpensive, but it has a low efficiency and a very low starting torque. Speed regulation is poor, and it must be fan-cooled during normal operation. Thus, its primary use is in shaft-mounted fans and blowers where the air is drawn over the motor. Some small pumps, toys, and intermittently used household items also employ shaded-pole motors because of their low cost.

21–8 AC MOTOR FRAME TYPES AND ENCLOSURES

Frame Types

The design of the equipment in which the motor is to be mounted determines the type of frame required. Some types are described below.

Foot-mounted.   The most widely used type for industrial machinery, the foot-mounted frame has integral feet with a standard hole pattern for bolting the motor to the machine (see Figure 21–12).

Cushion Base.   A foot mounting is provided with resilient isolation of the motor from the frame of the machine to reduce vibration and noise (see Figure 21–13).

C-Face Mounting.   A machined face is provided on the shaft end of the motor which has a standard pattern of tapped holes. Driven equipment is then bolted directly to the motor. The design of the face is standardized by the National Electrical Manufacturers Association (NEMA). See Figures 21–14 and 21–15.

FIGURE 21–12
Foot-mounted motors
with various enclosure
types (Rockwell
Automation/Reliance
Electric, Greenville, SC)

(a) Open protected, dripproof

(b) Totally enclosed, non-ventilated
lint-proof

(c) Totally enclosed, fan cooled

(d) Explosion proof, dust ignition
proof

FIGURE 21–13   Enclosed, cushion-base motor with
double-end shaft (A. O. Smith Electrical Products,
Tipp City, OH)

FIGURE 21–14   C-face motor. See Figure 8–25 for
an example of a reducer designed to mate with a
C-face motor. (Rockwell Automation/Reliance
Electric, Greenville, SC)

FIGURE 21–15
Rigid-base, dripproof,
NEMA C-face motor
for application to a
close-coupled pump
(A. O. Smith Electrical
Products, Tipp City, OH)

D-Flange Mounting.   A machined flange is provided on the shaft end of the motor with a standard pattern of through clearance holes for bolts for attaching the motor to the driven equipment. The flange design is controlled by NEMA.

Vertical Mounting.   Vertical mounting is a special design because of the effects of the vertical orientation on the bearings of the motor. Attachment to the driven equipment is usually through C-face or D-flange bolt holes as previously described (see Figure 21–16).

Unmounted.   Some equipment manufacturers purchase only the bare rotor and stator from the motor manufacturer and then build them into their machine. Compressors for refrigeration equipment are usually built in this manner.

Special-Purpose Mountings.   Many special designs are made for fans, pumps, oil burners, and so on.

Enclosures

The housings around the motor that support the active parts and protect them vary with the degree of protection required. Some of the enclosure types are shown in Figures 21–12 through 21–17 and are described next.

Open.   Typically a light-gage sheet-metal housing is provided around the stator with end plates to support the shaft bearings. The housing contains several holes or slots that permit cooling air to enter the motor. Such a motor must be protected by the housing of the machine itself (see Figure 21–17).

Protected.   Sometimes called *dripproof*, ventilating openings are provided only on the lower part of the housing so that liquids dripping on the motor from above cannot enter the motor. This is probably the most widely used type [see Figure 21–12(a)].

Totally Enclosed Nonventilated (TENV).   No openings at all are provided in the housing, and no special provisions are made for cooling the motor except for fins cast into the frame to promote convective cooling. The design protects the motor from harmful atmospheres [see Figure 21–12(b)].

Totally Enclosed Fan-cooled.   The totally enclosed fan-cooled (TEFC) design is similar to TENV design, except a fan is mounted to one end of the shaft to draw air over the finned housing [see Figure 21–12(c)].

FIGURE 21–16
Totally enclosed, nonventilated, C-face motor with dripcover for vertical operation (A. O. Smith Electrical Products, Tipp City, OH)

FIGURE 21–17
Open-frame motor, shown with cushion base. Mounting may also be by through-bolts, belly-band, or cushion ring (A. O. Smith Electrical Products, Tipp City, OH)

FIGURE 21–18 Key for NEMA standard motor frame dimensions listed in Table 21–3

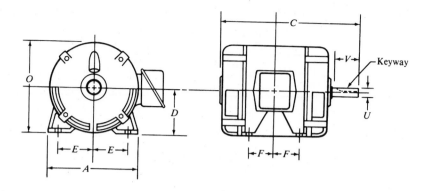

TABLE 21–3 Motor frame sizes

hp	Frame size	Dimensions (in)								
		A	C	D	E	F	O	U	V	Keyway
1/4	48	5.63	9.44	3.00	2.13	1.38	5.88	0.500	1.50	0.05 flat
1/2	56	6.50	10.07	3.50	2.44	1.50	6.75	0.625	1.88	3/16 × 3/32
1	143T	7.00	10.69	3.50	2.75	2.00	7.00	0.875	2.00	3/16 × 3/32
2	145T	7.00	11.69	3.50	2.75	2.50	7.00	0.875	2.00	3/16 × 3/32
5	184T	9.00	13.69	4.50	3.75	2.50	9.00	1.125	2.50	1/4 × 1/8
10	215T	10.50	17.25	5.25	4.25	3.50	10.56	1.375	3.13	5/16 × 5/32
15	254T	12.50	22.25	6.25	5.00	4.13	12.50	1.625	3.75	3/8 × 3/16
20	256T	12.50	22.25	6.25	5.00	5.00	12.50	1.625	3.75	3/8 × 3/16
25	284T	14.00	23.38	7.00	5.50	4.75	14.00	1.875	4.38	1/2 × 1/4
30	286T	14.00	24.88	7.00	5.50	5.50	14.00	1.875	4.38	1/2 × 1/4
40	324T	16.00	26.00	8.00	6.25	5.25	16.00	2.125	5.00	1/2 × 1/4
50	326T	16.00	27.50	8.00	6.25	6.00	16.00	2.125	5.00	1/2 × 1/4

Note: All motors are four-pole, three-phase, 60-Hz, AC induction motors. Refer to Figure 21–18 for description of dimensions.

TEFC-XP. The TEFC-XP (explosion-proof) design is similar to the TEFC housing, except special protection is provided for electrical connections to prohibit fire or explosion in hazardous environments [see Figure 21–12(d)].

Frame Sizes

The critical dimensions of motor frames are controlled by NEMA frame sizes. Included are the overall height and width; the height from the base to the shaft centerline; the shaft diameter, length, and keyway size; and mounting hole pattern dimensions. A few selected motor frame sizes for 1725-rpm, three-phase, induction, foot-mounted dripproof motors are listed in Table 21–3. For the descriptions of the dimensions, refer to Figure 21–18.

21–9 CONTROLS FOR AC MOTORS

Motor controls must perform several functions, as outlined in Figure 21–19. The complexity of the control depends on the size and the type of motor involved. Small fractional or subfractional motors may sometimes be started with a simple switch that connects the motor directly to the full line voltage. Larger motors, and some smaller motors on critical equipment, require greater protection.

FIGURE 21–19
Motor control block
diagram

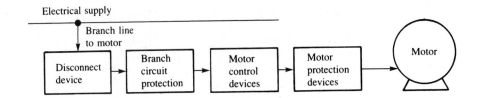

The functions of motor controls are as follows:

1. To start and stop the motor
2. To protect the motor from overloads that would cause the motor to draw dangerously high current levels
3. To protect the motor from overheating
4. To protect personnel from contact with hazardous parts of the electrical system
5. To protect the controls from the environment
6. To prohibit the controls from causing a fire or an explosion
7. To provide controlled torque, acceleration, speed, or deceleration of the motor
8. To provide for the sequential starting of a series of motors or other devices
9. To provide for the coordinated operation of different parts of a system, possibly
   including several motors
10. To protect the conductors of the branch circuit in which the motor is connected

The proper selection of a motor control system requires knowledge of at least the following factors:

1. The type of electrical service: voltage and frequency; single- or three-phase; current limitations
2. The type and size of motor: power and speed ratings; full-load current rating;
   locked-rotor current rating
3. Operation desired: duty cycle (continuous, start/stop, or intermittent); single
   or multiple discrete speeds, or variable-speed operation; one-direction or
   reversing
4. Environment: temperature; water (rain, snow, sleet, sprayed or splashed water); dust and dirt; corrosive gases or liquids; explosive vapors or dusts; oils or
   lubricants
5. Space limitations
6. Accessibility of controls
7. Noise or appearance factors

Starters

There are several classifications of motor starters: manual or magnetic; one-direction or reversing; two-wire or three-wire control; full-voltage or reduced-voltage starting; single-
speed or multiple-speed; normal stopping, braking, or plug stopping. All of these typically
include some form of overload protection, which will be discussed later.

Manual and Magnetic, Full-Voltage, One-Direction Starting

Figure 21–20 shows the schematic connection diagram for manual starters for single-phase and three-phase motors. The symbol *M* indicates a normally open contactor (switch) that is actuated manually, for example, by throwing a lever. The contactors are rated according to the motor power that they can safely handle. The power rating indirectly relates to the current drawn by the motor, and the contactor design must (1) safely make contact during the start-up of the motor, considering the high starting current; (2) carry the expected range of operating current without overheating; and (3) break contact without excessive arcing that could burn the contacts. The ratings are established by NEMA. Tables 21–4 and 21–5 show the ratings for selected NEMA starter sizes.

Note in Figure 21–20 that overload protection is required in all three lines for three-phase motors but in only one line of the single-phase motors.

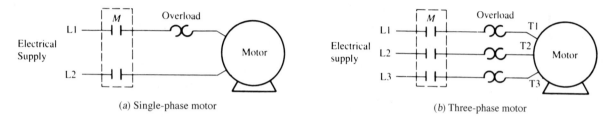

(a) Single-phase motor   (b) Three-phase motor

FIGURE 21–20   Manual starters. *M* = normally open contactors. All actuate together.

TABLE 21–4   Ratings of AC full-voltage starters for single-phase power

NEMA size number	Current rating (amperes)	Power rating at given voltages					
		110 V		220 V		440 and 550 V	
		(hp)	(kW)	(hp)	(kW)	(hp)	(kW)
00		1/2	0.37	3/4	0.56		
0	15	1	0.75	$1\frac{1}{2}$	1.12	$1\frac{1}{2}$	1.12
1	25	$1\frac{1}{2}$	1.12	3	2.24	5	3.73
2[a]	50	3	2.24	$7\frac{1}{2}$	5.60	10	7.46
3[a]	100	$7\frac{1}{2}$	5.60	15	11.19	25	18.65

[a]Applies to magnetically operated starters only.

TABLE 21–5   Ratings of AC full-voltage starters for three-phase power

NEMA size number	Current rating (amperes)	Power rating at given voltages					
		110 V		220 V		440 and 550 V	
		(hp)	(kW)	(hp)	(kW)	(hp)	(kW)
00		3/4	0.56	1	0.75	1	0.75
0	15	$1\frac{1}{2}$	1.12	2	1.49	2	1.49
1	25	3	2.24	5	3.73	$7\frac{1}{2}$	5.60
2	50	$7\frac{1}{2}$	5.60	15	11.19	25	18.65
3	100	15	11.19	30	22.38	50	37.30

FIGURE 21–21

Magnetic starters for
three-phase motors

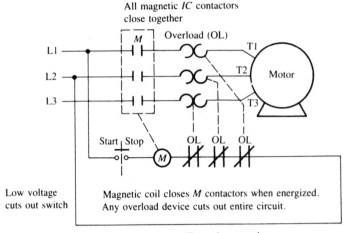

(a) Two-wire control

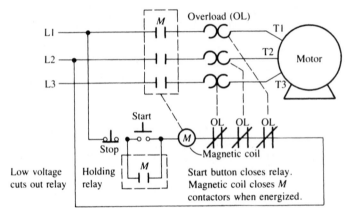

(b) Three-wire control

Figure 21–21 shows the schematic connection diagrams for magnetic starters using two-wire and three-wire controls. The "start" button in the three-wire control is a momentary contact type. As it is actuated manually, the coil in parallel with the switch is energized, and it magnetically closes the line contactors marked *M*. The contacts remain closed until the stop button is pushed or until the line voltage drops to a set low value. (Remember, a low line voltage causes the motor to draw excessive current.) Either case causes the magnetic contactors to open, stopping the motor. The start button must be manually pushed again to restart the motor.

The two-wire control has a manually operated start button that stays engaged after the motor starts. As a safety feature, the switch will open when a low-voltage condition occurs. But when the voltage rises again to an acceptable level, the contacts close, restarting the motor. You must ensure that this is a safe operating mode.

Reversing Starters

Figure 21–22 shows the connection for a reversing starter for a three-phase motor. You can reverse the direction of rotation of a three-phase motor by interchanging any two of the three power lines. The *F* contactors are used for the forward direction. The *R* contactors

FIGURE 21–22

Reversing control for a three-phase motor

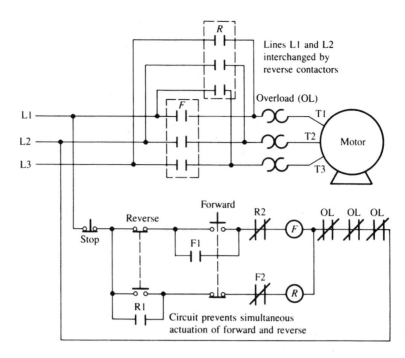

would interchange L1 and L3 to reverse the direction. The *Forward* and *Reverse* pushbuttons actuate only one of the sets of contactors.

Reduced-Voltage Starting

The motors discussed in the previous sections and the circuits shown in Figures 21–20 through 21–22 employ full-voltage starting. That is, when the system is actuated, the full line voltage is applied to the motor terminals. This will give the maximum starting effort, but in some cases it is not desirable. To limit jerk, to control the acceleration of a load, and to limit the starting current, reduced-voltage starting is sometimes used. This gentle start is used on some conveyors, hoists, pumps, and similar loads.

Figure 21–23 shows one method of providing a reduced voltage to the motor when starting. The first action is the closing of the contactors marked *A*. Thus, the power to the motor passes through a set of resistors that reduces the voltage at each motor terminal. A typical reduction would be to approximately 65% of normal line voltage. The peak line current would be reduced to 65% of normal locked-rotor current, and the starting torque would be 42% of normal locked-rotor torque. (See Reference 4.) After the motor is accelerated, the main contactors *M* are closed, and full voltage is applied to the motor. A timer is typically used to control the sequencing of the *A* and *M* contactors.

Dual-Speed Motor Starting

A dual-speed motor with two separate windings to produce the different speeds can be started with the circuit shown in Figure 21–24. The operator selectively closes either the *F* (fast) or *S* (slow) contacts to obtain the desired speed. The other features of starting circuits discussed earlier can also be applied to this circuit.

FIGURE 21–23
Reduced-voltage
starting by primary
resistor method

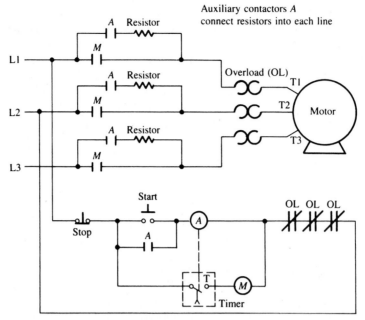

Auxiliary contactors *A*
connect resistors into each line

Auxiliary contactors engaged first.
Timer then actuates main contactors,
shorting around starting resistors.

FIGURE 21–24
Speed control for a
dual-winding, three-
phase motor

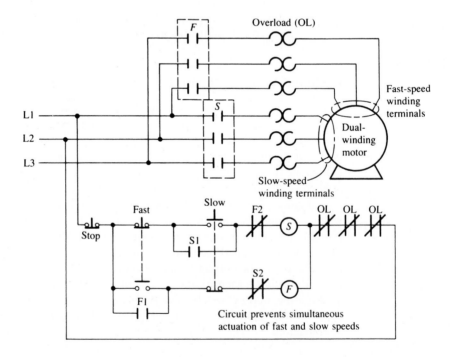

Circuit prevents simultaneous
actuation of fast and slow speeds

Stopping the Motor

Where no special conditions exist when the system is shut down, the motor can be permitted to coast to a stop after the power is interrupted. The time required to stop will depend on the inertia and the friction in the system. If controlled, rapid stopping is required, external brakes can be used. *Brake motors*, which have a brake integral with the motor, are available. Typically, the design is of the "fail-safe" nature, in which the brake is disengaged by an electromagnetic coil when the motor is energized. When the motor is de-energized, either on purpose or because of power failure, the brake is actuated by mechanical spring force.

On circuits with reversing starters, *plug stopping* can be used. When it is desired to stop the motor running in the forward direction, the control can be switched immediately to reverse. There would then be a decelerating torque applied to the rotor, stopping it quickly. Care must be exercised to cut out the reversing circuit when the motor is at rest to prevent it from continuing in the reverse direction.

Overload Protection

The chief cause of failure in electric motors is overheating of the wound coils due to excessive current. The current is dependent on the load on the motor. A short circuit, of course, would cause a virtually instantaneously high current of a damaging level.

The protection against a short circuit can be provided by *fuses*, but careful application of fuses to motors is essential. A fuse contains an element that literally melts when a particular level of current flows through it. Thus, the circuit is opened. Reactivating the circuit would require replacing the fuse. Time-delay fuses, or "slow-blowing" fuses, are needed for motor circuits to prevent the fuses from blowing when the motor starts, drawing the relatively high starting current that is normal and not damaging. After the motor starts, the fuse will blow at a set value of overcurrent.

Fuses are inadequate for larger or more critical motors because they provide protection at only one level of overcurrent. Each motor design has a characteristic *overheating curve*, as shown in Figure 21–25. This indicates that the motor could withstand different levels of overcurrent for different periods of time. For example, for the motor heating curve of Figure 21–25, a current twice as high as the full-load current (200%) could exist for up to 9 min before a damaging temperature was produced in the windings. But a 400% overload would cause damage in less than 2 min. An ideal overload protection device would parallel the overheating curve of the given motor, always cutting out the motor at a safe current level, as shown in Figure 21–25. Devices are available commercially to provide this protection. Some use special melting alloys, bimetallic strips similar to a thermostat, or magnetic coils that are sensitive to the current flowing in them. Most large motor starters include overload protection integral with the starter.

Another type of overload protection uses a temperature-sensitive device inserted in the windings of the motor when it is made. Then it opens the motor circuit when the windings reach a dangerous temperature, regardless of the reason.

FIGURE 21–25
Motor heating curve
and response curve of a
typical overload
protector (Square D
Company, Palatine, IL)

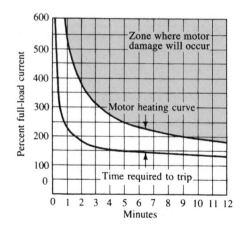

Solid State Overload Relay

Difficulties with thermal element or bimetallic overload devices can be overcome by using a solid-state overload relay. Thermal devices that use a melting element require the replacement of that element after tripping, resulting in extra cost for maintenance supplies and personnel. Both thermal element and bimetallic overload devices are affected by varying ambient temperatures that can change the actual current flow protection level. Temperature compensation devices are available, but they require careful setting and knowledge of expected conditions. Solid-state overload relays overcome these difficulties because only sensed current flow level is used to produce the tripping action. They are inherently insensitive to ambient temperature swings. Furthermore, they can sense the current flow in each of the three windings of three-phase motors and provide protection if any one of the phases experiences a failure or a given increase in current. This provides protection not only for the motor but also for associated equipment that may be damaged if the motor fails suddenly. See Internet site 5 for additional information.

Enclosures for Motor Controls

As stated before, one of the functions of a motor control system is to protect personnel from contact with dangerous parts of the electrical system. Also, protection of the system from the environment must be provided. These functions are accomplished by the enclosure.

NEMA has established standards for enclosures for the variety of environments encountered by motor controls. The most frequent types are described in Table 21–6.

AC Variable-Speed Drives

Standard AC motors operate at a fixed speed for a given load if powered by AC power at a fixed frequency, for example, 60 Hz. Variable-speed operation can be obtained with a control system that produces a variable frequency power. Two such control types are in common use: the *six-step method* and the *pulse-width modulation (PWM) method*. Either system takes the 60-Hz line voltage and first rectifies it to a DC voltage. The six-step method then uses an inverter to produce a series of square waves that provides a voltage to the motor winding which varies both voltage and frequency in six steps per cycle. In the PWM system,

TABLE 21–6 Motor control enclosures

NEMA design number	Description
1	General-purpose: indoor use; not dusttight
3	Dusttight, raintight: outdoor-weather-resistant
3R	Dusttight, rainproof, sleet-resistant
4	Watertight: can withstand a direct spray of water from a hose; used on ships and in food processing plants where washdown is required
4X	Watertight, corrosion-resistant
7	Hazardous locations, class I: can operate in areas where flammable gases or vapors are present
9	Hazardous locations, class II: combustible dust areas
12	Industrial use: resistant to dust, lint, oil, and coolants
13	Oiltight, dusttight

FIGURE 21–26
Six-step method of
variable-speed AC
motor control
(Rockwell
Automation/Allen-
Bradley, Milwaukee,
WI)

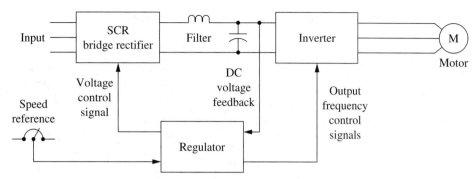

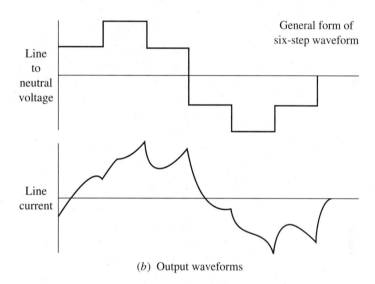

(a) Schematic diagram of variable voltage inverter

(b) Output waveforms

the DC voltage is input to an inverter that produces a series of pulses of variable width. The
rate of polarity reversals determines the frequency applied to the motor. See Figures 21–26
and 21–27.

Reasons for Applying Variable-Speed Drives

It is often desirable to vary the speed of mechanical systems to obtain operating characteristics that are more nearly optimum for the application. For example:

1. Conveyor speed can be varied to match the demand of production.
2. The delivery of bulk materials to a process can be varied continuously.
3. Automatic control can provide synchronization of two or more system components.
4. Dynamic control of system operation can be used during start-up and stopping se-
   quences, for torque control, or for in-process acceleration and deceleration control,
   often necessary when processing continuous webs such as paper or plastic film.
5. Spindle speeds of machine tools can be varied to produce optimum cutting for
   given materials, depth of cut, feeds, or cutting tools.
6. The speeds of fans, compressors, and liquid pumps can be varied in response to
   needs for cooling or for product delivery.

FIGURE 21–27
Pulse-width modulation
method of variable-
speed AC motor control
(Rockwell Automation/
Allen-Bradley,
Milwaukee, WI)

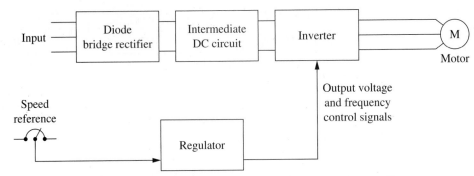

(*a*) Schematic diagram of pulse-width modulation (PWM) controller

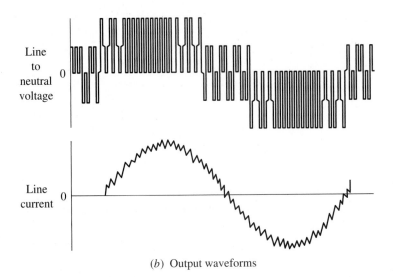

(*b*) Output waveforms

All of these situations allow for more flexible and better process control. Cost savings are also obtained, particularly for item 6. The difference in the power required to operate a pump at two speeds is proportional to the cube of the ratio of the speeds. For example, if the speed of the motor is reduced to one-half of its original speed, the power required to operate the pump is reduced to 1/8 of the original power. Matching the pump speed to the required delivery of the fluid can accumulate significant savings. Similar savings result for fans and compressors.

**21–10
DC POWER**

DC motors have several inherent advantages over AC motors, as discussed in the next section. A disadvantage of DC motors is that a source of DC power must be available. Most residential, commercial, and industrial locations have only AC power provided by the local utility. Three approaches are used to provide DC power:

1. *Batteries:* Typically batteries are available in voltages of 1.5, 6.0, 12.0, and 24.0 volts (V). They are used for portable devices or for mobile applications. The power is pure DC, but voltage varies with time as the battery discharges. The bulkiness, weight, and finite life are disadvantages.

2. *Generators:* Powered by AC electric motors, internal combustion engines, turbine engines, wind devices, water turbines, and so on; DC generators produce pure DC. The usual voltages are 115 and 230 V. Some industries maintain such generators to provide DC power throughout the plant.

TABLE 21–7   DC motor voltage ratings

Input AC voltage	DC motor rating	NEMA code
115 V AC, one-phase	90 V DC	K
230 V AC, one-phase	180 V DC	K
230 V AC, three-phase	240 V DC	C or D
460 V AC, three-phase	500 V DC or	C or D
	550 V DC	
460 V AC, three-phase	240 V DC	E

3. *Rectifiers: Rectification* is the process of converting AC power with its sinusoidal variation of voltage with time to DC power, which ideally is nonvarying. A readily available device is the *silicon-controlled rectifier* (*SCR*). One difficulty with rectification of AC power to produce DC power is that there is always some amount of "ripple," a small variation of voltage as a function of time. Excessive ripple can cause overheating of the DC motor. Most commercially available SCR devices produce DC power with an acceptably low ripple. Table 21–7 lists the commonly used DC voltage ratings for motors powered by rectified AC power as defined by NEMA.

21–11
DC MOTORS

The advantages of direct current motors are summarized here:

■ The speed is adjustable by use of a simple rheostat to adjust the voltage applied to the motor.

■ The direction of rotation is reversible by switching the polarity of the voltage applied to the motor.

■ Automatic control of speed is simple to provide for matching of the speeds of two or more motors or to program a variation of speed as a function of time.

■ Acceleration and deceleration can be controlled to provide the desired response time or to decrease jerk.

■ Torque can be controlled by varying the current applied to the motor. This is desirable in tension control applications, such as the winding of film on a spool.

■ Dynamic braking can be obtained by reversing the polarity of the power while the motor is rotating. The reversed effective torque slows the motor without the need for mechanical braking.

■ DC motors typically have quick response, accelerating quickly when voltage is changed, because they have a small rotor diameter, giving them a high ratio of torque to inertia.

DC motors have electric windings in the rotor, and each coil has two connections to the commutator on the shaft. The commutator is a series of copper segments through which the electric power is transferred to the rotor. The current path from the stationary part of the motor to the commutator is through a pair of brushes, usually made of carbon, which are held against the commutator by light coil or leaf springs. Maintenance of the brushes is one of the disadvantages of DC motors.

DC Motor Types

Four commonly used DC motor types are the *shunt-wound*, *series-wound*, *compound-wound*, and *permanent magnet* motors. They are described in terms of their speed/torque curves in a manner similar to that used for AC motors. One difference here is that the speed axis is expressed in percentage of *full-load rated speed*, rather than a percentage of synchronous speed, as that term does not apply to DC motors.

Shunt-Wound DC Motor.   The electromagnetic field is connected in parallel with the rotating armature, as sketched in Figure 21–28. The speed/torque curve shows relatively good speed regulation up to approximately two times full-load torque, with a rapid drop in speed after that point. The speed at no load is only slightly higher than the full-load speed. Contrast this with the series-wound motor described next. Shunt-wound motors are used mainly for small fans and blowers.

Series-Wound DC Motor.   The electromagnetic field is connected in series with the rotating armature, as shown in Figure 21–29. The speed/torque curve is steep, giving the motor a soft performance that is desirable in cranes, hoists, and traction drives for vehicles. The starting torque is very high, as much as 800% of full-load rated torque. A major difficulty, however, with series-wound motors is that the no-load speed is theoretically unlimited. The motor could reach a dangerous speed if the load were to be accidentally disconnected. Safety devices, such as overspeed detectors that shut off the power, should be used.

Compound-Wound DC Motors.   The compound-wound DC motor employs both a series field and a shunt field, as sketched in Figure 21–30. It has a performance somewhat between that of the series-wound and the shunt-wound motors. It has fairly high starting torque and a soft speed characteristic, but it has an inherently controlled no-load speed. This makes it good for cranes, which may suddenly lose their loads. The motor would run slowly when heavily loaded for safety and control, and fast when lightly loaded to improve productivity.

FIGURE 21–28
Shunt-wound DC
motor performance
curve

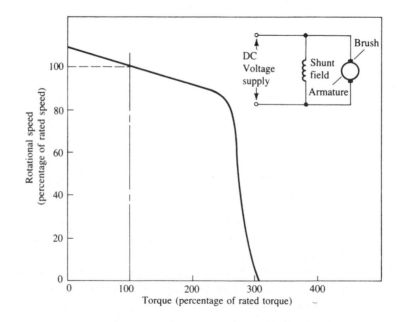

FIGURE 21–29
Series-wound DC
motor performance
curve

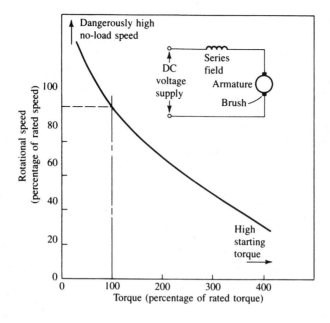

FIGURE 21–30
Compound-wound DC
motor performance
curve

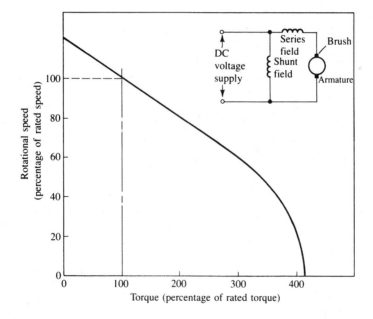

Permanent Magnet DC Motors. Instead of using electromagnets, the permanent magnet DC motor uses permanent magnets to provide the field for the armature. The direct current passes through the armature, as shown in Figure 21–31. The field is nearly constant at all times and results in a linear speed/torque curve. Current draw also varies linearly with torque. Applications include fans and blowers to cool electronics packages in aircraft, small actuators for control in aircraft, automotive power assists for windows and seats, and fans in automobiles for heating and air conditioning. These motors frequently have built-in gear-type speed reducers to produce a low-speed, high-torque output.

FIGURE 21–31
Permanent magnet DC
motor performance
curve

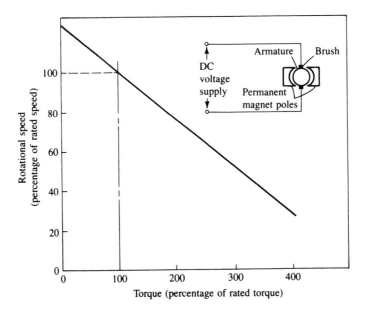

<table>
<tr><td>21–12</td><td></td></tr>
</table>

**21–12
DC MOTOR
CONTROL**

Starting DC motors presents essentially the same problems as discussed for AC motors in terms of limiting the starting current and the provision of switching devices and holding relays of sufficient capacity to handle the operating loads. The situation is made somewhat more severe, however, by the presence of the commutators in the rotor circuit which are more sensitive to overcurrent.

Speed control is provided by variation of the resistance in the lines containing the armature or the field of the motor. The details depend on whether the motor is a series, shunt, or compound type. The variable-resistance device, sometimes called a *rheostat*, can provide either stepwise variation in resistance or continuously varying resistance. Figure 21–32 shows the schematic diagrams for several types of DC motor speed controls.

**21–13
OTHER TYPES OF
MOTORS**

Torque Motors

As the name implies, *torque motors* are selected for their ability to exert a certain torque rather than for a rated power. Frequently this type of motor is operated at a stalled condition to maintain a set tension on a load. The continuous operation at slow speed or at zero speed causes heat generation to be a potential problem. In severe cases, external cooling fans may be required.

By special design, several of the AC and DC motor types discussed elsewhere in this chapter can be used as torque motors.

Servomotors

Either AC or DC *servomotors* are available to provide automatic control of position or speed of a mechanism in response to a control signal. Such motors are used in aircraft actuators, instruments, computer printers, and machine tools. Most have rapid response characteristics because of the low inertia of the rotating components and the relatively high torque exerted by the motor. See Internet sites 3, 7–10, and 12.

Figure 21–33 shows a schematic diagram of a servomotor controller system. Three control loops are shown: (1) the position loop, (2) the velocity loop, and (3) the current loop. Speed control is effected by sensing the motor speed with a tachometer and feeding the signal back through the velocity loop to the controller. Position is sensed by an optical encoder or a similar device on the driven load, with the signal fed back through the position loop to

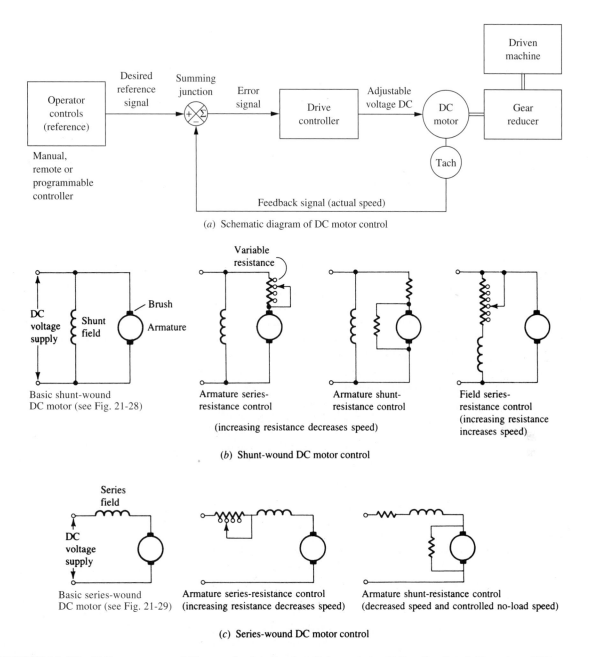

(a) Schematic diagram of DC motor control

(b) Shunt-wound DC motor control

(c) Series-wound DC motor control

FIGURE 21–32 DC motor control [Source for (a), Rockwell Automation/Allen-Bradley, Milwaukee, WI]

the controller. The controller sums the inputs, compares them with the desired value set by the control program, and generates a signal to control the motor. Thus, the system is a closed-loop servo-control. Typical uses for this type of control would be for numerical control machine tools, special-purpose assembly machines, and aircraft control surface actuators.

Stepping Motors

A stream of electronic pulses is delivered to a *stepping motor*, which then responds with a fixed rotation (step) for each pulse. Thus, a very precise angular position can be obtained by counting and controlling the number of pulses delivered to the motor. Several step angles are

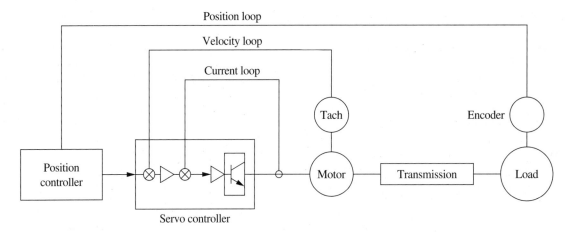

FIGURE 21–33   Servomotor controller system (Rockwell Automation/Allen-Bradley, Milwaukee, WI)

available in commercially provided motors, such as 1.8°, 3.6°, 7.5°, 15°, 30°, 45°, and 90°. When the pulses are stopped, the motor stops automatically and is held in position. Because many of these motors are connected through a gear-type speed reducer to the load, very precise positioning is possible to a small fraction of a step. Also, the reducer provides a torque increase. See Internet sites 8 and 10.

Brushless Motors

The typical DC motor requires brushes to make contact with the rotating commutator on the shaft of the motor. This is a major failure mode of such motors. In the *brushless DC motor*, the switching of the rotor coils is accomplished by solid-state electronic devices, resulting in very long life. The emission of electromagnetic interference is likewise reduced compared with brush-type DC motors.

Printed Circuit Motors

The rotor of the *printed circuit motor* is a flat disc that operates between two permanent magnets. The resulting design has a relatively large diameter and small axial length; sometimes it is called a *pancake motor*. The rotor has a very low inertia, so high acceleration rates are possible.

Linear Motors

Linear motors are electrically similar to rotary motors, except the components, stator and rotor, are laid flat instead of being formed into a cylindrical shape. Types include brush-type DC motors, brushless DC motors, stepping motors, and the single-phase type of motor. Capacity is measured in terms of the force that the motor can exert, ranging typically from a few pounds to 2500 lb. Speeds range from about 40 to 100 in/s. See Internet sites 3, 11, and 12.

REFERENCES

1. Avallone, Eugene P., and Theodore Baumeister III. *Marks' Handbook for Mechanical Engineers.* 10th ed. New York: McGraw-Hill, 1996.

2. Hubert, Charles I. I. *Electric Machines: Theory, Operating Applications, and Controls.* 2d ed. Upper Saddle River, NJ: Prentice Hall Professional Technical Reference, 2002.

3. Skvarenina, Timothy L., and William E. DeWitt. *Electrical Power and Controls.* Upper Saddle River, NJ: Prentice Hall Professional Technical Reference, 2001.

4. Wildi, Theodore. *Electrical Machines, Drives, and Power Systems.* 5th ed. Upper Saddle River, NJ: Prentice Hall Professional Technical Reference, 2002.

INTERNET SITES FOR ELECTRIC MOTORS AND CONTROLS

1. **Reliance Electric/Rockwell Automation.** *www.reliance.com* Manufacturer of AC and DC motors and the associated controls. The site provides a Motor Technical Reference resource giving motor performance, applications guidelines, motor construction information, and an online catalog.

2. **A. O. Smith Electrical Products Company.** *www.aosmithmotors.com* Manufacturer of electric motors from the subfractional (1/800 hp) to the large integral horsepower (800 hp) ranges, under the A. O. Smith, Universal, and Century brands.

3. **Baldor Electric Company.** *www.baldor.com* Manufacturer of AC and DC motors, gear motors, servomotors, linear motors, generators, linear motion products, and controls for a wide range of industrial and commercial applications. An online catalog, CAD drawings, and performance charts are included. An extensive, 124-page online product brochure provides much information about motor technology and frame sizes.

4. **U.S. Electric Motors Company.** *www.usmotors.com* Manufacturer of a wide variety of AC and DC electric motors from ¼ to 4000 hp for general and specific applications. An online catalog is included. Part of Emerson Electric Company, motors are offered under the U.S. Motors, Doerr, Emerson, and Hurst brand names. See also *www.emersonmotors.com*

5. **Square D.** *www.squared.com* Manufacturer of electric motor controls along with electrical distribution and industrial automation products, systems, and services, under the Square D, Modicon, Merlingerin, and Telemecanique brands. Included are adjustable frequency AC motor drives, motor contactors and starters, and motor control centers. Site provides technical data, CAD drawings, and an online catalog.

6. **Eaton/Cutler Hammer Company.** *www.ch.cutler-hammer.com* Manufacturer of a wide range of electrical control and power distribution products for industrial, commercial, and residential applications. Online product knowledge is available for motor control centers, circuit breakers, power conditioning equipment, switchgear, brakes, and many other products.

7. **Allen-Bradley/Rockwell Automation.** *www.ab.com* Manufacturer of a wide variety of controls for automation, including motor contactors, AC motor drives, motor control centers, servomotors, programmable logic controllers, sensors, relays, circuit protection devices, network communication, and control systems.

8. **GE Fanuc.** *www.gefanuc.com* Part of GE Industrial Systems, GE Fanuc Automation is a joint venture of General Electric Company and FANUC LTD of Japan. Manufacturer of servomotors, stepping motors, programmable logic controllers (PLCs), and other motion controls for automation.

9. **Parker Automation/Compumotor Division.** *www.compumotor.com* Manufacturer of brushless servomotors and controllers for a variety of industrial automation and commercial product applications.

10. **Oriental Motor U.S.A. Corporation.** *www.orientalmotor.com* Manufacturer of stepping motors with either AC or DC input power and low speed synchronous motors for a variety of industrial automation and commercial product applications.

11. **Trilogy Systems Company.** *www.trilogysystems.com* Manufacturer of linear electric motors and linear positioners for a variety of industrial automation and commercial product applications.

12. **Beckhoff Drive Technology.** *www.beckhoff.com* Manufacturer of brushless synchronous servomotors and synchronous linear motors for a variety of industrial automation and commercial product applications.

PROBLEMS

1. List six items that must be specified for electric motors.

2. List eight factors to be considered in the selection of an electric motor.

3. Define *duty cycle*.

4. How much variation in voltage will most AC motors tolerate?

5. State the relationship among torque, power, and speed.

6. What does the abbreviation AC stand for?

7. Describe and sketch the form of single-phase AC power.

8. Describe and sketch the form of three-phase AC power.

9. What is the standard frequency of AC power in the United States?

10. What is the standard frequency of AC power in Europe?

11. What type of electrical power is available in a typical American residence?

12. How many conductors are required to carry single-phase power? How many for three-phase power?

13. Assume that you are selecting an electric motor for a machine to be used in an industrial plant. The following types of AC power are available: 120-V, single-phase; 240-V, single-phase; 240-V, three-phase; 480-V, three-phase. In general, for which type of power would you specify your motor?

14. Define *synchronous speed* for an AC motor.

15. Define *full-load speed* for an AC motor.

16. What is the synchronous speed of a four-pole AC motor when running in the United States? In France?

17. A motor nameplate lists the full-load speed to be 3450 rpm. How many poles does the motor have? What would its approximate speed be at zero load?

18. If a four-pole motor operates on 400-Hz AC power, what will its synchronous speed be?

19. If an AC motor is listed as a normal-slip, four-pole/six-pole motor, what will its approximate full-load speeds be?

20. What type of control would you use to make an AC motor run at variable speeds?

21. Describe a C-face motor.

22. Describe a D-flange motor.

23. What does the abbreviation NEMA stand for?

24. Describe a protected motor.

25. Describe a TEFC motor.

26. Describe a TENV motor.

27. What type of motor enclosure would you specify to be used in a plant that manufactures baking flour?

28. What type of motor would you specify for a meat grinder if the motor is to be exposed?

29. Figure P21–29 shows a machine that is to be driven by a 5-hp, protected, foot-mounted AC motor having a 184T frame. The motor is to be aligned with the shaft of the driven machine. Make a complete dimensioned drawing, showing standard side and top views of the machine and the motor. Design a suitable mounting base for the motor showing the motor mounting holes.

30. Define *locked-rotor torque*. What is another term used for this parameter?

31. What is meant if one motor has a poorer speed regulation than another?

32. Define *breakdown torque*.

33. Name the four most common types of single-phase AC motors.

34. Refer to the AC motor performance curve in Figure P21–34.

    (a) What type of motor is the curve likely to represent?

    (b) If the motor is a six-pole type, rated at 0.75 hp, how much torque can it exert at the rated load?

    (c) How much torque can the motor develop to start a load?

    (d) What is the breakdown torque for the motor?

35. Repeat Parts (b), (c), and (d) of Problem 34 if the motor is a two-pole type, rated at 1.50 kW.

36. A cooling fan for a computer is to operate at 1725 rpm, direct-driven by an electric motor. The speed/torque curve for the fan is shown in Figure P21–36. Specify a suitable motor, giving the type of motor, horsepower, and number of poles.

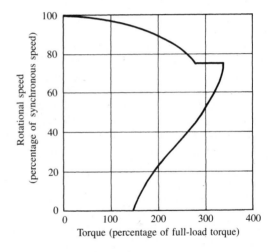

FIGURE P21–34 (Problems 34 and 35)

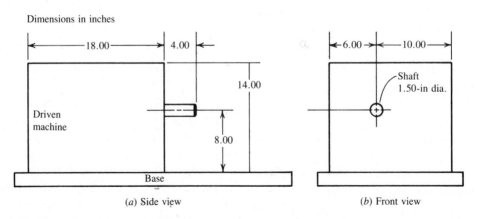

Dimensions in inches

(a) Side view

(b) Front view

FIGURE P21–29

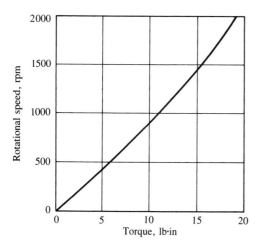

FIGURE P21–36

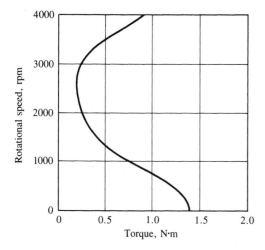

FIGURE P21–37

37. Figure P21–37 shows the speed/torque curve for a house-hold refrigeration compressor, designed to operate at 3450 rpm. Specify a suitable motor, giving the type, power rating in watts, and number of poles.

38. How is speed adjusted for a wound-rotor motor?

39. What is the full-load speed of a 10-pole synchronous motor?

40. What is meant by the term *pull-out torque*, as applied to a synchronous motor?

41. Discuss the reasons that universal motors are frequently used for hand-held tools and small appliances.

42. Why is the adjective *universal* used to describe a universal motor?

43. Name three ways to produce DC power.

44. List 12 common DC voltages.

45. What is an SCR control? For what is it used?

46. If a DC motor drive advertises that it produces *low ripple*, what does the term mean?

47. If you want to use a DC motor in your home, and your home has only standard 115 V AC, single-phase power, what will you need? What type of motor should you get?

48. List seven advantages of DC motors with respect to AC motors.

49. Discuss two disadvantages of DC motors.

50. Name four types of DC motors.

51. What happens to a series-wound DC motor if the load on the motor falls to nearly zero?

52. Assume that a permanent magnet DC motor can exert a torque of 15.0 N·m when operating at 3000 rpm. What torque could it exert at 2200 rpm?

53. List 10 functions of a motor control.

54. What size motor starter is required for a 10-hp, three-phase motor, operating on 220 V?

55. A single-phase, 110-V AC motor has a nameplate rating of 1.00 kW. What size motor starter is required?

56. What does the term *plug stopping* mean, and how is it accomplished?

57. Why is a fuse an inadequate protection device for an industrial motor?

58. What type of motor control enclosure would you specify for use in a car wash?

59. What could you do to the control circuit for a standard series DC motor to give it a controlled no-load speed?

60. What happens if you connect a resistance in series with the armature of a shunt-wound DC motor?

61. What happens if you connect a resistance in series with the shunt field of a shunt-wound DC motor?

22

Motion Control: Clutches and Brakes

The Big Picture

You Are the Designer

22–1 Objectives of This Chapter

22–2 Descriptions of Clutches and Brakes

22–3 Types of Friction Clutches and Brakes

22–4 Performance Parameters

22–5 Time Required to Accelerate a Load

22–6 Inertia of a System Referred to the Clutch Shaft Speed

22–7 Effective Inertia for Bodies Moving Linearly

22–8 Energy Absorption: Heat-Dissipation Requirements

22–9 Response Time

22–10 Friction Materials and Coefficient of Friction

22–11 Plate-Type Clutch or Brake

22–12 Caliper Disc Brakes

22–13 Cone Clutch or Brake

22–14 Drum Brakes

22–15 Band Brakes

22–16 Other Types of Clutches and Brakes

Motion Control: Clutches and Brakes

Discussion Map

- ☐ A *brake* is a device used to bring a moving system to rest, to slow its speed, or to control its speed to a certain value under varying conditions.

- ☐ A *clutch* is a device used to connect or disconnect a driven component from the prime mover of the system.

Discover

Where do you use brakes?

Consider the brakes for a car or a bicycle. Describe the components and the actuation cycle in as much detail as you can. Discuss your findings with your colleagues.

What kinds of equipment other than vehicles use clutches or brakes? Describe some scenarios.

Describe the physics of the operation of a clutch and a brake, considering energy and inertia effects.

This chapter helps you explore all of these issues, and many convenient design and analysis equations are developed. Many kinds of commercially available clutches and brakes are pictured and described. Look through the chapter now.

Machine systems require control whenever the speed or the direction of the motion of one or more components is to be changed. When a device is started initially, it must be accelerated from rest to the operating speed. As a function is completed, the system must frequently be brought to rest. In continuously operating systems, changing speeds to adjust to different operating conditions is often necessary. Safety sometimes dictates motion control, as with a load being lowered by a hoist or an elevator.

In this chapter we will be concerned mostly with control of rotary motion in systems driven by electric motors, engines, turbines, and the like. Ultimately, linear motion may be involved through linkages, conveyors, or other mechanisms.

The machine elements most frequently used for motion control are the clutch and the brake, defined as follows:

- A *clutch* is a device used to connect or disconnect a driven component from the prime mover of the system. For example, in a machine that must cycle frequently, the driving motor is left running continuously, and a clutch is interposed between the motor and the driven machine. Then the clutch is cycled on and off to connect and disconnect the load. This permits the motor to operate at an efficient speed, and it also permits the system to cycle more rapidly because there is no need to accelerate the heavy motor rotor with each cycle.

- A *brake* is a device used to bring a moving system to rest, to slow its speed, or to control its speed to a certain value under varying conditions.

Where do you use brakes? An obvious answer is in a car or a bicycle where safe operation requires fast, smooth stopping when emergency conditions occur or simply when you need to stop at a stop sign or in your driveway. And you do not always need to bring the car or bicycle to a full stop. Slowing down to conform to prevailing traffic flow or to round a curve requires braking to decrease the speed from a higher to a lower speed.

What really happens when you apply the brakes of a bicycle? Can you describe the essential elements of the braking system? With hand brakes, your actuation of the brake lever pulls on a cable that, in turn, pulls on a linkage at the brake assembly mounted above the rim of a wheel. The linkage causes the friction pad to press against the rim. When you pull harder on the lever, a higher force is developed between the pad and the rim. This is called the *normal force*. Recall from physics and statics that a frictional force is created between surfaces moving relative to each other when a normal force presses them together. The frictional force acts in a direction opposite to the relative motion, and thus it tends to slow the motion. With a sufficiently large frictional force applied for a sufficient time, the wheel stops. Notice also that the frictional force acts at a fairly large radius from the center of the wheel. Thus, the force causes a frictional *torque* to be developed, and what is really happening is the deceleration of the angular velocity of the wheel. But since this is directly proportional to the linear speed of the bicycle, you experience the stopping action as a linear deceleration.

But that is not all! Have you ever felt the brake pad after a hard stop? The fact that it gets warm is an indication that the brake absorbs energy during the stopping action. Where does the energy come from? Can you compute the amount of energy that must be absorbed? What are the parameters involved in this calculation? Compare the amount of energy that must be absorbed in stopping a bicycle that you are riding with that involved in stopping a large airliner landing at 120 mph with a full load of people and baggage in addition to the huge weight of the airplane itself. Imagine what those brakes look like in comparison with the bicycle brake!

What other kinds of equipment besides transportation vehicles require brakes? Consider elevators, escalators, hoists, and winches, which must stop and hold a load after lifting it. Machine tools, conveyors, and other manufacturing equipment must often be brought to a safe, quick stop.

But that equipment must also be *accelerated* to start a new cycle of operation. How can that be achieved? One way is to start and stop the motor or the engine that drives the equipment. However, that is inconvenient and time-consuming, and it may lead to early failure of the system.

What if you had to stop the engine of your car each time you came to a stop sign and then had to restart it to move on? How does the automotive system allow you to stop and then continue driving without shutting off the engine? Have you ever driven a standard shift car? A "stick shift"? A *clutch* is used to engage and disengage the drive train from the engine. In cars with automatic transmissions clutches are also built into the transmission.

What other kinds of equipment use clutches? Describe some examples from your personal experience, or create a scenario in which it would be desirable to employ a clutch.

Now, consider what the task of a clutch is. Some parts of a machine are rotating continuously, while others are temporarily stationary. Then you engage the clutch. What happens? Consider the physics of the situation. The stationary parts must be accelerated from zero speed to the desired speed according to the design of the drive system. Inertia must be overcome—sometimes rotational inertia; sometimes linear inertia; and sometimes both kinds in the same machine. How long do you want it to take to accelerate the load to its operating speed? You must realize that a shorter time requires a higher value of accelerating torque to be developed by the clutch, and it increases the technical demands on the system in terms of strength of its components, smoothness, and wear life of the frictional materials that actually accomplish the engagement.

This chapter helps you explore all of these issues, and many convenient design and analysis equations are developed. Several different styles of brakes and clutches are discussed, and photographs or detailed drawings of several commercially available designs are shown. Look through the chapter now to get a feel for the scope of the coverage of this chapter. It might be desirable to keep this book handy for future projects so that you can look up the details of different clutch and brake designs.

You Are the Designer

Your company manufactures conveyor systems for warehouses and truck terminals. The conveyors deliver cartons of materials to any of several stations where trucks are to be loaded. To save energy and to decrease the wear on operating parts of the conveyor system, it is decided to operate only the parts of the system that have a demand to deliver a carton. The system is to be operated automatically through a series of sensors, switches, programmable controllers, and an overall supervisory computer control. Your assignment is to recommend the type and the size of clutch and brake units to start and stop the various conveyors.

Some of the design decisions you will have to make are as follows:

1. How much time should be allowed to get the conveyors up to speed after the initial command to start has been given?
2. How fast should the conveyors come to a stop?
3. How many cycles per hour are expected?

4. How much room is available to install the clutch and brake units?
5. What means are available to actuate the clutch and brake units: electric power, compressed air, hydraulic pressure, or some other means?
6. What general type of clutch and brake should be used?
7. What size and model of units should be specified?

Along with these decisions, you will need information about the conveyor system itself, such as the following:

1. How much load will be on the conveyors when they are started and stopped?
2. What is the design of the conveyor system itself, and what are the weights, shapes, and dimensions of its components?
3. How is the conveyor driven: electric motor, hydraulic motor, or another means?
4. Are the products moved all on one level, or are there elevation changes in the system?

The information in this chapter will help you design such a system.

22–1
OBJECTIVES OF THIS CHAPTER

After completing this chapter, you will be able to:

1. Define the terms *clutch* and *brake*.
2. Differentiate between a clutch and a *clutch coupling*.
3. Describe a fail-safe brake.
4. Describe a clutch-brake module.
5. Specify the required capacity of a clutch or a brake to drive a given system reliably.
6. Compute the time required to accelerate a system or to bring it to a stop with the application of a given torque.
7. Define the inertia of a system in terms of its Wk^2 value.
8. Compute the energy-dissipation requirements of a clutch or a brake.
9. Determine the response time for a clutch-brake system.
10. Describe at least five types of clutches and brakes.
11. Name six actuation means used for clutches and brakes.
12. Perform the design and analysis of plate-type, caliper disc, cone, drum and shoe, and band brakes and clutches.
13. Name nine other types of clutches and brakes.

22–2
DESCRIPTIONS OF CLUTCHES AND BRAKES

Several arrangements of clutches and brakes are shown in Figure 22–1. By convention, the term *clutch* is reserved for the application in which the connection is made to a shaft parallel to the motor shaft, as illustrated in Figure 22–1(a). If the connection is to a shaft in-line with the motor, the term *clutch coupling* is used, as shown in Figure 22–1(b).

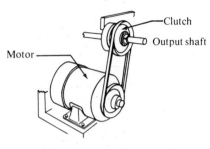

(a) Clutch: Transmits rotary motion to a parallel shaft only when coil is energized, by using sheaves, sprockets, gears, or timing pulleys.

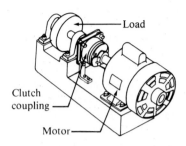

(b) Clutch coupling: Transmits rotary motion to an in-line shaft only when coil is energized. Split shaft applications.

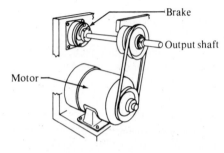

(c) Brake: Stops (brakes) load when coil is energized. Panel shows clutch imparting rotary motion to the output (load) shaft while brake is de-energized. Conversely, de-energizing clutch coil and energizing brake coil causes load to stop.

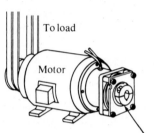

(d) Fail-safe brake: Stops load by de-energization of coil; power off—brake on.

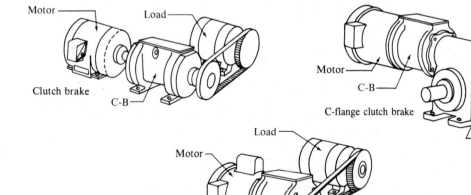

(e) Three types of mountings for clutch-brake modules: *Clutch brake* combines functions of clutch and brake in a complete preassembled package with input and output shafts. *C-flange clutch brake* performs same function but for use between NEMA "C" flange motor and speed reducer. *Motor clutch brake* is a preassembled module for mounting to a NEMA "C" flange motor and presenting an output shaft for connection to load.

FIGURE 22–1   Typical applications of clutches and brakes (Electroid Company, Springfield, NJ)

Also by convention, a brake [Figure 22–1(c)] is actuated by some overt action: the application of a fluid pressure, the switching on of an electric current, or the moving of a lever by hand. A brake that is spring-applied automatically in the absence of an overt action is called a *fail-safe brake* [Figure 22–1(d)]: When the power goes off, the brake goes on.

When the functions of both a clutch and a brake are required in a system, they are frequently arranged in the same unit, the *clutch-brake module*. When the clutch is activated, the brake is deactivated, and vice versa [see Figure 22–1(e)].

A *slip clutch* is a clutch that, by design, transmits only a limited torque; thus, it will slip at any higher torque. It is used to provide a controlled acceleration to a load that is smooth and that requires a smaller motor power. It is also used as a safety device, protecting expensive or sensitive components in the event that a system is jammed.

Most of the discussion in this chapter will concern clutches and brakes that transmit motion through friction at the interface between two rotating parts moving at different speeds. Other types are discussed briefly in the last section.

References 1-7 and Internet sites 1–8 provide additional information on the design, selection, and application of motion control systems, clutches, and brakes.

22–3 TYPES OF FRICTION CLUTCHES AND BRAKES

Clutches and brakes that use friction surfaces as the means of transmitting the torque to start or stop a mechanism can be classified according to the general geometry of the friction surfaces and by the means used to actuate them. In some cases, the same basic geometry can be used as either a clutch or a brake by selectively attaching the friction elements to the driver, the driven machine, or the stationary frame of the machine.

The following types of clutches and brakes are sketched in Figure 22–2:

1. *Plate clutch or brake:* Each friction surface is in the shape of an annulus on a flat plate. One or more friction plates move axially to contact a mating smooth plate, usually made of steel, to which the friction torque is transmitted.

2. *Caliper disc brake:* A disc-shaped rotor is attached to the machine to be controlled. Friction pads covering only a small portion of the disc are contained in a fixed assembly called a *caliper* and are forced against the disc by air pressure or hydraulic pressure.

3. *Cone clutch or brake:* A cone clutch or brake is similar to a plate clutch or brake except that the mating surfaces are on a portion of a cone instead of on a flat plate.

4. *Band brake:* Used only as a brake, the friction material is on a flexible band that nearly surrounds a cylindrical drum attached to the machine to be controlled. When braking is desired, the band is tightened on the drum, exerting a tangential force to stop the load.

5. *Block or shoe brake:* Curved, rigid pads faced with the friction material are forced against the surface of a drum, from either the outside or the inside, exerting a tangential force to stop the load.

Actuation

The following are the means used to actuate clutches or brakes. Each may be applied to several of the types described. Figures 22–3 through 22–9 show a variety of commercially available designs.

Manual.   The operator provides the force, usually through a lever arrangement to achieve force multiplication.

FIGURE 22–2 Types of friction clutches and brakes [(b), Tol-O-Matic, Hamel, MN]

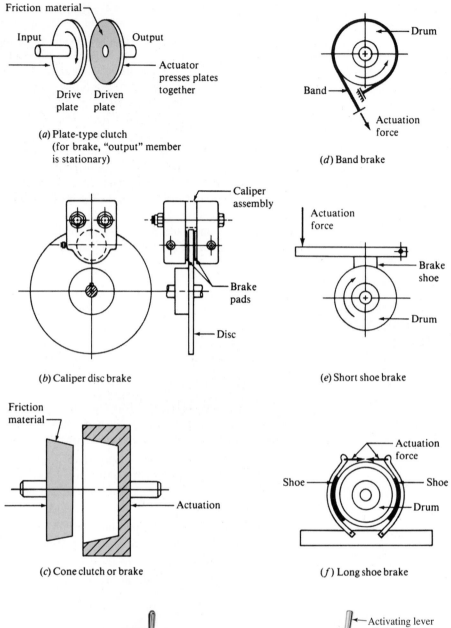

(a) Plate-type clutch (for brake, "output" member is stationary)

(b) Caliper disc brake

(c) Cone clutch or brake

(d) Band brake

(e) Short shoe brake

(f) Long shoe brake

FIGURE 22–3
Manually operated clutch (Rockford Division, Borg-Warner Corp., Rockford, IL)

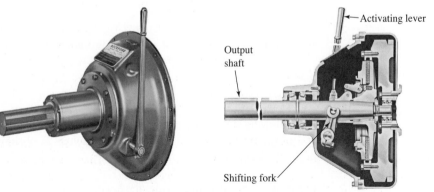

FIGURE 22–4
Spring-applied,
electrically released,
long shoe brake
(Eaton Corp., Cutler-
Hammer Products,
Milwaukee, WI)

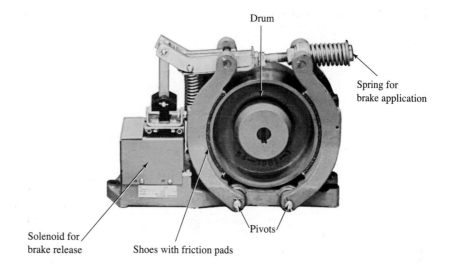

Drum

Spring for
brake application

Solenoid for
brake release

Pivots

Shoes with friction pads

FIGURE 22–5 Slip
clutch. Springs apply
normal pressure on the
friction plates. The
spring force is
adjustable to vary the
torque level at which
clutch will slip. (The
Hilliard Corp.,
Elmira, NY)

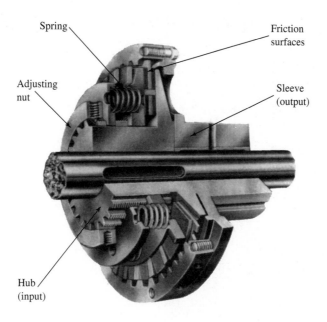

Spring

Friction
surfaces

Adjusting
nut

Sleeve
(output)

Hub
(input)

Spring-applied.   Sometimes called a *fail-safe* design when applied to a brake, the brake is applied automatically by springs unless there is some opposing force present. Thus, if power fails, or if air pressure or hydraulic pressure is lost, or if the operator is unable to perform the function, the springs apply the brake and stop the load. The concept may also be used to engage or to disengage a clutch.

Centrifugal.   A centrifugal clutch is sometimes employed to permit the driving system to accelerate without a connected load. Then, at a preselected speed, centrifugal force moves the clutch elements into contact to connect the load. As the system slows, the load will be automatically disconnected.

FIGURE 22–6 Air-actuated clutch or brake (Eaton Corp., Airflex Division, Cleveland, OH)

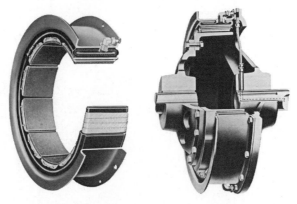

(a) Details of clutch design

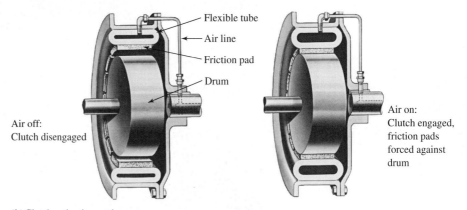

Flexible tube
Air line
Friction pad
Drum

Air off:
Clutch disengaged

Air on:
Clutch engaged, friction pads forced against drum

(b) Clutch activation cycle

FIGURE 22–7 Hydraulically actuated disc brake assembly (Tol-O-Matic, Hamel, MN)

Caliper assembly

Brake pads

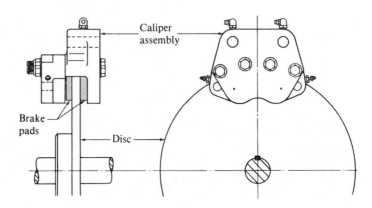

Caliper assembly

Brake pads

Disc

FIGURE 22–8
Electrically actuated
plate-type clutch or
brake (Warner Electric,
Inc., South Beloit, IL)

(a) Cutaway view of complete assembly

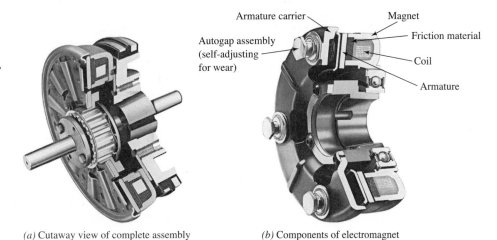

(b) Components of electromagnet

FIGURE 22–9
Electrically operated
clutch-brake module
(Electroid Company,
Springfield, NJ)

(a) External appearance

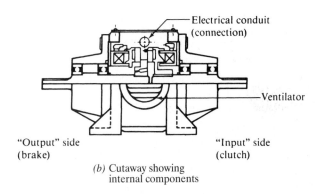

(b) Cutaway showing
internal components

Pneumatic. Compressed air is introduced into a cylinder or some other chamber. The force produced by the pressure on a piston or a diaphragm brings the friction surfaces into contact with the members connected to the load.

Hydraulic. Hydraulic brakes are similar to the pneumatic type, except that they use oil hydraulic fluids instead of air. The hydraulic actuator is usually applied where high actuation forces are required.

Electromagnetic. An electric current is applied to a coil, creating an electromagnetic flux. The magnetic force then attracts an armature attached to the machine that is to be controlled. The armature is usually of the plate type.

22–4
PERFORMANCE
PARAMETERS

The principles of physics tell us that whenever the speed or the direction of the motion of a body is changed, there must be a force exerted on the body. If the body is in rotation, a torque must be applied to the system to speed it up or to slow it down. When a change in speed occurs, a corresponding change in the kinetic energy of the system occurs. Thus, motion control inherently involves the control of energy, either adding energy to accelerate a system or absorbing energy to decelerate it.

The parameters involved in the rating of clutches and brakes are described in the following list:

1. Torque required to accelerate or decelerate the system
2. Time required to accomplish the speed change
3. The cycling rate: number of on/off cycles per unit time
4. The inertia of the rotating or translating parts
5. The environment of the system: temperature, cooling effects, and so on
6. Energy-dissipation capability of the clutch or the brake
7. Physical size and configuration
8. Actuation means
9. Life and reliability of the system
10. Cost and availability

Two basic methods are used to determine the torque capacity required of a clutch or a brake. One relates the capacity to the power of the motor driving the system. Recall that, in general, power = torque × rotational speed ($P = Tn$). The required torque capacity is then usually expressed in the form

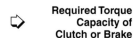

Required Torque
Capacity of
Clutch or Brake

$$T = \frac{CPK}{n} \qquad (22-1)$$

where C = conversion factor for units
K = service factor based on the application

More will be said about these later.

Note that the torque required is inversely proportional to the rotational speed. For this reason, it is advisable to locate the clutch or the brake on the highest-speed shaft in the system so that the required torque is a minimum. The size, cost, and response time are all typically lower when the torque is lower. One disadvantage is that the shaft to be accelerated or decelerated must undergo a larger change in speed, and the amount of slipping may be greater. This effect may generate more frictional heat, leading to thermal problems. However, it is offset by the increased cooling effect because of the faster motion of the clutch or brake parts.

The value for the K factor in the torque equation is largely a design decision. Some of the typical guidelines follow:

1. For brakes under average conditions, use $K = 1.0$.
2. For clutches in light duty where the output shaft does not assume its normal load until after it is up to speed, use $K = 1.5$.
3. For clutches in heavy duty where large attached loads must be accelerated, use $K = 3.0$.

4. For clutches in systems having varying loads, use a K factor at least equal to the factor by which the motor breakdown torque exceeds the full-load torque. This was discussed in Chapter 21, but for a typical industrial motor (design B), use $K = 2.75$. For a high starting torque motor (design C or capacitor-start motor), $K = 4.0$ may be required. This ensures that the clutch will be able to transmit at least as much torque as the motor and that it will not slip after getting up to speed.

5. For clutches in systems driven by gasoline engines, diesel engines, or other prime movers, consider the peak torque capability of the driver; $K = 5.0$ may be required.

The following list relates the value for C to typically used units for torque, power, and rotational speed. For example, if power is in hp and speed is in rpm, then to obtain torque in lb·ft, use $T = 5252(P/n)$.

Torque	Power	Speed	C
lb·ft	hp	rpm	5252
lb·in	hp	rpm	63 025
N·m	W	rad/s	1
N·m	W	rpm	9.549
N·m	kW	rpm	9549

Although the torque computation method of Equation (22–1) will produce generally satisfactory performance in typical applications, it does not provide a means of estimating the actual time required to accelerate the load with a clutch or to decelerate the load with a brake. The method described next should be used for systems with large inertia such as conveyors or presses having flywheels.

22–5 TIME REQUIRED TO ACCELERATE A LOAD

The basic principle involved is taken from dynamics:

$$T = I\alpha$$

where I = mass moment of inertia of the components being accelerated
α (alpha) = angular acceleration—that is, the rate of change of angular velocity

The usual objective of such an analysis is to determine the torque required to produce a change in rotational speed, Δn, of a system in a given amount of time, t. But $\Delta n/t = \alpha$. Also, it is more convenient to express the mass moment of inertia in terms of the *radius of gyration, k*. By definition,

$$k = \sqrt{I/m} \quad \text{or} \quad k^2 = I/m$$

where m = mass
$m = W/g$

Then

$$I = mk^2 = Wk^2/g$$

The equation for torque then becomes

**Torque Required
to Accelerate an
Inertia Load**

$$T = I\alpha = \frac{Wk^2}{g}\frac{(\Delta n)}{t} \tag{22–2}$$

The term Wk^2 is frequently called simply the *inertia* of the load, although that designation is not strictly correct. A large proportion of the components in a machine system to be accelerated are in the form of cylinders or discs. Figure 22–10 gives the relationships for the radius of gyration and Wk^2 for hollow discs. Solid discs are simply a special case with an inside radius of zero. We can analyze more complex objects by considering them to be made from a set of simpler discs. Example Problem 22–1 illustrates the process.

Now the torque required to accelerate the pulley can be computed. Equation (22–2) can be put into a more convenient form by noting that T is usually expressed in lb·ft, Wk^2 in lb·ft^2, n in rpm, and t in s. Using $g = 32.2$ ft/s^2 and converting units gives

$$T = \frac{Wk^2(\Delta n)}{308t}\ \text{lb}\cdot\text{ft} \tag{22–3}$$

FIGURE 22–10
Inertia properties of a
hollow disc

Radius of gyration:

$$k^2 = \frac{1}{2}(R_1^2 + R_2^2)$$

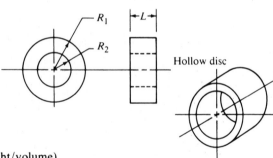

Hollow disc

Volume:

$$V = \pi(R_1^2 - R_2^2)L$$

Weight:

$$W = \delta_w V$$

δ_w = Weight density (weight/volume)

Inertia (Wk^2):

$$Wk^2 = \delta_w V k^2 = \delta_w \pi (R_1^2 - R_2^2)L(R_1^2 + R_2^2)/2$$

$$Wk^2 = \frac{\pi \delta_w L}{2}(R_1^4 - R_2^4)$$

Typical units: L, R_1, R_2 in inches

δ_w in lb/in^3

Wk^2 in lb·ft^2

$$Wk^2 = \frac{\pi}{2} \times \delta_w\frac{\text{lb}}{\text{in}^3} \times L(\text{in}) \times (R_1^4 - R_2^4)\ \text{in}^4 \times \frac{1\ \text{ft}^2}{144\ \text{in}^2}$$

$$Wk^2 = \frac{\delta_w L(R_1^4 - R_2^4)}{91.67}\ \text{lb}\cdot\text{ft}^2$$

Special case for steel: $\delta_w = 0.283$ lb/in^3

$$Wk^2 = \frac{L(R_1^4 - R_2^4)}{323.9}\ \text{lb}\cdot\text{ft}^2$$

FIGURE 22–11
Steel flat-belt pulley

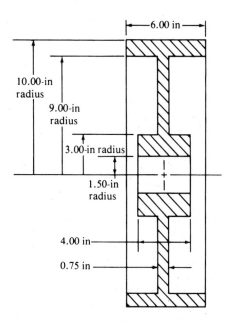

Example Problem 22–1   Compute the value of Wk^2 for the steel flat-belt pulley shown in Figure 22–11.

Solution   The pulley can be considered to be made up of three components, each of which is a hollow disc. The Wk^2 for the total pulley is the sum of that for each component.

Part 1
Using the formula from Figure 22–10 for a steel disc, we have

$$Wk^2 = \frac{(R_1^4 - R_2^4)(L)}{323.9} \text{ lb} \cdot \text{ft}^2 = \frac{[(10.0)^4 - (9.0)^4](6.0)}{323.9}$$

$$Wk^2 = 63.70 \text{ lb} \cdot \text{ft}^2$$

Part 2

$$Wk^2 = \frac{[(9.0)^4 - (3.0)^4](0.75)}{323.9} = 15.00 \text{ lb} \cdot \text{ft}^2$$

Part 3

$$Wk^2 = \frac{[(3.0)^4 - (1.5)^4](4.0)}{323.9} = 0.94 \text{ lb} \cdot \text{ft}^2$$

$$\text{Total } Wk^2 = 63.70 + 15.00 + 0.94 = 79.64 \text{ lb} \cdot \text{ft}^2$$

Example Problem 22–2   Compute the torque that a clutch must transmit to accelerate the pulley of Figure 22–11 from rest to 550 rpm in 2.50 s. From Example Problem 22–1, $Wk^2 = 79.64$ lb·ft^2.

Solution   Use Equation (22–3):

$$T = \frac{(79.64)(550)}{308(2.5)} = 56.9 \text{ lb} \cdot \text{ft}$$

In summary, if a clutch that is capable of exerting at least 56.9 lb·ft of torque engages a shaft carrying the pulley shown in Figure 22–11, the pulley can be accelerated from rest to 550 rpm in 2.50 s or less.

22–6
INERTIA OF A SYSTEM REFERRED TO THE CLUTCH SHAFT SPEED

In many practical machine systems, there are several elements on several shafts, operating at differing speeds. It is required that the effective inertia of the entire system *as it affects the clutch* be determined. The effective inertia of a connected load operating at a rotational speed different from that of the clutch is proportional to the square of the ratio of the speeds. That is,

Effective Inertia

$$Wk_e^2 = Wk^2 \left(\frac{n}{n_c}\right)^2 \tag{22–4}$$

where n = speed of the load of interest
n_c = speed of the clutch

Example Problem 22–3

Compute the total inertia of the system in Figure 22–12 as seen by the clutch. Then compute the time required to accelerate the system from rest to a motor speed of 550 rpm if the clutch exerts a torque of 24.0 lb·ft. The Wk^2 for the armature of the clutch, which must also be accelerated, is 0.22 lb·ft^2, including the 1.25-in shaft.

Solution

The clutch and gear A will be rotating at 550 rpm, but because of the gear reduction, gear B, its shaft, and the pulley rotate at

$$n_2 = 550 \text{ rpm}(24/66) = 200 \text{ rpm}$$

Now compute the inertia for each element referred to the clutch speed. Assume that the gears are discs having outside diameters equal to the pitch diameter of the gear and inside diameters equal to the shaft diameter. The equation in Figure 22–10 for a steel disc will be used to compute Wk^2.

FIGURE 22–12
Given system

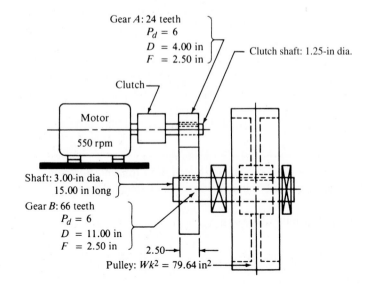

Gear A

$$Wk^2 = [(2.00)^4 - (0.625)^4](2.50)/323.9 = 0.122 \text{ lb·ft}^2$$

Gear B

$$Wk^2 = [(5.50)^4 - (1.50)^4](2.50)/323.9 = 7.02 \text{ lb·ft}^2$$

But, because of the speed difference, the effective inertia is

$$Wk_e^2 = 7.02(200/550)^2 = 0.93 \text{ lb·ft}^2$$

Pulley

From Example Problem 22–1, $Wk^2 = 79.64 \text{ lb·ft}^2$. The effective inertia is

$$Wk_e^2 = 79.64(200/550)^2 = 10.53 \text{ lb·ft}^2$$

Shaft

$$Wk^2 = (1.50)^4(15.0)/323.9 = 0.234 \text{ lb·ft}^2$$

The effective inertia is

$$Wk_e^2 = 0.234(200/550)^2 = 0.03 \text{ lb·ft}^2$$

The total effective inertia as seen by the clutch is

$$Wk_e^2 = 0.22 + 0.12 + 0.93 + 10.53 + 0.03 = 11.83 \text{ lb·ft}^2$$

Solving Equation (22–3) for the time gives

$$t = \frac{Wk_e^2(\Delta n)}{308T} = \frac{(11.83)(550)}{308(24.0)} = 0.88 \text{ s}$$

22–7
EFFECTIVE
INERTIA FOR
BODIES MOVING
LINEARLY

To this point, we have dealt only with components that rotate. Many systems include linear devices such as conveyors, hoist cables and their loads, or reciprocating racks driven by gears that also have inertia and must be accelerated. It would be convenient to represent these devices with an effective inertia measured by Wk^2 as we have for rotating bodies. We can accomplish this by relating the equations for kinetic energy for linear and rotary motion. The actual kinetic energy of a translating body is

$$KE = \frac{1}{2}mv^2 = \frac{1}{2}\frac{W}{g}v^2 = \frac{Wv^2}{2g}$$

where v = linear velocity of the body

We will use ft/min for the units of velocity. For a rotating body,

$$KE = \frac{1}{2}I\omega^2 = \frac{1}{2}\frac{Wk^2}{g}\omega^2 = \frac{Wk^2\omega^2}{2g}$$

Letting Wk^2 be the effective inertia, and equating these two formulas, gives

$$Wk_e^2 = W\left(\frac{v}{\omega}\right)^2$$

where ω must be in rad/min to be consistent

Using n rpm rather than ω rad/min, we must substitute $\omega = 2\pi n$. Then

**Effective Inertia for
a Load Moving
Linearly**

$$Wk_e^2 = W\left(\frac{v}{2\pi n}\right)^2 \qquad (22–5)$$

Example Problem 22–4   The conveyor in Figure 22–13 moves at 80 ft/min. The combined weight of the belt and the parts on it is 140 lb. Compute the equivalent Wk^2 inertia for the conveyor referred to the shaft driving the belt.

Solution   The rotational speed of the shaft is

$$\omega = \frac{v}{R} = \frac{80 \text{ ft}}{\text{min}} \frac{1}{5.0 \text{ in}} \frac{12 \text{ in}}{\text{ft}} = 192 \text{ rad/min}$$

Then the equivalent Wk^2 is

$$Wk_e^2 = W\left(\frac{v}{\omega}\right)^2 = (140 \text{ lb})\left(\frac{80 \text{ ft/min}}{192 \text{ rad/min}}\right)^2 = 24.3 \text{ lb}\cdot\text{ft}^2$$

**22–8
ENERGY
ABSORPTION:
HEAT-
DISSIPATION
REQUIREMENTS**

When you are using a brake to stop a rotating object or using a clutch to accelerate it, the clutch or brake must transmit energy through the friction surfaces as they slip in relation to each other. Heat is generated at these surfaces, tending to increase the temperature of the unit. Of course, heat is then dissipated from the unit, and for a given set of operating conditions an equilibrium temperature is achieved. That temperature must be sufficiently low to ensure a long life of the friction elements and other operating parts of the unit, such as electric coils, springs, and bearings.

FIGURE 22–13
Conveyor moving at
80 ft/min

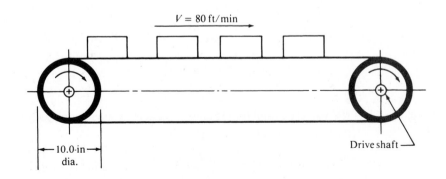

The energy to be absorbed or dissipated by the unit per cycle is equal to the change in kinetic energy of the components being accelerated or stopped. That is,

⇨ **Energy Absorption
by a Brake**

$$E = \Delta KE = \frac{1}{2} I\omega^2 = \frac{1}{2} mk^2\omega^2 = \frac{Wk^2\omega^2}{2g}$$

For typical units in the U.S. Customary System ($\omega = n$ rpm; Wk^2 in lb·ft; and $g = 32.2$ ft/s^2), we get

$$E = \frac{Wk^2 \, (\text{lb} \cdot \text{ft}^2)}{2(32.2 \text{ ft/s}^2)} \frac{n^2 \text{ rev}^2}{\text{min}^2} \frac{(2\pi)^2 \text{ rad}}{\text{rev}^2} \frac{1 \text{ min}^2}{60^2 \text{ s}^2}$$

⇨ **Energy Absorption
in U.S. Customary
Units**

$$E = 1.7 \times 10^{-4} Wk^2 n^2 \text{ lb} \cdot \text{ft} \tag{22–6}$$

In SI units, mass is in kilograms (kg), radius of gyration is in meters (m), and angular velocity is in radians per second (rad/s). Then

$$E = \frac{1}{2} I\omega^2 = \frac{1}{2} mk^2\omega^2 (\text{kg} \cdot \text{m}^2/\text{s}^2)$$

But the newton unit is equal to the kg·m/s^2. Then

⇨ **Energy Absorption
in SI Units**

$$E = \frac{1}{2} mk^2\omega^2 \text{ N} \cdot \text{m} \tag{22–7}$$

No further conversion factors are required.

If there are repetitive cycles of operation, the energy from Equation (22–6) or (22–7) must be multiplied by the cycling rate, usually in cycles/min in U.S. Customary units and cycles/s for SI units. The result would be the energy generation per unit time, which must be compared with the heat-dissipation capacity of the clutch or brake being considered for the application.

When the clutch-brake cycles on and off, part of its operation is at the full operating speed of the system, and part is at rest. The combined heat-dissipation capacity is the average of the capacity at each speed weighted by the proportion of the cycle at each speed (see Example Problem 22–5).

**22–9
RESPONSE TIME**

The term *response time* refers to the time required for the unit (clutch or brake) to accomplish its function after action is initiated by the application of an electric current, air pressure, spring force, or manual force. Figure 22–14 shows a complete cycle using a clutch-brake module. The straight-line curve is idealized while the curved line gives the general form of system motion. The actual response time will vary, even for a given unit, with variations in the load, environment, or other operating conditions.

Commercially available clutches and brakes for typical machine applications have response times from a few milliseconds (1/1000 s) for a small device, such as a paper transport, to approximately 1.0 s for larger machines, such as an assembly conveyor. Manufacturers' literature should be consulted. To give you a feel for the capabilities of commercially available clutches and brakes, Table 22–1 gives sample data for electrically powered units.

Example Problem 22–5   For the system in Figure 22–12, and using the data from Example Problem 22–3, estimate the total cycle time if the system is controlled by unit *G* from Table 22–1 and the system must stay on (at steady speed) for 1.50 s and off (at rest) for 0.75 s. Also estimate the response time of the clutch-brake and the acceleration and deceleration times. If the system

FIGURE 22–14
Typical cycle of
engagement and
disengagement for a
clutch

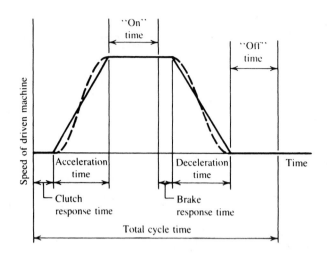

TABLE 22–1   Clutch-brake sample performance data

Unit size	Torque capacity (lb·ft)	Inertia, Wk^2 (lb·ft^2)	Heat dissipation (ft·lb/min) At rest	1800 rpm	Response time (s) Clutch	Brake
A	0.42	0.000 17	750	800	0.022	0.019
B	1.25	0.0014	800	1200	0.032	0.024
C	6.25	0.021	1050	2250	0.042	0.040
D	20.0	0.108	2000	6000	0.090	0.089
E	50.0	0.420	3000	13 000	0.110	0.105
F	150.0	1.17	9000	62 000	0.250	0.243
G	240.0	2.29	18 000	52 000	0.235	0.235
H	465.0	5.54	20 000	90 000	0.350	0.350
I	700.0	13.82	26 000	190 000	0.512	0.512

Note: Torque ratings are static. The torque capacity decreases as the speed difference between the parts being engaged increases. Interpolation may be used on heat-dissipation data.

cycles continuously, compute the heat-dissipation rate, and compare it with the capacity of the unit.

Solution   Figure 22–15 shows the estimated total cycle time for the system to be 2.896 s. From Table 22–1, we find that the clutch-brake exerts 240 lb·ft of torque and has a response time of 0.235 s for both the clutch and the brake.

Acceleration and Deceleration Time [Equation (22–3)]

$$t = \frac{Wk_e^2(\Delta n)}{308T} = \frac{(11.83)(550)}{308(240)} = 0.088 \text{ s}$$

Cycling Rate and Heat Dissipation [Equation (22–6)]
For a total cycle time of 2.896 s, the number of cycles per minute would be

$$C = \frac{1.0 \text{ cycle}}{2.896 \text{ s}} \frac{60 \text{ s}}{\text{min}} = 20.7 \text{ cycles/min}$$

FIGURE 22–15
Estimated total cycle
time

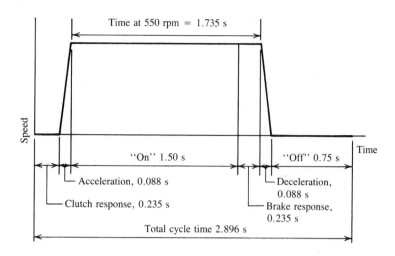

The energy generated with each engagement of either the clutch or the brake is

$$E = 1.7 \times 10^{-4} W k^2 n^2 = 1.7 \times 10^{-4} (11.83)(550)^2 = 608 \text{ lb·ft}$$

The energy generation per minute is

$$E_t = 2EC = (2)(608 \text{ lb·ft/cycle})(20.7 \text{ cycles/min}) = 25\ 200 \text{ lb·ft/min}$$

This is greater than the heat-dissipation capacity of unit G at rest (18 000 lb·ft/min). Then let's compute a weighted average capacity for this cycle. First, referring to Figure 22–15, approximately 1.735 s is "at speed," or 550 rpm. The balance of the cycle, 1.161 s, is at rest. From Table 22–1, and interpolating between zero speed and 1800 rpm, the heat-dissipation rate at 550 rpm is approximately 28 400 lb·ft/min. Then the weighted average capacity for unit G is

$$E_{avg} = \frac{t_0}{t_t} E_0 + \frac{t_{550}}{t_t} E_{550}$$

where t_t = total cycle time
t_0 = time at rest (0 rpm)
t_{550} = time at 550 rpm
E_0 = heat-dissipation capacity at rest
E_{550} = heat-dissipation capacity at 550 rpm

Then

$$E_{avg} = \frac{1.161}{2.896}(18\ 000) + \frac{1.735}{2.896}(28\ 400) = 24\ 230 \text{ lb·ft/min}$$

This is a little lower than required, and the design would be marginal. Fewer cycles per minute should be specified.

**22–10
FRICTION
MATERIALS AND
COEFFICIENT OF
FRICTION**

Many of the types of clutches and brakes discussed in this chapter employ mating surfaces driven through friction materials. The function of these materials is to develop a substantial friction force when a normal force is created by the brake actuation means. The friction force produces a force or a torque that retards existing motion if applied as a brake or that accelerates a member at rest or moving at a slower speed if applied as a clutch.

The desirable properties of friction materials are discussed here:

1. They should have a relatively high coefficient of friction when operating against the mating materials in the system. The highest coefficient of friction is not always the best choice, because smooth engagement is often aided by a more moderate frictional force or torque.

2. The coefficient of friction should be relatively constant over the range of operating pressures and temperatures so that reliable, predictable performance can be expected.

3. The materials should have good resistance to wear.

4. The materials must be chemically compatible with mating components.

5. Environmental hazards must be minimized.

Several different materials are used for friction elements in clutches and brakes, and many are proprietary to a particular manufacturer. Common in the past were various asbestos-based compounds having coefficients of friction in the range of 0.35 to 0.50. Asbestos has been shown to be a health hazard and is being replaced by molded compounds of polymers and rubber. Where flexibility is required, as in band brakes, the base material is woven into a fabric, sometimes strengthened with metal wire, saturated with a resin, and cured. Cork and wood are also sometimes used. Paper-based materials are used in some oil-filled clutches. In severe environments, cast iron, sintered iron or other metals, or graphite materials are employed. Table 22–2 gives approximate ranges of coefficients of friction and the pressure that the materials can withstand.

For automotive applications, standards are set by the Society of Automotive Engineers. In SAE J866 (Reference 7), a set of codes is defined to classify friction materials according to the coefficient of friction, regardless of the material used. Table 22–3 lists these codes.

In problems in this book that require a coefficient of friction, we will use a value of 0.25 unless otherwise stated. Based on the values reported here, this is a relatively low value that should yield conservative designs.

Materials for Discs and Drums

A variety of metals are applied in the manufacture of brake and clutch discs and drums. The material must have a sufficient strength, ductility, and stiffness to withstand the applied

TABLE 22–2   Coefficients of friction

Friction material	Dynamic friction coefficient		Pressure range	
	Dry	In oil	(psi)	(kPa)
Molded compounds	0.25–0.45	0.06–0.10	150–300	1035–2070
Woven materials	0.25–0.45	0.08–0.10	50–100	345–690
Sintered metal	0.15–0.45	0.05–0.08	150–300	1035–2070
Cork	0.30–0.50	0.15–0.25	8–15	55–100
Wood	0.20–0.45	0.12–0.16	50–90	345–620
Cast iron	0.15–0.25	0.03–0.06	100–250	690–1725
Paper-based		0.10–0.15		
Graphite/resin		0.10–0.14		

TABLE 22–3 Coefficient of friction classification codes of the Society of Automotive Engineers

Code letter	Coefficient of friction
C	Not over 0.15
D	Over 0.15 but not over 0.25
E	Over 0.25 but not over 0.35
F	Over 0.35 but not over 0.45
G	Over 0.45 but not over 0.55
H	Over 0.55
Z	Unclassified

forces while maintaining a precise geometry. It must also be able to absorb heat from the friction surface and dissipate it to the environment.

Some of the popular choices are gray cast iron, ductile iron, carbon steel, and copper alloys. Many discs and drums are cast for cost reasons and to obtain near net shape of the parts that require little machining after casting. Cast iron has low cost and high thermal conductivity as compared with ductile iron. However, ductile is better able to withstand shock or impact loading. Copper alloys have far higher thermal conductivity than the other materials, but they have poorer wear performance.

22–11 PLATE-TYPE CLUTCH OR BRAKE

Figure 22–2(a) shows a simple sketch of a plate-type clutch, and Figure 22–8 shows a cutaway view of a commercially available, electrically actuated clutch or brake. In Figure 22–8, an electromagnet exerts an axial force that draws the friction surfaces together. When two bodies are brought into contact with a normal force between them, a friction force that tends to resist relative motion is created. This is the principle on which the plate-type clutch or brake is based.

Friction Torque

As the friction plate rotates in relation to the mating plate with an axial force pressing them together, the friction force acts in a tangential direction, producing the brake or clutch torque. At any point, the local pressure times the differential area at the point is the *normal force*. The normal force times the coefficient of friction is the *friction force*. The friction force times the radius of the point is the *torque* produced at this point. The *total torque* is the sum of all of the torques over the entire area of the plate. We can find the sum by integrating over the area.

There is usually some variation of pressure over the surface of the friction plate, and some assumption about the nature of the variation must be made before the total torque can be computed. A conservative assumption that yields a useful result is that the friction surface will wear uniformly over the entire area as the brake or clutch operates. This assumption implies that the product of the local pressure, p, times the linear relative speed, v, between the plates is constant. Wear has been found to be approximately proportional to the product of p times v.

Taking all these factors into account and completing the analysis produces the following result for friction torque:

$$T_f = fN(R_o + R_i)/2$$

But the last part of this relation is the mean radius, R_m, of the annular plate. Then

⇨ **Friction Torque on Annular Plate**

$$T_f = fNR_m \qquad \text{(22–8)}$$

As stated before, this is a conservative result, meaning that the actual torque produced would be slightly larger than predicted.

Wear Rating

Note that the torque is proportional to the mean radius, but that no area relationship is involved in Equation (22–8). Thus, the completion of the design for final dimensions requires some other parameter. The missing factor in Equation (22–8) is the expected wear rate of the friction material. It should be obvious that even with the same mean radius, a brake with a larger area would wear less than one with a smaller area.

Manufacturers of friction materials can assist in the final determination of the relationship between wear and the area of the friction surface. However, the following guidelines allow the estimation of the physical size of brakes and will be used for problem solutions in this book.

The wear rating, WR, will be based on the frictional power, P_f, absorbed by the brake per unit area, A, where

Frictional Power
$$P_f = T_f \omega \qquad (22\text{–}9)$$

and ω is the angular velocity of the disc. In SI units, with torque in N·m and ω in rad/s, the frictional power is in N·m/s or watts. In the U.S. Customary System, with torque in lb·in and angular velocity expressed as n rpm, the frictional power is in hp, computed from

$$P_f = \frac{T_f n}{63\ 000} \text{ hp} \qquad (22\text{–}10)$$

For industrial applications, we will use

Wear Rating
$$WR = P_f / A \qquad (22\text{–}11)$$

where $WR = 0.04$ hp/in^2 for frequent applications, a conservative rating
$WR = 0.10$ hp/in^2 for average service
$WR = 0.40$ hp/in^2 for infrequently used brakes allowed to cool somewhat between applications

Example Problem 22–6

Compute the dimensions of an annular plate-type brake to produce a braking torque of 300 lb·in. Springs will provide a normal force of 320 lb between the friction surfaces. The coefficient of friction is 0.25. The brake will be used in average industrial service, stopping a load from 750 rpm.

Solution

Step 1. Compute the required mean radius. From Equation (22–8),

$$R_m = \frac{T_f}{fN} = \frac{300 \text{ lb·in}}{(0.25)(320 \text{ lb})} = 3.75 \text{ in}$$

Step 2. Specify a desired ratio of R_o/R_i, and solve for the dimensions. A reasonable value for the ratio is approximately 1.50. The range can be from 1.2 to about 2.5, at the designer's choice. Using 1.50, $R_o = 1.50R_i$, and

$$R_m = (R_o + R_i)/2 = (1.5R_i + R_i)/2 = 1.25\ R_i$$

Then

$$R_i = R_m/1.25 = (3.75 \text{ in})/1.25 = 3.00 \text{ in}$$
$$R_o = 1.50R_i = 1.50(3.00) = 4.50 \text{ in}$$

Step 3. Compute the area of the friction surface:

$$A = \pi(R_o^2 - R_i^2) = \pi[(4.50)^2 - (3.00)^2] = 35.3 \text{ in}^2$$

Step 4. Compute the frictional power absorbed:

$$P_f = \frac{T_f n}{63\,000} = \frac{(300)(750)}{63\,000} = 3.57 \text{ hp}$$

Step 5. Compute the wear ratio:

$$WR = \frac{P_f}{A} = \frac{3.57 \text{ hp}}{35.3 \text{ in}^2} = 0.101 \text{ hp/in}^2$$

Step 6. Judge the suitability of *WR*. If *WR* is too high, return to Step 2 and increase the ratio. If *WR* is too low, decrease the ratio. In this example, *WR* is acceptable.

A more compact unit can be designed if more than one friction plate is used. We multiply the friction torque for one plate by the number of plates to determine the total friction torque. One disadvantage of this approach is that the heat dissipation is relatively poorer than for the single plate.

Improving Wear Performance of Brakes

Actual wear in a given application depends on a combination of many variables. Friction materials are relatively softer and weaker than the metallic materials used for discs and drums. Wear is often characterized as adhesion. As the surface of the friction material rubs over the high spots of the metal, plastic deformation occurs at the surface and particles are sheared off, breaking the bond between particles or dislodging filler materials from the polymer bonding agents. This process accelerates when surface temperatures rise as the brake absorbs the energy required to stop the rotating system. The thermal behavior of the system is critical to good life. If temperatures rise above about 400°F (200°C), wear rate increase significantly and the coefficient of friction decreases, leading to poorer braking performance called *fade*.

It is difficult to predict the life of a given brake system analytically, and testing under real operating conditions is recommended for new designs. The following list gives general principles of improving wear performance.

- Specify friction materials that have relatively low adhesion when in contact with the disc or drum material.
- Specify friction materials that have high bonding strength between constituent particles.
- Provide high hardness on the surface of the disc or drum by heat treatment.
- Keep the pressure between the friction material and the material of the disc or drum as low as practical.
- Maintain the surface temperature at the interface between the friction material and the material of the disc or drum as low as practical by promoting heat transfer away from the system by conduction, convection, and radiation. Forced airflow or cooling with water is often applied in critical situations.
- Provide a smooth surface finish on the discs and drums.

- Provide lubricants such as oil or graphite at the friction interface.
- Exclude abrasive contaminants from the friction interface.
- Minimize slipping between the clutch or brake elements by promoting lockup of the engaging elements.

22–12
CALIPER DISC
BRAKES

The disc brake pads are brought into contact with the rotating disc by fluid pressure acting on a piston in the caliper. The pads are either round or in a short crescent shape to cover more of the disc surface [see Figure 22–2(b) and Figure 22–7]. However, one advantage of the disc brake is that the disc is exposed to the atmosphere, promoting the dissipation of heat. Because the disc rotates with the machine to be controlled, heat dissipation is further enhanced. The cooling effect improves the fade resistance of this type of brake relative to the shoe-type brake.

Designs for the friction torque and for the wear rating are similar to those explained for plate-type brakes.

22–13
CONE CLUTCH
OR BRAKE

The angle of inclination of the conical surface of the cone clutch or brake is typically 12°. A lower angle could be used with care, but there is a tendency for the friction surfaces to engage suddenly with a jerk. As the angle increases, the amount of axial force required to produce a given friction torque increases. Thus, 12° is a reasonable compromise.

Referring to Figure 22–16, we see that as an axial force F_a is applied by a spring, manually or by fluid pressure, a normal force N is created between the mating friction surfaces, all around the periphery of the cone. The desired friction force F_f is produced in the tangential direction, where $F_f = fN$. It is assumed that the friction force acts at the mean radius of the cone so that the friction torque is

$$T_f = F_f R_{\mathrm{m}} = fNR_m \tag{22–8}$$

In addition to the tangentially directed friction force, a friction force develops along the surface of the cone and opposes the tendency for the member having the internal cone surface to move axially away from the external cone. We will call this force F'_f and compute it also from

$$F'_f = fN$$

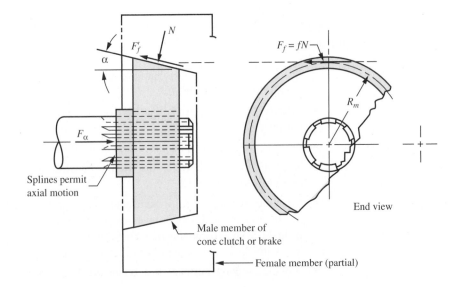

FIGURE 22–16
Cone clutch or brake

For the equilibrium condition of the external cone, the sum of the horizontal forces must be zero. Then

$$F_a = N \sin \alpha + F_f' \cos \alpha = N \sin \alpha + fN \cos \alpha = N(\sin \alpha + f \cos \alpha)$$

or

Frictional Torque on a Cone Clutch or Brake

$$N = \frac{F_a}{\sin \alpha + f \cos \alpha} \qquad (22\text{--}12)$$

Substituting this into Equation (22–8) gives

$$T_f = \frac{fR_m F_a}{\sin \alpha + f \cos \alpha} \qquad (22\text{--}13)$$

Example Problem 22–7   Compute the axial force required for a cone brake if it is to exert a braking torque of 50 lb·ft. The mean radius of the cone is 5.0 in. Use $f = 0.25$. Try cone angles of 10°, 12°, and 15°.

Solution   We can solve Equation (22–13) for the axial force F_a:

$$F_a = \frac{T_f(\sin \alpha + f \cos \alpha)}{fR_m} = \frac{(50 \text{ lb·ft})(\sin \alpha + 0.25 \cos \alpha)}{(0.25)(5.0/12) \text{ ft}}$$

$$F_a = 480(\sin \alpha + 0.25 \cos \alpha) \text{ lb}$$

Then the values of F_a as a function of the cone angle are as follows:

For $\alpha = 10°$
$F_a = 202$ lb

For $\alpha = 12°$
$F_a = 217$ lb

For $\alpha = 15°$
$F_a = 240$ lb

22–14
DRUM BRAKES

Short Shoe Drum Brakes

Figure 22–17 shows a sketch of a drum brake in which the actuating force W acts on the lever that pivots on pin A. This creates a normal force between the shoe and the rotating drum. The resulting friction force is assumed to act tangential to the drum if the shoe is short. The friction force times the radius of the drum is the friction torque that slows the drum.

The objectives of the analysis are to determine the relationship between the applied load and the friction force and to be able to evaluate the effect of design decisions such as the size of the drum, the lever dimensions, and the placement of the pivot A. The free-body

FIGURE 22–17
Short shoe drum brake

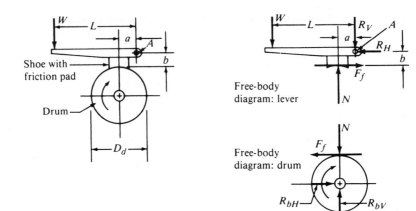

(a) Lever design 1

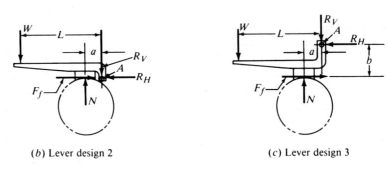

(b) Lever design 2   (c) Lever design 3

diagrams in Figure 22–17(a) support this analysis. For the lever, we can sum moments about the pivot A:

$$\Sigma M_a = 0 = WL - Na + F_f b \qquad (22\text{–}14)$$

But note that $F_f = fN$ or $N = F_f/f$. Then

$$0 = WL - F_f a/f + F_f b = WL - F_f(a/f - b)$$

Solving for W gives

$$W = \frac{F_f(a/f - b)}{L} \qquad (22\text{–}15)$$

Solving for F_f gives

▷ **Friction Force on Drum Brake**

$$F_f = \frac{WL}{(a/f - b)} \qquad (22\text{–}16)$$

We can use these equations for the friction torque by noting

▷ **Friction Torque**

$$T_f = F_f D_d/2 \qquad (22\text{–}17)$$

where D_d = diameter of the drum

Note the alternate positions of the pivot in Parts (b) and (c) of Figure 22–17. In (b), the dimension $b = 0$.

FIGURE 22–18
Results: actuating load
force vs. distance b

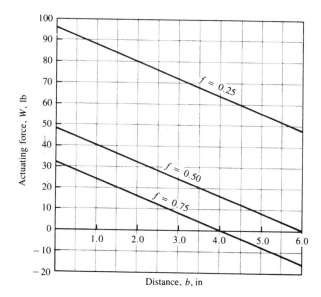

Distance, b, in

Example Problem 22–8

Compute the actuation force required for the short shoe drum brake of Figure 22–17 to produce a friction torque of 50 lb·ft. Use a drum diameter of 10.0 in, $a = 3.0$ in, and $L = 15.0$ in. Consider values of f of 0.25, 0.50, and 0.75, and different points of location of pivot A such that b ranges from 0 to 6.0 in.

Solution

The required friction force can be found from Equation (22–17):

$$F_f = 2T_f/D_d = (2)(50 \text{ lb·ft})/(10/12 \text{ ft}) = 120 \text{ lb}$$

In Equation (22–15), we can substitute for a, L, and F_f:

$$W = \frac{F_f(a/f - b)}{L} = \frac{120 \text{ lb}[(3.0 \text{ in})/f - b]}{15.0 \text{ in}} = 8(3/f - b) \text{ lb}$$

We can substitute the varying values of f and b into this last equation to compute the data for the curves of Figure 22–18, showing the actuating force versus the distance b for different values of f. Note that for some combinations, the value of W is *negative*. This means that the brake is *self-actuating* and that an upward force on the lever would be required to release the brake.

DESIGN

Long Shoe Drum Brakes

The assumption used for short shoe brakes, that the resultant friction force acted at the middle of the shoe, cannot be used in the case of shoes covering more than about 45° of the drum. In such cases, the pressure between the friction lining and the drum is very nonuniform, as is the moment of the friction force and of the normal force with respect to the pivot of the shoe.

The following equations govern the performance of a long shoe brake, using the terminology from Figure 22–19. (See Reference 4.)

1. ***Friction torque on drum:***

$$T_f = r^2 f w p_{max}(\cos \theta_1 - \cos \theta_2) \qquad \textbf{(22–18)}$$

FIGURE 22–19

Terminology for long shoe drum brake

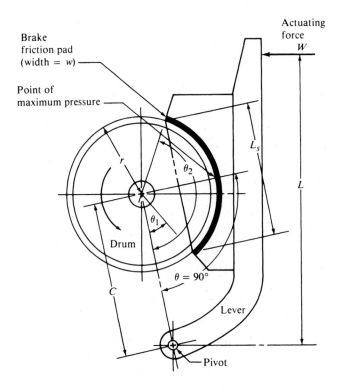

2. **Actuation force:**

$$W = (M_N + M_f)/L \qquad \text{(22–19)}$$

where M_N = moment of normal force with respect to the hinge pin

$$M_N = 0.25p_{max}wrC[2(\theta_2 - \theta_1) - \sin 2\theta_2 + \sin 2\theta_1] \qquad \text{(22–20)}$$

M_f = moment of friction force with respect to the hinge pin

$$M_f = fp_{max}wr[r(\cos \theta_1 - \cos \theta_2) + 0.25C(\cos 2\theta_2 - \cos 2\theta_1)] \qquad \text{(22–21)}$$

The sign of M_f is negative ($-$) if the drum surface is moving away from the pivot and positive ($+$) if it is moving toward the pivot.

3. **Friction power:**

$$P_f = T_f n/63\ 000 \text{ hp} \qquad \text{(22–22)}$$

where n = rotational speed in rpm

4. **Brake shoe area** (*Note*: Projected area is used):

$$A = L_s w = 2wr \sin[(\theta_2 - \theta_1)/2] \qquad \text{(22–23)}$$

5. **Wear ratio:**

$$WR = P_f/A \qquad \text{(22–24)}$$

The use of these relationships in the design and analysis of a long shoe brake is shown in Example Problem 22–9.

Example Problem 22–9   Design a long shoe drum brake to produce a friction torque of 750 lb·in to stop a drum from 120 rpm.

Solution   **Step 1.** Select a brake friction material, and specify the desired maximum pressure and the design value for the coefficient of friction. Table 22–2 lists some general properties for friction materials. Actual test values or specific manufacturer's data should be used where possible. The design value for p_{max} should be far less than the permissible pressure listed in Table 22–2 in order to improve wear life.

For this problem, let's choose a molded polymer compound and design for approximately 75 psi maximum pressure. Note, as shown in Figure 22–19, that the maximum pressure occurs at the section 90° from the pivot. If the shoe does not extend at least 90°, the equations used here are not valid. (See Reference 4.) Also, let's use $f = 0.25$ for the design.

Step 2. Propose trial values for the geometry of the brake drum and the brake pad. Several design decisions must be made here. The general arrangement shown in Figure 22–19 can be used as a guide. But your specific application and creativity may lead you to modify the arrangement.

Trial values are $r = 4.0$ in; $C = 8.0$ in; $L = 15$ in; $\theta_1 = 30°$; and $\theta_2 = 150°$.

Step 3. Solve for the required width of the shoe from Equation (22–18):

$$w = \frac{T_f}{r^2 f p_{max}(\cos \theta_1 - \cos \theta_2)}$$

For this problem,

$$w = \frac{750 \text{ lb} \cdot \text{in}}{(4.0 \text{ in})^2(0.25)(75 \text{ lb/in}^2)(\cos 30° - \cos 150°)} = 1.44 \text{ in}$$

For convenience, let $w = 1.50$ in. Because the maximum pressure is inversely proportional to the width, the actual maximum pressure will be

$$p_{max} = 75 \text{ psi}(1.44/1.50) = 72 \text{ psi}$$

Step 4. Compute M_N from Equation (22–20). The value of $\theta_2 - \theta_1$ must be in radians, with π radians = 180°. Then

$$\theta_2 - \theta_1 = 120°(\pi \text{ rad}/180°) = 2.09 \text{ rad}$$

The moment of the normal force on the shoe is

$$M_N = 0.25(72 \text{ lb/in}^2)(1.50 \text{ in})(4.0 \text{ in})(8.0 \text{ in})$$
$$[2(2.09) - \sin(300°) + \sin(60°)]$$
$$M_N = 5108 \text{ lb} \cdot \text{in}$$

Step 5. Compute the moment of the friction force on the shoe, M_f, from Equation (22–21):

$$M_f = 0.25(72 \text{ lb/in}^2)(1.50 \text{ in})(4.0 \text{ in})$$
$$[(4.0 \text{ in})(\cos 30° - \cos 150°)$$
$$+ 0.25(8.0 \text{ in})(\cos 300° - \cos 60°)]$$
$$M_f = 748 \text{ lb} \cdot \text{in}$$

Step 6. Compute the required actuation force, W, from Equation (22–19):

$$W = (M_N - M_f)/L = (5108 - 748)/(15) = 291 \text{ lb}$$

Note the minus sign because the drum surface is moving away from the pivot.

Step 7. Compute the frictional power from Equation (22–22):

$$P_f = T_f n/(63\,000) = (750)(120)/(63\,000) = 1.43 \text{ hp}$$

Step 8. Compute the projected area of the brake shoe from Equation (22–23).

$$A = L_s w = 2wr \sin[(\theta_2 - \theta_1)/2]$$
$$A = 2(1.50 \text{ in})(4.0 \text{ in})\sin(120°/2) = 10.4 \text{ in}^2$$

Step 9. Compute the wear ratio, WR:

$$WR = P_f/A = 1.43 \text{ hp}/10.4 \text{ in}^2 = 0.14 \text{ hp/in}^2$$

Step 10. Evaluate the suitability of the results. In this problem, we would need more information about the application to evaluate the results. However, the wear ratio seems reasonable for average service (see Section 22–11), and the geometry seems acceptable.

**22–15
BAND BRAKES**

Figure 22–20 shows the typical configuration of a *band brake*. The flexible band, usually made of steel, is faced with a friction material that can conform to the curvature of the drum. The application of a force to the lever puts tension in the band and forces the friction material against the drum. The normal force, thus created, causes the friction force tangential to the drum surface to be created, retarding the drum.

The tension in the band decreases from the value P_1 at the pivot side of the band to P_2 at the lever side. The net torque on the drum is then

$$T_f = (P_1 - P_2)r \qquad\qquad (22\text{–}25)$$

FIGURE 22–20
Band brake design

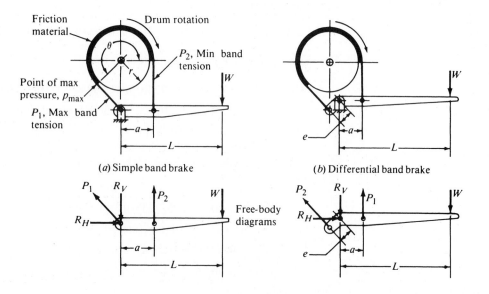

(a) Simple band brake   (b) Differential band brake

where r = radius of the drum

The relationship between P_1 and P_2 can be shown (see Reference 4) to be the logarithmic function

$$P_2 = P_1/e^{f\theta} \qquad (22\text{–}26)$$

where θ = total angle of coverage of the band in radians

The point of maximum pressure on the friction material occurs at the end nearest the highest tension, P_1, where

$$P_1 = p_{max}rw \qquad (22\text{–}27)$$

and w is the width of the band.

For the two types of band brakes shown in Figure 22–20, the free-body diagrams of the levers can be used to show the following relationships for the actuating force, W, as a function of the tensions in the band. For the simple band brake of Figure 22–20(a),

$$W = P_2a/L \qquad (22\text{–}28)$$

The style shown in Figure 22–20(b) is called a *differential band brake*, where the actuation force is

$$W = (P_2a - P_1e)/L \qquad (22\text{–}29)$$

The design procedure is presented in Example Problem 22–10.

Example Problem 22–10   Design a band brake to exert a braking torque of 720 lb·in while slowing the drum from 120 rpm.

Solution   *Step 1.* Select a material and specify a design value for the maximum pressure. A woven friction material is desirable to facilitate the conformance to the cylindrical drum shape. Let's use p_{max} = 25 psi and a design value of f = 0.25. See Section 22–10.

Step 2. Specify a trial geometry: r, θ, w. For this problem, let's try r = 6.0 in, θ = 225°, and w = 2.0 in. Note that 225° = 3.93 rad.

Step 3. Compute the maximum band tension, P_1, from Equation (22–27):

$$P_1 = p_{max}rw = (25 \text{ lb/in}^2)(6.0 \text{ in})(2.0 \text{ in}) = 300 \text{ lb}$$

Step 4. Compute tension P_2 from Equation (22–26):

$$P_2 = \frac{P_1}{e^{f\theta}} = \frac{300 \text{ lb}}{e^{(0.25)(3.93)}} = 112 \text{ lb}$$

Step 5. Compute the friction torque, T_f:

$$T_f = (P_1 - P_2)r = (300 - 112)(6.0) = 1128 \text{ lb·in}$$

Note: Repeat Steps 2–5 until you achieve a satisfactory geometry and friction torque. Let's try a smaller design, say, $r = 5.0$ in:

$$P_1 = (25)(5.0)(2.0) = 250 \text{ lb}$$

$$P_2 = \frac{250 \text{ lb}}{e^{(0.25)(3.93)}} = 93.7 \text{ lb}$$

$$T_f = (250 - 93.7)(5.0) = 782 \text{ lb} \cdot \text{in} \quad (\text{okay})$$

Step 6. Specify the geometry of the lever, and compute the required actuation force. Let's use $a = 5.0$ in and $L = 15.0$ in. Then

$$W = P_2(a/L) = 93.7 \text{ lb}(5.0/15.0) = 31.2 \text{ lb}$$

Step 7. Compute the average wear ratio from $WR = P_f/A$:

$$A = 2\pi rw(\theta/360) = 2(\pi)(5.0 \text{ in})(2.0 \text{ in})(225/360) = 39.3 \text{ in}^2$$

$$P_f = T_f n/63\,000 = (782)(120)/(63\,000) = 1.50 \text{ hp}$$

$$WR = P_f/A = (1.50 \text{ hp})/(39.3 \text{ in}^2) = 0.038 \text{ hp/in}^2$$

This should be conservative for average service.

22–16 OTHER TYPES OF CLUTCHES AND BRAKES

The chapter thus far has concentrated on clutches and brakes using friction materials to transmit the torque between rotating members, but many other types are available. Brief descriptions are given below, but specific design information is not given. Most are unique to the manufacturer, and application data are available through catalogs.

Jaw Clutch

The teeth of the mating sets of jaws are brought into engagement by sliding one or both members axially. The teeth may be straight-sided or triangular, or they may incorporate some smooth curve to facilitate engagement. Once the teeth are engaged, there is a positive transmission of torque. The jaw clutch is normally engaged while the system is stopped or is running very slowly.

Ratchet

Although not strictly a clutch, the familiar ratchet and pawl permits alternate engagement and disengagement of moving members, and thus it can be used in similar applications. Typically, the ratchet moves only a small fraction of a revolution per cycle.

Sprag, Roller, and Cam Clutches

There are differences in the specific geometry of sprag, roller, and cam clutches, but they all perform similar functions. When the input shaft is rotating in the driving direction, the internal elements (sprags, rollers, or cams) are wedged between the driving and driven members and thus transmit torque. But when the input member rotates in the opposite direction, the internal elements move out of engagement, and no torque is transmitted. Thus, they can be used for applications similar to ratchets, but with much smoother operation and with a virtually infinite amount of incremental motion. Another application is *backstopping*, in which the clutch runs free when the machine is being driven in the intended direction. But if the drive is disengaged and the load starts to reverse direction, the clutch locks up and prevents motion. This type of clutch is also used for *overrunning:* a positive drive as long as the load rotates no faster than the driver. If the load tends to run faster (overrun) than the driver, the clutch elements disengage. This protects equipment that might be damaged by overspeeding. See Internet site 6.

Fiber Clutch

A fiber clutch operates in a manner similar to the overrunning clutches described previously. But instead of driving through solid elements, the torque is transmitted through stiff fibers that have a preferred orientation. When rotated in a direction opposite the preferred direction, the fibers "lie down," and no torque is transmitted.

Wrap-Spring Clutch

Again used in cases similar to the overrunning clutches, the *wrap-spring clutch* is made from a rectangular wire and normally has an inside diameter slightly larger than that of the shaft on which it is installed. Thus, no torque is transmitted. But when one end of the spring is restrained, the spring "wraps down" tightly on the surface of the shaft, and torque is transmitted positively through the spring. See Internet site 4.

Single-Revolution Clutches

It is frequently desired to have a machine cycle one complete revolution and then come to a stop. The *single-revolution clutch* provides this feature. After it is tripped, it drives the output shaft until reaching a positive stop at the end of one revolution. Some types can be engaged for more than one revolution, but they will return to a fixed position, for example, at the top of the stroke of a press. See Internet site 6.

Fluid Clutches

The *fluid clutch* is made up of two separate parts with no mechanical connection between them. A fluid fills a cavity between the parts, and as one member rotates, it tends to shear the fluid, causing torque to be transmitted to the mating element. The resulting drive is smooth and soft because load peaks will simply cause one member to move relative to the other. In this situation, it is similar to the slip clutch described earlier.

Eddy Current Drive

When a conducting disc moves through a magnetic field, *eddy currents* are induced in the disc, causing a force to be exerted on the disc in a direction opposite to the direction of rotation. The force can be used to brake the disc or to transmit torque to a mating part, such as a clutch. An advantage of this type of unit is that there is no mechanical connection between the elements. The torque can be controlled by varying the current to the electromagnets.

Overload Clutches

The drive is positive, provided that the torque is below some set value. At higher torques, some element is disengaged automatically. One type uses a series of spherical balls positioned in detents and held under a spring force. When the tripping torque level is reached, the balls are forced out of the detents and disengage the drive. See Internet site 6.

Tensioners

The production of continuous products such as wire, webs of paper, or plastic film require that the drive system be carefully controlled to maintain a light tension without breaking the product. Similar control must be exerted when winding coils of paper, foil, or sheet metal as they are produced, or when unwinding them to feed a process such as printing or press forming. Drives for cranes and hoists must provide controlled braking while loads are lowered. In such cases, brakes are required to exert some braking action while allowing smooth motion. Many of the brake designs reviewed in this chapter can accomplish this function by moderating the force applied between the friction elements. An example is the air-actuated brake shown in Figure 22–6. The braking torque depends on the air pressure applied that can be controlled by an operator or automatically. See also Internet site 5.

REFERENCES

1. Avallone, Eugene A., and Theodore Baumeister III, eds. *Marks' Standard Handbook for Mechanical Engineers*. 10th ed. New York: McGraw-Hill, 1996.

2. Juvinall, Robert C., and Kurt M. Marshek. *Fundamentals of Machine Component Design*. 3rd ed. New York: John Wiley & Sons, 2000.

3. Orthwein, William C. *Clutches and Brakes: Design and Selection*. New York: Marcel Dekker, 1986.

4. Shigley, J. E., and C. R. Mischke. *Mechanical Engineering Design*. 6th ed. New York: McGraw-Hill, 2001.

5. Society of Automotive Engineers. *Standard J286 Clutch Friction Test Machine Guidelines*. Warrendale, PA: Society of Automotive Engineers, 1996.

6. Society of Automotive Engineers. *Standard J661 Brake Lining Quality Control Test Procedure*. Warrendale, PA: Society of Automotive Engineers, 1997.

7. Society of Automotive Engineers. *Standard J866 Friction Coefficient Identification System for Brake Linings*. Warrendale, PA: Society of Automotive Engineers, 2002.

INTERNET SITES FOR CLUTCHES AND BRAKES

1. **Rockford Powertrain, Inc.** *www.rockfordpowertrain. com* Manufacturer of clutches and other powertrain components for the automotive, truck, and off-road equipment markets.

2. **Eaton/Cutler Hammer.** *www.ch.cutler-hammer.com* Manufacturer of mechanical shoe brakes for industrial and commercial applications such as conveyors, machine tools, printing presses, cranes, and carnival rides. Also provides controls for electrical systems.

3. **BWD Automotive.** *www.bwdautomotive.com* Manufacturer of automotive clutch sets and other components under the Borg Warner brand.

4. **Warner Electric, Inc.** *www.warnernet.com* Manufacturer of clutch and brake systems for industrial, turf and garden, and vehicular applications. Includes a wide variety of disk type, wrap spring, and magnetic particle clutches and brakes for start, stop, hold, and tension control. Catalog information can be downloaded from the Literature section.

5. **Eaton/Airflex.** *www.airflex.com* Manufacturer of industrial clutches and brakes using the expanding tube principle actuated by pneumatic or hydraulic pressure. Applications include engines, paper making machinery, power presses, brakes, shears, marine drives, well drilling, and tensioning, winding, and unwinding equipment. Online catalog includes product data and descriptions, product application information, and technical information.

6. **Hilliard Corporation.** *www.hilliardcorp.com* Manufacturer of a wide variety of clutches and brakes for industry and commercial equipment applications. Included are centrifugal clutches, single-revolution clutches, overrunning clutches, ball detent overload release clutches, slip clutch torque limiters, intermittent drives, caliper disc brakes, and others.

7. **Tol-O-Matic, Inc.** *www.tolomatic.com*
Manufacturer of pneumatically and hydraulically operated caliper disc brakes and other automation and motion control products.

8. **Electroid Company.** *www.electroid.com*
Manufacturer of a wide variety of clutches, brakes, and tensioners for industrial, commercial, and aerospace applications.

PROBLEMS

1. Specify the required torque rating for a clutch to be attached to a motor shaft running at 1750 rpm. The motor is rated at 5.0 hp and is of the design B type.

2. Specify the required torque rating for a clutch to be attached to a diesel engine shaft running at 2500 rpm. The engine is rated at 75.0 hp.

3. Specify the required torque rating for a clutch to be attached to an electric motor shaft running at 1150 rpm. The motor is rated at 0.50 hp and drives a light fan.

4. An alternative design for the system described in Problem 1 is being considered. Instead of putting the clutch on the motor shaft, it is desired to place it on the output shaft of a speed reducer that rotates at 180 rpm. The power transmission is still approximately 5.0 hp. Specify the required torque rating for the clutch.

5. Specify the required brake torque rating for each of the conditions in Problems 1–4 for average industrial conditions.

6. Specify the required torque rating for a clutch in N · m if it is attached to a design B electric motor shaft rated at 20.0 kW and rotating at 3450 rpm.

7. A clutch-brake module is to be connected between a design C electric motor and a speed reducer. The motor is rated at 50.0 kW at 900 rpm. Specify the required torque ratings for the clutch and the brake portions of the module for average industrial service. The drive is to a large conveyor.

8. Compute the torque required to accelerate a solid steel disc from rest to 550 rpm in 2.0 s. The disc has a diameter of 24.0 in and is 2.5 in thick.

9. The assembly shown in Figure P22–9 is to be stopped by a brake from 775 rpm to zero in 0.50 s or less. Compute the required brake torque.

10. Compute the required clutch torque to accelerate the system shown in Figure P22–10 from rest to a motor speed of 1750 rpm in 1.50 s. Neglect the inertia of the clutch.

11. A winch, sketched in Figure P22–11, is lowering a load at the speed of 50 ft/min. Compute the required torque rating for the brake on the winch shaft to stop the system in 0.25 s.

12. Figure P22–12 shows a tumbling barrel being driven through a wormgear reduction unit. Evaluate the torque rating required for a clutch to accelerate the barrel to 38.0 rpm from rest in 2.0 s (a) if the clutch is placed on the motor shaft and (b) if it is placed at the output of the reducer. Neglect the inertia of the gear shafts, the bearing races,

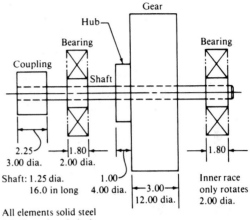

FIGURE P22–9 (Problems 9, 14, and 16)

All elements solid steel
All dimensions in inches

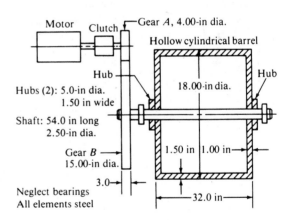

FIGURE P22–10

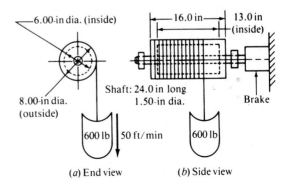

(*a*) End view   (*b*) Side view

FIGURE P22–11

and the clutch. Consider the worm and wormgear to be solid cylinders.

13. Compute the dimensions of an annular plate-type brake to produce a braking torque of 75 lb·in. Air pressure will develop a normal force of 150 lb between the friction surfaces. Use a coefficient of friction of 0.25. The brake will be used in average industrial service, stopping a load from 1150 rpm.

14. Design a plate-type brake for the application described in Problem 9. Specify the design coefficient of friction, the dimensions of the plate, and the axial force required.

15. Compute the axial force required for a cone clutch if it is to exert a driving torque of 15 lb·ft. The cone surface has a mean diameter of 6.0 in and an angle of 12°. Use $f = 0.25$.

16. Design a cone brake for the application described in Problem 9. Specify the design coefficient of friction, the mean diameter of the cone surface, and the axial force required.

17. Compute the actuation force required for the short shoe drum brake of Figure 22–17 to produce a friction torque of 150 lb·ft. Use a drum diameter of 12.0 in, $a = 4.0$ in, and $L = 24.0$ in. Use $f = 0.25$ and $b = 5.0$ in.

18. For all other data from Problem 17 being the same, determine the required dimension b for the brake to be self-actuating.

19. Design a short shoe drum brake to produce a torque of 100 lb·ft. Specify the drum diameter, the configuration of the actuation lever, and the actuation force.

20. Design a long shoe drum brake to produce a friction torque of 100 lb·ft to stop a load from 480 rpm. Specify the friction material, the drum size, the shoe configuration, pivot locations, and actuation force.

21. Design a band brake to exert a braking torque of 75 lb·ft while slowing a drum from 350 rpm to rest. Specify a material, the drum diameter, the band width, the angle of coverage of the friction material, the actuation lever configuration, and the actuation force.

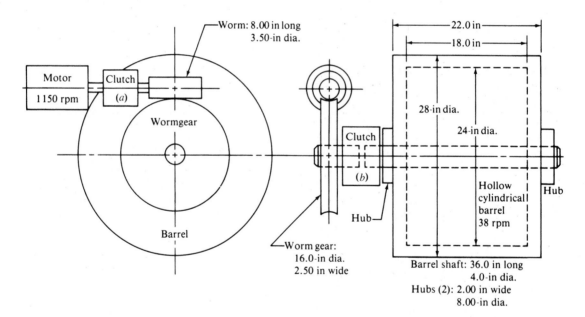

FIGURE P22–12

23

Design Projects

23–1 Objectives of This Chapter

23–2 Design Projects

**23–1
OBJECTIVES OF
THIS CHAPTER**

One of the primary focuses of this book has been to emphasize the integration of machine elements into complete mechanical designs. The interfaces between machine elements have been discussed for many examples. The forces exerted on one element by another have been computed. Commercially available components and complete devices have been shown in figures throughout the book.

Although these discussions and examples are helpful, one of the better ways to learn mechanical design is to *do* mechanical design. You should decide the detailed functions and design requirements for the design. You should conceptualize several approaches to a design. You should decide which approach to complete. You should complete the design of each element in detail. You should make assembly and detail drawings to communicate your design to others who may use it or be responsible for its fabrication. You should specify completely the purchased components that are part of the design.

Following are several projects that call for you to do these operations. You or your instructor may modify or amplify the projects to suit individual needs or the available time or information, such as manufacturers' catalogs. As in most design projects, many solutions are possible. Different solutions from several members of a class could be compared and critiqued to enhance learning. It may be helpful at this time to review Sections 1–4 and 1–5 about the functions and design requirements for design and a philosophy of design. Also, the problems at the end of Chapter 1 asked that you write a set of functions and design requirements for several of the same devices. If you have already done so, they can be used as a part of these exercises.

**23–2
DESIGN
PROJECTS**

Automobile Hood Latch

Design a hood latch for an automobile. The latch must be able to hold the hood securely closed during operation of the vehicle. But it should be easy to open for servicing the contents of the engine compartment. Theft-proofing is an important design goal. Attachment of the latch to the frame of the car and to the hood should be defined. Mass production should be a requirement.

Hydraulic Lift

Design a hydraulic lift to be used for car repair. Obtain pertinent dimensions from representative cars for initial height, extended height, design of the pads that contact the car, and so on. The lift will raise the entire car.

Car Jack

Design a floor jack for a car to lift either the entire front end or the entire rear end. The jack may be powered by hand, using mechanical or hydraulic actuation. Or it may be powered by pneumatic pressure or electrical power.

Portable Crane

Design a portable crane to be used in homes, small industries, warehouses, and garages. It should have a capacity of at least 1000 lb (4.45 kN). Typical uses would be to remove an engine from a car, lift machine components, or load trucks.

Can Crusher

Design a machine to crush soft-drink or beer cans. The crusher would be used in homes or restaurants as an aid to recycling efforts. It could be operated either by hand or electrically. It should crush the cans to approximately 20% of their original volume.

Transfer Device

Design an automatic transfer device for a production line. The parts to be handled are steel castings with the following characteristics:

Weight: 42.0 lb (187 N)

Size: Cylindrical; 6.75-in diameter and 10.0 in high. Exterior surface is free of projections or holes and has a reasonably smooth, as-cast finish.

Transfer rate: Continuous flow, 2.00 s between parts.

Parts enter at a 24.0-in elevation on a roller conveyor. They must be elevated to 48.0 inches in a space of 60.0 in horizontally. They leave on a separate conveyor.

Drum Dumper

Design a drum dumper. The machine is to raise a 55-gal drum of bulk material from floor level to a height of 60.0 in and dump the contents of the drum into a hopper.

Paper Feeder

Design a paper feed device for a copier. The paper must be fed at a rate of 120 sheets per min.

Gravel Conveyor

Design a conveyor to elevate gravel into a truck. The top edge of the truck bed is 8.0 ft (2.44 m) off the ground. The bed is 6.5 ft wide, 12.0 ft long, and 4.0 ft deep (1.98 m × 3.66 m × 1.22 m). It is desired to fill the truck in 5.0 min or less.

Construction Lift

Design a construction lift. The lift will raise building materials from ground level to any height up to 40.0 ft (12.2 m). The lift will be at the top of a rigid scaffold that is not a part of the design project. It will raise a load of up to 500 lb (2.22 kN) at the rate of 1.0 ft/s (0.30 m/s). The load will be on a pallet, 3.0 ft by 4.0 ft (0.91 m × 1.22 m). At the top of the lift, means must be provided to bring the load onto a platform that supports the lift.

Packaging Machine

Design a packaging machine. Toothpaste tubes are to be taken from a continuous belt and inserted into cartons. Any standard tube size may be chosen. The device may include the means to close the cartons after the tube is in place.

Carton Packer

Design a machine to insert 24 cartons of toothpaste into a shipping case.

Robot Gripper

Design a gripper for a robot to grasp a spare-tire assembly from a rack and insert it into the trunk of an automobile on an assembly line. Obtain dimensions from a particular car.

Weld Positioner

Design a weld positioner. A heavy frame is made of welded steel plate in the shape shown in Figure 23–1. The welding unit will be robot-guided, but it is essential that the weld line be horizontal as the weld proceeds. Design the device to hold the frame securely and move it to present the part to the robot. The plate has a thickness of 3/8 in (9.53 mm).

Garage Door Opener

Design a garage door opener.

Spur Gear Speed Reducer, Single-Reduction

Design a complete single-reduction, spur gear–type speed reducer. Specify the two gears, two shafts, four bearings, and a housing. Use any of the data from Chapter 9, Problems 60–70.

Spur Gear Speed Reducer, Double-Reduction

Design a complete double-reduction, spur gear–type speed reducer. Specify the four gears, three shafts, six bearings, and a housing. Use any of the data from Chapter 9, Problems 74–76.

Helical Gear Speed Reducer, Single-Reduction

Design a complete single-reduction, helical gear–type speed reducer. Use any of the data from Chapter 10, Problems 5–11.

Bevel Gear Speed Reducer, Single-Reduction

Design a complete single-reduction, bevel gear–type speed reducer. Use any of the data from Chapter 10, Problems 14–17.

Wormgear Speed Reducer, Single-Reduction

Design a complete single-reduction, wormgear–type speed reducer. Use any of the data from Chapter 10, Problems 18–24.

FIGURE 23–1
Frame to be handled by
a weld positioner

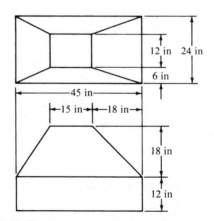

Weld all seams with weld line in horizontal plane

Lift Device Using Acme Screws

Design a device similar to that sketched in Figure 17–3. An electric motor drives the worm at a speed of 1750 rpm. The two Acme screws rotate and lift the yoke, which in turn lifts the hatch. See Example Problem 17–1 for additional details. Complete the entire unit, including the wormgear set, the chain drive, the Acme screws, the bearings, and their mountings. The hatch is 60 in (1524 mm) in diameter at its top surface. The screws should be nominally 30 in (762 mm) long. The total motion of the yoke will be 24 in, to be completed in 15.0 s or less.

Lift Device Using Ball Screws

Repeat the design of the lift device, using ball screws instead of Acme screws.

Brake for a Drive Shaft

Design a brake. A rotating load (as sketched in Figure 22–21) is to be stopped from 775 rpm in 0.50 s or less. Use any type of brake described in Chapter 22, and complete the design details, including the actuating means: springs, air pressure, manual lever, and so on. Show the brake attached to the shaft of Figure 22–21.

Brake for a Winch

Design a complete brake for the application shown in Figure 22–23 and described in Problem 11 in Chapter 22.

Indexing Drive

Design an indexing drive for an automatic assembly system. The items to be moved are mounted on a square steel fixture plate, 6.0 in (152 mm) on a side and 0.50 in (12.7 mm) thick. The total weight of each assembly is 10.0 lb (44.5 N). The center of each fixture (intersection of its diagonals) is to move 12.0 in (305 mm) with each index. The index is to be completed in 1.0 s or less, and the fixture must be held stationary at each station for a minimum of 2.0 s. Four assembly stations are required. The arrangement may be linear, rotary, or any other, provided that the fixtures move in a horizontal plane.

Child's Ferris Wheel

Design a child's Ferris wheel. It should be capable of holding one to four children weighing up to 80 lb (356 N) each. The rotational speed should be 1 rev in 6.0 s. It should be driven by an electric motor.

Merry-Go-Round

Design an amusement ride for small children, age 6 or younger, that they can sit in safely while enjoying an interesting motion pattern. At least two children at a time can ride the machine. The ride will be marketed to shopping centers and department stores to amuse customers' children.

Backyard Amusement Ride

Design a backyard amusement ride in which small coaster wagons are pulled along a circular path. The ride should be powered by an electric motor. Each wagon is 1.0 m long (39.4 in) and 0.50 m wide (19.7 in). The four wheels are 150 mm (6.0 in) in diameter. The wagon is to be attached to the drive rod at the point where the handle would normally be. The radial distance to the attachment point is to be 2.0 m (6.6 ft). The wagons are to make 1 rev in 8.0 s. Provide a means of starting and stopping the drive.

Transfer Device

Design a device to move automotive camshafts between processing stations. Each movement is to be 9.0 in (229 mm). The camshaft is to be supported on two unfinished bearing surfaces having a diameter of 3.80 in (96.5 mm) and an axial length of 0.75 in (19.0 mm). The spread between the bearing surfaces is 15.75 in (400.0 mm). Each camshaft weighs 16.3 lb (72.5 N). One motion cycle is to be completed each 2.50 s. Design the complete mechanism, including the drive from an electric motor.

Chain Conveyor

Design a powered, straight chain conveyor to move eight pallets along an assembly line. The pallets are 18 in long and 12 in wide. The maximum weight of each pallet and the product it carries is 125 lb. At the end of the conveyor, an external downward force of 500 lb is applied to the product which must be carried through the pallet to the conveyor structure. You may design the configuration of the sides and bottom of the pallet.

"You Are the Designer" Projects

At the beginning of each chapter in this book, a section called **You Are the Designer** appeared in which you were asked to imagine that you were the designer of some device or system. Choose any of those projects.

Appendices

APPENDIX 1   PROPERTIES OF AREAS

(a) Circle

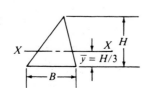

$$A = \pi D^2/4 \qquad r = D/4$$
$$I = \pi D^4/64 \qquad J = \pi D^4/32$$
$$S = \pi D^3/32 \qquad Z_p = \pi D^3/16$$

(b) Hollow circle (tube)

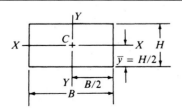

$$A = \pi(D^2 - d^2)/4 \qquad r = \sqrt{D^2 + d^2}/4$$
$$I = \pi(D^4 - d^4)/64 \qquad J = \pi(D^4 - d^4)/32$$
$$S = \pi(D^4 - d^4)/32D \qquad Z_p = \pi(D^4 - d^4)/16D$$

(c) Square

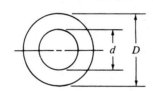

$$A = S^2 \qquad r = S/\sqrt{12}$$
$$I = S^4/12$$
$$S = S^3/6$$

(d) Rectangle

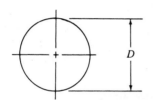

$$A = BH \qquad r_x = H/\sqrt{12}$$
$$I_x = BH^3/12 \qquad r_y = B/\sqrt{12}$$
$$S_x = BH^2/6$$

(e) Triangle

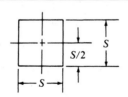

$$A = BH/2 \qquad r = H/\sqrt{18}$$
$$I = BH^3/36$$
$$S = BH^2/24$$

(*f*) Semicircle

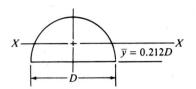

$A = \pi D^2/8$ $r = 0.132D$

$I = 0.007D^4$

$S = 0.024D^3$

(*g*) Regular hexagon

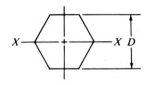

$A = 0.866D^2$ $r = 0.264D$

$I = 0.06D^4$

$S = 0.12D^3$

A = area

I = moment of inertia

S = section modulus

r = radius of gyration = $\sqrt{I/A}$

J = polar moment of inertia

Z_p = polar section modulus

APPENDIX 2 PREFERRED BASIC SIZES AND SCREW THREADS

TABLE A2–1 Preferred basic sizes

Fractional (in)				Decimal (in)			Metric (mm) First	Second	First	Second	First	Second
1/64	0.015 625	5	5.000	0.010	2.00	8.50	1		10		100	
1/32	0.031 25	5 1/4	5.250	0.012	2.20	9.00		1.1		11		110
1/16	0.0625	5 1/2	5.500	0.016	2.40	9.50	1.2		12		120	
3/32	0.093 75	5 3/4	5.750	0.020	2.60	10.00		1.4		14		140
1/8	0.1250	6	6.000	0.025	2.80	10.50	1.6		16		160	
5/32	0.156 25	6 1/2	6.500	0.032	3.00	11.00		1.8		18		180
3/16	0.1875	7	7.000	0.040	3.20	11.50	2		20		200	
1/4	0.2500	7 1/2	7.500	0.05	3.40	12.00		2.2		22		220
5/16	0.3125	8	8.000	0.06	3.60	12.50	2.5		25		250	
3/8	0.3750	8 1/2	8.500	0.08	3.80	13.00		2.8		28		280
7/16	0.4375	9	9.000	0.10	4.00	13.50	3		30		300	
1/2	0.5000	9 1/2	9.500	0.12	4.20	14.00		3.5		35		350
9/16	0.5625	10	10.000	0.16	4.40	14.50	4		40		400	
5/8	0.6250	10 1/2	10.500	0.20	4.60	15.00		4.5		45		450
11/16	0.6875	11	11.000	0.24	4.80	15.50	5		50		500	
3/4	0.7500	11 1/2	11.500	0.30	5.00	16.00		5.5		55		550
7/8	0.8750	12	12.000	0.40	5.20	16.50	6		60		600	
1	1.000	12 1/2	12.500	0.50	5.40	17.00		7		70		700
1 1/4	1.250	13	13.000	0.60	5.60	17.50	8		80		800	
1 1/2	1.500	13 1/2	13.500	0.80	5.80	18.00		9		90		900
1 3/4	1.750	14	14.000	1.00	6.00	18.50					1000	
2	2.000	14 1/2	14.500	1.20	6.50	19.00						
2 1/4	2.250	15	15.000	1.40	7.00	19.50						
2 1/2	2.500	15 1/2	15.500	1.60	7.50	20.00						
2 3/4	2.750	16	16.000	1.80	8.00							
3	3.000	16 1/2	16.500									
3 1/4	3.250	17	17.000									
3 1/2	3.500	17 1/2	17.500									
3 3/4	3.750	18	18.000									
4	4.000	18 1/2	18.500									
4 1/4	4.250	19	19.000									
4 1/2	4.500	19 1/2	19.500									
4 3/4	4.750	20	20.000									

Reprinted from ASME B4.1-1967, by permission of The American Society of Mechanical Engineers. All rights reserved.

TABLE A2–2   American Standard screw threads

A. American Standard thread dimensions, numbered sizes

Size	Basic major diameter, D (in)	Coarse threads: UNC		Fine threads: UNF	
		Threads per inch, n	Tensile stress area (in²)	Threads per inch, n	Tensile stress area (in²)
0	0.0600			80	0.001 80
1	0.0730	64	0.002 63	72	0.002 78
2	0.0860	56	0.003 70	64	0.003 94
3	0.0990	48	0.004 87	56	0.005 23
4	0.1120	40	0.006 04	48	0.006 61
5	0.1250	40	0.007 96	44	0.008 30
6	0.1380	32	0.009 09	40	0.010 15
8	0.1640	32	0.0140	36	0.014 74
10	0.1900	24	0.0175	32	0.0200
12	0.2160	24	0.0242	28	0.0258

B. American Standard thread dimensions, fractional sizes

Size	Basic major diameter, D (in)	Coarse threads: UNC		Fine threads: UNF	
		Threads per inch, n	Tensile stress area (in²)	Threads per inch, n	Tensile stress area (in²)
1/4	0.2500	20	0.0318	28	0.0364
5/16	0.3125	18	0.0524	24	0.0580
3/8	0.3750	16	0.0775	24	0.0878
7/16	0.4375	14	0.1063	20	0.1187
1/2	0.5000	13	0.1419	20	0.1599
9/16	0.5625	12	0.182	18	0.203
5/8	0.6250	11	0.226	18	0.256
3/4	0.7500	10	0.334	16	0.373
7/8	0.8750	9	0.462	14	0.509
1	1.000	8	0.606	12	0.663
$1\frac{1}{8}$	1.125	7	0.763	12	0.856
$1\frac{1}{4}$	1.250	7	0.969	12	1.073
$1\frac{3}{8}$	1.375	6	1.155	12	1.315
$1\frac{1}{2}$	1.500	6	1.405	12	1.581
$1\frac{3}{4}$	1.750	5	1.90		
2	2.000	$4\frac{1}{2}$	2.50		

TABLE A2–3   Metric sizes of screw threads

Basic major diameter, D (mm)	Coarse threads		Fine threads	
	Pitch (mm)	Tensile stress area (mm^2)	Pitch (mm)	Tensile stress area (mm^2)
1	0.25	0.460		
1.6	0.35	1.27	0.20	1.57
2	0.4	2.07	0.25	2.45
2.5	0.45	3.39	0.35	3.70
3	0.5	5.03	0.35	5.61
4	0.7	8.78	0.5	9.79
5	0.8	14.2	0.5	16.1
6	1	20.1	0.75	22.0
8	1.25	36.6	1	39.2
10	1.5	58.0	1.25	61.2
12	1.75	84.3	1.25	92.1
16	2	157	1.5	167
20	2.5	245	1.5	272
24	3	353	2	384
30	3.5	561	2	621
36	4	817	3	865
42	4.5	1121		
48	5	1473		

APPENDIX 3   DESIGN PROPERTIES OF CARBON AND ALLOY STEELS

Material designation (AISI number)	Condition	Tensile strength		Yield strength		Ductility (percent elongation in 2 inches)	Brinell hardness (HB)
		(ksi)	(MPa)	(ksi)	(MPa)		
1020	Hot-rolled	55	379	30	207	25	111
1020	Cold-drawn	61	420	51	352	15	122
1020	Annealed	60	414	43	296	38	121
1040	Hot-rolled	72	496	42	290	18	144
1040	Cold-drawn	80	552	71	490	12	160
1040	OQT 1300	88	607	61	421	33	183
1040	OQT 400	113	779	87	600	19	262
1050	Hot-rolled	90	620	49	338	15	180
1050	Cold-drawn	100	690	84	579	10	200
1050	OQT 1300	96	662	61	421	30	192
1050	OQT 400	143	986	110	758	10	321
1117	Hot-rolled	62	427	34	234	33	124
1117	Cold-drawn	69	476	51	352	20	138
1117	WQT 350	89	614	50	345	22	178
1137	Hot-rolled	88	607	48	331	15	176
1137	Cold-drawn	98	676	82	565	10	196
1137	OQT 1300	87	600	60	414	28	174
1137	OQT 400	157	1083	136	938	5	352
1144	Hot-rolled	94	648	51	352	15	188
1144	Cold-drawn	100	690	90	621	10	200
1144	OQT 1300	96	662	68	469	25	200
1144	OQT 400	127	876	91	627	16	277
1213	Hot-rolled	55	379	33	228	25	110
1213	Cold-drawn	75	517	58	340	10	150
12L13	Hot-rolled	57	393	34	234	22	114
12L13	Cold-drawn	70	483	60	414	10	140
1340	Annealed	102	703	63	434	26	207
1340	OQT 1300	100	690	75	517	25	235
1340	OQT 1000	144	993	132	910	17	363
1340	OQT 700	221	1520	197	1360	10	444
1340	OQT 400	285	1960	234	1610	8	578
3140	Annealed	95	655	67	462	25	187
3140	OQT 1300	115	792	94	648	23	233
3140	OQT 1000	152	1050	133	920	17	311
3140	OQT 700	220	1520	200	1380	13	461
3140	OQT 400	280	1930	248	1710	11	555
4130	Annealed	81	558	52	359	28	156
4130	WQT 1300	98	676	89	614	28	202
4130	WQT 1000	143	986	132	910	16	302
4130	WQT 700	208	1430	180	1240	13	415
4130	WQT 400	234	1610	197	1360	12	461
4140	Annealed	95	655	60	414	26	197
4140	OQT 1300	117	807	100	690	23	235
4140	OQT 1000	168	1160	152	1050	17	341
4140	OQT 700	231	1590	212	1460	13	461
4140	OQT 400	290	2000	251	1730	11	578

Material designation (AISI number)	Condition	Tensile strength (ksi)	Tensile strength (MPa)	Yield strength (ksi)	Yield strength (MPa)	Ductility (percent elongation in 2 inches)	Brinell hardness (HB)
4150	Annealed	106	731	55	379	20	197
4150	OQT 1300	127	880	116	800	20	262
4150	OQT 1000	197	1360	181	1250	11	401
4150	OQT 700	247	1700	229	1580	10	495
4150	OQT 400	300	2070	248	1710	10	578
4340	Annealed	108	745	68	469	22	217
4340	OQT 1300	140	965	120	827	23	280
4340	OQT 1000	171	1180	158	1090	16	363
4340	OQT 700	230	1590	206	1420	12	461
4340	OQT 400	283	1950	228	1570	11	555
5140	Annealed	83	572	42	290	29	167
5140	OQT 1300	104	717	83	572	27	207
5140	OQT 1000	145	1000	130	896	18	302
5140	OQT 700	220	1520	200	1380	11	429
5140	OQT 400	276	1900	226	1560	7	534
5150	Annealed	98	676	52	359	22	197
5150	OQT 1300	116	800	102	700	22	241
5150	OQT 1000	160	1100	149	1030	15	321
5150	OQT 700	240	1650	220	1520	10	461
5150	OQT 400	312	2150	250	1720	8	601
5160	Annealed	105	724	40	276	17	197
5160	OQT 1300	115	793	100	690	23	229
5160	OQT 1000	170	1170	151	1040	14	341
5160	OQT 700	263	1810	237	1630	9	514
5160	OQT 400	322	2220	260	1790	4	627
6150	Annealed	96	662	59	407	23	197
6150	OQT 1300	118	814	107	738	21	241
6150	OQT 1000	183	1260	173	1190	12	375
6150	OQT 700	247	1700	223	1540	10	495
6150	OQT 400	315	2170	270	1860	7	601
8650	Annealed	104	717	56	386	22	212
8650	OQT 1300	122	841	113	779	21	255
8650	OQT 1000	176	1210	155	1070	14	363
8650	OQT 700	240	1650	222	1530	12	495
8650	OQT 400	282	1940	250	1720	11	555
8740	Annealed	100	690	60	414	22	201
8740	OQT 1300	119	820	100	690	25	241
8740	OQT 1000	175	1210	167	1150	15	363
8740	OQT 700	228	1570	212	1460	12	461
8740	OQT 400	290	2000	240	1650	10	578
9255	Annealed	113	780	71	490	22	229
9255	Q&T 1300	130	896	102	703	21	262
9255	Q&T 1000	181	1250	160	1100	14	352
9255	Q&T 700	260	1790	240	1650	5	534
9255	Q&T 400	310	2140	287	1980	2	601

Note: Properties common to all carbon and alloy steels:

Poisson's ratio: 0.27

Shear modulus: 11.5×10^6 psi; 80 GPa

Coefficient of thermal expansion: 6.5×10^{-6} °F^{-1}

Density: 0.283 lb/in^3: 7680 kg/m^3

Modulus of elasticity: 30×10^6 psi; 207 GPa

APPENDIX 4   PROPERTIES OF HEAT-TREATED STEELS

FIGURE A4–1
Properties of heat-treated AISI 1040, water-quenched and tempered (*Modern Steels and Their Properties*, Bethlehem Steel Co., Bethlehem, PA)

Treatment: Normalized at 1650 F; reheated to 1550 F; quenched in water.
1-in. Round Treated; .505-in. Round Tested.   As-quenched HB 534.

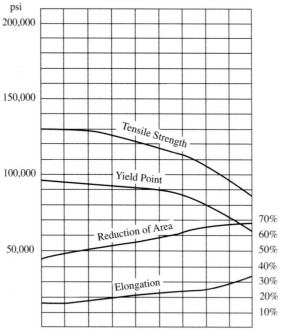

Temper, F	400	500	600	700	800	900	1000	1100	1200	1300
Hardness, HB	514	495	444	401	352	293	269	235	201	187

FIGURE A4–2
Properties of heat-treated AISI 1144, oil-quenched and tempered (*Modern Steels and Their Properties*, Bethlehem Steel Co., Bethlehem, PA)

Treatment: Normalized at 1650 F; reheated to 1550 F; quenched in oil.
1-in. Round Treated; .505-in. Round Tested.   As-quenched HB 285.

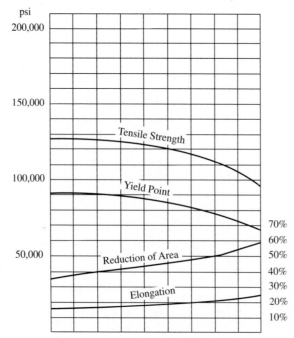

Temper, F	400	500	600	700	800	900	1000	1100	1200	1300
Hardness, HB	277	269	262	255	248	241	235	229	217	201

FIGURE A4–3
Properties of heat-treated AISI 1340, oil-quenched and tempered (*Modern Steels and Their Properties*, Bethlehem Steel Co., Bethlehem, PA)

Treatment: Normalized at 1600 F; reheated to 1525 F; quenched in agitated oil. .565-in. Round Treated; .505-in. Round Tested. As-quenched HB 601.

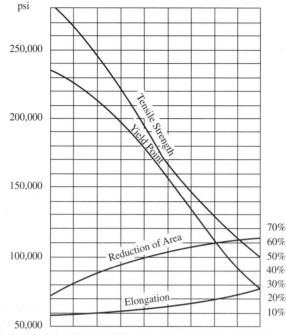

Temper, F	400	500	600	700	800	900	1000	1100	1200	1300
Hardness, HB	578	534	495	444	415	388	363	331	293	235

FIGURE A4–4
Properties of heat-treated AISI 4140, oil-quenched and tempered (*Modern Steels and Their Properties*, Bethlehem Steel Co., Bethlehem, PA)

Treatment: Normalized at 1600 F; reheated to 1550 F; quenched in agitated oil. .530-in. Round Treated; .505-in. Round Tested. As-quenched HB 601.

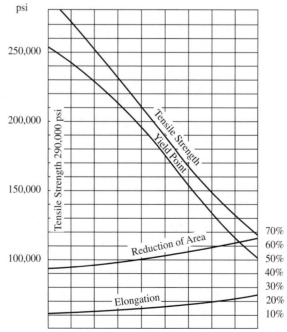

Temper, F	400	500	600	700	800	900	1000	1100	1200	1300
Hardness, HB	578	534	495	461	429	388	341	311	277	235

FIGURE A4–5

Properties of heat-treated AISI 4340, oil-quenched and tempered (*Modern Steels and Their Properties*, Bethlehem Steel Co., Bethlehem, PA)

Treatment: Normalized at 1600 F; reheated to 1475 F; quenched in agitated oil. .530-in. Round Treated; .505-in. Round Tested.   As-quenched HB 601.

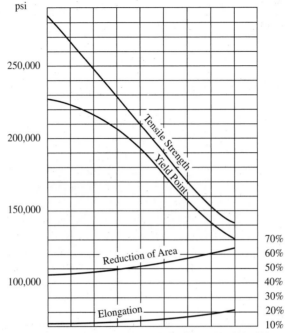

Temper, F	400	500	600	700	800	900	1000	1100	1200	1300
Hardness, HB	555	514	477	461	415	388	363	321	293	—

FIGURE A4–6

Properties of heat-treated AISI 6150, oil-quenched and tempered (*Modern Steels and Their Properties*, Bethlehem Steel Co., Bethlehem, PA)

Treatment: Normalized at 1600 F; reheated to 1550 F; quenched in agitated oil. .565-in. Round Treated; .505-in. Round Tested.   As-quenched HB 627.

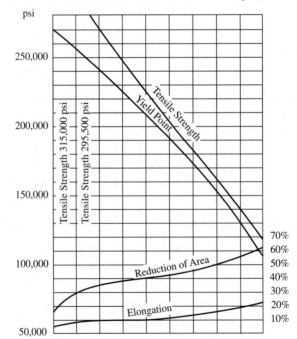

Temper, F	400	500	600	700	800	900	1000	1100	1200	1300
Hardness, HB	601	578	534	495	444	401	375	341	293	241

APPENDIX 5   PROPERTIES OF CARBURIZED STEELS

Material designation (AISI number)	Condition	Core properties					Brinell hardness (HB)	Case hardness (HRC)
		Tensile strength		Yield strength		Ductility (percent elongation in 2 inches)		
		(ksi)	(MPa)	(ksi)	(MPa)			
1015	SWQT 350	106	731	60	414	15	217	62
1020	SWQT 350	129	889	72	496	11	255	62
1022	SWQT 350	135	931	75	517	14	262	62
1117	SWQT 350	125	862	66	455	10	235	65
1118	SWQT 350	144	993	90	621	13	285	61
4118	SOQT 300	143	986	93	641	17	293	62
4118	DOQT 300	126	869	63	434	21	241	62
4118	SOQT 450	138	952	89	614	17	277	56
4118	DOQT 450	120	827	63	434	22	229	56
4320	SOQT 300	218	1500	178	1230	13	429	62
4320	DOQT 300	151	1040	97	669	19	302	62
4320	SOQT 450	211	1450	173	1190	12	415	59
4320	DOQT 450	145	1000	94	648	21	293	59
4620	SOQT 300	119	820	83	572	19	277	62
4620	DOQT 300	122	841	77	531	22	248	62
4620	SOQT 450	115	793	80	552	20	248	59
4620	DOQT 450	115	793	77	531	22	235	59
4820	SOQT 300	207	1430	167	1150	13	415	61
4820	DOQT 300	204	1405	165	1140	13	415	60
4820	SOQT 450	205	1410	184	1270	13	415	57
4820	DOQT 450	196	1350	171	1180	13	401	56
8620	SOQT 300	188	1300	149	1030	11	388	64
8620	DOQT 300	133	917	83	572	20	269	64
8620	SOQT 450	167	1150	120	827	14	341	61
8620	DOQT 450	130	896	77	531	22	262	61
E9310	SOQT 300	173	1190	135	931	15	363	62
E9310	DOQT 300	174	1200	139	958	15	363	60
E9310	SOQT 450	168	1160	137	945	15	341	59
E9310	DOQT 450	169	1170	138	952	15	352	58

Notes: Properties given are for a single set of tests on 1/2-in round bars.

SWQT: single water-quenched and tempered

SOQT: single oil-quenched and tempered

DOQT: double oil-quenched and tempered

300 and 450 are the tempering temperatures in °F. Steel was carburized for 8 h. Case depth ranged from 0.045 to 0.075 in.

APPENDIX 6   PROPERTIES OF STAINLESS STEELS

Material designation			Tensile strength		Yield strength		Ductility (percent elongation in 2 inches)
AISI number	UNS	Condition	(ksi)	(MPa)	(ksi)	(MPa)	
Austenitic steels							
201	S20100	Annealed	115	793	55	379	55
		1/4 hard	125	862	75	517	20
		1/2 hard	150	1030	110	758	10
		3/4 hard	175	1210	135	931	5
		Full hard	185	1280	140	966	4
301	S30100	Annealed	110	758	40	276	60
		1/4 hard	125	862	75	517	25
		1/2 hard	150	1030	110	758	15
		3/4 hard	175	1210	135	931	12
		Full hard	185	1280	140	966	8
304	S30400	Annealed	85	586	35	241	60
310	S31000	Annealed	95	655	45	310	45
316	S31600	Annealed	80	552	30	207	60
Ferritic steels							
405	S40500	Annealed	70	483	40	276	30
430	S43000	Annealed	75	517	40	276	30
446	S44600	Annealed	80	552	50	345	25
Martensitic steels							
410	S41000	Annealed	75	517	40	276	30
416	S41600	Q&T 600	180	1240	140	966	15
		Q&T 1000	145	1000	115	793	20
		Q&T 1400	90	621	60	414	30
431	S43100	Q&T 600	195	1344	150	1034	15
440A	S44002	Q&T 600	280	1930	270	1860	3
Precipitation-hardening steels							
17-4PH	S17400	H 900	200	1380	185	1280	14
		H 1150	145	1000	125	862	19
17-7PH	S17700	RH 950	200	1380	175	1210	10
		TH 1050	175	1210	155	1070	12

APPENDIX 7   PROPERTIES OF STRUCTURAL STEELS

Material designation (ASTM number)	Grade, product, or thickness	Tensile strength		Yield strength		Ductility (percent elongation in 2 inches)
		(ksi)	(MPa)	(ksi)	(MPa)	
A36	$t \leq 8$ in	58	400	36	250	21
A242	$t \leq 3/4$ in	70	480	50	345	21
A242	$t \leq 1\frac{1}{2}$ in	67	460	46	315	21
A242	$t \leq 4$ in	63	435	42	290	21
A500	Cold-formed structural tubing, round or shaped					
	Round, Grade A	45	310	33	228	25
	Round, Grade B	58	400	42	290	23
	Round, Grade C	62	427	46	317	21
	Shaped, Grade A	45	310	39	269	25
	Shaped, Grade B	58	400	46	317	23
	Shaped, Grade C	62	427	50	345	21
A501	Hot-formed structural tubing, round or shaped	58	400	36	250	23
A514	Quenched and tempered, $t \leq 2\frac{1}{2}$ in	110–130	760–895	100	690	18%
A572	42, $t \leq 6$ in	60	415	42	290	24
A572	50, $t \leq 4$ in	65	450	50	345	21
A572	60, $t \leq 1\frac{1}{4}$ in	75	520	60	415	18
A572	65, $t \leq 1\frac{1}{4}$ in	80	550	65	450	17
A588	$t \leq 4$ in	70	485	50	345	21
A992	W-shapes	65	450	50	345	21

Note: ASTM A572 is one of the high-strength, low-alloy steels (HSLA) and has properties similar to those of the SAE J410b steel specified by the SAE.

APPENDIX 8   DESIGN PROPERTIES OF CAST IRON

Material designation (ASTM number)	Grade	Tensile strength (ksi)	Tensile strength (MPa)	Yield strength (ksi)	Yield strength (MPa)	Ductility (percent elongation in 2 inches)	Modulus of elasticity (10^6 psi)	Modulus of elasticity (GPa)
Gray iron								
A48-94a	20	20	138			<1	12	83
	25	25	172			<1	13	90
	30	30	207			<1	15	103
	40	40	276			<1	17	117
	50	50	345			<1	19	131
	60	60	414			<1	20	138
Malleable iron								
A47-99	32510	50	345	32	221	10	25	172
	35018	53	365	35	241	18	25	172
A220-99	40010	60	414	40	276	10	26	179
	45006	65	448	45	310	6	26	179
	50005	70	483	50	345	5	26	179
	70003	85	586	70	483	3	26	179
	90001	105	724	90	621	1	26	179
Ductile iron								
A536-84	60-40-18	60	414	40	276	18	22	152
	80-55-06	80	552	55	379	6	22	152
	100-70-03	100	689	70	483	3	22	152
	120-90-02	120	827	90	621	2	22	152
Austempered ductile iron								
ASTM 897-90	1	125	850	80	550	10	22	152
	2	150	1050	100	700	7	22	152
	3	175	1200	125	850	4	22	152
	4	200	1400	155	1100	1	22	152
	5	230	1600	185	1300	<1	22	152

Notes: Strength values are typical. Casting variables and section size affect final values. Modulus of elasticity may also vary. Density of cast irons ranges from 0.25 to 0.27 lb/in^3 (6920 to 7480 kg/m^3). Compressive strength ranges 3 to 5 times higher than tensile strength.

APPENDIX 9   TYPICAL PROPERTIES OF ALUMINUM

Alloy and temper	Tensile strength (ksi)	(MPa)	Yield strength (ksi)	(MPa)	Ductility (percent elongation in 2 inches)	Shearing strength (ksi)	(MPa)	Endurance strength (ksi)	(MPa)
1060-O	10	69	4	28	43	7	48	3	21
1060-H14	14	97	11	76	12	9	62	5	34
1060-H18	19	131	18	124	6	11	121	6	41
1350-O	12	83	4	28	28	8	55		
1350-H14	16	110	14	97		10	69		
1350-H19	27	186	24	165		15	103	7	48
2014-O	27	186	14	97	18	18	124	13	90
2014-T4	62	427	42	290	20	38	262	20	138
2014-T6	70	483	60	414	13	42	290	18	124
2024-O	27	186	11	76	22	18	124	13	90
2024-T4	68	469	47	324	19	41	283	20	138
2024-T361	72	496	57	393	12	42	290	18	124
2219-O	25	172	11	76	18				
2219-T62	60	414	42	290	10			15	103
2219-T87	69	476	57	393	10			15	103
3003-O	16	110	6	41	40	11	121	7	48
3003-H14	22	152	21	145	16	14	97	9	62
3003-H18	29	200	27	186	10	16	110	10	69
5052-O	28	193	13	90	30	18	124	16	110
5052-H34	38	262	31	214	14	21	145	18	124
5052-H38	42	290	37	255	8	24	165	20	138
6061-O	18	124	8	55	30	12	83	9	62
6061-T4	35	241	21	145	25	24	165	14	97
6061-T6	45	310	40	276	17	30	207	14	97
6063-O	13	90	7	48		10	69	8	55
6063-T4	25	172	13	90	22				
6063-T6	35	241	31	214	12	22	152	10	69
7001-O	37	255	22	152	14				
7001-T6	98	676	91	627	9			22	152
7075-O	33	228	15	103	16	22	152		
7075-T6	83	572	73	503	11	48	331	23	159

Note: Common properties:
   Density: 0.095 to 0.102 lb/in^3 (2635 to 2829 kg/m^3)
   Modulus of elasticity: 10 to 10.6 × 10^6 psi (69 to 73 GPa)
   Endurance strength at 5 × 10^8 cycles

APPENDIX 10   TYPICAL PROPERTIES OF ZINC CASTING ALLOYS

	Alloy				
	Zamak 3	Zamak 5	ZA-8	ZA-12	ZA-27
Composition (% by weight)					
Aluminum content	4	4	8.4	11	27
Magnesium content	0.035	0.055	0.023	0.023	0.015
Copper content		1.0	1.0	0.88	2.25
Properties (die-cast form)					
Tensile strength [ksi (MPa)]	41 (283)	48 (331)	54 (374)	59 (404)	62 (426)
Yield strength [ksi (MPa)]	32 (221)	33 (228)	42 (290)	46 (320)	54 (371)
% elongation in 2 inches	10	7	8	5	2.5
Modulus of elasticity [10^6 psi (GPa)]	12.4 (85)	12.4 (85)	12.4 (85)	12.0 (83)	11.3 (78)

APPENDIX 11   PROPERTIES OF TITANIUM ALLOYS

Material designation	Condition	Tensile strength		Yield strength		Ductility (percent elongation in 2 inches)	Modulus of elasticity	
		(ksi)	(MPa)	(ksi)	(MPa)		(10^6 psi)	(GPa)
Commercially pure alpha titanium (density = 0.163 lb/in^3; 4515 kg/m^3)								
Ti-35A	Wrought	35	241	25	172	24	15.0	103
Ti-50A	Wrought	50	345	40	276	20	15.0	103
Ti-65A	Wrought	65	448	55	379	18	15.0	103
Alpha alloy (density = 0.163 lb/in^3; 4515 kg/m^3)								
Ti-0.2Pd	Wrought	50	345	40	276	20	14.9	103
Beta alloy (density = 0.176 lb/in^3; 4875 kg/m^3)								
Ti-3Al-13V-11Cr	Air-cooled from 1400°F	135	931	130	896	16	14.7	101
Ti-3Al-13V-11Cr	Air-cooled from 1400°F and aged	185	1280	175	1210	6	16.0	110
Alpha-beta alloy (density = 0.160 lb/in^3; 4432 kg/m^3)								
Ti-6Al-4V	Annealed	130	896	120	827	10	16.5	114
Ti-6Al-4V	Quenched and aged at 1000°F	160	1100	150	1030	7	16.5	114

APPENDIX 12   PROPERTIES OF BRONZES

Material	UNS number designation	Tensile strength (ksi)	Tensile strength (MPa)	Yield strength (ksi)	Yield strength (MPa)	Ductility (percent elongation in 2 inches)	Modulus of elasticity (10^6 psi)	Modulus of elasticity (GPa)
Leaded phosphor bronze	C54400	68	469	57	393	20	15	103
Silicon bronze	C65500	58	400	22	152	60	15	103
Manganese bronze	C67500	65	448	30	207	33	15	103
	C86200	95	655	48	331	20	15	103
Bearing bronze	C93200	35	241	18	124	20	14.5	100
Aluminum bronze	C95400	85	586	35	241	18	15.5	107
Copper-nickel alloy	C96200	45	310	25	172	20	18	124
Copper-nickel-zinc alloy (also called nickel silver)	C97300	35	241	17	117	20	16	110

APPENDIX 13   TYPICAL PROPERTIES OF SELECTED PLASTICS

Material	Type	Tensile strength (ksi)	Tensile strength (MPa)	Tensile modulus (ksi)	Tensile modulus (MPa)	Flexural strength (ksi)	Flexural strength (MPa)	Flexural modulus (ksi)	Flexural modulus (MPa)	Impact strength IZOD (ft·lb/in of notch)
Nylon	66 Dry	12.0	83	420	2900			410	2830	1.0
	66 50% R.H.	11.2	77					175	1210	2.1
ABS	Medium-impact	6.0	41	360	2480	11.5	79	310	2140	4.0
	High-impact	5.0	34	250	1720	8.0	55	260	1790	7.0
Polycarbonate	General-purpose	9.0	62	340	2340	11.0	76	300	2070	12.0
Acrylic	Standard	10.5	72	430	2960	16.0	110	460	3170	0.4
	High-impact	5.4	37	220	1520	7.0	48	230	1590	1.2
PVC	Rigid	6.0	41	350	2410			300	2070	0.4–20.0 (varies widely)
Polyimide	25% graphite powder filler	5.7	39			12.8	88	900	6210	0.25
	Glass-fiber filler	27.0	186			50.0	345	3250	22 400	17.0
	Laminate	50.0	345			70.0	483	4000	27 580	13.0
Acetal	Copolymer	8.0	55	410	2830	13.0	90	375	2590	1.3
Polyurethane	Elastomer	5.0	34	100	690	0.6	4			No break
Phenolic	General	6.5	45	1100	7580	9.0	62	1100	7580	0.3
Polyester with glass-fiber mat reinforcement (approx. 30% glass by weight)										
	Lay-up, contact mold	9.0	62			16.0	110	800	5520	
	Cold press molded	12.0	83			22.0	152	1300	8960	
	Compression molded	25.0	172			10.0	69	1300	8960	

APPENDIX 14 BEAM-DEFLECTION FORMULAS

TABLE A14–1 Beam-deflection formulas for simply supported beams

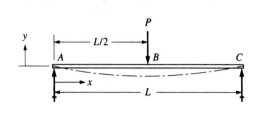

(a)

$$y_B = y_{max} = \frac{-PL^3}{48EI} \quad \text{at center}$$

Between A and B:

$$y = \frac{-Px}{48EI} (3L^2 - 4x^2)$$

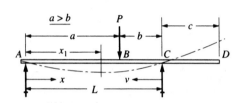

(b)

$$y_{max} = \frac{-Pab(L+b)\sqrt{3a(L+b)}}{27EIL}$$

$$\text{at } x_1 = \sqrt{a(L+b)/3}$$

$$y_B = \frac{-Pa^2b^2}{3EIL} \quad \text{at load}$$

Between A and B (the longer segment):

$$y = \frac{-Pbx}{6EIL} (L^2 - b^2 - x^2)$$

Between B and C (the shorter segment):

$$y = \frac{-Pav}{6EIL} (L^2 - v^2 - a^2)$$

At end of overhang at D:

$$y_D = \frac{Pabc}{6EIL} (L+a)$$

(c)

$$y_E = y_{max} = \frac{-Pa}{24EI} (3L^2 - 4a^2) \quad \text{at center}$$

$$y_B = y_C = \frac{-Pa^2}{6EI} (3L - 4a) \quad \text{at loads}$$

Between A and B:

$$y = \frac{-Px}{6EI} (3aL - 3a^2 - x^2)$$

Between B and C:

$$y = \frac{-Pa}{6EI} (3Lx - 3x^2 - a^2)$$

TABLE A14–1   *(continued)*

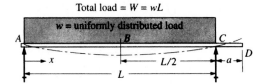

(d)

$$y_B = y_{max} = \frac{-5wL^4}{384EI} = \frac{-5WL^3}{384EI} \quad \text{at center}$$

Between *A* and *B*:

$$y = \frac{-wx}{24EI} (L^3 - 2Lx^2 + x^3)$$

At *D* at end:

$$y_D = \frac{wL^3a}{24EI}$$

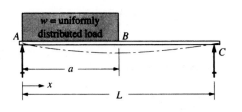

(e)

Between *A* and *B*:

$$y = \frac{-wx}{24EIL} [a^2(2L - a)^2 - 2ax^2(2L - a) + Lx^3]$$

Between *B* and *C*:

$$y = \frac{-wa^2(L - x)}{24EIL} (4Lx - 2x^2 - a^2)$$

(f)

$$M_B = \text{concentrated moment at } B$$

Between *A* and *B*:

$$y = \frac{-M_B}{6EI} \left[\left(6a - \frac{3a^2}{L} - 2L \right)x - \frac{x^3}{L} \right]$$

Between *B* and *C*:

$$y = \frac{M_B}{6EI} \left[3a^2 + 3x^2 - \frac{x^3}{L} - \left(2L + \frac{3a^2}{L} \right)x \right]$$

(g)

At *C* at end of overhang:

$$y_C = \frac{-Pa^2}{3EI} (L + a)$$

At *D*, maximum upward deflection:

$$y_D = 0.06415 \frac{PaL^2}{EI}$$

TABLE A14–1 *(continued)*

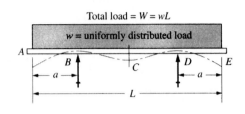 (h)	At C at center: $$y = \frac{-W(L-2a)^3}{384EI}\left[\frac{5}{L}(L-2a) - \frac{24}{L}\left(\frac{a^2}{L-2a}\right)\right]$$ At A and E at ends: $$y = \frac{-W(L-2a)^3 a}{24EIL}\left[-1 + 6\left(\frac{a}{L-2a}\right)^2 + 3\left(\frac{a}{L-2a}\right)^3\right]$$
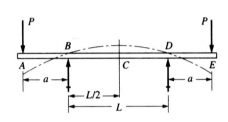 (i)	At C at center: $$y = \frac{PL^2 a}{8EI}$$ At A and E at ends at loads: $$y = \frac{-Pa^2}{3EI}\left(a + \frac{3}{2}L\right)$$
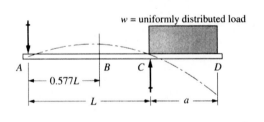 (j)	At B: $$y = 0.03208\frac{wa^2 L^2}{EI}$$ At D at end: $$y = \frac{-wa^3}{24EI}(4L + 3a)$$

Source: Engineering Data for Aluminum Structures (Washington, DC: The Aluminum Association, 1986), pp. 63–77.

TABLE A14–2 Beam-deflection formulas for cantilevers

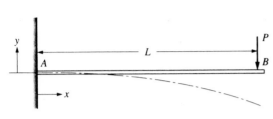

(a)

At B at end:

$$y_B = y_{max} = \frac{-PL^3}{3EI}$$

Between A and B:

$$y = \frac{-Px^2}{6EI}(3L - x)$$

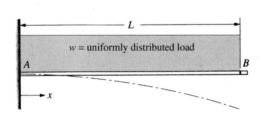

(b)

At B at load:

$$y_B = \frac{-Pa^3}{3EI}$$

At C at end:

$$y_C = y_{max} = \frac{-Pa^2}{6EI}(3L - a)$$

Between A and B:

$$y = \frac{-Px^2}{6EI}(3a - x)$$

Between B and C:

$$y = \frac{-Pa^2}{6EI}(3x - a)$$

(c)

W = total load = wL

At B at end:

$$y_B = y_{max} = \frac{-WL^3}{8EI}$$

Between A and B:

$$y = \frac{-Wx^2}{24EIL}[2L^2 + (2L - x)^2]$$

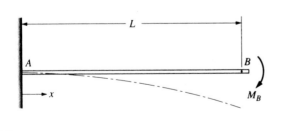

(d)

M_B = concentrated moment at end

At B at end:

$$y_B = y_{max} = \frac{-M_B L^2}{2EI}$$

Between A and B:

$$y = \frac{-M_B x^2}{2EI}$$

Source: Engineering Data for Aluminum Structures (Washington, DC: The Aluminum Association, 1986), pp. 63–77.

TABLE A14–3   Beam diagrams and beam-deflection formulas for statically indeterminate beams

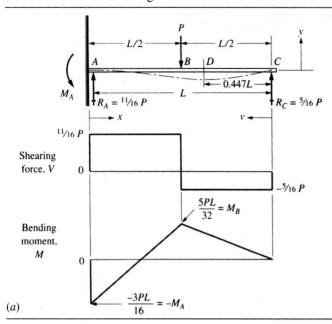

(a)

Deflections
At B at load:

$$y_B = \frac{-7}{768} \frac{PL^3}{EI}$$

y_{max} is at $v = 0.447L$ at D:

$$y_D = y_{max} = \frac{-PL^3}{107EI}$$

Between A and B:

$$y = \frac{-Px^2}{96EI}(9L - 11x)$$

Between B and C:

$$y = \frac{-Pv}{96EI}(3L^2 - 5v^2)$$

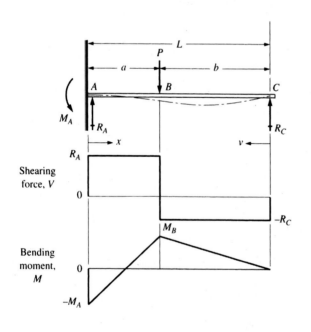

(b)

Reactions

$$R_A = \frac{Pb}{2L^3}(3L^2 - b^2)$$

$$R_C = \frac{Pa^2}{2L^3}(b + 2L)$$

Moments

$$M_A = \frac{-Pab}{2L^2}(b + L)$$

$$M_B = \frac{Pa^2b}{2L^3}(b + 2L)$$

Deflections
At B at load:

$$y_B = \frac{-Pa^3b^2}{12EIL^3}(3L + b)$$

Between A and B:

$$y = \frac{-Px^2b}{12EIL^3}(3C_1 - C_2x)$$

$$C_1 = aL(L + b); C_2 = (L + a)(L + b) + aL$$

Between B and C:

$$y = \frac{-Pa^2v}{12EIL^3}[3L^2b - v^2(3L - a)]$$

A–22

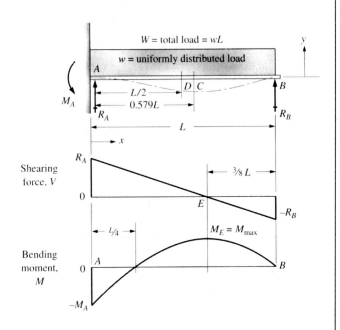

Reactions

$$R_A = \frac{5}{8}W$$

$$R_B = \frac{3}{8}W$$

Moments

$$M_A = -0.125WL$$
$$M_E = 0.0703WL$$

Deflections
At C at x = 0.579L:

$$y_C = y_{max} = \frac{-WL^3}{185EI}$$

At D at center:

$$y_D = \frac{-WL^3}{192EI}$$

Between A and B:

$$y = \frac{-Wx^2(L-x)}{48EIL}(3L - 2x)$$

(c)

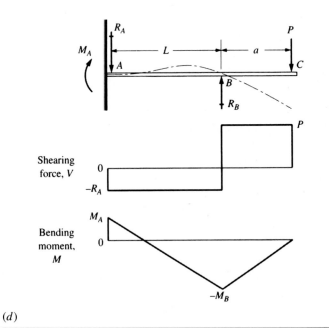

Reactions

$$R_A = \frac{-3Pa}{2L}$$

$$R_B = P\left(1 + \frac{3a}{2L}\right)$$

Moments

$$M_A = \frac{Pa}{2}$$

$$M_B = -Pa$$

Deflection
At C at end:

$$y_C = \frac{-PL^3}{EI}\left(\frac{a^2}{4L^2} + \frac{a^3}{3L^3}\right)$$

(d)

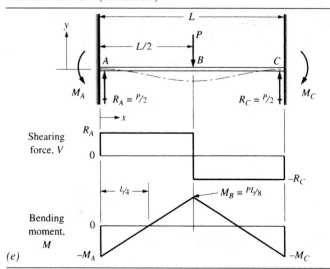

(e)

Moments

$$M_A = M_B = M_C = \frac{PL}{8}$$

Deflections
At *B* at center:

$$y_B = y_{max} = \frac{-PL^3}{192EI}$$

Between *A* and *B*:

$$y = \frac{-Px^2}{48EI}(3L - 4x)$$

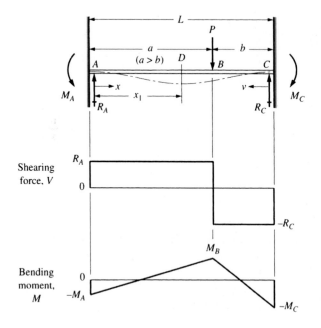

(f)

Reactions

$$R_A = \frac{Pb^2}{L^3}(3a + b)$$

$$R_C = \frac{Pa^2}{L^3}(3b + a)$$

Moments

$$M_A = \frac{-Pab^2}{L^2}$$

$$M_B = \frac{2Pa^2b^2}{L^3}$$

$$M_C = \frac{-Pa^2b}{L^2}$$

Deflections
At *B* at load:

$$y_B = \frac{-Pa^3b^3}{3EIL^3}$$

At *D* at $x_1 = \dfrac{2aL}{3a + b}$

$$y_D = y_{max} = \frac{-2Pa^3b^2}{3EI(3a + b)^2}$$

Between *A* and *B* (longer segment):

$$y = \frac{-Px^2b^2}{6EIL^3}[2a(L - x) + L(a - x)]$$

Between *B* and *C* (shorter segment):

$$y = \frac{-Pv^2a^2}{6EIL^3}[2b(L - v) + L(b - v)]$$

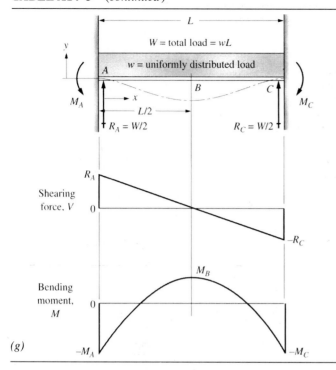

Moments

$$M_A = M_C = \frac{-WL}{12}$$

$$M_B = \frac{WL}{24}$$

Deflections
At B at center:

$$y_B = y_{max} = \frac{-WL^3}{384EI}$$

Between A and C:

$$y = \frac{-wx^2}{24EI}(L - x)^2$$

(g)

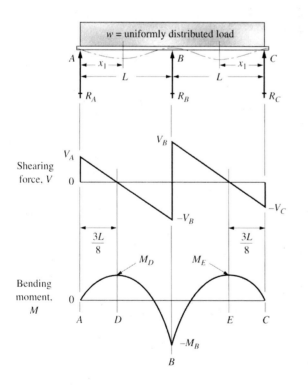

Reactions

$$R_A = R_C = \frac{3wL}{8}$$

$$R_B = 1.25wL$$

Shearing forces

$$V_A = V_C = R_A = R_C = \frac{3wL}{8}$$

$$V_B = \frac{5wL}{8}$$

Moments

$$M_D = M_E = 0.0703wL^2$$
$$M_B = -0.125wL^2$$

Deflections
At $x_1 = 0.4215L$ from A or C:

$$y_{max} = \frac{-wL^4}{185EI}$$

Between A and B:

$$y = \frac{-w}{48EI}(L^3x - 3Lx^3 + 2x^4)$$

(h)

TABLE A14–3 (*continued*)

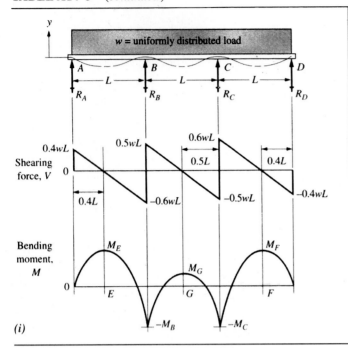

Reactions

$$R_A = R_D = 0.4wL$$
$$R_B = R_C = 1.10wL$$

Moments

$$M_E = M_F = 0.08wL^2$$
$$M_B = M_C = -0.10wL^2 = M_{max}$$
$$M_G = 0.025wL^2$$

(i)

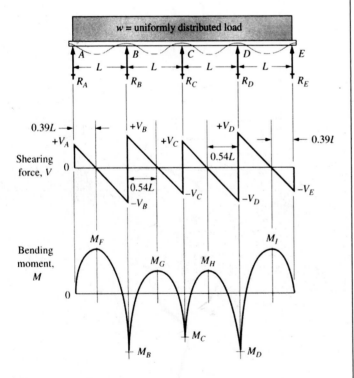

Reactions

$$R_A = R_E = 0.393wL$$
$$R_B = R_D = 1.143wL$$
$$R_C = 0.928wL$$

Shearing forces

$$V_A = +0.393wL$$
$$-V_B = -0.607wL$$
$$+V_B = +0.536wL$$
$$-V_C = -0.464wL$$
$$+V_C = +0.464wL$$
$$-V_D = -0.536wL$$
$$+V_D = +0.607wL$$
$$-V_E = -0.393wL$$

Moments

$$M_B = M_D = -0.1071wL^2 = M_{max}$$
$$M_F = M_I = 0.0772wL^2$$
$$M_C = -0.0714wL^2$$
$$M_G = M_H = 0.0364wL^2$$

(j)

Source: Engineering Data for Aluminum Structures (Washington, DC: The Aluminum Association, 1986), pp. 63–77.

APPENDIX 15   STRESS CONCENTRATION FACTORS

FIGURE A15–1
Stepped round shaft

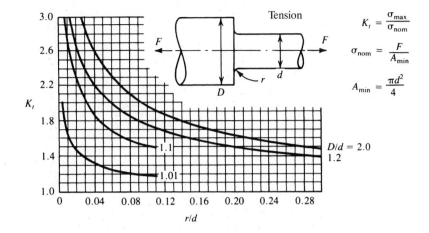

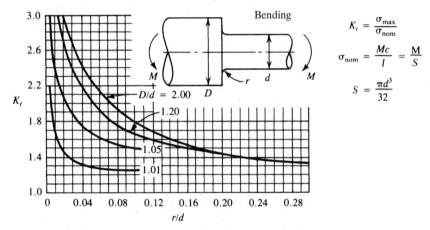

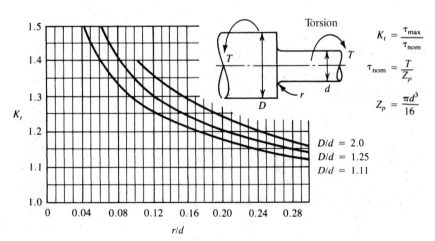

FIGURE A15–2
Stepped flat plate with
fillets

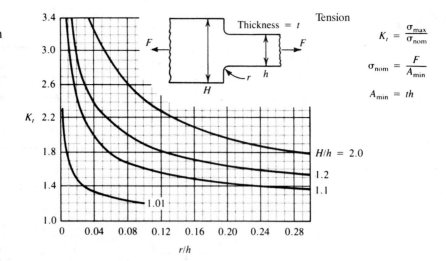

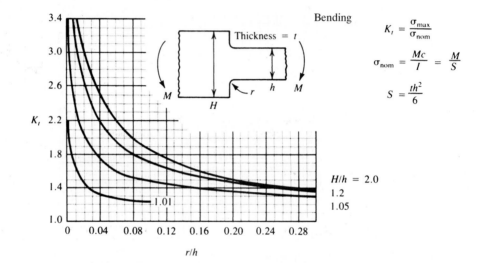

FIGURE A15–3
Flat plate with a
central hole

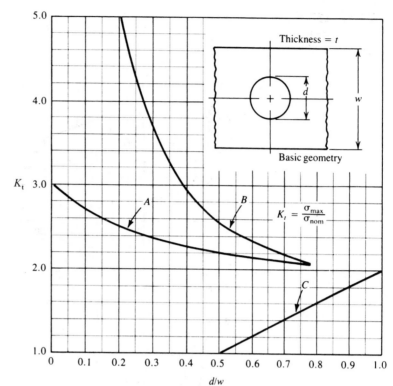

Curve A
Direct tension
on plate

$$\sigma_{nom} = \frac{F}{A_{net}} = \frac{F}{(w-d)t}$$

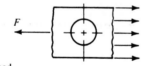

F F = total load

Curve B
Tension-load
applied through
a pin in the hole

$$\sigma_{nom} = \frac{F}{A_{net}} = \frac{F}{(w-d)t}$$

Curve C
Bending in
the plane of
the plate

$$\sigma_{nom} = \frac{Mc}{I_{net}} \qquad \frac{M}{S_{net}} = \frac{6Mw}{(w^3 - d^3)t}$$

Note : $K_t = 1.0$ for $d/w < 0.5$

FIGURE A15–4
Round bar with a
transverse hole

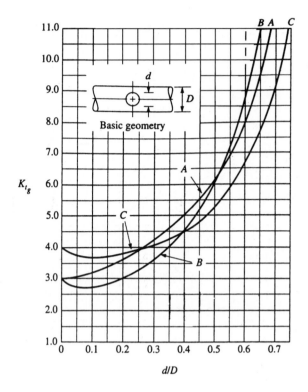

K_{t_g}

Basic geometry

d/D

Note: K_{t_g} is based on the nominal stress in a round
bar without a hole (gross section).

$$\sigma_{max} = K_{t_g}\,\sigma_{gross} \qquad\qquad \tau_{max} = K_{t_g}\,\tau_{gross}$$

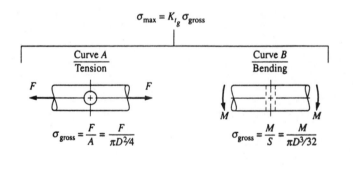

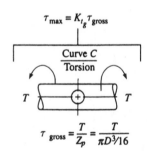

Curve A	Curve B	Curve C
Tension	Bending	Torsion

$$\sigma_{gross} = \frac{F}{A} = \frac{F}{\pi D^2/4}$$

$$\sigma_{gross} = \frac{M}{S} = \frac{M}{\pi D^3/32}$$

$$\tau_{gross} = \frac{T}{Z_p} = \frac{T}{\pi D^3/16}$$

APPENDIX 16   STEEL STRUCTURAL SHAPES

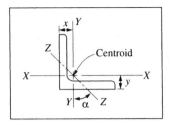
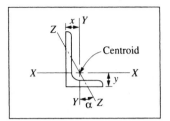

TABLE A16–1   Properties of steel angles, equal legs and unequal legs, L-shapes*

Designation	Area (in²)	Weight per foot (lb)	Axis X–X			Axis Y–Y			Axis Z–Z	
			I (in⁴)	S (in³)	y (in)	I (in⁴)	S (in³)	x (in)	r (in)	α (deg)
L8 × 8 × 1	15.0	51.0	89.0	15.8	2.37	89.0	15.8	2.37	1.56	45.0
L8 × 8 × 1/2	7.75	26.4	48.6	8.36	2.19	48.6	8.36	2.19	1.59	45.0
L8 × 4 × 1	11.0	37.4	69.6	14.1	3.05	11.6	3.94	1.05	0.846	13.9
L8 × 4 × 1/2	5.75	19.6	38.5	7.49	2.86	6.74	2.15	0.859	0.865	14.9
L6 × 6 × 3/4	8.44	28.7	28.2	6.66	1.78	28.2	6.66	1.78	1.17	45.0
L6 × 6 × 3/8	4.36	14.9	15.4	3.53	1.64	15.4	3.53	1.64	1.19	45.0
L6 × 4 × 3/4	6.94	23.6	24.5	6.25	2.08	8.68	2.97	1.08	0.860	23.2
L6 × 4 × 3/8	3.61	12.3	13.5	3.32	1.94	4.90	1.60	0.941	0.877	24.0
L4 × 4 × 1/2	3.75	12.8	5.56	1.97	1.18	5.56	1.97	1.18	0.782	45.0
L4 × 4 × 1/4	1.94	6.6	3.04	1.05	1.09	3.04	1.05	1.09	0.795	45.0
L4 × 3 × 1/2	3.25	11.1	5.05	1.89	1.33	2.42	1.12	0.827	0.639	28.5
L4 × 3 × 1/4	1.69	5.8	2.77	1.00	1.24	1.36	0.599	0.896	0.651	29.2
L3 × 3 × 1/2	2.75	9.4	2.22	1.07	0.932	2.22	1.07	0.932	0.584	45.0
L3 × 3 × 1/4	1.44	4.9	1.24	0.577	0.842	1.24	0.577	0.842	0.592	45.0
L2 × 2 × 3/8	1.36	4.7	0.479	0.351	0.636	0.479	0.351	0.636	0.389	45.0
L2 × 2 × 1/4	0.938	3.19	0.348	0.247	0.592	0.348	0.247	0.592	0.391	45.0
L2 × 2 × 1/8	0.484	1.65	0.190	0.131	0.546	0.190	0.131	0.546	0.398	45.0

*Data are taken from a variety of sources. Sizes listed represent a small sample of the sizes available.

Notes: Example designation: L4 × 3 × 1/2
4 = length of longer leg (in); 3 = length of shorter leg (in); 1/2 = thickness of legs (in)
Z–Z is axis of minimum moment of inertia (I) and radius of gyration (r).
I = moment of inertia; S = section modulus; r = radius of gyration.

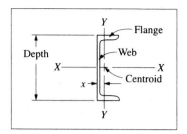

TABLE A16–2   Properties of American Standard steel channels, C-shapes*

				Flange		Axis X–X		Axis Y–Y		
Designation	Area (in^2)	Depth (in)	Web thickness (in)	Width (in)	Average thickness (in)	I (in^4)	S (in^3)	I (in^4)	S (in^3)	x (in)
C15 × 50	14.7	15.00	0.716	3.716	0.650	404	53.8	11.0	3.78	0.798
C15 × 40	11.8	15.00	0.520	3.520	0.650	349	46.5	9.23	3.37	0.777
C12 × 30	8.82	12.00	0.510	3.170	0.501	162	27.0	5.14	2.06	0.674
C12 × 25	7.35	12.00	0.387	3.047	0.501	144	24.1	4.47	1.88	0.674
C10 × 30	8.82	10.00	0.673	3.033	0.436	103	20.7	3.94	1.65	0.649
C10 × 20	5.88	10.00	0.379	2.739	0.436	78.9	15.8	2.81	1.32	0.606
C9 × 20	5.88	9.00	0.448	2.648	0.413	60.9	13.5	2.42	1.17	0.583
C9 × 15	4.41	9.00	0.285	2.485	0.413	51.0	11.3	1.93	1.01	0.586
C8 × 18.75	5.51	8.00	0.487	2.527	0.390	44.0	11.0	1.98	1.01	0.565
C8 × 11.5	3.38	8.00	0.220	2.260	0.390	32.6	8.14	1.32	0.781	0.571
C6 × 13	3.83	6.00	0.437	2.157	0.343	17.4	5.80	1.05	0.642	0.514
C6 × 8.2	2.40	6.00	0.200	1.920	0.343	13.1	4.38	0.693	0.492	0.511
C5 × 9	2.64	5.00	0.325	1.885	0.320	8.90	3.56	0.632	0.450	0.478
C5 × 6.7	1.97	5.00	0.190	1.750	0.320	7.49	3.00	0.479	0.378	0.484
C4 × 7.25	2.13	4.00	0.321	1.721	0.296	4.59	2.29	0.433	0.343	0.459
C4 × 5.4	1.59	4.00	0.184	1.584	0.296	3.85	1.93	0.319	0.283	0.457
C3 × 6	1.76	3.00	0.356	1.596	0.273	2.07	1.38	0.305	0.268	0.455
C3 × 4.1	1.21	3.00	0.170	1.410	0.273	1.66	1.10	0.197	0.202	0.436

*Data are taken from a variety of sources. Sizes listed represent a small sample of the sizes available.

Notes: Example designation: C15 × 50

15 = depth (in); 50 = weight per unit length (lb/ft)

I = moment of inertia; S = section modulus.

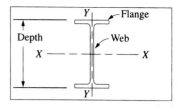

TABLE A16–3 Properties of steel wide-flange shapes, W-shapes*

			Web thickness (in)	Flange		Axis X–X		Axis Y–Y	
Designation	Area (in²)	Depth (in)		Width (in)	Thickness (in)	I (in⁴)	S (in³)	I (in⁴)	S (in³)
W24 × 76	22.4	23.92	0.440	8.990	0.680	2100	176	82.5	18.4
W24 × 68	20.1	23.73	0.415	8.965	0.585	1830	154	70.4	15.7
W21 × 73	21.5	21.24	0.455	8.295	0.740	1600	151	70.6	17.0
W21 × 57	16.7	21.06	0.405	6.555	0.650	1170	111	30.6	9.35
W18 × 55	16.2	18.11	0.390	7.530	0.630	890	98.3	44.9	11.9
W18 × 40	11.8	17.90	0.315	6.015	0.525	612	68.4	19.1	6.35
W14 × 43	12.6	13.66	0.305	7.995	0.530	428	62.7	45.2	11.3
W14 × 26	7.69	13.91	0.255	5.025	0.420	245	35.3	8.91	3.54
W12 × 30	8.79	12.34	0.260	6.520	0.440	238	38.6	20.3	6.24
W12 × 16	4.71	11.99	0.220	3.990	0.265	103	17.1	2.82	1.41
W10 × 15	4.41	9.99	0.230	4.000	0.270	69.8	13.8	2.89	1.45
W10 × 12	3.54	9.87	0.190	3.960	0.210	53.8	10.9	2.18	1.10
W8 × 15	4.44	8.11	0.245	4.015	0.315	48.0	11.8	3.41	1.70
W8 × 10	2.96	7.89	0.170	3.940	0.205	30.8	7.81	2.09	1.06
W6 × 15	4.43	5.99	0.230	5.990	0.260	29.1	9.72	9.32	3.11
W6 × 12	3.55	6.03	0.230	4.000	0.280	22.1	7.31	2.99	1.50
W5 × 19	5.54	5.15	0.270	5.030	0.430	26.2	10.2	9.13	3.63
W5 × 16	4.68	5.01	0.240	5.000	0.360	21.3	8.51	7.51	3.00
W4 × 13	3.83	4.16	0.280	4.060	0.345	11.3	5.46	3.86	1.90

*Data are taken from a variety of sources. Sizes listed represent a small sample of the sizes available.

Notes: Example designation: W14 × 43

14 = nominal depth (in); 43 = weight per unit length (lb/ft)

I = moment of inertia; S = section modulus.

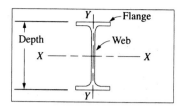

TABLE A16–4   Properties of American Standard steel beams, S-shapes*

Designation	Area (in²)	Depth (in)	Web thickness (in)	Flange Width (in)	Flange Average thickness (in)	Axis X–X I (in⁴)	Axis X–X S (in³)	Axis Y–Y I (in⁴)	Axis Y–Y S (in³)
S24 × 90	26.5	24.00	0.625	7.125	0.870	2250	187	44.9	12.6
S20 × 96	28.2	20.30	0.800	7.200	0.920	1670	165	50.2	13.9
S20 × 75	22.0	20.00	0.635	6.385	0.795	1280	128	29.8	9.32
S20 × 66	19.4	20.00	0.505	6.255	0.795	1190	119	27.7	8.85
S18 × 70	20.6	18.00	0.711	6.251	0.691	926	103	24.1	7.72
S15 × 50	14.7	15.00	0.550	5.640	0.622	486	64.8	15.7	5.57
S12 × 50	14.7	12.00	0.687	5.477	0.659	305	50.8	15.7	5.74
S12 × 35	10.3	12.00	0.428	5.078	0.544	229	38.2	9.87	3.89
S10 × 35	10.3	10.00	0.594	4.944	0.491	147	29.4	8.36	3.38
S10 × 25.4	7.46	10.00	0.311	4.661	0.491	124	24.7	6.79	2.91
S8 × 23	6.77	8.00	0.441	4.171	0.426	64.9	16.2	4.31	2.07
S8 × 18.4	5.41	8.00	0.271	4.001	0.426	57.6	14.4	3.73	1.86
S7 × 20	5.88	7.00	0.450	3.860	0.392	42.4	12.1	3.17	1.64
S6 × 12.5	3.67	6.00	0.232	3.332	0.359	22.1	7.37	1.82	1.09
S5 × 10	2.94	5.00	0.214	3.004	0.326	12.3	4.92	1.22	0.809
S4 × 7.7	2.26	4.00	0.193	2.663	0.293	6.08	3.04	0.764	0.574
S3 × 5.7	1.67	3.00	0.170	2.330	0.260	2.52	1.68	0.455	0.390

*Data are taken from a variety of sources. Sizes listed represent a small sample of the sizes available.

Notes: Example designation: S10 × 35

10 = nominal depth (in); 35 = weight per unit length (lb/ft)

I = moment of inertia; S = section modulus.

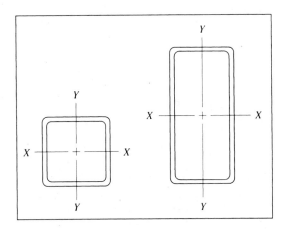

TABLE A16–5 Properties of steel structural tubing, square and rectangular*

Size	Area (in²)	Weight per foot (lb)	Axis X–X			Axis Y–Y		
			I (in⁴)	S (in³)	r (in)	I (in⁴)	S (in³)	r (in)
8 × 8 × 1/2	14.4	48.9	131	32.9	3.03	131	32.9	3.03
8 × 8 × 1/4	7.59	25.8	75.1	18.8	3.15	75.1	18.8	3.15
8 × 4 × 1/2	10.4	35.2	75.1	18.8	2.69	24.6	12.3	1.54
8 × 4 × 1/4	5.59	19.0	45.1	11.3	2.84	15.3	7.63	1.65
8 × 2 × 1/4	4.59	15.6	30.1	7.52	2.56	3.08	3.08	0.819
6 × 6 × 1/2	10.4	35.2	50.5	16.8	2.21	50.5	16.8	2.21
6 × 6 × 1/4	5.59	19.0	30.3	10.1	2.33	30.3	10.1	2.33
6 × 4 × 1/4	4.59	15.6	22.1	7.36	2.19	11.7	5.87	1.60
6 × 2 × 1/4	3.59	12.2	13.8	4.60	1.96	2.31	2.31	0.802
4 × 4 × 1/2	6.36	21.6	12.3	6.13	1.39	12.3	6.13	1.39
4 × 4 × 1/4	3.59	12.2	8.22	4.11	1.51	8.22	4.11	1.51
4 × 2 × 1/4	2.59	8.81	4.69	2.35	1.35	1.54	1.54	0.770
3 × 3 × 1/4	2.59	8.81	3.16	2.10	1.10	3.16	2.10	1.10
3 × 2 × 1/4	2.09	7.11	2.21	1.47	1.03	1.15	1.15	0.742
2 × 2 × 1/4	1.59	5.41	0.766	0.766	0.694	0.766	0.766	0.694

*Data are taken from a variety of sources. Sizes listed represent a small sample of the sizes available.

Notes: Example size: 6 × 4 × 1/4
6 = vertical depth (in); 4 = width (in); 1/4 = wall thickness (in).
I = moment of inertia; S = section modulus; r = radius of gyration.

TABLE A16–6 Properties of American National Standard Schedule 40 welded and seamless wrought steel pipe

Diameter (in)			Wall thickness (in)	Cross-sectional area of metal (in^2)	Properties of sections			
Nominal	Actual inside	Actual outside			Moment of inertia, I (in^4)	Radius of gyration (in)	Section modulus, S (in^3)	Polar section modulus, Z_p (in^3)
1/8	0.269	0.405	0.068	0.072	0.001 06	0.122	0.005 25	0.010 50
1/4	0.364	0.540	0.088	0.125	0.003 31	0.163	0.012 27	0.024 54
3/8	0.493	0.675	0.091	0.167	0.007 29	0.209	0.021 60	0.043 20
1/2	0.622	0.840	0.109	0.250	0.017 09	0.261	0.040 70	0.081 40
3/4	0.824	1.050	0.113	0.333	0.037 04	0.334	0.070 55	0.1411
1	1.049	1.315	0.133	0.494	0.087 34	0.421	0.1328	0.2656
$1\frac{1}{4}$	1.380	1.660	0.140	0.669	0.1947	0.539	0.2346	0.4692
$1\frac{1}{2}$	1.610	1.900	0.145	0.799	0.3099	0.623	0.3262	0.6524
2	2.067	2.375	0.154	1.075	0.6658	0.787	0.5607	1.121
$2\frac{1}{2}$	2.469	2.875	0.203	1.704	1.530	0.947	1.064	2.128
3	3.068	3.500	0.216	2.228	3.017	1.163	1.724	3.448
$3\frac{1}{2}$	3.548	4.000	0.226	2.680	4.788	1.337	2.394	4.788
4	4.026	4.500	0.237	3.174	7.233	1.510	3.215	6.430
5	5.047	5.563	0.258	4.300	15.16	1.878	5.451	10.90
6	6.065	6.625	0.280	5.581	28.14	2.245	8.496	16.99
8	7.981	8.625	0.322	8.399	72.49	2.938	16.81	33.62
10	10.020	10.750	0.365	11.91	160.7	3.674	29.91	59.82
12	11.938	12.750	0.406	15.74	300.2	4.364	47.09	94.18
16	15.000	16.000	0.500	24.35	732.0	5.484	91.50	183.0
18	16.876	18.000	0.562	30.79	1172	6.168	130.2	260.4

APPENDIX 17 ALUMINUM STRUCTURAL SHAPES

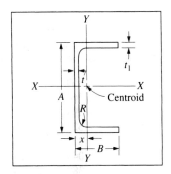

TABLE A17–1 Aluminum Association standard channels: dimensions, areas, weights, and section properties

| Size | | | | Flange thickness, | Web thickness, | Fillet radius, | Section properties[‡] | | | | | | |
| Depth, A (in) | Width, B (in) | Area* (in²) | Weight† (lb/ft) | t_1 (in) | t (in) | R (in) | Axis X–X | | | Axis Y–Y | | | |
							I (in⁴)	S (in³)	r (in)	I (in⁴)	S (in³)	r (in)	x (in)
2.00	1.00	0.491	0.577	0.13	0.13	0.10	0.288	0.288	0.766	0.045	0.064	0.303	0.298
2.00	1.25	0.911	1.071	0.26	0.17	0.15	0.546	0.546	0.774	0.139	0.178	0.391	0.471
3.00	1.50	0.965	1.135	0.20	0.13	0.25	1.41	0.94	1.21	0.22	0.22	0.47	0.49
3.00	1.75	1.358	1.597	0.26	0.17	0.25	1.97	1.31	1.20	0.42	0.37	0.55	0.62
4.00	2.00	1.478	1.738	0.23	0.15	0.25	3.91	1.95	1.63	0.60	0.45	0.64	0.65
4.00	2.25	1.982	2.331	0.29	0.19	0.25	5.21	2.60	1.62	1.02	0.69	0.72	0.78
5.00	2.25	1.881	2.212	0.26	0.15	0.30	7.88	3.15	2.05	0.98	0.64	0.72	0.73
5.00	2.75	2.627	3.089	0.32	0.19	0.30	11.14	4.45	2.06	2.05	1.14	0.88	0.95
6.00	2.50	2.410	2.834	0.29	0.17	0.30	14.35	4.78	2.44	1.53	0.90	0.80	0.79
6.00	3.25	3.427	4.030	0.35	0.21	0.30	21.04	7.01	2.48	3.76	1.76	1.05	1.12
7.00	2.75	2.725	3.205	0.29	0.17	0.30	22.09	6.31	2.85	2.10	1.10	0.88	0.84
7.00	3.50	4.009	4.715	0.38	0.21	0.30	33.79	9.65	2.90	5.13	2.23	1.13	1.20
8.00	3.00	3.526	4.147	0.35	0.19	0.30	37.40	9.35	3.26	3.25	1.57	0.96	0.93
8.00	3.75	4.923	5.789	0.41	0.25	0.35	52.69	13.17	3.27	7.13	2.82	1.20	1.22
9.00	3.25	4.237	4.983	0.35	0.23	0.35	54.41	12.09	3.58	4.40	1.89	1.02	0.93
9.00	4.00	5.927	6.970	0.44	0.29	0.35	78.31	17.40	3.63	9.61	3.49	1.27	1.25
10.00	3.50	5.218	6.136	0.41	0.25	0.35	83.22	16.64	3.99	6.33	2.56	1.10	1.02
10.00	4.25	7.109	8.360	0.50	0.31	0.40	116.15	23.23	4.04	13.02	4.47	1.35	1.34
12.00	4.00	7.036	8.274	0.47	0.29	0.40	159.76	26.63	4.77	11.03	3.86	1.25	1.14
12.00	5.00	10.053	11.822	0.62	0.35	0.45	239.69	39.95	4.88	25.74	7.60	1.60	1.61

Source: Aluminum Association, *Aluminum Standards and Data,* 11th ed., Washington, DC, © 1993, p. 187.

*Areas listed are based on nominal dimensions.

†Weights per foot are based on nominal dimensions and a density of 0.098 lb/in³, which is the density of alloy 6061.

‡I = moment of inertia; S = section modulus; r = radius of gyration.

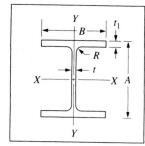

TABLE A17–2 Aluminum Association standard I-beams: dimensions, areas, weights, and section properties

Size							Section properties[‡]					
							Axis X–X			Axis Y–Y		
Depth, A (in)	Width, B (in)	Area* (in²)	Weight[†] (lb/ft)	Flange thickness, t_1 (in)	Web thickness, t (in)	Fillet radius, R (in)	I (in⁴)	S (in³)	r (in)	I (in⁴)	S (in³)	r (in)
3.00	2.50	1.392	1.637	0.20	0.13	0.25	2.24	1.49	1.27	0.52	0.42	0.61
3.00	2.50	1.726	2.030	0.26	0.15	0.25	2.71	1.81	1.25	0.68	0.54	0.63
4.00	3.00	1.965	2.311	0.23	0.15	0.25	5.62	2.81	1.69	1.04	0.69	0.73
4.00	3.00	2.375	2.793	0.29	0.17	0.25	6.71	3.36	1.68	1.31	0.87	0.74
5.00	3.50	3.146	3.700	0.32	0.19	0.30	13.94	5.58	2.11	2.29	1.31	0.85
6.00	4.00	3.427	4.030	0.29	0.19	0.30	21.99	7.33	2.53	3.10	1.55	0.95
6.00	4.00	3.990	4.692	0.35	0.21	0.30	25.50	8.50	2.53	3.74	1.87	0.97
7.00	4.50	4.932	5.800	0.38	0.23	0.30	42.89	12.25	2.95	5.78	2.57	1.08
8.00	5.00	5.256	6.181	0.35	0.23	0.30	59.69	14.92	3.37	7.30	2.92	1.18
8.00	5.00	5.972	7.023	0.41	0.25	0.30	67.78	16.94	3.37	8.55	3.42	1.20
9.00	5.50	7.110	8.361	0.44	0.27	0.30	102.02	22.67	3.79	12.22	4.44	1.31
10.00	6.00	7.352	8.646	0.41	0.25	0.40	132.09	26.42	4.24	14.78	4.93	1.42
10.00	6.00	8.747	10.286	0.50	0.29	0.40	155.79	31.16	4.22	18.03	6.01	1.44
12.00	7.00	9.925	11.672	0.47	0.29	0.40	255.57	42.60	5.07	26.90	7.69	1.65
12.00	7.00	12.153	14.292	0.62	0.31	0.40	317.33	52.89	5.11	35.48	10.14	1.71

Source: Aluminum Association, *Aluminum Standards and Data,* 11th ed., Washington, DC, © 1993, p. 187.

*Areas listed are based on nominal dimensions.

[†]Weights per foot are based on nominal dimensions and a density of 0.098 lb/in³, which is the density of alloy 6061.

[‡]I = moment of inertia; S = section modulus; r = radius of gyration.

APPENDIX 18   CONVERSION FACTORS

TABLE A18–1   Conversion of U.S. Customary units to SI units: basic quantities

Quantity	U.S. Customary unit		SI unit	Symbol	Equivalent units
Length	1 foot (ft)	= 0.3048	meter	m	
Mass	1 slug	= 14.59	kilogram	kg	
Time	1 second	= 1.0	second	s	
Force	1 pound (lb)	= 4.448	newton	N	$kg \cdot m/s^2$
Pressure	1 lb/in^2	= 6895	pascal	Pa	N/m^2 or $kg/m \cdot s^2$
Energy	1 ft-lb	= 1.356	joule	J	$N \cdot m$ or $kg \cdot m^2/s^2$
Power	1 ft-lb/s	= 1.356	watt	W	J/s

TABLE A18–2   Other convenient conversion factors

Length

1 ft	=	0.3048	m
1 in	=	25.4	mm
1 mi	=	5280	ft
1 mi	=	1.609	km
1 km	=	1000	m
1 cm	=	10	mm
1 m	=	1000	mm

Area

1 ft^2	=	0.0929	m^2
1 in^2	=	645.2	mm^2
1 m^2	=	10.76	ft^2
1 m^2	=	10^6	mm^2

Volume

1 ft^3	=	7.48	gal
1 ft^3	=	1728	in^3
1 ft^3	=	0.0283	m^3
1 gal	=	0.00379	m^3
1 gal	=	3.785	L
1 m^3	=	1000	L

Volume flow rate

1 ft^3/s	=	449	gal/min
1 ft^3/s	=	0.0283	m^3/s
1 gal/min	=	6.309×10^{-5}	m^3/s
1 gal/min	=	3.785	L/min
1 L/min	=	16.67×10^{-6}	m^3/s

Stress, pressure, or unit loading

1 lb/in^2	=	6.895	kPa
1 lb/ft^2	=	0.0479	kPa
1 kip/in^2	=	6.895	MPa

Section modulus

1 in^3	=	1.639×10^4	mm^3

Moment of inertia

1 in^4	=	4.162×10^5	mm^4

Density

1 slug/ft^3	=	515.4	kg/m^3

Specific weight

1 lb/ft^3	=	157.1	N/m^3

Energy

1 ft·lb	=	1.356	J
1 Btu	=	1.055	kJ
1 W·h	=	3.600	kJ

Torque or moment

1 lb·in	=	0.1130	N·m

Power

1 hp	=	550	ft·lb/s
1 hp	=	745.7	W
1 ft·lb/s	=	1.356	W
1 Btu/h	=	0.293	W

Temperature

$$T(°C) = [T(°F) - 32]5/9$$
$$T(°F) = \tfrac{9}{5}[T(°C)] + 32$$

APPENDIX 19   HARDNESS CONVERSION TABLE

Brinell		Rockwell		Steel: tensile strength, (1000 psi approx.)	Brinell		Rockwell		Steel: tensile strength, (1000 psi approx.)
Indent. dia. (mm)	No.*	B	C		Indent. dia. (mm)	No.*	B	C	
2.25	745		65.3		3.75	262	(103.0)	26.6	127
2.30	712				3.80	255	(102.0)	25.4	123
2.35	682		61.7		3.85	248	(101.0)	24.2	120
2.40	653		60.0		3.90	241	100.0	22.8	116
2.45	627		58.7		3.95	235	99.0	21.7	114
2.50	601		57.3		4.00	229	98.2	20.5	111
2.55	578		56.0		4.05	223	97.3	(18.8)	
2.60	555		54.7	298	4.10	217	96.4	(17.5)	105
2.65	534		53.5	288	4.15	212	95.5	(16.0)	102
2.70	514		52.1	274	4.20	207	94.6	(15.2)	100
2.75	495		51.6	269	4.25	201	93.8	(13.8)	98
2.80	477		50.3	258	4.30	197	92.8	(12.7)	95
2.85	461		48.8	244	4.35	192	91.9	(11.5)	93
2.90	444		47.2	231	4.40	187	90.7	(10.0)	90
2.95	429		45.7	219	4.45	183	90.0	(9.0)	89
3.00	415		44.5	212	4.50	179	89.0	(8.0)	87
3.05	401		43.1	202	4.55	174	87.8	(6.4)	85
3.10	388		41.8	193	4.60	170	86.8	(5.4)	83
3.15	375		40.4	184	4.65	167	86.0	(4.4)	81
3.20	363		39.1	177	4.70	163	85.0	(3.3)	79
3.25	352	(110.0)	37.9	171	4.80	156	82.9	(0.9)	76
3.30	341	(109.0)	36.6	164	4.90	149	80.8		73
3.35	331	(108.5)	35.5	159	5.00	143	78.7		71
3.40	321	(108.0)	34.3	154	5.10	137	76.4		67
3.45	311	(107.5)	33.1	149	5.20	131	74.0		65
3.50	302	(107.0)	32.1	146	5.30	126	72.0		63
3.55	293	(106.0)	30.9	141	5.40	121	69.8		60
3.60	285	(105.5)	29.9	138	5.50	116	67.6		58
3.65	277	(104.5)	28.8	134	5.60	111	65.7		56
3.70	269	(104.0)	27.6	130					

Source: Modern Steels and Their Properties, Bethlehem Steel Co., Bethlehem, PA.

Note: This is a condensation of Table 2, *Report J417b, SAE 1971 Handbook.* Values in ( ) are beyond normal range and are presented for information only.

*Values above 500 are for tungsten carbide ball; below 500, for standard ball.

APPENDIX 20   GEOMETRY FACTOR *I* FOR PITTING FOR SPUR GEARS

Section 9–10 introduced the geometry factor *I* for pitting resistance for spur gears as a factor that relates the gear tooth geometry to the radius of curvature of the teeth. The value of *I* is to be determined at the *lowest point of single tooth contact* (LPSTC). The AGMA defines *I* as

$$I = C_c C_x$$

where

C_c is the curvature factor at the pitch line
C_x is the factor to adjust for the specific height of the LPSTC

The variables involved are to be the pressure angle ϕ, the number of teeth in the pinion N_P, and the gear ratio $m_G = N_G/N_P$. Note that m_G is always greater than or equal to 1.0, regardless of which gear is driving. The value of C_c is easily calculated directly from

$$C_c = \frac{\cos \phi \sin \phi}{2} \frac{m_G}{m_G + 1}$$

The computation of the value of C_x requires the evaluation of several other terms.

$$C_x = \frac{R_1 R_2}{R_P R_G}$$

where each term is developed in the following equations in terms of ϕ, N_P, and m_G along with P_d. It will be shown that the diametral pitch appears in the denominator of each term, and it can then be cancelled. We also express each term in the form C/P_d for convenience.

R_P = Radius of curvature for pinion at the pitch point

$$R_P = \frac{D_P \sin \phi}{2} = \frac{N_P \sin \phi}{2 P_d} = \frac{C_1}{P_d}$$

R_G = Radius of curvature for gear at the pitch point

$$R_G = \frac{D_G \sin \phi}{2} = \frac{D_P m_G \sin \phi}{2} = \frac{N_P m_G \sin \phi}{2 P_d} = \frac{C_2}{P_d} = \frac{m_G C_1}{P_d}$$

R_1 = Radius of curvature of the pinion at the LPSTC = $R_P - Z_c$

R_2 = Radius of curvature of the gear at the LPSTC = $R_G + Z_c$

$$Z_c = p_b - Z_a$$

$$p_b = \text{Base pitch} = \frac{\pi \cos \phi}{P_d} = \frac{C_3}{P_d}$$

$$Z_a = 0.5\left[\sqrt{D_{oP}^2 - D_{bP}^2} - \sqrt{D_P^2 - D_{bP}^2}\right]$$

We now express all of the diameters in this equation in terms of ϕ, N_P, m_G, and P_d.

$$D_{oP} = \text{Outside diameter of the pinion} = (N_P + 2)/P_d$$

$$D_P = \text{Pinion diameter} = N_P/P_d$$

$$D_{bP} = \text{Base diameter for the pinion} = D_P \cos \phi = (N_P \cos \phi)/P_d$$

Note that each term has the diametral pitch P_d in the denominator. It can then be factored outside the square root sign. The resulting equation for Z_a is

$$Z_a = \frac{0.5}{P_d}\left[\sqrt{(N_P + 2)^2 - (N_P \cos \phi)^2} - \sqrt{N_P^2 - (N_P \cos \phi)^2}\right] = \frac{C_4}{P_d}$$

Now we can define Z_c.

$$Z_c = p_b - Z_a = \frac{C_3}{P_d} - \frac{C_4}{P_d} = \frac{C_3 - C_4}{P_d}$$

We can now complete the equations for R_1 and R_2.

$$R_1 = R_P - Z_c = \frac{C_1}{P_d} - \frac{C_3 - C_4}{P_d} = \frac{C_1 - C_3 + C_4}{P_d}$$

$$R_2 = R_G + Z_c = \frac{C_2}{P_d} + \frac{C_3 - C_4}{P_d} = \frac{C_2 + C_3 - C_4}{P_d}$$

Finally, all of these terms can be substituted into the equation for C_x.

$$C_x = \frac{R_1 R_2}{R_P R_G} = \frac{[(C_1 - C_3 + C_4)/P_d]\,[(C_2 + C_3 - C_4)/P_d]}{(C_1/P_d)(C_2/P_d)}$$

Now you can see that the diametral pitch P_d cancels out, resulting in the final form,

$$C_x = \frac{R_1 R_2}{R_P R_G} = \frac{(C_1 - C_3 + C_4)(C_2 + C_3 - C_4)}{(C_1)(C_2)}$$

The algorithm for computing I can now be stated. First compute each of the C terms.

$$C_1 = (N_P \sin \phi)/2$$

$$C_2 = (N_P m_G \sin\phi)/2 = (C_1)(m_G)$$

$$C_3 = \pi \cos \phi$$

$$C_4 = 0.5\left[\sqrt{(N_p + 2)^2 - (N_P \cos \phi)^2} - \sqrt{N_P^2 - (N_P \cos \phi)^2}\right]$$

$$C_x = \frac{R_1 R_2}{R_P R_G} = \frac{(C_1 - C_3 + C_4)(C_2 + C_3 - C_4)}{(C_1)(C_2)}$$

$$C_c = \frac{\cos \phi \sin \phi}{2}\,\frac{m_G}{m_G + 1}$$

Finally, $I = C_c C_x$

Example
Problem A20–1 Compute the value for the geometry factor I for pitting for the following data: Two spur gears in mesh with a pressure angle of $20°$, $N_P = 30$, $N_G = 150$.

Solution First compute: $m_G = N_G/N_P = 150/30 = 5.0$

Then

$C_1 = (N_P \sin \phi)/2 = (30)(\sin 20°)/2 = 5.1303$

$C_2 = (N_P\, m_G \sin \phi)/2 = (C_1)(m_G) = 25.652$

$C_3 = \pi \cos \phi = \pi \cos(20°) = 2.9521$

$C_4 = 0.5 \left[\sqrt{(N_P + 2)^2 - (N_P \cos \phi)^2} - \sqrt{N_P^2 - (N_P \cos \phi)^2} \right]$

$C_4 = 0.5 \left[\sqrt{(30 + 2)^2 - (30 \cos(20°))^2} - \sqrt{30^2 - (30 \cos(20°))^2} \right] = 2.4407$

$$C_x = \frac{R_1 R_2}{R_P R_G} = \frac{(C_1 - C_3 + C_4)(C_2 + C_3 - C_4)}{(C_1)(C_2)}$$

$$C_x = \frac{(5.1303 - 2.9521 + 2.4407)(25.652 + 2.9521 - 2.4407)}{(5.1303)(25.652)} = 0.91826$$

$$C_c = \frac{\cos \phi \sin \phi}{2}\, \frac{m_G}{m_G + 1} = \frac{\cos (20°) \sin (20°)(5)}{2(5 + 1)} = 0.13391$$

Finally, $I = C_c C_x = (0.13391)(0.91826) = 0.12297 \approx 0.123$

This process lends itself well to programming in a spreadsheet, MATLAB, BASIC, or any other convenient computational aid.

Answers to Selected Problems

Given here are the answers to problems for which there are unique solutions. Many of the problems for solution in this book are true design problems, and individual design decisions are required to arrive at the solutions. Others are of the review question form for which the answers are in the text of the associated chapter. It should also be noted that some of the problems require the selection of design factors and the use of data from charts and graphs. Because of the judgment and interpolation required, some of the answers may be slightly different from your solutions.

CHAPTER 1   The Nature of Mechanical Design

15. $D = 44.5$ mm
16. $L = 14.0$ m
17. $T = 1418$ N·m
18. $A = 2658$ mm^2
19. $S = 2.43 \times 10^5$ mm^3
20. $I = 3.66 \times 10^7$ mm^4
21. L 2 × 2 × 3/8
22. $P = 5.59$ kW
23. $s_u = 876$ MPa

24. Weight = 48.9 N
25. $T = 20.3$ N·m
    $\theta = 0.611$ rad
    Scale = 5.14 lb·in/deg
    Scale = 33.3 N·m/rad
26. Energy = 1.03×10^{11} lb·ft/yr
    Energy = 3.88×10^7 W·h/yr
27. $\mu = 540$ lb·s/ft^2
    $\mu = 25.9 \times 10^3$ N·s/m^2
28. 4.60×10^9 rev

CHAPTER 2   Materials in Mechanical Design

9. No. The percent elongation must be greater than 5.0 percent to be ductile.
11. $G = 42.9$ GPa
12. Hardness = 52.8 HRC
13. Tensile strength = 235 ksi (Approximately)

Questions 14–17 ask what is wrong with the given statements.

14. Annealed steels typically have hardness values in the range from 120 HB to 200 HB. A hardness of 750 HB is extremely hard and characteristic of as-quenched high alloy steels.
15. The HRB scale is normally limited to HRB 100.
16. The HRC hardness is normally no lower than HRC 20.
17. The given relationship between hardness and tensile strength is only valid for steels.
18. Charpy and Izod
19. Iron and carbon. Manganese and other elements are often present.
20. Iron, carbon, manganese, nickel, chromium, molybdenum.
21. Approximately 0.40 percent.
22. Low carbon: Less than 0.30 percent
    Medium carbon: 0.30 to 0.50 percent
    High carbon: 0.50 to 0.95 percent
23. Nominally 1.0 percent.

24. AISI 12L13 steel has lead added to improve machinability.
25. AISI 1045, 4140, 4640, 5150, 6150, 8650.
26. AISI 1045, 4140, 4340, 4640, 5150, 6150, 8650.
27. Wear resistance, strength, ductility. AISI 1080.
28. AISI 5160 OQT 1000 is a chromium steel, having nominally 0.80 percent chromium and 0.60 percent carbon, a high carbon alloy steel. It has fairly high strength and good ductility. It was through-hardened, quenched in oil, and tempered at 1000°F.
29. Yes, with careful specification of the quenching medium. A hardness of HRC 40 is equivalent to HB 375. Appendix 3 indicates that oil quenching would not produce an adequate hardness. However, Appendix 4–1 shows that a hardness of HB 400 could be obtained by quenching in water and tempering in 700°F while still having 20% elongation for good ductility.
33. AISI 200 and 300 series
34. Chromium
35. ASTM A992 structural steel
37. Gray iron, ductile iron, malleable iron
41. Stamping dies, punches, gages
43. Strain hardened
46. Alloy 6061
50. Gears and bearings
60. Auto and truck bodies; large housings

CHAPTER 3   Stress Analysis

1. $\sigma = 31.8$ MPa; $\delta = 0.12$ mm
2. $\sigma = 44.6$ MPa
3. $\sigma = 66.7$ MPa
4. $\sigma = 5375$ psi
5. $\sigma = 17\,200$ psi
6. For all materials, $\sigma = 34.7$ MPa
   Deflection:
   a. $\delta = 0.277$ mm   b. $\delta = 0.277$ mm
   c. $\delta = 0.377$ mm   d. $\delta = 0.830$ mm
   e. $\delta = 0.503$ mm   f. $\delta = 23.8$ mm
   g. $\delta = 7.56$ mm.

Note: The stress is close to the ultimate strength for (f) and (g).

7. Force = 2556 lb; σ = 2506 psi

8. σ = 595 psi

9. Force = 1061 lb

10. D = 0.274 in

13. $\sigma_{AD} = \sigma_{DE}$ = 6198 psi
    σ_{EF} = 7748 psi
    σ_{BD} = 0 psi
    σ_{BE} = 5165 psi
    $\sigma_{AB} = \sigma_{CE}$ = −4132 psi
    σ_{BC} = −3099 psi
    σ_{CF} = −5165 psi

15. σ = 144 MPa

16. At pins A and C: $\tau_A = \tau_C$ = 7958 psi
    At pin B: τ_B = 10 190 psi

19. τ = 98. 8 MPa

21. τ = 547 MPa

22. τ = 32.6 MPa

23. θ = 0.79°

25. τ = 32 270 psi

28. τ = 70.8 MPa; θ = 1.94°

30. T = 9624 lb·in; θ = 1.83°

31. Required section modulus = 2.40 in³. Standard nominal sizes given for each shape:

    a. Each side = 2.50 in

    c. Width = 5.00 in; height = 1.75 in

    e. S4 × 7.7

    g. 4-in schedule 40 pipe

32. Weights:

    a. 212 lb   c. 297 lb

    e. 77.0 lb   g. 107.9 lb

33. Maximum deflection   Deflection at loads

    a. 0.701 in   0.572 in

    c. 1.021 in   0.836 in

    e. 0.375 in   0.307 in

    g. 0.315 in   0.258 in

34. M_A = 330 N·m; M_B = 294 N·m; M_C = −40 N·m

36. a. y_A = 0.238 in; y_B = 0.688 in
    b. y_A = 0.047 in; y_B = 0.042 in

38. σ = 3480 psi; τ = 172 psi

For Problems 39 through 49, the complete solutions require drawings. Listed below are the maximum bending moments only:

39. 480 lb·in

41. 120 lb·in

43. 93 750 N·mm

45. 8640 lb·in

47. −11 250 N·mm

49. −1.49 kN·m

51. σ = 62.07 MPa

53. a. σ = 20.94 MPa tension on top of lever
    b. At section B, h = 35.1 mm; at C, h = 18 mm

55. σ = 84.6 MPa tension

57. Sides = 0.50 in

59. Maximum σ = −1.42 MPa compression on the top surface between A and C

61. σ = 86.2 MPa

63. Left: σ = 39 750 psi
    Middle: σ = 29 760 psi
    Right: σ = 31 700 psi

65. σ = 96.4 MPa

67. σ = 32 850 psi

69. Tension in member A–B: σ = 50.0 MPa
    Shear in pin: τ = 199 MPa
    Bending in A−C at B: σ = 14 063 MPa (Very high; redesign)

71. σ = 186 MPa

73. σ_{max} = 118.3 MPa at first step, 40 mm from either support.

75. σ = 108.8 MPa

77. With pivot at top hole: σ_{max} = 10 400 psi at fulcrum
    With pivot at bottom hole: σ_{max} = 5600 psi at fulcrum

CHAPTER 4   Combined Stresses and Mohr's Circle

Answers to Problems 1 to 35:

	Maximum Principal Stress	Minimum Principal Stress	Maximum Shear Stress
1.	24.14 ksi	−4.14 ksi	14.14 ksi
3.	50.0 ksi	−50.0 ksi	50.0 ksi
5.	124.7 ksi	0.0 ksi	62.4 ksi
7.	20.0 ksi	−40.0 ksi	30.0 ksi
9.	144.3 MPa	−44.3 MPa	94.3 MPa
11.	61.3 MPa	−91.3 MPa	76.3 MPa
13.	168.2 MPa	0.0 MPa	84.1 MPa
15.	250.0 MPa	−80.0 MPa	165.0 MPa
17.	453 MPa	−353 MPa	403 MPa
19.	42.2 MPa	−52.2 MPa	47.2 MPa
21.	40.0 ksi	0 ksi	20.0 ksi

Maximum Principal Stress	Minimum Principal Stress	Maximum Shear Stress
23. 42.8 ksi	−29.8 ksi	36.3 ksi
25. 23.9 ksi	−1.9 ksi	12.9 ksi
27. 328 MPa	0 MPa	164 MPa
29. 0 kPa	−868 kPa	434 kPa
31. 26.24 ksi	−5.70 ksi	15.97 ksi
33. 7730 psi	−4.0 psi	3867 psi
35. 398 psi	−6366 psi	3382 psi

CHAPTER 5 Design for Different Types of Loading

Stress Ratio

1. $\sigma_{max} = 44.6$ MPa
   $\sigma_{min} = 6.37$ MPa
   $\sigma_m = 25.5$ MPa
   $\sigma_a = 19.1$ MPa
   R = 0.143

3. $\sigma_{max} = 5375$ psi
   $\sigma_{min} = -750$ psi
   $\sigma_m = 2313$ psi
   $\sigma_a = 3063$ psi
   R = −0.140

5. $\sigma_{max} = 110.3$ MPa
   $\sigma_{min} = 50.9$ MPa
   $\sigma_m = 80.6$ MPa
   $\sigma_a = 29.7$ MPa
   R = 0.462

7. $\sigma_{max} = 9868$ psi
   $\sigma_{min} = 1645$ psi
   $\sigma_m = 5757$ psi
   $\sigma_a = 4112$ psi
   R = 0.167

9. $\sigma_{max} = 475$ MPa
   $\sigma_{min} = 297$ MPa
   $\sigma_m = 386$ MPa
   $\sigma_a = 89$ MPa
   R = 0.625

Endurance Strength

11. $s_n' = 219$ MPa

13. $s_n' = 29.0$ ksi

Design and Analysis

19. N = 1.57 (low)

23. N = 2.82 at right load. OK

25. N = 11.6

27. N = 9.60

29. D = 40 mm; a = 10.6 mm

31. $2\frac{1}{2}$ in pipe

33. b = 1.80 in

35. N = 9.11

36. σ = 34.7 MPa
    a. N = 5.96   c. N = 8.93
    e. N = 23.8   g. N = 1.30 (low)

37. N = 3.37

39. Force = 1061 lb; D = 5/16 in

41. N = 1.64 (low)

43. N = 5.17

45. D = 1.75 in

49. N = 8.18

51. N = 3.16

57. N = 2.05

61. Design a

64. N = 0.25 at pin holes (Failure)

67. N = 1.64 at B (low)

69. N = 1.74 (low)

73. N = 1.31 (low)

75. $r_{min} = 0.20$ in

CHAPTER 6 Columns

1. $P_{cr} = 4473$ lb

2. $P_{cr} = 14\,373$ lb

5. $P_{cr} = 4473$ lb

6. $P_{cr} = 32.8$ lb

8. a. Pinned ends: $P_{cr} = 7498$ lb
   b. Fixed ends: $P_{cr} = 12\,000$ lb
   c. Fixed-pinned ends: $P_{cr} = 10\,300$ lb
   d. Fixed-free ends: $P_{cr} = 1700$ lb

10. D = 1.45 in required. Use D = 1.50 in.

14. S = 1.423 in required. Use S = 1.500 in.

16. Use D = 1.50 in.

23. P = 1189 lb

25. P = 1877 lb

27. σ = 212 MPa; y = 25.7 mm

28. σ = 6685 psi; y = 0.045 in

30. P = 37 500 psi

31. $P_a = 22\,600$ lb

33. 4 × 4 × 1/2 is smallest outside dimension. $I = 12.3$ in^4.
    Weight = 21.6 lb/ft.
    6 × 4 × 1/4 is the lightest from those listed.
    $I_{min} = I_y = 11.7$ in^4. Weight = 15.6 lb/ft.

35. $P_a = 11\,750$ lb

37. $P_a = 18\,300$ lb

39. Piston rod is safe.

CHAPTER 7   Belts and Chains

V-belts

1. 3V Belt, 75 inches long

2. $C = 22.00$ in

3. $\theta_1 = 157°$; $\theta_2 = 203°$

10. $v_b = 2405$ ft/min

13. $P = 6.05$ hp

Roller Chain

25. No. 80 chain

28. Design power rating $= 18.08$ hp; Type B lubrication (bath)

29. Design power rating $= 45.2$ hp

34. $L = 96$ in, 128 pitches

35. $C = 35.57$ in

CHAPTER 8   Kinematics of Gears

Spur Gears

1. $N = 44$; $P_d = 12$

   a. $D = 3.667$ in   b. $p = 0.2618$ in

   c. $m = 2.117$ mm   d. $m = 2.00$ mm

   e. $a = 0.0833$ in   f. $b = 0.1042$ in

   g. $c = 0.0208$ in   h. $h_t = 0.1875$ in

   i. $h_k = 0.1667$ in   j. $t = 0.131$ in

   k. $D_o = 3.833$ in

3. $N = 45$; $P_d = 2$

   a. $D = 22.500$ in   b. $p = 1.571$ in

   c. $m = 12.70$ mm   d. $m = 12.0$ mm

   e. $a = 0.5000$ in   f. $b = 0.6250$ in

   g. $c = 0.1250$ in   h. $h_t = 1.1250$ in

   i. $h_k = 1.0000$ in   j. $t = 0.7854$ in

   k. $D_o = 23.500$ in

5. $N = 22$; $P_d = 1.75$

   a. $D = 12.571$ in   b. $p = 1.795$ in

   c. $m = 14.514$ mm   d. $m = 16.0$ mm

   e. $a = 0.5714$ in   f. $b = 0.7143$ in

   g. $c = 0.1429$ in   h. $h_t = 1.2857$ in

   i. $h_k = 1.1429$ in   j. $t = 0.8976$ in

   k. $D_o = 13.714$ in

7. $N = 180$; $P_d = 80$

   a. $D = 2.2500$ in   b. $p = 0.0393$ in

   c. $m = 0.318$ mm   d. $m = 0.30$ mm

   e. $a = 0.0125$ in   f. $b = 0.0170$ in

   g. $c = 0.0045$ in   h. $h_t = 0.0295$ in

   i. $h_k = 0.0250$ in   j. $t = 0.0196$ in

   k. $D_o = 2.2750$ in

9. $N = 28$; $P_d = 20$

   a. $D = 1.4000$ in   b. $p = 0.1571$ in

   c. $m = 1.270$ mm   d. $m = 1.25$ mm

   e. $a = 0.0500$ in   f. $b = 0.0620$ in

   g. $c = 0.0120$ in   h. $h_t = 0.1120$ in

   i. $h_k = 0.1000$ in   j. $t = 0.0785$ in

   k. $D_o = 1.5000$ in

11. $N = 45$; $m = 1.25$

   a. $D = 56.250$ mm   b. $p = 3.927$ mm

   c. $P_d = 20.3$ d. $P_d = 20$

   e. $a = 1.25$ mm   f. $b = 1.563$ mm

   g. $c = 0.313$ mm   h. $h_t = 2.813$ mm

   i. $h_k = 2.500$ mm   j. $t = 1.963$ mm

   k. $D_o = 58.750$ mm

13. $N = 22$; $m = 20$

   a. $D = 440.00$ mm   b. $p = 62.83$ mm

   c. $P_d = 1.270$ d. $P_d = 1.25$

   e. $a = 20.0$ mm   f. $b = 25.00$ mm

   g. $c = 5.000$ mm   h. $h_t = 45.00$ mm

   i. $h_k = 40.00$ mm   j. $t = 31.42$ mm

   k. $D_o = 480.00$ mm

15. $N = 180$; $m = 0.4$

   a. $D = 72.00$ mm   b. $p = 1.26$ mm

   c. $P_d = 63.5$ d. $P_d = 64$

   e. $a = 0.40$ mm   f. $b = 0.500$ mm

   g. $c = 0.100$ mm   h. $h_t = 0.90$ mm

   i. $h_k = 0.80$ mm   j. $t = 0.628$ mm

   k. $D_o = 72.80$ mm

17. $N = 28$; $m = 0.8$

   a. $D = 22.40$ mm   b. $p = 2.51$ mm

   c. $P_d = 31.75$ d. $P_d = 32$

   e. $a = 0.80$ mm   f. $b = 1.000$ mm

   g. $c = 0.200$ mm   h. $h_t = 1.800$ mm

   i. $h_k = 1.60$ mm   j. $t = 1.257$ mm

   k. $D_o = 24.00$ mm

19. Problem 1: $P_d = 12$; backlash $= 0.006$ to 0.009 in
     Problem 12: $m = 12$; backlash $= 0.52$ to 0.82 mm

21. a. $C = 14.000$ in   b. $VR = 4.600$

   c. $n_G = 48.9$ rpm   d. $v_t = 294.5$ ft/min

23. a. $C = 2.266$ in   b. $VR = 6.25$

   c. $n_G = 552$ rpm   d. $v_t = 565$ ft/min

25. a. $C = 90.00$ mm   b. $VR = 3.091$
   c. $n_G = 566$ rpm   d. $v_t = 4.03$ m/s

27. a. $C = 162.0$ mm   b. $VR = 1.250$
   c. $n_G = 120$ rpm   d. $v_t = 1.13$ m/s

For problems 29 through 32, the following are errors in the given statements:

29. The pinion and the gear cannot have different pitches.

30. The actual center distance should be 8.333 in.

31. There are too few teeth in the pinion; interference is to be expected.

32. The actual center distance should be 2.156 in. Apparently, the outside diameters were used instead of the pitch diameters to compute C.

33. $Y = 8.45$ in; $X = 10.70$ in

35. $Y = 44.00$ mm; $X = 58.40$ mm

37. Output speed = 111 rpm CCW

39. Output speed = 144 rpm CW

Helical Gears

41. $p = 0.3927$ in $p_n = 0.3401$ in
   $P_{nd} = 9.238$ $P_x = 0.680$ in
   $D = 5.625$ in $\phi_n = 12.62°$
   $F/P_x = 2.94$ axial pitches in the face width

42. $P_d = 8.485$ $p = 0.370$ in
   $p_c = 0.2618$ in $P_x = 0.370$ in
   $\phi_t = 27.2°$ $D = 5.657$ in
   $F/P_x = 4.05$ axial pitches in the face width

Bevel Gears

45. Selected results: $F = 1.25$ in specified.
   $d = 2.500$ in $D = 7.500$ in
   $\gamma = 18.435°$ $\Gamma = 71.565°$
   $A_o = 3.953$ in $F_{nom} = 1.186$ in
   $A_m = A_{mG} = 3.328$ in $h = 0.281$ in
   $c = 0.035$ in $h_m = 0.316$ in
   $a_P = 0.213$ in $a_G = 0.068$ in
   $d_o = 2.992$ in $D_o = 7.555$ in

49. Selected results: $F = 0.800$ in specified.
   $d = 1.500$ in $D = 6.000$ in
   $\gamma = 14.03°$ $\Gamma = 75.97°$
   $A_o = 3.092$ in $F_{nom} = 0.928$ in
   $A_m = A_{mG} = 2.692$ in $h = 0.145$ in
   $c = 0.018$ in $h_m = 0.163$ in

$a_P = 0.112$ in $a_G = 0.033$ in
$d_o = 1.755$ in $D_o = 6.020$ in

Wormgearing.

52. $L = 0.3142$ in $\lambda = 4.57°$
   $a = 0.100$ in $b = 0.1157$ in
   $D_{oW} = 1.450$ in $D_{RW} = 1.0186$ in
   $D_G = 4.000$ in $C = 2.625$ in
   $VR = 40$

59. 0.4067 rpm

61. 0.5074 rpm

63. $N_P = 22, N_G = 38$

65. $N_P = 19, N_G = 141$

68. One possible solution: Triple reduction; Layout as in Figure 8–31.
   $N_A = N_C = N_E = 17, N_B = 136, N_D = 119, N_F = 85$
   $TV = 280$ exactly; $n_{out} = 12$ rpm exactly; Used factoring method

69. One possible solution: Triple reduction with an idler
   $N_{P1} = 18, N_{G1} = 126, N_{P2} = 18, N_{G2} = 108, N_{P3} = 18,$
   $N_{G3} = 135, N_{idler} = 18$
   $n_{out} = 13.33$ rpm

72. One possible solution. Double reduction; Layout as in Figure 8–30.
   $N_A = N_C = 18, N_B = 75, N_D = 51$
   $n_{out} = 148.2$ rpm

CHAPTER 9   Spur Gear Design

1. a. $n_G = 486.1$ rpm
   b. $VR = m_G = 3.600$
   c. $D_P = 1.667$ in; $D_G = 6.000$ in
   d. $C = 3.833$ in
   e. $v_t = 764$ ft/min
   f. $T_P = 270$ lb·in; $T_G = 972$ lb·in
   g. $W_t = 324$ lb
   h. $W_r = 118$ lb
   i. $W_N = 345$ lb

3. a. $n_G = 752.7$ rpm
   b. $VR = m_G = 4.583$
   c. $D_P = 1.000$ in; $D_G = 4.583$ in
   d. $C = 2.792$ in
   e. $v_t = 903$ ft/min
   f. $T_P = 13.7$ lb·in; $T_G = 62.8$ lb·in
   g. $W_t = 27.4$ lb
   h. $W_r = 10.0$ lb
   i. $W_N = 29.2$ lb

5. a. $n_G = 304.4$ rpm

    b. $VR = m_G = 3.778$

    c. $D_P = 3.600$ in; $D_G = 13.600$ in

    d. $C = 8.600$ in

    e. $v_t = 1084$ ft/min

    f. $T_P = 2739$ lb·in; $T_G = 10\ 348$ lb·in

    g. $W_t = 1522$ lb

    h. $W_r = 710$ lb

    i. $W_N = 1680$ lb

8. $Q = 5$
    Pinion tol. = 0.0130 in; Gear tol. = 0.0150 in

9. $Q = 10$
    Pinion tol. = 0.0015 in; Gear tol. = 0.0017 in

11. $Q = 14$
    Pinion tol. = 0.00031 in; Gear tol. = 0.00035 in

15. $Q = 10$ (Interpolation used for tolerances)
    Pinion tol. = 0.0037 in; Gear tol. = 0.0039 in

26. a. $s_{at} = 28\ 200$ psi; $s_{ac} = 93\ 500$ psi

    c. $s_{at} = 43\ 700$ psi; $s_{ac} = 157\ 900$ psi

    e. $s_{at} = 36\ 800$ psi; $s_{ac} = 104\ 100$ psi

    g. $s_{at} = 57\ 200$ psi; $s_{ac} = 173\ 900$ psi

27. $HB = 300$ for Grade 1; $HB = 192$ for Grade 2

33. a. $s_{at} = 45\ 000$ psi; $s_{ac} = 170\ 000$ psi

    c. $s_{at} = 55\ 000$ psi; $s_{ac} = 180\ 000$ psi

    e. $s_{at} = 55\ 000$ psi; $s_{ac} = 180\ 000$ psi

    g. $s_{at} = 51\ 200$ psi; $s_{ac} = 168\ 000$ psi

    i. $s_{at} = 5000$ psi; $s_{ac} = 50\ 000$ psi

    k. $s_{at} = 27\ 000$ psi; $s_{ac} = 92\ 000$ psi

    m. $s_{at} = 23\ 600$ psi; $s_{ac} = 65\ 000$ psi

    o. $s_{at} = 9000$ psi; s_{ac} not listed

34. $h_e = 0.027$ in

35. $h_e = 0.90$ mm

The following three sets of answers are given in groups of four problems that all relate to the same basic set of design data.

37. $s_{tP} = 32\ 740$ psi; $s_{tG} = 26\ 940$ psi

43. $s_{atP} = 34\ 460$ psi; $s_{atG} = 28\ 100$ psi

49. $s_{cP} = 172\ 100$ psi; $s_{cG} = 172\ 100$ psi

55. $s_{acP} = 189\ 200$ psi; $s_{acG} = 185\ 100$ psi

39. $s_{tP} = 2300$ psi; $s_{tG} = 2000$ psi

45. $s_{atP} = 3700$ psi; $s_{atG} = 3100$ psi

51. $s_{cP} = 37\ 800$ psi; $s_{cG} = 37\ 800$ psi

57. $s_{acP} = 63\ 600$ psi; $s_{acG} = 62\ 200$ psi

41. $s_{tP} = 9458$ psi; $s_{tG} = 8134$ psi

47. $s_{atP} = 10\ 254$ psi; $s_{atG} = 8642$ psi

53. $s_{cP} = 78\ 263$ psi; $s_{cG} = 78\ 263$ psi

59. $s_{acP} = 87\ 531$ psi; $s_{acG} = 85\ 727$ psi

Problems 60–70 are design problems for which there are no unique solutions.

71. Power capacity = 12.9 hp based on gear contact stress at 15 000 hr life.

Problems 73–83 are design problems for which there is no unique solution.

CHAPTER 10 Helical Gears, Bevel Gears, and Wormgearing

1. $W_t = 89.6$ lb; $W_x = 51.7$ lb; $W_r = 23.2$ lb
    $Q = 6$; $s_t = 2778$ psi; $s_c = 36\ 228$ psi
    Cast iron Class 20

3. $W_t = 143$ lb; $W_x = 143$ lb; $W_r = 37.0$ lb
    $Q = 8$; $s_t = 9720$ psi; $s_c = 73\ 300$ psi
    Ductile iron 60–40–18 or Cast iron Class 40

14. $W_{tP} = W_{tG} = 599$ lb; $W_{xP} = W_{rG} = 69$ lb;
    $W_{rP} = W_{xG} = 207$ lb
    $Q = 8$; $s_{tP} = 34\ 100$ psi; $s_{tG} = 48\ 900$ psi; $s_c = 118\ 000$ psi
    Pinion: AISI 6150 OQT 1200, HB 293
    Gear: AISI 6150 OQT 1000, HB 375

18. $D_G = 4.000$ in; $C = 2.625$ in; $VR = 40$
    $W_{xW} = W_{tG} = 462$ lb; $W_{xG} = W_{tW} = 53$ lb;
    $W_{rG} = W_{rW} = 120$ lb
    Efficiency = 70.3%; Worm speed = 1200 rpm;
    $P_i = 0.626$ hp
    $\sigma_G = 24\ 223$ psi [Slightly high for phosphor bronze]
    Rated wear load = $W_{tr} = 659$ lb. [OK, > W_{tG}]

CHAPTER 11 Keys and Couplings

1. Use 1/2 in square key; AISI 1040 cold drawn steel; length = 3.75 in.

3. Use 3/8 in square key; AISI 1020 cold drawn steel; required length = 1.02 in; use $L = 1.50$ in to be just shorter than the 1.75 in hub length.

5. T = torque; D = shaft diameter; L = hub length. From Table 11–5, $K = T/(D^2L)$.

    a. Data from Problem 1: required $K = 1313$; too high for any spline in Table 11–5.

    c. Data from Problem 3: required $K = 208$; use 6 splines.

7. Sprocket: 1/2 in square key; AISI 1020 CD; $L = 1.00$ in
    Wormgear: 3/8 in square key; AISI 1020 CD; $L = 1.75$ in

13. $T = 2725$ lb·in

15. $T = 26\ 416$ lb·in

19. Data from Problem 16: $T = 313$ lb·in per inch of hub length

    Data from Problem 18: $T = 4300$ lb·in per inch of hub length

CHAPTER 12   Shaft Design

1. $T_B = 3436$ lb·in
   $F_{Bx} = W_{tB} = 430$ lb d
   $F_{By} = W_{rB} = 156$ lb T

3. $T_B = 656$ lb·in
   $F_{Bx} = W_{tB} = 437$ lb S
   $F_{By} = W_{rB} = 159$ lb c

5. $T_D = 3938$ lb·in
   $W_{tD} = 985$ lb Up at 30° left of vertical
   $W_{rD} = 358$ lb To right 30° above horizontal
   $F_{Dx} = 182$ lb d
   $F_{Dy} = 1032$ lb c

7. $T_C = 6563$ lb·in
   $F_{Cx} = W_{tC} = 1313$ lb S
   $F_{Cy} = W_{rC} = 478$ lb c

9. $T_C = 1432$ lb·in
   $F_{Cx} = W_{tC} = 477$ lb d
   $F_{Cy} = W_{rC} = 174$ lb T

11. $T_F = 1432$ lb·in
    $W_{tF} = 477$ lb Down at 45° left of vertical
    $W_{rF} = 174$ lb To right 45° below horizontal
    $F_{Fx} = 214$ lb d
    $F_{Fy} = 460$ lb T

13. $T_A = 3150$ lb·in
    $F_{Ax} = 0$
    $F_{Ay} = F_A = 630$ lb T

15. $T_C = 1444$ lb·in
    $F_C = 289$ lb Down at 15° left of vertical
    $F_{Cx} = 75$ lb d
    $F_{Cy} = 279$ lb T

17. $T_C = 2056$ lb·in
    $F_{Cx} = F_C = 617$ lb d
    $F_{Cy} = 0$

19. $T_C = 10\ 500$ lb·in
    $F_{Cx} = F_C = F_{Dx} = F_D = 1500$ lb d
    $F_{Cy} = F_{Dy} = 0$

21. $T_E = 1313$ lb·in
    $F_E = 438$ lb Up at 30° above horizontal
    $F_{Ex} = 379$ lb S
    $F_{Ey} = 219$ lb c

22. $T_B = 727$ lb·in
    $F_{Bx} = W_{tB} = 351$ lb S
    $F_{By} = W_{rB} = 132$ lb T
    $W_{xB} = 94$ lb exerts a CCW concentrated moment of 194.6 lb·in on shaft at B.

W_{xB} also places the shaft in compression from A to B if bearing A resists the thrust load.

23. $T_A = 270$ lb·in
    $F_{Ax} = 0$
    $F_{Ay} = F_A = 162$ lb T
    $F_{Cx} = W_{tW} = 265$ lb d

$F_{Cy} = W_{rW} = 352$ lb c
$W_{xW} = 962$ lb exerts a CW concentrated moment of 962 lb·in on shaft at worm.
W_{xW} also places the shaft in compression from bearing B to worm if bearing B resists the thrust load.

CHAPTER 13   Tolerances and Fits

1. RC8: Hole—3.5050/3.5000; shaft—3.4930/3.4895; clearance—0.0070 to 0.0155 in

3. RC8: Hole—0.6313/0.6285; shaft—0.6250/0.6234; clearance—0.0035 to 0.0079 in

5. RC8: Hole—1.2540/1.2500; pin—1.2450/1.2425; clearance—0.0050 to 0.0115 in

7. RC5: Hole—0.7512/0.7500; pin—0.7484/0.7476; clearance—0.0016 to 0.0036 in (tighter fit could be used)

10. FN5: Hole—3.2522/3.2500; shaft—3.2584/3.2570; interference—0.0048 to 0.0084 in;
    pressure = 26 350 psi; stress = 128 726 psi

12. FN5: Interference—0.0042 to 0.0072 in;
    pressure = 17 789 psi; stress = 37 800 psi at inner surface of aluminum cylinder; stress = $-17\ 789$ psi at outer surface of steel rod; stress in aluminum is very high

13. Maximum interference = 0.000 89 in.

14. Temperature = 567°F

15. Shrinkage = 0.003 8 in; $t = 284$°F

16. Final $ID = 3.4942$ in

CHAPTER 14   Rolling Contact Bearings

1. Life = 2.76×10^6 rev

2. $C = 12\ 745$ lb

For problems calling for the selection of suitable bearings for given applications, there are several possible solutions. The following are sample acceptable solutions using data in this chapter. Other possible solutions may be more optimum, particularly if a larger variety of manufacturers' data are available.

5. At B: Required $C = 47\ 637$ lb. Use bearing #6332
   At C: Required $C = 21\ 320$ lb. Use bearing #6318

7. At A: Required $C = 2519$ lb. Use bearing #6206
   (Carries radial load only)
   At C: Required $C = 8592$ lb. Use bearing #6213
   (Carries radial and thrust loads)

9. Required $C = 5066$ lb. Use bearing #6307 (Carries radial load only)

11. Required $C = 6412$ lb. Use bearing #6308 (Carries radial and thrust loads)

13. Required $C = 33\ 998$ lb. Use bearing #6322 (Carries radial and thrust loads)

14. Required $C = 12\ 462$ lb. Use bearing #6313 (Carries radial load only)

19. Life = 48 900 hours
21. Life = 25 300 hours
23. Life = 47 300 hours
25. $C = 17\,229$ lb
27. $C = 4580$ lb

CHAPTER 15   No practice problems

CHAPTER 16   Plain Surface Bearings

All problems in this chapter are design problems for which no unique solutions exist.

CHAPTER 17   Linear Motion Elements

5. 2 1/2–3 Acme thread
6. $L > 1.23$ in
7. $T = 6974$ lb·in
8. $T = 3712$ lb·in
11. Lead angle = 4.72°; self-locking
12. Efficiency = 35%
13. $n = 180$ rpm; $P = 0.866$ hp
17. 24.7 years

CHAPTER 18   Fasteners

4. Grade 2 bolts: 5/16–18; $T = 70.3$ lb·in
5. $F = 1190$ lb
6. $F = 4.23$ kN
7. Nearest metric thread is M24 × 2. Metric thread is 1.8 mm larger (8% larger).
8. Closest standard thread is M5 × 0.8. (#10–32 is also close.)
9. 3.61 kN
10. a. 1177 lb   c. 2385 lb
    e. 2862 lb   g. 1081 lb
    i. 2067 lb   k. 143 lb

CHAPTER 19   Springs

1. $k = 13.3$ lb/in
2. $L_f = 1.497$ in
3. $F_s = 47.8$ lb; $L_f = 1.25$ in
7. $ID = 0.93$ in; $D_m = 1.015$ in; $C = 11.94$; $N = 6.6$ coils
8. $C = 8.49$; $p = 0.241$ in; pitch angle = 8.70°; $L_s = 1.12$ in
9. $F_o = 10.25$ lb; stress = 74 500 psi
11. $OD = 0.583$ in when at solid length
12. $F_s = 26.05$ lb; stress = 189 300 psi (High)
31. Bending stress = 114 000 psi; torsion stress = 62 600 psi. Stresses are safe.
35. Torque = 0.91 lb·in to rotate spring 180°.
    Stress = 184 800 psi; OK for severe service

CHAPTER 20   Machine Frames, Bolted Connections and Welded Joints

Problems 1–16 are design problems for which there are no unique solutions.

17.

Material	Diameter (in)	Weight (lb per inch of length)
a. 1020 HR steel	0.638	0.0906
c. Aluminum 2014–T6	0.451	0.0160
e. Ti–6A1–4V (Annealed)	0.319	0.0128

CHAPTER 21   Electric Motors and Controls

13. 480V, 3-Phase because the current would be lower and the motor size smaller
16. $n_s = 1800$ rpm in the United States
    $n_s = 1500$ rpm in France
17. 2-Pole motor; $n = 3600$ rpm at zero load (approximate)
18. $n_s = 12\,000$ rpm
19. 1725 rpm and 1140 rpm
20. Variable frequency control
34. a. Single phase, split phase AC motor
    b. $T = 41.4$ lb·in   c. $T = 62.2$ lb·in
    d. $T = 145$ lb·in
35. b. $T = 4.15$ N·m   c. $T = 6.23$ N·m
    d. $T = 14.5$ N·m
39. Full load speed = synchronous speed = 720 rpm
47. Use a NEMA Type K SCR control to convert 115 VAC to 90 VDC; use a 90 VDC motor.
51. Speed theoretically increases to infinity
52. $T = 20.5$ N·m
54. NEMA 2 starter
55. NEMA 1 starter

CHAPTER 22   Motion Control: Clutches and Brakes

1. $T = 495$ lb·in
3. $T = 41$ lb·in
5. Data from Problem 1: $T = 180$ lb·in
    Data from Problem 3: $T = 27.4$ lb·in
7. Clutch: $T = 2122$ N·m
    Brake: $T = 531$ N·m
8. $T = 143$ lb·ft
9. $T = 60.9$ lb·ft
11. $T = 223.6$ lb·ft
15. $F_a = 109$ lb
17. $W = 138$ lb
18. $b > 16.0$ in

Index

Adhesives, 726
Allowance, 578
Aluminum, 57, A–5
Aluminum Association, 44
American Gear Manufacturers Association
    (AGMA), 308–17, 334–5, 344, 357,
    374–395, 400–4, 442–3, 455–60,
    470–2, 481–6, 488
American Iron and Steel Institute (AISI), 44
American National Standards Institute
    (ANSI), standards, 495, 548
American Society of Mechanical Engineers
    (ASME), standard, 495–6, 548
American Society for Testing and Materials
    (ASTM), 44, 54, 716, A–13–A–14
American standard beam shapes, 20, A–34
Angles, equal and unequal leg, A–31
Annealing, 50
Areas, properties of, A–1
Austempered ductile iron, 56, A–14

Babbitt, 667
Ball screws, 697, 704–7
    column buckling, 708
    efficiency, 706
    materials, 708
    performance, 706
    torque, 706
    travel life, 706
Basic sizes, preferred, 18, A–3
Beams
    bending stress, 105
    concentrated bending moments, 112
    deflections, 108, A–18–A–26
    flexural center, 107
    shear center, 108
    shapes, A–31–A–38
Bearings, plain surface, 660–83
    bearing characteristic number, 677
    bearing parameter, $\mu n/p$, 665
    boundary lubrication, 664, 668–74
    clearance, diametral, 661, 672, 676
    coefficient of friction, 680
    coefficient of friction variable, 679
    design of full film hydrodynamic
        bearings, 678–80
    film thickness, 675
    film thickness variable, 678

friction torque and power, 680
full-film (hydrodynamic) lubrication, 664,
    674–80
geometry, 661
grooving, 682
hydrostatic, 683–7
hydrostatic bearing performance, 684
journal, 661
length, 670, 676
materials, 666–8
mixed-film lubrication, 664
$\mu n/p$ parameter, 665
pressure, 665, 676
pV factor, 668–9
Sommerfeld number, 677
Striberk curve, 665
surface roughness, 675
temperature of lubricant, 670, 676
viscosity, 676–7
wear factor, 691
Bearings, rolling contact, 597–637
    brinelling, 610
    design life, 611–12
    dynamic load rating, 610–11
    equivalent load, 614, 618
    flange units, 605
    grease for, 622
    installation, 622
    life factor, 612
    load/life relationship, 606–9
    locknuts, 617–18
    lubrication, 621, 624
    manufacturers' data, 606
    materials, 606
    mean effective load, 626
    mounted bearings, 604
    mounting, 616
    oil film thickness, 624
    pillow blocks, 605
    preloading, 622
    rated life (L_{10}), 610
    reliability, 616
    rotation factor, 613
    sealing, 623
    selection, 613–15
    sizes, 610
    speed factor, 612
    speeds, limiting, 623

standards, 623
static load rating, 610
stiffness, 622
take-up bearings, 605
tapered roller bearings, 603, 618
thrust bearing, 604
thrust factor, 614
tolerances, 624
types, 600–4
varying loads, 623, 625–7
Belleville spring, 733–4
Belt pulleys, flat, 540
Belts and chains. *See* V-belt drives; Chain
    drives
Bolted connections, 719, 780–2
Brakes. *See* Clutches and brakes
Brass, 60
Brazing, 726
Brinell hardness, 38
Bronze, 60, 383, A–17
Bronze, sintered, 667
Buckling of columns, 230
Buckling of springs, 745–6
Bushing, split taper, 510

Carbo-nitriding, 52
Carburizing, 52, 382, A–11
Carburizing of gear teeth, 382
Cardan joint, 516
Cast iron, 54, 383, A–14
Chain drives, 283–96
    attachments, 285
    center distance formula, 290
    conveyor chain, 286
    design of, 285–96
    forces on shafts, 537–8
    length of chain formula, 290
    lubrication, 291
    multiple strands, factors, 286
    pitch, 283
    power ratings, 287–9
    roller chain, 283–4
    service factors, 290
    sizes, 284
    sprockets, 283
    styles, 284
Channel beam shapes, 22, A–32, A–37
Charpy test, 39

Clearance, 312, 576
Clearance fits, 581–5
Clutches and brakes, 830–64
   actuation, 835–9
   applications, typical, 834
   band brakes, 835, 860–2
   brake, defined, 831
   brake, fail-safe, 835
   clutch, defined, 831
   clutch coupling, 833
   clutch-brake module, 835
   coefficient of friction, 850–1
   cone clutch or brake, 835, 854
   disc brakes, 776, 835
   drum brakes, 855–9
   eddy current drive, 863
   energy absorption, 846–7
   fiber clutch, 863
   fluid clutch, 863
   friction materials, 849–50
   inertia, 841–6
   inertia, effective, 844–6
   jaw clutch, 862
   overload clutch, 863
   performance, 840–1
   plate type, 835, 851–3
   radius of gyration, 841
   ratchet, 862
   response time, 847–9
   single-revolution clutch, 863
   slip clutch, 835, 837
   sprag clutch, 862
   tensioners, 864
   types, 835
   wear, 852–3
   Wk^2, inertia, 842–6
   wrap spring clutch, 863
Coefficient of friction, 476, 680, 850
Coefficient of thermal expansion, 44
Collars, 519
Columns, 229–56
   buckling, 230
   column constant, 234
   crooked, 250
   design factors for, 238
   design of, 245
   eccentrically loaded, 251
   effective length, 233
   efficient shapes for column cross
      sections, 244
   end fixity, 232
   Euler formula, 235
   J. B. Johnson formula, 239
   radius of gyration, 232
   secant formula, 253
   slenderness ratio, 234
Combined stresses, 117, 138
Composite materials. See Materials,
      composites
Connections, keyless, 509
Contact stress, 401
Conversion factors, A–39
Copper, 60
Copper Development Assn., 44

Couplings, 513–6
   bellows, 515
   chain, 514
   Dynaflex, 515
   Ever-flex, 514
   flexible, 513
   Form-flex, 516
   gear, 514
   Grid-flex, 514
   jaw-type, 516
   Para-flex, 515
   rigid, 513–6
Creep, 40
Criteria for machine design, 11
Critical speed, 563, 707
Cyaniding, 52

Damage accumulation method, 214
Decision analysis, 634
Deformation, 92
Density, 43
Design analysis methods, 193
Design calculations, 17–19
Design details, power transmission,
      635–56
Design factor, 185
Design philosophy, 182
Design problem examples, 200–13
Design process example, 12–17
Design procedure, 197–200
Design projects, 867–92
Design requirements, 11
Design skills, 11
Direct stress element, 90
Distortion energy theory, 189, 195, 532, 546
Drawings, assembly, 656
   shaft details, 653–5
Drop weight test, 40
Ductile iron, 56, A–14
Ductility, 35

Elastic limit, 34
Electrical conductivity, 44
Electrical resistivity, 44
Endurance strength, 40, 172–81
   actual, 173
   graph vs. tensile strength, 175
Equivalent torque, 161
Euler formula for columns, 235
Evaluation criteria, 11

Factor of safety. See Design factor
Failure modes, 89, 186
Fasteners, 711–27
   adhesives, 726
   American Standard, 717
   bolt materials, 714–7
   bolted joints, 719, 780–2
   brazing, 726
   clamping load, 719
   coatings and finishes, 717
   head styles, 713
   locking devices, 725
   metric, 716–8

   screws, 713
   set screws, 725
   soldering, 726
   strength, 714–6
   thread designations, 717–8, A–4, A–5
   thread stripping, 723
   tightening torque, 719–21
   washers, 725
Fatigue, 44, 168
Fillets, shoulder, 541
Finite element analysis, 161
Finite life method, 214
Fit, 578
Fits, clearance, 581–5
   running and sliding, 581
   locational, 585
Fits, interference, 585–90
   force fits, 585
   locational, 585
   shrink, 585
   stresses for, 587–90
Fits, transition, 586
Fits for bearings, 645–6
Flame hardening, gears, 52, 381
Flexible couplings, 513
   effects on shafts, 540
Flexural center, 107
Flexural modulus, 37
Fluctuating stress, 168–70
Force, 26
Force fits, 585
   stresses for, 587–90
Forces exerted on shafts by machine
      elements, 535–40
Friction, 687
Function statements, 11

Garter springs, 733–4
Gear manufacture and quality, 370–8
   composite error, 374
   form milling, 371
   hobbing, 371
   measurement, 376
   quality numbers, 373–8
   shaping, 371
   total composite tolerance, 373
Gear-type speed reducers, 340, 431–4,
      451, 475
Gear trains
   train value, 323, 347–57
   velocity ratio, 322
Gearmotors, 431
Gears, bevel, 333–9
   bearing forces, 465–9
   bending moments on shafts, 468
   forces on, 463
   geometry, 334–9
   geometry factor, 471–2
   material selection for, 472
   miter gears, 335
   pitch cone angle, 335
   pitting resistance, 473
   reducers, 475
   stresses in teeth, 470

Gears, helical, 329–33
    crossed, 329
    design of, 460
    forces on shafts, 537
    forces on teeth, 330–1, 452
    geometry, 330–2
    geometry factor, 455–8
    helix angle, 829
    pitches, 331
    pitting resistance, 459
    reducers, 451
    stresses in teeth, 455
Gears, spur, 307–20, 363–442
    addendum, 312
    backlash, 315
    center distance, 316
    clearance, 312
    contact stress, 401
    dedendum, 312
    design of, 407–9, 441
    dynamic factor, 392
    efficiency, 370
    elastic coefficient, 400
    face width, 315
    forces on shafts, 535–7
    forces on teeth, 367–9
    geometry, 310–8, 437
    geometry factor, 386, 402
    Hertz stress on teeth, 400
    idler, 326
    internal, 326
    interference, 320
    involute tooth form, 307
    Lewis form factor, 385
    life factor, 395, 403
    load distribution factor, 389
    lubrication, 430–2
    manufacture, 370–2
    material specification, 399, 402
    materials, 379–85
    metric module, 312, 413
    overload factor, 388
    pitch, 309
    pitch, circular, 310
    pitch, diameter, 308
    pitch, diametral, 310
    pitting resistance, 399, A–41
    plastics gearing, 434–42
    power flow, 370
    power transmitting capacity, 428
    pressure angle, 316
    quality, 372–8
    rack, 327
    reliability factor, 396
    rim thickness factor, 391
    size factor, 389
    stresses in teeth, 385–92
    stresses, allowable, 378–85
    styles, 303, 306
    undercutting, 321
Gears, worm. See Wormgearing
Geometry factor, I, A–41
Gerber method, 192
Goodman method, 190–2

Gray iron, 55, 383, A–14
Grasses, 689

Hastelloy, 61
Hardness, 37
    conversions, 38, A–40
Heat treating of steels
    annealing, 50
    carbo-nitriding, 52
    carburizing, 52, 382, A–11
    case hardening, 51, 380
    cyaniding, 52
    flame hardening, 52, 381
    induction hardening, 52, 381
    nitriding, 52, 382
    normalizing, 51
    tempering, 51
    through-hardening, 51
Hertz stress, 400
Hollow structural shapes, 23
Hunting tooth, 350
Hydrodynamic lubrication, 674–80
Hydrostatic bearings, 683–7

I-beam shapes, 24, A–33–A–34, A–38
Idler gear, 326
Impact energy, 38
Inconel alloys, 61
Induction hardening, gears, 52, 381
Interference, 576
    fits, 585–90
    of spur gear teeth, 320
Involute tooth form, 307
Izod test, 39

J. B. Johnson formula for columns, 239
J-factor for helical gears, 455–8
J-factor for spur gears, 386
Jook screw, 695

Keys, 494–502
    chamfers, 496
    design of, 500
    forces on, 499
    gib-head, 495
    materials for, 498
    parallel, 494–5
    pin, 497
    sizes, 495
    stresses in, 499–503
    taper, 495
    tolerances, 652
    types, 494–8
    Woodruff, 497, 501–3
Keyseats and keyways
    dimensions, 496
    fillets, 496
    stress concentrations, 540

Law of gearing, 307
Leaf springs, 734
Linear motion elements, 695–7
Loading types, 664–5
Locknuts, 521, 617–8

Lubricants, 688–90
    solid, 690
Lubrication
    bearings, plain. See Bearings, plain
        surface
    bearings, rolling contact, 621, 624
    chain drives, 291–2
    gears, 430–2

Machinability, 38
Machine frames and structures, 776–9
    deflection limits, 777
    materials, 776
    torsion, 778–9
Malleable iron, 55, A–14
Mass, 26
Materials in mechanical design, 29–78
    aluminum, 57, A–15
    brass and bronze, 60, 384, A–17
    carbon and alloy steel, 46–9, 330,
        A–6–A–10
    cast iron, 54, 383, A–14
    composites, 65–78
        advantages of, 68
        construction of, 71
        design guidelines, 76
        filament winding, 68
        limitations of, 69
        preimpregnated materials, 67
        pultrusion, 67
        reinforcing fibers, 66–7
        sheet molding compound, 67
        wet processing, 67
    copper, 60
    gear, 379–85
    nickel alloys, 61
    plastics, 61, 435, A–17
    powdered metals, 56
    selection, 77
    stainless steels, 53, A–12
    structural steel, 54, A–13
    thermoplastics, 61
    thermosets, 62
    titanium, 60, A–16
    tool steels, 54
    zinc, 59, A–16
Maximum normal stress, 139, 187
Maximum shear stress, 139, 188
Metal Powder Industries Foundation
    (MPIF), 56
Metric units, 24
Miner's rule, 216, 625
Mises–Hencky theory, 189
Miter gears, 335
Modulus of elasticity
    in flexure, 37
    in shear, 36
    in tension, 34
    spring wire, 745
Mohr's circle, 145–8
    special stress conditions, 158
Mohr theory, 187
Monel alloys, 61
Moore, R.R. fatigue test device, 168

Motion control. *See* Clutches and brakes
Motor starters, 812–6
Motors, electric, 795–826
   AC motor types, 803–8
   AC motors, 800–11
   AC variable speed drive, 818–20
   AC voltages, 800
   brushless DC, 826
   capacitor start, 807
   compound-wound, DC, 822
   controls, AC, 811–20
   DC motor control, 824–5
   DC motor types, 821–3
   DC power, 820–1
   enclosures, motors, 810–1
     controls, 818
   frame sizes, 811
   frame types, 808–11
   induction motors, 800–2
   Linear motors, 826
   NEMA AC motor designs
     B, C, D, 803
   overload protection, 817
   performance curves, AC motors,
     802–6
   permanent magnet, DC, 823
   permanent split capacitor, 808
   printed circuit motors, 826
   rectifiers (SCR), 821
   selection factors, 798
   series-wound, DC, 822
   servomotors, 824
   shaded pole, 808
   shunt-wound, DC, 822
   single-phase motors, 806–8
   single-phase power, 799
   sizes, 798, 811
   speed control, AC, 818–20
   speeds, 800
   split-phase, 806
   squirrel cage motors, 801
   starters, 812–6
   stepping motors, 825
   synchronous motors, 805
   three-phase power, 799
   torque motors, 824
   universal motor, 805
   wound rotor motor, 804

National Electrical Manufacturers
   Association (NEMA), 803
Nickel-based alloys, 61
Ni-resist alloy, 61
Nitriding, 52, 179, 382
Normal stress element, 89
Normal stresses, combined, 117
  maximum, 139
Normalizing, 51
Notch sensitivity, 122

Oils, 688

Palmgren–Miner rule, 625
Percent elongation, 35
Pillow blocks, 605

Pinning, 508
Pipe, 24, A–36
Plastic gear materials, 435, 668
Plastics, 61, 435, 668, A–17
Poisson's ratio, 36
Polar section modulus, 97
Powdered metals, 56
Power, 94
Power screws, 699–702
  Acme thread, 697, 699, 702
  buttress thread, 697
  efficiency, 701
  lead angle, 701
  power required, 702
  self-locking, 701
  square thread, 697
  torque required, 700
Polygon connection, 510
Press fit, 512. *See also* Force fit; Shrink fit;
    Interference fit
Pressure angle
  spur gears, 316
  wormgearing, 343
Principal stresses, 139
Product realization process, 10
Properties of materials, 32
Proportional limit, 34
Pulleys
  flat belt, 540
  V-belt, 269–73, 538
pV factor, 668–9

Quality function deployment, 10

Rack, 327
Radius of gyration, 232
Ratchet, 862
Relaxation, 43
R.R. Moore fatigue test device, 168
Reliability factors, 175–6, 389
Residual stress, 179
Resistivity, electrical, 44
Retaining ring grooves, 542
Retaining rings, 518–9
Reyn, 676–7
Ringfeder, locking assemblies, 509
Robust product design, 592–4
Rockwell hardness, 37

Screw threads, 18, A–4, A–5
Seals, 521–6
  bearing, 524, 623
  elastomers, 525
  face, 524
  gaskets, 526
  materials, 525
  O-rings, 522
  packings, 526
  shaft, 523–4
  T-rings, 522
  types, 521–5
Section modulus, 106
  polar, 97
Set screws, 510

Shaft design, 530–64
  design stresses, 543–6
  dynamic considerations, 562
  equation for diameter, 548
  examples, 548–60
  flexible, 563
  forces exerted on shafts, 535–40
  preferred basic sizes, 552, A–3
  procedure, 532–5
  stress concentrations in, 540–3
Shaping of gears, 371
Shear center, 107
Shear pin, 508
Shear strength, 35
Shear stress
  direct, 92
  formulas, 104
  element, 90
  on keys, 92
  maximum, 139
  vertical, 102, 543
Sheaves, V-belt, 269
  forces on shafts, 538
Shoulders, shaft, 520, 541
Shrink fits, 585
  stresses for, 587–9
Size factor, 175–8, 389
SI units, 24
  prefixes, 25
  typical quantities in machine design, 25
Slenderness ratio for columns, 234
Society of Automotive Engineers (SAE), 44,
   272, 715, 850
Soderberg criterion, 193
Soldering, 726
Sommerfeld number, 677
Spacers, 520
Specific modulus, 68
Specific strength, 68
Specific weight, 44
Splines, 503–7
  fits, 504
  geometry, 504
  involute, 505
    modules, 507
    pitches, 506–7
  straight-sided, 503
  torque capacity, 505
Split taper bushings, 510
Spreadsheets as design aids
  chain design, 297
  columns, 242, 252, 256
  force fits, stresses, 590
  gear design, 415
  shaft design, 561
  springs, design, 752–7
Springs, helical compression, 735–57
  allowable stresses, 740–3
  analysis, 746–8
  buckling, 745–6
  deflection, 744
  design of, 749–57
  end treatments, 734
  materials, 740–1, 745
  number of coils, 739

pitch, 739
pitch angle, 739
spring index, 739
spring rate, 738
stresses, 744
Wahl factor, 744
wire diameters, 735–7
Springs, helical extension, 757–62
    allowable stresses, 760, 742–3
    end configurations, 734, 757
Springs, helical torsion, 762–9
    deflection, 763
    design procedure, 764, 768–9
    design stresses, 764–7
    number of coils, 764
    spring rate, 764
    stresses, 763
Springs, types, 732–5
    manufacturing, 770
    shot peening, 769
Sprockets, chain, 283, 537
Stainless steel, 53, A–12
Statical moment, 102
Statistical approaches to design, 213
Steel, 46–56
    alloy groups, 48
    bearing, 48
    carbon and alloy, 86–9, 379, A–6–A–10
    carbon content, 47
    carburized, 52, 382, A–11
    conditions, 49–53
    designation systems, 47
    heat treating, 49–53
    high-carbon, 47
    low-carbon, 47
    medium-carbon, 47
    stainless, 53, A–12
    structural, 54, A–13
    tool steels, 54
    uses for, 49
Strength
    endurance, 40, 172–81
    shear, 35
    tensile, 33
    yield, 34
Strength reduction factor, 122
Stress
    allowable for gears, 378–85
    amplitude, 166
    bending, 105
    combined bending and torsion on circular
        shafts, 159, 546
    combined normal stresses, 117
    combined stress, general, 138
    design, for shafts, 543–6
    direct shear, 92
    direct tension and compression, 90
    due to shrink or force fits, 587–9
    fluctuating, 168–70
    maximum shear stress, 139
    Mohr's circle, 145–8
    principal stresses, 139

ratio, 167, 186
    repeated and reversed, 167
    special shear stress formulas, 104
    torsional shear, 95
    vertical shearing stress, 102
Stress amplitude, 166
Stress concentration factors, 119, 540–3,
    A–27–A–30
Stress concentrations, 119
    in shaft design, 540–3
    keyseats, 541
Stress elements, 89
Stress ratio, 167, 180
Structural shapes, 20–24, A–31–A–38
    aluminum, A–37–A–38
    steel, A–31–A–36
Structural steel, 54, A–13
Surface finish, 174, 592, 675
Synchronous belts, 268, 280–3

Taguchi method, 594
Taper and screw, 512
Tapered roller bearings, 603, 618
Tempering, 51, A–6–A–10
Tensile strength, 33
Thermal conductivity, 44
Thermal expansion coefficient, 44
Thrust bearings, 604
Titanium, 60, A–16
Tolerance, 578–92
    geometric, 592
    grades, 579
Tool steels, 54
Torque, 94
    equivalent, 161
Torsion
    combined with tension, 159
    in closed thin-walled tubes, 100
    in noncircular cross sections, 98
    pure, Mohr's circle, 159
    stress distribution, 96
    torsional deformation, 97
    torsional shear stress formula, 95
Toughness, 38
Train value, 323, 347–8
Transmission, design of, 630–57
Transition fits, 585
Tubes, stresses in, 100, 104, 587–90
Tubing, structural
    Rectangular and square, A–35
    Round, A–36

Undercutting of gear teeth, 321
Unified numbering system, UNS, 44
Unit systems, 24
Universal joints, 516–8
U.S. customary units, 24
    typical quantities in machine design, 25

V-belt drives, 267–77
    angle of contact, 270
    angle of wrap correction factor, 277

belt construction, 268
    belt cross sections, 271–2
    belt length correction factor, 277
    belt length formula, 270
    belt lengths, standard, 277
    belt tension, 271, 280
    center distance formula, 270
    design of, 272–9
    forces on shafts, 538
    power rating charts, 275–6
    pulleys. See sheaves
    SAE standards, 272
    selection chart, 274
    service factors, 274
    sheaves, 269–73
    span length, 270
Velocity ratio, gears, 322
Viscosity, 676–7
Viscosity Index, 689
von Mises theory, 189

Wear, 690
Weight, 26
Welded joints, 783–91
    allowable stresses, 783, 785
    geometry factors, 786
    size of weld, 784, 787
    treating weld as a line, 785
    types of joints, 783–4
    types of welds, 784
Wide flange beam shapes, 22, A–33
Woodruff keys, 497, 501–3
Wormgearing, 339–46, 475–487
    coefficient of friction, 476
    design of, 481–7
    efficiency, 478–9
    face length of worm, 346
    forces on, 476
    geometry, 341–6
    lead, 341
    lead angle, 341
    pitches, 341
    pressure angle, 343
    reducer, 344–6
    self-locking, 343
    shell worm, 344
    stresses, 481
    surface durability, 482
    threads (teeth), 341
    tooth dimensions, 344
    types, 339–40
    velocity ratio, 342
    worm diameter, 344
    wormgear dimensions, 345–6

Yield point, 34
Yield strength, 34, 188

Zinc, 59, A–16